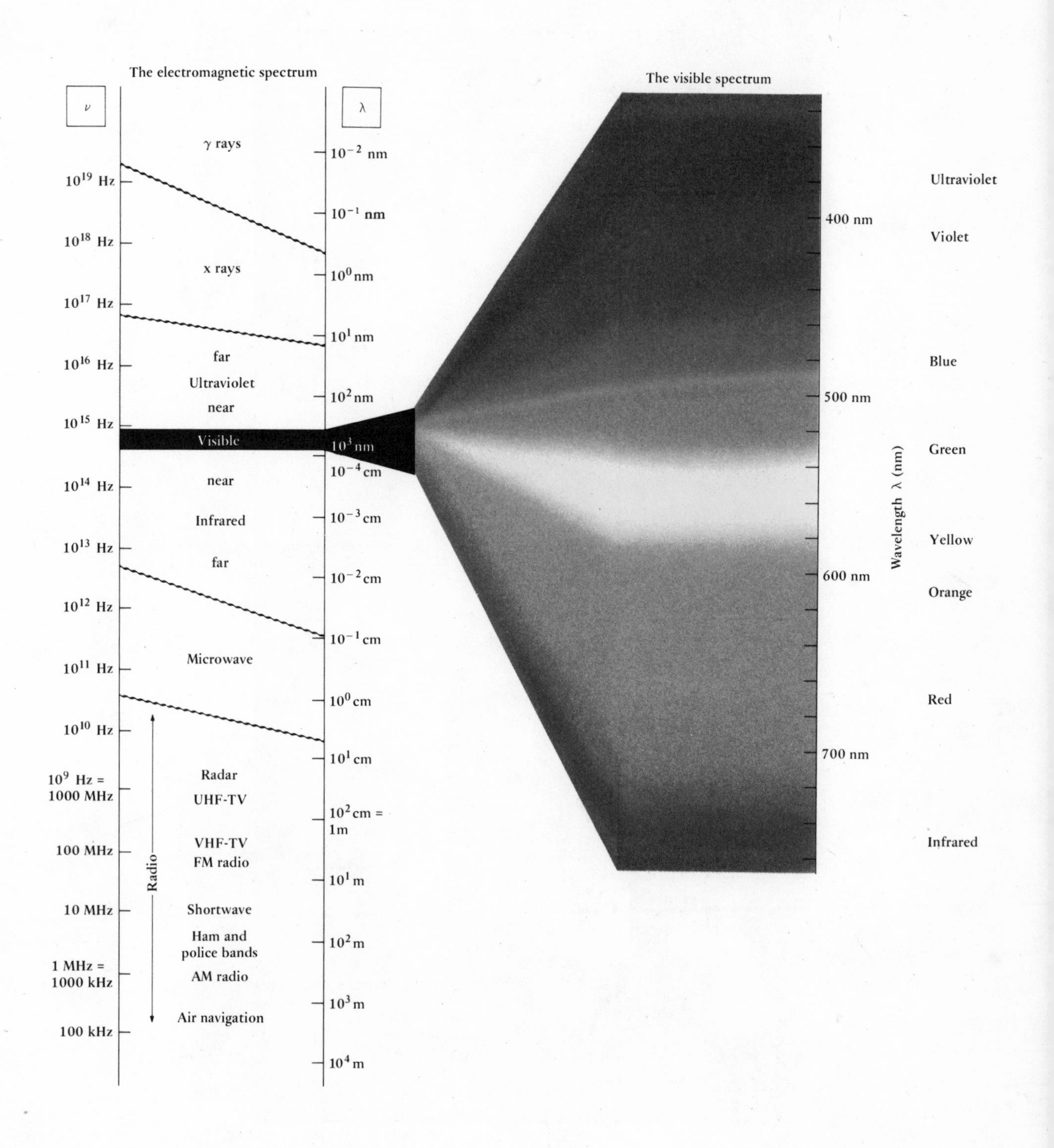

The electromagnetic spectrum
ν
λ
γ rays
x rays
far
Ultraviolet
near
Visible
near
Infrared
far
Microwave
Radar
UHF-TV
VHF-TV
FM radio
Shortwave
Ham and
police bands
AM radio
Air navigation
Radio
10¹⁹ Hz
10¹⁸ Hz
10¹⁷ Hz
10¹⁶ Hz
10¹⁵ Hz
10¹⁴ Hz
10¹³ Hz
10¹² Hz
10¹¹ Hz
10¹⁰ Hz
10⁹ Hz =
1000 MHz
100 MHz
10 MHz
1 MHz =
1000 kHz
100 kHz
10⁻² nm
10⁻¹ nm
10⁰ nm
10¹ nm
10² nm
10³ nm
10⁻⁴ cm
10⁻³ cm
10⁻² cm
10⁻¹ cm
10⁰ cm
10¹ cm
10² cm =
1m
10¹ m
10² m
10³ m
10⁴ m
The visible spectrum
400 nm
500 nm
600 nm
700 nm
Wavelength λ (nm)
Ultraviolet
Violet
Blue
Green
Yellow
Orange
Red
Infrared

FUNDAMENTALS OF Analytical Chemistry

FUNDAMENTALS OF

Analytical Chemistry

FIFTH EDITION

DOUGLAS A. SKOOG
Stanford University

DONALD M. WEST
San Jose State University

F. JAMES HOLLER
University of Kentucky

Saunders Golden Sunburst Series
Saunders College Publishing
New York Chicago San Francisco
Philadelphia Montreal Toronto
London Sydney Tokyo

Text Typeface: ITC New Baskerville
Compositor: General Graphic Services
Acquisitions Editor: John Vondeling
Project Editor: Merry Post
Copy Editor: Irene Nunes
Art Director: Carol Bleistine
Art Assistant: Doris Roessner
Text Designer: Tracy Baldwin
Cover Designer: Lawrence R. Didona
Text Artwork: Larry Ward, Tom Mallon
Production Manager: Harry Dean, Jr.
Assistant Production Manager: Jo Ann Melody

Cover credit: A blue (488.0 nm) beam from an argon-ion laser enters a cell containing a minute quantity of NO_2, a brown gas found in smog. The beam, which is directed through the sample several times by mirrors, is absorbed by the NO_2. The gas subsequently fluoresces to give the bright orange color characteristic of NO_2. The intensity of the color is a measure of the concentration of the pollutant. This is a dramatic illustration of the very sensitive technique of laser-induced fluorescence, which is discussed in Chapter 20. The photo is courtesy of Professor Dennis J. Clouthier of the University of Kentucky.

Printed in the United States of America

Fundamentals of Analytical Chemistry, 5th edition

ISBN 0-03-14828-6

Library of Congress Catalog Card Number: 87-24411

Preface

With the appearance of this edition—the fifth—*Fundamentals of Analytical Chemistry* enters its second quarter century as an introductory textbook on analytical chemistry designed primarily for a course for chemistry majors. During the lifetime of this text, the field of analytical chemistry has teemed with activity as never before: new instruments with previously unheard of sensitivities and selectivities have appeared; new methods for resolving the components of incredibly complex mixtures have been developed; and automation and computer control of the measurement process have become commonplace. Some of these developments have been incorporated in each new edition. Others have not, however, because the time alloted in most chemistry curricula for an introductory course in quantitative analysis is so limited.

In preparing each new edition, we have found ourselves facing the dilemma of what new developments to include and, equally important, what to delete from older editions to make space for the new. In making these decisions, we have adopted the general philosophy of avoiding superficiality by limiting the number of topics covered to those that we believe can be treated in sufficient depth so that the reader can develop a basic understanding of the principles upon which they are based.

Because disagreement is inevitable as to what topics should be included in an elementary analytical course, we have included more material than could possibly be covered in one or two semesters. We have tried to make chapters sufficiently independent so that some can be left out and the order of others changed without a loss in continuity. Thus, the text can be tailored to fit the tastes and prejudices of the individual instructor.

Much of the theoretical discussion in earlier editions of this text has centered on thermodynamics and the application of equilibrium calculations to analytical problems. With the appearance of automated instruments and the increasing importance of analytical methods for determining species of interest in medicine, biochemistry, and ecology, it is evident that a balanced analytical textbook must devote space not only to thermodynamic theory but also to the theory of reaction kinetics. Thus, the reader will find an entirely new chapter in this edition that deals with the kinetics and application of kinetic measurements in analytical chemistry.

Another significant change in this edition is the use of formula weights and molar concentrations in all volumetric calculations. The decision to abandon equivalent weights and normalities brings the text into line with current practice in most analytical journals. (We have, however, included a discussion of the use of normality and equivalent weight in Appendix 9.)

Other topics new to this edition include the operational definition of pH, modern voltammetric techniques including pulse polarography and stripping methods, diode array detectors and multichannel spectroscopic instruments, dc plasma sources for atomic spectroscopy, flow injection methods, fused silica columns for capillary gas chromatography, supercritical-fluid chromatography, and electronic balances. We have also introduced a short section and a laboratory experiment involving a very old technique—weight titrations. With modern top-loading balances and plastic reagent dispensers, a weight titration can be carried out more efficiently and accurately than one based on volumetric measurements.

To provide space for these new topics, we have had to condense or delete several parts of earlier editions. Thus, we have reduced the number of laboratory experiments from 53 to 34. In addition, we have eliminated the chapter devoted to nonaqueous titrations and substituted a brief discussion of this subject in the chapter on applications of acid/base titrations. We have also omitted the discussion of the use of mercury(II) for complex formation titrations and shortened the discussions devoted to the theory of membrane electrodes and to the applications of oxidation/reduction titrations.

In addition to updating the text, we have extensively reorganized the introductory chapters in order to remove redundancies and to present the material in a more logical and concise way. To this end, we have integrated the material dealing with stoichiometry and chemical calculations into two early chapters (3 and 4) dealing with gravimetry and titrimetry (including weight titrimetry). Chapter 5 is devoted to aqueous solution chemistry including simple equilibrium calculations of all types and the use of activities and activity coefficients in such calculations. Chapter 6 extends the discussion of equilibrium to complex systems involving several competing reactions. The chapters that follow dealing with precipitation, neutralization, and complex formation titrations retain the organizational pattern of earlier editions. Many parts of these chapters have been rewritten to improve their clarity and readability, however.

The introductory chapter on oxidation/reduction equilibrium and electrochemical theory (Chapter 12) has been completely rewritten. In addition, the material on polarization phenomena has been moved to the chapter on coulometry where it first becomes of importance. We believe that these changes provide a clearer and more logical presentation of electrochemical theory that will make it more readily absorbed and understood by students. Chapter 15 on potentiometric methods has also undergone reorganization and extensive rewriting with a particular emphasis on clarification of the sign conventions that are used for indicator and reference electrodes.

The organization of the chapters dealing with spectroscopy, chromatography, and preliminary steps in an analysis are substantially the same as in earlier editions. Much of the material in these chapters has also been rewritten for clarity and readability.

The problem sets at the ends of chapters have all been rewritten and, in addition, sets of questions have been introduced. Answers to approximately half of the questions and problems are found at the end of the text. A solutions manual is also available for instructors, and about 60 to 70 transparencies have been developed for use in lectures.

We wish to acknowledge with thanks the comments and suggestions of the following who have reviewed the manuscript for this edition at various stages in its production: Professor John Ganchoff of Elmhurst College; Professor Richard H. Hanson of the University of Arkansas at Little Rock; Professor T. J. Haupert of California State University, Sacramento; Professor John L. Plude of University of Wisconsin, Oshkosh; and Professor Joseph J. Topping of Towson State University. We also wish to thank Professor Elizabeth W. Kleppinger of Berea College, who was kind enough to read page proofs for the first 11 chapters. In addition, we want to thank Professor David K. Roe of Portland State University for bringing to our attention an inconsistency in the sign convention for electrodes that was present in our earlier editions as well as in other analytical textbooks. Finally, we offer particular thanks to Professor Alfred Armstrong of The College of William and Mary in Virginia for again reviewing the manuscript in detail and to Professor Peter F. Linde of San Francisco State University for his thoughtful and cogent comments.

Douglas A. Skoog
Donald M. West
F. James Holler

Contents Overview

Appendixes

Contents

CHAPTER 1

Introduction

Analytical chemistry is concerned with the separation, identification, and determination of the relative amounts of the components (the analytes) making up a sample of matter. A *qualitative analysis* provides information on the chemical identity of the analytes in the sample, whereas a *quantitative analysis* yields numerical information on the relative amount of one or more of these analytes. Generally, qualitative information is required before a quantitative analysis can be undertaken. A separation step is usually a necessary part of both a qualitative and a quantitative analysis.

The principal topics covered in this text are quantitative methods of analysis and methods of analytical separations, although references to qualitative methods appear from time to time.

1A The Role of Analytical Chemistry in the Sciences

Historically, analytical chemistry has played a vital role in the development of science. For example, in 1894 Wilhelm Ostwald wrote

> Analytical chemistry, or the art of recognizing different substances and determining their constituents, takes a prominent position among the applications of science, since the questions which it enables us to answer arise wherever chemical processes are employed for scientific or technical purposes. Its supreme importance has caused it to be assiduously cultivated from a very early period in the history of chemistry, and its records comprise a large part of the quantitative work which is spread over the whole domain of science.

Since 1894, analytical chemistry has evolved from an art to a science, due in no small part to the work of Ostwald himself, and its importance still spreads over all domains of science and technology. To cite but a few examples, consider the following: The effectiveness of smog-control devices is determined by measuring the parts per million of hydrocarbons, nitrogen oxides, and carbon monoxide in the exhaust gases of automobiles. Hyperparathyroidism in human patients is diagnosed by quantitative measurements of ionized calcium in blood serum. The protein content, and thus the nutritional value of foods, is ordinarily established by quantitative determination of their nitrogen content. Periodic analysis of steel during its production permits adjustment in the concentration of such elements as carbon, nickel, and chromium to give a product that has a desired strength, hardness, corrosion resistance, and ductility. Household gas supplies are continuously monitored for their mercaptan content in order to ensure sufficient levels of odorant to warn of leaks. Modern farmers tailor their fertilization and irrigation schedules to meet changing plant needs during the growing season; these needs are gauged from quantitative analyses of the plants and of the soil in which they grow.

In addition to everyday applications of the types just cited, quantitative analytical measurements play a vital role in many research areas in chemistry, biochemistry, biology, geology, and the other sciences. For example, chemists have learned much about mechanisms of chemical reactions through kinetic studies based upon periodic quantitative measurements that reveal the rates at which reactants are consumed or products are formed. Quantitative analyses for potassium, calcium, and sodium ions in the body fluids of animals have permitted physiologists to study the role these ions play in the conduction of nerve signals and the contraction and relaxation of muscles. Materials scientists, in their studies of the behavior of semiconductor devices, have relied heavily upon quantitative analyses of crystalline germanium and silicon for impurities in the concentration range from 1×10^{-6} to 1×10^{-10} percent. Archeologists have found it possible to identify sources of volcanic glasses (obsidian) based upon the concentrations of several minor elements in samples taken from various locations; this knowledge has made it possible to trace prehistoric trade routes for tools and weapons manufactured from obsidian.

Many chemists and biochemists devote a significant part of their time in the laboratory acquiring quantitative information about the systems in which they are interested. For such investigators, analytical chemistry serves as a tool in their scholarly efforts in much the same way that calculus and matrix algebra are tools of the theoretical physicist and ancient languages are tools of the classics scholar.

1B Classification of Quantitative Methods of Analysis

The results of a typical quantitative analysis are based upon two measurements. One is the weight or volume of sample to be analyzed. The second, which normally completes the analysis, is the measurement of some quantity that is proportional to the amount of analyte in that sample. Analytical methods are often classified according to the nature of this final measurement. In a *gravimetric method,* the mass of the analyte or that of some compound chemically related to the analyte is determined. In a *titrimetric method,* the quantity of reagent necessary to react completely with the analyte is measured. *Electroanalytical methods* involve the measurement of such properties as potential, current, resistance, and quantity of charge. *Spectroscopic methods* are based upon measurements of the interaction between electromagnetic radiation (including X-ray, ultraviolet, visible, infrared, microwave, and radio-frequency radiation) and analyte atoms or molecules or upon measurements of the amount of such radiation produced by analytes. Finally, there is a group of miscellaneous methods for completing analyses based upon measuring such properties as mass-to-charge ratio (mass spectrometry), rate of radioactive decay, heat of reaction, rate of reaction, thermal conductivity, optical activity, and refractive index.

1C Steps in a Typical Quantitative Analysis

A typical quantitative analysis involves a sequence of several steps:

1. Selecting a method of analysis
2. Sampling

3. Preparing a laboratory sample
4. Defining replicate samples
5. Preparing solutions of the sample
6. Eliminating interferences
7. Completing the analysis
8. Calculating results and estimating their reliability

In some instances, one or more of these steps can be dispensed with. Ordinarily, however, all play an important role in the success of an analysis.

The first 23 chapters of this text focus on the last two steps of this list. Step 7 involves measuring one of the physical properties mentioned in the previous section, preferably one that is proportional (in most cases) to the amount of analyte in a sample of known weight or volume. Step 8 consists of computing the relative amount of the analyte present in the samples and estimating the reliability of the results.

A brief description of each of these steps is provided at this juncture in order to give the reader an overall perspective on how quantitative chemical data are obtained.

1C-1 Selecting a Method of Analysis

Selecting which method will be used to solve an analytical problem is a vital first step in any quantitative analysis. The choice is sometimes difficult and requires experience as well as intuition on the part of the chemist. An important consideration in selection is the accuracy required. Unfortunately, high reliability nearly always entails a large expenditure of time; the method ultimately chosen may thus of necessity represent a compromise between accuracy and economics.

A second consideration, also related to economic factors, is the number of samples to be analyzed. If there are many, the chemist can afford to use a method that requires such preliminary operations as the assembling and calibrating of instrumental equipment and the preparing of standard solutions. On the other hand, with only a single sample or a few samples, it may be more expedient to select a procedure that avoids such preliminary steps.

Finally, the choice of method is always governed by the complexity of the sample being analyzed and by the number of components for which quantitative information is needed. Further details on choosing a method of analysis are given in Section 26A.

1C-2 Sampling

To produce meaningful information, an analysis must be performed on a sample whose composition faithfully reflects that of the bulk of material from which it is taken. Where the bulk is large and inhomogeneous, great effort is required to procure a representative sample. Consider, for example, a railroad car containing 25 tons of silver ore. Buyer and seller must come to agreement regarding the value of the shipment based primarily upon its silver content. The ore is inherently heterogeneous, consisting of lumps of various size as well as varying silver content. The assay of this shipment will be performed on a sample that has a mass of perhaps 1 g. For the analysis to have significance, it is essential that this small sample have a composition that is representative of the 25 tons (approximately 22,700,000 g) of ore in the shipment. The task of isolating 1 g with any confidence that its composition truly reflects the average composition of the nearly 23,000,000 g from which it is taken is clearly a nontrivial undertaking that requires a systematic manipulation of the entire shipment.

Many sampling problems are less formidable than the one just described. Nevertheless, the chemist must have some assurance that the laboratory sample is representative of the whole before proceeding with an analysis.

A detailed description of how various types of material are sampled is presented in Section 27A.

1C-3 Preparing a Laboratory Sample

After sampling, solid materials frequently are ground to decrease particle size, mixed to ensure homogeneity, and stored for various lengths of time before being analyzed. During each of these steps, absorption or desorption of water may occur, depending upon the humidity of the environment. Because a loss or gain of water changes the chemical composition of solids, such samples are frequently dried under carefully specified conditions just before analysis. Alternatively, the moisture content of the sample may be determined at the time of the analysis by a separate analytical procedure.

Further information on preparing samples for analysis and determining their moisture content is given in Chapter 27.

1C-4 Defining Replicate Samples

Most chemical analyses are performed on replicate samples whose weights or volumes have been determined by careful measurements with an analytical balance or with various volumetric devices. Replicate samples are used in order to improve the quality of the results and to provide a measure of their reliability.

Detailed instructions on techniques for measuring the weight or volume of samples are given in Sections 30D through 30H.

1C-5 Preparing Solutions of the Sample

Most (but certainly not all) analyses are performed on solutions of the sample. Ideally, the solvent should dissolve the entire sample (not just the analyte) rapidly and under sufficiently mild conditions that analyte loss cannot occur. Unfortunately, many of the materials of interest to scientists—such as silicate minerals, high-molecular-weight polymers, and specimens of animal tissue, to mention a few—are not directly soluble in common solvents. Conversion of the analyte in such materials to a soluble form can be a formidable and time-consuming task. Methods for decomposing and dissolving samples are described in Chapter 28.

1C-6 Eliminating Interferences

Few, if any, chemical or physical properties of importance in chemical analysis are unique to a single chemical species; instead, the reactions used and the properties measured are characteristic of a group of elements or compounds. This lack of truly specific reactions and properties adds greatly to the difficulties faced by the chemist undertaking an analysis because a scheme must be devised to isolate the species of interest from all others in the sample that can influence the final measurement. Substances that prevent the direct measurement of the analyte concentration are called *interferences;* their separation prior to the final measurement constitutes an important step in most analyses. No hard and fast rules can be given for the elimination of interferences, and the resolution of this problem is often the most demanding aspect of an analysis. Separation methods are treated in Chapters 23, 24, 25, and 29.

1C-7 Completing the Analysis; Calibration

The results of every analysis are based on a final measurement of some physical property X that varies in a known and reproducible way with the concentration C_A of the analyte. Ideally, a straight-line relationship exists between the two, that is,

$$C_A = kX$$

where k is a proportionality constant. With two exceptions, analytical methods require the empirical determination of k with chemical standards for which C_A is known (the two exceptions are gravimetric and coulometric methods, discussed in Chapters 3 and 16, respectively). The process of determining k is an important step in most analyses and is termed a *calibration*.

Smith and Parsons have assembled a list of chemical standards suitable for calibration.[1] This list is found in the end pages of this text.

1C-8 Calculating Results and Estimating Their Reliability

The computation of analyte concentrations from experimental data is ordinarily a simple and straightforward task, particularly with modern calculators and computers. The analytical result is incomplete, however, without an estimate of its reliability. The experimenter must therefore provide some measure of the uncertainties associated with computed results. Methods for obtaining such estimates are described in detail in Chapter 2.

[1]B. W. Smith and M. L. Parsons, *J. Chem. Educ.*, **1973,** *50*, 679.

CHAPTER 2

Evaluation of Analytical Data

It is not possible to perform a chemical analysis in such a way that the results are totally free of uncertainty. The best that can be hoped is that this uncertainty can be kept at a tolerable level and that its size can be estimated with acceptable accuracy.

The presence of uncertainty in analytical data is illustrated in Figure 2-1, which shows results for the quantitative determination of iron(III) based on the red color of the thiocyanate complex of that cation. Here, six identical portions of an aqueous solution known to contain exactly 20.00 ppm[1] of iron(III) were analyzed in exactly the same way. The data are typical of what might be obtained by a skilled and experienced technician working with reasonable care. Note that the results range from a low of 19.4 ppm to a high of 20.3 ppm of iron(III), with their average $\bar{x}$ being 19.8 ppm.

Every physical measurement is subject to a host of uncertainties that lead to a scatter of results similar to that in Figure 2-1. Because these uncertainties can never be completely eliminated, the true value for any measured quantity must always remain unknown. If the probable magnitude of the error in a measurement can be established, however, it becomes possible to define limits within which the true value of a measured quantity is likely to lie.

Estimating the reliability of experimental data is seldom easy. Nevertheless, such estimates must be made whenever laboratory results are collected *because data of unknown reliability are worthless.* In contrast, results that are not particularly accurate may be of great value if the limits of probable uncertainty can be set with reasonable confidence.

Unfortunately, no simple and generally applicable method exists for determining the reliability of data with absolute certainty. Indeed, the effort involved in establishing the quality of experimental results is frequently comparable to the effort expended in obtaining them. Evaluating the reliability of results may take several forms, including performing experiments specifically designed to reveal the presence of errors, analyzing standards whose composition is known from the way they were prepared, consulting the literature to profit from the experience of others, calibrating equipment, and applying statistical tests to the data. In the end, none of these recourses is infallible. Consequently, the scientist can only make a *judgment* as to the probable accuracy of a result. With experience, these judgments tend to become harsher and less optimistic.

One of the first questions that must be answered before an analysis is undertaken is, "What is the maximum allowable error in the result?" The answer to this question is of prime importance because it ordinarily determines the amount of time the scientist must expend in obtaining the result. For

[1]Parts per million, that is, 20.00 parts of iron(III) per million parts of solution.

FIGURE 2-1

Results from six replicate determinations for iron in aqueous samples of a standard solution containing 20.00 ppm of iron (III).

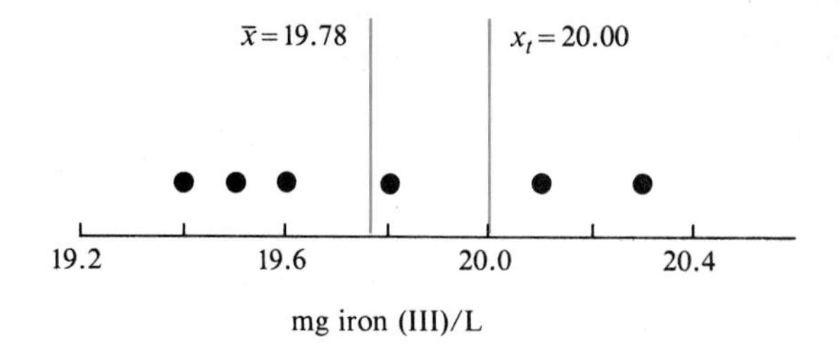

example, a tenfold increase in reliability may require hours, days, or even weeks of added labor. *It cannot be too strongly emphasized that a scientist cannot afford to waste time in the indiscriminate pursuit of greater reliability than is needed.*

This chapter is devoted to a consideration of the types of errors encountered in chemical analyses, methods for recognizing such errors, and techniques for estimating and reporting their magnitude.[2]

2A Definition of Terms

In performing an analysis, a chemist generally carries two to five portions of the sample through the entire procedure. The individual results from such a set of measurements are seldom identical (Figure 2-1), and selection of a central, or "best," value for the set becomes necessary. Intuitively, the added effort of replication can be justified in two ways. First, the central value of a set should be more reliable than any of the individual results. Second, the variation in the data should provide a measure of the uncertainty associated with the central result. Either the *mean* or the *median* may serve as the central value for a set of replicate measurements.

2A-1 The Mean and the Median

Mean, arithmetic mean, and *average* ($\bar{x}$) are synonymous terms for the quantity that is obtained by dividing the sum of replicate measurements by the number of results in the set:

$$\bar{x} = \frac{\sum_{i=1}^{N} x_i}{N} \qquad (2\text{-}1)$$

where x_i represent the individual values of x making up a set of N replicate measurements.

The *median* is that result about which all the others are equally distributed. For an odd number of results, the median can be evaluated directly. For an even number, the mean of the central pair is used.

EXAMPLE 2-1

Calculate the mean and the median for the data shown in Figure 2-1.

$$\text{Mean} = \bar{x} = \frac{19.4 + 19.5 + 19.6 + 19.8 + 20.1 + 20.3}{6} = 19.78 \simeq 19.8 \text{ ppm Fe}$$

[2]For detailed discussions of error analysis, see L. A. Currie, in *Treatise on Analytical Chemistry*, 2nd ed., I. M. Kolthoff and P. J. Elving, Eds., Part I, Vol. 1, Chapter 4, New York: Wiley, 1978; and J. Mandel, *ibid.*, Chapter 5.

Because the set contains an even number of measurements, the median is the average of the central pair:

$$\text{Median} = \frac{19.6 + 19.8}{2} = 19.7 \text{ ppm Fe}$$

Ideally, the mean and media are identical. Frequently they are not, however, particularly when the number of measurements in the set is small.

2A-2

Precision

Precision describes the agreement between two or more measurements that have been carried out *in exactly the same fashion.* Precision is expressed in several ways.

Standard Deviation (s)

Standard deviation is a term that is widely used in statistics as a measure of precision. For small sets of data, it is obtained by summing the squares of the individual deviations from the mean, taking the square root of this sum, and dividing by the *number of degrees of freedom,* which is one fewer than the number N of the results in the set:

$$s = \sqrt{\frac{\Sigma(x_i - \bar{x})^2}{N - 1}} \tag{2-2}$$

Variance (s^2)

The variance is simply the square of the standard deviation:

$$s^2 = \frac{\Sigma(x_i - \bar{x})^2}{N - 1} \tag{2-3}$$

Table 2-1 illustrates how the standard deviation and the variance for the data in Figure 2-1 are obtained. Here, the standard deviation is 0.35 ppm Fe, the variance is 0.13 (ppm Fe)^2, and the number of degrees of freedom is five. The significance of these three terms is discussed in detail in Section 2D-3.

Note that the standard deviation bears the same units as the data whereas the variance has units of the data squared. Chemists generally prefer to discuss precision of data in terms of standard deviation rather than variance because of this correspondence in units; statisticians, in contrast, use variances more often because they can be combined additively. That is, if n independent sources of indeterminate error exist in a system (or a calculation), the total sample variance s_t is

$$s_t^2 = s_1^2 + s_2^2 + \cdots + s_n^2 \tag{2-4}$$

where $s_1^2, s_2^2, \ldots, s_n^2$ are the individual variances.

An Alternative Method for Computing Standard Deviations

In calculating s with a handheld calculator that does not have a standard deviation key, the following algebraic identity for Equation 2-2 is somewhat more convenient to use:

$$s = \sqrt{\frac{\Sigma x_i^2 - (\Sigma x_i)^2/N}{N - 1}} \tag{2-5}$$

TABLE 2-1

Methods of Expressing Precision and Accuracy*

	Fe Concn, ppm x_i	Deviation from Mean $\lvert x_i - \bar{x} \rvert$	$(x_i - \bar{x})^2$
x_1	19.4	0.38	0.1444
x_2	19.5	0.28	0.0784
x_3	19.6	0.18	0.0324
x_4	19.8	0.02	0.0004
x_5	20.1	0.32	0.1024
x_6	20.3	0.52	0.2704
$\Sigma x_i =$	118.7	$\Sigma \lvert x_i - \bar{x} \rvert =$ 1.70	$\Sigma(x_i - \bar{x})^2 =$ 0.6284

Mean $= \bar{x} = 118.7/6 = 19.78 \simeq 19.8$ ppm Fe

$$\text{Standard deviation} = s = \sqrt{\frac{\Sigma(x_i - \bar{x})^2}{N - 1}} = \sqrt{\frac{0.6284}{6 - 1}}$$

$$= \sqrt{0.1257} = 0.354 \simeq 0.35 \text{ ppm Fe}$$

$$\text{Variance} = s^2 = \frac{\Sigma(x_i - \bar{x})^2}{N - 1} = \frac{0.6284}{6 - 1} = 0.1257 \simeq 0.13 \text{ (ppm Fe)}^2$$

Spread or range $= w = 20.3 - 19.4 = 0.9$ ppm Fe

Average deviation from mean $= \Sigma|x_i - \bar{x}|/N = 1.70/6 = 0.283 \simeq 0.28$ ppm Fe

$$\text{Relative standard deviation} = \frac{s}{\bar{x}} \times 1000 = \frac{0.354}{19.78} \times 1000 \text{ ppt} = 17.9 \simeq 18 \text{ ppt}$$

$$\text{Coefficient of variation} = \text{CV} = \frac{s}{\bar{x}} \times 100\% = \frac{0.354}{19.78} \times 100\% \simeq 1.8\%$$

Absolute error† $= 19.78 - 20.00 = -0.22 = -0.2$ ppm Fe

$$\text{Relative error} = \frac{19.78 - 20.00}{20.00} \times 100\% = -1.1\%$$

*For source of data, see Figure 2-1.

†Sample known to contain 20.00 ppm Fe.

EXAMPLE 2-2

Determine the mean and standard deviation for the five data making up trials 6 through 10 in Table 2-3, page 19.

Here, we compute Σx_i^2 and $(\Sigma x_i)^2/N$ in order to apply Equation 2-5.

Trial	x_i, mL	x_i^2
6	9.982	99.640324
7	9.986	99.720196
8	9.982	99.640324
9	9.981	99.620361
10	9.990	99.800100
	$\Sigma x_i = 49.921$	$\Sigma x_i^2 = 498.421305$

$$\bar{x} = 49.921/5 = 9.9842 \simeq 9.984 \text{ mL}$$

$$\frac{(\Sigma x_i)^2}{N} = \frac{(49.921)^2}{5} = 498.4212482$$

$$s = \sqrt{\frac{498.421305 - 498.4212482}{5 - 1}} = \sqrt{\frac{0.0000568}{4}} = 0.00377 \simeq 0.0038$$

Note that the difference between Σx_i^2 and $(\Sigma x_i)^2/N$ in Example 2-2 is so small that premature rounding would have led to a serious error in the computed value of s. Because of this source of error, Equation 2-5 should never be used to calculate the standard deviation for numbers containing five or more digits; instead, Equation 2-2 should be used.[3] It is also important to note that handheld calculators and small computers with a standard deviation function usually employ a version of Equation 2-5. Consequently, large errors in s are to be expected when these devices are applied to data having five or more significant figures.[4]

Relative Standard Deviation (RSD)

Chemists frequently quote standard deviations in relative rather than absolute terms. The relative standard deviation is obtained by dividing the standard deviation by the mean of the set and is often expressed in parts per thousand or in percent by multiplying this quotient by 1000 or by 100. For example,

$$\text{RSD} = (s/\bar{x}) \times 1000 \text{ ppt}$$

When 100% is the multiplier, the relative standard deviation is given the special name of *coefficient of variation* (CV). Thus,

$$\text{CV} = (s/\bar{x}) \times 100\% \tag{2-6}$$

Note that the relative standard deviation in Table 2-1 is reported as 18 parts per thousand (ppt), or 1.8%; the latter is also the coefficient of variation for the set.

Relative standard deviations often give a clearer picture of data quality than do absolute standard deviations. For example, the absolute standard deviation that causes a 1.8% coefficient of variation in a sample containing 20.0 ppm of analyte results in an 18% CV in a sample containing 2.00 ppm of analyte.

Spread or Range (w)

The *spread*, or *range*, of a set of data is obtained by subtracting the smallest value in the set from the largest. Thus, the spread of the data in Table 2-1 is 0.9 ppm Fe. In a sense, the spread is a measure of the scatter rather than the precision of a group of results.

Average Deviation from the Mean

In the past, the precision of experimental data was often reported in terms of *average deviation from the mean* because this quantity is so easy to compute. With the advent of handheld calculators and computers, however, it is equally easy to derive the two statistically significant measures of precision, namely, standard deviation and variance. As a consequence, the average deviation from the mean has become obsolete. Nevertheless, we include its definition for those who may encounter it in the older literature.

The *deviation from the mean* $(x_i - \bar{x})$ is simply the numerical difference

[3]In most cases, the first two or three digits in a set of data are identical to each other. Thus, as an alternative to using Equation 2-2, these identical digits can be dropped and the remaining digits used with Equation 2-5. For example, the standard deviation for the data in Example 2-2 could be based on 0.082, 0.086, 0.082, and so forth or even 82, 86, 82, and so forth.

[4]See H. E. Solbert, *Anal. Chem.*, **1983,** *55*, 1661; and P. M. Wanek *et al.*, *Anal. Chem.*, **1982,** *54*, 1877.

between an individual result x_i in a set of data and the mean of the set. The *average deviation from the mean* for a set is obtained by summing the individual deviations from the mean *without regard to sign* (if the signs are retained, the sum will equal zero) and dividing the sum by the number of data N in the set. Column 2 of Table 2-1 lists the individual deviations from the mean for the data plotted in Figure 2-1. Note that the sign of these differences has been ignored and that an extra digit has been retained in the computed mean to minimize rounding error. The average deviation is computed by dividing the sum of the deviations (1.70) by 6, giving an average deviation from the mean of 0.28 ppm Fe.

2A-3

Accuracy

The term *accuracy* denotes that closeness of a measurement or set of measurements to the accepted value and is expressed in terms of *error*. Note the fundamental difference between accuracy and precision. Accuracy measures agreement between a result and what is believed to be its true value, whereas precision describes the agreement among several results acquired in the same way. Precision can always be determined by simply replicating a measurement. In contrast, accuracy can never be determined exactly because the true value of a quantity can never be known exactly, and an accepted value must be used instead.

Absolute Error

The *absolute error* E in the measurement of a quantity x_i is given by the equation

$$E = x_i - x_t \tag{2-7}$$

where x_t is the true, or accepted, value of the quantity. Returning to our example in Table 2-1, we see that the absolute error is -0.2 ppm Fe. Note that the sign is retained (in contrast to measures of precision).[5] Thus, the negative sign in the example indicates that the experimental result is smaller than the accepted value.

Relative Error

Often, the *relative error* E_r is a more useful quantity than the absolute error. The percent relative error is given by the expression

$$E_r = \frac{x_i - x_t}{x_t} \times 100 \tag{2-8}$$

Relative error is also expressed in parts per thousand (ppt). Thus, the relative error for the mean of the data in Table 2-1 could be reported as -1.1% or as -11 ppt.

2A-4

Types of Errors in Experimental Data

The precision of a measurement is readily determined by comparing data from carefully replicated experiments. Unfortunately, an estimate of the accuracy is not equally available because this quantity requires sure knowledge of the very information that is being sought, namely, the true value. It is tempting to ascribe a direct relationship between precision and accuracy. The danger of this approach is illustrated in Figure 2-2, which summarizes the results for the determination of nitrogen in two pure compounds. The dots

[5] Note that the term absolute in this context has a meaning different from that used in mathematics.

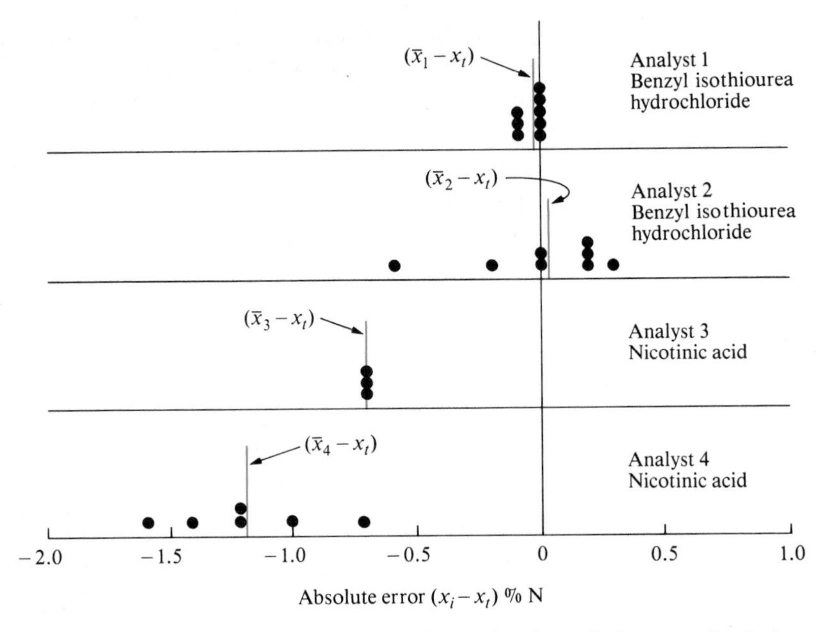

FIGURE 2-2 Absolute errors in the micro-Kjeldahl determination of nitrogen. Each dot represents the error associated with a single measurement. Each vertical line labeled $(\bar{x}_i - x_t)$ is the absolute average deviation of the set from the true value. (Data from C. O. Willits and C. L. Ogg, *J. Assoc. Offic. Anal. Chem.*, **1949**, *32*, 561. With permission.)

show the absolute errors of replicate results obtained by four analysts. Note that analyst 1 obtained relatively high precision and high accuracy. Analyst 2 had poor precision but good accuracy. The results of analyst 3 are of a kind that is by no means uncommon; the precision is excellent, but a significant error exists in the numerical average for the data. The scientist also encounters situations similar to that recorded by analyst 4, in which both precision and accuracy are poor.

The behavior illustrated by Figures 2-1 and 2-2 can be rationalized by assuming that a chemical analysis can be affected by at least two types of errors. One type, called *indeterminate* (or *random*) *error,* causes data to be scattered more or less symmetrically around a mean value. Referring again to Figure 2-2, notice that the indeterminate errors associated with the work of analysts 1 and 3 are significantly less than those of analysts 2 and 4. Generally, the indeterminate errors in an analysis are reflected in the precision of the data. For example, the relative standard deviation of 18 ppt Fe computed in Table 2-1 is a valid measure of the indeterminate errors associated with this method for determining iron.

A second type of error, called *determinate* (or *systematic*) *error*, causes the mean of a set of data to differ from the accepted value. For example, the data in Figure 2-1 appear to have a determinate error of about −0.22 ppm Fe. The results of analysts 1 and 2 in Figure 2-2 appear to be largely free of determinate errors, whereas the data of analysts 3 and 4 exhibit determinate errors of about −0.7 and −1.2% nitrogen.

A third type of error, which is less common than the two just described, is termed *gross error.* Gross errors differ from indeterminate and determinate errors in the respect that the former usually occur only occasionally, are often large, and may cause a result to be either high or low. Gross errors lead to *outliers*—results that appear to differ markedly from all other data in a set of replicate measurements. No evidence of a gross error is found in Figures 2-1 and 2-2.

2B Determinate Errors

Determinate errors have a definite source that can in principle be identified. Generally, they cause all the results from replicate measurements to be either high or low. Because of this unidirectional character, determinate errors are also called *systematic errors.* For example, the last two sets of data in Figure 2-2 appear to be affected by a negative determinate error. The source of this systematic error can be traced to the chemical nature of the sample, nicotinic acid. The analytical method used involves decomposition of the samples with hot concentrated sulfuric acid, which converts the nitrogen to ammonium sulfate; the amount of ammonia in the latter is then determined. It has been found, however, that compounds containing a pyridine ring (such as nicotinic acid) are incompletely decomposed by sulfuric acid unless special precautions are taken; low results are the consequence. It is highly likely that the negative errors ($\bar{x}_3 - x_t$ and $\bar{x}_4 - x_t$) in Figure 2-2 are determinate and attributable to incomplete decomposition.

2B-1 Sources of Determinate Errors

Determinate errors are of three types: (1) *instrument errors,* which are attributable to imperfections in measuring devices and instabilities in their power supplies, (2) *method errors,* which arise from nonideal chemical or physical behavior of analytical systems, and (3) *personal errors,* which are caused by the carelessness, inattention, or physical or psychological limitations of the experimenter.

Instrument Errors

All measuring devices are potential sources of determinate errors. For example, pipets, burets, and volumetric flasks frequently deliver or contain volumes slightly different from those indicated by their graduations. These differences may have such origins as use of glassware at a temperature that differs significantly from the calibration temperature, distortions in container walls due to heating while drying, errors in the original calibration, or contaminants on the inner surfaces of the containers. Most determinate errors of this type are readily eliminated by calibration.

Instruments powered by electricity are commonly subject to determinate errors. These uncertainties arise from decreased voltage of battery-operated power supplies with use, increased resistance in circuits because of dirty electrical contacts, temperature effects on resistors and standard potential sources, and currents induced from 110-V power lines. Again, these errors are detectable and correctable; most are unidirectional.

Method Errors

Determinate method errors are often introduced as a result of the nonideal chemical or physical behavior of the reagents and reactions upon which an analysis is based. Such sources of nonideality include the slowness of some reactions, the incompleteness of others, the instability of some species, the nonspecificity of most reagents, and the possible occurrence of side reactions that interfere with the measurement process. For example, a common method error in volumetric methods results from the small excess of reagent required to cause an indicator to undergo the color change that signals completion of the reaction. The ultimate accuracy of such an analysis is thus limited by the very phenomenon that makes the titration possible.

Another example of method error was described earlier in connection with the data in Figure 2-2c and d. Here, the source was the incompleteness of the reaction between the pyridine ring in nicotinic acid and the sulfuric acid.

Errors inherent in a method are frequently difficult to detect and are thus the most serious of the three types of determinate error.

Personal Errors

Many measurements require personal judgments. Examples include estimating the position of a pointer between two scale divisions, the color of a solution at the end point in a titration, or the level of a liquid with respect to a graduation in a pipet or buret. Judgments of this type are often subject to systematic, unidirectional errors. For example, one person may read a pointer consistently high, another may be slightly slow in activating a timer, and a third may be less sensitive to color changes and thus tend to employ excess reagent in a volumetric analysis. Color blindness or other physical handicaps often exacerbate personal determinate errors.

A near-universal source of personal error is prejudice, or *bias.* Most of us, no matter how honest, have a natural tendency to estimate scale readings in a direction that improves the precision in a set of results or causes the results to fall closer to a preconceived notion of the true value for the measurement. Number bias is another source of personal error that is widely encountered and varies considerably from person to person. The most common bias encountered in estimating the position of a needle on a scale involves a preference for the digits 0 and 5. Also prevalent is a prejudice favoring small digits over large and even numbers over odd.

Scientists must actively fight bias; it does not suffice to assume that problems occur only in others.

2B-2 The Effect of Determinate Errors upon Analytical Results

Determinate errors may be either *constant* or *proportional.* The magnitude of a constant error is independent of the size of the quantity measured, whereas proportional errors increase or decrease in proportion to the size of the sample taken for analysis.

Constant Errors

Constant errors become more serious as the size of the quantity measured decreases. This behavior is illustrated by the effect of solubility losses on the results of a gravimetric analysis.

EXAMPLE 2-3

Suppose that 0.50 mg of precipitate is lost as a result of being washed with 200 mL of wash liquid. If 500 mg of precipitate is involved, the relative error due to solubility loss is $-(0.50/500) \times 100\% = 0.1\%$. Loss of the same quantity from 50 mg of precipitate results in a relative error of -1.0%.

The excess of reagent required to bring about a color change during a titration is another example of constant error. This volume, usually small, remains the same regardless of the total volume of reagent required for the titration. Again, the relative error from this source becomes more serious as the total volume decreases. Clearly, one way of minimizing the effect of constant error is to use as large a sample as possible.

Proportional Errors

A common cause of proportional errors is the presence of interfering contaminants in the sample. For example, a widely used method for the determination of copper is based upon the reaction of copper(II) ion with potassium iodide to give iodine, the amount of which is then measured. Iron(III), if present, also liberates iodine from potassium iodide. Unless steps are taken to prevent this interference, high results are observed for the percentage of copper because the iodine produced will be a measure of the sum of the copper(II) and iron(III) in the sample. The magnitude of this error is fixed by the *fraction* of iron contamination, which is independent of the size of sample taken. If the sample size is doubled, for example, the amount of iodine liberated by both the copper and the iron contaminant is also doubled. Thus, the reported percentage of copper is independent of sample size.

2B-3 Detection of Determinate Instrument and Personal Errors

Determinate instrument errors are usually found and corrected by calibration. Indeed, periodic calibration of equipment is always desirable because the response of most instruments changes with time owing to wear, corrosion, or mistreatment.

Most personal errors can be minimized by care and self-discipline. Thus, most scientists develop the habit of systematically checking instrument readings, notebook entries, and calculations. Errors that result from a physical handicap can usually be avoided by a judicious choice of method—provided, of course, that the handicap is recognized.

2B-4 Detection of Determinate Method Errors

Determinate method errors are particularly difficult to detect. Identification and compensation for systematic errors of this type may take one or more of the courses described in the following paragraphs.

Analysis of Standard Samples

A method may be tested for determinate error by analysis of synthetic samples whose overall composition is known and closely approximates that of the material for which the analysis is intended. Great care must go into the preparation of standard samples to ensure that the concentration of the constituent to be determined is known with a high degree of certainty. Unfortunately, the preparation of a standard sample whose composition truly resembles that of a complex natural substance may be difficult, if not impossible. Moreover, the problem is compounded by the requirement that the standard must be prepared in such a way that the concentration of one of its components is known exactly. These problems are frequently so imposing as to prevent the use of this approach.

Several hundred common substances that have been carefully analyzed for one or more constituents are available from the National Bureau of Standards. These standard materials are valuable for testing analytical procedures for accuracy.[6] In addition, several commercial supply houses now offer a variety of analyzed materials for method testing.[7]

Independent Analysis

A valuable way of detecting method errors when samples of known composition are unavailable involves parallel analysis of typical samples by an independent method of established reliability. In general, the independent method should differ as much as possible from the one under study in order to minimize the possibility that some common factor in the sample has the same effect on both methods.

Blank Determinations

Blank determinations, in which all steps of the analysis are performed in the absence of a sample, are frequently useful for detecting certain types of constant errors. The results are then applied as a correction to the sample measurements. Blank determinations are of particular value in exposing errors that are due to interfering contaminants from the reagents and vessels employed in an analysis. Blanks also enable the analyst to correct titration data for the volume of reagent needed to cause an indicator to change color at the end point.

Variation in Sample Size

As demonstrated by Example 2-3, a constant error has a decreasing effect on a result as the size of the measurement increases. Thus, constant errors can often be detected by varying the sample size as widely as practicable.

2C Gross Errors

Most gross errors are personal and arise from carelessness, laziness, or ineptitude on the part of the experimenter. Gross errors are sometimes random but occur so infrequently that they are generally not included as indeterminate errors. Sources of gross errors include arithmetic mistakes, transposition of numbers in recording data, reading a scale backward, reversing a sign, using a wrong scale, spilling a solution, or just bad luck. Some gross errors affect only a single result; others, such as using the wrong scale of an instrument, affect an entire set of replicate measurements. Most gross personal errors are the consequence of carelessness and can be eliminated by self-discipline. Many scientists always follow the practice of re-reading an instrument after the information has been recorded and then checking the new reading against the one that has been recorded.

Gross errors are also encountered as a result of momentary interruptions in power or water supplies and other unexpected events of this kind.

[6]See U.S. Department of Commerce, *NBS Standard Reference Materials*, 1986–87, ed., NBS Special Publication 260. Washington: Government Printing Office, 1986. For a description of the reference material programs of the NBS, see R. A. Alvarez, S. D. Rasberry, and G. A. Uriano, *Anal. Chem.*, **1982,** *54,* 1226A; and G. A. Uriano, *ASTM Standardization News*, **1979,** *7,* 8.

[7]For sources of biological and environmental reference materials containing various elements, see C. Veillon, *Anal. Chem.*, **1986,** *58,* 851A.

2D

Indeterminate Errors

Indeterminate, or *random, errors* arise when a system of measurement is extended to its maximum sensitivity. They are the direct consequence of the many uncontrollable variables that inevitably exist in every physical or chemical measurement. While the sources of this type of error are many, none can be positively identified or measured because most are so small as to be undetectable individually. The accumulated effect of the individual indeterminate errors, however, causes the data from replicate measurements to fluctuate randomly around the mean of the set. For example, the scatter of data in Figures 2-1 and 2-2 is a direct consequence of an accumulation of small indeterminate errors. Note that, in Figure 2-2, there appears to be a larger uncertainty associated with the work of analysts 2 and 4 than with that of analysts 1 and 3.

2D-1

Sources of Indeterminate Error

A qualitative notion of the way small errors affect the outcome of replicate measurements can be developed by considering an imaginary situation in which just four small errors give rise to an overall uncertainty. We assume that each of these errors has an equal probability of occurring and that each can cause the final result to be in error by a fixed amount $\pm U$.

Table 2-2 shows all the possible ways the four errors can combine to give the indicated deviations from the mean (that is, the indeterminate uncertainty). Note that only one combination of errors leads to a deviation of $+4U$, whereas four combinations give a deviation of $+2U$ and six a deviation of $0U$. The same relationship exists for negative indeterminate errors. This ratio of 6:4:1 is a measure of the probability of a deviation of each magnitude; if we make a sufficiently large number of measurements, we can expect a frequency distribution like that shown in Figure 2-3a. Note that the ordinate in the plot is the relative frequency of occurrence of the five possible combinations.

TABLE 2-2

Possible Combinations of Four Equal-Sized Uncertainties

Combinations of Uncertainties	Magnitude of Indeterminate Error	Number of Combinations	Relative Frequency
$+U_1 + U_2 + U_3 + U_4$	$+4U$	1	0.0625
$-U_1 + U_2 + U_3 + U_4$ $+U_1 - U_2 + U_3 + U_4$ $+U_1 + U_2 - U_3 + U_4$ $+U_1 + U_2 + U_3 - U_4$	$+2U$	4	0.250
$-U_1 - U_2 + U_3 + U_4$ $+U_1 + U_2 - U_3 - U_4$ $+U_1 - U_2 + U_3 - U_4$ $-U_1 + U_2 - U_3 + U_4$ $-U_1 + U_2 + U_3 - U_4$ $+U_1 - U_2 - U_3 + U_4$	0	6	0.375
$+U_1 - U_2 - U_3 - U_4$ $-U_1 + U_2 - U_3 - U_4$ $-U_1 - U_2 + U_3 - U_4$ $-U_1 - U_2 - U_3 + U_4$	$-2U$	4	0.250
$-U_1 - U_2 - U_3 - U_4$	$-4U$	1	0.0625

FIGURE 2-3 Frequency distribution for measurements containing (a) four indeterminate uncertainties; (b) ten indeterminate uncertainties; (c) a very large number of indeterminate uncertainties.

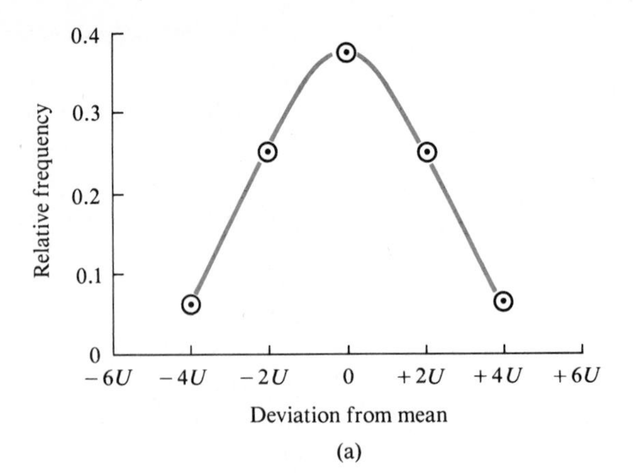

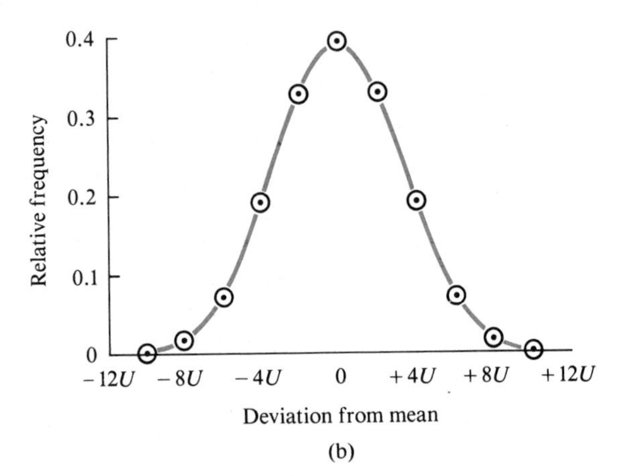

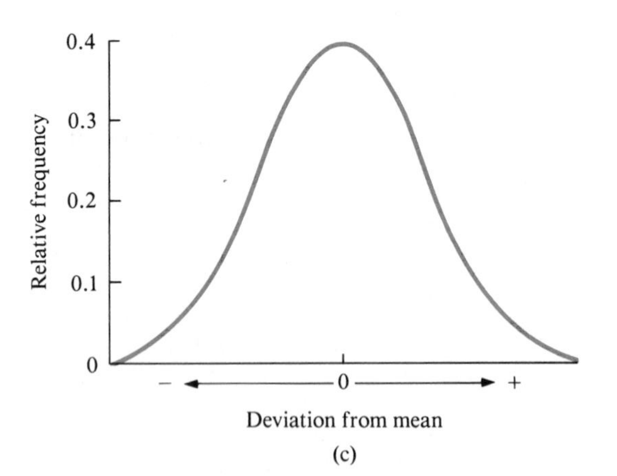

Figure 2-3b shows the theoretical distribution for ten equal-sized uncertainties; again we see that the most frequent occurrence is zero deviation from the mean, whereas a maximum deviation of $10U$ occurs only occasionally (about once in 500 measurements).

When the foregoing arguments are extended to a very large number of individual errors, a curve like that shown in Figure 2-3c results. This bell-shaped curve is called a *Gaussian curve* or a *normal error curve.*[8]

2D-2

Distribution of Experimental Data

It has been found empirically that the distribution of replicate data from most quantitative analytical experiments tends to approach that of the Gaussian curve shown in Figure 2-3c. For example, consider the data in Table 2-3 for the calibration of a pipet having a nominal volume of 10 mL. The experiment involved weighing a small flask and stopper, transferring 10 mL of water to the flask with the pipet, and weighing again. The temperature of the water was also measured to establish its density. The experimentally determined weight difference was then readily converted to the volume delivered by the pipet.

The data in Table 2-3 are typical of those that might be obtained by an experienced and competent worker weighing to the nearest milligram (which corresponds to 0.001 mL) on a top-loading balance and making every effort to recognize and eliminate determinate error. Even so, the standard deviation of the 50 measurements is 0.0056 mL and the spread is 0.025 mL. This

TABLE 2-3

Replicate Data on the Calibration of a 10-mL Pipet*

Trial	Vol, mL	Trial	Vol, mL	Trial	Vol, mL
1	9.988	18	9.975	35	9.976
2	9.973	19	9.980	36	9.990
3	9.986	20	9.994‡	37	9.998
4	9.980	21	9.992	38	9.971
5	9.975	22	9.984	39	9.986
6	9.982	23	9.981	40	9.978
7	9.986	24	9.987	41	9.986
8	9.982	25	9.978	42	9.982
9	9.981	26	9.983	43	9.977
10	9.990	27	9.982	44	9.977
11	9.980	28	9.991	45	9.986
12	9.989	29	9.981	46	9.978
13	9.978	30	9.969†	47	9.983
14	9.971	31	9.985	48	9.980
15	9.982	32	9.977	49	9.983
16	9.983	33	9.976	50	9.979
17	9.988	34	9.983		

Mean volume = 9.982 mL

Median volume = 9.982 mL

Spread = 0.025 mL

Standard deviation = 0.0056 mL

*Data listed in the order obtained.

†Minimum value.

‡Maximum value.

[8] We have developed our argument for an example in which all the uncertainties have the same magnitude. Such a restriction is not needed in deriving the equation for a Gaussian curve.

distribution of data about the mean is the direct consequence of indeterminate errors.

The information in Table 2-3 is more easily comprehended if the data are rearranged into frequency distribution groups, as in Table 2-4. Here, the number of data falling into a series of contiguous 0.003-mL *cells* are recorded as well as the percent of measurements falling into each cell. Note that 26% of the data reside in the cell that encompasses the mean and median value of 9.982 mL and that more than half the data are within ±0.004 mL of this mean. It is also of interest to note that 72% of the data are within ±0.0056 mL, or one standard deviation, of the mean (as shown later, the theoretical percent in this range is 68).

The frequency distribution data in Table 2-4 are plotted as a bar graph, or *histogram*, in Figure 2-4*A*. Also shown in this figure is a theoretical Gaussian curve (Figure 2-4*B*) derived for an infinite set of data having the same mean (9.982 mL), the same standard deviation (0.0056 mL), and the same area under the curve as the histogram. Note that, as the number of calibration experiments increases and the size of the cells decreases, the histogram approaches more and more closely the shape of the continuous curve.

Variations in replicate results such as those in Table 2-3 can be rationalized by assuming that each measurement is affected by numerous small and individually undetectable instrument, method, and personal indeterminate errors attributable to uncontrollable variables in the experiment. The cumulative effect of such errors is likewise indeterminate. Ordinarily, the small errors tend to cancel one another and thus exert a minimal effect. Occasionally, however, they act in concert to produce a relatively large positive or negative error.

Sources of uncertainty in the calibration of a pipet include such visual judgments as the level of the water with respect to the marking on the pipet and the mercury level in the thermometer (both personal indeterminate errors). Other sources are variation in the drainage time and in the angle of the pipet as it drains (both method errors). Instrument errors arise from temperature fluctuations, which affect (1) the volume of the pipet, (2) the viscosity of the liquid, and (3) the performance of the balance. Other sources of instrument error are vibrations and drafts that cause small variations in the balance reading. Numerous other sources of indeterminate error undoubtedly exist. Thus, it is clear that many small and uncontrollable variables affect even

TABLE 2-4

Frequency Distribution of Data from Table 2-3

Volume Range, mL	Number in Range	% in Range
9.969 to 9.971	3	6
9.972 to 9.974	1	2
9.975 to 9.977	7	14
9.978 to 9.980	9	18
9.981 to 9.983	13	26
9.984 to 9.986	7	14
9.987 to 9.989	5	10
9.990 to 9.992	4	8
9.993 to 9.995	1	2

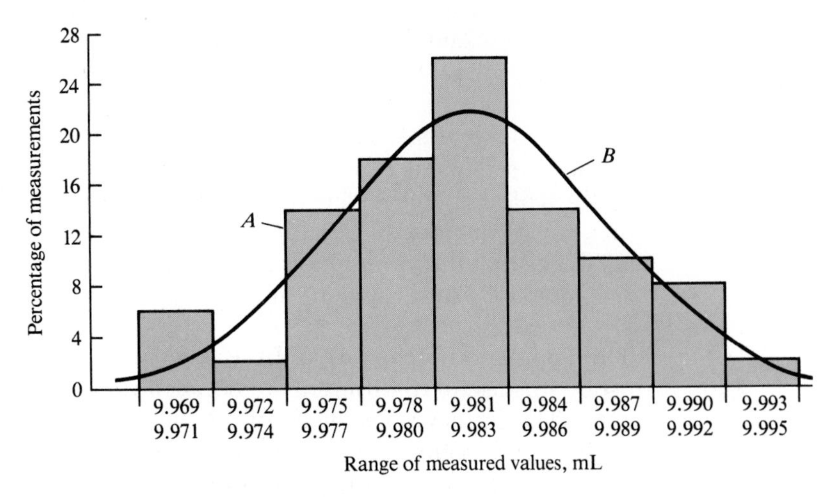

FIGURE 2-4

A Histogram showing distribution of the 50 results in Table 2-3. *B* Gaussian curve for data having the same mean and same standard deviation as the data in *A*.

as simple a process as calibrating a pipet. We are unable to account for the influence of any one of the indeterminate errors arising from these variables, but their cumulative effect is responsible for the scatter of data around the mean.

2D-3 The Statistical Treatment of Indeterminate Errors

Statistics is the mathematical science that deals with chance variations. As we have just shown, the magnitude of the indeterminate error associated with an individual measurement is determined by a chance combination of individual errors that may be positive or negative. As a consequence, the effects of indeterminate errors on replicate sets of experimental data are readily treated by the laws of statistics.[9]

It should be emphasized at the outset that statistics cannot reveal any information that is not already implicit in a set of data. It does, however, permit objective judgments that are otherwise difficult, particularly when large sets of data are at hand.

The Population and the Sample

In the statistical treatment of data, it is assumed that the handful of replicate experimental results the scientist obtains in the laboratory is a tiny fraction of an infinite number of results that could in theory be obtained, given infinite time for the measurements. Statisticians call this limited set of data a *sample* and view it as a subset of a *population* or a *universe* of data that in principle

[9]References dealing with application of statistics include R. Calcutt and R. Boddy, *Statistics for Analytical Chemists*. New York: Chapman and Hall, 1983; O. L. Davies and P. L. Goldsmith, *Statistical Methods in Research and Production*, 4th ed. New York: Longman, 1976; J. Mandel, *Treatise on Analytical Chemistry*, 2nd ed., I. M. Kolthoff and P. J. Elving, Eds., Part I, Vol. 1, Chapter 5. New York: Wiley, 1978.

exists.[10] For example, the data in Table 2-3 make up a sample of an infinite population of calibration experiments that can be imagined (but not performed).

The laws of statistics apply strictly to a population of data only. In applying these laws, it is necessary to assume that the handful of data making up the typical sample is truly representative of the infinite population of results that could, in principle, be derived in the same way. Because there is no assurance that this assumption is valid, statements about indeterminate errors are necessarily uncertain and must be couched in terms of probabilities.

The Population Mean (μ) and the Sample Mean ($\bar{x}$). In discussions of statistics, it is useful to differentiate between the *sample mean* and the *population mean.* The former is the mean of a limited sample drawn from a population of data. It is defined by Equation 2-1, where N is usually a relatively small number. The population mean, in contrast, is the true mean for the entire population (and the true value in the absence of determinate errors); here, N in Equation 2-1 is a number that approaches infinity. In order to emphasize the difference between the two means, the sample mean is symbolized by $\bar{x}$ and the population mean by μ. More often than not, particularly when N is small, $\bar{x}$ differs from μ because a small sample of data is unlikely to be truly representative of a population.

The Population Standard Deviation (σ) and the Sample Standard Deviation(s). Statisticians also differentiate between the *sample standard deviation* and the *population standard deviation,* which is the true standard deviation of an infinite set of data. The population standard deviation σ is defined by the equation

$$\sigma = \sqrt{\frac{\sum_{i=1}^{N}(x_i - \mu)^2}{N}} \tag{2-9}$$

In contrast, the sample standard deviation s is given by Equation 2-2:

$$s = \sqrt{\frac{\sum_{i=1}^{N}(x_i - \bar{x})^2}{N - 1}}$$

Note that this equation differs from Equation 2-9 in two regards. First, the sample mean $\bar{x}$ appears in the numerator of Equation 2-2 in place of the population mean μ in Equation 2-9. Second, the number of degrees of freedom $(N - 1)$ appears in the denominator of Equation 2-2 in place of the N term of Equation 2-9.

When Equation 2-9 is applied to several small samples of data drawn from the same population, it becomes evident that the calculated values for the standard deviation are on the average lower than the standard deviation

[10]Note that, in the context of statistics, the term *sample* refers to a set of replicate measurements. Thus, an analysis performed on four *chemical samples* represents a single *statistical sample.* Unfortunately, this dual use of *sample* sometimes leads to confusion in applying statistics to analytical chemistry.

for the population. That is, a *negative bias* develops.[11] This bias, which is attributable to the need to extract both a mean and a standard deviation from a small set of data, can be largely eliminated by substituting the *degrees of freedom* $N - 1$ for N in Equation 2-9. The efficacy of this substitution is demonstrated by the data in column 3 in footnote 11.

The rationale for the use of $N - 1$ in Equation 2-2 is as follows. When σ is unknown, two quantities must be extracted from the set of data: $\bar{x}$ and s. One degree of freedom is lost in establishing $\bar{x}$ because, with their signs retained, the sum of the individual deviations must add up to zero. Thus, when $N - 1$ deviations have been computed, the final one is known. Consequently, only $N - 1$ deviations provide an *independent* measure of the precision of the set.

Properties of the Normal Error Curve

Figure 2-5a shows two Gaussian curves in which the relative frequency of occurrence of various deviations from the mean is plotted as a function of deviation from the mean $(x - \mu)$ for two populations of data that differ only in standard deviation. The standard deviation for the population yielding the broader but lower curve (B) is twice that for the population yielding curve A.

Figure 2-5b shows another type of normal error curve, in which the abscissa is now a new variable z, which is defined as

$$z = (x - \mu)/\sigma \qquad (2\text{-}10)$$

Note that z is the standard deviation of the data stated *in units of standard deviation*. That is, when $x - \mu = \sigma$, z is equal to one standard deviation; when $x - \mu = 2\sigma$, z is equal to two standard deviations; and so forth. Since z is the deviation of the mean in standard deviation units, a plot of relative frequency versus this parameter yields a single Gaussian curve that describes *all* populations of data regardless of standard deviation. Thus, Figure 2-5b is the normal error curve for both sets of data used to plot curves A and B in Figure 2-5a.

General properties of normal error curves include (1) zero deviation from the mean occurring with maximum frequency, (2) symmetrical distribution of positive and negative deviations about this maximum, and (3) exponential decrease in frequency as the magnitude of the deviations increases. Thus, small indeterminate uncertainties are observed much more often than very large ones.

[11]The negative bias can be demonstrated by dividing the data in Table 2-3 into 10 samples of 5 data each, 5 samples of 10 data each, and 2 samples of 25 data each. When σ and s are calculated for each sample using Equations 2-9 and 2-2, the results are

Number of Samples and Size	Mean σ of Samples (Equation 2-9)	Mean s of Samples (Equation 2-2)
10 samples of 5	0.0051	0.0056
5 samples of 10	0.0052	0.0057
2 samples of 25	0.0055	0.0056
1 sample of 50	0.0056	0.0056

The negative bias that accompanies application of Equation 2-9 to small sets of data is reflected in the data in column 2. Note in column 3 that this bias disappears when s is calculated with Equation 2-2.

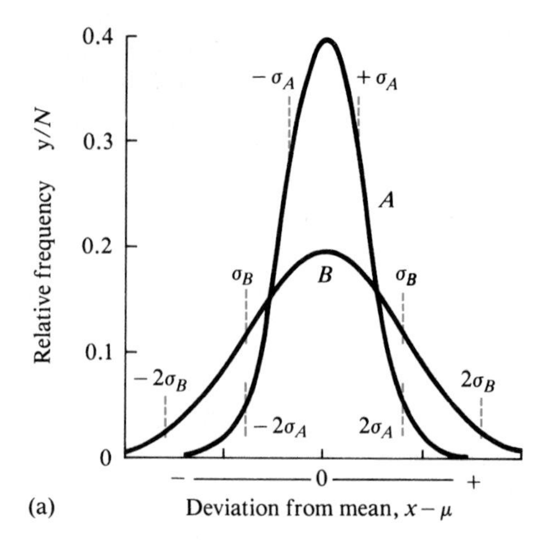

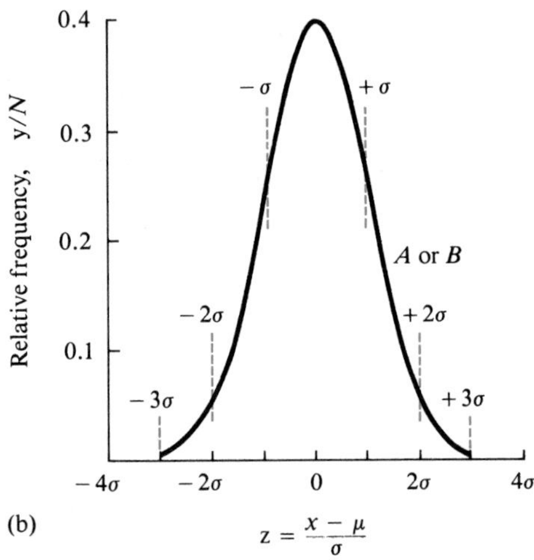

FIGURE 2-5 Normal error curves. The standard deviation for B is twice that for A, that is, $\sigma_B = 2\sigma_A$. (a) The abscissa is the deviation from the mean in the units of measurement. (b) The abscissa is the deviation from the mean in units of σ. Thus, A and B produce identical curves.

Areas Under a Normal Error Curve. It can be shown that, regardless of its breadth, 68.3% of the area beneath a normal error curve lies within one standard deviation ($\pm 1\sigma$) of the mean μ. Thus, 68.3% of the data making up the population lie within these bounds. Furthermore, approximately 95.5% of all data are within $\pm 2\sigma$ of the mean and 99.7% within $\pm 3\sigma$; vertical dashed lines show the areas encompassed by $\pm 1\sigma$ and $\pm 2\sigma$ in Figure 2-5a.

Because of area relationships such as these, the standard deviation of a population of data is a useful predictive tool. For example, one can assert that the chances are 68.3 in 100 that the indeterminate uncertainty of any single measurement is no more than $\pm 1\sigma$. Similarly, the chances are 95.5 in 100 that the error is less than $\pm 2\sigma$, and so forth.

Standard Error of a Mean. The figures on percentage distribution just quoted refer to the probable error for a *single* measurement. If a series of samples, each containing N data, are taken randomly from a population of data, the mean of each set will show less and less scatter as N increases. The standard deviation of each mean is known as the *standard error* of the mean and is given the symbol σ_m. It can be shown that the standard error is inversely proportional to the square root of the number of data N used to calculate the mean:

$$\sigma_m = \sigma/\sqrt{N} \tag{2-11}$$

where σ is defined by Equation 2-9. An analogous equation can be written for a sample standard deviation:

$$s_m = s/\sqrt{N} \tag{2-12}$$

Properties of the Standard Deviation

Effect of N on the Reliability of s. Uncertainty in the calculated value of s decreases as N in Equation 2-2 increases. In fact, when N is greater than about 20, s and σ can be assumed to be identical for all practical purposes. For example, if the 50 measurements in Table 2-3 are divided into 10 subgroups of 5 each, the value of s varies widely from one subgroup to another (0.0023 to 0.0079 mL) even though the average is that of the entire set (0.0056 mL). In contrast, the computed values of s for two subsets of 25 each are nearly identical (0.0054 and 0.0058 mL).

The rapid improvement in the reliability of s with increases in N makes it feasible to obtain a good approximation of σ when the method of measurement is not excessively time-consuming and when an adequate supply of sample is available. For example, if the pH of numerous solutions is to be measured in the course of an investigation, it might prove worthwhile to evaluate s in a series of preliminary experiments. This particular measurement is simple, requiring only that a pair of rinsed and dried electrodes be immersed in the test solution; the potential across the electrodes is proportional to pH. To determine s, 20 to 30 portions of a buffer solution of fixed pH could be measured with all steps of the procedure being followed exactly. Normally, it is safe to assume that the indeterminate error in this test is the same as that in subsequent measurements and that the value of s calculated by means of Equation 2-2 is a valid and accurate measure of the theoretical σ.

Pooling Data to Improve the Reliability of s. For analyses that are time-consuming, the foregoing procedure is not ordinarily practical. Here, however, precision data from a series of samples accumulated over time can often be pooled to provide an estimate of s that is superior to the value for any individual subset. Again, one must assume the same sources of indeterminate error in all the samples. This assumption is usually valid if the samples have similar compositions and have been analyzed identically.

To obtain a pooled estimate of s, deviations from the mean for each subset are squared; the squares of all subsets are then summed and divided by an appropriate number of degrees of freedom. The pooled s is obtained by extracting the square root of the quotient. One degree of freedom is lost for each subset. Thus, the number of degrees of freedom for the pooled s is equal to the total number of measurements minus the number of subsets.

EXAMPLE 2-4

The mercury in samples of seven fish taken from the Sacramento River was determined by a method based upon the absorption of radiation by gaseous elemental mercury. Calculate a pooled estimate of the standard deviation for the method, based upon the first three columns of data:

Specimen Number	Number of Samples Measured	Hg Content, ppm	Mean, ppm Hg	Sum of Squares of Deviations from Mean
1	3	1.80, 1.58, 1.64	1.673	0.0258
2	4	0.96, 0.98, 1.02, 1.10	1.015	0.0115
3	2	3.13, 3.35	3.240	0.0242
4	6	2.06, 1.93, 2.12, 2.16, 1.89, 1.95	2.018	0.0611
5	4	0.57, 0.58, 0.64, 0.49	0.570	0.0114
6	5	2.35, 2.44, 2.70, 2.48, 2.44	2.482	0.0685
7	4	1.11, 1.15, 1.22, 1.04	1.130	0.0170
Number of measurements = 28			Sum of squares =	0.2196

The values in the last two columns for sample 1 were computed as follows:

x_i	$\lvert(x_i - \bar{x})\rvert$	$(x_i - \bar{x})^2$
1.80	0.127	0.0161
1.58	0.093	0.0086
1.64	0.033	0.0011
5.02	Sum of squares =	0.0258

$$\bar{x} = \frac{5.02}{3} = 1.673$$

The other data in columns 4 and 5 were obtained similarly. Then

$$s_{\text{pooled}} = \sqrt{\frac{0.0258 + 0.0115 + 0.0242 + 0.0611 + 0.0114 + 0.0685 + 0.0170}{28 - 7}}$$

$$= 0.10 \text{ ppm Hg}$$

Note that one degree of freedom is lost for each of the seven samples. Because more than 20 degrees of freedom remain, however, the computed value of s can be considered a good approximation of σ; that is, $s \to \sigma \to 0.10$ ppm Hg.

2E The Uses of Statistics

Experimentalists employ statistical calculations to sharpen their judgment concerning the effects of indeterminate errors. The most common applications of statistics to analytical chemistry include:

1. Definition of the interval around the mean of a set within which the population mean can be expected to be found with a given probability.
2. Determination of the number of replicate measurements required to assure (with a certain probability) that an experimental mean will fall within a predetermined interval around the population mean.
3. Guidance concerning whether an outlying value in a set of replicate results should be retained or rejected in calculating the mean for the set.
4. Estimation of the probability that two samples analyzed by the same method are significantly different in composition, that is, whether a

difference in experimental results is likely to be a consequence of indeterminate error or real composition difference.

5. Estimation of the probability that a difference in precision exists between two sets of data obtained by different workers or by different methods.
6. Treatment of calibration data.

We shall examine each of these applications in the sections that follow.

2E-1 *Confidence Limits*

The exact value of the mean μ for a population of data can never be determined exactly because such a determination requires that an infinite number of measurements be made. With the aid of statistical theory, however, it is possible to set limits about an experimentally determined mean $\bar{x}$ wherein μ is expected to lie with a given degree of probability. The limits obtained in this manner are called *confidence limits*, and the interval they define is known as the *confidence interval*.

The magnitude of the confidence interval, which is derived from the sample standard deviation, depends on the certainty with which s is known. If there is reason to believe that s is a good approximation of σ, then the confidence interval can be significantly narrower than if the estimate of s is based upon only two or three measurements.

The Confidence Interval When s Is a Good Approximation of σ

Figure 2-6 is a normal error curve in which the abscissa is the quantity z, which is the deviation from the mean in units of the population standard deviation (Equation 2-10). The column of numbers in the center gives the percent of the total area under the curve encompassed by the indicated values of $-z$ and $+z$. For example, 50% of the area under any Gaussian curve is located between -0.67σ and $+0.67\sigma$, 80% lies between -1.29σ and $+1.29\sigma$, and so forth. From these limits then, we may assert, with 80 chances out of 100 of being

FIGURE 2-6 Confidence levels for various values of z.

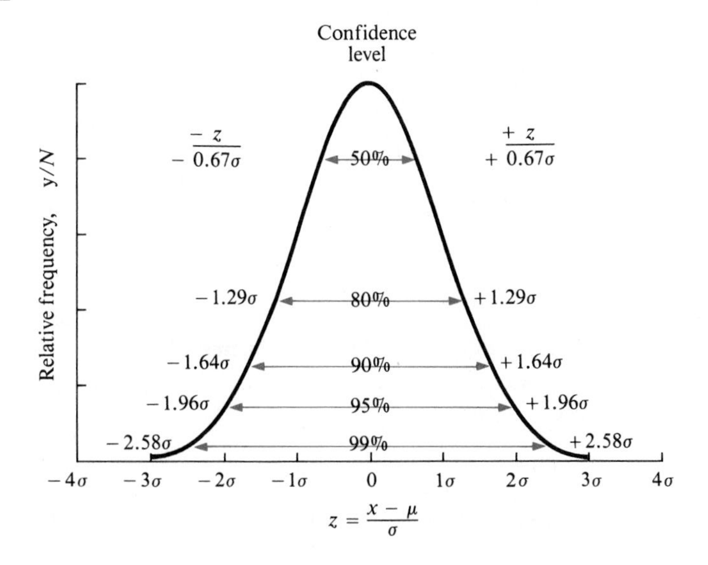

correct, that the population mean lies within $\pm 1.29\sigma$ of any single measurement we make. Here, the *confidence level* is 80% and the confidence interval is $\pm z\sigma = \pm 1.29\sigma$. A general statement for the confidence limits (CL) of a single measurement is obtained by rearranging Equation 2-10, remembering that z can take positive or negative values. Thus,

$$\text{CL for } \mu = x \pm z\sigma \tag{2-13}$$

Values for z at various confidence levels are found in Table 2-5.

EXAMPLE 2-5

Calculate the 50 and 95% confidence limits for the first entry (1.80 ppm Hg) in Example 2-4.

In that example, we calculated s to be 0.10 ppm Hg and had sufficient data to assume $s \rightarrow \sigma$. From Table 2-5, we see that $z = 0.67$ and 1.96 for the two confidence levels. Thus, from Equation 2-13,

$$50\% \text{ CL for } \mu = 1.80 \pm 0.67 \times 0.10 = 1.80 \pm 0.07$$

$$95\% \text{ CL for } \mu = 1.80 \pm 1.96 \times 0.10 = 1.80 \pm 0.20$$

From these calculations, we conclude that the chances are 50 in 100 that μ, the population mean (and, *in the absence of determinate error,* the true value), lies in the interval between 1.73 and 1.87 ppm Hg. Furthermore, there is a 95% chance that it lies in the interval between 1.60 and 2.00 ppm Hg.

Equation 2-13 applies to the result of *a single measurement.* Application of Equation 2-11 shows that the confidence interval is decreased by $\sqrt{N}$ for the average of N replicate measurements. Thus, a more general form of Equation 2-13 is

$$\text{CL for } \mu = \bar{x} \pm \frac{z\sigma}{\sqrt{N}} \tag{2-14}$$

EXAMPLE 2-6

Calculate the 50 and 95% confidence limits for the mean value (1.67 ppm Hg) for specimen 1 in Example 2-4. Again, $s \rightarrow \sigma = 0.10$.

For the three measurements,

$$50\% \text{ CL} = 1.67 \pm \frac{0.67 \times 0.10}{\sqrt{3}} = 1.67 \pm 0.04$$

$$95\% \text{ CL} = 1.67 \pm \frac{1.96 \times 0.10}{\sqrt{3}} = 1.67 \pm 0.11$$

Thus, the chances are 50 in 100 that the population mean is located in the

TABLE 2-5

Confidence Levels for Various Values of z

Confidence Level, %	z	Confidence Level, %	z
50	0.67	96	2.00
68	1.00	99	2.58
80	1.29	99.7	3.00
90	1.64	99.9	3.29
95	1.96		

interval from 1.63 to 1.71 ppm Hg and 95 in 100 that it lies between 1.56 and 1.78 ppm.

EXAMPLE 2-7

How many replicate measurements of specimen 1 in Example 2-4 are needed to decrease the 95% confidence interval to ±0.07 ppm Hg?

The pooled value of s is a good estimate of σ. For a confidence interval of ±0.07 ppm Hg, then, substitution into Equation 2-14 leads to

$$0.07 = \pm\frac{zs}{\sqrt{N}} = \pm\frac{1.96 \times 0.10}{\sqrt{N}}$$

$$\sqrt{N} = \pm\frac{1.96 \times 0.10}{0.07} = \pm 2.80$$

$$N = (\pm 2.8)^2 = 7.8$$

We thus conclude that eight measurements would provide a slightly better than 95% chance of the population mean lying within ±0.07 ppm of the experimental mean.

A consideration of Equation 2-14 indicates that the confidence interval for an analysis can be halved by employing the mean of four measurements. Sixteen measurements are required to narrow the limit by yet another factor of 2. It is apparent that a point of diminishing return is rapidly reached in acquiring additional data. Thus, the chemist ordinarily takes advantage of the relatively large gain afforded by averaging two to four measurements but can seldom afford the time required for further increases in confidence.

It is essential to keep in mind at all times that confidence intervals based on Equation 2-14 apply only *in the absence of determinate errors.*

The Confidence Limits When σ Is Unknown

The chemist is frequently faced with a situation in which limitations in time or amount of available sample preclude an accurate estimation of σ. Here, a single set of replicate measurements must provide not only a mean but also a precision estimate. As indicated earlier, s calculated from a small set of data may be subject to considerable uncertainty; thus, confidence limits are necessarily broader when a good estimate of σ is unavailable.

To account for the potential variability of s, use is made of the statistical parameter t, which is defined in a way analogous to z:

$$t = \frac{x - \mu}{s} \tag{2-15}$$

In contrast to z in Equation 2-10, the numerical value of t depends not only on the desired confidence level but also on the number of degrees of freedom available in the calculation of s. Table 2-6 provides values for t for a few degrees of freedom. More extensive tables are found in various mathematical and statistical handbooks. Note that t approaches z (Table 2-5) as the number of degrees of freedom becomes infinite.

The confidence limits for the mean $\bar{x}$ of N replicate measurements can be derived from t by an equation analogous to Equation 2-14:

$$\text{CL for } \mu = \bar{x} \pm \frac{ts}{\sqrt{N}} \tag{2-16}$$

TABLE 2-6

Values of t for Various Levels of Probability

Degrees of Freedom	Factor for Confidence Interval				
	80%	90%	95%	99%	99.9%
1	3.08	6.31	12.7	63.7	637
2	1.89	2.92	4.30	9.92	31.6
3	1.64	2.35	3.18	5.84	12.9
4	1.53	2.13	2.78	4.60	8.60
5	1.48	2.02	2.57	4.03	6.86
6	1.44	1.94	2.45	3.71	5.96
7	1.42	1.90	2.36	3.50	5.40
8	1.40	1.86	2.31	3.36	5.04
9	1.38	1.83	2.26	3.25	4.78
10	1.37	1.81	2.23	3.17	4.59
11	1.36	1.80	2.20	3.11	4.44
12	1.36	1.78	2.18	3.06	4.32
13	1.35	1.77	2.16	3.01	4.22
14	1.34	1.76	2.14	2.98	4.14
∞	1.29	1.64	1.96	2.58	3.29

EXAMPLE 2-8

A chemist obtained the following data for the alcohol content of a sample of blood: % C_2H_5OH = 0.084, 0.089, and 0.079. Calculate the 95% confidence limits for the mean assuming (a) no additional knowledge about the precision of the method and (b) that on the basis of previous experience, it is known that $s \rightarrow \sigma = 0.005\%$ C_2H_5OH.

(a) $\Sigma x_i = 0.084 + 0.089 + 0.079 = 0.252$

$$\Sigma x_i^2 = 0.007056 + 0.007921 + 0.006241 = 0.021218$$

$$s = \sqrt{\frac{0.021218 - (0.252)^2/3}{3 - 1}} = 0.0050\%\ C_2H_5OH$$

Here, $\bar{x} = 0.252/3 = 0.084$. Table 2-6 indicates that $t = 4.30$ for two degrees of freedom and 95% confidence. Thus,

$$95\%\ \text{CL} = \bar{x} \pm \frac{ts}{\sqrt{N}} = 0.084 \pm \frac{4.30 \times 0.0050}{\sqrt{3}} = 0.084 \pm 0.012\%\ C_2H_5OH$$

(b) Because a good value of σ is available,

$$95\%\ \text{CL} = \bar{x} \pm \frac{z\sigma}{\sqrt{N}} = 0.084 \pm \frac{1.96 \times 0.0050}{\sqrt{3}} = 0.084 \pm 0.006\%\ C_2H_5OH$$

Note that a sure knowledge of σ decreases the confidence interval by a significant amount.

2E-2

Rejection of Outliers

When a set of data contains an outlying result that appears to differ excessively from the average, the decision must be made whether to retain or reject it.[12] The choice of criterion for the rejection of a suspected result has its perils. If

[12]For further information on the treatment of outliers, see J. Mandel, in *Treatise on Analytical Chemistry,* 2nd ed., I. M. Kolthoff and P. J. Elving, Eds., Part I, Vol. 1, pp. 282–289. New York: Wiley, 1978.

we set a stringent standard that makes the rejection of a questionable measurement difficult, we run the risk of retaining results that are spurious and have an inordinate effect on the average of the data. If we set lenient limits on precision and thereby make the rejection of a result easy, we are likely to discard measurements that rightfully belong in the set, thus introducing a bias to the data. It is an unfortunate fact that no universal rule can be invoked to settle the question of retention or rejection.

Statistical Tests

Several statistical tests have been developed to provide criteria for rejection or retention of outliers. Generally, such tests assume that the distribution of the population data is normal, or Gaussian. Unfortunately, this condition cannot be proved or disproved for samples that have many fewer than 50 results. Consequently, statistical rules, which are perfectly reliable for normal distributions of data, should be *used with utmost caution* when applied to samples containing only a few data. J. Mandel, in discussing treatment of small sets of data, writes, "Those who believe that they can discard observations with statistical sanction by using statistical rules for the rejection of outliers are simply deluding themselves."[13] Thus, statistical tests for rejection should be used only as aids to common sense when small samples are involved.

The Q test is a simple, statistically based test that has been widely used.[14] Here, the difference between the questionable result x_q and its nearest neighbor x_n (without regard to sign) is divided by the spread w of the entire set to give the quantity Q_{exp}:

$$Q_{exp} = |x_q - x_n|/w \tag{2-17}$$

This ratio is then compared with rejection values Q_{crit} found in Table 2-7. If Q_{exp} is greater than Q_{crit}, the questionable result can be rejected with the indicated degree of confidence.

TABLE 2-7

Critical Values for Rejection Quotient Q*

	Q_{crit} (Reject if $Q_{exp} > Q_{crit}$)		
Number of Observations	90% Confidence	96% Confidence	99% Confidence
3	0.94	0.98	0.99
4	0.76	0.85	0.93
5	0.64	0.73	0.82
6	0.56	0.64	0.74
7	0.51	0.59	0.68
8	0.47	0.54	0.63
9	0.44	0.51	0.60
10	0.41	0.48	0.57

*Reproduced from W. J. Dixon, *Ann. Math. Stat.*, **1951,** *22,* 68.

[13] J. Mandel, in *Treatise on Analytical Chemistry,* 2nd ed., I. M. Kolthoff and P. J. Elving, Eds., Part I, Vol. 1, p. 282. New York: Wiley, 1978.

[14] R. B. Dean and W. J. Dixon, *Anal. Chem.*, **1951,** *23,* 636.

EXAMPLE 2-9

The analysis of a calcite sample yielded CaO percentages of 55.95, 56.00, 56.04, 56.08, and 56.23. The last value appears anomalous; should it be retained or rejected?

The difference between 56.23 and 56.08 is 0.15%. The spread (56.23 − 55.95) is 0.28%. Thus,

$$Q_{exp} = \frac{0.15}{0.28} = 0.54$$

For five measurements, Q_{crit} at the 90% confidence level is 0.64. Because $0.54 < 0.64$, retention is indicated.

In the American Society for Testing Materials (ASTM) T_n *test,* the quantity T_n serves as the rejection criterion,[15] where

$$T_n = |x_q - \bar{x}|/s$$

Here, x_q is the questionable result and $\bar{x}$ and s are the mean and standard deviations of the entire set *including the questionable result.* Rejection is indicated if the calculated T_n is greater than the critical values found in Table 2-8.

EXAMPLE 2-10

Apply the T_n test to the data in Example 2-9.

$$\Sigma x_i = 55.95 + 56.00 + 56.04 + 56.08 + 56.23 = 280.3$$

$$\Sigma x_i^2 = (55.95)^2 + (56.00)^2 + (56.04)^2 + (56.08)^2 + (56.23)^2 = 15713.6634$$

$$\bar{x} = 280.3/5 = 56.06$$

$$s = \sqrt{\frac{15713.6634 - (280.3)^2/5}{5 - 1}} = 0.107$$

$$T_n = |56.23 - 56.06|/0.107 = 1.59$$

TABLE 2-8 **Critical Values for Rejection Quotient T_n***

Number of Observations	T_n 95% Confidence	97.5% Confidence	99% Confidence
3	1.15	1.15	1.15
4	1.46	1.48	1.49
5	1.67	1.71	1.75
6	1.82	1.89	1.94
7	1.94	2.02	2.10
8	2.03	2.13	2.22
9	2.11	2.21	2.52
10	2.18	2.29	2.41

*Adapted from J. Mandel, in *Treatise on Analytical Chemistry,* 2nd ed., I. M. Kolthoff and P. J. Elving, Eds., Part I, Vol. 1, p. 284. New York: Wiley, 1978. With permission of John Wiley & Sons, Inc.

[15]For further discussion of this test, see J. Mandel, in *Treatise on Analytical Chemistry,* 2nd ed., I. M. Kolthoff and P. J. Elving, Eds., Part I, Vol. 1, pp. 283–285. New York: Wiley, 1978.

Table 2-8 indicates that the critical value of T_n for five measurements is greater than the experimental value at all confidence levels. Therefore, retention is also indicated by this test.

The blind application of statistical tests to the decision for retention or rejection of a suspect measurement in a small set of data is not likely to be much more fruitful than an arbitrary decision. Indeed, the application of good judgment based on broad experience with an analytical method is usually a sounder approach. In the end, the only entirely valid reason for rejecting a result from a small set of data is the sure knowledge that a mistake has been made in its acquisition. Lacking this knowledge, *a cautious approach to rejection of an outlier is wise.*

Recommendations for the Treatment of Outliers

Recommendations for the treatment of a small set of results that contains a suspect value follow:

1. Reexamine carefully all data relating to the outlying result to see if a gross error could have affected its value. This recommendation demands *a properly kept laboratory notebook containing careful notations of all observations.*
2. If possible, estimate the precision that can be reasonably expected from the procedure to be sure that the outlying result actually is questionable.
3. Repeat the analysis if sufficient sample and time are available. Agreement between the newly acquired data and those of the original set that appear to be valid will lend weight to the notion that the outlying result should be rejected. Furthermore, if retention is still indicated, the questionable result will have a smaller effect on the mean of the larger set of data.
4. If more data cannot be secured, apply the Q or the T_n test to the existing set to see if the doubtful result should be retained or rejected on statistical grounds.
5. If the statistical test indicates retention, consider reporting the median of the set rather than the mean. The median has the great virtue of allowing inclusion of all data in a set without undue influence from an outlying value. Moreover, it has been demonstrated that the median of a normally distributed set containing three measurements is likely to provide a more reliable estimate of the correct value than the mean of the set after the outlying value has been arbitrarily discarded.[16]

2E-3 Statistical Aids to Hypothesis Testing

Much of scientific and engineering endeavor is based upon hypothesis testing. Thus, in order to explain an observation, a hypothetical model is advanced and tested experimentally to determine its validity. If the results from these experiments do not support the model, it is rejected and a new hypothesis is sought. If agreement is found, the hypothetical model serves as the basis for further experiments. When the hypothesis is supported by sufficient experi-

[16]National Bureau of Standards, *Technical News Bulletin*, July 1949; *J. Chem. Educ.*, **1949,** *26*, 673.

mental data, it becomes recognized as a useful theory until such time as data that refute it are obtained.

Experimental results seldom agree exactly with those predicted from a theoretical model. As a consequence, scientists and engineers frequently must judge whether a numerical difference calls for rejection of a hypothesis or whether this difference is a manifestation of the indeterminate errors inevitable in all measurements. Certain statistical tests are useful in sharpening these judgments.

Tests of this kind make use of a *null hypothesis,* which assumes that the numerical quantities being compared are, in fact, the same. The probability of the observed differences appearing as a result of indeterminate error is then computed from statistical theory. Usually, if the observed difference is as large as or larger than the difference that would occur 5 times in 100 (the 5% probability level), the null hypothesis is considered questionable and the difference is judged to be significant. Other probability levels, such as 1 in 100 or 10 in 100, may also be adopted, depending upon the certainty desired in the judgment.

The kinds of testing that chemists use most often include the comparison of (1) the mean from an analysis $\bar{x}$ with what is believed to be the true value μ, (2) the means $\bar{x}_1$ and $\bar{x}_2$ from two sets of analyses, (3) the standard deviations s_1 and s_2 or σ_1 and σ_2 from two sets of measurements, and (4) the standard deviation s of a small set of data with the standard deviation σ of a larger set of measurements. The sections that follow consider some of the methods for dealing with these comparisons.

Comparison of an Experimental Mean with a True Value

A common way of testing for determinate errors in an analytical method is to use the method to analyze a sample whose composition is accurately known (page 6). In all probability, the experimental mean $\bar{x}$ will differ from the accepted value μ; the judgment must then be made whether this difference is the consequence of indeterminate error or of a determinate error.

In treating this type of problem statistically, the difference $\bar{x} - \mu$ is compared with the difference that could be caused by indeterminate error. If the observed difference is less than that computed for a chosen probability level, the null hypothesis that $\bar{x}$ and μ are the same cannot be rejected; that is, no significant determinate error has been demonstrated. It is important to realize, however, that this statement does not say that a determinate error does not exist; it says only that its presence has not been demonstrated. If $\bar{x} - \mu$ is significantly larger than either the expected or the critical value, it may be assumed that the difference is real and that a determinate error exists.

The critical value for the rejection of the null hypothesis can be obtained by rewriting Equation 2-16 in the form

$$\bar{x} - \mu = \pm \frac{ts}{\sqrt{N}} \tag{2-18}$$

where N is the number of replicate measurements employed in the test. If a good estimate of σ is available, Equation 2-18 can be modified by replacing t with z and s with σ.

EXAMPLE 2-11

A new procedure for the rapid determination of sulfur in kerosenes was tested on a sample known from its method of preparation to contain 0.123% S. The results were % S = 0.112, 0.118, 0.115, and 0.119. Do the data indicate the presence of a negative determinate error in the method?

$$\Sigma x_i = 0.112 + 0.118 + 0.115 + 0.119 = 0.464$$

$$\bar{x} = 0.464/4 = 0.116\% \text{ S}$$

$$\bar{x} - \mu = 0.116 - 0.123 = -0.007\% \text{ S}$$

$$\Sigma x_i^2 = 0.012544 + 0.013924 + 0.013225 + 0.014161 = 0.053854$$

$$s = \sqrt{\frac{(0.053854 - (0.464)^2/4}{4 - 1}} = \sqrt{\frac{0.000030}{3}} = 0.0032$$

From Table 2-6, we find that at the 95% confidence level, t has a value of 3.18 for three degrees of freedom. Thus,

$$\frac{ts}{\sqrt{4}} = \frac{3.18 \times 0.0032}{\sqrt{4}} = \pm 0.0051$$

An experimental mean can be expected to deviate by ± 0.0051 or greater no more frequently than 5 times in 100. Thus, if we conclude that $\bar{x} - \mu = -0.007$ is a significant difference and that a determinate error is present, we will, on the average, be wrong fewer than 5 times in 100.

If we make a similar calculation employing the value for t at the 99% confidence level, $ts/\sqrt{N}$ assumes a value of 0.0093. Thus, if we insist upon being wrong no more often than 1 time in 100, we must conclude that no difference between the results has been *demonstrated.* Note that this statement is different from saying that no determinate error exists.

Comparison of Two Experimental Means

The results of chemical analyses are frequently used to determine the identity or lack of identity of two materials. Here, the chemist must judge whether a difference in the means of two sets of measurements is real and constitutes evidence that the samples are different or whether the discrepancy is simply a consequence of indeterminate errors in the two sets of measurements. To illustrate, let us assume that N_1 replicate analyses of material 1 yielded a mean value of $\bar{x}_1$, and N_2 analyses by the same method of material 2 gave a mean of $\bar{x}_2$. If the data were obtained in an identical way, it is usually safe to assume that the variances of the two sets of measurements are the same and modify Equation 2-18 to take into account that one set of results is being compared with a second rather than with the true mean of the data μ.

In this case, as with the previous one, we invoke the null hypothesis that the samples are identical and that the observed difference in the results, $(\bar{x}_1 - \bar{x}_2)$, is the result of statistical fluctuations in the data arising from indeterminate errors. To test this hypothesis statistically, we make two changes in Equation 2-18. First, we substitute $\bar{x}_2$ for μ, thus making the left side of the equation the numerical difference between the two means. Second, we replace the standard error in the equation $(s/\sqrt{N})$ with a precision parameter based upon a pooled standard deviation for the two sets of data. In order to derive the latter, we recall that the variances of two sets of independent measurements are additive (see Equation 2-4). Thus, the variance of the difference in means $(\bar{x}_1 - \bar{x}_2)$ is given by the equation

$$\left(\frac{s}{\sqrt{N}}\right)^2 = \left(\frac{s}{\sqrt{N_1}}\right)^2 + \left(\frac{s}{\sqrt{N_2}}\right)^2 = s^2\left(\frac{N_1 + N_2}{N_1 N_2}\right)$$

and

$$\frac{s}{\sqrt{N}} = s\sqrt{\frac{N_1 + N_2}{N_1 N_2}}$$

where s is the pooled standard deviation for $(N_1 + N_2) = N$ measurements. Substituting this equation into Equation 2-8 (and also $\bar{x}_2$ for μ) gives

$$\bar{x}_1 - \bar{x}_2 = \pm ts\sqrt{\frac{N_1 + N_2}{N_1 N_2}} \tag{2-19}$$

The numerical value for the term on the right is computed employing t for the particular confidence level desired. The number of degrees of freedom for finding t in Table 2-6 is $N_1 + N_2 - 2$. If the experimental difference $\bar{x}_1 - \bar{x}_2$ is smaller than the computed value, the null hypothesis is not rejected and no significant difference between the means has been demonstrated. An experimental difference greater than the value computed from t indicates the existence of a significant difference.

If a good estimate of σ is available, Equation 2-19 can be modified by insertion of z for t and σ for s.

EXAMPLE 2-12

The composition of a flake of paint found on the clothes of a hit-and-run victim was compared with that of paint from the car suspected of causing the accident. Do the following data for the spectroscopic determination of titanium in the paint suggest a difference in composition between the two materials? From previous experience, the standard deviation for the method is known to be 0.35% Ti; that is, $s \rightarrow \sigma = 0.35\%$ Ti.

Paint from clothes % Ti = 4.0, 4.6

Paint from car % Ti = 4.5, 5.3, 5.5, 5.0, 4.9

$$\bar{x}_1 = \frac{4.6 + 4.0}{2} = 4.3$$

$$\bar{x}_2 = \frac{4.5 + 5.3 + 5.5 + 5.0 + 4.9}{5} = 5.0$$

$$\bar{x}_1 - \bar{x}_2 = 4.3 - 5.0 = -0.7\% \text{ Ti}$$

Modifying Equation 2-19 to take into account our knowledge that $s \rightarrow \sigma$ and taking values of z from Table 2-5, we calculate for the 95% confidence level

$$\pm z\sigma\sqrt{\frac{N_1 + N_2}{N_1 N_2}} = \pm 1.96 \times 0.35\sqrt{\frac{2 + 5}{2 \times 5}} = \pm 0.57$$

and for the 99% confidence level

$$\pm z\sigma\sqrt{\frac{N_1 + N_2}{N_1 N_2}} = \pm 2.58 \times 0.35\sqrt{\frac{2 + 5}{2 \times 5}} = \pm 0.76$$

Only 5 out of 100 data should differ by 0.57% Ti or greater, and only 1 out of 100 should differ by as much as 0.76% Ti. Thus, it seems reasonably probable (between 95 and somewhat less than 99% certain) that the observed difference of −0.7% does not arise from indeterminate error but in fact is caused, at least in part, by a real difference between the two paint samples. Hence, we conclude the suspected vehicle was probably not involved.

EXAMPLE 2-13

Two barrels of wine were analyzed for their alcohol content in order to determine whether they are from different sources. On the basis of six analyses, the average content of the first barrel was established to be 12.61% ethanol. Four analyses of the second barrel gave a mean of 12.53% alcohol. The ten analyses yielded a pooled value of $s = 0.070\%$. Is a difference between the wines indicated by the data?

Here we employ Equation 2-19, using t for eight degrees of freedom (10 − 2). At the 95% confidence level,

$$\pm ts\sqrt{\frac{N_1 + N_2}{N_1N_2}} = \pm 2.31 \times 0.070\sqrt{\frac{6 + 4}{6 \times 4}} = \pm 0.10\%$$

The observed difference is

$$\bar{x}_1 - \bar{x}_2 = 12.61 - 12.53 = 0.08\%$$

As often as 5 times in 100, indeterminate error will be responsible for a difference as great as 0.10%. At the 95% confidence level, then, no difference in the alcohol content of the wine has been established.

In Example 2-13, no significant difference between the two wines was detected at the 95% probability level. It should be noted that this statement is not equivalent to saying that $\bar{x}_1$ is equal to $\bar{x}_2$; nor do the tests prove that the wines come from the same source. Indeed, it is conceivable that one is a red and the other a white. To establish with a reasonable probability that the two wines are from the same source would require extensive testing of other characteristics, such as taste, color, odor, and refractive index, as well as tartaric acid, sugar, and trace element content. If no significant differences are revealed by all of these tests and by others, then it might be possible to judge the two as having a common genesis. In contrast, the finding of *one* significant difference in any test would clearly show that the two wines are different. Thus, the establishment of a significant difference by a single test is much more revealing than the establishment of an absence of difference.

Estimation of Detection Limits

Equation 2-19 is useful for estimating the detection limit for a measurement. Here, the standard deviation from several blank determinations is computed. The minimum detectable quantity Δx_{min} is

$$\Delta x_{\text{min}} = \bar{x}_1 - \bar{x}_b > ts_b\sqrt{\frac{N_1 + N_b}{N_1N_b}} \qquad (2\text{-}20)$$

where the subscript b refers to the blank determination.

EXAMPLE 2-14

A method for the analysis of DDT gave the following results when applied to pesticide-free foliage samples: μg DDT = 0.2, −0.5, −0.2, 1.0, 0.8, −0.6, 0.4, 1.2. Calculate the DDT-detection limit (at the 99% confidence level) of the method for (a) a single analysis and (b) the mean of five analyses.

Here we find

$$\Sigma x_i = 0.2 - 0.5 - 0.2 + 1.0 + 0.8 - 0.6 + 0.4 + 1.2 = 2.3$$

$$\Sigma x_i^2 = 0.04 + 0.25 + 0.04 + 1.0 + 0.64 + 0.36 + 0.16 + 1.44 = 3.93$$

$$s_b = \sqrt{\frac{3.93 - (2.3)^2/8}{8 - 1}} = \sqrt{\frac{3.26875}{7}} = 0.69\ \mu\text{g}$$

(a) For a single analysis, $N_1 = 1$ and the number of degrees of freedom is $1 + 8 - 2 = 7$. From Table 2-6, we find $t = 3.50$, and so

$$\Delta x_{\min} > 3.50 \times 0.68 \sqrt{\frac{1 + 8}{1 \times 8}} > 2.5 \ \mu\text{g DDT}$$

Thus, 99 times out of 100, a result greater than 2.5 μg DDT indicates the presence of the pesticide on the plant.

(b) Here, N_1 is 5 and the number of degrees of freedom is 11. Therefore, $t = 3.11$ and

$$\Delta x_{\min} > 3.11 \times 0.68 \sqrt{\frac{5 + 8}{5 \times 8}} = 1.2 \ \mu\text{g DDT}$$

Comparison of the Precision of Measurements

The *F* test provides a simple method for comparing the precision of two sets of measurements. The sets do not necessarily have to be obtained from the same sample as long as the samples are sufficiently alike that the sources of indeterminate error can be assumed to be the same. The *F* test is also based upon the null hypothesis and thus assumes that the precisions are identical. The quantity *F*, which is defined as the ratio of the variances of the two measurements, is computed and compared with the maximum values of *F* expected (at a certain probability level) if no difference in precision existed between the two sets of measurements. If the experimental *F* exceeds the critical value found in probability tables, a statistical basis exists for questioning the null hypothesis that the two standard deviations are the same. Table 2-9 provides critical values for *F* at the 5% probability level. These values will be exceeded only 5 times in 100 if the standard deviations of the two measurements are the same. Much more extensive tables of *F* values at various probability levels are found in most mathematics handbooks.

The *F* test may be used to provide insights into either of two questions: (1) whether method A is more precise than method B and (2) whether there is a difference in the precision of the two methods. For the first of these applications, the variance of the supposedly more precise procedure is always placed in the denominator and that of the less precise in the numerator; for the second application, the larger variance always appears in the numerator. This arbitrary placement of the larger variance in the numerator in the second case makes the outcome of the test less certain; thus, the uncertainty level of the *F* values in Table 2-9 is doubled from 5 to 10%.

TABLE 2-9 **Critical Values for *F* at the 5% Level**

Degrees of Freedom (Denominator)	Degrees of Freedom (Numerator)							
	2	3	4	5	6	12	20	∞
2	19.00	19.16	19.25	19.30	19.33	19.41	19.45	19.50
3	9.55	9.28	9.12	9.01	8.94	8.74	8.66	8.53
4	6.94	6.59	6.39	6.26	6.16	5.91	5.80	5.63
5	5.79	5.41	5.19	5.05	4.95	4.68	4.56	4.36
6	5.14	4.76	4.53	4.39	4.28	4.00	3.87	3.67
12	3.89	3.49	3.26	3.11	3.00	2.69	2.54	2.30
20	3.49	3.10	2.87	2.71	2.60	2.28	2.12	1.84
∞	3.00	2.60	2.37	2.21	2.10	1.75	1.57	1.00

EXAMPLE 2-15

A standard method for the determination of carbon monoxide in gaseous mixtures is known from many hundreds of measurements to have a standard deviation of 0.21 ppm CO. A modification of the method yielded a value for s of 0.15 ppm CO for a pooled set of data with 12 degrees of freedom. A second modification, also based on 12 degrees of freedom, has a standard deviation of 0.12 ppm CO. Is either modification significantly more precise than the original?

Because an improvement is claimed, the variances of the modifications are placed in the denominator. For the first,

$$F_1 = \frac{s_{\text{std}}^2}{s_1^2} = \frac{(0.21)^2}{(0.15)^2} = 1.96$$

and for the second,

$$F_2 = \frac{(0.21)^2}{(0.12)^2} = 3.06$$

For the standard procedure, $s \rightarrow \sigma$ and the number of degrees of freedom for the numerator can be taken as infinite. The critical value of F is thus 2.30 (Table 2-9).

The value of F for the first modification is less than 2.30, and so the null hypothesis has not been disproved at the 95% probability level. The second modification, however, does appear to have significantly greater precision.

It is interesting to note that if we ask whether the precision of the second modification is significantly better than that of the first, the F test indicates that such a difference has not been demonstrated. That is,

$$F = (s_1)^2/(s_2)^2 = (0.15)^2/(0.12)^2 = 1.56$$

In this instance, the critical value for F is 2.69.

2E-4 The Least-Squares Method for Deriving Calibration Plots

Most analytical methods are based on a calibration curve in which a measured quantity y is plotted as a function of the known concentration x of a series of standards. Figure 2-7 shows a typical calibration curve, which was derived for the determination of isooctane in a hydrocarbon sample. The ordinate is the area under the chromatographic peak for isooctane, and the abscissa is the mole percent of isooctane. As is typical (and desirable), the plot approximates a straight line. Note, however, that not all the data fall exactly on the line because of the indeterminate errors in the measuring process. Thus, the investigator must try to derive a "best" straight line from the points. A statistical technique called *regression analysis* provides the means for objectively obtaining such a line and also for specifying the uncertainties associated with its subsequent use. We consider here only the simplest regression procedure: the *method of least squares.*

Assumptions

When the method of least squares is used to generate a calibration curve, two assumptions are required. The first is that a linear relationship does in fact exist between the analyte concentration (x) and the measured variable (y); that is,

$$y = a + bx$$

where a is the intercept (the value of y when x is zero) and b is the slope of the line. It is also assumed that any deviation of individual points from the straight line is entirely the consequence of indeterminate error in the *meas-*

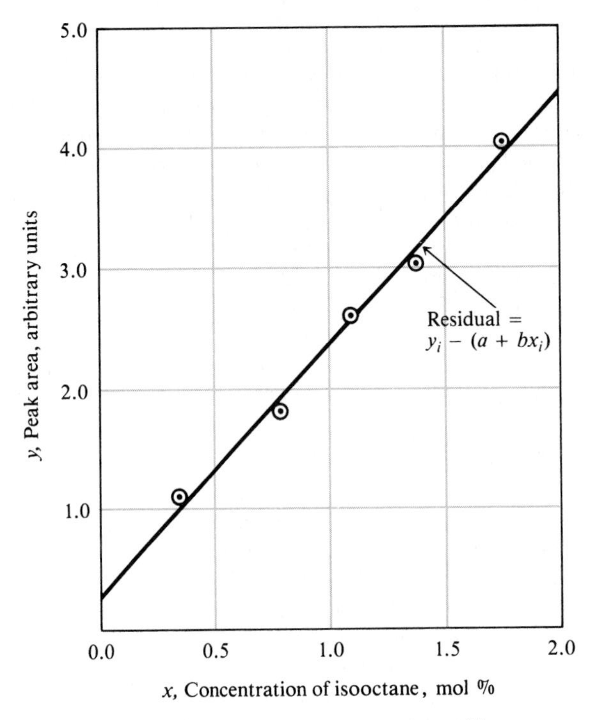

FIGURE 2-7 Calibration curve for the determination of isooctane in a hydrocarbon mixture.

urement—that is, the concentrations of the standards are known exactly. Both of these assumptions are appropriate for most analytical methods.

The Derivation of a Least-Squares Line

The line generated by the least-squares method is the one that minimizes the squares of the individual vertical displacements, or *residuals*, from that line. One of the residuals is indicated in Figure 2-7. In addition to providing the best fit between the experimental points and a straight line, the method provides the means for determining the standard deviations for a and b.[17]

For convenience, we define three quantities S_{xx}, S_{yy}, and S_{xy} as follows:

$$S_{xx} = \Sigma(x_i - \bar{x})^2 = \Sigma x_i^2 - (\Sigma x_i)^2/N \qquad (2\text{-}21)^{18}$$

$$S_{yy} = \Sigma(y_i - \bar{y})^2 = \Sigma y_i^2 - (\Sigma y_i)^2/N \qquad (2\text{-}22)^{18}$$

$$S_{xy} = \Sigma(x_i - \bar{x})(y_i - \bar{y}) = \Sigma x_i y_i - \Sigma x_i \Sigma y_i/N \qquad (2\text{-}23)$$

Here, x_i and y_i are individual pairs of data for x and y, N is the number of

[17] For detailed discussions of regression analysis and the least-squares method, see *Statistical Methods in Research and Production*, O. L. Davies and P. L. Goldsmith, Eds., Chapter 7. New York: Longman, 1972; and R. Calcutt and R. Boddy, *Statistics for Analytical Chemists*, Chapters 5 and 6. New York: Chapman and Hall, 1983.

[18] The reader should be careful to distinguish between Σx_i^2 and $(\Sigma x_i)^2$ [or Σy_i^2 and $(\Sigma y_i)^2$]. The term Σx_i^2 is obtained by first squaring the value of x_i and then summing. To obtain $(\Sigma x_i)^2$, the values of x_i are first summed and the sum is then squared.

pairs of data used in preparation of the calibration curve, and $\bar{x}$ and $\bar{y}$ are the average values for the variables, that is,

$$\bar{x} = \Sigma x_i/N \quad \text{and} \quad \bar{y} = \Sigma y_i/N$$

Note that S_{xx} and S_{yy} are simply the sum of the squares of the deviations from the mean for the individual values of x and y. The equivalent expressions shown to the far right in Equations 2-21 to 2-23 are more convenient when a handheld calculator is being used.

Five useful quantities can be derived from S_{xx}, S_{yy}, and S_{xy}:

1. The slope of the line b:

$$b = S_{xy}/S_{xx} \tag{2-24}$$

2. The intercept a:

$$a = \bar{y} - b\bar{x} \tag{2-25}$$

3. The standard deviation about regression s_r:

$$s_r = \sqrt{\frac{S_{yy} - b^2 S_{xx}}{N - 2}} \tag{2-26}$$

4. The standard deviation of the slope b:

$$s_b = \sqrt{s_r^2/S_{xx}} \tag{2-27}$$

5. The standard deviation for results obtained from the calibration curve s_c:

$$s_c = \frac{s_r}{b}\sqrt{\frac{1}{M} + \frac{1}{N} + \frac{(\bar{y}_c - \bar{y})^2}{b^2 S_{xx}}} \tag{2-28}$$

Equation 2-28 permits the calculation of the standard deviation from the mean $\bar{y}_c$ of a set of M replicate analyses when a calibration curve that contains N points is used; recall that $\bar{y}$ is the mean value of y for the n calibration data.

The standard deviation about regression s_r (Equation 2-26) is the standard deviation for y when the deviations are measured not from the mean of y (as is usually the case) but from the derived straight line:

$$s_r = \sqrt{\frac{[y_i - (a + bx_i)]^2}{N - 2}}$$

Here, the number of degrees of freedom is $N - 2$ since one degree is lost in the calculation of b and one in determining a.

TABLE 2-10

Calibration Data for a Chromatographic Method for the Determination of Isooctane in a Hydrocarbon Mixture

Mole Percent Isooctane, x_i	Peak Area, y_i	x_i^2	y_i^2	x_iy_i
0.352	1.09	0.12390	1.1881	0.38368
0.803	1.78	0.64481	3.1684	1.42934
1.08	2.60	1.16640	6.7600	2.80800
1.38	3.03	1.90440	9.1809	4.18140
1.75	4.01	3.06250	16.0801	7.01750
5.365	12.51	6.90201	36.3775	15.81992

EXAMPLE 2-16

Carry out a least-squares analysis of the experimental data provided in the first two columns in Table 2-10 and plotted in Figure 2-7.

Columns 3, 4, and 5 of the table contain computed values for x_i^2, y_i^2, and x_iy_i; their sums appear as the last entry of each column. Note that the number of digits carried in the computed values should be the *maximum allowed by the calculator;* that is, *rounding should not be performed until the end.*

We now substitute into Equations 2-21, 2-22, and 2-23 and obtain[18]

$$S_{xx} = \Sigma x_i^2 - (\Sigma x_i)^2/N = 6.90201 - (5.365)^2/5 = 1.14537$$

$$S_{yy} = \Sigma y_i^2 - (\Sigma y_i)^2/N = 36.3775 - (12.51)^2/5 = 5.07748$$

$$S_{xy} = \Sigma x_i y_i - \Sigma x_i \Sigma y_i/N = 15.81992 - 5.365 \times 12.51/5 = 2.39669$$

Substitution of these quantities into Equations 2-24 and 2-25 yields

$$b = 2.39669/1.14537 = 2.0925 = 2.09$$

$$a = \frac{12.51}{5} - 2.0925 \times \frac{5.365}{5} = 0.2567 = 0.26$$

Thus, the equation for the least-squares line is

$$y = 0.26 + 2.09x$$

Substitution into Equation 2-26 yields the standard deviation for the residuals:

$$s_r = \sqrt{\frac{S_{yy} - b^2 S_{xx}}{N - 2}} = \sqrt{\frac{5.07748 - (2.0925)^2 \times 1.14537}{5 - 2}} = 0.144 = 0.14$$

and substitution into Equation 2-27 gives the standard deviation of the slope:

$$s_b = \sqrt{s_r^2/S_{xx}} = \sqrt{(0.144)^2/1.14537} = 0.13$$

The confidence limits for the slope can be derived using t from Table 2-6. Here, the number of degrees of freedom is two fewer than the number of points because one degree of freedom is lost in calculating a and one in calculating b. The 90% confidence limit in this example is

$$90\%\ \text{CL} = 2.09 \pm ts_b = 2.09 \pm 2.355 \times 0.13 = 2.09 \pm 0.31$$

EXAMPLE 2-17

The calibration curve derived in Example 2-16 was used for the chromatographic determination of isooctane in a hydrocarbon mixture. A peak area of 2.65 was obtained. Calculate the mole percent of isooctane and the standard deviation for the result if the area was (a) the result of a single measurement and (b) the mean of four measurements.

In either case,

$$x = \frac{y - 0.26}{2.09} = \frac{2.65 - 0.26}{2.09} = 1.14\ \text{mol}\ \%$$

(a) Substituting into Equation 2-28, we obtain

$$s_c = \frac{0.14}{2.09}\sqrt{\frac{1}{1} + \frac{1}{5} + \frac{(2.65 - 12.51/5)^2}{(2.09)^2 \times 1.145}} = 0.074\ \text{mol}\ \%$$

(b) For the mean of four measurements,

$$s_c = \frac{0.14}{2.09}\sqrt{\frac{1}{4} + \frac{1}{5} + \frac{(2.65 - 12.51/5)^2}{(2.09)^2 \times 1.145}} = 0.046\ \text{mol}\ \%$$

2F

The Standard Deviation of Computed Results

Frequently it is necessary to estimate the standard deviation of a result that has been computed from two or more experimental data, each of which has a known standard deviation. The way in which such estimates are obtained depends upon the type of arithmetic calculation involved.

2F-1

The Standard Deviation of Sums and Differences

Consider the summation:

$$\begin{array}{r} +0.50\ (\pm 0.02) \\ +4.10\ (\pm 0.03) \\ \underline{-1.97}\ (\pm 0.05) \\ 2.63 \end{array}$$

where the numbers in parentheses are absolute standard deviations. Note that the standard deviation of the sum could be as large as ± 0.10 if the signs of the three individual standard deviations happen to be all positive or all negative. On the other hand, the three could combine fortuitously to give an accumulated value of zero. Neither of these possibilities is as probable as a combination leading to a standard deviation intermediate between these extremes. It can be shown from statistical theory[19] that the standard deviation of a sum or difference can be found by taking the square root of the sum of the squares of the individual *absolute* standard deviations. Thus, for the computation

$$y = a\ (\pm s_a) + b\ (\pm s_b) - c\ (\pm s_c)$$

the absolute standard deviation of the result s_y is

$$s_y = \sqrt{s_a^2 + s_b^2 + s_c^2} \tag{2-29}$$

where s_a, s_b, and s_c are the absolute standard deviations of the three terms in the sum. Substituting the standard deviations from the example gives

$$s_y = \sqrt{(\pm 0.02)^2 + (\pm 0.03)^2 + (\pm 0.05)^2} = \pm 0.06$$

and the sum could be reported as 2.63 (± 0.06).

2F-2

The Standard Deviation of Products and Quotients

Consider the following computation, where the numbers in parentheses are again absolute standard deviations:

$$\frac{4.10\ (\pm 0.02) \times 0.0050\ (\pm 0.0001)}{1.97\ (\pm 0.04)} = 0.010406\ (\pm\ ?)$$

In this instance, we note that the standard deviations of two of the numbers are larger than the result. It seems evident, then, that we cannot obtain the desired standard deviation by direct combination of absolute standard deviations the way we can in addition or subtraction. Indeed, it can be shown that the *relative* standard deviation of a product or quotient is determined by the *relative* standard deviations of the numbers forming the computed result.

[19]For a derivation of the various relationships shown in this section, see D. A. Skoog, *Principles of Instrumental Analysis*, 4th ed., pp. 14–18. Philadelphia: Saunders, 1985.

Thus, for the more general case of

$$y = a \times b/c$$

the relative standard deviation s_y/y of the result y is obtained by summing the squares of the relative standard deviations of a, b, and c and extracting the square root of the sum:

$$\frac{s_y}{y} = \sqrt{\left(\frac{s_a}{a}\right)^2 + \left(\frac{s_b}{b}\right)^2 + \left(\frac{s_c}{c}\right)^2} \tag{2-30}$$

Applying this equation to the numerical example gives

$$\frac{s_y}{y} = \sqrt{\left(\frac{\pm 0.02}{4.10}\right)^2 + \left(\frac{\pm 0.0001}{0.005}\right)^2 + \left(\frac{\pm 0.04}{1.97}\right)^2}$$

$$= \sqrt{(0.0048)^2 + (0.0200)^2 + (0.0203)^2} = \pm 0.0289$$

In order to complete the calculation, we must find the *absolute* standard deviation of the result. Thus,

$$s_y = y \times (\pm 0.0289) = 0.0104 \times (\pm 0.0289) = \pm 0.000301$$

and we can indicate the uncertainty of the answer as 0.0104 (±0.0003).

The following example demonstrates how the standard deviation of the result for a more complex calculation is determined.

EXAMPLE 2-18

Calculate the standard deviation of the result of

$$\frac{[14.3\ (\pm 0.2) - 11.6\ (\pm 0.2)] \times 0.050\ (\pm 0.001)}{[820\ (\pm 10) + 1030\ (\pm 5)] \times 42.3\ (\pm 0.4)} = 1.725\ (\pm\ ?) \times 10^{-6}$$

First we must calculate the standard deviation of the sum and the difference. For the difference in the numerator,

$$s_a = \sqrt{(\pm 0.2)^2 + (\pm 0.2)^2} = \pm 0.283$$

and for the sum in the denominator,

$$s_b = \sqrt{(\pm 10)^2 + (\pm 5)^2} = \pm 11.2$$

We may then rewrite the equation as

$$\frac{2.7\ (\pm 0.283) \times 0.050\ (\pm 0.001)}{1850\ (\pm 11.2) \times 42.3\ (\pm 0.4)} = 1.725 \times 10^{-6}$$

The equation now contains only products and quotients, and Equation 2-30 applies:

$$\frac{s_y}{y} = \sqrt{\left(\frac{\pm 0.283}{2.7}\right)^2 + \left(\frac{\pm 0.001}{0.050}\right)^2 + \left(\frac{\pm 11.2}{1850}\right)^2 + \left(\frac{\pm 0.4}{42.3}\right)^2} = \pm 0.107$$

To obtain the absolute standard deviation, we write

$$s_y = 1.725 \times 10^{-6} \times (\pm 0.107) = \pm 0.185 \times 10^{-6}$$

and the answer is rounded to $1.7\ (\pm 0.2) \times 10^{-6}$.

2F-3 The Standard Deviation in Exponential Calculations

Consider the relationship

$$y = a^x$$

where the exponent x can be considered to be free of uncertainty. It can be shown that the relative standard deviation in y resulting from the uncertainty in a is

$$s_y/y = x\, s_a/a \qquad (2\text{-}31)$$

Thus, the relative standard deviation of the square of a number is twice the relative standard deviation of the number, the relative standard deviation in the cube root of a number is simply one third that of the number, and so forth.

EXAMPLE 2-19

The standard deviation in measuring the diameter d of a sphere is ± 0.02 cm. What is the standard deviation in the calculated volume V of the sphere if $d = 2.15$ cm?

From the equation for the volume of a cube, we have

$$V = \frac{4}{3}\pi\left(\frac{d}{2}\right)^3 = \frac{4}{3}\pi\left(\frac{2.15}{2}\right)^3 = 5.20 \text{ cm}^3$$

Here we may write

$$\frac{s_V}{V} = 3 \times \frac{s_d}{d} = 3 \times \frac{0.02}{2.15} = 0.0279$$

The absolute standard deviation in V is then

$$s_V = 5.20 \times 0.0279 = 0.145$$

Thus,

$$V = 5.2\ (\pm 0.1) \text{ cm}^3$$

EXAMPLE 2-20

The solubility product K_{sp} for the silver salt AgX is $4.0\ (\pm 0.4) \times 10^{-8}$. The solubility of AgX in water is

$$\text{Solubility} = (K_{sp})^{1/2} = (4.0 \times 10^{-8})^{1/2} = 2.0 \times 10^{-4}$$

What is the uncertainty in the calculated solubility of AgX in water? Substituting y = solubility, $a = K_{sp}$, and $x = ½$ into Equation 2-31 gives

$$\frac{s_a}{a} = \frac{0.4 \times \cancel{10^{-8}}}{4.0 \times \cancel{10^{-8}}}$$

$$\frac{s_y}{y} = \frac{1}{2} \times \frac{0.4}{4.0} = 0.05$$

$$s_y = 2.0 \times 10^{-4} \times 0.05 = 0.1 \times 10^{-4}$$

$$\text{Solubility} = 2.0\ (\pm 0.1) \times 10^{-4} \text{ M}$$

It is important to note that the error propagation in taking a number to a power is different from the error propagation in multiplication. For example, consider the uncertainty in the square of $4.0\ (\pm 0.2)$. Here, the relative error in the result (16.0) is given by Equation 2-31:

$$s_y/y = 2 \times (0.2/4) = 0.1 \quad \text{or} \quad 10\%$$

Consider now the case where y is the product of *two independently measured* numbers that by chance happen to have values of $a_1 = 4.0\ (\pm 0.2)$ and $a_2 = 4.0\ (\pm 0.2)$. In this case, the relative error of the product $a_1a_2 = 16.0$ is given by Equation 2-30:

$$s_y/y = \sqrt{(0.2/4)^2 + (0.2/4)^2} = 0.07 \quad \text{or} \quad 7\%$$

The reason for this apparent anomaly lies in the fact that, in the second case, the sign associated with one error can be the same as or different from that of the other error. If they happen to be the same, the error is identical to that encountered in the first case, where the signs *must* be the same. On the other hand, the possibility exists that one sign could be positive and the other negative, in which case the relative errors tend to cancel. Thus, the probable error lies between the maximum (10%) and zero.

2F-4

The Standard Deviation of Logarithms and Antilogarithms

To show how errors are propagated when logarithms and antilogarithms are computed, we take the derivative of the expression

$$y = \log a$$

which gives

$$dy = 0.434 \frac{da}{a}$$

If we now assume that the small differences dy and da correspond to the absolute standard deviations in y and a, respectively, we obtain

$$s_y = 0.434 \frac{s_a}{a} \qquad (2\text{-}32)$$

Thus, the *absolute* standard deviation of the logarithm of a number is determined by the *relative* standard deviation of the number; conversely, the *relative* standard deviation of the antilogarithm of a number is determined by the *absolute* standard deviation of the number.

EXAMPLE 2-21

Calculate the absolute standard deviations of the results of the following computations. The absolute standard deviation for each quantity is given in parentheses.

(a) $y = \log [2.00\ (\pm 0.02) \times 10^{-4}] = -3.6990 \pm ?$

(b) $a = \text{antilog}\ [1.200\ (\pm 0.003)] = 15.849 \pm ?$

(c) $a = \text{antilog}\ [45.4\ (\pm 0.3)] = 2.5119 \times 10^{45} \pm ?$

(a) Referring to Equation 2-32, we see that we must multiply the *relative* standard deviation by 0.434:

$$s_y = \pm 0.434 \times \frac{0.02 \times 10^{-4}}{2.00 \times 10^{-4}} = \pm 0.004$$

Thus,

$$\log [2.00\ (\pm 0.02) \times 10^{-4}] = -3.699\ (\pm 0.004)$$

(b) Rearranging Equation 2-32, we have

$$\frac{s_a}{a} = \frac{s_y}{0.434} = \frac{\pm 0.003}{0.434} = \pm 0.0069$$

$$s_a = \pm 0.0069a = \pm 0.0069 \times 15.849 = \pm 0.11$$

Thus,

$$\text{antilog}\ [1.200\ (\pm 0.003)] = 15.8 \pm 0.1$$

(c) $$\frac{s_a}{a} = \frac{\pm 0.3}{0.434} = \pm 0.69$$

$$s_a = +0.69a = +0.69 \times 2.5119 \times 10^{45} = +1.7 \times 10^{45}$$

Thus,

$$\text{antilog } [45.4\ (\pm 0.3)] = 2.5\ (\pm 1.7) \times 10^{45}$$

Example 2-21c demonstrates that a large absolute error is associated with the antilogarithm of a number with few digits beyond the decimal point. This large uncertainty arises from the fact that the numbers to the left of the decimal (the characteristic) serve only to locate the decimal point. The large error in the antilogarithm results from the relatively large uncertainty in the *mantissa* of the number (that is, 0.4 ± 0.3).

2G Methods for Reporting Analytical Data

A numerical result is of little value unless something is known about its accuracy. Consequently, in reporting experimental results, responsible scientists always include some kind of indication of the likely reliability of the data. One of the best ways of indicating reliability is to give confidence limits at, say, the 90 or 95% confidence level. Alternatively, the absolute standard deviation or the coefficient of variation of the data may be reported. In this case, it is a good idea to indicate the number of data used in obtaining the standard deviation. The least satisfactory indicator of the quality of data (but one that is encountered frequently) is the *significant-figure convention.*

2G-1 The Significant-Figure Convention

A simple but crude way of indicating the probable uncertainty associated with an experimental measurement is to round the result so that it contains only *significant figures.* By definition, the significant figures in a number are all of the certain digits and *the first uncertain digit.* For example, in reading a 50-mL buret that has graduations every 0.1 mL, a chemist can easily tell that the liquid level is, let us say, greater than 30.2 mL and less than 30.3 mL. Furthermore, the position of the liquid between the graduations can be estimated to perhaps ±0.02 mL. Thus, by the significant-figure convention, the volume delivered would be reported as 30.24 mL, which corresponds to four significant figures; the first three digits are certain, and the 4 has some uncertainty associated with it.

A zero may or may not be significant, depending upon its location in a number. A zero surrounded by other digits (as in 30.24 mL) is always significant because it is read directly and with certainty from a scale or instrument readout. Zeros that serve only to place the decimal point are not significant. Thus, expression of the foregoing buret volume as 0.03204 L does not increase the number of significant figures because the zero before the 3 serves only to locate the decimal point and is therefore not significant. Terminal zeros may or may not be significant. For example, if the volume of a beaker is expressed as 2.0 L, the presence of the otherwise unnecessary zero implies that the volume is known to a few tenths of a liter; that is, both the 2 and the zero are significant figures. If this same volume is reported as 2000 mL, the last two zeros are not significant because the uncertainty here is still a few tenths of a liter, or a few hundred milliliters. In order to follow the significant-figure convention in a case such as this, scientific notation should be used and the volume reported as 2.0×10^3 mL.

An obvious limitation of the significant-figure convention as an indicator of data reliability is its ambiguity. For example, when a result is reported as 61.6, the uncertainty could range from a high of ± 0.5 to a low of ± 0.05.

2G-2

Significant Figures in Numerical Computations

Care is required in determining the appropriate number of significant figures in the result of an arithmetic combination of two or more numbers.[20]

Sums and Differences

For addition and subtraction, the number of significant figures can be seen by visual inspection. For example, in the expression

$$3.4 + 0.020 + 7.31 = 10.73 = 10.7$$

The second decimal place clearly cannot be significant because an uncertainty in the first decimal place is introduced by the 3.4. Note that the result contains three significant digits despite the fact that two of the numbers involved have but two.

Products and Quotients

A rounding rule sometimes encountered is that a product or quotient should be rounded so that it contains the same number of significant digits as the original number with the least number of significant digits. Unfortunately, this procedure often leads to incorrect rounding. For example, consider the calculations

$$\frac{24 \times 4.52}{100.0} = 1.08 \quad \text{and} \quad \frac{24 \times 4.02}{100.0} = 0.965$$

By this rule, the first answer would be rounded to 1.1 and the second to 0.96. If, however, we assume a unit uncertainty in the last digit of each number in the first quotient, the *relative* uncertainties associated with each of these numbers are 1/24, 1/452, and 1/1000. Because the first relative uncertainty is larger than the other two, we may assume the relative uncertainty in the result is also 1/24; the absolute uncertainty is then

$$1.08 \times 1/24 = 0.045 = 0.04$$

By the same argument, the absolute uncertainty of the second answer is

$$0.965 \times 1/24 = 0.040 = 0.04$$

Therefore, the first result should be rounded to *three* significant figures, or 1.08, whereas the second should be rounded to only two, or 0.96.

[20]For an extensive discussion of propagation of significant figures, see L. M. Schwartz, *J. Chem. Educ.*, **1985,** *62,* 693.

Logarithms and Antilogarithms

From our discussion of error propagation in logarithms and antilogarithms (pp. 46–47), it should be clear that particular care is needed in rounding the results in these types of calculations. The following rules are applicable to most situations[21]:

1. In the logarithm of a number, retain as many digits to the right of the decimal point as there are significant figures in the original number.
2. In the antilogarithm of a number, retain as many digits as there are digits to the right of the decimal point in the original number.

EXAMPLE 2-22

Round the following answers so that only significant digits are retained:

(a) $\log 4.000 \times 10^{-5} = -4.3979400$

(b) $\text{antilog } 12.5 = 3.162277 \times 10^{12}$

(a) Following rule 1, we retain four digits to the right of the decimal point:

$$\log 4.000 \times 10^{-5} = -4.3979$$

(b) Following rule 2, we retain only one digit:

$$\text{antilog } 12.5 = 3 \times 10^{12}$$

2G-3 Rounding Data

The computed results of a chemical analysis should always be rounded in an appropriate way. For example, consider the four replicate results 61.60, 61.46, 61.55, and 61.61. The mean of these data is 61.555, and the standard deviation is 0.069. In rounding the mean, the question of taking 61.55 or 61.56 must be considered, 61.555 being equally spaced between the two. A good guide to follow when rounding a 5 is always to round to the nearest even number. In this way, any tendency to round in a set direction is eliminated since, in any given situation, there is an equal likelihood that the nearest even number will be the higher or the lower. Accordingly, we might choose to report the result as 61.56 ± 0.07. If, on the other hand, we have reason to doubt the reliability of the estimated standard deviation, we might decide to report the result as 61.6 ± 0.1.

2G-4 Rounding the Results from Chemical Computations

Throughout this text and others, the reader is asked to perform calculations with data whose precision is indicated only by the significant-figure convention. In these circumstances, common-sense assumptions must be made as to the likely uncertainty in each number, and the techniques presented in Section 2F should be used to estimate the uncertainty of the result. Finally, the result should be rounded so that it contains only significant digits. It is of particular importance to note that rounding *should be postponed until the calculation is completed*—that is, at least one digit beyond the significant digits should be

[21] D. E. Jones, *J. Chem. Educ.*, **1971**, *49*, 753.

carried through all the computations in order to avoid a rounding error. This extra digit is sometimes called a "guard" digit. Modern calculators generally retain several extra digits that are not significant, and the user must become adept at rounding a final result properly so that only significant figures are included. Example 2-23 illustrates this procedure.

EXAMPLE 2-23

A 1.7421-g sample of a solid mixture containing benzoic acid, C_6H_5COOH (fw = 122.1247), was dissolved in water and titrated with base to a phenolphthalein end point; the acid consumed 41.36 mL of 0.1164 M NaOH. Calculate the percent benzoic acid (HBz) in the sample.

As is shown in Section 4C-4, the computation takes the following form:

$$\% \text{ HBz} = \frac{41.36 \cancel{\text{mL}} \times 0.1164 \frac{\cancel{\text{mmol NaOH}}}{\cancel{\text{mL NaOH}}} \times \frac{1 \cancel{\text{mmol HBz}}}{\cancel{\text{mmol NaOH}}} \times \frac{122.125 \text{ g HBz}}{1000 \cancel{\text{mmol HBz}}}}{1.7421 \text{ g sample}} \times 100\%$$

$$= 34.1631\%$$

Since the result is product and quotient, its *relative* uncertainty is determined by the *relative* uncertainties of the experiment data. Let us estimate what these uncertainties are.

(a) The position of the liquid level in a buret can be estimated to ±0.02 mL. Initial and final readings must be made, however, and so we assume that the maximum uncertainty is ±0.04 mL. The relative uncertainty is then

$$\frac{0.04}{41.36} \times 1000 \text{ ppt} = 0.97 \text{ ppt}$$

(b) Generally, the absolute uncertainty of a weight obtained with an analytical balance is on the order of ±0.0001 g. Thus the relative uncertainty of the denominator is

$$\frac{0.0001}{1.7421} \times 1000 \text{ ppt} = 0.057 \text{ ppt}$$

(c) We can usually assume that the absolute uncertainty in the molarity of a reagent solution is ±0.0001, and so

$$\frac{0.0001}{0.1164} \times 1000 \text{ ppt} = 0.86 \text{ ppt}$$

(d) The relative uncertainty in the formula weight of HBz is several orders of magnitude smaller than that of the three experimental data and is of no consequence. Note, however, that we should retain sufficient digits in the calculation so that the formula weight is given to at least one more digit (the guard digit) than any of the experimental data. Thus, we use 122.125 for the formula weight (here we are carrying two extra digits).

(e) No uncertainty is associated with the 100 and the 1000 since these are exact numbers.

In rounding a result of this kind, it ordinarily suffices to round the answer so that its relative uncertainty is of the *same order of magnitude* as the relative uncertainty of the number having the largest relative uncertainty. In practice, this means that the answer is rounded so that its relative uncertainty lies between 0.2 and 2 times the largest relative uncertainty of the input data.

Examining the relative uncertainties of the three input data, we see that the largest is 0.97 ppt. The answer should then be rounded to the same order of magnitude as 0.97, or about 1 ppt. If the answer is rounded to 34.2, the suggested relative uncertainty is $0.1 \times 1000/34.2 = 3$ ppt, which is well over $2 \times 1 = 2$ ppt. Rounding to 34.16 implies a relative uncertainty of $0.01 \times 1000/34.16 = 0.3$ ppt, which lies between 0.2×1 and 2×1. Thus the answer is written % HBz = 34.16.

With a little practice, rounding decisions such as that shown in Example 2-23 can be carried out quickly and usually in the head. For example, looking again at the equation in this example, one sees that the relative uncertainty of the volume measurement is about 4 in 4000, or 1 in 1000. Similarly, the uncertainty in the molarity is roughly 1 in 1000, and the uncertainty in the denominator is somewhat less than 1 in 2000. Thus, the uncertainty in the result is 1/1000 of 34, or about 0.03. Therefore we round to 34.16.

It is important to emphasize again that rounding decisions are an important part of *every calculation* and that such decisions *cannot* be based on the number of digits in the readout of a calculator.

Questions and Problems

2-1 Explain the difference between

*(a) accuracy and precision.
(b) indeterminate and determinate error.
*(c) confidence limits and confidence level.
(d) mean and median.
*(e) absolute and relative error.
(f) constant and proportional error.
*(g) population mean and sample mean.
(h) variance and standard deviation.

2-2 Define

*(a) range.
(b) coefficient of variation.
*(c) histogram.
(d) Gaussian distribution.
*(e) standard error of a mean.
(f) null hypothesis.
*(g) Q test.
(h) residual in the context of a least-squares line.
*(i) significant figures.

*2-3 Name three types of determinate errors.

2-4 How are determinate method errors detected?

*2-5 What kind of determinate errors are detected by varying the sample size?

2-6 What is the F test used for?

*2-7 What is the difference between s and σ? Under what circumstances are they likely to be the same?

2-8 Suggest some sources of indeterminate error in measuring the width of a 3-m table with a 1-m metal rule.

*Answers to the asterisked problems are given in the answer section at the back of the book.

2-9 Consider the following sets of replicate measurements:

*A	B	*C	D	*E	F
2.4	69.94	0.0902	2.3	69.65	0.624
2.1	69.92	0.0884	2.6	69.63	0.613
2.1	69.80	0.0886	2.2	69.64	0.596
2.3		0.1000	2.4	69.21	0.607
1.5			2.9		0.582

For each set, calculate the (a) mean, (b) median, (c) spread, or range, (d) standard deviation, and (e) coefficient of variation.

2-10 The accepted values for the sets of data in Problem 2-9 are: *set A, 2.0; set B, 69.75; *set C, 0.0930; set D, 3.0, *set E, 69.05; set F, 0.635. For each set, calculate (a) the absolute error and (b) the relative error in parts per thousand.

2-11 Calculate the 95% confidence interval for each set of data in Problem 2-9. What do these confidence intervals mean?

2-12 Calculate the 95% confidence interval for each set of data in Problem 2-9 if $s \rightarrow \sigma$ and has a value of: *set A, 0.20; set B, 0.050; *set C, 0.0070; set D, 0.50; set E, 0.15; set F, 0.015.

2-13 The last result in each set of data in Problem 2-9 may be an outlier. Apply (a) the Q test (96% confidence level) and (b) the T_n test (95% confidence level) to determine whether or not there is a statistical basis for rejection.

*2-14 Does the mean for set B in Problem 2-9 differ from that of set E

(a) at the 90% confidence level?
(b) at the 99% confidence level?

*2-15 Is the precision of the data of set B in Problem 2-9 better than that of set E at the 95% confidence level?

2-16 Do the means for sets A and D in Problem 2-9 differ from each other

(a) at the 80% confidence level?
(b) at the 99.9% confidence level?

*2-17 Is the precision of the data of set D better than that of set A in Problem 2-9 at the 95% confidence level?

2-18 A method of analysis yields weights for gold that are low by 0.3 mg. Calculate the percent relative error caused by this uncertainty if the weight of gold in the sample is

*(a) 800 mg. (b) 500 mg. *(c) 100 mg. (d) 25 mg.

2-19 The method described in Problem 2-18 is to be used for the analysis of ores that assay about 1.2% gold. What minimum sample weight should be taken if the relative error resulting from a 0.3-mg loss is not to exceed

*(a) −0.2%? (b) −0.5%? *(c) −0.8%? (d) −1.2%?

*2-20 Analysis of several plant-food preparations for potassium ion yielded the accompanying data:

Sample	Mean Percent K^+	Number of Observations	Deviation of Individual Results from Mean
1	4.80	5	0.13, 0.09, 0.07, 0.05, 0.06
2	8.04	3	0.09, 0.08, 0.12
3	3.77	4	0.02, 0.15, 0.07, 0.10
4	4.07	4	0.12, 0.06, 0.05, 0.11
5	6.84	5	0.06, 0.07, 0.13, 0.10, 0.09

(a) Evaluate the standard deviation s for each sample.
(b) Obtain a pooled estimate for s.

2-21 Six bottles of wine were analyzed for residual sugar, with the following results:

Bottle	Percent (w/v) Residual Sugar	Number of Observations	Deviation of Individual Results from Mean
1	0.94	3	0.050, 0.10, 0.08
2	1.08	4	0.060, 0.050, 0.090, 0.060
3	1.20	5	0.05, 0.12, 0.07, 0.00, 0.08
4	0.67	4	0.05, 0.10, 0.06, 0.09
5	0.83	3	0.07, 0.09, 0.10
6	0.76	4	0.06, 0.12, 0.04, 0.03

(a) Evaluate the standard deviation s for each set of data.
(b) Pool the data to establish an absolute standard deviation for the method.

*2-22 Nine samples of illicit heroin preparations were analyzed in duplicate by a gas-chromatographic technique. Pool the following data to establish an absolute standard deviation for the procedure:

Sample	Heroin, %	Sample	Heroin, %	Sample	Heroin, %
1	2.24, 2.27	4	11.9, 12.6	7	14.4, 14.8
2	8.4, 8.7	5	4.3, 4.2	8	21.9, 21.1
3	7.6, 7.5	6	1.07, 1.02	9	8.8, 8.4

2-23 Calculate a pooled estimate of s from the following spectrophotometric analysis for NTA (nitrilotriacetic acid) in water from the Ohio River:

Sample	NTA, ppb
1	13, 16, 14, 9
2	38, 37, 38
3	25, 29, 23, 29, 26

*2-24 An atomic-absorption method for the determination of the amount of iron present in used jet engine oil was found, from pooling 30 triplicate analyses, to have a standard deviation $s \rightarrow \sigma = 2.4$ μg/mL. Calculate the 80 and 95% confidence intervals for the result, 18.5 μg Fe/mL, if it was based upon (a) a single analysis, (b) the mean of two analyses, (c) the mean of four analyses.

2-25 The method described in Problem 2-24 yielded a pooled standard deviation of copper of $s \rightarrow \sigma = 0.32$ μg Cu/mL. The analysis of an oil from a reciprocating aircraft engine showed a copper content of 8.53 μg Cu/mL. Calculate the 90 and 99% confidence intervals for the result based upon (a) a single analysis, (b) the mean of 4 analyses, (c) the mean of 16 analyses.

*2-26 How many replicate measurements are needed to decrease the 95 and 99% confidence limits for the analysis described in Problem 2-24 to ±1.5 μg Fe/mL?

2-27 How many replicate measurements are necessary to decrease the 95 and 99% confidence limits for the analysis described in Problem 2-25 to ±0.2 μg Cu/mL?

*2-28 A volumetric calcium analysis on triplicate samples of the blood serum of a patient believed to be suffering from a hyperparathyroid condition produced the following data: meq Ca/L = 3.15, 3.25, 3.26. What is the 95% confidence interval for the mean of the data, assuming

(a) no prior information about the precision of the analysis?
(b) $s \rightarrow \sigma = 0.05$ meq Ca/L?

2-29 A chemist obtained the following data for percent lindane in the triplicate analysis of an insecticide preparation: 7.47, 6.98, 7.27. Calculate the 90% confidence interval for the mean of the three data, assuming that

(a) the only information about the precision of the method is the precision for the three data.

(b) on the basis of long experience with the method, it is believed that $s \rightarrow \sigma =$ 0.28% lindane.

2-30 A standard method for the determination of tetraethyl lead (TEL) in gasoline is reported to have a standard deviation of 0.040 mL TEL per gallon. If $s \rightarrow \sigma = 0.040$, how many replicate analyses should be made in order for the mean for the analysis of a sample to be within

*(a) ±0.03 mL/gal of the true mean 99% of the time?

(b) ±0.03 mL/gal of the true mean 95% of the time?

(c) ±0.02 mL/gal of the true mean 90% of the time?

*2-31 A spark-source mass-spectrometric method for the determination of various elements in steel was tested by analyzing several National Bureau of Standards samples. The results from three of the analyses are given below. Assume that the NBS analyses are correct and determine whether or not a determinate error in any of the analyses is indicated at the 95% confidence level.

Element	Number of Analyses	Mean, % (w/w)	Relative Standard Deviation, ppt	NBS Result, % (w/w)
(a) V	8	0.090	97	0.096
(b) Ni	5	0.36	55	0.39
(c) Cu	7	0.55	76	0.52

2-32 A spectrophotometric method for the determination of boron in animal tissue was tested by adding known amounts of boron as a borate-mannitol complex to samples of rat livers; the increase in boron concentration was then determined. The mean result from eight replicate analyses showed an increase in boron concentration of 1.49 μg/g, with a standard deviation of 0.064 μg/g. The samples contained 1.60 μg/g of added boron. Is a negative determinate error indicated at the 95% confidence level?

*2-33 A titrimetric method for the determination of calcium in limestone was tested by analysis of an NBS limestone containing 30.15% CaO. The mean result of four analyses was 30.26% CaO, with a standard deviation of 0.085%. By pooling data from several analyses, it was established that $s \rightarrow \sigma = 0.094\%$ CaO.

(a) Do the data indicate the presence of a determinate error at the 95% confidence level?

(b) Do the data indicate the presence of a determinate error at the 95% confidence level if no pooled value for σ was available?

2-34 In order to test the quality of the work of a commercial laboratory, duplicate analyses of a purified benzoic acid (68.8% C, 4.953% H) sample are requested. It is assumed that the relative standard deviation of the method is $s_r \rightarrow \sigma_r = 4$ ppt for carbon and 6 ppt for hydrogen. The means of the reported results are 68.5% C and 4.882% H. At the 95% confidence level, is there any indication of determinate error in either analysis?

*2-35 A prosecuting attorney in a criminal case presented as principal evidence small fragments of glass found imbedded in the coat of the accused. The attorney claimed that the fragments were identical in composition to a rare Belgian stained glass window broken during the crime. The averages of triplicate analyses for five elements in the glass are shown below. On the basis of these data, does the defendent have grounds for claiming reasonable doubt as to guilt? Employ the 99% confidence level as a criterion for doubt.

Element	Concentration, ppm		Standard Deviation
	From Clothes	From Window	$s \rightarrow \sigma$
As	129	119	9.5
Co	0.53	0.60	0.025
La	3.92	3.52	0.20
Sb	2.75	2.71	0.25
Th	0.61	0.73	0.043

2-36 The homogeneity of a standard chloride sample was tested by analyzing portions of the material from the top and the bottom of the container, with the following results:

% Chloride	
Top	Bottom
26.32	26.28
26.33	26.25
26.38	26.38
26.39	

(a) Is nonhomogeneity indicated at the 95% confidence level?
(b) Is nonhomogeneity indicated at the 95% level if it is known that $s \rightarrow \sigma =$ 0.03% Cl?

*2-37 A method for the analysis of codeine in prescription drugs yielded the following results when applied to a codeine-free blank: 0.1, −0.2, 0.3, 0.2, 0.0, −0.1 mg codeine. Calculate the detection limit (in terms of milligrams of codeine) at the 99% confidence level, based upon the mean of (a) two analyses, (b) four analyses, (c) six analyses.

2-38 A single alloy specimen was used to compare the results of two testing laboratories. The standard deviation s and degrees of freedom DF in pooled sets for four analyses are:

Element	Laboratory A		Laboratory B	
	s	DF	s	DF
*(a) Fe	0.10	6	0.12	12
(b) Ni	0.07	12	0.04	20
*(c) Cr	0.05	20	0.07	6
(d) Mn	0.020	20	0.035	6

Use the F test to determine whether the results from one laboratory are statistically more precise than those from the other.

*2-39 Estimate the absolute standard deviation and the coefficient of variation for the results of the following calculations. Round each result so that it contains only significant digits. The numbers in parentheses are absolute standard deviations.

(a) $y = 6.75\ (\pm 0.03) + 0.843\ (\pm 0.001) - 7.021\ (\pm 0.001) = 0.572$

(b) $y = 19.97\ (\pm 0.04) + 0.0030\ (\pm 0.0001) + 1.29\ (\pm 0.08) = 21.263$

(c) $y = 67.1\ (\pm 0.3) \times 1.03\ (\pm 0.02) \times 10^{-17} = 6.9113 \times 10^{-16}$

(d) $y = 243\ (\pm 1) \times \dfrac{760\ (\pm 2)}{1.006\ (\pm 0.006)} = 183578.5$

(e) $y = \dfrac{143\ (\pm 6) - 64\ (\pm 3)}{1249\ (\pm 1) + 77\ (\pm 8)} = 5.9578 \times 10^{-2}$

(f) $y = \dfrac{1.97\ (\pm 0.01)}{243\ (\pm 3)} = 8.106996 \times 10^{-3}$

(g) $y = [9.6\ (\pm 0.2)]^3 = 884.736$

(h) $y = [1.03\ (\pm 0.04) \times 10^{-16}]^{1/3} = 4.6875 \times 10^{-6}$

2-40 Estimate the absolute standard deviation and the coefficient of variation for the results of the following calculations. Round each result to include only significant figures. The numbers in parentheses are absolute standard deviations.

(a) $y = -1.02\ (\pm 0.02) \times 10^{-7} - 3.54\ (\pm 0.2) \times 10^{-8} = -1.374 \times 10^{-7}$

(b) $y = 100.20\ (\pm 0.08) - 99.62\ (\pm 0.06) + 0.200\ (\pm 0.004) = 0.780$

(c) $y = 0.0010\ (\pm 0.0005) \times 18.10\ (\pm 0.02) \times 200\ (\pm 1) = 3.62$

(d) $y = [33.33\ (\pm 0.03)]^3 = 37025.927$

(e) $y = \dfrac{1.73\ (\pm 0.03) \times 10^{-14}}{1.63\ (\pm 0.04) \times 10^{-16}} = 106.1349693$

(f) $y = \dfrac{100\ (\pm 1)}{2\ (\pm 1)} = 50$

(g) $y = \dfrac{1.43\ (\pm 0.02) \times 10^{-2} - 4.76\ (\pm 0.06) \times 10^{-3}}{24.3\ (\pm 0.7) + 8.06\ (\pm 0.08)} = 2.948 \times 10^{-4}$

(h) $y = [17.2\ (\pm 0.6)]^{1/4} = 2.036489$

2-41 Estimate the absolute standard deviation and the coefficient of variation for the results of the following calculations. Round each result to include only significant figures. The numbers in parentheses are absolute standard deviations.

*(a) $y = \log [1.73\ (\pm 0.030)] = 0.238046$

(b) $y = \log [0.0432\ (\pm 0.004)] = -1.364516$

*(c) $y = \log [6.02\ (\pm 0.02) \times 10^{23}] = 23.77960$

(d) $y = \text{antilog} [-3.47\ (\pm 0.05)] = 3.38844 \times 10^{-4}$

*(e) $y = \text{antilog} [5.7\ (\pm 0.5)] = 5.01187 \times 10^{5}$

(f) $y = \text{antilog} [0.99\ (\pm 0.05)] = 9.77237$

*2-42 Apply the Q test and the T_n test to the accompanying sets to determine whether the outlying result should be retained or rejected at the 95 to 96% confidence level.

(a) 41.27, 41.61, 41.84, 41.70

(b) 7.295, 7.284, 7.388, 7.292

2-43 Apply the Q test and the T_n test to the accompanying sets to determine whether the outlying result should be retained or rejected at the 95 to 96% confidence level.

(a) 85.10, 84.62, 84.70

(b) 85.10, 84.62, 84.65, 84.70

2-44 Verify the data in the table in footnote 11, page 23.

CHAPTER 3 Gravimetric Methods of Analysis

Gravimetric methods of analysis are based upon the measurement of the weight of the analyte or of a compound of known composition that contains the analyte.[1] Two types of gravimetric methods are encountered. In *precipitation methods,* the analyte (or a species chemically related to it) is isolated as a sparingly soluble precipitate that either has a known composition or can be converted to a product of known composition by suitable heat treatment. In *volatilization methods,* the analyte or its decomposition products are volatilized at a suitable temperature. The volatile product is either collected and weighed, or the weight of the product is determined indirectly from the loss in weight of the sample.

3A A Review of Chemical Stoichiometry

Stoichiometry is defined as the weight relationships among reacting chemical species. This section provides a brief review of stoichiometry and its applications to the treatment of gravimetric data.

3A-1 Empirical Formulas, Chemical Formulas, and the Mole

An *empirical formula* gives the simplest whole-number ratio of atoms in a chemical compound. In contrast, a *chemical formula* specifies the number of atoms in a molecule. Two or more substances may have the same empirical formula but different chemical formulas. For example, CH_2O is both the empirical and the chemical formula for formaldehyde; it is also the empirical formula for such diverse substances as acetic acid, $C_2H_4O_2$, glyceraldehyde, $C_3H_6O_3$, and glucose, $C_6H_{12}O_6$, as well as more than 50 other substances containing 6 or fewer carbon atoms. The empirical formula is obtained from the percent composition of a compound. The chemical formula requires, in addition, a knowledge of the molecular weight of the species.

A *molecular formula* is one written in such a way as to also provide structural information. For example, the chemically different ethanol and dimethyl ether share the same chemical formula, C_2H_6O. Their molecular formulas, C_2H_5OH and CH_3OCH_3, however, provide structural information about the molecules that cannot be discerned in their common chemical formulas.

[1]For an extensive treatment of gravimetric methods, see C. L. Rulfs, in *Treatise on Analytical Chemistry,* I. M. Kolthoff and P. J. Elving, Eds., Part I, Vol. 11, Chapter 13. New York: Wiley, 1975.

The Mole

The *mole* (mol) is the fundamental unit for describing amount of chemical species.[2] It is always associated with a chemical formula and represents one Avogadro's number (6.02×10^{23}) of atoms, ions, molecules, or electrons. The *formula weight* (fw) of a substance is the weight in grams of 1 mol of that substance. Formula weights are obtained by summing the atomic weights of all the atoms appearing in a chemical formula (*gram formula weight, molecular weight,* and *gram molecular weight* are all synonymous terms for the weight of a mole and are used in this text from time to time). For example, the formula weight of formaldehyde is

$$\text{fw } CH_2O = 12.0 \times 1 + 1.0 \times 2 + 16.0 \times 1 = 30.0 \text{ g/mol}$$

and that of glucose is

$$\text{fw } C_6H_{12}O_6 = 12.0 \times 6 + 1.0 \times 12 + 16.0 \times 6 = 180.0 \text{ g/mol}$$

Thus, 1 mol of formaldehyde weighs 30.0 g and 1 mol of glucose weighs 180.0 g. As further examples, 1 mol of sodium ion weighs 23.0 g, and 1 mol of chloride ion weighs 35.5 g. These numbers are the gram formula weights as well as the atomic weights of the two ions. The weight of 1 mol of sodium chloride is the sum of the atomic weights of sodium and chlorine, or 58.5 g.

The Millimole

Laboratory-size quantities are frequently more conveniently described in terms of *millimoles* (mmol) or *milliformula weights* (mfw), which are the mole or the formula weight divided by 1000.

EXAMPLE 3-1

How many moles and millimoles of benzoic acid (fw = 122.1 g) are contained in 2.00 g of the pure acid?

$$\text{Amount HBz} = \frac{2.00 \cancel{\text{ g HBz}}}{122.1 \cancel{\text{ g HBz}}\text{/mol HBz}} = 0.0164 \text{ mol HBz}$$

$$\text{Amount HBz} = \frac{2.00 \cancel{\text{ g HBz}}}{0.1221 \cancel{\text{ g HBz}}\text{/mmol HBz}} = 16.4 \text{ mmol HBz}$$

EXAMPLE 3-2

How many grams of Na^+ (fw = 23.00) are contained in 25.0 g of Na_2SO_4 (fw = 142.0)?

$$\text{Wt } Na^+ = \frac{25.0 \cancel{\text{ g } Na_2SO_4}}{0.1420 \cancel{\text{ g } Na_2SO_4\text{/mmol } Na_2SO_4}} \times \frac{2 \cancel{\text{ mmol } Na^+}}{\cancel{\text{mmol } Na_2SO_4}} \times 0.0230 \frac{\text{g } Na^+}{\cancel{\text{mmol } Na^+}}$$

$$= 8.10 \text{ g}$$

[2]In the International System (SI) of Units, proposed by the International Bureau of Weights and Measures, the only chemical unit for amount of substance is the mole, defined as that quantity of a material that contains as many elementary entities (these may be atoms, ions, electrons, ion-pairs, or molecules and must be explicitly defined) as there are atoms of carbon in exactly 0.012 kg of carbon-12 (that is, Avogadro's number).

3A-2

Stoichiometric Relationships

A balanced chemical equation provides the combining ratios (in moles) between reacting substances and their products. Thus, the equation

$$2NaI(aq) + Pb(NO_3)_2(aq) = PbI_2(s) + 2NaNO_3(aq)$$

indicates that 2 mol of aqueous sodium iodide combines with 1 mol of aqueous lead nitrate to produce 1 mol of solid lead iodide and 2 mol of aqueous sodium nitrate.[3,4] A statement such as this is called the *stoichiometry of the reaction.*

The chemist frequently needs to convert an experimentally determined weight of a substance to the weight of some other species that is chemically equivalent to the original substance. Generally, the experimental and calculated weights are in metric units, such as grams or milligrams.[5] A conversion of this type is a three-step process involving (1) transformation of the raw metric data to moles or millimoles, (2) multiplication by a factor that accounts for the stoichiometry, and (3) reconversion of the data in moles to the metric units called for in the answer. The process can be summarized by the equation

Quantity Measured	×	Conversion Factor	×	Stoichiometric Factor	×	Conversion Factor	=	Quantity Sought
metric units		metric to chemical				chemical to metric		metric units

The conversion factors in this sequence are derived from formula weights.

EXAMPLE 3-3

What weight of $AgNO_3$ (fw = 169.9) is needed to convert 2.33 g of Na_2CO_3 (fw = 106.0) to Ag_2CO_3?

$$2.33\ \cancel{\text{g Na}_2\text{CO}_3} \times \frac{1\ \cancel{\text{mol Na}_2\text{CO}_3}}{106.0\ \cancel{\text{g Na}_2\text{CO}_3}} \times \frac{2\ \cancel{\text{mol AgNO}_3}}{\cancel{\text{mol Na}_2\text{CO}_3}} \times \frac{169.9\ \text{g AgNO}_3}{\cancel{\text{mol AgNO}_3}} = 7.47\ \text{g AgNO}_3$$

metric quantity	conversion factor	stoichiometric factor	conversion factor	metric quantity

It is worthwhile emphasizing that the units on the two sides of an equation such as the foregoing must be the same. Therefore, it is wise to include these units as a check to see that the generated relationship is correct.

3A-3

Calculation of Results from Gravimetric Data

The results of a gravimetric analysis are generally computed from two experimental measurements: the weight of sample and the weight of a product

[3]Here it is advantageous to depict the reaction in terms of chemical compounds. When we wish to focus on reacting species, the *net ionic equation* is preferable:

$$2I^-(aq) + Pb^{2+}(aq) = PbI_2(s)$$

[4]Chemists frequently include information about the physical state of substances in equations; thus (g), (l), (s), and (aq) refer to gaseous, liquid, solid, and aqueous states, respectively. We shall follow this practice wherever the physical state of a reactant is pertinent to the discussion.

[5]The relationship among the various metric units of weight is

$$10^{-3}\ \text{kg} = 1\ \text{g} = 10^3\ \text{mg} = 10^6\ \mu\text{g (microgram)} = 10^9\ \text{ng (nanogram)} = 10^{12}\ \text{pg (picogram)}$$

of known composition. When the product is the analyte, the concentration is given by the equation

$$\% \text{ A} = \frac{\text{weight of A}}{\text{weight of sample}} \times 100\% \tag{3-1}$$

where A represents the analyte. When the product is not the analyte, the numerator in Equation 3-1 is obtained by means of a calculation similar to that shown in Example 3-3.

EXAMPLE 3-4

A 0.3516-g sample of a commercial phosphate detergent was ignited at a red heat to destroy the organic matter. The residue was then taken up in hot HCl, which converted the P to H_3PO_4. The phosphate was precipitated as $MgNH_4PO_4 \cdot 6H_2O$ by addition of Mg^{2+} followed by aqueous NH_3. After being filtered and washed, the precipitate was converted to $Mg_2P_2O_7$ (fw = 222.57) by ignition at 1000°C. This residue weighed 0.2161 g. Calculate the % P (fw = 30.974) in the sample.

The weight of analyte is

$$\text{Wt P} = \underbrace{\cancel{\text{g } Mg_2P_2O_7}}_{\text{metric quantity}} \times \underbrace{\frac{1 \cancel{\text{ mol } Mg_2P_2O_7}}{222.57 \cancel{\text{ g } Mg_2P_2O_7}}}_{\text{conversion factor}} \times \underbrace{\frac{2 \cancel{\text{ mol P}}}{1 \cancel{\text{ mol } Mg_2P_2O_7}}}_{\text{stoichiometric factor}} \times \underbrace{\frac{30.974 \text{ g P}}{\cancel{\text{mol P}}}}_{\text{conversion factor}}$$

$$= 0.2161 \times \frac{2 \times 30.974}{222.57} = 0.060147 \simeq 0.06015 \text{ g P}$$

Substituting into Equation 3-1 yields the percent analyte:

$$\% \text{ P} = \frac{0.2161 \times \dfrac{2 \times 30.974}{222.57}}{0.3516} \times 100\% = 17.107 \simeq 17.11\%$$

EXAMPLE 3-5

An iron ore was analyzed by dissolving a 1.1324-g sample in concentrated HCl. The resulting solution was diluted with water, and the iron(III) was precipitated as the hydrous oxide $Fe_2O_3 \cdot xH_2O$ by the addition of NH_3. After filtration and washing, the residue was ignited at a high temperature to give 0.5394 g of pure Fe_2O_3 (fw = 159.69). Calculate (a) the % Fe (fw = 55.847) and (b) the % Fe_3O_4 (fw = 231.54) in the sample.

(a) $$\text{Wt Fe} = \cancel{\text{g } Fe_2O_3} \times \frac{1 \cancel{\text{ mol } Fe_2O_3}}{159.69 \cancel{\text{ g } Fe_2O_3}} \times \frac{2 \cancel{\text{ mol Fe}}}{1 \cancel{\text{ mol } Fe_2O_3}} \times \frac{55.847 \text{ g Fe}}{\cancel{\text{mol Fe}}}$$

$$= 0.5394 \times \frac{2 \times 55.847}{159.69}$$

$$\% \text{ Fe} = \frac{0.5394 \times \dfrac{2 \times 55.847}{159.69}}{1.1324} \times 100\% = 33.317 \simeq 33.32\%$$

(b) In this calculation, we assume that 3 mol of Fe_2O_3 is equivalent to 2 mol of Fe_3O_4:

$$3Fe_2O_3 \longrightarrow 2\ Fe_3O_4 + \tfrac{1}{2}O_2$$

Then

$$\text{Wt } Fe_3O_4 = \cancel{g\ Fe_2O_3} \times \frac{1\ \cancel{mol\ Fe_2O_3}}{159.69\ \cancel{g\ Fe_2O_3}} \times \frac{2\ \cancel{mol\ Fe_3O_4}}{3\ \cancel{mol\ Fe_2O_3}} \times \frac{231.54\ g\ Fe_3O_4}{\cancel{mol\ Fe_3O_4}}$$

$$\%\ Fe_3O_4 = \frac{0.5394 \times \dfrac{2 \times 231.54}{3 \times 159.69}}{1.1324} \times 100\% = 46.043 \simeq 46.04\%$$

Note that the calculations in Examples 3-4 and 3-5 are similar in that the weight of analyte was obtained by multiplying the weight of the final product by a constant made up of the two conversion factors and the stoichiometric relationship between the analyte and the product weighed. This constant is sometimes called the *gravimetric factor* (GF). Thus, the gravimetric factor in Example 3-4 is

$$GF = \frac{2 \times \text{fw P}}{\text{fw } Mg_2P_2O_7} = \frac{2 \times 30.974}{222.57} = 0.27833$$

Similarly, the two gravimetric factors in Example 3-5 are

$$\text{(a) } GF = \frac{2 \times \text{fw Fe}}{\text{fw } Fe_2O_3} = \frac{2 \times 55.847}{159.69} = 0.69944$$

$$\text{(b) } GF = \frac{2 \times \text{fw } Fe_3O_4}{3 \times \text{fw } Fe_2O_3} = \frac{2 \times 231.54}{3 \times 159.69} = 0.96662$$

A general definition for the gravimetric factor is

$$GF = \frac{a}{b} \times \frac{\text{fw of substance sought}}{\text{fw of substance weighed}}$$

where a and b are small whole numbers that have values such that the number of formula weights in the numerator and the denominator are chemically equivalent.

A general equation for calculating the results of a gravimetric analysis is

$$\%\ A = \frac{\text{wt product} \times GF}{\text{wt sample}} \times 100\% \tag{3-2}$$

EXAMPLE 3-6

At elevated temperatures, $NaHCO_3$ is converted quantitatively to Na_2CO_3:

$$2NaHCO_3(s) = Na_2CO_3(s) + CO_2(g) + H_2O(g)$$

Ignition of a 0.3592-g sample containing $NaHCO_3$ and nonvolatile impurities yielded a residue weighing 0.2362 g. Calculate the percent purity of the sample.

The difference in weight before and after ignition represents the amount of CO_2 and H_2O evolved from the $NaHCO_3$ in the sample. The equation for the reaction indicates that

$$2 \text{ mol } NaHCO_3 \equiv 1 \text{ mol } CO_2 + 1 \text{ mol } H_2O$$

Thus,

$$\% \text{ NaHCO}_3 = \frac{(0.3592 - 0.2362) \times \dfrac{2 \times \text{fw NaHCO}_3}{\text{fw CO}_2 + \text{fw H}_2\text{O}}}{0.3592} \times 100\%$$

$$= \frac{0.1230 \times \dfrac{2 \times 84.01}{44.01 + 18.015}}{0.3592} \times 100\% = 92.76 \simeq 92.8\%$$

Note (1) that the denominator of the gravimetric factor in this example is the sum of the gram formula weights of the two volatile products and (2) that the combined weight of these products forms the basis for this analysis.

EXAMPLE 3-7

A 0.2795-g sample of an organic mixture containing only $C_6H_6Cl_6$ (fw = 290.83) and $C_{14}H_9Cl_5$ (fw = 354.49) was burned in a stream of oxygen in a quartz tube. The products (CO_2, H_2O, and HCl) were passed through a solution of $NaHCO_3$. After acidification, the chloride in this solution yielded 0.7161 g of AgCl (fw = 143.32). Calculate the percent of each halogen compound in the sample.

There are two unknowns, and we must therefore develop two independent equations that can be solved simultaneously. One equation is

$$\text{Wt C}_6\text{H}_6\text{Cl}_6 + \text{wt C}_{14}\text{H}_9\text{Cl}_5 = 0.2795 \text{ g}$$

A second equation is

$$\text{Wt AgCl from C}_6\text{H}_6\text{Cl}_6 + \text{wt AgCl from C}_{14}\text{H}_9\text{Cl}_5 = 0.7161 \text{ g}$$

After we insert the appropriate gravimetric factors, the second equation becomes

$$\text{Wt C}_6\text{H}_6\text{Cl}_6 \times \frac{6 \times \text{fw AgCl}}{\text{fw C}_6\text{H}_6\text{Cl}_6} + \text{wt C}_{14}\text{H}_9\text{Cl}_5 \times \frac{5 \times \text{fw AgCl}}{\text{fw C}_{14}\text{H}_9\text{Cl}_5} = 0.7161 \text{ g}$$

Substituting numerical values for the formula weights leads to

$$\text{Wt C}_6\text{H}_6\text{Cl}_6 \times 2.9568 + \text{wt C}_{14}\text{H}_9\text{Cl}_5 \times 2.0215 = 0.7161$$

The first equation can be rewritten as

$$\text{Wt C}_{14}\text{H}_9\text{Cl}_6 = 0.2795 - \text{wt C}_6\text{H}_6\text{Cl}_6$$

Substitution of this into the previous equation leads to

$$2.9568 \times \text{wt C}_6\text{H}_6\text{Cl}_6 + 2.0215(0.2795 - \text{wt C}_6\text{H}_6\text{Cl}_6) = 0.7161$$

Thus,

$$\text{Wt C}_6\text{H}_6\text{Cl}_6 = 0.16154 \text{ g}$$

$$\% \text{ C}_6\text{H}_6\text{Cl}_6 = \frac{0.16154}{0.2795} \times 100 = 57.80\%$$

$$\% \text{ C}_{14}\text{H}_9\text{Cl}_5 = 100 - 57.80\% = 42.20\%$$

3B Properties of Precipitates and Precipitating Reagents

The ideal precipitating reagent for a gravimetric analysis reacts specifically with the analyte to produce a solid that (1) has a sufficiently low solubility that losses from this source are negligible, (2) is readily filtered and washed free of contaminants, and (3) is unreactive and of known composition after drying or, if necessary, ignition. Few reagents produce precipitates that possess all these desirable properties.

The variables that influence the solubility of precipitates are discussed in Chapter 5. This section deals with methods for obtaining pure and easily filtered solids.[6]

3B-1 Particle Size and Filterability of Precipitates

Precipitates consisting of large particles are generally desirable in gravimetric work because such particles are readily retained by porous filtering media and are thus easily and rapidly filtered and washed. Coarse precipitates are also desirable because, more often than not, they carry down fewer contaminants during formation than do precipitates made up of fine particles.

Factors That Determine the Particle Size of Precipitates

Particle size depends not only upon the chemical composition of a precipitate but also upon the conditions that exist at the time of its formation. Enormous variations are observed. At one extreme are *colloidal suspensions,* whose individual particles are so small as to be invisible to the naked eye (10^{-6} to 10^{-4} mm in diameter). Colloidal particles show no tendency to settle from solution, nor are they retained upon common filtering media. At the other extreme are particles with dimensions on the order of several tenths of a millimeter or greater. The temporary dispersion of such particles in the liquid phase is called a *crystalline suspension.* The particles of a crystalline suspension tend to settle spontaneously and are readily filtered.

No sharp discontinuities in physical properties occur as the dimensions of the particles in a solid phase increase from colloidal to those typical of crystals; indeed, some precipitates possess characteristics between these defined extremes. The majority, however, are easily recognizable as being predominantly colloidal or predominantly crystalline. Thus, although imperfect, this classification can be usefully applied to most solid phases.

The phenomenon of precipitation has long attracted the attention of chemists, but the mechanism of the process is still not fully understood. It is certain, however, that the particle size of a precipitate is influenced in part by such experimental variables as precipitate solubility in the medium in which it is being formed, temperature, reactant concentrations, and rate at which reactants are mixed. The effect of these variables can be accounted for, at least qualitatively, by assuming that the particle size is related to a single property of the system called its *relative supersaturation,* where

$$\text{Relative supersaturation} = \frac{Q - S}{S} \tag{3-3}$$

In this equation, Q is the concentration of the solute at any instant and S is its equilibrium solubility.

[6]For a more detailed treatment of precipitates, see H. A. Laitinen and W. E. Harris, *Chemical Analysis,* 2nd ed., Chapters 8 and 9. New York: McGraw-Hill, 1975; A. E. Nielsen, in *Treatise on Analytical Chemistry,* 2nd ed., I. M. Kolthoff and P. J. Elving, Eds., Part I, Vol. 3, Chapter 27. New York: Wiley, 1983.

During the formation of a sparingly soluble precipitate, each addition of precipitating reagent presumably causes the solution to be momentarily supersaturated (that is, $Q > S$). Ordinarily, this unstable condition is relieved, usually after a brief period, by precipitate formation. Experimental evidence suggests, however, that the particle size of the resulting precipitate varies inversely with the average degree of relative supersaturation that exists after each addition of reagent. Thus, when $(Q - S)/S$ is large, the precipitate tends to be colloidal; when this parameter is small on the average, a crystalline solid is more likely.

Mechanism of Precipitate Formation

The effect of relative supersaturation on particle size can be rationalized by postulating two competing precipitation mechanisms: *nucleation* and *particle growth*. The particle size of a freshly formed precipitate is governed by the extent to which one of these processes predominates over the other.

Nucleation is a process by which some minimum number of ions, atoms, or molecules (perhaps as few as four or five) unite to form a stable second phase (most commonly, on the surface of suspended solid contaminants, such as dust particles). Further precipitation then occurs either by additional nucleation, by growth on existing nuclei (particle growth), or by a mixture of the two processes. If nucleation predominates, a precipitate containing a large number of small particles results; if growth predominates, a smaller number of larger particles is produced.

The rate of nucleation is believed to increase exponentially with relative supersaturation, whereas the rate of particle growth bears an approximately linear relationship to this parameter. Thus, when supersaturation is high, the nucleation rate far exceeds the particle-growth rate and is the predominant precipitation mechanism. At low relative supersaturations, the rate of particle growth may be the greater of the two, and deposition of solid on particles already present may occur to the exclusion of further nucleation.

Experimental Control of Particle Size

Experimental variables that minimize supersaturation and thus lead to crystalline precipitates include elevated temperatures (to increase S), dilute solutions (to minimize Q), and slow addition of the precipitating agent with good stirring (also to minimize Q).

Control of acidity can be used to enhance the particle size of precipitates whose solubilities are affected by this variable. For example, large, easily filtered crystals of calcium oxalate can be obtained by forming the bulk of the precipitate in a mildly acidic environment in which the salt is moderately soluble. The precipitation is then completed by slowly adding aqueous ammonia until the acidity is sufficiently low for quantitative removal of the calcium oxalate; the additional precipitate produced during this step forms on the existing solid.

Unfortunately, many solids do not precipitate as crystals under laboratory conditions that are practical. A noncrystalline solid is generally encountered when a precipitate has such a low solubility that S in Equation 3-3 always remains minute relative to Q. The relative supersaturation thus remains enormous throughout precipitate formation, and a colloidal suspension results. For example, under conditions feasible for an analysis, the hydrous oxides of iron(III), aluminum, and chromium(III) and the sulfides of most heavy-metal ions form only as colloids because of their very low solubilities.[7]

3B-2

Colloidal Precipitates

Individual colloidal particles are so small that they are not retained on ordinary filtering media; furthermore, Brownian motion prevents their settling from the solution under the influence of gravity. Fortunately, however, the individual particles of most colloids can be caused to coagulate or agglomerate to give a filterable, noncrystalline mass that rapidly settles from solution.

Stability of Colloids

Under some conditions, colloidal suspensions are indefinitely stable and have the appearance of true homogeneous solutions (although the particles are too small to be detected by the eye, they do scatter light so that the path of a light beam in colloidal suspension is visible). The stability of such suspensions is attributable to the like charge carried by each of the individual particles. The source of this charge is *adsorbed* ions that are firmly attached to the surface of the particles. The existence of this charge is readily demonstrated experimentally by observing that all the particles of a suspension migrate in the same direction under the influence of an electrical field.

Adsorption of ions upon the surface of an ionic solid has as its origin the normal bonding forces responsible for crystal growth. Thus, a lattice silver ion located at the surface of a silver chloride particle has a partially unsatisfied bonding capacity by virtue of its surface location. Negative ions from the solution are attracted to this site by the same forces that hold chloride ions in the silver chloride lattice. Lattice chloride ions on the surface of a silver chloride precipitate exert an analogous attraction for cations in the solvent.

The nature and magnitude of the charge on the solid particles of a colloidal suspension depend in a complex way on a number of variables. For colloidal suspensions of interest in analysis, however, the species adsorbed, and thus the charge on the particles, can be readily predicted from the empirical observation that ions identical to those already in the crystal lattice are generally more strongly adsorbed than any others. Thus, a silver chloride particle is positively charged in a solution containing an excess of silver ions, owing to the preferential adsorption of those ions; it has a negative charge in the presence of excess chloride ions for the same reason. The silver chloride particles formed in a gravimetric chloride analysis initially carry a negative charge but become positive as an excess of the precipitating agent is added.

The extent of adsorption increases rapidly as the concentration of the adsorbed ion increases. Ultimately, however, the surface of each particle becomes saturated, whereupon further increases in concentration have little or no effect.

Figure 3-1 is a schematic depiction of a colloidal silver chloride particle in a solution containing an excess of silver ions. Silver ions are attached directly to the solid surface in the *primary adsorption layer*. Surrounding the charged

[7]Silver chloride illustrates that the relative supersaturation concept is imperfect. This compound ordinarily forms as a colloid, and yet its molar solubility is not significantly different from that of other compounds, such as $BaSO_4$, which generally form as crystals.

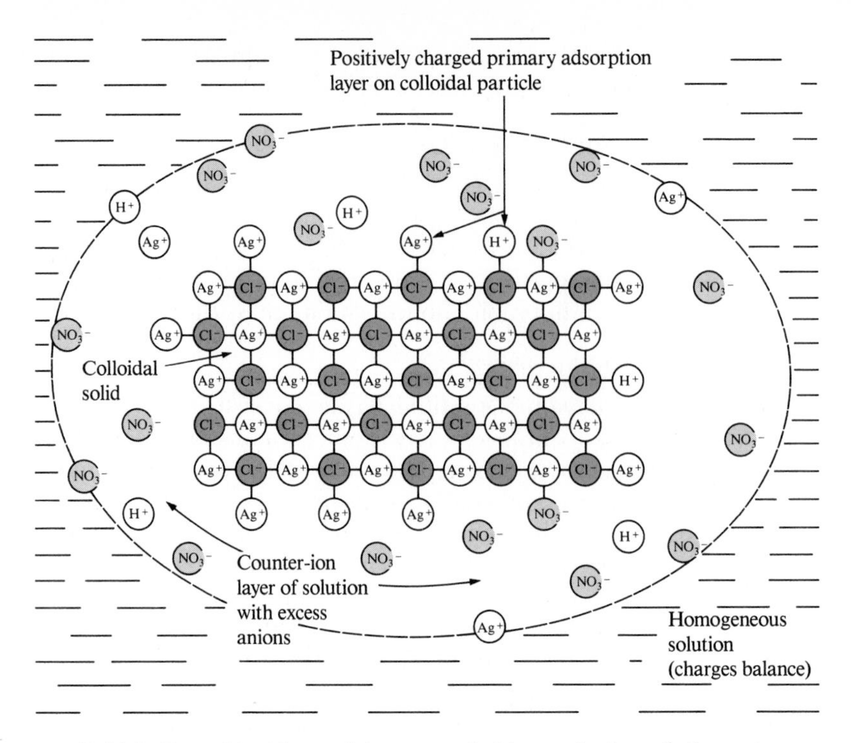

FIGURE 3-1 A colloidal silver chloride particle suspended in a solution of silver nitrate.

particle is a *region of solution* called the *counter-ion layer;* within this layer there exists a sufficient excess of negative ions to balance the charge of the adsorbed positive ions on the particle surface. The counter-ion layer forms as the result of electrostatic forces.

Figure 3-2a depicts the potential difference that develops across the counter-ion layer illustrated in Figure 3-1. The magnitude of the surface potential, E_0, is determined by the number of adsorbed silver ions, which in turn is determined by the concentration of silver ions in the solvent. Moving out from the surface, the potential decreases linearly to the potential E_1 owing to the high concentration of negative counter ions in the solution immediately adjacent to the solid surface. Beyond this point, the counter ions become more diffuse; here the potential decreases exponentially and ultimately approaches zero.

When an electric field is applied to such a colloidal suspension, the water molecules immediately surrounding each particle become highly oriented and move with the particle in the field. The interface between this sheath of water molecules and the bulk of the solution is called the *slipping plane;* the potential at this plane is termed the *zeta potential.* In order for a colloidal suspension to be stable, it is necessary that the zeta potentials of the particles be appreciably greater than zero and of the same sign as E_0. Under these circumstances, electrostatic repulsions among counter-ion layers prevent particles from approaching one another closely enough to agglomerate.

The precipitation of silver ions with chloride provides an example of the effect of the electrical double layer on stabilization of a colloid. With the initial addition of sodium chloride, silver chloride is formed in an environment that has a high silver ion concentration, which leads to colloidal particles that have

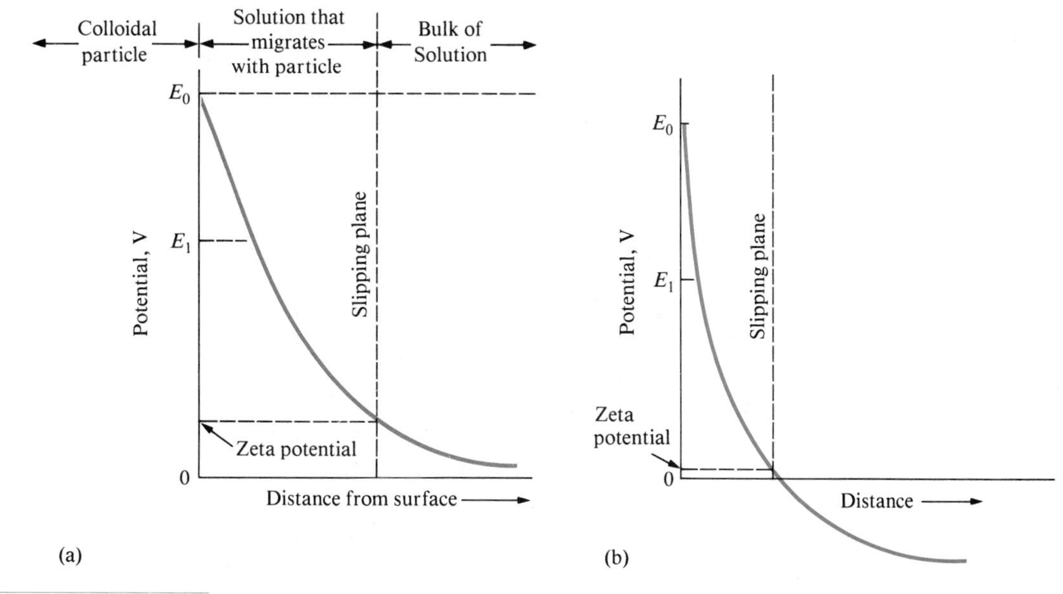

FIGURE 3-2 Potential of electric double layer for a silver chloride particle in the presence of excess silver ion. (a) Stable colloidal particle. (b) Unstable colloidal particle.

a high positive surface potential; that is, E_0 in Figure 3-2a is large and positive. The volume of the anionic counter-ion layer surrounding each particle must also be relatively large to contain enough negative ions (nitrate or hydroxide ions, for example) to neutralize the positive charge of the primary layer. Coagulation does not occur under these circumstances because the zeta potential is also quite positive. As more chloride ions are added, the charge per particle diminishes because the silver ion concentration is decreased, and the number of particles is increased; the double-layer potential curve in Figure 3-2a is thus moved downward to less positive potentials. With the approach of chemical equivalence, this shift is sufficient to cause the zeta potential to approach zero or even become slightly negative. The consequence is the sudden appearance of coagulated silver chloride. The agglomeration process is reversed when a large excess of chloride ions is added. Here, the charge of the double layer changes and its volume increases, which in turn reverses the coagulation process.

Coagulation of Colloids

Coagulation of a colloid is brought about by reducing the zeta potential of the particles by (1) heating the suspension for a brief period with good stirring and (2) increasing the electrolyte concentration of the solution.

When a colloid is heated, the number of adsorbed ions per particle decreases significantly as does its surface potential. The reduction in surface potential has the added effect of narrowing the width of the counter-ion layer and bringing the zeta potential down to a point where it approaches zero. Under this circumstance, little repulsion exists between counter-ion layers, and particles approach one another closely enough so that their water sheaths coalesce; agglomeration results.

An even more effective method of coagulation is to increase the electrolyte concentration of the solution by adding a suitable ionic compound.

The added electrolyte has the effect of reducing the volume of solution that contains enough ions of opposite charge to neutralize the charge on the particle. As a consequence, the counter-ion potential curve steepens (as in Figure 3-2b) until the zeta potential approaches zero. The particles then approach one another more closely and coagulation takes place.

Peptization of Colloids

Peptization is a process whereby a coagulated colloid reverts to its original dispersed state. Peptization frequently occurs when pure water is used to wash a colloidal precipitate. Although washing is not particularly effective in dislodging primarily adsorbed contaminants, it does tend to remove the electrolyte responsible for coagulation from the liquid present in the interstices of the coagulated solid. As the electrolyte is removed, the volumes of the counter-ion layers increase and the repulsive forces responsible for the original colloidal state are reestablished. Particles then detach themselves from the coagulated mass and return to their suspended state. The washings may become cloudy as the freshly dispersed particles pass through the filter.

The chemist is thus faced with a dilemma in handling coagulated colloids. Although washing is needed to minimize contamination, losses from peptization become a serious risk. This problem is commonly resolved by washing the agglomerated colloid with a solution containing a *volatile* electrolyte that can subsequently be removed from the solid by heating. For example, silver chloride precipitates are ordinarily washed with dilute nitric acid, giving a product that is contaminated with the acid. No harm results, however, because the nitric acid is volatilized when the precipitate is dried at 110°C.

Practical Treatment of Colloidal Precipitates

Colloids are ordinarily precipitated from hot, stirred solutions to which sufficient electrolyte has been added to assure coagulation. The filterability of a coagulated colloid is frequently improved by allowing it to stand for an hour or more in contact with the hot solution from which it was formed (the *mother liquor*). During this process, which is known as *digestion*, weakly bound water appears to be lost from the precipitate, and the result is a denser mass that is easier to filter. The coagulated mass is then washed with a solution of a volatile electrolyte to prevent peptization.

3B-3 *Crystalline Precipitates*

Crystalline precipitates are generally more easily filtered and purified than are coagulated colloids. In addition, the size of individual crystalline particles, and thus their filterability, can be controlled to a degree.

Methods of Improving Particle Size and Filterability

The particle size of crystalline solids can often be improved significantly by minimizing Q and/or maximizing S in Equation 3-3. Minimization of Q is generally accomplished by using dilute solutions and adding the precipitating reagent slowly and with good mixing. Often S is increased by precipitating from hot solution or by adjusting the pH of the precipitation medium.

Digestion of crystalline precipitates (without stirring) for some time after formation frequently yields a purer, more filterable product. The improvement in filterability undoubtedly results from the solution and recrystallization that occur continuously and at an enhanced rate at elevated temperatures.

Recrystallization is believed to result in bridging between adjacent particles, a process that yields larger and more easily filtered crystalline aggregates. This view is based upon the observation that little improvement in filtering characteristics occurs if the mixture is stirred during digestion.

3B-4 Coprecipitation

Coprecipitation is a phenomenon in which *otherwise soluble* compounds are removed from solution during precipitate formation. The reader should clearly understand that contamination of a precipitate by a second substance whose solubility limit has been exceeded *does not constitute coprecipitation.*

At least four sources of coprecipitation are recognized: *surface adsorption, mixed-crystal formation, occlusion,* and *mechanical entrapment.*[8] The first two of these sources are the result of equilibrium phenomena, whereas the third and fourth arise from the kinetics associated with the rapid growth of crystals.

Surface Adsorption

Adsorption is a common source of coprecipitation that is likely to cause significant contamination of precipitates with large specific surface areas[9]—that is, coagulated colloids. Although adsorption does occur in crystalline solids, its effects on purity are usually undetectable because of the relatively small specific surface area of these solids.

Coagulation of a colloid does not significantly decrease the amount of adsorption because the coagulated solid still contains large internal surface areas that remain exposed to the solvent (Figure 3-3). The coprecipitated

[8]Several systems of classification of coprecipitation phenomena have been suggested. We follow the simple system proposed by A. E. Nielsen, in *Treatise on Analytical Chemistry,* 2nd ed., I. M. Kolthoff and P. J. Elving, Eds., Part I, Vol. 3, p. 333. New York: Wiley, 1983.

[9]The specific surface area is defined as the surface area per unit weight of solid and is ordinarily expressed in terms of square centimeters per gram. Specific surface area increases dramatically as particle size decreases and becomes enormous for colloids. For example, the specific surface area of a 1-cm cube of a solid with a density of 2.0 g/cm^3 is 3 cm^2/g. This quantity increases to 30 cm^2/g when the cube is divided into 1000 cubes that are 0.1 cm on a side and to 3×10^5 cm^2/g (or 323 ft^2/g) when the cube is divided into 10^{15} cubes of colloidal dimensions (10^{-5} cm).

FIGURE 3-3 A coagulated colloid.

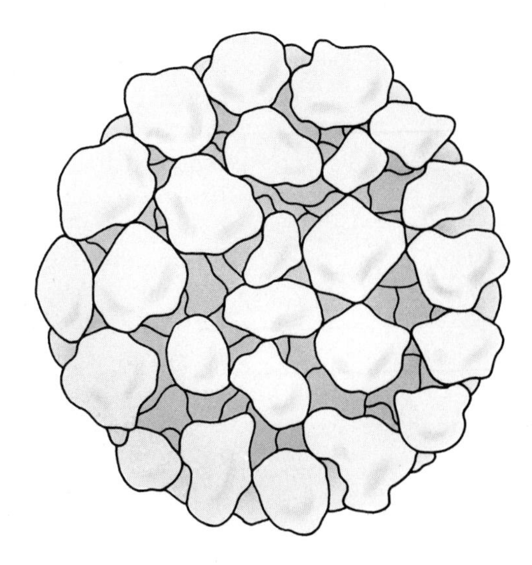

contaminant on the coagulated colloid consists of the ion originally adsorbed on the surface before coagulation and the counter ion of opposite charge held in the film of solution immediately adjacent to the particle. *The net effect of surface adsorption is therefore the carrying down of an otherwise soluble compound as a surface contaminant.* For example, the coagulated silver chloride precipitate formed in the gravimetric determination of chloride ion is contaminated with silver ions held on the surfaces of the particle and nitrate or other anions in the counter-ion layer. As a consequence, silver nitrate, a normally soluble compound, is coprecipitated with the silver chloride.

Methods for Minimizing Adsorbed Impurities on Colloids. The purity of many coagulated colloids is improved by digestion. During this process, water is expelled from the solid to give a denser mass that has a smaller specific surface area where adsorption can take place.

Washing a coagulated colloid with a solution containing a volatile electrolyte may also be helpful because any nonvolatile electrolyte added earlier to cause coagulation is displaced by the volatile species. Washing generally does not remove all primarily adsorbed ions because the attraction between these ions and the surface of the solid is too strong. Some exchange may occur, however, between existing counter ions and ions in the wash liquid. For example, in the determination of silver by precipitation with chloride ion, the primarily adsorbed species is chloride. Washing with an acidic solution converts the counter-ion layer largely to hydrogen ions so that both chloride and hydrogen ions (or hydrochloric acid) are retained by the solid. This compound is volatilized when the precipitate is dried.

Regardless of the method of treatment, it must be expected that a coagulated colloid will be contaminated to some degree, even after extensive washing. The error introduced into the analysis from this source can range from 1 or 2 ppt (as in the coprecipitation of silver nitrate on silver chloride) to an intolerable level (as in the coprecipitation of heavy-metal hydroxides on the hydrous oxides of trivalent iron or aluminum).

Reprecipitation. A drastic but effective way to minimize the effects of adsorption is *reprecipitation,* or *double precipitation.* Here, the filtered solid is redissolved and reprecipitated. The first precipitate ordinarily carries down only a fraction of the contaminant present in the original solvent. Thus, the solution containing the redissolved precipitate has a significantly lower contaminant concentration than the original, and consequently less adsorption occurs during the second precipitation. Reprecipitation adds substantially to the time required for an analysis but is often necessary for such precipitates as the hydrous oxides of iron(III) and aluminum, which possess extraordinary tendencies to adsorb the hydroxides of heavy-metal cations, such as zinc, cadmium, and manganese.

Mixed-Crystal Formation

In mixed-crystal formation, one of the ions in the crystal lattice of a solid is replaced by an ion of another element. For this exchange to occur, it is necessary that the two ions have the same charge and that their sizes differ by no more than about 5%. Furthermore, the two salts must belong to the same crystal class. For example, barium sulfate formed by adding barium chloride to a solution containing sulfate, lead, and acetate ions (acetate ions form a complex with lead ions and prevent precipitation of lead sulfate) is found to be severely contaminated by lead sulfate because the lead ions replace some of the barium ions in the barium sulfate crystals. Other examples of coprecipitation by mixed-crystal formation include $MgKPO_4$ in $MgNH_4PO_4$, $SrSO_4$ in $BaSO_4$, and MnS in CdS.

The extent of mixed-crystal contamination is governed by the law of mass action and increases as the ratio of contaminant concentration to analyte concentration increases. Mixed-crystal formation is a particularly troublesome type of coprecipitation because little can be done to ameliorate its effects when certain combinations of ions are present in a sample matrix. This phenomenon is encountered with both colloidal suspensions and crystalline precipitates. When mixed crystal formation becomes a problem, separation of the interfering ion may have to be carried out before the final precipitation step. Alternatively, the chemist may choose to use a different precipitating reagent that does not give mixed crystals with the ions in question.

Occlusion and Mechanical Entrapment

During crystal growth, the ion of the precipitating agent must displace similarly charged ions from the counter-ion layer so as to be in a position to be deposited on the solid surface. If a crystal is growing rapidly, some of the counter ions do not have time to escape from the solid surface and instead become trapped, or *occluded,* within the growing crystal. Because supersaturation, and thus growth rate, decrease as a precipitation progresses, the amount of occluded material is greatest in that part of a crystal that forms first.

Mechanical entrapment occurs when crystals lie close together during growth. Here, several crystals grow together and in so doing trap a portion of the solution in a tiny pocket.

Both occlusion and mechanical entrapment are minimized when the rate of precipitate formation is low—that is, under conditions of low supersaturation. In addition, digestion is often remarkably helpful in minimizing these types of coprecipitation. Undoubtedly, the rapid solution and reprecipitation that goes on at the elevated temperature of digestion opens up the pockets and permits the impurities to escape into the solution.

Coprecipitation Errors

Coprecipitated impurities may cause either negative or positive errors in an analysis. If the contaminant is not a compound of the ion being determined, positive errors always result. Thus, a positive error is observed when colloidal silver chloride adsorbs silver nitrate during a chloride analysis. In contrast, when the contaminant does contain the ion being determined, either positive or negative errors may be observed. For example, in the determination of barium by precipitation as barium sulfate, occlusion of other barium salts occurs. If the occluded contaminant is barium nitrate, a positive error is observed because this compound has a larger formula weight than the barium sulfate that would have formed had no coprecipitation occurred. If barium chloride is the contaminant, a negative error arises because its formula weight is less than that of the sulfate salt.

3B-5 Precipitation from Homogeneous Solution

Precipitation from homogeneous solution is performed by generating the precipitating agent by means of a slow chemical reaction in a solution of the analyte.[10] Local reagent excesses are avoided because the precipitating agent appears gradually and homogeneously throughout the solution. As a consequence, the relative supersaturation is kept low during the entire precipitation. In general, homogeneously formed precipitates, both colloidal and crystalline, are better suited for analysis than solids formed by direct addition of a precipitating reagent.

Urea is often employed for the homogeneous generation of hydroxide ion. The reaction can be expressed by the equation

$$(H_2N)_2CO + 3H_2O \longrightarrow CO_2 + 2NH_4^+ + 2OH^-$$

This hydrolysis proceeds slowly at temperatures just below boiling, and 1 to 2 hr is needed to produce enough reagent to complete a typical precipitation. Urea is particularly valuable for the precipitation of hydrous oxides or basic salts. For example, the hydrous oxides of iron(III) and aluminum are bulky, gelatinous masses that are heavily contaminated and difficult to filter when formed by the direct addition of base. In contrast, these same products are dense and readily filtered and have considerably higher purity when produced by the homogeneous generation of hydroxide ion, particularly in the presence of succinate ion.

Homogeneous precipitation of crystalline precipitates also results in marked increases in crystal size as well as improvements in purity.

Representative methods based upon precipitation by homogeneously generated reagents are given in Table 3-1.

3B-6 *Drying and Ignition of Precipitates*

After filtration, a gravimetric precipitate is heated until its weight becomes constant. Heating removes the solvent and any volatile species carried down with the precipitate. With some precipitates, heating is also necessary in order to decompose the solid to give a product of known composition, which is often called the *weighing form.*

The temperature required to produce a suitable product varies from precipitate to precipitate. Figure 3-4 shows weight loss as a function of temperature for several common analytical precipitates. These data were obtained with an automatic thermobalance,[11] an instrument that records the weight of a substance continuously as its temperature is increased at a constant rate in a furnace. Heating three of the precipitates—silver chloride, barium sulfate, and aluminum oxide—simply causes removal of water and perhaps volatile electrolytes. Note the vastly different temperatures required to produce an anhydrous precipitate of constant weight. Moisture is completely removed from silver chloride above 110 to 120°C, but dehydration of aluminum oxide is not complete until a temperature greater than 1000°C is achieved. It is of interest to note that aluminum oxide formed homogeneously with urea can be completely dehydrated at about 650°C.

The thermal curve for calcium oxalate is considerably more complex

[10]For a general reference on this technique, see L. Gordon, M. L. Salutsky, and H. H. Willard, *Precipitation from Homogeneous Solution*, New York: Wiley, 1959.

[11]For descriptions of thermobalances, see W. W. Wendlandt, *Thermal Methods of Analysis*, 2nd ed. New York: Wiley, 1974.

TABLE 3-1 **Methods for the Homogeneous Generation of Precipitants**

Precipitant	Reagent	Generation Reaction	Elements Precipitated
OH^-	Urea	$(NH_2)_2CO + 3H_2O \longrightarrow CO_2 + 2NH_4^+ + 2OH^-$	Al, Ga, Th, Bi, Fe, Sn
PO_4^{3-}	Trimethyl phosphate	$(CH_3O)_3PO + 3H_2O \longrightarrow 3CH_3OH + H_3PO_4$	Zr, Hf
$C_2O_4^{2-}$	Ethyl oxalate	$(C_2H_5)_2C_2O_4 + 2H_2O \longrightarrow 2C_2H_5OH + H_2C_2O_4$	Mg, Zn, Ca
SO_4^{2-}	Dimethyl sulfate	$(CH_3O)_2SO_2 + 4H_2O \longrightarrow 2CH_3OH + SO_4^{2-} + 2H_3O^+$	Ba, Ca, Sr, Pb
CO_3^{2-}	Trichloroacetic acid	$Cl_3CCOOH + 2OH^- \longrightarrow CHCl_3 + CO_3^{2-} + H_2O$	La, Ba, Ra
H_2S	Thioacetamide*	$CH_3CSNH_2 + H_2O \longrightarrow CH_3CONH_2 + H_2S$	Sb, Mo, Cu, Cd
DMG†	Biacetyl + hydroxylamine	$CH_3COCOCH_3 + 2H_2NOH \longrightarrow DMG + 2H_2O$	Ni
HOQ¶	8-Acetoxyquinoline§	$CH_3COOQ + H_2O \longrightarrow CH_3COOH + HOQ$	Al, U, Mg, Zn

*$CH_3C(=S)—NH_2$

†DMG = Dimethylglyoxime = $CH_3C(=N—OH)—C(=N—OH)—CH_3$

¶HOQ = 8-Hydroxyquinoline = [structure: quinoline ring with OH at position 8]

§ $CH_3C(=O)—O$– [attached at position 8 of quinoline ring]

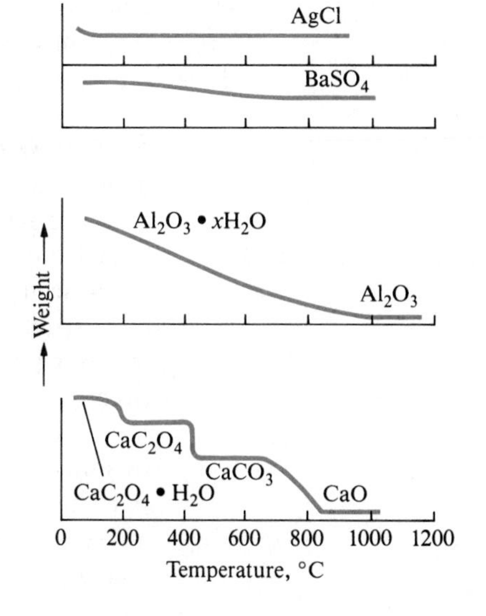

FIGURE 3-4 Effect of temperature on precipitate weight.

than the others shown in Figure 3-3. Below about 135°C, unbound water is eliminated to give the monohydrate $CaC_2O_4 \cdot H_2O$. This compound is then converted to the anhydrous oxalate CaC_2O_4 at 225°C. The abrupt change in weight at about 450°C signals the decomposition of calcium oxalate to calcium carbonate and carbon monoxide. The final step in the curve depicts the conversion of the carbonate to calcium oxide and carbon dioxide. It is evident that the compound finally weighed in a gravimetric calcium determination based upon precipitation as oxalate is highly dependent upon the ignition temperature.

3C A Critique of the Gravimetric Method

Some chemists are inclined to discount the present-day value of gravimetric methods on the grounds that they are inefficient and obsolete. We, on the other hand, believe that the gravimetric approach to an analytical problem, like all others, has strengths and weaknesses and that ample situations exist where it represents the best possible choice for the resolution of an analytical problem.

3C-1 Time Required for a Gravimetric Analysis

Gravimetric methods are purported to be slower than other analytical procedures. This assertion is usually true if it is based upon the difference in clock time between the start of an analysis and its end. On the other hand, in terms of operator time this difference often disappears because much of the elapsed time in a gravimetric procedure involves operations such as drying, igniting, digesting, and evaporating, that require little or no attention of the operator. Based on operator time, then, the gravimetric approach for a particular analysis may prove to be the most efficient, especially where only one or two samples are to be analyzed because no operator time is expended in calibration or standardization. All other analytical methods, with the exception of coulometry (Section 16D), require preparation of standard solutions that are then employed to either derive an empirical calibration curve or titrate a solution of the analyte. In contrast, a gravimetric method requires only the calculation of a gravimetric factor from data in a table of atomic weights. Thus, when only a few samples are to be analyzed, a gravimetric method, with its freedom from calibration or standardization, often requires less operator time, on a per sample basis, than procedures that require calibration.

3C-2 Sensitivity and Accuracy of Gravimetric Methods

The sensitivity and accuracy of many methods are limited by the device used for the analytical measurement. Such a limitation seldom, if ever, affects gravimetric analyses. Thus, with a suitable balance it is perfectly feasible to obtain the weight of a few micrograms of material to within a few parts per thousand of its true value; for larger masses, the weighing uncertainty can be decreased to a few parts per million. Few other analytical instruments exhibit such high precision.

The sensitivity and accuracy of a gravimetric method is usually limited by solubility losses, coprecipitation errors, and mechanical losses of precipitate, particularly when a small amount of precipitate must be isolated from a relatively large volume of solution that contains high concentrations of other constituents from the sample. Because of these problems, the chemist is wise to discard the idea of employing a gravimetric method for a constituent if its concentration is likely to be below 0.1%; for simple samples containing more than 1% of the analyte, however, gravimetric procedures are seldom surpassed in accuracy. Here, errors may often be decreased to 1 or 2 ppt. With increasing sample complexity, larger errors are inevitable unless a great deal of time is expended in circumventing them. With this type sample, the accuracy of a gravimetric method may be no better than, and may sometimes be poorer than, that of other analytical methods.

3C-3 Specificity of Gravimetric Methods

Gravimetric reagents are seldom specific but are instead *selective* in the sense that they tend to form precipitates with groups of ions. Each ion within a group interferes with the determination of any other ion in the group unless a preliminary separation is performed. In general, gravimetric procedures are less specific than some methods to be considered in later chapters.

3C-4 Equipment for Gravimetric Methods

In contrast to many analytical methods, the equipment required for a gravimetric procedure is simple, relatively inexpensive, reliable, and easy to maintain.

3D Applications of Gravimetric Methods

Gravimetric methods have been developed for most, if not all, inorganic anions and cations, as well as for such neutral species as water, sulfur dioxide, carbon dioxide, and iodine. A variety of organic substances can also be readily determined gravimetrically. Examples include lactose in milk products, salicylates in drug preparations, phenolphthalein in laxatives, nicotine in pesticides, cholesterol in cereals, and benzaldehyde in almond extracts. Indeed, gravimetric methods are among the most widely applicable of all analytical procedures.

3D-1 Inorganic Precipitating Agents

Table 3-2 lists common inorganic precipitating agents. These reagents typically cause formation of a slightly soluble salt or a hydrous oxide. The weighing form is either the salt or an oxide. The lack of specificity of most inorganic reagents is clear from the many entries in the table.

Detailed procedures for gravimetric determinations with inorganic reagents are given in Sections 31A-1 and 31A-2.

3D-2 Reducing Reagents

Table 3-3 lists several reagents that convert an analyte to its elemental form for weighing.

3D-3 Organic Precipitating Agents

Numerous organic reagents have been developed for the gravimetric determination of inorganic species. Some of these reagents are more selective in their reactions than many of the inorganic reagents listed in Table 3-2.

Two types of organic reagents are encountered. One forms slightly soluble nonionic products called *coordination compounds;* the other forms products in which the bonding between the inorganic species and the reagent is largely ionic.

TABLE 3-2

Some Inorganic Precipitating Agents*

Precipitating Agent	Element Precipitated†
$NH_3(aq)$	**Be** (BeO), **Al** (Al_2O_3), **Sc** (Sc_2O_3), Cr (Cr_2O_3),* **Fe** (Fe_2O_3), Ga (Ga_2O_3), Zr (ZrO_2), **In** (In_2O_3), Sn (SnO_2), U (U_3O_8)
H_2S	Cu (CuO),* **Zn** (ZnO, or $ZnSO_4$), **Ge** (GeO_2), As ($\underline{As_2O_3}$, or As_2O_5), Mo (MoO_3), Sn (SnO_2),* Sb ($\underline{Sb_2O_3}$, or Sb_2O_5), Bi (Bi_2S_3)
$(NH_4)_2S$	Hg ($\underline{HgS}$), Co (Co_3O_4)
$(NH_4)_2HPO_4$	**Mg** ($Mg_2P_2O_7$), Al ($AlPO_4$), Mn ($Mn_2P_2O_7$), Zn ($Zn_2P_2O_7$), Zr ($Zr_2P_2O_7$), Cd ($Cd_2P_2O_7$), Bi ($BiPO_4$)
H_2SO_4	Li, Mn, **Sr, Cd, Pb, Ba** (all as sulfates)
H_2PtCl_6	K (K_2PtCl_6, or Pt), Rb ($\underline{Rb_2PtCl_6}$), Cs ($\underline{Cs_2PtCl_6}$)
$H_2C_2O_4$	Ca (CaO), Sr (SrO), **Th** (ThO_2)
$(NH_4)_2MoO_4$	Cd ($CdMoO_4$),* Pb ($\underline{PbMoO_4}$)
HCl	**Ag** ($AgCl$), Hg (Hg_2Cl_2), Na (as $NaCl$ from butyl alcohol), Si (SiO_2)
$AgNO_3$	**Cl** ($AgCl$), Br ($\underline{AgBr}$), I($\underline{AgI}$)
$(NH_4)_2CO_3$	**Bi** (Bi_2O_3)
NH_4SCN	Cu [$Cu_2(SCN)_2$]
$NaHCO_3$	Ru, Os, Ir (precipitated as hydrous oxides; reduced with H_2 to metallic state)
HNO_3	Sn (SnO_2)
H_5IO_6	Hg ($Hg_5(IO_6)_2$)
$NaCl$, $Pb(NO_3)_2$	F ($PbClF$)
$BaCl_2$	$\mathbf{SO_4^{2-}}$ ($BaSO_4$)
$MgCl_2$, NH_4Cl	$\mathbf{PO_4^{3-}}$ ($Mg_2P_2O_7$)

*From W. F. Hillebrand, G. E. F. Lundell, H. A. Bright, and J. I. Hoffman, *Applied Inorganic Analysis.* New York: Wiley, 1953. By permission of John Wiley & Sons, Inc.

†Boldface type indicates that gravimetric analysis is the preferred method for the element or ion. The weighed form is indicated in parentheses. An asterisk indicates that the gravimetric method is seldom used. An underscore indicates the most reliable gravimetric method.

Organic reagents that yield sparingly soluble coordination compounds typically contain at least two functional groups, each of which is capable of bonding with a cation by donation of a pair of electrons. The functional groups are located in the molecule such that a five- or six-membered ring results from the reaction. Reagents that form compounds of this type are called *chelating agents,* and their products are called *chelates.*

Metal chelates are relatively nonpolar and, as a consequence, have solubilities that are low in water but high in organic liquids. Usually these compounds possess low densities and are often intensely colored. Because they are not wetted by water, coordination compounds are readily freed of moisture at low temperatures. Two widely used chelating reagents are described in the paragraphs that follow.

TABLE 3-3

Some Reducing Reagents Employed in Gravimetric Methods

Reducing Agent	Analyte
SO_2	Se, Au
$SO_2 + H_2NOH$	Te
H_2NOH	Se
$H_2C_2O_4$	Au
H_2	Re, Ir
HCOOH	Pt
$NaNO_2$	Au
$TiCl_2$	Rh
$SnCl_2$	Hg
Electrolytic reduction	Co, Ni, Cu, Zn, Ag, In, Sn, Sb, Cd, Re, Bi

8-Hydroxyquinoline

Approximately two dozen cations form sparingly soluble coordination compounds with 8-hydroxyquinoline, which is also known as *oxine.*

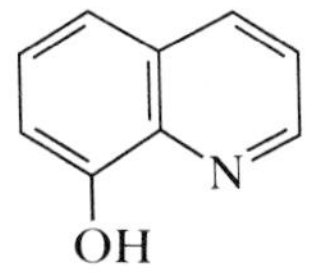

N
OH

The structure of magnesium 8-hydroxyquinolate is typical of these chelates:

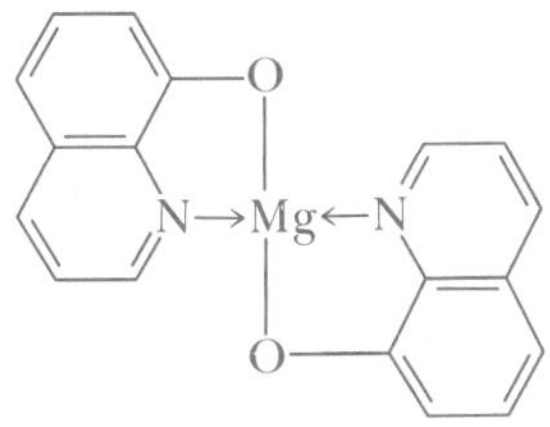

O
N→Mg←N
O

The solubilities of metal oxinates vary widely from cation to cation and are pH-dependent because 8-hydroxyquinoline is always deprotonated during chelation reaction. Therefore, a considerable degree of selectivity can be imparted to the action of 8-hydroxyquinoline through pH control.

Dimethylglyoxime

Dimethylglyoxime is an organic precipitating agent of unparalleled specificity.

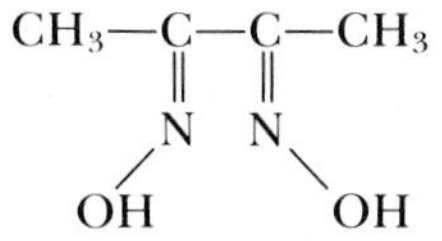

$$CH_3-\underset{\underset{\displaystyle N-OH}{\|}}{C}-\underset{\underset{\displaystyle N-OH}{\|}}{C}-CH_3$$

Its coordination compound with palladium is the only one that is sparingly soluble in acidic solution. Similarly, only the nickel compound precipitates from a weakly alkaline environment. Nickel dimethylglyoxime is bright red and has the structure

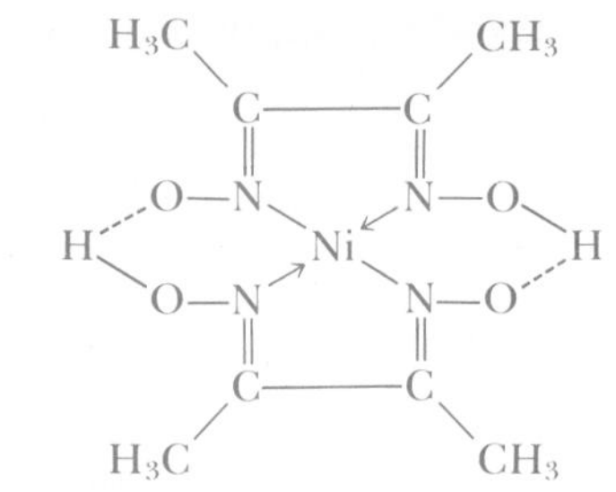

This precipitate is so bulky that only small amounts of nickel can be handled conveniently. It also has an exasperating tendency to creep up the sides of the container as it is filtered and washed. The solid is readily dried at 110°C and has the composition indicated by its formula. A procedure for the determination of nickel in a steel based on dimethylglyoxime is found in Section 31A-3.

Sodium Tetraphenylboron

Sodium tetraphenylboron, $(C_6H_5)_4B^-Na^+$, is an important example of an organic precipitating reagent that forms salt-like precipitates. In cold mineral-acid solutions, it is a near-specific precipitating agent for potassium and ammonium ions. The composition of the precipitates is stoichiometric and involve 1 mol of potassium or ammonium ion for each mole of tetraphenylboron ion; these ionic compounds are amenable to vacuum filtration and can be brought to constant weight at 105 to 120°C. Only mercury(II), rubidium, and cesium interfere and must be removed by prior treatment.

3D-4 Gravimetric Organic-Functional-Group Analysis

Several reagents react selectively with certain organic functional groups and thus can be used for the determination of most compounds containing these groups. A list of gravimetric functional-group reagents is given in Table 3-4. Many of the reactions shown can also be used for volumetric and spectrophotometric determinations. For the occasional analysis, the gravimetric procedure is often the method of choice since no calibration or standardization is required.

3D-5 Volatilization Procedures

The two most common gravimetric methods based on volatilization are those for water and carbon dioxide.

Water is quantitatively eliminated from many inorganic samples by ignition. In the direct determination, it is collected on any of several solid desiccants and its mass is determined from the weight gain of the desiccant.

Less satisfactory is the indirect method based upon the weight loss suffered by the sample as the result of ignition. Here, it must be assumed that water is the only component volatilized. This assumption is frequently unjustified, however, because ignition of many substances results in their decomposition and a consequent change in weight, irrespective of the presence of water. Nevertheless, the indirect method has found wide use for the determination of water in items of commerce. For example, a semiautomated instrument for the determination of moisture in cereal grains can be purchased.

TABLE 3-4 **Gravimetric Methods for Organic Functional Groups**

Functional Group	Basis for Method	Reaction and Product Weighed*
Carbonyl	Weight of precipitate with 2.4-dinitrophenylhydrazine	$RCHO + H_2NNHC_6H_3(NO_2)_2 \longrightarrow \underline{R{-}CH{=}NNHC_6H_3(NO_2)_2(s)} + H_2O$ (RCOR′ reacts similarly)
Aromatic carbonyl	Weight of CO_2 formed at 230°C in quinoline; CO_2 distilled, absorbed, and weighed	$ArCHO \xrightarrow[CuCO_3]{230°C} Ar + \underline{CO_2(g)}$
Methoxyl and ethoxyl	Weight of AgI formed after distillation and decomposition of CH_3I or C_2H_5I	$ROCH_3 + HI \longrightarrow ROH + CH_3I$; $RCOOCH_3 + HI \longrightarrow RCOOH + CH_3I$; $ROC_2H_5 + HI \longrightarrow ROH + C_2H_5I$; $CH_3I + Ag^+ + H_2O \rightarrow \underline{AgI(s)} + CH_3OH$
Aromatic nitro	Weight loss of Sn	$RNO_2 + \frac{3}{2}\underline{Sn(s)} + 6H^+ \longrightarrow RNH_2 + \frac{3}{2}Sn^{4+} + 2H_2O$
Azo	Weight loss of Cu	$RN{=}NR' + \underline{2Cu(s)} + 4H^+ \rightarrow RNH_2 + R'NH_2 + 2Cu^{2+}$
Phosphate	Weight of Ba salt	$ROP(=O)(OH)_2 + Ba^{2+} \longrightarrow \underline{ROP(=O)O_2Ba(s)} + 2H^+$
Sulfamic acid	Weight of $BaSO_4$ after oxidation with HNO_2	$RNHSO_3H + HNO_2 + Ba^{2+} \longrightarrow ROH + \underline{BaSO_4(s)} + N_2 + 2H^+$
Sulfinic acid	Weight of Fe_2O_3 after ignition of Fe^{3+} sulfinate	$3ROSOH + Fe^{3+} \longrightarrow (ROSO)_3Fe(s) + 3H^+$; $(ROSO)_3Fe \xrightarrow[O_2]{} CO_2 + H_2O + SO_2 + \underline{Fe_2O_3(s)}$

*The substance weighed is underlined.

It consists of a platform balance upon which a 10-g sample is heated with an infrared lamp. The percent residue is read directly.

Carbonates are ordinarily decomposed by acids to give carbon dioxide, which is readily evolved from solution by heat. As in the direct analysis for water, the weight of carbon dioxide is established from the increase in the weight of a solid absorbent. Ascarite II,[12] which consists of sodium hydroxide on a nonfibrous silicate, retains carbon dioxide by the reaction

$$2NaOH + CO_2 \longrightarrow Na_2CO_3 + H_2O$$

The absorption tube must also contain a desiccant to prevent loss of the evolved water.

Sulfides and sulfites can also be determined by volatilization. Hydrogen sulfide or sulfur dioxide evolved from the sample after treatment with acid is collected in a suitable absorbent.

Finally, the classical method for the determination of carbon and hydrogen in organic compounds is a gravimetric procedure in which the com-

[12]®Thomas Scientific, Swedesboro, NJ.

bustion products (H_2O and CO_2) are collected selectively on weighed absorbents. The increase in weight serves as the analytical parameter.

Questions and Problems

3-1 Explain the difference between
*(a) a colloidal and a crystalline precipitate.
(b) an empirical and a chemical formula.
*(c) precipitation and coprecipitation.
(d) peptization and coagulation.
*(e) occlusion and mixed-crystal formation.
(f) nucleation and particle growth.

3-2 Define
*(a) digestion
(b) adsorption.
*(c) reprecipitation.
(d) precipitation from homogeneous solution.
*(e) gravimetric factor.
(f) supersaturation.
*(g) counter-ion layer.
(h) mother liquor.

*3-3 What are the structural characteristics of a chelating agent?

3-4 How can the relative supersaturation during precipitate formation be varied?

*3-5 An aqueous solution contains $NaNO_3$ and KSCN. The thiocyanate ion is precipitated as AgSCN by addition of $AgNO_3$. After an excess of the precipitating reagent has been added,
(a) what is the charge on the surface of the coagulated colloidal particles?
(b) what is the source of the charge?
(c) what ions make up the counter-ion layer?

3-6 Suggest a method by which Cu^{2+} can be precipitated homogeneously as CuS.

*3-7 What is peptization and how is it avoided?

3-8 Suggest a precipitation method for separation of K^+ from Na^+ and Li^+.

3-9 Use chemical symbols to express the gravimetric factor for each of the following:

Sought	Weighed	Sought	Weighed
*(a) SO_3	$BaSO_4$	(f) Mn_2O	Mn_3O_4
(b) Zn	$Zn_2P_2O_7$	(g) Pb_3O_4	PbO_2
*(c) In	In_2O_3	(h) $U_2P_2O_{11}$	P_2O_5
(d) K	K_2PtCl_6	*(i) $Na_2B_4O_7 \cdot 10H_2O$	B_2O_3
*(e) CuO	$Cu_2(SCN)_2$	(j) Na_2O	$NaZn(UO_2)_3(C_2H_3O_2)_9 \cdot 6H_2O$

*3-10 Treatment of a 0.4000-g sample of impure potassium chloride with an excess of $AgNO_3$ resulted in the formation of 0.7332 g of AgCl. Calculate the percentage of KCl in the sample.

3-11 The aluminum in a 1.200-g sample of impure ammonium aluminum sulfate was precipitated with aqueous ammonia as the hydrous $Al_2O_3 \cdot xH_2O$. The precipitate was filtered and ignited at 1000°C to give anhydrous Al_2O_3, which weighed 0.1798 g.

Express the result of this analysis in terms of
(a) % $NH_4Al(SO_4)_2$. (b) % Al_2O_3. (c) % Al.

*3-12 What weight of $Cu(IO_3)_2$ can be formed from 0.400 g of $CuSO_4{\cdot}5H_2O$?

3-13 What weight of KIO_3 is needed to convert the copper in 0.4000 g of $CuSO_4{\cdot}5H_2O$ to $Cu(IO_3)_2$?

*3-14 What weight of AgI can be produced from a 0.240-g sample that assays 30.6% MgI_2?

3-15 Precipitates used in the gravimetric determination of uranium include $Na_2U_2O_7$ (fw 634.0), $(UO_2)_2P_2O_7$ (fw 714.0), and $V_2O_5{\cdot}2UO_3$ (fw 753.9). Which of these weighing forms provides the greatest weight of precipitate from a given quantity of uranium?

*3-16 A 0.7406-g sample of impure magnesite, $MgCO_3$, was decomposed with HCl; the liberated CO_2 was collected on calcium oxide and found to weigh 0.1881 g. Calculate the percentage of magnesium in the sample.

3-17 The hydrogen sulfide in a 50.0-g sample of crude petroleum was removed by distillation and collected in a solution of $CdCl_2$. The precipitated CdS was then filtered, washed, and ignited to $CdSO_4$. Calculate the percentage of H_2S in the sample if 0.108 g of $CdSO_4$ was recovered.

*3-18 A 0.1799-g sample of an organic compound was burned in a stream of oxygen, and the CO_2 produced was collected in a solution of barium hydroxide. Calculate the percentage of carbon in the sample if 0.5613 g of $BaCO_3$ was formed.

3-19 A 5.000-g sample of a pesticide was decomposed with metallic sodium in alcohol, and the liberated chloride ion was precipitated as AgCl. Express the results of this analysis in terms of percent DDT ($C_{14}H_9Cl_5$) based upon the recovery of 0.1606 g of AgCl.

*3-20 The mercury in a 0.7152-g sample was precipitated with an excess of para-periodic acid, H_5IO_6:

$$5Hg^{2+} + 2H_5IO_6 \longrightarrow Hg_5(IO_6)_2 + 10H^+$$

The precipitate was filtered, washed free of precipitating agent, dried, and weighed, 0.3408 g being recovered. Calculate the percentage of Hg_2Cl_2 in the sample.

3-21 The iodide in a sample that also contained chloride was converted to iodate by treatment with an excess of bromine:

$$3H_2O + 3Br_2 + I^- \longrightarrow 6Br^- + IO_3^- + 6H^+$$

The unused bromine was removed by boiling; an excess of barium ion was then added to precipitate the iodate:

$$Ba^{2+} + 2IO_3^- \longrightarrow Ba(IO_3)_2$$

In the analysis of a 2.72-g sample, 0.0720 g of barium iodate was recovered. Express the results of this analysis as percent potassium iodide.

*3-22 Ammoniacal nitrogen can be determined by treatment of the sample with chloroplatinic acid; the product is slightly soluble ammonium chloroplatinate:

$$H_2PtCl_6 + 2NH_4^+ \longrightarrow (NH_4)_2PtCl_6 + 2H^+$$

The precipitate decomposes upon ignition, yielding metallic platinum and gaseous products:

$$(NH_4)_2PtCl_6 \longrightarrow Pt + 2Cl_2 + 2NH_3 + 2HCl$$

Calculate the percentage of ammonia in a sample if 0.2213 g gave rise to 0.5881 g of platinum.

3-23 A 0.6447-g portion of manganese dioxide was added to an acidic solution in which 1.1402 g of a chloride-containing sample was dissolved. After the evolution of chlorine was complete,

$$MnO_2(s) + 2Cl^- + 4H^+ \longrightarrow Mn^{2+} + Cl_2 + 2H_2O$$

the excess MnO_2 was collected by filtration, washed, and weighed, 0.3521 g being recovered. Express the results of this analysis in terms of percent aluminum chloride in the chloride-containing sample.

*3-24 A series of sulfate samples is to be analyzed by precipitation as $BaSO_4$. If it is known that the sulfate content in these samples ranges between 20 and 55%, what minimum sample weight should be taken to ensure that a precipitate weight no smaller than 0.300 g is produced? What is the maximum precipitate weight to be expected if this quantity of sample is taken?

3-25 The addition of dimethylglyoxime, $H_2C_4H_6O_2N_2$, to a solution containing nickel(II) ion gives rise to a precipitate:

$$Ni^{2+} + 2H_2C_4H_6O_2N_2 \longrightarrow 2H^+ + Ni(HC_4H_6O_2N_2)_2$$

Nickel dimethylglyoxime is a bulky precipitate that is inconvenient to manipulate in amounts greater than 175 mg. The amount of nickel in a type of permanent-magnet alloy ranges between 24 and 35%. Calculate the sample size that should not be exceeded when analyzing these alloys for nickel.

*3-26 The success of a particular catalyst is highly dependent upon its zirconium content. The starting material for this preparation is received in batches that assay between 68 and 84% $ZrCl_4$. Routine analysis based upon precipitation of AgCl is feasible, it having been established that there are no sources of chloride ion other than the $ZrCl_4$ in the sample.

(a) What sample weight should be taken to ensure a AgCl precipitate that weighs at least 0.400 g?
(b) If this sample weight is used, what is the maximum weight of AgCl that can be expected in this analysis?
(c) To simplify calculations, what sample weight should be taken in order to have the percentage of $ZrCl_4$ exceed the weight of AgCl produced by a factor of 100?

3-27 A 0.8720-g sample of a mixture consisting solely of sodium bromide and potassium bromide yields 1.505 g of silver bromide. What are the percentages of the two salts in the sample?

*3-28 A 0.6407-g sample containing chloride and iodide ions gave a silver halide precipitate weighing 0.4430 g. This precipitate was then strongly heated in a stream of Cl_2 gas to convert the AgI to AgCl; upon completion of this treatment, the precipitate weighed 0.3181 g. Calculate the percentage of chloride and iodide in the sample.

3-29 The phosphorus in a 0.2374-g sample was precipitated as the slightly soluble $(NH_4)_3PO_4 \cdot 12MoO_3$. This precipitate was filtered, washed, and then redissolved in acid. Treatment of the resulting solution with an excess of Pb^{2+} resulted in the formation of 0.2752 g of $PbMoO_4$. Express the results of this analysis in terms of percent P_2O_5.

*3-30 How many grams of CO_2 are evolved from a 1.204-g sample that is 36.0% $MgCO_3$ and 44.0% K_2CO_3 by weight?

3-31 A 6.881-g sample containing magnesium chloride and sodium chloride was dissolved in sufficient water to give 500 mL of solution. Analysis for the chloride content of a 50.0-mL aliquot resulted in the formation of 0.5923 g of AgCl. The magnesium in a second 50.0-mL aliquot was precipitated as $MgNH_4PO_4$; upon ignition,

0.1796 g of $Mg_2P_2O_7$ was found. Calculate the percentage of $MgCl_2 \cdot 6H_2O$ and of NaCl in the sample.

*3-32 A 50.0-mL portion of a solution containing 0.200 g of $BaCl_2 \cdot 2H_2O$ is mixed with 50.0 mL of a solution containing 0.300 g of $NaIO_3$. Assume that the solubility of $Ba(IO_3)_2$ in water is negligibly small and calculate

(a) the weight of the precipitated $Ba(IO_3)_2$.

(b) the weight of the unreacted compound that remains in solution.

3-33 When 100.0 mL of a solution containing 0.500 g of $AgNO_3$ is mixed with 100.0 mL of a solution containing 0.300 g of K_2CrO_4, a bright red precipitate of Ag_2CrO_4 forms.

(a) Assuming the solubility of Ag_2CrO_4 is negligible, calculate the weight of the precipitate.

(b) Calculate the weight of the unreacted component that remains in solution.

CHAPTER 4 Titrimetric Methods of Analysis

Titrimetric methods of analysis are a large and powerful group of quantitative procedures that are based upon measuring the combining capacity of an analyte for a reagent. Three types of titrimetry are encountered, namely, *volumetric titrimetry, weight* or *gravimetric titrimetry,* and *coulometric titrimetry.* In the first, the volume of a liquid reagent of known concentration required to react essentially completely with the analyte is determined. In the second, the weight of the reagent is measured instead. In coulometric titrimetry, the "reagent" is a constant direct electrical current of known magnitude that reacts directly or indirectly with the analyte; here, the quantity measured is the time required for quantitative oxidation or reduction of the analyte.

This chapter provides an introduction to volumetric titrimetry as well as weight titrimetry; the several chapters that follow deal exclusively with titrimetric methods based on volume measurements, although much of the material applies equally well to weight titrimetry. Coulometric titrimetry is considered in Section 16D-5. All these titrimetric methods are used for routine analyses because they are generally rapid, convenient, accurate, and readily automated.

4A Some General Aspects of Volumetric Titrimetry[1]

4A-1 Definition of Some Terms

The reagent of known concentration upon which a volumetric method is based is called a *standard solution* (or a *standard titrant*). A *titration* is a process by which a standard solution is slowly added from a buret (a device for delivering accurately known volumes of solutions) to a solution of the analyte until the reaction between the two is judged to be complete. The volume needed to complete the titration is determined from the difference between the initial and final buret readings.

The point in a titration where the amount of added titrant is chemically equivalent to the amount of analyte in the sample is called the *equivalence point.* For example, the equivalence point in the titration of sodium chloride with silver nitrate occurs after exactly 1 mol of silver ions has been introduced for each mol of chloride ion in the sample. Likewise, the equivalence point in the titration of sulfuric acid with sodium hydroxide is reached after introduction of 2 mol of base for each mole of acid.

[1] For a detailed discussion of volumetric methods, see J. I. Watters, in *Treatise on Analytical Chemistry,* I. M. Kolthoff and P. J. Elving, Eds., Part I, Vol. 11, Chapter 114. New York: Wiley, 1975.

It is sometimes necessary to add an excess of the standard titrant and then determine the excess amount by *back-titration* with a second standard titrant. Here, the equivalence point corresponds to the point where the amount of initial titrant is chemically equivalent to the amount of analyte plus the amount of back-titrant.

4A-2 Equivalence Points and End Points

It should be understood that the equivalence point of a titration is a theoretical point that cannot be determined experimentally. It can only be estimated by observing some physical change associated with the condition of equivalence. This change signals an *end point* for the titration. Every effort is made to ensure that any volume difference between the equivalence point and the end point is small. Such differences do exist, however, owing to inadequacies in the physical changes and in our ability to observe them; the result is a *titration error*.

A common method of detecting end points involves the introduction of a supplementary substance called an *indicator,* which produces an observable physical change in the solution at or near the equivalence point. This change is caused by a large change in the relative concentration of analyte or titrant that occurs in the equivalence-point region. Typical indicator changes include the appearance or disappearance of a color, a change in color, and the appearance or disappearance of turbidity.

End points are also obtained with instruments that respond to certain properties of the solution that change in a characteristic way during the titration. Among such properties are electric potential, current, and conductance; optical absorbance; and refractive index.

4A-3 Primary Standards

Every volumetric method is based upon a *primary standard,* which is a highly purified compound that is used—directly or indirectly—to establish the concentration of the standard solution. The accuracy of a volumetric method is critically dependent on the properties of this compound. Important requirements for primary-standard substances include the following:

1. High purity. Established methods for confirming purity should also be available.
2. Stability toward air.
3. Absence of hydrate water so that compositional uncertainties that accompany the gain or loss of water as a function of relative humidity are minimized.
4. Ready availability at modest cost.
5. Reasonable solubility in the titration medium.
6. Reasonably large formula weight so that the relative error associated with weighing the standard is minimized.

The number of compounds that meet or even approach these criteria is so small that only a limited number of primary-standard substances are available to the chemist. As a consequence, it is occasionally necessary to employ a less pure substance in lieu of a primary standard. The *assay* (that is, the percent purity) of such a *secondary standard* must be established by careful analysis.

4B Standard Solutions

Standard solutions play a central role in all volumetric methods of analysis. It is therefore worthwhile examining the properties such solutions should possess, how they are prepared, and how their concentrations are expressed.

4B-1 Desirable Properties of Standard Solutions

The ideal standard solution for a volumetric method should (1) be sufficiently stable so that its concentration remains unchanged indefinitely, thus eliminating the need to reestablish its concentration periodically, (2) react rapidly with the analyte so that the time required between additions of reagent is minimized, (3) react more or less completely with the analyte so that satisfactory end points are realized, and (4) undergo a selective reaction with the analyte that can be described by a simple balanced equation. The number of reagents that meet all these ideals is relatively small.

4B-2 Methods for Establishing the Concentration of Standard Solutions

The concentration of the standard solutions employed in volumetric methods must be known with a high degree of accuracy because the ultimate reliability of a volumetric method depends directly on the quality of this parameter.

Two basic methods are used for establishing the concentration of standard solutions. The first is the *direct method,* in which a carefully weighed quantity of a primary standard is dissolved in a suitable solvent and diluted to an exactly known volume in a volumetric flask. The second is by *standardization,* in which the titrant to be standardized is used to titrate (1) a weighed quantity of a primary standard, (2) a weighed quantity of a secondary standard, or (3) a measured volume of another standard solution. A titrant that is standardized against a secondary standard or against another standard solution is sometimes referred to as a *secondary-standard solution.* A secondary-standard solution is less desirable than a primary-standard solution because its concentration is subject to greater uncertainty. Ordinarily, the best standard solutions are those prepared by the direct method. Unfortunately, this procedure is often impossible to use because of a lack of reagents that possess the properties of a primary standard.

4B-3 Methods for Expressing the Concentration of a Standard Solution

The concentrations of standard solutions are generally expressed in units of either *molarity* c or *normality* c_N. The first gives the number of moles of reagent contained in 1 L of solution, and the second gives the number of chemical equivalents of reagent in the same volume (the term *equivalent* is defined in Appendix 9).

Many chemists (perhaps a majority) feel that the chemical literature is overly cluttered with redundant terms and believe that the terms *normality* and *equivalent* offer so few real advantages that they can be abandoned without serious loss.[2] We are sympathetic to this view and base most of the calculations in this text on molarities. We also recognize, however, that normalities and equivalents abound in the chemical and biochemical literature of the last century and are also still found in various monographs and compilations of

[2]For example, *Analytical Chemistry,* a leading scientific journal of analytical chemistry, no longer permits the use of either *normality* or *titer* (Section 4D-4) in the papers it accepts for publication.

analytical methods of considerable value to present-day chemists.[3] For this reason, we have included a section in the appendix (Appendix 9) that demonstrates how volumetric computations are performed with normalities and equivalents. Readers may find this appendix useful when they have occasion to consult the analytical literature.

4C Volumetric Calculations Based on Molar Concentrations

The term *molar* M is a concentration unit that is equal to the number of moles of a solute contained in 1 L of solution or the number of millimoles of a solute in 1 mL. For example, a 0.1 M NaCl solution contains 0.1 mol of the reagent (NaCl) in each liter of solution, or 0.1 mmol in each milliliter of solution.[4]

4C-1 Molar Concentration

The molar concentration of a solution of a chemical species A is given by the relationships

$$c_A = \frac{\text{no. mmol A}}{\text{no. mL soln}} \qquad (4\text{-}1a)$$

$$c_A = \frac{\text{no. mol A}}{\text{no. L soln}} \qquad (4\text{-}1b)$$

where c_A is the molar concentration, or *molarity.*

EXAMPLE 4-1

Calculate the molar concentration of an aqueous solution of ethanol that contains 2.30 g of C_2H_5OH (fw = 46.07) in 3.50 L of solution.

Substituting into Equation 4-1b yields

$$c_{C_2H_5OH} = \frac{2.30\ \cancel{\text{g } C_2H_5OH}}{3.50\text{ L} \times 46.07\ \cancel{\text{g } C_2H_5OH}/\text{mol } C_2H_5OH}$$
$$= 0.0143\text{ mol } C_2H_5OH/\text{L} = 0.0143\text{ M}$$

It is important to note that two types of molarity are encountered: *analytical* molarity and *species,* or *equilibrium,* molarity.

Analytical Molarity

The analytical, or total, molarity of a solution gives the *total* number of moles of a solute in 1 L of the solution (or the total number of millimoles in 1 mL). That is, the analytical molarity specifies a recipe by which the solution can be prepared. For example, a sulfuric acid solution that has an analytical concen-

[3] For example, The Association of Official Analytical Chemists employs normality in its publications, which are useful sources of tested methods for the determination of a wide variety of species in materials of agriculture and commerce. See *Official Methods of Analysis,* 14th ed., Washington, D.C.: Association of Official Analytical Chemists, 1984.

[4] For many years, the fundamental unit of volume was the liter, which was defined as the volume occupied by 1 kg of water at 4°C. In the SI system, the liter is an acceptable secondary unit that is redefined as 1 dm^3. The difference between the two definitions is small: the liter based on 1 kg of water has a volume of 1.000028 dm^3. Generally, in speaking of the liter and the milliliter, we are referring to the new definitions—that is, 1 dm^3 and 1 cm^3, respectively.

tration of 1.0 M can be prepared by dissolving 1.0 mol, or 98 g, of H_2SO_4 in water and diluting to exactly 1.0 L.

Equilibrium, or Species, Molarity

The equilibrium, or species, molarity expresses the molar concentration of a particular species that exists in a solution at equilibrium. In order to state the species molarity, it is necessary to know something about the behavior of a solute when it is dissolved in a solvent. For example, the species molarity of H_2SO_4 in a solution with an analytical concentration of 1.0 M is 0.0 M since the sulfuric acid is entirely dissociated into a mixture of H_3O^+, HSO_4^- and SO_4^{2-} ions. The equilibrium concentrations of these three ions can be shown to be 1.01, 0.99, and 0.01 M, respectively.

Equilibrium molar concentrations are often symbolized by placing square brackets around the chemical formula for the species. Thus, for a solution of H_2SO_4 with an analytical concentration of 1.0 M, we can write

$$[H_2SO_4] = 0.00 \text{ M} \qquad [H_3O^+] = 1.01 \text{ M}$$

$$[HSO_4^-] = 0.99 \text{ M} \qquad [SO_4^{2-}] = 0.01 \text{ M}$$

EXAMPLE 4-2

Calculate the analytical and equilibrium molar concentrations of the solute species in an aqueous solution that contains 285 mg of trichloroacetic acid, Cl_3CCOOH (fw = 163.4), in 10.0 mL (the acid is 73% ionized in water).

Employing HA as the symbol for Cl_3CCOOH, we substitute into Equation 4-1a to obtain the analytical, or total, concentration of the acid:

$$c_{HA} = \frac{285 \cancel{\text{mg HA}}}{10.0 \text{ mL} \times 163.4 \cancel{\text{mg HA}}/\text{mmol HA}} = 0.174 \text{ mmol HA/mL} = 0.174 \text{ M}$$

Because all but 27% of the acid is dissociated into H_3O^+ and A^-, the species concentration of HA is

$$[HA] = \frac{(0.27 \times 285) \cancel{\text{mg HA}}}{10.0 \text{ mL} \times 163.4 \cancel{\text{mg HA}}/\text{mmol HA}} = 0.047 \text{ mmol/mL} = 0.047 \text{ M}$$

The molarity of H_3O^+ as well as that of A^- is equal to the analytical concentration of the acid minus the species concentration of undissociated acid:

$$[H_3O^+] = [A^-] = 0.174 - 0.047 = 0.127 \text{ mmol/mL} = 0.127 \text{ M}$$

Note that the analytical concentration of HA is the sum of the species concentrations of HA and A^-:

$$c_{HA} = [HA] + [A^-] = [HA] + [H_3O^+]$$

Some chemists prefer to distinguish between species and analytical concentrations by restricting the use of *molar concentration* to the former and using the term *formal concentration* (F) for the latter.

4C-2 Some Useful Algebraic Relationships

Most volumetric calculations are based on two pairs of simple equations. The first pair is

$$\text{No. mmol A} = \frac{\text{wt A (g)}}{\text{mfw A (g/mmol)}} \qquad (4\text{-}2a)$$

$$\text{No. mol A} = \frac{\text{wt A (g)}}{\text{fw A (g/mol)}} \qquad (4\text{-}2b)$$

where A refers to a chemical species. The second pair is obtained by rearranging Equations 4-1a and 4-1b:

$$\text{No. mmol A} = V\text{ (mL)} \times c_A\text{(mmol A/mL)} \qquad (4\text{-}3a)$$

$$\text{No. mol A} = V\text{ (L)} \times c_A\text{ (mol A/L)} \qquad (4\text{-}3b)$$

where V is the volume of the solution.

Equations 4-2a and 4-3a are more convenient to use when volumes are measured in milliliters, whereas the other two are better suited when liters are used.

4C-3

Calculation of the Molarity of Standard Solutions

Chemists prepare a standard solution by diluting to a known volume (1) a weighed quantity of a solid reagent, (2) a measured volume of a solution whose molarity is known, or (3) a measured volume of a commercial liquid reagent whose composition is given in terms of weight percentages and specific gravities.

EXAMPLE 4-3

Describe the preparation of 5.000 L of 0.1000 M Na_2CO_3 (fw = 105.99) from the primary-standard solid.

Since the volume is in liters, we employ Equation 4-3b to compute the amount of Na_2CO_3 needed:

$$\text{No. mol } Na_2CO_3 = V_{soln}\text{ (L)} \times c_{Na_2CO_3}\text{ (mol/L)}$$

$$= 5.000 \cancel{L} \times \frac{0.1000 \text{ mol } Na_2CO_3}{\cancel{L}} = 0.5000 \text{ mol } Na_2CO_3$$

To obtain the weight of Na_2CO_3, we rearrange Equation 4-2b to give

$$\text{Wt } Na_2CO_3 = 0.5000 \cancel{\text{mol } Na_2CO_3} \times \frac{105.99 \text{ g } Na_2CO_3}{\cancel{\text{mol } Na_2CO_3}} = 53.00 \text{ g } Na_2CO_3$$

Therefore the solution is prepared by dissolving 53.00 g of Na_2CO_3 in water and diluting to exactly 5.000 L.

EXAMPLE 4-4

A standard 0.0100 M solution of Na^+ is required for calibrating a flame-photometric method for determining the element. Describe how 500 mL of this solution can be prepared from primary-standard Na_2CO_3.

We wish to compute the weight of reagent required to give a species molarity of 0.0100. Equation 4-3a is preferable here, since V is in milliliters. Because Na_2CO_3 is completely dissociated into two Na^+ ions and one CO_3^{2-} ion, we can write

$$\text{No. mmol } Na_2CO_3 = 500 \cancel{\text{mL}} \times 0.0100 \frac{\cancel{\text{mmol } Na^+}}{\cancel{\text{mL}}} \times \frac{1 \text{ mmol } Na_2CO_3}{2 \cancel{\text{mmol } Na^+}} = 2.50 \text{ mmol}$$

Substituting into Equation 4-2a, we get

$$\text{Wt } Na_2CO_3 = 2.50 \cancel{\text{mmol } Na_2CO_3} \times 0.10599 \frac{\text{g } Na_2CO_3}{\cancel{\text{mmol } Na_2CO_3}} = 0.265 \text{ g}$$

The solution is therefore prepared by dissolving 0.265 g of Na_2CO_3 in water and diluting to 500 mL.

EXAMPLE 4-5

Describe how the solution in Example 4-4 could be used to prepare 50.0 mL of calibration standards that are 0.00500 M, 0.00200 M, and 0.00100 M in Na^+.

The number of millimoles of Na_2CO_3 taken from the concentrated solution must equal the number in the diluted solution:

$$\text{No. mmol } Na_2CO_3 \text{ from concd soln} = \text{no. mmol } Na_2CO_3 \text{ in dil soln}$$

Substituting the right-hand term of Equation 4-3a into both sides of this equation yields

$$V_{\text{concd}} \times c_{\text{concd}} = V_{\text{dil}} \times c_{\text{dil}}$$

For the 0.00500 M solution,

$$V_{\text{concd}} \times 0.0100 \cancel{\text{mmol/mL}} = 50.0 \text{ mL} \times 0.00500 \cancel{\text{mmol/mL}}$$

$$V_{\text{concd}} = 25.0 \text{ mL}$$

Thus 25.0 mL of the concentrated solution should be diluted to exactly 50.0 mL.

Repeating the calculation for the other two molarities reveals that diluting 10.0 and 5.00 mL of the concentrated solution to 50.0 mL produces the desired solutions.

EXAMPLE 4-6

Describe the preparation of 100 mL of approximately 6.0 M HCl from the commercial concentrated reagent. The label on the bottle states that the reagent is 37% HCl and has a specific gravity of 1.18.

Generally, the percentages employed in describing commercial reagents are weight-weight; that is, the reagent in question contains about 37 g of HCl per 100 g of the concentrated reagent. Furthermore, we assume (Section 4D-2) that the reagent has a density of 1.18 g/mL. Therefore,

$$\frac{\text{Wt HCl}}{\text{Vol HCl}} = \frac{1.18 \cancel{\text{g concd soln}}}{\text{mL concd soln}} \times \frac{37 \text{ g HCl}}{100 \cancel{\text{g concd soln}}} = 0.437 \text{ g/mL}$$

To obtain the weight of HCl required, we combine Equations 4-3a and 4-2a:

$$\text{Wt HCl required} = 100 \cancel{\text{mL dil soln}} \times \frac{6.00 \cancel{\text{mmol HCl}}}{\cancel{\text{mL dil soln}}} \times \frac{0.0365 \text{ g}}{\cancel{\text{mmol HCl}}} = 21.9 \text{ g}$$

and the volume of concentrated reagent required is

$$V = \frac{21.9 \cancel{\text{g HCl}}}{0.437 \cancel{\text{g HCl}}/\text{mL concd soln}} = 50.2 \text{ mL}$$

The solution is prepared by diluting 50 mL of the concentrated reagent to a volume of about 100 mL.

4C-4 *Treatment of Titration Data*

Two types of calculations are considered in this section. The first involves computing the molarity of solutions that have been standardized against either a primary standard or another standard solution. The second involves calculating the amount of analyte in a sample from titration data. Both types are based on three simple algebraic relationships. Two of these are Equations 4-2a and 4-3a, both of which are based on millimoles and milliliters. The third relationship is the stoichiometric ratio of the number of millimoles of the species being titrated to the number of millimoles of titrant.

In order to become familiar with the use of these three equations, we suggest that readers employ a systematic five-step approach:

1. Derive a stoichiometric relationship between the number of millimoles of the species whose concentration is being determined and the number of millimoles of the species whose concentration is known.
2. Calculate the number of millimoles of standard used in the titration by means of either Equation 4-2a or 4-3a, whichever is appropriate.
3. Convert the number of millimoles found in step 2 to the number of millimoles of the species titrated by multiplying by the stoichiometric ratio derived in step 1.
4. Equate the result from step 3 to either Equation 4-2a or 4-3a, whichever contains the quantity sought.
5. Solve the resulting equation.

Calculation of Molarities from Standardization Data

Examples 4-7 and 4-8 illustrate how this approach is applied to standardization data.

EXAMPLE 4-7

Exactly 50.00 mL of an HCl solution required 29.71 mL of 0.01963 M $Ba(OH)_2$ to reach an end point with bromocresol green indicator. Calculate the molarity of the HCl.

Step 1. In the titration, 1 mmol of $Ba(OH)_2$ reacts with 2 mmol of HCl, and thus the stoichiometric ratio is

$$\text{Stoichiometric ratio} = \frac{2 \text{ mmol HCl}}{1 \text{ mmol Ba(OH)}_2}$$

Step 2. The number of millimoles of the standard is obtained by substituting into Equation 4-3a:

$$\text{No. mmol Ba(OH)}_2 = 29.71 \cancel{\text{mL Ba(OH)}_2} \times 0.01963 \frac{\text{mmol Ba(OH)}_2}{\cancel{\text{mL Ba(OH)}_2}}$$

Step 3. To obtain the number of millimoles of HCl, we multiply the result from step 2 by the ratio derived in step 1:

$$\text{No. mmol HCl} = (29.71 \times 0.01963) \cancel{\text{mmol Ba(OH)}_2} \times \frac{2 \text{ mmol HCl}}{1 \cancel{\text{mmol Ba(OH)}_2}}$$

Step 4. Equating the result from step 3 to Equation 4-3a yields

$$\text{No. mmol HCl} = 50.00 \text{ mL HCl} \times c_{\text{HCl}} = (29.71 \times 0.01963 \times 2) \text{ mmol HCl}$$

Step 5.

$$c_{\text{HCl}} = \frac{(29.71 \times 0.01963 \times 2) \text{ mmol HCl}}{50.00 \text{ mL HCl}} = 0.02333 \frac{\text{mmol HCl}}{\text{mL HCl}}$$

$$= 0.02333 \text{ M}$$

EXAMPLE 4-8

A solution of $KMnO_4$ was standardized against 0.2121 g of pure $Na_2C_2O_4$ (fw = 134.00). If 43.31 mL of $KMnO_4$ was required to obtain an end point, what is the molarity of the $KMnO_4$ solution? The chemical reaction is

$$2MnO_4^- + 5C_2O_4^{2-} + 16H^+ \longrightarrow 2Mn^{2+} + 10CO_2 + 8H_2O$$

Step 1. From the chemical equation, we see that the stoichiometric ratio is

$$\text{Stoichiometric ratio} = \frac{2 \text{ mmol KMnO}_4}{5 \text{ mmol Na}_2\text{C}_2\text{O}_4}$$

Step 2. The number of millimoles of standard $Na_2C_2O_4$ is given by Equation 4-2a:

$$\text{No. mmol } Na_2C_2O_4 = \frac{0.2121 \cancel{\text{g } Na_2C_2O_4}}{0.13400 \cancel{\text{g } Na_2C_2O_4}/\text{mmol } Na_2C_2O_4}$$

Step 3. To obtain the number of millimoles of $KMnO_4$, we multiply the result of step 2 by the stoichiometric factor:

$$\text{No. mmol } KMnO_4 = \frac{0.2121}{0.13400} \cancel{\text{mmol } Na_2C_2O_4} \times \frac{2 \text{ mmol } KMnO_4}{5 \cancel{\text{mmol } Na_2C_2O_4}}$$

Step 4. Equating this relationship with Equation 4-3a gives

$$\text{No. mmol } KMnO_4 = 43.31 \text{ mL } KMnO_4 \times c_{KMnO_4} = \frac{0.2121}{0.13400} \times \frac{2}{5} \text{ mmol } KMnO_4$$

$$\textit{Step 5. } c_{KMnO_4} = \frac{\frac{0.2121}{0.13400} \times \frac{2}{5} \text{ mmol } KMnO_4}{43.31 \text{ mL } KMnO_4} = 0.01462 \text{ M}$$

Note that units are carried through all calculations as a check on the correctness of the relationships used in Examples 4-7 and 4-8.

Calculation of Quantity of Analyte from Titration Data

As shown by the examples that follow, the same systematic approach just described is also used to compute analyte concentrations from titration data.

EXAMPLE 4-9

A 0.8040-g sample of an iron ore was dissolved in acid. The iron was then reduced to Fe^{2+} and titrated with 47.22 mL of 0.02242 M $KMnO_4$ solution. Calculate the results of this analysis in terms of (a) % Fe (fw = 55.847) and (b) % Fe_3O_4 (fw = 231.54). The reaction of the analyte with the reagent is described by the equation

$$MnO_4^- + 5Fe^{2+} + 8H^+ \longrightarrow Mn^{2+} + 5Fe^{3+} + 4H_2O$$

(a) *Step 1.* Stoichiometric ratio $= \dfrac{5 \text{ mmol } Fe^{2+}}{1 \text{ mmol } KMnO_4}$

Step 2. $\text{No. mmol } KMnO_4 = \dfrac{47.22 \cancel{\text{mL } KMnO_4} \times 0.02242 \text{ mmol } KMnO_4}{\cancel{\text{mL } KMnO_4}}$

Step 3. $\text{No. mmol } Fe^{2+} = (47.22 \times 0.02242) \cancel{\text{mmol } KMnO_4} \times \dfrac{5 \text{ mmol } Fe^{2+}}{1 \cancel{\text{mmol } KMnO_4}}$

Step 4. Substitution of the numerical value from step 3 into Equation 4-2a gives

$$\text{No. mmol } Fe^{2+} = \frac{\text{wt } Fe^{2+} \text{ (g)}}{\text{mfw } Fe^{2+} \text{ (g/mmol)}}$$
$$= (47.22 \times 0.02242 \times 5) \text{ mmol } Fe^{2+}$$

which rearranges to

$$\text{Wt } Fe^{2+} = (47.22 \times 0.02242 \times 5) \cancel{\text{mmol } Fe^{2+}} \times 0.055847 \frac{\text{g } Fe^{2+}}{\cancel{\text{mmol } Fe^{2+}}}$$

The percent Fe^{2+} is

$$\% \; Fe^{2+} = \frac{(47.22 \times 0.02242 \times 5 \times 0.055847) \text{ g } Fe^{2+}}{0.8040 \text{ g sample}} \times 100\% = 36.77\%$$

(b) In order to derive a stoichiometric factor, we note that 5 Fe^{2+} ≡ 1 MnO_4^-. Therefore, $5Fe_3O_4$ ≡ $15Fe^{2+}$ ≡ $3MnO_4^-$, and

$$\text{Stoichiometric ratio} = \frac{5 \text{ mmol } Fe_3O_4}{3 \text{ mmol } KMnO_4}$$

As in part (a),

$$\text{No. mmol } KMnO_4 = 47.22 \text{ mL } KMnO_4 \times 0.02242 \text{ mmol } KMnO_4/\text{mL } KMnO_4$$

$$\text{No. mmol } Fe_3O_4 = (47.22 \times 0.02242) \text{ mmol } KMnO_4 \times \frac{5 \text{ mmol } Fe_3O_4}{3 \text{ mmol } KMnO_4}$$

Substituting this last equation into Equation 4-2a gives

$$\text{No. mmol } Fe_3O_4 = \frac{\text{wt } Fe_3O_4 \text{ (g)}}{\text{mfw } Fe_3O_4 \text{ (g/mmol)}} = \frac{47.22 \times 0.02242 \times 5}{3} \text{ mmol } Fe_3O_4$$

$$\text{Wt } Fe_3O_4 = (47.22 \times 0.02242 \times \tfrac{5}{3}) \text{ mmol } Fe_3O_4 \times 0.23154 \text{ g } Fe_3O_4/\text{mmol } Fe_3O_4$$

$$\% \; Fe_3O_4 = \frac{(47.22 \times 0.02242 \times \frac{5}{3}) \times 0.23154 \text{ g } Fe_3O_4}{0.8040 \text{ g sample}} \times 100\% = 50.81\%$$

EXAMPLE 4-10

The organic matter in a 3.776-g sample of a mercuric ointment was decomposed with HNO_3. After dilution, the Hg^{2+} was titrated with 21.30 mL of a 0.1144 M solution of NH_4SCN. Calculate the percent Hg (fw = 200.59) in the ointment.

This titration involves the formation of a stable neutral complex, $Hg(SCN)_2$:

$$Hg^{2+} + 2SCN^- \longrightarrow Hg(SCN)_2(aq)$$

At the equivalence point,

$$\text{Stoichiometric ratio} = \frac{1 \text{ mmol } Hg^{2+}}{2 \text{ mmol } NH_4SCN}$$

$$\text{No. mmol } NH_4SCN = 21.30 \text{ mL} \times 0.1144 \text{ mmol } NH_4SCN/\text{mL}$$

$$\text{No. mmol } Hg^{2+} = (21.30 \times 0.1144) \text{ mmol } NH_4SCN \times \frac{1 \text{ mmol } Hg^{2+}}{2 \text{ mmol } NH_4SCN}$$

$$\text{Wt } Hg^{2+} = (21.30 \times 0.1144 \times \tfrac{1}{2}) \text{ mmol } Hg^{2+} \times \frac{0.20059 \text{ g } Hg^{2+}}{\text{mmol } Hg^{2+}}$$

$$\% \text{ Hg} = \frac{(21.30 \times 0.1144 \times \frac{1}{2}) \times 0.20059 \text{ g } Hg^{2+}}{3.776 \text{ g sample}} \times 100\% = 6.47\%$$

EXAMPLE 4-11

A 0.4755-g sample containing $(NH_4)_2C_2O_4$ and inert compounds was dissolved in H_2O and made strongly alkaline with KOH, which converted NH_4^+ to NH_3. The liberated NH_3 was distilled into exactly 50.00 mL of 0.05035 M H_2SO_4. The excess H_2SO_4 was back-titrated with 11.13 mL of 0.1214 M NaOH. Calculate (a) the % N (fw = 14.007) and (b) the % $(NH_4)_2C_2O_4$ (fw = 124.10) in the sample.

(a) The H_2SO_4 reacts with both NH_3 and NaOH, and the stoichiometric relationships are

$$\text{Stoichiometric ratios} = \frac{2 \text{ mmol } NH_3}{1 \text{ mmol } H_2SO_4} \quad \text{and} \quad \frac{1 \text{ mmol } H_2SO_4}{2 \text{ mmol NaOH}}$$

$$\text{Total no. mmol } H_2SO_4 = 50.00 \text{ mL } H_2SO_4 \times 0.05035 \text{ mmol } H_2SO_4/\text{mL } H_2SO_4$$
$$= 2.5175 \text{ mmol } H_2SO_4$$

The number of millimoles of H_2SO_4 consumed by the NaOH in the back-titration is

$$\text{No. mmol } H_2SO_4 = (11.13 \times 0.1214)\ \cancel{\text{mmol NaOH}} \times \frac{1 \text{ mmol } H_2SO_4}{2\ \cancel{\text{mmol NaOH}}}$$
$$= 0.6756 \text{ mmol } H_2SO_4$$

The number of millimoles of H_2SO_4 that reacted with NH_3 is then

$$\text{No. mmol } H_2SO_4 = (2.5175 - 0.6756) = 1.8419 \text{ mmol } H_2SO_4$$

The number of millimoles of NH_3, which is equal to the number of millimoles of N, is

$$\text{No. mmol N} = \text{no. mmol } NH_3 = 1.8419\ \cancel{\text{mmol } H_2SO_4} \times \frac{2 \text{ mmol N}}{1\ \cancel{\text{mmol } H_2SO_4}}$$
$$= 3.6838 \text{ mmol N}$$

$$\%\ \text{N} = \frac{3.6838\ \cancel{\text{mmol N}} \times 0.014007 \text{ g N}/\cancel{\text{mmol N}}}{0.4755 \text{ g sample}} \times 100\% = 10.85\%$$

(b) Since each millimole of $(NH_4)_2C_2O_4$ produces 2 mmol of NH_3, which reacts with 1 mmol of H_2SO_4,

$$\text{Stoichiometric ratio} = \frac{1 \text{ mmol } (NH_4)_2C_2O_4}{1 \text{ mmol } H_2SO_4}$$

$$\text{No. mmol } (NH_4)_2C_2O_4 = 1.8419\ \cancel{\text{mmol } H_2SO_4} \times \frac{1 \text{ mmol } (NH_4)_2C_2O_4}{1\ \cancel{\text{mmol } H_2SO_4}}$$

$$\%\ (NH_4)_2C_2O_4 = \frac{1.8419\ \cancel{\text{mmol } (NH_4)_2C_2O_4} \times 0.12410 \dfrac{\text{g } (NH_4)_2C_2O_4}{\cancel{\text{mmol } (NH_4)_2C_2O_4}}}{0.4755 \text{ g sample}} \times 100\%$$
$$= 48.07\%$$

EXAMPLE 4-12

The CO concentration in a sample of gas was obtained by passing 20.3 L of the gas over iodine pentoxide heated to 150°C. The reaction is

$$I_2O_5(s) + 5CO(g) \longrightarrow 5CO_2(g) + I_2(g)$$

The iodine distilled at this temperature and was collected in an absorber containing 8.25 mL of 0.01101 M $Na_2S_2O_3$:

$$I_2(aq) + 2S_2O_3^{2-}(aq) \longrightarrow 2I^-(aq) + S_4O_6^{2-}(aq)$$

The excess $Na_2S_2O_3$ was back-titrated with 2.16 mL of 0.00947 M I_2 solution. Calculate the number of milligrams of CO (fw = 28.01) per liter of sample.

Based on the two reactions, the stoichiometric ratios are

$$\text{Stoichiometric ratios} = \frac{5 \text{ mmol CO}}{1 \text{ mmol } I_2} \quad \text{and} \quad \frac{2 \text{ mmol } Na_2S_2O_4}{1 \text{ mmol } I_2}$$

Dividing the first ratio by the second yields a third ratio that is useful:

$$\text{Stoichiometric ratio} = \frac{5 \text{ mmol CO}}{2 \text{ mmol } Na_2S_2O_3}$$

The total number of millimoles of $Na_2S_2O_3$ is

$$\text{No. mmol } Na_2S_2O_3 = 8.25\ \cancel{\text{mL } Na_2S_2O_3} \times 0.01101 \text{ mmol } Na_2S_2O_3/\cancel{\text{mL } Na_2S_2O_3}$$
$$= 0.09083 \text{ mmol } Na_2S_2O_3$$

The number of millimoles of $Na_2S_2O_3$ consumed in the back-titration is

$$\text{No. mmol Na}_2\text{S}_2\text{O}_3 = 2.16 \cancel{\text{mL I}_2} \times 0.00947 \frac{\cancel{\text{mmol I}_2}}{\cancel{\text{mL I}_2}} \times \frac{2 \text{ mmol Na}_2\text{S}_2\text{O}_3}{\cancel{\text{mmol I}_2}}$$

$$= 0.04091 \text{ mmol Na}_2\text{S}_2\text{O}_3$$

The number of millimoles of CO can then be obtained by employing the third stoichiometric ratio:

$$\text{No. mmol CO} = (0.09083 - 0.04091) \cancel{\text{mmol Na}_2\text{S}_2\text{O}_3} \times \frac{5 \text{ mmol CO}}{2 \cancel{\text{mmol Na}_2\text{S}_2\text{O}_3}}$$

$$\frac{\text{Wt CO}}{\text{Vol sample}} = \frac{0.04992 \cancel{\text{mmol Na}_2\text{S}_2\text{O}_3} \times \frac{5 \cancel{\text{mmol CO}}}{2 \cancel{\text{mmol Na}_2\text{S}_2\text{O}_3}} \times \frac{28.01 \text{ mg CO}}{\cancel{\text{mmol CO}}}}{20.3 \text{ L sample}}$$

$$= 0.172 \text{ mg CO/L}$$

4D Other Methods for Expressing Concentration

Several other methods for expressing the concentrations of solutions are encountered. The most common of these are described in this section.

4D-1 p-Functions

Scientists frequently express the concentration of a species in a dilute solution in terms of its *p-function,* or *p-value,* which is defined as the negative logarithm (to the base 10) of the molar concentration of that species. Thus, for the species X,

$$\text{pX} = -\log [\text{X}]$$

As shown by the following examples, p-values offer the advantage of allowing concentrations that vary over ten or more orders of magnitude to be expressed in terms of small numbers that are usually positive.

EXAMPLE 4-13

Calculate the p-value for each ion in a solution that is 2.00×10^{-3} M in NaCl and 5.4×10^{-4} M in HCl.

$$\text{pH} = -\log [\text{H}_3\text{O}^+] = -\log (5.4 \times 10^{-4})$$
$$= -\log 5.4 - \log 10^{-4} = -0.73 - (-4) = 3.27$$

To obtain pNa, we write

$$\text{pNa} = -\log (2.00 \times 10^{-3}) = -\log 2.00 - \log 10^{-3}$$
$$= -0.301 - (-3.00) = 2.699$$

The total Cl^- concentration is given by the sum of the two Cl^- concentrations:

$$[\text{Cl}^-] = 2.00 \times 10^{-3} + 5.4 \times 10^{-4} = 2.00 \times 10^{-3} + 0.54 \times 10^{-3} = 2.54 \times 10^{-3}$$

$$\text{pCl} = -\log 2.54 \times 10^{-3} = 2.595$$

Note that in Example 4-13 and in the one that follows, the results are rounded according to the rules listed on page 49.

EXAMPLE 4-14

Calculate the molar concentration of Ag^+ in a solution that has a pAg of 6.372.

$$\text{pAg} = -\log[\text{Ag}^+] = 6.372$$

$$\log[\text{Ag}^+] = -6.372 = -7.000 + 0.628$$

$$[\text{Ag}^+] = \text{antilog}(-7.000) \times \text{antilog}(0.628) = 10^{-7} \times 4.246 = 4.25 \times 10^{-7}$$

4D-2 Density and Specific Gravity

Density and specific gravity are concentration-related terms often encountered in the analytical literature. The *density* of a substance measures its mass per unit volume, and the *specific gravity* of a substance is the ratio of its mass to the mass of an equal volume of water at 4°C. Density has units of kilograms per liter or grams per milliliter in the metric system. Specific gravity is dimensionless and thus is not tied to any particular system of units; this term is widely used in describing items of commerce. Since the density of water is approximately 1.00 g/mL and since we employ the metric system throughout this text, *density* and *specific gravity* are used interchangeably.

4D-3 Parts per Million and Parts per Billion

For very dilute solutions, parts per million (ppm) is a convenient way to express concentration:

$$c_{\text{ppm}} = \frac{\text{weight of solute}}{\text{weight of solution}} \times 10^6 \text{ ppm}$$

where c_{ppm} is the concentration in parts per million. Obviously, the units of weight in the numerator and denominator must agree. For even more dilute solutions, 10^9 ppb rather than 10^6 ppm is employed in the foregoing equation to give the results in parts per billion (ppb). The term *parts per thousand* (ppt) is also encountered, especially in oceanography.

If the solvent is water and the quantity of solute is so small that the density of the solution is still essentially 1.00 g/mL, then

$$c_{\text{ppm}} = \frac{\text{wt solute (mg)}}{\text{vol soln (L)}} \tag{4-4}$$

EXAMPLE 4-15

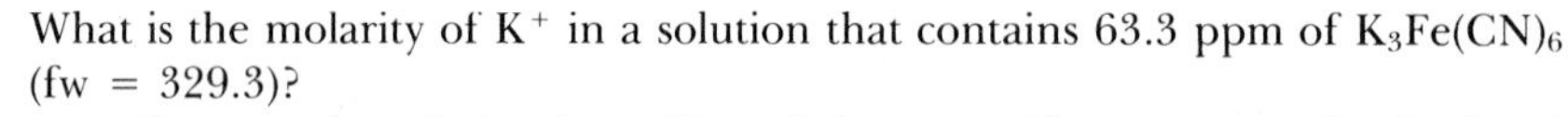

What is the molarity of K^+ in a solution that contains 63.3 ppm of $K_3Fe(CN)_6$ (fw = 329.3)?

Because the solution is so dilute, it is reasonable to assume that its density is 1.00 g/mL. Therefore, according to Equation 4-4,

$$63.3 \text{ ppm } K_3Fe(CN)_6 = 63.3 \text{ mg } K_3Fe(CN)_6/L$$

$$[K^+] = \frac{63.3 \cancel{\text{mg } K_3Fe(CN)_6}}{L} \times \frac{1 \cancel{\text{mmol } K_3Fe(CN)_6}}{329.3 \cancel{\text{mg } K_3Fe(CN)_6}} \times \frac{3 \cancel{\text{mmol } K^+}}{\cancel{\text{mmol } K_3Fe(CN)_6}} \times \frac{1 \text{ mol}}{10^3 \cancel{\text{mmol}}}$$

$$= 5.77 \times 10^{-4} \text{ mol/L}$$

4D-4 Titer

Titer defines the concentration of a solution in terms of the mass (often in milligrams) of some species that is equivalent to 1 mL of the solution.

EXAMPLE 4-16

Describe the preparation of 1.000 L of a solution of $K_2Cr_2O_7$ (fw = 294.19) that has a titer of 5.00 mg Fe_2O_3/mL, assuming that the Fe_2O_3 (fw = 159.69) is to be converted to Fe^{2+} and titrated with $Cr_2O_7^{2-}$:

$$Cr_2O_7^{2-} + 6Fe^{2+} + 14H^+ \longrightarrow 2Cr^{3+} + 6Fe^{3+} + 7H_2O$$

Note that 1 mmol of $K_2Cr_2O_7$ reacts with 6 mmol of Fe^{2+}, which is equivalent to 3 mmol of Fe_2O_3.

$$c_{K_2Cr_2O_7} = \frac{5.00\ \cancel{\text{mg Fe}_2\text{O}_3}/\text{mL K}_2\text{Cr}_2\text{O}_7}{159.69\ \cancel{\text{mg Fe}_2\text{O}_3/\text{mmol Fe}_2\text{O}_3}} \times \frac{1\ \text{mmol K}_2\text{Cr}_2\text{O}_7}{3\ \cancel{\text{mmol Fe}_2\text{O}_3}}$$

$$= 0.010437\ \text{mmol K}_2\text{Cr}_2\text{O}_7/\text{mL K}_2\text{Cr}_2\text{O}_7$$

$$\text{Wt K}_2\text{Cr}_2\text{O}_7 \text{ required} = 0.010437 \frac{\cancel{\text{mmol K}_2\text{Cr}_2\text{O}_7}}{\cancel{\text{mL K}_2\text{Cr}_2\text{O}_7}} \times 1000\ \cancel{\text{mL K}_2\text{Cr}_2\text{O}_7}$$

$$\times\ 0.29419 \frac{\text{g K}_2\text{Cr}_2\text{O}_7}{\cancel{\text{mmol K}_2\text{Cr}_2\text{O}_7}} = 3.070\ \text{g}$$

The solution is thus prepared by dissolving 3.070 g of $K_2Cr_2O_7$ in water and diluting to 1.000 L.

Titer is a useful unit for a standard solution that is to be used for the repetitive analysis of a large number of samples.

4D-5 Percent Concentration

Chemists frequently express concentrations in terms of percent (parts per hundred). Unfortunately, this practice can be a source of ambiguity because of the many ways in which the percent composition of a solution can be expressed. Three common methods are

$$\text{Weight percent (w/w)} = \frac{\text{wt solute}}{\text{wt soln}} \times 100\%$$

$$\text{Volume percent (v/v)} = \frac{\text{vol solute}}{\text{vol soln}} \times 100\%$$

$$\text{Weight-volume percent (w/v)} = \frac{\text{wt solute, g}}{\text{vol soln, mL}} \times 100\%$$

Note that the denominator in each of these expressions refers to the *solution* rather than to the solvent. Moreover, the first two expressions do not depend on the units employed (provided, of course, that there is consistency between numerator and denominator), whereas units must be defined for the third. Of the three expressions, only weight percent has the virtue of being temperature-independent.

Weight percent is frequently used to express the concentration of commercial aqueous reagents. For example, nitric acid is sold as a 70% solution, which means that the reagent contains 70 g of HNO_3 per 100 g of solution (see Example 4-6).

Weight-volume percent is often employed to indicate the composition of dilute aqueous solutions of solid reagents. For example, 5% aqueous silver nitrate *usually* refers to a solution prepared by dissolving 5 g of silver nitrate in sufficient water to give 100 mL of solution.

To avoid uncertainty, it is necessary to specify explicitly the type of percent composition used. If this information is lacking, the user is forced to decide intuitively which of the several types is involved. The potential error resulting from a wrong choice is considerable. For example, commercial 50% (w/w) sodium hydroxide contains 763 g of the reagent per liter, which corresponds to 76.3% (w/v) sodium hydroxide.

4D-6 Solution-Diluent Volume Ratios

The composition of a dilute solution is sometimes specified in terms of the volume of a more concentrated solution and the volume of solvent to be used in diluting it. The volume of the former is separated from that of the latter by a colon. Thus, a 1:4 HCl solution contains four volumes of water for each volume of concentrated hydrochloric acid. This method of notation is frequently ambiguous in that the concentration of the original solution is not always obvious to the reader; indeed, in some usages 1:4 means dilute one volume with three volumes. Because of such uncertainties, solution-diluent ratios should be avoided.

4E Weight Titrimetry

Weight or *gravimetric titrimetry* differs from its volumetric counterpart in the respect that the *mass* of titrant is measured rather than the volume. Thus, in a weight titration, a balance and a solution dispenser are substituted for a buret and its markings. Weight titrimetry actually predates volumetric titrimetry by more than 50 years.[5] With the advent of reliable burets, however, weight titrations were largely supplanted by volumetric methods because the former required relatively elaborate equipment and were tedious and time consuming. The recent advent of sensitive top-loading balances and convenient plastic solution dispensers has changed this situation completely, and weight titrations can now be performed more easily and more rapidly than volumetric titrations.

4E-1 Calculations Associated with Weight Titrations

The most convenient unit of concentration for weight titrations is *weight molarity* M_w, which is the number of moles of a reagent in one kilogram of solution or the number of millimoles in one gram of solution. Thus aqueous 0.1 M_w NaCl contains 0.1 mol of the salt in 1 kg of solution or 0.1 mmol in 1 g of the solution.

The weight molarity $c_{w(\mathrm{A})}$ of a solution of a solute A is computed by means of either of two equations that are analogous to Equations 4-1a and 4-1b:

$$c_{w(\mathrm{A})} = \frac{\text{no. mmol A}}{\text{no. g solution}} \tag{4-5a}$$

$$c_{w(\mathrm{A})} = \frac{\text{no. mol A}}{\text{no. kg solution}} \tag{4-5b}$$

Weight titration data can then be treated by the methods illustrated in Sections

[5] For a brief history of gravimetric and volumetric titrimetry, see B. Kratochvil and C. Maitra, *Amer. Lab.*, **1983** (*1*), 22.

4C-3 and 4C-4 after substitution of weight molarity for molarity and grams and kilograms for milliliters and liters.

4E-2

Advantages of Weight Titrations

In addition to greater speed and convenience, weight titrations offer certain other advantages over their volumetric counterparts:

1. Calibration of glassware and tedious cleaning to ensure proper drainage is avoided.
2. Temperature corrections are unnecessary because weight molarity does not change with temperature in contrast to volume molarity. This advantage is particularly important in nonaqueous titrations because of the high coefficients of expansion of most organic liquids (about ten times that of water).
3. Weight measurements can be made with considerably greater precision and accuracy than can volume measurements. For example, 50 or 100 g of an aqueous solution can be readily measured to ± 1 mg, which corresponds to ± 0.001 mL. This greater sensitivity makes it possible to choose sample sizes that lead to significantly smaller consumption of standard reagents.
4. Weight titrations are more easily automated than are volumetric titrations.

Questions and Problems

Unless stated otherwise, the molarity in all the following problems is analytical molarity (see Section 4C-1).

4-1 Define
- *(a) millimole.
- *(b) milliformula weight.
- (c) parts per million.
- (d) titration error.

*4-2 Distinguish between the end point and the equivalence point in a titration.

4-3 What is the difference between species molarity and analytical molarity?

*4-4 What is a primary standard?

4-5 What is a secondary standard? How does it differ from a primary standard?

4-6 How many moles and how many millimoles are contained in
- *(a) 27.3 g of Mn_3O_4?
- (b) 163 μg of BF_3?
- *(c) 6.92 L of 0.0400 M $Na_2B_4O_7$?
- (d) 10.0 mL of 2.00×10^{-3} M $HgCl_2$?
- *(e) 10.0 mL of an aqueous solution containing 143 ppm SO_2?
- (f) a solution of H_2SO_4 having a specific gravity of 1.218 and containing 30% H_2SO_4?

4-7 How many moles and how many millimoles are contained in
- *(a) 23.4 mg of $Mg_2P_2O_7$?
- (b) 100 g of dry ice (CO_2)?
- *(c) 1.00 lb of NaCl?
- (d) 7.50 L of 0.0525 M H_2CO_3?

*(e) 0.500 mL of a 0.0300 M solution of ascorbic acid (fw = 176)?
(f) 200 mL of 3.4 M KCl?

4-8 How many grams are contained in
*(a) 2.00 mol of CO_2?
(b) 1.84 mmol of benzene (fw = 78.11)?
*(c) 40.0 fw of NaOH?
(d) 6.24 mL of 0.121 M sucrose (fw = 342.3)?
*(e) 3.33 L of 12.2 M HCl?
(f) 750 mL of 4.50 M KNO_3?

*4-9 Describe the preparation of 3.00 L of 0.0800 M H_2SO_4 from
(a) 4.00 M H_2SO_4.
(b) 13.0% (w/w) H_2SO_4 solution.
(c) the concentrated reagent (specific gravity = 1.84 g/mL; % H_2SO_4 = 95).

4-10 Describe the preparation of 800 mL of 0.0500 M KOH from
(a) a 6.00 M solution.
(b) a 3.61% (w/w) KOH solution.
(c) a concentrated reagent (specific gravity = 1.505 g/mL; % KOH = 50.0).

*4-11 A solution of $HClO_4$ was standardized by dissolving 0.3745 g of primary-standard-grade HgO in a solution of KBr:

$$HgO(s) + 4Br^- + H_2O \longrightarrow HgBr_4^{2-} + 2OH^-$$

The liberated OH^- required 37.79 mL of the acid to be neutralized. Calculate the molarity of the $HClO_4$.

4-12 A 0.3367-g sample of primary-standard-grade Na_2CO_3 required 28.66 mL of a H_2SO_4 solution to reach the end point in the reaction

$$CO_3^{2-} + 2H^+ \longrightarrow H_2O + CO_2(g)$$

What is the molarity of the H_2SO_4?

*4-13 What is the molarity of an $AgNO_3$ solution that has a titer of 5.63 mg $BaCl_2 \cdot 2H_2O$/mL?

4-14 What is the molarity of a solution of $KMnO_4$ that has a titer of 11.0 mg Fe_2O_3/mL? The reaction is

$$MnO_4^- + 5Fe^{2+} + 8H^+ \longrightarrow Mn^{2+} + 5Fe^{3+} + 4H_2O$$

*4-15 Exactly 40.00 mL of a solution of $HClO_4$ was added to a solution containing 0.4793 g of primary-standard-grade Na_2CO_3. The solution was boiled to remove CO_2, and the excess $HClO_4$ was back-titrated with 8.70 mL of a NaOH solution. In a separate experiment, 25.00 mL of the NaOH neutralized 27.43 mL of $HClO_4$. Calculate the molarity of the $HClO_4$ and that of the NaOH.

4-16 A 0.3396-g sample of Na_2SO_4 that had an assay of 96.4% Na_2SO_4 was titrated with a solution of $BaCl_2$:

$$Ba^{2+} + SO_4^{2-} \longrightarrow BaSO_4(s)$$

What is the molarity of the $BaCl_2$ solution if the end point was observed when 35.70 mL of the reagent was added? What is the SO_4^{2-} titer of the solution?

*4-17 The arsenic in a 1.223-g sample of a pesticide was converted to H_3AsO_4 by suitable treatment. The acid was then neutralized, and exactly 40.00 mL of 0.07891 M $AgNO_3$ was added to precipitate the arsenic quantitatively as Ag_3AsO_4. The excess Ag^+ in the filtrate and washings from the precipitate was titrated with 11.27 mL of 0.1000 M KSCN; the reaction was

$Ag^+ + SCN^- \longrightarrow AgSCN(s)$

Calculate the percent As_2O_3 in the sample.

4-18 The ethyl acetate concentration in an alcoholic solution was determined by diluting a 10.00-mL sample to exactly 100 mL. A 20.00-mL portion of the diluted solution was refluxed with 40.00 mL of 0.04672 M KOH:

$CH_3COOC_2H_5 + OH^- \longrightarrow CH_3COO^- + C_2H_5OH$

After cooling, the excess OH^- was back-titrated with 3.41 mL of 0.05042 M H_2SO_4. Calculate the number of grams of ethyl acetate (fw = 76.10) per 100 mL of the original sample.

*4-19 The thiourea in a 1.455-g sample of organic material was extracted into a dilute H_2SO_4 solution and titrated with 37.31 mL of 0.009372 M Hg^{2+} via the reaction

$4(NH_2)_2CS + Hg^{2+} \longrightarrow [(NH_2)_2CS]_4Hg^{2+}$

Calculate the percent $(NH_2)_2CS$ in the sample.

4-20 Calculate the p-functions for each ion in a solution that is:
*(a) 0.0100 M in NaBr.
(b) 0.0100 M in $BaBr_2$.
*(c) 3.5×10^{-3} M in $Ba(OH)_2$.
(d) 0.040 M in HCl and 0.020 M in NaCl.
*(e) 5.2×10^{-3} M in $CaCl_2$ and 3.6×10^{-3} M in $BaCl_2$.
(f) 4.8×10^{-8} M in $Zn(NO_3)_2$ and 5.6×10^{-7} M $Cd(NO_3)_2$.

4-21 Convert the following p-functions to molar concentrations:
*(a) pH = 8.67
(b) pOH = 0.125
*(c) pBr = 0.034
(d) pCa = 12.35
*(e) pLi = −0.321
(f) pNO_3 = 7.77
*(g) pMn = 0.0025
(h) pCl = 1.020

*4-22 A solution of $Ba(OH)_2$ was standardized against 0.1016 g of primary-standard-grade benzoic acid, C_6H_5COOH (fw = 122.12). An end point was observed after addition of 44.42 mL of base.
(a) Calculate the molarity of the base.
(b) Calculate the standard deviation of the molarity if the standard deviation for weighing was ±0.2 mg and that for the volume measurement was ±0.03 mL.
(c) Assuming an error of −0.3 mg in the weighing, calculate the absolute and relative determinate error in the molarity.

4-23 A 0.1475 M solution of $Ba(OH)_2$ was used to titrate the acetic acid (fw = 60.05) in a dilute aqueous solution. The following results were obtained:

Sample	Sample Volume, mL	$Ba(OH)_2$ Volume, mL
1	50.00	43.17
2	49.50	42.68
3	25.00	21.47
4	50.00	43.33

(a) Calculate the mean w/v percentage of acetic acid in the sample.
(b) Calculate the standard deviation for the results.
(c) At the 90% confidence level, could any of the results be discarded?

(d) Assume that the buret used to measure out the acetic acid had a determinate error of -0.05 mL at all volumes delivered. Calculate the determinate error in the mean result.

*4-24 Calculate the molar concentration of a solution that is 25.0% in H_2SO_4 (w/w) and has a specific gravity of 1.19.

4-25 Calculate the molar concentration of a 12.0% solution (w/w) of $CuSO_4$ that has a specific gravity of 1.13.

*4-26 Describe the preparation of

(a) 200 mL of a 10.0% (w/v) aqueous glucose solution.
(b) 200 g of a 10.0% (w/w) aqueous glucose solution.
(c) 200 mL of a 10.0% (v/v) aqueous methanol solution.

4-27 Describe the preparation of

(a) 500 mL of a solution that is 1.0% (w/v) I_2 in ethanol.
(b) 500 g of a solution that is 1.0% (w/w) I_2 in ethanol.
(c) 500 mL of a 1.0% (v/v) aqueous ethanol solution.

*4-28 The sulfur in a petroleum product was determined by combusting a 4.476 g sample in a tube furnace and bubbling the combustion products through a 3% solution of H_2O_2. The SO_2 was converted to H_2SO_4:

$$SO_2(g) + H_2O_2 \longrightarrow H_2SO_4$$

A 25.00-mL portion of 0.00923 M NaOH was added to the solution and the excess base back-titrated with 13.33 mL of 0.01007 M HCl. Calculate the parts per million S in the sample.

4-29 A 100.0-mL sample of a spring water was analyzed for its Fe content by acidifying and reducing all the Fe present to Fe^{2+}. A 25.00-mL aliquot of a 0.002107-mL solution of $K_2Cr_2O_7$ was added, which resulted in the reaction

$$6Fe^{2+} + Cr_2O_7^{2-} + 14H^+ \longrightarrow 6Fe^{3+} + 2Cr^{3+} + 7H_2O$$

The excess $K_2Cr_2O_7$ was back-titrated with 7.47 mL of 0.00979 M solution of Fe^{2+}. Calculate the parts per million Fe in the sample.

*4-30 (a) A 0.1752-g sample of primary standard $AgNO_3$ was dissolved in 502.3 g of distilled water. Calculate the weight molarity of Ag^+ in this solution.
(b) The standard solution described in part (a) was used to titrate a 25.171-g sample of a KSCN solution. An end point was obtained after adding 23.765 g of the $AgNO_3$ solution. Calculate the weight molarity of the KSCN solution.
(c) The solutions described in parts (a) and (b) were used to determine the $BaCl_2 \cdot 2H_2O$ in a 0.7120-g sample. A 20.102-g portion of the $AgNO_3$ was added to a solution of the sample and the excess $AgNO_3$ was back-titrated with 7.543 g of the KSCN solution. Calculate the % $BaCl_2 \cdot 2H_2O$ in the sample.

4-31 (a) A 0.3147-g sample of primary standard $Na_2C_2O_4$ was dissolved in dilute H_2SO_4 and titrated with $KMnO_4$; 31.672 g of the $KMnO_4$ was consumed. What is the weight molarity of the $KMnO_4$ solution if the reaction was

$$2MnO_4^- + 5C_2O_4^{2-} + 16H^+ \longrightarrow 2Mn^{2+} + 10CO_2 + 8H_2O$$

(b) A 0.6656-g sample of an iron ore was dissolved and the Fe reduced quantitatively to Fe^{2+}. The resulting solution required 26.753 g of the $KMnO_4$ to reach an end point. Calculate the % Fe_2O_3 in the sample.

CHAPTER 5

A Review of Aqueous-Solution Chemistry

Most analyses are performed on solutions of an analyte with water being the most common solvent. For this reason, any discussion of the fundamental principles upon which many analytical methods are based requires a familiarity with aqueous-solution chemistry as well as an understanding of the mass relationships between reactants and products in aqueous media. This chapter provides a review of these topics.

5A The Chemical Composition of Aqueous Solutions

Water is the most plentiful solvent available on earth and as such has served for millennia as a medium for chemical processes important to human beings. In this section, we consider some important characteristics of water and aqueous solutions.

5A-1 Solutions of Electrolytes

Electrolytes are substances that dissolve in water to produce enough free ions to enhance significantly the electrical conductivity of the medium. *Strong electrolytes* are ionized essentially completely, whereas *weak electrolytes* are only partially ionized and thus impart less conductivity to the solvent. Table 5-1 is a compilation of solutes that act as strong and weak electrolytes in aqueous solutions. Among the strong electrolytes are *salts,* which are the product (along with water) of the reaction of acids with bases. Three examples of common salts are sodium chloride, barium nitrate, and potassium acetate.

TABLE 5-1 **Classification of Electrolytes**

Strong Electrolytes	Weak Electrolytes
1. The inorganic acids HNO_3, $HClO_4$, H_2SO_4,* HCl, HI, HBr, $HClO_3$, $HBrO_3$	1. Many inorganic acids, such as H_2CO_3, H_3BO_3, H_3PO_4, H_2S, H_2SO_3
2. Alkali and alkaline-earth hydroxides	2. Most organic acids
3. Most salts	3. Ammonia and most organic bases
	4. Halides, cyanides, and thiocyanates of Hg, Zn, and Cd

*H_2SO_4 is completely dissociated into HSO_4^- and H_3O^+ ions and for this reason is classified as a strong electrolyte. However, it should be noted that the HSO_4^- ion is a weak electrolyte, being only partially dissociated.

5A-2

Acids and Bases

The concept of acid/base behavior proposed independently by Brønsted and Lowry in 1923 is of particular importance to the analytical chemist.[1] According to this view, *an acid is a substance capable of donating a proton, and a base is a substance that can accept a proton.* It is important to appreciate that the proton-donating capacity of an acid is observed only in the presence of a proton acceptor, or base. Similarly, a substance manifests its proton-accepting character only in the presence of a proton donor, or acid.

An important feature of the Brønsted-Lowry concept is that each acid has associated with it a *conjugate base,* which is the entity that remains following the donation of a proton. Similarly, every base produces a *conjugate acid* as a result of accepting a proton. Examples of conjugate acid/base relationships are shown in Equations 5-1 through 5-4.

Many solvents are themselves proton donors or acceptors and can thus induce basic or acidic behavior in solutes dissolved in them. For example, in an aqueous solution of ammonia, the solvent donates a proton and thus acts as an acid with respect to the solute:

$$\underset{base_1}{NH_3} + \underset{acid_2}{H_2O} \rightleftharpoons \underset{acid_1}{NH_4^+} + \underset{base_2}{OH^-} \quad (5\text{-}1)$$

Ammonium ion is the conjugate acid of the base NH_3, and OH^- is the conjugate base of the acid H_2O. In contrast, water acts as a proton acceptor, or base in an aqueous solution of nitrous acid:

$$\underset{base_1}{H_2O} + \underset{acid_2}{HNO_2} \rightleftharpoons \underset{acid_1}{H_3O^+} + \underset{base_2}{NO_2^-} \quad (5\text{-}2)$$

Nitrite ion is the conjugate base of the acid HNO_2; H_3O^+ is the conjugate acid of the base H_2O. Neither NH_3 nor HNO_2 reacts completely with H_2O; therefore, ammonia and nitrous acid are both termed weak electrolytes.

Water is the classic example of an *amphiprotic* solvent, which is one that acts either as a donor (Equation 5-1) or as an acceptor (Equation 5-2) of protons, depending upon the solute. Other common amphiprotic solvents are methanol, ethanol, and anhydrous acetic acid. In methanol, for example, the equilibria analogous to those shown in Equations 5-1 and 5-2 are

$$\underset{base_1}{NH_3} + \underset{acid_2}{CH_3OH} \rightleftharpoons \underset{acid_1}{NH_4^+} + \underset{base_2}{CH_3O^-} \quad (5\text{-}3)$$

$$\underset{base_1}{CH_3OH} + \underset{acid_2}{HNO_2} \rightleftharpoons \underset{acid_1}{CH_3OH_2^+} + \underset{base_2}{NO_2^-} \quad (5\text{-}4)$$

It is important to recognize that an acid that has donated a proton becomes a conjugate base capable of accepting a proton to re-form the original acid; the converse holds equally well. Thus, nitrite ion, the species produced by the loss of a proton from nitrous acid, is a potential acceptor of a proton from a suitable donor. It is this reaction that causes an aqueous solution of

[1]For a thorough treatment of the various acid/base concepts, see I. M. Kolthoff, in *Treatise on Analytical Chemistry,* 2nd ed., I. M. Kolthoff and P. J. Elving, Eds., Part I, Vol. 2, Chapter 17. New York: Wiley, 1979.

sodium nitrite to be slightly basic:

$$\underset{\text{base}_1}{NO_2^-} + \underset{\text{acid}_2}{H_2O} \rightleftharpoons \underset{\text{acid}_1}{HNO_2} + \underset{\text{base}_2}{OH^-}$$

Autoprotolysis

Amphiprotic solvents undergo self-ionization, or *autoprotolysis*, to form a pair of ionic species. Autoprotolysis is yet another example of an acid/base reaction, as illustrated by the following equations:

$$\begin{array}{llll} \text{base}_1 & + \text{ acid}_2 & \rightleftharpoons \text{ acid}_1 & + \text{ base}_2 \\ H_2O & + H_2O & \rightleftharpoons H_3O^+ & + OH^- \\ CH_3OH & + CH_3OH & \rightleftharpoons CH_3OH_2^+ & + CH_3O^- \\ HCOOH & + HCOOH & \rightleftharpoons HCOOH_2^+ & + HCOO^- \\ NH_3 & + NH_3 & \rightleftharpoons NH_4^+ & + NH_2^- \end{array}$$

The product H_3O^+, called the *hydronium ion,* consists of a proton covalently bonded to the parent molecule by one of the unshared electron pairs of the oxygen. Higher hydrates, such as $H_5O_2^+$ and $H_9O_4^+$, also exist, but they are orders of magnitude less stable than H_3O^+. Essentially no unhydrated protons appear to exist in aqueous solutions.[2]

To emphasize the extraordinary stability of the singly hydrated proton, many chemists use the notation H_3O^+ when writing equations containing the proton. Others use H^+ because this notation simplifies the balancing of equations in which the proton is a participant. Ordinarily, we shall use H_3O^+ in equilibrium calculations but the simpler H^+ notation otherwise.

The extent to which water undergoes autoprotolysis is slight at room temperature, with the hydronium and hydroxide ion concentrations in pure water being only about 10^{-7} M. Nevertheless, this dissociation reaction is of utmost importance in understanding the behavior of aqueous solutions.

Strengths of Acids and Bases

Figure 5-1 shows the dissociation reaction of a few common acids in water. The first two are *strong acids* because reaction with the solvent is sufficiently complete as to leave no undissociated solute molecules in aqueous solution. The remainder are *weak acids,* which react incompletely with water to give

FIGURE 5-1 Dissociation reactions and relative strengths of some common weak acids and their conjugate bases.

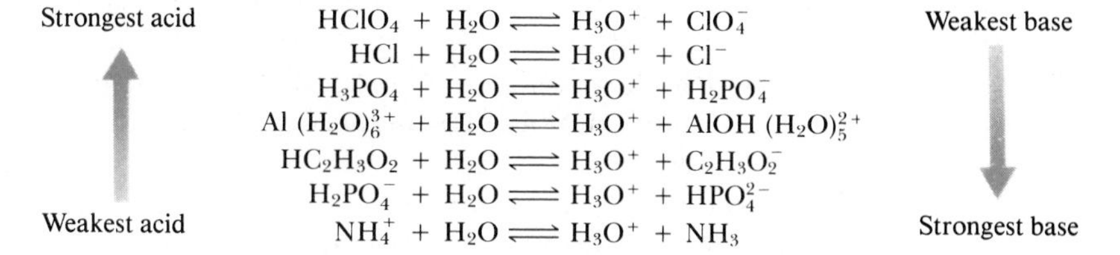

[2]See P. A. Giguère, *J. Chem. Educ.*, **1979,** *56,* 571.

solutions that contain significant quantities of both the parent acid and its conjugate base. Note that acids can be cationic, anionic, or electrically neutral.

The acids in Figure 5-1 become progressively weaker from top to bottom. Perchloric acid and hydrochloric acid are completely dissociated, whereas ammonium ion is dissociated only to a few thousandths of a percent. Another generality illustrated in Figure 5-1 is that the weakest acid forms the strongest conjugate base; that is, ammonia has a much stronger affinity for protons than any base above it.

The tendency of a solvent to accept or donate protons determines in large measure the strength of a solute acid or base dissolved in it. For example, perchloric, hydrobromic, and hydrochloric acids are all strong acids in water. If anhydrous acetic acid, a poorer proton acceptor than water, is substituted *as the solvent,* none of these acids undergoes complete dissociation; instead, equilibria such as the following are established:

$$\underset{\text{base}_1}{CH_3COOH} + \underset{\text{acid}_2}{HClO_4} \rightleftharpoons \underset{\text{acid}_1}{CH_3COOH_2^+} + \underset{\text{base}_2}{ClO_4^-}$$

It is noteworthy that perchloric acid is considerably stronger than hydrochloric or hydrobromic acid in this solvent, its dissociation being about 5000 times greater than that of hydrochloric acid. Acetic acid thus acts as a *differentiating* solvent toward these acids in the sense that its use reveals inherent differences in their acidities. Water, on the other hand, is a *leveling* solvent for these three acids in that all are completely ionized in it and thus exhibit no differences in strength.

5B Chemical Equilibrium

Few ideas are more pervasive throughout the field of analytical chemistry than the concept of chemical equilibrium. The notion that a reaction is never complete but rather is characterized by a condition in which the ratio between reactants and products is fixed crops up time and again throughout this text.

Equilibrium constant expressions are *algebraic* equations that relate the molar concentrations of reactants and products to one another by means of a numerical quantity called an *equilibrium constant.* The ability to extract useful information from equilibrium constants is essential to the study of analytical chemistry.

5B-1 The Equilibrium State

For purposes of discussion, consider the equilibrium

$$H_3AsO_4 + 3I^- + 2H^+ \rightleftharpoons H_3AsO_3 + I_3^- + H_2O \qquad (5\text{-}5)$$

The rate of this reaction and the extent to which it proceeds to the right can be readily judged by observing the orange-red color imparted to the solution by the triiodide ion (the other participants in the reaction are colorless). If, for example, 1 mmol of arsenic acid is added to 100 mL of a solution containing 3 mmol of potassium iodide, color appears almost immediately and, within a few seconds, the color intensity becomes constant with time, showing that the triiodide concentration has become invariant.

A solution of identical color intensity (and hence identical triiodide concentration) can be produced by adding 1 mmol of arsenous acid to 100 mL of a solution containing 1 mmol of triiodide ion. Here, the color intensity is initially greater than in the first solution but rapidly decreases as a result of the reaction

$$H_3AsO_3 + I_3^- + H_2O \rightleftharpoons H_3AsO_4 + 3I^- + 2H^+$$

Ultimately the color of the two solutions will be identical. Many other combinations of the four reactants could be employed to yield solutions indistinguishable from the two just described.

The foregoing examples illustrate that the concentration relationship at chemical equilibrium (that is, the *position of equilibrium*) is independent of the route by which the equilibrium state is achieved. However, it is readily shown that this relationship is altered by the application of stress to the system—for example, by changes in temperature, in pressure (if one of the reactants or products is a gas), or in total concentration of a reactant or product. These effects can be predicted qualitatively from the *principle of Le Châtelier,* which states that the position of chemical equilibrium always shifts in a direction that tends to relieve the effect of an applied stress. Thus, an increase in temperature alters the concentration relationship in the direction that tends to absorb heat, and an increase in pressure favors those participants that occupy the smaller total volume. Of particular importance in an analysis is the effect of introducing an additional amount of a participating species to the reaction mixture. Here, the resulting stress is relieved by a shift in equilibrium in the direction that partially uses up the added substance. Thus, for the equilibrium we have been considering (Equation 5-5), the addition of arsenic acid or hydrogen ions causes an increase in color as more triiodide ion and arsenous acid are formed; the addition of arsenous acid has the reverse effect. An equilibrium shift brought about by changing the amount of one of the participating species is called a *mass-action effect.*

If it were possible to examine the system under discussion at the molecular level, it would be found that interactions among the species present continue unabated even after equilibrium is achieved. The observed constant concentration relationship is thus the consequence of an equality in the rates of the forward and reverse reactions; in other words, chemical equilibrium is a dynamic state.

5B-2 Equilibrium-Constant Expressions

The influence of concentration (or pressure if the species are gases) on the position of a chemical equilibrium is conveniently described in quantitative terms by means of an equilibrium-constant expression. Such expressions are readily derived from thermodynamic theory and are of great practical importance because they permit the chemist to predict the direction and completeness of a chemical reaction. It is important to appreciate, however, that an equilibrium-constant expression yields no information concerning the *rate* at which equilibrium is approached. Thus, we encounter on occasion reactions that have highly favorable equilibrium constants but are of little analytical use because their rates are so very small. To be sure, this limitation can often be overcome by the use of a catalyst, which speeds the attainment of equilibrium without changing its position.

Let us consider a generalized equation for a chemical equilibrium:

$$mM + nN \rightleftharpoons pP + qQ \qquad (5\text{-}6)$$

where the capital letters represent the formulas of participating chemical species and the lower-case italic letters are the small whole numbers required to balance the equation. Thus, the equation states that *m* mol of M reacts with *n* mol of N to form *p* mol of P and *q* mol of Q. The equilibrium-constant expression for this reaction is

$$\frac{[P]^p[Q]^q}{[M]^m[N]^n} = K \qquad (5\text{-}7)$$

where the letters in brackets represent the molar concentrations of dissolved solutes or partial pressures (in atmospheres) if the reacting substances are gases.

The constant *K* in Equation 5-7 is a temperature-dependent numerical quantity called the *equilibrium constant.* By convention, the concentrations of the products, *as the equation is written,* are always placed in the numerator and the concentrations of the reactants in the denominator. Note that each concentration is raised to a power identical to the integer stoichiometric coefficient of the formula of that species in the balanced equation describing the equilibrium.

Equation 5-7 is a statement of the *mass law,* which is a *limiting law* in the sense that it applies strictly only when the solution is so dilute that the behavior of any participant in the equilibrium is unaffected by the presence of other participants or *other solute species.* That is, Equation 5-7 describes *ideal* solute behavior in *infinitely dilute* solutions. Unfortunately, the solutions encountered in analytical chemistry are usually far from ideal because they typically contain moderate to high concentrations of electrolytes; the resulting charged environment influences the position of any equilibrium that involves ionic species. As shown in Section 5C, Equation 5-7 can be corrected so that it is applicable to solutions containing moderate electrolyte concentrations. The corrections are somewhat tedious to apply, however, and are usually dispensed with unless failure to use them creates serious errors.

5B-3 Common Types of Equilibrium Constants Encountered in Analytical Chemistry

Table 5-2 summarizes the types of chemical equilibria that are of importance in analytical chemistry. Simple applications of these constants are illustrated in the paragraphs that follow.

The Ion-Product Constant for Water

Aqueous solutions contain small amounts of hydronium and hydroxide ions as a consequence of the dissociation reaction

$$2H_2O \rightleftharpoons H_3O^+ + OH^-$$

An equilibrium constant for this reaction can be formulated as shown in Equation 5-7:

$$\frac{[H_3O^+][OH^-]}{[H_2O]^2} = K \qquad (5\text{-}8)$$

In dilute aqueous solutions, however, the concentration of water is large compared with the concentration of its ions and can be considered to be invariant. That is, the quantity of H_2O molecules in each liter of a dilute aqueous solution (approximately 55 mol) is enormous compared with the amount formed or

TABLE 5-2 **Equilibria and Equilibrium Constants of Importance to Analytical Chemistry**

Type of Equilibrium	Name and Symbol of Equilibrium Constant	Typical Example	Equilibrium-Constant Expression
Dissociation of water	Ion-product constant, K_w	$2H_2O \rightleftharpoons H_3O^+ + OH^-$	$K_w = [H_3O^+][OH^-]$
Heterogeneous equilibrium between a slightly soluble substance and its ions in a saturated solution	Solubility product, K_{sp}	$BaSO_4(s) \rightleftharpoons Ba^{2+} + SO_4^{2-}$	$K_{sp} = [Ba^{2+}][SO_4^{2-}]$
Dissociation of a weak acid or base	Dissociation constant, K_a or K_b	$CH_3COOH + H_2O \rightleftharpoons H_3O^+ + CH_3COO^-$	$K_a = \frac{[H_3O^+][CH_3COO^-]}{[CH_3COOH]}$
		$CH_3COO^- + H_2O \rightleftharpoons OH^- + CH_3COOH$	$K_b = \frac{[OH^-][CH_3COOH]}{[CH_3COO^-]}$
Formation of a complex ion	Formation constant, K_f	$Ni^{2+} + 4CN^- \rightleftharpoons Ni(CN)_4^{2-}$	$K_f = \beta_4 = \frac{[Ni(CN)_4^{2-}]}{[Ni^{2+}][CN^-]^4}$
Oxidation/reduction equilibrium	K_{redox}	$MnO_4^- + 5Fe^{2+} + 8H^+ \rightleftharpoons Mn^{2+} + 5Fe^{3+} + 4H_2O$	$K_{redox} = \frac{[Mn^{2+}][Fe^{3+}]^5}{[MnO_4^-][Fe^{2+}]^5[H^+]^8}$
Distribution equilibrium between immiscible solvents	K_d	$I_2(aq) \rightleftharpoons I_2(org)$	$K_d = \frac{[I_2]_{org}}{[I_2]_{aq}}$

lost through any equilibrium shift. Therefore, $[H_2O]$ in Equation 5-8 can be taken as constant and we can write

$$[H_3O^+][OH^-] = K[H_2O]^2 = K_w \qquad (5\text{-}9)$$

where the new constant K_w is given a special name, the *ion-product constant for water.*

At 25°C, the ion-product constant for water is 1.008×10^{-14} (for convenience, we normally use the approximation that $K_w \simeq 1.00 \times 10^{-14}$). Table 5-3 shows the dependence of this constant upon temperature.

The ion-product constant for water permits the ready calculation of the hydronium and hydroxide ion concentrations in aqueous solutions.

TABLE 5-3

Variation of K_w with Temperature

Temperature, °C	K_w
0	0.114×10^{-14}
25	1.01×10^{-14}
50	5.47×10^{-14}
100	49×10^{-14}

EXAMPLE 5-1

Calculate the hydronium and hydroxide ion concentrations of pure water at 25 and 100°C.

Because OH^- and H_3O^+ are formed only from the dissociation of water, their concentrations must be equal:

$$[H_3O^+] = [OH^-]$$

Substitution into Equation 5-9 gives

$$[H_3O^+]^2 = [OH^-]^2 = K_w$$

$$[H_3O^+] = [OH^-] = \sqrt{K_w}$$

At 25°C,

$$[H_3O^+] = [OH^-] = \sqrt{1.00 \times 10^{-14}} = 1.00 \times 10^{-7}$$

At 100°C, from Table 5-3,

$$[H_3O^+] = [OH^-] = \sqrt{49 \times 10^{-14}} = 7.0 \times 10^{-7}$$

EXAMPLE 5-2

Calculate the hydronium and hydroxide ion concentrations in 0.200 M aqueous NaOH.

Sodium hydroxide is a strong electrolyte, and its contribution to the hydroxide ion concentration in this solution is 0.200 mol/L. As in Example 5-1, hydroxide ions and hydronium ions from the dissociation of water are formed *in equal amounts.* Therefore, we write

$$[OH^-] = 0.200 + [H_3O^+]$$

where $[H_3O^+]$ accounts for the hydroxide ions contributed by the solvent. The concentration of OH^- from the water is certainly insignificant, however, when compared with 0.200, and we can therefore write

$$[OH^-] \simeq 0.200$$

Equation 5-9 can then be used to calculate the hydronium ion concentration:

$$[H_3O^+] = \frac{K_w}{[OH^-]} = \frac{1.00 \times 10^{-14}}{0.200} = 5.00 \times 10^{-14}$$

Note that the approximation

$$[OH^-] = 0.200 + 5.00 \times 10^{-14} \simeq 0.200$$

causes no significant error.

The Solubility-Product Constant

Most sparingly soluble salts are essentially completely dissociated in saturated aqueous solution. For example, when an excess of silver carbonate is equilibrated with water, the dissociation process is adequately described by the equation

$$Ag_2CO_3(s) \rightleftharpoons 2Ag^+ + CO_3^{2-}$$

Application of Equation 5-7 leads to

$$\frac{[Ag^+]^2[CO_3^{2-}]}{[Ag_2CO_3(s)]} = K$$

The denominator represents the molar concentration of Ag_2CO_3 *in the solid*, which exists in a phase that is separate from the saturated solution. The concentration of a compound in its solid state is, however, invariant. In other words, the number of moles of Ag_2CO_3 divided by the *volume* of the Ag_2CO_3 is constant no matter how much excess solid is present. Therefore, the foregoing equation can be rewritten in the form

$$[Ag^+]^2[CO_3^{2-}] = K[Ag_2CO_3(s)] = K_{sp} \qquad (5\text{-}10)$$

where the new constant is called the *solubility-product constant* or the *solubility product.* It is important to appreciate that Equation 5-10 states in effect that the position of this equilibrium is independent of the *amount* of Ag_2CO_3 present, be it a few milligrams or several grams. For Equation 5-10 to obtain, it is necessary only that *some solid be present. If no $Ag_2CO_3(s)$ is present, Equation 5-10 is not applicable.*

A table of solubility products for numerous inorganic salts is found in Appendix 3. The examples that follow demonstrate some typical uses of solubility-product expressions. Other applications are considered in Chapters 6 and 7.

The Solubility of a Precipitate in Pure Water. The solubility-product expression permits the ready calculation of the solubility of a sparingly soluble substance that ionizes in water.

EXAMPLE 5-3

How many grams of $Ba(IO_3)_2$ (fw = 487) can be dissolved in 500 mL of water at 25°C?

The solubility-product constant for $Ba(IO_3)_2$ is 1.57×10^{-9} (Appendix 3). The equilibrium between the solid and its ions in solution is described by the equation

$$Ba(IO_3)_2(s) \rightleftharpoons Ba^{2+} + 2IO_3^-$$

and so

$$[Ba^{2+}][IO_3^-]^2 = 1.57 \times 10^{-9} = K_{sp}$$

The equation describing the equilibrium reveals that 1 mol of Ba^{2+} is formed for each mole of $Ba(IO_3)_2$ that dissolves. Therefore,

$$\text{Molar solubility of } Ba(IO_3)_2 = [Ba^{2+}]$$

The iodate concentration is clearly twice that for barium ion:

$$[IO_3^-] = 2[Ba^{2+}]$$

Substituting this last equation into the equilibrium-constant expression gives

$$[Ba^{2+}](2[Ba^{2+}])^2 = 1.57 \times 10^{-9}$$

$$[Ba^{2+}] = \left(\frac{1.57 \times 10^{-9}}{4}\right)^{1/3} = 7.32 \times 10^{-4}$$

Since 1 mol Ba^{2+} is produced for every mole of $Ba(IO_3)_2$

$$\text{Solubility} = 7.32 \times 10^{-4} \text{ M}$$

To obtain the weight of $Ba(IO_3)_2$ dissolved in 500 mL of solution, we write

$$\text{Wt } Ba(IO_3)_2 = 7.32 \times 10^{-4} \frac{\cancel{\text{mmol } Ba(IO_3)_2}}{\cancel{\text{mL}}} \times 500 \cancel{\text{mL}} \times 0.487 \frac{\text{g } Ba(IO_3)_2}{\cancel{\text{mmol } Ba(IO_3)_2}}$$

$$= 0.178 \text{ g}$$

The Solubility of a Precipitate in the Presence of a Common Ion. The common-ion effect predicted from the Le Châtelier principle is demonstrated by the following examples.

EXAMPLE 5-4

Calculate the molar solubility of $Ba(IO_3)_2$ in a solution that is 0.0200 M in $Ba(NO_3)_2$.

The solubility is no longer equal to $[Ba^{2+}]$ because $Ba(NO_3)_2$ is also present. We know, however, that it is related to $[IO_3^-]$ by the relationship

$$\text{Molar solubility of } Ba(IO_3)_2 = \tfrac{1}{2}[IO_3^-]$$

There are two sources of barium ions: $Ba(NO_3)_2$ and $Ba(IO_3)_2$. The contribution from the former is 0.0200 M, and that from the latter is equal to the molar solubility, or $\frac{1}{2}[IO_3^-]$. Thus,

$$[Ba^{2+}] = 0.0200 + \tfrac{1}{2}[IO_3^-]$$

Substitution of these quantities into the solubility-product expression yields

$$(0.0200 + \tfrac{1}{2}[IO_3^-])[IO_3^-]^2 = 1.57 \times 10^{-9}$$

Since the exact solution for $[IO_3^-]$ requires solving a cubic equation, we seek an approximation that simplifies the algebra. The small numerical value of K_{sp} suggests that the solubility of $Ba(IO_3)_2$ is not large, and this is confirmed by the result obtained in Example 5-3. Therefore, it is reasonable to suppose that the barium ion concentration arising from the solubility of $Ba(IO_3)_2$ is small relative to that from the $Ba(NO_3)_2$. That is, $\frac{1}{2}[IO_3^-] \ll 0.0200$, and so

$$[Ba^{2+}] = 0.0200 + \tfrac{1}{2}[IO_3^-] \simeq 0.0200$$

The original equation then simplifies to

$$0.0200[IO_3^-]^2 = 1.57 \times 10^{-9}$$

$$[IO_3^-] = \sqrt{7.85 \times 10^{-8}} = 2.80 \times 10^{-4}$$

The assumption that $(0.0200 + \frac{1}{2} \times 2.80 \times 10^{-4}) \simeq 0.0200$ does not appear to cause serious error because the second term, representing the amount of Ba^{2+} arising from the dissociation of $Ba(IO_3)_2$, is only about 0.7% of 0.0200. Ordinarily, we consider an assumption of this type to be satisfactory if the discrepancy is less than 10%. Finally, then,

$$\text{Solubility of } Ba(IO_3)_2 = \tfrac{1}{2}[IO_3^-] = \tfrac{1}{2} \times 2.80 \times 10^{-4} = 1.40 \times 10^{-4}\ M$$

If we compare this result with the solubility of barium iodate in pure water (Example 5-3), we see that the presence of a small concentration of the common ion has lowered the molar solubility of $Ba(IO_3)_2$ by a factor of about 5.

EXAMPLE 5-5

Calculate the solubility of $Ba(IO_3)_2$ in the solution that results when 200 mL of 0.0100 M $Ba(NO_3)_2$ is mixed with 100 mL of 0.100 M $NaIO_3$.

We must first establish whether either reactant is present in excess at equilibrium. The amounts taken are

$$\text{No. mmol } Ba^{2+} = 200\ mL \times 0.0100\ mmol/mL = 2.00$$

$$\text{No. mmol } IO_3^- = 100\ mL \times 0.100\ mmol/mL = 10.0$$

If formation of $Ba(IO_3)_2$ is complete,

$$\text{No. mmol excess } NaIO_3 = 10.0 - 2(2.00) = 6.0$$

Thus,

$$[IO_3^-] = \frac{6.0\ mmol}{300\ mL} = 0.0200\ M$$

As in Example 5-3,

$$\text{Molar solubility of } Ba(IO_3)_2 = [Ba^{2+}]$$

Here, however,

$$[c_{NaIO_3}] = 0.0200 + 2[Ba^{2+}]$$

where $2[Ba^{2+}]$ represents the iodate contributed by the sparingly soluble $Ba(IO_3)_2$. We can obtain a provisional answer after making the assumption that $[IO_3^-] \simeq 0.0200$ and so

$$\text{Solubility of } Ba(IO_3)_2 = [Ba^{2+}] = \frac{K_{sp}}{[IO_3^-]^2} = \frac{1.57 \times 10^{-9}}{(0.0200)^2} = 3.93 \times 10^{-6}\ mol/L$$

The approximation appears to be reasonable.

Note that the results from the last two examples demonstrate that an excess of iodate ions is more effective in decreasing the solubility of $Ba(IO_3)_2$ than is the same excess of barium ions.

Acid and Base Dissociation Constants

When a weak acid or a weak base is dissolved in water, partial dissociation occurs. Thus, for nitrous acid, we can write

$$HNO_2 + H_2O \rightleftharpoons H_3O^+ + NO_2^- \qquad \frac{[H_3O^+][NO_2^-]}{[HNO_2]} = K_a$$

where K_a is the *acid dissociation constant* for nitrous acid. In an analogous way, the *base dissociation constant* for ammonia is

$$NH_3 + H_2O \rightleftharpoons NH_4^+ + OH^- \qquad \frac{[NH_4^+][OH^-]}{[NH_3]} = K_b$$

Note that $[H_2O]$ does not appear in the denominator of either equation because the concentration of water is so large relative to the concentration of the weak acid or base that the dissociation does not alter $[H_2O]$ appreciably. Thus, as in the derivation of the ion-product constant for water, $[H_2O]$ is incorporated in the equilibrium constants K_a and K_b. Dissociation constants for weak acids and weak bases are found in Appendixes 4 and 5.

Relationship Between Dissociation Constants for Conjugate Acid/Base Pairs. Consider the dissociation-constant expressions for ammonia and its conjugate acid, ammonium ion:

$$NH_3 + H_2O \rightleftharpoons NH_4^+ + OH^- \qquad \frac{[NH_4^+][OH^-]}{[NH_3]} = K_b$$

$$NH_4^+ + H_2O \rightleftharpoons NH_3 + H_3O^+ \qquad \frac{[NH_3][H_3O^+]}{[NH_4^+]} = K_a$$

Multiplication of one equilibrium-constant expression by the other gives

$$\frac{\cancel{[NH_3]}[H_3O^+]}{\cancel{[NH_4^+]}} \times \frac{\cancel{[NH_4^+]}[OH^-]}{\cancel{[NH_3]}} = [H_3O^+][OH^-] = K_aK_b$$

but

$$[H_3O^+][OH^-] = K_w$$

and therefore

$$K_aK_b = K_w \qquad (5\text{-}11)$$

This relationship is general for all conjugate acid/base pairs. Most tables of dissociation constants do not list both the acid and the base dissociation constants for conjugate pairs since it is so easy to calculate one from the other with Equation 5-11.

EXAMPLE 5-6

What is K_b for the equilibrium

$$CN^- + H_2O \rightleftharpoons HCN + OH^-$$

Examination of Appendix 5 (dissociation constants for bases) reveals no entry for CN^-. Appendix 4, however, lists a K_a value of 2.1×10^{-9} for HCN. Thus,

$$\frac{[HCN][OH^-]}{[CN^-]} = K_b = \frac{K_w}{K_{HCN}}$$

$$K_b = \frac{1.00 \times 10^{-14}}{2.1 \times 10^{-9}} = 4.8 \times 10^{-6}$$

Hydronium Ion Concentration in Solutions of Weak Acids. When the weak acid HA is dissolved in water, two equilibria are established that yield hydronium ions:

$$HA + H_2O \rightleftharpoons H_3O^+ + A^- \qquad \frac{[H_3O^+][A^-]}{[HA]} = K_a$$

$$2H_2O \rightleftharpoons H_3O^+ + OH^- \qquad [H_3O^+][OH^-] = K_w$$

Ordinarily, the hydronium ions produced from the first reaction suppress the dissociation of water to such an extent that the contribution of hydronium ions from the second equilibrium is negligible. Under these circumstances, one H_3O^+ ion is formed for each A^- ion and we write

$$[A^-] \simeq [H_3O^+] \qquad (5\text{-}12)$$

Furthermore, the sum of the molar concentrations of the weak acid and its conjugate base must equal the analytical concentration of the acid c_{HA} because the solution contains no other source of A^- ions. Thus,

$$c_{HA} = [A^-] + [HA] \qquad (5\text{-}13)$$

When we substitute $[H_3O^+]$ for $[A^-]$ (Equation 5-12) into Equation 5-13, we obtain, after rearranging,

$$[HA] = c_{HA} - [H_3O^+] \qquad (5\text{-}14)$$

When $[A^-]$ and $[HA]$ are replaced by their equivalent terms from Equations 5-12 and 5-14, the equilibrium constant expression becomes

$$\frac{[H_3O^+]^2}{c_{HA} - [H_3O^+]} = K_a \qquad (5\text{-}15)$$

which rearranges to

$$[H_3O^+]^2 + K_a[H_3O^+] - K_a c_{HA} = 0$$

The positive solution to this quadratic equation is

$$[H_3O^+] = \frac{-K_a + \sqrt{K_a^2 + 4K_a c_{HA}}}{2} \qquad (5\text{-}16)$$

As an alternative to using Equation 5-16, the quadratic equation may be solved by successive approximations as shown in Appendix 11. Equation 15-15 can frequently be simplified by making the assumption that dissociation does not appreciably decrease the molar concentration of HA. Thus, provided $[H_3O^+] \ll c_{HA}$, $c_{HA} - [H_3O^+] \cong c_{HA}$ and Equation 5-15 reduces to

$$\frac{[H_3O^+]^2}{c_{HA}} = K_a$$

$$[H_3O^+] = \sqrt{K_a c_{HA}} \qquad (5\text{-}17)$$

The magnitude of the error introduced by the assumption that $[H_3O^+] \ll c_{HA}$ increases as the molar concentration of acid becomes smaller and as the dissociation constant for the acid becomes larger. This statement is supported by the data in Table 5-4. Note that the error introduced by the assumption is about 0.5% when the ratio c_{HA}/K_a is 10^4. The error increases to about 1.6% when the ratio is 10^3, to about 5% when it is 10^2, and to about 17% when it is 10. Figure 5-2 illustrates the effect graphically. It is noteworthy that the hydronium ion concentration from the approximate solution becomes

TABLE 5-4

Error Introduced by Assuming H_3O^+ Concentration Is Small Relative to c_{HA} in Equation 5-15

K_a	c_{HA}	$[H_3O^+]$ Using Assumption	$[H_3O^+]$ Using More Exact Equation	Percent Error
1.00×10^{-2}	1.00×10^{-3}	3.16×10^{-3}	0.92×10^{-3}	244
	1.00×10^{-2}	1.00×10^{-2}	0.62×10^{-2}	61
	1.00×10^{-1}	3.16×10^{-2}	2.70×10^{-2}	17
1.00×10^{-4}	1.00×10^{-4}	1.00×10^{-4}	0.62×10^{-4}	61
	1.00×10^{-3}	3.16×10^{-4}	2.70×10^{-4}	17
	1.00×10^{-2}	1.00×10^{-3}	0.95×10^{-3}	5.3
	1.00×10^{-1}	3.16×10^{-3}	3.11×10^{-3}	1.6
1.00×10^{-6}	1.00×10^{-5}	3.16×10^{-6}	2.70×10^{-6}	17
	1.00×10^{-4}	1.00×10^{-5}	0.95×10^{-5}	5.3
	1.00×10^{-3}	3.16×10^{-5}	3.11×10^{-5}	1.6
	1.00×10^{-2}	1.00×10^{-4}	9.95×10^{-5}	0.5
	1.00×10^{-1}	3.16×10^{-4}	3.16×10^{-4}	0.0

equal to or greater than the molar concentration of the acid when the ratio is unity or smaller, which is clearly a meaningless result.

In general, it is good practice to make the simplifying assumption and obtain a trial value for $[H_3O^+]$ that can be compared with c_{HA} in Equation 5-14. If the trial value alters [HA] by an amount smaller than the allowable error in the calculation, the solution may be considered satisfactory. Otherwise, the quadratic equation must be solved to obtain a better value for $[H_3O^+]$. Alternatively, the method of successive approximations (Appendix 11) may be employed.

FIGURE 5-2

Relative error resulting from the assumption that $c_{HA} - [H_3O^+] = c_{HA}$ in Equation 5-15.

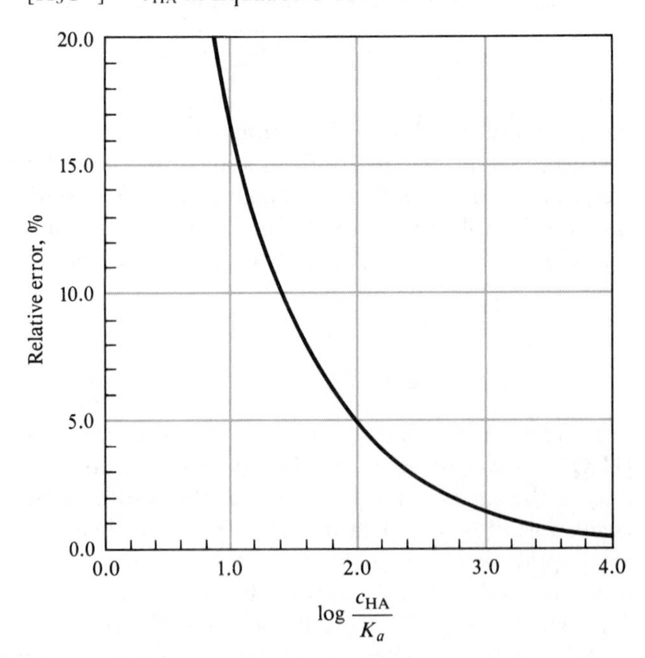

EXAMPLE 5-7

Calculate the hydronium ion concentration in an aqueous 0.120 M nitrous acid solution. The principal equilibrium is

$$HNO_2 + H_2O \rightleftharpoons H_3O^+ + NO_2^-$$

for which (Appendix 4)

$$\frac{[H_3O^+][NO_2^-]}{[HNO_2]} = K_a = 5.1 \times 10^{-4}$$

Substitution into Equations 5-12 and 5-14 gives

$$[H_3O^+] = [NO_2^-]$$

$$[HNO_2] = 0.120 - [H_3O^+]$$

When these relationships are introduced into the expression for K_a, we obtain

$$\frac{[H_3O^+]^2}{0.120 - [H_3O^+]} = 5.1 \times 10^{-4}$$

If we now assume $[H_3O^+] \ll 0.120$, we find

$$[H_3O^+] = \sqrt{0.120 \times 5.1 \times 10^{-4}} = 7.8 \times 10^{-3}$$

We now examine the assumption that $0.120 - 0.0078 \approx 0.120$ and see that a difference of about 7% is involved. The relative error in $[H_3O^+]$ is actually smaller than this figure, however, as can be seen by calculating $\log c_{HA}/K_a = 2.4$, which, from Figure 5-2, suggests an error of about 3%. If a more accurate figure is needed, solution of the quadratic equation yields 7.6×10^{-3} M for the hydronium ion concentration.

EXAMPLE 5-8

Calculate the hydronium ion concentration in a solution that is 2.0×10^{-4} M in aniline hydrochloride, $C_6H_5NH_3Cl$.

In aqueous solution, dissociation of the salt to Cl^- and $C_6H_5NH_3^+$ is complete. The weak acid $C_6H_5NH_3^+$ dissociates as follows:

$$C_6H_5NH_3^+ + H_2O \rightleftharpoons C_6H_5NH_2 + H_3O^+ \qquad \frac{[H_3O^+][C_6H_5NH_2]}{[C_6H_5NH_3^+]} = K_a$$

Inspection of Appendix 4 reveals no entry for $C_6H_5NH_3^+$, but Appendix 5 gives a base constant for aniline, $C_6H_5NH_2$:

$$C_6H_5NH_2 + H_2O \rightleftharpoons C_6H_5NH_3^+ + OH^- \qquad K_b = 3.94 \times 10^{-10}$$

The acid dissociation constant is then obtained from Equation 5-11:

$$K_a = \frac{1.00 \times 10^{-14}}{3.94 \times 10^{-10}} = 2.54 \times 10^{-5}$$

Proceeding as in Example 5-7, we have

$$[H_3O^+] = [C_6H_5NH_2]$$

$$[C_6H_5NH_3^+] = 2.0 \times 10^{-4} - [H_3O^+]$$

Let us assume that $[H_3O^+] \ll 2.0 \times 10^{-4}$ and substitute the simplified value for $[C_6H_5NH_3^+]$ into the dissociation-constant expression:

$$\frac{[H_3O^+]^2}{2.0 \times 10^{-4}} = 2.54 \times 10^{-5}$$

$$[H_3O^+] = 7.1 \times 10^{-5}$$

Comparison of 7.1×10^{-5} with 2.0×10^{-4} suggests that a significant error has been introduced by the assumption that $[H_3O^+] \ll c_{C_6H_5NH_3^+}$ (Figure 5-2 indicates that this error is about 20%). Thus, unless only an approximate value for $[H_3O^+]$ is needed, it is necessary to use the more nearly exact expression

$$\frac{[H_3O^+]^2}{2.0 \times 10^{-4} - [H_3O^+]} = 2.54 \times 10^{-5}$$

which rearranges to

$$[H_3O^+]^2 + 2.54 \times 10^{-5} [H_3O^+] - 5.08 \times 10^{-9} = 0$$

$$[H_3O^+] = \frac{-2.54 \times 10^{-5} + \sqrt{(2.54 \times 10^{-5})^2 + 4 \times 5.08 \times 10^{-9}}}{2} = 6.0 \times 10^{-5}$$

This quadratic equation can also be solved by the iterative method described in Appendix 11.

Hydronium Ion Concentration in Solutions of Weak Bases. The techniques discussed in previous sections are readily adapted to the calculation of the hydroxide or hydronium ion concentration in solutions of weak bases.

Aqueous ammonia is basic by virtue of the reaction

$$NH_3 + H_2O \rightleftharpoons NH_4^+ + OH^-$$

Here, the predominant species has been clearly demonstrated to be NH_3. Nevertheless, such solutions are sometimes called ammonium hydroxide, the terminology being vestigial from the time when NH_4OH rather than NH_3 was believed to be the undissociated form of the base. Application of the mass law to the equilibrium as written yields

$$\frac{[NH_4^+][OH^-]}{[NH_3]} = K_b$$

EXAMPLE 5-9

Calculate the hydronium ion concentration in a 0.075 M NH_3 solution. The predominant equilibrium is

$$NH_3 + H_2O \rightleftharpoons NH_4^+ + OH^-$$

From Appendix 5,

$$\frac{[NH_4^+][OH^-]}{[NH_3]} = 1.76 \times 10^{-5} = K_b$$

The equation for the equilbrium indicates that

$$[NH_4^+] = [OH^-]$$

$$[NH_4^+] + [NH_3] = c_{NH_3} = 0.075 \text{ M}$$

If we substitute $[OH^-]$ for $[NH_4^+]$ in the second of these equations and rearrange, we find that

$$[NH_3] = 0.075 - [OH^-]$$

Substituting these quantities into the dissociation-constant expression yields

$$\frac{[OH^-]^2}{7.5 \times 10^{-2} - [OH^-]} = 1.76 \times 10^{-5}$$

Provided that $[OH^-] \ll 7.5 \times 10^{-2}$, this equation simplifies to

$$[OH^-]^2 \simeq 7.5 \times 10^{-2} \times 1.76 \times 10^{-5}$$

$$[OH^-] = 1.15 \times 10^{-3}$$

Upon comparing the calculated value for $[OH^-]$ with 7.5×10^{-2}, we see that the error in $[OH^-]$ is less than 2%. If needed, a better value for $[OH^-]$ can be obtained by solving the quadratic equation.

Finally, then,

$$[H_3O^+] = \frac{K_w}{[OH^-]} = \frac{1.00 \times 10^{-14}}{1.15 \times 10^{-3}} = 8.7 \times 10^{-12}$$

EXAMPLE 5-10

Calculate the hydronium ion concentration in a 0.010 M sodium hypochlorite solution.

The equilibrium between OCl^- and water is

$$OCl^- + H_2O \rightleftharpoons HOCl + OH^-$$

for which

$$\frac{[HOCl][OH^-]}{[OCl^-]} = K_b$$

Appendix 5 does not contain a value for K_b, but Appendix 4 reveals that the acid dissociation constant for HOCl is 3.0×10^{-8}. Therefore, employing Equation 5-11, we write

$$K_b = \frac{K_w}{K_a} = \frac{1.00 \times 10^{-14}}{3.0 \times 10^{-8}} = 3.3 \times 10^{-7}$$

Proceeding as in Example 5-9, we have

$$[OH^-] = [HOCl]$$

$$[OCl^-] + [HOCl] = 0.010$$

$$[OCl^-] = 0.010 - [OH^-] \simeq 0.010$$

Here we have assumed that $[OH^-] \ll 0.010$. Substitution into the equilibrium-constant expression gives

$$\frac{[OH^-]^2}{0.010} = 3.3 \times 10^{-7}$$

$$[OH^-] = 5.7 \times 10^{-5}$$

The error resulting from the approximation is clearly small. To obtain the hydronium ion concentration, we write

$$[H_3O^+] = \frac{1.00 \times 10^{-14}}{5.7 \times 10^{-5}} = 1.8 \times 10^{-10}$$

Formation Constants for Complex Ions

An analytically important class of reactions involves the formation of soluble complex ions. Two examples are

$$Fe^{3+} + SCN^- \rightleftharpoons Fe(SCN)^{2+} \qquad \frac{[Fe(SCN)^{2+}]}{[Fe^{3+}][SCN^-]} = K_f$$

$$Zn(OH)_2(s) + 2OH^- \rightleftharpoons Zn(OH)_4^{2-} \qquad \frac{[Zn(OH)_4^{2-}]}{[OH^-]^2} = K_f$$

where K_f is called the *formation constant* for the complex.[3] Note that the second constant applies only to a saturated solution that is in contact with the sparingly soluble zinc hydroxide. Note also that no concentration term for the zinc hydroxide appears in the formation-constant expression because its concentration in the solid is invariant and is included in K_f. In this regard, the treatment is analogous to that described for solubility-product expressions.

Formation constants for numerous complex ions appear in Appendix 6, and an application is given in the following example.

EXAMPLE 5-11

A person with average eyesight can see the red color $FeSCN^{2+}$ imparts to an aqueous solution when the concentration of the complex is 6.4×10^{-6} M or greater. What minimum concentration of KSCN is required to make it possible to detect 1 ppm of iron(III) in water from a spring if the formation constant for the complex is 1.4×10^2?

$$Fe^{3+} + SCN^- \rightleftharpoons FeSCN^{2+} \qquad K_f = 1.4 \times 10^2$$

$$\frac{[FeSCN^{2+}]}{[Fe^{3+}][SCN^-]} = 1.4 \times 10^2$$

To convert the minimum detectable concentration of iron(III) to molarity, we recall (page 96) that 1 ppm corresponds to 1 mg/L. Thus,

$$c_{Fe^{3+}} = \frac{1 \text{ mg}}{\text{L}} \times \frac{1 \text{ g}}{1000 \text{ mg}} \times \frac{1 \text{ mol}}{55.8 \text{ g}} = 1.8 \times 10^{-5} \text{ mol/L}$$

This detectable concentration is distributed between two species, Fe^{3+} and $FeSCN^{2+}$, and we may write

$$1.8 \times 10^{-5} = [Fe^{3+}] + [FeSCN^{2+}]$$

Since $[FeSCN^{2+}]$ must be 6.4×10^{-6} M if this species is to be seen, we have that

$$[Fe^{3+}] = 1.8 \times 10^{-5} - 6.4 \times 10^{-6} = 1.16 \times 10^{-5}$$

We now substitute these concentrations into the formation-constant expression in order to determine the SCN^- concentration required:

$$\frac{6.4 \times 10^{-6}}{(1.16 \times 10^{-5})[SCN^-]} = 1.4 \times 10^2$$

$$[SCN^-] = 0.0039 \simeq 0.004$$

[3]Equilibria involving complex ions are sometimes described in terms of dissociation reactions and *instability constants;* the latter, which are found in the older literature, are the reciprocals of formation constants. For example,

$$Fe(SCN)^{2+} \rightleftharpoons Fe^{3+} + SCN^- \qquad \frac{[Fe^{3+}][SCN^-]}{[FeSCN^{2+}]} = K_{inst} = \frac{1}{K_f}$$

Thus, if the test solution is made 0.004 M or greater in KSCN, a detectable amount of $FeSCN^{2+}$ will form, provided the total iron(III) concentration is 1 ppm or greater. In practice, higher concentrations of KSCN are desirable in order to convert a larger fraction of the Fe^{3+} to the complex and thus enhance the sensitivity.

Oxidation/Reduction Equilibrium Constants

Equilibrium constants for oxidation/reduction reactions can be formulated in the usual way. For example,

$$6Fe^{2+} + Cr_2O_7^{2-} + 14H_3O^+ \rightleftharpoons 6Fe^{3+} + 2Cr^{3+} + 21H_2O$$

$$\frac{[Fe^{3+}]^6[Cr^{3+}]^2}{[Fe^{2+}]^6[Cr_2O_7^{2-}][H_3O^+]^{14}} = K_{redox}$$

As in earlier examples, no term for the concentration of water is needed.

Tables of equilibrium constants for oxidation/reduction reactions are not generally available because these constants are readily derived from more fundamental constants called *standard electrode potentials.* Calculations of this kind are treated in Chapter 12.

Distribution Equilibrium Constants

Another important type of equilibrium involves the distribution of a solute between two immiscible liquid phases. For example, if an aqueous solution of iodine is shaken with an immiscible organic solvent such as hexane or chloroform, a portion of the iodine is extracted into the organic layer. Ultimately, an equilibrium is established between the two phases that can be described by

$$I_2(aq) \rightleftharpoons I_2(org) \qquad \frac{[I_2]_{org}}{[I_2]_{aq}} = K_d$$

The equilibrium constant K_d for this reaction is often called a *distribution* or *partition coefficient.*

We shall see in Chapters 23 and 29 that distribution equilibria are of vital importance in many separation processes.

Stepwise Equilibrium Constants

Many electrolytes associate or dissociate in a stepwise manner, and an equilibrium constant can be written for each step. For example, when ammonia is added to a solution containing zinc ions, the following equilibria are established:

$$Zn^{2+} + NH_3 \rightleftharpoons ZnNH_3^{2+} \qquad K_1 = \frac{[ZnNH_3^{2+}]}{[Zn^{2+}][NH_3]} = 2.5 \times 10^4$$

$$ZnNH_3^{+} + NH_3 \rightleftharpoons Zn(NH_3)_2^{2+} \qquad K_2 = \frac{[Zn(NH_3)_2^{2+}]}{[ZnNH_3^{2+}][NH_3]} = 2.5 \times 10^4$$

$$Zn(NH_3)_2^{2+} + NH_3 \rightleftharpoons Zn(NH_3)_3^{2+} \qquad K_3 = \frac{[Zn(NH_3)_3^{2+}]}{[Zn(NH_3)_2^{2+}][NH_3]} = 3.2 \times 10^4$$

$$Zn(NH_3)_3^{2+} + NH_3 \rightleftharpoons Zn(NH_3)_4^{2+} \qquad K_4 = \frac{[Zn(NH_3)_4^{2+}]}{[Zn(NH_3)_3^{2+}][NH_3]} = 1.3 \times 10^4$$

where K_1, K_2, K_3, and K_4 are *stepwise formation constants* for the four complexes.

Overall formation constants are obtained by multiplying stepwise constants together. These are symbolized by β_n, where n corresponds to the number of moles of complexing species that combine with 1 mol of cation. For the foregoing example,

$$\beta_2 = K_1K_2 = 2.5 \times 10^4 \times 2.5 \times 10^4 = 6.2 \times 10^8$$

where β_2 is the overall formation constant for the reaction

$$Zn^{2+} + 2NH_3 \rightleftharpoons Zn(NH_3)_2^{2+} \qquad \beta_2 = \frac{[Zn(NH_3)_2^{2+}]}{[Zn^{2+}][NH_3]^2} = 6.2 \times 10^8$$

Similarly,

$$\beta_3 = K_1K_2K_3 = 6.2 \times 10^8 \times 3.2 \times 10^4 = 2.0 \times 10^{13}$$

where β_3 is the equilibrium constant for the reaction.

$$Zn^{2+} + 3NH_3 \rightleftharpoons Zn(NH_3)_2^{2+} \qquad \beta_3 = \frac{[Zn(NH_3)_3^{2+}]}{[Zn^{2+}][NH_3]^3} = 2.0 \times 10^{13}$$

Obviously β_1 and K_1 are identical.

Many common acids and bases undergo stepwise dissociation. For example, the following equilibria exist in an aqueous solution of phosphoric acid:

$$H_3PO_4 + H_2O \rightleftharpoons H_2PO_4^- + H_3O^+ \qquad \frac{[H_3O^+][H_2PO_4^-]}{[H_3PO_4]} = K_1 = 7.11 \times 10^{-3}$$

$$H_2PO_4^- + H_2O \rightleftharpoons HPO_4^{2-} + H_3O^+ \qquad \frac{[H_3O^+][HPO_4^{2-}]}{[H_2PO_4^-]} = K_2 = 6.34 \times 10^{-8}$$

$$HPO_4^{2-} + H_2O \rightleftharpoons PO_4^{3-} + H_3O^+ \qquad \frac{[H_3O^+][PO_4^{3-}]}{[HPO_4^{2-}]} = K_3 = 4.2 \times 10^{-13}$$

As illustrated by H_3PO_4, numerical values of K_n for an acid or a base become smaller with each successive dissociation step.

5C The Effect of Electrolyte Concentration on Chemical Equilibria

It is found experimentally that the position of many solution equilibria depends upon the electrolyte concentration of the medium. This dependency is observed even when the added electrolyte contains no ion in common with those of the equilibrium. For example, consider again the equilibrium established when a solution containing a small concentration of arsenic acid is mixed with a dilute solution of potassium iodide:

$$H_3AsO_4 + 3I^- + 2H^+ \rightleftharpoons H_3AsO_3 + I_3^- + H_2O$$

When an electrolyte, such as barium nitrate, potassium sulfate, or sodium perchlorate, is added to this solution, the position of equilibrium shifts to the left, as shown by a decrease in the triiodide color.

Figure 5-3 illustrates the effect of added electrolyte on the first three equilibrium constants shown in Table 5-2. Curve A is a plot of the product of the molar hydronium and hydroxide ion concentrations ($\times$ 10^{14}) as a function of molarity of sodium chloride. This product is designated K'_w. At low concentrations of sodium chloride, K'_w becomes independent of the electrolyte

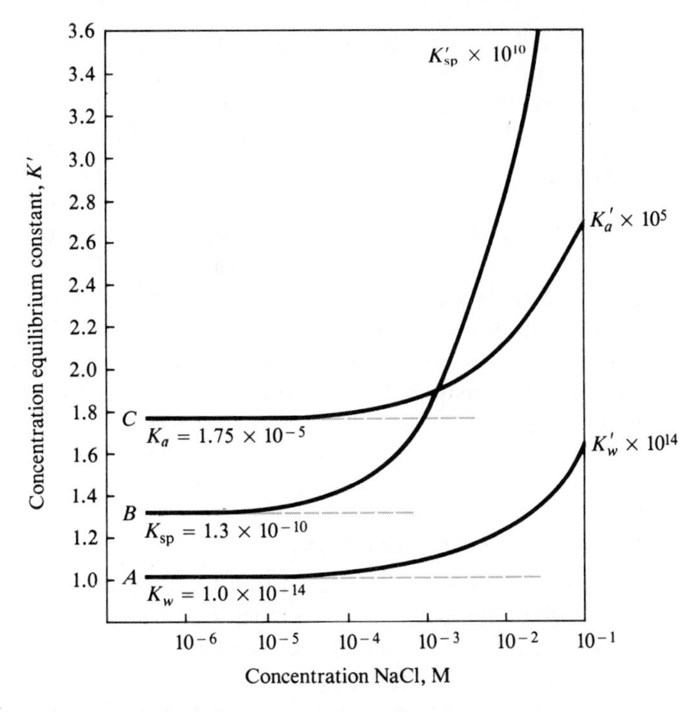

FIGURE 5-3 Effect of electrolyte concentration on concentration-based equilibrium constants.

Curve A: $2H_2O \rightleftharpoons H_3O^+ + OH^-$;

$$K_w = [H_3O^+][OH^-]$$

Curve B: $BaSO_4(s) \rightleftharpoons Ba^{2+} + SO_4^{2-}$;

$$K_{sp} = [Ba^{2+}][SO_4^{2-}]$$

Curve C: $HOAc + H_2O \rightleftharpoons H_3O^+ + OAc^-$;

$$K_a = \frac{[H_3O^+][OAc^-]}{[HOAc]}$$

concentration and assumes a value of 1.00×10^{-14}, which is of course the ion product constant for water K_w. A relationship, such as the ion product constant expression, where the numerical value for the relationship reaches a constant value as some concentration parameter approaches zero, is often called a *limiting law;* the numerical constant obtained at the limit is referred to as a *limiting value.*

The ordinate of Curve B in Figure 5-3 is the product of the molar concentrations of barium and sulfate ions ($\times 10^{10}$) in saturated solutions of barium sulfate. This concentration product is designated as K'_{sp}. At low electrolyte concentrations, K'_{sp} reaches a limiting value of 1.3×10^{-10}, which is by definition the solubility product K_{sp} for barium sulfate.

Curve C is a plot of K'_a ($\times 10^5$), the concentration quotient for the equilibrium involving the dissociation of acetic acid as a function electrolyte molarity. Here again, the ordinate function approaches a limiting value K_a, which is the acid dissociation constant for acetic acid.

The dotted lines in Figure 5-3 represent ideal behavior of the solutes.

Note that departures from ideality can be significant. For example, the product of the molar concentrations of hydrogen and hydroxide ion increase from 1.0×10^{-14} in pure water to about 1.7×10^{-14} in a solution that is 0.1 M in sodium chloride. The effect is even more pronounced with barium sulfate; here K'_{sp} in 0.1 M sodium chloride is more than double that of its limiting value.

The electrolyte effect shown in Figure 5-3 is not peculiar to sodium chloride. Indeed, similar curves are obtained with potassium nitrate or sodium perchlorate. In each case, the effect has as its origin the electrostatic attraction between the ions of the electrolyte and the ions of opposite charge from the reacting species. Since the electrostatic forces associated with all singly charged ions are approximately the same, the three salts exhibit similar effects on equilibria.

We must now consider how Equation 5-7 can be modified to take into account the electrolyte effect.

5C-1 *Some Empirical Observations*

Extensive studies have revealed that the magnitude of the electrolyte effect is highly dependent upon the charges of the participants in an equilibrium. When only neutral species are involved, the position of equilibrium is, to a first approximation, independent of electrolyte concentration. With ionic species, the magnitude of the electrolyte effect increases with charge. This generality is demonstrated by the three solubility curves in Figure 5-4. Note, for example, that in a 0.02 M solution of potassium nitrate, the solubility of barium sulfate, with its pair of doubly charged ions, is a factor of 2 larger than it is in pure water. This same change in electrolyte concentration increases the solubility of barium iodate by a factor of only 1.25 and that of silver chloride by 1.2. The enhanced effect due to doubly charged ions is also reflected in the greater slope of Curve *B* in Figure 5-3.

A second important observation is that, over a considerable electrolyte concentration range, the electrolyte effect depends only upon a concentration

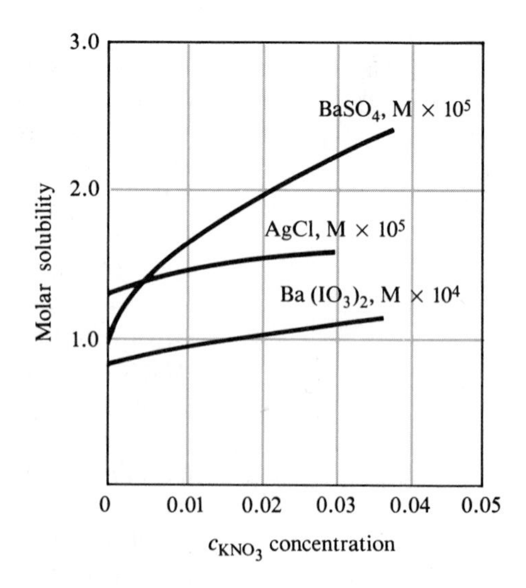

FIGURE 5-4 Effect of electrolyte concentration on the solubility of some salts.

parameter called the *ionic strength.* This quantity is defined as

$$\text{Ionic strength} = \mu = \tfrac{1}{2}([A]Z_A^2 + [B]Z_B^2 + [C]Z_C^2 + \cdots) \qquad (5\text{-}18)$$

where [A], [B], [C], . . . represent the molar concentration of ions A, B, C, . . . and Z_A, Z_B, Z_C, . . . are their charges.

EXAMPLE 5-12

Calculate the ionic strength of (a) a 0.1 M solution of KNO_3 and (b) a 0.1 M solution of Na_2SO_4.

(a) For the KNO_3 solution, $[K^+]$ and $[NO_3^-]$ are 0.1 and

$$\mu = \tfrac{1}{2}(0.1 \times 1^2 + 0.1 \times 1^2) = 0.1$$

(b) For the Na_2SO_4 solution, $[Na^+] = 0.2$ and $[SO_4^{2-}] = 0.1$. Therefore,

$$\mu = \tfrac{1}{2}(0.2 \times 1^2 + 0.1 \times 2^2) = 0.3$$

EXAMPLE 5-13

What is the ionic strength of a solution that is 0.05 M in KNO_3 and 0.1 M in Na_2SO_4?

$$\mu = \tfrac{1}{2}(0.05 \times 1^2 + 0.05 \times 1^2 + 0.2 \times 1^2 + 0.1 \times 2^2) = 0.35$$

It is apparent from these examples that the ionic strength of a solution of a strong electrolyte consisting solely of singly charged ions is identical with the total molar salt concentration. If the solution contains ions with multiple charges, however, the ionic strength is greater than the molar concentration.

For solutions with ionic strengths of 0.1 or less, the electrolyte effect is independent of the *kind* of ions and dependent *only upon the ionic strength.* Thus, the degree of dissociation of acetic acid is the same in the presence of sodium chloride, potassium nitrate, or barium iodide, provided the concentrations of these species are such that the ionic strength is fixed. It should be noted that this independence with respect to electrolyte species disappears at high ionic strengths.

5C-2 Activity and Activity Coefficients

In order to describe the effect of ionic strength on equilibria in quantitative terms, chemists adopt a concentration parameter called the *activity,* defined as

$$a_A = [A]f_A \qquad (5\text{-}19)$$

where a_A is the activity of the species A, [A] is its molar concentration, and f_A is a dimensionless quantity called the *activity coefficient.* The activity coefficient and thus the activity of A vary with ionic strength such that substitution of a_A for [A] in an equilibrium-constant expression frees the numerical value of the constant from dependence on the ionic strength. To illustrate, for the dissociation of acetic acid, we write

$$K_a = \frac{a_{H_3O^+} \cdot a_{OAc^-}}{a_{HOAc}} = \frac{[H_3O^+][OAc^-]}{[HOAc]} \times \frac{f_{H_3O^+} \cdot f_{OAc^-}}{f_{HOAc}}$$

where $f_{H_3O^+}$, f_{OAc^-}, and f_{HOAc} vary with ionic strength to keep K_a numerically constant over a wide range of ionic strengths (in contrast to the concentration constant K'_a shown in curve C in Figure 5-3).

Properties of Activity Coefficients

Activity coefficients have the following properties:

1. The activity coefficient of a species is a measure of the effectiveness with which that species influences an equilibrium in which it is a participant. In very dilute solutions, where ionic strength is minimal, this effectiveness becomes constant, and the activity coefficient acquires a value of unity. Under such circumstances, the activity and the molar concentration are identical. As the ionic strength increases, however, an ion loses some of its effectiveness, and its activity coefficient decreases. We may summarize this behavior in terms of Equation 5-19. At moderate ionic strengths, $f_A < 1$; as the solution approaches infinite dilution, however, $f_A \rightarrow 1$ and thus $a_A \rightarrow [A]$.

   At high ionic strengths ($\mu > 0.1$), activity coefficients often increase and may even become greater than unity. Because interpretation of the behavior of solutions in this region is difficult, we confine our discussion to regions of low or moderate ionic strength (that is, where $\mu < 0.1$).

   The variation of typical activity coefficients as a function of ionic strength is shown in Figure 5-5.
2. In solutions that are not too concentrated, the activity coefficient for a given species is independent of the nature of the electrolyte and dependent only upon the ionic strength.
3. For a given ionic strength, the activity coefficient of an ion departs farther from unity as the charge carried by the species increases. This effect is shown in Figure 5-5. The activity coefficient of an uncharged molecule is approximately unity, regardless of ionic strength.
4. At any given ionic strength, the activity coefficients of ions of the same charge are approximately equal. The small variations that do exist can be correlated with the effective diameter of the hydrated ions.
5. The activity coefficient of a given ion describes its effective behavior in all equilibria in which it participates. For example, at a given ionic

FIGURE 5-5 Effect of ionic strength on activity coefficients.

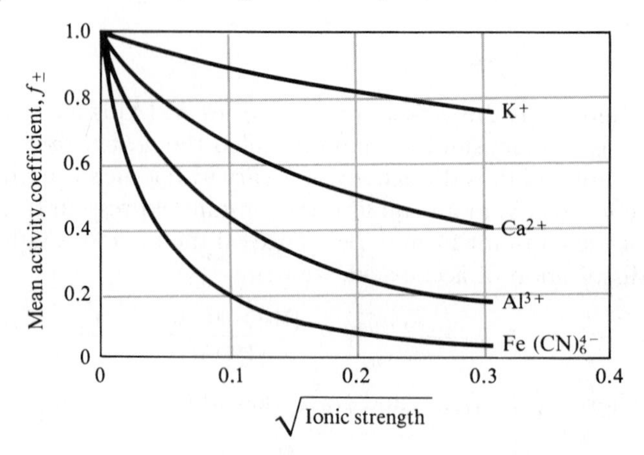

strength, a single activity coefficient for cyanide ion describes the influence of that species upon any of the following equilibria:

$$HCN + H_2O \rightleftharpoons H_3O^+ + CN^-$$

$$AgCN(s) \rightleftharpoons Ag^+ + CN^-$$

$$Ni(CN)_4^{2-} \rightleftharpoons Ni^{2+} + 4CN^-$$

The Debye-Hückel Equation

As noted earlier, the electrolyte effect results from the electrostatic attractive and repulsive forces that exist between the ions of an electrolyte and the ions involved in an equilibrium. These forces cause each ion from the dissociated reactant to be surrounded by a sheath of solution that contains a slight excess of electrolyte ions of opposite charge. For example, when a barium sulfate precipitate is equilibrated with a sodium chloride solution, each dissolved barium ion is surrounded by an ionic atmosphere that, because of electrostatic attraction and repulsion, carries a net negative charge on the average. Similarly, each sulfate ion is surrounded by an ionic atmosphere that tends to be slightly positive. These charged layers make the barium ions somewhat less positive and the sulfate ions somewhat less negative than they would be in the absence of electrolyte. The consequence of this effect is a decrease in the overall attraction between barium and sulfate ions and an increase in solubility, which becomes greater as the number of electrolyte ions in the solution becomes larger.

In 1923, Debye and Hückel used the foregoing model to derive a theoretical expression that permits the calculation of activity coefficients of ions from their charge and average size.[4] This equation, which has become known as the *Debye-Hückel equation*, takes the form

$$-\log f_A = \frac{0.51 Z_A^2 \sqrt{\mu}}{1 + 0.33 \alpha_A \sqrt{\mu}} \tag{5-20}$$

where

f_A = activity coefficient of the species A

Z_A = charge on the species A

μ = ionic strength of the solution

α_A = effective diameter of the hydrated ion A in ångström units ($1\ \text{Å} = 10^{-8}$ cm)

The constants 0.51 and 0.33 are applicable to aqueous solutions at 25°C; other values must be employed at other temperatures.

Unfortunately, considerable uncertainty exists regarding the magnitude of α_A in Equation 5-20. Its value appears to be approximately 3 Å for most singly charged ions; for these species, then, the denominator of the Debye-Hückel equation simplifies to approximately $1 + \sqrt{\mu}$. For ions with higher charge, α_A may be as large as 10 Å. It should be noted that the second term of the denominator becomes small with respect to the first when the ionic

[4]P. Debye and E. Hückel, *Physik. Z.*, **1923,** *24,* 185.

strength is less than 0.01, so that at these ionic strengths, uncertainties in α_A are of little significance in calculating activity coefficients.

Kielland[5] has derived values of α_A for numerous ions from a variety of experimental data. His best values for effective diameters are given in Table 5-5. Also presented are activity coefficients calculated from Equation 5-20 using these values for the size parameter.

Experimental determination of single-ion activity coefficients such as those shown in Table 5-5 is unfortunately impossible because all experimental methods give only a mean activity coefficient for the positively and negatively charged ions in a solution.[6] In other words, it is impossible to measure the

TABLE 5-5 **Activity Coefficient for Ions at 25°C***

Ion	α_A, Å	Activity Coefficient at Indicated Ionic Strength				
		0.001	0.005	0.01	0.05	0.1
H_3O^+	9	0.967	0.933	0.914	0.86	0.83
Li^+, $C_6H_5COO^-$	6	0.965	0.929	0.907	0.84	0.80
Na^+, IO_3^-, HSO_3^-, HCO_3^-, $H_2PO_4^-$, $H_2AsO_4^-$, OAc^-	4–4.5	0.964	0.928	0.902	0.82	0.78
OH^-, F^-, SCN^-, HS^-, ClO_3^-, ClO_4^-, BrO_3^-, IO_4^-, MnO_4^-	3.5	0.964	0.926	0.900	0.81	0.76
K^+, Cl^-, Br^-, I^-, CN^-, NO_2^-, NO_3^-, $HCOO^-$	3	0.964	0.925	0.899	0.80	0.76
Rb^+, Cs^+, Tl^+, Ag^+, NH_4^+	2.5	0.964	0.924	0.898	0.80	0.75
Mg^{2+}, Be^{2+}	8	0.872	0.755	0.69	0.52	0.45
Ca^{2+}, Cu^{2+}, Zn^{2+}, Sn^{2+}, Mn^{2+}, Fe^{2+}, Ni^{2+}, Co^{2+}, $Phthalate^{2-}$	6	0.870	0.749	0.675	0.48	0.40
Sr^{2+}, Ba^{2+}, Cd^{2+}, Hg^{2+}, S^{2-}	5	0.868	0.744	0.67	0.46	0.38
Pb^{2+}, CO_3^{2-}, SO_3^{2-}, $C_2O_4^{2-}$	4.5	0.868	0.742	0.665	0.46	0.37
Hg_2^{2+}, SO_4^{2-}, $S_2O_3^{2-}$, CrO_4^{2-}, HPO_4^{2-}	4.0	0.867	0.740	0.660	0.44	0.36
Al^{3+}, Fe^{3+}, Cr^{3+}, La^{3+}, Ce^{3+}	9	0.738	0.54	0.44	0.24	0.18
PO_4^{3-}, $Fe(CN)_6^{3-}$	4	0.725	0.50	0.40	0.16	0.095
Th^{4+}, Zr^{4+}, Ce^{4+}, Sn^{4+}	11	0.588	0.35	0.255	0.10	0.065
$Fe(CN)_6^{4-}$	5	0.57	0.31	0.20	0.048	0.021

*From J. Kielland, *J. Am. Chem. Soc.*, 1937, *59*, 1675. By Courtesy of the American Chemical Society.

[5]J. Kielland, *J. Amer. Chem. Soc.*, **1937**, *59*, 1675.

[6]The mean activity of the electrolyte A_mB_n is defined as

$$f_{\pm} = \text{mean activity coefficient} = (f_A^m \cdot f_B^n)^{1/(m+n)}$$

The mean activity coefficient can be measured in any of several ways, but it is impossible experimentally to resolve this term into the individual activity coefficients for f_A and f_B. For example, if A_mB_n is a precipitate, we can write

$$K_{sp} = [A]^m[B]^n \cdot f_A^m \cdot f_B^n = [A]^m[B]^n \cdot f_{\pm}^{m+n}$$

By measuring the solubility of A_mB_n in a solution in which the electrolyte concentration approaches zero (that is, where both f_A and $f_B \rightarrow 1$), we can obtain K_{sp}. A second solubility measurement at some ionic strength μ_1 gives values for [A] and [B]. These data then permit the calculation of $f_A^m \cdot f_B^n = f_{\pm}^{m+n}$ for ionic strength μ_1. It is important to understand that this procedure does not provide enough experimental data to permit the calculation of the *individual* quantities f_A and f_B and that there appears to be no additional experimental information that would permit evaluation of these quantities. This situation is general, and the *experimental* determination of individual activity coefficients appears to be impossible.

properties of individual ions in the presence of counter ions of opposite charge and solvent molecules. It should be pointed out, however, that mean activity coefficients calculated from the data in Table 5-5 agree satisfactorily with the experimental values.

The Debye-Hückel relationship and the data in Table 5-5 give satisfactory activity coefficients for ionic strengths up to about 0.1. Beyond this value, the equation fails, and experimentally determined mean activity coefficients must be used.

Equilibrium Calculations Employing Activity Coefficients

The use of activities rather than molar concentrations in equilibrium-constant calculations yields information that is more accurate. Unless otherwise specified, equilibrium constants found in tables are generally based upon activities (activity-based equilibrium constants are sometimes called *thermodynamic equilibrium constants*). The examples that follow illustrate how activity coefficients from Table 5-5 are applied to such data.

EXAMPLE 5-14

Use activities to calculate the solubility of $Ba(IO_3)_2$ in a 0.033 M solution of $Mg(IO_3)_2$. The thermodynamic solubility product for $Ba(IO_3)_2$ is 1.57×10^{-9} (Appendix 3).

At the outset, we write the solubility-product expression in terms of activities:

$$a_{Ba^{2+}} \times a^2_{IO_3^-} = K_{sp} = 1.57 \times 10^{-9}$$

where $a_{Ba^{2+}}$ and $a_{IO_3^-}$ are the activities of barium and iodate ions. Replacing activities in this equation by activity coefficients and concentrations from Equation 5-19 yields

$$[Ba^{2+}]f_{Ba^{2+}} \times [IO_3^-]^2 f^2_{IO_3^-} = K_{sp}$$

where $f_{Ba^{2+}}$ and $f_{IO_3^-}$ are the activity coefficients for the two ions. Rearranging this expression gives

$$[Ba^{2+}][IO_3^-]^2 = \frac{K_{sp}}{f_{Ba^{2+}} \times f^2_{IO_3^-}} = K'_{sp} \qquad (5\text{-}21)$$

where K'_{sp} is the *concentration-based solubility product.*

The ionic strength of the solution is obtained by substituting into Equation 5-18:

$$\mu = \tfrac{1}{2}([Mg^{2+}] \times 2^2 + [IO_3^-] \times 1^2)$$

$$= \tfrac{1}{2}(0.033 \times 4 + 0.066 \times 1) = 0.099 \simeq 0.1$$

In calculating μ, we have assumed that the Ba^{2+} and IO_3^- ions from the precipitate do not significantly affect the ionic strength of the solution. This simplification seems justified, considering the low solubility of barium iodate and the relatively high concentration of $Mg(IO_3)_2$. In situations where it is not possible to make such an assumption, the concentrations of the two ions can be approximated by solubility calculation in which activities and concentrations are assumed to be identical (as in Examples 5-3 and 5-4). These concentrations can then be introduced to give a better value for μ.

Turning now to Table 5-5, we find that at an ionic strength of 0.1,

$$f_{Ba^{2+}} = 0.38 \qquad f_{IO_3^-} = 0.78$$

If the calculated ionic strength did not match that of one of the columns in the table, $f_{Ba^{2+}}$ and $f_{IO_3^-}$ could be calculated from Equation 5-20.

Substituting into the thermodynamic solubility-product expression gives

$$\frac{1.57 \times 10^{-9}}{(0.38)(0.78)^2} = 6.8 \times 10^{-9} = K'_{sp}$$

$$[Ba^{2+}][IO_3^-]^2 = 6.8 \times 10^{-9}$$

Proceeding now as in earlier solubility calculations,

$$\text{Solubility} = [Ba^{2+}]$$

$$[IO_3^-] = 0.066$$

$$[Ba^{2+}](0.066)^2 = 6.8 \times 10^{-9}$$

$$[Ba^{2+}] = \text{solubility} = 1.6 \times 10^{-6}\ M$$

It is of interest to note that the calculated solubility is 3.6×10^{-7} M when the effects of ionic strength are neglected, which is less than one fourth as large as the value obtained here.

EXAMPLE 5-15

Using activities, calculate the hydronium ion concentration in a 0.120 M solution of HNO_2 that is also 0.050 M in NaCl.

The ionic strength of this solution is

$$\mu = \tfrac{1}{2}(0.0500 \times 1^2 + 0.0500 \times 1^2) = 0.0500$$

In Table 5-5, at ionic strength 0.050, we find

$$f_{H_3O^+} = 0.86 \qquad f_{NO_2^-} = 0.80$$

Also, from rule 3 (page 126), we can write

$$f_{HNO_2} = 1.0$$

These three values for f permit the calculation of a concentration-based dissociation constant from the thermodynamic constant of 5.1×10^{-4} (Appendix 4):

$$\frac{[H_3O^+][NO_2^-]}{[HNO_2]} = \frac{K_a \times f_{HNO_2}}{f_{H_3O^+} \times f_{NO_2^-}} = \frac{5.1 \times 10^{-4} \times 1.0}{0.86 \times 0.81} = K'_a = 7.3 \times 10^{-4}$$

Proceeding as in Example 5-7, we write

$$[H_3O^+] = \sqrt{0.120 \times 7.3 \times 10^{-4}} = 9.4 \times 10^{-3}$$

Note that neglecting the activity coefficient gives $[H_3O^+] = 7.8 \times 10^{-3}$.

Omission of Activity Coefficients in Equilibrium Calculations

We shall ordinarily neglect activity coefficients and simply use molar concentrations in applications of the equilibrium law. This recourse simplifies the calculations and greatly decreases the amount of data needed. For most purposes, the error introduced by the assumption of unity for the activity coefficient is not large enough to lead to false conclusions. It should be apparent from the preceding examples, however, that disregard of activity coefficients may introduce a significant numerical error in calculations of this kind. Note, for example, that an error greater than −75% was incurred in Example 5-14. The reader should be alert to the conditions under which the substitution of concentration for activity is likely to lead to the largest error. Significant discrepancies occur when the ionic strength is large (0.01 or larger) or when the ions involved have multiple charges (Table 5-5). With dilute solutions (ionic strength < 0.01) of nonelectrolytes or of singly charged ions, the use of concentrations in a mass-law calculation often provides reasonably accurate results.

It is also important to note that the decrease in solubility resulting from the presence of an ion common to the precipitate is in part counteracted by the larger electrolyte concentration associated with the presence of the salt containing the common ion. This effect is illustrated in Example 5-14.

Questions and Problems

*5-1 Differentiate between activity and activity coefficient.

5-2 Predict how the position of each of the following equilibria is affected when the ionic strength of the solution is increased from 1.0×10^{-3} to 1.0×10^{-1}:

*(a) $Fe^{3+} + SCN^- \rightleftharpoons Fe(SCN)^{2+}$
(b) $H_2PO_4^- + H_2O \rightleftharpoons HPO_4^{2-} + H_3O^+$
*(c) $PO_4^{3-} + H_2O \rightleftharpoons HPO_4^{2-} + OH^-$
(d) $Hg^{2+} + 2Cl^- \rightleftharpoons HgCl_2$

5-3 Define

*(a) salt.
(b) electrolyte.
*(c) hydronium ion.
(d) conjugate base.
*(e) base.
(f) mass law.
*(g) Le Châtelier principle.
(h) autoprotolysis.
*(i) amphiprotic solvent.
(j) common-ion effect.

*5-4 Explain why water is termed a leveling solvent toward mineral acids, whereas ethanol is a differentiating solvent.

5-5 What is a thermodynamic equilibrium-constant expression?

*5-6 The solubility product for Ag_2CO_3 is 8.1×10^{-12} and that for AgCl is 1.8×10^{-10}. Why is it wrong to conclude that Ag_2CO_3 is less soluble than silver chloride because its K_{sp} is smaller?

5-7 Appendix 6 is a tabulation of logarithms of stepwise formation constants for a number of complexes. From these data, calculate numerical values for β_2, β_3, β_4, β_5, and β_6 for the

*(a) NH_3 complexes of Ni^{2+}.
(b) F^- complexes of Al^{3+}.

5-8 Write equilibrium-constant expressions for the various constants derived in Problem 5-7.

*5-9 Describe the conditions that lead to the largest discrepancies between thermodynamic and concentration-based equilibrium calculations.

5-10 Write the equilibrium-constant expressions, and give the numerical values for each constant in

*(a) the basic dissociation of ethylamine, $C_2H_5NH_2$.
(b) the acidic dissociation of hydrogen cyanide, HCN.
*(c) the acidic dissociation of pyridine hydrochloride, C_5H_5NHCl.
(d) the basic dissociation of NaCN.
*(e) the formation of $AgCl_2^-$ from AgCl.
(f) the formation of CuI_2^- from CuI.
*(g) the formation of $Cd(NH_3)_2^{2+}$ from NH_3 and Cd^{2+}.
(h) the formation of AlF_6^{3-} from Al^{3+} and F^-.
*(i) the dissociation of H_3AsO_4 to H_3O^+ and AsO_4^{3-}.
(j) the reaction of CO_3^{2-} with H_2O to give H_2CO_3 and OH^-.

5-11 Calculate the solubility-product constant for each of the following substances, given that the molar concentrations of their saturated solutions are as indicated:

*(a) AgSeCN (2.0×10^{-8} mol/L; products are Ag^+ and $SeCN^-$).
(b) $RaSO_4$ (6.6×10^{-6} mol/L).
*(c) $Pb(BrO_3)_2$ (1.7×10^{-1} mol/L).
(d) $PbBr_2$ (2.1×10^{-2} mol/L).
*(e) $Ce(IO_3)_3$ (1.9×10^{-3} mol/L).
(f) BiI_3 (1.3×10^{-5} mol/L).

*5-12 Calculate the weight in grams of PbI_2 that dissolves in 100 mL of

(a) H_2O.
(b) 2.00×10^{-2} M KI.
(c) 2.00×10^{-2} M $Pb(NO_3)_2$.
(d) 0.100 M $NaClO_4$ (use activities here).

5-13 Calculate the weight of $La(IO_3)_3$ that dissolves in 100 mL of

(a) H_2O.
(b) 2.00×10^{-2} M $NaIO_3$.
(c) 2.00×10^{-2} M $La(NO_3)_3$.
(d) 0.0500 M $NaNO_3$ (use activities here).

*5-14 The solubility product for Tl_2CrO_4 is 9.8×10^{-13}. What CrO_4^{2-} concentration is required to

(a) initiate precipitation of Tl_2CrO_4 from a solution that is 2.12×10^{-3} M in Tl^+?
(b) lower the concentration of Tl^+ in a solution to 1.00×10^{-6} M?

5-15 What hydroxide concentration is required to

(a) initiate precipitation of Fe^{3+} from a 1.00×10^{-3} M solution of $Fe_2(SO_4)_3$?
(b) lower the Fe^{3+} concentration in the foregoing solution to 1.00×10^{-9} M?

*5-16 The solubility-product constant for $Ce(IO_3)_3$ is 3.2×10^{-10}. What is the Ce^{3+} concentration in a solution prepared by mixing 50.0 mL of 0.0500 M Ce^{3+} with

(a) 50.0 mL of water?
(b) 50.0 mL of 0.050 M IO_3^-?
(c) 50.0 mL of 0.150 M IO_3^-?
(d) 50.0 mL of 0.300 M IO_3^-?

5-17 The solubility-product constant for K_2PtCl_6 is 1.1×10^{-5} ($K_2PtCl_6 \rightleftharpoons 2K^+ + PtCl_6^{2-}$). What is the K^+ concentration of a solution prepared by mixing 50.0 mL of 0.400 M KCl with

(a) 50.0 mL of 0.100 M $PtCl_6^{2-}$?
(b) 50.0 mL of 0.200 M $PtCl_6^{2-}$?
(c) 50.0 mL of 0.400 M $PtCl_6^{2-}$?

5-18 At 25°C, what are the molar H_3O^+ and OH^- concentrations in

*(a) 0.0200 M HOCl?
(b) 0.0800 M propanoic acid?
*(c) 0.200 M methylamine?
(d) 0.100 M trimethylamine?
*(e) 0.120 M NaOCl?
(f) 0.0860 M CH_3COONa?
*(g) 0.100 M hydroxylamine hydrochloride?
(h) 0.0500 M ethanolamine hydrochloride?

*5-19 What is the hydronium ion concentration in water at 0°C?

5-20 At 25°C, what is the hydronium ion concentration in

*(a) 0.100 M chloroacetic acid?
*(b) 0.100 M sodium chloroacetate?
(c) 0.0100 M methylamine?
(d) 0.0100 M methylamine hydrochloride?

*(e) 1.00×10^{-3} M aniline hydrochloride?
(f) 0.200 M HIO_3?

*5-21 Mercury(II) forms the soluble neutral complex $Hg(SCN)_2$ with SCN^-. The formation constant β_2 of the complex has a value of 2.0×10^{17}. Calculate the Hg^{2+} and SCN^- concentrations in a solution prepared by mixing 10 mL of 0.0200 M Hg^{2+} with
(a) 10 mL of 0.0200 M SCN^-.
(b) 10 mL of 0.0400 M SCN^-.
(c) 10 mL of 0.0600 M SCN^-.

*5-22 Given that the formation constant for the reaction $CuI(s) + I^- \rightleftharpoons CuI_2^-$ is 8×10^{-4}, calculate the molar concentration of CuI_2^- in a solution that is saturated with CuI and has a molar KI concentration of
(a) 0.00100. (b) 0.0100. (c) 0.100.

5-23 The formation constant for $Pb(OH)_3^-$ is

$$Pb(OH)_2(s) + OH^- \rightleftharpoons Pb(OH)_3^- \qquad \beta_2 = 5 \times 10^{-2}$$

Calculate the concentration of $Pb(OH)_3^-$ in a saturated $Pb(OH)_2$ solution that is
(a) 0.00100 M in NaOH. (b) 0.100 M in NaOH.

*5-24 The solubility products for a series of iodides are

TlI $K_{sp} = 6.5 \times 10^{-8}$

AgI $K_{sp} = 8.3 \times 10^{-17}$

PbI_2 $K_{sp} = 7.1 \times 10^{-9}$

BiI_3 $K_{sp} = 8.1 \times 10^{-19}$

List these four compounds in order of decreasing molar solubility in
(a) water.
(b) 0.10 M NaI.
(c) a 0.10 M solution of the solute cation.

5-25 The solubility products for a series of iodates are

$AgIO_3$ $K_{sp} = 3.0 \times 10^{-8}$

$Sr(IO_3)_2$ $K_{sp} = 3.3 \times 10^{-7}$

$La(IO_3)_3$ $K_{sp} = 6.2 \times 10^{-12}$

$Ce(IO_3)_4$ $K_{sp} = 4.7 \times 10^{-17}$

List these four compounds in order of decreasing molar solubility in
(a) water.
(b) 0.10 M $NaIO_3$.
(c) a 0.10 M solution of the solute cation.

*5-26 Calculate the solubilities of the following compounds in a 0.0333 M solution of $Mg(ClO_4)_2$ employing (1) activities and (2) molar concentrations:
(a) AgSCN.
(b) PbI_2.
(c) $BaSO_4$.
(d) $Cd_2Fe(CN)_6$ [$Cd_2Fe(CN)_6(s) \rightleftharpoons 2Cd^{2+} + Fe(CN)_6^{4-}$; $K_{sp} = 3.2 \times 10^{-17}$].

5-27 Calculate the solubilities of the following compounds in a 0.0167 M solution of $Ba(NO_3)_2$ employing (1) activities and (2) molar concentrations:
(a) $AgIO_3$.
(b) $Mg(OH)_2$.
(c) $BaSO_4$.
(d) $La(IO_3)_3$.

5-28 Calculate the hydronium ion concentration in the following solutions based upon (1) activities and (2) molar concentrations:

*(a) 0.0200 M HNO_2 that is 0.0010 M in $NaNO_3$.
*(b) 0.0200 M HNO_2 that is 0.10 M in $NaNO_3$.
(c) 0.0500 M NH_3 that is 0.0010 M in NaCl.
(d) 0.0500 M NH_3 that is 0.100 M in NaCl.
(e) 0.05 M NH_4^+ that is 0.001 M in NaCl.
(f) 0.05 M NH_4^+ that is 0.100 M in NaCl.

5-29 Mercury(II) forms a soluble neutral complex with Cl^-:

$$Hg^{2+} + 2\ Cl^- \rightleftharpoons HgCl_2 \qquad K_f = 1.6 \times 10^{13}$$

Calculate the Hg^{2+} concentration of each of the following solutions based upon (1) activities and (2) molarities:

*(a) A solution prepared by dissolving 0.0100 mol of $HgCl_2$ in 1.00 L of water.
(b) A solution prepared by dissolving 0.0100 mol of $HgCl_2$ in 1.00 L of 0.0500 M $NaNO_3$.
*(c) A solution prepared by dissolving 0.0100 mol of $HgCl_2$ in 1.00 L of 0.0500 M NaCl.
(d) A solution prepared by dissolving 0.0100 mol of $HgCl_2$ in 1.00 L of 0.0333 M $Hg(NO_3)_2$.
*(e) A solution prepared by mixing 50.0 mL of a solution that is 0.0100 M in $Hg(NO_3)_2$ with 50.0 mL of a solution that is 0.0400 M in NaCl and 0.0600 M in $NaNO_3$.
(f) A solution prepared by mixing 50.0 mL of a solution that is 0.0100 M in $Hg(NO_3)_2$ with 50.0 mL of a solution that is 0.0400 M in $BaCl_2$ and 0.0500 M in $NaNO_3$.

CHAPTER 6

The Application of Equilibrium Calculations to Complex Systems

Many systems encountered in the laboratory are chemically complex in the sense that they are made up of several species that interact with each other and with the solvent to give two or more equilibria that operate simultaneously. For example, consider a system prepared by dissolving 1.0 mmol of sodium hydrogen carbonate in 100 mL of water. Since this salt is a strong electrolyte, the solution has a species sodium ion concentration of 0.010 M. Its species hydrogen carbonate ion concentration is less than the analytical concentration of the salt, however, because of the reaction of its anion with water:

$$HCO_3^- + H_2O \rightleftharpoons CO_3^{2-} + H_3O^+ \qquad (6\text{-}1)$$

$$HCO_3^- + H_2O \rightleftharpoons H_2CO_3 + OH^- \qquad (6\text{-}2)$$

Note that the hydrogen carbonate ion behaves as a weak acid in the first reaction and as a weak base in the second.

Any equilibrium involving H_3O^+ or OH^- is necessarily influenced by the autoprotolysis of water; thus in order to describe this system in quantitative terms, it is necessary to consider yet a third equilibrium:

$$2H_2O \rightleftharpoons H_3O^+ + OH^- \qquad (6\text{-}3)$$

We see from these three equations that a 0.010 M solution of sodium hydrogen carbonate contains, in addition to 0.010 mol/L of Na^+, five other species: HCO_3^-, CO_3^{2-}, H_2CO_3, H_3O^+, and OH^-. The determination of the concentration of one or more of these species requires that five independent algebraic expressions be generated and solved simultaneously. As is shown shortly, generation of such algebraic equations is relatively easy if approached systematically. In fact, developing the necessary equations is often far easier than finding their numerical solutions (unless a suitable computer program is available).

In the sections that follow, a systematic approach for attacking any problem involving multiple equilibria is presented. This approach is then illustrated with a number of typical examples that involve computation of the solubility of precipitates in the presence of species that interact with the ions of the precipitates and thus enhance their solubility. Throughout the discussion, the reader should constantly bear in mind that *the validity and form of a particular equilibrium-constant expression are in no way affected by the existence of additional equilibria in the solution.* For example, if barium chloride is added to the sodium hydrogen carbonate solution just described, barium carbonate precipitates and a new equilibrium is established:

$$BaCO_3(s) \rightleftharpoons Ba^{2+} + CO_3^{2-} \qquad (6\text{-}4)$$

This additional equilibrium in no way affects the concentration *relationships* among the other carbonate species and the hydronium and hydroxide ion. That is, the following two equilibrium-constant expressions, which describe the equilibria shown in Equations 6-1 and 6-2, are valid regardless of whether or not barium carbonate is present:

$$\frac{[H_3O^+][CO_3^{2-}]}{[HCO_3^-]} = K_a \tag{6-5}$$

$$\frac{[OH^-][H_2CO_3]}{[HCO_3^-]} = K_b \tag{6-6}$$

To be sure, except for the sodium ion, the concentrations of all the species in the original solution are markedly altered by the addition of barium ion. Nevertheless, the relationships among concentrations of species remain those given by Equations 6-5 and 6-6.

6A A Systematic Method for Deriving Algebraic Equations Describing Multiequilibrium Systems

In order to determine the equilibrium concentrations of one or more species in a system in which several equilibria exist, it must be possible to write as many independent algebraic equations as there are participants in the equilibria. Thus, in order to obtain the equilibrium concentrations of one or more of the five species in a solution of sodium hydrogen carbonate, it is necessary to develop five independent algebraic equations that contain concentration terms for these five constituents. For the solution that also contains barium carbonate, a sixth independent equation is required because of the presence of barium ions in the equilibrium mixture.

Three types of algebraic equations can be developed for solving multiequilibrium problems: (1) two or more equilibrium-constant expressions, (2) one or more *mass-balance expressions,* and (3) a single *charge-balance* equation. We have already shown how equilibrium-constant expressions are written; methods for obtaining the other two types of equations are given in the next two sections.

6A-1 Mass-Balance Equations

Mass-balance equations are algebraic expressions that relate the equilibrium concentrations of various species in a solution to one another and to the analytical concentrations of the various substances present. They are derived from information about how the solution was prepared and from a knowledge of the kinds of equilibria established in the solution.

EXAMPLE 6-1

Write mass-balance expressions for the system formed when a 0.010 M NH_3 solution is saturated with AgBr.

In order to develop mass-balance expressions, it is necessary to write chemical equations describing all pertinent equilibria in the solution. Here these equations are

$$AgBr(s) \rightleftharpoons Ag^+ + Br^-$$

$$Ag^+ + 2NH_3 \rightleftharpoons Ag(NH_3)_2^+$$

$$NH_3 + H_2O \rightleftharpoons NH_4^+ + OH^-$$

Because the only source of Br^-, Ag^+, and $Ag(NH_3)_2^+$ is AgBr and because silver and bromide ions exist in a 1:1 ratio in the starting material, it follows that one mass-balance equation is

$$[Ag^+] + [Ag(NH_3)_2^+] = [Br^-]$$

where the bracketed terms are molar species concentrations. Also, we know that the only source of ammonia-containing species is the 0.010 M NH_3. Therefore,

$$c_{NH_3} = [NH_3] + [NH_4^+] + 2[Ag(NH_3)_2^+] = 0.010$$

6A-2

Charge-Balance Equations

A charge-balance equation is based on the observation that solutions of electrolytes carry no net charge because the concentrations of anions and cations are such that the total number of moles of positive charge in the solution is equal to the total number of moles of negative charge. In deriving charge-balance equations, keep in mind that *the number of moles of charge contributed by an ion is equal to the number of charges carried by that ion multiplied by its molar concentration.* For example, the number of moles of positive charge contributed to a solution by $MgCl_2$ is twice the molar concentration of magnesium ion (that is, $2[Mg^{2+}]$) since one mole of Mg^{2+} contributes two moles of positive charge. The chloride ion contributes but one mole of negative charge per mole of ion. A charge-balance equation for such a solution is then

$$2[Mg^{2+}] + [H_3O^+] = [Cl^-] + [OH^-]$$

This equation states that the number of moles of positive charge equals the number of moles of negative charge when the sum of the molar concentrations of OH^- and Cl^- just equals the molar concentration of H_3O^+ ion plus *two times* the molar concentration of Mg^{2+}. Note that when a solution is neutral, the concentrations of hydronium and hydroxide ions cancel and need not be included in the charge-balance expression. Thus, in the case at hand, we can write simply

$$2[Mg^{2+}] = [Cl^-]$$

Note that this equation implies that charge balance is realized when the chloride ion concentration is *twice* the magnesium ion concentration.

EXAMPLE 6-2

Neglecting the dissociation of water, write a charge-balance equation for a solution that contains NaCl, $Mg(NO_3)_2$, and $Al_2(SO_4)_3$.

$$[Na^+] + 2[Mg^{2+}] + 3[Al^{3+}] = [Cl^-] + [NO_3^-] + 2[SO_4^{2-}]$$

6A-3

Steps for Solving Problems Involving Several Equilibria

1. Write a set of balanced chemical equations for all pertinent equilibria.
2. State in terms of equilibrium concentrations what quantity is being sought.
3. Write equilibrium-constant expressions for all equilibria developed in step 1, and find numerical values for the constants in tables of equilibrium constants.
4. Write mass-balance expressions for the system.
5. If possible, write a charge-balance expression for the system.

6. Count the number of unknown concentrations in the equations developed in steps 3, 4, and 5, and compare this number with the number of independent equations. If the number of equations is equal to the number of unknowns, proceed to step 7. If they are not, seek additional equations. If enough equations cannot be developed, try to eliminate unknowns by suitable approximations regarding the concentration of one or more of the unknowns. If such approximations cannot be found, the problem cannot be solved.
7. Make suitable approximations to simplify the algebra.
8. Solve the algebraic equations for the equilibrium concentrations needed to give a provisional answer as defined in step 2.
9. Check the validity of the approximations made in step 7 using the provisional concentrations computed in step 8.

Step 6 is particularly important because it shows whether or not an exact solution to the problem can be obtained. If the number of unknowns is identical to the number of equations, the problem has been reduced to one of *algebra* alone. That is, answers can be obtained given sufficient perseverance. On the other hand, if a sufficient number of equations cannot be found even when approximations are made, the problem should be abandoned. Time expended in attempting to solve a problem for which there are insufficient data is time wasted.

6A-4 *The Use of Approximations in Solving Equilibrium Calculations*

Upon completing step 6 of the systematic approach described in Section 6A-3, the chemist is faced with the mathematical problem of solving several nonlinear simultaneous equations—a task that is often formidable, tedious, and time-consuming unless a suitable computer program is available or unless the equations can be simplified by approximations that decrease the number of unknowns and equations. In this section, we consider in general terms how equations describing equilibrium relationships can be simplified by suitable approximations.

It should be borne in mind whenever step 7 of the systematic approach is undertaken that the *only* equations that can be simplified are the mass-balance and charge-balance equations because it is only in these equations that concentration terms appear as sums or differences rather than as products or quotients. It is always possible to assume that one or more of the terms in a sum or difference are so much smaller than the others that they can be ignored without destroying the equality. When any concentration term in an equilibrium-constant expression is assumed to be zero, the expression becomes meaningless.

Many students find step 7 to be the most troublesome because they fear that their lack of experience will lead to invalid approximations and thus to serious errors in their computed results. Such fears are groundless. Experienced scientists are often as puzzled as neophytes when making an approximation that simplifies an equilibrium calculation. Nonetheless, they make such approximations without trepidation because they know that the effects of an invalid assumption will become obvious by the time a computation is completed. Generally, questionable assumptions should always be tried at the outset and provisional answers computed. If the assumption leads to an intolerable error, a recalculation without the faulty approximation is then performed. Usually, it is more efficient to try a questionable assumption at the outset than to undertake a more time-consuming and tedious calculation without the assumption.

6B The Calculation of Solubility by the Systematic Method

In this section, the use of the systematic method is illustrated by examples involving the solubility of precipitates under various conditions.

6B-1 Metal Hydroxides

Examples 6-3 and 6-4 involve calculating the solubilities of two metal hydroxides, one of which is relatively soluble compared with the second. These examples illustrate the importance of steps 7 and 8 in the systematic procedure for solving mass-law problems.

EXAMPLE 6-3

Calculate the molar solubility of $Mg(OH)_2$ in water.

Step 1. Pertinent equilibria:
Two equilibria that need to be considered are

$$Mg(OH)_2(s) \rightleftharpoons Mg^{2+} + 2OH^-$$

$$2H_2O \rightleftharpoons H_3O^+ + OH^-$$

Step 2. Definition of unknown:
Since 1 mol of Mg^{2+} is formed for each mole of $Mg(OH)_2$ dissolved,

$$\text{Solubility} = [Mg^{2+}]$$

Step 3. Equilibrium-constant expressions:

$$[Mg^{2+}][OH^-]^2 = 1.7 \times 10^{-11} \quad (6\text{-}7)$$

$$[H_3O^+][OH^-] = 1.00 \times 10^{-14} \quad (6\text{-}8)$$

Step 4. Mass-balance expression:

$$[OH^-] = 2[Mg^{2+}] + [H_3O^+] \quad (6\text{-}9)$$

The first term on the right-hand side of Equation 6-9 represents the hydroxide ion concentration resulting from dissolved $Mg(OH)_2$ in the solution, and the second term is the hydroxide ion concentration resulting from the dissociation of water.

Step 5. Charge-balance expression:

$$2[Mg^{2+}] + [H_3O^+] = [OH^-]$$

Note that this equation is identical to Equation 6-9. We shall find that equations derived from mass-balance and charge-balance considerations are often the same.

Step 6. Number of independent equations and unknowns:
We have developed three independent algebraic equations (Equations 6-7, 6-8, and 6-9) and have three unknowns ($[Mg^{2+}]$, $[OH^-]$, and $[H_3O^+]$). Therefore, a rigorous solution to the problem can, if necessary, be realized.

Step 7. Approximations:
Equation 6-9 is the only expression wherein approximations can be made. Since the solubility-product constant for $Mg(OH)_2$ is relatively large, the solution will be somewhat basic. Therefore, it seems reasonable to assume that $[H_3O^+] \ll [Mg^{2+}]$, whereupon Equation 6-9 simplifies to

$$2[Mg^{2+}] \simeq [OH^-] \quad (6\text{-}10)$$

Step 8. Solution to equations:

Substitution of Equation 6-10 into Equation 6-7 gives

$$[Mg^{2+}](2[Mg^{2+}])^2 = 1.7 \times 10^{-11}$$

$$[Mg^{2+}] = \text{Solubility} = 1.6 \times 10^{-4} \text{ mol/L}$$

Step 9. Check of assumptions:

Substitution into Equation 6-10 yields

$$[OH^-] = 2 \times 1.6 \times 10^{-4} = 3.2 \times 10^{-4}$$

From the ion-product constant for water, we obtain

$$[H_3O^+] = (1.00 \times 10^{-14})/(3.2 \times 10^{-4}) = 3.1 \times 10^{-11}$$

Thus our assumption that $3.1 \times 10^{-11} \ll 1.6 \times 10^{-4}$ is certainly valid.

EXAMPLE 6-4

Calculate the solubility of $Fe(OH)_3$ in water.

Proceeding as in Example 6-3, we write the pertinent equilibria

$$Fe(OH)_3(s) \rightleftharpoons Fe^{3+} + 3OH^-$$

$$2H_2O \rightleftharpoons H_3O^+ + OH^-$$

and

$$\text{Solubility} = [Fe^{3+}]$$

We then write three algebraic equations as in the previous example:

$$[Fe^{3+}][OH^-]^3 = 4 \times 10^{-38}$$

$$[H_3O^+][OH^-] = 1.00 \times 10^{-14}$$

$$3[Fe^{3+}] + [H_3O^+] = [OH^-]$$

We again assume that $[H_3O^+] \ll 3[Fe^{3+}]$, so that

$$3[Fe^{3+}] \simeq [OH^-]$$

Substituting this equation into the solubility-product expression gives

$$[Fe^{3+}](3[Fe^{3+}])^3 = 4 \times 10^{-38}$$

$$[Fe^{3+}] = \left(\frac{4 \times 10^{-38}}{27}\right)^{1/4} = 2 \times 10^{-10}$$

$$\text{Solubility} = [Fe^{3+}] = 2 \times 10^{-10} \text{ mol/L}$$

We have assumed, however, that

$$[OH^-] \simeq 3[Fe^{3+}] = 3 \times 2 \times 10^{-10} = 6 \times 10^{-10}$$

which means that

$$[H_3O^+] = \frac{1.00 \times 10^{-14}}{6 \times 10^{-10}} = 1.7 \times 10^{-5}$$

Clearly, $[H_3O^+]$ is *not* much smaller than $3[Fe^{3+}]$; indeed, the reverse appears to be the case. That is,

$$3[Fe^{3+}] \ll [H_3O^+]$$

and the mass-balance equation reduces to

$$[H_3O^+] = [OH^-] = 1.00 \times 10^{-7}$$

Substitution for $[OH^-]$ in the solubility-product expression yields

$$[Fe^{3+}] = \frac{4 \times 10^{-38}}{(1.00 \times 10^{-7})^3} = 4 \times 10^{-17}$$

$$\text{Solubility} = 4 \times 10^{-17} \text{ mol/L}$$

The assumption that $3[Fe^{3+}] \ll [H_3O^+]$ is clearly valid. Note the very large error that arises with the faulty assumption that $[H_3O^+] \ll 3[Fe^{3+}]$.

From this example, it is apparent that solubility calculations for metal hydroxides can frequently be made simpler by the proper choice of one of two assumptions. It is to be expected that there is a range of solubility products for which neither assumption is valid and for which equations similar to Equations 6-7, 6-8, and 6-9 must be solved for all three variables.

6B-2 The Solubility of Precipitates in the Presence of Complexing Agents

The solubility of a precipitate often increases dramatically in the presence of reagents that form complexes with the anion or the cation of the precipitate. For example, fluoride ions prevent the quantitative precipitation of aluminum hydroxide even though the solubility product of this precipitate is remarkably small (2×10^{-32}). The cause of the increase in solubility is shown by the equations

$$\begin{array}{l} Al(OH)_3(s) \rightleftharpoons Al^{3+} + 3OH^- \\ \qquad\qquad\qquad\quad + \\ \qquad\qquad\qquad\quad 6F^- \\ \qquad\qquad\qquad\quad \upharpoonleft\downharpoonright \\ \qquad\qquad\qquad\quad AlF_6^{3-} \end{array}$$

The fluoride complex is sufficiently stable to permit fluoride ions to compete successfully with hydroxide ions for aluminum ions.

EXAMPLE 6-5

The solubility product of CuI is 1.1×10^{-12}. The formation constant K_2 for the reaction of CuI with I^- to give CuI_2^- is 7.9×10^{-4}. Calculate the molar solubility of CuI in a 1.0×10^{-4} M solution of KI.

Step 1. Pertinent equilibria:

From the input data, we assume that the following equilibria are present in a solution of KI that is saturated with CuI:

$$CuI(s) \rightleftharpoons Cu^+ + I^-$$

$$CuI(s) + I^- \rightleftharpoons CuI_2^-$$

Since neither H_3O^+ nor OH^- reacts to any significant extent with any of the participants in these equilibria, we need not include the dissociation of water in our list.

Step 2. Definition of unknown:

From inspection of the two equations, we see that the dissolved copper(I) is present as either Cu^+ or CuI_2^-. Therefore,

$$\text{Solubility} = [Cu^+] + [CuI_2^-]$$

Step 3. Equilibrium-constant expressions

$$[Cu^+][I^-] = K_{sp} = 1.1 \times 10^{-12} \qquad (6\text{-}11)$$

$$\frac{[CuI_2^-]}{[I^-]} = K_2 = 7.9 \times 10^{-4} \qquad (6\text{-}12)$$

Step 4. Mass-balance expression:
Here, we may write

$$[I^-] = c_{KI} + [Cu^+] - [CuI_2^-] \qquad (6\text{-}13)$$

The first term on the right-hand side of this equation represents the concentration of iodide from the KI, and the second term corresponds to the contribution of iodide from the dissolution of CuI. The third term gives the concentration of iodide needed to form the complex from CuI.

Step 5. Charge-balance equation:

$$[Cu^+] + [K^+] = [I^-] + [CuI_2^-] \qquad (6\text{-}14)$$

The K^+ concentration is known to be 1.00×10^{-4} M. Substituting this figure into Equation 6-14 gives, after rearranging,

$$[I^-] = 1.0 \times 10^{-4} + [Cu^+] - [CuI_2^-] \qquad (6\text{-}15)$$

Note, however, that as in Example 6-4, the charge-balance and mass-balance equations are identical. Therefore, we have only three independent equations (Equations 6-11, 6-12, and 6-15).

Step 6. Number of independent equations and unknowns:
We have three unknowns, $[Cu^+]$, $[I^-]$, and $[CuI_2^-]$, and as noted in step 5, three algebraic equations. Therefore, an exact solution to the equations is possible.

Step 7. Approximations:
Two approximations are possible in Equation 6-15. The first is that $[Cu^+]$ is so much smaller than 1.0×10^{-4} that the former can be eliminated from the equation. The second is that $[CuI_2^-]$ is also small enough to be ignored. The first assumption is surely valid in light of the very small numerical value for the solubility-product constant for CuI. The validity of the second assumption is less obvious. As noted earlier, however, it is usually wise to make the assumption, calculate a provisional value for the concentration, and see whether the provisional value is indeed much smaller than the number with which it is being compared. If it is not, a recalculation must be made with the questionable concentration term retained in the equation.

Following this procedure, we assume $([Cu^+] - [CuI_2^-]) \ll 1.0 \times 10^{-4}$, whereupon Equation 6-15 simplifies to

$$[I^-] = 1.0 \times 10^{-4}$$

Step 8. Solution to equations:
Substituting for $[I^-]$ in Equation 6-11 gives, after rearranging,

$$[Cu^+] = (1.1 \times 10^{-12})/(1.0 \times 10^{-4}) = 1.1 \times 10^{-8}$$

Substituting this value and the value for $[I^-]$ into Equation 6-12 and rearranging yield

$$[CuI_2^-] = 1.0 \times 10^{-4} \times 7.9 \times 10^{-4} = 7.9 \times 10^{-8}$$

When these concentrations are substituted into the relationship developed in step 2, we obtain

$$\text{Solubility} = 1.1 \times 10^{-8} + 7.9 \times 10^{-8} = 9.0 \times 10^{-8} \text{ mol/L}$$

Step 9. Check of assumptions:

We see that the assumption $(1.1 \times 10^{-8} - 7.9 \times 10^{-8}) \ll 1.0 \times 10^{-4}$ is valid, and the provisional values for $[Cu^+]$ and $[CuI_2^-]$ are therefore acceptable.

EXAMPLE 6-6

Calculate the solubility of AgBr in 0.0200 M NH_3, given that the formation constant for $Ag(NH_3)_2^+$ is 1.3×10^7, the solubility product for AgBr is 5.2×10^{-13}, and the dissociation constant for NH_3 is 1.76×10^{-5}.

Step 1. Pertinent equilibria:

$$AgBr \rightleftharpoons Ag^+ + Br^-$$

$$Ag^+ + 2NH_3 \rightleftharpoons Ag(NH_3)_2^+$$

$$NH_3 + H_2O \rightleftharpoons NH_4^+ + OH^-$$

Step 2. Definition of unknown:

It is apparent from these equations that 1 mol of AgBr produces 1 mol of Br^- and 1 mol of silver-containing species. Therefore,

$$\text{Solubility} = [Br^-] = [Ag^+] + [Ag(NH_3)_2^+]$$

Step 3. Equilibrium-constant expressions:

$$[Ag^+][Br^-] = K_{sp} = 5.2 \times 10^{-13} \qquad (6\text{-}16)$$

$$\frac{[Ag(NH_3)_2^+]}{[Ag^+][NH_3]^2} = \beta_2 = 1.3 \times 10^7 \qquad (6\text{-}17)$$

$$\frac{[NH_4^+][OH^-]}{[NH_3]} = K_b = 1.76 \times 10^{-5} \qquad (6\text{-}18)$$

Step 4. Mass-balance expressions:

As shown in Example 6-1 and in step 2,

$$[Br^-] = [Ag^+] + [Ag(NH_3)_2^+] \qquad (6\text{-}19)$$

Since the analytical concentration of NH_3 is 0.0200,

$$0.0200 = [NH_3] + [NH_4^+] + 2[Ag(NH_3)_2^+] \qquad (6\text{-}20)$$

Furthermore, reaction of NH_3 with water produces one NH_4^+ for each OH^-. Therefore,

$$[OH^-] = [NH_4^+] \qquad (6\text{-}21)$$

Step 5. Charge-balance expression:

$$[Ag^+] + [NH_4^+] + [Ag(NH_3)_2^+] = [Br^-] + [OH^-] \qquad (6\text{-}22)[1]$$

Step 6. Number of equations and unknowns:

We count seven equations (6-16 through 6-22) but only six unknowns ($[OH^-]$, $[Br^-]$, $[Ag^+]$, $[Ag(NH_3)_2^+]$, $[NH_4^+]$, and $[OH^-]$). This discrepancy suggests that the algebraic equations are not all independent. Close examination shows that Equation 6-22 is in fact the sum of Equations 6-19 and 6-21 and therefore not independent. Thus, we remove Equation 6-22 from consideration and now have six independent equations and an equal number of unknowns.

[1]We do not include $[H_3O^+]$ in this equation because its concentration is negligible in a solution made basic with NH_3.

Step 7. Approximations:

To locate possible approximations, we again turn to those equations in which concentrations appear as sums or differences—that is, Equations 6-19 and 6-20.

1. Examining Equation 6-19 first, we speculate that $[Ag^+]$ may be considerably smaller than $[Ag(NH_3)_2^+]$ because the formation constant for the complex is so very large. Therefore, we assume provisionally that

$$[Ag^+] \ll [Ag(NH_3)_2^+]$$

2. The value for K_b for ammonia is small, which suggests that the concentration of NH_3 is significantly greater than the concentration of NH_4^+. We therefore assume that in Equation 6-20

$$[NH_4^+] \ll [NH_3] + 2[Ag(NH_3)_2^+]$$

3. A second possible assumption regarding Equation 6-20 is that $[Ag(NH_3)_2^+]$ is also significantly smaller than $[NH_3]$. The reason for this assumption is not immediately obvious from the equations at hand. If, however, Equations 6-16 and 6-17 are multiplied together, one obtains

$$\frac{[Ag(NH_3)_2^+][Br^-]}{[NH_3]^2} = \beta_2 K_{sp} = 6.8 \times 10^{-6} \tag{6-23}$$

This equation suggests that the numerator is significantly smaller than $[NH_3]^2$. Therefore, we shall try the assumption that

$$2[Ag(NH_3)_2^+] \ll [NH_3]$$

Each of these three assumptions is open to some doubt, but no harm is done in making them because a lack of validity in any one will become obvious by the time the calculation is completed. With the assumptions, Equations 6-18 and 6-21 are no longer needed and Equations 6-19 and 6-20 simplify to

$$[Br^-] = [Ag(NH_3)_2^+] \tag{6-24}$$

$$0.0200 = [NH_3]$$

We now have just three equations (6-24, 6-16, and 6-17) and three unknowns ($[Ag^+]$, $[Ag(NH_3)_2^+]$, and $[Br^-]$).

Step 8. Solution to equations:

Substituting $[NH_3] = 0.0200$ into Equation 6-17 gives, after rearranging,

$$[Ag(NH_3)_2^+] = (1.3 \times 10^7)(0.0200)^2[Ag^+] = (5.2 \times 10^3)[Ag^+]$$

Equation 6-24 then becomes

$$[Br^-] = (5.2 \times 10^3)[Ag^+]$$

Replacing $[Ag^+]$ with $K_{sp}/[Br^-]$ (Equation 6-16) gives

$$[Br^-] = \frac{5.2 \times 10^3 \times 5.2 \times 10^{-13}}{[Br^-]}$$

$$[Br^-] = \sqrt{5.2 \times 10^3 \times 5.2 \times 10^{-13}} = 5.2 \times 10^{-5}$$

Referring to step 2, we see

$$\text{Solubility} = [Br^-] = 5.2 \times 10^{-5} \text{ mol/L}$$

Step 9. Check of assumptions:

To check assumption 1, we turn to Equation 6-24 and find

$$[Ag(NH_3)_2^+] = [Br^-] = 5.2 \times 10^{-5}$$

We assumed that this concentration is much smaller than $[NH_3]$, that is, $5.2 \times 10^{-5} \ll 0.0100$. Essentially no error is introduced by this assumption.

To check assumption 2, we substitute $[NH_3] = 0.0200$ and Equation 6-21 into Equation 6-18, which gives

$$\frac{[NH_4^+]^2}{0.0200} = 1.76 \times 10^{-5}$$

$$[NH_4^+] = \sqrt{1.76 \times 10^{-5} \times 0.0200} = 5.9 \times 10^{-4}$$

We have assumed that

$$5.9 \times 10^{-4} = 0.00059 \ll 0.0100$$

The resulting error is then less than $(0.00059/0.0100) \times 100\% = 3.0\%$, which is acceptable in most cases.

To check assumption 3, we must calculate $[Ag^+]$ by substituting $[Br^-] = 5.2 \times 10^{-5}$ into the expression for K_{sp}:

$$[Ag^+] = (5.2 \times 10^{-13})/(5.2 \times 10^{-5}) = 1.0 \times 10^{-8}$$

We know from Equation 6-24 that

$$[Ag(NH_3)_2^+] = [Br^-] = 5.2 \times 10^{-5}$$

and we have assumed that $1.0 \times 10^{-8} \ll 5.2 \times 10^{-5}$. The percent error introduced by this assumption is less than 0.1%.

6B-3

The Effect of pH on Solubility

The solubility of precipitates containing an anion with basic properties, a cation with acidic properties, or both is dependent upon pH. For example, when barium sulfate is equilibrated with a solution containing hydrochloric acid, the following equilibria are established:

$$BaSO_4(s) \rightleftharpoons Ba^{2+} + SO_4^{2-}$$

$$SO_4^{2-} + H_3O^+ \rightleftharpoons HSO_4^- + H_2O$$

If more acid is added to this system, the sulfate ion concentration decreases by the common-ion effect. This decrease causes the first equilibrium to shift to the right, thus increasing the solubility. The examples that follow illustrate how the effect of pH on solubility can be treated in quantitative terms.

Solubility Calculations When the pH Is Fixed and Known

Analytical precipitations are frequently performed in solutions in which the hydronium ion concentration is fixed at some predetermined and known value. The calculation of solubility under this circumstance is a relatively straightforward process when the systematic approach is used, as illustrated by the following example.

EXAMPLE 6-7

Calculate the molar solubility of calcium oxalate in a solution that has a constant pH of 4.00.

Step 1. Pertinent equilibria:

$$CaC_2O_4(s) \rightleftharpoons Ca^{2+} + C_2O_4^{2-} \quad (6\text{-}25)$$

Since both oxalate and hydrogen oxalate ions are conjugate bases of weak acids, they react with water to produce hydroxide ions:

$$C_2O_4^{2-} + H_2O \rightleftharpoons HC_2O_4^- + OH^- \quad (6\text{-}26)$$

$$HC_2O_4^- + H_2O \rightleftharpoons H_2C_2O_4 + OH^- \quad (6\text{-}27)$$

Thus, a solution that is saturated with calcium oxalate will become basic unless an auxiliary reagent that maintains the hydroxide concentration at a constant level is added. Such a reagent is called a *buffer,* and presumably a pH 4.0 buffer was used in this case. Buffers are discussed in detail in Section 8D.

Step 2. Definition of the unknown:

Since CaC_2O_4 is a strong electrolyte, its solubility is equal to the molar concentration of calcium ions and to the sum of the equilibrium concentrations of the oxalate species in the solution:

$$\text{Solubility} = [Ca^{2+}] = [C_2O_4^{2-}] + [HC_2O_4^-] + [H_2C_2O_4]$$

Step 3. Equilibrium-constant expression:

$$[Ca^{2+}][C_2O_4^{2-}] = K_{sp} = 2.3 \times 10^{-9} \quad (6\text{-}28)$$

Recall that for a conjugate acid pair, $K_w = K_aK_b$ (Equation 5-11). Application of Equation 5-11 to the reactions of the two oxalate conjugate bases gives

$$\frac{[HC_2O_4^-][OH^-]}{[C_2O_4^{2-}]} = \frac{K_w}{K_2} = \frac{1.00 \times 10^{-14}}{5.42 \times 10^{-5}} = 1.85 \times 10^{-10} \quad (6\text{-}29)$$

$$\frac{[H_2C_2O_4][OH^-]}{[HC_2O_4^-]} = \frac{K_w}{K_1} = \frac{1.00 \times 10^{-14}}{5.36 \times 10^{-2}} = 1.87 \times 10^{-13} \quad (6\text{-}30)$$

where K_1 and K_2 are the first and second dissociation constants for oxalic acid, $H_2C_2O_4$.

Step 4. Mass-balance expressions:

Because the only source of Ca^{2+} and the various oxalate species is the dissolved CaC_2O_4, it follows that

$$[Ca^{2+}] = [C_2O_4^{2-}] + [HC_2O_4^-] + [H_2C_2O_4] \quad (6\text{-}31)$$

Furthermore, the solution is buffered to a pH of 4.00. Therefore,

$$[H_3O^+] = \text{antilog}(-4.00) = 1.0 \times 10^{-4}$$

$$[OH^-] = (1.00 \times 10^{-14})/(1.0 \times 10^{-4}) = 1.0 \times 10^{-10} \quad (6\text{-}32)$$

Step 5. Charge-balance expressions:

In order to maintain the solution at pH 4.00, a buffer is required. In all probability, this buffer consists of some weak acid HA and its conjugate base NaA. The nature and concentrations of these two species have not been specified; thus we are unable to write a charge-balance equation. As shown in step 6, however, this equation is not needed.

Step 6. Number of independent equations and unknowns:

We have four unknowns ($[Ca^{2+}]$, $[C_2O_4^{2-}]$, $[HC_2O_4^-]$, and $[H_2C_2O_4]$) as well as four independent algebraic relationships (Equations 6-28, 6-29, 6-30, and 6-31). Therefore, an exact solution is possible, and the problem becomes one of algebra.

Step 7. Approximations:

An exact solution is so readily obtained that we dispense with approximations.

Step 8. Solution of the equations:

A convenient way to solve for the four unknowns is to make suitable substitutions into Equation 6-31 and thereby establish a relationship between $[Ca^{2+}]$ and $[C_2O_4^{2-}]$. In order to do this, we first derive expressions for $[HC_2O_4^-]$ and $[H_2C_2O_4]$ in terms of $[C_2O_4^{2-}]$. Substitution of 1.0×10^{-10} for $[OH^-]$ in Equation 6-29 yields

$$\frac{[HC_2O_4^-](1.0 \times 10^{-10})}{[C_2O_4^{2-}]} = 1.85 \times 10^{-10}$$

$$[HC_2O_4^-] = 1.85[C_2O_4^{2-}]$$

Upon substituting this relationship and the hydroxide ion concentration into Equation 6-30, we obtain

$$\frac{[H_2C_2O_4](1.0 \times 10^{-10})}{1.85[C_2O_4^{2-}]} = 1.87 \times 10^{-13}$$

$$[H_2C_2O_4] = \frac{1.87 \times 10^{-13} \times 1.85[C_2O_4^{2-}]}{1.0 \times 10^{-10}} = 0.0035[C_2O_4^{2-}]$$

When these values for $[H_2C_2O_4]$ and $[HC_2O_4^-]$ are substituted into Equation 6-31, we obtain

$$[Ca^{2+}] = [C_2O_4^{2-}] + 1.85[C_2O_4^{2-}] + 0.0035[C_2O_4^{2-}] = 2.85[C_2O_4^{2-}]$$

Substitution for $[C_2O_4^{2-}]$ in Equation 6-28 gives

$$\frac{[Ca^{2+}][Ca^{2+}]}{2.85} = 2.3 \times 10^{-9}$$

$$[Ca^{2+}] = \sqrt{6.56 \times 10^{-9}} = 8.1 \times 10^{-5}$$

Thus, from step 2 we conclude that

$$\text{Solubility} = 8.1 \times 10^{-5} \text{ mol/L}$$

Solubility Calculations When the pH is Variable

Saturating an unbuffered solution with a sparingly soluble salt made up of a basic anion or an acidic cation causes the pH of the solution to change. For example, pure water saturated with barium carbonate is basic as a consequence of the reactions

$$BaCO_3(s) \rightleftharpoons Ba^{2+} + CO_3^{2-}$$

$$CO_3^{2-} + H_2O \rightleftharpoons HCO_3^- + OH^-$$

$$HCO_3^- + H_2O \rightleftharpoons H_2CO_3 + OH^-$$

In contrast to Example 6-7, the hydroxide ion concentration now becomes an unknown, and an additional algebraic equation must therefore be developed if the solubility of barium carbonate is to be calculated.

In many instances, the reaction of a precipitate with water cannot be neglected without introducing an error in the calculation. As shown by the data in Table 6-1, the magnitude of the error depends upon the solubility of the precipitate as well as on the base dissociation constant of the anion. The solubilities of the hypothetical precipitate MA, shown in column 4, were obtained by taking into account the reaction of A^- with water. Column 5 gives the calculated results when the basic properties of A^- are neglected; here, the solubility is simply the square root of the solubility product. Two solubility products, 1.0×10^{-10} and 1.0×10^{-20}, have been assumed for these calculations as well as several values for K_b (column 3). It is apparent that ne-

TABLE 6-1

Calculated Solubility of MA from Various Assumed Values of K_{sp} and K_b

Assumed K_{sp} for MA	Assumed K_{HA}	Dissociation Constant for A^- $K_b = K_w/K_{HA}$	Calculated Solubility of MA, M	Calculated Solubility of MA Neglecting Reaction of A^- with Water, M
1.0×10^{-10}	1.0×10^{-6}	1.0×10^{-8}	1.02×10^{-3}	1.0×10^{-5}
	1.0×10^{-8}	1.0×10^{-6}	1.2×10^{-5}	1.0×10^{-5}
	1.0×10^{-10}	1.0×10^{-4}	2.4×10^{-5}	1.0×10^{-5}
	1.0×10^{-12}	1.0×10^{-2}	10×10^{-5}	1.0×10^{-5}
1.0×10^{-20}	1.0×10^{-6}	1.0×10^{-8}	1.05×10^{-10}	1.0×10^{-10}
	1.0×10^{-8}	1.0×10^{-6}	3.3×10^{-10}	1.0×10^{-10}
	1.0×10^{-10}	1.0×10^{-4}	32×10^{-10}	1.0×10^{-10}
	1.0×10^{-12}	1.0×10^{-2}	290×10^{-10}	1.0×10^{-10}

glecting the reaction of the anions with water leads to a negative error that becomes more pronounced both as the solubility of the precipitate decreases (smaller K_{sp}) and as the conjugate base becomes stronger. Note that the error becomes insignificant for anions derived from acids with dissociation constants greater than about 10^{-6}.

It is not difficult to write the algebraic relationships needed to calculate the solubility of such a precipitate. Solving the equations, however, is tedious. Fortunately, it is ordinarily possible to invoke one of two simplifying assumptions to decrease the algebraic labor:

1. The first simplification is applicable to moderately soluble compounds containing an anion that reacts extensively with water. It is assumed that sufficient hydroxide ions are formed to make it unnecessary to consider the hydronium ion concentration in the calculations. Another way of stating this assumption is to say that the hydroxide ion concentration of the solution is determined exclusively by the reaction of the anion with water and that the contribution of hydroxide ions from the dissociation of water is negligible by comparison.
2. The second simplification is applicable to precipitates of very low solubility, particularly those containing an anion that does not react extensively with water. In such a system, it can be assumed that dissolution of the precipitate does not significantly change the hydronium or hydroxide ion concentrations and that, at room temperature, these concentrations remain essentially 10^{-7} mol/L. The solubility calculation then follows the course shown in Example 6-7.

EXAMPLE 6-8

Calculate the solubility of $BaCO_3$ in water.

Step 1. Pertinent equilibria:

$$BaCO_3(s) \rightleftharpoons Ba^{2+} + CO_3^{2-} \quad (6\text{-}33)$$

$$CO_3^{2-} + H_2O \rightleftharpoons HCO_3^- + OH^- \quad (6\text{-}34)$$

$$HCO_3^- + H_2O \rightleftharpoons H_2CO_3 + OH^- \quad (6\text{-}35)$$

$$2H_2O \rightleftharpoons H_3O^+ + OH^- \quad (6\text{-}36)$$

Step 2. Definition of the unknown:

$$\text{Solubility} = [Ba^{2+}] = [CO_3^{2-}] + [HCO_3^-] + [H_2CO_3]$$

Step 3. Equilibrium-constant expressions:

$$[Ba^{2+}][CO_3^{2-}] = K_{sp} = 5.1 \times 10^{-9} \quad (6\text{-}37)$$

$$\frac{[HCO_3^-][OH^-]}{[CO_3^{2-}]} = \frac{K_w}{K_2} = \frac{1.00 \times 10^{-14}}{4.7 \times 10^{-11}} = 2.13 \times 10^{-4} \quad (6\text{-}38)$$

$$\frac{[H_2CO_3][OH^-]}{[HCO_3^-]} = \frac{K_w}{K_1} = \frac{1.00 \times 10^{-14}}{4.45 \times 10^{-7}} = 2.25 \times 10^{-8} \quad (6\text{-}39)$$

and

$$[H_3O^+][OH^-] = 1.00 \times 10^{-14} \quad (6\text{-}40)$$

Step 4. Mass-balance expression:

$$[Ba^{2+}] = [CO_3^{2-}] + [HCO_3^-] + [H_2CO_3] \quad (6\text{-}41)$$

Step 5. Charge-balance expression:

$$2[Ba^{2+}] + [H_3O^+] = 2[CO_3^{2-}] + [HCO_3^-] + [OH^-] \quad (6\text{-}42)$$

Step 6. Number of equations and unknowns:

We have developed six equations (Equations 6-37 through 6-42), which are sufficient to solve for the six unknowns ($[Ba^{2+}]$, $[CO_3^{2-}]$, $[HCO_3^-]$, $[H_2CO_3]$, $[OH^-]$, and $[H_3O^+]$).

Step 7. Approximations:

We examine Equations 6-41 and 6-42 with the goal of eliminating one or more terms on the grounds that their elimination does not create a significant error. One candidate for elimination is $[H_3O^+]$ in Equation 6-42. Because the solution is basic as a consequence of reactions 6-34 and 6-35, $[H_3O^+]$ must be smaller than 10^{-7}. Also, $[Ba^{2+}]$ must be considerably larger than 10^{-7} because, if no reaction occurs between CO_3^{2-} and H_2O, $[Ba^{2+}]$ is simply the square root of K_{sp}, or 7×10^{-5} M. The fact that reaction does occur means that $[CO_3^{2-}]$ is smaller than 7×10^{-5} and therefore $[Ba^{2+}] > 7 \times 10^{-5}$ M. Thus, the assumption that $[H_3O^+] \ll 2[Ba^{2+}]$ in Equation 6-42 appears entirely reasonable. With this assumption, we no longer need Equation 6-40.

A second possible assumption is that $[H_2CO_3]$ is so much smaller than $[HCO_3^-]$ that the former can be deleted from Equation 6-41. We base this assumption on our knowledge that the solution is basic and therefore $[OH^-] > 10^{-7}$. If $[OH^-]$ is 10^{-6} M, substitution into Equation 6-39 reveals that

$$\frac{[H_2CO_3]}{[HCO_3^-]} = \frac{2.25 \times 10^{-8}}{10^{-6}} = 0.0225$$

If $[OH^-]$ is 10^{-5}, this ratio is about 0.002. Although these calculations suggest that $[H_2CO_3] \ll [HCO_3^-]$, we cannot be sure that this assumption is valid. It is surely worth trying, however.

As a result of the first assumption, Equation 6-42 becomes

$$2[Ba^{2+}] = 2[CO_3^{2-}] + [HCO_3^-] + [OH^-] \quad (6\text{-}43)$$

Further, insofar as $[HCO_3^-] \gg [H_2CO_3]$, the mass-balance expression simplifies to

$$[Ba^{2+}] = [CO_3^{2-}] + [HCO_3^-] \quad (6\text{-}44)$$

Equations 6-39 and 6-40 are now no longer needed. Thus, we have reduced the number of equations to four and the number of unknowns to four.

Step 8. Solution of the equations:

If we multiply Equation 6-44 by 2 and subtract the product from Equation 6-43, we obtain, upon rearrangement,

$$[OH^-] = [HCO_3^-] \tag{6-45}$$

Substitution of $[HCO_3^-]$ for $[OH^-]$ in Equation 6-38 gives

$$\frac{[HCO_3^-]^2}{[CO_3^{2-}]} = \frac{K_w}{K_2}$$

$$[HCO_3^-] = \sqrt{\frac{K_w}{K_2}[CO_3^{2-}]}$$

This expression permits elimination of $[HCO_3^-]$ from Equation 6-44:

$$[Ba^{2+}] = [CO_3^{2-}] + \sqrt{\frac{K_w}{K_2}[CO_3^{2-}]} \tag{6-46}$$

From Equation 6-37, we have

$$[CO_3^{2-}] = \frac{K_{sp}}{[Ba^{2+}]}$$

Substituting for $[CO_3^{2-}]$ in Equation 6-46 yields

$$[Ba^{2+}] = \frac{K_{sp}}{[Ba^{2+}]} + \sqrt{\frac{K_w K_{sp}}{K_2[Ba^{2+}]}}$$

It is convenient to multiply through by $[Ba^{2+}]$ and rearrange:

$$[Ba^{2+}]^2 - \sqrt{\frac{K_w}{K_2} K_{sp}[Ba^{2+}]} - K_{sp} = 0$$

Finally, after numerical values are supplied for the constants, we obtain

$$[Ba^{2+}]^2 - (1.04 \times 10^{-6})[Ba^{2+}]^{1/2} - 5.1 \times 10^{-9} = 0$$

In order to solve this equation by successive approximations (Appendix 11), we let $x = [Ba^{2+}]$ and rewrite the equation in the form

$$x = (1.04 \times 10^{-6}\sqrt{x} + 5.1 \times 10^{-9})^{1/2}$$

If we let $x_1 = 0$ and solve for x_2, we obtain

$$x_2 = (0 + 5.1 \times 10^{-9})^{1/2} = 7.14 \times 10^{-5}$$

Substituting this value for x in the original equation gives

$$x_3 = (1.04 \times 10^{-6} \times \sqrt{7.14 \times 10^{-5}} + 5.1 \times 10^{-9})^{1/2} = 1.18 \times 10^{-4}$$

Further iteration leads to $x_4 = 1.28 \times 10^{-4}$, $x_5 = 1.30 \times 10^{-4}$, and $x_6 = 1.30 \times 10^{-4}$. Thus

$$\text{Solubility} = [Ba^{2+}] = 1.30 \times 10^{-4} \simeq 1.3 \times 10^{-4}\ M$$

Step 9. Check of approximations:

To check the two assumptions that were made, we must calculate the concentrations of most of the other ions in the solution. We can evaluate $[CO_3^{2-}]$ from Equation 6-37:

$$[CO_3^{2-}] = \frac{5.1 \times 10^{-9}}{1.3 \times 10^{-4}} = 3.9 \times 10^{-5}$$

From Equation 6-44,

$$[HCO_3^-] = 13.0 \times 10^{-5} - 3.9 \times 10^{-5} = 9.1 \times 10^{-5}$$

From Equation 6-45,

$$[OH^-] = [HCO_3^-] = 9.1 \times 10^{-5}$$

From Equation 6-39,

$$\frac{[H_2CO_3](9.1 \times 10^{-5})}{9.1 \times 10^{-5}} = 2.25 \times 10^{-8}$$

$$[H_2CO_3] = 2.2 \times 10^{-8}$$

Finally, from Equation 6-40,

$$[H_3O^+] = \frac{1.00 \times 10^{-14}}{9.1 \times 10^{-5}} = 1.1 \times 10^{-10}$$

We see that the two assumptions do not lead to large errors; $[HCO_3^-]$ is about 4000 times greater than $[H_2CO_3]$, and $[H_3O^+]$ is clearly much smaller than any of the species in Equation 6-42.

Failure to take into account the basic reaction of CO_3^{2-} would have yielded a solubility of 7.1×10^{-5}, which is only about one half the value yielded by the more rigorous method.

An example of the second type of calculation referred to on page 148 follows. Here, it is assumed that the hydronium and hydroxide ion concentrations are 10^{-7} M after the solid has dissolved.

EXAMPLE 6-9

Calculate the solubility of silver sulfide in pure water.

Step 1. Pertinent equilibria:

$$Ag_2S(s) \rightleftharpoons 2Ag^+ + S^{2-} \quad (6\text{-}47)$$

$$S^{2-} + H_2O \rightleftharpoons HS^- + OH^- \quad (6\text{-}48)$$

$$HS^- + H_2O \rightleftharpoons H_2S + OH^- \quad (6\text{-}49)$$

$$2H_2O \rightleftharpoons H_3O^+ + OH^- \quad (6\text{-}50)$$

Step 2. Definition of unknown:

$$\text{Solubility} = \tfrac{1}{2}[Ag^+] = [S^{2-}] + [HS^-] + [H_2S]$$

Step 3. Equilibrium-constant expressions:

$$[Ag^+]^2[S^{2-}] = 6 \times 10^{-50} \quad (6\text{-}51)$$

$$\frac{[HS^-][OH^-]}{[S^{2-}]} = \frac{K_w}{K_2} = \frac{1.0 \times 10^{-14}}{1.2 \times 10^{-15}} = 8.3 \quad (6\text{-}52)$$

$$\frac{[H_2S][OH^-]}{[HS^-]} = \frac{K_w}{K_1} = \frac{1.0 \times 10^{-14}}{5.7 \times 10^{-8}} = 1.8 \times 10^{-7} \quad (6\text{-}53)$$

$$[H_3O^+][OH^-] = 1.00 \times 10^{-14}$$

Step 4. Mass-balance expression:

$$\tfrac{1}{2}[Ag^+] = [S^{2-}] + [HS^-] + [H_2S] \quad (6\text{-}54)$$

Step 5. Charge-balance expression:

$$[Ag^+] + [H_3O^+] = 2[S^{2-}] + [HS^-] + [OH^-] \quad (6\text{-}55)$$

Step 6. Comparison of equations and unknowns:

We have six unknowns and six equations. Thus, an exact solution is feasible.

Step 7. Approximations:

The solubility product for Ag_2S is very small; therefore, it is probable that there is little change in hydroxide ion concentration as the precipitate dissolves. As a consequence, we can assume tentatively that

$$[OH^-] \simeq [H_3O^+] = 1.0 \times 10^{-7}$$

This assumption will be correct if, in Equation 6-55,

$$[Ag^+] \ll [H_3O^+] \qquad \text{and} \qquad 2[S^{2-}] + [HS^-] \ll [OH^-]$$

Step 8. Solution of equations:

We now proceed exactly as we did in Example 6-7. Substitution of 1.0×10^{-7} for $[OH^-]$ in Equations 6-52 and 6-53 gives

$$\frac{[HS^-]}{[S^{2-}]} = \frac{8.3}{1.0 \times 10^{-7}} = 8.3 \times 10^7$$

$$[H_2S] = \frac{(1.8 \times 10^{-7})[HS^-]}{1.0 \times 10^{-7}} = 1.8[HS^-]$$

Substituting the first equation into the second yields

$$[H_2S] = (1.8)(8.3 \times 10^7)[S^{2-}] = (14.9 \times 10^7)[S^{2-}]$$

When these relationships are substituted into Equation 6-54, we obtain

$$\tfrac{1}{2}[Ag^+] = [S^{2-}] + (8.3 \times 10^7)[S^{2-}] + (14.9 \times 10^7)[S^{2-}]$$

$$[S^{2-}] = (2.3 \times 10^{-9})[Ag^+]$$

Substituting this relationship into the solubility-product expression gives

$$(2.3 \times 10^{-9})[Ag^+]^3 = 6 \times 10^{-50}$$

$$[Ag^+] = 3.0 \times 10^{-14}$$

$$\text{Solubility} = \tfrac{1}{2}[Ag^+] = 1.5 \times 10^{-14} \simeq 2 \times 10^{-14} \text{ mol/L}$$

Step 9. Check of approximations:

The assumption that $[Ag^+]$ is much smaller than $[H_3O^+]$ is clearly valid. We can readily calculate a value for $2[S^{2-}] + [HS^-]$ and confirm that this sum is likewise much smaller than $[OH^-]$. Therefore, we conclude that the assumptions made are reasonable and that the approximate solution is satisfactory.

6B-4 The Effect of Undissociated Solute on Solubility

Thus far, we have considered only solutes that dissociate completely when dissolved in aqueous media. There are some inorganic substances, however, such as calcium sulfate and the silver halides, that act as weak electrolytes and only partially dissociate in water. For example, a saturated solution of silver chloride contains significant amounts of undissociated silver chloride molecules as well as silver and chloride ions. Here, two equilibria are required to adequately describe the system:

$$AgCl(s) \rightleftharpoons AgCl(aq) \qquad (6\text{-}56)$$

$$AgCl(aq) \rightleftharpoons Ag^+ + Cl^- \qquad (6\text{-}57)$$

The equilibrium constant for the first reaction takes the form

$$\frac{[AgCl(aq)]}{[AgCl(s)]} = K$$

where the numerator is the concentration of the undissociated species *in the solution* and the denominator is the concentration of silver chloride *in the solid phase*. The latter term is a constant, however (page 111), and so the equation can be written

$$[AgCl(aq)] = K[AgCl(s)] = K_s \tag{6-58}$$

where K_s is the constant for the equilibrium shown in Equation 6-56. It is evident from this equation that, at a given temperature, the concentration of the undissociated silver chloride is constant and *independent* of the chloride and silver ion concentrations.

The equilibrium constant for the dissociation reaction K_d is

$$\frac{[Ag^+][Cl^-]}{[AgCl(aq)]} = K_d = 3.9 \times 10^{-4} \tag{6-59}$$

The product of these two constants is equal to the solubility product:

$$[Ag^+][Cl^-] = K_d K_s = K_{sp}$$

As shown by Example 6-10, both reaction 6-56 and reaction 6-57, as well as two others, contribute significantly to the solubility of silver chloride.

6B-5

The Solubility of Silver Chloride as a Function of Chloride Ion Concentration

The reaction between silver ions and chloride ions finds widespread use in analytical chemistry for the gravimetric and volumetric determination of both species. The effect of excess chloride ion on the solubility of the silver chloride precipitate is complex, as revealed by the following set of equations that describe the chemistry of the system:

$$AgCl(s) \rightleftharpoons AgCl(aq) \tag{6-60}$$

$$AgCl(aq) \rightleftharpoons Ag^+ + Cl^- \tag{6-61}$$

$$AgCl(s) + Cl^- \rightleftharpoons AgCl_2^- \tag{6-62}$$

$$AgCl_2^- + Cl^- \rightleftharpoons AgCl_3^{2-} \tag{6-63}$$

Note that equilibrium 6-61 and thus equilibrium 6-60 shift to the left with added chloride ion, whereas equilibria 6-62 and 6-63 shift to the right under the same circumstance. The consequence of these opposing effects is that a plot of silver chloride solubility as a function of concentration of added chloride exhibits a minimum. Example 6-10 illustrates how this behavior can be described in quantitative terms.

EXAMPLE 6-10

Derive an equation that describes the effect of the analytical concentration of KCl on the solubility of AgCl in an aqueous solution. Calculate the concentration of KCl at which the solubility is a minimum.

Step 1. Pertinent equilibria:

Equations 6-60 through 6-63 describe the pertinent equilibria.

Step 2. Definition of unknown:

The molar solubility S of AgCl is equal to the sum of the concentrations of the silver-containing species:

$$\text{Solubility} = S = [AgCl(aq)] + [Ag^+] + [AgCl_2^-] + [AgCl_3^{2-}] \tag{6-64}$$

Step 3. Equilibrium-constant expressions:

Equilibrium constants available in the literature include

$$[Ag^+][Cl^-] = K_{sp} = 1.82 \times 10^{-10} \tag{6-65}$$

$$\frac{[Ag^+][Cl^-]}{[AgCl(aq)]} = K_d = 3.9 \times 10^{-4} \tag{6-66}$$

$$\frac{[AgCl_2^-]}{[Cl^-]} = K_2 = 2.0 \times 10^{-5} \tag{6-67}$$

$$\frac{[AgCl_3^{2-}]}{[AgCl_2^-][Cl^-]} = K_3 = 1 \tag{6-68}$$

Step 4. Mass-balance equation:

$$[Cl^-] = c_{KCl} + [Ag^+] - [AgCl_2^-] - 2[AgCl_3^{2-}] \tag{6-69}$$

The second term on the right-hand side of this equation gives the chloride ion concentration produced by the dissolution of the precipitate, and the next two terms correspond to the *decrease* in chloride ion concentration resulting from the formation of the two chloro complexes from AgCl.

Step 5. Charge-balance equation:

As in some of the earlier examples, the charge-balance equation is identical to the mass-balance equation.

Step 6. Number of equations and unknowns:

We have five equations (6-65 through 6-69) and five unknowns ($[Ag^+]$, $[AgCl(aq)]$, $[AgCl_2^-]$, $[AgCl_3^{2-}]$, and $[Cl^-]$).

Step 7. Assumptions:

We assume that, over a considerable range of chloride ion concentration, the solubility of AgCl is so small that Equation 6-69 can be greatly simplified by the assumption that

$$[Ag^+] - [AgCl_2^-] - 2[AgCl_3^{2-}] \ll c_{KCl}$$

It is not certain that this is a valid assumption, but it is worth trying because it simplifies the problem so much. With this assumption, then, Equation 6-69 reduces to

$$[Cl^-] = c_{KCl} \tag{6-70}$$

Step 8. Solution of equations:

For convenience, we multiply Equations 6-67 and 6-68 together to give

$$\frac{[AgCl_3^{2-}]}{[Cl^-]^2} = K_2K_3 = 1 \times 2.0 \times 10^{-5} = 2.0 \times 10^{-5} \tag{6-71}$$

To calculate [AgCl(aq)], we divide Equation 6-65 by Equation 6-66 and rearrange:

$$[AgCl(aq)] = \frac{K_{sp}}{K_d} = \frac{1.82 \times 10^{-10}}{3.9 \times 10^{-4}} = 4.7 \times 10^{-7} \tag{6-72}$$

Note that the concentration of this species is *constant and independent of the chloride concentration.*

Substitution of Equations 6-72, 6-66, 6-67, and 6-71 into Equation 6-64 permits us to express the solubility in terms of the chloride ion concentration and the several constants:

$$S = \frac{K_{sp}}{K_d} + \frac{K_{sp}}{[Cl^-]} + K_2[Cl^-] + K_2K_3[Cl^-]^2 \tag{6-73}$$

Substitution of Equation 6-70 yields the desired relationship between the solubility and the analytical concentration of KCl:

$$S = \frac{K_{sp}}{K_d} + \frac{K_{sp}}{c_{KCl}} + K_2 c_{KCl} + K_2 K_3 c_{KCl}^2 \qquad (6\text{-}74)$$

To find the minimum in S, we set the derivative of S with respect to c_{KCl} equal to zero:

$$\frac{dS}{dc_{KCl}} = 0 = -\frac{K_{sp}}{c_{KCl}^2} + K_2 + 2K_2K_3c_{KCl}$$

$$2K_2K_3c_{KCl}^3 + c_{KCl}^2K_2 - K_{sp} = 0$$

Substituting numerical values gives

$$(4.0 \times 10^{-5})c_{KCl}^3 + (2.0 \times 10^{-5})c_{KCl}^2 - 1.82 \times 10^{-10} = 0$$

Following the procedure shown in Appendix 11, we can solve this equation by successive approximations to obtain

$$c_{KCl} = 0.0030 = [Cl^-]$$

In order to check the assumption made earlier, we calculate the concentration of the various species. Substitutions into Equations 6-65, 6-67, and 6-69 yield

$$[Ag^+] = (1.82 \times 10^{-10})/0.0030 = 6.1 \times 10^{-8}$$

$$[AgCl_2^-] = 2.0 \times 10^{-5} \times 0.0030 = 6.0 \times 10^{-8}$$

$$[AgCl_3^{2-}] = 2.0 \times 10^{-5} \times (0.0030)^2 = 1.8 \times 10^{-10}$$

Thus our assumption that c_{KCl} is much larger than the concentrations of the ions of the precipitate is reasonable. The minimum solubility is obtained by substitution of these concentrations and [AgCl(aq)] into Equation 6-64:

$$S = 4.7 \times 10^{-7} + 6.1 \times 10^{-8} + 6.0 \times 10^{-8} + 1.8 \times 10^{-10} = 5.9 \times 10^{-7}\ M$$

The solid curve in Figure 6-1 illustrates the effect of chloride ion concentration on the solubility of silver chloride; data for the curve were obtained by substituting various chloride concentrations into Equation 6-74. Note that at high concentrations of the common ion, the solubility becomes greater than that in pure water. The broken lines represent the equilibrium concentrations of the various silver-containing species as a function of c_{KCl}. Note that at the solubility minimum, undissociated silver chloride, AgCl(aq), is the major silver species in the solution, representing about 80% of the total dissolved silver. Its concentration is invariant, as has been demonstrated.

Unfortunately, reliable equilibrium data regarding undissociated species such as AgCl(aq) and complex species such as $AgCl_2^-$ are not abundant; consequently, solubility calculations are often, of necessity, based on solubility-product equilibria alone. Example 6-10 shows that, under some circumstances, such neglect of other equilibria can lead to serious error.

6C Separation of Ions by Control of the Concentration of a Precipitating Reagent

When two different ions react with a reagent to form precipitates of different solubilities, the less soluble compound will form at a lower reagent concentration. If the solubilities are sufficiently different, quantitative removal of the

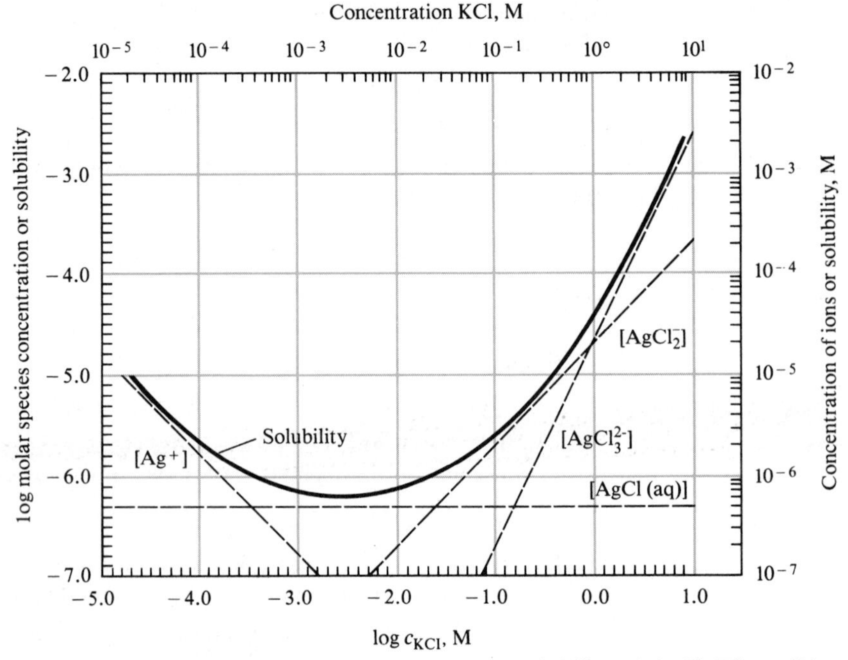

FIGURE 6-1 The effect of chloride ion concentration on the solubility of AgCl. The solid curve shows the total concentration of dissolved AgCl. The broken lines show the concentrations of the various silver-containing species.

first species from solution may be achieved without precipitation of the second. Such separations require careful control of the precipitating reagent concentration at some suitable predetermined level. Many important analytical separations, notably those involving sulfide ion, hydroxide ion, and organic reagents, are based on this concept.

6C-1 Calculation of the Feasibility of Separations

The following examples illustrate how solubility-product calculations are used to determine the feasibility of separations based upon solubility differences.

EXAMPLE 6-11

Can Fe^{3+} and Mg^{2+} be separated quantitatively as hydroxides from a solution that is 0.10 M in each cation? If the separation is possible, what range of OH^- concentration is permissible? Solubility-product constants for the two precipitates are

$$[Fe^{3+}][OH^-]^3 = 4 \times 10^{-38}$$

$$[Mg^{2+}][OH^-]^2 = 1.8 \times 10^{-11}$$

The K_{sp} for $Fe(OH)_3$ is so much smaller than that for $Mg(OH)_2$ that it appears likely that the former will precipitate at a lower OH^- concentration.[2] We can answer the questions posed in this problem by (1) calculating the OH^- concentration required to achieve quantitative precipitation of Fe^{3+} and (2) computing the OH^-

[2]The reader should be aware, however, that it is only the enormous numerical difference between the constants that permits this judgment. The solubility products are not strictly comparable inasmuch as the hydroxide concentration appears as a squared factor in one and as a cubed factor in the other.

concentration at which $Mg(OH)_2$ just begins to precipitate. If (1) is smaller than (2), a separation is feasible in principle, and the range of permissible OH^- concentrations is defined by the two values.

To determine (1), we must first specify what constitutes a quantitative removal of Fe^{3+} from the solution. The decision here is arbitrary and depends upon the purpose of the separation. In this example and the next, we shall consider a precipitation to be quantitative when all but 1 part in 10,000 of the ion has been removed from the solution—that is, when $[Fe^{3+}] \ll 1 \times 10^{-5}$.

We can readily calculate the OH^- concentration in equilibrium with 1×10^{-1} M Fe^{3+} by substituting directly into the solubility-product expression:

$$(1.0 \times 10^{-5})[OH^-]^3 = 4 \times 10^{-38}$$

$$[OH^-] = [(4 \times 10^{-38})/(1.0 \times 10^{-5})]^{1/3} = 2 \times 10^{-11} \text{ M}$$

Thus, if we maintain the OH^- concentration at about 2×10^{-11} mol/L, the Fe^{3+} concentration will be lowered to 1×10^{-5} mol/L. Note that quantitative precipitation of $Fe(OH)_3$ is achieved in a distinctly acidic medium.

To determine what maximum OH^- concentration can exist in the solution without causing formation of $Mg(OH)_2$, we note that precipitation cannot occur until the product $[Mg^{2+}][OH^-]^2$ exceeds the solubility product, 1.8×10^{-11}. Substitution of 0.1 (the molar Mg^{2+} concentration of the solution) into the solubility-product expression permits the calculation of the *maximum* OH^- concentration that can be tolerated:

$$0.10[OH^-]^2 = 1.8 \times 10^{-11}$$

$$[OH^-] = 1.3 \times 10^{-5}$$

When the OH^- concentration exceeds this level, the solution will be supersaturated with respect to $Mg(OH)_2$ and precipitation can begin.

From these calculations, we conclude that quantitative separation of $Fe(OH)_3$ can be achieved if the OH^- concentration is greater than 2×10^{-11} mol/L and that $Mg(OH)_2$ will not precipitate until a OH^- concentration of 1.3×10^{-5} mol/L is reached. Therefore, it is possible, in principle, to separate Fe^{3+} from Mg^{2+} by maintaining the OH^- concentration between these levels. In practice, the concentration of OH^- is kept as low as practical—often about 10^{-10} M.

6C-2 *Sulfide Separations*

A number of important methods for the separation of metallic ions involve controlling the concentration of the precipitating anion by regulating the hydronium ion concentration of the solution. Such methods are particularly attractive because of the relative ease with which the hydronium ion concentration can be maintained at some predetermined level by the use of a suitable buffer.[3] Perhaps the best known of these methods makes use of hydrogen sulfide as the precipitating reagent. Hydrogen sulfide is a weak acid, dissociating as follows:

$$H_2S + H_2O \rightleftharpoons H_3O^+ + HS^- \qquad \frac{[H_3O^+][HS^-]}{[HS^-]} = K_1 = 5.7 \times 10^{-8}$$

$$HS^- + H_2O \rightleftharpoons H_3O^+ + S^{2-} \qquad \frac{[H_3O^+][S^{2-}]}{[HS^-]} = K_2 = 1.2 \times 10^{-15}$$

[3]The preparation and properties of buffer solutions are considered in Section 8D. An important property of a buffer is that it maintains the hydronium ion concentration at an approximately fixed and predetermined level.

These equations may be combined to give an expression for the overall dissociation of hydrogen sulfide to sulfide ion:

$$H_2S + 2H_2O \rightleftharpoons 2H_3O^+ + S^{2-} \qquad \frac{[H_3O^+]^2[S^{2-}]}{[H_2S]} = K_1K_2 = 6.8 \times 10^{-23}$$

The constant for this overall reaction is simply the product of K_1 and K_2.

In sulfide separations, the solution is continuously saturated with hydrogen sulfide so that the molar concentration of the reagent is essentially constant throughout the precipitation. Because hydrogen sulfide is such a weak acid, its species concentration corresponds closely to its solubility in water, which is about 0.1 M. It is thus permissible to assume that, throughout any sulfide precipitation,

$$[H_2S] \simeq 0.10 \text{ mol/L}$$

Substituting this value into the overall dissociation-constant expression gives

$$\frac{[H_3O^+]^2[S^{2-}]}{0.10} = 6.8 \times 10^{-23}$$

Thus,

$$[S^{2-}] = \frac{6.8 \times 10^{-24}}{[H_3O^+]^2} \qquad (6\text{-}75)$$

Note that the molar concentration of the sulfide ion varies inversely as the square of the hydronium ion concentration of the solution. This relationship is useful for calculating the optimum conditions for the separation of cations by sulfide precipitation.

EXAMPLE 6-12

Find the conditions under which Cd^{2+} and Tl^+ can, in theory, be separated quantitatively with H_2S from a solution that is 0.1 M in each cation.

The constants for the two solubility equilibria are:

$$CdS(s) \rightleftharpoons Cd^{2+} + S^{2-} \qquad [Cd^{2+}][S^{2-}] = 2 \times 10^{-28}$$

$$Tl_2S(s) \rightleftharpoons 2Tl^+ + S^{2-} \qquad [Tl^+]^2[S^{2-}] = 1 \times 10^{-22}$$

Since CdS precipitates at a lower $[S^{2-}]$ than does Tl_2S, we first compute the sulfide ion concentration necessary for quantitative removal of Cd^{2+} from solution. We again define quantitative removal as lowering $[Cd^{2+}]$ to 1×10^{-5} M or smaller and substitute this value into the solubility-product expression:

$$10^{-5}[S^{2-}] = 2 \times 10^{-28}$$

$$[S^{2-}] = 2 \times 10^{-23}$$

This value is then compared with the $[S^{2-}]$ needed to initiate precipitation of Tl_2S from a 0.1 M solution:

$$(0.1)^2[S^{2-}] = 1 \times 10^{-22}$$

$$[S^{2-}] = 1 \times 10^{-20}$$

We see that $[S^{2-}]$ must be kept between 2×10^{-23} and 1×10^{-20} M to achieve separation. Equation 6-75, upon rearrangement, permits calculation of the $[H_3O^+]$ necessary to hold the $[S^{2-}]$ within these confines. Substitution of the two limiting values for $[S^{2-}]$ into this equation gives

$$[H_3O^+]^2 = \frac{6.8 \times 10^{-24}}{2 \times 10^{-23}} = 0.34$$

$$[H_3O^+] = 0.58 \simeq 0.6$$

and

$$[H_3O^+]^2 = \frac{6.8 \times 10^{-24}}{1 \times 10^{-20}} = 6.8 \times 10^{-4}$$

$$[H_3O^+] \simeq 0.026 = 0.03$$

By maintaining $[H_3O^+]$ between 0.03 and 0.6 M, we can, in principle, separate CdS quantitatively from Tl_2S. From a practical standpoint, however, it is questionable whether conditions could be controlled closely enough to give a clean separation.

Questions and Problems

Unless otherwise directed, base all calculations in this set on molar concentrations rather than activities.

6-1 Write the mass-balance expression for a solution that is
*(a) 0.10 M in H_3PO_4.
(b) 0.10 M in Na_2HPO_4.
*(c) 0.100 M in HNO_2 and 0.0500 M in $NaNO_2$.
(d) 0.025 M in NaF and saturated with CaF_2.
*(e) 0.100 M in NaOH and saturated with $Zn(OH)_2$ (which undergoes the reaction $Zn(OH)_2 + 2OH^- \rightleftharpoons Zn(OH)_4^{2-}$).
(f) saturated with $MgCO_3$.
*(g) saturated with CaF_2.
(h) 0.010 M in NaF and saturated with $Al(OH)_3$. (Al^{3+} forms a series of six F^- complexes having formulas of AlF^{2+}, AlF_2^+, . . . , AlF_6^{3-}.)
*(i) 0.0100 M in NH_3 and saturated with $Cd(OH)_2$. (Cd^{2+} forms a series of ammine complexes having formulas of $Cd(NH_3)_2^{2+}$, $Cd(NH_3)_2^{2+}$, . . . , $Cd(NH_3)_6^{2+}$.)

6-2 Write the charge-balance equations for the solutions in Problem 6-1.

6-3 Generate the solubility-product expression for
*(a) AgSCN.
(b) Ag_2CrO_4.
*(c) $PbCrO_4$.
(d) PbI_2.
*(e) BiI_3.
(f) $In_4[Fe(CN)_6]_3$.
*(g) PbClF.
(h) $Ce(IO_3)_4$.

6-4 Express the solubility-product constant for each substance in Problem 6-3 in terms of its molar solubility S.

6-5 Calculate the solubility-product constant for each of the following substances, given that the molar analytical concentrations of their saturated solutions are as indicated:
*(a) $AlPO_4$ (7.62×10^{-10} M).
(b) Tl_2S (1.08×10^{-7} M).
*(c) Ag_3PO_4 (4.68×10^{-6} M).
(d) Ce_2S_3 (3.51×10^{-5} M).
*(e) $NaUO_2AsO_4$ (5.06×10^{-8} M). (Products: Na^+, UO_2^{2+}, and AsO_4^{3-})

6-6 Tabulated with each of the following substances is the amount of solute contained in 1.00 L of a saturated solution. Use this information to evaluate the solubility-product constant.
*(a) TlBr (0.524 g)
(b) $Co(OH)_2$ (3.42×10^{-1} mg)
*(c) BiI_3 (7.76 mg)
(d) La_2S_3 (4.23×10^{-3} g)
*(e) $MgNH_4PO_4$ (9.19 mg)

*6-7 Calculate the solubility-product constant for $Fe(OH)_2$ if a saturated solution is 1.17×10^{-5} M in OH^-.

6-8 A 0.040 M KI solution saturated with PbI_2 is found to be 4.44×10^{-6} M in Pb^{2+}. Calculate K_{sp} for PbI_2.

*6-9 Calculate the solubility-product constant for $La(IO_3)_3$ if a saturated solution is 2.08×10^{-3} M in IO_3^-.

6-10 What is the Pb^{2+} concentration in a saturated aqueous solution of
(a) $PbSO_4$? (b) PbI_2? (c) $Pb(OH)_2$? (d) $PbCl_2$?

*6-11 What is the I^- concentration in a saturated aqueous solution of
(a) AgI? (b) PbI_2? (c) BiI_3 ($K_{sp} = 8.1 \times 10^{-19}$)?

6-12 Calculate the weight of PbI_2 that dissolves in 100 mL of
(a) H_2O. (b) 5.0×10^{-3} M KI. (c) 2.3×10^{-5} M $Pb(NO_3)_2$.

*6-13 Calculate the weight of TlCl that dissolves in 200 mL of
(a) H_2O. (b) 1.0×10^{-1} M NaCl. (c) 1.0×10^{-3} M NaCl.

6-14 What SO_4^{2-} concentration is required to lower the Pb^{2+} concentration of a solution to 1.0 mg/L?

*6-15 What I^- concentration is required to lower the Pb^{2+} concentration of a solution to 1.0×10^{-6} M?

6-16 What OH^- concentration must be achieved in order to diminish that of Pb^{2+} to 1.0×10^{-5} M?

*6-17 What Pb^{2+} concentration is needed to
(a) initiate precipitation of $PbSO_4$ from a solution that is 0.050 M in SO_4^{2-}?
(b) lower the SO_4^{2-} concentration to 1.0×10^{-6} M?

6-18 What IO_3^- concentration is needed to
(a) initiate precipitation of $Cu(IO_3)_2$ ($K_{sp} = 7.4 \times 10^{-8}$) from a solution that is 5.0×10^{-3} M in Cu^{2+}?
(b) lower the Cu^{2+} concentration of a solution to 2.0×10^{-6} M?

*6-19 What is the Cu^{2+} concentration of a
(a) saturated aqueous solution of $Cu(IO_3)_2$ ($K_{sp} = 7.4 \times 10^{-8}$)?
(b) 0.100 M $CuCl_2$ solution saturated with $Cu(IO_3)_2$?
(c) 0.100 M KIO_3 solution saturated with $Cu(IO_3)_2$?
(d) 1.0×10^{-3} M $CuCl_2$ solution saturated with $Cu(IO_3)_2$?

6-20 What is the Ag^+ concentration in the solution that results from the addition of 32.4 mL of 0.117 M $AgNO_3$ solution of 67.4 mL of
(a) distilled water?
(b) 0.0562 M $NaNO_3$?
(c) 0.0562 M NaBr?
(d) 0.0562 M $MgBr_2$?
(e) 0.00562 M $MgBr_2$?

*6-21 Calculate the equilibrium concentration of each ion in the solution that results when 0.180 g of $Mg(OH)_2$ is added to 45.0 mL of (a) 0.0204 M HCl and (b) 0.204 M HCl.

6-22 Calculate the equilibrium concentration of each ion in the solution that results when 16.3 mL of 0.144 M $MgSO_4$ is
(a) diluted to 85.5 mL with distilled water.
(b) mixed with 69.2 mL of 0.0300 M KBr solution.
(c) mixed with 69.2 mL of 0.0300 M KOH solution.
(d) mixed with 69.2 mL of 0.0300 M $Ba(OH)_2$ solution.
(e) mixed with 69.2 mL of 0.0300 M BaI_2 solution.

*6-23 The solubility products for a series of hydroxides are

BiOOH $K_{sp} = 4.0 \times 10^{-10} = [BiO^+][OH^-]$

$Be(OH)_2$ $K_{sp} = 7.0 \times 10^{-22}$

$Tm(OH)_3$ $K_{sp} = 3.0 \times 10^{-24}$

$Hf(OH)_4$ $K_{sp} = 4.0 \times 10^{-26}$

Which hydroxide has

(a) the lowest molar solubility in H_2O?
(b) the lowest molar solubility in a solution that is 0.10 M in NaOH?

6-24 The solubility products for a series of IO_3^- salts are

$TlIO_3$ $K_{sp} = 3.1 \times 10^{-6}$

$Cu(IO_3)_2$ $K_{sp} = 7.4 \times 10^{-8}$

$In(IO_3)_3$ $K_{sp} = 3.3 \times 10^{-11}$

$Ce(IO_3)_4$ $K_{sp} = 4.7 \times 10^{-17}$

Arrange these iodate salts in an order that depicts

(a) decreasing molar solubility in water.
(b) decreasing solubility (grams per liter) in a 0.10 M solution of the solute cation.
(c) decreasing molar solubility in a 0.10 M solution of the solute cation.
(d) decreasing molar solubility in a 0.10 M IO_3^- solution.

*6-25 Silver ion is being considered as a reagent for separating I^- from SCN^- in a solution that is 0.060 M in KI and 0.070 M in NaSCN.

(a) What Ag^+ concentration is needed to lower the I^- concentration to 1.0×10^{-6} M?
(b) What is the Ag^+ concentration of the solution when AgSCN begins to precipitate?
(c) What is the ratio of SCN^- to I^- ion when AgSCN begins to precipitate?
(d) What is the ratio of SCN^- to I^- when the Ag^+ concentration is 1.0×10^{-3} M?

6-26 A solution is 0.040 M in Na_2SO_4 and 0.050 M in NaOH. To this is added a dilute solution containing Pb^{2+}.

(a) Which compound precipitates first, $PbSO_4$ or $Pb(OH)_2$?
(b) What is the Pb^{2+} concentration as the first precipitate appears?
(c) What Pb^{2+} concentration is required to initiate precipitation of the more soluble substance?
(d) What is the concentration of the first anion when the more soluble precipitate begins to form?

*6-27 Using 1.0×10^{-6} M as the criterion for quantitative removal, determine whether it is feasible to use

(a) SO_4^{2-} to separate Ba^{2+} and Sr^{2+} in a solution that is initially 0.10 M in Sr^{2+} and 0.25 M in Ba^{2+}.
(b) SO_4^{2-} to separate Ba^{2+} and Ag^+ in a solution that is initially 0.040 M in each cation. For Ag_2SO_4, $K_{sp} = 1.6 \times 10^{-5}$.
(c) OH^- to separate Be^{3+} and Hf^{4+} in a solution that is initially 0.020 M in Be^{2+} and 0.010 M in Hf^{4+}. See Problem 6-23 for K_{sp} data.
(d) IO_3^- to separate In^{3+} and Tl^+ in a solution that is initially 0.11 M in In^{3+} and 0.060 M in Tl^+. See Problem 6-24 for K_{sp} data.

6-28 Dilute NaOH is introduced into a solution that is 0.050 M in Cu^{2+} and 0.040 M in Mn^{2+}.

(a) Which hydroxide precipitates first?
(b) What OH^- concentration is needed to initiate precipitation of the first hydroxide?
(c) What is the concentration of the cation forming the less soluble hydroxide when the more soluble hydroxide begins to form?

6-29 Calculate the molar solubility of Ag_2CO_3 in a solution that has a H_3O^+ concentration of

*(a) 1.0×10^{-6} M. *(c) 1.0×10^{-9} M.
(b) 1.0×10^{-7} M. (d) 1.0×10^{-11} M.

6-30 Calculate the molar solubility of $BaSO_4$ in a solution having an $[H_3O^+]$ of

*(a) 2.0 M. (b) 1.0 M. *(c) 0.50 M. (d) 0.10 M.

6-31 For a solution in which μ is 5.0×10^{-2}, calculate K'_{sp} for

*(a) AgSCN. *(c) $La(IO_3)_3$.
(b) PbI_2. (d) $MgNH_4PO_4$.

6-32 Use activities to calculate the molar solubility of $Mn(OH)_2$ in

(a) 0.0100 M $NaNO_3$.
*(b) 0.0167 M K_2SO_4.
(c) the solution that results when 60.0 mL of 0.0333 M $MnCl_2$ is mixed with 40.0 mL of 0.0500 M KOH.

*6-33 Calcium sulfate is only partially dissociated in aqueous solution:

$$CaSO_4(aq) \rightleftharpoons Ca^{2+} + SO_4^{2-} \qquad K_d = 5.2 \times 10^{-3}$$

The solubility-product constant for $CaSO_4$ is 2.6×10^{-5}. Calculate the solubility of $CaSO_4$ in (a) water and (b) 0.0100 Na_2SO_4. In each case, calculate the percent of undissociated $CaSO_4$ present in solution.

*6-34 The solubility-product constant for $CdCO_3$ is 2.5×10^{-14}. Calculate the equilibrium solubility of $CdCO_3$ in

(a) water.
(b) a solution that is 0.10 M in CO_3^{2-}.
(c) a solution in which the H_3O^+ concentration is 1.0×10^{-8} M.

6-35 Calculate the molar solubility of $MgCO_3$ in

(a) water.
(b) a solution in which $[H_3O^+]$ is 1.0×10^{-8} M.
(c) 0.10 M Na_2CO_3.

*6-36 Calculate the molar solubility of CuS in a solution in which the H_3O^+ concentration is held constant at (a) 1.0×10^{-1} M and (b) 1.0×10^{-4} M.

6-37 Calculate the concentration of CdS in a solution in which the $[H_3O^+]$ is held constant at (a) 1.0×10^{-1} M and (b) 1.0×10^{-4} M.

*6-38 What is the equilibrium solubility of MnS in a solution in which the $[H_3O^+]$ is held constant at (a) 1.0×10^{-5} M and (b) 1.0×10^{-8} M?

6-39 What weight of AgBr dissolves in 200 mL of 0.100 M NaCN?
($Ag^+ + 2CN^- \rightleftharpoons Ag(CN)_2^-$; $K_f = 1.3 \times 10^{21}$)

*6-40 The equilibrium constant for formation of $CuCl_2^-$ is given by

$$Cu^+ + 2Cl^- \rightleftharpoons CuCl_2^- \qquad K_f = \frac{[CuCl_2^-]}{[Cu^+][Cl^-]^2} = 7.9 \times 10^4$$

What is the solubility of CuCl in solutions having the following analytical NaCl concentrations:

(a) 1.0 M?
(b) 1.0×10^{-1} M?
(c) 1.0×10^{-2} M?
(d) 1.0×10^{-3} M?
(e) 1.0×10^{-4} M?

6-41 The equilibrium constant for the formation of $Al(OH)_4^-$ is given by

$$Al(OH)_3(s) + OH^- \rightleftharpoons Al(OH)_4^- \qquad K_f = 10$$

How many milliliters of 1.0 M NaOH is required to completely redissolve 1.00 g of $Al(OH)_3$ suspended in 100.0 mL of water?

*6-42 What concentration of OH^- must be maintained in order to dissolve 0.200 g of $Pb(OH)_2$ in 200 mL of solution?

$$Pb(OH)_2(s) + OH^- \rightleftharpoons Pb(OH)_3^- \qquad K_f = 5.0 \times 10^{-2}$$

6-43 Formation constants for the reaction of Ag^+ with $S_2O_3^{2-}$ are

$$Ag^+ + S_2O_3^{2-} \rightleftharpoons AgS_2O_3^- \qquad K_1 = 6.6 \times 10^8$$

$$AgS_2O_3^- + S_2O_3^{2-} \rightleftharpoons Ag(S_2O_3)_2^{3-} \qquad K_2 = 4.4 \times 10^3$$

Calculate the solubility of AgI in 0.200 M $Na_2S_2O_3$ (assume that $S_2O_3^{2-}$ does not combine with H_3O^+).

CHAPTER 7 Precipitation Titrations

In principle, any precipitation reaction that is rapid and reasonably complete can serve as the basis for a titration, provided an indicator for the reaction exists. Unfortunately, however, few precipitation reactions proceed rapidly enough to make them practical for titrimetry. Furthermore, few satisfactory indicators exist for precipitation reactions. Consequently, only a handful of precipitating reagents are useful for titrimetric methods. By far the most important such reagent is silver nitrate, and most of the discussion in this chapter is devoted to *argentometric methods,* which are those that are based on this reagent. Argentometric methods have found widespread use in the routine determination of such species as halides, several divalent anions, mercaptans, and certain fatty acids.

7A Titration Curves in Titrimetric Methods

As noted in Section 4A-2, an end point consists of an observable physical change that occurs at or near the equivalence point of a titration. The two most widely used involve (1) changes in color due to the reagent, the analyte, or an indicator and (2) a change in potential of an electrode that responds to the concentration of the reagent or the analyte.

In order to understand the theoretical basis of end points and the sources of titration errors, it is often helpful to derive a *titration curve* for the system under consideration. Generally, titration curves consist of a plot of reagent volume as the abscissa and some function of the analyte or reagent concentration as the ordinate.

7A-1 Types of Titration Curves

Two general types of titration curves (and thus two general types of end points) are employed in titrimetric methods. In the first, the important observations are confined to a small region (typically ± 1 or ± 2 mL) surrounding the equivalence point. In the second, measurements are made on both sides of but well away from the equivalence point; indeed, measurements near equivalence are avoided. The former offers the advantages of speed and convenience. The latter is advantageous for reactions that are complete only in the presence of a goodly excess of the reagent or analyte. The focus here, and in the several chapters that follow, will be on the first type of end point; the second type is discussed in Chapters 17 and 20.

7A-2 Concentration Changes During Titrations

The equivalence point in a titration is characterized by major shifts in the *relative* concentrations of reagent and analyte. Table 7-1 illustrates this phe-

TABLE 7-1

Concentration Changes During a Titration*

Volume of 0.1000 M R	[A], mol/L	Volume of R That Causes a Tenfold Decrease in [A]	pA	pR
0.00	1.0×10^{-1}	—	1.00	—
40.90	1.0×10^{-2}	40.9	2.00	8.00
49.00	1.0×10^{-3}	8.1	3.00	7.00
49.90	1.0×10^{-4}	0.9	4.00	6.00
50.00	1.0×10^{-5}	0.1	5.00	5.00
50.10	1.0×10^{-6}	0.1	6.00	4.00
51.00	1.0×10^{-7}	0.9	7.00	3.00
61.10	1.0×10^{-8}	10.1	8.00	2.00

*Reaction: A(aq) + R(aq) → AR(s)
For AR(s), $K_{sp} = 1.00 \times 10^{-10}$

nomenon. The data in the second column of the table show the changes in concentration of an analyte species A as a 50.00 mL aliquot of a 0.1000 M solution of A is titrated with a 0.1000 M solution of a reagent R. The chemical reaction produces the precipitate AR, which has a solubility product of 1.0×10^{-10}. In order to emphasize the changes in *relative* concentration that occur in the equivalence region, the volume increments selected are those required to cause tenfold decrease in the concentration of A. Thus, we see in the third column that an addition of 40.90 mL of R is needed to decrease the concentration of A by one order of magnitude from 0.10 M to 0.010 M. An additional 8.1 mL portion is required to lower the concentration to 0.0010 M; 0.9 mL causes yet another tenfold decrease, and so on (it can be shown that corresponding increases in [R] occur simultaneously). End point detection, then, is based upon this large change in *relative* concentration of the analyte (or the reagent) that occurs at the equivalence point for every type of titration.

In order to depict clearly the large relative concentration changes that occur in the region of chemical equivalence, it is common practice to plot the p-function of the analyte or the reagent as a function of reagent volume as has been done in Figure 7-1. The data for these plots are found in the fourth and fifth columns of Table 7-1. Titration curves for reactions involving precipitate formation, neutralization, complex formation, and oxidation/reduction all exhibit the same sharp increase or decrease in p-function as those shown in Figure 7-1. Typical examples are discussed in this and later chapters. Titration curves are of considerable importance in titrimetric studies because they define the properties required of an indicator and permit estimation of the titration error associated with the analytical method.

7B Titration Curves for Precipitation Reactions

Titration curves for precipitation reactions in which silver nitrate is the standard reagent are illustrated in this section. Such curves ordinarily consist of a plot of pAg as a function of reagent volume, although occasionally the p-value of the analyte serves as the ordinate.

The generation of an argentometric titration curve requires three types of calculation, which correspond to three distinct stages in the titration: (1) preequivalence point, (2) equivalence point, and (3) postequivalence point. In

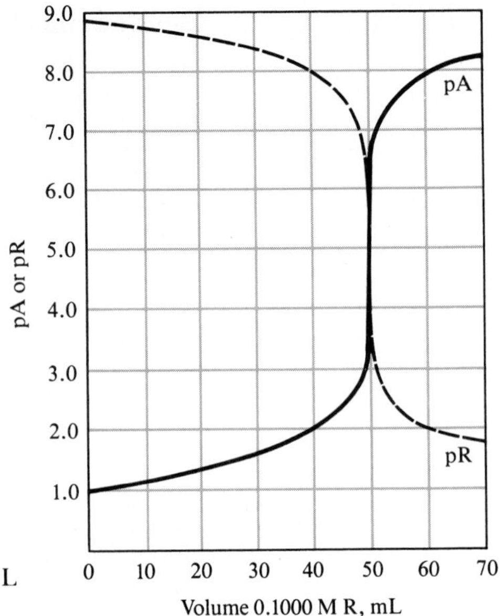

FIGURE 7-1 Precipitation titration curve for 50.00 mL of 0.1000 M A with 0.1000 M R.

the preequivalence stage, the analytical concentration of the analyte is readily computed from its starting concentration and the volumetric data. The equilibrium analyte concentration is then assumed to be identical to its analytical concentration and is substituted into the solubility-product expression to obtain the silver ion concentration. At the equivalence point, neither silver ion nor analyte ion is present in excess, and the silver ion concentration is derived directly from the solubility product constant. In the postequivalence stage, the analytical concentration of the excess silver nitrate is computed and assumed to be identical to its equilibrium concentration.

EXAMPLE 7-1

Derive a titration curve for the reaction of 50.00 mL of 0.00500 M NaBr with 0.01000 M $AgNO_3$.

We shall calculate both pBr and pAg, although the latter more commonly serves as the ordinate in titration curves.

Initial point

At the outset, the solution is 0.00500 M in Br^- and 0.000 M in Ag^+. Thus, pBr $= -\log(5.00 \times 10^{-3}) = 2.301 = 2.30$. Since no silver nitrate has been introduced, pAg is indeterminate.

After addition of 5.00 mL of reagent

The bromide ion concentration is decreased as a result of both precipitate formation and dilution. Thus,

$$c_{NaBr} = \frac{\text{no. mmol NaBr after addition of AgNO}_3}{\text{total volume soln}}$$

$$= \frac{\text{original no. mmol NaBr} - \text{no. mmol AgNO}_3\text{ added}}{\text{total volume soln}}$$

$$= \frac{(50.00\text{ mL} \times 0.00500\text{ M}) - (5.000\text{ mL} \times 0.01000\text{ M})}{50.00\text{ mL} + 5.00\text{ mL}}$$

$$= \frac{(0.2500\text{ mmol} - 0.0500\text{ mmol})}{55.00\text{ mL}} = 3.64 \times 10^{-3}\text{ M}$$

The first term in the numerator of these equations is the number of millimoles of NaBr originally in the sample, and the second term is the number of millimoles of $AgNO_3$ added and hence the number of millimoles of Br^- consumed. The denominator takes into account the dilution of the solution by the added reagent.

Both the unreacted NaBr and the slightly soluble AgBr contribute to the species concentration of bromide ion. Thus, the equilibrium concentration of Br^- is larger than the analytical concentration of NaBr by an amount equal to the molar solubility of the precipitate:

$$[Br^-] = 3.64 \times 10^{-3} + [Ag^+] \qquad (7\text{-}1)$$

The contribution of the silver bromide to the equilibrium bromide ion concentration is equal to $[Ag^+]$ because one silver ion is formed for each bromide ion from this source. Unless the concentration of NaBr is very small, this term can be neglected. That is, if $[Ag^+] \ll 3.64 \times 10^{-3}$,

$$[Br^-] \simeq 3.64 \times 10^{-3}$$

$$pBr = -\log(3.64 \times 10^{-3}) = 2.439 = 2.44$$

A convenient way to find pAg is to take the negative logarithm of the solubility-product expression for AgBr:

$$-\log([Ag^+][Br^-]) = -\log K_{sp} = -\log(5.2 \times 10^{-13})$$

$$-\log[Ag^+] - \log[Br^-] = -\log K_{sp} = 12.28$$

or

$$pAg + pBr = pK_{sp} = 12.28 \qquad (7\text{-}2)$$

$$pAg = 12.28 - 2.44 = 9.84$$

This relationship applies to any solution containing silver and bromide ions in contact with solid silver bromide. Note that the calculated pAg corresponds to $[Ag^+] = 1.4 \times 10^{-10}$, which, as we assumed at the outset, is certainly much smaller than 3.4×10^{-4}.

Other data points in the region before chemical equivalence can be derived in this same way. Data for several such points are found in columns 4 and 5 of Table 7-2.

TABLE 7-2

Changes in pAg and pBr During Titration with Solutions of Different Concentrations

Volume $AgNO_3$, mL	50.00 mL 0.0500 M Br^- with 0.1000 M $AgNO_3$		50.00 mL 0.00500 M Br^- with 0.01000 M $AgNO_3$		50.00 mL 0.000500 M Br^- with 0.001000 M $AgNO_3$	
	pAg	pBr	pAg	pBr	pAg	pBr
0.00	—	1.30	—	2.30	—	3.30
10.00	10.68	1.60	9.68	2.60	8.68	3.60
20.00	10.13	2.15	9.13	3.15	8.13	4.15
23.00	9.72	2.56	8.72	3.56	7.72	4.56
24.90	8.41	3.87	7.41	4.87	6.50	5.78*
24.95	8.10	4.18	7.10	5.18	6.33	5.95*
25.00	6.14	6.14	6.14	6.14	6.14	6.14
25.05	4.18	8.10	5.18	7.10	5.95	6.33†
25.10	3.88	8.40	4.88	7.40	5.78	6.50†
27.00	2.58	9.70	3.58	8.70	4.58	7.70
30.00	2.20	10.08	3.20	9.08	4.20	8.08

*Approximation $[Ag^+] \ll c_{NaBr}$ not valid.

†Approximation $[Br^-] \ll c_{AgNO_3}$ not valid.

Equivalence point

At the equivalence point, neither NaBr nor $AgNO_3$ is in excess and so the concentrations of silver and bromide ions must be equal. Substitution of this equality into the solubility-product expression yields

$$[Ag^+] = [Br^-] = \sqrt{5.2 \times 10^{-13}} = 7.21 \times 10^{-7}$$

$$pAg = pBr = -\log(7.21 \times 10^{-7}) = 6.14$$

After addition of 25.10 mL of reagent

The solution now contains an excess of $AgNO_3$, and we can write

$$c_{AgNO_3} = \frac{\text{total no. mmol } AgNO_3 - \text{original no. mmol NaBr}}{\text{total volume of solution}}$$

$$= \frac{25.10 \text{ mL} \times 0.01000 \text{ M} - 50.00 \text{ mL} \times 0.00500 \text{ M}}{(50.00 + 25.10) \text{ mL}} \simeq 1.33 \times 10^{-5} \text{ M}$$

and the equilibrium concentration of silver ion is

$$[Ag^+] = 1.33 \times 10^{-5} + [Br^-] \simeq 1.33 \times 10^{-5}$$

In this equation, $[Br^-]$ is a measure of the Ag^+ concentration resulting from the slight solubility of AgBr; it can ordinarily be neglected. Thus,

$$pAg = -\log(1.33 \times 10^{-5}) = 4.876 = 4.88$$

$$pBr = 12.28 - 4.88 = 7.40$$

Additional points defining the titration curve beyond the equivalence point can be obtained in an analogous way and are found in Table 7-2.

7B-1

Significant Figures in Titration-Curve Derivations

The concentration data associated with the equivalence-point region of a titration curve are usually of low precision because they are based upon small differences between large numbers. For example, in the calculations of c_{AgNO_3} following the introduction of 25.10 mL of 0.01000 M $AgNO_3$, the numerator (0.2510 − 0.2500 = 0.0010) contains only two significant figures; at best then, c_{AgNO_3} is known to two significant figures. To minimize the rounding error, however, three digits (1.33×10^{-5}) were retained in this calculation and rounding was postponed until after pAg was computed.

In rounding the calculated value for pAg, it is important to recall (page 49) that it is the *mantissa of a logarithm (that is, the number to the right of the decimal point) that should be rounded to include only significant figures* because the characteristic serves merely to locate the decimal point. Thus, in Example 7-1 we rounded pAg to two figures to the right of the decimal point; that is, pAg = 4.88.

Note that the pAg values for points well away from the equivalence point contain one additional significant figure. For example, the initial pBr could be reported as 2.301. Nothing is gained by the added precision, however, because our primary interest is in the equivalence-point region. For this reason, in deriving data for titration curves, we shall generally round p-functions to two digits to the right of the decimal point. The large changes in p-functions characteristic of most equivalence points are not obscured by the limited precision in the data in this region.

7B-2

Factors Influencing the Sharpness of End Points

A sharp and easily located end point is observed when small additions of titrant cause large changes in p-function. It is therefore of interest to examine the variables that influence the magnitude of such changes during a titration.

Reagent Concentration

Table 7-2 is a compilation of data computed by the method shown in Example 7-1. Three concentrations of titrant and analyte, each differing by a factor of 10, are assumed. The data are plotted in Figure 7-2. It is apparent that an increase in analyte and reagent concentrations enhances the change in pAg in the equivalence-point region; an analogous effect is observed when pBr is plotted rather than pAg.

These effects have practical significance for the titration of bromide ion. If the analyte concentration is sufficient to permit the use of a silver nitrate solution that is 0.1 M or stronger, easily detected end points are observed, and the titration error is minimal. On the other hand, with standard solutions that are 0.001 M or less, the change in pAg or pBr is so small that end-point detection is difficult and a large titration error is expected. It is noteworthy that the effect of titrant concentration on the sharpness of end points is observed in neutralization and complex-formation titrations as well.

Completeness of Reaction

Figure 7-3 demonstrates the effect of product solubility on the sharpness of end points for titrations in which 0.1 M silver nitrate is the titrant. Clearly, the greatest change in pAg occurs in the titration of iodide ion, which, of all the anions considered, forms the least soluble silver salt and hence reacts most completely with silver ion. The smallest change in pAg is observed for the reaction product that is most soluble—that is, in the titration of bromate ion.

FIGURE 7-2 Effect of titrant concentration on titration curves: *A*, 50.00 mL of 0.0500 M NaBr with 0.1000 M $AgNO_3$; *B*, 50.00 mL of 0.00500 M NaBr with 0.01000 M $AgNO_3$; *C*, 50.00 mL of 0.000500 M NaBr with 0.001000 M $AgNO_3$.

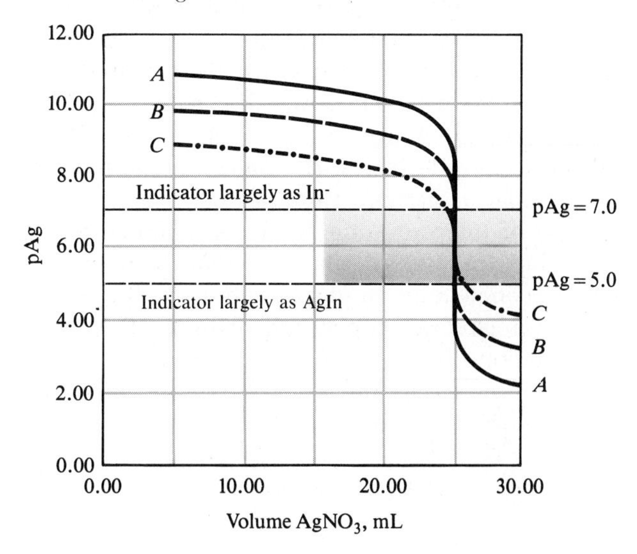

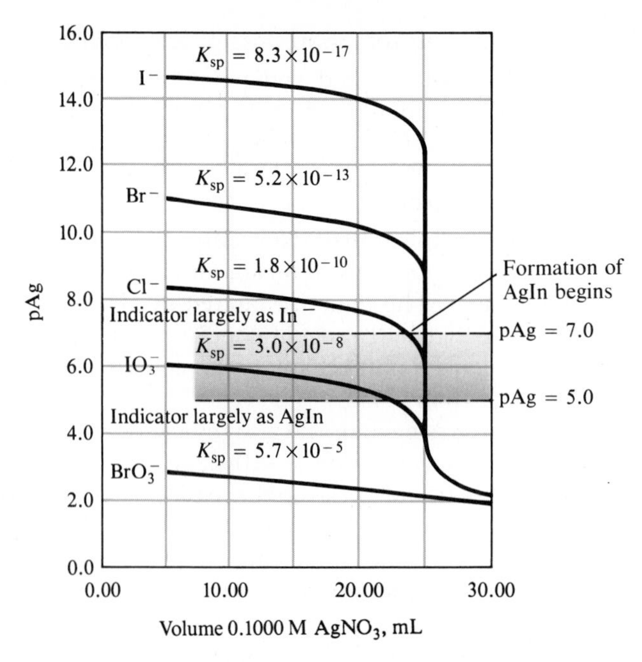

FIGURE 7-3 Effect of completeness of reaction on titration curves. For each curve, 50.00 mL of a 0.0500 M solution of the anion was titrated with 0.1000 M $AgNO_3$.

Reactions that produce silver salts with solubilities intermediate between these extremes yield titration curves with end-point changes that are also intermediate in magnitude. This effect is common to all reaction types.

7B-3 *Titration Curves for Mixtures*

The methods developed in the previous section for deriving titration curves can be extended to mixtures that form precipitates of different solubilities. To illustrate, consider the titration of 50.00 mL of a solution that is 0.0500 M in iodide ion and 0.0800 M in chloride ion with 0.1000 M silver nitrate. The curve for the initial stages of this titration is identical to the curve shown for iodide in Figure 7-3 because silver chloride, with its much larger solubility product, does not begin to precipitate until well into the titration. That is, until silver chloride forms, the curve for the mixture is derived from calculations similar to those in Example 7-1.

It is of interest to determine how much iodide is precipitated before appreciable amounts of silver chloride form. With the appearance of the smallest possible amount of silver chloride, the solubility-product expressions for both precipitates apply, and division of one by the other provides the useful relationship

$$\frac{\cancel{[Ag^+]}[I^-]}{\cancel{[Ag^+]}[Cl^-]} = \frac{8.3 \times 10^{-17}}{1.82 \times 10^{-10}} = 4.56 \times 10^{-7}$$

$$[I^-] = (4.56 \times 10^{-7})[Cl^-]$$

It is apparent from this relationship that the iodide concentration is decreased to a tiny fraction of the chloride ion concentration prior to the onset of silver chloride precipitation. Thus, for all practical purposes, formation of silver chloride will occur only after 25.00 mL of titrant have been added in this titration. At this point, the chloride ion concentration, because of dilution, is approximately

$$c_{Cl} = [Cl^-] = \frac{50.00 \times 0.0800}{50.00 + 25.00} = 0.0533 \text{ M}$$

Substituting into the previous equation yields

$$[I^-] = 4.56 \times 10^{-7} \times 0.0533 = 2.43 \times 10^{-8}$$

The percentage of iodide unprecipitated at this point can be calculated as follows:

$$\text{No. mmol I}^- \text{ remaining} = (75.00 \text{ mL})(2.43 \times 10^{-8} \text{ mmol I}^-/\text{mL}) = 1.82 \times 10^{-6}$$

$$\text{Original no. mmol I}^- = (50.00 \text{ mL})(0.0500 \text{ mmol/mL}) = 2.50$$

$$\text{I}^- \text{ unprecipitated} = \frac{1.82 \times 10^{-6}}{2.50} \times 100\% = 7.3 \times 10^{-5}\%$$

Thus, to within about 7.3×10^{-5} percent of the equivalence point for iodide, no silver chloride forms, and up to this point, the titration curve is indistinguishable from that for the iodide alone (Figure 7-3). The first part of the titration curve shown by the solid line in Figure 7-4 was derived on this basis.

As chloride ion begins to precipitate, the rapid decrease in pAg is terminated abruptly at a level that can be calculated from the solubility-product constant for silver chloride and the computed chloride concentration:

$$[Ag^+] = \frac{1.82 \times 10^{-10}}{0.0533} = 3.41 \times 10^{-9}$$

$$\text{pAg} = -\log(3.41 \times 10^{-9}) = 8.47$$

Further additions of silver nitrate decrease the chloride ion concentration, and the curve then becomes that for the titration of chloride by itself. For example, after 30.00 mL of titrant has been added,

$$c_{Cl} = [Cl^-] = \frac{50.00 \times 0.0800 + 50.00 \times 0.0500 - 30.00 \times 0.100}{50.00 + 30.00}$$

Here, the first two terms in the numerator give the number of millimoles of chloride and iodide, respectively, and the third is the number of millimoles of titrant added. Thus,

$$[Cl^-] = 0.0438$$

$$[Ag^+] = \frac{1.82 \times 10^{-10}}{0.0438} = 4.16 \times 10^{-9}$$

$$\text{pAg} = 8.38$$

The remainder of the curve can be derived in the same way as a curve for chloride by itself.

Curve *A* in Figure 7-4, which is the titration curve for the chloride/iodide mixture just considered, is a composite of the individual curves for the two

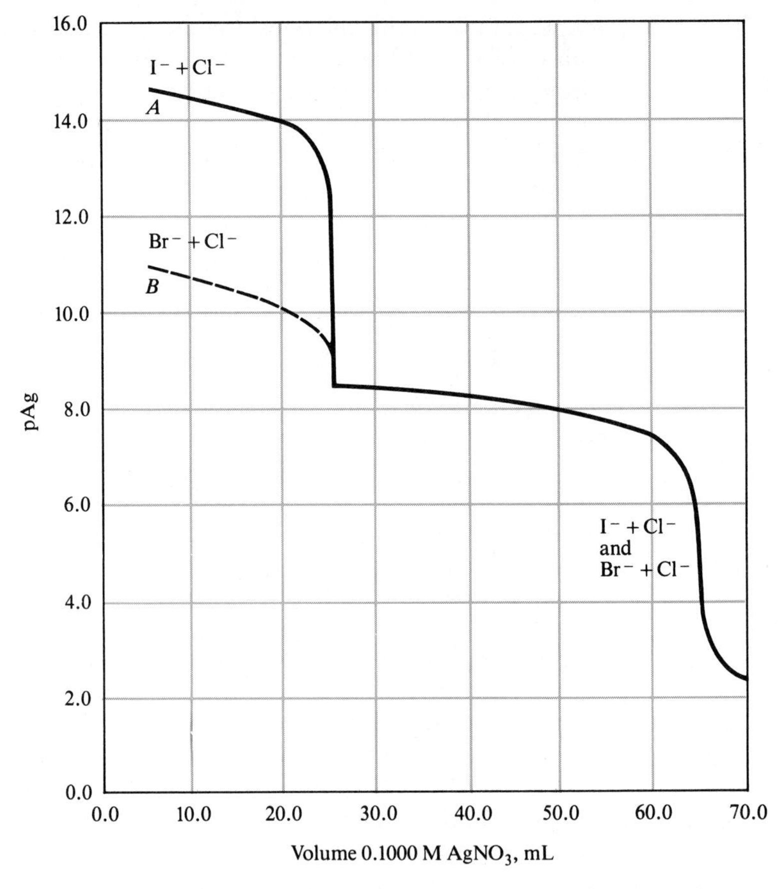

FIGURE 7-4 Titration curves for 50.00 mL of a solution 0.0800 M in Cl^- and 0.0500 M in I^- or Br^-.

anionic species. Two equivalence points are evident. Curve *B* is the titration curve for a mixture of bromide and chloride ions. Clearly, the change associated with the first equivalence point becomes less distinct as the solubilities of the two precipitates approach one another. In the bromide/chloride titration, the initial pAg values are lower than they are in the iodide/chloride titration because the solubility of silver bromide exceeds that of silver iodide. Beyond the first equivalence point, however, where chloride ion is being titrated, the two titration curves are identical.

It is possible to obtain experimental curves similar to those shown in Figure 7-4 by measuring the potential between a silver electrode and a reference electrode immersed in the solution. This technique, which is discussed in Chapter 15, permits the analysis of the individual components in certain halide-ion mixtures.

7B-4 Chemical Indicators for Precipitation Titrations

The end point produced by a chemical indicator usually consists of a color change or, occasionally, the appearance or disappearance of turbidity in the solution being titrated. Chemical indicators function by reacting competitively with one of the reactants of the titration. For example, consider the titration of the analyte A with a titrant R in the presence of an indicator In that reacts with R. A chemical description of this system throughout the titration is

$$A + R \rightleftharpoons AR(s)$$

$$In + R \rightleftharpoons InR$$

For indicator action, it is necessary that InR impart a significantly different appearance to the mixture than does In. In addition, the amount of InR required for an observable change should be so small that no appreciable consumption of R occurs as InR is formed. Finally, the equilibrium constant for the indicator reaction must be such that the ratio [InR]/[In] is shifted from a small to a large value as a consequence of the change in [R] (or pR) that occurs in the equivalence-point region. The ratio [InR]/[In] for most indicators must shift by one or two orders of magnitude in order to produce a color change detectable to the human eye. Such a change corresponds to a change of 1 or 2 in pR.

To illustrate, consider the application of a hypothetical indicator to the three bromide titrations described in Table 7-2 and Figure 7-2. We shall assume that the equilibrium constant for the reaction of the indicator In^- with silver ion is such that a full color change occurs when pAg shifts from 7 to 5. It is clear that each of the three titrations requires a different volume of titrant to encompass this range. Thus, the data in the second column of Table 7-2 indicate that, with 0.1000 M $AgNO_3$, the color change begins after 24.95 mL of titrant has been added and is complete before 25.05 mL has been added. An abrupt change in color and a minimal titration error can be expected. In contrast, with 0.001 M $AgNO_3$, the color starts to change at about 24.5 mL and is complete at 25.8 mL. Exact location of the end point in this titration is impossible. The pAg change for the titration with 0.01 M reagent is such that somewhat less than 0.2 mL is required to bring about a color change; here, the indicator is usable but the titration error is significant.

Consider the applicability and effectiveness of this same indicator for the titrations represented by the curves in Figure 7-3. Here the indicator exists largely as AgIn throughout the titration of bromate and iodate ions, and no color change is observed after the first addition of silver nitrate. In contrast, the solubilities of silver bromide and silver iodide are small enough to prevent formation of significant amounts of AgIn until the equivalence-point region is reached. Note that pAg remains above 7 until just before the equivalence point for the bromide titration and until just beyond the equivalence point for the iodide. In both cases, the color change occurs over a range of less than 0.01 mL of reagent and so titration errors are negligible.

Turning to the titration curve for chloride ion in Figure 7-3, we see that an indicator with a pAg range of 5 to 7 is not satisfactory because formation of appreciable amounts of AgIn begins approximately 1 mL short of the equivalence point and extends over a range of about 1 mL, thus making exact location of the end point impossible. In contrast, an indicator with a pAg range of 4 to 6 would be perfectly satisfactory. No chemical indicator exists for the iodate and bromate titrations because the changes in pAg in the equivalence-point region are far too small.

The Formation of a Second Precipitate: The Mohr Method

Chromate ion serves as the indicator for the *Mohr method,* which is widely used for the argentometric determination of chloride and bromide ions. The end point in this titration is signaled by the appearance of brick-red silver chromate, Ag_2CrO_4. The molar solubility of silver chromate is several times greater than that of silver chloride. Consequently, the latter tends to form first in the titration mixture. By adjustment of the chromate concentration to a suitable level, formation of silver chromate can be retarded until the silver ion concentration in the solution reaches the theoretical equivalence-point concentration. As shown on page 168, at equivalence we can write

$$[Ag^+] = [Cl^-] = \sqrt{K_{sp}}$$

$$[Ag^+] = \sqrt{1.82 \times 10^{-10}} = 1.35 \times 10^{-5}$$

The chromate concentration required to initiate precipitation of silver chromate under these conditions can be computed from the solubility-product constant for silver chromate:

$$[CrO_4^{2-}] = \frac{K_{sp}}{[Ag^+]^2} = \frac{1.1 \times 10^{-12}}{(1.35 \times 10^{-5})^2} = 6 \times 10^{-3}$$

In principle, then, at this concentration of chromate ion, the red color of silver chromate appears with the first excess of silver ion over its equivalence concentration. In practice, however, end-point detection is difficult if not impossible in solutions with chromate ion concentrations as large as 6×10^{-3} M because the yellow color imparted by this concentration of chromate ion masks the red color of the silver chromate. Empirically, it has been found that the optimum concentration of the indicator is about 2.5×10^{-3} M. At this chromate ion concentration, precipitation of silver chromate is delayed until the silver ion concentration becomes about 2.1×10^{-5} M instead of the theoretical 1.35×10^{-5} M. In addition, a finite amount of silver nitrate must be added to produce a detectable quantity of precipitate. Both factors cause an overconsumption of titrant. The problem is serious with dilute solutions but small with 0.1 M solutions. A correction can be made by obtaining an *indicator blank*—that is, by determining the silver ion consumption of a suspension of chloride-free calcium carbonate in about the same volume and with the same quantity of indicator as the sample. The blank titration mixture serves as a convenient color standard for subsequent titrations. A better method, which largely eliminates the indicator error, is to use the Mohr method to standardize the silver nitrate solution against pure sodium chloride. The "working molarity" obtained for the solution compensates not only for the overconsumption of reagent but also for the ability of the analyst to detect the color change.

Attention must be paid to the acidity of the medium in the Mohr method because the equilibrium

$$2CrO_4^{2-} + 2H^+ \rightleftharpoons Cr_2O_7^{2-} + H_2O$$

is displaced to the right as the hydrogen ion concentration increases. Because silver dichromate is considerably more soluble than silver chromate, the indicator reaction in acidic solution requires substantially higher silver ion concentrations—if indeed it occurs at all. If the medium is made strongly alkaline, there is the danger that silver will precipitate as its oxide:

$$2Ag^+ + 2OH^- \rightleftharpoons 2AgOH(s) \rightleftharpoons Ag_2O(s) + H_2O$$

Thus, the determination of chloride by the Mohr method must be carried out in a medium that is neutral or nearly so (pH 7 to 10). A convenient way of maintaining a suitable pH is to add an excess of solid sodium hydrogen carbonate or borax to the solution being titrated.

The Mohr method is not applicable to the determination of iodide because chromate ion oxidizes iodide ion to iodine.

The Formation of a Colored Complex: The Volhard Method

The Volhard method employs a standard solution of thiocyanate ion to titrate silver ion:

$$Ag^+ + SCN^- \rightleftharpoons AgSCN(s)$$

Iron(III) serves as the indicator, imparting a red color to the solution with the first slight excess of thiocyanate ion:

$$Fe^{3+} + SCN^- \rightleftharpoons \underset{\text{red}}{FeSCN^{2+}} \qquad K_f = 1.4 \times 10^2$$

The titration must be carried out in acidic solution to prevent precipitation of iron(III) as the hydrated oxide.

The most important application of the Volhard method is for the indirect determination of halide ions. A measured excess of standard silver nitrate solution is added to the sample, and the excess silver ion is determined by back-titration with a standard thiocyanate solution. The strong acid environment required for the Volhard procedure represents a distinct advantage over other methods of halide analysis because such ions as carbonate, oxalate, and arsenate (which form slightly soluble silver salts in neutral media but not in acidic media) do not interfere.

In contrast to the other silver halides, silver chloride is more soluble than silver thiocyanate. As a consequence, in chloride determinations, the reaction

$$AgCl(s) + SCN^- \rightleftharpoons AgSCN(s) + Cl^-$$

occurs to a significant extent near the end of the back-titration of the excess silver ion. This reaction causes the end point to fade and results in an overconsumption of thiocyanate ion, which leads to low values for the chloride analysis. This error can be circumvented by filtering the silver chloride before undertaking the back-titration.

Adsorption Indicators

An *adsorption indicator* is an organic compound that tends to be adsorbed onto the surface of the solid formed during a precipitation titration. Ideally, the adsorption (or desorption) occurs near the equivalence point and results not only in a color change but also in a transfer of color from the solution to the solid (or the reverse). Analytical procedures based upon adsorption indicators are sometimes called *Fajans methods* in honor of the scientist who was most active in their development.

A typical adsorption indicator is the organic dye *fluorescein,* which is useful for the titration of chloride ion with silver nitrate. In aqueous solution, fluorescein partially dissociates into hydrogen ions and negatively charged fluoresceinate ions that impart a yellow-green color to the medium. The fluoresceinate ion forms a highly colored silver salt of limited solubility. Whenever this dye is used as an indicator, however, *its concentration is never large enough to exceed the solubility product of silver fluoresceinate.*

In the early stages of the titration of chloride ion with silver nitrate, the colloidal silver chloride particles are negatively charged owing to adsorption of excess chloride ions (page 65). The dye anions are repelled from this surface by electrostatic forces and impart a yellow-green color to the solution. Beyond the equivalence point, however, the silver chloride particles strongly adsorb silver ions and thereby acquire a positive charge. Fluoresceinate anions are now attracted *into the counter-ion layer.* The net result is the appearance of the red color of silver fluoresceinate *on the surface of the precipitate.* It is important to emphasize that the color change is an *adsorption* (not a precipitation) process inasmuch as the solubility product of the silver fluoresceinate is never exceeded. The adsorption is reversible, the dye being desorbed upon back-titration with chloride ion.

The successful application of an adsorption indicator requires that the precipitate and the indicator have the following properties:

1. The particles of the precipitate must be of colloidal dimensions in order to maximize the quantity of indicator adsorbed.
2. The precipitate must strongly adsorb its own ions. We have seen (Section 3B-2) that this property is characteristic of colloidal precipitates.
3. The indicator dye must be strongly held in the counter-ion layer by the primarily adsorbed ion. In general, this type of adsorption occurs when the solubility of the salt formed by the dye and the lattice ion is low. At the same time, this salt must be sufficiently soluble so that it does not precipitate.
4. The pH of the solution must be such as to ensure that the ionic form of the indicator predominates. Because the active constituent of most adsorption indicators is an ion that is the conjugate acid or base of the dye molecule, the concentration of that ion is pH-dependent.

Titrations involving adsorption indicators are rapid, accurate, and reliable, but their application is limited to the relatively few precipitation reactions in which a colloidal precipitate is formed rapidly. High electrolyte concentrations must be avoided with adsorption indicators because electrolytes tend to cause coagulation of the precipitate, a process that decreases the surface area on which adsorption occurs. Addition of a few milliliters of a 2% solution of chloride-free dextrin has been shown to delay the coagulation of silver precipitates, which leads to sharper end points.

Most adsorption indicators are weak acids. Their use is thus confined to basic, neutral, or slightly acidic solutions, where the indicator exists predominantly as the anion. A few cationic adsorption indicators suitable for titrations in strongly acidic solutions are known, however. For such indicators, adsorption of the dye and coloration of the precipitate occur in the presence of an excess of the precipitate anion (that is, when the precipitate particles possess a negative charge).

Finally, many adsorption indicators sensitize silver-containing precipitates toward photodecomposition, which may cause difficulties.

7B-5

Other Methods for End-Point Detection

Electroanalytical methods, which can be applied for the detection of end points in some precipitation reactions, are described in Chapters 15 and 17.

7C

Applications of Precipitation Titrations

As noted at the outset of this chapter, most applications of precipitation titrations are based upon the use of a standard silver nitrate solution. Table 7-3 lists some typical applications of argentometry. In many of these methods, the analyte is precipitated with a measured excess of silver nitrate, and this step is then followed by a Volhard titration with standard potassium thiocyanate.

Both silver nitrate and potassium thiocyanate are obtainable in primary-standard quality. The latter is, however, somewhat hygroscopic, and thiocyanate solutions are ordinarily standardized against silver nitrate. Both silver nitrate and potassium thiocyanate solutions are stable indefinitely.

Silver nitrate is a remarkably useful reagent for both titrimetric and gravimetric methods of analysis. Its greatest drawback is its cost, which fluctuates widely with the world price of silver. The high cost requires that considerable care be exercised to minimize the quantity of silver nitrate used. The reagent can be conserved by the use of small samples so that dilute solutions

TABLE 7-3 **Typical Argentometric Precipitation Methods**

Substance Determined	End Point	Remarks
AsO_4^{3-}, Br^-, I^-, CNO^-, SCN^-	Volhard	Removal of silver salt not required
CO_3^{2-}, CrO_4^{2-}, CN^-, Cl^-, $C_2O_4^{2-}$, PO_4^{3-}, S^{2-}, NCN^{2-}	Volhard	Removal of silver salt required before back-titration of excess Ag^+
BH_4^-	Modified Volhard	Titration of excess Ag^+ following $BH_4^- + 8Ag^+ + 8OH^- \rightarrow 8Ag(s) + H_2BO_3^- + 5H_2O$
Epoxide	Volhard	Titration of excess Cl^- following hydrohalogenation
K^+	Modified Volhard	Precipitation of K^+ with known excess of $B(C_6H_5)_4^-$, addition of excess Ag^+ giving $AgB(C_6H_5)_4(s)$, and back-titration of the excess
Br^-, Cl^-	Mohr	
Br^-, Cl^-, I^-, SeO_3^{2-}	Adsorption indicator	
$V(OH)_4^+$, fatty acids, mercaptans	Electroanalytical	Direct titration with Ag^+
Zn^{2+}	Modified Volhard	Precipitation as $ZnHg(SCN)_4$, filtration, dissolution in acid, addition of excess Ag^+, back-titration of excess Ag^+
F^-	Modified Volhard	Precipitation as PbClF, filtration, dissolution in acid, addition of excess Ag^+, back-titration of excess Ag^+

TABLE 7-4 **Miscellaneous Volumetric Precipitation Methods**

Reagent	Ion Determined	Reaction Product	Indicator
$K_4Fe(CN)_6$	Zn^{2+}	$K_2Zn_3[Fe(CN)_6]_2$	Diphenylamine
$Pb(NO_3)_2$	SO_4^{2-}	$PbSO_4$	Erythrosin B
	MoO_4^{2-}	$PbMoO_4$	Eosin A
$Pb(OAc)_2$	PO_4^{3-}	$Pb_3(PO_4)_2$	Dibromofluorescein
	$C_2O_4^{2-}$	PbC_2O_4	Fluorescein
$BaCl_2$	SO_4^{2-}	$BaSO_4$ (50% methanol solvent)	Alizarin red S
$Th(NO_3)_4$	F^-	ThF_4	Alizarin red
$Hg_2(NO_3)_2$	Cl^-, Br^-	Hg_2Cl_2, Hg_2Br_2	Bromophenol blue
NaCl	Hg_2^{2+}	Hg_2Cl_2	Bromophenol blue

or small volumes of silver nitrate are needed. In addition, the silver salt formed in the analysis and any reagent remaining afterward should be collected for subsequent processing to recover the silver or perhaps to convert it to silver nitrate for reuse. Several methods have been recommended for treating silver wastes to yield silver or silver nitrate.[1]

Table 7-4 lists some miscellaneous volumetric methods based on reagents other than silver nitrate.

Specific directions for the argentometric titration of chloride ion with an adsorption indicator are found in Section 31B.

Questions and Problems

*7-1 Why does a Volhard determination of chloride ion require more steps than a Volhard determination of bromide ion?

7-2 Explain how a cationic adsorption indicator InH^+ works in the titration of
(a) Br^- with Ag^+.
(b) Ag^+ with Cl^-.

*7-3 Under what circumstances can a cationic adsorption indicator be used to advantage?

7-4 Give equations for the volumetric determination of $C_2O_4^{2-}$ by the Volhard method. Would it be necessary to isolate the $Ag_2C_2O_4$ before back-titration with KSCN? Why or why not?

*7-5 Why must the titration of Cl^- with $AgNO_3$ using fluorescein as an indicator be carried out in subdued light?

7-6 In the determination of Cl^- by a Mohr titration, why is it desirable to standardize the $AgNO_3$ against primary-standard NaCl rather than prepare the reagent directly from primary-standard $AgNO_3$?

*7-7 Outline a method for the determination of K^+ based on argentometry. Write balanced equations for the reactions.

7-8 Suggest an argentometric method for the determination of F^-. Write balanced equations for the reactions.

[1] K. J. Bush and H. Diehl, *J. Chem. Educ.*, 1979, *56*, 54 and J. P. Rawat and L. Bellows, *ibid.*, **1986,** *63*, 357.

7-9 Calculate the molar concentration of an $AgNO_3$ solution if a 26.12-mL portion is needed to react with

*(a) 0.2124 g of KSCN.
(b) 0.6120 g of $BaCl_2 \cdot 2H_2O$.
*(c) 222.4 mg of Na_3PO_4.
(d) 50.00 mL of 0.01375 M H_2S.
*(e) 50.00 mL of 0.05451 M H_3AsO_4.
(f) 50.00 mL of 0.02756 M K_2CrO_4.

7-10 The As in a 9.13-g sample of pesticide was converted to AsO_4^{3-} and precipitated as Ag_3AsO_4 with 50.00 mL of 0.02105 M $AgNO_3$. The excess Ag^+ was then titrated with 4.75 mL of 0.04321 M KSCN. Calculate the percentage of As_2O_3 in the sample.

*7-11 Lead(II) can be titrated with standard K_2CrO_4 (product: $PbCrO_4$); an adsorption indicator signals the end point. A 1.1622-g mineral sample consisting mainly of Pb_3O_4 was decomposed by treatment with acid and subsequently titrated with 34.47 mL of 0.04176 M K_2CrO_4. Calculate the percentage of Pb_3O_4 in the sample.

7-12 A 50.00-mL aliquot of 0.1011 M $AgNO_3$ was added to an ammoniacal solution containing 0.2005 g of a sample containing an unknown amount of propargyl alcohol. The reaction is

$$2Ag^+ + NO_3^- + HC{\equiv}C{-}CH_2OH \rightleftharpoons AgC{\equiv}C{-}CH_2OH{\cdot}AgNO_3(s) + H^+$$

When reaction was complete, the excess Ag^+ was titrated with 6.797 mL of 0.08143 M KSCN. Calculate the percentage of propargyl alcohol (fw = 56.065) in the sample.

*7-13 Gallium(III) is quantitatively precipitated with $Fe(CN)_6^{4-}$:

$$4Ga^{3+} + 3Fe(CN)_6^{4-} \longrightarrow Ga_4[Fe(CN)_6]_3(s)$$

Calculate the percentage of gallium(III) sulfate in a 1.9671-g sample that required a 31.39-mL titration with 0.07208 M $K_4Fe(CN)_6$.

7-14 A $K_4Fe(CN)_6$ solution was standardized against a 0.1997-g sample of pure ZnO; 31.90 mL was required for the titration. The same solution was then used to determine the percentage of $Zn_2P_2O_7$ in a 0.2360-g sample of a mineral. This titration required 37.76 mL. The same reaction was involved in both the standardization and the titration:

$$2Fe(CN)_6^{4-} + 3Zn^{2+} + 2K^+ \longrightarrow K_2Zn_3[Fe(CN)_6]_2(s)$$

What is the percentage of $Zn_2P_2O_7$ in the sample?

*7-15 The monochloroacetic acid ($ClCH_2COOH$) preservative in 100.0 mL of a carbonated beverage was extracted into diethyl ether and then returned to aqueous solution as $ClCH_2COO^-$ by extraction with 1 M NaOH. This aqueous extract was acidified and treated with 50.00 mL of 0.04521 M $AgNO_3$. The reaction is

$$ClCH_2COOH + Ag^+ + H_2O \longrightarrow HOCH_2COOH + H^+ + AgCl(s)$$

After filtration of the AgCl, titration of the filtrate and washings required 10.43 mL of an NH_4SCN solution. Titration of a blank taken through the entire process used 22.98 mL of the NH_4SCN. Calculate the weight (in milligrams) of $ClCH_2COOH$ in the sample.

*7-16 A 0.1064-g sample of a pesticide was decomposed by the action of sodium biphenyl in toluene. The liberated Cl^- was extracted with dilute HNO_3 and titrated with 23.28 mL of 0.03337 M $AgNO_3$ by the Mohr method. Express the results of this analysis in terms of percent aldrin, $C_{12}H_8Cl_6$ (fw = 364.92).

7-17 A carbonate fusion was needed to free the Bi from a 0.6423-g sample containing the mineral eulytite, $2Bi_2O_3{\cdot}3SiO_2$. The fused mass was dissolved in dilute acid, following which the Bi^{3+} was titrated with 27.36 mL of 0.03369 M NaH_2PO_4 solution. The reaction is

$$Bi^{3+} + H_2PO_4^- \longrightarrow BiPO_4(s) + 2H^+$$

Calculate the percentage purity of eulytite (fw = 1112) in the sample.

*7-18 A 20-tablet sample of soluble saccharin was treated with 20.00 mL of 0.08181 M $AgNO_3$. The reaction is

$$C_6H_4(CO)(SO_2)NNa + Ag^+ \longrightarrow C_6H_4(CO)(SO_2)NAg(s) + Na^+$$

After removal of the solid, titration of the filtrate and washings required 2.81 mL of 0.04124 M KSCN. Calculate the average number of milligrams of saccharin (fw = 205.17) in each tablet.

7-19 The Fajans method is to be used in the routine analysis of solids for their chloride content. It is desired that the volume of standard $AgNO_3$ used in these titrations be numerically equal to the percent Cl^- when 0.2500-g samples are used. What should the molarity of the $AgNO_3$ solution be?

*7-20 The Association of Official Analytical Chemists recommends a Volhard titration for analysis of the insecticide heptachlor, $C_{10}H_5Cl_7$. The percentage of heptachlor is given by

$$\%\text{ heptachlor} = \frac{(\text{mL}_{Ag} \times c_{Ag} - \text{mL}_{SCN} \times c_{SCN}) \times 37.33}{\text{wt sample}}$$

What does this calculation reveal concerning the stoichiometry of this titration?

7-21 An analysis for borohydride ion is based upon its reaction with Ag^+:

$$BH_4^- + 8Ag^+ + 8OH^- \longrightarrow H_2BO_3^- + 8Ag(s) + 5H_2O$$

The purity of a quantity of KBH_4 for use in an organic synthesis was established by diluting 0.3213 g of the material to exactly 500.0 mL, treating a 100.0-mL aliquot with 50.00 mL of 0.2221 M $AgNO_3$, and titrating the excess silver ion with 3.36 mL of 0.0397 M KSCN. Calculate the percent purity of the KBH_4.

7-22 What volume of 0.04642 M KSCN would be needed if the analysis in Problem 7-21 were completed by filtering off the metallic Ag, dissolving it in acid, diluting to 250.0 mL, and titrating a 50.00-mL aliquot?

*7-23 A 100-mL sample of brackish water was made ammoniacal and the sulfide it contained titrated with 8.47 mL of 0.01310 M $AgNO_3$. The net reaction is

$$2Ag^+ + S^{2-} \longrightarrow Ag_2S(s)$$

Calculate the parts per million of H_2S in the water.

7-24 A 2.000-L water sample was evaporated to a small volume and treated with an excess of sodium tetraphenylboron, $NaB(C_6H_5)_4$. The precipitated $KB(C_6H_5)_4$ was filtered and then redissolved in acetone. The analysis was completed by a Mohr titration, with 37.90 mL of 0.03981 M $AgNO_3$ being used. The net reaction is

$$KB(C_6H_5)_4 + Ag^+ \longrightarrow AgB(C_6H_5)_4 + K^+$$

Express the results of this analysis in terms of parts per million of K (that is, mg K/L).

*7-25 The action of an alkaline I_2 solution upon the rodenticide warfarin, $C_{19}H_{16}O_4$ (fw = 308.34), results in the formation of 1 mol of iodoform, CHI_3 (fw = 393.73), for each mole of the parent compound reacted. Analysis for warfarin can then be based upon the reaction between CHI_3 and Ag^+:

$$CHI_3 + 3Ag^+ + H_2O \longrightarrow 3AgI(s) + 3H^+ + CO(g)$$

The CHI_3 produced from a 13.96-g sample was treated with 25.00 mL of 0.02979 M $AgNO_3$, and the excess Ag^+ was then titrated with 2.85 mL of 0.05411 M KSCN. Calculate the percentage of warfarin in the sample.

*7-26 A 1.998-g sample containing Cl^- and ClO_4^- was dissolved in sufficient water to give 250.0 mL of solution. A 50.00-mL aliquot required 13.97 mL of 0.08551 M $AgNO_3$ to titrate the Cl^-. A second 50.00-mL aliquot was treated with $V_2(SO_4)_3$ to reduce the ClO_4^- to Cl^-:

$$ClO_4^- + 4V_2(SO_4)_3 + 4H_2O \longrightarrow Cl^- + 12SO_4^{2-} + 8VO^{2+} + 8H^+$$

Titration of the reduced sample required 40.12 mL of the $AgNO_3$ solution. Calculate the percentages of Cl^- and ClO_4^- in the sample.

7-27 A 2.4414-g sample containing KCl, K_2SO_4, and inert materials was dissolved in sufficient water to give 250.0 mL of solution. A Mohr titration of a 50.00-mL aliquot required 41.36 mL of 0.05818 M $AgNO_3$. A second 50.00-mL aliquot was treated with 40.00 mL of 0.1083 M $NaB(C_6H_5)_4$. The reaction is

$$NaB(C_6H_5)_4 + K^+ \longrightarrow KB(C_6H_5)_4(s) + Na^+$$

The solid was filtered, redissolved in acetone, and titrated with 49.98 mL of the $AgNO_3$ solution (see Problem 7-24). Calculate the percentages of KCl and K_2SO_4 in the sample.

*7-28 A 0.2185-g sample containing only KCl and K_2SO_4 yielded a precipitate of $KB(C_6H_5)_4$ that—after isolation and solution in acetone—required a Fajans titration involving 25.02 mL of 0.1126 M $AgNO_3$ for the reaction

$$KB(C_6H_5)_4 + Ag^+ \longrightarrow AgB(C_6H_5)_4(s) + K^+$$

Calculate the percentages of KCl and K_2SO_4 in the sample.

7-29 For each of the following precipitation titrations, calculate the cation and anion concentrations at equivalence as well as at reagent volumes corresponding to ±20.00 mL, ±10.00 mL, and ±1.00 mL of equivalence. Construct a titration curve from the data, plotting the p-function of the cation versus reagent volume.

*(a) 25.00 mL of 0.05000 M $AgNO_3$ with 0.02500 M NH_4SCN
(b) 20.00 mL of 0.06000 M $AgNO_3$ with 0.03000 M KI
*(c) 30.00 mL of 0.07500 M $AgNO_3$ with 0.07500 M NaCl
(d) 35.00 mL of 0.4000 M Na_2SO_4 with 0.2000 M $Pb(NO_3)_2$
*(e) 40.00 mL of 0.02500 M $BaCl_2$ with 0.05000 M Na_2SO_4
(f) 50.00 mL of 0.2000 M NaI with 0.4000 M $TlNO_3$
(K_{sp} for TlI $= 6.5 \times 10^{-8}$)

7-30 Calculate the silver ion concentration after the addition of 5.00*, 15.00, 25.00, 30.00, 35.00, 39.00, 40.00*, 45.00*, and 50.00 mL of 0.05000 M $AgNO_3$ to 50.0 mL of 0.0400 M KBr. Construct a titration curve from these data, plotting pAg as a function of titrant volume.

7-31 Calculate the Hg_2^{2+} concentration after the addition of 0, 10.00*, 20.00, 30.00*, 35.00, 39.00, 40.00*, 41.00, 45.00, and 50.00* mL of 0.2000 M NaCl to 80.0 mL of a solution that is 0.0500 M in Hg_2^{2+}. For the process,

$$Hg_2Cl_2 \rightleftharpoons Hg_2^{2+} + 2Cl^- \qquad K_{sp} = 1.3 \times 10^{-18}$$

Construct a titration curve from these data, plotting pHg_2 as a function of titrant volume. *Note:* No evidence exists for the intermediate species Hg_2Cl^+.

CHAPTER 8

Titration Curves for Simple Acid/Base Systems

In this chapter, we describe the properties of aqueous solutions of simple acids and bases—that is, acids and bases that react with water to produce a single titratable hydronium or hydroxide ion. Consideration is first given to those simple acids and bases that possess indicator properties. The discussion then turns to the derivation of titration curves for strong acids and strong bases. Next, the pH of solutions containing conjugate acid/base pairs is considered, as well as the buffer properties of these solutions. Finally, methods for the derivation of titration curves for simple weak acids and bases are presented.[1]

In preparation for the discussion that follows, the reader may find it helpful to review the material on acids and bases in Sections 5A-2 and 5B-3.

8A Solutions and Indicators for Neutralization Titrations

8A-1 Standard Solutions

The standard solutions employed in neutralization titrations are usually strong acids or strong bases because these substances react more completely with an analyte than do their weaker counterparts. Standard acids are prepared from hydrochloric, perchloric, and sulfuric acids. Nitric acid is seldom used because its property as an oxidizing agent offers the potential for undesirable side reactions. It should be noted that hot concentrated solutions of sulfuric and perchloric acids are also potent oxidizing agents and thus hazardous. Fortunately, however, dilute solutions of these reagents are relatively benign and can be used in the analytical laboratory without any special precautions except eye protection.

Standard basic solutions are ordinarily prepared from sodium, potassium, and occasionally barium hydroxides. Again, eye protection should always be used when handling these reagents or their solutions.

8A-2 The Theory of Indicator Behavior

Many substances, both naturally occurring and synthetic, display colors that depend upon the pH of the solution in which they are dissolved. Some of these substances, which have been used for centuries to indicate the acidity or alkalinity of water, find current application as acid/base indicators. Generally, acid/base indicators are weak organic acids or bases that, upon disso-

[1]For a general reference on these topics, see D. R. Rosenthal and P. Zuman, in *Treatise on Analytical Chemistry*, 2nd ed., I. M. Kolthoff and P. J. Elving, Eds., Part I, Vol. 2, Chapter 18. New York: Wiley, 1979.

ciation or association, undergo internal structural changes that give rise to color differences. Color changes for acid/base indicators can be ascribed to such equilibria as

$$\underset{\text{acid color}}{HIn} + H_2O \rightleftharpoons H_3O^+ + \underset{\text{base color}}{In^-}$$

$$\underset{\text{base color}}{In} + H_2O \rightleftharpoons \underset{\text{acid color}}{InH^+} + OH^-$$

In both cases, the color of the molecular form of the indicator is different from the color of the ionic form. The structural changes that account for these differences in the most common types of indicators are described in Section 8G-1.

Equilibrium-constant expressions for the two dissociations are

$$\frac{[H_3O^+][In^-]}{[HIn]} = K_a \tag{8-1}$$

$$\frac{[InH^+][OH^-]}{[In]} = K_b \tag{8-2}$$

Rearranging Equation 8-1 reveals that the ratio of the concentration of the basic form of the indicator to that of the acid form is inversely proportional to the hydronium ion concentration:

$$\frac{[In^-]}{[HIn]} = \frac{K_a}{[H_3O^+]} \tag{8-3}$$

The human eye is not very sensitive to color differences in a solution containing a mixture of In^- and HIn, particularly when the ratio $[In^-]/[HIn]$ is greater than about 10 or smaller than about 0.1. Consequently, the color imparted to a solution by a typical indicator appears to the average observer to change rapidly only within the limited concentration ratio of 10 to 0.1. At greater or smaller ratios, the color becomes essentially constant to the human eye and independent of the ratio. Therefore, we can write that the indicator HIn exhibits its pure acid color to the average observer when

$$\frac{[In^-]}{[HIn]} \le \frac{1}{10}$$

and its base color when

$$\frac{[In^-]}{[HIn]} \ge \frac{10}{1}$$

The color appears to be intermediate for ratios between these two values. These ratios vary considerably from indicator to indicator, of course. Furthermore, people differ significantly in their ability to distinguish between colors, with a color-blind person representing one extreme.

If the two concentration ratios are substituted into Equation 8-1, the range of hydronium ion concentrations needed to effect the complete indicator color change can be evaluated. Thus, for the full acid color,

$$\frac{[H_3O^+][In^-]}{[HIn]} = \frac{[H_3O^+]1}{10} = K_a$$

$$[H_3O^+] = 10K_a$$

and similarly for the full base color,

$$\frac{[H_3O^+]10}{1} = K_a$$

$$[H_3O^+] = \frac{1}{10} K_a$$

To obtain the indicator range, we take the negative logarithms of the two expressions:

$$\text{Indicator pH range} = -\log 10K_a \text{ to } -\log \frac{K_a}{10}$$

$$= -1 + pK_a \text{ to } -(-1) + pK_a$$

$$= pK_a \pm 1$$

Thus, the typical indicator with an acid dissociation constant of 1×10^{-5} ($pK_a = 5$) shows a complete color change when the pH of the solution in which it is dissolved changes from 4 to 6. A similar relationship is easily derived for a basic indicator.

The list of compounds possessing acid/base indicator properties is large and includes a number of organic structures. An indicator covering almost any desired pH range can ordinarily be found. A few common indicators and their properties are listed in Table 8-1. Note that most of the transition ranges cover something less than two pH units.

TABLE 8-1

Some Important Acid/Base Indicators*

Common Name	Transition Range, pH	pK_a**	Color Change†	Indicator Type‡
Thymol blue	1.2–2.8	1.65	R–Y	1
	8.0–9.6	8.90	Y–B	
Methyl yellow	2.9–4.0		R–Y	2
Methyl orange	3.1–4.4	3.46§	R–O	2
Bromocresol green	3.8–5.4	4.66	Y–B	1
Methyl red	4.2–6.3	5.00§	R–Y	2
Bromocresol purple	5.2–6.8	6.12	Y–P	1
Bromothymol blue	6.2–7.6	7.10	Y–B	1
Phenol red	6.8–8.4	7.81	Y–R	1
Cresol purple	7.6–9.2		Y–P	1
Phenolphthalein	8.3–10.0		C–R	1
Thymolphthalein	9.3–10.5		C–B	1
Alizarin yellow GG	10–12		C–Y	2

*From C. A. Streuli, in *Handbook of Analytical Chemistry*, L. Meites, Ed., pp. **3**-35 and **3**-36. New York: McGraw-Hill, 1963.

**At ionic strength of 0.1.

†B = blue; C = colorless; O = orange; P = purple; R = red; Y = yellow.

‡(1) Acidic type: $HIn + H_2O \rightleftharpoons H_3O^+ + In^-$
(2) Basic type: $In + H_2O \rightleftharpoons InH^+ + OH^-$

§For reaction $InH^+ + H_2O \rightleftharpoons H_3O^+ + In$

8B

Titration Curves for Strong Acids and Strong Bases

When both reagent and analyte are strong, the net neutralization reaction is

$$H_3O^+ + OH^- \rightleftharpoons 2H_2O$$

Titration curves based on this reaction are derived by methods that are entirely analogous to those shown in Section 7A for precipitation titrations. In the present case, however, K_w is substituted for K_{sp} whenever needed.

8B-1

The Titration of a Strong Acid with a Strong Base

The hydronium ions in an aqueous solution of a strong acid arise from two sources: (1) the reaction of the solute with water and (2) the dissociation of water. In all but the most dilute solutions, however, the contribution from the solute far exceeds that from the solvent. Thus, for a solution of HCl having a concentration greater than about 1×10^{-6} M, we can write

$$[H_3O^+] = c_{HCl} + [OH^-] \simeq c_{HCl} \qquad (8\text{-}4)$$

where $[OH^-]$ represents the contribution of water dissociation to the hydronium ion concentration. Note the similarity between Equations 8-4 and 7-1 on page 167.

For a solution of a strong base, such as sodium hydroxide, an analogous situation exists:

$$[OH^-] = c_{NaOH} + [H_3O^+] \simeq c_{NaOH} \qquad (8\text{-}5)$$

A useful relationship for calculating the pH of basic solutions can be obtained by taking the negative logarithm of each side of the ion-product-constant expression for water:

$$-\log K_w = -\log ([H_3O^+][OH^-]) = -\log [H_3O^+] - \log [OH^-]$$

$$pK_w = pH + pOH$$

At 25°C, pK_w is 14.00. Note the similarity of this expression to Equation 7-2, page 167.

As seen in the following example, the derivation of a curve for the titration of a strong acid with a strong base is simple because the concentration of the hydronium ion or the hydroxide ion is obtained directly from stoichiometric calculations.

EXAMPLE 8-1

Derive a titration curve for the reaction of 50.00 mL of 0.05000 M HCl with 0.1000 M NaOH. Round pH data to two places to the right of the decimal point.

Initial point

The solution is 5.00×10^{-2} M in HCl. Since HCl is completely dissociated,

$$[H_3O^+] = 5.00 \times 10^{-2}$$

$$pH = -\log (5.00 \times 10^{-2}) = -\log 5.00 - \log 10^{-2}$$

$$= -0.699 + 2 = 1.301 = 1.30$$

pH after addition of 10.00 mL of NaOH
The volume of the solution is now 60.00 mL, and part of the HCl has been neutralized. Thus,

$$[H_3O^+] = \frac{50.00 \times 0.0500 - 10.00 \times 0.1000}{60.00} = 2.50 \times 10^{-2}$$

$$pH = 2 - \log 2.50 = 1.60$$

Additional data to define the curve in the region before the equivalence point are derived in the same way. The results of such calculations are given in column 2 of Table 8-2.

pH after addition of 25.00 mL of NaOH
At the equivalence point, the solution contains neither an excess of HCl nor an excess of NaOH and hydronium ions arise only from the dissociation of water. Therefore,

$$[H_3O^+] = [OH^-] = \sqrt{K_w} = 1.00 \times 10^{-7}$$

$$pH = 7.00$$

pH after addition of 25.10 mL of NaOH
The concentration of excess base is

$$c_{NaOH} = \frac{25.10 \times 0.1000 - 50.0 \times 0.0500}{75.10} = 1.33 \times 10^{-4} \text{ M}$$

Provided the concentration of OH^- from the dissociation of water is negligible with respect to c_{NaOH},

$$[OH^-] = c_{NaOH} = 1.33 \times 10^{-4} \text{ M}$$

$$pOH = -\log(1.33 \times 10^{-4}) = 3.88$$

$$pH = 14.00 - 3.88 = 10.12$$

Additional data for this titration, calculated in the same way, are given in column 2 of Table 8-2.

The Effect of Concentration

The effects of reagent and analyte concentration on the neutralization titration curves for strong acids are shown by the two sets of data in Table 8-2 and the

TABLE 8-2 **Changes in pH During the Titration of a Strong Acid with a Strong Base**

Volume of NaOH, mL	50.00 mL of 0.0500 M HCl with 0.1000 M NaOH	50.00 mL of 0.000500 M HCl with 0.001000 M NaOH
	pH	pH
0.00	1.30	3.30
10.00	1.60	3.60
20.00	2.15	4.15
24.00	2.87	4.87
24.90	3.87	5.87
25.00	7.00	7.00
25.10	10.12	8.12
26.00	11.12	9.12
30.00	11.80	9.80

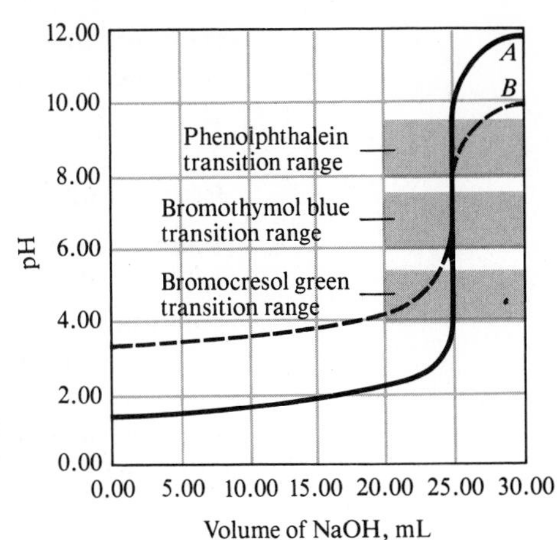

FIGURE 8-1 Curve for the titration of HCl with standard NaOH. *A*: 50.00 mL of 0.0500 M HCl with 0.1000 M NaOH. *B*: 50.00 mL of 0.000500 M HCl with 0.001000 M NaOH.

plots in Figure 8-1. Note the similarity between the curves in this figure and those for precipitation titrations shown in Figure 7-2 (Section 7A-2).

With 0.1 M NaOH as the titrant (curve *A*), the change in pH in the equivalence-point region is large. With 0.001 M NaOH, the change is markedly less but still pronounced.

Indicator Choice

Figure 8-1 shows that the selection of an indicator is not critical when the reagent concentration is approximately 0.1 M. Here, the volume differences in titrations with the three indicators shown are of the same magnitude as the uncertainties associated with reading the buret and are thus negligible. Note, however, that bromocresol green is clearly unsuited for a titration involving the 0.001 M reagent because not only does the color change continuously over a substantial range of titrant volumes but also the transition to the alkaline form is essentially complete before the equivalence point is reached. Therefore, a significant determinate error would result. The use of phenolphthalein is subject to similar objections. Of the three indicators, then, only bromothymol blue provides a satisfactory end point with a minimal determinate titration error when the more dilute solutions are used.

8B-2 The Titration of a Strong Base with a Strong Acid

Normally, pH serves as the ordinate in titration curves for strong bases as well as strong acids since the output of a pH meter, which is widely used in deriving experimental acid/base titration curves, is in terms of this parameter. Derivation of a titration curve for a strong base is analogous to that for strong acids, which is illustrated by Example 8-1. In the region before the equivalence point, the solution is highly alkaline, the hydroxide ion concentration being numerically equal to the molarity of the base. The solution is neutral at the equivalence point for precisely the same reason noted previously. Finally, the solution becomes acidic in the region beyond the equivalence point; here, the hydronium ion concentration is equal to the concentration of the excess strong acid. A curve for the titration of a strong base with 0.1 M hydrochloric acid is shown in Figure 8-5. Indicator selection is based upon the considerations described for the titration of a strong acid with a strong base.

8C Properties of Weak Acid and Weak Base Systems

Four types of computations are required to generate a titration curve for a solution of a weak acid. These four include computing the pH of a solution containing (1) a weak acid alone, (2) the conjugate base of that acid alone, (3) a mixture of the weak acid and its conjugate base, and (4) a mixture of the conjugate base and a strong base. An analogous set of calculations is required to obtain titration curves for weak bases. Techniques applicable to solutions (1) and (2) were described in Section 5B-3. In this section, we will consider methods for the calculation of the pH of the solutions described in (3) and (4).

8C-1 The pH of Solutions Containing Conjugate Acid/Base Pairs

A solution containing a conjugate acid/base pair may be acidic, neutral, or basic, depending upon the position of two competitive equilibria. For a solution of the weak acid HA and its sodium salt NaA, these equilibria are

$$HA + H_2O \rightleftharpoons H_3O^+ + A^- \qquad \frac{[H_3O^+][A^-]}{[HA]} = K_a \tag{8-6}$$

$$A^- + H_2O \rightleftharpoons OH^- + HA \qquad \frac{[OH^-][HA]}{[A^-]} = K_b = \frac{K_w}{K_a} \tag{8-7}$$

If the first equilibrium lies farther to the right than the second, the solution is acidic. If the second equilibrium is more favorable, the solution is basic. From the two equilibrium-constant expressions, it is evident that the relative concentrations of the hydronium and hydroxide ions depend not only upon the magnitudes of K_a and K_b but also upon the ratio of the concentrations of the acid and its conjugate base.

A similar situation prevails for a weak base and its conjugate acid. Thus, for a solution containing both ammonia and ammonium chloride, the equilibria are

$$NH_3 + H_2O \rightleftharpoons NH_4^+ + OH^-$$

$$\frac{[NH_4^+][OH^-]}{[NH_3]} = K_b = 1.76 \times 10^{-5}$$

$$NH_4^+ + H_2O \rightleftharpoons NH_3 + H_3O^+$$

$$\frac{[H_3O^+][NH_3]}{[NH_4^+]} = \frac{K_w}{K_b} = K_a$$

$$K_a = \frac{1.00 \times 10^{-14}}{1.76 \times 10^{-5}} = 5.68 \times 10^{-10}$$

We see that K_b is much larger than K_a; thus, the first equilibrium ordinarily predominates, and the solutions are basic. When the ratio of $[NH_4^+]$ to $[NH_3]$ becomes greater than about 200, however, the difference in equilibrium constants is offset, and such solutions are acidic.

Calculation of the pH of a Solution Containing a Weak Acid and Its Conjugate Base

In showing how the pH of a solution of a weak acid and its conjugate base can be computed, we shall follow the systematic approach developed in Section 6A and consider a solution that has a molar analytical concentration of weak acid c_{HA} and a molar analytical concentration of conjugate base c_{NaA}. The important equilibria as well as their equilibrium constants are given as part of Equations 8-6 and 8-7, and the quantity sought is $[H_3O^+]$.

Mass- and Charge-Balance Equations. Since the A-containing species in the solution have been introduced from two sources, the following mass-balance relationship exists:

$$c_{HA} + c_{NaA} = [HA] + [A^-]$$

Electrical neutrality considerations require that

$$[Na^+] + [H_3O^+] = [A^-] + [OH^-]$$

but

$$[Na^+] = c_{NaA}$$

Therefore, the charge-balance equation is

$$c_{NaA} + [H_3O^+] = [A^-] + [OH^-]$$

which rearranges to

$$[A^-] = c_{NaA} + [H_3O^+] - [OH^-] \tag{8-8}$$

Subtraction of the first equation from the fourth gives, upon rearrangement,

$$[HA] = c_{HA} - [H_3O^+] + [OH^-] \tag{8-9}$$

It is of interest to note that the two equations developed from mass- and charge-balance considerations can be arrived at by logical deductions from the two equilibria operating in the solution. Thus, the reaction shown by Equation 8-6 causes $[A^-]$ to increase by an amount equal to $[H_3O^+]$, whereas reaction 8-7 consumes $[A^-]$ in an amount corresponding to $[OH^-]$. Therefore, as shown by Equation 8-8, the resulting equilibrium concentration of A^- is the analytical concentration of NaA plus the hydronium ion concentration and minus the hydroxide ion concentration. Similarly, the reaction shown in Equation 8-6 tends to decrease the concentration of HA by an amount equal to $[H_3O^+]$. Reaction 8-7, conversely, increases the concentration of HA by an amount equal to $[OH^-]$. Thus, the equilibrium concentration of HA, as shown by Equation 8-9, is equal to its analytical concentration minus the hydronium ion concentration plus the hydroxide ion concentration.

Number of Equations and Unknowns. We have four equations (8-6, 8-7, 8-8, and 8-9) as well as four unknowns ($[HA]$, $[A^-]$, $[H_3O^+]$, and $[OH^-]$). Therefore, an exact solution for $[H_3O^+]$ is feasible.

Approximations. Because of the inverse relationship between $[H_3O^+]$ and $[OH^-]$, it is *always* possible to eliminate one or the other from Equations 8-8 and 8-9. Moreover, the *difference* in concentration between these two species is often so small relative to the molar concentrations of acid and conjugate base that Equations 8-8 and 8-9 simplify to

$$[A^-] = c_{NaA} \quad (8\text{-}10)$$

$$[HA] = c_{HA} \quad (8\text{-}11)$$

Solution of Equations. Substitution of Equations 8-10 and 8-11 into the dissociation-constant expression and rearrangement yield

$$[H_3O^+] = K_a \frac{c_{HA}}{c_{NaA}} \quad (8\text{-}12)^2$$

The assumption leading to Equations 8-10 and 8-11 sometimes breaks down with acids or bases that have dissociation constants greater than about 10^{-3} or when the molar concentration of either the acid or its conjugate base (or both) is very small. In these circumstances, either $[OH^-]$ or $[H_3O^+]$ must be retained in Equations 8-8 and 8-9, depending upon whether the solution is acidic or basic. In any case, Equations 8-10 and 8-11 should always be used initially. The provisional values for $[H_3O^+]$ and $[OH^-]$ can then be employed to test the assumptions.

Within the limits imposed by the assumptions made in its derivation, Equation 8-12 states that the hydronium ion concentration of a solution containing a weak acid and its conjugate base is dependent only upon the *ratio* between the molar concentrations of these two solutes. Furthermore, this ratio remains *independent of dilution* because the concentrations of each component changes in a proportionate manner upon a change in volume. Thus, the hydronium ion concentration of a solution containing appreciable quantities of a weak acid and its conjugate base tends to be independent of dilution and depends only upon the ratio of molar concentrations between the two solutes. That pH is independent of dilution is one manifestation of the *buffering* properties of such solutions.

EXAMPLE 8-2

What is the pH of a solution that is 0.400 M in formic acid and 1.00 M in sodium formate?

[2] An alternative form of Equation 8-12 is frequently encountered in the biological literature and biochemical texts. It is obtained by expressing each term in the form of its negative logarithm and inverting the concentration ratio to keep all signs positive:

$$-\log [H_3O^+] = -\log K_a - \log \frac{c_{HA}}{c_{NaA}}$$

Therefore,

$$\text{pH} = \text{p}K_a + \log \frac{c_{NaA}}{c_{HA}} \quad (8\text{-}13)$$

This expression is known as the *Henderson-Hasselbalch equation.*

The equilibrium governing the hydronium ion concentration in this solution is

$$H_2O + HCOOH \rightleftharpoons H_3O^+ + HCOO^-$$

for which

$$K_a = \frac{[H_3O^+][HCOO^-]}{[HCOOH]} = 1.77 \times 10^{-4}$$

$$[HCOO^-] \simeq c_{HCOO^-} = 1.00$$

$$[HCOOH] \simeq c_{HCOOH} = 0.400$$

Substitution into Equation 8-12 gives

$$[H_3O^+] = 1.77 \times 10^{-4} \times \frac{0.400}{1.00} = 7.08 \times 10^{-5}$$

Note that the assumption that $[H_3O^+] \ll$ the molarity of HCOOH and $HCOO^-$ is valid. Thus,

$$pH = -\log (7.08 \times 10^{-5}) = 4.15$$

If we compare the pH of this solution with that calculated for a 0.400 M formic acid solution (Equation 5-17), we find that the addition of sodium formate raises the pH from 2.08 to 4.15.

Calculation of the pH of a Solution of a Weak Base and Its Conjugate Acid

The calculation of pH for a solution of a weak base and its conjugate acid is completely analogous to that discussed in the preceding section.

EXAMPLE 8-3

Calculate the pH of a solution that is 0.280 M in NH_4Cl and 0.0700 M in NH_3.

The equilibrium of interest is

$$NH_3 + H_2O \rightleftharpoons NH_4^+ + OH^-$$

for which $K_b = 1.76 \times 10^{-5}$. As before, we assume that

$$[NH_3] \simeq c_{NH_3} = 0.0700$$

$$[NH_4^+] \simeq c_{NH_4^+} = 0.280$$

A provisional value for $[OH^-]$ is then obtained by substituting these values into the equilibrium-constant expression:

$$\frac{0.280[OH^-]}{0.0700} = 1.76 \times 10^{-5}$$

$$[OH^-] = 4.40 \times 10^{-6}$$

Clearly, the approximation is justified and

$$pOH = -\log (4.40 \times 10^{-6}) = 5.36$$

$$pH = 14.00 - 5.36 = 8.64$$

8C-2 The pH of Conjugate Base/Strong Base and Conjugate Acid/Strong Acid Mixtures

Beyond the equivalence point in the titration of the weak acid HA with standard sodium hydroxide, the solution contains a relatively large concentration of the salt NaA and a small amount of the strong base NaOH. The hydroxide ion concentration is then made up of the contributions from the two bases:

$$[OH^-] = c_{NaOH} + [HA]$$

where the second term accounts for the concentration of hydroxide ion arising from the reaction

$$A^- + H_2O \rightleftharpoons HA + OH^-$$

In most titrations, it is possible to assume that the excess strong base represses the foregoing equilibrium to the extent that $[HA] \ll c_{NaOH}$ and

$$[OH^-] \simeq c_{NaOH}$$

For the titration of the weak base B with standard hydrochloric acid, the solution beyond equivalence consists of a mixture of the hydrochloric acid and the conjugate acid BH^+; thus, an analogous set of equations can be written:

$$[H_3O^+] = c_{HCl} + [B]$$

where [B] accounts for the hydronium ion concentration arising from the reaction

$$BH^+ + H_2O \rightleftharpoons B + H_3O^+$$

Usually it is possible to assume that $[B] \ll c_{HCl}$, so that

$$[H_3O^+] = c_{HCl}$$

EXAMPLE 8-4

At 0.10 mL beyond the equivalence point in the titration of acetic acid with sodium hydroxide, a solution was 1.0×10^{-4} M in NaOH and 0.05 M in NaOAc. Calculate the pH of this solution.

Initially we assume that

$$[OH^-] = 1.00 \times 10^{-4} + [HOAc] \simeq 1.00 \times 10^{-4}$$

$$pH = 14.00 - (-\log 1.00 \times 10^{-4}) = 10.0$$

To check the validity of the approximations, we use the dissociation-constant expression for the conjugate base OAc^-:

$$\frac{[HOAc][OH^-]}{[OAc^-]} = \frac{K_w}{K_a} = \frac{1.00 \times 10^{-14}}{1.75 \times 10^{-5}} = 5.71 \times 10^{-10}$$

$$[OAc^-] = 0.050 - [HOAc] \simeq 0.050$$

Substituting,

$$\frac{[HOAc] \times 1.0 \times 10^{-4}}{0.050} = 5.71 \times 10^{-10}$$

$$[HOAc] = 2.9 \times 10^{-7}$$

We see then that [HOAc] is indeed smaller than c_{NaOH} and $[OAc^-]$.

EXAMPLE 8-5

Repeat the calculation in Example 8-4 but assume that the acid being titrated is HOCl.

Again, we assume

$$[OH^-] = 1.00 \times 10^{-4} + [HOCl] \simeq 1.00 \times 10^{-4}$$

$$pH = 10.00$$

and to check the assumption, we write

$$\frac{[HOCl][OH^-]}{[OCl^-]} = \frac{[HOCl] \times 1.0 \times 10^{-4}}{0.050} = \frac{1.00 \times 10^{-14}}{3.0 \times 10^{-8}} = 3.33 \times 10^{-7}$$

$$[HOCl] = 1.7 \times 10^{-4}$$

which is certainly *not* much smaller than $[OH^-] = 1.0 \times 10^{-4}$.

To obtain a more exact solution, let us write the mass-balance equation in the form

$$[OCl^-] = c_{NaOCl} - [HOCl] = 0.050 - [HOCl]$$

and assume that $[HOCl] \ll 0.050$, which leads to

$$[OCl^-] \simeq 0.050$$

Substituting into the base dissociation constant expression for OCl^- gives

$$\frac{[HOCl][OH^-]}{0.050} = 3.37 \times 10^{-7}$$

$$[HOCl] = 3.33 \times 10^{-7} \times 0.050/[OH^-] = 1.67 \times 10^{-8}/[OH^-]$$

Substituting into the first equation leads to

$$[OH^-] = 1.0 \times 10^{-4} + 1.67 \times 10^{-8}/[OH^-]$$

$$[OH^-]^2 - (1.0 \times 10^{-4})[OH^-] - 1.67 \times 10^{-8} = 0$$

$$[OH^-] = 1.89 \times 10^{-4}$$

$$pOH = 14.00 - (-\log 1.89 \times 10^{-4}) = 10.27$$

If we calculate [HOCl], we find that it is indeed much smaller than 0.050, as we have assumed.

8D Properties of Buffer Solutions

A *buffer solution* is defined as a solution that resists changes in pH either when it is diluted or when small amounts of acid or base are added to it. The most effective buffer solutions contain large and approximately equal concentrations of a weak acid and its conjugate base.

8D-1 The Effect of Dilution

The pH of a buffer solution remains essentially independent of dilution until the concentrations of the species it contains are decreased to the point where the approximations used to develop Equations 8-10 and 8-11 become invalid. Example 8-6 illustrates the behavior of a typical buffer during dilution.

EXAMPLE 8-6

Calculate the pH of the buffer described in Example 8-2 (original pH 4.15) upon dilution by a factor of (a) 50 and (b) 10,000.

(a) Upon dilution by a factor of 50,

$$c_{HCOOH} = \frac{0.400}{50} = 8.00 \times 10^{-3}\ M$$

$$c_{HCOONa} = \frac{1.00}{50} = 2.00 \times 10^{-2}\ M$$

If we assume, as before, that $[H_3O^+] - [OH^-]$ is small relative to the two molarities (page 189), we obtain

$$\frac{[H_3O^+] \times 2.00 \times 10^{-2}}{8.00 \times 10^{-3}} = 1.77 \times 10^{-4}$$

$$[H_3O^+] = 7.08 \times 10^{-5}$$

$$pH = 4.15$$

The error introduced by the assumption $7.08 \times 10^{-5} \ll 8.00 \times 10^{-3}$ and 2.00×10^{-2} is acceptable. (If the assumption is not made, one obtains $[H_3O^+] = 6.99 \times 10^{-5}$ and pH = 4.16.)

(b) Upon dilution by a factor of 10,000,

$$c_{HCOOH} = 4.00 \times 10^{-5}\ M$$

$$c_{HCOONa} = 1.00 \times 10^{-4}\ M$$

In this solution, the solute concentrations are of the same magnitude as $[H_3O^+]$, and so the more exact statements given by Equations 8-8 and 8-9 must be used:

$$[HCOOH] = 4.00 \times 10^{-5} - [H_3O^+] + \cancel{[OH^-]}$$

$$[HCOO^-] = 1.00 \times 10^{-4} + [H_3O^+] - \cancel{[OH^-]}$$

Because the solution is acidic, $[OH^-] \ll [H_3O^+]$. Substitution of the foregoing relationships into the dissociation-constant expression gives

$$\frac{[H_3O^+](1.00 \times 10^{-4} + [H_3O^+])}{4.00 \times 10^{-5} - [H_3O^+]} = 1.77 \times 10^{-4}$$

This equation rearranges to the quadratic form

$$[H_3O^+]^2 + (2.77 \times 10^{-4})[H_3O^+] - 7.08 \times 10^{-9} = 0$$

the solution of which is

$$[H_3O^+] = 2.36 \times 10^{-5}$$

$$pH = 4.63$$

Thus, a 10,000-fold dilution causes the pH to increase from 4.15 to 4.63, whereas a 50-fold dilution has essentially no effect.

Figure 8-2 contrasts the behavior of buffered and unbuffered solutions with dilution. For each, the initial solute concentration is 1.00 M. The resistance of the buffered solution to changes in pH during dilution is clear.

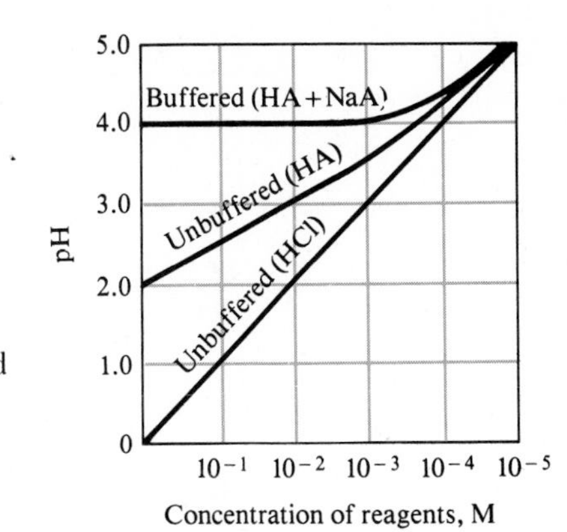

FIGURE 8-2 The effect of dilution on the pH of buffered and unbuffered solutions. The dissociation constant for HA is assumed to be 1.00×10^{-4}. Initial solute concentrations are 1.00 M.

8D-2 *Addition of Acids and Bases*

The following example illustrates the second property of buffer solutions, namely, their resistance to pH change after addition of small amounts of strong acids or bases.

EXAMPLE 8-7

Calculate the pH change that takes place when a 100-mL portion of (a) 0.0500 M NaOH and (b) 0.0500 M HCl is added to 400 mL of a buffer solution that is 0.200 M in NH_3 and 0.300 M in NH_4Cl.

The initial pH of the solution is obtained by assuming that

$$[NH_3] \simeq c_{NH_3} = 0.200$$

$$[NH_4^+] \simeq c_{NH_4Cl} = 0.300$$

and substituting into the dissociation-constant expression for NH_3:

$$\frac{0.300[OH^-]}{0.200} = 1.76 \times 10^{-5}$$

$$[OH^-] = 1.17 \times 10^{-5}$$

$$pH = 14.00 - (-\log 1.17 \times 10^{-5}) = 9.07$$

(a) Addition of NaOH converts part of the NH_4^+ in the buffer to NH_3:

$$NH_4^+ + OH^- \longrightarrow NH_3 + H_2O$$

The analytical concentrations of NH_3 and NH_4Cl then become

$$c_{NH_3} = \frac{400 \times 0.200 + 100 \times 0.0500}{500} = \frac{85.0}{500} = 0.170 \text{ M}$$

$$c_{NH_4Cl} = \frac{400 \times 0.300 - 100 \times 0.0500}{500} = \frac{115}{500} = 0.230 \text{ M}$$

When substituted into the dissociation-constant expression, these values yield

$$[OH^-] = \frac{1.76 \times 10^{-5} \times 0.170}{0.230} = 1.30 \times 10^{-5}$$

$$pH = 14.00 - (-\log 1.30 \times 10^{-5}) = 9.11$$

and the change in pH is

$$\Delta pH = 9.11 - 9.07 = 0.04$$

(b) Addition of HCl converts part of the NH_3 to NH_4^+; thus,

$$NH_3 + H_3O^+ \longrightarrow NH_4^+ + H_2O$$

$$c_{NH_3} = \frac{400 \times 0.200 - 100 \times 0.0500}{500} = \frac{75}{500} = 0.150 \text{ M}$$

$$c_{NH_4^+} = \frac{400 \times 0.300 + 100 \times 0.0500}{500} = \frac{125}{500} = 0.250 \text{ M}$$

$$[OH^-] = 1.76 \times 10^{-5} \times \frac{0.150}{0.250} = 1.06 \times 10^{-5}$$

$$pH = 14.00 - (-\log 1.06 \times 10^{-5}) = 9.02$$

$$\Delta pH = 9.02 - 9.07 = -0.05$$

It is of interest to contrast the behavior of an unbuffered solution with a pH of 9.07 to that of the buffered one just considered. It is readily shown that adding the same quantity of base to the unbuffered solution would increase the pH to 12.00—a pH change of 2.93 units. Adding the acid would decrease the pH by slightly more than 7 units.

8D-3

Buffer Capacity

Example 8-7 clearly demonstrates that a solution containing a conjugate acid/base pair possesses remarkable resistance to changes in pH. The ability of a buffer to prevent a significant change in pH is directly related to the total concentration of the buffering species as well as to their concentration ratio. For example, the pH of a 400-mL portion of a buffer formed by diluting the solutions described in Example 8-7 by 10 would change by about 0.4 to 0.5 unit when treated with the same amounts of sodium hydroxide or hydrochloric acid. In contrast, the change is about 0.04 to 0.05 unit for the more concentrated buffer.

The *buffer capacity* of a solution is defined as the number of equivalents of strong acid or base that causes 1.00 L of the buffer to undergo a 1.00-unit change in pH. The capacity of a buffer depends not only on the total concentration of the two buffer components but also on their concentration ratio. In fact, it is readily shown that the maximum buffer capacity is realized when the ratio of the molar concentration of the weak acid to the molar concentration conjugate base is unity.[3] That is, maximum buffer capacity occurs when $[HA] = [A^-]$. Note that the pH of such a buffer is then equal to the pK_a of the weak acid. Buffer capacity falls off moderately rapidly as the ratio of the molar concentrations of the acid and conjugate base becomes larger or smaller than unity. For this reason, the pK_a of the acid chosen for a given application should lie within ± 1 unit of the desired pH in order for the buffer to have a reasonable capacity.

[3]See, for example, D. R. Rosenthal and P. Zuman, in *Treatise on Analytical Chemistry,* 2nd ed., I. M. Kolthoff and P. J. Elving, Eds., Part I, Vol. 2, p. 208. New York: Wiley, 1979.

8D-4

Preparation of Buffers

In principle, a buffer solution of any desired pH can be prepared by combining calculated quantities of a suitable conjugate acid/base pair. In practice, however, the pH values of buffers prepared from theoretically generated recipes differ somewhat from the predicted values. These differences arise in part from the uncertainties that exist in the numerical values of many dissociation constants and in part from the simplifications used in calculations. More important, however, is the fact that the ionic strength of a buffer is usually so high that good values for the activity coefficients of the ions in the solution cannot be obtained from the Debye-Hückel relationship. Thus, the ionic strength of the NH_3/NH_4^+ buffer considered in Example 8-7 is about 0.30; the concentration equilibrium constant K_b' (page 130) is therefore significantly larger than 1.76×10^{-5} and quite uncertain (about 4×10^{-5}).

Because of these uncertainties, the chemist does not attempt to prepare buffers from carefully measured quantities of reagents but instead makes up a solution of approximately the desired pH and then adjusts by adding acid or conjugate base until the required pH is indicated by a pH meter.

Alternatively, empirically derived recipes for preparing buffer solutions of known pH are available in chemical handbooks and reference works.[4] Because of their widespread applications, two buffer systems deserve specific mention. McIlvaine buffers cover a pH range from about 2 to 8 and are prepared by mixing solutions of citric acid and disodium hydrogen phosphate. Clark and Lubs buffers, which encompass a pH range from 2 to 10, make use of three systems: phthalic acid/potassium hydrogen phthalate, potassium dihydrogen phosphate/dipotassium hydrogen phosphate, and boric acid/sodium borate. Table 15-4 gives the composition of buffers recommended by the National Bureau of Standards.

Buffers are of tremendous importance in biological and biochemical studies where a low but constant concentration of hydronium ions (10^{-6} to 10^{-10} M) must be maintained throughout experiments. Several biological supply houses offer a variety of such buffers.[5]

8E

Titration Curves for Weak Acids

8E-1

Qualitative Features of a Weak Acid Titration Curve

Before demonstrating how titration curves for weak acids are derived, let us deduce what qualitative differences are expected between these curves and the curves for strong acids having the same concentration. One obvious difference is the initial pH, which is greater for the weak acid HA because of its incomplete dissociation. Furthermore, the weak acid solution exhibits a much faster rise in pH with the first small addition of reagent (say, 1 mL) than the strong acid solution. The reason for this difference is that a profound increase in the *relative* concentration of A^- accompanies this addition because the concentration of A^- at the outset is vanishingly small. In contrast, the analytical

[4]See, for example, L. Meites, Ed., *Handbook of Analytical Chemistry*, pp. 5-112 and 11-3 to 11-8. New York: McGraw-Hill, 1963.

[5]For example, see *Aldrichimica Acta*, **1983**, *16* (2) 35.

concentrations of the weak or strong acids are changed minimally by this addition because their initial concentrations are high. As a consequence, the ratio [HA]/[A^-], which determines the pH of the weak acid solution (Equation 8-12), changes dramatically at the outset, as does the pH. In contrast, the pH of the strong acid solution is not affected greatly at this point because the change in its relative concentration is minimal. These effects are demonstrated in the example that follows.

EXAMPLE 8-8

Calculate the percent change of hydronium ion concentration that occurs when 1.0 mL of 0.1 M NaOH is added to 50 mL of 0.10 M (a) HCl and (b) acetic acid (HOAc).

(a)

$$\text{Initial } [H_3O^+] = c_{HCl} = 0.10$$

$$\text{New } [H_3O^+] = \frac{50 \times 0.10 - 1.0 \times 0.10}{50 + 1.0} = 0.096$$

$$\text{Change in } [H_3O^+] = \frac{0.096 - 0.10}{0.10} \times 100\% = -4.0\%$$

(b) Substituting into Equation 5-17 gives

$$\text{Initial } [H_3O^+] = \sqrt{K_a c_{HOAc}} = \sqrt{1.75 \times 10^{-5} \times 0.10} = 1.32 \times 10^{-3}$$

To obtain the new [H_3O^+], we must calculate [HOAc] and [OAc^-]:

$$c_{HOAc} = \frac{50 \times 0.10 - 1.0 \times 0.10}{51} = 0.096$$

$$c_{NaOAc} = \frac{1.0 \times 0.10}{51} = 1.96 \times 10^{-3}$$

Substituting into Equation 8-12 yields

$$\text{New } [H_3O^+] = \frac{(1.75 \times 10^{-5})c_{HOAc}}{c_{NaOAc}} = \frac{1.75 \times 10^{-5} \times 0.096}{1.96 \times 10^{-3}} = 8.6 \times 10^{-4}$$

$$\text{Change in } [H_3O^+] = \frac{8.6 \times 10^{-4} - 1.32 \times 10^{-3}}{1.32 \times 10^{-3}} \times 100\% = -35\%$$

Note that the addition of 0.10 mmol of base to the strong acid solution causes a 4% decrease in hydronium ion concentration while the same addition to the weak acid solution results in a 35% decrease.

After the initial surge in pH, the weak acid solution enters the buffer region, in which the analytical concentrations of HA and A^- are both relatively large, and their ratio changes only slowly with additions of base. As we showed earlier, such solutions are remarkably resistant to changes in pH with additions of strong base. Thus, the curve in this region is relatively flat and above that for a strong acid until most of the weak acid has been consumed.

Additions of base when the titration is within 1 mL of the equivalence point cause relatively small changes in the concentration of A^-, which by now is large, but large changes in the concentrations of HA, which is small. The consequence is an acceleration in the rate of change of pH with respect to volume of added base.

At the equivalence point, the pH of the strong acid solution is 7.0 but that of the weak acid solution is higher because the conjugate base dissociates to give hydroxide ions. Slightly beyond equivalence, the excess base represses this dissociation so that the titration curves for the weak and strong acid become identical.

8E-2

The Derivation of Titration Curves

As noted earlier, four distinctly different types of calculations are employed in deriving a titration curve for a weak acid (or weak base). (1) At the outset, the solution contains only a weak acid or a weak base, and the pH is calculated from the concentration of that solute and the corresponding dissociation constant. (2) After various increments of titrant have been added (in quantities up to, but not including, an equivalent amount), the solution consists of a series of buffers; the pH of each can be calculated from the analytical concentrations of the conjugate base or acid and the residual concentrations of the weak acid or base. (3) At the equivalence point, the solution contains only the conjugate of the weak acid or base being titrated (that is, a salt), and the pH is calculated from the concentration of this product. (4) Beyond the equivalence point, the excess of strong acid or base titrant represses the basic or acidic character of the reaction product to such an extent that the pH is governed largely by the concentration of the excess titrant.

EXAMPLE 8-9

Derive a curve for the titration of 50.00 mL of 0.1000 M acetic acid ($K_a = 1.75 \times 10^{-5}$) with 0.1000 M sodium hydroxide.

Initial pH

At the outset, we must calculate the pH of a 0.1000 M solution of HOAc using Equation 5-17:

$$[H_3O^+] = \sqrt{c_{HOAc}K_a} = \sqrt{0.1000 \times 1.75 \times 10^{-5}} = 1.32 \times 10^{-3}$$

$$pH = -\log 1.32 \times 10^{-3} = 2.88$$

pH after addition of 10.00 mL of reagent

A buffer solution consisting of NaOAc and HOAc has now been produced. The analytical concentrations of the two constituents are

$$c_{HOAc} = \frac{50.00 \text{ mL} \times 0.1000 \text{ M} - 10.00 \text{ mL} \times 0.1000 \text{ M}}{60.00 \text{ mL}} = \frac{4.000}{60.00} \text{ M}$$

$$c_{NaOAc} = \frac{10.00 \text{ mL} \times 0.1000 \text{ M}}{60.00 \text{ mL}} = \frac{1.000}{60.00} \text{ M}$$

Upon substituting these concentrations into the dissociation-constant expression for acetic acid, we obtain

$$\frac{[H_3O^+](1.000/\cancel{60.00})}{4.000/\cancel{60.00}} = K_a = 1.75 \times 10^{-5}$$

$$[H_3O^+] = 7.00 \times 10^{-5}$$

$$pH = 4.16$$

Calculations similar to this delineate the curve throughout the buffer region. Data from such calculations are given in column 2 of Table 8-3.

Equivalence point pH

At the equivalence point, all the acetic acid has been converted to sodium acetate. The solution is therefore similar to one formed by dissolving that base in water, and the pH calculation is identical to that shown in Example 5-10 for a weak base. In the present example, the NaOAc concentration is 0.0500 M. Thus,

$$OAc^- + H_2O \rightleftharpoons HOAc + OH^-$$

$$[OH^-] = [HOAc]$$

$$[OAc^-] = 0.0500 - [OH^-] \simeq 0.0500$$

Substituting in the base-dissociation-constant expression for OAc^- gives

$$\frac{[OH^-]^2}{0.0500} \simeq \frac{K_w}{K_a} = \frac{1.00 \times 10^{-14}}{1.75 \times 10^{-5}} = 5.71 \times 10^{-10}$$

$$[OH^-] = \sqrt{0.0500 \times 5.71 \times 10^{-10}} = 5.34 \times 10^{-6}$$

$$pH = 14.00 - (-\log 5.34 \times 10^{-6}) = 8.73$$

pH after addition of 50.10 mL of base

After the addition of 50.10 mL of NaOH, both the excess base and the acetate ion contribute to the hydroxide ion concentration. The contribution of the latter is vanishingly small, however, because the excess of strong base represses the reaction. This fact becomes evident when we consider that the hydroxide ion concentration is only 5.34×10^{-6} M at the equivalence point; once an excess of strong base is added, the contribution from the reaction of the acetate is even smaller. Thus,

$$[OH^-] \simeq c_{NaOH} = \frac{50.10\text{ mL} \times 0.1000\text{ M} - 50.00\text{ mL} \times 0.1000\text{ M}}{100.1\text{ mL}} = 1.00 \times 10^{-4}\text{ M}$$

$$pH = 14.00 - (-\log 1.00 \times 10^{-4}) = 10.00$$

Note that the titration curve for a weak acid with a strong base becomes identical with that for a strong acid with a strong base in the region slightly beyond the equivalence point.

TABLE 8-3

Changes in pH During the Titration of a Weak Acid with a Strong Base

	50.00 mL of 0.1000 M HOAc with 0.1000 M NaOH	50.00 mL of 0.001000 M HOAc with 0.001000 M NaOH
Volume of NaOH, mL	**pH**	**pH**
0.00	2.88	3.91
10.00	4.16	4.30
25.00	4.76	4.80
40.00	5.36	5.38
49.00	6.45	6.46
49.90	7.46	7.47
50.00	8.73	7.73
50.10	10.00	8.09
51.00	11.00	9.00
60.00	11.96	9.96
75.00	12.30	10.30

It is also noteworthy that the analytical concentrations of acid and conjugate base are identical when an acid has been half neutralized (in Example 8-9, after the addition of exactly 25.00 mL of base). Thus, these terms cancel in the equilibrium-constant expression, and the hydronium ion concentration is numerically equal to the dissociation constant. Likewise, in the titration of a weak base, the hydroxide ion concentration is numerically equal to the dissociation constant of the base at the midpoint in the titration curve. In addition, the buffer capacities of each of the solutions is at a maximum at the midpoint of the titration.

8E-3 The Effect of Concentration

The second and third columns of Table 8-3 contain pH data for the titration of 0.1000 M and of 0.001000 M acetic acid with sodium hydroxide solutions of the same concentration. In deriving the data for the more dilute acid, none of the approximations shown in Example 8-9 were valid, and solution of a quadratic equation was necessary throughout.

Figure 8-3 is a plot of the data in Table 8-3. Note that the initial pH values are higher and the equivalence-point pH is lower for the more dilute solution. At intermediate titrant volumes, however, the pH values differ only slightly because of the buffering action of the acetic acid/sodium acetate system that exists in this region. Figure 8-3 is graphical confirmation of the fact that the pH of buffers is largely independent of dilution.

8E-4 The Effect of Reaction Completeness

Titration curves for 0.1000 M solutions of acids with different dissociation constants are shown in Figure 8-4. Note that the pH change in the equivalence-point region becomes smaller as the acid becomes weaker—that is, as the reaction between the acid and the base becomes less complete. The relation between completeness of reaction and reagent concentration illustrated by Figures 8-3 and 8-4 is analogous to these effects on precipitation titration curves (Section 7A-2).

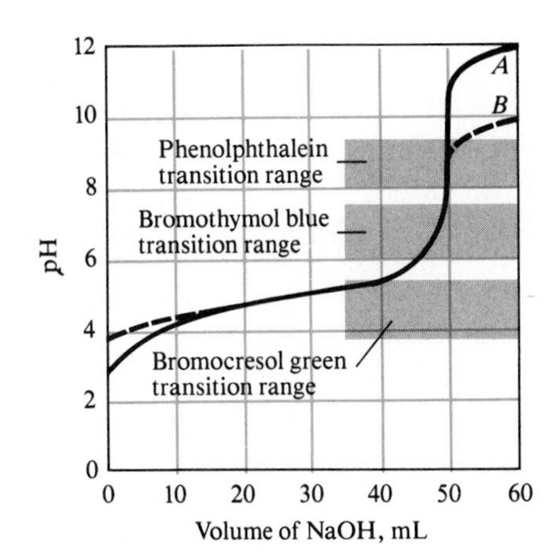

FIGURE 8-3 Curve for the titration of acetic acid with NaOH. *A*: 0.1000 M acid with 0.1000 M base. *B*: 0.001000 M acid with 0.001000 M base.

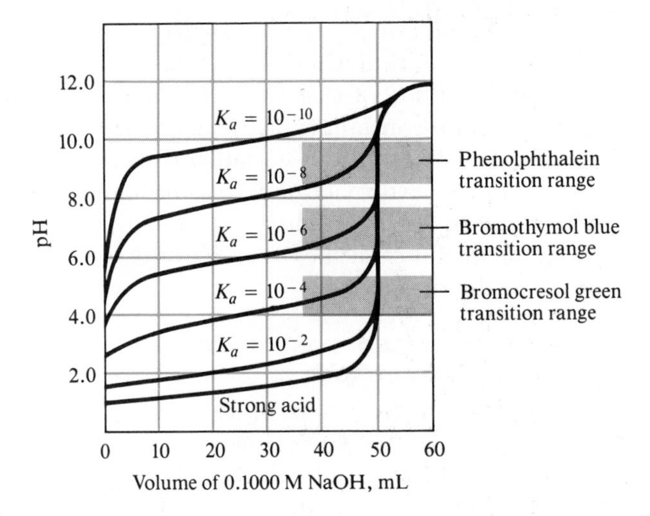

FIGURE 8-4 The effect of acid strength on titration curves. Each curve represents the titration of 50.0 mL of 0.1000 M acid with 0.1000 M NaOH.

8E-5 Indicator Choice: The Feasibility of Titration

Comparison of Figures 8-3 and 8-4 with Figure 8-1 shows clearly that the choice of indicator for the titration of a weak acid is more limited than that for a strong acid. For example, from Figure 8-3, it is obvious that bromocresol green is totally unsuited for titration of 0.1000 M acetic acid. Bromothymol blue is also unsatisfactory because its full color change occurs over a range from about 47 mL to 50 mL of 0.1000 M base. On the other hand, an indicator exhibiting a color change in the basic region, such as phenolphthalein, should provide a sharp end point with a minimal titration error.

The end-point pH change associated with the titration of 0.001000 M acetic acid (curve *B*, Figure 8-3) is so small that a significant titration error is likely to be introduced regardless of indicator. However, use of an indicator with a transition range between that of phenolphthalein and that of bromothymol blue in conjunction with a suitable color comparison standard makes it possible to establish the end point in this titration with a reproducibility of a few percent relative.

Figure 8-4 illustrates that similar problems exist as the strength of the acid being titrated decreases. Precision on the order of ± 2 ppt can be achieved in the titration of a 0.1000 M acid solution with a dissociation constant of 10^{-8} provided a suitable color comparison standard is available. With more concentrated solutions, somewhat weaker acids can be titrated with reasonable precision.

8F Titration Curves for Weak Bases

The derivation of a curve for the titration of a weak base is analogous to that of a weak acid.

EXAMPLE 8-10

A 50.00-mL aliquot of 0.0500 M NaCN is titrated with 0.1000 M HCl. The reaction is

$$CN^- + H_3O^+ \rightleftharpoons HCN + H_2O$$

Calculate the pH after the addition of (a) 0.00, (b) 10.00, (c) 25.00, and (d) 26.00 mL of acid.

(a) 0.00 mL of reagent:
The pH of a solution of NaCN can be derived by the method shown in Example 5-10:

$$CN^- + H_2O \rightleftharpoons HCN + OH^-$$

$$\frac{[OH^-][HCN]}{[CN^-]} = K_b = \frac{K_w}{K_a} = \frac{1.00 \times 10^{-14}}{2.1 \times 10^{-9}} = 4.76 \times 10^{-6}$$

$$[OH^-] = [HCN]$$

$$[CN^-] = c_{NaCN} - [OH^-] \approx c_{NaCN} = 0.0500$$

Substitution into the dissociation-constant expression gives, after rearrangement,

$$[OH^-] = \sqrt{K_b\, c_{NaCN}} = \sqrt{4.76 \times 10^{-6} \times 0.0500} = 4.88 \times 10^{-4}$$

$$pH = 14.00 - (-\log 4.88 \times 10^{-4}) = 10.69$$

(b) 10.00 mL of reagent:
Addition of acid produces a buffer with a composition given by

$$c_{NaCN} = \frac{50.00 \times 0.0500 - 10.00 \times 0.1000}{60.00} = \frac{1.500}{60.00}\ M$$

$$c_{HCN} = \frac{10.00 \times 0.1000}{60.00} = \frac{1.000}{60.00}\ M$$

These values are then substituted into the expression for the acid dissociation constant of HCN to give

$$[H_3O^+] = \frac{2.1 \times 10^{-9} \times (1.000/\cancel{60.00})}{1.500/\cancel{60.00}} = 1.4 \times 10^{-9}$$

$$pH = 8.85$$

(c) 25.00 mL of reagent:
This volume corresponds to the equivalence point, where the principal solute species is the weak acid HCN. Thus,

$$c_{HCN} = \frac{25.00 \times 0.1000}{75.00} = 0.03333\ M$$

Applying Equation 5-17 gives

$$[H_3O^+] = \sqrt{K_a\, c_{HCN}} = \sqrt{2.1 \times 10^{-9} \times 0.03333} = 8.37 \times 10^{-6}$$

$$pH = 5.08$$

(d) 26.00 mL of reagent:
The excess of strong acid now present represses the dissociation of the HCN to the point where its contribution to the pH is negligible. Thus,

$$[H_3O^+] = c_{HCl} = \frac{26.00 \times 0.1000 - 50.00 \times 0.0500}{76.00} = 1.32 \times 10^{-3}$$

$$pH = 2.88$$

Figure 8-5 shows theoretical curves for a series of weak bases of different strengths. Clearly, indicators with *acidic* transition ranges must be employed for weak bases.

8G Common Types of Acid/Base Indicators

Numerous organic compounds serve as indicators for neutralization titrations.

8G-1 Common Indicator Structures

The majority of acid/base indicators possess structural properties that permit classification into perhaps half a dozen categories.[6] Three of these classes are described in the following paragraphs.

Phthalein Indicators

Most phthalein indicators are colorless in moderately acidic solutions and exhibit a variety of colors in alkaline media. These colors tend to fade slowly in strongly alkaline solutions, which is an inconvenience in some applications. As a group, the phthaleins are sparingly soluble in water but readily dissolve in ethanol to give dilute solutions of the indicator.

FIGURE 8-5 The effect of base strength on titration curves. Each curve represents the titration of 50.0 mL of 0.1000 M base with 0.1000 M HCl.

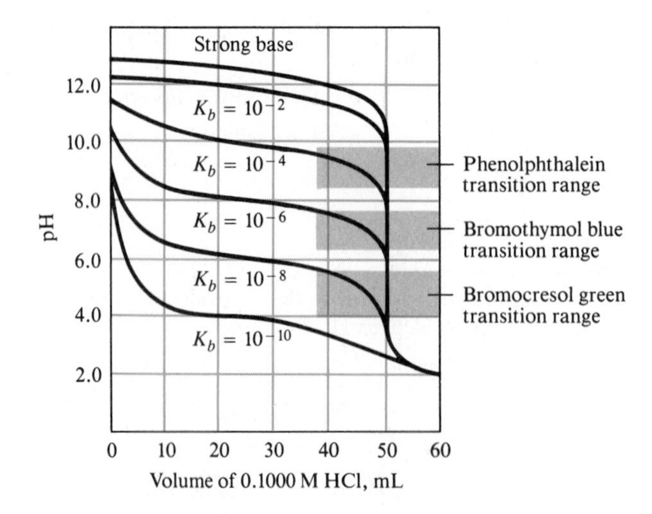

[6] See E. Banyai, in *Indicators*, E. Bishop, Ed., Chapter 3. New York: Pergamon, 1972.

The best-known phthalein indicator is *phenolphthalein*, whose structures can be represented as

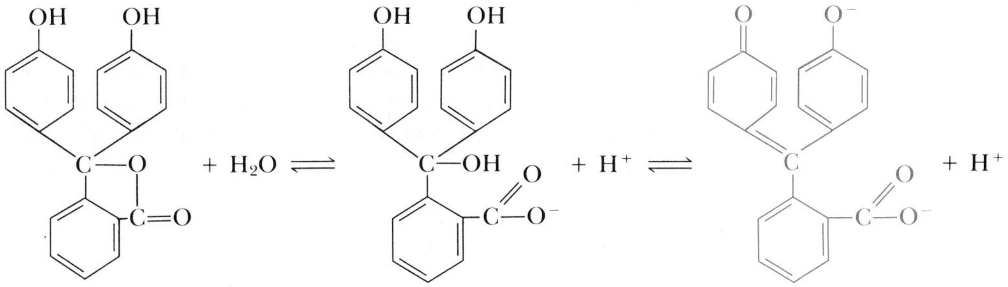

colorless $+ H_2O \rightleftharpoons$ colorless $+ H^+ \rightleftharpoons$ red $+ H^+$

Note that a quinoid ring, which imparts color to most organic compounds, is formed in the second reaction. The pH at which the pink color of this quinoid structure first becomes detectable depends on the concentration of the indicator and on the visual acuity of the observer. For most people, however, the pink appears in the pH range of 8.0 to 8.2.

The other phthalein indicators have various functional groups substituted on the phenolic rings. For example, *thymolphthalein* contains two alkyl groups on each ring. The structural alterations associated with the color change of this indicator are similar to those of phenolphthalein.

Sulfonphthalein Indicators

Many of the sulfonphthaleins exhibit two useful color-change ranges; one in somewhat acidic solutions and the other in neutral or moderately basic media. In contrast to the phthaleins, the basic color shows good stability toward strong alkali.

The sodium salts of the sulfonphthaleins are ordinarily used for the preparation of indicator solutions, owing to the appreciable acidity of the parent molecule. Solutions can be prepared directly from the sodium salt or indirectly by dissolving the sulfonphthalein in its acidic form in an appropriate volume of dilute aqueous sodium hydroxide.

The simplest sulfonphthalein indicator is *phenolsulfonphthalein*, known also as *phenol red*. The principal equilibria for a solution of the sodium salt of this compound are

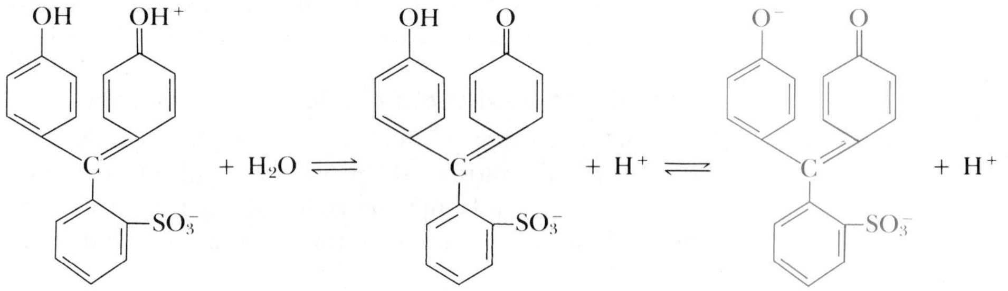

red $+ H_2O \rightleftharpoons$ yellow $+ H^+ \rightleftharpoons$ red $+ H^+$

Only the second color change, which occurs in the pH range between 6.4 and 8.0, is useful.

Substitution of halogens or alkyl groups for the hydrogens in the phenolic rings of the parent compound yields sulfonphthaleins that differ in color and pH range.

Azo Indicators

Most azo indicators exhibit a color change from red to yellow with increasing basicity; their transition ranges are generally on the acidic side of neutrality. The most commonly encountered examples are *methyl orange* and *methyl red*. The behavior of the former is described by the equations

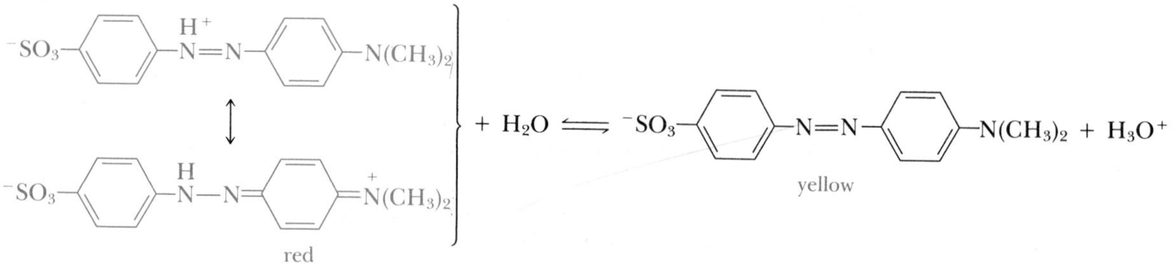

$$\left.\begin{matrix} {}^{-}SO_3-C_6H_4-\overset{H^+}{N}=N-C_6H_4-N(CH_3)_2 \\ \updownarrow \\ {}^{-}SO_3-C_6H_4-\overset{H}{N}-N=C_6H_4=\overset{+}{N}(CH_3)_2 \\ \text{red} \end{matrix}\right\} + H_2O \rightleftharpoons \underset{\text{yellow}}{{}^{-}SO_3-C_6H_4-N=N-C_6H_4-N(CH_3)_2} + H_3O^+$$

Methyl red contains a carboxylic acid group in place of the sulfonic acid group. Variations in the substituents on the amino nitrogen and in the rings give rise to a series of indicators with slightly different properties.

8G-2 Titration Errors with Acid/Base Indicators

Two types of titration errors are encountered in neutralization titrations. The first is a determinate error that occurs when the pH at which the indicator changes color differs from the pH at chemical equivalence. This type of error can usually be minimized by judicious indicator selection or by a blank correction.

The second type is an indeterminate error that originates from the limited ability of the eye to distinguish reproducibly the intermediate color of the indicator. The magnitude of this error depends on the change in pH per milliliter of reagent at the equivalence point, on the concentration of the indicator, and on the sensitivity of the eye to the two indicator colors. On the average, the visual uncertainty with an acid/base indicator is in the range of ± 0.5 to ± 1 pH unit. This uncertainty can often be decreased to as little as ± 0.1 pH unit by matching the color of the solution being titrated with that of a reference standard containing a similar amount of indicator at the appropriate pH. These uncertainties are of course approximations that vary considerably from indicator to indicator as well as from person to person.

8G-3 Variables That Influence the Behavior of Indicators

The pH interval over which a given indicator exhibits a color change is influenced by temperature, by the ionic strength of the medium, and by the presence of organic solvents and colloidal particles. Some of these effects, particularly the last two, can cause the transition range to shift by one or more pH units.[7]

[7]For a discussion of these effects, see H. A. Laitinen and W. E. Harris, *Chemical Analysis*, 2nd ed., pp. 48–51. New York: McGraw-Hill, 1975.

Questions and Problems

*8-1 Give a qualitative explanation of how the titration curves for 0.10 M NaOH and 0.1 M NH_3 differ from each other. Account for the difference.

8-2 What is a buffer solution and what are its properties?

*8-3 Why are buffers important in physiological fluids?

8-4 Define buffer capacity.

*8-5 Which has the greater buffer capacity, (a) a mixture containing 0.100 mol of NH_3 and 0.200 mol of NH_4Cl or (b) a mixture containing 0.0500 mol of NH_3 and 0.100 mol of NH_4Cl?

8-6 Would a buffer prepared by mixing 1 mol of acetic acid and 0.5 mol of NaOH differ from one prepared by mixing 1 mol of sodium acetate with 0.5 mol of HCl? Explain.

*8-7 Explain the fundamental difference between the equivalence point in a titration and the end point.

8-8 What variables can cause the pH range of an indicator to shift?

*8-9 Why are the standard reagents used in neutralization titrations generally strong acids and bases rather than weak acids and bases?

8-10 Before glass electrodes and pH meters became so widely used, pH was often determined by measuring the concentration of the acid and base forms of the indicator colorimetrically. If bromothymol blue is introduced into a solution and the concentration ratio of acid to base form is found to be 1.43, what is the pH of the solution?

*8-11 The procedure described in Question 8-10 was used to determine pH with methyl orange as the indicator. The concentration ratio of the acid to base form of the indicator was 1.64. Calculate the pH of the solution.

8-12 Which is expected to yield a sharper end point in a titration with 0.10 M HCl or 0.1 M NaOH:

*(a) 0.10 M NaOCl or 0.10 M hydroxylamine?

(b) 0.10 M anilinium hydrochloride ($C_6H_5NH_3^+Cl^-$) or benzoic acid?

*(c) 0.10 M NaOCl or 0.10 M hydrazine?

(d) 0.10 M sodium phenolate ($NaOC_6H_5$) or 0.100 M NH_3?

8-13 Consult Appendixes 4 and 5 and pick out a suitable acid/base pair to prepare a buffer of pH

*(a) 3.5. (b) 7.6. *(c) 9.3. (d) 5.1.

*8-14 Values for K_w at 0, 50, and 100°C are 1.14×10^{-15}, 5.47×10^{-14}, and 4.9×10^{-13}, respectively. Calculate the pH for a neutral solution at each of these temperatures.

8-15 What is the pH of a 1.00×10^{-2} M NaOH solution at 0°C?

*8-16 What is the pH of an aqueous solution that is 39.9% HCl by weight and has a density of 1.200 g/mL?

8-17 Calculate the pH of a solution that contains 1.00% (w/w) NaOH and has a density of 1.011 g/mL.

*8-18 Calculate the pH of the solution that results upon mixing 20.0 mL of 0.200M HCl with 25.0 mL of

(a) distilled water.
(b) 0.132 M $AgNO_3$.
(c) 0.132 M NaOH.
(d) 0.132 M NH_3.
(e) 0.232 M NaOH.

8-19 What is the pH of the solution that results when 0.102 g of $Mg(OH)_2$ is mixed with

(a) 75.0 mL of 0.0600 M HCl?
(b) 15.0 mL of 0.0600 M HCl?
(c) 30.0 mL of 0.0600 M HCl?
(d) 30.0 mL of 0.0600 M $MgCl_2$?

*8-20 What is the pH of a 2.0×10^{-7} M HCl solution?

8-21 What is the pH of a solution that is 0.0100 M in H_2SO_4?

*8-22 Calculate the hydronium ion concentration and pH of a solution that is 0.0500 M in HCl

(a) neglecting activities.
(b) using activities.

8-23 A solution is 0.050 M in NH_4Cl and 0.0300 M in NH_3. Calculate its OH^- concentration and its pH

(a) neglecting activities.
(b) taking activities into account.

*8-24 Calculate the pH of a solution prepared by

(a) dissolving 43.0 g of lactic acid in water and diluting to 500 mL.
(b) diluting 25.0 mL of the solution in (a) to 250 mL.
(c) diluting 10.0 mL of the solution in (b) to 1.00 L.

8-25 Calculate the pH of an iodic acid solution that is (a) 1.00×10^{-1} M, (b) 1.00×10^{-2} M, (c) 1.00×10^{-4} M.

*8-26 Calculate the pH of a solution prepared by

(a) dissolving 25.0 g of picric acid, $(NO_2)_3C_6H_2OH$, in 100 mL of water.
(b) diluting 10.0 mL of the solution in (a) to 100 mL.
(c) diluting 10.0 mL of the solution in (b) to 1.00 L.

*8-27 Calculate the pH of an ammonia solution that is (a) 1.00×10^{-1} M, (b) 1.00×10^{-2} M, (c) 1.00×10^{-4} M.

*8-28 Calculate the pH of an NH_4Cl solution that is (a) 1.00×10^{-1} M, (b) 1.00×10^{-2} M, (c) 1.00×10^{-4} M.

8-29 Calculate the pH of a HOCl solution that is (a) 1.00×10^{-1} M, (b) 1.00×10^{-2} M, (c) 1.00×10^{-4} M.

*8-30 Calculate the pH of a NaOCl solution that is (a) 1.00×10^{-1} M, (b) 1.00×10^{-2} M, (c) 1.00×10^{-4} M.

8-31 Calculate the pH of a solution in which the concentration of the piperidine is (a) 1.00×10^{-1} M, (b) 1.00×10^{-2} M, (c) 1.00×10^{-4} M.

*8-32 Calculate the pH of the solution that results when 20.0 mL of 0.200 M formic acid is

(a) diluted to 45.0 mL with distilled water.
(b) mixed with 25.0 mL of 0.160 M NaOH solution.

(c) mixed with 25.0 mL of 0.200 M NaOH solution.
(d) mixed with 25.0 mL of 0.200 sodium formate solution.

8-33 Calculate the pH of the solution that results when 40.0 mL of 0.100 M NH_3 is

(a) diluted to 60 mL with distilled water.
(b) mixed with 20.0 mL of 0.200 M HCl solution.
(c) mixed with 20.0 mL of 0.250 M HCl solution.
(d) mixed with 20.0 mL of 0.200 M NH_4Cl solution.
(e) mixed with 20.0 mL of 0.100 M HCl solution.

*8-34 What is the pH of a solution that is

(a) prepared by dissolving 9.20 g of lactic acid and 11.15 g of sodium lactate in water and diluting to 1.00 L?
(b) 0.055 M in acetic acid and 0.011 M in sodium acetate?
(c) prepared by dissolving 3.00 g of salicylic acid, $C_6H_4(OH)COOH$, in 50.0 mL of 0.1130 M NaOH and diluting to 500.0 mL?
(d) 0.010 M in picric acid and 0.100 M in sodium picrate?

8-35 What is the pH of a solution that is

(a) prepared by dissolving 3.30 g of $(NH_4)_2SO_4$ in water, adding 125.0 mL of 0.1011 M NaOH, and diluting to 500.0 mL?
(b) 0.120 M in piperidine and 0.080 M in its chloride salt?
(c) 0.050 M in ethylamine and 0.167 M in its chloride salt?
(d) prepared by dissolving 2.32 g of aniline in 100 mL of 0.0200 M HCl and diluting to 250.0 mL?

*8-36 Calculate the change in pH that occurs in each of the solutions listed below as a result of a tenfold dilution with water:

(a) H_2O.
(b) 0.0500 M HCl.
(c) 0.0500 M NaOH.
(d) 0.0500 M NH_3.
(e) 0.0500 M NH_4Cl.
(f) 0.0500 M NH_3 + 0.0500 M NH_4Cl.
(g) 0.500 M NH_3 + 0.500 M NH_4Cl.

*8-37 Calculate the change in pH that occurs when 1.00 mmol of a strong acid is added to 100 mL of the solutions listed in Problem 8-36.

*8-38 Calculate the change in pH that occurs when 1.00 mmol of a strong base is added to 100 mL of the solutions listed in Problem 8-36.

8-39 Calculate the change in pH that occurs when 0.50 mmol of a strong acid is added to 100 mL of

(a) 0.0200 M lactic acid + 0.0800 M sodium lactate.
(b) 0.0800 M lactic acid + 0.0200 M sodium lactate.
(c) 0.0500 M lactic acid + 0.0500 M sodium lactate.

*8-40 What weight of sodium formate must be added to 400 mL of 1.00 M formic acid to produce a buffer solution that has a pH of 3.50?

8-41 What weight of sodium glycolate should be added to 300 mL of 1.00 M glycolic acid to produce a buffer solution with a pH of 4.00?

*8-42 What volume of 0.200 M HCl must be added to 250 mL of 0.300 M sodium mandelate to produce a buffer solution with a pH of 3.37?

8-43 What volume of 2.00 M NaOH must be added to 300 mL of 1.00 M glycolic acid to produce a buffer solution having a pH of 4.00?

*8-44 In a titration of 50.00 mL of 0.05000 M formic acid with 0.1000 M KOH, the titration error must be smaller than ±0.05 mL. What indicator can be chosen to realize this goal?

8-45 In a titration of 50.00 mL of 0.1000 M ethylamine with 0.1000 M $HClO_4$, the titration error must be no more than ±0.05 mL. What indicator can be chosen to realize this goal?

*8-46 A 50.00-mL aliquot of 0.1000 M NaOH is titrated with 0.1000 M HCl. Calculate the pH of the solution after the addition of 0.00, 10.00, 25.00, 40.00, 45.00, 49.00, 50.00, 51.00, 55.00, and 60.00 mL of acid and prepare a titration curve from the data.

8-47 Calculate the pH after addition of 0.00, 5.00, 15.00, 25.00, 40.00, 45.00, 49.00, 50.00, 51.00, 55.00, and 60.00 mL of 0.1000 M NaOH in the titration of 50.00 mL of

*(a) 0.1000 M HNO_2.
(b) 0.1000 M lactic acid.
*(c) 0.1000 M pyridinium chloride.

8-48 Calculate the pH after addition of 0.00, 5.00, 15.00, 25.00, 40.00, 45.00, 49.00, 50.00, 51.00, 55.00, and 60.00 mL of 0.1000 M HCl in the titration of 50.00 mL of

*(a) 0.1000 M ammonia.
(b) 0.1000 M hydrazine.
(c) 0.1000 M sodium cyanide.

8-49 Calculate the pH after addition of 0.00, 5.00, 15.00, 25.00, 40.00, 49.00, 50.00, 51.00, 55.00, and 60.00 mL of reagent in the titration of

*(a) 0.1000 M anilinium chloride with 0.1000 M NaOH.
(b) 0.01000 M picric acid with 0.01000 M NaOH.
*(c) 0.1000 M hypochlorous acid with 0.1000 M NaOH.
(d) 0.1000 M hydroxylamine with 0.1000 M HCl.

Construct titration curves from the data.

CHAPTER 9

Titration Curves for Complex Acid/Base Systems

In this chapter, methods for deriving titration curves for complex acid/base systems are described. For the purpose of this discussion, complex systems are defined as solutions made up of (1) two acids or two bases of different strengths, (2) an acid or base that has two or more acidic or basic functional groups, or (3) an amphiprotic substance, which is capable of acting as both an acid and a base. Equations for more than one equilibrium are required to describe the characteristics of any of these systems.

9A Mixtures of Strong and Weak Acids or Strong and Weak Bases

To illustrate the derivation of titration curves for mixtures of a strong and a weak acid, consider the titration of a solution containing hydrochloric acid and a weak acid HA that has a dissociation constant of 1.00×10^{-4}. The molar hydronium ion concentration in the early stages of this titration is

$$[H_3O^+] = c_{HCl} + [A^-]$$

The terms on the right account for the contributions of the two solute acids to the hydronium ion concentration of the solution. A term for hydronium ions resulting from dissociation of water is omitted because it is vanishingly small. This assumption is surely valid for any system that contains an appreciable amount of either hydrochloric acid or HA.

Example 9-1 demonstrates that hydrochloric acid represses the dissociation of the weak acid in the early stages of the titration to such an extent that we can assume that $[A^-] \ll c_{HCl}$. The hydronium ion concentration is then simply the molar concentration of the strong acid.

EXAMPLE 9-1

Calculate the pH of a mixture that is 0.1200 M in hydrochloric acid and 0.0800 M in the weak acid HA ($K_a = 1.00 \times 10^{-4}$).

$$[H_3O^+] = 0.1200 + [A^-]$$

If we assume that $[A^-] \ll 0.1200$, then $[H_3O^+] \simeq 0.1200$ and the pH is 0.92. To check this assumption, the provisional value for $[H_3O^+]$ is substituted into the dissociation-constant expression for HA, which upon rearrangement gives

$$\frac{[A^-]}{[HA]} = \frac{K_a}{[H_3O^+]} = \frac{1.00 \times 10^{-4}}{0.1200} = 8.33 \times 10^{-4}$$

which can be rearranged to

$$[HA] = [A^-]/(8.33 \times 10^{-4})$$

From mass-balance considerations, we can write

$$0.0800 = c_{HA} = [HA] + [A^-]$$

Substituting the value of [HA] from the previous equation gives

$$0.0800 = [A^-]/(8.33 \times 10^{-4}) + [A^-] = (1.20 \times 10^3)[A^-]$$

$$[A^-] = 6.7 \times 10^{-5}$$

We see that $[A^-]$ is indeed much smaller than 0.1200 M, as assumed.

The approximation employed in Example 9-1 can be shown to apply until most of the hydrochloric acid has been neutralized by the titrant. Therefore, the curve in this region *is identical to the titration curve for a 0.1200 M solution of a strong acid by itself.*

As shown by the following example, the presence of HA must be taken into account as the first end point in the titration is approached.

EXAMPLE 9-2

Calculate the pH of the solution that results when 29.00 mL of 0.1000 M NaOH is added to 25.00 mL of the solution described in Example 9-1.

Here,

$$c_{HCl} = \frac{25.00 \times 0.1200 - 29.00 \times 0.1000}{54.00} = 1.85 \times 10^{-3}\ M$$

$$c_{HA} = \frac{25.00 \times 0.0800}{54.00} = 3.70 \times 10^{-2}\ M$$

A provisional result based (as in the previous example) on the assumption that $[H_3O^+] = 1.85 \times 10^{-3}$ yields a value of 1.90×10^{-3} for $[A^-]$. Clearly, $[A^-]$ is no longer much smaller than $[H_3O^+]$, and we must write

$$[H_3O^+] = c_{HCl} + [A^-] = 1.85 \times 10^{-3} + [A^-] \qquad (9\text{-}1)$$

In addition, from mass-balance considerations, we know that

$$[HA] + [A^-] = c_{HA} = 3.70 \times 10^{-2} \qquad (9\text{-}2)$$

We rearrange the acid dissociation constant expression for HA and obtain

$$[HA] = \frac{[H_3O^+][A^-]}{1.00 \times 10^{-4}}$$

Substitution of this expression into Equation 9-2 yields

$$\frac{[H_3O^+][A^-]}{1.00 \times 10^{-4}} + [A^-] = 3.70 \times 10^{-2}$$

$$[A^-] = \frac{3.70 \times 10^{-6}}{[H_3O^+] + 1.00 \times 10^{-4}}$$

Substitution for $[A^-]$ and c_{HCl} in Equation 9-1 yields

$$[H_3O^+] = 1.85 \times 10^{-3} + \frac{3.70 \times 10^{-6}}{[H_3O^+] + 1.00 \times 10^{-4}}$$

$$[H_3O^+]^2 + (1.00 \times 10^{-4})[H_3O^+] = (1.85 \times 10^{-3})[H_3O^+] + 1.85 \times 10^{-7} + 3.7 \times 10^{-6}$$

Collecting terms gives

$$[H_3O^+]^2 - (1.75 \times 10^{-3})[H_3O^+] - 3.885 \times 10^{-6} = 0$$

$$[H_3O^+] = 3.03 \times 10^{-3}$$

$$pH = 2.52$$

Note that the contributions to the hydronium ion concentration from HCl (1.85×10^{-3} M) and HA (3.03×10^{-3} M $- 1.85 \times 10^{-3}$ M) are of comparable magnitude.

When the amount of base added is equivalent to the amount of hydrochloric acid originally present, the solution is identical in all respects to one prepared by dissolving appropriate quantities of the weak acid and sodium chloride in a suitable amount of water. The sodium chloride, however, has no effect on the pH (neglecting the influence of increased ionic strength); thus, the remainder of the titration curve is identical to that for a dilute solution of HA.

The shape of the curve for a mixture of weak and strong acids, and hence the information obtainable from it, depend in large measure upon the strength of the weak acid. Figure 9-1 depicts the pH changes that occur during the titration of mixtures containing hydrochloric acid and several weak acids. Note that the rise in pH at the first equivalence point is small or essentially nonexistent when the weak acid has a relatively large dissociation constant (curves *A* and *B*). In cases such as these, only the total number of millimoles of weak and strong acid can be ascertained. Conversely, when the weak acid has a very small dissociation constant, only the strong acid content can be determined. For weak acids of intermediate strength (K_a somewhat less than 10^{-4} but greater than 10^{-8}), two useful end points usually exist.

The determination of the amount of each component in a mixture that contains a strong base and a weak base is also possible, subject to the constraints just described for the strong acid/weak acid system. The derivation of a curve for such a titration is analogous to that for a mixture of acids.

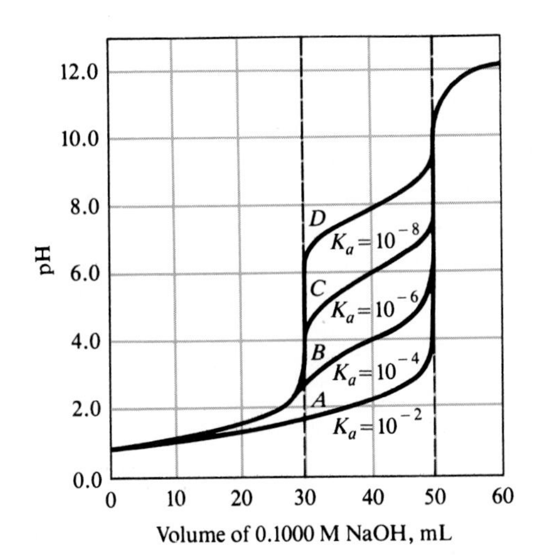

FIGURE 9-1 Curves for the titration of strong acid/weak acid mixtures with 0.1000 M NaOH. Each titration involves 25.00 mL of a solution that is 0.1200 M in HCl and 0.0800 M in HA.

9B

Equilibrium Calculations for Compounds with Multiple Acidic or Basic Functional Groups

In this section, consideration is given to pH calculations for solutions of compounds that contain two acidic functional groups, two basic functional groups, or an acidic and a basic functional group. In addition, the properties of buffer systems prepared from these compounds are described.

9B-1

Solutions of Polyprotic Acids

The dibasic weak acid H_2A dissociates in two steps:

$$H_2A + H_2O \rightleftharpoons H_3O^+ + HA^-$$

$$HA^- + H_2O \rightleftharpoons H_3O^+ + A^{2-}$$

Equilibrium constant expressions for the two reactions are:

$$\frac{[H_3O^+][HA^-]}{[H_2A]} = K_1 \qquad (9\text{-}3)$$

$$\frac{[H_3O^+][A^{2-}]}{[HA^-]} = K_2 \qquad (9\text{-}4)$$

Mass-balance considerations demand that

$$c_{H_2A} = [H_2A] + [HA^-] + [A^{2-}] \qquad (9\text{-}5)$$

and charge-balance requires that

$$[H_3O^+] \simeq [HA^-] + 2[A^{2-}] + \cancel{[OH^-]} \qquad (9\text{-}6)$$

Note that the concentration of hydroxide ion in the acidic solution is assumed to be negligible in Equation 9-6.

These four independent algebraic expressions can be solved to give $[H_3O^+]$ (as well as $[H_2A]$, $[HA^-]$, and $[A^{2-}]$). The mathematical manipulations required are somewhat awkward and tedious without the aid of a computer, however. Fortunately, simplification is possible provided the ratio K_1/K_2 is sufficiently large ($>10^3$). Under this circumstance, the hydronium ions produced in the first dissociation can be assumed to repress the inherently less favorable second dissociation to such an extent that the latter can be neglected. That is, the system is treated as if the second equilibrium does not exist and as if only the first equilibrium serves as a source of hydronium ions. In other words, we assume that $[A^{2-}]$ is much smaller than $[HA^-]$ and $[H_2A]$ in Equations 9-5 and 9-6 so that the two equations reduce to

$$c_{H_2A} \simeq [H_2A] + [HA^-] \qquad (9\text{-}7)$$

$$[H_3O^+] = [HA^-] \qquad (9\text{-}8)$$

Substitution of Equations 9-7 and 9-8 into Equation 9-3 yields

$$\frac{[H_3O^+]^2}{c_{H_2A} - [H_3O^+]} = K_1 \qquad (9\text{-}9)$$

Equation 9-9 is of course identical to the one used previously to calculate the hydronium ion concentration for a simple weak acid (Equation 5-15, page 115). As before, it can be solved either by the quadratic formula, by successive approximations, or by further assuming that $[H_3O^+]$ is small relative to c_{H_2A}.

The propriety of neglecting the contribution of the second dissociation to the hydronium ion concentration can be judged by substituting Equation 9-8 into Equation 9-4, which gives

$$[A^{2-}] = K_2$$

The trial values for $[H_3O^+]$, or $[HA^-]$, can then be compared with this approximate concentration for $[A^{2-}]$. In general, the second dissociation can be neglected except when K_1/K_2 is small or when the solution is very dilute. When the assumption is not valid, the four equations must be solved rigorously.

EXAMPLE 9-3

Calculate the pH of a 0.100 M maleic acid solution.

If we symbolize the acid by H_2M, we can write

$$H_2M + H_2O \rightleftharpoons H_3O^+ + HM^- \qquad \frac{[H_3O^+][HM^-]}{[H_2M]} = K_1 = 1.20 \times 10^{-2}$$

$$HM^- + H_2O \rightleftharpoons H_3O^+ + M^{2-} \qquad \frac{[H_3O^+][M^{2-}]}{[HM^-]} = K_2 = 5.96 \times 10^{-7}$$

Neglecting the effects of the second dissociation and employing Equation 9-9, we obtain

$$\frac{[H_3O^+]^2}{0.100 - [H_3O^+]} = K_1 = 1.20 \times 10^{-2}$$

The first dissociation constant for maleic acid is so large that the quadratic equation must be used to calculate $[H_3O^+]$:

$$[H_3O^+]^2 + (1.20 \times 10^{-2})[H_3O^+] - 1.20 \times 10^{-3} = 0$$

$$[H_3O^+] = 2.92 \times 10^{-2} = [HM^-]$$

$$pH = -\log (2.92 \times 10^{-2}) = 1.53$$

To check the assumption that $[M^{2-}]$ is much smaller than $[HM^-]$, we substitute the relationship $[H_3O^+] = [HM^-]$ into the expression for K_2:

$$\frac{[\cancel{H_3O^+}][M^{2-}]}{[\cancel{HM^-}]} = 5.96 \times 10^{-7}$$

$$[M^{2-}] = 5.96 \times 10^{-7}$$

Clearly, the approximation does not cause a significant error.

EXAMPLE 9-4

Calculate the hydronium ion concentration of a 0.0400 M H_2SO_4 solution.

One proton in H_2SO_4 is completely dissociated, but the other is not. If the dissociation of HSO_4^- is assumed to be negligible, it follows that

$$[H_3O^+] = [HSO_4^-] = 0.0400$$

However, an estimate of $[SO_4^{2-}]$ based upon this approximation and the expression for K_2 reveals that

$$\frac{[\cancel{H_3O^+}][SO_4^{2-}]}{[\cancel{HSO_4^-}]} = 1.20 \times 10^{-2}$$

Clearly, $[SO_4^{2-}]$ is *not* small relative to $[HSO_4^-]$, and a more rigorous solution is required.

From stoichiometric considerations, it is necessary that

$$[H_3O^+] = 0.0400 + [SO_4^{2-}]$$

The first term on the right is the concentration of H_3O^+ from dissociation of the H_2SO_4 to HSO_4^-. The second term is the contribution of the dissociation of HSO_4^-. Rearrangement yields

$$[SO_4^{2-}] = [H_3O^+] - 0.0400$$

Mass-balance considerations require that

$$c_{H_2SO_4} = 0.0400 = [HSO_4^-] + [SO_4^{2-}]$$

Combining the last two equations and rearranging yield

$$[HSO_4^-] = 0.0800 - [H_3O^+]$$

Introduction of these equations for $[SO_4^{2-}]$ and $[HSO_4^-]$ into the expression for K_2 yields

$$\frac{[H_3O^+]([H_3O^+] - 0.0400)}{0.0800 - [H_3O^+]} = 1.20 \times 10^{-2}$$

$$[H_3O^+]^2 - (0.0280)[H_3O^+] - 9.60 \times 10^{-4} = 0$$

$$[H_3O^+] = 0.0480$$

9B-2

Solutions of the Conjugate Bases of Polyprotic Acids

Calculation of the pH of a solution of the conjugate base of a polyprotic acid is analogous to that for the weak acid in Example 9-3.

EXAMPLE 9-5

Calculate the pH of a 0.100 M Na_2CO_3 solution.

The equilibria that must be considered are

$$CO_3^{2-} + H_2O \rightleftharpoons HCO_3^- + OH^- \qquad \frac{[HCO_3^-][OH^-]}{[CO_3^{2-}]} = \frac{K_w}{K_2} = 2.13 \times 10^{-4}$$

$$HCO_3^- + H_2O \rightleftharpoons H_2CO_3 + OH^- \qquad \frac{[H_2CO_3][OH^-]}{[HCO_3^-]} = \frac{K_w}{K_1} = 2.25 \times 10^{-8}$$

To obtain a provisional solution to the problem, we assume that $[OH^-]$ is determined by the first equilibrium only, in other words, that $[H_2CO_3] \ll [HCO_3^-]$ and $[CO_3^{2-}]$. The problem then reduces to one of determining the pH of a simple weak base. Thus,

$$[HCO_3^-] = [OH^-]$$

$$[CO_3^{2-}] = c_{Na_2CO_3} - [OH^-] \simeq c_{Na_2CO_3}$$

Note that we have further assumed that $[OH^-] \ll c_{Na_2CO_3}$ in the last expression. Substitution of these two equations into the first equilibrium-constant expression gives, upon rearrangement,

$$[OH^-]^2 = 2.13 \times 10^{-4} \times 0.100 = 2.13 \times 10^{-5}$$

$$[OH^-] = 4.62 \times 10^{-3}$$

$$pH = 14.00 - (-\log 4.62 \times 10^{-3}) = 11.66$$

To test the assumption that $[H_2CO_3]$ is negligible, we turn to the second equilibrium-constant expression and write

$$\frac{[H_2CO_3][OH^-]}{[HCO_3^-]} = 2.25 \times 10^{-8}$$

Clearly, $[H_2CO_3]$ is much smaller than the concentrations of the other two carbonate species. Furthermore, the assumption that $[CO_3^{2-}] \simeq c_{Na_2CO_3}$ introduces an error of about 2% in the hydrogen ion concentration.

9B-3

Buffer Solutions Involving Polyprotic Acids

Two buffer systems can be prepared from a weak dibasic acid and its salts. The first consists of free acid H_2A and its conjugate base NaHA, and the second makes use of the acid NaHA and its conjugate base Na_2A. The pH of the latter system is higher than that of the former because the acid dissociation constant for HA^- is always less than that for H_2A.

Sufficient independent equations are readily written to permit a rigorous evaluation of the hydronium ion concentration for either of these systems. Ordinarily, however, it is permissible to introduce the simplifying assumption that only one of the equilibria is important in determining the hydronium ion concentration of the solution. Thus, for a buffer prepared from H_2A and NaHA, the dissociation of HA^- to yield A^{2-} is neglected, and the calculation is based on the first dissociation only. With this simplification, the hydronium ion concentration is calculated by the method described in Section 8C-1 for a simple buffer solution. As before, it is an easy matter to check the validity of the assumption by calculating an approximate concentration of A^{2-} and comparing this value with the concentrations of H_2A and HA^-.

EXAMPLE 9-6

Calculate the hydronium ion concentration of a buffer solution that is 2.00 M in phosphoric acid and 1.50 M in potassium dihydrogen phosphate.

The principal equilibrium in this solution is the dissociation of H_3PO_4:

$$H_3PO_4 + H_2O \rightleftharpoons H_3O^+ + H_2PO_4^- \qquad \frac{[H_3O^+][H_2PO_4^-]}{[H_3PO_4]} = K_1 = 7.11 \times 10^{-3}$$

The dissociation of $H_2PO_4^-$ is assumed to be negligible, that is, $[HPO_4^{2-}]$ and $[PO_4^{3-}] \ll [H_2PO_4^-]$ or $[H_3PO_4]$. Then,

$$[H_3PO_4] \simeq c_{H_3PO_4} = 2.00$$

$$[H_2PO_4^-] \simeq c_{KH_2PO_4} = 1.50$$

$$[H_3O^+] = \frac{7.11 \times 10^{-3} \times 2.00}{1.50} = 9.48 \times 10^{-3}$$

We now use the equilibrium-constant expression for K_2 to show that $[HPO_4^{2-}]$ can be neglected:

$$\frac{[H_3O^+][HPO_4^{2-}]}{[H_2PO_4^-]} = 6.34 \times 10^{-8}$$

and our assumption is valid. Note that $[PO_4^{3-}]$ is even smaller than $[HPO_4^{2-}]$.

For a buffer prepared from NaHA and Na_2A, the second dissociation is assumed to predominate, and the reaction

$$HA^- + H_2O \rightleftharpoons H_2A + OH^-$$

is disregarded. The concentration of H_2A is ordinarily negligible compared with that of HA^- or A^{2-}; the hydronium ion concentration can then be calculated from the second dissociation constant, again employing the techniques for a simple buffer solution. To test the assumption, an estimate of the H_2A concentration is compared with the concentrations of HA^- and A^{2-}.

EXAMPLE 9-7

Calculate the hydronium ion concentration of a buffer that is 0.0500 M in potassium hydrogen phthalate (KHP) and 0.150 M in potassium phthalate (K_2P).

$$HP^- + H_2O \rightleftharpoons H_3O^+ + P^{2-} \qquad \frac{[H_3O^+][P^{2-}]}{[HP^-]} = K_2 = 3.91 \times 10^{-6}$$

Provided the concentration of H_2P in this solution is negligible,

$$[HP^-] \simeq c_{KHP} \simeq 0.0500$$

$$[P^{2-}] \simeq c_{K_2P} = 0.150$$

$$[H_3O^+] = \frac{3.91 \times 10^{-6} \times 0.0500}{0.150} = 1.30 \times 10^{-6}$$

To check the first assumption, an approximate value for $[H_2P]$ is calculated by substituting numerical values for $[H_3O^+]$ and $[HP^-]$ into the expression for K_1:

$$\frac{(1.30 \times 10^{-6})(0.0500)}{[H_2P]} = 1.12 \times 10^{-3}$$

$$[H_2P] = 6 \times 10^{-5}$$

This result justifies the assumption that $[H_2P] \ll [HP^-]$ and $[P^{2-}]$, that is, that the dissociation of HP^- as a base can be neglected.

In all but a few situations, the assumption of a single principal equilibrium, as invoked in Examples 9-6 and 9-7, provides a satisfactory estimate of the pH of buffer mixtures derived from polybasic acids. Appreciable errors occur, however, when the concentration of the acid or the salt is very low or when the two dissociation constants are numerically close to one another. A more laborious and rigorous calculation is then required.

9B-4 *Solutions of Amphiprotic Substances*

Many substances exhibit both acidic and basic character when dissolved in water. For example, both acidic and basic dissociation equilibria are established in a solution of the amphiprotic compound ammonium formate:

$$NH_4^+ + H_2O \rightleftharpoons H_3O^+ + NH_3 \qquad K_a = \frac{K_w}{K_b} = \frac{1.00 \times 10^{-14}}{1.76 \times 10^{-5}} = 5.68 \times 10^{-10}$$

$$A^- + H_2O \rightleftharpoons HA + OH^- \qquad K_b = \frac{K_w}{K_a} = \frac{1.00 \times 10^{-14}}{1.77 \times 10^{-4}} = 5.65 \times 10^{-11}$$

where A^- represents the formate anion. Note that a solution of ammonium formate is slightly acidic because the first reaction is somewhat more favorable than the second.

An acid salt of the type NaHA is another amphiprotic substance. Thus, its dissociation as an acid can be formulated as

$$HA^- + H_2O \rightleftharpoons A^{2-} + H_3O^+$$

whereas its reaction as a base is

$$HA^- + H_2O \rightleftharpoons H_2A + OH^-$$

One of these reactions produces hydronium ions and the other hydroxide ions. Whether a solution NaHA is acidic or basic is determined by the relative magnitude of the equilibrium constants for these processes:

$$K_2 = \frac{[H_3O^+][A^{2-}]}{[HA^-]} \qquad (9\text{-}10)$$

$$K_b = \frac{K_w}{K_1} = \frac{[H_2A][OH^-]}{[HA^-]} \qquad (9\text{-}11)$$

If K_b is greater than K_2, the solution is basic; otherwise, it is acidic.

A solution of NaHA can be described in terms of mass balance:

$$c_{\text{NaHA}} = [HA^-] + [H_2A] + [A^{2-}] \qquad (9\text{-}12)$$

and charge balance:

$$[Na^+] + [H_3O^+] = [HA^-] + 2[A^{2-}] + [OH^-]$$

Since the sodium ion concentration is equal to the molar concentration of the salt, the last equation can be rewritten as

$$c_{\text{NaHA}} = [HA^-] + 2[A^{2-}] + [OH^-] - [H_3O^+] \qquad (9\text{-}13)$$

One additional algebraic equation is needed to solve for the five unknowns rigorously. The ion-product constant for water serves this purpose:

$$K_w = [H_3O^+][OH^-]$$

The derivation of a rigorous expression for the hydronium ion concentration from these five equations is difficult. However, a reasonable approximation, applicable to solutions of most acid salts, can be obtained as follows.

Let us subtract Equation 9-13 from Equation 9-12, and rearrange to give

$$[H_2A] = [A^{2-}] + [OH^-] - [H_3O^+]$$

We can then express $[H_2A]$ and $[A^{2-}]$ in this equation in terms of $[HA^-]$ by substituting Equations 9-10 and 9-11:

$$\frac{K_w[HA^-]}{K_1[OH^-]} = \frac{K_2[HA^-]}{[H_3O^+]} + [OH^-] - [H_3O^+]$$

Replacement of $[OH^-]$ by the equivalent expression $K_w/[H_3O^+]$ gives

$$\frac{[H_3O^+][HA^-]}{K_1} = \frac{K_2[HA^-]}{[H_3O^+]} + \frac{K_w}{[H_3O^+]} - [H_3O^+]$$

Multiplication of both sides of this equation by $[H_3O^+]$ and rearrangement yield

$$[H_3O^+]^2\left(\frac{[HA^-]}{K_1} + 1\right) = K_2[HA^-] + K_w$$

Finally, this equation rearranges to

$$[H_3O^+] = \sqrt{\frac{K_2[HA^-] + K_w}{1 + [HA^-]/K_1}} \qquad (9\text{-}14)$$

Under most circumstances, it can be assumed that

$$[HA^-] \simeq c_{NaHA} \qquad (9\text{-}15)$$

Introduction of this relationship into Equation 9-14 gives

$$[H_3O^+] = \sqrt{\frac{K_2c_{NaHA} + K_w}{1 + c_{NaHA}/K_1}} \qquad (9\text{-}16)$$

It is important to understand that the approximation shown as Equation 9-15 requires that $[HA^-]$ be much larger than any of the other equilibrium concentrations in Equations 9-12 and 9-13. This assumption is not valid for very dilute solutions of NaHA or when K_2 or K_w/K_1 is relatively large.

Frequently, the ratio c_{NaHA}/K_1 is much larger than unity and K_2c_{NaHA} is considerably greater than K_w. With these assumptions, Equation 9-16 simplifies to

$$[H_3O^+] \simeq \sqrt{K_1K_2} \qquad (9\text{-}17)$$

Note that Equation 9-17 does not contain c_{NaHA}, which implies that the pH of solutions of this type remains constant over a considerable range of solute concentrations.

EXAMPLE 9-8

Calculate the hydronium ion concentration of a 0.100 M $NaHCO_3$ solution.

We first examine the assumptions leading to Equation 9-17. The dissociation constants for H_2CO_3 are $K_1 = 4.45 \times 10^{-7}$ and $K_2 = 4.7 \times 10^{-11}$. Clearly, c_{NaHA}/K_1 is much larger than unity; in addition, K_2c_{NaHA} has a value of 4.7×10^{-12}, which is substantially greater than K_w. Thus Equation 9-17 applies and

$$[H_3O^+] = \sqrt{4.45 \times 10^{-7} \times 4.7 \times 10^{-11}} = 4.6 \times 10^{-9}$$

EXAMPLE 9-9

Calculate the hydronium ion concentration of a 1.0×10^{-3} M Na_2HPO_4 solution.

The pertinent dissociation constants are K_2 and K_3, which both contain $[HPO_4^{2-}]$. Their values are $K_2 = 6.34 \times 10^{-8}$ and $K_3 = 4.2 \times 10^{-13}$. Considering again the assumptions implicit in Equation 9-17, we find that $(1.0 \times 10^{-3})/(6.34 \times 10^{-8})$ is large enough so that the denominator can again be simplified. The product $K_2c_{Na_2HPO_4}$ is by no means much larger than K_w, however. We therefore use a partially simplified version of Equation 9-16:

$$[H_3O^+] = \sqrt{\frac{4.2 \times 10^{-13} \times 1.0 \times 10^{-3} + 1.0 \times 10^{-14}}{(1.0 \times 10^{-3})/(6.34 \times 10^{-8})}} = 8.1 \times 10^{-10}$$

Use of Equation 9-17 yields a value of 1.6×10^{-10} M.

EXAMPLE 9-10

Find the hydronium ion concentration of a 0.0100 M NaH_2PO_4 solution.

The two dissociation constants of importance (those containing $[H_2PO_4^-]$) are $K_1 = 7.11 \times 10^{-3}$ and $K_2 = 6.34 \times 10^{-8}$. We see that the denominator of Equation 9-16 cannot be simplified, but the numerator reduces to $K_2 c_{NaH_2PO_4}$. Thus, Equation 9-16 becomes

$$[H_3O^+] = \sqrt{\frac{6.34 \times 10^{-8} \times 1.0 \times 10^{-2}}{1.00 + (1.0 \times 10^{-2})/(7.11 \times 10^{-3})}} = 1.6 \times 10^{-5}$$

9C Titration Curves for Polyprotic Acids and Their Conjugate Bases

Compounds with two or more acid or base functional groups yield multiple end points in a titration provided the functional groups differ sufficiently in strengths as acids or bases. The computational techniques described thus far permit the derivation of reasonably accurate theoretical titration curves for polyprotic acids or bases if the ratio K_1/K_2 is somewhat greater than 10^3. If the ratio is smaller, the error, particularly in the region of the first equivalence point, becomes excessive, and a more rigorous treatment of the equilibrium relationships is required.

9C-1 Titration Curves for Acids

In the example that follows, we derive data for obtaining the titration curve of maleic acid, a weak dibasic organic acid with the formula $C_2H_2(COOH)_2$. The two dissociation equilibria are

$$H_2M + H_2O \rightleftharpoons H_3O^+ + HM^- \qquad K_1 = 1.20 \times 10^{-2}$$

$$HM^- + H_2O \rightleftharpoons H_3O^+ + M^{2-} \qquad K_2 = 5.96 \times 10^{-7}$$

where H_2M symbolizes the free acid. Because the ratio K_1/K_2 is large (2×10^4), the second dissociation can be neglected when deriving points in the early part of the curve; that is, we assume that $[M^{2-}] \ll [HM^-]$ and $[H_2M]$ in this region. It can be shown that to within a few tenths of a milliliter of the first equivalence point, this assumption does not lead to serious error. Shortly beyond the first equivalence point, the second equilibrium is sufficiently dominant so that the basic reaction of HM^-,

$$HM^- + H_2O \rightleftharpoons OH^- + H_2M$$

does not significantly influence the pH. Here, we assume that $[H_2M] \ll [HM^-]$ and $[M^{2-}]$.

EXAMPLE 9-11

Derive a curve for the titration of 25.00 mL of 0.1000 M maleic acid with 0.1000 M NaOH.

Initial pH

Example 9-3 reveals that the initial pH is 1.53.

First buffer region

The addition of 5.00 mL of base results in the formation of a buffer consisting of the weak acid H_2M and its conjugate base HM^-. To the extent that dissociation of

HM^- to give M^{2-} is negligible, the solution can be treated as a simple buffer system. Thus applying Equations 8-10 and 8-11 (page 190) gives

$$c_{NaHM} \simeq [HM^-] = \frac{5.00 \times 0.1000}{30.00} = 1.67 \times 10^{-2}\ M$$

$$c_{H_2M} \simeq [H_2M] = \frac{25.00 \times 0.1000 - 5.00 \times 0.1000}{30.00} = 6.67 \times 10^{-2}\ M$$

Substitution of these values into the equilibrium-constant expression for K_1 yields a tentative value of 4.8×10^{-2} M for $[H_3O^+]$. It is clear, however, that the approximation $[H_3O^+] \ll c_{H_2M}$ or c_{HM^-} is not valid; therefore Equations 8-8 and 8-9 must be used, and

$$[HM^-] = 1.67 \times 10^{-2} + [H_3O^+] - [OH^-]$$

$$[H_2M] = 6.67 \times 10^{-2} - [H_3O^+] + [OH^-]$$

Because the solution is quite acidic, the approximation that $[OH^-]$ is very small is surely justified. Substitution of these expressions into the dissociation-constant relationship gives

$$\frac{[H_3O^+](1.67 \times 10^{-2} + [H_3O^+])}{6.67 \times 10^{-2} - [H_3O^+]} = 1.20 \times 10^{-2} = K_1$$

$$[H_3O^+]^2 + (2.87 \times 10^{-2})[H_3O^+] - 8.00 \times 10^{-4} = 0$$

$$[H_3O^+] = 1.74 \times 10^{-2}$$

$$pH = 1.76$$

Additional points in the first buffer region can be computed in a similar way.

First equivalence point

At the first equivalence point,

$$[HM^-] \simeq \frac{2.500}{50.00} = 5.00 \times 10^{-2}$$

Simplification of the numerator in Equation 9-16 is clearly justified. On the other hand, the concentration of HM^- is relatively close to the value of K_1. Hence,

$$[H_3O^+] \simeq \sqrt{\frac{K_2 c_{HM}}{1 + c_{HM^-}/K_1}} = \sqrt{\frac{5.96 \times 10^{-7} \times 5.00 \times 10^{-2}}{1 + (5.00 \times 10^{-2})/(1.20 \times 10^{-2})}} = 7.60 \times 10^{-5}$$

$$pH = 4.12$$

Second buffer region

Further additions of base to the solution create a new buffer system consisting of HM^- and M^{2-}. When enough base has been added so that the reaction of HM^- with water to give OH^- can be neglected (a few tenths of a milliliter beyond the first equivalence point), the pH of the mixture is readily obtained from K_2. With the introduction of 25.50 mL of NaOH, for example,

$$c_{Na_2M} \simeq \frac{(25.50 - 25.00)(0.1000)}{50.50} = \frac{0.050}{50.50}\ M$$

and the molar concentration of NaHM is

$$c_{NaHM} \simeq \frac{(25.00 \times 0.1000) - (25.50 - 25.00)(0.1000)}{50.50} = \frac{2.45}{50.50}\ M$$

Substituting these values into the expression for K_2 gives

$$\frac{[H_3O^+](0.050/\cancel{50.50})}{2.45/\cancel{50.50}} = 5.96 \times 10^{-7}$$

$$[H_3O^+] = 2.92 \times 10^{-5}$$

The assumption that $[H_3O^+]$ is small relative to the two molar concentrations is valid and pH = 4.54.

Second equivalence point

After the addition of 50.00 mL of 0.1000 M sodium hydroxide, the solution is 0.0333 M in Na_2M. Reaction of the base M^{2-} with water is the predominant equilibrium in the system and the only one that must be taken into account. Thus,

$$M^{2-} + H_2O \rightleftharpoons OH^- + HM^-$$

$$\frac{[OH^-][HM^-]}{[M^{2-}]} = \frac{K_w}{K_2} = \frac{1.00 \times 10^{-14}}{5.96 \times 10^{-7}} = 1.68 \times 10^{-8}$$

$$[OH^-] \simeq [HM^-]$$

$$[M^{2-}] = 0.0333 - [OH^-] \simeq 0.0333$$

$$\frac{[OH^-]^2}{0.0333} = \frac{1.00 \times 10^{-14}}{5.96 \times 10^{-7}}$$

$$[OH^-] = 2.36 \times 10^{-5}$$

$$pH = 14.00 - (-\log 2.36 \times 10^{-5}) = 9.37$$

pH beyond the second equivalence point

Further additions of sodium hydroxide repress the basic dissociation of M^{2-}. The pH is calculated from the concentration of NaOH added in excess of that required for the complete neutralization of H_2M.

Figure 9-2 is the titration curve for 0.1000 M maleic acid derived by the techniques shown in Example 9-11. Two end points are apparent, either of which could in principle be used as a measure of the concentration of the

FIGURE 9-2 Titration curve for 25.00 mL of 0.1000 M maleic acid, H_2M with 0.1000 M NaOH.

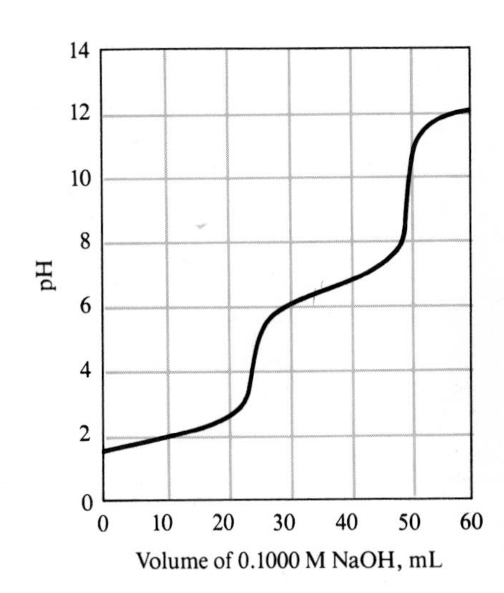

acid. The second end point is clearly more satisfactory, however, inasmuch as the pH change is more pronounced.

Figure 9-3 shows titration curves for three other polyprotic acids. These curves illustrate that a well-defined end point corresponding to the first equivalence point is observed only when the degree of dissociation of the two acids is sufficiently different. The ratio of K_1 to K_2 for oxalic acid (curve B) is approximately 1000. The curve for this titration shows an inflection corresponding to the first equivalence point. However, the magnitude of the pH change is too small to permit precise location of equivalence with an indicator. The second end point, however, provides a means for the accurate determination of the amount of oxalic acid.

Curve A is the theoretical titration curve for triprotic phosphoric acid. Here, the ratio K_1/K_2 is approximately 10^5, which is about 100 times greater than that for oxalic acid. As a result, two well-defined end points are observed, either of which is satisfactory for analytical purposes. If an indicator with an acidic transition range is used, 1 mol of base will be consumed per mole of acid. With an indicator exhibiting a color change in the basic region, 2 mol of base will be used. The third hydrogen of phosphoric acid is so slightly dissociated ($K_3 = 4.2 \times 10^{-13}$) that no practical end point is associated with its neutralization. The buffering effect of the third dissociation is noticeable, however, and causes the pH for curve A to be lower than the pH for the other two curves in the region beyond the second equivalence point.

Curve C is the titration curve for sulfuric acid, a substance that has one fully dissociated proton and one that is dissociated to a relatively large extent ($K = 1.2 \times 10^{-2}$). Because of the similarity in strengths of the two acids, only a single end point, corresponding to the titration of both protons, is observed.

In general, the titration of acids or bases that have two reactive groups yields individual end points that are of practical value only when the ratio between the two dissociation constants is at least 10^4. If the ratio is much smaller than this, the pH change at the first equivalence point will be too small for accurate detection—only the second end point will prove satisfactory for analysis.

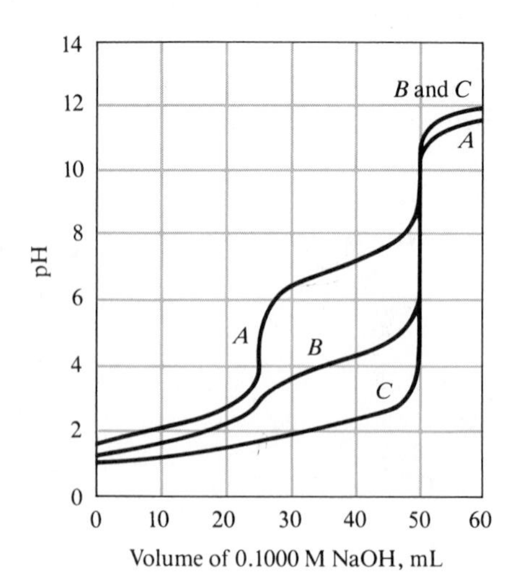

FIGURE 9-3 Curves for the titration of polybasic acids. A 0.1000 M NaOH solution is used to titrate 25.00 mL of 0.1000 M H_3PO_4 (A), 0.1000 M oxalic acid (B), and 0.1000 M H_2SO_4 (C).

9C-2

Titration Curves for Substances with Two Base Functional Groups

The derivation of a titration curve for a polyfunctional base involves no new principles. To illustrate, consider the titration of a sodium carbonate solution with standard hydrochloric acid. The important equilibrium constants are

$$CO_3^{2-} + H_2O \rightleftharpoons OH^- + HCO_3^- \qquad K_{b1} = \frac{K_w}{K_{a2}} = 2.13 \times 10^{-4}$$

$$HCO_3^- + H_2O \rightleftharpoons OH^- + H_2CO_3 \qquad K_{b2} = \frac{K_w}{K_{a1}} = 2.25 \times 10^{-8}$$

The reaction of carbonate ion with water governs the initial pH of the solution, which can be computed by the method shown in Example 9-5. With the first additions of acid, a carbonate/hydrogen carbonate buffer is established. In this region, the pH can be derived from *either* the hydroxide ion concentration calculated from K_{b1} *or* the hydronium ion concentration calculated from K_{a2}.

Sodium hydrogen carbonate is the principal solute species at the first equivalence point, and Equation 9-17 is used to compute the hydronium ion concentration. With the addition of more acid, a new buffer consisting of sodium hydrogen carbonate and carbonic acid is formed. The pH of this buffer is readily obtained from either K_{b2} or K_{a1}.

At the second equivalence point, the solution consists of carbonic acid and sodium chloride. The carbonic acid can be treated as a simple weak acid having a dissociation constant K_{a1}. Finally, when excess hydrochloric acid has been introduced, the dissociation of the weak acid is repressed to a point where the hydronium ion concentration is essentially that of the molar concentration of the strong acid.

Figure 9-4 illustrates that two end points are observed in the titration of sodium carbonate, the second being appreciably sharper than the first. It is apparent that the individual components in mixtures of sodium carbonate and sodium hydrogen carbonate can be determined by neutralization methods. Thus, the number of millimoles of acid required to reach a phenolphthalein end point gives the number of millimoles of carbonate originally present in the sample. In contrast, the number of millimoles of acid consumed in a

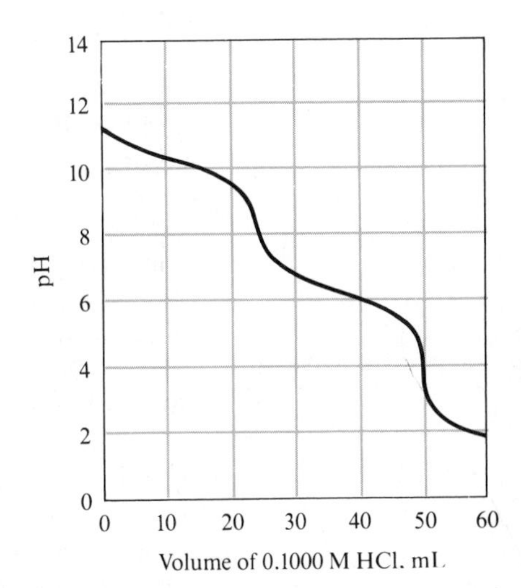

FIGURE 9-4 Curve for the titration of 25.00 mL of 0.1000 M Na_2CO_3 with 0.1000 M HCl.

titration to the acidic end point equals twice the number of millimoles of carbonate plus the number of millimoles of bicarbonate in the sample; that is,

$$\text{no. mmol acid} = 2 \times \text{no. mmol } CO_3^{2-} + \text{no. mmol } HCO_3^-$$

9C-3

Titration Curves for Amphiprotic Species

As noted earlier, an amphiprotic substance, when dissolved in a suitable solvent, behaves both as a weak acid and as a weak base. If either its acidic or its basic character predominates sufficiently, titration of the species with a strong base or a strong acid may be feasible. For example, in sodium dihydrogen phosphate solution, the following equilibria exist:

$$H_2PO_4^- + H_2O \rightleftharpoons H_3O^+ + HPO_4^- \qquad K_{a2} = 6.34 \times 10^{-8}$$

$$H_2PO_4^- + H_2O \rightleftharpoons OH^- + H_3PO_4 \qquad K_{b3} = \frac{K_w}{K_{a1}} = 1.41 \times 10^{-12}$$

Note that K_{b3} is much too small to permit titration of $H_2PO_4^-$ with an acid, but K_{a2} is large enough for a successful titration of the ion with a standard base solution.

A different situation prevails in solutions containing disodium hydrogen phosphate, for which the analogous equilibria are

$$HPO_4^{2-} + H_2O \rightleftharpoons H_3O^+ + PO_4^{3-} \qquad K_{a3} = 4.2 \times 10^{-13}$$

$$HPO_4^{2-} + H_2O \rightleftharpoons OH^- + H_2PO_4^- \qquad K_{b2} = \frac{K_w}{K_{a2}} = 1.58 \times 10^{-7}$$

The magnitude of the constants indicates that HPO_4^{2-} can be titrated with standard acid but not with standard base.

The simple amino acids are an important class of amphiprotic compounds that contain both a weak acid and a weak base functional group. In an aqueous solution of a typical amino acid, such as glycine, three important equilibria operate:

$$NH_2CH_2COOH \rightleftharpoons NH_3^+CH_2COO^- \tag{9-18}$$

$$NH_3^+CH_2COO^- + H_2O \rightleftharpoons NH_2CH_2COO^- + H_3O^+ \quad K_a = 2 \times 10^{-10} \tag{9-19}$$

$$NH_3^+CH_2COO^- + H_2O \rightleftharpoons NH_3^+CH_2COOH + OH^- \quad K_b = 2 \times 10^{-12} \tag{9-20}$$

The first reaction constitutes a kind of internal acid/base reaction and is analogous to the reaction one would observe between a carboxylic acid and an amine:

$$R_1NH_2 + R_2COOH \rightleftharpoons R_1NH_3^+ + R_2COO^- \tag{9-21}$$

The typical aliphatic amine has a base dissociation constant of 10^{-4} to 10^{-5} (Appendix 5), while many carboxylic acids have acid dissociation constants of about the same magnitude. The consequence is that both reaction 9-18 and reaction 9-21 proceed far to the right, with the product or products being the predominant species in the solution.

The amino acid species in Equation 9-18, bearing both a positive and a negative charge, is called a *zwitterion.* As shown by Equations 9-19 and 9-20, the zwitterion of glycine is slightly stronger as an acid than as a base. Thus, an aqueous solution of glycine is slightly acidic.

The zwitterion of an amino acid, containing as it does a positive and a negative charge, has no tendency to migrate to an electric field, whereas the singly charged anionic and cationic species are attracted to electrodes of opposite charge. No *net* migration of the amino acid occurs in an electric field when the pH of the solvent is such that the concentrations of the anionic and cationic forms are identical. The pH at which no net migration occurs is called the *isoelectric point* and is an important physical constant for characterizing amino acids. The isoelectric point is readily related to the ionization constants for the species. Thus, for glycine

$$\frac{[H_3O^+][NH_2CH_2COO^-]}{[NH_3^+CH_3COO^-]} = K_a$$

$$\frac{[OH^-][NH_3^+CH_2COOH]}{[NH_3^+CH_2COO^-]} = K_b$$

At the isoelectric point,

$$[NH_2CH_2COO^-] = [NH_3^+CH_2COOH]$$

Thus, division of K_a by K_b gives

$$\frac{[H_3O^+][\cancel{NH_2CH_2COO^-}]}{[OH^-][\cancel{NH_3^+CH_2COOH}]} = \frac{[H_3O^+]}{[OH^-]} = \frac{K_a}{K_b}$$

Substitution of $K_w/[H_3O^+]$ for $[OH^-]$ and rearrangement yield

$$[H_3O^+] = \sqrt{\frac{K_aK_w}{K_b}}$$

The isoelectric point for glycine occurs at a pH of 6.0. That is,

$$[H_3O^+] = \left(\frac{2 \times 10^{-10}}{2 \times 10^{-12}} \times 1 \times 10^{-14}\right)^{1/2} = 1 \times 10^{-6}$$

For simple amino acids, K_a and K_b are generally so small that their determination by direct neutralization titration is impossible. Addition of formaldehyde removes the amine functional group, however, and leaves the carboxylic acid available for titration with a standard base. For example, with glycine,

$$NH_3^+CH_2COO^- + CH_2O \longrightarrow CH_2{=}NCH_2COOH + H_2O$$

The titration curve for the product is that of a typical carboxylic acid.

9D The Composition of a Solution of a Polyprotic Acid as a Function of pH

In order to visualize the compositional changes that occur in the course of a titration involving a polyprotic acid, it is instructive to plot the *relative* concentration of the free acid as well as that of each of its anions as a function of solution pH. These relative concentrations are called alpha values. For example, if we let c_T be the sum of the molar concentrations of the maleate-containing species in the solution throughout the titration described in Example 9-11, the alpha value for the free acid α_0 is defined as

$$\alpha_0 = \frac{[H_2M]}{c_T}$$

where

$$c_T = [H_2M] + [HM^-] + [M^{2-}]$$

The alpha values for HM^- and M^{2-} are given by similar equations:

$$\alpha_1 = \frac{[HM^-]}{c_T}$$

$$\alpha_2 = \frac{[M^{2-}]}{c_T}$$

Alpha values are unitless ratios of the concentration of individual species to the total analytical concentration of all the related species. By the nature of their definitions, the sum of the alpha values for a system must equal unity:

$$\alpha_0 + \alpha_1 + \alpha_2 = 1$$

The alpha values for the maleic acid system are readily expressed in terms of $[H_3O^+]$, K_1, and K_2 by rearranging the dissociation-constant expressions to give

$$[HM^-] = \frac{K_2[H_2M]}{[H_3O^+]}$$

$$[M^{2-}] = \frac{K_1K_2[H_2M]}{[H_3O^+]^2}$$

After substituting these quantities, the mass-balance equation becomes

$$c_T = [H_2M] + \frac{K_1[H_2M]}{[H_3O^+]} + \frac{K_1K_2[H_2M]}{[H_3O^+]^2}$$

which can be converted to

$$[H_2M] = \frac{c_T[H_3O^+]^2}{[H_3O^+]^2 + K_1[H_3O^+] + K_1K_2}$$

Substituting this value for $[H_2M]$ into the equation defining α_0 gives

$$\alpha_0 = \frac{[H_3O^+]^2}{[H_3O^+]^2 + K_1[H_3O^+] + K_1K_2} \quad (9\text{-}22)$$

By similar manipulations, it is easily shown that

$$\alpha_1 = \frac{K_1[H_3O^+]}{[H_3O^+]^2 + K_1[H_3O^+] + K_1K_2} \quad (9\text{-}23)$$

$$\alpha_2 = \frac{K_1K_2}{[H_3O^+]^2 + K_1[H_3O^+] + K_1K_2} \quad (9\text{-}24)$$

Note that the denominator is the same for each expression, that the numerator for α_0 is the first term in the denominator, and that each successive alpha value has as its numerator the next term from the denominator. Generation of expressions for alpha values is thus a simple matter,[1] as is the calculation

[1]For the weak acid H_nA, the denominator takes the form:

$$[H^+] + K_1[H^+]^{(n-1)} + K_1K_2[H^+]^{(n-2)} + \cdots K_1K_2 \cdots K_n$$

Alpha values for polyfunctional bases are generated in an analogous way, with the equations being written in terms of base dissociation constants and $[OH^-]$.

FIGURE 9-5

Composition of H_2M solutions as a function of pH.

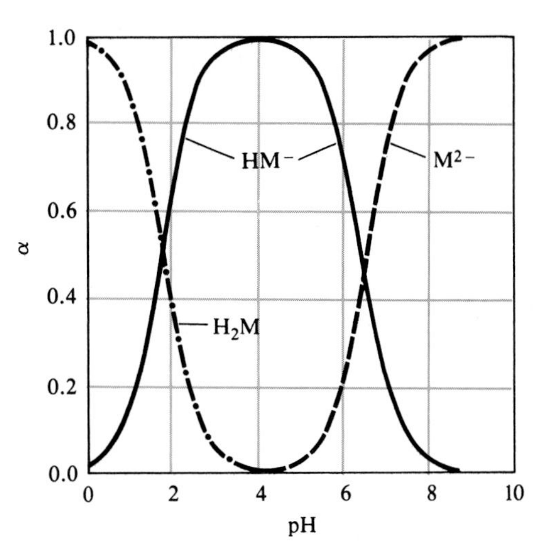

of their numerical values. Note that the fractional amount of each species is fixed at any pH and is *independent* of the total concentration, c_T.

The three curves plotted in Figure 9-5 show the alpha value for each maleate-containing species as a function of pH. In contrast, the solid curves in Figure 9-6 depict the same alpha values but now plotted as a function of volume of sodium hydroxide as the acid is titrated. The titration curve also appears as a dashed line in the latter figure. Consideration of these curves gives a clear picture of all concentration changes that occur during the titration. For example, Figure 9-6 reveals that before the addition of any base, α_0 for H_2M is roughly 0.7 and α_1 for HM^- is approximately 0.3. For all practical

FIGURE 9-6

Titration of 25.00 mL of 0.1000 M maleic acid with 0.1000 M NaOH. Solid curves are plots of alpha values versus volume. Broken curve is a plot of pH versus volume.

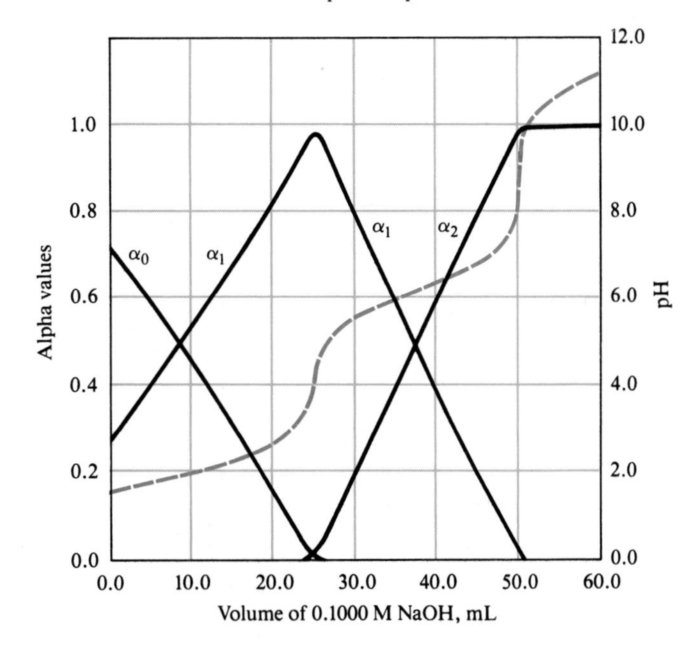

purposes, α_2 is zero. Thus, approximately 70% of the maleic acid exists as H_2M and 30% as HM^-. With addition of base, the pH rises, as does the fraction of HM^-. At the first equivalence point (pH = 4.12), essentially all of the maleate is present as HM^- ($\alpha_1 \rightarrow 1$). Beyond the first equivalence point, HM^- decreases and M^{2-} increases. At the second equivalence point (pH = 9.37) and beyond, essentially all of the maleate exists as M^{2-}.

Questions and Problems

*9-1 What is an amphiprotic species?

9-2 What is a zwitterion?

9-3 Indicate whether an aqueous solution of the following compounds is acidic, neutral, or basic:

*(a) NH_4OAc.
(b) NH_4NO_2.
*(c) $(NH_4)_2C_2O_4$.
(d) $NaHC_2O_4$.
*(e) $Na_2C_2O_4$.
(f) Na_2HCit (where Cit^{3-} is the citrate anion)
*(g) NaH_2Cit.
(h) Na_3Cit.

9-4 Suggest an indicator that could be used to provide an end point for the titration of the first proton in H_3AsO_4.

*9-5 Suggest an indicator that would give an end point when two of the protons in H_3AsO_4 have been titrated.

9-6 Suggest a method for the determination of the amounts of H_3PO_4 and NaH_2PO_4 in an aqueous solution.

*9-7 Give equations that define α_0, α_1, α_2, and α_3 for the acid H_3AsO_4.

9-8 Formulate equilibrium constants for the following equilibria, giving numerical values for the constants:

*(a) $H_2AsO_4^- + H_2AsO_4^- \rightleftharpoons H_3AsO_4 + HAsO_4^{2-}$
(b) $HAsO_4^{2-} + HAsO_4^{2-} \rightleftharpoons AsO_4^{3-} + H_2AsO_4^-$

*9-9 Derive a numerical value for the equilibrium constant for the reaction:

$NH_4^+ + OAc^- \rightleftharpoons NH_3 + HOAc$

9-10 Suggest a suitable indicator for a titration based upon the following reactions:

*(a) $H_2CO_3 + NaOH \longrightarrow NaHCO_3 + H_2O$
(b) $H_2P + 2NaOH \longrightarrow Na_2P + 2H_2O$ (H_2P = *o*-phthalic acid)
*(c) $H_2T + 2NaOH \longrightarrow Na_2T + 2H_2O$ (H_2T = tartaric acid)
(d) $NH_2C_2H_4NH_2 + HCl \longrightarrow NH_2C_2H_4NH_3Cl$
*(e) $NH_2C_2H_4NH_2 + 2HCl \longrightarrow ClNH_3C_2H_4NH_3Cl$
(f) $H_2SO_3 + NaOH \longrightarrow NaHSO_3 + H_2O$
*(g) $H_2SO_3 + 2NaOH \longrightarrow Na_2SO_3 + 2H_2O$

*9-11 Calculate the pH of a solution that is 0.0600 M in

(a) hydrogen sulfide.
(b) sulfuric acid.
(c) malonic acid.
(d) sodium sulfide.
(e) ethylenediamine.
(f) sodium oxalate.

9-12 Calculate the pH of a solution that is 0.0600 M in

(a) phosphoric acid.
(b) oxalic acid.
(c) phosphorous acid.
(d) sodium sulfite.
(e) trisodium phosphate.
(f) sodium sulfate.

*9-13 Calculate the pH of a solution that is 0.0400 M in
(a) sodium hydrogen sulfide.
(b) sodium hydrogen oxalate.
(c) sodium hydrogen sulfite.
(d) ethylenediamine hydrochloride ($NH_2C_2H_4NH_3Cl$).

9-14 Calculate the pH of a solution that is 0.0400 M in
(a) sodium hydrogen fumarate.
(b) sodium hydrogen sulfate.
(c) disodium hydrogen arsenate.
(d) sodium dihydrogen arsenate.

*9-15 Calculate the pH of a solution that is
(a) 0.0100 M in HCl and 0.0200 M in picric acid.
(b) 0.0100 M in HCl and 0.0200 M in benzoic acid.
(c) 0.0100 M in NaOH and 0.100 M in Na_2CO_3.
(d) 0.0100 M in NaOH and 0.100 M in NH_3.

9-16 Calculate the pH of a solution that is
(a) 0.0100 M in $HClO_4$ and 0.0300 M in monochloroacetic acid.
(b) 0.0100 M in HCl and 0.0150 M in H_2SO_4.
(c) 0.0100 M in NaOH and 0.0300 M in Na_2S.
(d) 0.0100 M in NaOH and 0.0300 M in sodium acetate.

*9-17 Identify the principal conjugate acid/base pair and calculate the ratio between them in a solution that is buffered to pH 6.00 and contains
(a) H_2SO_3.
(b) citric acid.
(c) malonic acid.
(d) tartaric acid.

9-18 Identify the principal conjugate acid/base pair and calculate the ratio between them in a solution that is buffered to pH 9.00 and contains
(a) H_2S.
(b) ethylenediamine dihydrochloride.
(c) H_3AsO_4.
(d) H_2CO_3.

*9-19 Calculate the pH of a solution made up to contain the following analytical concentrations:
(a) 0.0500 M in H_3AsO_4 and 0.0200 M in NaH_2AsO_4.
(b) 0.0300 M in NaH_2AsO_4 and 0.0500 M in Na_2HAsO_4.
(c) 0.0600 M in Na_2CO_3 and 0.0300 M in $NaHCO_3$.
(d) 0.0400 M in H_3PO_4 and 0.0200 M in Na_2HPO_4.
(e) 0.0500 M in $NaHSO_4$ and 0.0400 M in Na_2SO_4.

9-20 Calculate the pH of a solution made up to contain the following analytical concentrations:
(a) 0.240 M in H_3PO_3 and 0.480 M in NaH_2PO_3.
(b) 0.0670 M in Na_2SO_3 and 0.0315 M in $NaHSO_3$.
(c) 0.640 M in $HOC_2H_4NH_2$ and 0.750 M in $HOC_2H_4NH_3Cl$.
(d) 0.240 M in $H_2C_2O_4$ (oxalic acid) and 0.360 M in $Na_2C_2O_4$.
(e) 0.0100 M in $Na_2C_2O_4$ and 0.0400 M in $NaHC_2O_4$.

*9-21 What is the pH of the buffer formed when 50.0 mL of 0.200 M NaH_2PO_4 is mixed with
(a) 50.0 mL of 0.120 M HCl?
(b) 50.0 mL of 0.120 M NaOH?

9-22 What is the pH of the buffer formed by adding 100 mL of 0.150 M potassium hydrogen phthalate to
(a) 100 mL of 0.0800 M NaOH?
(b) 100 mL of 0.0800 M HCl?

*9-23 Describe the preparation of 1.00 L of a buffer of pH 9.60 from 0.300 M Na_2CO_3 and 0.200 M HCl.

9-24 Describe how to prepare 1.00 L of a buffer of pH 7.00 from 0.200 M H_3PO_4 and 0.160 M NaOH.

9-25 Describe the preparation of 1.00 L of a buffer of pH 6.00 from 0.500 M Na_3AsO_4 and 0.400 M HCl.

*9-26 How many grams of $Na_2HPO_4 \cdot 2H_2O$ must be added to 400 mL of 0.200 M H_3PO_4 to give a buffer of pH 7.30?

9-27 How many grams of dipotassium phthalate must be added to 750 mL of 0.0500 M phthalic acid to give a buffer of pH 5.75?

9-28 Derive a curve for the titration of 50.00 mL of a 0.1000 M solution of compound A with a 0.2000 M solution of compound B in the following list. For each titration, calculate the pH after the addition of 0.00, 12.50, 20.00, 24.00, 25.00, 26.00, 37.50, 45.00, 49.00, 50.00, 51.00, and 60.00 mL of compound B:

	A	B
*(a)	Na_2CO_3	HCl
(b)	ethylenediamine	HCl
*(c)	H_2SO_4	NaOH
(d)	$H_2C_2O_4$	NaOH

*9-29 Generate a titration curve for 50.00 mL of a solution that has a NaOH analytical concentration of 0.1000 M and a hydrazine analytical concentration of 0.08000 M. Calculate the pH after addition of 0.00, 10.00, 20.00, 24.00, 25.00, 26.00, 35.00, 44.00, 45.00, 46.00, and 50.00 mL of 0.2000 M $HClO_4$.

9-30 Generate a titration curve for 50.00 mL of a solution that has an $HClO_4$ analytical concentration of 0.1000 M and a formic acid analytical concentration of 0.08000 M. Calculate the pH after addition of 0.00, 10.00, 20.00, 24.00, 25.00, 26.00, 35.00, 44.00, 45.00, 46.00, and 50.00 mL of 0.2000 M KOH.

9-31 For pH values of 2.00, 6.00, and 10.00, calculate the alpha value for each species in an aqueous solution of

*(a) phthalic acid.
(b) phosphoric acid.
*(c) citric acid.
(d) arsenic acid.
*(e) phosphorous acid.
(f) oxalic acid.

CHAPTER 10

Applications of Neutralization Titrations

Neutralization titrations are widely used for determining the concentration of analytes that react directly or indirectly with a solvent to form hydrogen or hydroxide ions. For most applications, water is chosen as the solvent because of its ready availability, low cost, and low toxicity. Some analytes, however, are not titratable in aqueous media because their solubilities are too low or because their strengths as acids or bases are not sufficiently great to provide satisfactory end points. The concentration of such substances can often be determined by substituting a suitable nonaqueous solvent. Nonaqueous titrations are discussed briefly at the end of this chapter.

10A Reagents for Neutralization Reactions

In Chapter 8, it was shown that in a neutralization titration, strong acids and bases cause the most pronounced change in pH at the equivalence point. For this reason, standard solutions for such titrations are always prepared from strong acids or strong bases.

10A-1 Preparation of Standard Acid Solutions

Hydrochloric acid is the most commonly used standard acid for the titration of bases. Dilute solutions of the reagent are stable indefinitely and can be used in the presence of most cations without causing troublesome precipitation reactions. It is reported that 0.1 M solutions of HCl can be boiled for as long as 1 h without loss of acid, provided that the water lost by evaporation is periodically replaced; 0.5 M solutions can be boiled for at least 10 min without significant loss.

Solutions of perchloric acid and sulfuric acid are also stable and are useful for titrations where chloride ion interferes by forming precipitates. Standard solutions of nitric acid are seldom used because of their oxidizing properties.

Standard acid solutions are ordinarily prepared by diluting an approximate volume of the concentrated reagent and subsequently standardizing the diluted solution against a primary-standard base. Less frequently, the composition of the concentrated acid is established through careful density measurement; a weighed quantity is then diluted to an exact volume. (Tables relating reagent density to composition are found in most chemistry and chemical engineering handbooks.) A stock solution with an exactly known hydrochloric acid concentration can also be prepared by dilution of a quantity of the concentrated reagent with an equal volume of water followed by distillation. Under controlled conditions, the final quarter of the distillate, which is known as *constant-boiling* HCl, has a fixed and known composition, its acid content being dependent only upon atmospheric pressure. For a pressure P

between 670 and 780 torr, the weight in air of the distillate that contains exactly one mole of H_3O^+ is[1]

$$\frac{\text{wt constant-boiling HCl in g}}{\text{mol } H_3O^+} = 164.673 + 0.02039P \qquad (10\text{-}1)$$

Standard solutions are prepared by diluting weighed quantities of this acid to accurately known volumes.

10A-2 The Standardization of Acids

Sodium Carbonate

Acids are frequently standardized by titration of weighed quantities of sodium carbonate. Primary-standard-grade sodium carbonate is available commercially or can be prepared by heating purified sodium hydrogen carbonate between 270 to 300°C for 1 h:

$$2NaHCO_3(s) \longrightarrow Na_2CO_3(s) + H_2O(g) + CO_2(g)$$

As shown in Figure 9-4, two end points are observed in the titration of sodium carbonate. The first, corresponding to the conversion of carbonate to hydrogen carbonate, occurs at about pH 8.3; the second, involving the formation of carbonic acid, is observed at about pH 3.8. The second end point is always used for standardization because the change in pH is greater than that of the first.

An even sharper end point can be achieved by boiling the solution briefly to eliminate the reaction product, carbonic acid. The sample is titrated to the first appearance of the acid color of the indicator (such as bromocresol green or methyl orange). At this point, the solution contains a large amount of carbonic acid and a small amount of unreacted hydrogen carbonate. Boiling effectively destroys this buffer by eliminating the carbonic acid:

$$H_2CO_3(aq) \longrightarrow CO_2(g) + H_2O(l)$$

As a result, the solution again acquires an alkaline pH owing to the residual hydrogen carbonate ion. The titration is completed after the solution has cooled. Now, however, a substantially larger decrease in pH attends the final additions of acid, thus giving a more abrupt color change.

As an alternative, the acid can be introduced in an amount slightly in excess of that needed to convert the sodium carbonate to carbonic acid. The solution is boiled as before to remove carbon dioxide and cooled; the excess acid is then back-titrated with a dilute solution of base. Any indicator suitable for a strong acid/strong base titration can be employed. The volume ratio of acid to base must of course be established by an independent titration.

Directions for the standardization of hydrochloric acid solutions against sodium carbonate are found in Section 31C-6.

Other Primary Standards for Acids

Tris-(hydroxymethyl)aminomethane, $(HOCH_2)_3CNH_2$, known also as TRIS or THAM, is available in primary standard purity from commercial sources.

[1] *Official Methods of Analysis of the AOAC,* 14th ed., p. 1004. Washington, D.C.: Association of Official Analytical Chemists, 1984.

It possesses the advantage of a substantially greater equivalent weight[2] (121.1) than sodium carbonate (53.0).

Sodium tetraborate decahydrate, mercury(II) oxide, and calcium oxide have also been recommended as primary standards. The reaction of an acid with the tetraborate is

$$B_4O_7^{2-} + 2H_3O^+ + 3H_2O \longrightarrow 4H_3BO_3$$

10A-3 Preparation of Standard Solutions of Base

Sodium hydroxide is the most common base for preparing standard solutions although potassium hydroxide and barium hydroxide are also encountered. None of these is obtainable in primary-standard purity, and so standardization is required after preparation.

The Effect of Carbon Dioxide Upon Standard Base Solutions

In solution as well as in the solid state, the hydroxides of sodium, potassium, and barium react avidly with atmospheric carbon dioxide to produce the corresponding carbonate:

$$CO_2(g) + 2OH^- \longrightarrow CO_3^{2-} + H_2O$$

Although this reaction consumes hydroxide ions, absorption of carbon dioxide by a solution of standardized base does not necessarily alter the number of moles of acid that a milliliter of the solution will consume. For example, if potassium or sodium hydroxide solution is employed for a titration in which an acid-range indicator is appropriate (bromocresol green, for example), each carbonate ion has consumed two hydronium ions of the analyte by the time the end point is reached (Figure 9-4):

$$CO_3^{2-} + 2H_3O^+ \longrightarrow H_2CO_3 + 2H_2O$$

Since the amount of hydronium ion consumed by this reaction is identical to the amount of hydroxide lost during formation of the carbonate ion, no error is incurred.

Unfortunately, most applications of standard base require an indicator with a basic transition range (phenolphthalein, for example). Here, each carbonate ion has consumed only one hydronium ion when the color change of the indicator is observed:

$$CO_3^{2-} + H_3O^+ \longrightarrow HCO_3^- + H_2O$$

The effective concentration of the base is thus diminished by absorption of carbon dioxide, and a determinate error (called a *carbonate error*) results.

[2]The equivalent weight of an acid is that weight that contains one mole of titratable protons; the equivalent weight of a base is that weight that consumes one mole of protons. Thus, the equivalent weight of H_2SO_4 is one-half of its formula weight. The equivalent weight of Na_2CO_3 is usually one-half of its formula weight because in most applications its reaction is

$$Na_2CO_3 + 2H_3O^+ \longrightarrow H_2O + H_2CO_3 + 2Na^+$$

When titrated with some indicators, however, it consumes but a single proton:

$$Na_2CO_3 + H_3O^+ \longrightarrow NaHCO_3 + Na^+$$

In this case, the equivalent weight and the formula weight of Na_2CO_3 are identical. A high equivalent weight is desirable in a primary standard because a large amount of reagent is needed thus reducing the relative error in weighing.

When carbon dioxide is absorbed by standard barium hydroxide, precipitation of barium carbonate occurs:

$$CO_2(g) + Ba^{2+} + 2OH^- \longrightarrow BaCO_3(s) + H_2O$$

The acid titer is thus decreased regardless of the indicator employed in the titration; a carbonate error is the inevitable consequence.

The solid reagents used to prepare standard solutions of base are always contaminated by significant amounts of carbonate ion. As a result, even freshly prepared solutions of base are likely to contain substantial quantities of carbonate. The presence of this contaminant does not cause a carbonate error provided the same indicator is used for both standardization and analysis. This restriction causes the reagent to lose much of its versatility, however.

Several methods exist for the preparation of carbonate-free hydroxide solutions. One is based upon the use of barium hydroxide to which an excess of barium chloride has been added to diminish the solubility of barium carbonate further. Barium salts are also added to solutions of potassium or sodium hydroxide to precipitate carbonate ion. The presence of barium ion is frequently undesirable, however, owing to its tendency to form slightly soluble salts with anions that may be present in the sample.

The preferred method for preparing carbonate-free sodium hydroxide solutions takes advantage of the very low solubility of sodium carbonate in concentrated solutions of the alkali. An approximately 50% aqueous solution of sodium hydroxide is prepared (or purchased from commercial sources). The solid sodium carbonate is allowed to settle to give a clear liquid that is decanted and diluted to give the desired concentration (alternatively the solid is removed by vacuum filtration). Details for this procedure are given in Section 31C-3.

A carbonate-free solution of base must be prepared from water that contains no carbon dioxide. Distilled water, which is sometimes supersaturated with carbon dioxide, should be boiled briefly to eliminate the gas. The water should then be allowed to cool to room temperature before the introduction of base because hot alkali solutions rapidly absorb carbon dioxide.

Standard solutions of base are reasonably stable as long as they are protected from contact with the atmosphere. Figure 10-1 shows an arrange-

FIGURE 10-1 Arrangement for the storage of standard base solutions.

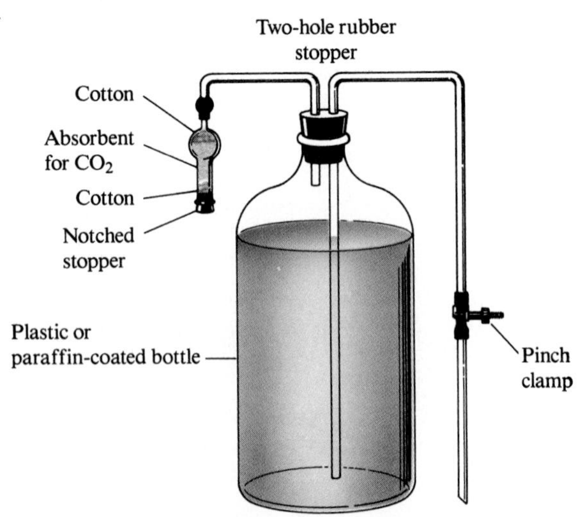

ment for preventing the uptake of atmospheric carbon dioxide during storage and when the reagent is dispensed. Air entering the vessel is passed over a solid absorbent for CO_2, such as soda lime or Ascarite II.[3] The contamination that occurs as the solution is transferred from this storage bottle to the buret is ordinarily negligible.

A tightly capped polyethylene bottle usually provides sufficient short-term protection against the uptake of atmospheric carbon dioxide. Before capping, the bottle is squeezed to minimize the interior air space. Care should also be taken to keep the bottle closed except during the brief periods when the contents are being transferred to a buret.

Sodium hydroxide reacts slowly with glass to form silicates. For this reason, standard solutions of base should not be stored for extended periods (longer than 1 or 2 weeks) in glass containers.[4] In addition, bases should never be kept in glass-stoppered containers because the reaction between the base and the stopper may cause the latter to "freeze" after a brief period. Finally, to avoid the same type of freezing, burets with glass stopcocks should be promptly drained and thoroughly rinsed with water after use with standard base solutions. This problem is avoided with burets equipped with Teflon stopcocks.

10A-4 The Standardization of Bases

Several excellent primary standards are available for the standardization of bases. Most are weak organic acids that require the use of an indicator with a basic transition range.

Potassium Hydrogen Phthalate, $KHC_8H_4O_4$

Potassium hydrogen phthalate is an ideal primary standard. It is a nonhygroscopic crystalline solid with a high equivalent weight (204.2). For most purposes, the commercial analytical-grade salt can be used without further purification. For the most exacting work, potassium hydrogen phthalate of certified purity is available from the National Bureau of Standards.

Directions for the standardization of sodium hydroxide with potassium hydrogen phthalate are given in Section 31C-7.

Other Primary Standards for Bases

Benzoic acid is obtainable in primary-standard purity and can be used for the standardization of bases. Because its solubility in water is limited, this reagent is ordinarily dissolved in ethanol prior to dilution with water and titration. A blank should always be carried through this procedure because commercial alcohol is sometimes slightly acidic.

Potassium hydrogen iodate, $KH(IO_3)_2$, is an excellent primary standard with a high equivalent weight. It is also a strong acid that can be titrated using virtually any indicator with a transition range between pH 4 and 10.

[3]Thomas Scientific, Swedesboro, NJ. Ascarite II consists of sodium hydroxide deposited on a nonfibrous silicate structure.

[4]It has been shown that the concentration of a 0.1 M NaOH solution changes by 0.12 to 0.29% per week during storage in a glass bottle. See A. A. Smith, *J. Chem. Educ.*, **1986**, *63*, 85. In addition, A. R. Armstrong of The College of William and Mary has related to us that 50-mL burets used for a few weeks each year in a quantitative analysis laboratory increased in volume at a rate of 1 to 3 ppt per year over a ten-year period owing to solution of the glass in standard base solutions.

10B

Typical Applications of Neutralization Titrations

Neutralization titrations are used for determining the innumerable inorganic, organic, and biological species that possess inherent acidic or basic properties. Equally important, however, are the many applications that involve conversion of an analyte to an acid or base by suitable chemical treatment followed by titration with a standard strong base or acid.

Two major types of end points find widespread use in neutralization titrations. The first is a visual end point based on indicators such as those described in Section 8G. The second is a *potentiometric* end point in which the potential of a glass/calomel electrode system is determined with a voltage-measuring device. The measured potential is directly proportional to pH. Potentiometric end points are described in Chapter 15.

10B-1

Elemental Analysis

Several important elements that occur in organic and biological systems are conveniently determined by methods that involve an acid/base titration as the final step. Generally, the elements susceptible to this type of analysis are nonmetallic and include carbon, nitrogen, chlorine, bromine, fluorine, and a few other less common species. In each instance, the element is converted to an inorganic acid or base that is then titrated. A few examples follow.

Nitrogen

Nitrogen is found in a wide variety of substances of interest in research, industry, and agriculture, including amino acids, proteins, synthetic drugs, fertilizers, explosives, soils, potable water supplies, and dyes. Thus analytical methods for the determination of nitrogen, particularly in organic substrates, are of singular importance.

Nitrogen in organic materials is determined by one of two principal methods: the *Dumas* method and the *Kjeldahl* method. The former is discussed in Chapter 28. The Kjeldahl method, which was first developed in 1883, is considered in this chapter because the analysis is usually completed by an acid/base titration. The procedure is straightforward, requires no special equipment, and is readily adapted to the routine analysis of large numbers of samples. It (or one of its modifications) is the standard means for determining the protein content of grains, meats, and other biological materials. Since most proteins contain approximately the same percentage of nitrogen, multiplication of this percentage by a suitable factor (6.25 for meats, 6.38 for dairy products, and 5.7 for cereals) gives the percentage of protein in a sample.

In the Kjeldahl method, the sample is decomposed in hot, concentrated sulfuric acid to convert the bound nitrogen to ammonium ion. The solution is then cooled, diluted, and made basic with an excess of strong base. The liberated ammonia is distilled, collected in an acidic solution, and determined by a neutralization titration.

The critical step in the Kjeldahl method is the decomposition with sulfuric acid, which oxidizes the carbon and hydrogen in the sample to carbon dioxide and water. The fate of the nitrogen, however, depends upon its state of combination in the original sample. Amine and amide nitrogens are quantitatively converted to ammonium ion. In contrast, nitro, azo, and azoxy nitrogens are likely to yield elemental nitrogen or various oxides of nitrogen, all of which are lost from the hot acidic medium. This loss can be avoided by pretreatment of the sample with a reducing agent to form reduced products that behave as amide or amine nitrogen. In one such prereduction scheme, salicylic acid and sodium thiosulfate are added to the concentrated sulfuric acid solution containing the sample. After a brief period, the digestion is performed in the usual way.

Certain aromatic heterocyclic compounds, such as pyridine and its derivatives, are particularly resistant to complete decomposition by sulfuric acid. Such compounds yield low results as a consequence (Figure 2-2) unless special precautions are taken.

The decomposition step in a Kjeldahl determination is slow, with 1 h or more required for refractory samples (that is, samples that are resistant to high temperature decomposition). Numerous modifications of the original procedure have been proposed with the aim of shortening the digestion time. One method, proposed by Gunning, is now almost universally employed. A neutral salt, such as potassium sulfate, is added to increase the boiling point of the sulfuric acid solution and thus the temperature at which the decomposition occurs. Care is needed, however, because oxidation of the ammonium ion may occur if excessive evaporation of the acid occurs during digestion.

Many substances catalyze the decomposition of organic compounds by sulfuric acid. Mercury, copper, and selenium, either combined or in the elemental state, are effective. Mercury(II), if present, must be precipitated with hydrogen sulfide prior to distillation to prevent retention of ammonia as a mercury(II) ammine complex.

Figure 10-2 illustrates typical equipment for a Kjeldahl distillation. The long-necked container, which is used for both digestion and distillation, is called a *Kjeldahl flask.* After the decomposition is judged complete, the contents

FIGURE 10-2 Kjeldahl distillation apparatus.

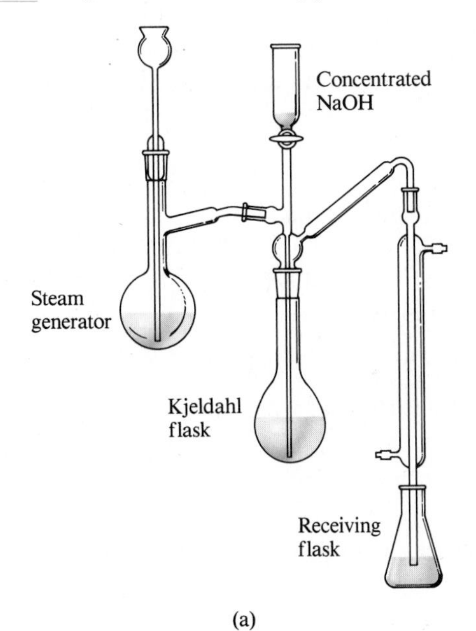

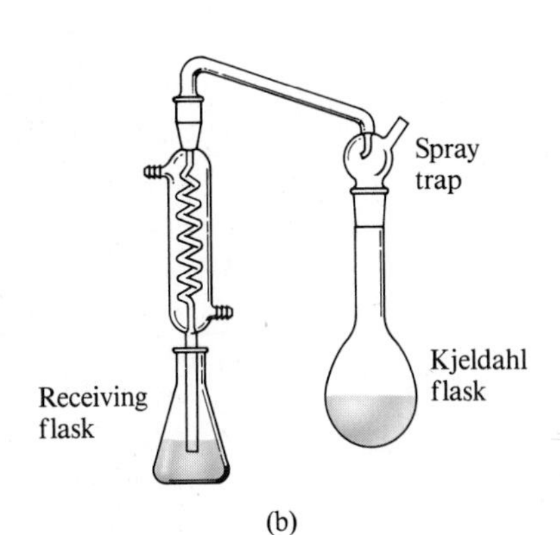

of the flask are cooled, diluted with water, and made basic to liberate the ammonia:

$$NH_4^+ + OH^- \longrightarrow NH_3(g) + H_2O$$

In the apparatus shown in Figure 10-2a, the base is added slowly by partially opening the stopcock from the storage vessel; the liberated ammonia is then carried to the receiving flask by steam distillation.

In an alternative method, a dense, concentrated sodium hydroxide solution is carefully poured down the side of the flask to form a second, lower layer. The flask is then quickly connected to a spray trap (Figure 10-2b) and an ordinary condenser before loss of ammonia can occur. Only then are the two layers mixed by gentle swirling of the flask.

Note that in both types of apparatus shown in Figure 10-2, the tip of the condenser extends into the acid in the receiving flask during the distillation. The tip of the condenser must be removed when heating is discontinued, however, to prevent the contents of the flask from being drawn back into the Kjeldahl flask.

Two methods are commonly used for collecting and determining the ammonia liberated from the sample. In one, the ammonia is distilled into a measured volume of standard acid. After the distillation is complete, the excess acid is back-titrated with standard base. An indicator with an acidic transition range is required because of the acidity of the ammonium ions present at equivalence. A convenient alternative, which requires only one standard solution, involves the collection of the ammonia in an unmeasured excess of boric acid, which retains the ammonia by the reaction

$$H_3BO_3 + NH_3 \longrightarrow NH_4^+ + H_2BO_3^-$$

The dihydrogen borate ion produced is a reasonably strong base that can be titrated with a standard solution of hydrochloric acid:

$$H_2BO_3^- + H_3O^+ \longrightarrow H_3BO_3 + H_2O$$

At the equivalence point, the solution contains boric acid and ammonium ions; an indicator with an acidic transition interval (such as bromocresol green) is again required.

Details of the Kjeldahl method are found in Section 31C-11.

Sulfur

Sulfur in organic and biological materials is conveniently determined by burning the sample in a stream of oxygen. The sulfur dioxide (as well as the sulfur trioxide) formed during the oxidation is collected by distillation into a dilute solution of hydrogen peroxide:

$$SO_2(g) + H_2O_2 \longrightarrow H_2SO_4$$

The sulfuric acid is then titrated with standard base.

Other Elements

Table 10-1 lists other elements that can be determined by neutralization methods.

TABLE 10-1 **Elemental Analysis Based on Neutralization Titrations**

Element	Converted to	Absorption or Precipitation Products	Titration
N	NH_3	$NH_3(g) + H_3O^+ \longrightarrow NH_4^+ + H_2O$	Excess HCl with NaOH
S	SO_2	$SO_2(g) + H_2O_2 \longrightarrow H_2SO_4$	NaOH
C	CO_2	$CO_2(g) + Ba(OH)_2 \longrightarrow BaCO_3(s) + H_2O$	Excess $Ba(OH)_2$ with HCl
Cl(Br)	HCl	$HCl(g) + H_2O \longrightarrow Cl^- + H_3O^+$	NaOH
F	SiF_4	$3SiF_4(g) + 2H_2O \longrightarrow 2H_2SiF_6 + SiO_2(s)$	NaOH
P	H_3PO_4	$12H_2MoO_4 + 3NH_4^+ + H_3PO_4 \longrightarrow (NH_4)_3PO_4 \cdot 12MoO_3(s) + 12H_2O + 3H^+$	
		$(NH_4)_3PO_4 \cdot 12MoO_3(s) + 26OH^- \longrightarrow HPO_4^{2-} + 12MoO_4^{2-} + 14H_2O + 3NH_3(g)$	Excess NaOH with HCl

10B-2 The Determination of Inorganic Substances

Numerous inorganic species can be determined by titration with strong acids or bases. A few examples follow.

Ammonium Salts

Ammonium salts are conveniently determined by conversion to ammonia with strong base followed by distillation in the Kjeldahl apparatus shown in Figure 10-2. The ammonia is collected and titrated as in the Kjeldahl method.

Nitrates and Nitrites

The method just described for ammonium salts can be extended to the determination of inorganic nitrate or nitrite. These ions are first reduced to ammonium ion by Devarda's alloy (50% Cu, 45% Al, 5% Zn). Granules of the alloy are introduced into a strongly alkaline solution of the sample in a Kjeldahl flask. The ammonia is distilled after reaction is complete. Arnd's alloy (60% Cu, 40% Mg) has also been used as the reducing agent.

Carbonate and Carbonate Mixtures

The qualitative and quantitative determination of the constituents in a solution containing sodium carbonate, sodium hydrogen carbonate, and sodium hydroxide, either alone or admixed, provides interesting examples of how neutralization titrations can be employed to analyze mixtures. No more than two of these three constituents can exist in appreciable amount in any solution because reaction eliminates the third. Thus, mixing sodium hydroxide with sodium hydrogen carbonate results in the formation of sodium carbonate until one or the other (or both) of the original reactants is exhausted. If the sodium hydroxide is used up, the solution will contain sodium carbonate and sodium hydrogen carbonate; if sodium hydrogen carbonate is depleted, sodium carbonate and sodium hydroxide will remain; if equimolar amounts of sodium hydrogen carbonate and sodium hydroxide are mixed, the principal solute species will be sodium carbonate.

The analysis of such mixtures requires two titrations: one employing an alkaline-range indicator, such as phenolphthalein, and the other an acid-range indicator, such as bromocresol green. The composition of the solution can then be deduced from the relative volumes of acid needed to titrate equal volumes of the sample (Table 10-2 and Figure 9-4). Once the composition of the solution has been established, the volume data can be used to determine the concentration of each component in the sample.

EXAMPLE 10-1

A solution contains $NaHCO_3$, Na_2CO_3, and NaOH, either alone or in permissible combination. Titration of a 50.0-mL portion to a phenolphthalein end point requires 22.1 mL of 0.100 M HCl. A second 50.0-mL aliquot requires 48.4 mL of the HCl when titrated to a bromocresol green end point. Deduce the composition, and calculate the molar solute concentrations of the original solution.

If the solution contains only NaOH, the volume of acid required would be the same regardless of indicator (that is, $V_{ph} = V_{bg}$). Similarly, we can rule out the presence of Na_2CO_3 alone because titration of this compound to a bromocresol green end point would consume just twice the volume of acid required to reach the phenolphthalein end point. In fact, however, the second titration requires 48.4 mL. Because less than half of this amount is involved in the first titration, the solution must contain some $NaHCO_3$ in addition to Na_2CO_3. We can now calculate the concentration of the two constituents.

When the phenolphthalein end point is reached, the CO_3^{2-} originally present is converted to HCO_3^-. Thus,

$$\text{no. mmol } Na_2CO_3 = 22.1 \cancel{\text{mL}} \times 0.100 \text{ mmol}/\cancel{\text{mL}} = 2.21$$

The titration from the phenolphthalein to the bromocresol green end point (48.4 − 22.1 = 26.3 mL) involves both the hydrogen carbonate originally present and that formed by titration of the carbonate. Thus,

$$\text{no. mmol } NaHCO_3 + \text{no. mmol } Na_2CO_3 = 26.3 \times 0.100 = 2.63$$

Hence,

$$\text{no. mmol } NaHCO_3 = 2.63 - 2.21 = 0.42$$

The molar concentrations are readily calculated from these data:

$$c_{Na_2CO_3} = \frac{2.21 \text{ mmol}}{50.0 \text{ mL}} = 0.0442 \text{ M}$$

$$c_{NaHCO_3} = \frac{0.42 \text{ mmol}}{50.0 \text{ mL}} = 0.0084 \text{ M}$$

The method described in Example 10-1 is not entirely satisfactory because the pH change corresponding to the hydrogen carbonate equivalence

TABLE 10-2

Volume Relationship in the Titration of Mixtures Containing CO_3^{2-}, HCO_3^-, and OH^-

Constituents	Relationship Between Acid to Reach Phenolphthalein End Point, V_{ph}, and Bromocresol Green End Point, V_{bg}
NaOH	$V_{ph} = V_{bg}$
Na_2CO_3	$2V_{ph} = V_{bg}$
$NaHCO_3$	$V_{ph} = 0$, $V_{bg} > 0$
NaOH, Na_2CO_3	$2V_{ph} > V_{bg}$
Na_2CO_3, $NaHCO_3$	$2V_{ph} < V_{bg}$

point is not sufficient to give a sharp color change with a chemical indicator (Figure 9-4). Relative errors of 1% or more must be expected as a consequence.

The accuracy of methods for analyzing solutions containing mixtures of carbonate and hydrogen carbonate ions or carbonate and hydroxide ions can be greatly improved by taking advantage of the limited solubility of barium carbonate in neutral and basic solutions. For example, in the *Winkler method* for the analysis of carbonate/hydroxide mixtures, both components are titrated with a standard acid to the end point with an acid-range indicator, such as bromocresol green (the end point is established after the solution is boiled to remove carbon dioxide). An unmeasured excess of neutral barium chloride is then added to a second aliquot of the sample solution to precipitate the carbonate ion, following which the hydroxide ion is titrated to a phenolphthalein end point. The presence of the sparingly soluble barium carbonate does not interfere as long as the concentration of barium ion is greater than 0.1 M.

Carbonate and hydrogen carbonate ions can be accurately determined in mixtures by first titrating both ions with standard acid to an end point with an acid-range indicator (with boiling to eliminate carbon dioxide). The hydrogen carbonate in a second aliquot is converted to carbonate by the addition of a known excess of standard base. After a large excess of barium chloride has been introduced, the excess base is titrated with standard acid to a phenolphthalein end point.

The presence of solid barium carbonate does not hamper end-point detection in either of these methods.

10B-3

The Determination of Organic Functional Groups

Neutralization titrations provide convenient methods for the direct or indirect determination of several organic functional groups. Brief descriptions of methods for the more common groups follow.

Carboxylic and Sulfonic Acid Groups

Carboxylic and sulfonic acid groups are the two most common structures that impart acidity to organic compounds. Most carboxylic acids have dissociation constants that range between 10^{-4} and 10^{-6}, and thus these compounds are readily titrated. An indicator that changes color in the basic range, such as phenolphthalein, is required.

Many carboxylic acids are not sufficiently soluble in water to permit direct titration in this medium. Where this problem exists, the acid can be dissolved in ethanol and titrated with aqueous base. Alternatively, the acid can be dissolved in an excess of standard base followed by back-titration with standard acid.

Sulfonic acids are generally strong acids and readily dissolve in water. Their titration with a base is therefore straightforward.

Neutralization titrations are often employed to determine the equivalent weight of purified organic acids (that is, the weight of acid that contains 1 mol of titratable protons). Equivalent weights serve as an aid in qualitative identification of organic acids.

Amine Groups

Aliphatic amines generally have base dissociation constants on the order of 10^{-5} and can thus be titrated directly with a solution of a strong acid. Aromatic amines, such as aniline and its derivatives, in contrast are usually too weak for titration in aqueous medium ($K_b \cong 10^{-10}$). The same is true for cyclic amines with aromatic character, such as pyridine and its derivatives. Many saturated cyclic amines, such as piperidine, tend to resemble aliphatic amines in their acid/base behavior and thus can be titrated in aqueous media.

Many amines that are too weak to be titrated as bases in water are readily titrated in nonaqueous solvents, such as anhydrous acetic acid, which enhance their basicity. These procedures are discussed in Section 10C.

Ester Groups

Esters are commonly determined by *saponification* with a measured quantity of standard base:

$$R_1COOR_2 + OH^- \longrightarrow R_1COO^- + HOR_2$$

The excess base is then titrated with standard acid.

Esters vary widely in their rate of saponification. Some require several hours of heating with a base to complete the saponification. A few react rapidly enough to permit direct titration with the base. Typically, the ester is refluxed with standard 0.5 M KOH for 1 to 2 h. After cooling, the excess base is determined with standard acid.

Hydroxyl Groups

Hydroxyl groups in organic compounds can be determined by esterification with various carboxylic acid anhydrides or chlorides; the two most common reagents are acetic anhydride and phthalic anhydride. With acetic anhydride, the reaction is

$$(CH_3CO)_2O + ROH \longrightarrow CH_3COOR + CH_3COOH$$

The acetylation is ordinarily carried out by mixing the sample with a carefully measured volume of acetic anhydride in pyridine. After heating, water is added to hydrolyze the unreacted anhydride:

$$(CH_3CO)_2O + H_2O \longrightarrow 2CH_3COOH$$

The acetic acid is then titrated with a standard solution of alcoholic sodium or potassium hydroxide. A blank is carried through the analysis to establish the original amount of anhydride.

Amines, if present, are converted quantitatively to amides by acetic anhydride; a correction for this source of interference is frequently possible by a direct titration of another portion of the sample with standard acid.

Carbonyl Groups

Many aldehydes and ketones can be determined with a solution of hydroxylamine hydrochloride. The reaction, which produces an oxime, is

$$\begin{matrix} R_1 \\ \quad\diagdown \\ \quad\quad C{=}O \\ \quad\diagup \\ R_2 \end{matrix} + NH_2OH{\cdot}HCl \longrightarrow \begin{matrix} R_1 \\ \quad\diagdown \\ \quad\quad C{=}NOH \\ \quad\diagup \\ R_2 \end{matrix} + HCl + H_2O$$

where R_2 may be an atom of hydrogen. The liberated hydrochloric acid is titrated with base. Here again, the conditions necessary for quantitative reaction vary. Typically, 30 min suffices for aldehydes. Many ketones require refluxing with the reagent for 1 h or more.

10B-4

The Determination of Salts

The total salt content of a solution can be accurately and readily determined by an acid/base titration. The salt is converted to an equivalent amount of an acid or a base by passage through a column packed with an ion-exchange resin. (This application is considered in more detail in Section 29E-3).

Standard acid or base solutions can also be prepared with ion-exchange resins. Here, a solution containing a known weight of a pure compound, such as sodium chloride, is washed through the resin column and diluted to a known volume. The salt liberates an equivalent amount of acid or base, permitting calculation of the molarity of the reagent in a straightforward way.

10C

Application of Neutralization Titrations in Nonaqueous Media

Two types of compounds that are not titratable in aqueous media can be determined by neutralization titration in suitable nonaqueous solvents. The first are high-molecular-weight organic acids and bases that have limited solubility in water. The second are inorganic and organic compounds that are such weak acids or bases (K_a or $K_b < 10^{-8}$) that they do not yield satisfactory end points in aqueous solutions. Examples of the latter include aromatic amines, phenols, and the salts of a variety of inorganic and carboxylic acids. Often compounds that yield unsatisfactory end points in water exhibit sharp end points in solvents that enhance their acidic or basic character.[5]

Although nonaqueous titrations make possible the determination of species that cannot be titrated in water, several disadvantages attend their use. Generally, the solvents are expensive and in addition are often volatile and toxic. Furthermore, most have significantly larger coefficients of expansion than water and require a much closer control of reagent temperature to avoid determinate errors in the measurement of volume.

10C-1

Solvents for Nonaqueous Titrations

A wide variety of organic solvents have been employed for nonaqueous acid/base titrations. The most common, and the only ones considered here, are *amphiprotic solvents,* which possess both acidic and basic properties and undergo self-dissociation or *autoprotolysis* to yield an acid and a base. Although water is the most common amphiprotic solvent, many other substances exhibit analogous behavior. Three amphiprotic solvents that have found considerable use in nonaqueous titrimetry are anhydrous acetic acid, ethanol, and ethylenediamine. Their autoprotolysis equilibria are:

$$2CH_3COOH \rightleftharpoons CH_3COOH_2^+ + CH_3COO^- \qquad (10\text{-}2)$$

$$2C_2H_5OH \rightleftharpoons C_2H_5OH_2^+ + C_2H_5O^- \qquad (10\text{-}3)$$

$$2NH_2CH_2CH_2NH_2 \rightleftharpoons NH_2CH_2CH_2NH_3^+ + NH_2CH_2CH_2NH^- \qquad (10\text{-}4)$$

Other examples of autoprotolytic reactions are found on page 105.

[5]For more extensive discussions of nonaqueous neutralization titrations, see J. S. Fritz, *Titrations in Nonaqueous Solvents.* Boston: Allyn and Bacon, 1973; I. M. Kolthoff *et al., Treatise on Analytical Chemistry,* 2nd ed., I. M. Kolthoff and P. J. Elving, Eds., Part I, Vol. 2, Chapter 19, New York: Wiley, 1979.

10C-2

The Completeness of Acid/Base Reactions in Amphiprotic Solvents

The completeness of a neutralization reaction in an amphiprotic solvent depends not only upon the strength of the analyte as an acid or base but also upon the autoprotolysis constant, the inherent acidity or basicity, and the dielectric constant of the solvent.

The Effect of Solvent Autoprotolysis Constant

In water, the titration of a weak base B with a strong acid can be formulated as

$$B + H_3O^+ \rightleftharpoons BH^+ + H_2O \tag{10-5}$$

The magnitude of the equilibrium constant, which provides a measure of the completeness of this reaction, is readily shown to be the ratio of base dissociation constant for B divided by the autoprotolysis constant for water:

$$K_{equil} = \frac{K_b}{K_w} = \frac{[BH^+]\cancel{[OH^-]}}{[B][H_3O^+]\cancel{[OH^-]}} = \frac{[BH^+]}{[B][H_3O^+]} \tag{10-6}$$

Note that the completeness of the reaction is determined in part by the magnitude of K_w, *the autoprotolysis constant for the solvent.*

By analogy, the completeness of the reaction between a weak acid HA and a strong base in water can be expressed by an equilibrium constant that is numerically equal to K_a/K_w.

Similar relationships can be derived for reactions in nonaqueous solvents. For example, when the weak base B reacts with perchloric acid in anhydrous acetic acid, the reaction is

$$B + CH_3COOH_2^+ \rightleftharpoons BH^+ + CH_3COOH \tag{10-7}$$

Here, $CH_3COOH_2^+$ is the solvated proton formed when perchloric acid is dissolved in anhydrous acetic acid:

$$HClO_4 + CH_3COOH \rightleftharpoons CH_3COOH_2^+ + ClO_4^-$$

The solvated proton $CH_3COOH_2^+$ is analogous to H_3O^+ in an aqueous solution. The equilibrium constant for the neutralization reaction (Equation 10-7) is given by

$$K_{equil} = \frac{[BH^+]}{[B][CH_3COOH_2^+]} = \frac{K'_b}{K_s} \tag{10-8}$$

where K'_b is the dissociation constant for the base *in acetic acid;* that is,

$$B + CH_3COOH \rightleftharpoons BH^+ + CH_3COO^- \qquad K'_b = \frac{[BH^+][CH_3COO^-]}{[B]} \tag{10-9}$$

The constant K_s in Equation 10-8 is the autoprotolysis constant for the equilibrium shown as Equation 10-2 and is analogous to the ion-product constant for water:

$$K_s = [CH_3COOH_2^+][CH_3COO^-] \tag{10-10}$$

As with the aqueous equilibrium constants, the concentration of the solvent CH_3COOH is essentially invariant and is thus included in K'_b and K_s.

A similar set of relationships can be developed for the titration of a weak acid with a strong base in an anhydrous medium.

Equations 10-6 and 10-8 show that the completeness of an acid/base

reaction is directly related to the dissociation constant of the weak acid but is inversely related *to the autoprotolysis constant of the solvent in which the titration is carried out.* These equations can be understood by considering a neutralization reaction as a competition for protons between solvent and base. Thus, for example, the extent of reaction 10-7 is governed by the success with which base molecules B compete with solvent molecules CH_3COOH for a stoichiometrically limited number of hydrogen ions. The effectiveness of each participant in this competition is measured by its dissociation constant K_b' and its autoprotolysis constant K_s respectively.

Effect of Acid or Base Characteristics of the Solvent

The behavior of solutes as acids or bases is strongly influenced by the strength of the solvent as an acid or a base. A number of amphiprotic solvents, including formic acid, acetic acid, and sulfuric acid, are considerably better proton donors than proton acceptors and are therefore classified as acidic solvents. In such media, the basic properties of a solute are magnified while its acidic properties are attenuated. Thus, for example, aniline, $C_6H_5NH_2$, cannot be titrated in water because its base dissociation constant is only about 10^{-10}. In anhydrous acetic acid, however, aniline is an appreciably stronger base because the solvent gives up protons more readily than does water. Thus, the equilibrium constant, K_b' for the reaction

$$C_6H_5NH_2 + CH_3COOH \rightleftharpoons C_6H_5NH_3^+ + CH_3COO^-$$

is significantly larger than K_b for the analogous reaction in water:

$$C_6H_5NH_2 + H_2O \rightleftharpoons C_6H_5NH_3^+ + OH^-$$

Solvents such as ethylenediamine and liquid ammonia have a strong affinity for protons and are therefore classified as basic solvents. In these media, the acidic character of a solute is enhanced. For example, phenol, which has an acid dissociation constant of about 10^{-10} in water, is strong enough in ethylenediamine to be readily titrated with a standard base. The strengths of bases are of course diminished in solvents of this type and solutes that are strong bases in water may only partially dissociate in solvents with basic properties.

Effect of Solvent Dielectric Constant

The dielectric constant of a solvent measures its capacity to cause particles of opposite charge to separate from one another and thus behave as independent entities. For example, in a solvent with a high dielectric constant, such as water (D_{H_2O} = 78.5), a minimum amount of work is required to separate a positively charged ion from one with a negative charge. In contrast, a much greater amount of energy is expended in accomplishing this process in a solvent with a low dielectric constant, such as acetic acid (D_{HOAc} = 6.2). Methanol and ethanol, with dielectric constants of 33 and 24, respectively, are intermediate in their behavior.

Solvents with a low dielectric constant suffer a disadvantage when used as a titration medium involving acids or bases that dissociate to give two oppositely charged ions because the extent of dissociation is markedly decreased by the limited ability of the solvent to cause ion separation. For example, perchloric acid, which behaves as a strong acid in solvents with high dielectric constants, has a dissociation constant of only 1.1×10^{-4} in *tert*-butyl alcohol, a solvent with a dielectric constant of 12.5. That is,

$$HClO_4 + C_4H_9OH \rightleftharpoons C_4H_9OH_2^+ + ClO_4^- \qquad K_a' = 1.1 \times 10^{-4}$$

It is important to note that the dielectric constant of a solvent has little effect on the degree of dissociation of an acid or a base when the dissociation does not require a separation of charged particles. For example, the dielectric effect is minimal for dissociation reactions such as

$$NH_4^+ + C_2H_5OH \rightleftharpoons C_2H_5OH_2^+ + NH_3$$

10C-3 Choice of Amphiprotic Solvents for Neutralization Titrations

In light of the discussion in Section 10C-2, it is possible to conclude that the best solvent for a given acid/base titration hinges upon three interrelated properties:

1. Its autoprotolysis constant; a numerically small value is desirable.
2. Its properties as a proton donor or acceptor. For the titration of a weak base, a solvent with strong proton donor tendencies (that is, an acidic solvent) is helpful; for the determination of a weak acid, a solvent that is a good proton acceptor is desirable.
3. Its dielectric constant; a high value is most useful.

In addition, of course, the solute must be reasonably soluble in the solvent.

Anhydrous acetic acid (often called glacial acetic acid[6]) is frequently chosen as solvent for titration of very weak bases because it tends to donate protons and thus enhances the strength of a dissolved base. Also, its autoprotolysis constant (3.6×10^{-15}) is somewhat more favorable than that of water. On the other hand, its low dielectric constant partially offsets these advantages. The two favorable properties outweigh the single disadvantage, however, and acetic acid is a generally superior solvent for the titration of weak bases; it is clearly inferior to water for the titration of weak acids because of its weakness as a proton acceptor.

Ethanol, which has been widely applied for nonaqueous titrations, is classified as a neutral solvent because its proton donor and acceptor properties do not differ markedly. The autoprotolysis constant for ethanol is far superior to that of water (8×10^{-20}). On the other hand, its low dielectric constant compared with water frequently offsets the autoprotolysis advantage. For example, the dissociation constants of most uncharged acids, such as benzoic acid, are about 10^{-6} as great in ethanol as in water. At the same time, the ratio of autoprotolysis constants is smaller by nearly the same factor (8×10^{-6}). Thus the ratio K_a'/K_s is only slightly more favorable in ethanol than in water, and the improvement in end points gained by use of this solvent is modest. In contrast, significant gain is realized by the use of ethanol for the titration of a charged weak acid such as the ammonium ion because no charge separation is involved in the dissociation:

$$NH_4^+ + C_2H_5OH \rightleftharpoons NH_3 + C_2H_5OH_2^+$$

[6]Glacial acetic acid derives its name from the fact that at about 16°C large ice-like crystals of acetic acid form in the medium.

As a consequence, dissociation of the acid NH_4^+ is not significantly less in ethanol than in water. The reaction of NH_4^+ with a strong base is much more complete in ethanol, however, because of the low autoprotolysis constant of the solvent. Thus, NH_4^+ can be titrated successfully in ethanol but not in water.

10C-4 End-Point Detection in Nonaqueous Titrations

End points in nonaqueous titrations are most commonly based on potential measurements with a glass electrode. This type of end point is discussed in detail in Chapter 15.

Many of the acid/base indicators developed for aqueous titrations are also applicable in nonaqueous media. To be sure, their behavior in an aqueous environment cannot be extrapolated to predict their properties in nonaqueous solutions. The limited information available with respect to these properties in solvents other than water makes the choice of indicator largely a matter of experience and empirical observation.

10C-5 Applications of Nonaqueous Acid/Base Titrations

Innumerable combinations of solvents, titrants, and end points are discussed in the literature of nonaqueous titrimetry. The choice among them is usually based upon such considerations as solubility, convenience, reagent cost, toxicity, and availability of suitable equipment.[7]

Titration of Bases in Glacial Acetic Acid

Glacial acetic acid is the most widely used acidic solvent. The titrant is a standard perchloric acid solution prepared in the anhydrous medium. A standard solution of sodium acetate in the same solvent serves as a base for back-titrations when needed. Solutions of the acid are generally standardized against sodium carbonate or potassium acid phthalate (in acetic acid, the hydrogen phthalate ion is a strong enough base to make it an excellent primary standard for *acids*).

Some Typical Applications

Standard solutions of perchloric acid in glacial acetic acid are useful for the determination of aromatic amines, amides, urea, and other very weak nitrogen bases. An important application of acetic acid is to the direct titration of most amino acids with a standard acid. It was noted earlier (page 226) that in aqueous media, these compounds exist largely as zwitterions which are not strong enough acids or bases for titration. In glacial acetic acid, however, the dissociation of the carboxylic acid group is essentially completely repressed, leaving the amine group available for titration with perchloric acid.

Many inorganic and organic salts that are neutral or very weak bases in aqueous solvent, are easily determined by titration with perchloric acid in anhydrous acetic acid. For example, the sodium salts of inorganic anions such as chloride, bromide, iodide, nitrate, chlorate, and sulfate have all been titrated as bases in glacial acetic acid. The ammonium and alkali metal salts of most carboxylic acids can also be determined in this medium; typical examples

[7]Reviews of applications of nonaqueous titrations are found in B. Kratochvil, *Anal. Chem.*, 1982, *54,* 105R; 1980, *52,* 151R; 1978, *50,* 153R.

include ammonium benzoate, sodium salicylate, sodium acetate, potassium tartrate, and sodium citrate.

10C-6

Titration of Acids

Several basic solvents are useful for determining acids that are too weak to be titrated in water. Some of these solvents are ethylenediamine, dimethylformamide, pyridine, dimethylsulfoxide, and butylamine. In addition, aliphatic alcohols, acetone, and acetonitrile find use in the determination of acids.

A variety of inorganic and organic compounds including amine salts, inorganic salts, carboxylic acids, phenols, enols, and imides are soluble in ethylenediamine and exhibit enhanced acidic characteristics therein. Tetrabutylammonium hydroxide $[(C_4H_9)_4NOH]$ often serves as titrant.

Questions and Problems

*10-1 The boiling points of HCl and CO_2 are nearly the same (−85 and −78°C). Explain why CO_2 can be removed from an aqueous solution by boiling briefly while essentially no HCl is lost even after boiling for 1 h or more.

10-2 What is constant-boiling HCl?

*10-3 Explain how Na_2CO_3 of primary standard grade can be prepared from a $NaHCO_3$ primary standard.

10-4 Suggest two advantages of $KH(IO_3)_2$ as a primary standard for bases.

*10-5 What is the carbonate error in acid/base titrations? How can it be avoided?

10-6 A sample of alfalfa was found to contain 2.83% nitrogen when analyzed by the Kjeldahl procedure. What was the protein content?

*10-7 Outline a method for the determination of $NaNO_2$ in a mixture of inorganic halides.

10-8 Outline a method for the determination of SO_2 in the effluent gases from a paper mill.

*10-9 List the solvent properties that play a part in determining the completeness of acid/base reactions.

10-10 What is the autoprotolysis constant for a solvent? Are large or small values for this constant desirable in titration reactions? Explain.

*10-11 What is the pH of a 0.010 M solution of a strong acid in ethylenediamine? Of a 0.01 M solution of a strong base? Assume that both are fully dissociated. The autoprotolysis constant for ethylenediamine is 5×10^{-16}.

10-12 What is the pH of a 0.010 M solution of a strong acid in ethanol? Of a 0.01 M solution of a strong base? Assume that both are fully dissociated. The autoprotolysis constant for ethanol is 8×10^{-20}.

10-13 Define the following terms and illustrate with an example:

*(a) autoprotolysis.
(b) amphiprotic solvent.
*(c) acidic amphiprotic solvent.
*(d) basic amphiprotic solvent.
*(e) dielectric constant.

10-14 Write autoprotolysis equations and expressions for

*(a) H_2O.
(b) C_2H_5OH.
*(c) H_2SO_4.
(d) CH_3OH.
*(e) HCOOH.
(f) $NH_2C_2H_4NH_2$.
*(g) HCN.
(h) HF.

*10-15 If 1.000 L of 0.1500 M NaOH was unprotected from the air after standardization and absorbed 11.2 mmol of CO_2, what is its new molarity when it is standardized against a standard solution of HCl using

(a) phenolphthalein?
(b) bromocresol green?

10-16 A NaOH solution was 0.1019 M immediately after standardization. Exactly 500.0 mL of the reagent was left exposed to air for several days and absorbed 0.652 g of CO_2. Calculate the relative carbonate error in the determination of acetic acid with this solution if the titrations were performed with phenolphthalein.

*10-17 Describe the preparation of 2.00 L of

(a) 0.15 M KOH from the solid.
(b) 0.015 M $Ba(OH)_2 \cdot 8H_2O$ from the solid.
(c) 0.200 M HCl from a reagent that has a density of 1.0579 g/mL and is 11.50% HCl (w/w).
(d) 0.150 M reagent from constant-boiling HCl that was distilled at 750 mm Hg.

10-18 Describe the preparation of 500 mL of

(a) 0.250 M H_2SO_4 from a reagent that has a density of 1.1539 g/mL and is 21.8% H_2SO_4 (w/w).
(b) 0.30 M NaOH from the solid.
(c) 0.500 M HCl from constant-boiling HCl that was distilled at 770 mm Hg.
(d) 0.0800 M Na_2CO_3 from the pure solid.

*10-19 The following data were obtained for the standardization of HCl against samples of sodium tetraborate, $Na_2B_4O_7 \cdot 10H_2O$:

$$B_4O_7^{2-} + 2H_3O^+ + 3H_2O \longrightarrow 4H_3BO_3$$

$Na_2B_4O_7 \cdot 10H_2O$, g	HCl, mL
0.6442	33.74
0.7102	37.56
0.5934	31.26

(a) Calculate the mean molarity for the set.
(b) Calculate the standard deviation for the molarity.

10-20 (a) Calculate the mean molarity of a $Ba(OH)_2$ solution from the following standardization data:

$KH(IO_3)_2$, g	$Ba(OH)_2$, mL
0.2574	26.77
0.2733	28.45
0.2885	30.11

(b) Calculate the standard deviation for the molarity.

10-21 Suggest a range of sample weights for the indicated primary standard if it is desired to use between 35 and 45 mL of titrant:

*(a) 0.150 M $HClO_4$ titrated against Na_2CO_3 (CO_2 product).
(b) 0.075 M HCl titrated against $Na_2C_2O_4$:

$$Na_2C_2O_4 \longrightarrow Na_2CO_3 + CO$$

$$CO_3^{2-} + 2H^+ \longrightarrow H_2O + CO_2$$

*(c) 0.20 M NaOH titrated against benzoic acid.
(d) 0.030 M $Ba(OH)_2$ titrated against $KH(IO_3)_2$.
*(e) 0.040 M $HClO_4$ titrated against THAM.
(f) 0.080 M H_2SO_4 titrated against $Na_2B_4O_7 \cdot 10H_2O$ (see Problem 10-19).

*10-22 A 25.00-mL aliquot of dilute H_2SO_4 yielded 0.3472 g of $BaSO_4$. Calculate the molarity of the acid.

10-23 A 50.00-mL aliquot of dilute HCl yielded 0.4771 g of AgCl. What was the molarity of the acid?

*10-24 A 50.00-mL sample of a white dinner wine required 21.48 mL of 0.03776 M NaOH to achieve a phenolphthalein end point. Express the acidity of the wine in terms of grams of tartaric acid ($H_2C_4H_4O_6$; fw = 150.09) per 100 mL. (Assume that two hydrogens of the acid are titrated.)

10-25 A 25.00-mL sample of a household cleaning solution was diluted to 250.0 mL in a volumetric flask. A 50.00-mL aliquot of this solution required 40.38 mL of 0.2506 M HCl to reach a bromocresol green end point. Calculate the weight-volume percentage of NH_3 in the sample. (Assume that all the alkalinity results from the ammonia.)

*10-26 A 0.2296-g sample of a recrystallized organic acid required a 29.83-mL titration with 0.1000 M NaOH to reach a phenolphthalein end point. What is the equivalent weight of the acid (that is, the number of grams of acid per mole of titratable protons)?

10-27 In order to establish the identity of the cation in a pure carbonate, a 0.1401-g sample was dissolved in 50.00 mL of 0.1140 M HCl and boiled to remove CO_2. The excess HCl was back-titrated with 24.21 mL of 0.09802 M NaOH. Identify the carbonate.

*10-28 The active ingredient in Antabuse, a drug used for the treatment of chronic alcoholism, is tetraethylthiuram disulfide,

(fw = 296.54). The sulfur in a 0.4329-g sample of an Antabuse preparation was oxidized to SO_2, which was absorbed in H_2O_2 to give H_2SO_4. The acid was titrated with 22.13 mL of 0.03736 M base. Calculate the percentage of active ingredient in the preparation.

10-29 Neohetramine, $C_{16}H_{21}ON_4$ (fw = 285.37), is a common antihistamine. A 0.1247 g sample containing this compound was analyzed by the Kjeldahl method. The ammonia produced was collected in H_3BO_3; the resulting $H_2BO_3^-$ was titrated with 26.13 mL of 0.01477 M HCl. Calculate the percentage of neohetramine in the sample.

*10-30 To obtain the percentage of protein in a wheat product, the percentage of nitrogen present is generally multiplied by 5.70. A 0.9092-g sample of a wheat flour was analyzed by the Kjeldahl procedure. The ammonia formed was distilled into 50.00 mL of 0.05063 M HCl; a 7.46-mL back-titration with 0.04917 M base was required. Calculate the percentage of protein in the flour.

10-31 The formaldehyde content of a pesticide preparation was determined by weighing 0.3124 g of the liquid sample into a flask containing 50.0 mL of 0.0996 M NaOH and 50 mL of 3% H_2O_2. Upon heating, the following reaction took place:

$$OH^- + HCHO + H_2O_2 \longrightarrow HCOO^- + 2H_2O$$

After cooling, the excess base was titrated with 23.3 mL of 0.05250 M H_2SO_4. Calculate the percentage of HCHO in the sample.

*10-32 A 3.00-L sample of urban air was bubbled through a solution containing 50.0 mL of 0.0116 M $Ba(OH)_2$, and $BaCO_3$ precipitated. The excess base was back-titrated to a phenolphthalein end point with 23.6 mL of 0.0108 M HCl. Calculate the parts per million of CO_2 in the air (that is, mL $CO_2/10^6$ mL air) if the density of CO_2 is 1.98 g/L.

10-33 Air was bubbled at a rate of 30.0 L/min through a trap containing 75 mL of 1% H_2O_2 ($H_2O_2 + SO_2 \longrightarrow H_2SO_4$). After 10.0 min, the H_2SO_4 was titrated with 11.1 mL of 0.00204 M NaOH. Calculate the parts per million of SO_2 (that is, mL $SO_2/10^6$ mL air) if the density of SO_2 is 0.00285 g/mL.

*10-34 What is the molarity of a $Ba(OH)_2$ solution if its titer is 1.00 mg HCl/mL?

10-35 What is the H_3AsO_4 titer of a 0.0676 M NaOH solution if phenolphthalein is to serve as the indicator?

*10-36 A 0.8160-g sample containing dimethylphthalate, $C_6H_4(COOCH_3)_2$ (fw = 194.19), and unreactive species was saponified by refluxing with 50.00 mL of 0.1031 M NaOH. After reaction was complete, the excess NaOH was back-titrated with 24.27 mL of 0.1644 M HCl. Calculate the percentage of dimethylphthalate in the sample.

10-37 A 50.00-mL sample containing methylethylketone, $CH_3COC_2H_5$ (fw = 72.108), and unreactive components was treated with an excess of hydroxylamine hydrochloride, $NH_2OH{\cdot}HCl$. After oxime formation was complete, the liberated HCl was titrated with 19.15 mL of 0.01123 M NaOH. Calculate the number of milligrams of ketone per liter of sample.

*10-38 A 50.00-mL sample containing acetaldehyde, ethyl acetate, and unreactive substances was diluted to 250.0 mL with H_2O. A 50.00-mL aliquot of the diluted sample was mixed with 40.00 mL of 0.05454 M NaOH and 20 mL of 3% H_2O_2. Upon refluxing for 30 min, quantitative saponification of the ester occurred; at the same time, the acetaldehyde was oxidized to sodium acetate:

$$H_2O_2 + CH_3CHO + OH^- \longrightarrow CH_3COO^- + 2H_2O$$

The unreacted NaOH consumed 10.05 mL of 0.02513 M HCl. A 25.00-mL aliquot of the sample solution was treated with $NH_2OH{\cdot}HCl$ to convert the acetaldehyde to the corresponding oxime:

$$CH_3CHO + NH_2OH{\cdot}HCl \longrightarrow CH_3CH{=}NOH + HCl + H_2O$$

The liberated HCl consumed 10.97 mL of the standard base. Calculate the weight-volume percentage of acetaldehyde and ethyl acetate in the sample.

10-39 A 1.219-g sample containing $(NH_4)_2SO_4$, NH_4NO_3, and nonreactive substances was diluted to 200 mL in a volumetric flask. A 50.00-mL aliquot was made basic with strong alkali, and the liberated NH_3 was distilled into 30.00 mL of 0.08421 M HCl. The excess HCl required 10.17 mL of 0.08802 M NaOH. A 25.00-mL aliquot of the sample was made alkaline after the addition of Devarda's alloy, and the NO_3^- was reduced to NH_3. The NH_3 from both NH_4^+ and NO_3^- was then distilled into 30.00 mL of the standard acid and back-titrated with 14.16 mL of the base. Calculate the percentage of $(NH_4)_2SO_4$ and NH_4NO_3 in the sample.

*10-40 The digestion of a 0.1417-g sample of a phosphorus-containing compound in a mixture of HNO_3 and H_2SO_4 resulted in the formation of CO_2, H_2O, and H_3PO_4. Addition of ammonium molybdate yielded a solid having the composition $(NH_4)_3PO_4{\cdot}12MoO_3$. This precipitate was filtered, washed, and dissolved in 50.00 mL of 0.2000 M NaOH:

$$(NH_4)_3PO_4{\cdot}12MoO_3(s) + 26OH^- \longrightarrow HPO_4^{2-} + 12MoO_4^{2-} + 14H_2O + 3NH_3(g)$$

After the solution was boiled to remove the NH_3, the excess NaOH was titrated with 14.17 mL of 0.1741 M HCl. Calculate the percentage of phosphorus in the sample.

*10-41 A 1.217-g sample of commercial KOH contaminated by K_2CO_3 was dissolved in water, and the resulting solution was diluted to 500.0 mL. A 50.00-mL aliquot of this solution was treated with 40.00 mL of 0.05304 M HCl and boiled to remove CO_2. The excess acid consumed 4.74 mL of 0.04983 M NaOH (phenolphthalein indicator). An excess of neutral $BaCl_2$ was added to another 50.00-mL aliquot to precipitate the carbonate as $BaCO_3$. The solution was then titrated with 28.56 mL of the acid to a phenolphthalein end point. Calculate the percentage KOH, K_2CO_3, and H_2O in the sample, assuming that these are the only compounds present.

10-42 A 0.5000-g sample containing $NaHCO_3$, Na_2CO_3, and H_2O was dissolved and diluted to 250.0 mL. A 25.00-mL aliquot was then boiled with 50.00 mL of 0.01255 M HCl. After cooling, the excess acid in the solution required 2.34 mL of 0.01063 M NaOH when titrated to a phenolphthalein end point. A second 25.00-mL aliquot was then treated with an excess of $BaCl_2$ and 25.00 mL of the base; precipitation of all the carbonate resulted, and 7.63 mL of the HCl was required to titrate the excess base. Calculate the composition of the mixture.

*10-43 Calculate the volume of 0.06122 M HCl needed to titrate

(a) 20.00 mL of 0.05555 M Na_3PO_4 to a thymolphthalein end point.
(b) 25.00 mL of 0.05555 M Na_3PO_4 to a bromocresol green end point.
(c) 40.00 mL of a solution that is 0.02102 M in Na_3PO_4 and 0.01655 M in Na_2HPO_4 to a bromocresol green end point.
(d) 20.00 mL of a solution that is 0.02102 M in Na_3PO_4 and 0.01655 M in NaOH to a thymolphthalein end point.

10-44 Calculate the volume of 0.07731 M NaOH needed to titrate

(a) 25.00 mL of a solution that is 0.03000 M in HCl and 0.01000 M in H_3PO_4 to a bromocresol green end point.
(b) the solution in (a) to a thymolphthalein end point.
(c) 30.00 mL of 0.06407 M NaH_2PO_4 to a thymolphthalein end point.
(d) 25.00 mL of a solution that is 0.02000 M in H_3PO_4 and 0.03000 M in NaH_2PO_4 to a thymolphthalein end point.

*10-45 A series of solutions containing NaOH, Na_2CO_3, and $NaHCO_3$, alone or in compatible combination, was titrated with 0.1202 M HCl. Tabulated below are the volumes of acid needed to titrate 25.00-mL portions of each solution to a (1) phenolphthalein and (2) bromocresol green end point. Use this information to deduce the composition of the solutions. In addition, calculate the number of milligrams of each solute per milliliter of solution.

	(1)	(2)
(a)	22.42	22.44
(b)	15.67	42.13
(c)	29.64	36.42
(d)	16.12	32.23
(e)	0.00	33.33

10-46 A series of solutions containing NaOH, Na_3AsO_4, and Na_2HAsO_4, alone or in compatible combination, was titrated with 0.08601 M HCl. Tabulated below are the volumes of acid needed to titrate 25.00-mL portions of each solution to a (1) phenolphthalein and (2) bromocresol green end point. Use this information to deduce the composition of the solutions. In addition, calculate the number of milligrams of each solute per milliliter of solution.

	(1)	(2)
(a)	0.00	18.15
(b)	21.00	28.15
(c)	19.80	39.61
(d)	18.04	18.03
(e)	16.00	37.37

*10-47 A series of solutions can contain HCl, H_3PO_4, or NaH_2PO_4, alone or in any compatible combination. Tabulated below are the volumes of 0.1200 M NaOH needed to titrate 25.00-mL portions of each solution to a (1) bromocresol green and (2) thymolphthalein end point. Use this information to deduce the composition of the solutions. In addition, calculate the number of milligrams of each solute per milliliter of solution.

	(1)	(2)
(a)	18.72	23.60
(b)	7.93	7.95
(c)	0.00	16.77
(d)	13.12	35.19
(e)	13.33	26.65

10-48 A series of solutions can contain HCl, maleic acid, and sodium hydrogen maleate, alone or in any compatible combination. Tabulated below are the volumes of 0.0994 M NaOH needed to titrate 25.00-mL portions of each solution to a (1) bromocresol green and (2) thymolphthalein end point. Use this information to deduce the composition of the solutions. In addition, calculate the number of milligrams of each solute per milliliter of solution.

	(1)	(2)
(a)	0.00	27.67
(b)	7.34	23.34
(c)	7.34	14.68
(d)	9.99	29.00
(e)	12.70	12.70

*10-49 (a) Derive a curve for the titration of 50.0 mL of 0.0500 M $HClO_4$ with 0.100 M C_2H_5ONa when both are dissolved in anhydrous ethanol. Assume that both the acid and the base are completely dissociated in this solvent and that the autoprotolysis constant of the solvent is 8×10^{-20}. Calculate the pH ($-\log [C_2H_5OH_2^+]$) after the addition of 0.00, 12.5, 24.0, 24.9, 25.0, 25.1, 26.0, and 30.0 mL of the base.
(b) Compare the pH change from 24.9 to 25.1 mL in (a) with the pH change for the same range with water as the solvent and NaOH as the base.

*10-50 Calculate the pH of each of the following solutions in water and in ethanol (pH = $-\log [C_2H_5OH_2^+]$), given that the dissociation constant for acetic acid in anhydrous ethanol is 5.6×10^{-11}, and the autoprotolysis constant for C_2H_5OH is 8×10^{-20}:
(a) 0.0500 M acetic acid.
(b) a solution that is 0.0500 M in acetic acid and 0.0500 M in sodium acetate.
(c) 0.0500 M sodium acetate.

*10-51 Calculate the pH of each of the following solutions in water and in anhydrous ethanol (pH = $-\log [C_2H_5OH_2^+]$), given that the base dissociation constant of aniline, $C_6H_5NH_2$, is 4.0×10^{-14} in ethanol, and the autoprotolysis constant for ethanol is 8×10^{-20}:
(a) 0.0100 M $C_6H_5NH_2$.
(b) a solution that is 0.0100 M in $C_6H_5NH_3^+$ and 0.0200 M in $C_6H_5NH_2$.
(c) 0.0100 M $C_6H_5NH_3^+$.

10-52 A 0.100 M solution of sodium ethoxide, C_2H_5ONa, in anhydrous ethanol was employed to titrate 50.00 mL of a 0.1000 M ethanolic solution of acetic acid:
(a) Derive a curve for this titration, calculating the pH ($-\log [C_2H_5OH_2^+]$) after the addition of 0.00, 10.00, 25.00, 40.00, 49.00, 49.90, 50.00, 50.10, 51.00, and 60.00 mL of base. (See Problem 10-50 for the necessary equilibrium constant.)
(b) Compare the data in (a) with the analogous data for the titration of 0.1000 M aqueous acetic acid with 0.1000 M aqueous NaOH.

*10-53 A 0.1000 M solution of sodium ethoxide, C_2H_5ONa, in anhydrous ethanol was employed to titrate 50.00 mL of a 0.1000 M ethanolic solution of anilinium chloride, $C_6H_5NH_3^+Cl^-$).

(a) Derive a curve for this titration, calculating the pH ($-\log [C_2H_5OH_2^+]$) after the addition of 0.00, 10.00, 25.00, 40.00, 49.00, 49.90, 50.00, 50.10, 51.00, and 60.00 mL of base. (See Problem 10-51 for the necessary equilibrium constant.)

(b) Compare the data in (a) with the analogous data for the titration of 0.1000 M aqueous anilinium chloride with 0.1000 M aqueous NaOH.

CHAPTER 11

Complex-Formation Titrations

Complex-formation titrations are widely used for determining metallic elements. For most applications, the reagents are organic compounds that contain several electron-donor groups that form stable covalent bonds with metal ions.

11A Complex-Formation Reactions

Most metal ions react with electron-pair donors to form coordination compounds or complex ions. The donor species, or *ligand,* must have at least one pair of unshared electrons available for bond formation. Water, ammonia, and halide ions are common inorganic ligands.

The *coordination number* of a cation is the number of covalent bonds that a cation tends to form with electron donor groups. Typical values for the coordination number are two, four, and six. The species formed as a result of coordination can be electrically positive, neutral, or negative. For example, copper(II), which has a coordination number of four, forms a cationic ammine complex, $Cu(NH_3)_4^{2+}$; a neutral complex with glycine, $Cu(NH_2CH_2COO)_2$; and an anionic complex with chloride ion, $CuCl_4^{2-}$.

Titrimetric methods based upon complex formation (sometimes called *complexometric methods*) have been used for at least a century. The truly remarkable growth in their analytical application is of more recent origin, however, and is based upon a particular class of coordination compounds called *chelates.* A chelate is produced when a metal ion coordinates with two (or more) donor groups of a single ligand to form a five- or six-membered heterocyclic ring. The copper complex of glycine just mentioned is an example. Here, copper bonds to both the oxygen of the carboxyl group and the nitrogen of the amine group:

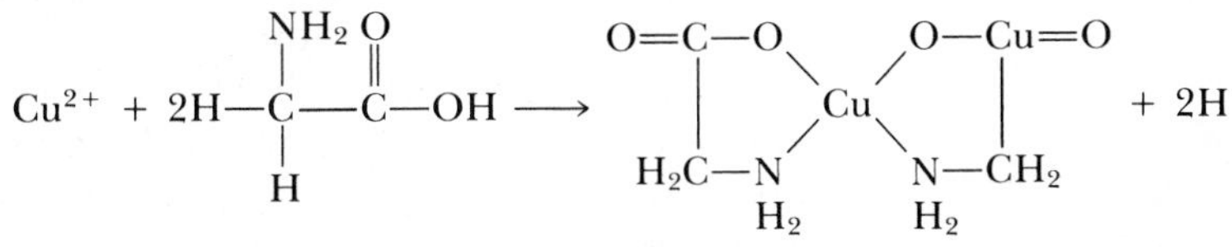

$$Cu^{2+} + 2H-\underset{H}{\overset{NH_2}{C}}-\overset{O}{\overset{\|}{C}}-OH \longrightarrow \text{[chelate: } O{=}C{-}O{-}Cu{-}O{-}C{=}O \text{ rings closed through } H_2C{-}N(H_2) \text{ and } N(H_2){-}CH_2\text{]} + 2H^+$$

A ligand that has a single donor group, such as ammonia, is called *unidentate* (single-toothed), whereas one such as glycine, which has two groups available for coordination bonding, is called *bidentate. Tridentate, tetradentate, pentadentate,* and *hexadentate* chelating agents are also known.

As titrants, multidentate ligands have two advantages over their unidentate counterparts. First, they generally react more completely with cations and thus provide sharper end points. Second, they ordinarily react with metal ions in a single-step process, whereas complex formation with unidentate ligands usually involves two or more intermediate species. Consider, for example, the equilibrium that exists between the metal ion M with a coordination number of four and the tetradentate ligand D.[1]

$$M + D \rightleftharpoons MD$$

The equilibrium expression for this process is

$$K_f = \frac{[MD]}{[M][D]}$$

where K_f is the *formation constant.*

Similarly, the equilibrium between M and the bidentate ligand B can be represented by

$$M + 2B \rightleftharpoons MB_2$$

This reaction, however, is the sum of a two-step process that involves formation of the intermediate MB:

$$M + B \rightleftharpoons MB$$

$$MB + B \rightleftharpoons MB_2$$

for which

$$K_1 = \frac{[MB]}{[M][B]} \qquad K_2 = \frac{[MB_2]}{[MB][B]}$$

The product K_1K_2 yields the equilibrium constant for the overall process.

$$\beta_2 = K_1K_2 = \frac{\cancel{[MB]}}{[M][B]} \times \frac{[MB_2]}{\cancel{[MB]}[B]} = \frac{[MB_2]}{[M][B]^2}$$

In a like manner, the reaction between M and the unidentate ligand A involves the overall equilibrium

$$M + 4A \rightleftharpoons MA_4$$

and the equilibrium constant β_4 for the formation of MA_4 from M and A is numerically equal to the product of the equilibrium constants for the four constituent processes.

Each of the titration curves depicted in Figure 11-1 involves a reaction that has an overall equilibrium constant of 10^{20}. Curve *A* is derived for the formation of MD in a single step. Curve *B* is based upon the formation of MB_2 in two steps, for which K_1 is 10^{12} and K_2 is 10^8. Curve *C* was obtained by assuming that MA_4 forms in four steps having successive formation constants of 10^8, 10^6, 10^4, and 10^2. These curves demonstrate that a ligand that combines with a metal ion in a 1:1 ratio is clearly superior to one that forms one or more intermediates because the change in pM in the equivalence-point region is largest with the 1:1 system. Thus, polydentate ligands, which combine in lower ratios with metal ions, are ordinarily preferred for complex-formation titrations.

11B Titrations with Aminopolycarboxylic Acids

Tertiary amines that also contain carboxylic acid groups form remarkably

[1]The electrostatic charges associated with M and D determine the charge of the product but are not important in terms of the present argument and are therefore not shown.

FIGURE 11-1

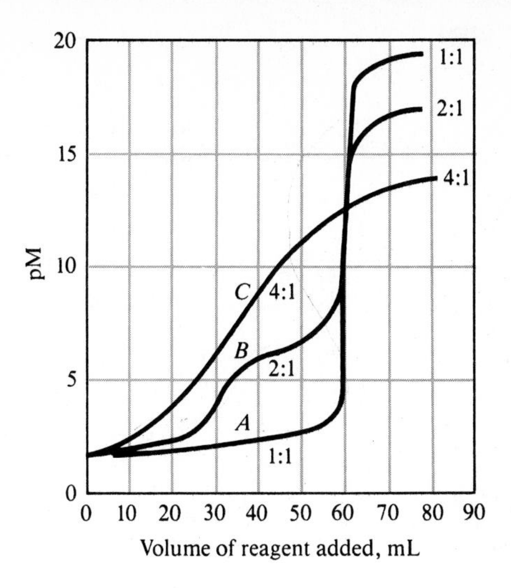

Curves for complex-formation titrations. Titration of 60.0 mL of a solution that is 0.020 M in M with (*A*) a 0.020 M solution of the tetradentate ligand D to give MD as product, (*B*) a 0.040 M solution of the bidentate ligand B to give MB_2, and (*C*) a 0.080 M solution of the unidentate ligand A to give MA_4. The overall formation constant for each product is 1.0×10^{20}.

stable chelates with many metal ions.[2] Schwarzenbach first recognized their potential as analytical reagents in 1945. Since this original work, investigators throughout the world have described applications of these compounds to the volumetric determination of most of the metals in the periodic table.

The discussion that follows is based largely on ethylenediaminetetraacetic acid, a reagent that has become one of the most widely used titrants in analytical chemistry.

11B-1 *Ethylenediaminetetraacetic Acid*

Ethylenediaminetetraacetic acid [also called (ethylenedinitrilo)tetraacetic acid], which is commonly shortened to EDTA, has the structure

```
HOOC—CH2                      CH2—COOH
        \                    /
         :N—CH2—CH2—N:
        /                    \
HOOC—CH2                      CH2—COOH
```

The dissociation constants for the four carboxylic acid groups are $K_1 = 1.02 \times 10^{-2}$, $K_2 = 2.14 \times 10^{-3}$, $K_3 = 6.92 \times 10^{-7}$, and $K_4 = 5.50 \times 10^{-11}$. It is of interest that the first two constants are of the same order of magnitude, which suggests that the two protons involved dissociate from opposite ends of the rather long molecule. As a consequence of their physical separation, the negative charge created by the first dissociation does not greatly inhibit the removal of the second proton. The same cannot be said for the dissociation of the other two protons, however, which are much closer to the negatively charged carboxylate ions created by the initial dissociations.

Note that the molecule has six potential sites for bonding a metal ion: the four carboxyl groups and the two amino groups, each of the latter with an unshared pair of electrons. Thus, EDTA is a hexadentate ligand.

The various EDTA species are often abbreviated H_4Y, H_3Y^-, H_2Y^{2-}, HY^{3-}, and Y^{4-}.

The free acid, H_4Y, and the dihydrate of the sodium salt, $Na_2H_2Y \cdot 2H_2O$, are available in reagent quality. The former can serve as a primary standard

[2]These reagents are the subject of several excellent monographs. See, for example, R. Pribil, *Applied Complexometry*. New York: Pergamon, 1982; A. Ringbom and E. Wänninen, in *Treatise on Analytical Chemistry*, 2nd ed., I. M. Kolthoff and P. J. Elving, Eds., Part I, Vol. 2, Chapter 11. New York: Wiley, 1979.

after it has been dried for several hours at 130 to 145°C. It is then dissolved in the minimum amount of base required for complete solution.

Under normal atmospheric conditions, the dihydrate $Na_2H_2Y \cdot 2H_2O$ contains 0.3% moisture in excess of the stoichiometric amount. For all but the most exacting work, this excess is sufficiently reproducible to permit use of a corrected weight of the salt in the direct preparation of a standard solution. If necessary, the pure dihydrate can be prepared by drying at 80°C in an atmosphere of 50% relative humidity for several days.

Another common reagent is nitrilotriacetic acid (abbreviated NTA), which has the structure

$$\mathrm{HOOC{-}CH_2{-}\ddot{N}(CH_2{-}COOH)_2}$$

Aqueous solutions of this tetradentate ligand are prepared from the free acid.

Other related substances have also been investigated but have not been as widely applied as titrants. We shall thus confine our discussion to the properties and applications of EDTA.

11B-2 Complexes of EDTA and Metal Ions

Solutions of EDTA are particularly valuable as titrants because the reagent combines with metal ions in a 1:1 ratio regardless of the charge on the cation. For example, formation of the silver and aluminum complexes is described by the equations

$$Ag^+ + Y^{4-} \rightleftharpoons AgY^{3-}$$

$$Al^{3+} + Y^{4-} \rightleftharpoons AlY^-$$

EDTA is a remarkable reagent not only because it forms chelates with all cations but also because most of these chelates are sufficiently stable to form the basis for a titrimetric method. This great stability undoubtedly results from the several complexing sites within the molecule that give rise to a cagelike structure in which the cation is effectively surrounded and isolated from solvent molecules. One form of the complex is depicted in Figure 11-2. Note that all six ligand groups are involved in bonding the divalent metal ion.

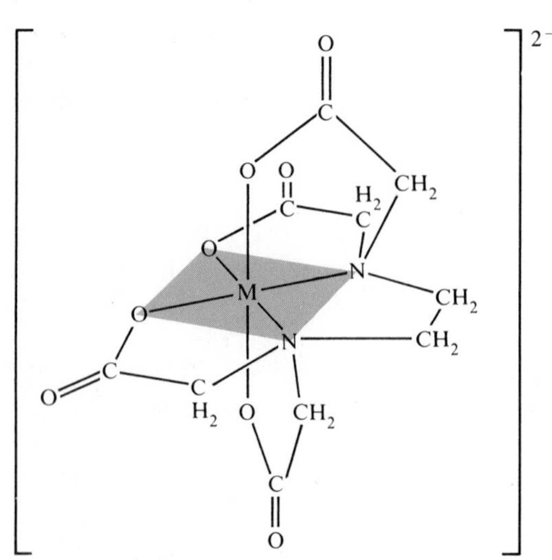

FIGURE 11-2 Structure of a metal/EDTA chelate.

Table 11-1 lists formation constants K_{MY} for common EDTA complexes. Note that the constant refers to the equilibrium involving the species Y^{4-} with the metal ion:

$$M^{n+} + Y^{4-} \rightleftharpoons MY^{(n-4)+} \qquad \frac{[MY^{(n-4)+}]}{[M^{n+}][Y^{4-}]} = K_{MY} \tag{11-1}$$

11B-3

Equilibrium Calculations Involving EDTA

The ordinate for EDTA titration curves is generally pM because most of the indicators used in these titrations are metal ion indicators, which respond to changes in the concentration of the analyte cation M^{n+}. Values for pM are readily computed in the early stage of an EDTA titration because the equilibrium concentration of M^{n+} can be assumed to be the same as its analytical concentration, which in turn is readily derived from stoichiometric data.

In order to calculate the concentration of M^{n+} at and beyond the equivalence point, however, Equation 11-1, which contains three unknown terms—$[Y^{4-}]$, $[MY^{(n-4)+}]$, and $[M^{n+}]$—must be used. The first of these terms is directly dependent upon pH because Y^{4-} is a conjugate base. This pH sensitivity makes evaluating the equilibrium concentrations of the three species troublesome and time-consuming when hydrogen ion concentration is an additional unknown variable. Fortunately, however, EDTA titrations are always performed on solutions that are buffered to a known pH because interferences are often avoided by exercising pH control. Calculating the concentration of M^{n+} in a buffered solution containing EDTA is a relatively straightforward procedure provided the pH is known. In this procedure, use is made of the alpha value for Y^{4-}. Recall from Section 9D that α_4 for Y^{4-} is defined as

$$\alpha_4 = \frac{[Y^{4-}]}{c_T} \tag{11-2}$$

where c_T is the total molar concentration of *uncomplexed* EDTA:

$$c_T = [Y^{4-}] + [HY^{3-}] + [H_2Y^{2-}] + [H_3Y^{-}] + [H_4Y]$$

Conditional Formation Constants

Conditional, or *effective, formation constants* are equilibrium constants that are pH-dependent and apply *at a single pH only.* To obtain the conditional constant

TABLE 11-1

Formation Constants for EDTA Complexes*

Cation	K_{MY}	log K_{MY}	Cation	K_{MY}	log K_{MY}
Ag^{+}	2.1×10^{7}	7.32	Cu^{2+}	6.3×10^{18}	18.80
Mg^{2+}	4.9×10^{8}	8.69	Zn^{2+}	3.2×10^{16}	16.50
Ca^{2+}	5.0×10^{10}	10.70	Cd^{2+}	2.9×10^{16}	16.46
Sr^{2+}	4.3×10^{8}	8.63	Hg^{2+}	6.3×10^{21}	21.80
Ba^{2+}	5.8×10^{7}	7.76	Pb^{2+}	1.1×10^{18}	18.04
Mn^{2+}	6.2×10^{13}	13.79	Al^{3+}	1.3×10^{16}	16.13
Fe^{2+}	2.1×10^{14}	14.33	Fe^{3+}	1.3×10^{25}	25.1
Co^{2+}	2.0×10^{16}	16.31	V^{3+}	7.9×10^{25}	25.9
Ni^{2+}	4.2×10^{18}	18.62	Th^{4+}	1.6×10^{23}	23.2

*Data from G. Schwarzenbach, *Complexometric Titrations,* p. 8. London: Chapman and Hall, 1957. With permission. Constants valid at 20°C and an ionic strength of 0.1.

for the equilibrium shown in Equation 11-1, we use Equation 11-2 to eliminate $[Y^{4-}]$ from the formation-constant expression (Equation 11-1):

$$M^{n+} + Y^{4-} \rightleftharpoons MY^{(n-4)+} \qquad \frac{[MY^{(n-4)+}]}{[M^{n+}]\alpha_4 c_T} = K_{MY} \tag{11-3}$$

Combining the two constants yields a new constant:

$$\frac{[MY^{(n-4)+}]}{[M^{n+}]c_T} = \alpha_4 K_{MY} = K'_{MY} \tag{11-4}$$

where K'_{MY}, the conditional formation constant, describes equilibrium relationships *only at the pH for which α_4 is applicable.*

Conditional constants are readily computed and provide a simple means by which the equilibrium concentrations of the metal ion and the complex can be calculated at any point in a titration curve. Note that the expression for the conditional constant differs from that for the formation constant only in the respect that, in the former, the term c_T replaces the equilibrium concentration of the completely dissociated anion, $[Y^{4-}]$. This difference is significant, however, because c_T is readily determined from the reaction stoichiometry, whereas $[Y^{4-}]$ is not.

The Computation of α_4 Values for EDTA Solutions

An expression for calculating α_4 at a given hydrogen ion concentration is readily derived by the method demonstrated in Section 9D. Thus, α_4 for EDTA is

$$\alpha_4 = \frac{K_1K_2K_3K_4}{[H^+]^4 + K_1[H^+]^3 + K_1K_2[H^+]^2 + K_1K_2K_3[H^+] + K_1K_2K_3K_4} \tag{11-5}$$

$$\alpha_4 = \frac{K_1K_2K_3K_4}{D} \tag{11-6}$$

where K_1, K_2, K_3, and K_4 are the four dissociation constants for H_4Y and D is the denominator of Equation 11-5.[3]

Alpha values for the other EDTA species are readily obtained in a similar way and are found to be

$$\alpha_0 = [H^+]^4/D \qquad \alpha_2 = K_1K_2[H^+]^2/D$$

$$\alpha_1 = K_1[H^+]^3/D \qquad \alpha_3 = K_1K_2K_3[H^+]/D$$

Figure 11-3 illustrates how the relative amounts of the five EDTA species vary as a function of pH. It is apparent that H_2Y^{2-} predominates in a moderately acidic medium (pH 3 to 6). Only at pH values greater than 10 does Y^{4-} become a major component of the solution.

[3]In this chapter and those that follow, we shall revert to the use of H^+ as a convenient shorthand notation for H_3O^+; from time to time, we shall refer to the species represented by H^+ as the hydrogen ion.

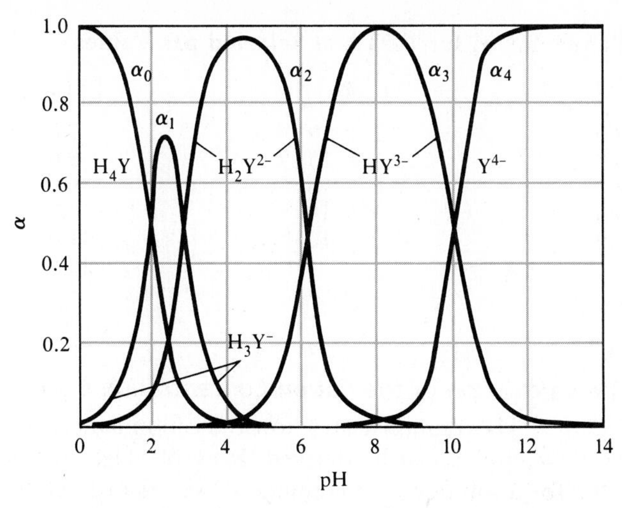

FIGURE 11-3 Composition of EDTA solutions as a function of pH.

EXAMPLE 11-1

Calculate α_4 and the mole percent of Y^{4-} in a solution of EDTA that is buffered to pH 10.20.

$$[H^+] = \text{antilog}(-10.20) = 6.31 \times 10^{-11} \simeq 6.3 \times 10^{-11}$$

From the values for the dissociation constants for H_4Y (page 259), we obtain

$$K_1 = 1.02 \times 10^{-2} \qquad K_1K_2K_3 = 1.51 \times 10^{-11}$$

$$K_1K_2 = 2.18 \times 10^{-5} \qquad K_1K_2K_3K_4 = 8.31 \times 10^{-22}$$

Numerical values for the several terms in the denominator in Equation 11-5 are

$$[H^+]^4 = (6.31 \times 10^{-11})^4 = 1.58 \times 10^{-41}$$

$$K_1[H^+]^3 = (1.02 \times 10^{-2})(6.31 \times 10^{-11})^3 = 2.56 \times 10^{-33}$$

$$K_1K_2[H^+]^2 = (2.18 \times 10^{-5})(6.31 \times 10^{-11})^2 = 8.68 \times 10^{-26}$$

$$K_1K_2K_3[H^+] = (1.51 \times 10^{-11})(6.31 \times 10^{-11}) = 9.53 \times 10^{-22}$$

$$K_1K_2K_3K_4 = \underline{8.31 \times 10^{-22}}$$

$$D = 1.78 \times 10^{-21}$$

and Equation 11-6 becomes

$$\alpha_4 = \frac{K_1K_2K_3K_4}{D} = \frac{8.31 \times 10^{-22}}{1.78 \times 10^{-21}} = 0.466 \simeq 0.47$$

$$\text{mol \% } Y^{4-} = 0.47 \times 100\% = 47\%$$

Note that only the last two terms in the denominator contribute significantly to the sum at pH 10.20. At low pH values, in contrast, only the first two or three terms are important.

Table 11-2 lists α_4 at selected pH values. Note that only about 4×10^{-12} percent of the EDTA exists as Y^{4-} at pH 2.00.

TABLE 11-2

Values for α_4 for EDTA at Selected pH Values

pH	α_4	pH	α_4
2.0	3.7×10^{-14}	7.0	4.8×10^{-4}
3.0	2.5×10^{-11}	8.0	5.4×10^{-3}
4.0	3.6×10^{-9}	9.0	5.2×10^{-2}
5.0	3.5×10^{-7}	10.0	3.5×10^{-1}
6.0	2.2×10^{-5}	11.0	8.5×10^{-1}
		12.0	9.8×10^{-1}

The Calculation of the Cation Concentration in EDTA Solutions

Example 11-2 demonstrates how the cation concentration in a solution of an EDTA complex can be derived. Example 11-3 shows how the calculation is done for a solution that contains an excess of EDTA.

EXAMPLE 11-2

Calculate the equilibrium concentration of Ni^{2+} in a solution with an analytical NiY^{2-} concentration of 0.0150 M at pH (a) 3.0 and (b) 8.0.

From Table 11-1,

$$Ni^{2+} + Y^{4-} \rightleftharpoons NiY^{4-} \qquad \frac{[NiY^{4-}]}{[Ni^{2+}][Y^{4-}]} = K_{MY} = 4.2 \times 10^{18}$$

The equilibrium concentration of NiY^{4-} is equal to the analytical concentration of the complex minus the concentration lost by dissociation. The latter is given by the equilibrium nickel ion concentration. Thus,

$$[NiY^{4-}] = 0.0150 - [Ni^{2+}]$$

If we assume $[Ni^{2+}] \ll 0.0150$, an assumption that is almost certainly valid in light of the large formation constant of the complex, the foregoing equation simplifies to

$$[NiY^{4-}] \simeq 0.0150$$

Since the complex is the only source of both Ni^{2+} and the EDTA species,

$$[Ni^{2+}] = [Y^{4-}] + [HY^{3-}] + [H_2Y^{2-}] + [H_3Y^{-}] + [H_4Y] = c_T$$

Substitution of this equality into Equation 11-4 gives

$$\frac{[NiY^{2-}]}{[Ni^{2+}]c_T} = \frac{[NiY^{2-}]}{[Ni^{2+}]^2} = \alpha_4 K_{MY} = K'_{MY} \qquad (11\text{-}7)$$

(a) Table 11-2 indicates that α_4 is 2.5×10^{-11} at pH 3.0. Substitution of this value and the concentration of NiY^{2-} into Equation 11-7 gives

$$\frac{0.0150}{[Ni^{2+}]^2} = 2.5 \times 10^{-11} \times 4.2 \times 10^{18} = 1.05 \times 10^{8}$$

$$[Ni^{2+}] = \sqrt{1.43 \times 10^{-10}} = 1.2 \times 10^{-5}$$

Note that $[Ni^{2+}] \ll 0.0150$, as assumed.

(b) At pH 8.0, the conditional constant is much larger. Thus,

$$K'_{MY} = 5.4 \times 10^{-3} \times 4.2 \times 10^{18} = 2.27 \times 10^{16}$$

and substitution into Equation 11-7 followed by rearrangement gives

$$[Ni^{2+}] = \sqrt{0.0150/(2.27 \times 10^{16})} = 8.1 \times 10^{-10}$$

EXAMPLE 11-3

Calculate the concentration of Ni^{2+} in a solution of pH 3.00 that was prepared by mixing 50.0 mL of 0.0300 M Ni^{2+} with 50.0 mL of 0.0500 M EDTA.

Here, an excess of EDTA is present and the analytical concentration of the complex is determined by the amount of Ni^{2+} originally present. Thus,

$$c_{NiY^{2-}} = \frac{50.0 \times 0.0300}{100} = 0.0150 \text{ M}$$

$$c_{EDTA} = \frac{50.0 \times 0.0500 - 50.0 \times 0.0300}{100} = 0.0100 \text{ M}$$

Again let us assume that $[Ni^{2+}] \ll [NiY^{2-}]$ so that

$$[NiY^{2-}] = 0.0150 - [Ni^{2+}] \simeq 0.0150$$

At this point, the total concentration of uncomplexed EDTA is given by its molarity:

$$c_T = 0.0100 \text{ M}$$

Applying Equation 11-4 gives

$$\frac{0.0150}{[Ni^{2+}]0.0100} = \alpha_4 K_{MY} = K'_{MY} = 2.5 \times 10^{-11} \times 4.2 \times 10^{18} = 1.05 \times 10^{8}$$

$$[Ni^{2+}] = \frac{0.0150}{0.0100 \times 1.05 \times 10^{8}} = 1.4 \times 10^{-8}$$

11B-4 The Derivation of Curves for EDTA Titrations

Example 11-4 demonstrates how an EDTA titration curve for a metal ion is derived when the pH of the analyte solution is fixed and known.

EXAMPLE 11-4

Derive a curve (pCa as a function of volume of EDTA) for the titration of 50.0 mL of 0.00500 M Ca^{2+} with 0.0100 M EDTA in a solution buffered to a constant pH of 10.0.

Calculation of a conditional constant

The conditional formation constant for the calcium-EDTA complex at pH 10 can be obtained from the formation constant of the complex (Table 11-1) and the α_4 value for EDTA at pH 10 (Table 11-2). Thus, substitution into Equation 11-4 gives

$$\frac{[CaY^{2-}]}{[Ca^{2+}]c_T} = \alpha_4 K_{CaY} = K'_{CaY}$$

$$= 0.35 \times 5.0 \times 10^{10} = 1.75 \times 10^{10}$$

Preequivalence-point values for pCa

Before the equivalence point is reached, the equilibrium concentration of Ca^{2+} is equal to the sum of the contributions from the untitrated excess of the cation and from the dissociation of the complex, the latter being numerically equal to c_T. It is ordinarily reasonable to assume that c_T is small relative to the analytical concentration of the uncomplexed calcium ion. Thus, for example, after the addition of 10.0 mL of reagent,

$$[Ca^{2+}] = \frac{50.0 \times 0.00500 - 10.0 \times 0.0100}{60.0} + c_T \simeq 2.50 \times 10^{-3}$$

$$pCa = -\log 2.50 \times 10^{-3} = 2.60$$

The equivalence-point pCa

Following the method shown in Example 11-2, we first compute the analytical concentration of CaY^{2-}:

$$c_{CaY^{2-}} = \frac{50.0 \times 0.00500}{50.0 + 25.0} = 3.33 \times 10^{-3}$$

The only source of Ca^{2+} ions is the dissociation of this complex. It also follows that the calcium ion concentration must be identical to the sum of the concentrations of the uncomplexed EDTA ions, c_T. Thus,

$$[Ca^{2+}] = c_T$$

$$[CaY^{2-}] = 0.00333 - [Ca^{2+}] \simeq 0.00333$$

Substituting into the conditional formation-constant expression gives

$$\frac{0.00333}{[Ca^{2+}]^2} = 1.75 \times 10^{10}$$

$$[Ca^{2+}] = \sqrt{\frac{0.00333}{1.75 \times 10^{10}}} = 4.36 \times 10^{-7}$$

$$pCa = -\log 4.36 \times 10^{-7} = 6.36$$

Postequivalence-point pCa

Beyond the equivalence point, analytical concentrations of CaY^{2-} and EDTA are obtained directly from the stoichiometric data. Here, a calculation similar to that in Example 11-3 is performed. Thus, after 35.0 mL of reagent has been added.

$$c_{CaY^{2-}} = \frac{50.0 \times 0.00500}{85.0} = 2.94 \times 10^{-3}\ M$$

$$c_{EDTA} = \frac{35.0 \times 0.0100 - 50.0 \times 0.00500}{85.0} = 1.18 \times 10^{-3}\ M$$

As an approximation, we can write

$$[CaY^{2-}] = 2.94 \times 10^{-3} - [Ca^{2+}] \simeq 2.94 \times 10^{-3}$$

$$c_T = 1.18 \times 10^{-3} + [Ca^{2+}] \simeq 1.18 \times 10^{-3}\ M$$

and substitution into the conditional formation-constant expression gives

$$\frac{2.94 \times 10^{-3}}{[Ca^{2+}] \times 1.18 \times 10^{-3}} = 1.75 \times 10^{10} = K'_{CaY}$$

$$[Ca^{2+}] = \frac{2.94 \times 10^{-3}}{1.18 \times 10^{-3} \times 1.75 \times 10^{10}} = 1.42 \times 10^{-10}$$

$$pCa = -\log 1.42 \times 10^{-10} = 9.85$$

Curve A in Figure 11-4 is a plot of data derived by the methods illustrated in Example 11-4. Curve B is the titration curve for a solution of magnesium ion under identical conditions. The formation constant for the EDTA complex of magnesium is smaller than that of the calcium complex. Consequently, the reaction of calcium ion with the EDTA is more complete, which leads to a larger change in p-function in the equivalence region. The effect here is analogous to the effects seen earlier for precipitation and neutralization titrations.

Figure 11-5 provides titration curves for calcium ion in solutions buffered to various pH levels. Obviously, end points are less sharp at lower pH values. The cause of the poorer end points is the diminished concentration of active

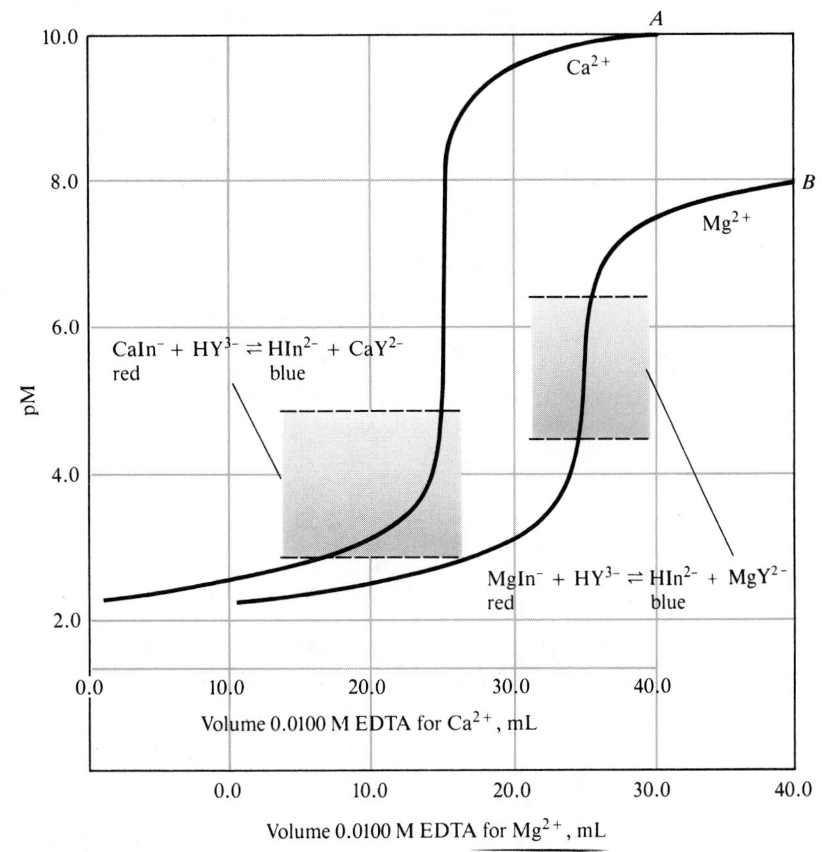

FIGURE 11-4 EDTA titration curves for 50.00 mL of 0.00500 M Ca^{2+} (K' for CaY^{2-} = 1.75×10^{10}) and Mg^{2+} (K' for MgY^{2-} = 1.72×10^{8}) at pH 10.0. The shaded areas show the transition range for Eriochrome Black T.

reagent (Y^{4-}) that accompanies lower pH values. It is apparent from Figure 11-5 that an adequate end point in the titration of calcium requires a pH of about 8 or greater. As shown in Figure 11-6, however, cations with larger formation constants provide good end points even in acidic media. Figure 11-7 shows the minimum permissible pH for a satisfactory end point in the

FIGURE 11-5 Influence of pH upon the titration of 0.0100 M Ca^{2+} with 0.0100 M EDTA.

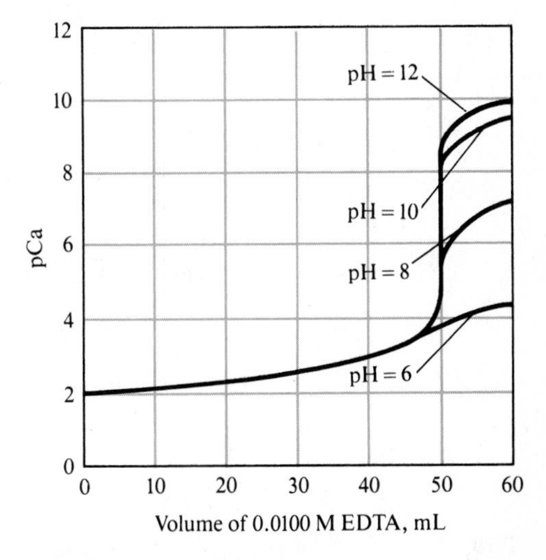

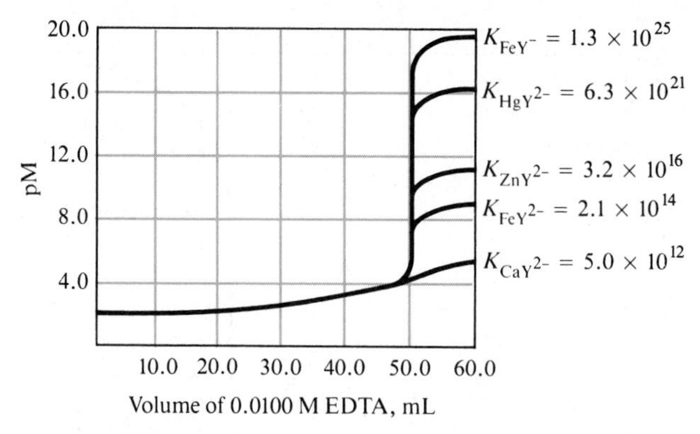

FIGURE 11-6 Titration curves for 50.0 mL of 0.0100 M cation solutions at pH 6.0.

FIGURE 11-7 Minimum pH needed for satisfactory titration of various cations with EDTA. (From C. N. Reilley and R. W. Schmid, *Anal. Chem.*, **1958,** *30,* 947. With permission of the American Chemical Society.)

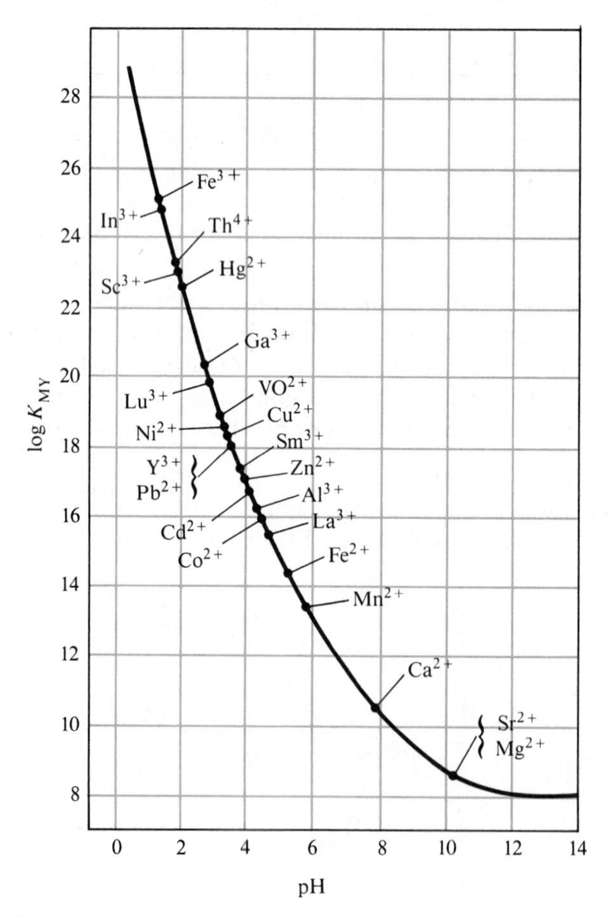

titration of various metal ions in the absence of competing complexing agents. Note that a moderately acidic environment is satisfactory for many divalent heavy-metal cations and that a strongly acidic medium can be tolerated in the titration of such ions as iron(III) and indium(III).

The Effect of Other Complexing Agents on EDTA Titrations

Many metal ions tend to form hydroxides or basic oxides at the pH levels required for their titration with EDTA. To avoid the problems created by formation of these precipitates, it is common practice to incorporate an auxiliary complexing agent when the solution is being buffered. For example, the titration of zinc(II) is ordinarily performed in the presence of moderate concentrations of ammonia and ammonium chloride. The ammonia serves not only as part of a buffer but also as a complexing agent that forms four ammine complexes that prevent precipitation of zinc hydroxide: $Zn(NH_3)^{2+}$, $Zn(NH_3)_2^{2+}$, $Zn(NH_3)_3^{2+}$, and $Zn(NH_3)_4^{2+}$. As expected, the added complexing agent competes with EDTA for the analyte ions and thus has the unfortunate effect of making the analytical reaction less complete.

A quantitative description of the effects of an auxiliary complexing reagent can be derived by a procedure similar to that used to determine the influence of pH on EDTA titration curves. In this instance, a quantity α_M is defined that is analogous to α_4:

$$\alpha_M = \frac{[M^{n+}]}{c_M} \tag{11-8}$$

where c_M is the sum of the concentrations of species containing the metal ion *exclusive* of that combined with EDTA. For solutions containing zinc(II) and ammonia, then

$$c_M = [Zn^{2+}] + [Zn(NH_3)^{2+}] + [Zn(NH_3)_2^{2+}] + [Zn(NH_3)_3^{2+}] + [Zn(NH_3)_4^{2+}] \tag{11-9}$$

The value of α_M can be readily expressed in terms of the ammonia concentration and the formation constants for the various ammine complexes. To arrive at such an expression, we write

$$K_1 = \frac{[Zn(NH_3)^{2+}]}{[Zn^{2+}][NH_3]}$$

$$[Zn(NH_3)^{2+}] = K_1[Zn^{2+}][NH_3]$$

Similarly, it is readily shown that

$$[Zn(NH_3)_2^{2+}] = K_1K_2[Zn^{2+}][NH_3]^2$$

$$[Zn(NH_3)_3^{2+}] = K_1K_2K_3[Zn^{2+}][NH_3]^3$$

$$[Zn(NH_3)_4^{2+}] = K_1K_2K_3K_4[Zn^{2+}][NH_3]^4$$

Substitution of these expressions into Equation 11-9 gives

$$c_M = [Zn^{2+}](1 + K_1[NH_3] + K_1K_2[NH_3]^2 + K_1K_2K_3[NH_3]^3 + K_1K_2K_3K_4[NH_3]^4)$$

Substituting this expression for c_M in Equation 11-8 (here, $[M^{n+}] = [Zn^{2+}]$) yields

$$\alpha_M = \frac{1}{1 + K_1[NH_3] + K_1K_2[NH_3]^2 + K_1K_2K_3[NH_3]^3 + K_1K_2K_3K_4[NH_3]^4} \tag{11-10}$$

Finally, a conditional constant for the equilibrium between EDTA and zinc(II) in an ammonia/ammonium chloride buffer is obtained by substituting Equation 11-8 into Equation 11-4 and rearranging:

$$\frac{[ZnY^{2-}]}{c_M c_T} = \alpha_4 \alpha_M K_{ZnY} = K''_{ZnY} \tag{11-11}$$

where K''_{ZnY} is a new conditional constant that applies at a single pH as well as a single concentration of ammonia. The following example shows how this conditional constant is employed in the derivation of titration curves.

EXAMPLE 11-5

Calculate the pZn of solutions prepared by adding 20.0, 25.0, and 30.0 mL of 0.0100 M EDTA to 50.0 mL of 0.00500 M Zn^{2+}. Assume that both the Zn^{2+} and EDTA solutions are 0.100 M in NH_3 and 0.176 M in NH_4Cl to provide a constant pH of 9.0.

In Appendix 6, we find that the logarithms of the stepwise formation constants for the four zinc complexes with ammonia are 2.4, 2.4, 2.5, and 2.1. Thus,

$$K_1 = \text{antilog } 2.4 = 2.51 \times 10^2$$

$$K_1K_2 = \text{antilog } (2.4 + 2.4) = 6.31 \times 10^4$$

$$K_1K_2K_3 = \text{antilog } (2.4 + 2.4 + 2.5) = 2.00 \times 10^7$$

$$K_1K_2K_3K_4 = \text{antilog } (2.4 + 2.4 + 2.5 + 2.1) = 2.51 \times 10^9$$

Calculation of a conditional constant

A value for α_M can be obtained from Equation 11-10 by assuming that the molar and analytical concentrations of ammonia are essentially the same; thus, for $[NH_3] = 0.100$,

$$\alpha_M = \frac{1}{1 + 25 + 631 + 2.00 \times 10^4 + 2.51 \times 10^5} = 3.68 \times 10^{-6}$$

A value for K_{ZnY} is found in Table 11-1, and α_4 for pH 9.0 is given in Table 11-2. Substituting into Equation 11-11, we find

$$K''_{ZnY} = 5.2 \times 10^{-2} \times 3.68 \times 10^{-6} \times 3.2 \times 10^{16} = 6.12 \times 10^9 \simeq 6.1 \times 10^9$$

Calculation of pZn after addition of 20.0 mL of EDTA

At this point, only part of the zinc has been complexed by EDTA. The remainder is present as Zn^{2+} and the four ammine complexes. By definition, the sum of the concentrations of these five species is c_M. Therefore,

$$c_M = \frac{50.0 \times 0.00500 - 20.0 \times 0.0100}{70.0} = 7.14 \times 10^{-4}\text{ M}$$

Substitution of this value into Equation 11-9 gives

$$[Zn^{2+}] = c_M \alpha_M = (7.14 \times 10^{-4})(3.68 \times 10^{-6}) = 2.63 \times 10^{-9}\text{ M}$$

$$\text{pZn} = 8.58$$

Calculation of pZn after addition of 25.0 mL of EDTA

At the equivalence point, the analytical concentration of ZnY^{2-} is

$$c_{ZnY^{2-}} = \frac{50.0 \times 0.00500}{50.0 + 25.0} = 3.33 \times 10^{-3}\text{ M}$$

The sum of the concentrations of the various zinc species not combined with EDTA equals the sum of the concentrations of the uncomplexed EDTA species:

$$c_M = c_T$$

and

$$[ZnY^{2-}] = 3.33 \times 10^{-3} - c_M \approx 3.33 \times 10^{-3}\ M$$

Substituting into Equation 11-11, we have

$$\frac{3.33 \times 10^{-3}}{c_M^2} = 6.12 \times 10^9 = K''_{ZnY}$$

$$c_M = 7.38 \times 10^{-7}\ M$$

Employing Equation 11-8, we obtain

$$[Zn^{2+}] = c_M\alpha_M = (7.38 \times 10^{-7})(3.68 \times 10^{-6}) = 2.72 \times 10^{-12}$$

$$pZn = 11.57$$

Calculation of pZn after addition of 30.0 mL of EDTA
The solution now contains an excess of EDTA; thus,

$$c_{EDTA} = c_T = \frac{30.0 \times 0.0100 - 50.0 \times 0.00500}{80.0} = 6.25 \times 10^{-4}\ M$$

and since essentially all the original Zn^{2+} is now complexed,

$$c_{ZnY^{2+}} = [ZnY^{2-}] = \frac{50.0 \times 0.00500}{80.0} = 3.12 \times 10^{-3}\ M$$

Rearranging Equation 11-11 gives

$$c_M = \frac{[ZnY^{2-}]}{c_T K''_{ZnY}} = \frac{3.12 \times 10^{-3}}{(6.25 \times 10^{-4})(6.12 \times 10^9)} = 8.16 \times 10^{-10}\ M$$

and, from Equation 11-8,

$$[Zn^{2+}] = c_M\alpha_M = (8.16 \times 10^{-10})(3.68 \times 10^{-6}) = 3.00 \times 10^{-15}$$

$$pZn = 14.52$$

Figure 11-8 shows two theoretical curves for the titration of zinc(II) with EDTA at pH 9.00. The equilibrium concentration of ammonia was 0.100 M for one titration and 0.0100 M for the other. Note that the presence of ammonia decreases the change in pZn near the equivalence point. For this reason, the concentration of auxiliary complexing reagents should always be kept to the minimum required to prevent precipitation of the analyte. Note that the magnitude of α_M does not affect pZn beyond the equivalence point. On the other hand, it will be recalled (Figure 11-5) that α_4, and thus pH, plays an important role in defining this part of the titration curve.

11B-5 Indicators for EDTA Titrations

Reilley and Barnard[4] have listed nearly 200 organic compounds that have been suggested as indicators for metal ions in EDTA titrations. In general, these indicators are organic dyes that form colored chelates with metal ions in a pM range that is characteristic of the particular cation and dye. The

[4]C. N. Reilley and A. J. Barnard Jr., in *Handbook of Analytical Chemistry*, L. Meites, Ed., p. **3**-77. New York: McGraw-Hill, 1963.

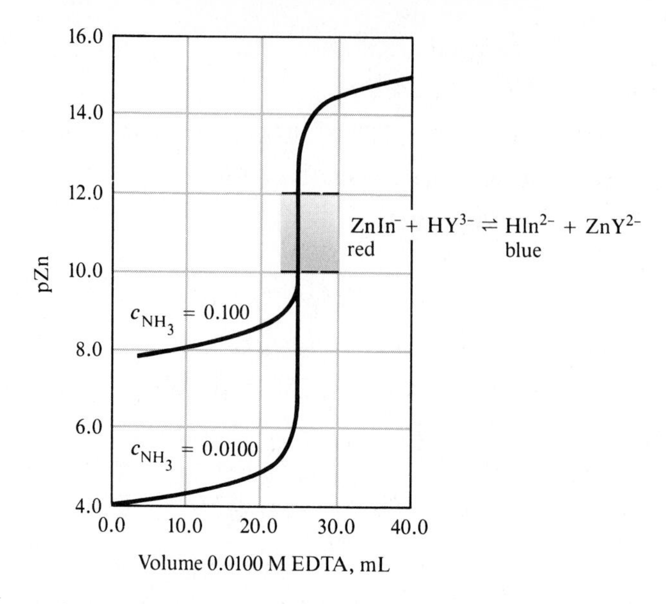

FIGURE 11-8 Influence of ammonia concentration upon the end point for the titration of 50.0 mL of 0.00500 M Zn^{2+}. Solutions buffered to pH 9.00. The shaded area shows the transition range for Eriochrome Black T.

complexes are often intensely colored, being discernible to the eye in the range of 10^{-6} to 10^{-7} M.

The majority of the indicators listed in the reference in footnote 4 contain an aromatic azo functional group. One of the first and most widely used of these compounds has the trivial name *Eriochrome Black T*. Its structure is

OH OH

^-O_3S —N=N—

NO_2

The proton associated with the sulfonic acid group is completely dissociated in an aqueous medium, but the phenolic protons are only partially dissociated:

$$\underset{\text{red}}{H_2O + H_2In^-} \rightleftharpoons \underset{\text{blue}}{HIn^{2-}} + H_3O^+ \qquad K_1 = 5 \times 10^{-7}$$

$$\underset{\text{blue}}{H_2O + HIn^{2-}} \rightleftharpoons \underset{\text{orange}}{In^{3-}} + H_3O^+ \qquad K_2 = 2.8 \times 10^{-12}$$

Note that the acids and their conjugate bases differ in color. Thus, Eriochrome Black T behaves as an acid/base indicator as well as a metal ion indicator.

The metal complexes of Eriochrome Black T are generally red. Thus, to observe a color change with this indicator, it is necessary to adjust the pH to 7 or above so that the blue form of the species HIn^{2-} predominates in the absence of a metal ion. The end-point reaction is

$$\underset{\text{red}}{MIn^-} + HY^{3-} \rightleftharpoons \underset{\text{blue}}{HIn^{2-}} + MY^{2-}$$

Eriochrome Black T forms red complexes with more than two dozen metal ions, but the formation constants of only a few are appropriate for end-point detection. As shown in the following example, the applicability of a given indicator to an EDTA titration can be determined from the change in pM in the equivalence-point region, provided the formation constant for the metal-indicator complex is known.[5]

EXAMPLE 11-6

Determine the transition ranges for Eriochrome Black T in titrations of Mg^{2+} and Ca^{2+} at pH 10.0, given (a) that the second acid dissociation constant for the indicator is

$$HIn^{2-} + H_2O \rightleftharpoons In^{3-} + H_3O^+ \qquad K_2 = 2.8 \times 10^{-12} = \frac{[H_3O^+][In^{3-}]}{[HIn^{2-}]}$$

(b) that the formation constant for $MgIn^-$ is

$$Mg^{2+} + In^{3-} \rightleftharpoons MgIn^- \qquad K_f = 1.0 \times 10^7 = \frac{[MgIn^-]}{[Mg^{2+}][In^{3-}]}$$

and (c) that the analogous constant for Ca^{2+} is 2.5×10^5.

We shall assume, as we did earlier (Section 8A-2), that a detectable color change requires a tenfold excess of one or the other of the colored species; that is, a detectable color change is observed when $[MgIn^-]/[HIn^{2-}]$ changes from 10 to 0.10.

Multiplication of K_2 for the indicator by K_f for $MgIn^-$ gives an expression that contains the foregoing ratio:

$$\frac{[MgIn^-][H_3O^+]}{[HIn^{2-}][Mg^{2+}]} = 2.8 \times 10^{-12} \times 1.0 \times 10^7 = 2.8 \times 10^{-5}$$

which rearranges to

$$[Mg^{2+}] = \frac{[MgIn^-]}{[HIn^{2-}]} \times \frac{[H_3O^+]}{2.8 \times 10^{-5}}$$

Substitution of 1.0×10^{-10} for $[H_3O^+]$ and 10 and 0.10 for the ratio yields the range of $[Mg^{2+}]$ over which the color change occurs:

$$[Mg^{2+}] = 3.6 \times 10^{-5} \text{ to } 3.6 \times 10^{-7}$$

$$pMg = 5.4 \pm 1.0$$

Proceeding in the same way, we find the range for pCa to be 3.8 ± 1.0.

The ranges for magnesium and calcium are indicated on the titration curves in Figure 11-4. Eriochrome Black T is clearly an ideal indicator for magnesium but a totally unsatisfactory one for calcium. Note that the formation constant for $CaIn^-$ is only about one fortieth that for magnesium. As

[5]For a complete discussion of the principles of indicator choice in complex-formation titrations, see C. N. Reilley and R. W. Schmid, *Anal. Chem.*, **1959**, *31*, 887.

a consequence, significant conversion of $CaIn^-$ to HIn^{2-} occurs well before equivalence.

A similar calculation reveals that Eriochrome Black T is also well suited for the titration of zinc with EDTA (Figure 11-8).

A limitation of Eriochrome Black T is that its solutions decompose slowly with standing. It is claimed that solutions of Calmagite, an indicator that for all practical purposes is identical in behavior to Eriochrome Black T, do not suffer this disadvantage. The structure of Calmagite is similar to that of Eriochrome Black T:

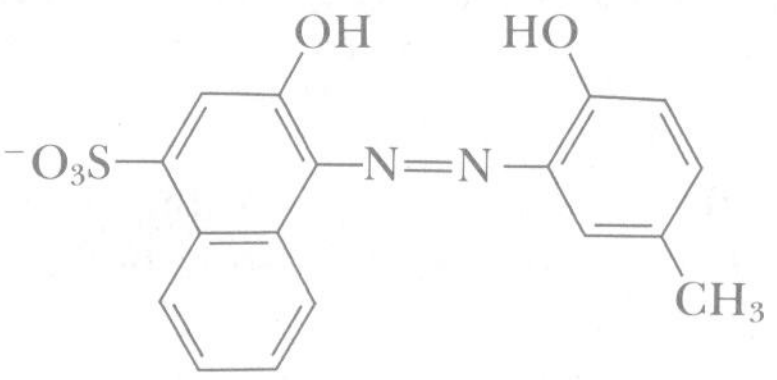

OH HO

^-O_3S—[naphthalene]—N=N—[benzene]

CH_3

Many other metal indicators have been developed for EDTA titrations.[6] In contrast to Eriochrome Black T, some of these indicators can be employed in strongly acidic media.

11B-6 Titration Methods Employing EDTA

The paragraphs that follow describe several ways in which EDTA is employed for the determination of cations.

Direct Titration

Methods Based on Indicators for the Analyte Ion. Reilley and Barnard[7] list 40 elements that can be determined by direct titration with EDTA using metal ion indicators. The direct method is not applicable to certain cations, however, either because no good indicators have been developed or because the reaction between the metal ion and EDTA is so slow as to make titration impractical. Such cations can usually be determined by one of the several methods described in the paragraphs that follow.

Methods Based on Indicators for an Added Metal Ion. It is often convenient to introduce into an EDTA solution a small amount of a cation that forms an EDTA complex that is less stable than the analyte complex and for which a good indicator exists. For example, indicators for calcium ion are generally less satisfactory than those we have described for magnesium ion. Consequently, a small amount of magnesium chloride is often added to an EDTA solution that is to be used for the determination of calcium with Eriochrome Black T. In the initial stages in the titration, magnesium ions are displaced from the EDTA complex by calcium ions and are free to combine with the Eriochrome Black T, thus imparting a red color to the solution. When all of

[6]See, for example, L. Meites, *Handbook of Analytical Chemistry*, pp. **3**-101 to **3**-165. New York: McGraw-Hill, 1963.

[7]C. N. Reilley and A. J. Barnard Jr., in *Handbook of Analytical Chemistry*, L. Meites, Ed., pp. **3**-166 to **3**-200. New York: McGraw-Hill, 1963.

the calcium ions have been complexed, however, the liberated magnesium ions again combine with the EDTA until the end point is observed. This procedure requires standardization of the EDTA solution against primary-standard calcium carbonate.

Potentiometric Methods. Potential measurements can be used for end-point detection in the EDTA titration of those metal ions for which specific ion electrodes are available. Electrodes of this type are described in Section 15B-4. In addition, a mercury electrode can be made sensitive to EDTA ions and used in titrations with this reagent. This application is discussed in Section 15E-3.

Back-Titration Methods

Back-titration procedures are useful for the determination of cations that form stable EDTA complexes and for which a satisfactory indicator is not available. The method is also useful for cations that react only slowly with EDTA. A measured excess of standard EDTA solution is added to the analyte solution. After the reaction is judged complete, the excess EDTA is back-titrated with a standard magnesium or zinc ion solution to an Eriochrome Black T or Calmagite end point.[8] For this procedure to be successful, it is necessary that the magnesium or zinc ions form an EDTA complex that is less stable than the corresponding analyte complex.

Back-titration is also useful for analyzing samples that contain anions that would otherwise form sparingly soluble precipitates with the analyte under the analytical conditions. Here, the excess EDTA prevents precipitate formation.

Displacement Methods

In displacement titrations, an unmeasured excess of a solution containing the magnesium or zinc complex of EDTA is introduced into the analyte solution. If the analyte forms a more stable complex than that of magnesium or zinc, the following displacement reaction occurs:

$$MgY^{2-} + M^{2+} \longrightarrow MY^{2-} + Mg^{2+}$$

where M^{2+} represents the analyte cation. The liberated magnesium or zinc is then titrated with a standard EDTA solution.

Displacement methods are useful where no satisfactory indicator is available for the metal ion being determined.

11B-7

The Scope of EDTA Titrations

Complexometric titrations with EDTA have been applied to the determination of virtually every metal cation with the exception of the alkali metal ions. Because of the ubiquity of EDTA chelates, the reagent might appear at first glance to be totally lacking in selectivity. In fact, however, considerable control over interferences can be realized by pH regulation. For example, trivalent cations can usually be titrated without interference from divalent species in

[8]For a recent analysis of the back-titration procedure, see C. Macca and M. Fiorana, *J. Chem. Educ.*, **1986**, *63*, 121.

solutions that have a pH of about 1 (Figure 11-7). At this pH, the less stable divalent chelates do not form to any significant extent, but the trivalent ions are quantitatively complexed.

Similarly, ions such as cadmium and zinc, which form more stable EDTA chelates than does magnesium, can be determined in the presence of the latter ion by buffering the mixture to pH 7 before titration. Eriochrome Black T serves as an indicator for the cadmium or zinc end points without interference from magnesium because the indicator chelate with magnesium is not formed at this pH.

Finally, interference from a particular cation can sometimes be eliminated by adding a suitable *masking agent,* an auxiliary ligand that preferentially forms highly stable complexes with the potential interference.[9] For example, cyanide ion is often employed as a masking agent to permit the titration of magnesium and calcium ions in the presence of ions such as cadmium, cobalt, copper, nickel, zinc, and palladium. All of the latter form sufficiently stable cyanide complexes to prevent reaction with EDTA.

Specific directions for the preparation and use of EDTA solutions are given in Section 31D.

11B-8 The Determination of Water Hardness

Historically, water "hardness" was defined in terms of the capacity of cations in the water to replace the sodium or potassium ions in soaps and form sparingly soluble products. Most multiply charged cations share this undesirable property. In natural waters, however, the concentration of calcium and magnesium ions generally far exceeds that of any other metal ion. Consequently, hardness is now expressed in terms of the concentration of calcium carbonate that is equivalent to the total concentration of all the multivalent cations in the sample.

The determination of hardness is a useful analytical test that provides a measure of the quality of water for household and industrial uses. The test is important to industry because hard water, upon being heated, precipitates calcium carbonate, which then clogs boilers and pipes.

Water hardness is ordinarily determined by an EDTA titration after the sample has been buffered to pH 10. Magnesium, which forms the least stable EDTA complex of all of the common multivalent cations in typical water samples, is not titrated until enough reagent has been added to complex all of the other cations in the sample. Therefore, a magnesium ion indicator, such as Calmagite or Eriochrome Black T, can serve as indicator in water-hardness titrations. Often, a small concentration of the EDTA complex of magnesium is incorporated in the buffer or in the titrant to ensure the presence of sufficient magnesium ions for satisfactory indicator action.

Test kits for determining the hardness of household water are available commercially. They consist of a vessel calibrated to contain a known volume of water, a measuring scoop to deliver an appropriate amount of a solid buffer mixture, an indicator solution, and a bottle of standard EDTA, which is equipped with a medicine dropper. The volume of standard reagent consumed is obtained by counting drops to the end point. The concentration of the

[9]For further information, see D. D. Perrin, *Masking and Demasking of Chemical Reactions.* New York: Wiley-Interscience, 1970; and C. N. Reilley and A. J. Barnard Jr., in *Handbook of Analytical Chemistry,* L. Meites, Ed., pp. **3**-208 to **3**-225. New York: McGraw-Hill, 1963.

TABLE 11-3

Typical Inorganic Complex-Formation Titrations*

Titrant	Analyte	Remarks
$Hg(NO_3)_2$	Br^-, Cl^-, SCN^-, CN^-, thiourea	Products are neutral mercury(II) complexes; various indicators used
$AgNO_3$	CN^-	Product is $Ag(CN)_2^-$; indicator is I^-; titrate to first turbidity of AgI
$NiSO_4$	CN^-	Product is $Ni(CN)_4^{2-}$; indicator is AgI; titrate to first turbidity of AgI
KCN	Cu^{2+}, Hg^{2+}, Ni^{2+}	Products are $Cu(CN)_4^{2-}$, $Hg(CN)_2$, $Ni(CN)_4^{2-}$; various indicators used

*For further applications and selected references, see L. Meites, *Handbook of Analytical Chemistry*, pp. **3**-226. New York: McGraw-Hill, 1963.

EDTA solution is ordinarily such that one drop corresponds to one grain (about 0.065 g) of calcium carbonate per gallon of water.

11C Titrations with Inorganic Complexing Agents

Complexometric titrations with inorganic reagents are among the oldest volumetric methods.[10] For example, the titration of iodide ion with mercury(II) ions,

$$Hg^{2+} + 4I^- \rightleftharpoons HgI_4^{2-}$$

was first described in 1833. Table 11-3 lists the common inorganic complexing agents as well as some of their applications.

Questions and Problems

11-1 Define

*(a) chelate.
(b) tetradendate chelating agent.
*(c) ligand.
(d) coordination number.
*(e) conditional formation constant.
(f) NTA.
*(g) water hardness.
(h) EDTA displacement titration.

*11-2 Describe three general methods for performing EDTA titrations. What are the advantages of each?

11-3 Propose a complexometric method for the determination of the individual components in a solution containing In^{3+}, Zn^{2+}, and Mg^{2+}.

*11-4 Why are multidentate ligands preferable to unidentate ligands for complexometric titrations?

11-5 Write chemical equations and equilibrium-constant expressions for the stepwise formation of

*(a) $Ag(S_2O_3)_2^{3-}$.
(b) AlF_6^{3-}.
*(c) $Cd(NH_3)_4^{2+}$.
(d) $Ni(SCN)_3^-$.

[10]For further information, see I. M. Kolthoff and V. A. Stenger, *Volumetric Analysis*, Vol. 2, pp. 282–331. New York: Interscience, 1947.

11-6 Explain how stepwise and overall formation constants are related.

*11-7 Why is a small amount of MgY^{2-} often added to a solution that is to be titrated for hardness?

*11-8 A potassium cyanide solution contains 1.62 g of KCN in 550 mL. Express the concentration of this solution in terms of
(a) molarity.
(b) titer as mg Ag^+/mL [reaction product: $Ag(CN)_2^-$].
(c) titer as mg Cu_2O/mL [reaction product: $Cu(CN)_2^-$].

11-9 Thiourea, $(H_2N)_2CS$ (fw = 76.12), can be used for the complexometric titration of mercury (II):

$$Hg^{2+} + 4tu \rightleftharpoons Hg(tu)_4^{2+}$$

where tu represents the thiourea molecule. A solution contains 1.917 g of thiourea in 500.0 mL. Calculate the concentration of this solution in terms of
(a) molarity.
(b) titer as mg Hg^{2+}/mL.
(c) titer as mg HgO/mL.

*11-10 An EDTA solution has a titer of 1.08 mg of CaO/mL. Express the concentration of this solution in terms of
(a) molarity.
(b) titer as mg MgO/mL.
(c) titer as mg Fe_2O_3/mL.

11-11 An EDTA solution was prepared by dissolving 8.642 g of pure H_4Y (fw = 292.24) in a small volume of sodium hydroxide and diluting to 500.0 mL. Express the concentration of this solution in terms of
(a) molarity.
(b) titer as mg Ca^{2+}/mL.
(c) titer as mg $MgCO_3$/mL.

*11-12 An EDTA solution was prepared by dissolving 3.853 g of purified and dried $Na_2H_2Y \cdot 2H_2O$ in sufficient water to give 1.000 L. Calculate the molar concentration, given that the solute contained 0.3% excess moisture (page 260).

11-13 An EDTA solution was prepared by dissolving about 3.0 g of $Na_2H_2Y \cdot 2H_2O$ in approximately 1 L of water. Calculate the molar concentration of this solution if an average of 32.22 mL was required to titrate 50.00-mL aliquots of 0.004517 M Mg^{2+}.

*11-14 Calculate the concentration of an EDTA solution from the accompanying titration data:

	Buret Containing	
	EDTA Soln	0.01470 M Mg^{2+} Soln
Final volume, mL	46.39	31.69
Initial volume, mL	0.04	0.00

11-15 A 50.00-mL aliquot of a solution containing iron(II) and iron(III) required 13.73 mL of 0.01200 M EDTA when titrated at pH 2.0 and 29.62 mL when titrated at pH 6.0. Express the concentration of the solution in terms of the parts per million of each solute.

*11-16 A 24-hr urine specimen was diluted to 2.000 L. After being buffered to pH 10, a 10.00-mL aliquot was titrated with 26.81 mL of 0.003474 M EDTA. The calcium in a second 10.00-mL aliquot was isolated as $CaC_2O_4(s)$, redissolved in acid, and titrated with 11.63 mL of the EDTA solution. Assuming that 15 to 300 mg of magnesium and 50 to 400 mg of calcium per day are normal, did this specimen fall within these ranges?

11-17 An EDTA solution was prepared by dissolving approximately 4 g of the disodium salt in approximately 1 L of water. An average of 42.35 mL of this solution was required to titrate 50.00-mL aliquots of a standard that contained 0.7682 g of $MgCO_3$ per liter. Titration of a 25.00-mL sample of mineral water at pH 10 required 18.81 mL of the EDTA solution. A 50.00-mL aliquot of the mineral water was rendered strongly alkaline to precipitate the magnesium as $Mg(OH)_2$. Titration with a calcium-specific indicator required 31.54 mL of the EDTA solution. Calculate

(a) the molarity of the EDTA solution.
(b) the ppm of $CaCO_3$ in the mineral water.
(c) the ppm of $MgCO_3$ in the mineral water.

*11-18 The sulfate in a 1.515-g sample was homogeneously precipitated as $BaSO_4$ by adding an excess of BaY^{2-} solution and slowly increasing the acid concentration to liberate Ba^{2+}. The precipitate was filtered and washed, and the filtrate and washings were collected in a 250-mL volumetric flask. A pH-10.0 buffer was added and the solution diluted to the mark. A 25.00-mL aliquot required a 28.73-mL titration with 0.01545 M Mg^{2+} solution. Express the results of this analysis in terms of percent $Na_2SO_4 \cdot 10H_2O$.

11-19 Calamine, which is used for relief of skin irritations, is a mixture of zinc and iron oxides. A 1.022-g specimen of dried calamine was dissolved in acid and diluted to 250.0 mL. Potassium fluoride was added to a 10.00-mL aliquot of the diluted solution to mask the iron; after suitable adjustment of the pH, Zn^{2+} consumed 38.78 mL of 0.01294 M EDTA. A second 50.00-mL aliquot was suitably buffered and titrated with 2.40 mL of 0.002727 M ZnY^{2-} solution:

$$Fe^{3+} + ZnY^{2-} \longrightarrow FeY^- + Zn^{2+}$$

Calculate the percentages of ZnO and Fe_2O_3 in the sample.

*11-20 A 3.650-g sample containing bromate and bromide was dissolved in sufficient water to give 250.0 mL. After acidification, silver nitrate was introduced to a 25.00-mL aliquot to precipitate AgBr, which was filtered, washed, and then redissolved in an ammoniacal solution of potassium tetracyanonickelate(II):

$$Ni(CN)_4^{2-} + 2AgBr(s) \longrightarrow 2Ag(CN)_2^- + Ni^{2+} + 2Br^-$$

The liberated nickel ion required 26.73 mL of 0.02089 M EDTA. The bromate in a 10.00-mL aliquot was reduced to bromide with arsenic(III) prior to the addition of silver nitrate; 21.94 mL of the EDTA solution was needed to titrate the nickel ion subsequently released. Calculate the percentages of NaBr and $NaBrO_3$ in the sample.

11-21 The chromium (d = 7.10 g/cm^3) plated on a 9.75-cm^2 surface was dissolved with hydrochloric acid and diluted to 100.0 mL. A 25.00-mL aliquot was buffered to pH 5, and 50.00 mL of 0.00862 M EDTA was added. Titration of the excess chelating reagent required 7.36 mL of 0.01044 M Zn^{2+}. Calculate the average thickness of the chromium plating.

*11-22 The silver ion in a 25.00-mL sample was converted to dicyanoargentate(I) ion by the addition of an excess of a solution containing $Ni(CN)_4^{2-}$:

$$Ni(CN)_4^{2-} + 2Ag^+ \longrightarrow 2Ag(CN)_2^- + Ni^{2+}$$

The liberated nickel ion was titrated with 43.77 mL of 0.02408 M EDTA. Calculate the molar concentration of the silver solution.

11-23 The potassium ion in a 250.0-mL sample of mineral water was precipitated with sodium tetraphenylboron:

$$K^+ + B(C_6H_4)_4^- \longrightarrow KB(C_6H_5)_4(s)$$

The precipitate was filtered, washed, and redissolved in an organic solvent. An excess of the mercury(II)/EDTA chelate was added:

$$4HgY^{2-} + B(C_6H_4)_4^- + 4H_2O \longrightarrow H_3BO_3 + 4C_6H_5Hg^+ + 4HY^{3-} + OH^-$$

The liberated EDTA was titrated with 29.64 mL of 0.05581 M Mg^{2+}. Calculate the potassium ion concentration in parts per million.

*11-24 A 0.4085-g sample containing lead, magnesium, and zinc was dissolved and treated with cyanide to complex and mask the zinc:

$$Zn^{2+} + 4CN^- \longrightarrow Zn(CN)_4^{2-}$$

Titration of the lead and magnesium required 42.22 mL of 0.02064 M EDTA. The lead was next masked with BAL (2,3-dimercaptopropanol), and the released EDTA was titrated with 19.35 mL of a 0.007657 M magnesium solution. Finally, formaldehyde was introduced to demask the zinc:

$$Zn(CN)_4^{2-} + 4HCHO + 4H_2O \longrightarrow Zn^{2+} + 4HOCH_2CN + 4OH^-$$

which was titrated with 28.63 mL of 0.02064 M EDTA. Calculate the percentages of the three metals in the sample.

11-25 Chromel is an alloy composed of nickel, iron, and chromium. A 0.6472-g sample was dissolved and diluted to 250.0 mL. When a 50.00-mL aliquot of 0.05180 M EDTA was mixed with an equal volume of the diluted sample, all three ions were chelated, and a 5.11-mL back-titration with 0.06241 M copper(II) was required. The chromium in a second 50.0-mL aliquot was masked through the addition of hexamethylenetetramine; titration of the Fe and Ni required 36.28 mL of 0.05182 M EDTA. Iron and chromium were masked with pyrophosphate in a third 50.0-mL aliquot, and the nickel was titrated with 25.91 mL of the EDTA solution. Calculate the percentages of nickel, chromium, and iron in the alloy.

*11-26 A 0.3284-g sample of brass (containing lead, zinc, copper, and tin) was dissolved in nitric acid. The sparingly soluble $SnO_2 \cdot 4H_2O$ was removed by filtration, and the combined filtrate and washings were then diluted to 500.0 mL. A 10.00-mL aliquot was suitably buffered; titration of the lead, zinc, and copper in this aliquot required 37.56 mL of 0.002500 M EDTA. The copper in a 25.00-mL aliquot was masked with thiosulfate; the lead and zinc were then titrated with 27.67 mL of the EDTA solution. Cyanide ion was used to mask the copper and zinc in a 100-mL aliquot; 10.80 mL of the EDTA solution was needed to titrate the lead ion. Determine the composition of the brass sample; evaluate the percentage of tin by difference.

*11-27 Calculate conditional constants for the formation of the EDTA complex of Mn^{2+} at pH (a) 6.0, (b) 8.0, and (c) 10.0.

11-28 Calculate conditional constants for the formation of the EDTA complex of Sr^{2+} at pH (a) 7.0, (b) 9.0, and (c) 11.0.

*11-29 Formation constants for the successive ammine complexes of cadmium are 320, 91, 20, and 6.2. Calculate the conditional constant for the reaction of Cd^{2+} with Y^{4-} when

(a) the molar concentration of NH_3 is 0.050 and the pH is 9.0.
(b) the molar concentration of NH_3 is 0.050 and the pH is 11.0.
(c) the molar concentration of NH_3 is 0.50 and the pH is 9.0.
(d) the molar concentration of NH_3 is 0.50 and the pH is 11.0.

11-30 Formation constants for the successive ethylenediamine (en) complexes of nickel are 4.6×10^7, 2.5×10^6, and 3.4×10^4. Calculate the conditional constant for the reaction of Ni^{2+} with Y^{4-} when

(a) the concentration of en is 0.025 M and the pH is 9.0.
(b) the concentration of en is 0.025 M and the pH is 11.0.
(c) the concentration of en is 0.25 M and the pH is 9.0.
(d) the concentration of en is 0.25 M and the pH is 11.0.

*11-31 Derive a titration curve for 50.00 mL of 0.01000 M Sr^{2+} with 0.02000 M EDTA in a solution buffered to pH 11.0. Calculate pSr values after the addition of 0.00, 10.00, 24.00, 24.90, 25.00, 25.10, 26.00, and 30.00 mL of titrant.

11-32 Derive a titration curve for 50.00 mL of 0.0150 M Fe^{2+} with 0.0300 M EDTA in a solution buffered to pH 7.0. Calculate pFe values after the addition of 0.00, 10.00, 24.00, 24.90, 25.00, 25.10, 26.00, and 30.00 mL of titrant.

*11-33 Derive a curve for the titration of 25.00 mL of 0.040000 M Co^{2+} with 0.05000 M Na_2H_2Y in a solution maintained at pH 9.00 with an NH_3/NH_4^+ buffer. Assume that the NH_3 concentration is constant at 0.04000 M throughout. Calculate values for pCo after the addition of 0.00, 5.00, 10.00, 19.00, 20.00, 21.00, and 30.00 mL of reagent.

11-34 Derive a curve for the titration of 25.00 mL of 0.02000 M Ni^{2+} with 0.01000 M Na_2H_2Y in a solution maintained at pH 10.00 with an NH_3/NH_4^+ buffer. Assume that the concentration of NH_3 remains 0.1000 M throughout. Calculate values for pNi after the addition of 0.00, 10.00, 25.00, 40.00, 45.00, 49.00, 50.00, 51.00, 55.00, and 60.00 mL of reagent.

CHAPTER 12

An Introduction to Oxidation/Reduction Equilibria and Electrochemical Theory

Several important and widely used analytical methods are based upon *oxidation/reduction,* or *redox,* reactions. These methods, which are discussed in Chapters 13 through 17, include oxidation/reduction titrimetry, potentiometry, coulometry, electrogravimetry, and voltammetry. This chapter provides a theoretical background for understanding these methods.[1]

12A Oxidation/Reduction Processes

In an oxidation/reduction reaction, electrons are transferred from one reactant to another. A substance that has a strong affinity for electrons and thus tends to extract them from other species is called an *oxidizing agent* or an *oxidant.* A *reducing agent,* or *reductant,* is a reagent that readily donates electrons to another species.

12A-1 A Comparison of Oxidation/Reduction and Acid/Base Reactions

Oxidation/reduction reactions resemble acid/base reactions in several regards. In both, one or more charged particles are transferred from a donor to an acceptor—the particles being electrons in the former case and protons in the latter. Furthermore, the relationships between reactants and products in the two types of reactions are analogous. Recall that according to the Brønsted-Lowry concept, discussed in Section 5A-2, an acid, after donating a proton in an acid/base reaction, becomes a base (a conjugate base); similarly, a base is converted to its conjugate acid after accepting a proton in the reaction. In the same way, the products of an oxidation/reduction reaction are a conjugate oxidant and a conjugate reductant (although the term *conjugate* is seldom, if ever, used in this context). For example, consider the redox reaction

$$\text{ox}_1 + \text{red}_2 \rightleftharpoons \text{red}_1 + \text{ox}_2$$

The oxidant ox_1 accepts electrons from reductant red_2 to form a new reductant, red_1. At the same time, reductant red_2, having given up electrons, becomes an oxidizing agent, ox_2. If equilibrium in this process favors the products, ox_1 is a more effective electron acceptor (and thus a stronger oxidizing agent) than ox_2, the species that results from the loss of electrons from red_2. By the same reasoning, red_2 is a more effective electron donor (and hence a stronger reducing agent) than red_1.

[1]For further reading on oxidation/reduction equilibria, see J. A. Goldman, in *Treatise on Analytical Chemistry,* 2nd ed., I. M. Kolthoff and P. J. Elving, Eds., Part I, Vol. 3, Chapter 24. New York: Wiley, 1983.

12A-2

Half-Reactions

It is often helpful to separate an oxidation/reduction equation into two *half-reactions*, one of which describes the oxidation and the other the reduction. Consider, for example, the equation for the oxidation of iron(II) ions by permanganate ions in an acidic solution:

$$5Fe^{2+} + MnO_4^- + 8H^+ \rightleftharpoons 5Fe^{3+} + Mn^{2+} + 4H_2O$$

The two half-reactions are

$$MnO_4^- + 8H^+ + 5e \rightleftharpoons Mn^{2+} + 4H_2O$$

$$5Fe^{2+} \rightleftharpoons 5Fe^{3+} + 5e$$

These equations show clearly that the permanganate ion is the electron acceptor and therefore the oxidizing agent while the iron(II) ion is the electron donor and reducing agent. Note that before combining the two half-reactions, it is necessary to multiply the second one by 5 in order to eliminate electrons from the final equation.[2]

12A-3

Oxidation/Reduction Reactions in Electrochemical Cells

Most oxidation/reduction reactions can be carried out in two ways. In the first, the oxidant and reductant are brought together in a medium wherein the reacting molecules, atoms, or ions approach one another closely enough so that a direct transfer of electrons can take place. In the second, the reactants are made a part of an electrochemical cell in which the two half-reactions are physically separated; electron transfer then takes place via a metallic conductor of electricity.

As an example of the first method, metallic zinc can be oxidized to zinc ions by immersing a piece of the metal in a solution containing copper sulfate. Copper(II) ions migrate to the surface of the zinc and are reduced:

$$Cu^{2+} + 2e \rightleftharpoons Cu(s)$$

while a chemically equivalent quantity of zinc is oxidized:

$$Zn(s) \rightleftharpoons Zn^{2+} + 2e$$

The equation for the overall process is obtained by adding the equations for the two half-reactions:

$$\underset{ox_1}{Cu^{2+}} + \underset{red_2}{Zn(s)} \rightleftharpoons \underset{red_1}{Cu(s)} + \underset{ox_2}{Zn^{2+}}$$

This same reaction can be carried out in an electrochemical cell, such as that shown in Figure 12-1. The compartment on the left is made up of a strip of zinc immersed in a solution of yM zinc sulfate; the one on the right consists of a strip of copper in an xM copper sulfate solution. The reactions at the two metal-solution interfaces are then

$$Zn(s) \rightleftharpoons Zn^{2+} + 2e$$

$$Cu^{2+} + 2e \rightleftharpoons Cu(s)$$

[2]The experimental observation that this reaction proceeds essentially to completion makes it possible to state that iron(II) is a more effective electron donor than manganese(II) and that permanganate ion is a more effective electron acceptor than iron(III).

In this case, an external conductor provides the means by which electrons are transferred from the zinc to the copper.

The porous disk in Figure 12-1 prevents extensive mixing of the two electrolyte solutions but offers a path by which electricity is carried in the cell by ion migration. Mixing is undesirable because it would lead to direct reaction between the zinc metal and the copper ions, thus decreasing the cell efficiency.

The equilibrium established in the electrochemical cell is identical in every respect to the equilibrium established when a piece of zinc is immersed in a solution of copper sulfate. In the cell, however, electrons are transferred from one species to the other as an electric current. Here, electron transfer continues until the concentrations of copper(II) and zinc(II) achieve levels corresponding to equilibrium for the reaction

$$Zn(s) + Cu^{2+} \rightleftharpoons Zn^{2+} + Cu(s)$$

When this condition is reached, no further net flow of electrons occurs, and so the current drops to zero. It is essential to recognize that the overall process and the concentrations that exist at equilibrium are totally independent of the route that led to equilibrium, be it by direct contact between the reactants or by indirect reaction as in Figure 12-1.

12B Electrochemical Cells

The current that develops in the cell in Figure 12-1 when the switch *S* is closed arises from the strong tendency of zinc to be oxidized at the surface of one electrode and of copper(II) to be reduced at the surface of the other. The magnitude of the potential that develops between the two electrodes provides a measure of the tendency of the two half-reactions to proceed toward equilibrium. As is shown in Section 12D-2, this potential—which is readily measured by the voltage-measuring device, *V*, shown in Figure 12-1—is directly

FIGURE 12-1 A galvanic cell with a liquid junction. Arrows show the direction of charge flow when the switch is closed.

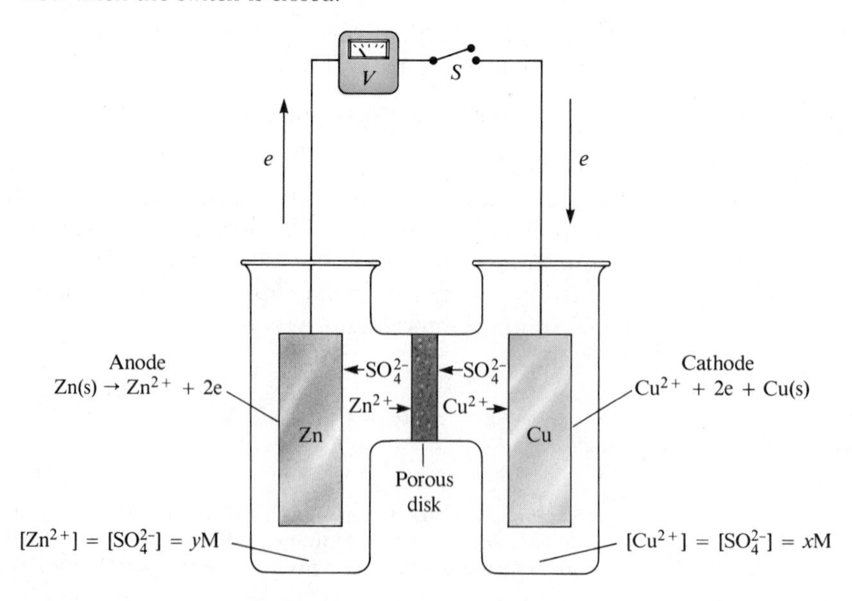

related to the equilibrium constant for the oxidation-reduction process involved as well as to the extent to which the existing concentrations of the participants differ from their equilibrium values. Such potentials, in fact, are important experimental sources of numerical equilibrium constants for redox reactions.

12B-1

Types of Electrochemical Cells

Electrochemical cells fall into two categories, depending upon whether they are operated in such a way as to produce or to consume electrical energy. They can be further subdivided into reversible and irreversible cells.

Galvanic and Electrolytic Cells

Electrochemical cells are classified as *galvanic* (or *voltaic*) when they act as a source of electrical power and *electrolytic* when they consume electrical energy. Many cells can be converted from one mode to the other. For example, the cell in Figure 12-1 is a galvanic cell that develops a potential of roughly 1 V. It could operate as an electrolytic cell, however, if a battery having a potential somewhat greater than 1 V were incorporated into the circuit in such a way as to force electrons to flow in the opposite direction. Under these circumstances, zinc would deposit and copper would dissolve. Because these processes are nonspontaneous, they would consume energy from the external source.

Reversible and Irreversible Cells

The cell in Figure 12-1 is an example of a *reversible* cell, in which reversing the current direction simply results in a reversal of the chemical processes that take place at the two electrodes. Changing the direction of the current in an irreversible cell causes entirely different reactions at one or both electrodes. For example, the cell in Figure 12-1 becomes irreversible if the solution in the zinc compartment is made acidic. If this is done, zinc will not deposit when a potential is applied to the cell; instead, hydrogen will form at the electrode:

$$2H^+ + 2e \rightleftharpoons H_2(g)$$

The overall process now becomes

$$Cu(s) + 2H^+ \rightarrow Cu^{2+} + H_2(g)$$

12B-2

Conduction in Electrochemical Cells

Three separate mechanisms are involved in the transport of electricity through an electrochemical cell:

1. Electrons serve as carriers in the electrodes as well as in the external conductor.
2. Anionic and cationic migration is responsible for transport in the solutions. Referring again to Figure 12-1, copper ions, zinc ions, and other positively charged species tend to move toward the copper electrode and negatively charged ions (SO_4^{2-}, HSO_4^-, and OH^-) are attracted to the zinc electrode.
3. An oxidation at one electrode surface and a reduction at the other provide the mechanism whereby the ionic conduction of the solution is coupled with the electronic conduction in the electrodes and in the external circuit.

12B-3

The Components of Cells

Electrochemical cells are made up of two electrodes, one called an *anode* and the other a *cathode;* each is immersed in an electrolyte solution. In most cells, the composition or the concentration of the electrolyte solutions differ, and the interface between the two is termed a *liquid junction.* A few cells operate without a liquid junction; here the two electrolyte solutions are identical.

Anodes and Cathodes

The anode of an electrochemical cell is the electrode at which oxidation occurs. The cathode is the electrode at which reduction occurs. Typical cathodic processes are

$$Ag^+ + e \rightleftharpoons Ag(s)$$

$$2H^+ + 2e \rightleftharpoons H_2(g)$$

$$Fe^{3+} + e \rightleftharpoons Fe^{2+}$$

$$IO_4^- + 2H^+ + 2e \rightleftharpoons IO_3^- + H_2O$$

The last three reactions take place at the surface of an inert conductor, such as platinum, that serves as the cathode. Note that anions as well as cations can undergo reduction at a platinum cathode. Hydrogen is frequently the cathode product in solutions that contain no species that is more easily reduced.

Typical anodic reactions are

$$Zn(s) \rightleftharpoons Zn^{2+} + 2e$$

$$2Cl^- \rightleftharpoons Cl_2(g) + 2e$$

$$Fe^{2+} \rightleftharpoons Fe^{3+} + e$$

$$2H_2O \rightleftharpoons O_2(g) + 4H^+ + 4e$$

The first reaction requires a zinc anode, whereas the others occur at an inert electrode. The last reaction commonly occurs in solutions that do not contain any species that is more easily oxidized than water.

Liquid Junctions and Salt Bridges

As noted earlier, the interface between two electrolyte solutions, such as in the porous disk in Figure 12-1, is termed a liquid junction. Liquid junctions are of considerable importance in certain types of electrochemical measurements because associated with them is a small potential called a *liquid-junction potential* that contributes to the overall potential of a cell.

The source of the liquid-junction potential is described in Section 15D-1. At this point, it suffices to mention that the effect of a liquid-junction potential on the overall potential of a cell can be minimized by interposing a *salt bridge* between the two cell compartments. This device consists of a U-shaped tube that contains a concentrated solution of a salt such as potassium chloride (Figure 12-3). A salt bridge creates two liquid-junction potentials, one at each end of the bridge. The effects of these two potentials are largely self-canceling.

Cells Without Liquid Junctions

Occasionally, useful cells can be constructed in which the electrodes share a common electrolyte. An example of a cell without a liquid junction is shown

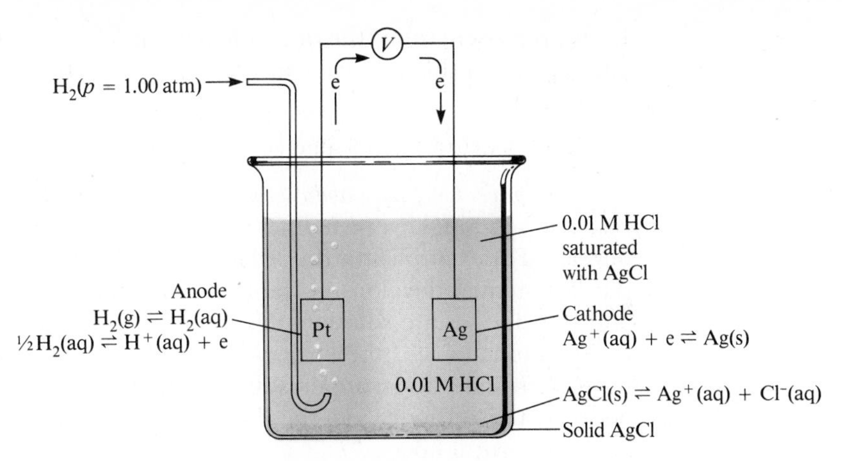

FIGURE 12-2 A galvanic cell without a liquid junction.

in Figure 12-2. Here, the reaction at the silver cathode can be considered a two-step process described by the equations

$$AgCl(s) \rightleftharpoons Ag^+(aq) + Cl^-(aq)$$

$$Ag^+(aq) + e \rightleftharpoons Ag(s)$$

The overall reaction at the cathode is then

$$AgCl(s) + e \rightleftharpoons Ag(s) + Cl^-(aq)$$

The reaction at the anode can also be considered to be a two-step process described by the equations

$$H_2(g) \rightleftharpoons H_2(aq)$$

$$H_2(aq) \rightleftharpoons 2H^+(aq) + e$$

The overall anodic reaction involves the oxidation of hydrogen gas to give an aqueous solution of hydrogen ions:

$$H_2(g) \rightleftharpoons 2H^+(aq) + 2e$$

The overall cell reaction is obtained by multiplying each term in the cathodic reaction by 2 and then adding the cathodic and anodic equations, which yields

$$2AgCl(s) + H_2(g) \rightleftharpoons 2Ag(s) + 2H^+(aq) + 2Cl^-(aq)$$

The direct reaction between molecular hydrogen and solid silver chloride is sufficiently slow that a common electrolyte can be employed without a significant loss of cell efficiency.

12B-4 *The Schematic Representation of Cells*

Chemists frequently use a shorthand notation to describe electrochemical cells. For example, the cell in Figure 12-1 can be represented as

$$Zn|ZnSO_4(yM)|CuSO_4(xM)|Cu$$

where yM is the molar concentration of the $ZnSO_4$ and xM is the molar concentration of the $CuSO_4$. An alternative way of describing this cell is

$$Zn|Zn^{2+}(yM)|Cu^{2+}(xM)|Cu$$

In this representation (the one more commonly encountered), only the active participants in the cell reaction are indicated. The cell in Figure 12-2 is represented as

$$\mathrm{Pt,H_2}(p = 1.00\ \mathrm{atm})|\mathrm{H^+}(0.01\ M),\mathrm{Cl^-}(0.01\ M),\mathrm{AgCl(sat'd)}|\mathrm{Ag}$$

By convention, the anodic process is always displayed to the left in these representations. Single vertical lines signify phase boundaries at which potentials develop. For example, for the cell in Figure 12-1, the first vertical line indicates that a potential develops at the phase boundary between the zinc electrode and the zinc sulfate solution. As noted earlier, small potentials also develop at liquid junctions; thus, another vertical line is inserted between symbols for the zinc sulfate and copper sulfate solutions. Finally, the phase boundary between the cathode and the copper sulfate solution is symbolized by yet another vertical line.

The cell in Figure 12-2 contains only two phase boundaries because the electrolyte is common to the two half-cells; thus, only two vertical lines are used. An equally correct representation of this cell is

$$\mathrm{Pt,H_2(sat'd)}|\mathrm{HCl}(0.01\ M),\mathrm{Ag^+}(1.8 \times 10^{-8}\ M)|\mathrm{Ag}$$

Here, the molecular hydrogen concentration is that of a saturated solution (in the absence of partial pressure data, 1.00 atm is implied). The indicated silver ion concentration was computed from the solubility product for silver chloride.

Two vertical lines are used to symbolize the presence of a salt bridge, implying that a potential develops at each of the two interfaces formed by this device. For example, the cell shown in Figure 12-3 is represented as

$$\mathrm{Pt,H_2}(p = 1.00\ \mathrm{atm})|\mathrm{H^+}(a_{\mathrm{H_3O^+}} = 1.00)||\mathrm{Cu^{2+}}(a_{\mathrm{Cu^{2+}}} = 1.00)|\mathrm{Cu}$$

12C Electrode Potentials

It is both convenient and reasonable to think of a cell potential as being the difference between two electrode potentials, or half-cell potentials, one of which is associated with the cathode of the cell and the other with the anode. Unfortunately, no direct method exists by which the absolute potential of a single electrode can be determined because all voltage-measuring devices measure only *differences* in potential. One conductor from such a device is connected to the electrode in question. A second conductor must then make contact—directly or indirectly—with the electrolyte solution of the half-cell. The latter contact inevitably involves a solid-solution interface that acts as a second half-cell at which chemical change *must occur* if charge is to flow. A potential is associated with this second reaction. Thus, an absolute value for the desired half-cell potential is not realized; instead, the measured potential is a combination of the potential of interest and the one that develops between the second contact and the solution.

It turns out that the lack of absolute half-cell potentials is not a serious handicap because *relative* half-cell potentials measured against a common reference half-cell are just as useful. The common reference electrode that has been used by scientists for decades is the *standard hydrogen electrode* (SHE), which is illustrated in the left part of Figure 12-3. By now, potential data relative to the standard hydrogen electrode have been compiled for several

FIGURE 12-3 A hypothetical cell illustrating the definition of electrode potential for the half-reaction $Cu^{2+} + 2e \rightleftharpoons Cu(s)$.

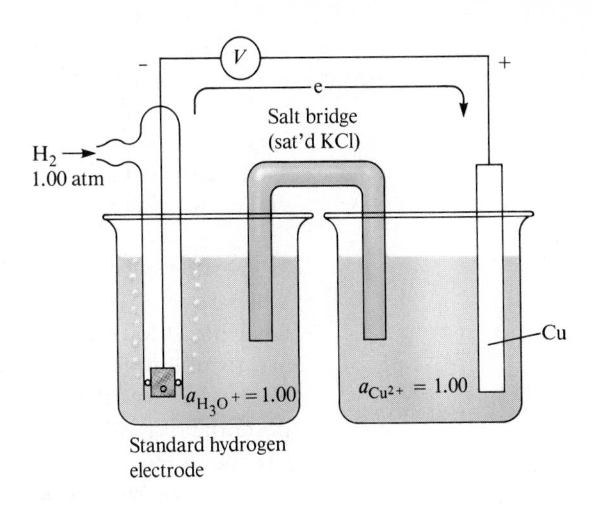

hundred half-reactions. These data are useful for calculating cell potentials and computing equilibrium constants for oxidation/reduction processes.

12C-1 Reference Electrodes

The standard hydrogen electrode is the universal reference electrode for reporting relative half-cell potentials. In addition, however, several secondary reference electrodes that are more convenient to use have been developed. Potentials relative to these secondary electrodes are readily converted to standard potentials in which the standard hydrogen electrode is the reference.

The Standard Hydrogen Electrode

The *hydrogen-gas electrode,* which was used extensively in early electrochemical studies, consists of a platinum electrode immersed in an electrolyte solution that is kept saturated with hydrogen gas. In order to ensure that the half-reaction $2H^+ + 2e \rightleftharpoons H_2(g)$ proceeds rapidly and reversibly, the platinum electrode must be *platinized*—that is, coated with a layer of finely divided platinum to maximize its specific surface area.

The hydrogen electrode can serve as an anode or a cathode, depending upon the half-cell with which it is coupled. Hydrogen is oxidized to hydrogen ions when the electrode acts as an anode; the reverse occurs when it acts as a cathode. The purpose of the gas stream shown in Figure 12-3 is to ensure that the solution is continuously saturated with molecular hydrogen. Thus, when a hydrogen electrode operates as an anode, the half-cell process is the sum of two reactions:

$$H_2(g) \rightleftharpoons H_2(\text{aq sat'd})$$

$$H_2(\text{aq sat'd}) \rightleftharpoons 2H^+(aq) + 2e$$

The overall reaction is

$$H_2(g) \rightleftharpoons 2H^+(aq) + 2e$$

it being understood that sufficient $H_2(g)$ is available to maintain saturation. The potential of a hydrogen electrode depends upon the temperature, the hydrogen ion concentration (or, more correctly, its activity), and the pressure of the hydrogen gas at the electrode surface. Values for these parameters must be carefully defined in order for the half-cell process to serve as a reference.

The *standard hydrogen electrode* is a hydrogen-gas electrode for which the hydrogen ion activity is specified as unity and the partial pressure of the hydrogen gas is exactly 1.00 atm. *By convention, the potential of this electrode is assigned a value of exactly zero at all temperatures.* Figure 12-3 is a diagram showing a standard hydrogen electrode coupled to a copper/copper ion half-cell system via a salt bridge.

The Saturated Calomel Electrode

The stream of gas needed for the operation of a hydrogen electrode is somewhat hazardous, and preparation and maintenance of the platinized surface are troublesome. As a consequence, more convenient secondary reference electrodes are often substituted for the standard hydrogen electrode. The potentials of these secondary reference electrodes relative to the standard hydrogen electrode have been carefully measured so that data collected with such electrodes can be converted to a standard hydrogen electrode basis.

Perhaps the most widely used secondary electrode is the saturated calomel electrode (SCE). When coupled to a standard hydrogen electrode, it forms a galvanic cell that is represented by the diagram

$$Pt,H_2|H^+(a = 1.00)||(KCl(sat'd),Hg_2Cl_2(sat'd)|Hg$$

$$E_{SCE} = +0.244\ V$$

Note that the saturated calomel electrode behaves as a cathode with respect to the standard hydrogen electrode and has a potential of 0.244 V at 25°C.[3] The saturated calomel electrode as well as several other secondary reference electrodes is described in detail in Section 15A.

12C-2

The Definition of Electrode Potential

An *electrode potential* is defined as the potential of a cell consisting of the electrode in question *acting as cathode* and the standard hydrogen electrode *acting as anode.* It should be emphasized that, despite its name, *an electrode potential is in fact the potential of an electrochemical cell involving a carefully defined reference electrode.* It could be more properly called a "relative electrode potential" (but seldom is).

The cell in Figure 12-3 illustrates the definition of the electrode potential for the half-reaction

$$Cu^{2+} + 2e \rightleftharpoons Cu(s)$$

Here, the half-cell on the right consists of a strip of pure copper in contact with a solution that has a copper(II) ion activity of 1.00; the electrode on the left is the standard hydrogen electrode. This cell develops a potential of 0.334 V with the copper electrode functioning *as the cathode;* that is, the spontaneous cell reaction is

$$Cu^{2+} + H_2(g) \rightleftharpoons Cu(s) + 2H^+$$

Because the copper electrode serves as the cathode, the measured potential is, *by definition,* the electrode potential for the copper half-reaction (or the

[3]Calomel is the common name for Hg_2Cl_2, and *saturated* refers *not* to the calomel but to the KCl, which is present in excess. There are also 1 M and 0.1 M calomel electrodes in which the KCl concentrations are 1 and 0.1 M, respectively.

copper *couple*). Note that the copper electrode bears a positive charge with respect to the hydrogen electrode. The electrode potential is therefore assigned a positive sign, and we write

$$Cu^{2+} + 2e \rightleftharpoons Cu(s) \qquad E_{Cu^{2+}} = +0.334 \text{ V}$$

Replacement of the Cu/Cu^{2+} couple with a zinc electrode immersed in a solution with a zinc ion activity of unity results in a potential of -0.763 V. In contrast to the copper electrode, the zinc electrode acts as the anode and the electric potential of the cell is negative. Here, the spontaneous cell reaction is

$$Zn(s) + 2H^{+} \rightleftharpoons H_2(g) + Zn^{2+}$$

That is, the zinc electrode bears a negative charge with respect to the standard hydrogen electrode. In order to reverse this reaction so that the zinc electrode acts as a cathode, a potential more negative than -0.763 V must be applied to the cell. Consequently, the electrode potential of the Zn/Zn^{2+} couple is *by convention* given a negative sign so that it is equal to -0.763 V.

A cadmium electrode immersed in a solution having a cadmium ion activity of unity develops a potential of -0.403 V when paired with a standard hydrogen electrode. Therefore, the electrode potential for the Cd/Cd^{2+} system is -0.403 V.

The electrode potentials for the four half-cells just described can be arranged in the order

Half-Reaction	Electrode Potential, V
$Cu^{2+} + 2e \rightleftharpoons Cu(s)$	+0.334
$2H^{+} + 2e \rightleftharpoons H_2(g)$	0.000
$Cd^{2+} + 2e \rightleftharpoons Cd(s)$	−0.403
$Zn^{2+} + 2e \rightleftharpoons Zn(s)$	−0.763

The magnitudes of these electrode potentials indicate the relative strength of the four ionic species as electron acceptors (oxidizing agents); that is, in decreasing strength, $Cu^{2+} > H^{+} > Cd^{2+} > Zn^{2+}$.

Sign Conventions for Electrode Potentials

Historically, electrochemists have not always used the sign convention just described. Indeed, disagreements regarding the conventions to be used in specifying signs for half-cell processes caused much controversy and confusion in the development of electrochemistry. The International Union of Pure and Applied Chemistry (IUPAC) addressed itself to this problem at its 1953 meeting in Stockholm. The usages adopted at that meeting are collectively referred to as the *Stockholm Convention* or the *IUPAC Convention* and are by now generally accepted. The sign convention described in the previous section and in the paragraphs that follow is based upon the IUPAC recommendations.

Any sign convention must be based upon expressing half-cell processes in a single way—that is, either as oxidations or as reductions. According to the IUPAC convention, the term *electrode potential* (or, more exactly, *relative electrode potential*) *is reserved exclusively to describe half-reactions written as reductions.* There is no objection to the use of the term *oxidation potential* to connote a process written in the opposite sense, but it is not proper to refer to such a potential as an electrode potential.

The sign of an electrode potential is determined by the electric potential of a galvanic cell consisting of the half-cell in question and the standard hydrogen electrode. When the former behaves as a cathode, the cell potential is positive, and so the electrode potential is positive also. When the half-cell of interest behaves as an anode, the cell potential and the electrode potential are both negative.

It is important to emphasize that the electrode potential refers to a half-cell process written *as a reduction.* For the zinc and cadmium electrodes we have been considering, the spontaneous reactions are oxidations. *It is evident, then, that the sign of an electrode potential indicates whether the reduction is spontaneous with respect to the standard hydrogen electrode.* The positive sign associated with the electrode potential for copper indicates that the process

$$Cu^{2+} + H_2(g) \rightleftharpoons Cu(s) + 2H^+$$

favors the products under ordinary conditions. Similarly, the negative sign of the electrode potential for zinc means that the analogous reaction

$$Zn^{2+} + H_2(g) \rightleftharpoons Zn(s) + 2H^+$$

is nonspontaneous; indeed, equilibrium favors the reactants over the products.

Reference works, particularly those published before 1953, often contain tabulations of electrode potentials that are not in accord with the IUPAC recommendations. For example, in a classic source of oxidation-potential data compiled by Latimer,[4] one finds

$$Zn(s) \rightleftharpoons Zn^{2+} + 2e \qquad E = +0.76\ V$$

$$Cu(s) \rightleftharpoons Cu^{2+} + 2e \qquad E = -0.34\ V$$

In converting these oxidation potentials to electrode potentials as defined by the IUPAC convention, one must mentally (1) express the half-reactions as reductions and (2) change the signs of the potentials.

The sign convention used in a tabulation of electrode potentials may not be explicitly stated. This information can be readily deduced, however, by noting the direction and sign of the potential for a half-reaction with which one is familiar. If the sign agrees with the IUPAC convention, the table can be used as is; if not, the signs of all of the data must be reversed. For example, the reaction

$$O_2(g) + 4H^+ + 4e \rightleftharpoons 2H_2O \qquad E = +1.229\ V$$

occurs spontaneously with respect to the standard hydrogen electrode and thus carries a positive sign. If the potential for this half-reaction is negative in a tabulation, it and all the other potentials should be multiplied by -1.

12C-3 The Effect of Concentration on Electrode Potentials: The Nernst Equation

The magnitude of an electrode potential is a measure of the force that drives the half-reaction to equilibrium (with respect to the standard hydrogen electrode). It follows, then, that numerical values for electrode potentials are concentration-dependent. Thus, copper ions in a concentrated solution of copper(II) tend to be more readily reduced to the elemental state than those

[4]W. M. Latimer, *The Oxidation States of the Elements and Their Potentials in Aqueous Solutions,* 2nd ed. Englewood Cliffs, NJ,: Prentice-Hall, 1952.

in a more dilute solution. Consequently, the electrode potential for copper is correspondingly greater in a more concentrated solution.

A quantitative relationship between electrode potential and concentration was first enunciated in 1889 by Hermann Walther Nernst, a German chemist, and it is fitting that his name should be associated with this important relationship.

Consider the generalized reversible half-reaction

$$aA + bB + \cdots + ne \rightleftharpoons cC + dD + \cdots$$

where the capital letters are formulas for species (whether charged or uncharged) participating in the electron transfer, e is the electron, and *a, b, c, d,* and *n* are the numbers of moles of participants involved in the half-cell process as written. It can be shown theoretically as well as experimentally that the electrode potential for this process is described by the equation

$$E = E^0 - \frac{RT}{nF} \ln \frac{[C]^c[D]^d \ldots}{[A]^a[B]^b \ldots} \tag{12-1}$$

where

E^0 = a constant called the *standard electrode potential,* which is characteristic for each half-reaction
R = the gas constant, 8.314 J K^{-1} mol^{-1}
T = temperature in kelvins
n = number of moles of electrons that appear in the half-reaction for the electrode process as written
F = the faraday = 96 485 C
ln = the natural logarithm = 2.303 log

Substitution of numerical values for the several constants, conversion to base 10 logarithms, and specification of 25°C for the temperature give

$$E = E^0 - \frac{0.0592}{n} \log \frac{[C]^c[D]^d \ldots}{[A]^a[B]^b \ldots} \tag{12-2}$$

The letters in brackets strictly represent activities, but we shall ordinarily follow our practice of substituting molar concentrations for activities in most calculations. Thus, if some participating species A is a solute,

$$[A] = \text{concentration of A in moles per liter}$$

If A is a gas, [A] in Equation 12-2 is replaced by p_A, the partial pressure of A in atmospheres. If A exists as a pure liquid in excess as a second phase, a pure solid present in excess as a second phase, or the solvent water, we assume that [A] has a value of 1.00. The rationale for this assumption is the same as that described in Section 5B-3, which deals with equilibrium-constant expressions.

EXAMPLE 12-1

Generate the Nernst expression for the following half-reactions.

(a) $Zn^{2+} + 2e \rightleftharpoons Zn(s)$ $\quad E = E^0 - \frac{0.0592}{2} \log \frac{1}{[Zn^{2+}]}$

Since the equilibrium is unaffected by the amount of elemental zinc (as long as some is present), [Zn(s)] has a value of 1.00, and the electrode potential varies with the logarithm of the reciprocal of the molar Zn^{2+} concentration.

(b) $Fe^{3+} + e \rightleftharpoons Fe^{2+} \qquad E = E^0 - \frac{0.0592}{1} \log \frac{[Fe^{2+}]}{[Fe^{3+}]}$

The potential for this half-reaction can be measured with an inert electrode immersed in a solution containing Fe(II) and Fe(III) ions. The potential that develops is dependent upon the logarithm of the ratio of the molar concentrations of these ions.

(c) $2H^+ + 2e \rightleftharpoons H_2(g) \qquad E = E^0 - \frac{0.0592}{2} \log \frac{p_{H_2}}{[H^+]^2}$

The term p_{H_2} represents the partial pressure of H_2, in atmospheres, at the surface of the electrode. Ordinarily, p_{H_2} is very close to atmospheric pressure.

(d) $Cr_2O_7^{2-} + 6e + 14H^+ \rightleftharpoons 2Cr^{3+} + 7H_2O$

$$E = E^0 - \frac{0.0592}{6} \log \frac{[Cr^{3+}]^2}{[Cr_2O_7^{2-}][H^+]^{14}}$$

Note that the potential of this half-cell depends not only upon the concentrations of the two chromium-containing species but also upon the hydrogen ion concentration. The potential is independent of the concentration of water, however, which is constant for all practical purposes.

(e) $AgCl(s) + e \rightleftharpoons Ag(s) + Cl^- \qquad E = E^0 - \frac{0.0592}{1} \log [Cl^-]$

This half-reaction describes the behavior of a silver electrode immersed in a chloride solution that has been *saturated* with AgCl; to ensure this condition, an excess of the solid is always present. Note that this electrode reaction is the sum of two reactions, namely,

$$AgCl(s) \rightleftharpoons Ag^+ + Cl^-$$

$$Ag^+ + e \rightleftharpoons Ag(s)$$

As long as both are present, the activities of metallic Ag and AgCl are equal to unity. Therefore, the logarithmic term in the Nernst equation contains the Cl^- concentration only.

12C-4

The Standard Electrode Potential E^0

Examination of Equation 12-2 reveals that the constant E^0 is equal to the half-cell potential when the logarithmic term is zero. This condition prevails whenever the activity quotient is unity; one such circumstance is when the activities of all reactants and products are equal to unity. Thus, *the standard electrode potential can be defined as the electrode potential of a half-cell reaction (versus the standard hydrogen electrode) when all reactants and products exist at unit activity.*

The standard electrode potential is an important physical constant that provides a quantitative description of the relative driving force for a half-cell reaction.[5] A number of facts concerning this constant and the half-cell potentials calculated from it should be kept in mind:

[5]For further reading on standard electrode potentials, see R. G. Bates, in *Treatise on Analytical Chemistry*, 2nd ed., I. M. Kolthoff and P. J. Elving, Eds., Part I, Vol. 1, Chapter 13. New York: Wiley, 1978.

1. The standard electrode potential is a relative quantity in the sense that it is the potential of an electrochemical cell in which the anode is the standard hydrogen electrode, whose potential has been arbitrarily assigned a value of zero.
2. Standard electrode potentials refer exclusively to half-cell processes that are written as reductions; that is, it is a relative reduction potential.
3. The standard electrode potential measures the relative intensity of the driving force for a half-reaction; as such, its numerical value is independent of the notation used to express the half-reaction. Thus, the potential for the process

$$Ag^+ + e \rightleftharpoons Ag(s) \qquad E^0 = +0.799 \text{ V}$$

does not change if for some reason we find it convenient to express this half-reaction as

$$100Ag^+ + 100e \rightleftharpoons 100Ag(s) \qquad E^0 = +0.799 \text{ V}$$

To be sure, the corresponding Nernst expression must be consistent with the half-reaction *as written*. For the first of these, it will be

$$E = 0.799 - \frac{0.0592}{1} \log \frac{1}{[Ag^+]}$$

and for the second

$$E = 0.799 - \frac{0.0592}{100} \log \frac{1}{[Ag^+]^{100}}$$

4. The sign of a standard electrode potential provides an indication of the tendency of the species involved to be reduced relative to the tendency of hydrogen ions to be reduced. A positive sign indicates that the oxidant involved (with all reactants and products at unit activity) is more easily reduced than are hydrogen ions and that the half-cell behaves as a cathode when coupled with a standard hydrogen electrode. Conversely, a negative sign indicates that the species involved is a weaker oxidizing agent than hydrogen ion and that the electrode behaves as the anode in a galvanic cell containing the standard hydrogen electrode.

Compilations of standard electrode potentials provide the chemist with qualitative information regarding the extent and direction of electron-transfer reactions between the tabulated species. Table 12-1 contains data needed for the examples that follow; a more extensive listing is given in Appendix 2.[6]

Entries in Table 12-1 are tabulated according to decreasing numerical values; those in Appendix 2, on the other hand, are arranged alphabetically in order to make it easier to find the potential of a given system. Proceeding down the left side of the tabulation in Table 12-1, each succeeding species is

[6]Comprehensive sources for standard electrode potentials include *Standard Electrode Potentials in Aqueous Solutions*, A. J. Bard, R. Parsons, and J. Jordan, Eds. New York: Marcel Dekker, 1985; G. Milazzo and S. Caroli, *Tables of Standard Electrode Potentials*. New York: Wiley-Interscience, 1977; M. S. Antelman and F. J. Harris, *Chemical Electrode Potentials*. New York: Plenum Press, 1982. Some compilations are arranged alphabetically by element; others are tabulated according to the numerical value of E^0.

a less effective electron acceptor (or oxidizing agent) than the one above it. The half-reactions at the bottom of the table have little tendency to take place, as written, but their reverse half-reactions have a strong tendency to occur. The most effective reducing agents, then, are species that appear in the lower right-hand side of the equations in a table of standard electrode potentials arranged in order of decreasing potentials.

Inspection of Table 12-1 indicates that zinc is more readily oxidized than cadmium, which makes it possible to conclude that the reaction

$$\underset{\text{red}_1}{Zn(s)} + \underset{\text{ox}_2}{Cd^{2+}} \rightleftharpoons \underset{\text{ox}_1}{Zn^{2+}} + \underset{\text{red}_2}{Cd(s)}$$

is spontaneous as written, and nonspontaneous in the opposite sense. Similarly, iron(III) is a more effective electron acceptor than is triiodide ion. Thus, iron(II) and triiodide ions predominate in the equilibrium

$$3I^- + 2Fe^{3+} \rightleftharpoons I_3^- + 2Fe^{2+}$$

12C-5 The Calculation of Electrode Potentials from Standard Electrode Potential Data

Typical applications of the Nernst equation are illustrated in Examples 12-2 through 12-5.

EXAMPLE 12-2

Calculate the potential of a cadmium electrode immersed in a 0.0100 M Cd^{2+} solution.

In Table 12-1, we find

$$Cd^{2+} + 2e \rightleftharpoons Cd(s) \qquad E^0 = -0.403 \text{ V}$$

The Nernst equation for this reaction is

$$E = E^0 - \frac{0.0592}{2} \log \frac{1}{[Cd^{2+}]}$$

TABLE 12-1 Standard Electrode Potentials*

Reaction	E^0 at 25°C, V
$Cl_2(g) + 2e \rightleftharpoons 2Cl^-$	+1.359
$O_2(g) + 4H^+ + 4e \rightleftharpoons 2H_2O$	+1.229
$Br_2(aq) + 2e \rightleftharpoons 2Br^-$	+1.087
$Br_2(l) + 2e \rightleftharpoons 2Br^-$	+1.065
$Ag^+ + e \rightleftharpoons Ag(s)$	+0.799
$Fe^{3+} + e \rightleftharpoons Fe^{2+}$	+0.771
$I_3^- + 2e \rightleftharpoons 3I^-$	+0.536
$Cu^{2+} + 2e \rightleftharpoons Cu(s)$	+0.337
$Hg_2Cl_2(s) + 2e \rightleftharpoons 2Hg(l) + 2Cl^-$	+0.268
$AgCl(s) + e \rightleftharpoons Ag(s) + Cl^-$	+0.222
$Ag(S_2O_3)_2^{3-} + e \rightleftharpoons Ag(s) + 2S_2O_3^{2-}$	+0.017
$2H^+ + 2e \rightleftharpoons H_2(g)$	0.000
$AgI(s) + e \rightleftharpoons Ag(s) + I^-$	−0.151
$PbSO_4(s) + 2e \rightleftharpoons Pb(s) + SO_4^{2-}$	−0.350
$Cd^{2+} + 2e \rightleftharpoons Cd(s)$	−0.403
$Zn^{2+} + 2e \rightleftharpoons Zn(s)$	−0.763

*See Appendix 2 for a more extensive list.

Substitution of 0.0100 for $[Cd^{2+}]$ and -0.403 for E^0 yields

$$E = -0.403 - \frac{0.0592}{2} \log \frac{1}{0.0100}$$

$$= -0.403 - \frac{0.0592}{2} \log 100$$

$$= -0.403 - \frac{0.0592}{2} \times 2 = -0.462 \text{ V}$$

The negative sign for this potential indicates that the reduction of cadmium ion is nonspontaneous with respect to the standard hydrogen electrode. Note also that the potential for the half-cell is more negative than the standard electrode potential. The direction of this difference follows from mass-law considerations, since the half-reaction *as written* has less tendency to occur when the cadmium ion concentration is 0.0100 M than when it is 1.00 M.

EXAMPLE 12-3

Calculate the potential of a platinum electrode immersed in a solution prepared by saturating 0.0100 M KBr with Br_2.

The pertinent half-reaction is

$$Br_2(l) + 2e \rightleftharpoons 2Br^-(aq) \qquad E^0 = 1.065 \text{ V}$$

Note that the symbol (l) in this half-reaction indicates that liquid Br_2 is present in excess and that the aqueous solution of KBr is at all times saturated with Br_2; then, by definition, the activity of Br_2 is constant and assigned a value of 1.00. The overall process can be considered the sum of the two equilibria

$$Br_2(l) \rightleftharpoons Br_2(aq)$$

$$Br_2(aq) + 2e \rightleftharpoons 2Br^-(aq)$$

The Nernst equation for the overall process is therefore

$$E = 1.065 - \frac{0.0592}{2} \log \frac{[Br^-]^2}{1.00}$$

$$= 1.065 - \frac{0.0592}{2} \log \frac{(1.00 \times 10^{-2})^2}{1.00}$$

$$= 1.065 - \frac{0.0592}{2} (-4.00) = 1.183 \text{ V}$$

EXAMPLE 12-4

Calculate the potential of a platinum electrode immersed in a solution that is 0.0100 M in KBr and 1.00×10^{-3} M in Br_2.

The standard electrode potential used in Example 12-3 is not applicable here because *the solution is no longer saturated with* Br_2. Table 12-1, however, has an entry for the half-reaction

$$Br_2(aq) + 2e \rightleftharpoons 2Br^-(aq) \qquad E^0 = 1.087 \text{ V}$$

The term (aq) refers to elemental Br_2 *in solution*, and 1.087 V is the potential for this half-cell when both Br_2 and Br^- have unit activity. It turns out, however, that the solubility of Br_2 in water is only about 0.18 M at 25°C. Thus, the recorded potential of 1.087 V is based on a system that—in terms of our definition of E^0—cannot be realized experimentally. Nevertheless, this hypothetical potential is useful because it permits the calculation of potentials for half-cells involving solutions that are undersaturated with respect to elemental bromine. Thus,

$$E = 1.087 - \frac{0.0592}{2} \log \frac{[Br^-]^2}{[Br_2]}$$

$$= 1.087 - \frac{0.0592}{2} \log \frac{(1.00 \times 10^{-2})^2}{1.00 \times 10^{-3}}$$

$$= 1.087 - \frac{0.0592}{2} \log 0.100 = 1.117 \text{ V}$$

Note that the activity of the Br_2 is taken as 1.00×10^{-3} rather than the 1.00 used in Example 12-3, where the solution was saturated and in contact with excess liquid bromine.

12C-6 *Standard Electrode Potentials for Half-Reactions Involving Precipitation or Complex Formation*

Example 12-5 demonstrates that reagents that compete for a participant in an electrode process can markedly influence potential.

EXAMPLE 12-5

Calculate the potential of a silver electrode in a solution that is saturated with AgCl and has a Cl^- activity of 1.00.

The standard electrode potential for the half-cell process occurring at the silver electrode (Table 12-1) is

$$Ag^+ + e \rightleftharpoons Ag(s) \qquad E^0 = 0.799 \text{ V}$$

and

$$E = 0.799 - \frac{0.0592}{1} \log \frac{1}{[Ag^+]}$$

Replacement of $[Ag^+]$ with $K_{sp}/[Cl^-]$ in the Nernst equation gives

$$E = 0.799 - \frac{0.0592}{1} \log \frac{[Cl^-]}{K_{sp}}$$

which can be rewritten as

$$E = 0.799 + 0.0592 \log K_{sp} - 0.0592 \log [Cl^-] \qquad (12\text{-}3)$$

Substitution of 1.82×10^{-10} for K_{sp} (Appendix 3) and 1.00 for $[Cl^-]$ gives

$$E = 0.799 + 0.0592 \log 1.82 \times 10^{-10} - 0.0592 \log 1.00$$

$$= 0.799 - 0.0592\,(-9.740) - 0 = 0.222 \text{ V}$$

Example 12-5 demonstrates that the half-cell potential for a silver electrode is profoundly altered by the presence of chloride ion, which lowers the silver ion concentration and thus diminishes the tendency for reduction of that ion.

Equation 12-3 relates the potential of a silver electrode to the chloride ion concentration of a solution that is also saturated with silver chloride. The overall process can be considered the resultant of the two reactions

$$AgCl(s) \rightleftharpoons Ag^+ + Cl^-$$

$$Ag^+ + e \rightleftharpoons Ag(s)$$

and the potential is related to the constants associated with these processes. When the chloride ion activity is unity, the second logarithmic term in Equation 12-3 becomes zero, and the resulting potential is then, *by definition, the standard electrode potential* for the half-reaction

$$AgCl(s) + e \rightleftharpoons Ag(s) + Cl^- \qquad E^0 = 0.222 \text{ V}$$

where

$$E^0_{AgCl} = 0.799 + 0.0592 \log K_{sp} = 0.222 \text{ V} \tag{12-4}$$

The potential for a silver electrode immersed in a solution saturated with silver chloride can thus be described *either* in terms of its silver ion concentration (with E^0 for the reduction of silver ion) *or* in terms of the chloride ion concentration (with E^0 for the silver/silver chloride half-reaction).

The potential of a silver electrode in a solution containing an ion that forms a soluble complex with silver ion can be treated analogously. For example, in a solution containing thiosulfate ion, the half-reaction can be written as

$$Ag(S_2O_3)_2^{3-} + e \rightleftharpoons Ag(s) + 2S_2O_3^{2-}$$

A silver electrode immersed in a solution in which the activities of the complex ion and the thiosulfate ion are both unity has a potential that is numerically equal to the standard electrode potential for this half-cell process; here,

$$E^0 = +0.799 - 0.0592 \log K_f$$

where K_f is the formation constant for the complex.

Data for the potential of a silver electrode in contact with solutions that contain ions that form complexes or sparingly soluble compounds with silver ion can be found in the literature (Appendix 2). Similar information is available for other electrode systems. Such data simplify the calculation of many half-cell potentials.

12D Applications of Electrode Potentials

In this section, we demonstrate how standard electrode potentials are applied to problems in analytical chemistry.

12D-1 The Calculation of Cell Potentials

A valuable application of standard electrode potentials is the calculation of the potential obtainable from a galvanic cell or the potential required to operate an electrolytic cell. These calculated potentials (sometimes called either *reversible* or *thermodynamic potentials*) are theoretical in the sense that they apply only to cells in the absence of current. As shown in Section 16A-3, additional factors need to be taken into account when a current exists in a cell.

The potential of an electrochemical cell arises from the difference in the potentials of its two half-cells. For example, the two half-reactions in a cell shown in Figure 12-1 can be written as

$$Cu^{2+} + 2e \rightleftharpoons Cu(s) \qquad E_{Cu^{2+}} = 0.3 \text{ V}$$

$$Zn^{2+} + 2e \rightleftharpoons Zn(s) \qquad E_{Zn^{2+}} = -0.8 \text{ V}$$

where we have assumed that the concentrations of copper and zinc ions (xM and yM, respectively) are such that the electrode potentials for the two half-reactions are as given. We can obtain an equation describing the cell reaction by subtracting the anodic half-reaction from the cathodic and rearranging:

$$Cu^{2+} + Zn(s) \rightleftharpoons Cu(s) + Zn^{2+}$$

Similarly, we obtain the potential for the cell by subtracting the anodic electrode potential from the cathodic:

$$E_{cell} = E_{Cu^{2+}} - E_{Zn^{2+}} = 0.3 - (-0.8) = 1.1 \text{ V}$$

The positive sign of the cell potential indicates that the reaction as written is spontaneous; the cell is therefore galvanic.

In general, the potential of a cell, be it galvanic or electrolytic, is the difference between the electrode potential of the cathode and that of the anode:

$$E_{cell} = E_{cathode} - E_{anode} \quad (12\text{-}5)$$

It should be stressed that *both* $E_{cathode}$ and E_{anode} are *electrode potentials* and are therefore the potentials of the half-reactions *written as reductions.*

Consider now the cell represented by the diagram

$$Cu|CuSO_4(x\text{M})||ZnSO_4(y\text{M})|Zn$$

This cell differs from the one in Figure 12-1 only in that the copper electrode is now the anode and the zinc electrode is now the cathode. The cell potential is obtained by substituting numerical values for the two electrode potentials into Equation 12-5:

$$E_{cell} = -0.8 - 0.3 = -1.1 \text{ V}$$

The negative sign signifies that the process

$$Cu(s) + Zn^{2+} \rightleftharpoons Cu^{2+} + Zn(s)$$

is nonspontaneous and that application of an external potential greater than -1.1 V is needed to cause this reaction to occur.

EXAMPLE 12-6

Calculate the theoretical potential for the cell

$$Ag|AgCl(\text{sat'd}),HCl(0.0200 \text{ M})|H_2(0.800 \text{ atm}),Pt$$

Note that this cell does not require two compartments (and a salt bridge) because molecular H_2 has little tendency to react directly with the low concentration of Ag^+ in the electrolyte solution.

The two half-reactions and their corresponding standard electrode potentials are (Table 12-1)

$$2H^+ + 2e \rightleftharpoons H_2(g) \qquad E^0 = 0.000 \text{ V}$$

$$AgCl(s) + e \rightleftharpoons Ag(s) + Cl^- \qquad E^0 = 0.222 \text{ V}$$

Provided the molar concentrations are reasonable approximations of activities, the two electrode potentials are

$$E_{AgCl} = 0.222 - \frac{0.0592}{1} \log 0.0200 = 0.323 \text{ V}$$

$$E_{Pt,H_2} = 0.000 - \frac{0.0592}{2} \log \frac{0.800}{(0.0200)^2} = -0.098 \text{ V}$$

The cell diagram specifies the silver electrode as the anode and the hydrogen electrode as the cathode. Thus,

$$E_{cell} = -0.098 - 0.323 = -0.421 \text{ V}$$

The negative sign indicates that the cell reaction

$$2H^+ + 2Ag(s) + 2Cl^- \rightleftharpoons H_2(g) + 2AgCl(s)$$

is nonspontaneous, and thus the cell is electrolytic.

EXAMPLE 12-7

The shorthand notation for the cell produced when a saturated calomel electrode ($E_{SCE} = 0.244$ V) is substituted for the hydrogen electrode in Example 12-6 is

$$Ag|AgCl(sat'd),HCl(0.0200 \text{ M})||KCl(sat'd),Hg_2Cl_2(sat'd)Hg(l)$$

Note that a salt bridge is required in this case.[7] The potential for this cell is

$$E_{cell} = 0.244 - 0.323 = -0.079 \text{ V}$$

The negative sign indicates that the reaction

$$2Ag(s) + Hg_2Cl_2(s) + 2Cl^- \rightleftharpoons 2Hg(l) + 2AgCl(s) + 2Cl^-$$

is also nonspontaneous and that the cell is thus electrolytic.

EXAMPLE 12-8

Calculate the potential for the following cell employing (a) concentrations and (b) activities:

$$Zn|ZnSO_4(x\text{M}),PbSO_4(sat'd)|Pb$$

where $x = 5.00 \times 10^{-4}$, 2.00×10^{-3}, 1.00×10^{-2}, 2.00×10^{-2}, and 5.00×10^{-2}.

(a) In a neutral solution, little HSO_4^- is formed and we can assume that

$$[SO_4^{2-}] = c_{ZnSO_4} = x = 5.00 \times 10^{-4}$$

The half-reactions and standard potentials are

$$PbSO_4(s) + 2e \rightleftharpoons Pb(s) + SO_4^{2-} \qquad E^0 = -0.350 \text{ V}$$

$$Zn^{2+} + 2e \rightleftharpoons Zn \qquad E^0 = -0.763 \text{ V}$$

The potential of the lead electrode is

$$E_{PbSO_4} = -0.350 - \frac{0.0592}{2} \log (5.00 \times 10^{-4}) = -0.252 \text{ V}$$

For the zinc half-reaction,

$$[Zn^{2+}] = 5.00 \times 10^{-4}$$

$$E_{Zn^{2+}} = -0.763 - \frac{0.0592}{2} \log \frac{1}{5.00 \times 10^{-4}} = -0.860 \text{ V}$$

[7]We shall sometimes abbreviate the cell diagram even further, as

$$Ag|AgCl(sat'd),HCl(0.0200 \text{ M})||SCE$$

where SCE is a shorthand representation for the saturated calomel electrode.

Since the lead electrode is specified as the cathode,

$$E_{cell} = -0.252 - (-0.860) = 0.608 \text{ V}$$

Cell potentials at the other concentrations can be derived in the same way. Their values are given in Table 12-2.

(b) To obtain activity coefficients for Zn^{2+} and SO_4^{2-}, we must first calculate the ionic strength with the aid of Equation 5-18:

$$\mu = [5.00 \times 10^{-4} \times (2)^2 + 5.00 \times 10^{-4} \times (2)^2]/2 = 2.00 \times 10^{-3}$$

In Table 5-4, we find $\alpha_A = 4.0$ for SO_4^{2-} and $\alpha_A = 6.0$ for Zn^{2+}. Substituting these values into Equation 5-20 gives for sulfate ion

$$-\log f_{SO_4^{2-}} = \frac{0.51 \times (2)^2 \times \sqrt{2.00 \times 10^{-3}}}{1 + 0.33 \times 4.0 \sqrt{2.00 \times 10^{-3}}} = 8.59 \times 10^{-2}$$

$$f_{SO_4^{2-}} = 0.820$$

Repeating the calculations for Zn^{2+}, for which $\alpha_A = 6.0$, yields

$$f_{Zn^{2+}} = 0.825$$

The Nernst equation for the lead electrode now becomes

$$E_{PbSO_4} = -0.350 - \frac{0.0592}{2} \log (0.820 \times 5.00 \times 10^{-4}) = -0.250$$

Similarly, for the zinc electrode

$$E_{Zn^{2+}} = -0.763 - \frac{0.0592}{2} \log \frac{1}{0.825 \times 5.00 \times 10^{-4}} = -0.863$$

Thus,

$$E_{cell} = -0.250 - (-0.863) = 0.613 \text{ V}$$

Values for other concentrations are found in Table 12-2 as well as experimentally determined potentials for the cell.

It is evident from the data in Table 12-2 that neglect of activity coefficients causes significant errors in cell potentials. It is also evident from the data in the fifth column of this table that potentials computed with activities agree reasonably well with experiment.

TABLE 12-2 **Effect of Ionic Strength on the Potential of a Galvanic Cell***

Concentration, $ZnSO_4$	**Ionic Strength, μ**	**(a) *E*, Based on Concentrations**	**(b) *E*, Based on Activities**	***E*, Experimental Values†**
5.00×10^{-4}	2.00×10^{-3}	0.608	0.613	0.611
2.00×10^{-3}	8.00×10^{-3}	0.573	0.582	0.583
1.00×10^{-2}	4.00×10^{-2}	0.531	0.550	0.553
2.00×10^{-2}	8.00×10^{-2}	0.513	0.537	0.542
5.00×10^{-2}	2.00×10^{-1}	0.490	0.521	0.529

*Cell described in Example 12-8.

†Experimental data from I. A. Cowperthwaite and V. K. LaMer, *J. Amer. Chem. Soc.*, **1931**, *53*, 4333.

EXAMPLE 12-9

Calculate the potential required to initiate deposition of copper from a solution that is 0.010 M in $CuSO_4$ and contains sufficient H_2SO_4 to give an H^+ concentration of 1.00×10^{-4} M.

The deposition of copper necessarily occurs at the cathode. Since there is no more easily oxidizable species than water in the system, evolution of O_2 occurs at the anode. The required standard electrode potentials are (Table 12-1)

$$O_2(g) + 4H^+ + 4e \rightleftharpoons 2H_2O \qquad E^0 = +1.229 \text{ V}$$

$$Cu^{2+} + 2e \rightleftharpoons Cu(s) \qquad E^0 = +0.337 \text{ V}$$

The potential for the Cu electrode is given by

$$E = +0.337 - \frac{0.0592}{2} \log \frac{1}{0.010} = +0.278 \text{ V}$$

If O_2 is evolved at 1.00 atm, the potential for the oxygen electrode is

$$E = +1.229 - \frac{0.0592}{4} \log \frac{1}{(1.00)(1.00 \times 10^{-4})^4} = +0.992 \text{ V}$$

and the cell potential is

$$E_{cell} = +0.278 - 0.992 = -0.714 \text{ V}$$

Thus, initiation of the reaction

$$2Cu^{2+} + 2H_2O \rightleftharpoons O_2(g) + 4H^+ + 2Cu(s)$$

requires the application of a potential greater than -0.714 V.

12D-2 The Calculation of Equilibrium Constants for Oxidation/Reduction Reactions from Standard Electrode Potentials

Consider the galvanic cell

$$Cu|Cu^{2+}(x\text{M})||Ag^+(y\text{M})|Ag$$

for which the spontaneous cell reaction is

$$Cu(s) + 2Ag^+ \rightleftharpoons Cu^{2+} + 2Ag(s)$$

The equilibrium-constant expression for this reaction is

$$K_{eq} = \frac{[Cu^{2+}]}{[Ag^+]^2}$$

The cell potential is at all times given by

$$E_{cell} = E_{cathode} - E_{anode}$$

and at all times,

$$E_{cathode} = E_{Ag+} = E^0_{Ag} - \frac{0.0592}{1} \log \frac{1}{[Ag^+]}$$

$$E_{anode} = E_{Cu^{2+}} = E^0_{Cu} - \frac{0.0592}{2} \log \frac{1}{[Cu^{2+}]}$$

If the two electrodes in this cell are connected by a wire, a current will develop. With the passage of time, the concentration of Cu(II) ions will increase and the concentration of Ag(I) ions will decrease. These changes make the potential of the copper electrode more positive and that of the silver electrode less positive. The net effect is a decrease in the potential of the cell as it is discharged. The concentrations of Cu(II) and Ag(I) ultimately attain values such that there remains no further tendency for the transfer of electrons to occur. Under these conditions, *the potential of the cell becomes zero, and the system is at equilibrium.* Thus,

$$E_{cell} = 0 = E_{cathode} - E_{anode}$$

or, at equilibrium,

$$E_{cathode} = E_{anode} \qquad (12\text{-}6)$$

Equation 12-6 is an important and general relationship; *the electrode potentials for all half-reactions of an oxidation/reduction system are equal when that system is at equilibrium.* Note that this generalization applies regardless of the number of half-reactions involved in the system because interactions among *all* must take place until all the electrode potentials are identical. In terms of the galvanic cell we have been considering, equilibrium is attained when

$$E_{Ag^+} = E_{Cu^{2+}}$$

This equality can be expressed in terms of the Nernst equations for the two half-cell processes:

$$E^0_{Ag^+} - \frac{0.0592}{2}\log\frac{1}{[Ag^+]^2} = E^0_{Cu^{2+}} - \frac{0.0592}{2}\log\frac{1}{[Cu^{2+}]} \qquad (12\text{-}7)$$

It is important to note that the Nernst equation is applied to the silver half-reaction as it appears in the balanced equation:

$$2Ag^+ + 2e \rightleftharpoons 2Ag(s) \qquad E^0 = 0.799\text{ V}$$

Rearrangement gives

$$E^0_{Ag^+} - E^0_{Cu^{2+}} = \frac{0.0592}{2}\log\frac{1}{[Ag^+]^2} - \frac{0.0592}{2}\log\frac{1}{[Cu^{2+}]}$$
$$= \frac{0.0592}{2}\log\frac{1}{[Ag^+]^2} + \frac{0.0592}{2}\log\frac{[Cu^{2+}]}{1}$$

Finally, then,

$$\frac{2(E^0_{Ag^+} - E^0_{Cu^{2+}})}{0.0592} = \log\frac{[Cu^{2+}]}{[Ag^+]^2} = \log K_{eq}$$

Note that the concentration terms in this particular relationship are *equilibrium concentrations; the quotient in the logarithmic term is thus the equilibrium constant for the reaction.*

EXAMPLE 12-10

A piece of copper is placed in 0.050 M $AgNO_3$. Calculate (a) the equilibrium constant for the chemical reaction that occurs and (b) the equilibrium composition of this solution.

(a) The required standard electrode potentials are (Table 12-1)

$$Ag^+ + e \rightleftharpoons Ag(s) \qquad E^0 = +0.799\text{ V}$$

$$Cu^{2+} + 2e \rightleftharpoons Cu(s) \qquad E^0 = +0.337\text{ V}$$

and, as just shown,

$$\log K_{eq} = \log \frac{[Cu^{2+}]}{[Ag^+]^2} = \frac{2(E^0_{Ag^+} - E^0_{Cu^{2+}})}{0.0592} = \frac{2(0.799 - 0.337)}{0.0592} = 15.61$$

$$K_{eq} = \frac{[Cu^{2+}]}{[Ag^+]^2} = \text{antilog}\ (15.61) = 4.1 \times 10^{15}$$

(b) The magnitude of the equilibrium constant suggests that nearly all the Ag^+ is reduced. The concentration of Cu^{2+} is therefore

$$[Cu^{2+}] = (0.050 - [Ag^+])/2 = 0.025 - [Ag^+]/2$$

It is worthwhile to obtain a provisional solution based on the assumption that $[Ag^+]$ is small with respect to 0.025, that is,

$$[Cu^{2+}] \simeq 0.025$$

and

$$\frac{0.025}{[Ag^+]^2} = 4.1 \times 10^{15}$$

$$[Ag^+] = 2.5 \times 10^{-9}$$

The assumption is clearly justified.

In order to derive a general relationship for computing equilibrium constants from standard potential data, let us consider a reaction in which a species A_{red} reacts with a species B_{ox} to yield A_{ox} and B_{red}. The two electrode reactions are

$$A_{ox} + me \rightleftharpoons A_{red}$$

$$B_{ox} + ne \rightleftharpoons B_{red}$$

In order to obtain a balanced equation for the desired reaction, we must multiply the first equation through by n and the second through by m to give

$$nA_{ox} + nme \rightleftharpoons nA_{red}$$

$$mB_{ox} + nme \rightleftharpoons mB_{red}$$

Subtraction of the first equation from the second yields

$$nA_{red} + mB_{ox} \rightleftharpoons nA_{ox} + mB_{red}$$

When this system is in equilibrium, the two electrode potentials E_A and E_B are numerically identical, that is,

$$E_A = E_B$$

Substitution of the Nernst expressions into this equation reveals that, *at equilibrium*,

$$E^0_A - \frac{0.0592}{nm} \log \frac{[A_{red}]^n}{[A_{ox}]^n} = E^0_B - \frac{0.0592}{nm} \log \frac{[B_{red}]^m}{[B_{ox}]^m}$$

which rearranges to

$$E^0_B - E^0_A = \frac{0.0592}{nm} \log \frac{[A_{ox}]^n[B_{red}]^m}{[A_{red}]^n[B_{ox}]^m} = \frac{0.0592}{nm} \log K_{eq}$$

Finally, then,

$$\log K_{eq} = \frac{nm(E^0_B - E^0_A)}{0.0592} \qquad (12\text{-}8)$$

EXAMPLE 12-11

Calculate the equilibrium constant for the reaction

$$MnO_4^- + 5Fe^{2+} + 8H^+ \rightleftharpoons Mn^{2+} + 5Fe^{3+} + 4H_2O$$

The standard electrode potentials are (Appendix 2)

$$MnO_4^- + 5e + 8H^+ \rightleftharpoons Mn^{2+} + 4H_2O \qquad E^0_{MnO_4^-} = +1.51 \text{ V}$$

$$Fe^{3+} + e \rightleftharpoons Fe^{2+} \qquad E^0_{Fe^{3+}} = +0.771 \text{ V}$$

The balanced net-ionic equation for this reaction involves 5 mol of Fe for 1 mol of Mn; it is therefore necessary to write the half-reaction for the Fe^{3+}/Fe^{2+} couple as

$$5Fe^{3+} + 5e \rightleftharpoons 5Fe^{2+} \qquad E^0_{Fe^{3+}} = +0.771 \text{ V}$$

Note that multiplication of this half-reaction by 5 does not alter the value of E^0 (page 295).

When the system is in equilibrium,

$$E_{Fe^{3+}} = E_{MnO_4^-}$$

or

$$E^0_{Fe^{3+}} - \frac{0.0592}{5}\log\frac{[Fe^{2+}]^5}{[Fe^{3+}]^5} = E^0_{MnO_4^-} - \frac{0.0592}{5}\log\frac{[Mn^{2+}]}{[MnO_4^-][H^+]^8}$$

This equality can be rearranged to provide the logarithm of the equilibrium constant:

$$\frac{0.0592}{5}\log\frac{[Mn^{2+}][Fe^{3+}]^5}{[MnO_4^-][Fe^{2+}]^5[H^+]^8} = \frac{0.0592}{5}\log K_{eq} = E^0_{MnO_4^-} - E^0_{Fe^{3+}}$$

Finally, then

$$\log K_{eq} = \frac{5(1.51 - 0.771)}{0.0592} = 62.52 \simeq 62.5$$

$$K_{eq} = 10^{0.52} \times 10^{62} = 3 \times 10^{62}$$

Note that this answer is rounded to one digit because $\log K_{eq}$ contains only one significant digit to the right of the decimal point (page 49).

12D-3

The Derivation of Equilibrium Constants for Acid/Base, Precipitation, and Complex-Formation Reactions from Electrode Potential Measurements

Numerical values for solubility-product constants, dissociation constants, and formation constants are conveniently evaluated through the measurement of cell potentials. One important virtue of this technique is that the measurement can be made without appreciably affecting any equilibria that may exist in the solution. For example, the potential of a silver electrode in a solution containing silver ion, cyanide ion, and the complex formed between them depends upon the activities of the three species. It is possible to measure this potential with negligible current. Since the activities of the participants are not sensibly altered during the measurement, the position of the equilibrium

$$Ag^+ + 2CN^- \rightleftharpoons Ag(CN)_2^-$$

is likewise undisturbed.

EXAMPLE 12-12

Calculate the formation constant K_f for $Ag(CN)_2^-$:

$$Ag^+ + 2CN^- \rightleftharpoons Ag(CN)_2^-$$

if the cell

$$Ag|Ag(CN)_2^-(7.50 \times 10^{-3}\ M),CN^-(0.0250\ M)||SCE$$

develops a potential of 0.625 V.

Proceeding as in the earlier examples, we have

$$Ag^+ + e \rightleftharpoons Ag(s) \qquad E^0 = +0.799\ V$$

$$0.625 = E_{cathode} - E_{anode} = 0.244 - E_{Ag+}$$

$$E_{Ag+} = 0.244 - 0.625 = -0.381\ V$$

Application of the Nernst equation for the silver electrode gives

$$-0.381 = 0.799 - \frac{0.0592}{1} \log \frac{1}{[Ag^+]}$$

$$\log [Ag^+] = \frac{-0.381 - 0.799}{0.0592} = -20.0$$

$$[Ag^+] = 1 \times 10^{-20}$$

The result must be rounded to one significant figure:

$$K_f = \frac{[Ag(CN)_2^-]}{[Ag^+][CN^-]^2} = \frac{7.50 \times 10^{-3}}{(1 \times 10^{-20})(2.50 \times 10^{-2})^2} = 1.2 \times 10^{21} \approx 1 \times 10^{21}$$

In theory, any electrode system in which hydrogen ions are participants can be used to evaluate dissociation constants for acids and bases. To be sure, relatively few of these systems have been used for determining such constants.

EXAMPLE 12-13

Calculate the dissociation constant for the weak acid HP if the cell

$$Pt,H_2(1.00\ atm)|HP(0.010\ M),NaP(0.040\ M)||SCE$$

develops a potential of 0.591 V.

The diagram for this cell indicates that the saturated calomel electrode is the cathode. Thus,

$$E_{cell} = 0.244 - E_{anode}$$

$$E_{anode} = 0.244 - 0.591 = -0.347\ V$$

Application of the Nernst equation for the hydrogen electrode gives

$$-0.347 = 0.000 - \frac{0.0592}{2} \log \frac{1.00}{[H^+]^2} = 0.000 + \frac{2 \times 0.0592}{2} \log [H^+]$$

$$\log [H^+] = \frac{-0.347 - 0.000}{0.0592} = -5.87$$

$$[H^+] = 1.34 \times 10^{-6}$$

Substitution of this value for $[H^+]$ as well as the concentrations of the weak acid and its conjugate base into the dissociation-constant expression gives

$$K_{HP} = \frac{[H^+][P^-]}{[HP]} = \frac{(1.34 \times 10^{-6})(0.040)}{0.010} = 5.4 \times 10^{-6}$$

12D-4

Limitations to the Use of Standard Electrode Potentials

The preceding examples suggest that the Nernst equation can provide answers to questions that are of importance to the analytical chemist. The reader should be aware, however, that significant differences between calculated and experimental potentials are sometimes encountered. Some of the sources of such differences are described in the paragraphs that follow.

Use of Concentrations Instead of Activities

It is ordinarily expedient to use molar concentrations instead of activities in calculations even though differences in these quantities tend to increase with increasing ionic strength. Because the ionic strength of the typical electrochemical cell is large, potentials based on molar concentrations are likely to differ appreciably from those obtained by direct measurement. Table 12-2 illustrates the effect of ionic strength on typical calculated cell potentials.

The Effect of Other Equilibria

In many systems of interest, the application of electrode-potential data is further complicated by solvolysis, dissociation, association, and complex-formation equilibria that involve species appearing in the Nernst equation. Accounting for these phenomena requires knowledge of their existence and availability of the appropriate equilibrium-constant data.

Formal Potentials

Swift[8] proposed substitution of *formal potentials* (also called *conditional potentials*) in place of standard electrode potentials to compensate for activity effects as well as errors due to the existence of competing equilibria. The formal potential of a system is the potential of the half-cell (with respect to the standard hydrogen electrode) when the concentration of each solute participating in the half-reaction is 1 M and the concentrations of all other solutes are carefully specified.

Formal potentials for many half-reactions are listed in Appendix 2. Note that large differences exist between the formal and standard potentials for some half-reactions. For example, the standard potential for the reduction of iron(III) to iron(II) is 0.771 V. In 1 M perchloric acid, the formal potential for the same half-reaction is 0.731 V. This difference is attributable to the fact that the activity coefficient of iron(III) is considerably smaller than that of iron(II) at the high ionic strength of the 1M perchloric acid medium. As a consequence, the ratio of activities of the two species is less than unity, a condition that leads to a decrease in the electrode potential. In 1 M hydrochloric acid, the formal potential for this couple is only 0.700 V. Here, the iron(III)/iron(II) activity ratio is even smaller because the chloro complexes of the former are more stable than those of iron(II); a change in potential larger than that seen in perchloric acid is the consequence.

Substitution of formal potentials for standard electrode potentials in the Nernst equation yields better agreement between calculated and experimental results—provided, of course, that the electrolyte concentration of the solution approximates that for which the formal potential is applicable. Not surprisingly, attempts to apply formal potentials to systems that differ substantially in type and in concentration of electrolyte can result in errors that

[8]E. H. Swift, *A System of Chemical Analysis,* p. 50. San Francisco: Freeman, 1939.

are larger than those associated with the use of standard electrode potentials. Henceforth, we shall use whichever is the more appropriate.

Reaction Rates

Standard potentials reveal whether or not a reaction proceeds far enough toward completion to be useful in a particular analytical problem, but they provide no information about the rate at which the equilibrium state is approached. Consequently, a reaction that appears extremely favorable from equilibrium considerations may be totally unacceptable from a kinetic standpoint. The oxidation of arsenic(III) with cerium(IV) in dilute sulfuric acid is a typical example. The reaction is

$$H_3AsO_3 + 2Ce^{4+} + H_2O \rightleftharpoons H_3AsO_4 + 2Ce^{3+} + 2H^+$$

The formal potentials E^f for these two systems are

$$Ce^{4+} + e \rightleftharpoons Ce^{3+} \qquad E^f = +1.4 \text{ V}$$

$$H_3AsO_4 + 2H^+ + 2e \rightleftharpoons H_3AsO_3 + H_2O \qquad E^f = +0.56 \text{ V}$$

and an equilibrium constant of about 10^{28} can be deduced from these data. Even though this reaction is highly favorable from an equilibrium standpoint, titration of solutions of arsenic(III) with cerium(IV) is impossible without a catalyst because several hours are needed for the attainment of equilibrium. Fortunately, several substances catalyze the reaction and thus make the titration feasible.

12D-5 The Experimental Determination of Standard Potentials

Although standard electrode potentials for hundreds of half-reactions are found in compilations of electrochemical data, it is noteworthy that neither the standard hydrogen electrode nor any other electrode in its standard state can be realized in the laboratory. That is, the standard hydrogen electrode is a *hypothetical electrode,* as is any other electrode system in which the reactants and products are at unit activity or pressure. The reason such electrode systems cannot be prepared experimentally is that chemists lack the knowledge to produce solutions having ionic activities of exactly unity. That is, no adequate theory exists that permits calculation of the *concentration* of a compound needed to give a solution of unit ionic activity. At such high ionic strengths, the Debye-Hückel relationship (Section 5C-2) is not valid, and no independent experimental method exists for the determination of activity coefficients in such solutions. Thus, for example, the *concentration* of HCl or other acids required to give the unit hydrogen ion activity specified in the standard hydrogen electrode *cannot* be calculated or determined experimentally. Notwithstanding, data taken in solutions of low ionic strengths can be extrapolated to give valid measures of standard electrode potentials as theoretically defined. An example of how hypothetical standard electrode potentials might be obtained from experimental data follows.

EXAMPLE 12-14

D. A. MacInnes[9] found that a cell similar to that shown in Figure 12-2 developed a potential of 0.52053 V. The cell is described by

$$\text{Pt},H_2(1.00 \text{ atm})|HCl(3.215 \times 10^{-3} \text{ M}),AgCl(\text{sat'd})|Ag$$

[9]D. A. MacInnes, *The Principles of Electrochemistry,* p. 187. New York: Reinhold, 1939.

Calculate the standard electrode potential for the half-reaction

$$AgCl(s) + e \rightleftharpoons Ag(s) + Cl^-$$

Here, the electrode potential for the cathode is

$$E_{cathode} = E^0_{AgCl} - 0.0592 \log c_{HCl} f_{Cl^-}$$

where f_{Cl^-} is the activity coefficient of Cl^-. The second half-cell reaction is

$$H^+ + e \rightleftharpoons \tfrac{1}{2} H_2(g)$$

and

$$E_{anode} = E^0_{H_2} = \frac{0.0592}{1} \log \frac{p^{1/2}_{H_2}}{c_{HCl} f_{H^+}}$$

The measured potential is the difference between these potentials (Equation 12-5):

$$E_{cell} = (E^0_{AgCl} - 0.0592 \log c_{HCl} f_{Cl^-}) - \left(0.000 - 0.0592 \log \frac{p^{1/2}_{H_2}}{c_{HCl} f_{H^+}}\right)$$

Combining the two logarithmic terms gives

$$E_{cell} = E^0_{AgCl} - 0.0592 \log \frac{c^2_{HCl} f_{H^+} f_{Cl^-}}{p^{1/2}_{H_2}}$$

The activity coefficients for H^+ and Cl^- can be derived from Equation 5-20 employing 3.215×10^{-3} for the ionic strength μ; these values are 0.945 and 0.939, respectively. Substitution of these activity coefficients and the experimental data into the foregoing equation gives, upon rearrangement,

$$E^0_{AgCl} = 0.52053 + 0.0592 \log \frac{(3.215 \times 10^{-3})^2(0.945)(0.939)}{1.00^{1/2}}$$

$$= 0.2223 \simeq 0.222 \text{ V}$$

(The mean for this and similar measurements at other concentrations was 0.222 V.)

Questions and Problems

12-1 Define

*(a) oxidation.
(b) reducing agent.
*(c) galvanic cell.
(d) electrolytic cell.
*(e) anode.
(f) cathode.
*(g) liquid junction.
(h) irreversible cell.
*(i) standard hydrogen electrode.
(j) electrode potential.
*(k) standard electrode potential.
(l) salt bridge.
*(m) formal potential.
(n) Nernst equation.

12-2 What is meant by a redox conjugate pair? How does this differ from an acid/base conjugate pair?

*12-3 Why is the standard hydrogen electrode never used in the laboratory?

12-4 Employing activities, calculate the electrode potential of a hydrogen electrode in which the electrolyte is 0.0100 M HCl and the activity of hydrogen gas is 1.00 atm.

*12-5 The standard electrode potential for the reduction of Ni^{2+} to Ni is −0.23 V. Would the potential of a nickel electrode immersed in a 1.00 M NaOH solution saturated with $Ni(OH)_2$ be more negative than $E^0_{Ni^{2+}}$ or less? Explain.

12-6 Why is it necessary to bubble hydrogen gas through the electrolyte in a hydrogen-gas electrode?

*12-7 The following two entries are found in a table of standard electrode potentials:

$I_2(s) + 2e \rightleftharpoons 2I^-$ $E^0 = 0.536$ V

$I_2(aq) + 2e \rightleftharpoons 2I^-$ $E^0 = 0.615$ V

What is the significance of the difference between these two?

12-8 Quinhydrone is a crystalline solid consisting of 1 mol of quinone and 1 mol of hydroquinone. These two species react reversibly at a platinum electrode according to the reaction

$$\text{quinone} + 2H^+ + 2e \rightleftharpoons \text{hydroquinone} \qquad E^0 = 0.699 \text{ V}$$

Before the invention of the glass electrode, quinhydrone was often used for the potentiometric determination of pH.

(a) Draw a diagram of a cell that might be used for the determination of pH with quinhydrone.
(b) Derive an equation relating pH to the potential of a cell containing a quinhydrone electrode.

*12-9 Draw a diagram of a cell that could be used for the determination of the formation constant for $Cu(NH_3)_2^{2+}$. Derive an equation relating K_f to the cell potential.

*12-10 Complete and balance the following equations, adding H^+, OH^-, or H_2O as required:

(a) $Tl^{3+} + Ag(s) + Br^- \rightleftharpoons TlBr(s) + AgBr(s)$

(b) $Fe^{2+} + UO_2^{2+} \rightleftharpoons Fe^{3+} + U^{4+}$

(c) $N_2(g) + H_2(g) \rightleftharpoons N_2H_5^+$

(d) $Cr_2O_7^{2-} + I^- \rightleftharpoons Cr^{3+} + I_3^-$

(e) $H_2O_2 + Ce^{4+} \rightleftharpoons O_2 + Ce^{3+}$

(f) $IO_3^- + I^- \rightleftharpoons I_2(s)$

12-11 Complete and balance the following equations, adding H^+, OH^-, or H_2O as required:

(a) $Cu^{2+} + I^- \rightleftharpoons CuI(s) + I_3^-$

(b) $I_3^- + S_2O_3^{2-} \rightleftharpoons I^- + S_4O_6^{2-}$

(c) $MnO_4^- + H_2SO_3 \rightleftharpoons Mn^{2+} + SO_4^{2-}$

(d) $MnO_4^- + Mn^{2+} \rightleftharpoons MnO_2(s)$

(e) $IO_3^- + H_3AsO_3 + Cl^- \rightleftharpoons ICl_2^- + H_3AsO_4$

(f) $S_2O_8^{2-} + Mn^{2+} \rightleftharpoons SO_4^{2-} + MnO_4^-$

*12-12 Identify the oxidizing agent and the reducing agent on the left side of each equation in Problem 12-10; write a balanced equation for each half-reaction.

12-13 Identify the oxidizing agent and the reducing agent on the left side of each equation in Problem 12-11; write a balanced equation for each half-reaction.

*12-14 Calculate the equilibrium constant for each reaction in Problem 12-10. (For $Tl^{3+} + Br^- + 2e \rightleftharpoons TlBr(s)$, $E^0 = 1.44$ V.)

12-15 Calculate the equilibrium constant for each reaction in Problem 12-11.

*12-16 Calculate the electrode potential of a mercury electrode immersed in
(a) 0.0400 M $Hg(NO_3)_2$.
(b) 0.0400 M $Hg_2(NO_3)_2$.
(c) 0.0400 M KCl saturated with Hg_2Cl_2.
(d) 0.0400 M $Hg(SCN)_2$ ($Hg^{2+} + 2SCN^- \rightleftharpoons Hg(SCN)_2$; $K_f = 1.8 \times 10^{17}$).

12-17 Calculate the electrode potential for a copper electrode immersed in
(a) 0.0200 M Cu^{2+}.
(b) 0.0200 M Cu^+.
(c) 0.0300 M KI saturated with CuI.
(d) 0.01 M NaOH saturated with $Cu(OH)_2$.

*12-18 Calculate the electrode potential for a platinum electrode immersed in a solution that is
(a) 0.075 M in $Fe_2(SO_4)_3$ and 0.060 M in $FeSO_4$.
(b) 0.244 M in V^{3+}, 0.414 M in VO^{2+}, and 1.00×10^{-5} M in NaOH.
(c) 0.111 M in KI and 0.200 M in KI_3.
(d) 0.117 M in $K_4Fe(CN)_6$ and 0.333 M in $K_3Fe(CN)_6$.
(e) 0.0731 M in $SbONO_3$, 0.0100 M in HNO_3, and saturated with Sb_2O_5.
(f) 0.0731 M in $SbONO_3$, 1.00×10^{-5} M in HNO_3, and saturated with Sb_2O_5.

12-19 Calculate the electrode potential for a platinum electrode immersed in a solution that is
(a) 0.313 M in $Tl_2(SO_4)_3$ and 0.209 M in Tl_2SO_4.
(b) saturated with hydrogen at 1.00 atm and has a pH of 3.50.
(c) 0.0774 M in UO_2^{2+}, 0.0507 M in U^{4+}, and 1.00×10^{-4} M in $HClO_4$.
(d) 0.0627 M in $S_2O_3^{2-}$ and 0.0714 M in $S_4O_6^{2-}$.
(e) 0.0540 M in $Cr_2O_7^{2-}$, 0.149 M in Cr^{3+}, and 0.100 M in $HClO_4$.

*12-20 Indicate whether each of the following half-cells behaves as anode or cathode when coupled with a standard hydrogen electrode in a galvanic cell, and calculate the potential of the cell:
(a) $Pb|Pb^{2+}(2.00 \times 10^{-4}\ M)$.
(b) $Pt|Sn^{4+}(0.200\ M),Sn^{2+}(0.100\ M)$.
(c) $Pt|Sn^{4+}(1.0 \times 10^{-6}\ M),Sn^{2+}(0.50\ M)$.
(d) $Pt|Ti^{3+}(0.300\ M),TiO^{2+}(0.100\ M),H^+(0.200\ M)$.
(e) $Ag|AgBr(sat'd),KBr(1.00 \times 10^{-4}\ M)$.
(f) $Ag|Ag(CN)_2^-(0.200\ M),CN^-(1.00 \times 10^{-6}\ M)$.

12-21 Indicate whether each of the following half-cells behaves as anode or cathode when coupled with a standard hydrogen electrode in a galvanic cell and calculate the potential of the cell:
(a) $Pt|V^{3+}(0.50\ M),V^{2+}(1.00 \times 10^{-6}\ M)$.
(b) $Ag|AgNO_3(0.0100\ M),Na_2S_2O_3(0.0800\ M)$.
(c) $Ag|AgNO_3(0.0100\ M),Na_2S_2O_3(0.200\ M)$.
(d) $Bi|BiCl_4^-(0.010\ M),Cl^-(0.500\ M)$.
(e) $Ag|Ag(CN)_2^-(0.400\ M),CN^-(1.00 \times 10^{-4}\ M)$.

*12-22 Indicate in which direction the reactions in Problem 12-10 proceed if all species are initially at unit activity.

$$Tl^{3+} + Br^- + 2e \rightleftharpoons TlBr(s), \qquad E^0 = 1.44\ V$$

12-23 Indicate in which direction the reactions in Problem 12-11 proceed if all species are initially at unit activity.

*12-24 Calculate the theoretical cell potential for each of the following. Is the cell as written galvanic or electrolytic?

(a) $Pb|PbSO_4$(sat'd),SO_4^{2-}(0.200 M)||Sn^{2+}(0.150 M),Sn^{4+}(0.250 M)|Pt
(b) Pt|Fe^{3+}(0.0100 M),Fe^{2+}(0.00100 M)||Ag^+(0.0350 M)|Ag
(c) Cu|CuI(sat'd),KI(0.0100 M)||KI(0.200 M),CuI(sat'd)|Cu
(d) Pt|UO_2^{2+}(0.100 M),U^{4+}(0.0100 M),H^+(1.00×10^{-6} M)|| AgCl(sat'd),KCl(1.00×10^{-4} M)|Ag
(e) Hg|Hg_2Cl_2(sat'd),Cl^-(0.0500 M)||V^{2+}(0.200 M),V^{3+}(0.300 M)|Pt
(f) Pt|VO^{2+}(0.250 M),V^{3+}(0.100 M),H^+(1.00×10^{-3} M)|| Tl^{3+}(0.100 M),Tl^+(0.0500 M)|Pt

12-25 Calculate the theoretical cell potential for each of the following. Is the cell as written galvanic or electrolytic?

(a) Ag|AgBr(sat'd),Br^-(0.0400 M)||H^+(1.00×10^{-4} M)|H_2(0.90 atm),Pt
(b) Pt|Cr^{3+}(0.0500 M),Cr^{2+}(0.0250 M)||Ni^{2+}(0.0100 M)|Ni
(c) Ag|$Ag(CN)_2^-$(0.240 M),CN^-(0.100 M)||Br_2(1.00×10^{-3} M),KBr(0.200 M)|Pt
(d) Ag|AgCl(sat'd),HCl(5.00×10^{-3} M)|H_2(0.300 atm),Pt
(e) Hg|Hg_2Cl_2(sat'd),HCl(0.0050 M)||HCl(1.50 M),Hg_2Cl_2(sat'd)|Hg
(f) Pt|TiO^{2+}(0.200 M),Ti^{3+}(0.100 M),H^+(2.00×10^{-3} M)|| SO_4^{2-}(0.200 M),$PbSO_4$(sat'd)|Pb

*12-26 The solubility-product constant for Ag_2SO_3 is 1.5×10^{-14}. Calculate E^0 for the process

$$Ag_2SO_3(s) + 2e \rightleftharpoons 2Ag(s) + SO_3^{2-}$$

12-27 The solubility-product constant for $Ni_2P_2O_7$ is 1.7×10^{-13}. Calculate E^0 for the process

$$Ni_2P_2O_7(s) + 4e \rightleftharpoons 2Ni(s) + P_2O_7^{4-}$$

*12-28 Calculate the solubility product of Hg_2SO_4, given the standard potentials

$$Hg_2SO_4(s) + 2e \rightleftharpoons 2Hg(l) + SO_4^{2-} \qquad E^0 = 0.615 \text{ V}$$

$$Hg_2^{2+} + 2e \rightleftharpoons Hg(l) \qquad E^0 = 0.789 \text{ V}$$

12-29 Calculate the solubility product of Ag_2MoO_4, given the standard potentials

$$Ag_2MoO_4(s) + 2e \rightleftharpoons 2Ag(s) + MoO_4^{2-} \qquad E^0 = 0.486 \text{ V}$$

$$Ag^+ + e \rightleftharpoons Ag(s) \qquad E^0 = 0.799 \text{ V}$$

*12-30 Compute E^0 for the process

$$ZnY^{2-} + 2e \rightleftharpoons Zn(s) + Y^{4-}$$

where Y^{4-} is the completely deprotonated anion of EDTA. The formation constant for ZnY^{2-} is 3.2×10^{16}.

12-31 Calculate E^0 for the process

$$VY^- + e \longrightarrow VY^{2-}$$

if the formation constant for the EDTA complex of V^{2+} is 5.0×10^{12} and that for the V^{3+} complex is 7.9×10^{25}.

*12-32 A silver electrode immersed in 1.00×10^{-2} M Na_2SeO_3 saturated with Ag_2SeO_3 acts as a cathode when coupled with a standard hydrogen electrode. Calculate K_{sp} for Ag_2SeO_3 if this cell develops a potential of 0.450 V.

12-33 A lead electrode is immersed in a solution of pH 8.00 that is 2.00×10^{-3} M in KBr and saturated with PbOHBr. This electrode acts as an anode when coupled with a standard hydrogen electrode. Calculate K_{sp} for PbOHBr if this cell develops a potential of 0.303 V ($PbOHBr \rightleftharpoons Pb^{2+} + OH^- + Br^-$).

*12-34 The potential of the following cell is 0.361 V:

$SHE\|Hg(OAc)_2(2.50 \times 10^{-3}\ M),OAc^-(0.0500\ M)|Hg$

where $Hg(OAc)_2$ is the neutral acetate complex of Hg^{2+}. Calculate its formation constant.

12-35 The potential for the following cell

$Zn|X^-(0.150\ M),ZnX_4^{2-}(6.00 \times 10^{-2}\ M)\|SHE$

is 1.072 V. Calculate the formation constant for ZnX_4^-.

*12-36 The cell

$Cu|CuCit^-(0.0400\ M),Na_3Cit(0.100\ M),H^+(1.00 \times 10^{-6}\ M)\|SHE$

was employed to determine the formation constant of the citrate (Cit^{3-}) complex of Cu(II). (Note that Cit^{3-} is the conjugate base of $HCit^{2-}$, H_2Cit^-, and H_3Cit.) The cell potential was 0.091 V. Calculate the formation constant for $CuCit^-$.

12-37 To determine the formation constant for the EDTA complex MY^{2-}, a 25.0 mL portion of 0.100 M MCl_2 was mixed with 25.0 mL of 0.200 M Na_2H_2Y. The solution was diluted to 100.0 mL with a pH 9.00 buffer. An electrode constructed from the metal M behaved as a cathode in this solution and developed a potential of 0.373 V when coupled with a standard hydrogen electrode. Calculate the formation constant for MY^{2-} ($E^0_{M^{2+}} = 0.889$ V).

*12-38 The cell

$Pt,H_2(1.00\ atm)|NaA(0.250\ M),HA(0.150\ M)\|SHE$

was employed to determine the dissociation constant of the weak acid HA. The potential was 0.470 V. Calculate K_a.

12-39 The cell

$Pt,H_2(1.00\ atm)|RNH_2(0.0540\ M),RNH_3Cl(0.0750\ M)\|SHE$

was employed to determine the dissociation constant of the amine RNH_2, where RNH_3Cl is the chloride salt of the amine. The potential of the cell was 0.481 V. Calculate K_b.

CHAPTER 13

Theory of Oxidation/Reduction Titrations

This chapter deals with titration curves and indicators for oxidation/reduction titrimetry. Because most redox indicators respond to changes in electrode potential, the ordinate in oxidation/reduction titration curves is an electrode potential rather than a p-function as has been the case for titration curves we have thus far considered. Note, however, that an electrode potential, like a p-function, bears a logarithmic relationship to the concentration of the analyte or the titrant. Thus, redox titration curves are much like those for neutralization, precipitation, or complex formation titrations.

13A Electrode Potentials for Redox Titration Systems

In order to demonstrate the nature of the electrode potential used as the ordinate in redox titrations curves, let us consider the oxidation of iron(II) by cerium(IV), which can be represented by the equation

$$Fe^{2+} + Ce^{4+} \rightleftharpoons Fe^{3+} + Ce^{3+}$$

This reaction is rapid and reversible, so that the system is at equilibrium throughout the titration. Consequently, the electrode potentials for the two half-reactions are always identical (Section 12D-2), that is,

$$E_{Ce^{4+}} = E_{Fe^{3+}} = E_{system}$$

where E_{system} is termed *the potential of the system.* If a redox indicator is present in this solution, the ratio of the concentrations of its oxidized and reduced forms must adjust so that the electrode potential for the indicator is equal to the system potential, that is,

$$E_{In} = E_{Ce^{4+}} = E_{Fe^{3+}} = E_{system}$$

The electrode potential of a system is readily derived from standard-potential data. Thus, for the reaction under consideration, the titration mixture can be treated as being part of the hypothetical cell

$$SHE \| Ce^{4+}, Ce^{3+}, Fe^{3+}, Fe^{2+} | Pt$$

where SHE symbolizes the standard hydrogen electrode. The potential of the platinum electrode with respect to the standard hydrogen electrode is determined by the affinities of iron(III) and cerium(IV) for electrons—that is, by the tendency of the following reactions to occur:

$$Fe^{3+} + e \rightleftharpoons Fe^{2+}$$

$$Ce^{4+} + e \rightleftharpoons Ce^{3+}$$

At equilibrium, the concentration ratios of the oxidized and reduced forms of the two species are such that their electron affinities (and thus their electrode potentials) are identical. Note that these concentration ratios vary continuously throughout the titration; so also must E_{system}. It is this characteristic variation in E_{system} that forms the basis for end-point detection.

From the foregoing discussion, it is clear that E_{system} data for a titration curve can be generated through the use of the Nernst equation for *either* the cerium(IV) half-reaction or the iron(III) half-reaction. It turns out, however, that one or the other is more convenient, depending upon the stage of the titration. For example, the iron(III) potential is easier to compute in the region short of the equivalence point because here the concentrations of iron(II) and iron(III) are appreciable and readily deduced from stoichiometric considerations. In contrast, the concentration of cerium(IV), which is negligible prior to equivalence because of the large excess of iron(II), can be obtained at this stage only by calculations based upon the equilibrium constant for the reaction. Beyond the equivalence point, the concentrations of cerium(IV) and cerium(III) are readily computed directly from the volumetric data, whereas that for iron(II) is not. In this region, then, the cerium(IV) electrode potential is the easier to use.

13A-1

The Derivation of Equivalence-Point Potentials

At the equivalence point, the concentration of cerium(IV) and iron(II) are minute and cannot be obtained from the stoichiometry of the reaction. Fortunately, a simple expression for the equivalence-point potential is readily obtained by taking advantage of the fact that the two reactant species and the two product species are present in known concentration ratios at chemical equivalence.

At the equivalence point in the titration of iron(II) with cerium(IV), the potential of the system E_{eq} is given by both

$$E_{eq} = E^0_{Ce^{4+}} - \frac{0.0592}{1} \log \frac{[Ce^{3+}]}{[Ce^{4+}]}$$

and

$$E_{eq} = E^0_{Fe^{3+}} - \frac{0.0592}{1} \log \frac{[Fe^{2+}]}{[Fe^{3+}]}$$

Adding these two expressions gives

$$2E_{eq} = E^0_{Ce^{4+}} + E^0_{Fe^{3+}} - \frac{0.0592}{1} \log \frac{[Ce^{3+}][Fe^{2+}]}{[Ce^{4+}][Fe^{3+}]}$$

Note that the concentration quotient in this expression is *not* the usual ratio of product concentrations and reactant concentrations that appear in equilibrium constant expressions.

The stoichiometry of the reaction requires that, *at the equivalence point,*

$$[Fe^{3+}] = [Ce^{3+}]$$

$$[Fe^{2+}] = [Ce^{4+}]$$

Substitution of these equalities into the previous equation results in elimination of the logarithmic term:

$$2E_{eq} = E^0_{Ce^{4+}} + E^0_{Fe^{3+}} - \frac{0.0592}{1} \log \frac{\cancel{[Ce^{3+}][Ce^{4+}]}}{\cancel{[Ce^{4+}][Ce^{3+}]}}$$

$$E_{eq} = \frac{E^0_{Ce^{4+}} + E^0_{Fe^{3+}}}{2} \tag{13-1}$$

Examples 13-1 and 13-2 illustrate how equivalence-point potentials are derived for more complex reactions.

EXAMPLE 13-1

Derive an equation for the equivalence point potential in the titration of Sn^{2+} with MnO_4^-. The reaction is

$$5Sn^{2+} + 2MnO_4^- + 16H^+ \rightleftharpoons 5Sn^{4+} + 2Mn^{2+} + 8H_2O$$

The half-reactions are

$$MnO_4^- + 5e + 8H^+ \rightleftharpoons Mn^{2+} + 4H_2O \qquad E^0 = +1.51 \text{ V}$$

$$Sn^{4+} + 2e \rightleftharpoons Sn^{2+} \qquad E^0 = +0.154 \text{ V}$$

The equivalence-point potential of this system is given by both

$$E_{eq} = E^0_{Sn^{4+}} - \frac{0.0592}{2} \log \frac{[Sn^{2+}]}{[Sn^{4+}]}$$

and

$$E_{eq} = E^0_{MnO_4^-} - \frac{0.0592}{5} \log \frac{[Mn^{2+}]}{[MnO_4^-][H^+]^8}$$

In order to add the two log terms, as we did in deriving Equation 13-1, it is necessary to multiply the first equation through by 2 and the second by 5, which gives

$$2E_{eq} = 2E^0_{Sn^{4+}} - 0.0592 \log \frac{[Sn^{2+}]}{[Sn^{4+}]}$$

$$5E_{eq} = 5E^0_{MnO_4^-} - 0.0592 \log \frac{[Mn^{2+}]}{[MnO_4^-][H^+]^8}$$

Addition of the two expressions leads to

$$7E_{eq} = 2E^0_{Sn^{4+}} + 5E^0_{MnO_4^-} - \frac{0.0592}{1} \log \frac{[Sn^{2+}][Mn^{2+}]}{[Sn^{4+}][MnO_4^-][H^+]^8}$$

At the equivalence point,

$$[Sn^{2+}] = \frac{5}{2}[MnO_4^-]$$

$$[Sn^{4+}] = \frac{5}{2}[Mn^{2+}]$$

Substitution and rearrangement give

$$E_{eq} = \frac{2E^0_{Sn^{4+}} + 5E^0_{MnO_4^-}}{7} - \frac{0.0592}{7} \log \frac{(5/2)[MnO_4^-][Mn^{2+}]}{(5/2)[Mn^{2+}][MnO_4^-][H^+]^8}$$

$$= \frac{2 \times 0.154 + 5 \times 1.51}{7} + \frac{0.0592}{7} \log [H^+]^8$$

$$= 1.12 + \frac{8 \times 0.0592}{7} \log [H^+] = 1.12 + 0.0677 \log [H^+]$$

$$= 1.12 - 0.068 \text{ pH}$$

Note that the equivalence-point potential for this titration is pH-dependent.

EXAMPLE 13-2

Derive an expression for the equivalence-point potential for the reaction

$$6Fe^{2+} + Cr_2O_7^{2-} + 14H^+ \rightleftharpoons 6Fe^{3+} + 2Cr^{3+} + 7H_2O$$

Proceeding as before, we obtain the expression

$$7E_{eq} = E^0_{Fe^{3+}} + 6E^0_{Cr_2O_7^{2-}} - 0.0592 \log \frac{[Fe^{2+}][Cr^{3+}]^2}{[Fe^{3+}][Cr_2O_7^{2-}][H^+]^{14}}$$

At the equivalence point

$$[Fe^{2+}] = 6[Cr_2O_7^{2-}]$$

$$[Fe^{3+}] = 3[Cr^{3+}]$$

Substitution of these quantities into the previous equation reveals that

$$E_{eq} = \frac{E^0_{Fe^{3+}} + 6E^0_{Cr_2O_7^{2-}}}{7} - \frac{0.0592}{7} \log \frac{2[Cr^{3+}]}{[H^+]^{14}}$$

Note that the equivalence-point potential in Example 13-2 is dependent not only upon the concentration of hydrogen ion but also upon that of a product ion (Cr^{3+}). In general, the equivalence-point potential depends upon the concentration of one of the participants in a reaction whenever the stoichiometric ratio of a reactant and its product has a value other than unity.

13A-2 The Derivation of Titration Curves

Examples 13-3 and 13-4 illustrate how data for titration curves are computed from standard-potential data.

EXAMPLE 13-3

Generate a curve for the titration of 0.0500 M Fe^{2+} with 0.1000 M Ce^{4+} in a medium that is 1.0 M in H_2SO_4 at all times.

Formal-potential data for both half-cell processes are available in Appendix 2 and are used for these calculations.

Initial potential

The solution contains no cerium species at the outset. In all likelihood, a small but unknown amount of Fe^{3+} is present due to air oxidation of Fe^{2+}. In any event, we lack sufficient information to calculate an initial potential.

Potential after the addition of 5.00 mL of cerium(IV)

With the introduction of oxidant, the solution acquires an appreciable and known concentration of three of the participants; that of the fourth, Ce^{4+}, is vanishingly small. Therefore, it is more convenient to use the concentrations of the two iron species to calculate the electrode potential of the system.

The concentration of Fe(III) is equal to its molar concentration less the equilibrium concentration of the unreacted Ce(IV):

$$[Fe^{3+}] = \frac{5.00 \times 0.100}{50.00 + 5.000} - [Ce^{4+}] \simeq \frac{0.500}{55.00}$$

Similarly, the Fe^{2+} concentration is given by its molarity plus $[Ce^{4+}]$:

$$[Fe^{2+}] = \frac{50.00 \times 0.0500 - 5.00 \times 0.1000}{55.00} + [Ce^{4+}] \simeq \frac{2.00}{55.00}$$

The validity of the indicated assumptions is readily confirmed by computing the equilibrium constant for the reaction between Fe(II) and Ce(IV) using the technique described in Section 12D-2. The large value for this constant (7×10^{12}) indicates clearly that the concentration of Ce(IV) must indeed be inconsequential relative to the concentrations of the two iron species.

Substitution for $[Fe^{2+}]$ and $[Fe^{3+}]$ in the Nernst equation gives

$$E = +0.68 - \frac{0.0592}{1} \log \frac{2.00/\cancel{55.00}}{0.500/\cancel{55.00}} = 0.64 \text{ V}$$

Note that the volumes in the numerator and denominator cancel, which indicates that the potential is independent of dilution. This independence persists until the solution becomes so dilute that the two assumptions made in the calculation become invalid.

It is worth emphasizing again that the use of the Nernst equation for the Ce(IV)/Ce(III) system would yield the same value for *E*, but to use the standard potential for Ce(IV) would require computation of the equilibrium constant for the reaction in order to obtain $[Ce^{4+}]$.

Additional potentials needed to define the titration curve short of the equivalence point can be obtained similarly. Such data are given in Table 13-1; the reader is encouraged to confirm one or two of these values.

Equivalence-point potential

Substitution of the two standard potentials into Equation 13-1 yields

$$E_{eq} = \frac{E^0_{Ce^{4+}} + E^0_{Fe^{3+}}}{2} = \frac{1.44 + 0.68}{2} = 1.06 \text{ V}$$

Potential after the addition of 25.10 mL of cerium(IV)

The molar concentrations of Ce(III), Ce(IV), and Fe(III) are readily computed at this point, but that for Fe(II) is not. Therefore, computations based on the cerium half-reaction are more convenient. The concentrations of the two cerium ion species are

$$[Ce^{3+}] = \frac{25.00 \times 0.100}{75.10} - [Fe^{2+}] \simeq \frac{2.500}{75.10}$$

$$[Ce^{4+}] = \frac{25.10 \times 0.1000 - 50.00 \times 0.0500}{75.10} + [Fe^{2+}] \simeq \frac{0.010}{75.10}$$

TABLE 13-1

Electrode Potentials versus SHE in the Titration of 50.0 mL of 0.0500 M Iron(II) Solutions

Reagent Volume, mL	Potential, V: Titration with 0.1000 M Ce^{4+}		Potential, V: Titration with 0.02000 M MnO_4^-*
5.00	0.64		0.64
15.00	0.69		0.69
20.00	0.72		0.72
24.00	0.76		0.76
24.90	0.82		0.82
25.00	1.06	← Equivalence point →	1.37
25.10	1.30		1.48
26.00	1.36		1.49
30.00	1.40		1.50

*H_2SO_4 concentration is such that $[H^+] = 1.0$ M throughout.

The indicated approximations should be reasonable in view of the favorable equilibrium constant. Substitution of these concentrations into the Nernst equation for the cerium couple gives

$$E = +1.44 - \frac{0.0592}{1} \log \frac{[Ce^{3+}]}{[Ce^{4+}]} = +1.44 - \frac{0.0592}{1} \log \frac{2.500/\cancel{75.10}}{0.010/\cancel{75.10}} = +1.30 \text{ V}$$

The additional postequivalence potentials in Table 13-1 were derived in a similar fashion.

The titration of iron(II) with cerium(IV) appears as curve *A* in Figure 13-1. Its general shape resembles the curves encountered in neutralization, precipitation, and complex-formation titrations, the equivalence point being signaled by a rapid change in the ordinate function. A titration involving 0.00500 M iron(II) and 0.01000 M cerium(IV) yields a curve that is, for all practical purposes, identical to the one derived, since the electrode potential of the system is independent of dilution. Note that the cerium(IV) titration is symmetric about the equivalence point, a consequence of the equimolar combining ratio that exists between oxidant and reductant. Example 13-4 demonstrates that an asymmetric curve results when this ratio differs from unity.

EXAMPLE 13-4

Derive a curve for the titration of 50.00 mL of 0.05000 M Fe^{2+} with 0.02000 M $KMnO_4$. For convenience, assume that the concentration of H_2SO_4 in the solution is sufficient to make $[H^+] = 1.00$ throughout the titration and that the formal electrode potential for Fe(III) in this medium is 0.68 V. The reaction is

$$5Fe^{2+} + MnO_4^- + 8H^+ \rightleftharpoons 5Fe^{3+} + Mn^{2+} + 4H_2O$$

Since no formal potential is available for the reduction of MnO_4^-, we use the standard potential of 1.51 V instead.

Potential after the addition of 5.00 mL of MnO_4^-
The stoichiometry at this point is

$$\text{No. mmol Fe}^{2+} \text{ taken} = 50.00 \text{ mL Fe}^{2+} \times \frac{0.05000 \text{ mmol Fe}^{2+}}{\text{mL Fe}^{2+}} = 2.500 \text{ mmol Fe}^{2+}$$

$$\text{No. mmol KMnO}_4 \text{ added} = 5.00 \text{ mL KMnO}_4 \times \frac{0.02000 \text{ mmol KMnO}_4}{\text{mL KMnO}_4}$$

$$= 0.1000 \text{ mmol KMnO}_4$$

$$[Fe^{3+}] = \frac{0.1000 \text{ mmol KMnO}_4 \times 5 \text{ mmol Fe}^{3+}/\text{mmol KMnO}_4}{(50.00 + 5.00) \text{ mL}} = \frac{0.5000 \text{ mmol Fe}^{3+}}{55.00 \text{ mL}}$$

$$[Fe^{2+}] = \frac{2.500 \text{ mmol Fe}^{2+} - 0.1000 \text{ mmol KMnO}_4 \times 5 \text{ mmol Fe}^{2+}/\text{mmol KMnO}_4}{(50.00 + 5.00) \text{ mL}}$$

$$= \frac{2.000 \text{ mmol Fe}^{2+}}{55.00 \text{ mL}}$$

Substituting into the Nernst expression for the iron couple leads to

$$E = 0.68 - \frac{0.0592}{1} \log \frac{2.000/\cancel{55.00}}{0.5000/\cancel{55.00}} = 0.64 \text{ V}$$

Note in Table 13-1 that the preequivalence potentials for the titration with MnO_4^- are identical to those for the titration with Ce^{4+}.

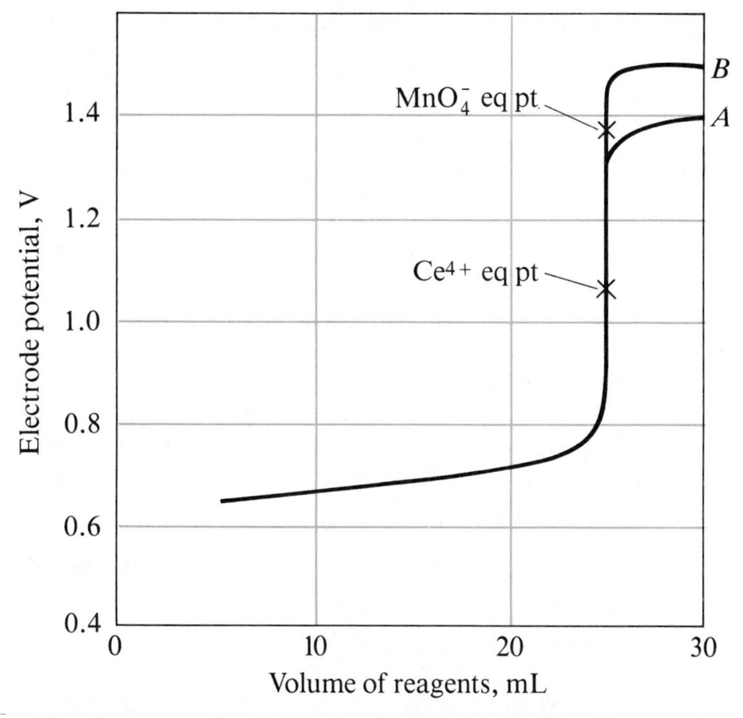

FIGURE 13-1 Titration curves for 50.00 mL of 0.05000 M Fe(II) with 0.1000 M Ce(IV) (*A*) and with 0.02000 M $KMnO_4$ (*B*).

Equivalence-point potential

Following the procedure in Example 13-1, we obtain for the equivalence-point potential

$$E_{eq} = \frac{E^0_{Fe^{3+}} + 5E^0_{MnO_4^-}}{6} + \frac{0.0592}{6}\log[H^+]^8$$

Substituting the formal potential for $E^0_{Fe^{3+}}$, we obtain

$$E_{eq} = \frac{0.68 + 5 \times 1.51}{6} + \frac{0.0592 \times 8}{6}\log(1.00) = 1.37\text{ V}$$

Potential after the addition of 25.10 mL of $KMnO_4$

It is now advantageous to derive data using the standard potential for the manganese system:

$$\text{No. mmol Fe}^{2+}\text{ taken} = 50.00\text{ mL Fe}^{2+} \times 0.05000\,\frac{\text{mmol Fe}^{2+}}{\text{mL Fe}^{2+}} = 2.500\text{ mmol Fe}^{2+}$$

$$\text{No. mmol KMnO}_4^-\text{ added} = 25.10\text{ mL KMnO}_4 \times 0.02000\,\frac{\text{mmol KMnO}_4}{\text{mL KMnO}_4}$$

$$= 0.5020\text{ mmol KMnO}_4$$

$$[\text{Mn}^{2+}] = \frac{2.500\text{ mmol Fe}^{2+} \times 1\text{ mmol Mn}^{2+}/5\text{ mmol Fe}^{2+}}{(50.00 + 25.10)\text{ mL}} = \frac{0.5000\text{ mmol Mn}^{2+}}{75.10\text{ mL}}$$

$[MnO_4^-]$

$$= \frac{0.5020\text{ mmol KMnO}_4 - 2.500\text{ mmol Fe}^{2+} \times 1\text{ mmol KMnO}_4/5\text{ mmol Fe}^{2+}}{(50.00 + 25.10)\text{ mL}}$$

$$= \frac{0.0020\text{ mmol MnO}_4^-}{75.10\text{ mL}}$$

Substituting into the Nernst equation for MnO_4^- gives

$$E = 1.51 - \frac{0.0592}{5} \log \frac{0.0020/\cancel{75.10}}{(1.00)^8 \times 0.5000/\cancel{75.10}} = 1.51 - (-0.028) = 1.48 \text{ V}$$

Additional data for the titration of iron(II) with permanganate ion are tabulated in Table 13-1 and are plotted as curve *B* in Figure 13-1.

Figure 13-1 demonstrates that the curves for the titration of iron(II) with the two oxidants are indistinguishable to within about 99.9% of the titrant volume needed for equivalence. Note, however, that the equivalence-point potentials are quite different. Moreover, the curve involving permanganate is markedly asymmetric, the potential increasing only slightly beyond the equivalence point. Finally, note that the change in potential associated with the equivalence-point region is somewhat greater with permanganate as titrant because the equilibrium constant for the reaction with this reagent is somewhat more favorable.

13A-3 The Effect of Reactant Concentration on Titration Curves

The quantity plotted as the ordinate for an oxidation/reduction titration is the electrode potential of the system, which is ordinarily unaffected by dilution. Consequently, titration curves for oxidation/reduction reactions are usually independent of analyte and reagent concentrations.[1] This behavior is in distinct contrast to that observed in the other types of titration curves we have encountered.

13A-4 The Effect of Reaction Completeness on Titration Curves

The change in ordinate function in the equivalent-point region of an oxidation/reduction titration becomes larger as the reaction becomes more complete. This effect is demonstrated in Figure 13-2, which depicts curves for the titration of a hypothetical reductant having a standard electrode potential of 0.20 V with several hypothetical oxidants with standard potentials ranging from 0.40 to 1.20 V; the corresponding equilibrium constants lie between about 2×10^3 and 8×10^{16}. Clearly, the greatest change in potential of the system is associated with the reaction that is most nearly complete; in this respect, oxidation/reduction titration curves are no different from those involving other types of reactions.

[1]Electrode potentials become dependent upon dilution when the number of moles of the reactant and product of a half-reaction differ. An example is the reaction

$$I_3^- + 2e \rightleftharpoons 3I^-$$

Application of the Nernst equation gives

$$E = E^0 - \frac{0.0592}{2} \log \frac{[I^-]^3}{[I_3^-]}$$

Here, the concentration term in the numerator bears units of $(mol/L)^3$ whereas the units in the denominator are mol/L. Consequently, the ratio of the two terms is related to the square of the iodide concentration. The result is that, in a titration in which the standard reagent is triiodide ion, potentials at and beyond the equivalence point depend upon dilution.

As mentioned earlier, electrode potentials also become concentration-dependent when it can no longer be assumed that the molar concentrations of the various species are equal to the concentrations derived from stoichiometry.

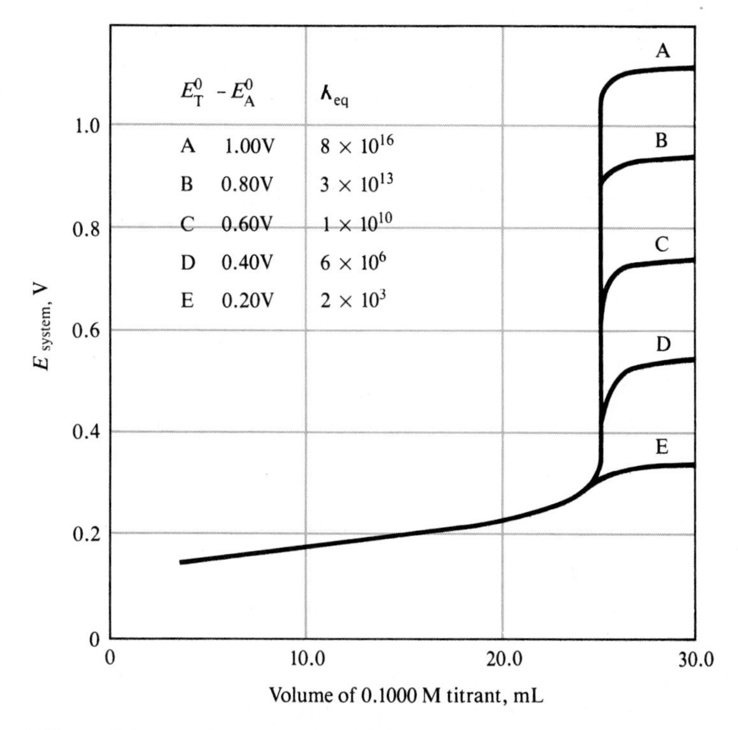

FIGURE 13-2

Effect of titrant electrode potential upon completeness of reaction. The standard electrode potential for the analyte (E^0_A) is 0.20 V; starting with curve A, standard electrode potentials for the titrant (E^0_T) are 1.20, 1.00, 0.80, 0.60, and 0.40 V. Both analyte and titrant undergo a one-electron change.

The curves in Figure 13-2 were derived for reactions in which both oxidant and reductant undergo a one-electron change. When one of the participants undergoes a two-electron change, the corresponding increase in potential in the region between 24.9 and 25.1 mL is larger by about 0.14 V.

13A-5 The Titration of Mixtures

Solutions containing two oxidizing agents or two reducing agents yield titration curves that contain two inflection points, provided the standard potentials involved are sufficiently different from each other. If this difference is greater than about 0.2 V, the end points are usually distinct enough to permit determination of each component. This situation is quite comparable to the titration of two acids having different dissociation constants or to that of two ions forming precipitates that have different solubilities.

In addition, the behavior of a few redox systems is analogous to that of polyprotic acids. For example, consider the two half-reactions

$$VO^{2+} + 2H^+ + e \rightleftharpoons V^{3+} + H_2O \qquad E^0 = +0.359\ V$$

$$V(OH)_4^+ + 2H^+ + e \rightleftharpoons VO^{2+} + 3H_2O \qquad E^0 = +1.00\ V$$

The curve for the titration of V^{3+} with a strong oxidizing agent, such as permanganate, has two inflection points, the first corresponding to the oxidation of V^{3+} to VO^{2+} and the second to the oxidation of VO^{2+} to $V(OH)_4^+$. The stepwise oxidation of molybdenum(III), first to the +5 oxidation state and subsequently to the +6 state, is another common example. Here again, satisfactory inflections occur in the curves because the difference in the standard potentials of the pertinent half-reactions is about 0.4 V.

The derivation of titration curves for either of these types of reactions is not difficult if the difference in standard potentials is sufficiently great. An example is the titration of a solution containing iron(II) and titanium(III) ions with potassium permanganate. The standard potentials are

$$TiO^{2+} + 2H^+ + e \rightleftharpoons Ti^{3+} + H_2O \qquad E^0 = +0.099\text{ V}$$

$$Fe^{3+} + e \rightleftharpoons Fe^{2+} \qquad E^0 = +0.77\text{ V}$$

The first additions of permanganate are used up by the more readily oxidized titanium(III) ion because, as long as an appreciable concentration of this species remains in solution, the potential of the system cannot become high enough to change the concentration of iron(II) ions appreciably. Thus, points defining the first part of the titration curve can be obtained by substituting the stoichiometric concentrations of titanium(III) and titanium(IV) ion into the equation

$$E = +0.099 - 0.0592 \log \frac{[Ti^{3+}]}{[TiO^{2+}][H^+]^2}$$

For all practical purposes then, the first part of this curve is identical to the titration curve for titanium(III) ion by itself. Beyond the first equivalence point, the solution contains both iron(II) and iron(III) ions in significant concentrations and so points on the curve can be most conveniently obtained from the relationship

$$E = E^0_{Fe^{3+}} - 0.0592 \log \frac{[Fe^{2+}]}{[Fe^{3+}]}$$

Throughout this region and beyond the second equivalence point, the curve is essentially identical to that for the titration of iron(II) ion alone (Figure 13-1). Such calculations leave undefined only the potential at the first equivalence point. A convenient way of estimating its value is to add the Nernst equations for the iron(II) and titanium(III) potentials. Since the electrode potentials for the two systems are identical at equilibrium, we can write

$$2E = +0.099 + 0.771 - 0.0592 \log \frac{[Ti^{3+}][Fe^{2+}]}{[TiO^{2+}][Fe^{3+}][H^+]^2}$$

Iron(III) and titanium(III) ions exist in small and equal amounts as a result of the equilibrium

$$2H^+ + TiO^{2+} + Fe^{2+} \rightleftharpoons Fe^{3+} + Ti^{3+} + H_2O$$

Thus, we can write

$$[Fe^{3+}] = [Ti^{3+}]$$

Substitution of this equality into the previous equation for the potential yields

$$E = \frac{+0.87}{2} - \frac{0.0592}{2} \log \frac{[Fe^{2+}]}{[TiO^{2+}][H^+]^2}$$

Finally, if $[TiO^{2+}]$ and $[Fe^{2+}]$ are assumed to be essentially identical to their analytical concentrations, we can compute the equivalence-point potential.

A curve for a mixture of iron(II) and titanium(III) ions titrated with a standard permanganate solution is shown in Figure 13-3.

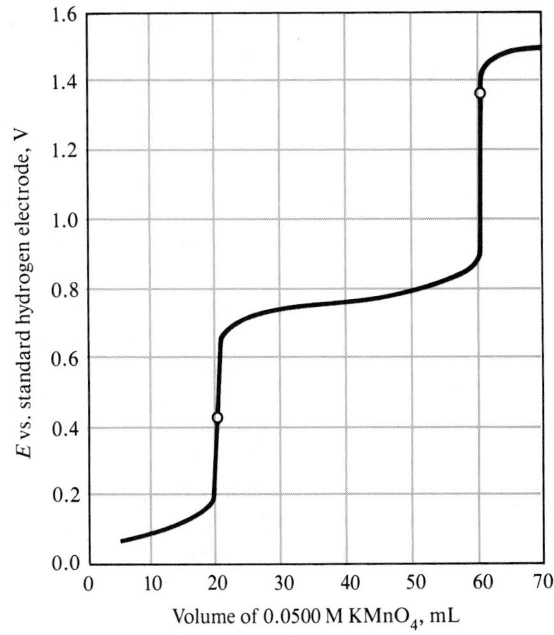

FIGURE 13-3 Curve for the titration of 50.0 mL of a solution that is 0.0500 M in Ti^{3+} and 0.200 M in Fe^{2+} with 0.0500 M $KMnO_4$. The concentration of H^+ is 1.0 M throughout.

13B Oxidation/Reduction Indicators

Two types of indicators are used for obtaining end points for oxidation/reduction titrations: *general redox indicators* and *specific indicators.*

13B-1 General Redox Indicators

General oxidation/reduction indicators are substances that change color upon being oxidized or reduced. In contrast to specific indicators, the color changes of true redox indicators are largely independent of the chemical nature of the analyte and titrant and depend instead upon the changes in the electrode potential of the system that occur as the titration progresses.

The half-reaction responsible for color change in a typical general oxidation/reduction indicator can be written as

$$\mathrm{In_{ox}} + ne \rightleftharpoons \mathrm{In_{red}}$$

If the indicator reaction is reversible, we can write

$$E = E^0_{\mathrm{In}} - \frac{0.0592}{n} \log \frac{[\mathrm{In_{red}}]}{[\mathrm{In_{ox}}]} \tag{13-2}$$

Typically, a change from the color of the oxidized form of the indicator to the color of the reduced form requires a change of about 100 in the ratio of reactant concentrations; that is, a color change is seen when

$$\frac{[\mathrm{In_{red}}]}{[\mathrm{In_{ox}}]} \leq \frac{1}{10}$$

changes to

$$\frac{[\mathrm{In_{red}}]}{[\mathrm{In_{ox}}]} \geq 10$$

The potential change required to produce the full color change of a typical general indicator can be found by substituting these two values into Equation 13-2, which gives

$$E = E^0_{\text{In}} \pm \frac{0.0592}{n}$$

This equation shows that a typical general indicator exhibits a detectable color change when a titrant causes the system potential to shift from $E^0_{\text{In}} + 0.0592/n$ to $E^0_{\text{In}} - 0.0592/n$, or about $(0.118/n)$ V. For many indicators, $n = 2$ and a change of 0.059 V is thus sufficient.[2]

Table 13-2 lists transition potentials for several redox indicators. Note that indicators functioning in any desired potential range up to about +1.25 V are available. Structures for and reactions of a few of the indicators listed in the table are considered in the paragraphs that follow.

Iron(II) Complexes of Orthophenanthrolines

A class of organic compounds known as 1,10-phenanthrolines (or orthophenanthrolines) form stable complexes with iron(II) and certain other ions. The parent compound has a pair of nitrogen atoms located in such positions that each can form a covalent bond with the iron(II) ion. Three orthophenanthroline molecules combine with each iron ion to yield a complex with the structure

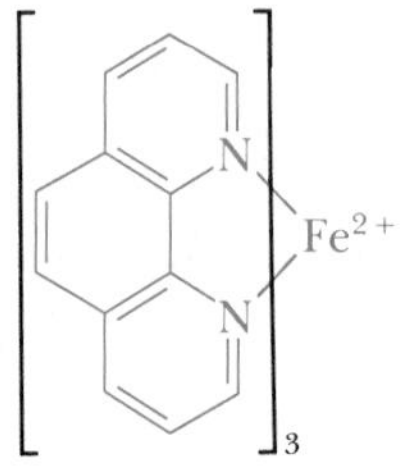

This complex, which is sometimes called "ferroin," is conveniently formulated as $(\text{Phen})_3\text{Fe}^{2+}$.

The complexed iron in the ferroin undergoes a reversible oxidation/reduction reaction that can be written

$$\underset{\text{pale blue}}{(\text{Phen})_3\text{Fe}^{3+}} + e \rightleftharpoons \underset{\text{red}}{(\text{Phen})_3\text{Fe}^{2+}} \qquad E^0 = +1.06\text{ V}$$

In practice, the color of the oxidized form is so slight as to go undetected, and the color change associated with this reduction appears to be from nearly colorless to red. Because of the difference in color intensity, the end point is usually taken when only about 10% of the indicator is in the iron(II) form. The transition potential is thus approximately +1.11 V in 1 M sulfuric acid.

Of all the oxidation/reduction indicators, ferroin approaches most closely the ideal substance. It reacts rapidly and reversibly, the color change is pronounced, and solutions of it are stable and readily prepared. In contrast to many indicators, the oxidized form of ferroin is remarkably inert toward strong oxidizing agents. At temperatures above 60°C, ferroin decomposes.

[2]Protons are involved in the reduction of many indicators, and so the range of potentials over which a color change occurs (the *transition potential*) is often pH-dependent.

TABLE 13-2 **Selected General Oxidation/Reduction Indicators***

Indicator	Color: Oxidized	Color: Reduced	Transition Potential, V	Conditions
5-Nitro-1,10-phenanthroline iron(II) complex	Pale blue	Red-violet	+1.25	1 M H_2SO_4
2,3′-Diphenylamine dicarboxylic acid	Blue-violet	Colorless	+1.12	7–10 M H_2SO_4
1,10-Phenanthroline iron(II) complex	Pale blue	Red	+1.11	1 M H_2SO_4
Erioglaucin A	Blue-red	Yellow-green	+0.98	0.5 M H_2SO_4
Diphenylamine sulfonic acid	Red-violet	Colorless	+0.85	Dilute acid
Diphenylamine	Violet	Colorless	+0.76	Dilute acid
p-Ethoxychrysoidine	Yellow	Red	+0.76	Dilute acid
Methylene blue	Blue	Colorless	+0.53	1 M acid
Indigo tetrasulfonate	Blue	Colorless	+0.36	1 M acid
Phenosafranine	Red	Colorless	+0.28	1 M acid

*Data taken in part from I. M. Kolthoff and V. A. Stenger, *Volumetric Analysis*, 2d ed., Vol. 1, p. 140. New York: Interscience, 1942.

A number of substituted phenanthrolines have been investigated for their indicator properties, and some have proved to be as useful as the parent compound. Among these, the 5-nitro and 5-methyl derivatives are noteworthy, with transition potentials of +1.25 V and +1.02 V, respectively.

Diphenylamine and Its Derivatives

Diphenylamine, $C_{12}H_{11}N$, which was one of the first redox indicators discovered, was used by Knop in 1924 for the titration of iron(II) with potassium dichromate. In the presence of a strong oxidizing agent, diphenylamine is believed to undergo the reactions

$$2\ \text{diphenylamine (colorless)} \longrightarrow \text{diphenylbenzidine (colorless)} + 2H^+ + 2e$$

$$\text{diphenylbenzidine (colorless)} \rightleftharpoons \text{diphenylbenzidine violet (violet)} + 2H^+ + 2e$$

The first reaction is irreversible, but the second can be reversed and constitutes the actual indicator reaction.

The reduction potential for the second reaction is about +0.76 V. Despite the fact that hydrogen ions appear in the equation, variations in acidity have little effect upon the magnitude of this potential, perhaps because the colored product is a protonated species.

Diphenylamine is not very soluble in water, and indicator solutions must be prepared in rather concentrated sulfuric acid. In addition, the indicator is not applicable to solutions containing tungstate ion, which precipitates the violet reaction product. Mercury(II) ions also interfere by inhibiting the indicator reaction.

The sulfonic acid derivative of diphenylamine, which has the structure

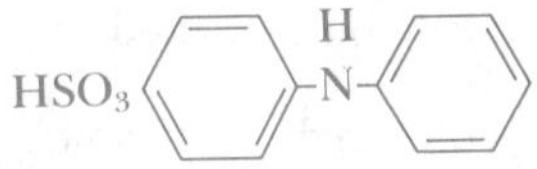

$HSO_3-C_6H_4-NH-C_6H_5$

does not suffer from these disadvantages. The barium or sodium salt of this acid is used to prepare aqueous indicator solutions, which behave in essentially the same manner as the parent substance. The color change is somewhat sharper, however, passing from colorless through green to deep violet. The transition potential is about +0.8 V and again is independent of the acid concentration. The sulfonic acid derivative is now widely used in redox titrations.

Diphenylbenzidine, the intermediate in the oxidation of diphenylamine, should behave in redox reactions just as diphenylamine does but should consume less oxidizing agent. Unfortunately, the low solubility of diphenylbenzidine in water and in sulfuric acid has precluded its widespread use. As might be expected, the sulfonic acid derivative of diphenylbenzidine has proved to be a satisfactory indicator.

Starch/Iodine Solutions

Starch, which forms a blue complex with triiodide ion, is a widely used specific indicator in oxidation/reduction reactions involving iodine as an oxidant or iodide ion as a reductant. A starch solution containing a little triiodide or iodide ion can also function as a true redox indicator, however. In the presence of excess oxidizing agent, the concentration ratio of iodine to iodide is high, giving a blue color to the solution. With excess reducing agent, on the other hand, iodide ion predominates, and the blue color is absent. Thus, the indicator system changes from colorless to blue in the titration of many reducing agents with various oxidizing agents. This color change is quite independent of the chemical composition of the reactants, depending only upon the potential of the system at the equivalence point.

The Choice of Redox Indicator

It is apparent from Figure 13-2 that all the indicators in Table 13-2 except for the first and the last could be employed with reagent A. In contrast, with reagent D, only indigo tetrasulfonate could be employed. The change in potential with reagent E is too small to be satisfactorily detected by an indicator.

13B-2 Specific Indicators

Perhaps the best-known specific indicator is starch, which forms a dark blue complex with triiodide ion. This complex signals the end point in titrations in which iodine is either produced or consumed.

Another specific indicator is potassium thiocyanate, which may be employed, for example, in the titration of iron(III) with solutions of titanium(III) sulfate. The end point involves the disappearance of the red color of the iron(III)/thiocyanate complex as a result of the marked decrease in the iron(III) concentration at the equivalence point.

13C Potentiometric End Points

End points for many oxidation/reduction titrations are readily observed by making the solution of the analyte part of the cell:

reference electrode||analyte solution|Pt

By measuring the potential of this cell during a titration, data for curves analogous to those shown in Figures 13-1 and 13-2 can be generated. End points are readily estimated from such curves. Potentiometric end points are considered in detail in Chapter 15.

Questions and Problems

*13-1 Why are the transition potentials of many redox indicators pH-dependent?

13-2 Why are titration curves for most redox reactions independent of the concentrations of analyte and standard?

*13-3 Under what conditions are oxidation/reduction titration curves symmetric about the equivalence point?

13-4 Differentiate between a specific and a general redox indicator.

*13-5 A solution of V^{2+} is titrated with Ce^{4+}. Indicate the qualitative features of the titration curve.

13-6 The half-reaction for a general redox indicator is

$$In + 2H^+ + 2e \rightleftharpoons InH_2 \qquad E^0 = 0.600\ V$$

Calculate the transition potential range of this indicator in a solution having a pH of

*(a) 2.00. (b) 4.00. *(c) 6.00. (d) 8.00.

13-7 Suggest a general redox indicator suitable for the following (assume $[H_3O^+] = 1.00$)

*(a) the titration of quinone, $C_6H_4O_2$, with a standard solution of $V(OH)_4^+$.
(b) the titration of selenous acid, H_2SeO_3, with a standard solution of Ce^{4+}.
*(c) the titration of Fe^{3+} with a standard solution of U^{4+}.
(d) the titration of $Fe(CN)_6^{4-}$ with a standard solution of Tl^{3+}.

*13-8 Calculate the electrode potential of the system at the equivalence point for each of the following reactions. Where necessary, assume that $[H^+] = 0.100$ at equivalence.

(a) $2Ti^{2+} + Sn^{4+} \rightleftharpoons 2Ti^{3+} + Sn^{2+}$

(b) $2Ce^{4+} + H_2SeO_3 + H_2O \rightleftharpoons SeO_4^{2-} + 2Ce^{3+} + 4H^+$ (in 1 M H_2SO_4)

(c) $2MnO_4^- + 5HNO_2 + H^+ \rightleftharpoons 2Mn^{2+} + 5NO_3^- + 3H_2O$

13-9 Calculate the electrode potential of the system at the equivalence point for each of the following reactions. Where necessary, assume that $[H^+] = 0.100$ at equivalence.

(a) $Cr^{2+} + Fe(CN)_6^{3-} \rightleftharpoons Cr^{3+} + Fe(CN)_6^{4-}$

(b) $Tl^{3+} + 2V^{3+} + 2H_2O \rightleftharpoons Tl^+ + 2VO^{2+} + 4H^+$

(c) $5U^{4+} + 2MnO_4^- + 2H_2O \rightleftharpoons 5UO_2^{2+} + 2Mn^{2+} + 4H^+$

*13-10 Calculate equilibrium constants for the reactions in Problem 13-8.

13-11 Calculate equilibrium constants for the reactions in Problem 13-9.

13-12 Calculate the equivalence-point concentration of each of the first-mentioned species in Problem 13-8 if 0.100 M solutions were employed.

13-13 Calculate the equivalence-point concentration of each of the first-mentioned species in Problem 13-9 if 0.100 M solutions were employed.

13-14 Construct curves for the following titrations. Calculate potentials after the addition of 10.00, 25.00, 49.00, 49.90, 50.00, 50.10, 51.00, and 60.00 mL of the reagent. Where necessary, assume that $[H^+] = 1.00$ throughout.

*(a) 50.00 mL of 0.1000 M V^{2+} with 0.05000 M Sn^{4+}

(b) 50.00 mL of 0.1000 M $Fe(CN)_6^{3-}$ with 0.1000 M Cr^{2+}

*(c) 50.00 mL of 0.1000 M $Fe(CN)_6^{4-}$ with 0.05000 M Tl^{3+}

(d) 50.00 mL of 0.1000 M Fe^{3+} with 0.05000 M Sn^{2+}

*(e) 50.00 mL of 0.05000 M U^{4+} with 0.02000 M MnO_4^-

(f) 50.00 mL of 0.02000 M Sn^{2+} with 0.02000 M I_3^- (assume that $[I^-] = 0.500$ M throughout the titration)

CHAPTER 14

Applications of Oxidation/Reduction Titrations

This chapter is concerned with the preparation of standard solutions of oxidants and reductants and with their applications in analytical chemistry. In addition, auxiliary reagents that convert an analyte to a single oxidation state are described.[1]

14A Auxiliary Oxidizing and Reducing Reagents

To obtain meaningful results from an oxidation/reduction titration, the analyte must be entirely in a single oxidation state at the outset. Often, however, the steps that precede the titration (dissolution of the sample and separation of interferences) convert the analyte to a mixture of oxidation states. Consequently, an auxiliary reagent is required to ensure that the analyte is in a single oxidation state before the titration is undertaken.[2] For example, the solution formed when an iron-containing sample is dissolved usually contains a mixture of iron(II) and iron(III). This solution must be treated with a reducing agent to convert all the iron to the +2 state or, alternatively, with an oxidizing agent to form iron(III) quantitatively.

To be useful as a preoxidant or a prereductant, a reagent must react quantitatively with the analyte. In addition, excesses of the reagent must be readily removable because such excesses will inevitably interfere by consuming standard solution. For example, a reagent capable of quantitatively converting iron from the +3 to the +2 state would certainly consume some of the standard oxidant used subsequently to titrate the iron(II).

14A-1 Auxiliary Reducing Reagents

A number of metals are good reducing agents and have been used for the prereduction of analytes. Included among these are zinc, aluminum, cadmium, lead, nickel, copper, and silver (in the presence of chloride ion). Sometimes, sticks or coils of the metal are immersed directly in the analyte solution. After reduction is judged complete, the solid is removed manually and rinsed with water. Filtration of the analyte solution is needed to remove granular or powdered forms of the metal. An alternative to filtration is the use of a *reductor,*

[1]For further reading on redox titrimetry, see J. A. Goldman and V. A. Stenger, in *Treatise on Analytical Chemistry,* I. M. Kolthoff and P. J. Elving, Eds., Part I, Vol. 11, Chapter 119. New York: Wiley, 1975; and I. M. Kolthoff and R. Belcher, *Volumetric Analysis,* Vol. 2. New York: Interscience, 1957.

[2]For a brief summary of auxiliary reagents, see J. A. Goldman and V. A. Stenger, in *Treatise on Analytical Chemistry,* I. M. Kolthoff and P. J. Elving, Eds., Part I, Vol. 11, pp. 7204–7206. New York: Wiley, 1975.

such as that shown in Figure 14-1.[3] Here, the finely divided metal is held in a vertical glass tube through which the solution is drawn under a mild vacuum. The metal in a reductor is ordinarily sufficient for hundreds of reductions.

A typical *Jones reductor* has a diameter of about 2 cm and is packed with a 40- to 50-cm column of amalgamated zinc. Amalgamation is accomplished by allowing zinc granules to stand briefly in a solution of mercury(II) chloride:

$$2Zn(s) + Hg^{2+} \longrightarrow Zn^{2+} + Zn(Hg)(s)$$

Zinc amalgam is nearly as effective a reducing agent as the pure metal and has the important virtue of inhibiting the reduction of hydrogen ions by zinc, a parasitic reaction that not only needlessly uses up the reducing agent but also heavily contaminates the sample solution with zinc(II) ions. Solutions that are quite acidic can be passed through a Jones reductor without significant hydrogen formation.

Table 14-1 lists the principal applications of the Jones reductor. Also listed in this table are reductions that can be accomplished with a *Walden reductor,* in which granular metallic silver held in a narrow glass column is the reductant. Silver is not a good reducing agent unless chloride or some other ion that forms a silver salt of low solubility is present. For this reason, prereductions with a Walden reductor are generally carried out from hydrochloric acid solutions of the analyte. The coating of silver chloride produced on the metal is removed periodically by dipping a zinc rod into the solution that covers the packing.

Table 14-1 suggests that the Walden reductor is somewhat more selective in its action than is the Jones reductor.

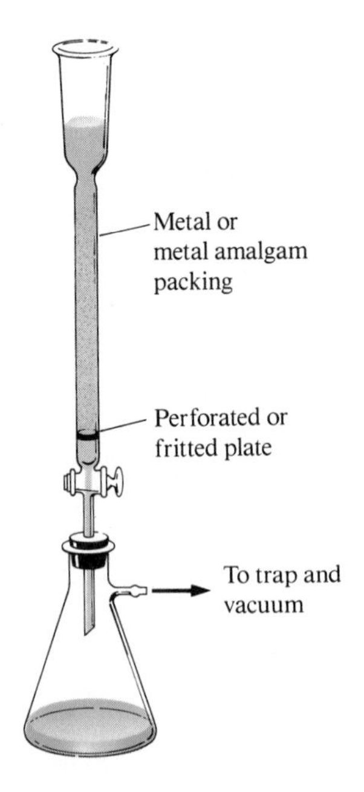

FIGURE 14-1 A metal or metal amalgam reductor.

[3]For a discussion of reductors, see F. Hecht, in *Treatise on Analytical Chemistry,* I. M. Kolthoff and P. J. Elving, Eds., Part I, Vol. 11, pp. 6703–6707. New York: Wiley, 1975.

TABLE 14-1

Uses of Walden and Jones Reductors*

Walden $Ag(s) + Cl^- \rightarrow AgCl(s) + e$	Jones $Zn(Hg)(s) \rightarrow Zn^{2+} + Hg + 2e$
$Fe^{3+} + e \rightarrow Fe^{2+}$	$Fe^{3+} + e \rightarrow Fe^{2+}$
$Cu^{2+} + e \rightarrow Cu^+$	Cu^{2+} reduced to metallic Cu
$H_2MoO_4 + 2H^+ + e \rightarrow MoO_2^+ + 2H_2O$	$H_2MoO_4 + 6H^+ + 3e \rightarrow Mo^{3+} + 4H_2O$
$UO_2^{2+} + 4H^+ + 2e \rightarrow U^{4+} + 2H_2O$	$UO_2^{2+} + 4H^+ + 2e \rightarrow U^{4+} + 2H_2O$
	$UO_2^{2+} + 4H^+ + 3e \rightarrow U^{3+} + 2H_2O$†
$V(OH)_4^+ + 2H^+ + e \rightarrow VO^{2+} + 3H_2O$	$V(OH)_4^+ + 4H^+ + 3e \rightarrow V^{2+} + 4H_2O$
TiO^{2+} not reduced	$TiO^{2+} + 2H^+ + e \rightarrow Ti^{3+} + H_2O$
Cr^{3+} not reduced	$Cr^{3+} + e \rightarrow Cr^{2+}$

*Information taken from I. M. Kolthoff and R. Belcher, *Volumetric Analysis*, Vol. 3, p. 12. New York: Interscience, 1957. With permission.

†A mixture of oxidation states is obtained. The Jones reductor may still be used for the analysis of uranium, however, because any U^{3+} formed can be converted to U^{4+} by shaking the solution with air for a few minutes.

14A-2 Auxiliary Oxidizing Reagents

Sodium Bismuthate

Sodium bismuthate is an extremely powerful oxidizing agent capable, for example, of converting manganese(II) quantitatively to permanganate ion. This bismuth salt is a sparingly soluble solid with a formula that is usually written $NaBiO_3$, although its exact composition is somewhat uncertain. Oxidations are performed by suspending the bismuthate in the analyte solution and boiling for a brief period. The unused reagent is then removed by filtration.

Ammonium Peroxydisulfate

Ammonium peroxydisulfate, $(NH_4)_2S_2O_8$, is also a powerful oxidizing agent. In acidic solution, it converts chromium(III) to dichromate, cerium(III) to its tetravalent state, and manganese(II) to permanganate. The half-reaction is

$$S_2O_8^{2-} + 2e \rightleftharpoons 2SO_4^{2-} \qquad E^0 = 2.01\text{ V}$$

The oxidations are catalyzed by traces of silver ion. The excess reagent is readily decomposed by a brief period of boiling:

$$2S_2O_8^{2-} + 2H_2O \longrightarrow 4SO_4^{2-} + O_2(g) + 4H^+$$

Sodium Peroxide and Hydrogen Peroxide

Peroxide is a convenient oxidizing agent either as the solid sodium salt or as a dilute solution of the acid. The half-reaction for hydrogen peroxide in acidic solution is

$$H_2O_2 + 2H^+ + 2e \rightleftharpoons 2H_2O \qquad E^0 = 1.78\text{ V}$$

After oxidation is complete, the solution is freed of excess reagent by boiling:

$$2H_2O_2 \longrightarrow 2H_2O + O_2(g)$$

14B Applications of Standard Oxidants

Table 14-2 summarizes the properties of the four most widely used volumetric oxidizing reagents. Note that the standard potentials for these reagents vary from 0.5 to 1.5 V. The choice among them depends upon the strength of the analyte as a reducing agent, the rate of reaction between oxidant and analyte, the stability of the standard oxidant solutions, the cost, and the availability of a satisfactory indicator.

14B-1 The Strong Oxidants—Potassium Permanganate and Cerium(IV)

Solutions of permanganate ion and cerium(IV) ion are strong oxidizing reagents whose applications closely parallel one another. Half-reactions for the two are

$$MnO_4^- + 8H^+ + 5e \rightleftharpoons Mn^{2+} + 4H_2O \qquad E^0 = 1.51 \text{ V}$$

$$Ce^{4+} + e \rightleftharpoons Ce^{3+} \qquad E^f = 1.44 \text{ V}$$

The formal potential shown for the reduction of cerium(IV) is for solutions that are 1 M in sulfuric acid. In 1 M perchloric acid and 1 M nitric acid, the potentials are 1.70 and 1.61 V, respectively. Solutions of cerium(IV) in the latter two acids are not very stable and thus find limited application.

The half-reaction shown for permanganate ion obtains only in solutions that are 0.1 M or greater in strong acid. In less acidic media, the product may be Mn(III), Mn(IV), or Mn(VI), depending upon conditions.

Comparison of the Two Reagents

For all practical purposes, the oxidizing strengths of permanganate and cerium(IV) solutions are comparable. Solutions of cerium(IV) in sulfuric acid, however, are stable indefinitely, whereas permanganate solutions decompose slowly and thus require occasional restandardization. Furthermore, cerium(IV) solutions do not oxidize chloride ion and thus can be used to titrate hydrochloric acid solutions of analytes; in contrast, permanganate ion cannot be used with hydrochloric acid solutions unless special precautions are taken to prevent the slow oxidation of chloride ion that leads to overconsumption

TABLE 14-2 **Common Oxidants Employed for Standard Solutions**

Reagent and Formula	Reduction Product	Standard Potential, V	Primary Standard for	Indicator*	Stability†
Potassium permanganate, $KMnO_4$	Mn^{2+}	1.51	$Na_2C_2O_4$, Fe	MnO_4^-	(b)
Cerium(IV), Ce^{4+}	Ce^{3+}	1.44‡	$Na_2C_2O_4$, Fe	(1)	(a)
Potassium dichromate, $K_2Cr_2O_7$	Cr^{3+}	1.33 1.0§	$K_2Cr_2O_7$, Fe	(2)	(a)
Iodine, I_2	I^-	0.536	$BaS_2O_3 \cdot H_2O$, $Na_2S_2O_3$, I_2	Starch	(c)

*(1) 1,10-phenanthroline iron(II) complex (ferroin); (2) diphenylamine sulfonic acid.

†(a) Indefinitely stable; (b) moderately stable, requires periodic standardization; (c) somewhat unstable, requires frequent standardization.

‡Formal potential in 1 M H_2SO_4.

§Formal potential in 1 M HCl and 1 M H_2SO_4.

of the standard reagent. A further advantage of cerium(IV) is that a primary-standard-grade salt of the reagent is available, thus making possible the direct preparation of standard solutions.

Despite these several advantages of cerium over potassium permanganate, solutions of the latter appear to be more widely used than are their cerium(IV) counterparts. One reason is the color of permanganate solutions, which is intense enough to serve as an indicator in titrations. A second reason for the popularity of permanganate solutions is their modest cost. The cost of 1 L of 0.02 M solution is about $0.08, whereas 1 L of cerium(IV) of comparable titer costs about $2.20 ($4.40 if reagent of primary-standard grade is employed). Another disadvantage of cerium(IV) solutions arises from their tendency to form precipitates of basic cerium(IV) salts in solutions that are less than 0.1 M in strong acid.

End Points

An obvious property of a potassium permanganate solution is its intense purple color, which is sufficient to serve as an indicator for most titrations. As little as 0.01 to 0.02 mL of a 0.02 M solution imparts a perceptible color to 100 mL of water. If the permanganate solution is very dilute, diphenylamine sulfonic acid or the 1,10-phenanthroline complex of iron(II) (Table 13-2) provides a sharper end point.

The permanganate end point is not permanent because excess permanganate ions react slowly with the relatively large concentration of manganese(II) ions present at the end point:

$$2MnO_4^- + 3Mn^{2+} + 2H_2O \longrightarrow 5MnO_2(s) + 4H^+$$

The equilibrium constant for this reaction is about 10^{47}, which indicates that the *equilibrium* concentration of permanganate ion is vanishingly small even in highly acidic media. Fortunately, the rate at which this equilibrium is approached is so slow that the end point fades only gradually over a period of perhaps 30 s.

Solutions of cerium(IV) are yellow-orange, but the color is not intense enough to act as an indicator in titrations. Several oxidation/reduction indicators are available for titrations with standard solutions of cerium(IV). The most widely used of these is the iron(II) complex of 1,10-phenanthroline or one of its substituted derivatives (Table 13-2).

The Preparation and Stability of Standard Solutions

Aqueous solutions of permanganate are not entirely stable because the ion tends to oxidize water:

$$4MnO_4^- + 2H_2O \longrightarrow 4MnO_2(s) + 3O_2(g) + 4OH^-$$

Although the equilibrium constant for this reaction indicates that the products are favored, permanganate solutions, when properly prepared, are reasonably stable because the decomposition reaction is slow. It is catalyzed by light, heat, acids, bases, manganese(II), and manganese dioxide.

Moderately stable solutions of permanganate ion can be prepared if the effects of these catalysts, particularly manganese dioxide, are minimized. Manganese dioxide is a contaminant in even the best grade of solid potassium permanganate. Furthermore, this compound forms in freshly prepared solutions of the reagent as a consequence of the reaction of permanganate ion with organic matter and dust present in the water used to prepare the solution. Removal of manganese dioxide by filtration before standardization markedly enhances the stability of standard permanganate solutions. Before filtration, the reagent solution is allowed to stand for about 24 h or is heated for a brief period to hasten oxidation of the organic species generally present in small amounts in distilled and deionized water. Paper cannot be used for filtering because permanganate ion reacts with it to form additional manganese dioxide.

Standardized permanganate solutions should be stored in the dark. Filtration and restandardization are required if any solid is detected in the solution or on the walls of the storage bottle. In any event, restandardization every 1 or 2 weeks is a good precautionary measure.

The most widely used compounds for the preparation of solutions of cerium(IV) are listed in Table 14-3. Primary-standard-grade cerium ammonium nitrate is available commercially and can be used to prepare standard solutions of the cation directly by weight. More commonly, less expensive reagent-grade cerium(IV) ammonium nitrate or ceric hydroxide is used to give solutions that are subsequently standardized. In either case, the reagent is dissolved in a solution that is at least 0.1 M in sulfuric acid to prevent the precipitation of basic salts.

Sulfuric acid solutions of tetravalent cerium are remarkably stable and can be stored for months or heated at 100°C for prolonged periods without change in concentration.

Primary Standards

Several excellent primary standards are available for the standardization of solutions of potassium permanganate and tetravalent cerium.

Sodium Oxalate. Sodium oxalate is widely used for standardizing permanganate and cerium(IV) solutions. In acidic solutions, the oxalate ion is converted to the undissociated acid. Thus, its reaction with the permanganate ion can be depicted as

$$2MnO_4^- + 5H_2C_2O_4 + 6H^+ \rightleftharpoons 2Mn^{2+} + 10CO_2(g) + 8H_2O$$

The same oxidation products are formed with cerium(IV).

The reaction between permanganate ion and oxalic acid is complex and proceeds slowly even at elevated temperature unless manganese(II) is present as a catalyst. Thus, when the first few milliliters of standard permanganate are added to a hot solution of oxalic acid, several seconds are required before the color of the permanganate ion disappears. As the concentration of manganese(II) builds up, however, the reaction proceeds more and more rapidly as a result of autocatalysis.

TABLE 14-3 **Analytically Useful Cerium(IV) Compounds**

Name	Formula	Formula Weight
Cerium(IV) ammonium nitrate	$Ce(NO_3)_4 \cdot 2NH_4NO_3$	548.2
Cerium(IV) ammonium sulfate	$Ce(SO_4)_2 \cdot 2(NH_4)_2SO_4 \cdot 2H_2O$	632.6
Cerium(IV) hydroxide	$Ce(OH)_4$	208.1
Cerium(IV) hydrogen sulfate	$Ce(HSO_4)_4$	528.4

It has been found that when solutions of sodium oxalate are titrated at 60 to 90°C, the consumption of permanganate is from 0.1 to 0.4% less than theoretical, probably due to the air-oxidation of a fraction of the oxalic acid. This small error can be avoided by adding 90 to 95% of the required permanganate to a cool solution of the oxalate. After the added permanganate is completely consumed as indicated by the disappearance of color, the solution is heated to about 60°C and titrated to a pink color that persists for about 30 s. The disadvantage of this procedure is that it requires a knowledge of the approximate concentration of the permanganate solution so that a proper initial volume of it can be added. For most purposes, the direct titration of the hot oxalic acid solution provides perfectly adequate data (usually 0.2 to 0.3% high). If greater accuracy is required, a direct titration of the hot solution of one portion of the primary standard can be followed by titration of two or three portions in which the solution is not heated until the end.

Tetravalent cerium standardizations against sodium oxalate are usually performed at 50°C in a hydrochloric acid solution containing iodine monochloride as a catalyst.

Other Primary Standards. Permanganate and cerium(IV) solutions can also be standardized against Oesper's salt [iron(II) ethylenediamine sulfate], electrolytic iron wire, and potassium ferrocyanide.

Applications of Potassium Permanganate and Cerium(IV) Solutions

Table 14-4 lists some of the many applications of permanganate and cerium(IV) solutions to the volumetric determination of inorganic species. Both

TABLE 14-4 **Some Applications of Potassium Permanganate and Cerium(IV) Solutions**

Substance Sought	Half-Reaction	Conditions
Sn	$Sn^{2+} \rightleftharpoons Sn^{4+} + 2e$	Prereduction with Zn
H_2O_2	$H_2O_2 \rightleftharpoons O_2(g) + 2H^+ + 2e$	
Fe	$Fe^{2+} \rightleftharpoons Fe^{3+} + e$	Prereduction with $SnCl_2$ or with Jones or Walden reductor
$Fe(CN)_6^{4-}$	$Fe(CN)_6^{4-} \rightleftharpoons Fe(CN)_6^{3-} + e$	
V	$VO^{2+} + 3H_2O \rightleftharpoons V(OH)_4^+ + 2H^+ + e$	Prereduction with Bi amalgam or SO_2
Mo	$Mo^{3+} + 4H_2O \rightleftharpoons MoO_4^{2-} + 8H^+ + 3e$	Prereduction with Jones reductor
W	$W^{3+} + 4H_2O \rightleftharpoons WO_4^{2-} + 8H^+ + 3e$	Prereduction with Zn or Cd
U	$U^{4+} + 2H_2O \rightleftharpoons UO_2^{2+} + 4H^+ + 2e$	Prereduction with Jones reductor
Ti	$Ti^{3+} + H_2O \rightleftharpoons TiO^{2+} + 2H^+ + e$	Prereduction with Jones reductor
$H_2C_2O_4$	$H_2C_2O_4 \rightleftharpoons 2CO_2 + 2H^+ + 2e$	
Mg, Ca, Zn, Co, Pb, Ag	$H_2C_2O_4 \rightleftharpoons 2CO_2 + 2H^+ + 2e$	Sparingly soluble metal oxalates filtered, washed, and dissolved in acid; liberated oxalic acid titrated
HNO_2	$HNO_2 + H_2O \rightleftharpoons NO_3^- + 3H^+ + 2e$	15-min reaction time; excess $KMnO_4$ back-titrated
K	$K_2NaCo(NO_2)_6 + 6H_2O \rightleftharpoons Co^{2+} + 6NO_3^- + 12H^+ + 2K^+ + Na^+ + 11e$	Precipitated as $K_2NaCo(NO_2)_6$; filtered and dissolved in $KMnO_4$; excess $KMnO_4$ back-titrated
Na	$U^{4+} + 2H_2O \rightleftharpoons UO_2^{2+} + 4H^+ + 2e$	Precipitated as $NaZn(UO_2)_3(OAc)_9$; filtered, washed, dissolved; U determined as above

reagents have also been applied to the determination of organic compounds with oxidizable functional groups.

Directions for the preparation, standardization, and use of solutions of potassium permanganate are found in Section 31E.

14B-2 Potassium Dichromate

In its analytical applications, dichromate ion is reduced to green chromium(III) ion:

$$Cr_2O_7^{2-} + 14H^+ + 6e \rightleftharpoons 2Cr^{3+} + 7H_2O \qquad E^0 = 1.33 \text{ V}$$

Dichromate titrations are generally carried out in solutions that are about 1 M in hydrochloric or sulfuric acid. In these media, the formal potential for the half-reaction is 1.0 to 1.1 V.

Potassium dichromate solutions are indefinitely stable, can be boiled without decomposition, and do not react with hydrochloric acid. Moreover, the reagent is available in high purity and at a modest cost. The disadvantages of potassium dichromate compared with cerium(IV) and permanganate ion are its lower electrode potential and the slowness of its reaction with certain reducing agents.

The Preparation, Properties, and Uses of Dichromate Solutions

For most purposes, reagent-grade potassium dichromate is sufficiently pure to permit the direct preparation of standard solutions; the solid is simply dried at 150 to 200°C before being weighed. A product of even higher quality can be obtained by recrystallizing potassium dichromate two or three times from water.

The orange color of a dichromate solution is not intense enough for use in end-point detection. However, diphenylamine sulfonic acid (Table 13-2) is an excellent indicator for titrations with this reagent. The oxidized form of the indicator is violet, and its reduced form is essentially colorless; thus, the color change observed in a direct titration is from the green of chromium(III) to violet.

Applications of Potassium Dichromate Solutions

The principal use of dichromate is for the volumetric titration of iron(II) based upon the reaction

$$Cr_2O_7^{2-} + 6Fe^{2+} + 14H^+ \longrightarrow 2Cr^{3+} + 6Fe^{3+} + 7H_2O$$

Often, this titration is performed in the presence of moderate concentrations of hydrochloric acid.

The reaction of dichromate with iron(II) has been widely used for the indirect determination of a variety of oxidizing agents. In these applications, a measured excess of a solution of iron(II) is added to an acidic solution of the analyte. The excess iron(II) is then back-titrated with standard potassium dichromate. Standardization of the iron(II) solution by titration with the dichromate is performed concurrently when the analysis is performed because solutions of iron(II) tend to be air-oxidized. This method has been applied to the determination of nitrate, chlorate, permanganate, and dichromate ions as well as organic peroxides and several other oxidizing agents.

14B-3

Iodine

Solutions of iodine are weak oxidizing agents that are used for the determination of strong reductants. The most accurate description of the half-reaction for iodine in these applications is

$$I_3^- + 2e \rightleftharpoons 3I^- \qquad E^0 = 0.536 \text{ V}$$

where I_3^- is the triiodide ion.

Standard iodine solutions have relatively limited application compared with the other oxidants we have described because of their significantly smaller electrode potential. Occasionally, however, this low potential is advantageous because it imparts a degree of selectivity that makes possible the determination of strong reducing agents in the presence of weak ones. An important advantage of iodine is the availability of a sensitive and reversible indicator for the titrations. Unfortunately, iodine solutions lack stability and must be restandardized regularly.

The Preparation and Properties of Iodine Solutions

Aqueous solutions of iodine are generally prepared by dissolving the element in a relatively concentrated solution of potassium iodide because the solubility of iodine in water is low ($\simeq 0.001$ M). More concentrated iodine solutions can then be obtained as a consequence of the reaction

$$I_2(s) + I^- \rightleftharpoons I_3^- \qquad K = 7.1 \times 10^2$$

These solutions should be called *triiodide solutions* since I_3^- is the principal iodine-containing solute species. In practice, however, they are called *iodine solutions* because this terminology suffices to account for the stoichiometric behavior of the reagent ($I_2 + 2e \longrightarrow 2I^-$).

Iodine dissolves only slowly in solutions of potassium iodide, particularly when the iodide concentration is low. To ensure complete solution, the iodine is always dissolved in a small volume of concentrated iodide solution, care being taken to avoid dilution of the concentrated solution until the last trace of solid iodine has disappeared. Otherwise, the molarity of the diluted solution gradually increases with time. This problem is sometimes avoided by filtering the solution through a sintered glass crucible before standardization.

Iodine solutions lack stability for several reasons, one being the volatility of the solute. Losses of iodine from an open vessel occur in a relatively short time even in the presence of an excess of iodide ion. In addition, iodine slowly attacks most organic materials. Consequently, cork or rubber stoppers are never used to close containers of the reagent, and precautions must be taken to protect standard solutions from contact with organic dusts and fumes.

Air-oxidation of iodide ion also causes changes in the molarity of an iodine solution:

$$4I^- + O_2(g) + 4H^+ \longrightarrow 2I_2 + 2H_2O$$

In contrast to the other effects, this reaction causes the molarity of the iodine to increase. Air-oxidation is promoted by acids, heat, and light.

The End Points of Iodine Titrations

Several end points are available for titrations involving iodine. For example, the color of triiodide ion can be discerned in a solution that is about 5×10^{-6} M in I_3^-, which corresponds to an overtitration of less than one drop of a 0.05 M iodine solution. Thus, provided the analyte solution is colorless, the reagent can serve as its own indicator.

A greater sensitivity can be obtained, at the sacrifice of convenience, by adding a few milliliters of chloroform or carbon tetrachloride to the solution. Shaking causes the bulk of any iodine to collect in the immiscible organic layer and imparts an intense violet color to it.

The deep blue complex that is formed between starch and iodine is the most widely used indicator for iodine titrations. The color is believed to arise from the absorption of iodine into the helical chain of β-amylose, a macromolecular component of most starches. The closely related α-amylose forms a red adduct with iodine. This reaction is not readily reversible and is thus undesirable. So-called *soluble starch,* which is available from commercial sources, consists principally of β-amylose, the alpha fraction having been removed; indicator solutions are readily prepared from this product.

Aqueous starch suspensions decompose within a few days, primarily because of bacterial action. The decomposition products tend to interfere with the indicator properties of the preparation and may also be oxidized by iodine. The rate of decomposition can be inhibited by preparing and storing the indicator under sterile conditions and by adding mercury(II) iodide or chloroform as a bacteriostat. Perhaps the simplest alternative is to prepare a fresh suspension of the indicator, which requires only a few minutes, on the day it is to be used.

Starch decomposes irreversibly in solutions containing large concentrations of iodine. Addition of the indicator to solutions containing an excess of iodine should therefore be postponed until the titration is nearly complete as shown by a change in color from deep red to faint yellow.

The Standardization of Iodine Solutions

Iodine solutions can be standardized against anhydrous sodium thiosulfate, barium thiosulfate monohydrate, or potassium antimony(III) tartrate.[4]

Sodium thiosulfate pentahydrate can be rendered anhydrous by refluxing in methanol for a short time.[5] The product is readily soluble in water, takes up moisture from the atmosphere only slowly, and is an excellent primary standard for triiodide solutions. Anhydrous sodium thiosulfate is available commercially.

Barium thiosulfate monohydrate is also a satisfactory primary standard for iodine solutions.[6] Although the salt is soluble to the extent of only about 0.01 M, the solid reacts so rapidly with iodine solutions that a direct titration is entirely feasible. The reaction is

$$I_2 + Ba_2S_2O_3 \cdot H_2O(s) \longrightarrow S_4O_6^{2-} + Ba^{2+} + 2I^-$$

It is reported that $BaS_2O_3 \cdot H_2O$ with a purity of 99.85% is readily prepared and stable at room temperature. Barium thiosulfate monohydrate begins to lose water above 50°C; the anhydrous salt, however, is unsuitable for stand-

[4]Arsenic(III) oxide, for many years the preferred primary standard for iodine solutions, has been identified as a carcinogen; its use is now subject to strict regulation.

[5]A. A. Woolf, *Anal. Chem.*, **1982**, *54*, 2134.

[6]W. M. MacNevin and O. H. Kriege, *Anal. Chem.*, **1953**, *25*, 767.

ardization owing to its low solubility. Therefore, the monohydrate should be used without drying.

The reaction between thiosulfate and iodine is discussed in Section 14C-2.

Applications of Standard Iodine Solutions

Iodine titrations are generally performed in neutral or acidic media. Basic solutions are avoided because hypoiodite ion forms according to the equation

$$I_2 + OH^- \rightleftharpoons IO^- + I^- + H^+$$

The hypoiodite may subsequently disproportionate to iodide and iodate:

$$3IO^- \rightleftharpoons IO_3^- + 2I^-$$

The occurrence of these reactions can cause serious errors by disturbing the stoichiometry of the reaction between the standardized reagent and the analyte. Thus, solutions to be titrated with standardized iodine cannot have pH values higher than 9. Occasionally, a pH greater than 7.0 is detrimental.

Table 14-5 summarizes methods that use iodine as an oxidizing agent. Directions for the preparation, standardization, and application of standard iodine solutions are given in Section 31F.

14C Volumetric Applications of Reductants

Standard solutions of most reducing agents tend to react with atmospheric oxygen. For this reason, standard solutions of reductants are seldom used for the direct titration of oxidizing analytes; indirect methods are used instead. The two most common indirect methods, discussed in the paragraphs that follow, are based upon iron(II) and iodide ions as reductants.

14C-1 Reductions with Iron(II) Solutions

Solutions of iron(II) are readily prepared from iron(II) ammonium sulfate, $Fe(NH_4)_2(SO_4)_2 \cdot 6H_2O$ (Mohr's salt), or from the closely related iron(II) ethylenediamine sulfate, $FeC_2H_4(NH_3)_2(SO_4)_2 \cdot 4H_2O$ (Oesper's salt). Air-oxidation of iron(II) takes place rapidly in neutral solutions but is inhibited in the presence of acids, with the most stable preparations being about 0.5 M in H_2SO_4. Such solutions are stable for no longer than one day, if that long.

Numerous oxidizing agents are conveniently determined by treatment of the sample solution with a measured excess of standard iron(II) followed

TABLE 14-5

Analysis with Standard Iodine Solutions

Substance Analyzed	Half-Reaction
As	$H_3AsO_3 + H_2O \leftrightarrows H_3AsO_4 + 2H^+ + 2e$
Sb	$H_3SbO_3 + H_2O \leftrightarrows H_3SbO_4 + 2H^+ + 2e$
Sn	$Sn^{2+} \leftrightarrows Sn^{4+} + 2e$
H_2S	$H_2S \leftrightarrows S(s) + 2H^+ + 2e$
SO_2	$SO_3^{2-} + H_2O \leftrightarrows SO_4^{2-} + 2H^+ + 2e$
$S_2O_3^{2-}$	$2S_2O_3^{2-} \leftrightarrows S_4O_6^{2-} + 2e$
N_2H_4	$N_2H_4 \leftrightarrows N_2(g) + 4H^+ + 4e$
Ascorbic acid*	$C_6H_8O_6 \rightarrow C_6H_6O_6 + 2H^+ + 2e$

*For the structure of ascorbic acid, see Section 31H-4.

by immediate titration of the excess with a standard solution of potassium dichromate or cerium(IV). Just before or after the samples are titrated, the concentration of the iron(II) solution is established by titrating two or three aliquots of the reagent in the absence of sample.

This procedure has been applied to the determination of organic peroxides; hydroxylamine; chromium(VI); cerium(IV); molybdenum(VI); nitrate, chlorate, and perchlorate ions; and numerous other reductants.

14C-2 Reductions with Iodide Ion; Titrations with Sodium Thiosulfate

Numerous oxidizing agents are commonly determined by a procedure in which iodide ion, the primary reducing agent, is converted to iodine according to the reaction

$$2I^- \rightleftharpoons I_2 + 2e$$

Solutions of iodide ion can never be used for the direct titration of oxidants, however, because the intense color of the product precludes the possibility of visual end points. Furthermore, iodide solutions are quite unstable because of the air-oxidation of the reagent, a factor that limits their use even when nonvisual end points are available.

Indirect methods involving iodine are carried out by adding an unmeasured excess of potassium iodide to a solution of the analyte. The iodine produced is then determined by titration with a standard solution of sodium thiosulfate, $Na_2S_2O_3$, one of the few reducing agents that is stable to air-oxidation. The analytically important half-reaction is

$$2S_2O_3^{2-} \rightleftharpoons S_4O_6^{2-} + 2e$$

where the product is the tetrathionate ion. Note that a one-electron change is associated with each thiosulfate ion.

The Reaction of Iodine with Thiosulfate Ion

The reaction between iodine and thiosulfate ion is described by the equation

$$I_2 + 2S_2O_3^{2-} \rightarrow 2I^- + S_4O_6^{2-}$$

The quantitative conversion of thiosulfate ion to tetrathionate ion is unique with iodine; other oxidizing agents tend to carry the oxidation further, to sulfate ion or to a mixture of tetrathionate and sulfate ions.

Thiosulfate titrations of iodine are best performed in neutral or slightly acidic media. If strongly acidic solutions must be titrated, air-oxidation of the excess iodide must be prevented by blanketing the solution with an inert gas, such as carbon dioxide or nitrogen. One simple way of providing a blanket of carbon dioxide is to introduce a quantity of solid sodium hydrogen carbonate into the solution, which reacts to form a layer of carbon dioxide that excludes oxygen from the titration vessel.

The reaction shown by Equation 14-1 is no longer quantitative in an iodine solution with a pH somewhat greater than 7 because the hypoiodite ion that forms under these conditions (page 341) converts the thiosulfate to sulfate ion. Because of this effect, it is recommended that a pH of less than 7.6 be maintained when 0.05 M solutions of iodine are being titrated. For 0.005 and 0.0005 M solutions, the maximum pH should be 6.5 and 5, respectively.

The titration of highly acidic iodine solutions with thiosulfate ion yields quantitative results, provided care is taken to prevent air-oxidation of the iodide ion. The end point for this titration is readily established by means of a starch solution. It is worth noting again that when titrating solutions of iodine with thiosulfate ion, addition of the starch indicator must be delayed until most of the iodine has been consumed—that is, until the solution has changed from red-brown to pale yellow. This precaution is necessary to avoid decomposition of the starch by the high concentration of iodine.

The Stability of Sodium Thiosulfate Solutions

Although sodium thiosulfate solutions are resistant to air-oxidation, they do tend to decompose to give sulfur and hydrogen sulfite ion:

$$S_2O_3^{2-} + H^+ \rightleftharpoons HSO_3^- + S(s)$$

Variables that influence the rate of this reaction include pH, presence of microorganisms, concentration of the solution, presence of copper(II) ions, and exposure to sunlight. These variables may cause the concentration of a thiosulfate solution to change by several percent over a period of a few weeks. On the other hand, proper attention to detail will yield solutions that need only occasional restandardization.

The rate of the decomposition reaction increases markedly as the solution becomes acidic. In fact, when sodium thiosulfate is added to a strongly acidic medium, a cloudiness develops almost immediately as a consequence of the precipitation of elemental sulfur. Even in neutral solution, this reaction proceeds at such a rate that standard sodium thiosulfate must be restandardized periodically.

The most important single cause for the instability of thiosulfate solutions can be traced to bacteria that metabolize thiosulfate ion to sulfite and sulfate ions as well as elemental sulfur. To minimize this problem, standard solutions of the reagent are prepared under reasonably sterile conditions. Bacterial activity appears to be at a minimum at a pH between 9 and 10, which accounts, at least in part, for the reagent's greater stability in slightly basic solutions. The presence of a bactericide, such as chloroform, sodium benzoate, or mercury(II) iodide, also slows decomposition.

The decomposition reaction may cause the concentration of a thiosulfate solution to increase or decrease. Increases in concentration arise because each mole of hydrogen sulfite ion formed by the reaction consumes 1 mol of iodine whereas the thiosulfate ion from which it was derived consumes just half this amount. Decreases in concentration are sometimes encountered when the hydrogen sulfite ion is air-oxidized to the unreactive sulfate ion. As noted earlier, strongly acidic solutions of iodine can be titrated with standard sodium thiosulfate without reagent decomposition, provided care is taken to introduce the thiosulfate slowly and with good mixing. Under these circumstances, the reagent is oxidized by the iodine so rapidly that the slower decomposition reaction is precluded.

The Standardization of Thiosulfate Solutions

Potassium iodate is an excellent primary standard for thiosulfate solutions. In this application, weighed amounts of primary-standard-grade reagent are dissolved in water containing an excess of potassium iodide. When this mixture is acidified with a strong acid, the reaction

$$IO_3^- + 5I^- + 6H^+ \rightleftharpoons 3I_2 + 2H_2O$$

occurs instantaneously. The liberated iodine is then titrated with the thiosulfate solution. The stoichiometry of the reactions is

$$1 \text{ mol } IO_3^- \equiv 3 \text{ mol } I_2 \equiv 6 \text{ mol e} \equiv 6 \text{ mol } S_2O_3^{2-}$$

The sole disadvantage of potassium iodate as a primary standard for thiosulfate solutions is the small weight required for a standardization. For example, the amount needed for the standardization of a 0.05 M thiosulfate solution is only about 0.07 g, which is small enough to lead to an unacceptable weighing uncertainty in some applications. This problem is often circumvented by dissolving a larger quantity of potassium iodate in a known volume of water and taking aliquots of this solution for the standardization. This approach suffers from the disadvantage of providing no duplicate check on the precision of the weighing process unless two or more iodate solutions are prepared.

Other primary standards for sodium thiosulfate are potassium dichromate, potassium bromate, potassium hydrogen iodate, potassium ferricyanide, and metallic copper. All these compounds liberate stoichiometric amounts of iodine when treated with excess potassium iodide.

Applications of Sodium Thiosulfate Solutions

Numerous substances can be determined by the indirect method involving titration with sodium thiosulfate; typical applications are summarized in Table 14-6. Directions for the preparation and standardization of thiosulfate solutions and examples of their applications are found in Section 31G.

14D Some Specialized Oxidants

In this section, we describe three oxidizing agents that are used primarily for determining certain special groups of compounds. Potassium bromate is used for the determination of organic compounds that contain olefinic and certain types of aromatic functional groups; periodic acid reacts selectively with organic compounds having hydroxyl, carbonyl, or amine groups on adjacent carbon atoms; and Karl Fischer reagent is widely employed for the determination of water in a variety of organic and inorganic samples.

TABLE 14-6 **Applications of Indirect Methods Involving Titration with Sodium Thiosulfate**

Substance	Half-Reaction	Special Conditions
IO_4^-	$IO_4^- + 8H^+ + 7e \rightarrow \frac{1}{2}I_2 + 4H_2O$	Acidic solution
	$IO_4^- + 2H^+ + 2e \rightarrow IO_3^- + H_2O$	Neutral solution
IO_3^-	$IO_3^- + 6H^+ + 5e \rightarrow \frac{1}{2}I_2 + 3H_2O$	Strong acid
BrO_3^-, ClO_3^-	$XO_3^- + 6H^+ + 6e \rightarrow X^- + 3H_2O$	Strong acid
Br_2, Cl_2	$X_2 + 2I^- \rightarrow I_2 + 2X^-$	
NO_2^-	$HNO_2 + H^+ + e \rightarrow NO(g) + H_2O$	
Cu^{2+}	$Cu^{2+} + I^- + e \rightarrow CuI(s)$	
O_2	$O_2 + 4Mn(OH)_2(s) + 2H_2O \rightarrow 4Mn(OH)_3(s)$	Basic solution
	$Mn(OH)_3(s) + 3H^+ + e \rightarrow Mn^{2+} + 3H_2O$	Acidic solution
O_3	$O_3(g) + 2H^+ + 2e \rightarrow O_2(g) + H_2O$	
Organic peroxide	$ROOH + 2H^+ + 2e \rightarrow ROH + H_2O$	

14D-1

Potassium Bromate as a Source of Bromine

Primary-standard potassium bromate is available from commercial sources and can be used directly to prepare standard solutions that are stable indefinitely. Direct titrations with potassium bromate are relatively few. The reagent is a convenient and widely used stable source of bromine, however.[7] In this application, an unmeasured excess of potassium bromide is added to an acidic solution of the analyte. Upon introduction of a measured volume of standard potassium bromate, a stoichiometric quantity of bromine is produced as a result of the reaction

$$\underset{\text{standard solution}}{BrO_3^-} + \underset{\text{excess}}{5Br^-} + 6H^+ \longrightarrow 3Br_2 + 3H_2O$$

This indirect generation circumvents the problems associated with the use of standard bromine solutions, which lack stability. Note that each bromate is responsible for the formation of three bromine molecules, which in turn require six electrons for reduction to bromide:

$$1 \text{ mol } BrO_3^- \equiv 3 \text{ mol } Br_2 \equiv 6 \text{ mol e}$$

Applications of Standard Potassium Bromate Solutions

Potassium bromate is a convenient source of bromine for the determination of compounds containing certain organic functional groups. Few of these groups react rapidly enough with bromine to make direct titration feasible. Instead, a measured excess of standard bromate is added to the solution that contains the sample plus an excess of potassium bromide. After acidification, the mixture is allowed to stand in a glass-stoppered vessel until reaction of the bromine with the analyte is judged complete. To determine the excess bromine, an excess of potassium iodide is introduced to convert the bromine to iodine:

$$2I^- + Br_2 \longrightarrow I_2 + 2Br^-$$

The liberated iodine is then titrated with standard sodium thiosulfate (Equation 14-1).

Bromine is incorporated into an organic molecule either by substitution or by addition.

Substitution Reactions. Halogen substitution involves the replacement of hydrogen in an aromatic ring by a halogen. For example, the bromination of phenol involves replacement of three hydrogen atoms:

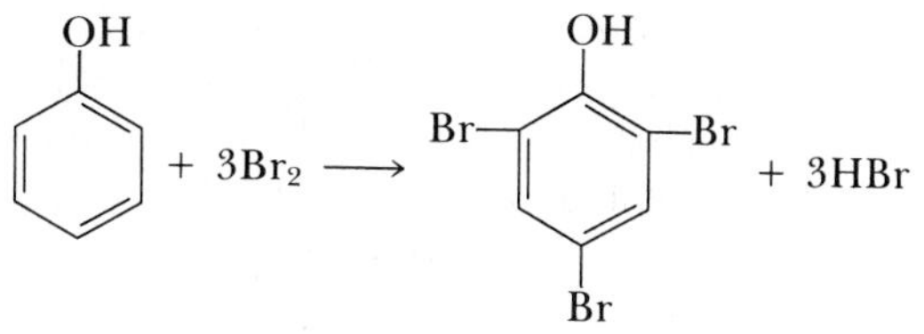

$$C_6H_5OH + 3Br_2 \longrightarrow C_6H_2Br_3OH + 3HBr$$

(phenol + 3Br$_2$ → 2,4,6-tribromophenol + 3HBr)

[7]For a discussion of bromate solutions and their applications, see M.R.F. Ashworth, *Titrimetric Organic Analysis,* Part I, pp. 118–130. New York: Interscience, 1964.

Substitution methods have been successfully applied to the determination of aromatic compounds that contain strong ortho-para-directing groups, particularly amines and phenols. An important example of this application is found in the determination of 8-hydroxyquinoline:

$$\text{8-hydroxyquinoline } (C_9H_6N\text{–OH}) + 2Br_2 \rightleftharpoons \text{5,7-dibromo-8-hydroxyquinoline} + 2HBr$$

In contrast to most bromine substitutions, this reaction takes place rapidly enough in hydrochloric acid solution to make direct titration feasible.

The titration of 8-hydroxyquinoline with bromine is particularly significant because the former is an excellent precipitating reagent for cations (Section 3D-3). For example, aluminum can be determined according to the sequence

$$Al^{3+} + 3HOC_9H_6N \xrightarrow{\text{pH 4–9}} Al(OC_9H_6N)_3(s) + 3H^+$$

$$Al(OC_9H_6N)_3(s) \xrightarrow{\text{hot 4 M HCl}} 3HOC_9H_6N + Al^{3+}$$

$$3HOC_9H_6N + 6Br_2 \longrightarrow 3HOC_9H_4NBr_2 + 6HBr$$

The stoichiometric relationships in this case are

$$1 \text{ mol } Al^{3+} \equiv 3 \text{ mol } HOC_9H_6N \equiv 6 \text{ mol } Br_2 \equiv 12 \text{ mol e}$$

Addition Reactions. Addition reactions involve the opening of an olefinic double bond. For example, 1 mol of ethylene reacts with 1 mol of bromine in the reaction

$$H_2C{=}CH_2 + Br_2 \longrightarrow H_2CBr{-}CBrH_2$$

The literature contains numerous references to the use of bromine for the estimation of olefinic unsaturation in fats, oils, and petroleum products.

A procedure for the determination of phenol in waste water by bromine substitution is found in Section 31H-3, and a direct-oxidation method for the determination of ascorbic acid in vitamin C tablets is given in Section 31H-4.

14D-2 *Periodic Acid*

Aqueous solutions of iodine in the +7 oxidation state are highly complex. In the presence of strong acid, paraperiodic acid, H_5IO_6, and its conjugate base predominate, although the metaperiodates HIO_4 and IO_4^- are undoubtedly present as well. The reduction of periodic acid to iodate ion is best described by the half-reaction

$$H_5IO_6 + H^+ + 2e \rightleftharpoons IO_3^- + 3H_2O \qquad E^0 = 1.6 \text{ V}$$

The Preparation and Properties of Periodic Acid Solutions

Several periodates are available for the preparation of standard solutions. Among these is paraperiodic acid itself, a crystalline, readily soluble, hygroscopic solid. An even more useful compound is sodium metaperiodate, $NaIO_4$, which is soluble in water to the extent of 0.06 M at 25°C. Sodium paraperiodate, Na_5IO_6, is not sufficiently soluble for the preparation of standard solutions; it is, however, readily converted to the more soluble metaperiodate by recrystallization from hot concentrated nitric acid. Potassium metaperiodate can be used as a primary standard for the preparation of periodate solutions. Although its solubility is only about 5 g/L at room temperature, it readily dissolves at elevated temperatures and can be subsequently converted to a more soluble form by the addition of base.

Periodate solutions vary considerably in stability, depending on their mode of preparation and storage. A solution prepared by dissolving sodium metaperiodate in water decomposes at the rate of several percent per week. By way of contrast, a solution of potassium metaperiodate in excess alkali was found to change no more than 0.3 to 0.4% in 100 days. The most stable periodate solutions appear to be those containing an excess of sulfuric acid; such solutions decrease in molarity by less than 0.1% in four months.

The Standardization of Periodate Solutions

Periodate solutions are most conveniently standardized by buffering aliquots of the reagent with solid borax or sodium hydrogen carbonate to ensure that they remain slightly alkaline. An excess of potassium iodide is then introduced, which results in the formation of 1 mol of iodine for each mole of periodate:

$$H_4IO_6^- + 2I^- \longrightarrow IO_3^- + I_2 + 2OH^- + H_2O$$

As long as the solution is kept slightly alkaline, further reduction of iodate does not occur, and the liberated iodine can be titrated directly with a standard solution of sodium thiosulfate or sodium arsenite.

Applications of Periodic Acid

The reason periodic acid is widely used is that it reacts remarkably selectively with organic compounds containing certain combinations of functional groups.[8] Ordinarily, these oxidations are performed at room temperature in the presence of a measured excess of the periodate; most reactions are complete in 0.5 to 1 h. After oxidation, the excess periodate is determined by the method described for standardization. Alternatively, a reaction product such as ammonia, formaldehyde, or a carboxylic acid may be determined. In this case, the exact quantity of periodate used need not be known.

Periodate oxidations are usually carried out in aqueous solution, although solvents such as methanol, ethanol, or dioxane may be added to enhance the solubility of the sample.

Compounds Attacked by Periodate. At room temperature, organic compounds containing aldehyde, ketone, or alcohol groups *on adjacent carbon atoms* are rapidly oxidized by periodic acid. Primary and secondary α-hydroxyamines are also readily attacked; α-diamines are not. With few exceptions, other organic compounds do not react at a significant rate. Thus, compounds containing isolated aldehydes, ketone, alcohol, or amine groups are not affected

[8]The use of periodic acid for this purpose was first reported by L. Malaprade, *Compt. rend.*, **1928,** *186,* 382. See also M.R.F. Ashworth, *Titrimetric Organic Analysis,* Part II, pp. 724–744. New York: Interscience, 1965.

by periodic acid; nor are compounds with a carboxylic acid group either isolated from or adjacent to any of the reactive groups. At elevated temperatures, the extraordinary selectivity of periodic acid tends to disappear.

Periodate oxidations of organic compounds follow a regular and predictable set of rules:

1. Attack of adjacent functional groups always results in rupture of the carbon-to-carbon bond between the groups.
2. A carbon atom containing a hydroxyl group is oxidized to an aldehyde or ketone.
3. A carbonyl group is converted to a carboxylic-acid group.
4. A carbon atom containing an amine group loses ammonia (or a substituted amine) and is itself converted to an aldehyde.

The following half-reactions illustrate these rules:

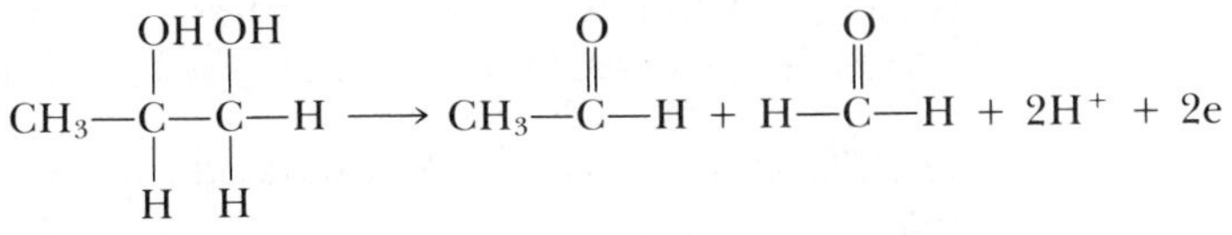

$$\mathrm{CH_3{-}CH(OH){-}CH(OH){-}H \longrightarrow CH_3{-}C({=}O){-}H + H{-}C({=}O){-}H + 2H^+ + 2e}$$

propylene glycol

$$\mathrm{H{-}CH(OH){-}CH(OH){-}CH(OH){-}H + H_2O \longrightarrow 2H{-}C({=}O){-}H + H{-}C({=}O){-}OH + 4H^+ + 4e}$$

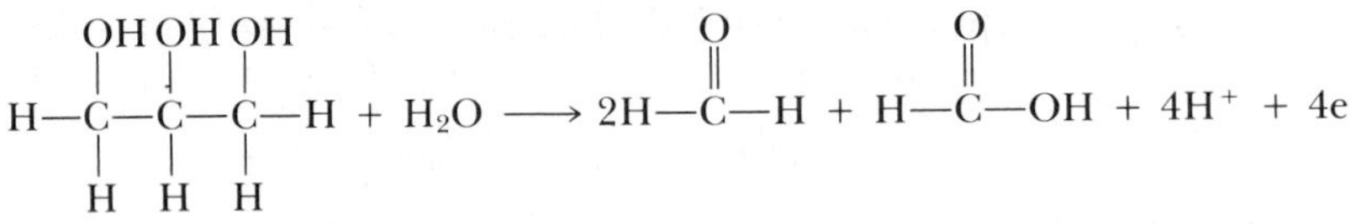

glycerol

For predicting the reaction products of the oxidation of glycerol, the first step can be thought of as producing 1 mol of formaldehyde and 1 mol of an α-hydroxyaldehyde (glycolic aldehyde). The latter is then further oxidized and by the second rule produces a second mole of formaldehyde plus 1 mol of formic acid.

Additional examples of the four rules include:

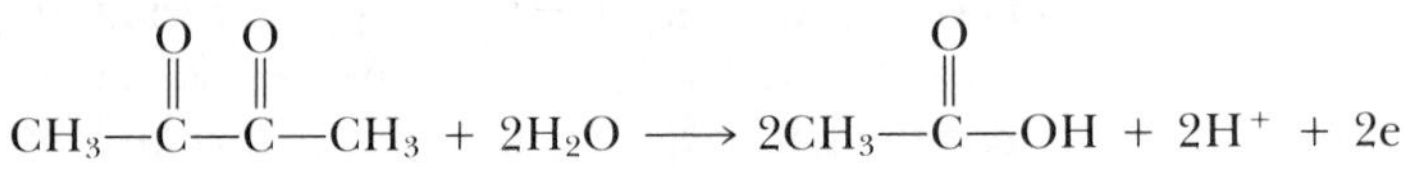

$$\mathrm{CH_3{-}C({=}O){-}C({=}O){-}CH_3 + 2H_2O \longrightarrow 2CH_3{-}C({=}O){-}OH + 2H^+ + 2e}$$

biacetyl

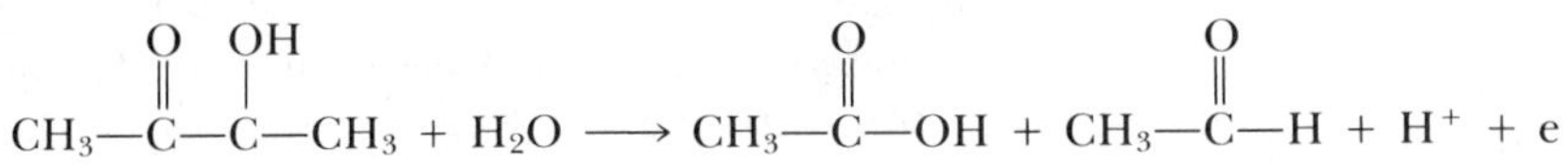

$$\mathrm{CH_3{-}C({=}O){-}C(OH){-}CH_3 + H_2O \longrightarrow CH_3{-}C({=}O){-}OH + CH_3{-}C({=}O){-}H + H^+ + e}$$

acetoin

$$\mathrm{H{-}CH(OH){-}CH(NH_2){-}H + H_2O \longrightarrow 2H{-}C({=}O){-}H + NH_3 + 2H^+ + 2e}$$

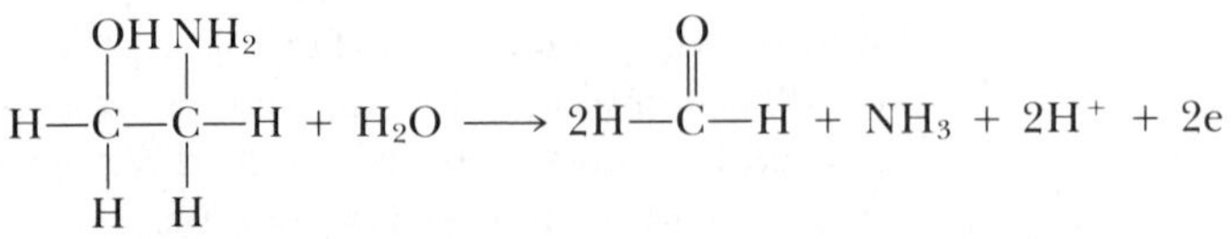

ethanolamine

The Determination of α-Hydroxyamines. As mentioned earlier, the periodate oxidation of compounds having hydroxyl and amino groups on adjacent carbon atoms results in the formation of aldehydes and the liberation of ammonia (rule 4). The latter is readily distilled from the alkaline oxidation mixture and determined by a neutralization titration. This procedure is particularly useful in the analysis of mixtures of the various amino acids that occur in proteins. Because only serine, threonine, β-hydroxyglutamic acid, and hydroxylysine have the requisite structure for the liberation of ammonia, the method is selective for these species.

14D-3 Karl Fischer Reagent for Water Determination

A number of chemical methods for the determination of water in solids and organic solvents have been devised (Section 27C). Unquestionably, the most important of these involves the use of Karl Fischer reagent, which is relatively specific for water.[9]

The Reaction and Stoichiometry

Karl Fischer reagent is composed of iodine, sulfur dioxide, pyridine, and methanol. This mixture reacts with water according to the equation

$$C_5H_5N{\cdot}I_2 + C_5H_5N{\cdot}SO_2 + C_5H_5N + H_2O \longrightarrow 2C_5H_5N{\cdot}HI + C_5H_5N{\cdot}SO_3 \qquad (14\text{-}2)$$

$$C_5H_5N{\cdot}SO_3 + CH_3OH \longrightarrow C_5H_5N(H)SO_4CH_3 \qquad (14\text{-}3)$$

Note that only the first step, which involves the oxidation of sulfur dioxide by iodine to give sulfur trioxide and hydrogen iodide, consumes water. In the presence of a large amount of pyridine, C_5H_5N, all reactants and products exist as complexes, as indicated in the equations. The second step, which occurs when an excess of methanol is present, is important to the success of the titration because the pyridine/sulfur trioxide complex is also capable of consuming water:

$$C_5H_5N{\cdot}SO_3 + H_2O \longrightarrow C_5H_5NHSO_4H \qquad (14\text{-}4)$$

This last reaction is undesirable because it is not as specific for water as the reaction shown in Equation 14-2; it can be prevented completely by having a large excess of methanol present.

Properties of the Reagent

Equation 14-2 indicates that the stoichiometry of the Karl Fischer titration involves the consumption of 1 mol of iodine, 1 mol of sulfur dioxide, and 3 mol of pyridine for each mole of water. In practice, excesses of both sulfur dioxide and pyridine are employed so that the reagent's combining capacity for water is determined by its iodine content. For typical applications, the titer is between 2 and 5 mg of water per milliliter of reagent; a twofold excess of sulfur dioxide and a threefold to fourfold excess of pyridine are provided.

The titer of Karl Fischer reagent decreases with standing. Because decomposition is particularly rapid immediately after preparation, it is common practice to prepare the reagent a day or two before it is to be used. Ordinarily, its titer must be established at least daily against a standard solution of water in methanol. A proprietary commercial Karl Fischer reagent reported to require only occasional restandardization is now available.

[9]For a monograph on this subject, see J. Mitchell and D. M. Smith, *Aquametry*, 2nd ed. New York: Interscience, 1977.

It is obvious that great care must be exercised to keep atmospheric moisture from contaminating the Karl Fischer reagent and the sample. All glassware must be carefully dried before use, and the standard solution must be stored out of contact with air. It is also necessary to minimize contact between the atmosphere and the solution during the titration.

End-Point Detection

The end point in a Karl Fischer titration is signaled by the appearance of the first excess of the pyridine/iodine complex when all water has been consumed. The color of the reagent is intense enough for a visual end point; the change is from the yellow of the reaction products to the brown of the excess reagent. With some practice, and in the absence of other colored materials, the end point can be established with a reasonable degree of certainty (that is, to perhaps ± 0.2 mL).

Various electrometric end points are also employed for Karl Fischer titrations, the most widely used being the amperometric technique with twin microelectrodes discussed in Section 17F-4. Several instrument manufacturers offer fully automated instruments for performing such titrations.

Applications

Karl Fischer reagent has been applied to the determination of water in numerous types of samples.[10] There are numerous variations of the basic technique, depending upon the solubility of the material, the state in which the water is retained, and the physical state of the sample. If the sample can be dissolved completely in methanol, a direct and rapid titration is usually feasible. This method has been applied to the determination of water in many organic acids, alcohols, esters, ethers, anhydrides, and halides. The hydrated salts of most organic acids, as well as the hydrates of a number of inorganic salts that are soluble in methanol, can also be determined by direct titration.

Direct titration of samples that are only partially dissolved in the reagent usually leads to incomplete recovery of the water. Satisfactory results with this type of sample are often obtained, however, by the addition of an excess of reagent and back-titration with a standard solution of water in methanol after a suitable reaction time. An often effective alternative is to extract the water from the sample by refluxing with anhydrous methanol or other organic solvents. The resulting solution is then titrated directly with the Karl Fischer solution.

Difficulty is also encountered in the analysis of sorbed moisture and tightly bound hydrate water. For these, the extraction techniques just described are frequently effective.

Certain substances interfere with the Karl Fischer method. Among these are compounds that react with one of the components of the reagent to produce water. For example, carbonyl compounds combine with methanol to give acetals:

$$RCHO + 2CH_3OH \longrightarrow R{-}CH(OCH_3)_2 + H_2O$$

[10]For a complete discussion of the applications of the reagent, see J. Mitchell and D. M. Smith, *Aquametry*, 2nd ed. New York: Interscience, 1977.

The result is a fading end point. Many metal oxides react with the hydrogen iodide formed in the titration to give water:

$$MO + 2HI \rightleftharpoons MI_2 + H_2O$$

Again, erroneous data result. Preliminary treatment of the sample can sometimes prevent these interferences.

Oxidizing or reducing substances frequently interfere with the Karl Fischer titration by reoxidizing the iodide produced or reducing the iodine in the reagent.

Questions and Problems

*14-1 Why is a Walden reductor always used with solutions that contain appreciable concentrations of HCl?

14-2 Why is zinc amalgam rather than pure zinc used in a Jones reductor?

*14-3 Write a balanced equation for the reduction of UO_2^{2+} in a Walden reductor.

14-4 Write a balanced equation for the reduction of TiO^{2+} in a Jones reductor.

*14-5 Why are standard $KMnO_4$ solutions seldom used for the titration of solutions containing HCl?

14-6 Why are Ce^{4+} solutions never used for the titration of reductants in basic solutions?

*14-7 Write a balanced equation showing why $KMnO_4$ end points fade.

14-8 Why are $KMnO_4$ solutions filtered before they are standardized?

*14-9 What is the primary use of standard solutions of $K_2Cr_2O_7$?

14-10 Why are iodine solutions prepared by dissolving I_2 in concentrated KI?

*14-11 A standard solution of I_2 increased in molarity with standing. Write a balanced equation that accounts for the increase.

14-12 Write equations that describe what occurs when I_2 is introduced into a basic solution.

*14-13 Why are standard solutions of reductants less often used for titrations than standard solutions of oxidants?

14-14 Write a balanced equation describing the titration of hydrazine with standard iodine.

*14-15 Write a balanced equation for the reaction of H_3AsO_3 with I_2. In order for this reaction to be used for the determination of H_3AsO_3, it is necessary that the analyte solution be buffered to a pH slightly greater than 8. Suggest an explanation.

14-16 When a solution of $Na_2S_2O_3$ is introduced into a solution of HCl, a cloudiness develops almost immediately. Write a balanced equation explaining this phenomenon.

*14-17 In the titration of I_2 solutions with $Na_2S_2O_3$, starch indicator is never added until just before chemical equivalence. Why?

14-18 Why are solutions of $KMnO_4$ and $Na_2S_2O_3$ generally stored in dark reagent bottles?

*14-19 When a solution of $KMnO_4$ was left standing in a buret for 3 h, a brownish ring formed at the surface of the liquid. Write a balanced equation that explains this observation.

14-20 Suggest a way in which a solution of KIO_3 could be used as a source of known quantities of I_2.

*14-21 Write balanced equations showing how $KBrO_3$ could be used for standardizing solutions of $Na_2S_2O_3$.

14-22 Write balanced equations showing how $K_2Cr_2O_7$ could be used for standardizing solutions of $Na_2S_2O_3$.

*14-23 Write balanced equations showing how a solution of H_5IO_6 is standardized against a standard solution of $Na_2S_3O_3$.

14-24 List the components of Karl Fischer reagent.

*14-25 Suggest how Karl Fischer reagent could be used for determining the bound water in $BaCl_2{\cdot}2H_2O$.

14-26 What types of species interfere in the determination of water with Karl Fischer reagent?

*14-27 Write balanced equations to describe

(a) the oxidation of Mn^{2+} to MnO_4^- by ammonium peroxydisulfate.
(b) the oxidation of Ce^{3+} to Ce^{4+} by sodium bismuthate.
(c) the oxidation of U^{4+} to UO_2^{2+} by H_2O_2.
(d) the reaction of $V(OH)_4^+$ in a silver reductor.
(e) the titration of H_2O_2 with $KMnO_4$.
(f) the reaction between KI and ClO_3^- in acidic solution.

14-28 Write balanced equations to describe

(a) the reduction of Fe^{3+} to Fe^{2+} by SO_2.
(b) the reaction of H_2MoO_4 in a Jones reductor.
(c) the oxidation of HNO_2 by a solution of MnO_4^-.
(d) the reaction of aniline ($C_6H_4NH_2$) with a mixture of $KBrO_3$ and KBr in acidic solution.
(e) the air-oxidation of $HAsO_3^{2-}$ to $HAsO_4^{2-}$.
(f) the reaction of KI with HNO_2 in acidic solution.

*14-29 Indicate the number of moles of each product formed from 1 mol of the following compounds when oxidized by periodic acid at room temperature:

(a) $CH_2OH(CHOH)_4CH_2OH$

```
        O   O
        ‖   ‖
(b) H—C—C—H
```

```
        H  NH₂ O
        |   |   ‖
(c) H—C—C—C—H
        |   |
        H   H
```

14-30 Indicate the number of moles of each product formed from 1 mol of the following compounds when oxidized by periodic acid at room temperature:

```
        H  NH₂ O
        |   |   ‖
(a) H—C—C—C—CH₂
        |   |
        H   H
```

```
             O   O
             ‖   ‖
(b) CH₃—C—C—H
```

$$\text{(c) } CH_3—\underset{\underset{H}{|}}{\overset{\overset{OH}{|}}{C}}—\overset{\overset{O}{\|}}{C}—H$$

*14-31 Describe the preparation of 2.500 L of 0.03500 M $K_2Cr_2O_7$ from the pure salt.

14-32 Describe the preparation of 500 mL of approximately 0.025 M $KMnO_4$ to be used for the titration of a reducing agent.

*14-33 Describe the preparation of 750.0 mL of 0.02500 M $KBrO_3$ to be used as a source of Br_2.

14-34 Describe the preparation of 2.000 L of 0.1700 M I_3^- from pure I_2 and KI.

*14-35 What is the Fe_2O_3 titer of the solution described in Problem 14-31?

14-36 What is the phenol titer of the solution described in Problem 14-33?

*14-37 To standardize a solution of $Na_2S_2O_3$, 0.1518 g of $K_2Cr_2O_7$ was dissolved in dilute HCl. An excess of KI was added, following which the liberated I_2 was titrated with 46.13 mL of the reagent. Calculate the molarity of the $Na_2S_2O_3$.

14-38 To standardize a solution of $Na_2S_2O_3$, 0.1017 g of $KBrO_3$ was dissolved in dilute HCl. An excess of KI was added, following which the liberated I_2 was titrated with 39.75 mL of the reagent. Calculate the molarity of the $Na_2S_2O_3$.

*14-39 Arsenic(III) oxide occurs in nature as the mineral claudetite. A 0.2104-g sample of impure claudetite required 29.36 mL of 0.02643 M iodine solution. Calculate the percentage of As_2O_3 in the sample.

14-40 The $KClO_3$ in a 0.1342-g sample of an explosive was determined by reaction with 50.00 mL of 0.09601 M Fe^{2+}:

$$ClO_3^- + 6Fe^{2+} + 6H^+ \longrightarrow Cl^- + 3H_2O + 6Fe^{3+}$$

When the reaction was complete, the excess Fe^{2+} was back-titrated with 12.99 mL of 0.08362 M Ce^{4+}. Calculate the percentage of $KClO_3$ in the sample.

*14-41 The Sb(III) in a 1.080-g stibnite ore required a 41.67-mL titration with 0.03134 M I_2. Express the results of this analysis in terms of (a) percentage of Sb and (b) percentage of stibnite (Sb_2S_3).

14-42 A 0.2236-g sample of limestone was dissolved in dilute HCl. After $(NH_4)_2C_2O_4$ was introduced and the pH of the resulting solution adjusted to permit the quantitative precipitation of CaC_2O_4, the solid was isolated by filtration, washed free of excess $C_2O_4^{2-}$, and dissolved in dilute H_2SO_4. Titration of the liberated $H_2C_2O_4$ required 26.77 mL of 0.02356 M $KMnO_4$. Calculate the percentage of CaO in the sample.

*14-43 An 8.13-g sample of an ant-control preparation was decomposed by wet-ashing with H_2SO_4 and HNO_3. The As in the residue was reduced to the trivalent state with hydrazine. After removal of the excess reducing agent, the As(III) required a 23.77-mL titration with 0.02425 M I_2 in a faintly alkaline medium. Express the results of this analysis in terms of percentage of As_2O_3 in the original sample.

14-44 A 4.971-g sample containing the mineral tellurite was dissolved and then treated with 50.00 mL of 0.03114 M $K_2Cr_2O_7$:

$$3TeO_2 + Cr_2O_7^{2-} + 8H^+ \longrightarrow 3H_2TeO_4 + 2Cr^{3+} + H_2O$$

Upon completion of the reaction, the excess $Cr_2O_7^{2-}$ required a 10.05-mL back-titration with 0.1135 M Fe^{2+}. Calculate the percentage of TeO_2 in the sample.

*14-45 A 25.0-mL aliquot of a solution containing Tl(I) ion was treated with K_2CrO_4. The Tl_2CrO_4 was filtered, washed free of excess precipitating agent, and dissolved in dilute H_2SO_4. The $Cr_2O_7^{2-}$ produced was titrated with 40.60 mL of 0.1004 M Fe^{2+} solution. What was the weight of Tl in the sample? The reactions are

$$2Tl^+ + CrO_4^{2-} \longrightarrow Tl_2CrO_4(s)$$

$$2Tl_2CrO_4(s) + 2H^+ \longrightarrow 4Tl^+ + Cr_2O_7^{2-} + H_2O$$

$$Cr_2O_7^{2-} + 6Fe^{2+} + 14H^+ \longrightarrow 6Fe^{3+} + 2Cr^{3+} + 7H_2O$$

14-46 A 0.306-g sample of an impure aluminum salt was dissolved in dilute acid, treated with an excess of $(NH_4)_2C_2O_4$, and slowly made alkaline through the addition of aqueous NH_3. The precipitated $Al_2(C_2O_4)_3$ was filtered, washed, dissolved in dilute acid, and titrated with 36.08 mL of 0.02413 M $KMnO_4$. Calculate the percentage of Al in the sample.

*14-47 The chromium in a 1.876-g sample of impure chromite ($FeO \cdot Cr_2O_3$) was oxidized to the +6 state by fusion with Na_2O_2. The fused mass was treated with water and boiled to destroy the excess peroxide. After acidification, the sample was treated with 50.00 mL of 0.1603 M Fe^{2+}. A back-titration of 2.97 mL of 0.01000 M $K_2Cr_2O_7$ was required to oxidize the excess iron(II). What is the percentage of (a) chromite and (b) chromium in the sample?

14-48 A 0.240-g sample of impure pyrolusite, MnO_2, was reduced to Mn^{2+} with 25.00 mL of 0.05000 M Na_2AsO_3. The excess AsO_3^{3-} was oxidized to AsO_4^{3-} with 3.125 mL of 0.02172 M $KMnO_4$ in the presence of a catalytic quantity of iodate. What is the percentage of pyrolusite in the original sample?

*14-49 In the presence of F^-, Mn^{2+} can be titrated with MnO_4^-, both reactants being converted to a complex of Mn(III). A 0.5451-g sample containing Mn_3O_4 was dissolved, and all the manganese was converted to Mn^{2+}. Titration in the presence of F^- consumed 31.16 mL of 0.02007 M $KMnO_4$.

(a) Write a balanced equation for the reaction, assuming that the complex is MnF_4^-.

(b) What is the percentage of Mn_3O_4 in the sample?

14-50 A 50.0-mL aliquot containing UO_2^{2+} was passed through a Jones reductor to reduce the U to the +3 state. Aeration converted the U^{3+} to U^{4+}, following which the latter was titrated with 29.37 mL of 0.01119 M $K_2Cr_2O_7$. What weight of U was contained in each liter of the sample solution?

*14-51 A 1.065-g sample of stainless steel was dissolved in HCl (this treatment converts the Cr present to Cr^{3+}) and diluted to 500.0 mL in a volumetric flask. One 50.00-mL aliquot was passed through a Walden reductor and then titrated with 13.72 mL of 0.01920 M $KMnO_4$. A 100.0-mL aliquot was passed through a Jones reductor into 50 mL of 0.10 M Fe^{3+}. Titration of the resulting solution required 36.43 mL of the $KMnO_4$ solution. Calculate the percentages of Fe and Cr in the alloy.

14-52 A 2.559-g sample containing both Fe and V was dissolved under conditions that converted the elements to Fe(III) and V(V). The solution was diluted to 500.0 mL, and a 50.00-mL aliquot was passed through a Walden reductor and titrated with 17.74 mL of 0.1000 M Ce^{4+}. A second 50.00-mL aliquot was passed through a Jones reductor and required 44.67 mL of the same Ce^{4+} solution to reach an end point. Calculate the percentage of Fe_2O_3 and V_2O_5 in the sample.

*14-53 A 0.6490-g sample of alkali metal sulfates was dissolved and diluted to 100.0 mL. The Na^+ in a 25.00-mL aliquot was precipitated as $NaZn(UO_2)_3(OAc)_9 \cdot 6H_2O$. After filtration and washing, the precipitate was dissolved in acid and the uranium reduced to U^{3+} in a Jones reductor. After aeration to convert the U^{3+} to U^{4+}, the solution was titrated with 33.33 mL of 0.1086 M Ce^{4+}. Calculate the percentage of Na_2SO_4 in the sample.

14-54 A 10.95-g sample of an insecticide preparation was wet-ashed with a sulfuric/nitric acid mixture to destroy its organic components. The Cu in the sample was then precipitated with an excess of K_2CrO_4 as the basic chromate, $CuCrO_4 \cdot 2CuO \cdot 2H_2O$. The solid was filtered, washed free of excess reagent, and dissolved in acid:

$$2(CuCrO_4 \cdot 2CuO \cdot 2H_2O)(s) + 10H^+ \longrightarrow 6Cu^{2+} + Cr_2O_7^{2-} + 9H_2O$$

Titration of the liberated $Cr_2O_7^{2-}$ required 31.47 mL of 0.1162 M Fe^{2+}. Express the results of this analysis in terms of the percentage of copper(II) oleate, $Cu(C_{18}H_{33}O_2)_2$ (fw = 626.47), in the sample.

*14-55 The Ba^{2+} in a 1.500-g sample was precipitated with 50.00 mL of 0.05000 M KIO_3. The $Ba(IO_3)_2$ was filtered, and the filtrate and washings were treated with an excess of KI. The I_2 liberated required 10.75 mL of 0.03962 M $Na_2S_2O_3$. Express the results of this analysis in terms of percentage of BaO.

14-56 When a slightly basic solution of Mn^{2+} is titrated with standard permanganate solution, the product is MnO_2:

$$2MnO_4^- + 3Mn^{2+} + 4OH^- \longrightarrow 5MnO_2(s) + 2H_2O$$

Calculate the percentage of Mn in a mineral specimen if a 0.4675-g sample required 36.24 mL of 0.03135 M $KMnO_4$.

*14-57 The H_2S and SO_2 concentrations of a gas were determined by passage of the gas through three absorber solutions connected in series. The first contained an ammoniacal solution of Cd^{2+} to trap the sulfide as CdS. The second contained 10.00 mL of 0.02006 M I_2 to oxidize the SO_2 to SO_4^{2-}. The third contained 2.00 mL of 0.03457 M $Na_2S_2O_3$ to retain any I_2 carried over from the second absorber. A 25.0-L gas sample was passed through the apparatus followed by pure N_2 to sweep the last traces of SO_2 from the first absorber to the second. The solution from the first absorber was made acidic, and 20.00 mL of the 0.02006 M I_2 was added, which converted the CdS to Cd^{2+} and S. The excess I_2 was back-titrated with 7.45 mL of the $Na_2S_2O_3$ solution. The solutions in the second and third absorbers were combined, and the residual I_2 was titrated with 2.44 mL of the $Na_2S_2O_3$ solution. Calculate the concentrations of SO_2 and H_2S in milligrams per liter.

14-58 A 25.00-mL sample of a household bleach was diluted to 250.0 mL in a volumetric flask. Calculate the weight/volume percentage of NaClO in the sample if 36.30 mL of 0.09611 M $Na_2S_2O_3$ was needed to titrate the I_2 liberated when excess KI was introduced to 50.00 mL of the diluted sample.

*14-59 A sensitive method for I^- in the presence of Cl^- and Br^- entails oxidation of the I^- to IO_3^- with Br_2. The Br_2 is then removed by boiling or by reduction with formate ion. The IO_3^- is determined by addition of excess I^- and titration of the resulting I_2. A 1.204-g sample of mixed halides was dissolved and analyzed by the foregoing procedure; 20.66 mL of 0.05551 M thiosulfate was required. Calculate the percentage of KI in the sample.

14-60 A solution containing $NaIO_3$ and $NaIO_4$ was analyzed by buffering a 25.00-mL sample with borax and adding an excess of KI. In this slightly alkaline medium, IO_4^- was reduced to IO_3^-, liberating an equivalent quantity of I_2 that consumed 15.52 mL of 0.08029 M $Na_2S_2O_3$. A 10.0-mL portion of the sample was made strongly acidic with HCl after the addition of excess KI; 36.73 mL of the $Na_2S_2O_3$ was needed to titrate the I_2 liberated by both the IO_4^- and the IO_3^-. Calculate the milligrams of $NaIO_3$ and $NaIO_4$ per milliliter of sample.

*14-61 The Winkler method for determining the amount of dissolved O_2 in water is based upon the rapid oxidation of solid $Mn(OH)_2$ to $Mn(OH)_3$ in an alkaline medium. When acidified, the Mn(III) readily releases I_2 from I^-. A 250-mL water sample in a stoppered vessel was treated with 1.00 mL of a concentrated solution of NaI and NaOH and 1.00 mL of a Mn^{2+} solution. Oxidation of the $Mn(OH)_2$ was complete in about 1 min. The precipitates were then dissolved by addition of 2.00 mL of concentrated H_2SO_4, whereupon an amount of I_2 equivalent to the $Mn(OH)_3$ (and hence to the dissolved O_2) was liberated. A 25.0-mL aliquot (of the 254 mL) was titrated with 6.35 mL of 0.00962 M $Na_2S_2O_3$. Calculate the milligrams of O_2 per liter of sample. (Assume that the concentrated reagents are free of O_2 and take their dilutions of the sample into account.)

14-62 The CO concentration in a 3.21-L sample of air was obtained by passage over iodine pentoxide at 150°C:

$$I_2O_5 + 5CO \rightleftharpoons 5CO_2 + I_2$$

The I_2 distilled at this temperature and was collected in a solution of KI. The resulting I_3^- was titrated with 7.76 mL of 0.00221 M $Na_2S_2O_3$. Calculate the parts per million of CO in the gas, assuming an air density of 1.20×10^{-3} g/mL.

*14-63 Titration of a 3.066-g sample of a hair-setting preparation required 24.7 mL of 0.05565 M I_2. Calculate the percentage of thioglycolic acid (fw = 92.12) in the sample. The reaction is

$$2HSCH_2COOH + I_2 \longrightarrow HOOCCH_2SSCH_2COOH + 2HI$$

14-64 A square of photographic film 2.00 cm on an edge was suspended in a 5% solution of $Na_2S_2O_3$ to dissolve the silver halides. After removal and washing of the film, the solution was treated with an excess of Br_2 to oxidize the I^- to IO_3^- and destroy the excess $S_2O_3^{2-}$. The solution was boiled to remove the Br_2, and an excess of KI was added. The liberated I_2 was titrated with 13.7 mL of 0.0352 M $Na_2S_2O_3$ solution.

(a) Write balanced equations for the reactions involved.
(b) Calculate the milligrams of AgI per square centimeter of film.

*14-65 The ethyl mercaptan concentration in a mixture was determined by shaking a 1.657-g sample with 50.0 mL of 0.01194 M I_2 in a tightly stoppered flask:

$$2C_2H_5SH + I_2 \longrightarrow C_2H_5SSC_2H_5 + 2I^- + 2H^+$$

The excess I_2 was back-titrated with 16.77 mL of 0.01325 M $Na_2S_2O_3$. Calculate the percentage of C_2H_5SH (fw = 62.13).

14-66 A 0.6467-g sample containing $BaCl_2 \cdot 2H_2O$ was dissolved, and an excess of a K_2CrO_4 solution was added. After a suitable period, the $BaCrO_4$ was filtered, washed, and dissolved in HCl to convert the CrO_4^{2-} to $Cr_2O_7^{2-}$. An excess of KI was added, and the liberated I_2 was titrated with 48.75 mL of 0.1370 M thiosulfate. Calculate the percentage of $BaCl_2 \cdot 2H_2O$.

CHAPTER 15

Potentiometric Methods

In Chapter 12, it was shown that the relative potential of an electrode is determined by the concentration (strictly, the activity) of one or more of the species in the solution in which the electrode is immersed. This chapter is concerned with the way in which this concentration sensitivity is employed to provide analytical information.[1]

An electrode employed for the determination of analyte concentration is termed an *indicator electrode;* it is used in conjunction with a *reference electrode,* whose potential is independent of the concentration of the analyte or, for that matter, of the concentration of any other ions in the solution under study. A potentiometric analysis thus involves determining the voltage of a cell consisting of the two electrodes and the solution containing the analyte.

15A Reference Electrodes

The ideal reference electrode has a potential that is known (relative to the standard hydrogen electrode), constant, and completely insensitive to the composition of the analyte solution. In addition, a reference electrode should be rugged and easy to assemble and should maintain a constant potential in the presence of small currents.

15A-1 Calomel Electrodes

A calomel electrode can be represented schematically as

$$Hg|Hg_2Cl_2(\text{sat'd}),KCl(xM)||$$

where x represents the molar concentration of potassium chloride in the solution.[2] The electrode potential for a calomel electrode is based upon the half-reaction

$$Hg_2Cl_2(s) + 2e \rightleftharpoons 2Hg(l) + 2Cl^-$$

Table 15-1 lists the composition and electrode potential for the three most common calomel electrodes. Note that these half-cells differ only in their potassium chloride concentrations; all are saturated with mercury(I) chloride.

In the saturated calomel electrode (SCE), the electrolyte solution is also

[1]For further reading on potentiometric methods, see E. P. Serjeant, *Potentiometry and Potentiometric Titrations.* New York: Wiley, 1984.

[2]By convention, a reference electrode is *always* treated *as an anode,* as it is in this diagram for the calomel electrode. This practice is consistent with the IUPAC convention for electrode potentials, which is discussed in Section 12C-2 and in which the reference electrode is the standard hydrogen electrode acting as an anode or the half-cell on the left in a cell diagram.

TABLE 15-1

Electrode Potentials for Reference Electrodes as a Function of Composition and Temperature

Temperature, °C	Potential (vs. SHE), V				
	0.1 M Calomel*	3.5 M Calomel†	Sat'd Calomel*	3.5 M Ag/AgCl†	Sat'd Ag/AgCl†
12	0.3362		0.2528		
15	0.3362	0.254	0.2511	0.212	0.209
20	0.3359	0.252	0.2479	0.208	0.204
25	0.3356	0.250	0.2444	0.205	0.199
30	0.3351	0.248	0.2411	0.201	0.194
35	0.3344	0.246	0.2376	0.197	0.189

*From R. G. Bates, in *Treatise on Analytical Chemistry,* 2nd ed., I. M. Kolthoff and P. J. Elving, Eds., Part I, Vol. 1, p. 793. New York: Wiley, 1978. With permission.

†From D. T. Sawyer and J. L. Roberts Jr., *Experimental Electrochemistry for Chemists,* p. 42. New York: Wiley, 1974. With permission.

calomel electrodes because of the ease with which it can be prepared. Its temperature coefficient is somewhat larger than that for electrodes with lower chloride ion concentrations, which is a disadvantage only in those rare circumstances where substantial temperature changes occur during a measurement. The *electrode* potential of the saturated calomel electrode is 0.2444 V at 25°C.

The saturated calomel electrode shown in Figure 15-1 is typical of electrodes obtainable from commercial sources. It consists of a 5- to 15-cm tube that is 0.5 to 1.0 cm in diameter. A mercury/mercury(I) chloride paste in saturated potassium chloride is contained in an inner tube and connected to the saturated potassium chloride solution in the outer tube through a small opening. Contact with the analyte solution is made through a fritted disk or a porous fiber sealed in the end of the outer tubing.

Figure 15-2 represents a saturated calomel electrode that is simple to construct from materials available in most laboratories. A salt bridge (Section 12B-3) provides electrical contact with the analyte solution. A fritted disk or

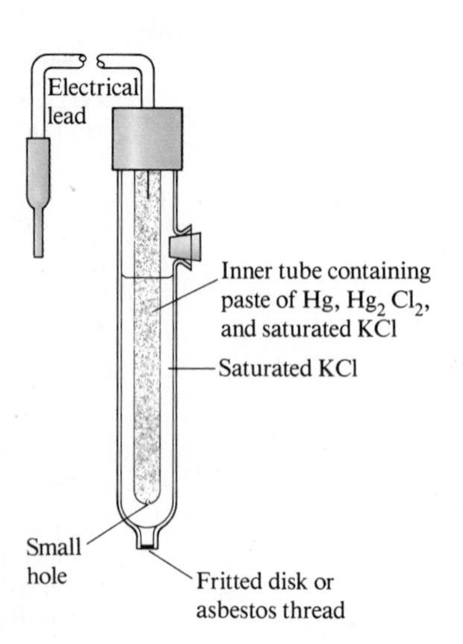

FIGURE 15-1 Diagram of a typical commercial saturated calomel electrode.

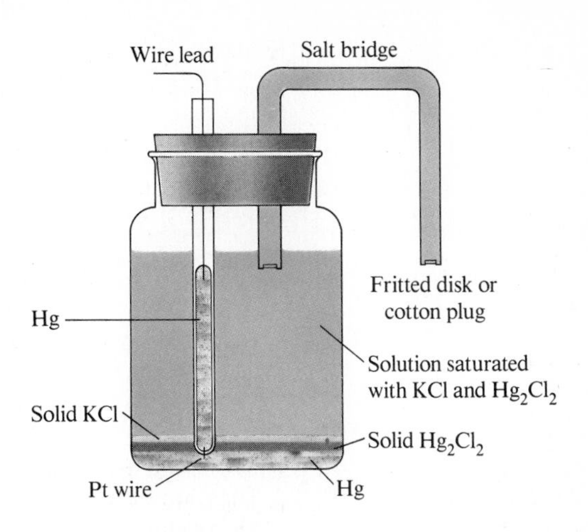

FIGURE 15-2 Diagram of an easily constructed saturated calomel electrode.

a wad of cotton at one end of the bridge prevents siphoning of liquids in either direction. Alternatively, an agar gel made with potassium chloride can be used as the conducting medium in the bridge.[3] This electrode has a lower electrical resistance than the one shown in Figure 15-1.

15A-2 Silver/Silver Chloride Electrodes

A system analogous to the calomel electrode consists of a silver electrode immersed in a solution of potassium chloride that has been saturated with silver chloride:

$$\mathrm{Ag}|\mathrm{AgCl(sat'd)},\mathrm{KCl}(x\mathrm{M})||$$

The half-reaction is

$$\mathrm{AgCl(s)} + \mathrm{e} \rightleftharpoons \mathrm{Ag(s)} + \mathrm{Cl}^-$$

This electrode is ordinarily prepared with a saturated solution of potassium chloride; its electrode potential is 0.199 V at 25°C (Table 15-1).

A simple and easily constructed silver/silver chloride electrode is shown in Figure 15-3. It consists of a tube fitted with a fritted glass disk. A layer of agar gel saturated with potassium chloride (footnote 3) is formed on the disk to prevent loss of solution from the tube. A layer of potassium chloride is introduced, and the tube is then filled with a saturated solution of that salt. A few drops of silver nitrate are added to ensure that the solution is also saturated with silver chloride. Finally, a heavy-gauge (1 to 2 mm in diameter) silver wire is inserted to provide electrical contact.

15B Indicator Electrodes

An ideal indicator electrode responds rapidly and reproducibly to changes in the concentration of a single analyte ion (or group of ions). Although no

[3]A conducting gel can be prepared by heating about 5 g of agar in 100 mL of water containing about 35 g of potassium chloride. The resulting liquid is poured into the bridge and allowed to cool.

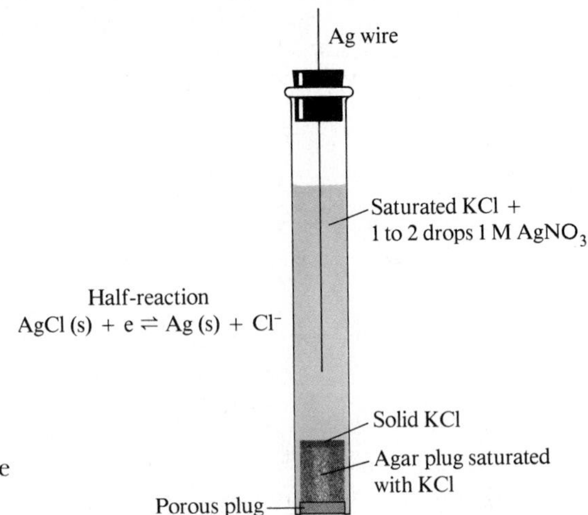

FIGURE 15-3 Diagram of a silver/silver chloride electrode.

electrode has yet been developed that is entirely specific in its response, a few are now available that are remarkably close in their approach to ideal behavior.

Two types of indicator electrodes are encountered: metallic and membrane.

15B-1 Metallic Indicator Electrodes

It is convenient to classify metallic indicator electrodes as *electrodes of the first kind, electrodes of the second kind,* and *redox electrodes.*

Electrodes of the First Kind

An electrode of the first kind is a pure metal electrode that is in direct equilibrium with the cation derived from the metal. A single reaction is involved. For example, the equilibrium between a metal X and its cation X^{n+} is

$$X^{n+} + ne \rightleftharpoons X(s)$$

for which

$$E_{ind} = E^0_{X^{n+}} - \frac{0.0592}{n} \log \frac{1}{[X^{n+}]}$$

where E_{ind} is the electrode potential of the metal electrode and $[X^{n+}]$ is the concentration of the ion (or more exactly, its activity a_X). The electrode potential of the indicator electrode is often expressed in terms of the p-function of the cation. Thus, substituting the definition of pX into the foregoing equation gives

$$E_{ind} = E^0_{X^{n+}} - \frac{0.0592}{n} \text{pX} \qquad (15\text{-}1)$$

$$\text{pX} = n(E^0_{X^{n+}} - E_{ind})/0.0592 \qquad (15\text{-}2)$$

When a metallic indicator electrode is to be used to determine the pX of a solution, it is typically made part of the cell

$$\text{SCE}\|X^{n+}\ (x\text{M})|X$$

Note that, by convention, the saturated calomel electrode is written as the anode and the measured cell potential is

$$E_{cell} = E_{ind} - E_{SCE} \qquad (15\text{-}3)$$

Substituting Equation 15-1 into 15-3 gives, upon rearranging,

$$pX = n(E^0_{X^{n+}} - E_{SCE} - E_{cell})/0.0592 \qquad (15\text{-}4)$$

Equation 15-4 permits calculation of the pX of a solution from potentiometric data.

EXAMPLE 15-1

The cell

$$SCE||Cd^{2+}(xM)|Cd$$

developed a potential of −0.723 V. Calculate the pCd of the solution using −0.403 V as E^0 for Cd^{2+}.

Substituting into Equation 15-4 gives

$$pCd = 2[-0.403 - 0.244 - (-0.723)]/0.0592 = 2.57$$

Equation 15-1 accurately describes the behavior of a number of common metals that are used as indicator electrodes of the first kind. In contrast, certain harder metals—notably iron, chromium, tungsten, cobalt, and nickel—do not provide reproducible potentials. Moreover, plots of the potential of a metal of this latter kind as a function of pX often yield slopes that differ significantly from the theoretical $0.0592/n$. The nonideal behavior of this type of electrode is attributed to strains and deformations in the crystal structure of the metal or to the presence of oxide films on the surface.

Electrodes of the Second Kind

Metals not only serve as indicator electrodes for their own cations but also respond to the concentration of anions that form sparingly soluble precipitates or highly stable complexes with such cations. The potential of a silver electrode, for example, correlates reproducibly with the concentration of chloride ion in a solution saturated with silver chloride. Here, two equilibria are involved:

$$AgCl(s) \rightleftharpoons Ag^+ + Cl^-$$

$$Ag^+ + e \rightleftharpoons Ag(s) \qquad E^0_{Ag^+} = 0.799 \text{ V}$$

Combination of these equations (Example 12-5, page 298) gives

$$AgCl(s) + e \rightleftharpoons Ag(s) + Cl^-$$

The Nernst expression for this process is

$$E_{ind} = E^0_{AgCl} - 0.0592 \log [Cl^-] = E^0_{AgCl} + 0.0592 \text{ pCl} \qquad (15\text{-}5)$$

where (Equation 12-4)

$$E^0_{AgCl} = E^0_{Ag^+} + 0.0592 \log K_{sp} = +0.222 \text{ V}$$

Equation 15-5 shows that the potential of a silver electrode is proportional to pCl, the negative logarithm of the chloride ion concentration. Thus, in a solution saturated with silver chloride, a silver electrode can serve as an indicator electrode of the second kind for chloride ion. Note that the sign of the log term for an electrode of this type is opposite that for an electrode of the first kind. Thus, when Equation 15-5 is substituted into Equation 15-3, we obtain, upon rearrangement,

$$pCl = (E^0_{cell} + E_{SCE} - E^0_{AgCl})/0.0592$$

Mercury serves as an indicator electrode of the second kind for the EDTA anion Y^{4-}. For example, when a small amount of HgY^{2-} is added to a solution containing Y^{4-}, the half-reaction at a mercury cathode is

$$HgY^{2-} + 2e \rightleftharpoons Hg(l) + Y^{4-} \qquad E^0 = 0.21\ V$$

for which

$$E_{ind} = 0.21 - \frac{0.0592}{2} \log \frac{[Y^{4-}]}{[HgY^{2-}]}$$

The formation constant for HgY^{2-} is very large (6.3×10^{21}), and so the concentration of the complex remains essentially constant over a large range of Y^{4-} concentrations. The Nernst equation for the process can therefore be written as

$$E = K - \frac{0.0592}{2} \log [Y^{4-}] = K + \frac{0.0592}{2} pY \qquad (15\text{-}6)$$

where

$$K = 0.21 - \frac{0.0592}{2} \log \frac{1}{[HgY^{2-}]}$$

The mercury electrode is thus a valuable electrode of the second kind for EDTA titrations.

Indicator Electrodes for Oxidation/Reduction Systems

As noted in Chapter 12, an inert metal—such as platinum, gold, palladium, or carbon—responds to the potential of redox systems with which it is in contact. For example, the potential of a platinum electrode immersed in a solution containing cerium(III) and cerium(IV) is

$$E_{ind} = E^0_{Ce(IV)} - 0.0592 \log \frac{[Ce^{3+}]}{[Ce^{4+}]}$$

A platinum electrode is thus a convenient indicator electrode for titrations involving standard cerium(IV) solutions.

15B-2

Membrane Electrodes[4]

For many years, the most convenient method for determining pH has involved measurement of the potential developed across a thin glass membrane that separates two solutions with different hydrogen ion concentrations. The phenomenon upon which the measurement is based was first reported in 1906 and by now has been extensively studied by many investigators. As a result, the sensitivity and selectivity of glass membranes toward hydrogen ions are reasonably well understood. Furthermore, this understanding has led to the development of other types of membranes that respond selectively to more than two dozen other ions.

Membrane electrodes are sometimes called *p-ion electrodes* because the data obtained from them are usually presented as p-functions, such as pH, pCa, or pNO_3. The membranes used in constructing these electrodes are

[4]Some suggested sources for additional information on this topic are *Ion-Selective Electrodes in Analytical Chemistry,* H. Freiser, Ed. New York: Plenum Press, 1978; J. Vesely, D. Weiss, and K. Stulik, *Analysis with Ion-Selective Electrodes.* New York: Wiley, 1979; *Ion-Selective Methodology,* A. K. Covington, Ed. Boca Raton, FL: CRC Press, 1979.

classified as crystalline or noncrystalline. The latter can be further subdivided into glass, liquid, and immobilized liquid. In this section, we consider all these types of p-ion membranes. In addition, a gas-sensing probe based on a noncrystalline membrane is described. It, however, is sensitive to molecules rather than ions.

It is important to note at the outset of this discussion that membrane electrodes are *fundamentally different* from metal electrodes both in design and in principle. We shall use the glass electrode for pH measurements to illustrate these differences.

15B-3

The Glass Electrode for pH Measurements

Figure 15-4 shows a typical cell for measuring pH. The cell consists of a glass indicator electrode and a saturated calomel reference electrode immersed in the solution whose pH is to be determined. The indicator electrode consists of a thin, pH-sensitive glass membrane sealed onto one end of a heavy-walled glass or plastic tube. A small volume of dilute hydrochloric acid saturated with silver chloride is contained in the tube (the inner solution in some electrodes is a buffer containing chloride ion). A silver wire in this solution forms a silver/silver chloride reference electrode, which is connected to one of the terminals of a potential-measuring device. A calomel electrode is connected to the other terminal.

The schematic representation of this cell in Figure 15-5 reveals that it contains *two* reference electrodes: (1) the *external* calomel electrode and (2) the *internal* silver/silver chloride electrode, which, while a part of the glass electrode, is not the pH-sensing element. Instead, *it is the thin glass membrane at the tip of the electrode that responds to pH.*

The Composition of Glass Membranes

Much systematic investigation has been devoted to the effects of glass composition on the sensitivity of membranes to protons and other cations, and a number of formulations are now used for the manufacture of electrodes. Early

FIGURE 15-4 Typical electrode system for measuring pH.

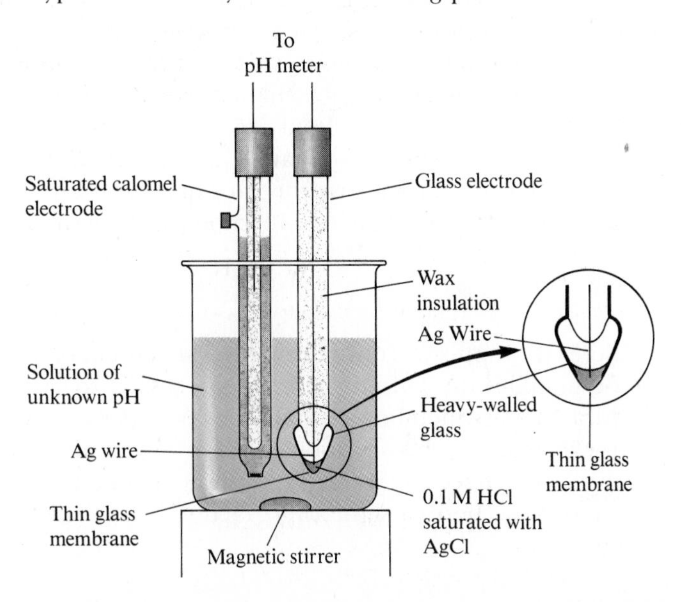

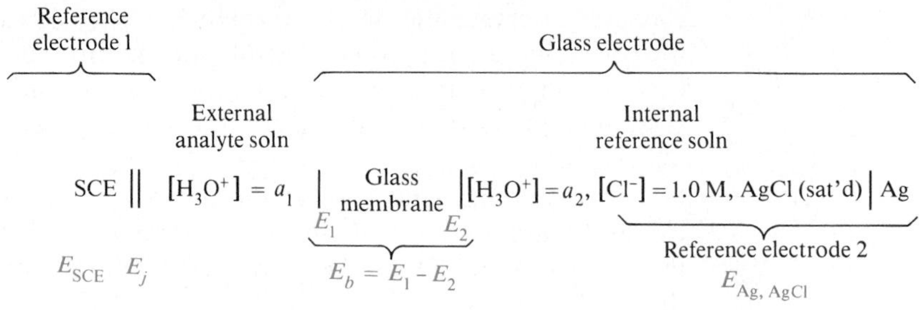

Reference electrode 1 — Glass electrode

External analyte soln — Internal reference soln

SCE || $[H_3O^+] = a_1$ | Glass membrane | $[H_3O^+] = a_2$, $[Cl^-] = 1.0$ M, AgCl (sat'd) | Ag

E_{SCE} E_j — E_1, E_2: $E_b = E_1 - E_2$ — Reference electrode 2: $E_{Ag, AgCl}$

FIGURE 15-5 Diagram of a glass/calomel cell for measurement of pH.

membranes were fabricated from a glass consisting of approximately 22% Na_2O, 6% CaO, and 72% SiO_2. These membranes showed excellent specificity toward hydrogen ions up to a pH of about 9. At higher pH values, however, they became somewhat responsive to sodium, as well as to other singly charged cations. Other glass formulations are now used, in which sodium and calcium ions are replaced to various degree by barium and lithium ions. These membranes have superior selectivity and lifetime.

The Structure of Membrane Glasses

A silicate glass used for membranes consists of an infinite three-dimensional network of SiO_4^{4-} groups in which each silicon is bonded to four oxygens and each oxygen is shared by two silicons. Within the interstices of this structure are sufficient cations to balance the negative charge of the silicate groups. Singly charged cations, such as sodium and lithium, are mobile in the lattice and are responsible for electrical conduction within the membrane.

The Hygroscopicity of Glass Membranes

The surface of a glass membrane must be hydrated before it will function as a pH electrode. The amount of water involved is approximately 50 mg per cubic centimeter of glass. Nonhygroscopic glasses show no pH function. Even hygroscopic glasses lose their pH sensitivity after dehydration by storage over a desiccant. The effect is reversible, however, and the response of a glass electrode is restored by soaking it in water.

It has also been demonstrated that the hydration of a pH-sensitive glass membrane involves an ion-exchange reaction between singly charged cations in the glass lattice and protons from the solution. The process involves univalent cations exclusively because di- and trivalent cations are too strongly held within the silicate structure to exchange with ions in the solution. Typically, then, the ion-exchange reaction can be written as

$$\underset{\text{soln}}{H^+} + \underset{\text{glass}}{Na^+Gl^-} \rightleftharpoons \underset{\text{soln}}{Na^+} + \underset{\text{glass}}{H^+Gl^-} \qquad (15\text{-}7)$$

where Gl^- represents one of many negatively charged sites in the glass surface. The equilibrium constant for this process is so large that the surface of a hydrated glass membrane ordinarily consists entirely of silicic acid (H^+Gl^-). An exception to this situation exists in highly alkaline media, where the hydrogen ion concentration is vanishingly small and the sodium ion concentration is large; here, a significant fraction of the sites are occupied by sodium ions.

Electrical Conduction in Membranes

In order to serve as an indicator for cations, a glass membrane must be capable of conducting electricity. Conduction within the hydrated glass membrane involves the movement of sodium and hydrogen ions, the former in the dry interior of the membrane and the latter in the gel layer. Conduction across the solution/gel interfaces occurs by the reactions

$$\underset{\text{soln}_1}{H^+} + \underset{\text{glass}_1}{Gl^-} \rightleftharpoons \underset{\text{glass}_1}{H^+Gl^-} \tag{15-8}$$

$$\underset{\text{glass}_2}{H^+Gl^-} \rightleftharpoons \underset{\text{soln}_2}{H^+} + \underset{\text{glass}_2}{Gl^-} \tag{15-9}$$

where subscript 1 refers to the interface between the glass and the analyte solution and subscript 2 refers to the interface between the internal solution and the glass. The positions of these two equilibria are determined by the hydrogen ion concentrations in the solutions on the two sides of the membrane. Where these positions differ, the surface at which the greater dissociation has occurred is negative with respect to the other surface. A boundary potential thus develops across the membrane, the magnitude of which depends upon the *difference* in the hydrogen ion concentrations of the two solutions. It is this potential difference that serves as the analytical parameter in a potentiometric pH measurement.

The membrane of a typical glass electrode (with a thickness of 0.03 to 0.1 mm) has an electrical resistance of 50 to 500 MΩ.

Membrane Potentials

The lower part of Figure 15-5 shows four potentials that develop in a cell when pH is being determined with a glass electrode. Two of these, $E_{Ag,AgCl}$ and E_{SCE}, are reference electrode potentials. A third potential exists across the salt bridge that separates the calomel electrode from the analyte solution. This junction and its associated *junction potential,* E_j, are found in all cells used for the potentiometric measurement of ion concentration. The source and effect of the junction potential are discussed later in this chapter. The fourth, and most important, potential shown in Figure 15-5 is the *boundary potential,* E_b, *which varies with the pH of the analyte solution.* The two reference electrodes simply provide a means of measuring the magnitude of the boundary potential.

Figure 15-5 reveals that the potential of a glass electrode has two components: the fixed potential of a silver/silver chloride electrode $E_{Ag,AgCl}$ and the pH-dependent boundary potential E_b. Not shown is a third potential, called the *asymmetry potential,* which is found in most membrane electrodes and which changes slowly with time. The source and characteristics of the boundary potential and the asymmetry potential are discussed in the two sections that follow.

The Boundary Potential

As shown in Figure 15-5, the boundary potential consists of two potentials, E_1 and E_2, each of which is associated with one of the two gel/solution interfaces. The boundary potential is simply the difference between these potentials:

$$E_b = E_1 - E_2 \tag{15-10}$$

The potential E_1 is determined by the ratio of the hydrogen ion activity in the analyte solution to the hydrogen ion activity in the gel surface and can be considered a measure of the driving force for the reaction shown in Equation 15-8. Similarly, E_2 is related to the ratio of the hydrogen ion activities in the internal solution and the corresponding gel surface and is related to the driving force for the reaction shown in Equation 15-9.

The relationship between the two hydrogen ion activities and the potentials E_1 and E_2 is given by the equations

$$E_1 = j_1 + 0.0592 \log \frac{a_1}{a_1'} \qquad (15\text{-}11)$$

$$E_2 = j_2 + 0.0592 \log \frac{a_2}{a_2'} \qquad (15\text{-}12)$$

where a_1 and a_2 are the activities of the hydrogen ion *in the two solutions* and a_1' and a_2' are the corresponding activities *in the surface layers of the two gels.* If the surfaces of the two gels have the same number of sites available to accommodate protons, then the constants j_1 and j_2 are identical. In addition, if all the sodium ions on both surfaces have been replaced by protons, the two activities a_1' and a_2' are also the same. Assuming these equalities and subtracting Equation 15-12 from 15-11 gives

$$E_b = E_1 - E_2 = 0.0592 \log \frac{a_1}{a_2} \qquad (15\text{-}13)$$

Thus, *provided the two gel surfaces are identical,* the boundary potential E_b depends only upon the hydrogen ion activities of the solutions on either side of the membrane. For a glass pH electrode, the hydrogen ion activity of the internal solution a_2 is held constant so that Equation 15-13 simplifies to

$$E_b = L' + 0.0592 \log a_1 = L' - 0.0592 \text{ pH} \qquad (15\text{-}14)$$

where

$$L' = -0.0592 \log a_2$$

Thus, the potential of the electrode becomes a measure of the hydrogen ion activity of the external solution.

The Asymmetry Potential

When identical solutions and reference electrodes are placed on the two sides of a glass membrane, a potential of zero is expected. In fact, however, a small *asymmetry potential* that changes gradually with time is frequently encountered.

The sources of the asymmetry potential are obscure but undoubtedly include such causes as differences in strain on the two surfaces of the membrane imparted during manufacture, mechanical abrasion on the outer surface during use, and chemical etching of the outer surface. In order to eliminate the determinate errors caused by the asymmetry potential, all membrane electrodes must be calibrated against one or more standard analyte solutions. Such calibrations should be carried out at least daily and more often when the electrode receives heavy use.

The Potential of the Glass Electrode

As noted earlier, the potential of a glass indicator electrode E_{ind} has three components: (1) the boundary potential, given by Equation 15-14, (2) the potential of the internal Ag/AgCl reference electrode, and (3) a small asymmetry potential. In equation form,

$$E_{ind} = E_b + E_{Ag/AgCl} + E_{asy}$$

Substitution of Equation 15-14 for E_b gives

$$E_{ind} = L' + 0.0592 \log a_1 + E_{Ag/AgCl} + E_{asy}$$

or

$$E_{ind} = L + 0.0592 \log a_1 = L - 0.0592 \text{ pH} \quad (15\text{-}15)$$

where L is a combination of the three constant terms. Note the similarity between Equations 15-15 and 15-1.

The Alkaline Error

Glass electrodes respond to the concentration of both hydrogen ion and alkali metal ions in basic solution, where the former is necessarily much smaller than the latter. The magnitude of this *alkaline error* for four different glass membranes is shown in Figure 15-6 (curves C to F). These curves refer to solutions in which the sodium ion concentration was held constant at 1 M while the pH was varied. Note that the error is negative (that is, the measured pH values are lower than the true values), which suggests that the electrode is responding to sodium ions as well as to protons. This observation is confirmed by data obtained for solutions containing different sodium ion concentrations. Thus at pH 12, the electrode with a Corning 015 membrane (curve C in Figure 15-6) registered a pH of 11.3 when immersed in a solution having a sodium ion concentration of 1 M but 11.7 in a solution that was 0.1 M in this ion. All singly charged cations induce an alkaline error whose magnitude depends upon both the cation in question and the composition of the glass membrane.

The alkaline error can be satisfactorily explained by assuming an exchange equilibrium between the hydrogen ions on the glass surface and the cations in solution. This process is simply the reverse of that shown in Equation 15-7:

$$\underset{\text{glass}}{H^+Gl^-} + \underset{\text{soln}}{B^+} \rightleftharpoons \underset{\text{glass}}{B^+Gl^-} + \underset{\text{soln}}{H^+}$$

where B^+ represents some singly charged cation, such as sodium ion.

The equilibrium constant for this reaction is

$$K_{ex} = \frac{a_1 b_1'}{a_1' b_1} \quad (15\text{-}16)$$

where a_1 and b_1 represent the activities of H^+ and B^+ in solution and a_1' and b_1' are the activities of these ions on the gel surface. Equation 15-16 can be rearranged to give the ratio of the activities B^+ to H^+ on the glass surface:

$$\frac{b_1'}{a_1'} = \frac{b_1}{a_1} K_{ex}$$

For the glasses used for pH electrodes, K_{ex} is so small that the activity ratio b_1'/a_1' is ordinarily minuscule. The situation differs in strongly alkaline media, however. For example, b_1'/a_1' for an electrode immersed in a pH 11 solution that is 1 M in sodium ions (Figure 15-6) is $10^{11}K_{ex}$. In this case, the activity

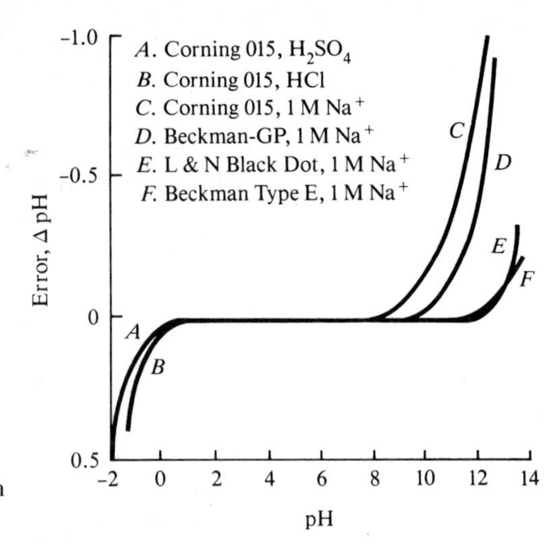

FIGURE 15-6 Acid and alkaline error for selected glass electrodes at 25°C. (From R. G. Bates, *Determination of pH*, 2nd ed., p. 365. New York: Wiley, 1973. With permission.)

of the sodium ions relative to that of the hydrogen ions becomes so large that the electrode responds to both species.

Selectivity Coefficients

The effect of an alkali metal ion on the potential across a membrane can be accounted for by inserting an additional term in Equation 15-14 to give

$$E_b = L + 0.0592 \log (a_1 + k_{H,B}b_1) \qquad (15\text{-}17)$$

where $k_{H,B}$ is the *selectivity coefficient* for the electrode. Equation 15-17 applies not only to glass indicator electrodes for hydrogen ion but also to all other types of membrane electrodes. Selectivity coefficients range from zero (no interference) to values greater than unity. Thus, if an electrode for ion A responds 20 times more strongly to ion B than to ion A, $k_{A,B}$ has a value of 20. If the response of the electrode to ion C is 0.001 of its response to A (a much more desirable situation), $k_{A,C}$ is 0.001.[5]

The product $k_{H,B}b_1$ for a glass pH electrode is ordinarily small relative to a_1 provided the pH is less than 9; under these conditions, Equation 15-17 simplifies to Equation 15-14. At high pH values and at high concentrations of a singly charged ion, however, the second term in Equation 15-17 assumes a more important role in determining E, and an alkaline error is encountered. For electrodes specifically designed for work in highly alkaline media (curve E in Figure 15-6), the magnitude of $k_{H,B}b_1$ is appreciably smaller than for ordinary glass electrodes.

The Acid Error

As shown in Figure 15-6, the typical glass electrode exhibits an error, opposite in sign to the alkaline error, in solutions of pH less than about 0.5; pH readings tend to be too high in this region. The magnitude of the error depends upon a variety of factors and is generally not very reproducible. The causes of the acid error are not well understood.

[5]The numerical value of k for a given electrode is affected by the total electrolyte concentration of the solution and by the concentration ratio of the analyte to the interfering species. Consequently, as a criterion for the response of an electrode to the presence of an interfering ion, the selectivity coefficient is reliable to about one order of magnitude only.

15B-4

Glass Electrodes for Cations Other Than Protons

The existence of the alkaline error in early glass electrodes led to investigations concerning the effect of glass composition upon the magnitude of this error. One consequence has been the development of glasses for which the alkaline error is negligible below about pH 12. Other studies have been directed toward discovering glass compositions that permit the determination of cations other than hydrogen. This application requires that the hydrogen ion activity a_1 in Equation 15-17 be negligible relative to $k_{H,B}b_1$; under such circumstances, the potential is independent of pH and is a function of pB instead. Several investigators have demonstrated that incorporation of Al_2O_3 or B_2O_3 in the glass has the desired effect. Glass electrodes that permit the direct potentiometric measurement of such singly charged species as Na^+, K^+, NH_4^+, Rb^+, Cs^+, Li^+, and Ag^+ have been developed. Some of these glasses are reasonably selective toward particular singly charged cations. Glass electrodes for Na^+, Li^+, NH_4^+, and total concentration of univalent cations are now available from commercial sources.

15B-5

Liquid-Membrane Electrodes

Liquid-membrane electrodes owe their response to the potential that develops across the interface between the solution containing the analyte and a liquid ion exchanger that selectively bonds with the analyte ion. These electrodes have been developed for the direct potentiometric measurement of numerous polyvalent cations as well as certain anions.

Figure 15-7 is a schematic diagram of a liquid-membrane electrode for calcium. It consists of a conducting membrane that selectively bonds calcium ions, an internal solution containing a fixed concentration of calcium chloride, and a silver electrode that is coated with silver chloride to form an internal reference electrode. The similarity to the glass electrode is obvious (Figure 15-8). The active membrane ingredient is an ion exchanger that consists of a calcium dialkyl phosphate of limited solubility in water. In the electrode shown in Figure 15-7, the ion exchanger is dissolved in an immiscible organic liquid that is forced by gravity into the pores of a hydrophobic porous disk. This disk then serves as the membrane that separates the internal solution from

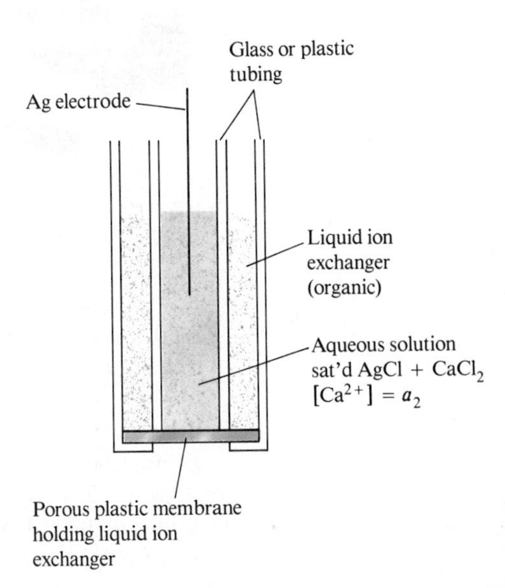

FIGURE 15-7 Diagram of a liquid-membrane electrode for Ca^{2+}.

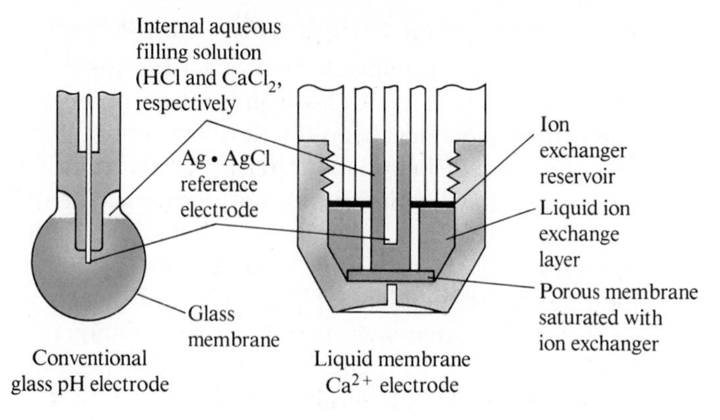

FIGURE 15-8 Comparison of a liquid-membrane calcium ion electrode with a glass electrode. (Courtesy of Orion Research, Boston, MA.)

the analyte solution. In a more recent design, the ion exchanger is immobilized in a tough polyvinyl chloride gel cemented to the end of a tube that holds the internal solution and reference electrode. In either design, there is set up at each membrane interface a dissociation equilibrium that is analogous to Equation 15-9:

$$\underset{\text{organic}}{[(RO)_2POO]_2Ca} \rightleftharpoons \underset{\text{organic}}{2(RO)_2POO^-} + \underset{\text{aqueous}}{Ca^{2+}}$$

where R is a high-molecular-weight aliphatic group. As with the glass electrode, a potential develops across the membrane when the extent of dissociation of the ion exchanger dissociation at one surface differs from the extent at the other surface as a consequence of differences in the calcium ion activity of the internal and external solutions. The relationship between this potential and the calcium ion activities is given by an equation that is similar to Equation 15-13:

$$E_b = E_1 - E_2 = \frac{0.0592}{2} \log \frac{a_1}{a_2} \tag{15-18}$$

where a_1 and a_2 are the activities of calcium ion in the external and internal solutions, respectively. Since the calcium ion activity of the internal solution is constant,

$$E_b = N + \frac{0.0592}{2} \log a_1 = N - \frac{0.0592}{2} \text{pCa} \tag{15-19}$$

where N is a constant (compare Equations 15-19 and 15-14). Note that, because calcium is divalent, a 2 appears in the denominator of the coefficient of the log term.

Figure 15-8 compares the structural features of a glass-membrane electrode and a commercially available liquid-membrane electrode for calcium ion. The sensitivity of the latter for calcium ion is reported to be 50 times greater than for magnesium ion and 1000 times greater than for sodium or potassium ions. Calcium ion activities as low as 5×10^{-7} M can be measured. Performance of the electrode is said to be independent of pH in the range between 5.5 and 11. At lower pH levels, hydrogen ions undoubtedly replace some of the calcium ions on the exchanger; the electrode then becomes sensitive to pH as well as to pCa.

The calcium ion liquid-membrane electrode is a valuable tool for physiological investigations because this ion plays important roles in such processes as nerve conduction, bone formation, muscle contraction, cardiac expansion and contraction, renal tubular function, and perhaps hypertension. At least some of these processes are more influenced by the activity than the concentration of the calcium ion; activity, of course, is the parameter measured by the membrane electrode.

A liquid-membrane electrode specific for potassium ion is also of great value for physiologists because the transport of neural signals appears to involve movement of this ion across nerve membranes. Investigation of this process requires an electrode that can detect small concentrations of potassium ion in media that contain much larger concentrations of sodium ion. Several liquid-membrane electrodes show promise in meeting this requirement. One is based upon the antibiotic valinomycin, a cyclic ether that has a strong affinity for potassium ion. Of equal importance is the observation that a liquid membrane consisting of valinomycin in diphenyl ether is about 10^4 times as responsive to potassium ion as to sodium ion.[6]

Table 15-2 lists liquid-membrane electrodes available from commercial sources. The anion-sensitive electrodes shown make use of a solution containing an anion-exchange resin in an organic solvent. Liquid-membrane electrodes in which the exchange liquid is held in a polyvinyl chloride gel have been developed for Ca^{2+}, K^+, NO_3^-, and BF_4^-. These have the appearance of crystalline electrodes, which are considered in the following section.

15B-6 Crystalline-Membrane Electrodes

Considerable work has been devoted to the development of solid membranes that are selective toward anions in the way that some glasses respond to cations. We have seen that the existence of anionic sites on a glass surface accounts for the selectivity of a membrane toward certain cations. By analogy, a membrane with cationic sites might be expected to respond selectively toward anions. Attempts have been made to prepare membranes containing a sparingly soluble salt of the anion of interest because such salts will bind that anion selectively. It has been difficult, however, to find methods of fabricating such membranes so that they possess adequate physical strength, conductivity, and resistance to abrasion and corrosion.

TABLE 15-2 **Characteristics of Liquid-Membrane Electrodes***

Analyte Ion	Concentration Range, M	Major Interferences
Ca^{2+}	10^0 to 5×10^{-7}	Pb^{2+}, Fe^{2+}, Ni^{2+}, Hg^{2+}, Sr^{2+}
Cl^-	10^0 to 5×10^{-6}	I^-, OH^-, SO_4^{2-}
NO_3^-	10^0 to 7×10^{-6}	ClO_4^-, I^-, ClO_3^-, CN^-, Br^-
ClO_4^-	10^0 to 7×10^{-6}	I^-, ClO_3^-, CN^-, Br^-
K^+	10^0 to 1×10^{-6}	Cs^+, NH_4^+, Tl^+
Water hardness ($Ca^{2+} + Mg^{2+}$)	10^0 to 6×10^{-6}	Cu^{2+}, Zn^{2+}, Ni^{2+}, Sr^{2+}, Fe^{2+}, Ba^{2+}

*From *Orion Guide to Ion Analysis*. Boston, MA: Orion Research, 1983. With permission.

[6] M. S. Frant and J. W. Ross Jr., *Science*, **1970**, *167*, 987.

Membranes prepared from cast pellets of silver halides have been successfully used in electrodes for the selective determination of chloride, bromide, and iodide ions. In addition, an electrode based upon a polycrystalline Ag_2S membrane is offered by one manufacturer for the determination of sulfide ion. In both types of membranes, silver ions are sufficiently mobile to conduct electricity through the solid medium. Mixtures of PbS, CdS, and CuS with Ag_2S provide membranes that are selective for Pb^{2+}, Cd^{2+}, and Cu^{2+}, respectively. Silver ion again conducts electricity in these solid membranes. The potential that develops across crystalline solid-state electrodes is described by a relationship similar to Equation 15-19.

A crystalline electrode for fluoride ion is available from commercial sources. The membrane consists of a slice of a single crystal of lanthanum fluoride that has been doped with europium(II) fluoride to improve its conductivity. The membrane, supported between a reference solution and the solution to be measured, shows a theoretical response to changes in fluoride ion activity in the range from 10^0 to 10^{-6} M. The electrode is reported to be selective for fluoride ion over other common anions by several orders of magnitude; only hydroxide ion appears to offer serious interference. Solid-state electrodes available from commercial sources are listed in Table 15-3.

15B-7 Gas-Sensing Probes

Figure 15-9 illustrates the essential features of a gas-sensing probe, which consists of a tube containing a reference electrode, a specific ion electrode, and an electrolyte solution. A thin, replaceable, gas-permeable membrane attached to one end of the tube serves as a barrier between the internal and analyte solutions. As can be seen from Figure 15-9, this device is a complete electrochemical cell and is more properly referred to as a probe rather than an electrode.

Membrane Composition

A *microporous membrane* is fabricated from a hydrophobic polymer. As the name implies, the membrane is highly porous (the average pore size is less than 1 μm) and allows the free passage of gases; at the same time, the water-repellent polymer prevents water and solute ions from entering the pores. The thickness of the membrane is about 0.1 mm.

TABLE 15-3

Analyte Ion	Concentration Range, M	Major Interferences
Br^-	10^0 to 5×10^{-6}	CN^-, I^-, S^{2-}
Cd^{2+}	10^{-1} to 1×10^{-7}	Fe^{2+}, Pb^{2+}, Hg^{2+}, Ag^+, Cu^{2+}
Cl^-	10^0 to 5×10^{-5}	CN^-, I^-, Br^-, S^{2-}
Cu^{2+}	10^{-1} to 1×10^{-8}	Hg^{2+}, Ag^+, Cd^{2+}
CN^-	10^{-2} to 1×10^{-6}	S^{2-}
F^-	Sat'd to 1×10^{-6}	OH^-
I^-	10^0 to 5×10^{-8}	
Pb^{2+}	10^{-1} to 1×10^{-6}	Hg^{2+}, Ag^+, Cu^{2+}
Ag^+/S^{2-}	Ag^+: 10^0 to 1×10^{-7} S^{2-}: 10^0 to 1×10^{-7}	Hg^{2+}
SCN^-	10^0 to 5×10^{-6}	I^-, Br^-, CN^-, S^{2-}

*From *Orion Guide to Ion Analysis*. Boston, MA: Orion Research, 1983. With permission.

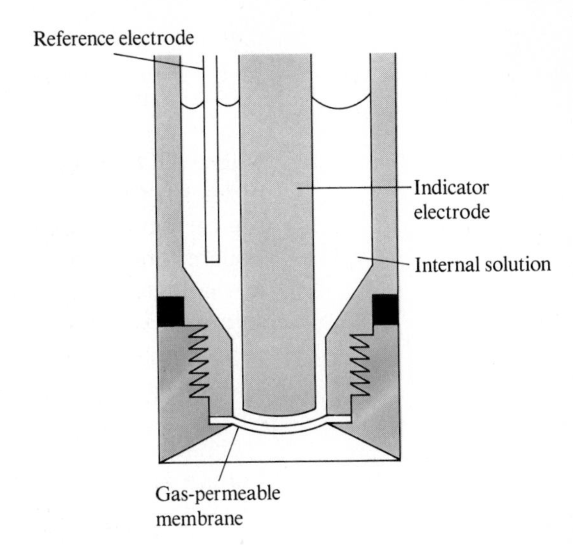

FIGURE 15-9 Diagram of a gas-sensing probe.

Nonporous *homogeneous membranes* are also used to fabricate gas-sensing probes. With these membranes, the gas from the external solution dissolves in the membrane and is subsequently extracted by the internal solution. Homogeneous membranes are usually fabricated from silicone rubber and range in thickness from 0.01 to 0.03 mm.

The Mechanism of Response

Using carbon dioxide as an example, we can represent the transfer of gas to the pores of the membrane by the equation

$$\underset{\text{analyte solution}}{CO_2(aq)} \rightleftharpoons \underset{\text{membrane pores}}{CO_2(g)}$$

Because the membrane contains a large number of pores, this equilibrium is rapidly established. The CO_2 in the pores is in contact with the internal solution as well, and a second equilibrium,

$$\underset{\text{membrane pores}}{CO_2(g)} \rightleftharpoons \underset{\text{internal solution}}{CO_2(aq)}$$

is also rapidly established. These two processes result in equilibrium between the analyte solution and the film of internal liquid in contact with the membrane. Here, however, yet another equilibrium,

$$\underset{\text{internal solution}}{CO_2(aq)} + 2H_2O \rightleftharpoons \underset{\text{internal solution}}{HCO_3^-} + H_3O^+$$

causes the pH of the internal surface film to change. This change is then detected by the internal glass/calomel electrode system. A description of the overall process is obtained by adding the equations for the three individual equilibria to give

$$\underset{\text{analyte solution}}{CO_2(aq)} + 2H_2O \rightleftharpoons H_3O^+ + \underset{\text{internal solution}}{HCO_3^-}$$

The net equilibrium constant is

$$K = \frac{[H_3O^+][HCO_3^-]}{[CO_2(aq)]_{ext}}$$

where $[CO_2(aq)]_{ext}$ is the concentration of the gas in the analyte solution. In order for the measured cell potential to vary linearly with the logarithm of the carbon dioxide concentration of the external solution, the hydrogen carbonate concentration of the internal solution must be sufficiently large so that it is not altered significantly by the carbon dioxide entering from the external solution. Assuming then that $[HCO_3^-]$ is constant, we can rearrange the previous equation to

$$\frac{[H_3O^+]}{[CO_2(aq)]_{ext}} = \frac{K}{[HCO_3^-]} = K_g$$

Letting a_1 be the hydrogen ion activity of the internal solution, we can write

$$a_1 = [H_3O^+] = K_g[CO_2(aq)]_{ext} \tag{15-20}$$

Substitution of Equation 15-20 for a_1 in Equation 15-15 yields

$$E_{ind} = L + 0.0592 \log K_g[CO_2(aq)]_{ext}$$

or

$$E_{ind} = L' + 0.0592 \log [CO_2(aq)]_{ext} \tag{15-21}$$

where

$$L' = L + 0.0592 \log K_g$$

Thus the potential between the glass electrode and the reference electrode in the internal solution is determined by the CO_2 concentration in the external solution. *Note that no electrode comes in direct contact with the analyte solution.* Thus, as noted earlier, these devices are gas-sensing *cells,* or *probes,* rather than gas-sensing electrodes. Nevertheless, they continue to be called electrodes in some literature and many advertising brochures.

It is also noteworthy that the only species that interfere are other dissolved gases that permeate the membrane and then affect the pH of the internal solution.

The possibility exists for increasing the selectivity of the gas-sensing probe through the use of an indicator electrode that is responsive to some species other than hydrogen ion. For example, a nitrate-sensing electrode can be used to provide a cell that is sensitive exclusively to oxides of nitrogen; the response of this electrode is determined by such equilibria as

$$\underset{\text{external solution}}{2NO_2(aq)} + H_2O \rightleftharpoons NO_2^- + \underset{\text{internal solution}}{NO_3^-} + 2H^+$$

Gas-sensing probes for CO_2, NO_2, H_2S, SO_2, HF, HCN, and NH_3 are now available from commercial sources.

15C Instruments for the Measurement of Cell Potentials

Potentiometric measurements must be made in the absence of significant currents in the cell being studied. One reason for this requirement is that a current causes changes in reactant concentrations in the cell and thus changes in its potential. Even more important are the effects of IR drop and polarization phenomena (Section 16A) upon the measured potential.

As shown by the example that follows, the influence of IR drop is particularly significant with specific ion electrodes, which may have resistances of 100 MΩ or more. With such electrodes, currents must be limited to 10^{-12} A or smaller; this limitation requires that the potential-measuring device have an internal resistance of 10^{12} Ω or more.

EXAMPLE 15-2

The true potential of a glass/calomel electrode system is 0.800 V; its internal resistance is 120 MΩ. What is the relative error in the measured potential if the measuring device has a resistance of 600 MΩ?

Here, the circuit can be considered as consisting of a potential source E_S and two resistors in series, R_S being that of the source and R_M that of the measuring device:

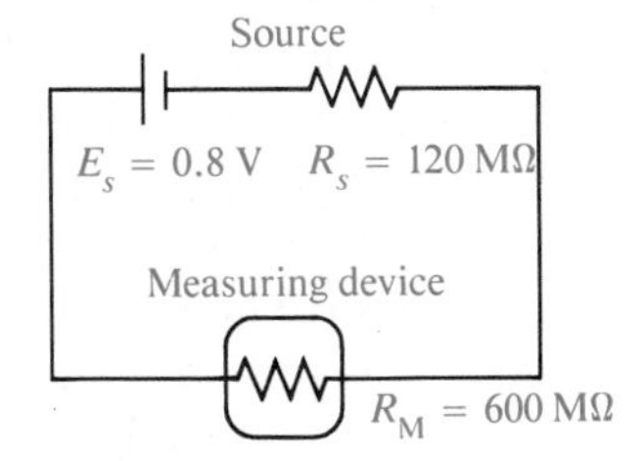

From Ohm's law, we write

$$E_S = IR_S + IR_M$$

where I is the current in the circuit, which is given by

$$I = \frac{0.800 \text{ V}}{(120 + 600) \times 10^6 \Omega} = 1.11 \times 10^{-9} \text{ A}$$

The potential drop across the measuring device (which is the potential it indicates) is IR_M. Thus,

$$\text{Indicated potential} = 1.11 \times 10^{-9}\text{A} \times 600 \times 10^6 \Omega = 0.667 \text{ V}$$

$$\text{Relative error} = \frac{0.667 - 0.800}{0.800} \times 100\% = -17\%$$

It is important to appreciate that an error in potential such as that shown in Example 15-2 (−0.133 V) would have an enormous effect on the accuracy of a concentration measurement based upon that potential. Thus, as shown in Section 15D-3, a 0.001 V uncertainty in potential leads to a relative error of about 4% when hydrogen ion concentration is determined with a glass electrode. An error such as that in Example 15-2 would result in a concentration error of two orders of magnitude or more.

Historically, potential measurements were performed with a potentiometer, a null-point instrument in which the unknown potential is just balanced by a standard reference potential. At null, no charge is drawn from the cell whose potential is being measured (see Appendix 10 for a description of a potentiometer). Potentiometers have now been almost entirely supplanted by electronic voltmeters, which have sufficiently high internal resistance to avoid the type of error illustrated in Example 15-2. Instruments of this type are commonly called *pH meters* but could more properly be referred to as *pIon meters* or *ion meters* since they are frequently used for the measurement of concentrations of other ions as well.

Numerous high-resistance, direct-reading pH meters are now on the market. The readout is either digital or analog (in the latter, a needle sweeps a range on a scale from 0 to 14 pH units). Many of the analog meters are equipped with the capability of scale expansion, which provide full-scale ranges from 0.5 to 2 pH units; precision on the order of 0.001 to 0.005 pH unit can thus be realized. It should be appreciated, however, that it is seldom, if ever, possible to measure a pH with a comparable degree of *accuracy*. Indeed, inaccuracies of ± 0.02 to ± 0.03 pH unit are typical.

15D Direct Potentiometric Measurements

Direct potentiometric measurements are used to establish the concentration of species for which an indicator electrode is available. The technique is simple, requiring only a comparison of the potential developed by the indicator electrode in the analyte solution with its potential when immersed in one or more solutions of known analyte concentration. Insofar as the response of the electrode is specific for the analyte and independent of matrix effects (that is, to the presence of other species in the sample), no preliminary separation steps are required. Direct potentiometric measurements are also readily adapted to the continuous and automatic recording of analytical data.

15D-1 The Liquid-Junction Potential

Notwithstanding these advantages, the user of direct potentiometric measurements must be alert to limitations inherent to the method. Principal among these is the existence of a *liquid-junction potential* that affects most potentiometric measurements. The junction potential is inconsequential in most electroanalytical methods and can be neglected; its existence, however, places a limitation upon the accuracy that can be attained from a direct potentiometric measurement.

A liquid-junction potential is caused by an unequal distribution of cations and anions across the boundary between two dissimilar electrolyte solutions. This inhomogeneity is the result of differences in the rates at which the various charged species diffuse across the boundary between the solutions. Consider, for example, the situation at the interface between 1 M and 0.01 M solutions of hydrochloric acid. This interface can be symbolized as

$$\text{HCl(1 M)}|\text{HCl(0.01 M)}$$

Both hydrogen ions and chloride ions tend to diffuse across this boundary from the more concentrated to the more dilute solution, the driving force being proportional to the concentration difference. The two species move at different rates, however, under the influence of this force. In the present example, hydrogen ions are substantially more mobile than chloride ions. Thus, hydrogen ions outstrip chloride ions during diffusion, and a separation of charge results (Figure 15-10). The more dilute side of the boundary becomes positively charged, owing to the more rapid diffusion of hydrogen ions, and the concentrated side therefore acquires a negative charge from the excess of slower moving chloride ions. The charge developed tends to counteract the differences in diffusion rates of the two ions so that a condition of equilibrium

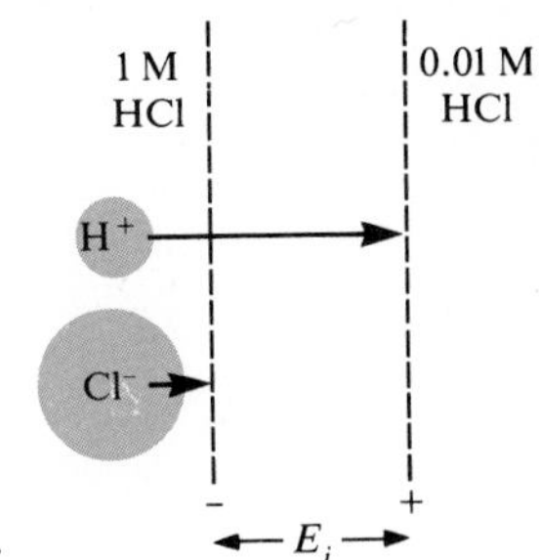

FIGURE 15-10 Schematic representation of a liquid junction showing the source of the junction potential E_j. The length of the arrows corresponds to the relative mobility of the two ions.

is soon attained. The potential difference resulting from this charge separation may amount to several hundredths of a volt.

The magnitude of the potential for a junction as simple as the one under consideration can be calculated from knowledge of the mobilities of the ions involved. However, seldom (if ever) does a cell of analytical importance possess a sufficiently simple composition to permit such a computation.

It is an experimental fact that the magnitude of the liquid-junction potential can be greatly decreased by interposition of a concentrated electrolyte (a *salt bridge*) between the two solutions. The effectiveness of this device improves as the mobilities of the ions in the bridge approach one another and as the concentrations increase. A saturated solution of potassium chloride is good from both standpoints, its concentration being somewhat greater than 4 M at room temperature and the diffusion rate of its ions differing from each other by only 4%. A junction potential with such a bridge typically amounts to a few millivolts and is negligible for most electroanalytical procedures; this potential appreciably affects direct potentiometric measurements, however.

15D-2 The Sign Convention and Equations for Direct Potentiometry

In the past, there was considerable controversy regarding the sign convention for potentiometry. Recently, however, the situation has been clarified and the convention has been made consistent with the convention described in Chapter 12 for standard electrode potential.[7] In the newly endorsed convention, the indicator electrode is *always* treated as the *cathode* and the reference electrode as the *anode*.[8] For direct potentiometric measurements, the potential of a cell can then be expressed in terms of the potentials developed by the indicator electrode, the reference electrode, and a junction potential:

$$E_{cell} = E_{ind} - E_{ref} + E_j \tag{15-22}$$

The junction potential E_j has two components: the first at the interface between the analyte solution and one end of the salt bridge and the second

[7]According to Bates, the convention being described here has been endorsed by standardizing groups in the United States and Great Britain as well as IUPAC. See R. G. Bates, in *Treatise on Analytical Chemistry,* 2nd ed., I. M. Kolthoff and P. J. Elving, Eds., Part I, Vol. 1, pp. 831–832. New York: Wiley, 1978.

[8]In effect, the sign convention for electrode potentials described in Section 12C-2 also designates the indicator electrode as the cathode by stipulating that half-reactions always be written as reductions; the standard hydrogen electrode, which is the reference electrode in this case, then becomes the anode.

between the reference electrode solution and the other end of the bridge. We have noted that these two potentials tend to cancel one another but seldom do so completely.

The potential of the indicator electrode is ideally related to the activity a_1 of the analyte by a version of the Nernst equation (Equation 15-1, 15-5, 15-15, and 15-19). Thus, for the cation X^{n+} at 25°C,

$$E_{ind} = L - \frac{0.0592}{n}\text{pX} = L + \frac{0.0592}{n}\log a_1 \tag{15-23}$$

where L is a constant. For metallic indicator electrodes, L is ordinarily the standard electrode potential; for membrane electrodes, L is the summation of several constants, including the time-dependent asymmetry potential (page 366) of uncertain magnitude.

Combination of Equation 15-22 with Equation 15-23 and rearrangement yield

$$\text{pX} = -\log a_1 = -\frac{E_{cell} - (E_j - E_{ref} + L)}{0.0592/n}$$

The constant terms in parentheses can be combined to give a new constant K.

$$\text{pX} = -\log a_1 = -\frac{E_{cell} - K}{0.0592/n} \tag{15-24}$$

Note that K includes at least one term (E_j) with a magnitude that cannot be evaluated from theory. Thus, before Equation 15-24 can be used for the determination of pX, K must be evaluated experimentally with a standard solution of the analyte.

For an anion A^{n-}, the signs of Equation 15-24 are reversed:

$$\text{pA} = \frac{E_{cell} - K}{0.0592/n} \tag{15-25}$$

All direct potentiometric methods are based upon Equation 15-24 or 15-25. It is noteworthy that the difference in sign in the two equations has a subtle but important consequence in the way that ion-selective electrodes are connected to pH meters and pIon meters. When the two equations are solved for E_{cell} we find that for cations

$$E_{cell} = K - (n/0.0592)\text{pX} \tag{15-26}$$

and for anions

$$E_{cell} = K + (n/0.0592)\text{pA} \tag{15-27}$$

Equation 15-26 shows that for a cation-selective electrode, an increase in pX results in a *decrease* in E_{cell}. Thus, when a high-resistance voltmeter is connected to the cell in the usual way, with the indicator electrode attached to the positive terminal, the meter reading decreases as pX increases. In order to alleviate this problem, instrument manufacturers generally reverse the leads so that cation-sensitive electrodes are connected to the *negative* terminal of the voltage measuring device; meter readings then increase with increases of pX. Anion-selective electrodes, on the other hand, are connected to the *positive* terminal of the meter so that increases in pA also yield larger readings.

15D-3

The Electrode-Calibration Method

In the electrode-calibration method, K in Equation 15-24 is determined by measuring E_{cell} for one or more standard solutions of known pX. The assumption is then made that K is unchanged when the standard is replaced by the analyte solution. The calibration is ordinarily performed at the time pX for the unknown is determined; recalibration may be required if measurements extend over several hours.

The electrode-calibration method offers the advantages of simplicity, speed, and applicability to the continuous monitoring of pX. Two important sources of difficulty attend its use, however. One is that the results of an analysis are in terms of activities rather than concentrations (to be sure, this factor may be an advantage rather than a disadvantage in some situations).[9] The other difficulty is the uncertainty introduced by the need to assume that the junction potential remains unchanged when the analyte solution replaces the standard; unfortunately, this uncertainty can never be totally eliminated.

Activity Versus Concentration

Electrode response is related to analyte activity rather than analyte concentration. The scientist is ordinarily interested in concentration, however, and the determination of this quantity from a potentiometric measurement requires activity coefficient data. More often than not, activity coefficients are not available because the ionic strength of the solution is either unknown or else so large that the Debye-Hückel equation is not applicable.

The difference between activity and concentration is illustrated by Figure 15-11, in which the response of a calcium ion electrode is plotted against a logarithmic function of calcium chloride *concentration*. The nonlinearity is due to the increase in ionic strength—and the consequent decrease in the activity of calcium ion—with increasing electrolyte concentration. The upper curve is obtained when these concentrations are converted to activities; note that this straight line possesses the theoretical slope of 0.0296 (0.0592/2).

Activity coefficients for singly charged species are less affected by changes in ionic strength than are the coefficients for ions with multiple charges. Thus,

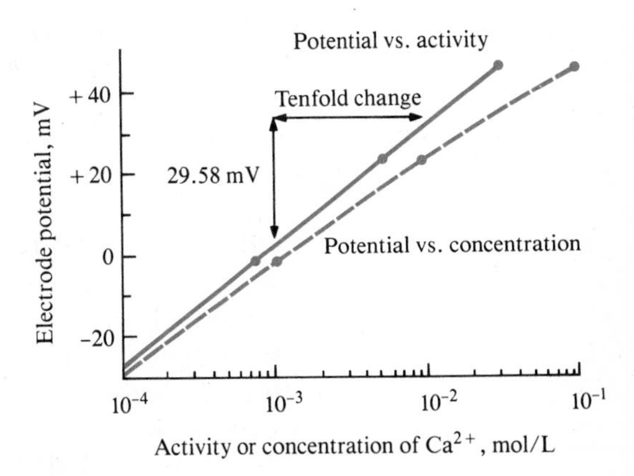

FIGURE 15-11 Response of a liquid-membrane electrode to variations in the concentration and activity of calcium ion. (Courtesy of Orion Research, Boston, MA.)

[9] Many chemical reactions of physiological importance depend upon the activity of metal ions rather than their concentration. For studies of such reactions, a potentiometric measurement with a reversible indicator electrode is an ideal analytical tool.

the effect shown in Figure 15-11 is less pronounced for electrodes that respond to H^+, Na^+, and other univalent ions.

In potentiometric pH measurements, the pH of the standard buffer employed for calibration is generally based on the activity of hydrogen ions. Thus, the results are also on an activity scale. If the unknown sample has a high ionic strength, the hydrogen ion *concentration* will differ appreciably from the activity measured.

Inherent Error in the Electrode-Calibration Procedure

A serious disadvantage of the electrode-calibration method is the inherent error that results from the assumption that K in Equations 15-24 and 15-25 remains constant after calibration. This assumption can seldom, if ever, be exactly true because the electrolyte composition of the unknown almost inevitably differs from that of the solution employed for calibration. The junction potential term contained in K varies slightly as a consequence, even when a salt bridge is used. This error is frequently on the order of 1 mV or more; unfortunately, because of the nature of the potential/activity relationship, such an uncertainty has an amplified effect on the inherent accuracy of the analysis. The magnitude of the error in analyte concentration can be estimated by differentiating Equation 15-24 while holding E_{cell} constant:

$$-\log_{10} e \frac{da_1}{a_1} = -0.434 \frac{da_1}{a_1} = -\frac{dK}{0.0592/n}$$

$$\frac{da_1}{a_1} = \frac{ndK}{0.0257}$$

Upon replacing da_1 and dK with finite increments and multiplying both sides of the equation by 100%, we obtain

$$\frac{\Delta a_1}{a_1} \times 100\% = 3.89 \times 10^3\, n\Delta K\% = \%\text{ relative error}$$

The quantity $\Delta a_1/a_1$ is the relative error in a_1 associated with an absolute uncertainty ΔK in K. If, for example, ΔK is ± 0.001 V, a relative error in activity of about $\pm 4n\%$ can be expected. *It is important to appreciate that this error is characteristic of all measurements involving cells that contain a salt bridge and that this error cannot be eliminated by even the most careful measurements of cell potentials or the most sensitive and precise measuring devices.* Furthermore it does not appear possible to devise a method for completely eliminating the error in K that is the source of this error.

15D-4 **Calibration Curves for Direct Potentiometry**

An obvious way to convert potentiometric measurements from activity to concentration is to make use of an empirical calibration curve, such as the lower plot in Figure 15-11. For this approach to be successful, it is necessary to make the ionic composition of the standards essentially the same as that of the analyte solution. Matching the ionic strength of standards to that of samples is often difficult, particularly for samples that are chemically complex.

Where electrolyte concentrations are not too great, it is often useful to swamp both samples and standards with a measured excess of an inert electrolyte. The added effect of the electrolyte from the sample matrix becomes negligible under these circumstances, and the empirical calibration curve yields results in terms of concentration. This approach has been used, for example, in the potentiometric determination of fluoride ion in drinking water. Both samples and standards are diluted with a solution that contains sodium chloride, an acetate buffer, and a citrate buffer; the diluent is sufficiently concentrated so that the samples and standards have essentially identical ionic strengths. This method provides a rapid means for measuring fluoride concentrations in the part-per-million range with an accuracy of about 5% relative.

15D-5 The Standard-Addition Method

The standard-addition method involves determining the potential of the electrode system before and after a measured volume of a standard has been added to a known volume of the analyte solution. Often an excess of an electrolyte is incorporated into the analyte solution at the outset to prevent any major shift in ionic strength that might accompany the addition of standard. It is also necessary to assume that the junction potential remains constant during the two measurements.

EXAMPLE 15-3

A cell consisting of a saturated calomel electrode and a lead ion electrode developed a potential of −0.4706 V when immersed in 50.00 mL of a sample. A 5.00-mL addition of standard 0.02000 M lead solution caused the potential to shift to −0.4490 V. Calculate the molar concentration of lead in the sample.

We shall assume that the activity of Pb^{2+} is approximately equal to $[Pb^{2+}]$ and apply Equation 15-24. Thus,

$$\text{pPb} = -\log [\text{Pb}^{2+}] = -\frac{E'_{\text{cell}} - K}{0.0592/2}$$

where E'_{cell} is the initial measured potential (−0.4706 V).

After the standard solution is added, the potential becomes E''_{cell} (−0.4490 V), and

$$-\log \frac{50.00 \times [\text{Pb}^{2+}] + 5.00 \times 0.0200}{50.00 + 5.00} = -\frac{E''_{\text{cell}} - K}{0.0592/2}$$

$$-\log (0.9091\,[\text{Pb}^{2+}] + 1.818 \times 10^{-3}) = -\frac{E''_{\text{cell}} - K}{0.0592/2}$$

Subtracting this equation from the first leads to

$$-\log \frac{[\text{Pb}^{2+}]}{0.9091[\text{Pb}^{2+}] + 1.818 \times 10^{-3}} = \frac{2(E''_{\text{cell}} - E'_{\text{cell}})}{0.0592}$$

$$= \frac{2[-0.4490 - (-0.4706)]}{0.0592} = 0.7297$$

$$\frac{[\text{Pb}^{2+}]}{0.9091[\text{Pb}^{2+}] + 1.818 \times 10^{-3}} = 0.1863$$

$$[\text{Pb}^{2+}] = 4.08 \times 10^{-4}$$

15D-6 Potentiometric pH Measurements with a Glass Electrode[10]

The glass electrode is unquestionably the most important indicator electrode for hydrogen ion. It is convenient to use and subject to few of the interferences that affect other pH-sensing electrodes.

[10]For a detailed discussion of potentiometric pH measurements, see R. G. Bates, *Determination of pH*, 2nd ed. New York: Wiley, 1973.

The glass/calomel electrode system is a remarkably versatile tool for the measurement of pH under many conditions. It can be used without interference in solutions containing strong oxidants, strong reductants, proteins, and gases; the pH of viscous or even semisolid fluids can be determined. Electrodes for special applications are available. Included among these are small electrodes for pH measurements in one drop (or less) of solution, in a tooth cavity, or in the sweat on the skin; microelectrodes that permit the measurement of pH inside a living cell; rugged electrodes for insertion in a flowing liquid stream to provide a continuous monitoring of pH; and small electrodes that can be swallowed to indicate the acidity of the stomach contents (the calomel electrode is kept in the mouth).

The Operational Definition of pH

The utility of pH as a measure of the acidity and alkalinity of aqueous media, the wide availability of commercial glass electrodes, and the relatively recent proliferation of inexpensive solid-state pH meters have made the potentiometric measurement of pH perhaps the most common analytical technique in all of science. It is therefore imperative that pH be defined in a manner that is easily duplicated at various times and in various laboratories throughout the world. To meet this requirement, it is necessary to define pH in operational terms—that is, by the way the measurement is made. Only then will the pH measured by one worker be the same as that by another.

An operational definition of pH endorsed by the National Bureau of Standards (NBS), similar organizations in other countries, and the IUPAC is based upon the direct calibration of the meter with carefully prescribed standard buffers followed by potentiometric determination of the pH of unknown solutions.

Consider, for example, the glass/calomel system in Figures 15-4 and 15-5. When these electrodes are immersed in a standard buffer, Equation 15-24 applies and we can write

$$\text{pH}_S = -\frac{E_S - K}{0.0592}$$

where E_S is the cell potential when the electrodes are immersed in the buffer. Similarly, if the cell potential is E_U when the electrodes are immersed in a solution of unknown pH, we have

$$\text{pH}_U = -\frac{E_U - K}{0.0592}$$

By subtracting the first equation from the second and solving for pH_U, we find

$$\text{pH}_U = \text{pH}_S - \frac{(E_U - E_S)}{0.0592} \qquad (15\text{-}28)$$

Equation 15-28 has been adopted throughout the world as the *operational definition of pH.*

Workers at the NBS and elsewhere have used cells without liquid junctions, similar to the one shown in Figure 12-2, to study primary-standard buffers extensively. Some of the properties of these buffers are presented in Table 15-4 and are discussed in detail elsewhere.[11] Note that the NBS buffers

[11]R. G. Bates, *Determination of pH*, 2nd ed., Chapter 4. New York: Wiley, 1973.

are described by their *molal* concentrations (mmol solute/g solvent) for accuracy and precision of preparation. For general use, the buffers can be prepared from relatively inexpensive laboratory reagents; for careful work, however, certified buffers can be purchased from the NBS. As Table 15-4 shows, the buffers form an internally consistent series over the pH range from 3.5 to 10.3, and their pH values are presented as a function of temperature. The buffer capacity listed for each solution indicates its pH stability with the addition of acid or base.

It should be emphasized that the strength of the operational definition of pH is that it provides a coherent scale for the determination of acidity or alkalinity. However, measured pH values cannot be expected to yield a detailed picture of solution composition that is entirely consistent with solution theory. This uncertainty stems from our fundamental inability to measure single ion activities. That is, the operational definition of pH does not yield the exact pH as defined by the equation

$$pH = -\log [H^+]f_{H^+}$$

In addition to this fundamental limitation of pH measurements, certain practical limitations apply as well.

A Summary of Errors That Affect pH Measurements with the Glass Electrode

The ubiquity of the pH meter and the general applicability of the glass electrode tend to lull the chemist into the attitude that any measurement obtained with such equipment is surely correct. The reader must be alert to the fact

TABLE 15-4 **Values of NBS Primary-Standard pH Solutions from 0 to 60°C***

Temperature, °C	Sat'd (25°C) KH tartrate	0.05 *m* KH_2 citrate†	0.05 *m* KHphthalate	0.025 *m* KH_2PO_4/ 0.025 *m* Na_2HPO_4	0.008695 *m* KH_2PO_4/ 0.03043 *m* Na_2HPO_4	0.01 *m* $Na_2B_4O_7$	0.025 *m* $NaHCO_3$/ 0.025 *m* Na_2CO_3
0	—	3.863	4.003	6.984	7.534	9.464	10.317
5	—	3.840	3.999	6.951	7.500	9.395	10.245
10	—	3.820	3.998	6.923	7.472	9.332	10.179
15	—	3.802	3.999	6.900	7.448	9.276	10.118
20	—	3.788	4.002	6.881	7.429	9.225	10.062
25	3.557	3.776	4.008	6.865	7.413	9.180	10.012
30	3.552	3.766	4.015	6.853	7.400	9.139	9.966
35	3.549	3.759	4.024	6.844	7.389	9.102	9.925
40	3.547	3.753	4.035	6.838	7.380	9.068	9.889
45	3.547	3.750	4.047	6.834	7.373	9.038	9.856
50	3.549	3.749	4.060	6.833	7.367	9.011	9.828
55	3.554	—	4.075	6.834	—	8.985	—
60	3.560	—	4.091	6.836	—	8.962	—
Buffer capacity‡ (mol/pH unit)	0.027	0.034	0.016	0.029	0.016	0.020	0.029
$\Delta pH_{1/2}$ for 1:1 dilution§	0.049	0.024	0.052	0.080	0.07	0.01	0.079

*Adapted from R. G. Bates, *Determination of pH*, 2nd ed., p. 73. New York: Wiley, 1973.

†*m* = molality (mol solute/kg H_2O).

‡See page 196.

§Change in pH that occurs when one volume of buffer is diluted with one volume of H_2O.

that there are distinct limitations to the electrode, some of which were discussed in earlier sections:

1. *The alkaline error.* The ordinary glass electrode becomes somewhat sensitive to alkali metal ions and gives low readings at pH values greater than 9. It should be noted that a greater tolerance (to pH 11 to 12) is achieved with membranes in which lithium and barium have replaced sodium and calcium to some extent.
2. *The acid error.* Values registered by the glass electrode tend to be somewhat high when the pH is less than about 0.5.
3. *Dehydration.* Dehydration may cause erratic electrode performance.
4. *Error in unbuffered neutral solutions.* Because equilibrium between the bulk of the solution and the layer of solution at the surface of a membrane is achieved only slowly in poorly buffered, approximately neutral solutions, time must be allowed for equilibrium between the two to be established. Before being used to determine the pH of such solutions, the glass electrode should be thoroughly rinsed with water. Then both electrodes should be immersed in successive portions of the unknown until a constant pH reading is obtained. Good stirring is also helpful, and several minutes should be allowed for the attainment of steady readings.
5. *Variation in junction potential.* A fundamental source of uncertainty for which a correction cannot be applied is the junction-potential variation resulting from differences in the composition of the standard and the unknown solution. Absolute pH values more reliable than 0.01 unit are generally unobtainable; even reliability to 0.03 unit requires considerable care. To be sure, it is often possible to detect pH *differences* between similar solutions or pH *changes* in a single solution as small as 0.001 unit. Many pH meters are designed to permit readings in increments smaller than 0.01 unit for this reason.
6. *Error in the pH of the standard buffer.* Any inaccuracies in the preparation of the buffer used for calibration or any changes in its composition during storage cause an error in subsequent pH measurements. The action of bacteria on organic buffer components is a common cause for deterioration.

15E Potentiometric Titrations

A *potentiometric titration* involves measurement of the potential of a suitable indicator electrode as a function of titrant volume. The information provided by a potentiometric titration is not the same as that obtained from a direct potentiometric measurement. For example, the direct measurement of 0.100 M solutions of hydrochloric and acetic acids would yield two substantially different hydrogen ion concentrations because the latter is only partially dissociated. In contrast, the potentiometric titration of equal volumes of the two acids would require the same amount of standard base because both solutes have the same number of titratable protons.

Potentiometric titrations provide data that are inherently more reliable than data from titrations that use chemical indicators and are particularly useful with colored or turbid solutions and for detecting the presence of unsuspected species. They suffer from the disadvantage of being more time-consuming than those involving indicators; on the other hand, they are readily automated.

Figure 15-12 illustrates a typical apparatus for performing a manual potentiometric titration. Its use involves measuring and recording the cell potential (in units of millivolts or pH, as appropriate) after each addition of reagent. The titrant is added in large increments at the outset and in smaller and smaller increments as the end point is approached (as indicated by larger changes in response per unit volume).

Sufficient time must be allowed for the attainment of equilibrium after each addition of reagent. Precipitation reactions may require several minutes for equilibration, particularly in the vicinity of the equivalence point. A close approach to equilibrium is indicated by the disappearance of drift in the potential. Effective stirring is often helpful in hastening the achievement of equilibrium.

The first two columns of Table 15-5 show typical potentiometric titration data obtained with the apparatus illustrated in Figure 15-12. The data in the vicinity of the end point are plotted in Figure 15-13a; note that this experimental plot closely resembles the titration curves derived from theoretical considerations.

15E-1 End-Point Detection

Several methods can be used to determine the end point of a potentiometric titration. The most straightforward involves a direct plot of potential as a function of reagent volume, as in Figure 15-13a; the midpoint in the steeply rising portion of the curve is estimated visually and taken as the end point. Various graphical methods have been proposed to aid in the establishment of the midpoint, but it is doubtful that these procedures significantly improve its determination.

FIGURE 15-12 Apparatus for a potentiometric titration.

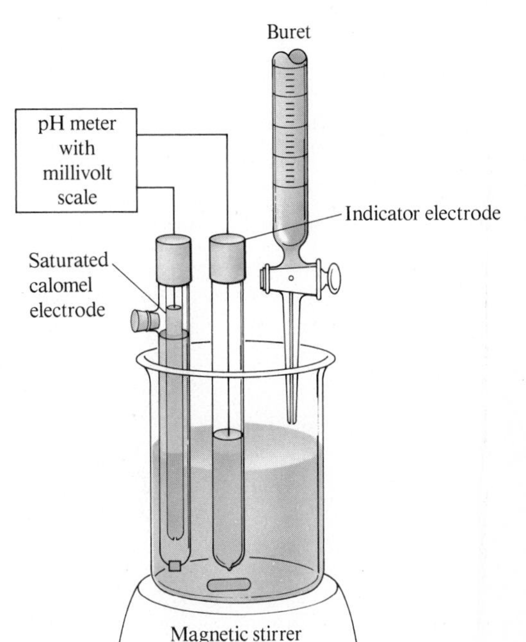

A second approach to end-point detection is to calculate the change in potential per unit volume of titrant (that is, $\Delta E/\Delta V$), as in column 3 of Table 15-5. A plot of these data as a function of the average volume V produces a curve with a maximum that corresponds to the point of inflection (Figure 15-13b). Alternatively, this ratio can be evaluated during the titration and recorded in lieu of the potential. Inspection of column 3 of Table 15-5 reveals that the maximum is located between 24.30 and 24.40 mL; selection of 24.35 mL would be adequate for most purposes.

Column 4 of Table 15-5 and Figure 15-13c show that the second derivative for the data changes sign at the point of inflection. This change is used as the analytical signal in some automatic titrators.

All the foregoing methods of end-point evaluation are predicated on the assumption that the titration curve is symmetric about the equivalence point and that the inflection in the curve corresponds to this point. This assumption is perfectly valid, provided the participants in the titration react with one another in an equimolar ratio and also provided the electrode reaction is perfectly reversible. The former condition is lacking in many oxida-

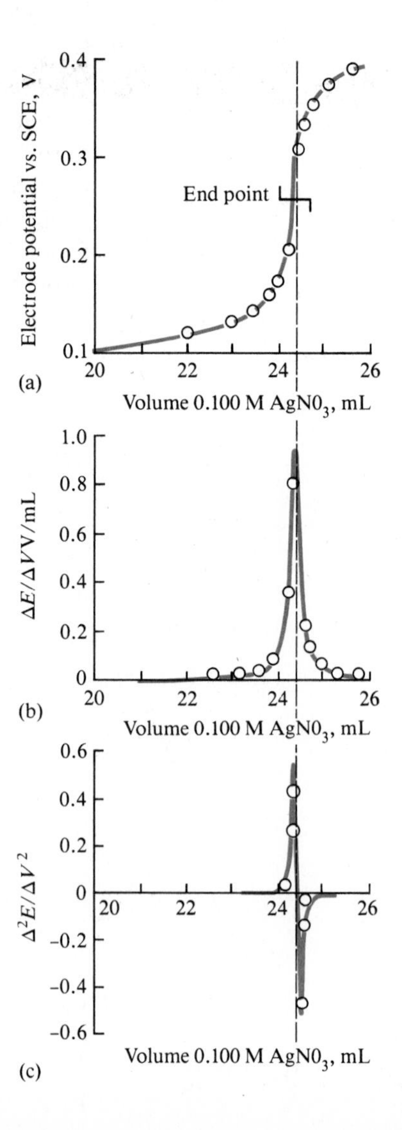

FIGURE 15-13 Titration of 2.433 mmol of chloride ion with 0.1000 M silver nitrate. (a) Titration curve. (b) First-derivative curve. (c) Second-derivative curve.

TABLE 15-5

Potentiometric Titration Data for 2.433 mmol of Chloride with 0.1000 M Silver Nitrate

Volume $AgNO_3$, mL	E vs. SCE, V	$\Delta E/\Delta V$, V/mL	$\Delta^2 E/\Delta V^2$, V^2/mL^2
5.0	0.062		
		0.002	
15.0	0.085		
		0.004	
20.0	0.107		
		0.008	
22.0	0.123		
		0.015	
23.0	0.138		
		0.016	
23.50	0.146		
		0.050	
23.80	0.161		
		0.065	
24.00	0.174		
		0.09	
24.10	0.183		
		0.11	
24.20	0.194		2.8
		0.39	
24.30	0.233		4.4
		0.83	
24.40	0.316		−5.9
		0.24	
24.50	0.340		−1.3
		0.11	
24.60	0.351		−0.4
		0.07	
24.70	0.358		
		0.050	
25.00	0.373		
		0.024	
25.5	0.385		
		0.022	
26.0	0.396		
		0.015	
28.0	0.426		

tion/reduction titrations; the titration of iron(II) with permanganate (Figure 13-1b) is an example. The curve for such titrations is ordinarily so steep, however, that failure to account for asymmetry results in a vanishingly small titration error.

15E-2 *Potentiometric Precipitation Titrations*

Electrode Systems

The indicator electrode for a precipitation titration is often the metal from which the reacting cation is derived. A membrane electrode responsive to this cation or to the anion in the titration can also be used.

Silver nitrate is without question the most versatile reagent for precipitation titrations. A silver wire serves as the indicator electrode. For reagent and analyte concentrations of 0.1 M or greater, a calomel reference electrode can be located directly in the titration vessel without serious error from the slight leakage of chloride ions from the salt bridge. This leakage can be a source of significant error in titrations that involve very dilute solutions or require high precision, however. The difficulty is eliminated by immersing the calomel electrode in a potassium nitrate solution that is connected to the analyte solution by a salt bridge containing potassium nitrate. Reference electrodes with bridges of this type can be purchased from laboratory supply houses.

Titration Curves

A theoretical curve for a potentiometric titration is readily derived. For example, the potential of a silver electrode in the argentometric titration of chloride can be described by

$$E_{Ag} = E^0_{AgCl} - 0.0592 \log [Cl^-] = 0.222 - 0.0592 \log [Cl^-]$$

where E^0_{AgCl} is the standard potential for the reduction of AgCl to silver. Alternatively, the standard potential for the reduction of silver ion can be used:

$$E_{Ag} = E^0_{Ag^+} - 0.0592 \log \frac{1}{[Ag^+]} = 0.799 - 0.0592 \log \frac{1}{[Ag^+]}$$

The former is convenient for calculating the potential of the silver electrode when an excess of chloride exists, whereas the latter is preferable for solutions containing an excess of silver ion.

Potentiometric measurements are particularly useful for titrations of mixtures of anions with standard silver nitrate. For example, Figure 7-4a shows a theoretical curve for the titration of a chloride/iodide mixture. An experimental curve has the same general appearance, although the ordinate units are different.

15E-3 Complex-Formation Titrations

Both metallic and membrane electrodes have been used to detect end points in potentiometric titrations involving complex formation. The mercury electrode[12] (Figure 15-14) is particularly useful for EDTA titrations of cations that form complexes that are less stable than HgY^{2-} (see page 362 for the half-reactions involved).

15E-4 Neutralization Titrations

Experimental neutralization curves closely approximate the theoretical curves described in Chapter 8. Ordinarily, however, the experimental curves are somewhat displaced from the theoretical along the pH axis because concen-

FIGURE 15-14 A typical mercury electrode.

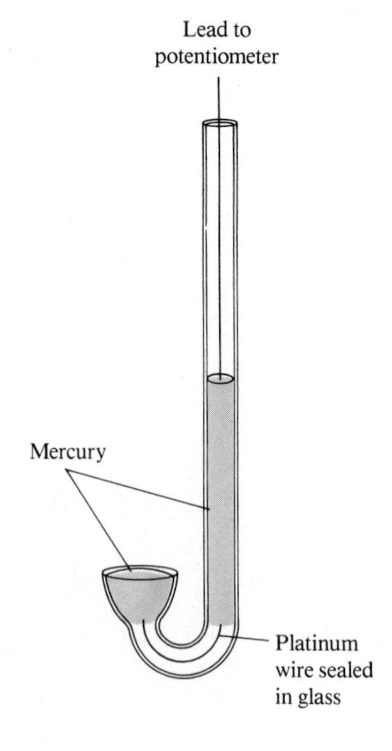

[12]These electrodes can be obtained from Kontes Manufacturing Corp., Vineland, NJ 08360.

trations rather than activities are used in their derivation; this displacement is of no consequence in locating end points. Potentiometric neutralization titrations are particularly valuable for the analysis of mixtures of acids or polyprotic acids. The same considerations apply to bases.

The Determination of Dissociation Constants

An approximate numerical value for the dissociation constant of a weak acid or base can be estimated from potentiometric titration curves. This quantity can be computed from the pH at any point along the curve, but as a practical matter, the pH at half-neutralization is most convenient. Here,

$$[HA] \simeq [A^-]$$

Therefore,

$$K_a = \frac{[H_3O^+]\cancel{[A^-]}}{\cancel{[HA]}} = [H_3O^+]$$

$$pK_a = pH$$

It is important to note that the use of concentrations instead of activities may cause the value for K_a to differ from its published value by a factor of 2 or more. The more exact form of the dissociation constant for HA is

$$K_a = \frac{a_{H_3O^+}a_{A^-}}{a_{HA}} = \frac{a_{H_3O^+}\cancel{[A^-]}f_{A^-}}{\cancel{[HA]}f_{HA}}$$

$$K_a = \frac{a_{H_3O^+}f_{A^-}}{f_{HA}} \quad (15\text{-}29)$$

Since the glass electrode provides a good approximation of $a_{H_3O^+}$, the value of K_a measured differs from the thermodynamic value by the ratio of the two activity coefficients. The activity coefficient in the denominator of Equation 15-29 does not change significantly as ionic strength increases because HA is a neutral species. The activity coefficient for A^-, on the other hand, decreases as the electrolyte concentration increases. Therefore, the observed hydrogen ion activity is larger than the thermodynamic dissociation constant.

EXAMPLE 15-4

In order to determine K_1 and K_2 for H_3PO_4 from titration data, careful pH measurements are made after 0.5 and 1.5 mol of base are added for each mole of acid. It is then assumed that the hydrogen ion activities computed from these data are identical to the desired dissociation constants. Calculate the relative error incurred by this assumption if the ionic strength is 0.1 at the time of each measurement. (From Appendix 4, K_1 and K_2 for H_3PO_4 are 7.11×10^{-3} and 6.34×10^{-8}.)

Rearrangement of Equation 15-29 gives

$$K_a(\text{exptl}) = a_{H_3O^+} = K_a(f_{HA}/f_{A^-})$$

The activity coefficient for H_3PO_4 is approximately unity since this species is uncharged. In Table 5-4, we find that the activity coefficient for $H_2PO_4^-$ is 0.78 and that for HPO_4^{2-} is 0.36. Substituting these values into the equations for K_1 and K_2 gives

$$K_1(\text{exptl}) = 7.11 \times 10^{-3} \times 1.0/0.78 = 9.1 \times 10^{-3}$$

$$\text{Error} = \frac{9.1 \times 10^{-3} - 7.11 \times 10^{-3}}{7.11 \times 10^{-3}} \times 100\% = 28\%$$

$$K_2(\text{exptl}) = 6.34 \times 10^{-8} \times 0.78/0.36 = 1.37 \times 10^{-7}$$

$$\text{Error} = \frac{1.37 \times 10^{-7} - 6.34 \times 10^{-8}}{6.34 \times 10^{-8}} \times 100\% = 116\%$$

A single titration of a pure acid can provide sufficient data (equivalent weight and dissociation constant) to make identification of the acid possible.

15E-5

Oxidation/Reduction Titrations

An inert indicator electrode constructed of platinum is ordinarily used to detect end points in oxidation/reduction titrations. Occasionally, other inert metals, such as silver, palladium, gold, and mercury, are used instead. Titrations curves similar to those derived in Chapter 13 are ordinarily obtained, although they may be displaced along the ordinate axis as a consequence of the effects of high ionic strengths. End points are determined by the methods described earlier in this chapter.

Questions and Problems

*15-1 Differentiate between an electrode of the first kind and an electrode of the second kind.

15-2 What occurs when a newly manufactured glass electrode is immersed in water?

15-3 What is the source of

(a) the asymmetry potential in a membrane electrode?
*(b) the boundary potential in a membrane electrode?
(c) a junction potential in a glass/calomel electrode system?
*(d) the potential of a crystalline membrane electrode used to determine the concentration of F^-?

*15-4 What is the alkaline error in pH measurement with a glass electrode?

15-5 List the advantages and disadvantages of a potentiometric titration relative to a titration with chemical indicators.

*15-6 A solution of ethylamine is titrated with HCl using a glass/calomel electrode system. Show how K_b can be obtained from the pH at the point of half-neutralization.

*15-7 (a) Calculate the standard potential for the reaction

$$CuSCN(s) + e \rightleftharpoons Cu(s) + SCN^-$$

(b) Sketch a cell having a copper indicator electrode as a cathode and a saturated calomel electrode as an anode that could be used for the determination of SCN^-.
(c) Derive an equation relating the measured potential of the cell in (b) to pSCN. (Assume the junction potential is zero.)
(d) Calculate the pSCN of a thiocyanate-containing solution that is saturated with CuSCN and employed in conjunction with a copper electrode in the cell sketched in (b) if the resulting potential is −0.076 V.

15-8 (a) Calculate the standard potential for the reaction

$$Ag_2S(s) + 2e \rightleftharpoons 2Ag(s) + S^{2-}$$

(b) Sketch a cell having a silver indicator electrode as the cathode and a standard calomel electrode as an anode that could be used for determining S^{2-}.

(c) Derive an equation relating the measured potential of the cell in (b) to pS. (Assume the junction potential is zero.)
(d) Calculate the pS of a solution that is saturated with Ag_2S and then employed with the cell sketched in (b) if the resulting potential is 0.538 V.

*15-9 Give a schematic representation of each of the following cells. Derive an equation relating cell potential to p-function. Assume the junction potential is negligible; treat the indicator electrode as the cathode; and specify any necessary concentrations as 1.00×10^{-4} M.

(a) A cell with a mercury indicator electrode for the determination of pCl.
(b) A cell with a silver indicator electrode for the determination of pCO_3.
(c) A cell with a platinum electrode for the determination of pSn(IV).

15-10 Give a schematic representation of each of the following cells. Derive an equation relating cell potential and p-function. Assume the junction potential is negligible; treat the indicator electrode as the cathode; and specify any necessary concentrations as 1.00×10^{-4} M.

(a) A cell with a lead electrode for the determination of $pCrO_4$.
(b) A cell with a silver indicator electrode for the determination of $pAsO_4$.
(c) A cell with a platinum electrode for the determination of pTl(III).

*15-11 The cell

$SCE||Ag_2CrO_4(sat'd),CrO_4^{2-}(x M)|Ag$

is employed for the determination of $pCrO_4$. Calculate $pCrO_4$ when the cell potential is 0.402 V.

15-12 Calculate the potential of the cell

SCE||aqueous solution|Hg

when the aqueous solution is

(a) 7.40×10^{-3} M Hg^{2+}.
(b) 7.40×10^{-3} M Hg_2^{2+}.
(c) $Hg_2SO_4(sat'd),SO_4^{2-}(0.0250\ M)$.
(d) $Hg^{2+}(2.00 \times 10^{-3}),OAc^-(0.100\ M)$:

$$Hg^{2+} + 2OAc^- \rightleftharpoons Hg(OAc)_2(aq) \qquad K_f = 2.7 \times 10^8$$

*15-13 The standard potential for the reduction of the EDTA complex of Hg(II) is

$$HgY^{2-} + 2e \rightleftharpoons Hg(l) + Y^{4-} \qquad E^0 = 0.210\ V$$

Calculate the potential of the cell

$SCE||HgY^{4-}(2.00 \times 10^{-4}),Z|Hg$

when Z is

(a) $H^+(1.00 \times 10^{-4}\ M),EDTA(0.0200\ M)$.
(b) $H^+(1.00 \times 10^{-8}\ M),EDTA(0.0200\ M)$.
(c) $H^+(1.00 \times 10^{-10}\ M),CaCl_2(0.0200\ M),EDTA(0.0200\ M)$.

15-14 Calculate the potential of the cell described in Problem 15-13 when Z is

(a) $H^+(1.00 \times 10^{-6}\ M),EDTA(0.0100\ M)$.
(b) $H^+(1.00 \times 10^{-10}\ M),EDTA(0.0100\ M)$.
(c) $H^+(1.00 \times 10^{-7}\ M),Zn(NO_3)_2(0.0100\ M),EDTA(0.0100\ M)$.

*15-15 The cell

$SCE||H^+(a = x)|$glass electrode

has a potential of 0.2094 V when the solution in the right-hand compartment is a buffer of pH 4.006. The following potentials are obtained when the buffer is replaced with unknowns: (a) −0.3011 V and (b) +0.1163 V. Calculate the pH and the hydrogen ion activity of each unknown. (c) Assuming an uncertainty of ±0.002 V in the junction potential, what is the range of hydrogen ion activities within which the true value might be expected to lie?

15-16 The following cell

$SCE||MgA_2(a_{Mg^{2+}} = 9.62 \times 10^{-3})|$membrane electrode for Mg^{2+}

has a potential of 0.367 V.

(a) When the solution of known magnesium activity is replaced with an unknown solution, the potential is −0.544 V. What is the pMg of this unknown solution?

(b) Assuming an uncertainty of ±0.002 V in the junction potential, what is the range of Mg^{2+} activities within which the true value might be expected?

*15-17 The cell

$SCE||CdA_2$(sat'd),A^-(0.0250 M)|Cd

has a potential of −0.721 V. Calculate the solubility product of CdA_2, neglecting the junction potential.

15-18 The cell

SCE||HA(0.250 M),NaA(0.180 M)|H_2(1.00 atm),Pt

has a potential of −0.797 V. Calculate the dissociation constant of HA, neglecting the junction potential.

15-19 A 40.00-mL aliquot of 0.05000 M HNO_2 is diluted to 75.00 mL and titrated with 0.08000 M Ce^{4+}. The pH of the solution is maintained at 1.00 throughout the titration, and the formal potential of the cerium system is 1.44 V.

*(a) Calculate the potential of the indicator electrode with respect to a saturated calomel reference electrode after the addition of 5.00, 10.00, 15.00, 25.00, 40.00, 49.00, 50.00, 51.00, 55.00, and 60.00 mL of cerium(IV).

(b) Draw a titration curve for these data.

15-20 Calculate the potential of a silver cathode versus the standard calomel electrode after the addition of 5.00, 15.00, 25.00, 30.00, 35.00, 39.00, 40.00, 41.00, 45.00, and 50.00 mL of 0.1000 M $AgNO_3$ to 50.00 mL of 0.0800 M KSeCN. Construct a titration curve from these data. (K_{sp} for AgSeCN $= 4.20 \times 10^{-16}$.)

15-21 Quinhydrone is an equimolar mixture of quinone (Q) and hydroquinone (H_2Q). These two compounds react reversibly at a platinum electrode:

$$Q + 2H^+ + 2e \rightleftharpoons H_2Q \qquad E^0 = 0.699 \text{ V}$$

The pH of a solution can be determined by saturating it with quinhydrone and making it a part of the cell

SCE||quinhydrone(sat'd),H^+(xM)|Pt

If such a cell has a potential of 0.313 V, what is the pH of the solution assuming the junction potential is zero?

CHAPTER 16

Electrogravimetric and Coulometric Methods

In this chapter, two related electroanalytical methods are described: *electrogravimetry* and *coulometry*. Each is based upon an electrolysis that is carried out for a sufficient length of time to ensure quantitative oxidation or reduction of the analyte to a single product of known composition. In electrogravimetric methods, the product is weighed as a deposit on one of the electrodes (*the working electrode*). In coulometric procedures, the quantity of electricity needed to complete the electrolysis serves as a measure of the amount of analyte present.[1]

Electrogravimetry and coulometry are moderately sensitive, rapid, and among the most accurate and precise techniques available to the chemist. Thus, in many applications, relative uncertainties of a few parts per thousand are readily obtained. In common with gravimetry but in contrast to the other procedures discussed in this text, these methods require no preliminary calibration against standards because the functional relationship between the quantity measured and the analyte concentration can be derived from theory and atomic-weight data.

Electrogravimetry and coulometry differ from potentiometry in the respect that they require the presence of a significant current throughout the analytical process. In contrast, potentiometric measurements are performed under conditions of minimal current. When electricity flows in an electrochemical cell, the cell potential is no longer simply the difference between the electrode potentials of the anode and the cathode because of the effects of *IR drop, concentration polarization,* and *kinetic polarization.* Before proceeding, we must examine these three phenomena in detail.

16A The Effect of Current on Cell Potentials

In order to develop a current in an electrolytic cell, it is necessary to apply a potential that is more negative than the thermodynamic, or equilibrium, cell potential. Similarly, when electricity is drawn from a galvanic cell, a decrease in output potential of the cell is observed. These effects are the direct consequence of the three phenomena just mentioned.

16A-1 Ohmic Potential: IR Drop

Electrochemical cells, like metallic conductors, offer resistance to the flow of charge. In both types of conduction, the effect of this resistance is described

[1]For further information concerning the methods in this chapter, see J. A. Plambeck, *Electroanalytical Chemistry,* New York: Wiley, 1982; *Laboratory Techniques in Electroanalytical Chemistry,* P. T. Kissinger and W. R. Heineman, Eds., New York: Marcel Dekker, 1984.

by Ohm's law. The product of the resistance R of a cell in ohms (Ω) and the current I in amperes (A) is called the *ohmic potential* or the *IR drop* of the cell. In order to develop a current of I amperes in an electrolytic cell having an electrical resistance of R ohms, it is necessary to apply a potential that exceeds the theoretical cell potential by an amount equal to $-IR$ volts. Similarly, when a current of I amperes is produced by a galvanic cell with a resistance of R ohms, the output potential of the cell is reduced by $-IR$ volts. Thus, in the presence of a current, Equation 12-5 for a cell potential E_{cell} must be modified by addition of the term $-IR$:

$$E_{cell} = E_{cathode} - E_{anode} - IR \qquad (16\text{-}1)^2$$

where $E_{cathode}$ and E_{anode} are thermodynamic potentials computed with the techniques discussed in Section 12D-5.

EXAMPLE 16-1

Consider a cell consisting of a copper electrode in contact with 1.00 M Cu^{2+}, a cadmium electrode in contact with 1.00 M Cd^{2+}, and a connecting salt bridge. The cell has a resistance of 4.00 Ω.

(a) Calculate the potential needed to develop a current of 0.0200 A in the electrolytic cell

$$Cu|Cu^{2+}(1.00\text{ M})||Cd^{2+}(1.00\text{ M})|Cd$$

Since both cation concentrations are 1.00 M, the electrode potentials and the standard electrode potentials are numerically equal. Substituting into Equation 16-1 gives

$$E_{cell} = E^0_{Cd} - E^0_{Cu} - IR$$
$$= -0.403 - 0.337 - 0.0200 \times 4.00 = -0.740 - 0.080 = -0.820\text{ V}$$

Thus a potential that is 0.08 V more negative than theoretical is needed to deposit cadmium at the rate required to maintain a current of 0.0200 A.

(b) Calculate the cell potential when 0.0200 A is developed in the galvanic cell

$$Cd|Cd^{2+}(1.00\text{ M})||Cu^{2+}(1.00\text{ M})|Cu$$

Here,

$$E_{cell} = E^0_{Cu} - E^0_{Cd} - IR$$
$$= 0.337 - (-0.403) - 0.0200 \times 4.00 = 0.740 - 0.080 = 0.660\text{ V}$$

Note that the production of a current in this cell causes the output potential to be markedly less than theoretical owing to the effect of the IR drop.

16A-2 Polarization Effects

Equation 16-1 can be rearranged to give

$$I = -\frac{1}{R}E_{cell} + \frac{1}{R}(E_{cathode} - E_{anode})$$

For small currents and brief periods of time, $E_{cathode}$ and E_{anode} remain relatively constant during an electrolysis. The cell behavior can then be approximated by the relationship

[2]The junction potential is unimportant in this discussion.

$$I = -\frac{1}{R}E_{\text{cell}} + k \tag{16-2}$$

where k is a constant.

Equation 16-2 reveals that a plot of current in an electrolytic cell as a function of applied potential should be a straight line with a slope that equals the negative reciprocal of the resistance (note that, by convention, E_{applied} is negative; thus, the current in an electrolytic cell is positive). As shown in Figure 16-1a, the expected linearity is observed at small currents; marked departures occur as currents become larger, however. Figure 16-1b reveals that galvanic cells behave in an analogous way. Here again, a linear relationship is observed at small currents, but significant departures from a straight line occur as the current becomes greater.

Cells that exhibit nonlinear behavior are said to be *polarized,* and the degree of polarization is given by an *overvoltage,* or *overpotential,* which is symbolized by Π in the figure. Note that polarization requires the application of a potential that is greater than theoretical to give a current of the expected magnitude. Thus, the overpotential required to achieve a current of 0.06 A in the electrolytic cell in Figure 16-1a is about -0.04 V. For the galvanic cell in Figure 16-1b, the existence of polarization resulting from a 0.06-A current causes the output cell potential to decrease by about 0.03 V—that is, the overvoltage is -0.03 V. Note that in each case the overvoltage carries a negative sign.

Polarization is an electrode phenomenon that can affect either or both of the electrodes in a cell. The degree of polarization of an electrode can vary

FIGURE 16-1 Current-voltage curves for

(a) $Cu|Cu^{2+}(1.0\ M)\ \|\ Cd^{2+}(1.0\ M)|Cd$

(b) $Cd|Cd^{2+}(1.0\ M)\ \|\ Cu^{2+}(1.0\ M)|Cd$

In both cells, the overvoltage at 0.06 A is symbolized by Π. For cell (a), $\Pi = -0.040$ V; for cell (b), $\Pi = -0.03$ V.

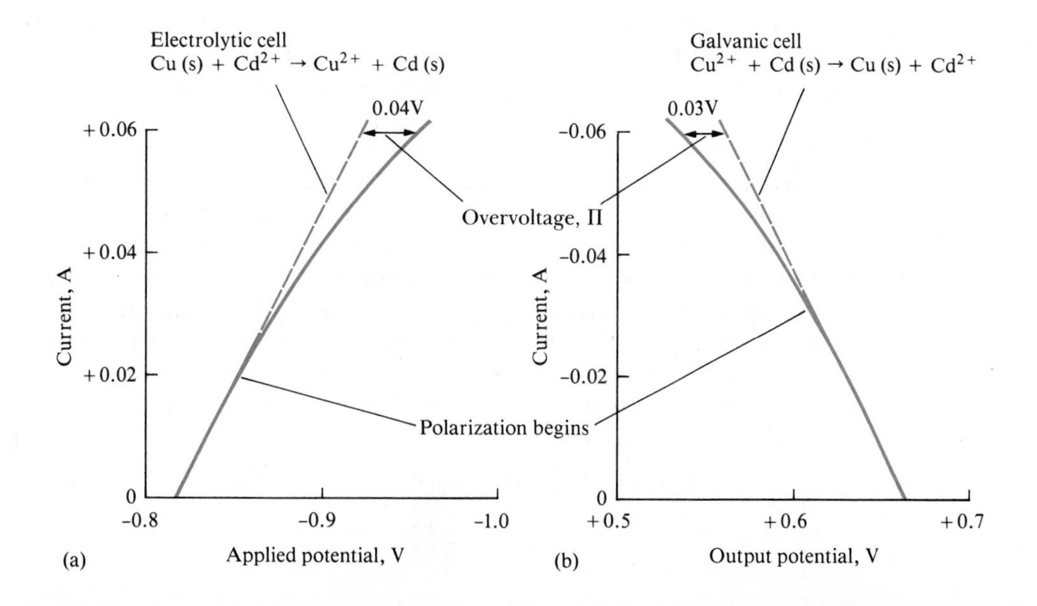

widely. In some instances it approaches zero, whereas in others it can be so nearly complete that the current in the cell becomes independent of potential. Several factors influence the extent of polarization: (1) the size, shape, and composition of the electrode, (2) the composition of the electrolyte solution, its temperature, and the rate at which it is stirred, (3) the magnitude of the current, and (4) the physical state of the species involved in the cell reaction. While some of these factors are understood sufficiently to permit their quantitative description, others can be accounted for only on an empirical basis.

Polarization phenomena are conveniently divided into two categories: *concentration polarization* and *kinetic polarization.*

Concentration Polarization

Electron transfer between a reactive species in a solution and an electrode can take place only from a thin film of solution located immediately adjacent to the surface of the electrode; this film is only a few angstrom units in thickness and contains a limited number of reactive ions or molecules. In order for a steady current to exist in a cell, it is necessary that this film be continuously replenished with reactant from the bulk of the solution. That is, as reactant ions or molecules are consumed by the electrochemical reaction, more must be transported into the surface film at a rate that is sufficient to sustain the current at the desired level. For example, in order to maintain a current of 0.01 A in the cell described in Example 16-1a, it is necessary to transport cadmium ions to the electrode surface at a rate of about 5×10^{-8} mol/s or 3×10^{16} cadmium ions per second. (Similarly, copper ions must be removed from the surface film of the anode at this same rate.)

Concentration polarization is encountered when the mechanisms of mass transport are inadequate to supply reactant to (or remove product from) the surface of an electrode at a rate commensurate with the desired current. Under this circumstance, the magnitude of the current is limited by the mass-transport rate; currents less than those predicted by Equation 16-2 are then observed.

Reactants are transported to the surface film of an electrode by three mechanisms: (1) *diffusion,* (2) *migration,* and (3) *convection.* Products are removed from electrode surfaces in the same way. In discussing these mechanisms, we shall focus on mass-transport processes as they occur in the cell compartment of a cathode. The conclusions reached apply equally well, however, to anodes.

Diffusion. When a concentration difference develops between two regions of a solution, as it does when a species is reduced at a cathode surface, ions or molecules move from the more concentrated region to the more dilute; this process is called *diffusion* and ultimately leads to a disappearance of the concentration difference. The rate of diffusion is directly proportional to the concentration difference. For example, when cadmium ions are deposited at a cathode by a current, a concentration gradient is formed that causes cadmium ions to diffuse from the bulk of the solution to the surface film. The rate of diffusion is given by

$$\text{Rate of diffusion to cathode surface} = k'([Cd^{2+}] - [Cd^{2+}]_0) \qquad (16\text{-}3)$$

where $[Cd^{2+}]$ is the reactant concentration in the bulk of the solution, $[Cd^{2+}]_0$

is its equilibrium concentration at the surface of the cathode, and k' is a proportionality constant. The value of $[Cd^{2+}]_0$ at any instant is fixed by *the potential of the electrode* and can be calculated through use of the Nernst equation. In this case, for example, the surface cadmium ion concentration can be obtained from the relationship

$$E_{\text{applied}} = E^0_{Cd^{2+}} - \frac{0.0592}{2} \log \frac{1}{[Cd^{2+}]_0}$$

where E_{applied} is the potential applied to the cathode. As the applied potential becomes more and more negative, $[Cd^{2+}]_0$ becomes smaller and smaller; as a consequence, the rate of diffusion becomes correspondingly larger as does the current.

Migration. The process by which ions move under the influence of an electric field is called *migration*. It is the primary process by which mass transfer occurs in the bulk of the solution in a cell. The rate at which ions migrate to or away from an electrode surface generally increases as the electrode potential increases. This charge movement constitutes a current, which also increases with potential. The larger the number of different kinds of ions in a given solution, the smaller is the fraction of the total charge that is carried by a particular species. For example, if a solution contains only cadmium ions (and their corresponding counter ions) the total current in the solution is carried by these ions. If a second salt such as sodium chloride is added to the solution, the fraction of the total current carried by the cadmium ions decreases. When the concentration of sodium ions exceeds that of cadmium ions by a factor of 50 to 100, the fraction of the total current carried by cadmium approaches zero. As a result, the rate of migration of cadmium toward the cathode becomes essentially independent of applied potential.

Convection. Reactants can also be transferred to or from an electrode by mechanical means. Thus, forced convection, such as stirring or agitation, will tend to decrease concentration polarization. Natural convection resulting from temperature or density differences also contributes to material transport.

The Importance of Concentration Polarization in Electroanalytical Chemistry. As noted earlier, concentration polarization sets in when the effects of diffusion, electrostatic migration, and mechanical mixing are insufficient to transport a reactant to or from an electrode surface at a rate that produces a current of the magnitude given by Equation 16-2. Concentration polarization requires applied potentials that are larger than theoretical to maintain a given current in an electrolytic cell (Figure 16-1a); similarly, the phenomenon causes a galvanic cell potential to be smaller than the value predicted on the basis of the theoretical potential and the IR drop (Figure 16-1b).

Concentration polarization is important in several electroanalytical methods. In some applications, its effects are deleterious, and steps are taken to eliminate it; in others, it is essential to the analytical method, and every effort is made to promote its occurrence. The experimental variables that influence the degree of concentration polarization are (1) reactant concentration, (2) total electrolyte concentration, (3) mechanical agitation, and (4) electrode size (as the area toward which a reactant can be transported increases, polarization effects become smaller).

Kinetic Polarization

In kinetic polarization, the magnitude of the current is limited by the rate of one or both of the electrode reactions. In order to offset kinetic polarization, an additional potential, or overvoltage, is required to overcome the energy barrier to the half-reaction. Note that the *current in a kinetically polarized cell is governed by the rate of electron transfer* rather than the rate of mass transfer.

Kinetic polarization is most pronounced for electrode processes that yield gaseous products and is often negligible for reactions that involve the deposition or solution of a metal. Kinetic effects usually decrease with increasing temperature and decreasing current density.[3] These effects also depend upon the composition of the electrode and are most pronounced with softer metals, such as lead, zinc, and particularly mercury. The magnitude of overvoltage effects cannot be predicted from present theory and can only be estimated from empirical information in the literature.[4] In common with *IR* drop, overvoltage effects cause the potential of a galvanic cell to be smaller than theoretical and to require potentials greater than theoretical to operate an electrolytic cell at a desired current.

The overvoltages associated with the formation of hydrogen and oxygen are often 1 V or more and are of considerable importance because these molecules are frequently produced by electrochemical reactions. Of particular interest is the high overvoltage of hydrogen on such metals as copper, zinc, and mercury, which permits deposition of these metals and several others without interference from hydrogen evolution. Thus, in theory, it is not possible to deposit zinc from a neutral aqueous solution because hydrogen forms at a potential that is considerably less than that required for zinc deposition. In fact, the metal can be deposited on a copper electrode with no significant hydrogen formation because the rate at which the gas forms on both zinc and copper is negligible, as shown by the high hydrogen overvoltage associated with these metals.

16A-3 The Potential of Electrolytic Cells in the Presence of a Current

The potential relationships in an electrolytic cell through which electricity is being passed are given by the equation

$$E_{\text{applied}} = E_{\text{cathode}} - E_{\text{anode}} - IR - \Pi_{\text{cathode}} - \Pi_{\text{anode}} \tag{16-4}$$

where Π_{cathode} and Π_{anode} are the kinetic and concentration overvoltages associated with the reactions at the cathode and the anode. In this equation, the junction potential has been neglected because it is generally so small as to be of no consequence in electrolytic methods.

16B The Potential Selectivity of Electrolytic Methods

In principle, electrolytic methods offer a reasonably selective means for separating and determining a number of ions. The feasibility of and theoretical

[3]Current density is defined as amperes per square centimeter (A/cm^2) of electrode surface.

[4]Overvoltage data for various gaseous species on different electrode surfaces have been compiled by J. A. Page, in *Handbook of Analytical Chemistry*, L. Meites, Ed., p. 5–184. New York: McGraw-Hill, 1963.

conditions for accomplishing a given separation can be readily derived from the standard electrode potentials for the species of interest.

EXAMPLE 16-2

Is a quantitative separation of Cu^{2+} and Pb^{2+} by electrolytic deposition feasible in principle? If so, what range of cathode potentials can be used? Assume that the sample solution is initially 0.1000 M in each ion and that quantitative removal of an ion is realized when only 1 part in 10,000 remains undeposited.

In Appendix 2, we find

$$Cu^{2+} + 2e \rightleftharpoons Cu(s) \qquad E^0 = 0.337 \text{ V}$$

$$Pb^{2+} + 2e \rightleftharpoons Pb(s) \qquad E^0 = -0.126 \text{ V}$$

It is apparent that copper will begin to deposit before lead. Let us first calculate the potential required to reduce the Cu^{2+} concentration to 10^{-4} of its original concentration (that is, to 1.00×10^{-5} M). Substituting into the Nernst equation, we obtain

$$E = 0.337 - \frac{0.0592}{2} \log \frac{1}{1.00 \times 10^{-5}} = 0.189 \text{ V}$$

Similarly, we can derive the cathode potential at which lead begins to deposit:

$$E = -0.126 - \frac{0.0592}{2} \log \frac{1}{0.100} = -0.156 \text{ V}$$

Therefore, if the cathode potential is maintained between 0.189 and −0.156 V, a quantitative separation should in theory occur.

Calculations such as the foregoing make it possible to compute the differences in standard electrode potentials theoretically needed for determining one ion without interference from another; these differences range from about 0.04 V for triply charged ions to about 0.24 V for singly charged ions.

These theoretical separation limits can be approached only by maintaining the potential of the *working electrode* (usually the cathode, at which metal deposition occurs) at a level required by theory. The potential of this electrode can *be controlled only by variation of the potential applied to the cell,* however. From Equation 16-4, it is evident that variations in $E_{applied}$ affect not only the cathode potential but also the anode potential, the *IR* drop, and any overvoltages associated with the electrode processes. All these potentials vary continuously as the electrolysis proceeds, and it is not feasible to calculate the fraction of the applied potential attributable to the working electrode. As a consequence, the only *practical* way of achieving the separation of species whose electrode potentials differ by only a few tenths of a volt is to *measure* the cathode potential continuously against a reference electrode whose potential is known; the cell potential can then be adjusted to maintain the potential of the working electrode within the desired range. An analysis performed in this way is called a *controlled-cathode-potential electrolysis* or a *controlled-anode-potential electrolysis.*

16C Electrogravimetric Methods of Analysis

Electrolytic precipitation has been used for over a century for the gravimetric determination of metals. In most applications, the metal is deposited on a weighed platinum cathode, and the increase in weight is determined. Important exceptions to this procedure include the anodic deposition of lead as lead dioxide on platinum and of chloride as silver chloride on silver.

16C-1 Types of Electrogravimetry

Two types of electrogravimetric methods are encountered. In one, no control of the potential of the working electrode is exercised, and the applied cell potential is held at a more or less constant level that provides a large enough current to complete the electrolysis in a reasonable length of time. The second type of electrogravimetric method is the controlled-cathode- or anode-potential method mentioned in the previous section. This method is also called a *potentiostatic method.*

Electrogravimetry Without Potential Control

Electrolytic procedures in which the potential applied to a cell is held approximately constant throughout the electrolysis offer the advantage that the equipment required is simple and inexpensive and requires little operator attention. To understand the limitations of this procedure, it is necessary to consider how the current and the potential of the working electrode vary as a function of time when a constant potential is applied to a cell.

In order to illustrate the current-voltage relationships that apply during an electrolysis at constant cell potential, it is convenient to consider the deposition of copper from a solution that is 0.1 M in copper(II) ions and 1 M in acid. The electrode reactions are

$$\text{Cathode} \qquad Cu^{2+} + 2e \longrightarrow Cu(s)$$

$$\text{Anode} \qquad H_2O \longrightarrow \frac{1}{2}O_2(g) + 2H^+ + 2e$$

Employing the technique demonstrated in Example 12-9, we can show that the theoretical potential of this cell is -0.92 V. We shall assume that an initial current of 1.5 A is desired and that the cell resistance is 0.5 Ω.

To obtain an estimate of the applied potential required to generate a current of 1.5 A during the electrolysis, we apply Equation 16-4, taking into account both kinetic and concentration overvoltages at the two electrodes. At the anode, concentration polarization should be negligible throughout the electrolysis because an enormous excess of the reactant (H_2O) is always present; therefore, its concentration at the electrode surface should be sufficient to sustain the desired current. As indicated earlier, however, kinetic polarization often accompanies the formation of gases at electrodes. Consultation of the reference in footnote 4 suggests that the resulting overvoltage required to bring about the formation of oxygen at the desired rate is about 0.7 V (that is, $\Pi_{\text{anode}} = 0.7$ V).

At the outset, the cathodic overvoltage Π_{cathode} should be negligible because the concentration of copper is high, thus making concentration polarization unlikely. Furthermore, the reaction of metal ions at a platinum electrode is generally rapid enough so that kinetic polarization does not occur. Thus, the applied potential required to produce a current of 1.5 A is obtained by substituting into Equation 16-4:

$$E_{\text{applied}} = -0.92 - 1.5 \times 0.5 - 0.7 - 0.0 = -2.4 \text{ V}$$

The variation in current as a function of time for the cell under consideration is shown in Figure 16-2a. The exponential decrease in current seen here arises from concentration polarization, which begins at the cathode almost immediately after application of the potential. That is, after a few seconds, mechanical mixing and diffusion are not sufficient to transport copper ions to the cathode surface at a rate great enough to maintain a current of 1.5 A. At this point, then, the current is limited by the *rate of mass transport* of copper ions from the bulk of the solution to the electrode surface. This rate decreases rapidly as the copper concentration in the bulk of the solution becomes smaller.

Figure 16-2b shows that the decrease in current brought about by concentration polarization results in a positive shift in both IR and the kinetic overpotential Π_{anode}. Since $E_{applied}$ in Equation 16-4 is fixed, however, the remaining terms—$E_{cathode}$, E_{anode}, and $\Pi_{cathode}$—must become more negative to offset these changes. The reversible anode potential is

$$E_{anode} = +1.229 - \frac{0.0592}{2} \log \frac{1}{p_{O_2}^{1/2}[H^+]^2}$$

Because the initial hydrogen ion concentration is high and the partial pressure of oxygen over the solution is constant, E_{anode} changes only slightly as the reaction proceeds. Thus, as shown in Figure 16-2b, the positive shift in the IR drop results in a corresponding negative shift in the potential of the working electrode ($E_{cathode} - \Pi_{cathode}$).

The rapid change in cathode potential that accompanies concentration polarization often leads to the codeposition of other species and the loss of selectivity. For example, lead, if present in a concentration roughly equal to that of copper, begins to codeposit at point A in Figure 16-2b. In the absence of lead, the evolution of hydrogen commences at about point B (this process was not taken into account in deriving the curve in Figure 16-2a).

FIGURE 16-2 Changes in (a) current and (b) potential during the electrolytic deposition of copper and oxygen at an applied potential of -2.4 V.

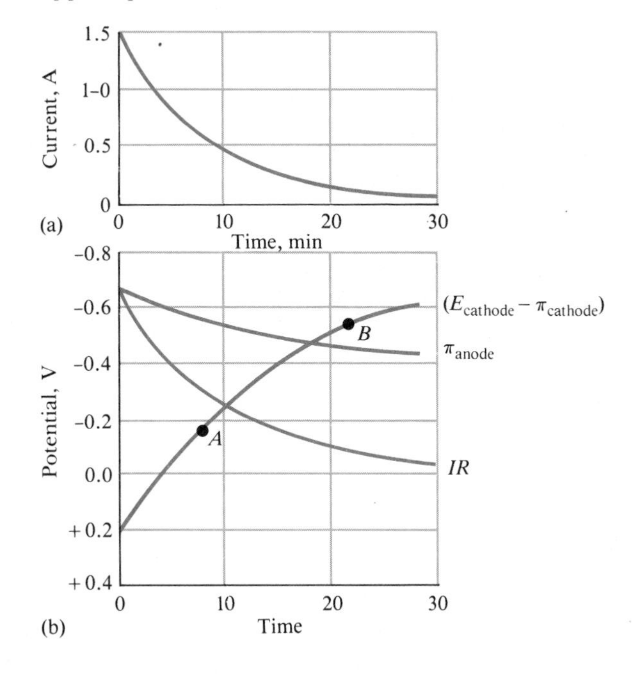

The lead interference just described can be avoided by decreasing the initial applied potential by several tenths of a volt to limit the negative drift of the cathode potential. The consequence, however, is a decrease in the initial currents and, ordinarily, an enormous increase in the time required for the analysis.

In practice, electrolysis at a constant cell potential is limited to the separation of an easily reduced cation from cations that are more difficult to reduce than hydrogen ion. Here, the reduction of hydrogen ions or some other species such as nitrate ions at the cathode stabilizes the cathode potential so that it cannot become negative enough to cause deposition of other cations.

Constant-Cathode-Potential Electrogravimetry

In the discussion that follows, the working electrode is assumed to be a cathode, at which an analyte is deposited as a metal. The remarks are readily extended to an anodic working electrode, however, or to products that do not form as metallic deposits.

In order to separate species having electrode potentials that differ by only a few tenths of a volt, it is necessary to employ a considerably more sophisticated technique than the one just described because concentration polarization at the cathode, if unchecked, causes the potential of that electrode to become so negative that codeposition of the other species present begins before the analyte is completely deposited. A large negative drift in the cathode potential can be avoided by employing a three-electrode system, such as that shown in Figure 16-3.

The controlled-potential apparatus shown in Figure 16-3 is made up of two independent electrical circuits that share a common electrode, the *working electrode* at which the analyte is deposited. The *electrolysis circuit* consists of a dc source, a potential divider (ACB) that permits continuous variation in the potential applied across the working electrode, a *counter electrode,* and an ammeter. The *control circuit* is made up of a saturated calomel electrode, a potentiometer (Appendix 10), and the working electrode. The electrical resist-

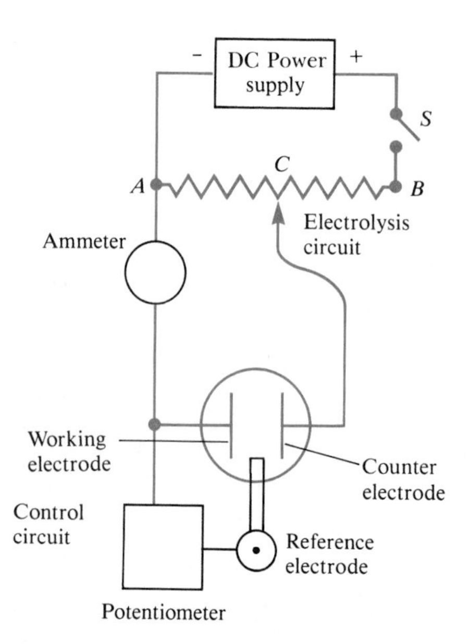

FIGURE 16-3 Apparatus for a controlled-cathode-potential or controlled-anode-potential electrolysis. Contact C is adjusted continuously to maintain the working electrode (the cathode) at a constant potential with respect to the reference electrode. Note that the current in the reference electrode circuit is negligible at all times.

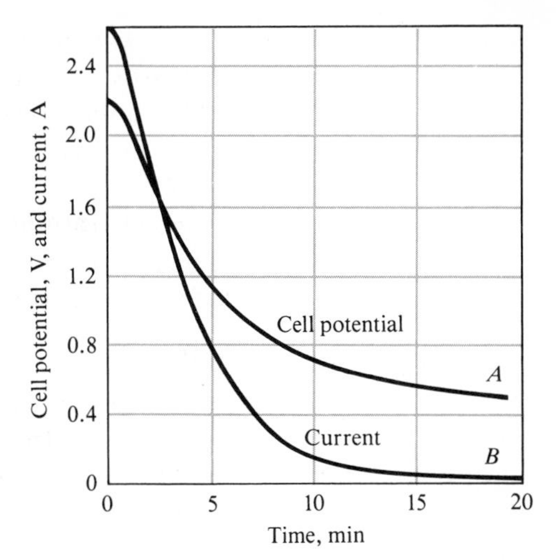

FIGURE 16-4 Changes in applied potential and current during a controlled-cathode-potential deposition of copper. The cathode is maintained at −0.36 V versus SCE throughout the experiment. (Data from J. J. Lingane, *Anal. Chem. Acta,* **1948,** *2,* 590. With permission.)

ance of the control circuit is so much greater than that of the electrolysis circuit that the latter supplies essentially all of the electrolysis current.

The sole purpose of the control circuit is to continuously monitor the potential between the working electrode and the saturated calomel electrode. When this potential reaches a level at which codeposition of an interference is about to begin, the potential across the working and counter electrodes is decreased by moving contact *C* to the left. Since the potential of the counter electrode remains constant during this change, the cathode potential becomes smaller, thus preventing interference from codeposition.

The current and applied voltage changes that occur in a typical constant-cathode-potential electrolysis are depicted in Figure 16-4. Note that the applied cell potential has to be decreased continuously almost immediately after the electrolysis is begun. As a consequence, the full attention of the chemist is required throughout the analysis. To avoid such waste of operator time, controlled-cathode-potential electrolyses are generally performed with automated instruments called *potentiostats,* which maintain a constant cathode potential electronically.

16C-2 The Physical Properties of Electrolytic Precipitates

Ideally, an electrolytically deposited metal should be strongly adherent, dense, and smooth so that it can be washed, dried, and weighed without mechanical loss or reaction with the atmosphere. Good metallic deposits are fine-grained and have a metallic luster; spongy, powdery, or flaky precipitates are likely to be less pure and less adherent.

The principal factors that influence the physical characteristics of deposits are current density, temperature, and the presence of complexing agents. Ordinarily, the best deposits are formed at current densities that are less than 0.1 A/cm^2. Stirring generally improves the quality of a deposit. The effects of temperature are unpredictable and must be determined empirically.

Many metals form smoother and more adherent films when deposited from solutions in which their ions exist primarily as complexes. Cyanide and ammonia complexes often provide the best deposits. The reasons for this effect are not obvious.

Codeposition of hydrogen during electrolysis is likely to cause the formation of nonadherent deposits, which are unsatisfactory for analytical purposes. The evolution of hydrogen can be avoided by introduction of a *cathode depolarizer*—a substance that is reduced at a lower potential than hydrogen ion. Nitrate functions in this manner, being reduced to ammonium ion:

$$NO_3^- + 10H^+ + 8e \rightleftharpoons NH_4^+ + 3H_2O \qquad (16\text{-}5)$$

16C-3

Instrumentation

The apparatus for an analytical electrodeposition without cathode potential control consists of a suitable cell and a direct-current power supply.

Cells

Figure 16-5 shows a typical cell for the deposition of a metal on a solid electrode. Tall-form beakers are ordinarily employed, and mechanical stirring is provided to minimize concentration polarization; frequently, the anode serves as a mechanical stirrer.

Electrodes

Ordinarily, the working electrode in electrogravimetry is a metallic gauze cylinder 2 or 3 cm in diameter and perhaps 6 cm in length. Electrodes are usually constructed of platinum, although copper, brass, and other metals find occasional use. Platinum electrodes have the advantage of being relatively nonreactive and can be ignited to remove any grease, organic matter, or gases that could have a deleterious effect on the physical properties of the deposit.

Certain metals (notably bismuth, zinc, and gallium) cannot be deposited directly onto platinum without causing permanent damage to the electrode. Consequently, a protective coating of copper is always deposited on a platinum electrode before the electrolysis of these metals is undertaken.

The Mercury Cathode

A mercury cathode, such as that shown in Figure 16-6, is particularly useful for removing easily reduced elements as a preliminary step in an analysis. For

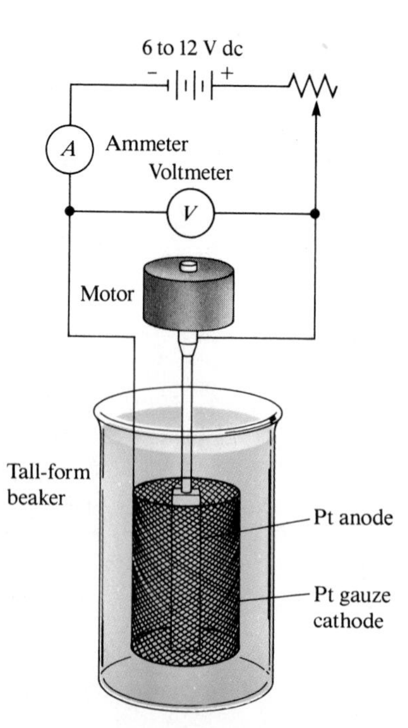

FIGURE 16-5 Apparatus for the electrodeposition of metals without cathode-potential control.

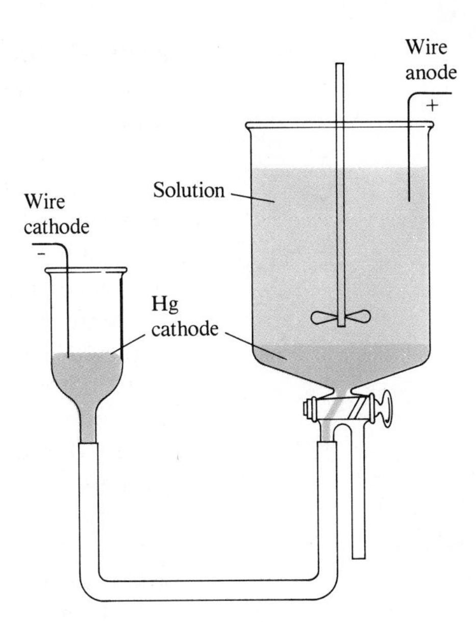

FIGURE 16-6 A mercury cathode for the electrolytic removal of metal ions from solution.

example, copper, nickel, cobalt, silver, and cadmium are readily separated at this electrode from such ions as aluminum, titanium, the alkali metals, sulfates, and phosphates. The precipitated elements dissolve in the mercury with little hydrogen evolution because even at high applied potentials, formation of the gas is prevented by the high overvoltage associated with its evolution on mercury. Ordinarily, the deposited metals are not determined after electrolysis, the goal being simply the removal of interfering ions from a solution of an analyte.

Power Supplies

The apparatus shown in Figure 16-5 is typical for most electrolytic analyses. The dc power supply usually consists of an ac rectifier, although a 6-V storage battery or a dc generator can also be employed. The voltage applied to the cell is controlled by a rheostat, with an ammeter and a voltmeter being used to indicate the approximate current and applied voltage.

Potentiostats for Controlled-Electrode-Potential Methods

Several instrument manufacturers sell potentiostats, which are devices that automatically vary the potential applied to a working/auxiliary electrode system in a cell to maintain the working electrode at a constant potential with respect to the third reference electrode. In one type of potentiostat, the small out-of-balance current in the potentiometer shown in Figure 16-3 is amplified and used to drive a reversible *servomotor* that moves contact C so as to zero the out-of-balance current. In newer instruments, potential control is achieved by all-electronic devices.

16C-4 Applications of Electrogravimetry

We have already noted that an electrogravimetric method can be performed in one of two ways: (1) by operating at a more or less constant applied cell potential with no attempt to control the potential of the working electrode and (2) by varying the applied cell potential in such a way that the potential of the working electrode is maintained at a constant predetermined level. The former is simpler to carry out but lacks the selectivity of the latter.

Methods Without Working-Electrode-Potential Control

Electrolytic methods performed without electrode-potential control, while somewhat limited by their lack of selectivity, do have several applications of practical importance. In these applications, a depolarizer is often introduced in excess; a depolarizer not only prevents formation of hydrogen but also stabilizes the electrode potential at a point where other cations do not codeposit. For example, copper can be determined in the presence of lead, zinc, tin, nickel, and manganese by carrying out the deposition in the presence of moderate concentrations of nitrate ion. When the copper concentration is depleted enough to cause concentration polarization, the resulting negative drift in the cathode potential is terminated by the half-reaction shown as Equation 16-5. This reaction occurs at a potential less negative than that required to deposit the other cations, and so their interference is prevented. Cations that are more easily reduced than the depolarizer interfere, of course. In such cases, a preliminary chemical separation may be needed or alternatively, a complexing agent that reacts selectively with the interfering species and thus prevents its codeposition with the analyte can be added.

Constant-cell-potential deposition with a mercury cathode is also useful in removing easily reduced ions from solution prior to completion of an analysis by some other method. The deposition of interfering heavy metals prior to the quantitative determination of the alkali metals is an example of this application.

Table 16-1 lists the common elements that can be determined by electrogravimetric procedures in which control over the potential of the working electrode is not required.

16C-5 Methods with Electrode-Potential Control

The controlled-cathode-potential method is a potent tool for separating and determining metallic species having standard potentials that differ by only a few tenths of a volt. For example, copper, bismuth, lead, cadmium, zinc, and tin in mixtures have been separated and determined by successive deposition on a weighed platinum cathode. The first three elements are deposited from a nearly neutral solution containing tartrate ion to complex the tin(IV) and prevent its deposition. Copper is first reduced quantitatively by maintaining the cathode potential at -0.2 V with respect to a saturated calomel electrode. After being weighed, the copper-plated cathode is returned to the solution,

TABLE 16-1

Typical Applications of Electrogravimetric Methods Without Potential Control

Analyte	Weighed as	Cathode	Anode	Conditions
Ag^+	Ag	Pt	Pt	Alkaline CN^- solution
Br^-	AgBr(on anode)	Pt	Ag	
Cd^{2+}	Cd	Cu on Pt	Pt	Alkaline CN^- solution
Cu^{2+}	Cu	Pt	Pt	H_2SO_4/HNO_3 solution
Mn^{2+}	MnO_2 (on anode)	Pt	Pt dish	HCOOH/HCOONa solution
Ni^{2+}	Ni	Cu on Pt	Pt	Ammoniacal solution
Pb^{2+}	PbO_2 (on anode)	Pt	Pt	Strong HNO_3 solution
Zn^{2+}	Zn	Cu on Pt	Pt	Acidic citrate solution

TABLE 16-2 **Some Applications of Controlled-Cathode-Potential Electrolysis**

Element Determined	Other Elements That Can Be Present
Ag	Cu and heavy metals
Cu	Bi, Sb, Pb, Sn, Ni, Cd, Zn
Bi	Cu, Pb, Zn, Sb, Cd, Sn
Sb	Pb, Sn
Sn	Cd, Zn, Mn, Fe
Pb	Cd, Sn, Ni, Zn, Mn, Al, Fe
Cd	Zn
Ni	Zn, Al, Fe

and bismuth is removed at a potential of -0.4 V. Lead is then deposited quantitatively by increasing the cathode potential to -0.6 V. When lead deposition is complete, the solution is made strongly ammoniacal, and cadmium and zinc are deposited successively at -1.2 and -1.5 V. Finally, the solution is acidified in order to decompose the tin complex by the formation of undissociated tartaric acid. Tin is then deposited at a cathode potential of -0.65 V. A fresh cathode must be used here because the zinc redissolves under these conditions.

A procedure such as this is particularly attractive for use with a potentiostat because little operator time is required for the complete analysis.

Table 16-2 lists some other separations performed by the controlled-cathode-potential method.

16D Coulometric Methods of Analysis

Coulometric methods are based upon the measurement of the quantity of electricity required to convert an analyte quantitatively to a different oxidation state. Coulometric and gravimetric methods share the common advantage that the proportionality constant between the quantity measured and the analyte weight can be derived from accurately known physical constants, thus eliminating the need for calibration standards. In contrast to gravimetric methods, coulometric procedures are usually rapid and do not require that the product of the electrochemical reaction be a weighable solid. Coulometric methods are as accurate as conventional gravimetric and volumetric procedures and in addition are readily automated.[5]

16D-1 The Quantity of Electricity

Units for the quantity of electricity include the coulomb (C) and the faraday (F). *The coulomb is the quantity of electrical charge transported by a constant current of one ampere in one second.* Thus, the number of coulombs (Q) resulting from a constant current of I amperes operated for t seconds is

$$Q = It \tag{16-6}$$

[5]For additional information about coulometric methods, see E. Bishop, in *Comprehensive Analytical Chemistry,* C. L. Wilson and D. W. Wilson, Eds., Vol. 11D. New York: Elsevier Scientific, 1975; J. A. Plambeck, *Electroanalytical Chemistry,* Chapter 12. New York: Wiley, 1982; G.W.C. Milner and G. Phillips, *Coulometry in Analytical Chemistry.* New York: Pergamon Press, 1967.

For a variable current i, the number of coulombs is given by the integral

$$Q = \int_0^t i\,dt \tag{16-7}$$

The faraday is the quantity of electricity that produces one equivalent of chemical change at an electrode. Since the equivalent in an oxidation/reduction reaction is that amount of a substance that donates or accepts 1 mol of electrons (Appendix 9), the faraday is equivalent to 6.02×10^{23} electrons. The faraday also equals 96,485 C. As shown in Example 16-3, these definitions make it possible to calculate the weight of a chemical species that is formed at an electrode by a current of known magnitude.

EXAMPLE 16-3

A constant current of 0.800 A is used to deposit copper at the cathode and oxygen at the anode of an electrolytic cell. Calculate the number of grams of each product formed in 15.2 min, assuming no other redox reaction.

The equivalent weights of copper and oxygen are determined from consideration of the two half-reactions

$$Cu^{2+} + 2e \rightarrow Cu(s)$$

$$2H_2O \rightarrow 4e + O_2(g) + 4H^+$$

Thus 1 mol of copper contains 2 eq and 1 mol of oxygen represents 4 eq of that species.

Substituting into Equation 16-6 yields

$$Q = 0.800\text{ A} \times 15.2\text{ min} \times 60\text{ s/min} = 729.6\text{ A·s} = 729.6\text{ C}$$

$$\text{No. } F = \frac{729.6\text{ C}}{96{,}485\text{ C}/F} = 7.56 \times 10^{-3} \equiv 7.56 \times 10^{-3}\text{ eq Cu and O}_2$$

From the definition of the faraday, 7.56×10^{-3} equivalent of copper is deposited on the cathode; a similar quantity of oxygen is evolved at the anode. Therefore,

$$\text{Wt Cu} = 7.56 \times 10^{-3}\text{ eq Cu} \times \frac{63.54\text{ g Cu/mol}}{2\text{ eq Cu/mol}} = 0.240\text{ g Cu}$$

$$\text{Wt O}_2 = 7.56 \times 10^{-3}\text{ eq O}_2 \times \frac{32.00\text{ g O}_2\text{/mol}}{4\text{ eq O}_2\text{/mol}} = 0.0605\text{ g O}_2$$

16D-2 Types of Coulometric Methods

Two types of methods have been developed that are based on the quantity of electricity: *potentiostatic coulometry* and *amperostatic coulometry* or *coulometric titrimetry*. Potentiostatic methods are performed in much the same way as controlled-potential gravimetric methods, with the potential of the working electrode relative to a reference electrode being maintained at a constant level throughout the electrolysis. In this case, however, the electrolysis current is recorded as a function of time to give a curve similar to curve A in Figure 16-4. The analysis is then completed by integrating the current-time curve to obtain the number of coulombs and thus the number of equivalents of analyte. As an alternative, the quantity of electricity consumed by the analyte can be determined with a *chemical coulometer* connected in series with the analytical cell. An example of a chemical coulometer is described in Example 16-4.

Coulometric titrations are similar to other titrimetric methods in that analyses are based on measuring the combining capacity of the analyte with a standard reagent. In the coulometric procedure, the reagent is electrons and the standard solution is a constant current of known magnitude. Electrons are added to the analyte (in this case via the direct current) or to some species that immediately reacts with the analyte until an end point is reached, whereupon the electrolysis is discontinued. The amount of analyte is determined from the magnitude of the current and the time required to complete the titration; the magnitude of the current in amperes is analogous to the molarity of a standard solution, and the time measurement is analogous to the volume measurement in conventional titrimetry.

16D-3

Current-Efficiency Requirements

A fundamental requirement for all coulometric methods is 100% current efficiency; that is, each faraday of electricity must bring about one equivalent of chemical change in the analyte. Note that 100% current efficiency can be achieved without direct participation of the analyte in electron transfer at an electrode. For example, chloride ions are readily determined by either coulometric method by the generation of silver ions at a silver anode. These ions then react with the analyte to form a precipitate or deposit of silver chloride. The quantity of electricity required to complete the silver chloride formation serves as the analytical parameter. In this instance, 100% current efficiency is realized because the number of moles of electrons is exactly equal to the number of moles of chloride ion in the sample despite the fact that these ions do not react directly at the electrode surface.

16D-4

Controlled-Potential Coulometry

In controlled-potential coulometry, the potential of the working electrode is maintained at a constant level such that only the analyte is responsible for the conduction of charge across the solution/electrode interface. The number of coulombs of electricity required to convert the analyte to its reaction product is then determined by recording and integrating the current-versus-time curve during the electrolysis or by means of a chemical coulometer.

Instrumentation

The instrumentation for potentiostatic coulometry consists of an electrolysis cell, a potentiostat, and a device for determining the number of coulombs of electricity consumed by the analyte.

Cells. Figure 16-7 illustrates two types of cells that are used for potentiostatic coulometry. The first consists of a platinum-gauze working electrode, a platinum-wire counter electrode, and a saturated calomel reference electrode. The counter electrode is separated from the analyte solution by a salt bridge that usually contains the same electrolyte as the solution being analyzed. This bridge is present to prevent the reaction products formed at the counter electrode from diffusing into the analyte solution and interfering. For example, hydrogen gas is a common product at a cathodic counter electrode. Unless this species is physically isolated from the analyte solution by the bridge, it will react directly with many of the analytes that are determined by oxidation at the working anode.

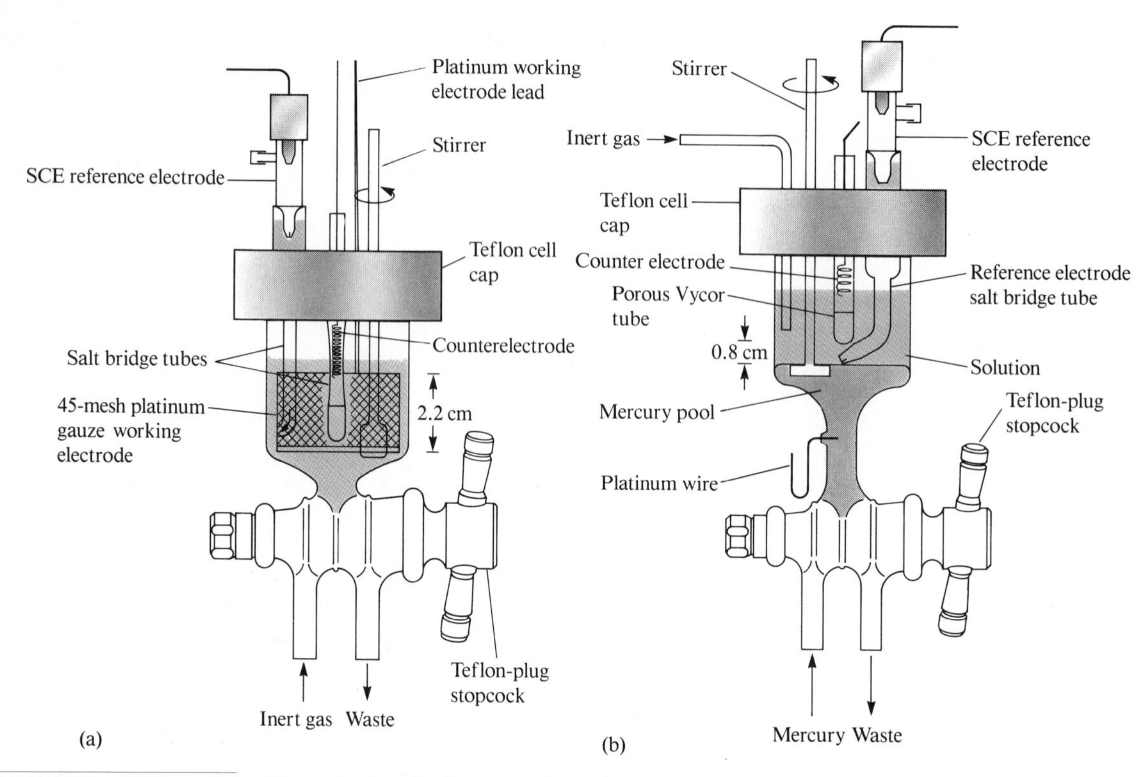

FIGURE 16-7 Electrolysis cells for potentiostatic coulometry. Working electrode: (a) platinum gauze, (b) mercury pool. (Reprinted with permission from J. E. Harrar and C. L. Pomernacki, *Anal. Chem.*, **1973,** *45,* 57. Copyright 1973 American Chemical Society.)

The second type of cell, shown in Figure 16-7b, is a mercury-pool type. As we have already noted, a mercury cathode is particularly useful for separating easily reduced elements as a preliminary step in an analysis. In addition, however, it has found considerable use for the coulometric determination of several metallic cations that form metals that are soluble in mercury. In these applications, little or no hydrogen evolution occurs even at high applied potentials because of the large overvoltage effects. A coulometric cell such as that shown in Figure 16-7b is also useful for the coulometric determination of certain types of organic compounds.

Potentiostats and Coulometers. For controlled-potential coulometry, an automatic potentiostat similar to that described on page 402 is required. Generally, the potentiostat is equipped with a current recorder that provides a plot of current as a function of time; the plot is ordinarily similar to curve *A* in Figure 16-4. The quantity of analyte is then determined from the area under the current-time curve. This area is usually evaluated with an electronic integrator, although as shown in the following example, a chemical coulometer can be employed.

EXAMPLE 16-4

The Fe(III) in a 0.8202-g sample was determined by coulometric reduction to Fe(II) at a platinum cathode. Calculate the percentage of $Fe_2(SO_4)_3$ (fw = 399.88) in the sample if a hydrogen/oxygen coulometer arranged in series with the cell containing the sample evolved 19.37 mL of gas ($H_2 + O_2$) at 23°C and 765 torr (after correction for water vapor).

Converting the gas volume to standard conditions gives

$$V = 19.37 \text{ mL} \times \frac{765 \cancel{\text{torr}}}{760 \cancel{\text{torr}}} \times \frac{273 \cancel{\text{K}}}{296 \cancel{\text{K}}} = 17.982 \text{ mL (STP)}$$

The reactions in the coulometer are

$$4H^+ + 4e \rightarrow 2H_2(g)$$

$$2H_2O \rightarrow O_2(g) + 4H^+ + 4e$$

Thus 3 mol of gas is produced by the passage of 4 mol of electrons, which is equivalent to 0.750 mol of gas per faraday.

$$\text{No. eq } Fe_2(SO_4)_3 = \frac{17.982 \cancel{\text{mL gas}}}{22{,}400 \cancel{\text{mL gas/mol gas}}} \times \frac{1 \cancel{F}}{0.750 \cancel{\text{mol gas}}} \times \frac{1 \text{ eq } Fe_2(SO_4)_3}{\cancel{F}}$$

$$= 1.0704 \times 10^{-3}$$

Since 1 mol of $Fe_2(SO_4)_3$ consumes 2 mol of electrons, there are 2 eq of $Fe_2(SO_4)_3$ in each mole and

$$\text{Wt } Fe_2(SO_4)_3 = 1.0704 \times 10^{-3} \cancel{\text{eq } Fe_2(SO_4)_3} \times \frac{1 \cancel{\text{mol } Fe_2(SO_4)_3}}{2 \cancel{\text{eq } Fe_2(SO_4)_3}} \times \frac{399.88 \text{ g } Fe_2(SO_4)_3}{\cancel{\text{mol } Fe_2(SO_4)_3}}$$

$$= 0.21401 \text{ g } Fe_2(SO_4)_3$$

$$\% \ Fe_2(SO_4)_3 = \frac{0.21401 \text{ g } Fe_2(SO_4)_3}{0.8202 \text{ g sample}} \times 100\% = 26.09\%$$

Applications of Controlled-Potential Coulometry

Controlled-potential coulometric methods have been applied to the determination of some 55 elements in inorganic compounds.[6] Mercury appears to be favored as the cathode, and methods for the deposition at this electrode of two dozen or more metals have been described. The method has found widespread use in the nuclear-energy field for the relatively interference-free determination of uranium and plutonium.

The controlled-potential coulometric procedure also offers possibilities for the electrolytic determination (and synthesis) of organic compounds. For example, trichloroacetic acid and picric acid are quantitatively reduced at a mercury cathode whose potential is suitably controlled:

$$Cl_3CCOO^- + H^+ + 2e \longrightarrow Cl_2HCCOO^- + Cl^-$$

$$\text{2,4,6-}(NO_2)_3C_6H_2OH + 18H^+ + 18e \longrightarrow \text{2,4,6-}(NH_2)_3C_6H_2OH + 6H_2O$$

[6]For a summary of the applications, see J. E. Harrar, in *Electroanalytical Chemistry*, A. J. Bard, Ed., Vol. 8. New York: Marcel Dekker, 1975; E. Bishop, in *Comprehensive Analytical Chemistry*, C. L. Wilson and D. W. Wilson, Eds., Vol. 11D, Chapter XV. New York: Elsevier, 1975.

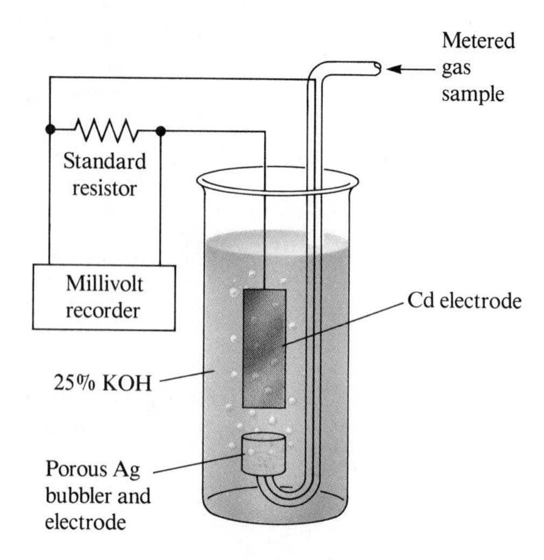

FIGURE 16-8 An instrument for continuously recording the oxygen content of a gas stream.

Coulometric measurements permit the determination of these compounds with a relative error of a few tenths of a percent.

Variable-current coulometric methods are frequently used to monitor continuously and automatically the concentrations of constituents in gas or liquid streams. An important example is the determination of low concentrations of oxygen.[7] A schematic diagram of the apparatus is shown in Figure 16-8. The porous silver cathode breaks up the incoming gas into small bubbles, with the reduction of oxygen then taking place quantitatively within the pores:

$$O_2(g) + 2H_2O + 4e \longrightarrow 4OH^-$$

The anode is a heavy cadmium sheet that reacts to form cadmium hydroxide:

$$2Cd(s) + 4OH^- \longrightarrow 2Cd(OH)_2(s) + 4e$$

Note that a galvanic cell is formed so that no external power supply is required; nor is a potentiostat necessary because the potential of the working anode can never become great enough to cause oxidation of other species. The current passes through a standard resistor, and the potential drop is recorded. The oxygen concentration is proportional to this potential, and the chart paper can be made to display the instantaneous oxygen concentration directly. The instrument is reported to provide oxygen concentration data in the range from 1 ppm to 1%.

16D-5 Coulometric Titrations[8]

Coulometric titrations are carried out with a constant-current source called an *amperostat,* which senses decreases in current in a cell and responds by increasing the potential applied to the cell until the current is restored to its original level. Because of the effects of concentration polarization, 100% current efficiency with respect to the analyte can be maintained only by having present in large excess an auxiliary reagent that is oxidized or reduced at the electrode to give a product that reacts with the analyte. As an example, consider

[7]F. A. Keidel, *Ind. Eng. Chem.*, **1960,** *52,* 491.

[8]For further details on this technique, see D. J. Curran, in *Laboratory Techniques in Electroanalytical Chemistry,* P. T. Kissinger and W. R. Heineman, Eds., Chapter 20. New York: Marcel Dekker, 1984.

the coulometric titration of iron(II) at a platinum anode. At the beginning of the titration, the primary anodic reaction is undoubtedly

$$Fe^{2+} \longrightarrow Fe^{3+} + e$$

As the concentration of iron(II) decreases, however, the requirement of a constant current results in an increase in the applied cell potential. Because of concentration polarization, this increase in potential causes the anode potential to increase to the point where the decomposition of water becomes a competing process:

$$2H_2O \longrightarrow O_2(g) + 4H^+ + 4e$$

The quantity of electricity required to complete the oxidation of iron(II) then exceeds that demanded by theory, and the current efficiency is less than 100%. The lowered current efficiency is avoided, however, by introducing at the outset an unmeasured quantity of cerium(III), which is oxidized at a lower potential than is water:

$$Ce^{3+} \longrightarrow Ce^{4+} + e$$

With stirring, the cerium(IV) produced is rapidly transported from the surface of the electrode to the bulk of the solution, where it oxidizes an equivalent amount of iron(II):

$$Ce^{4+} + Fe^{2+} \longrightarrow Ce^{3+} + Fe^{3+}$$

The net effect is an electrochemical oxidation of iron(II) with 100% current efficiency, even though only a fraction of that species is directly oxidized at the electrode surface.

End Points in Coulometric Titrations

Coulometric titrations, like their volumetric counterparts, require a means for determining when the reaction between analyte and reagent is complete. Generally, the end points described in the chapters on volumetric methods are applicable to coulometric titrations as well. Thus, for the titration of iron(II) just described, an oxidation/reduction indicator, such as 1,10-phenanthroline, can be used; as an alternative, the end point can be established potentiometrically. Similarly, an adsorption indicator or a potentiometric end point can be employed in the coulometric titration of chloride ion by the silver ions generated at a silver anode.

Instrumentation

As shown in Figure 16-9, the equipment required for a coulometric titration includes a source of constant current having a range of one to several hundred milliamperes, a titration vessel, a switch, an electric timer, and a device for monitoring current. Movement of the switch to position 1 simultaneously starts the timer and initiates a current in the titration cell. When the switch is moved to position 2, the electrolysis and the timing are discontinued. With the switch in this position, however, electricity continues to be drawn from the source and passes through a dummy resistor R_D that has about the same electrical resistance as the cell. This arrangement ensures continuous operation of the source, which aids in maintaining the current at a constant level.

The constant-current source for a coulometric titration is often an *amperostat,* an electronic device capable of maintaining a current of 200 mA or

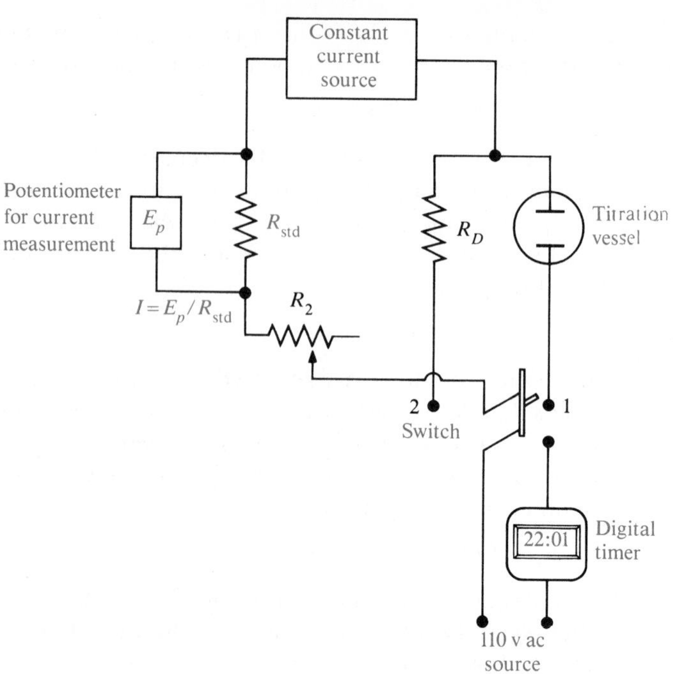

FIGURE 16-9 Diagram of a coulometric titration apparatus.

more that is constant to a few hundredths percent of the current. Amperostats are available from several instrument manufacturers. A much less expensive source that produces reasonably constant currents can be constructed from several heavy-duty batteries connected in series to give an output potential of 100 to 300 V. This output is applied to the cell that is in series with a resistor whose resistance is large relative to the resistance of the cell. Small changes in the cell conductance then have a negligible effect on the current in the resistor and cell.

An ordinary motor-driven electric clock is unsatisfactory for the measurement of the electrolysis time because the rotor of such a device tends to coast when stopped and lag when started. Modern digital electronic timers eliminate this problem.

As shown in Figure 16-9, the currents in a coulometric titration are often measured by determining the potential drop across the standard resistor R_{std} by means of a potentiometer.

Cells for Coulometric Titrations. Figure 16-10 shows a typical coulometric titration cell consisting of a generator electrode at which the reagent is produced and a counter electrode to complete the circuit. The generator electrode—ordinarily a platinum rectangle, a coil of wire, or a gauze cylinder—has a relatively large surface area to minimize polarization effects. In most instances, the counter electrode is isolated from the reaction medium by a sintered disk or some other porous medium in order to prevent interference by the reaction products from this electrode. For example, hydrogen is often evolved at the cathode as an oxidizing agent is being generated at an anode. Hydrogen reacts rapidly with most oxidizing agents, however, which leads to a positive determinate error unless the gas is generated in a separate compartment.

An alternative to isolation of the auxiliary electrode is a device in which

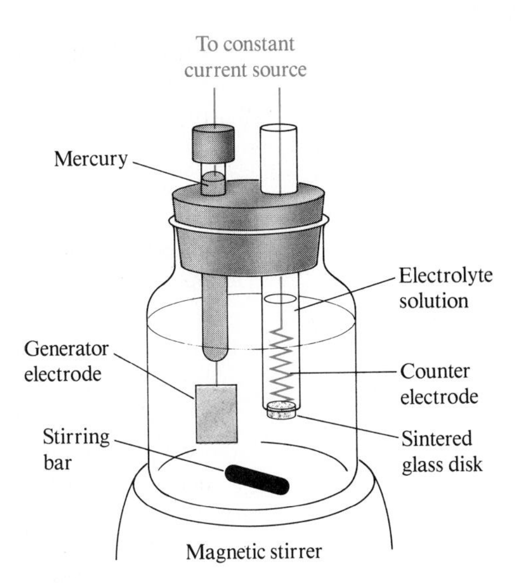

FIGURE 16-10 A typical coulometric titration cell.

the reagent is generated externally, such as that shown in Figure 16-11. The apparatus is so arranged that electrolyte flow continues briefly after the current is discontinued, thus flushing the residual reagent into the titration vessel. Note that the apparatus shown in Figure 16-11 provides either hydrogen or hydroxide ions, depending upon which arm is used. The apparatus has also been used for the generation of other reagents, such as iodine produced by oxidation of iodide at the anode.

A Comparison of Coulometric and Conventional Titrations

The various components of the titrator in Figure 16-9 have their counterparts in the reagents and apparatus required for a volumetric titration. The constant-current source of known magnitude serves the same function as the standard solution in a volumetric method. The electronic timer and switch correspond to the buret and stopcock, respectively. Electricity is passed through the cell for relatively long periods of time at the outset of a coulometric

FIGURE 16-11 A cell for the external coulometric generation of acid and base.

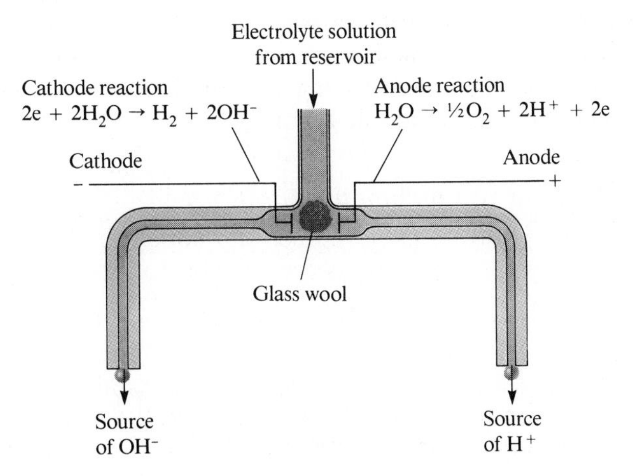

titration, but the time intervals are made smaller and smaller as chemical equivalence is approached. Note that these steps are analogous to the way a buret is operated in a conventional titration.

A coulometric titration offers several significant advantages over a conventional volumetric procedure. Principal among these is the elimination of the problems associated with the preparation, standardization, and storage of standard solutions. This advantage is particularly significant with labile reagents such as chlorine, bromine, and titanium(III) ion, which are sufficiently unstable in aqueous solution to seriously limit their use as volumetric reagents. In contrast, their utilization in a coulometric determination is straightforward because they are consumed as soon as they are generated.

Coulometric methods also excel when small amounts of analyte have to be titrated because tiny quantities of reagent are generated with ease and accuracy through the proper choice of current. In contrast, the use of very dilute solutions and the accurate measurement of small volumes are inconvenient at best.

A further advantage of the coulometric procedure is that a single constant-current source provides reagents for precipitation, complex formation, neutralization, or oxidation/reduction titrations. Finally, coulometric titrations are more readily automated since current control is more readily accomplished than is control of liquid flow.

Coulometric titrations are subject to five principal sources of error: (1) fluctuations in the current during electrolysis, (2) departures from 100% current efficiency, (3) errors in current measurement, (4) errors in time measurement, and (5) titration errors due to the difference between the equivalence point and the end point. The last of these difficulties is encountered in conventional volumetric methods as well. Where indicator error is the limiting factor, the two methods possess comparable reliability.

Currents constant to 0.2 and 0.5% relative are easily realized with simple battery-operated current sources. Control to 0.01% is readily obtainable with most modern amperostatic apparatus. Although generalizations concerning the magnitude of the uncertainty associated with electrode processes are difficult, current efficiencies exceeding 99.5% are reported regularly in the literature. Similarly, modern electronic digital timers permit the measurement of time to within $\pm 0.1\%$ relative (or better).

To summarize, the current-time measurements required for a coulometric titration are inherently as accurate as or more accurate than the comparable volume/molarity measurements of a conventional volumetric method, particularly where small quantities of reagent are involved. When the accuracy of a titration is limited by the sensitivity of the end point, the two titration methods have comparable accuracies.

Applications of Coulometric Titrations

Coulometric titrations have been developed for all types of volumetric reactions.[9] Selected applications are described in this section.

[9]For additional information, see E. Bishop, in *Comprehensive Analytical Chemistry*, C. L. Wilson and D. W. Wilson, Eds., Vol. 11D, Chapters XVIII to XXIV. New York: Elsevier, 1975; J. T. Stock, *Anal. Chem.*, **1984,** *56,* 1R, and **1980,** *52,* 1R.

Neutralization Titrations. Hydroxide ion can be generated at the surface of a platinum cathode immersed in a solution containing the analyte acid:

$$2H_2O + 2e \longrightarrow 2OH^- + H_2(g)$$

The platinum anode must be isolated by some sort of diaphragm to eliminate potential interference from the hydrogen ions simultaneously produced by anodic oxidation of water (Figure 16-10). As a convenient alternative, a silver wire can be substituted for the platinum anode, provided chloride or bromide ions are added to the analyte solution. The anode reaction then becomes

$$Ag(s) + Br^- \longrightarrow AgBr(s) + e$$

Silver bromide does not interfere with the neutralization reaction.

Both strong and weak acids can be titrated coulometrically with a high degree of accuracy. Either a potentiometric or an indicator end point is used. The problems associated with the estimation of the equivalence point are identical with those encountered in a conventional neutralization titration. The coulometric method, however, has the advantage that interference from carbonate ion (Section 10A-3) is far less troublesome. The only measure required to avoid a carbonate error is to remove the carbon dioxide from the solvent by boiling it or by bubbling an inert gas, such as nitrogen, through the solution for a brief period (the latter process is called *sparging*).

Hydrogen ions generated at the surface of a platinum anode can be used for the coulometric titration of strong as well as weak bases:

$$2H_2O \longrightarrow O_2 + 4H^+ + 4e$$

Here, the cathode must be isolated from the analyte solution to prevent interference from hydroxide ion.

Precipitation and Complex-Formation Reactions. Coulometric titrations with EDTA are carried out by reduction of the ammine mercury(II) EDTA chelate at a mercury cathode:

$$HgNH_3Y^{2-} + NH_4^+ + 2e \longrightarrow Hg(l) + 2NH_3 + HY^{3-} \qquad (16\text{-}8)$$

Because the mercury chelate is more stable than the corresponding complexes of such cations as calcium, zinc, lead, or copper, complexation of these ions occurs only after the ligand has been freed by the electrode process.

As shown in Table 16-3, several precipitating reagents can be generated

TABLE 16-3 **Summary of Applications of Coulometric Titrations Involving Neutralization, Precipitation, and Complex-Formation Reactions**

Species Determined	Generator Electrode Reaction	Secondary Analytical Reaction
Acids	$2H_2O + 2e \rightleftharpoons 2OH^- + H_2$	$OH^- + H^+ \rightleftharpoons H_2O$
Bases	$H_2O \rightleftharpoons 2H^+ + \frac{1}{2}O_2 + 2e$	$H^+ + OH^- \rightleftharpoons H_2O$
Cl^-, Br^-, I^-	$Ag \rightleftharpoons Ag^+ + e$	$Ag^+ + Cl^- \rightleftharpoons AgCl(s)$, etc.
Mercaptans	$Ag \rightleftharpoons Ag^+ + e$	$Ag^+ + RSH \rightleftharpoons AgSR(s) + H^+$
Cl^-, Br^-, I^-	$2Hg \rightleftharpoons Hg_2^{2+} + 2e$	$Hg_2^{2+} + 2Cl^- \rightleftharpoons Hg_2Cl_2(s)$, etc.
Zn^{2+}	$Fe(CN)_6^{3-} + e \rightleftharpoons Fe(CN)_6^{4-}$	$3Zn^{2+} + 2K^+ + 2Fe(CN)_6^{4-} \rightleftharpoons K_2Zn_3[Fe(CN)_6]_2(s)$
Ca^{2+}, Cu^{2+}, Zn^{2+}, Pb^{2+}	See Equation 16-8	$HY^{3-} + Ca^{2+} \rightleftharpoons CaY^{2-} + H^+$, etc.

TABLE 16-4 **Summary of Applications of Coulometric Titrations Involving Oxidation/Reduction Reactions**

Reagent	Generator Electrode Reaction	Substance Determined
Br_2	$2Br^- \rightleftharpoons Br_2 + 2e$	As(III), Sb(III), U(IV), Tl(I), I^-, SCN^-, NH_3, N_2H_4, NH_2OH, phenol, aniline, mustard gas, mercaptans, 8-hydroxyquinoline, olefins
Cl_2	$2Cl^- \rightleftharpoons Cl_2 + 2e$	As(III), I^-, styrene, fatty acids
I_2	$2I^- \rightleftharpoons I_2 + 2e$	As(III), Sb(III), $S_2O_3^{2-}$, H_2S, ascorbic acid
Ce^{4+}	$Ce^{3+} \rightleftharpoons Ce^{4+} + e$	Fe(II), Ti(III), U(IV), As(III), I^-, $Fe(CN)_6^{4-}$
Mn^{3+}	$Mn^{2+} \rightleftharpoons Mn^{3+} + e$	$H_2C_2O_4$, Fe(II), As(III)
Ag^{2+}	$Ag^+ \rightleftharpoons Ag^{2+} + e$	Ce(III), V(IV), $H_2C_2O_4$, As(III)
Fe^{2+}	$Fe^{3+} + e \rightleftharpoons Fe^{2+}$	Cr(VI), Mn(VII), V(V), Ce(IV)
Ti^{3+}	$TiO^{2+} + 2H^+ + e \rightleftharpoons Ti^{3+} + H_2O$	Fe(III), V(V), Ce(IV), U(VI)
$CuCl_3^{2-}$	$Cu^{2+} + 3Cl^- + e \rightleftharpoons CuCl_3^{2-}$	V(V), Cr(VI), IO_3^-
U^{4+}	$UO_2^{2+} + 4H^+ + 2e \rightleftharpoons U^{4+} + 2H_2O$	Cr(VI), Ce(IV)

coulometrically. The most widely used of these is silver ion, which is generated at a silver anode.

Oxidation/Reduction Titrations. Table 16-4 reveals that a variety of redox reagents can be generated coulometrically. Of particular interest is bromine, the coulometric generation of which forms the basis for a large number of methods. Of interest as well are reagents not ordinarily encountered in conventional volumetric analysis owing to the instability of their solutions; silver(II), manganese(III), and the chloride complex of copper(I) are examples.

Automatic Coulometric Titrators

A number of instrument manufacturers offer automatic coulometric titrators, most of which employ a potentiometric end point. Some of these instruments are multipurpose and can be used for the determination of a variety of species. Others are designed for a single type of analysis. Examples of the latter are chloride titrators, in which silver ion is generated coulometrically; sulfur dioxide monitors, where anodically generated bromine oxidizes the analyte to sulfate ions; carbon dioxide monitors, in which the gas, absorbed in monoethanolamine, is titrated with coulometrically generated base; and water titrators, in which Karl Fischer reagent is generated electrolytically.

Questions and Problems

16-1 Define

*(a) coulometric titration.
(b) kinetic polarization.
*(c) concentration polarization.
(d) potentiostat.
*(e) faraday.
(f) overvoltage.
*(g) controlled-cathode-potential electrolysis.
(h) ohmic potential.

*(i) cathode depolarizer.
(j) coulomb.
*(k) working electrode.
(l) diffusion.

*16-2 Differentiate between amperostatic coulometry and potentiostatic coulometry.

16-3 Differentiate between kinetic polarization and concentration polarization.

*16-4 Describe three phenomena that cause ions to migrate to an electrode surface from the bulk of a solution.

16-5 Describe variables that tend to decrease concentration polarization.

*16-6 Under what circumstances is kinetic polarization likely to occur?

16-7 List variables that affect the physical properties of electrolytic deposits.

*16-8 Compare a coulometric and a volumetric titration.

16-9 Why is an auxiliary reagent always required in a coulometric titration?

*16-10 How is a constant-cathode-potential electrolysis performed?

*16-11 Nickel is to be deposited from a solution that is 0.200 M in Ni^{2+} and buffered to pH 2.00. Oxygen is evolved at a partial pressure of 1.00 atm at a platinum anode. The cell has a resistance of 3.15 Ω; the temperature is 25°C. Calculate

(a) the thermodynamic potential needed to initiate the deposition of nickel.
(b) the *IR* drop for a current of 1.10 A.
(c) the initial applied potential, given that the oxygen overvoltage is 0.85 V.
(d) the applied potential needed when $[Ni^{2+}]$ is 0.00020, assuming that all other variables remain unchanged.

16-12 Bismuth is to be deposited from a solution in which the analytical concentration of $BiCl_4^-$ is 0.0800 M and that of KCl is 0.400 M. Oxygen is evolved at the anode at a partial pressure of 765 torr. The cell has a resistance of 1.80 Ω, the temperature is 25°C, and the solution is buffered to pH 1.50. Calculate

(a) the thermodynamic potential needed to initiate the deposition of bismuth.
(b) the *IR* drop for a current of 0.42 A.
(c) the initial applied potential, given that the oxygen overvoltage is 0.72 V.
(d) the applied potential when the analytical concentration of the undeposited bismuth is 1.00×10^{-5} M.

16-13 Calculate the minimum difference in standard electrode potentials needed to lower the concentration of the metal M_1 to 1.00×10^{-4} M in a solution that is 0.200 M in the less reducible metal M_2, where

*(a) M_2 is univalent and M_1 is divalent.
(b) M_1 and M_2 are both divalent.
*(c) M_2 is trivalent and M_1 is univalent.
(d) M_2 is divalent and M_1 is univalent.
(e) M_2 is divalent and M_1 is trivalent.

*16-14 A solution is 0.150 M in Co^{2+} and 0.0750 M in Cd^{2+}. Calculate

(a) the Co^{2+} concentration in the solution as the first cadmium starts to deposit.
(b) the cathode potential needed to lower the Co^{2+} concentration to 1×10^{-5} M.

16-15 A solution is 0.0500 M in BiO^+ and 0.0400 M in Co^{2+} and has a pH of 2.50.

(a) What is the concentration of the more readily reduced cation at the onset of deposition of the less reducible one?
(b) What is the potential of the cathode when the concentration of the more easily reduced species is 1.00×10^{-6} M?

*16-16 Electrodeposition is to be used to separate the cations in a solution that is buffered to pH 4.00 and is 5.00×10^{-2} M in Cu^{2+} and 8.00×10^{-3} M in Ag^{+}. Oxygen is evolved at a platinum anode at a pressure of 0.80 atm, the oxygen overvoltage is 0.80 V, and the cell has a resistance of 2.40 Ω.

(a) Which cation deposits first?
(b) Estimate the initial potential that must be applied in order to operate the cell at 0.50 A.
(c) Taking 1.00×10^{-6} M as a reasonable estimate for quantitative removal, calculate the range (versus SCE) within which it is necessary to maintain the cathode potential.

16-17 Electrodeposition is proposed as the means for separating the cations in a solution that is buffered to pH 3.00 and is 1.00×10^{-2} M in Sn^{2+} and 5.00×10^{-2} M in Tl^{+}. Oxygen is evolved at a platinum anode at a pressure of 750 torr. The cell has a resistance of 1.50 Ω; the oxygen overvoltage is 0.75 V.

(a) Estimate the initial potential that must be applied in order to operate the cell at 0.20 A.
(b) Within what range (versus SCE) should the cathode potential be maintained in order to lower the Sn^{2+} concentration to at least 1.00×10^{-6} M without interference from Tl^{+}?

16-18 Electrogravimetric analysis involving control of the cathode potential is proposed as a means for separating Bi^{3+} and Sn^{2+} in a solution that is 0.200 M in each ion and buffered to pH 1.50.

(a) Calculate the theoretical cathode potential at the start of deposition of the more readily reduced ion.
(b) Calculate the residual concentration of the more readily reduced species at the outset of the deposition of the less readily reduced species.
(c) Propose a range (versus SCE), if such exists, within which the cathode potential should be maintained; consider a residual concentration less than 10^{-6} M as constituting quantitative removal.

*16-19 Halide ions can be deposited on a silver anode via the reaction

$$Ag(s) + X^- \longrightarrow AgX(s) + e$$

(a) If 1.00×10^{-5} M is used as the criterion for quantitative removal, is it theoretically feasible to separate Br^- from I^- through control of the anode potential in a solution that is initially 0.250 M in each ion?
(b) Is a separation of Cl^- and I^- theoretically feasible in a solution that is initially 0.250 M in each ion?
(c) If a separation is feasible in either (a) or (b), what range of anode potential (versus SCE) should be used?

*16-20 What cathode potential (versus SCE) is required to lower the total Hg(II) concentration of the following solutions to 1.00×10^{-6} M (assume the reaction product in each case is elemental mercury):

(a) an aqueous solution of Hg^{2+}?
(b) a solution with an equilibrium SCN^- concentration of 0.100 M?

$$Hg^{2+} + 2SCN^- \rightleftharpoons Hg(SCN)_2(aq) \qquad K = 1.8 \times 10^7$$

(c) a solution with an equilibrium Br^- concentration of 0.250 M?

$$HgBr_4^- + 2e \rightleftharpoons Hg(l) + 4Br^- \qquad E^0 = 0.223 \text{ V}$$

16-21 What cathode potential (versus SCE) is needed to lower the analytical concentration of a Ni(II)-containing species to 1.00×10^{-6} M in a solution that is

(a) 0.0010 M in HCl (assume no chloride complex forms)?

(b) 0.212 M in CN^-?

$$Ni^{2+} + 4CN^- \longrightarrow Ni(CN)_4^{2-} \qquad \beta_4 = 1.0 \times 10^{22}$$

(c) 0.215 M in NH_3 (analytical concentration) and has a pH of 9.0?

$$Ni(NH_3)_4^{2-} + 2e \longrightarrow Ni(s) + 4NH_3 \qquad E^0 = -0.530 \text{ V}$$

(d) 0.0612 M in EDTA and buffered to pH 5.00?

*16-22 Calculate the time needed for a constant current of 0.961 A to deposit 0.500 g of Co(II) as

(a) elemental cobalt on the surface of a cathode.
(b) Co_3O_4 on an anode.

16-23 Calculate the time needed for a constant current of 1.20 A to deposit 0.500 g of

(a) Tl(III) as the element on a cathode.
(b) Tl(I) as Tl_2O_3 on an anode.
(c) Tl(I) as the element on a cathode.

*16-24 The cadmium and zinc in a 1.06-g sample were dissolved and subsequently deposited from an ammoniacal solution with a mercury cathode. When the cathode potential was maintained at −0.95 V (versus SCE), only the cadmium deposited. When the current ceased at this potential, a hydrogen/oxygen coulometer in series with the cell had evolved 44.6 mL of gas (corrected for water vapor) at 21.0°C and a barometric pressure of 773 mm Hg. The potential was raised to about −1.3 V, whereupon Zn^{2+} ion was reduced. Upon completion of this electrolysis, an additional 31.3 mL of gas was produced under the same conditions. Calculate the percentage of cadmium and zinc in the ore.

16-25 A 1.74-g sample of a solid containing $BaBr_2$, KI, and inert species was dissolved, made ammoniacal, and placed in a cell equipped with a silver anode. When the potential was maintained at −0.06 V (versus SCE), I^- was quantitatively precipitated as AgI without interference from Br^-. The volume of H_2 and O_2 formed in a gas coulometer in series with the cell was 39.7 mL (corrected for water vapor) at 21.7°C and 748 mm Hg. After precipitation of I^- was complete, the solution was acidified, and the Br^- was removed from solution as AgBr at a potential of 0.016 V. The volume of gas formed under the same conditions was 23.4 mL. Calculate the percentage of $BaBr_2$ and KI in the sample.

*16-26 An excess of $HgNH_3Y^{2-}$ was introduced to 25.00 mL of well water. Express the hardness of the water in terms of ppm $CaCO_3$ if the EDTA needed for the titration was generated at a mercury cathode (Equation 16-8) in 2.02 min by a constant current of 31.6 mA.

16-27 A 0.1516-g sample of a purified organic acid was neutralized by the hydroxide ion produced in 5 min and 24 s by a constant current of 0.401 A. Calculate the equivalent weight of the acid.

*16-28 The nitrobenzene in 210 mg of an organic mixture was reduced to phenylhydroxylamine at a constant potential of −0.96 V (versus SCE) applied to a mercury cathode:

$$C_6H_5NO_2 + 4H^+ + 4e \longrightarrow C_6H_5NHOH + H_2O$$

The sample was dissolved in 100 mL of methanol; after electrolysis for 30 min, the reaction was judged complete. An electronic coulometer in series with the cell indicated that the reduction required 26.74 C. Calculate the percentage of $C_6H_5NO_2$ in the sample.

16-29 Electrolytically generated I_2 was used to determine the amount of H_2S in 100.0 mL of brackish water. Following addition of excess KI, a titration required a constant current of 36.32 mA for 10.12 min. The reaction was

$$H_2S + I_2 \longrightarrow S(s) + 2H^+ + 2I^-$$

Express the results of the analysis in terms of ppm H_2S.

*16-30 At a potential of −1.0 V (versus SCE), CCl_4 in methanol is reduced to $CHCl_3$ at a mercury cathode:

$$2CCl_4 + 2H^+ + 2e + 2Hg(l) \longrightarrow 2CHCl_3 + Hg_2Cl_2(s)$$

At −1.80 V, the $CHCl_3$ further reacts to give CH_4:

$$2CHCl_3 + 6H^+ + 6e + 6Hg(l) \longrightarrow 2CH_4 + 3Hg_2Cl_2(s)$$

A 0.750-g sample containing CCl_4, $CHCl_3$, and inert organic species was dissolved in methanol and electrolyzed at −1.0 V until the current approached zero. A coulometer indicated that 11.63 C was required to complete the reaction. The potential of the cathode was adjusted to −1.8 V. Completion of the titration at this potential required an additional 68.6 C. Calculate the percentage of CCl_4 and $CHCl_3$ in the mixture.

16-31 A 0.1309-g sample containing only $CHCl_3$ and CH_2Cl_2 was dissolved in methanol and electrolyzed in a cell containing a mercury cathode; the potential of the cathode was held constant at −1.80 V (versus SCE). Both compounds were reduced to CH_4 (see Problem 16-30 for the reaction type). Calculate the percentage of $CHCl_3$ and CH_2Cl_2 if 306.7 C was required to complete the reduction.

*16-32 The iron in 0.854-g of ore was converted to the +2 state by suitable treatment and then oxidized quantitatively at a platinum anode maintained at −1.0 V (versus SCE). The quantity of electricity required to complete the oxidation was determined with a chemical coulometer equipped with a platinum anode immersed in an excess of I^- ions. The I_2 liberated by the current through the cell required 26.3 mL of 0.0197 M $Na_2S_2O_3$ to reach a starch end point. What was the percentage of Fe_3O_4 in the sample?

16-33 The phenol content of water downstream from a coking furnace was determined by coulometric analysis. A 100-mL sample was rendered slightly acidic, and an excess of KBr was introduced. To produce Br_2 for the reaction

$$C_6H_5OH + 3Br_2 \longrightarrow Br_3C_6H_2OH(s) + 3HBr$$

a steady current of 0.0313 A for 7 min and 33 s was required. Express the results of this analysis in terms of parts of C_6H_5OH per million parts of water. (Assume that the density of water is 1.00 g/mL.)

*16-34 The CN^- concentration in 10.0 mL of a plating solution was determined by titration with electrogenerated hydrogen ion to a methyl orange end point. A color change occurred after 3 min and 22 s with a current of 43.4 mA. Calculate the number of grams of NaCN per liter of solution.

16-35 Traces of $C_6H_5NH_2$ can be determined by reaction with an excess of electrolytically generated Br_2:

$$3Br_2 + C_6H_5NH_2 \longrightarrow Br_3C_6H_2NH_2 + 3H^+ + 3Br^-$$

The polarity of the working electrode is then reversed, and the excess Br_2 is determined by a coulometric titration involving the generation of Cu(I):

$$Br_2 + 2Cu^+ \longrightarrow 2Br^- + 2Cu^{2+}$$

Suitable quantities of KBr and $CuSO_4$ were added to a 25.0-mL sample containing aniline. Calculate the number of micrograms of $C_6H_5NH_2$ in the sample from the data:

Working Electrode Functioning As	Generation Time with a Constant Current of 1.51 mA, min
Anode	3.76
Cathode	0.270

16-36 Quinone can be reduced to hydroquinone with an excess of electrolytically generated Sn(II):

O O (quinone ring) $+ Sn^{2+} + 2H^+ \rightleftharpoons$ OH OH (hydroquinone ring) $+ Sn^{4+}$

The polarity of the working electrode is then reversed, and the excess Sn(II) is oxidized with Br_2 generated in a coulometric titration:

$$Sn^{2+} + Br_2 \rightleftharpoons Sn^{4+} + 2Br^-$$

Appropriate quantities of $SnCl_4$ and KBr were added to a 50.0-mL sample. Calculate the weight of $C_6H_4O_2$ in the sample from the data:

Working Electrode Functioning As	Generation Time with a Constant Current of 1.062 mA, min
Cathode	8.34
Anode	0.691

CHAPTER 17

Voltammetry and Polarography

Voltammetry comprises a group of electroanalytical methods in which information about the analyte is derived from the measurement of current as a function of applied potential during an electrolysis that is carried out under conditions that encourage polarization of the indicator, or working, electrode. Generally, working electrodes in voltammetry are tiny (usually with surface areas of a few square millimeters) in order to enhance polarization and are called *microelectrodes* as a consequence.

At the outset, it is worthwhile pointing out the basic differences between voltammetry and the two types of electrochemical methods (coulometry and potentiometry) discussed in earlier chapters. Voltammetry is based upon the measurement of a current that develops in an electrochemical cell under conditions of complete concentration polarization. In contrast, potentiometric measurements are made at currents that approach zero and where polarization is absent. Voltammetry differs from coulometry in the respect that with the latter, measures are taken to either minimize or compensate for the effects of concentration polarization. Furthermore, in voltammetry (and also potentiometry), a minimal consumption of analyte takes place, whereas in coulometry essentially all the analyte is converted from one state to another.

Historically, voltammetry developed from the discovery of *polarography* by the Czechoslovakian chemist Jaroslav Heyrovsky[1] in the early 1920s. Polarography, which is still the most widely used of all voltammetric methods, differs from the others in the respect that a special type of microelectrode called a *dropping mercury electrode* is used.[2]

17A Polarographic Measurements

Polarographic data are obtained by recording the current generated in a special type of electrolytic cell as the applied potential is continuously increased.

[1] J. Heyrovsky, *Chem. Listy,* **1922**, *16*, 256. Heyrovsky was awarded the 1959 Nobel Prize in Chemistry for his discovery and development of polarography.

[2] The principles and applications of polarography are considered in detail in a number of monographs; see, for example, I. M. Kolthoff and J. J. Lingane, in *Polarography,* 2nd ed. New York: Interscience, 1952; L. Meites, *Polarographic Techniques,* 2nd ed. New York: Interscience, 1965; J. Heyrovsky and J. Kůta, *Principles of Polarography.* New York: Academic Press, 1966; P. Zuman, *Topics in Organic Polarography.* New York: Pergamon Press, 1970; A. M. Bond, *Modern Polarographic Methods in Analytical Chemistry.* New York: Marcel Dekker, 1980.

The resulting current-voltage curve is called a *polarogram.* Both qualitative and quantitative information are contained in a polarogram.

17A-1 Equipment for Polarographic Measurements

Figure 17-1 is a block diagram of an instrument used for polarographic measurements. The components include a dc source with an output that can be varied from about 0.0 to ± 3.0 V in increments of a few hundredths of a volt, a direct current measuring device that is sensitive to 0.01 μA and that has a range of about 0.0 to ± 100 μA, and a cell containing a dropping mercury electrode and a counter electrode. For convenience, most modern polarographs are automated so that the potential is varied continuously over a desired range and the current is recorded on a strip-chart recorder whose sensitivity can be set at various levels.

Figure 17-2 shows the components of the simplest type of polarographic cell. It consists of a dropping mercury electrode (DME), a mercury-pool electrode, and a vessel having two side arms for bubbling nitrogen through the analyte solution or over its surface. Bubbling nitrogen through the solution (sparging) prior to the analysis removes dissolved oxygen, which interferes with most determinations; reabsorption of oxygen is prevented by maintaining a blanket of nitrogen above the solution throughout the electrolysis.

The dropping mercury electrode, which is used in all polarographic measurements, usually consists of a 5- to 20-cm length of fine capillary tubing (inside diameter approximately 0.05 mm) through which mercury is forced by a head of mercury having a height of about 50 cm (Figure 17-8). This device produces a continuous stream of mercury droplets, with highly reproducible diameters and lifetimes; typically, diameters range from 0.1 to 1 mm and lifetimes from 2 to 6 s.

The second electrode shown in Figure 17-2 consists of a large pool of mercury. Often, however, a saturated calomel electrode is substituted for the pool. In either case, it is important that the size of the second electrode be sufficiently great to make concentration polarization unlikely at this electrode.

In most applications, the dropping electrode, at which the analytical reaction occurs, serves as the cathode, where the analyte is reduced. Anodic applications of the mercury electrode are limited because of the ease with which mercury is oxidized. As a consequence, it can be used to determine only the relatively few species that are more readily oxidized than mercury.

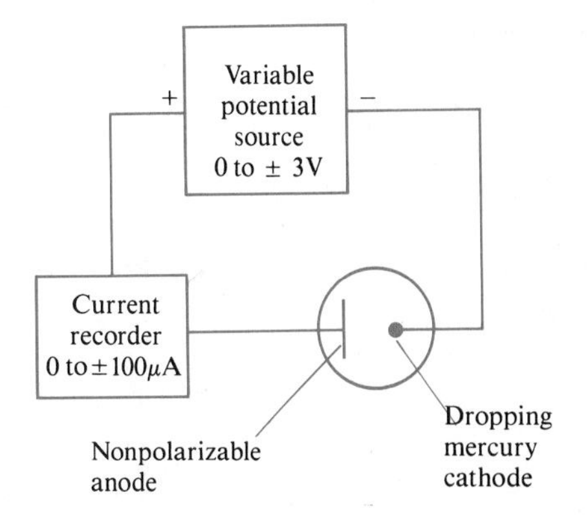

FIGURE 17-1 The equipment for performing polarographic measurements.

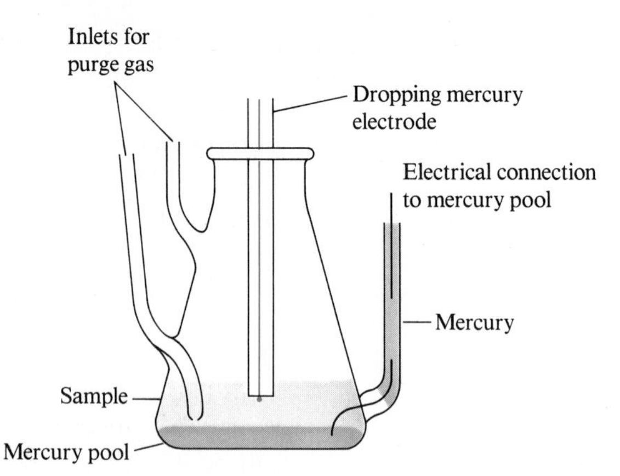

FIGURE 17-2 A simple cell for polarographic measurements.

17A-2 Current Variations with a Dropping Mercury Electrode

As shown in Figure 17-3, the current in a cell containing a dropping electrode undergoes periodic fluctuations corresponding in frequency to the drop rate. These fluctuations occur because the electrode surface area towards which the analyte diffuses varies widely during the lifetime of a drop. After a drop falls from the capillary, the area is zero and so is the current. As a new drop forms and begins to grow, the surface area increases dramatically but then slows and becomes nearly constant near the end of the lifetime of the drop. The current in the cell reflects these changes.

Polarographic currents are measured in one of two ways. In the classical method, the large oscillations in current are reduced in magnitude by damping with an electronic filter, which leads to polarograms such as that shown as curve *A* in Figure 17-4. The average current or the maximum current is then used for analytical purposes. Note in the upper part of the curve, the occasional irregularity in the current, probably caused by vibrations that shortened the drop times of a few of the drops.

Many modern polarographic instruments avoid the problem of large current oscillations by measuring the current only during the last few milliseconds of the lifetime of each drop. From Figure 17-3, it is apparent that current variations at this time are minimal.

FIGURE 17-3 Current changes in a cell containing a dropping mercury electrode.

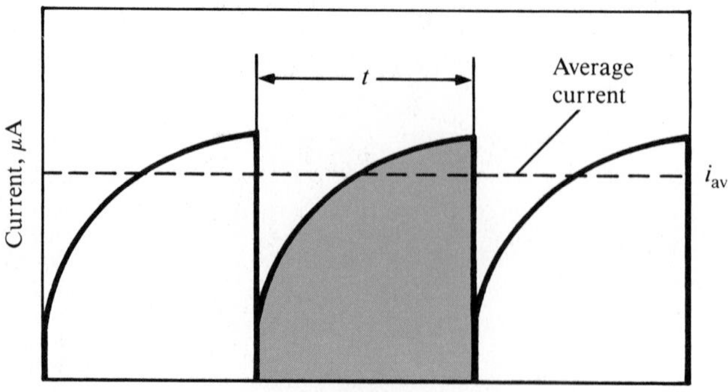

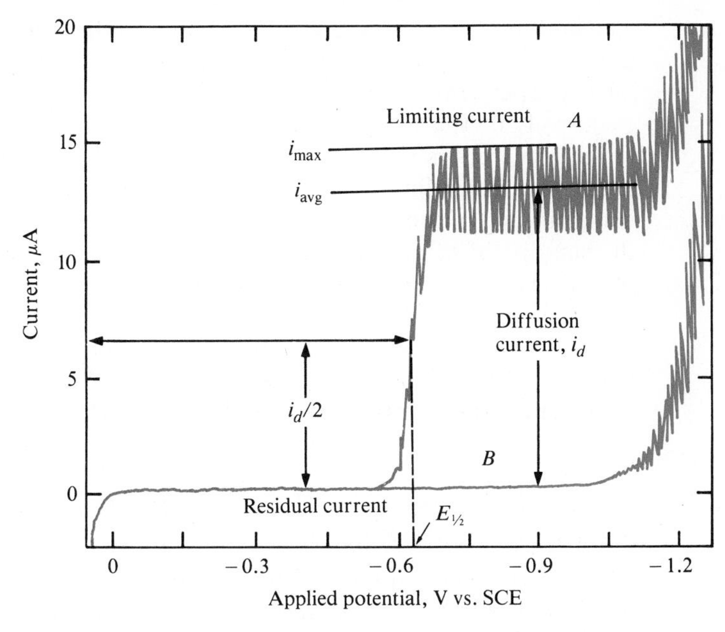

FIGURE 17-4 Polarograms for (*A*) a 1 M solution of HCl that is 5×10^{-4} M in Cd^{2+} and (*B*) a 1 M solution of HCl. (From D. T. Sawyer and J. L. Roberts Jr., *Experimental Electrochemistry for Chemists.* New York: Wiley, 1974. Reprinted by permission of John Wiley & Sons, Inc.)

17A-3 Polarograms

Figure 17-4 shows two typical polarograms, one for a solution that is 1.0 M in hydrochloric acid and 5×10^{-4} M in cadmium ion (curve *A*) and the other for the 1.0 M acid alone (curve *B*). As is usually the case, the microelectrode in these experiments acted as the cathode of an electrolytic cell; that is, it was connected to the negative terminal of the power supply. By convention, the applied potential is given a negative sign under this circumstance. Also by convention, currents are designated as positive when the flow of electrons is from the power supply into the microelectrode.

The step-shaped part of curve *A* in Figure 17-4, called a *polarographic wave,* arises from the reaction

$$Cd^{2+} + 2e + Hg \rightleftharpoons Cd(Hg) \quad (17\text{-}1)$$

where Cd(Hg) represents an amalgam of elemental cadmium dissolved in mercury. The sharp increase in current at about −1.2 V in both polarograms is caused by the reduction of hydrogen ions to hydrogen.

For reasons to be considered presently, a polarographic wave suitable for analysis is obtained only in the presence of a large excess of a *supporting electrolyte;* hydrochloric acid serves this function in the present example. Examination of the polarogram produced by the supporting electrolyte alone reveals that a small current called the *residual current* exists in the cell even in the absence of cadmium ions.

A characteristic feature of a polarographic wave is the *limiting current,* which is the current plateau observed immediately following the sharp current rise. Here the current becomes essentially independent of the applied voltage.

The limiting current is the result of a limit in the rate at which the participant in the electrode process can be brought to the surface of the microelectrode. With proper control over experimental conditions, this rate is determined exclusively by the rate at which the reactant diffuses. A diffusion-controlled limiting current is given a special name, the *diffusion current,* and is assigned the symbol i_d. Ordinarily, the diffusion current is directly proportional to the concentration of the reactive constituent and is thus of prime importance from the standpoint of analysis. As shown in Figure 17-4, the diffusion current is the difference between the limiting and residual currents.

The *half-wave potential,* which is shown in Figure 17-4, is the potential at the point on the polarogram where the current is equal to one half the diffusion current. This potential is usually given the symbol $E_{1/2}$ and is related to the standard electrode potential for the half-reaction responsible for the wave.

17B Polarographic Currents

As just mentioned, limiting currents are observed when the current is limited by the rate at which the reactive species is brought to the electrode surface. We have seen in Section 16A-2 that three mechanisms supply an electrode surface with reactant during an electrolysis: diffusion, migration, and convection. In polarography, every effort is made to minimize the last two and arrange conditions so that the current is totally diffusion-controlled.

To minimize migration, an inactive supporting electrolyte is added to the solution. To be effective, the concentration of the supporting electrolyte must exceed that of the analyte by a factor of 50 to 100. Under this circumstance, only a very small fraction of the migration current involves migration of the analyte ion. The transport of ions to the electrode surface by convection is prevented by avoiding cell vibrations and temperature differentials. With these measures taken, the current becomes limited by the diffusion rate alone and is termed a diffusion current.

17B-1 The Diffusion Current

Polarographic waves are readily interpreted provided the reaction at the dropping mercury electrode is sufficiently rapid that the concentrations of reactants and products at the solution/mercury interface are determined at any instant by the electrode potential alone. Thus, for the reversible reduction of cadmium(II), the concentrations of Cd^{2+} and Cd(Hg) at the interface are always those needed to satisfy the equation

$$E_{\text{applied}} = E_{\text{A}}^{0} - \frac{0.0592}{2} \log \frac{[\text{Cd}]_0}{[\text{Cd}^{2+}]_0} - E_{\text{ref}} \qquad (17\text{-}2)$$

where $[\text{Cd}]_0$ is the concentration of metallic cadmium dissolved in the surface film of the mercury, and $[\text{Cd}^{2+}]_0$ is the concentration of the ion in the aqueous phase. Note that the subscript zero has been employed for the concentration terms to emphasize that this relationship *applies to the surface films of the two media only;* the concentration of cadmium ion in the bulk of the solution and of elemental cadmium in the interior of the mercury drop *are ordinarily quite different from the surface concentrations.* The films we are concerned with are no more than a few atoms or molecules thick.

The term $E_{applied}$ in Equation 17-2 is the potential applied to the cell consisting of the dropping electrode and a reference electrode whose potential is E_{ref}; the term E^0_A is the standard potential for the half-reaction in which a saturated cadmium amalgam is the product. The difference between E^0_A and the standard electrode potential for the reduction to elemental cadmium is about +0.05 V.

Consider what occurs when $E_{applied}$ is sufficiently negative to cause appreciable reduction of cadmium ion. Because the reaction is reversible, the concentration of cadmium ion in the film surrounding the electrode decreases, and the concentration of cadmium in the outer layer of the mercury drop increases *instantaneously* to the levels demanded by Equation 17-2; a surge of current results. This current would rapidly decay to zero were it not for the fact that cadmium ions are mobile in the aqueous medium and can move to the surface of the mercury. The result is a current, *the magnitude of which depends upon the rate at which the cadmium ions move from the bulk of the solution to the surface where reaction occurs;* that is,

$$i = k'v_{Cd^{2+}}$$

where i is the current at an applied potential $E_{applied}$, $v_{Cd^{2+}}$ is the rate of migration of cadmium ions, and k' is a proportionality constant.

As mentioned earlier in this section, every effort is made to minimize the contribution of migration and convection to the mass transport of ions or molecules to the electrode surface. Under these circumstances, diffusion becomes the sole means of movement of cadmium ions to the electrode surface. Because the rate of diffusion is directly proportional to the concentration difference between the two parts of a solution, we can write

$$v_{Cd^{2+}} = k''([Cd^{2+}] - [Cd^{2+}]_0)$$

where $[Cd^{2+}]$ is the concentration *in the bulk of the solution* from which ions are diffusing, $[Cd^{2+}]_0$ is the concentration in the aqueous film surrounding the electrode, and k'' is a proportionality constant. As long as diffusion is the only process bringing cadmium ions to the surface, it follows that

$$i = k'v_{Cd^{2+}} = k'k''([Cd^{2+}] - [Cd^{2+}]_0) = k([Cd^{2+}] - [Cd^{2+}]_0)$$

where k is a new constant. Note that $[Cd^{2+}]_0$ becomes smaller as $E_{applied}$ is made more negative (Equation 17-2). Thus, the rate of diffusion as well as the current increases with increases in applied potential. This potential ultimately becomes so negative that essentially every cadmium ion reaching the drop is reduced, and the concentration of that ion in the surface film approaches zero; the rate of diffusion, and thus the current, then become constant, and the expression for the current becomes

$$i_d = k[Cd^{2+}]$$

where i_d is the potential-independent diffusion current. Note that *the magnitude of the diffusion current is directly proportional to the reactant concentration in the bulk of the solution.* Quantitative polarography is based upon this fact.

A state of *complete concentration polarization* is said to exist when the current in a cell is limited by the rate at which a reactant can be brought to the surface of an electrode. The current required to reach this condition with a microelectrode is small—typically, 3 to 10 μA for a 10^{-3} M solution. It is important to note that deposition by such current levels does not significantly alter the reactant concentration in the time required to obtain a polarogram (see Problem 17-8).

17B-2

Variables Affecting the Magnitude of Diffusion Currents

It has been shown[3] that the magnitude of the average diffusion current is approximated by the relationship

$$(i_d)_{ave} = 607nD^{1/2}m^{2/3}t^{1/6}C \tag{17-3}$$

where n is the number of moles of electrons shown in the half-reaction, m is the mass flow rate of mercury in mg/s, D is the diffusion coefficient of the reactive species in cm^2/s, and C is the analyte concentration in mmol/L. Equation 17-3 is known as the *Ilkovic equation* in honor of the man who first derived it. To obtain an expression for the maximum current, the constant in the Ilkovic equation becomes 706 instead of 607. Note that either the average or the maximum current can be used in quantitative polarography.

The Effect of Capillary Characteristics on Diffusion Currents

The product $m^{2/3}t^{1/6}$ in the Ilkovic equation, called the *capillary constant,* describes the influence of dropping-electrode characteristics upon the diffusion current. Since both m and t are readily evaluated experimentally, comparison of diffusion currents from different capillaries is possible.

The Effect of Temperature on Diffusion Currents

Temperature affects several of the variables that govern the diffusion current for a given species, and its overall influence is thus complex. The most temperature-sensitive factor in the Ilkovic equation is the diffusion coefficient, which ordinarily can be expected to change by about 2.5%/°C. As a consequence, temperature control to a few tenths of a degree is needed for accurate polarographic measurements.

17B-3

Residual Currents

Figure 17-5 shows a residual-current curve (obtained at high sensitivity) for a 0.1 M solution of hydrogen chloride. This current has two sources. The first is the reduction of trace impurities that are almost inevitably present in the

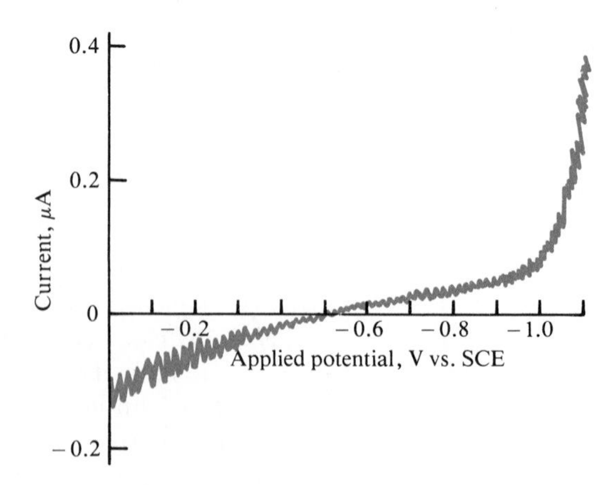

FIGURE 17-5 Residual current for a 0.1 M solution of HCl.

[3] I. M. Kolthoff and J. J. Lingane, *Polarography,* Vol. I, pp. 18–46. New York: Interscience, 1952.

blank solution; contributors here include small amounts of dissolved oxygen, heavy metal ions from the distilled or deionized water, and impurities present in the supporting electrolyte.

A second component of the residual current is the so-called *charging*, or *nonfaradaic, current* resulting from a flow of electrons that charge the mercury drops with respect to the solution; this current may be either negative or positive. At potentials more negative than about −0.4 V, an excess of electrons from the dc source provides the surface of each droplet with a negative charge. These excess electrons are carried down with the drop when it breaks. Since each new drop is charged as it forms, a small but significant current results. At applied potentials smaller than about −0.4 V, the mercury tends to be positive with respect to the solution. Thus, as each drop is formed, electrons are repelled from its surface toward the bulk of the mercury, and a negative current is the result. At about −0.4 V, the mercury surface is uncharged and the charging current is zero. The charging current is a type of nonfaradaic *current* in the sense that electricity is carried across an electrode-solution interface without an accompanying oxidation/reduction process.

The accuracy and sensitivity of the polarographic method ultimately depend upon the magnitude of the nonfaradaic residual current and the accuracy with which a correction for its effect can be determined.

17C Half-Wave Potentials

For electrode reactions that are rapid and reversible, the half-wave potential is a reference point on a polarographic wave that is independent of the reactant concentration but directly related to the standard electrode potential for the half-reaction. In practice, the half-wave potential can be a useful quantity for identification of the species responsible for a given polarographic wave. In this section, we examine some of the factors that affect half-wave potentials.

17C-1 The Effect of Complex Formation on Half-Wave Potentials

We saw in Section 12C-6 that the potential for the oxidation or reduction of a metallic ion is greatly affected by the presence of species that form complexes with that ion. It is not surprising, therefore, that similar effects are observed with polarographic half-wave potentials. The data in Table 17-1 show clearly that the half-wave potential for the reduction of a metal complex is generally more negative than that for reduction of the corresponding simple metal ion. In fact, this negative shift in potential permits the elucidation of the composition of the complex ion and the determination of its formation constant *provided the electrode reaction is reversible.* Thus, for the reactions

$$M^{n+} + Hg + ne \rightleftharpoons M(Hg)$$

and

$$M^{n+} + xA^- \rightleftharpoons MA_x^{(n-x)+}$$

Lingane[4] derived the following relationship between the molar concentrations of the ligand c_L and the shift in half-wave potential brought about by its presence:

$$(E_{1/2})_c - E_{1/2} = -\frac{0.0592}{n}\log K_f - \frac{0.0592x}{n}\log c_L \quad (17\text{-}4)$$

[4] J. J. Lingane, *Chem. Rev.*, **1941**, *29*, 1.

where $(E_{1/2})_c$ and $E_{1/2}$ are the half-wave potentials for the complexed and uncomplexed cations, respectively, K_f is the formation constant for the complex, and x is the molar combining ratio of complexing agent to cation.

Equation 17-4 makes it possible to evaluate the formula for the complex. Thus, a plot of the half-wave potential against log c_L for several ligand concentrations gives a straight line, the slope of which is $0.0592x/n$. If n is known, the combining ratio of ligand to metal ion is readily calculated. Equation 17-4 can then be employed to calculate K_f.

17C-2

The Effect of pH on Half-Wave Potentials

Half-wave potentials become pH-dependent when the analyte consumes or produces hydrogen ions in its reduction or oxidation. For example, quinone is reduced to hydroquinone rapidly and reversibly at a dropping mercury electrode, as shown by the reaction

$$C_6H_4O_2(aq) + 2H^+(aq) + 2e \rightleftharpoons C_6H_4(OH)_2(aq) \qquad E^0 = 0.699\ V$$

Applying the Nernst equation to this half-reaction gives

$$E = 0.699 - \frac{0.0592}{2}\log\frac{[C_6H_4(OH)_2]_0}{[C_6H_4O_2]_0[H^+]_0^2}$$

$$= 0.699 - \frac{0.0592}{2}\log\frac{[C_6H_4(OH)_2]_0}{[C_6H_4O_2]_0} - \frac{0.0592}{2}\log\frac{1}{[H^+]_0^2}$$

$$= 0.699 - \frac{0.0592}{2}\log\frac{[C_6H_4(OH)_2]_0}{[C_6H_4O_2]_0} - 0.0592\ pH_0$$

Thus, the electrode potential becomes more negative as the pH increases; the half-wave potential for quinone has a similar pH dependence.

The reduction of most organic groups involves the consumption of hydrogen ions, as do the electrode reactions of several inorganic anions, such as iodate, nitrite, and sulfite ions. The half-wave potentials of these species are also pH-dependent in a way that is analogous to quinone.

TABLE 17-1

Effect of Complexing Agents on Polarographic Half-Wave Potentials at the Dropping Mercury Electrode

Ion	Noncomplexing Media	1 M KCN	1 M KCl	1 M NH_3, 1 M NH_4Cl
Cd^{2+}	−0.59	−1.18	−0.64	−0.81
Zn^{2+}	−1.00	NR*	−1.00	−1.35
Pb^{2+}	−0.40	−0.72	−0.44	−0.67
Ni^{2+}	−1.01	−1.36	−1.20	−1.10
Co^{2+}	—	−1.45	−1.20	−1.29
Cu^{2+}	+0.02	NR*	+0.04 and −0.22†	−0.24 and −0.51†

*No reduction occurs before involvement of the supporting electrolyte.

†Reduction occurs in two steps having different electrode potentials:

$$Cu^{2+} + 2Cl^- + e \rightleftharpoons CuCl_2^-$$

$$CuCl_2^- + Hg + e \rightleftharpoons Cu(Hg) + 2Cl^-$$

It is important to appreciate that an electrode process that consumes or produces hydrogen ions alters the pH_0 of the solution *at the electrode surface,* often drastically, unless the solution is well buffered. These changes alter the reduction potential of the analyte continuously and cause drawn-out and poorly defined waves. Moreover, pH_0 changes sometimes alter the reaction product, in which case the relationship between diffusion current and concentration may change markedly. Thus, good buffering is *essential* whenever the half-reaction at the dropping electrode consumes or produces hydrogen ions.

17C-3 The Effect of Reversibility on Half-Wave Potentials

Many polarographic electrode processes, particularly those associated with organic systems, are not reversible and cause drawn-out and poorly defined waves. The quantitative description of such waves requires an additional term (involving the activation energy of the reaction) to account for the kinetics of the electrode process. Although half-wave potentials for irreversible reactions ordinarily show a dependence upon concentration, diffusion currents remain linear relative to this variable; such processes are therefore readily adapted to quantitative analysis. The wave for the reduction of oxygen shown in Figure 17-11 is an example of a nonreversible wave.

17C-4 The Effect of Cell Resistance on Half-Wave Potentials

The relationship between the potential applied to a polarographic cell and the potential of the dropping mercury electrode E_{DME} includes a term for IR drop:

$$E_{applied} = E_{DME} - E_{ref} - IR$$

Under many circumstances, the IR drop is negligible with respect to the other two terms on the right-hand side of the equation and E_{DME} responds linearly to changes in $E_{applied}$. When the solution resistance R is large, however, increases in I require a greater and greater fraction of the applied potential to be used in overcoming the IR drop, and linearity is lost. The effect of IR drop on polarograms is illustrated in Figure 17-6. Note that when R is 100 Ω, IR is small enough to have no effect on the slope of the wave. At the higher resistances, however, the influence of the IR drop becomes greater and greater, which leads to drawn-out, poorly defined waves.

The use of potentiostatic control permits the extension of polarography to organic solvents having high electrical resistances. Here, three electrodes are used: a dropping mercury electrode, a small counter electrode, and a

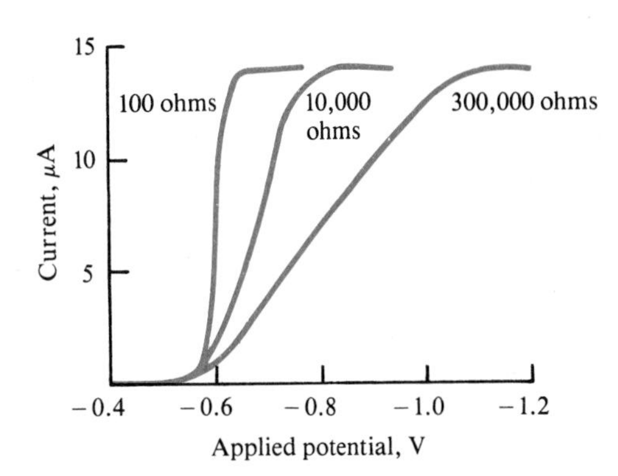

FIGURE 17-6 The effect of cell resistance on a reversible polarographic wave.

reference electrode such as a saturated calomel electrode. The arrangement is similar to that described for potentiostatic coulometry in Section 16D-4 and Figure 16-3. The reference electrode, which is located as close as possible to the dropping electrode, controls the applied potential such that the change in E_{DME} is *linear with time.* The abscissa for the polarogram now becomes E_{DME} rather than $E_{applied}$. The effect is to produce sharply defined waves similar to the steepest wave in Figure 17-6 even when the cell resistance is orders of magnitude greater.

17C-5 Polarograms for Mixtures of Reactants

Ordinarily, the reactants of a mixture behave independently of one another at a microelectrode; a polarogram for a mixture is thus simply the sum of the waves for the individual components. Figure 17-7 shows the polarogram of a mixture of five cations. Clearly, a single polarogram may permit the quantitative determination of several elements. Success depends upon the existence of a sufficient difference between succeeding half-wave potentials to permit evaluation of individual diffusion currents. Approximately 0.2 V is required if the more reducible species undergoes a two-electron reduction; a minimum of about 0.3 V is needed if the first reduction is a one-electron process.

FIGURE 17-7 Polarograms of (*A*) approximately 0.1 mmol each of silver(I), thallium(I), nickel(II), and zinc(II), listed in the order in which their waves appear, in 1 M ammonia/1 M ammonium chloride containing 0.002% Triton X-100 and (*B*) the supporting electrolyte alone. (From L. Meites, *Polarographic Techniques,* 2nd ed., p. 164. New York: Wiley, 1967. Reprinted by permission of John Wiley & Sons, Inc.)

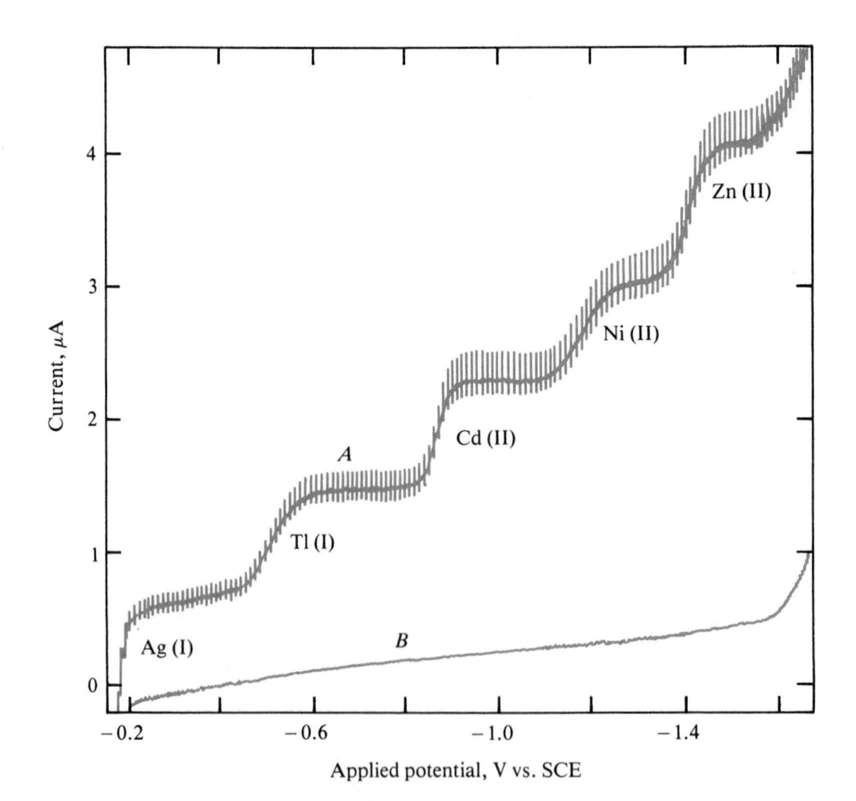

17D Instrumentation

Instrumentation for classical polarography can be relatively simple and inexpensive, costing a few hundred dollars. On the other hand, a modern instrument capable of performing the various types of modified polarography described in later sections is complex and often computer-controlled; and the cost of such instruments may be several thousand dollars.

17D-1 Cells

The simple cell with a mercury-pool anode shown in Figure 17-2 is perfectly adequate for routine quantitative work in aqueous solution. When maximum sensitivity and precision are required, however, and for solutions having high electric resistances, a more sophisticated cell, such as that shown in Figure 17-8a, is more useful. The cell is a heavy-walled glass container that is threaded at the top so that it can be screwed into the polypropylene cap that holds the electrodes. The cell walls are tapered so that as much as 30 mL and as little as 2 mL of the analyte solution can be studied. As shown in Figure 17-8b, the cap is fitted with five O-ring adapters to hold electrodes and other accessories.

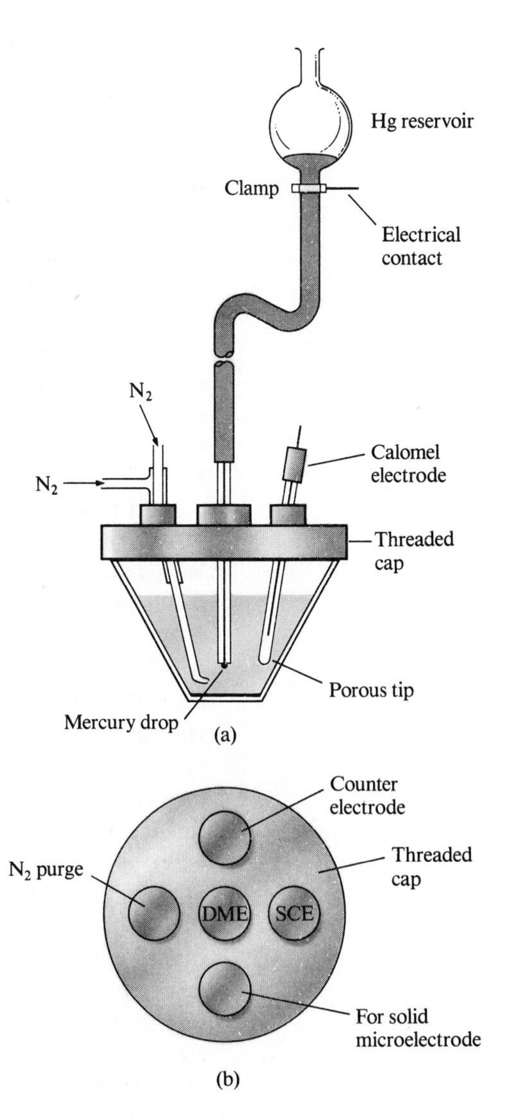

FIGURE 17-8 A dropping mercury electrode and polarographic cell: (a) cross-sectional view; (b) top view of cap.

As shown in (a), one accessory always present is a purge tube fitted to a three-way stopcock (not shown) that permits nitrogen to be bubbled through the solution for deaeration or over the liquid surface to prevent reabsorption of oxygen.

Modern polarography is generally carried out with three electrodes: a dropping mercury electrode, a reference electrode, and a counter electrode. The last is not shown in Figure 17-8a because it is behind the plane of the paper; its position is indicated in Figure 17-8b. The counter electrode can be any inert conducting electrode—a mercury pool, a platinum wire, a graphite surface, or a host of other conductors. Generally, the counter electrode does not have to be isolated from the test solution, as in coulometry, because the quantity of products formed here is small and the time scale of the experiment is such that the likelihood of these products reaching the microelectrode area is negligible.

The anode in a three-electrode arrangement does not need to be large, as it must in the two-electrode arrangement, because any polarization effects at this electrode do not affect the shape of the polarogram since the abscissa is the potential of the dropping electrode versus the reference electrode, rather than the applied potential.

17D-2 Dropping Electrodes

A dropping electrode, such as that shown in Figure 17-8, can be purchased from commercial sources. A 10-cm capillary ordinarily has a drop time of 3 to 6 s under a mercury head of about 50 cm. The tip of the capillary should be as nearly perpendicular to the length of the tube as possible, and care should be taken to ensure a vertical mounting of the electrode; otherwise, erratic and nonreproducible drop times and sizes will be observed.

With reasonable care, a capillary can be used for several months or even years. Such performance, however, requires the use of scrupulously clean mercury and the maintenance of a mercury head, no matter how slight, whenever the electrode is immersed in a solution. If solution comes into contact with the inner surface of the tip, malfunction of the electrode is inevitable. For this reason, the head of mercury should always be increased to provide a good flow before the tip is immersed in a solution. Cleaning of a malfunctioning electrode is usually not very successful in restoring its performance. When not in use, the electrode is immersed in clean mercury (after immersion, the head can be decreased).

Mechanically Controlled Dropping Electrodes

As will be seen in later sections, it is often advantageous to be able to dislodge the drop from a mercury electrode at a precisely controllable time. Various types of hammers, or drop knockers, have been devised for this purpose and are available commercially. An electrode system equipped with one of these devices is shown in Figure 17-9. The mercury is contained in a plastic-lined reservoir about 25 cm above the upper end of the capillary. A compression spring forces the polyurethane-tipped plunger against the head of the capillary, thus preventing a flow of mercury. This plunger is lifted upon activation of the solenoid by a signal from the control system. The capillary is much larger in diameter (0.15 mm) than the typical one. As a result, the formation of the drop is extremely rapid. This system has the advantage that the full-sized drop forms quickly and current measurements can be delayed until the surface area is stable and constant. This procedure largely eliminates the

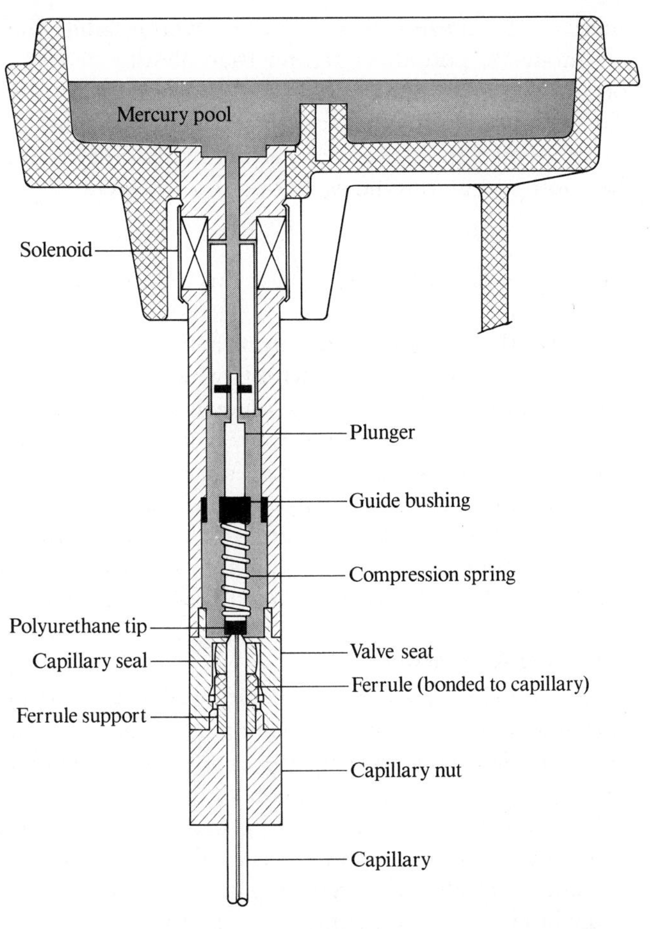

FIGURE 17-9 A modern dropping mercury electrode with mechanical control of drop size and time. (Courtesy EG&G Princeton Applied Research, Princeton, NJ.)

residual, or charging, current that limits the sensitivity of classical polarography.

Advantages and Disadvantages of the Dropping Mercury Electrode

The dropping mercury electrode is the most widely used microelectrode for voltammetry because of several unique features. The first is the unusually high overvoltage associated with the reduction of hydrogen ions. As a consequence, metal ions, such as zinc and cadmium, can be deposited from acidic solution even though their thermodynamic potentials suggest that their deposition without hydrogen formation is impossible. A second advantage is that a new metal surface is generated continuously; thus the behavior of the electrode is independent of its past history. In contrast, solid metal electrodes are notorious for their irregular behavior, which is related to adsorbed or deposited impurities. A third unusual feature of the dropping electrode is that reproducible average currents are *immediately* realized at any given potential regardless of whether this potential is approached from lower or higher settings.

One serious limitation of the dropping electrode is the ease with which mercury is oxidized; this property severely limits the use of the electrode as an anode. At potentials greater than about +0.4 V, formation of mercury(I) occurs and gives a wave that masks the curves of other oxidizable species. In the presence of ions that form precipitates or complexes with mercury(I), this behavior occurs at even lower potentials. For example, in Figure 17-4, the beginning of an anodic wave due to the reaction

$$2Hg + 2Cl^- \longrightarrow Hg_2Cl_2(s) + 2e$$

can be seen at 0 V. Incidentally, this anodic wave can be used for the determination of chloride ion.

Another important disadvantage of the dropping mercury electrode is the nonfaradaic residual current, which limits the sensitivity of the *classical* method to concentrations of somewhat less than 10^{-5} M. At lower concentrations, the residual current is likely to be greater than the diffusion current, a situation that prohibits accurate measurement of the latter. As will be shown later, methods are now available for avoiding this disadvantage.

Finally, the dropping mercury electrode is cumbersome to use and tends to malfunction as a result of clogging. In addition, spilled mercury can create a safety problem because of the volatility and toxicity of the element.

17D-3 Electrical Apparatus

The electrical apparatus for classical polarographic measurements can be relatively simple, consisting of a voltage divider that permits continuous variation of an applied voltage from 0 to about ±3 V known to about ±0.01 V. In addition, provision must be made to measure currents over the range between 0.01 and perhaps 100 μA with a precision of ±0.01 μA. A manual instrument that meets these requirements is easily constructed from equipment available in most laboratories. Recording instruments are more convenient and are available from several commercial suppliers.

A circuit for a simple polarographic instrument is shown in Figure 17-10. Two 1.5-V batteries provide a potential across R_1, a 100-Ω voltage divider that permits variation in the potential applied to the cell. A potentiometer not only measures this potential but also monitors the currents by measuring the potential drop across a precision 10,000-Ω resistor. For convenience, a high-resistance voltmeter can be substituted for the potentiometer.

17E Applications of Polarography

Polarography has been used for the quantitative determination of numerous inorganic and organic species, including molecules of biological and biochemical interest. The applicability of polarography to organic species has been greatly extended in recent years by the development of three-electrode polarography because this modification makes it possible to obtain sharply defined polarographic waves from nonaqueous solvents that have low electrical conductivities.

17E-1 Experimental Details

Oxygen Removal

Figure 17-11 is a polarogram for the reduction of oxygen in an air-saturated solution. Two waves with equal diffusion currents are apparent. The first,

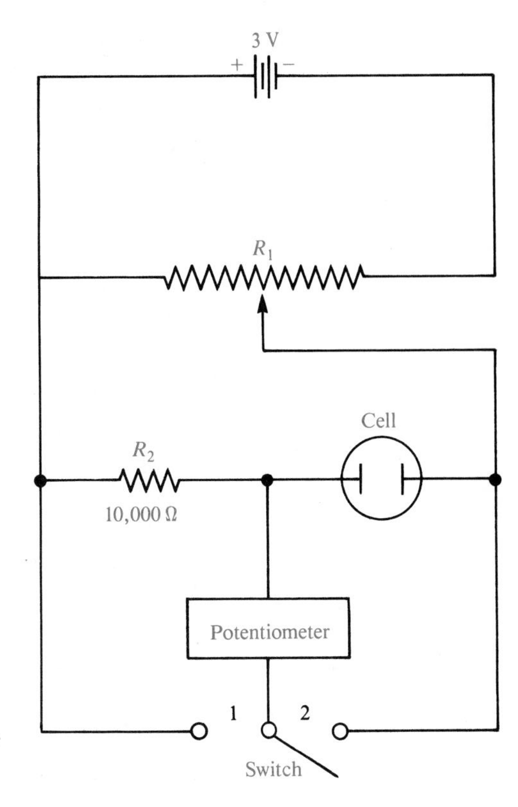

FIGURE 17-10 A simple polarographic circuit. (J. J. Lingane, *Anal. Chem.*, **1949,** *21,* 47. (Reprinted with permission of the American Chemical Society.)

which has a half-wave potential of −0.14 V (versus SCE), results from the reduction of molecular oxygen to hydrogen peroxide; the second, at −0.9 V, arises from the reduction of the hydrogen peroxide to water. Because both reactions are somewhat irreversible, the two waves are drawn out over a region of 0 to −1 V (or greater) and thus interfere with the determination of other species reduced in this useful range. To avoid this interference, removal of oxygen by sparging with nitrogen gas for several minutes generally precedes every polarographic measurement. Blanketing the solution with nitrogen is also necessary to prevent reabsorption.

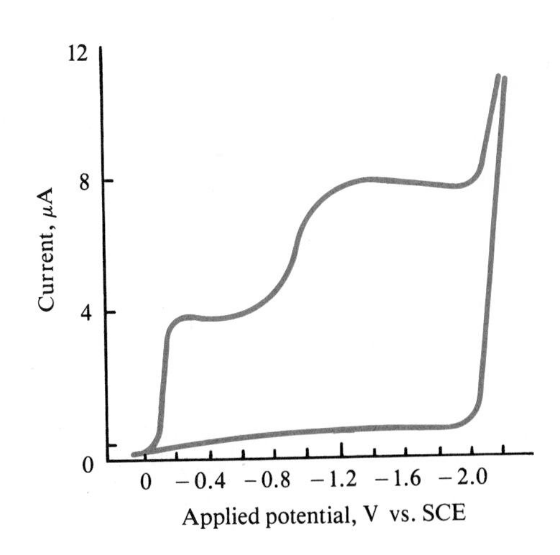

FIGURE 17-11 Polarogram for the reduction of oxygen in an air-saturated 0.1 M KCl solution. The lower curve is for oxygen-free 0.1 M KCl.

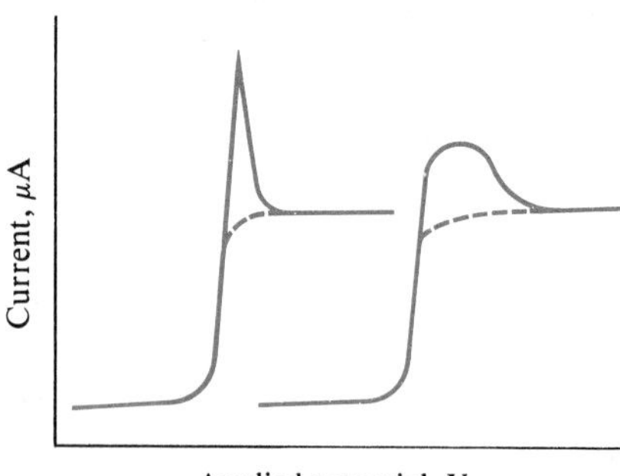

FIGURE 17-12 Typical current maxima.

Current Maxima

Polarograms are frequently distorted by so-called *current maxima* (Figure 17-12), which are troublesome because they interfere with the accurate evaluation of diffusion currents and half-wave potentials. Although the cause or causes of maxima are not fully understood, there is considerable empirical knowledge of methods for eliminating them. Generally, the addition of traces of such high-molecular-weight substances as gelatin, Triton X-100 (a commercial surface-active agent), methyl red, or other dyes will cause a maximum to disappear. Care must be taken to avoid large amounts of these reagents, however, because the excess may decrease the magnitude of the diffusion current. The proper amount of suppressor must be determined by trial and error; the amount required varies widely from one analyte to another.

Evaluation of Diffusion Currents

Limiting currents must be corrected for the residual current, as shown in Figure 17-4. If it is known that the residual current curve increases linearly with applied potential (as it usually does), the diffusion current can be evaluated by extrapolation of the residual current portion of the curve for the sample, as shown in Figure 17-13.

FIGURE 17-13 Extrapolation method for measuring diffusion current. The solution was 0.0010 M in Pb^{2+} in 1 M in $NaClO_4$.

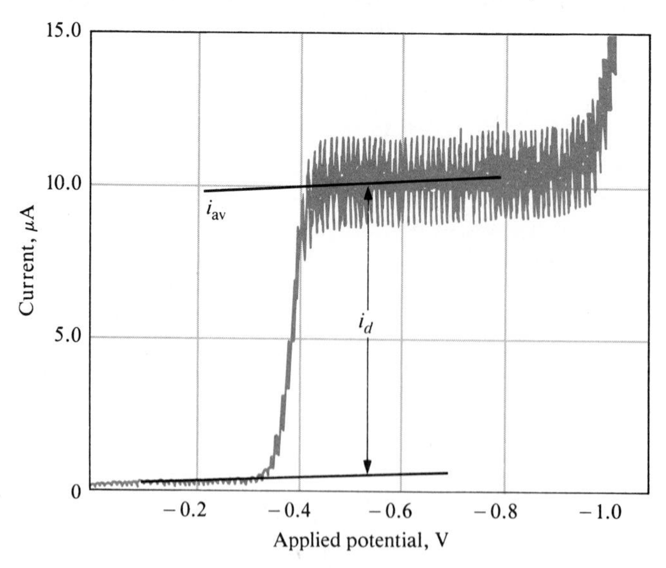

Concentration Determination

Polarographic analyses are generally based on the linear calibration curve produced by the measurement of solutions containing known amounts of analyte. As nearly as possible, these standards should closely resemble the samples to be analyzed in overall composition and encompass a range within which the concentration of the analyte is anticipated. An analysis can of course be carried out even if the calibration curve is nonlinear.

The standard addition method (page 381) is applicable to polarographic analysis. This approach is particularly useful where the diffusion current is sensitive to other components in the sample.

17E-2 The Sensitivity, Accuracy, and Precision of Polarographic Procedures

Classical polarography is most commonly performed on aqueous solutions that are 10^{-2} to 10^{-5} M in analyte. A polarographic analysis is easily performed on 1 to 2 mL of solution and, with a little effort, on volumes as small as a single drop.

The accuracy and precision of polarography depend upon the shape of the analyte wave. For a well-defined wave such as that shown in Figure 17-13, current measurements accurate to 1 to 2% relative can be achieved. Uncertainties from other sources, such as temperature, drop-time variations, or instrumental noise, lead to overall deviations of perhaps 3% relative. For nonreversible and ill-defined waves, the uncertainty may lie in the 5 to 20% range.

17E-3 Inorganic Polarographic Analysis

The polarographic method is widely applicable to the analysis of inorganic species. Most cations, for example, are reduced at the dropping mercury electrode, including ions of alkali and alkaline earth metals. In order to reach the high reduction potentials characteristic of the latter two groups, it is necessary to employ one of the tetraalkyl ammonium halides as supporting electrolyte; these compounds are reduced at even more negative potentials than are the alkali and alkaline earth cations.

The polarographic method is also applicable to the determination of such anions as bromate, iodate, dichromate, vanadate, and nitrite. Generally, these determinations must be carried out in buffered solutions since the half-reactions all consume hydronium ions (Section 17C-2).

17E-4 Organic Polarographic Analysis

Almost from its inception, polarography has been used for the study and determination of organic species, with many papers being devoted to this subject. Several common functional groups are oxidized or reduced at the dropping electrode, making possible the determination of a wide variety of organic compounds.[5]

In general, the reactions of organic compounds at a microelectrode are slower and more complex than those for inorganic species. Consequently, theoretical interpretation of the data is more difficult and often impossible;

[5] For a detailed discussion of organic polarographic analysis, see P. Zuman, *Organic Polarographic Analysis.* Oxford: Pergamon Press, 1964; *Polarography of Molecules of Biological Significance,* W. F. Smyth, Ed. New York: Academic Press, 1979; *Topics in Organic Polarography,* P. Zuman, Ed. New York: Plenum Press, 1970.

moreover, a much stricter adherence to detail is required for quantitative work. Despite these handicaps, organic polarography has proved fruitful for the determination of structure, for quantitative analysis of mixtures, and occasionally for the qualitative identification of compounds.

Solvents for Organic Polarography

Solubility considerations frequently dictate the use of solvents other than pure water for organic polarography; aqueous mixtures containing various amounts of such miscible solvents as glycols, dioxane, acetonitrile, alcohols, Cellosolve, or acetic acid have been employed. Anhydrous media such as acetic acid, formamide, diethylamine, and ethylene glycol have also been investigated. Supporting electrolytes are often lithium salts or tetraalkyl ammonium salts.

Reactive Functional Groups

Organic compounds containing any of the following functional groups can be expected to produce one or more polarographic waves:

1. *Carbonyl groups* in aldehydes, ketones, and quinones produce polarographic waves. In general, aldehydes are reduced at lower potentials than ketones; conjugation of the carbonyl double bond also results in lower half-wave potentials.
2. *Certain carboxylic acids* are reduced polarographically, although simple aliphatic and aromatic monocarboxylic acids are not. Dicarboxylic acids such as fumaric, maleic, or phthalic acid, in which the carboxyl groups are conjugated with one another, give characteristic polarograms; the same is true of certain keto and aldehydo acids.
3. *Most peroxides and epoxides* yield polarograms.
4. *Nitro, nitroso, amine oxide, and azo groups* are generally reduced at the dropping electrode.
5. *Most organic halogen groups* produce a polarographic wave that results from the replacement of the halogen group with an atom of hydrogen.
6. *The carbon/carbon double bond* is reduced when it is conjugated with another double bond, an aromatic ring, or an unsaturated group.
7. *Hydroquinones and mercaptans* produce anodic waves.

In addition, a number of other organic groups cause catalytic hydrogen waves that can be used for analysis. These include amines, mercaptans, acids, and heterocyclic nitrogen compounds. Numerous applications to biological systems have been reported.[6]

17F Modified Voltammetric Methods

As mentioned in Section 17E-2, classical polarography with a dropping mercury electrode is limited to solutions with concentrations greater than about 10^{-5} M. This limitation results from the nonfaradaic current associated with the charging of each mercury drop as it forms. Thus, when the ratio of the faradaic current (from the reduction of the analyte) to nonfaradaic current

[6]M. Brezina and P. Zuman, *Polarography in Medicine, Biochemistry and Pharmacy.* New York: Interscience, 1958; *Polarography of Molecules of Biological Significance,* W. F. Smyth, Ed. New York: Academic Press, 1979.

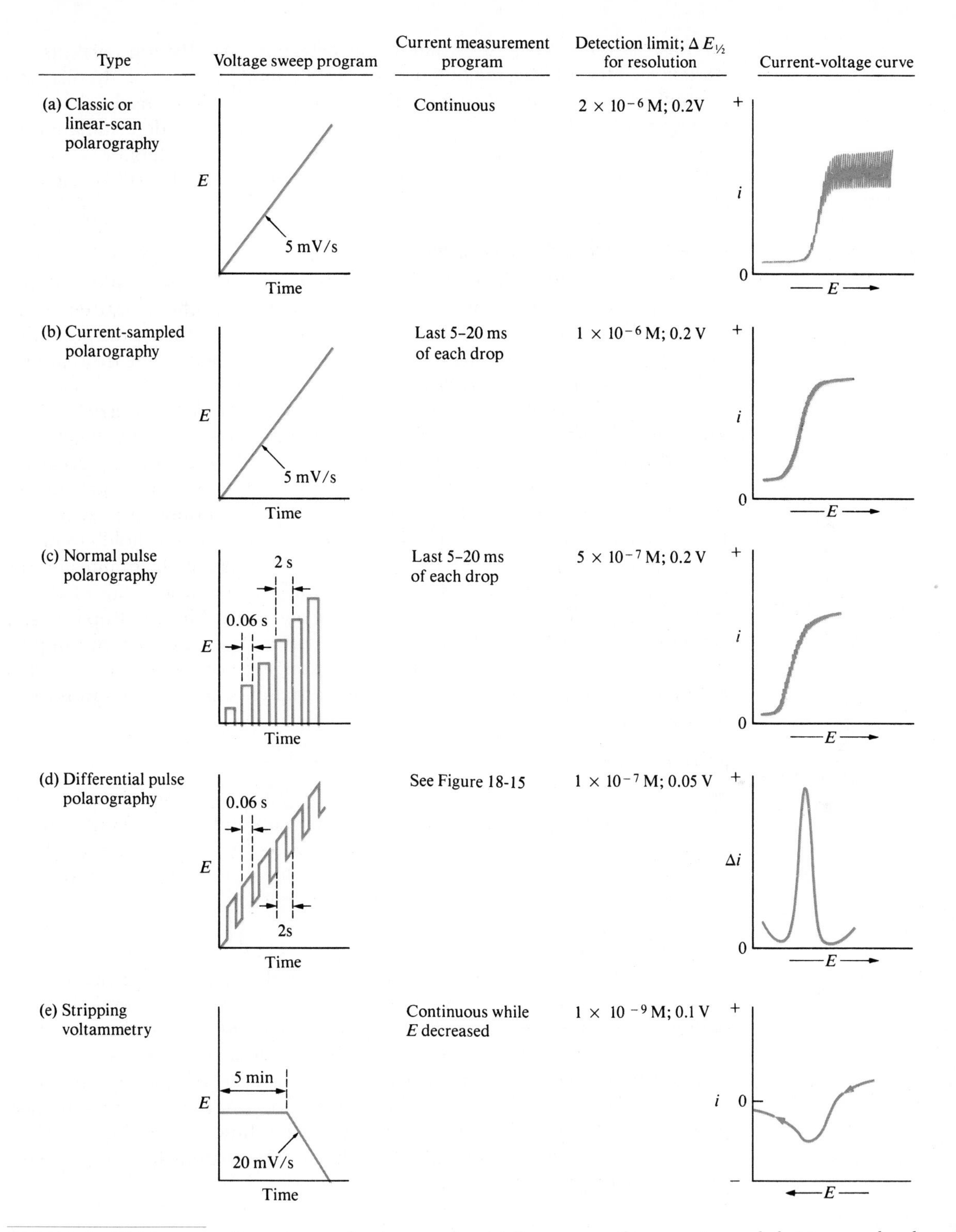

Type
Voltage sweep program
Current measurement program
Detection limit; Δ E½ for resolution
Current-voltage curve
(a) Classic or linear-scan polarography
E
5 mV/s
Time
Continuous
2 × 10⁻⁶ M; 0.2V
+
i
0
E
(b) Current-sampled polarography
E
5 mV/s
Time
Last 5–20 ms of each drop
1 × 10⁻⁶ M; 0.2 V
+
i
0
E
(c) Normal pulse polarography
2 s
0.06 s
E
Time
Last 5–20 ms of each drop
5 × 10⁻⁷ M; 0.2 V
+
i
0
E
(d) Differential pulse polarography
0.06 s
E
2s
Time
See Figure 18-15
1 × 10⁻⁷ M; 0.05 V
+
Δi
0
E
(e) Stripping voltammetry
5 min
E
20 mV/s
Time
Continuous while E decreased
1 × 10⁻⁹ M; 0.1 V
+
i 0
–
E

FIGURE 17-14 Some types of polarography: (a) linear-scan; (b) current-sampled; (c) normal-pulse; (d) differential-pulse; (e) stripping.

approaches unity, large uncertainties in determining diffusion currents are inevitable. One of the major goals of recent modifications of the classical method has been to increase the ratio of the faradaic to nonfaradaic currents by suppressing the latter, thus permitting the quantitative determination of species at lower concentrations. Some of these modern adaptations of the classical method are summarized in Figure 17-14 and are described briefly in the paragraphs that follow.[7]

17F-1

Current-Sampled Polarography

A simple modification of the classical polarographic technique, and one that is incorporated into most modern instruments, involves the measurement of current only for a period near the end of the lifetime of each drop. In this procedure, a mechanical knocker is generally used to detach the drop after a highly reproducible time interval (usually 0.5 to 5 s).

Figure 17-14b illustrates the sequence of events in recording a dc current-sampled polarogram. As in classical polarography, the applied voltage is increased linearly at perhaps 5 mV/s. Rather than the current being recorded continuously, however, it is instead sampled for a 5- to 20-ms period just before termination of each drop. Between sampling periods, the recorder is maintained at its last current level by means of a sample and hold circuit.

The advantage of current sampling is that it substantially diminishes the large current fluctuations due to the continuous growth and fall of drops. Note in Figure 17-3 that the current near the end of the life of a drop is nearly constant, and it is this current only that is recorded in the current-sampled technique. The result is a smoothed curve consisting of a series of steps that are significantly smaller than the current fluctuations encountered in normal polarography.

17F-2

Pulse Polarography

Several types of pulse polarography have been developed since about 1960. The voltage-sweep programs for the two most common types of pulse methods, *normal* and *differential,* are shown in Figure 17-14c and d. Here, dc voltage pulses are applied for 60 ms (or sometimes less) just before each drop is detached from the capillary by a mechanical knocker.

Normal-Pulse Polarograms

As shown in Figure 17-14c, in normal-pulse polarography, the size of successive pulses is increased linearly with time. As in current-sampled polarography, the current is sampled during the last 5 to 20 ms of the drop, where increases in current due to increases in the electrode surface area are minimal. Here again, the large current variations of a classical polarogram are eliminated and a curve similar to that obtained with the normal current-sampled procedure is recorded. An additional advantage, however, is the enhanced sensitivity that accrues to the pulse techniques. The reasons for this improvement become apparent in the next section.

[7]For additional details, see J. B. Flato, *Anal. Chem.,* **1972,** *44* (11), 75A; A. M. Bond, *Modern Polarographic Methods in Analytical Chemistry,* Chapters 4–9. New York: Marcel Dekker, 1980.

Differential-Pulse Polarograms

In differential-pulse polarography, a dc potential that increases linearly with time is applied to the polarographic cell (Figure 17-14d). As in classical polarography, the rate of increase is perhaps 5 mV/s. In contrast, however, a dc pulse of an additional 20 to 100 mV is applied for 60 ms just before detachment of the mercury drop from the electrode. Here again, to synchronize the pulse with the drop, the latter is detached at an appropriate time by a mechanical means.

As shown in Figure 17-15, two current measurements are made alternately—one just prior to the dc pulse and one near the end of the pulse. The *difference in current per pulse* (Δi) is recorded as a function of the linearly increasing voltage. The differential curve that results consists of a peak (Figure 17-16a) whose height is directly proportional to analyte concentration.

One advantage of the derivative type of polarogram is that individual peak maxima can be observed for substances with half-wave potentials differing by as little as 0.04 V; in contrast, classical and normal-pulse polarography require a potential difference of at least 0.2 V for wave resolution. More important, however, is the fact that differential-pulse polarography increases the sensitivity of the polarographic method significantly. This enhancement is illustrated in Figure 17-16. Note that a classical polarogram for a solution containing 180 ppm of the antibiotic tetracycline (Figure 17-16b) gives two barely discernible waves; differential-pulse polarography, in contrast, provides well-defined peaks at a concentration level that is 2×10^{-3} that for the classic wave, or 0.36 ppm. Note also that the current scale for Δi is in nA (nanoamperes), or 10^{-3} μA.

The greater sensitivity of pulse polarography, both normal and differential, can be attributed to two sources. The first is an enhancement of the faradaic current, and the second is a decrease in the nonfaradaic charging current. To account for the former, let us consider the events that must occur in the surface layer around an electrode as the potential is suddenly increased by 20 to 100 mV. If a reactive species is present in this layer, there is a surge of current that lowers the reactant concentration to that demanded by the new potential. As the equilibrium concentration for that potential is approached, however, the current decays to a level just sufficient to counteract

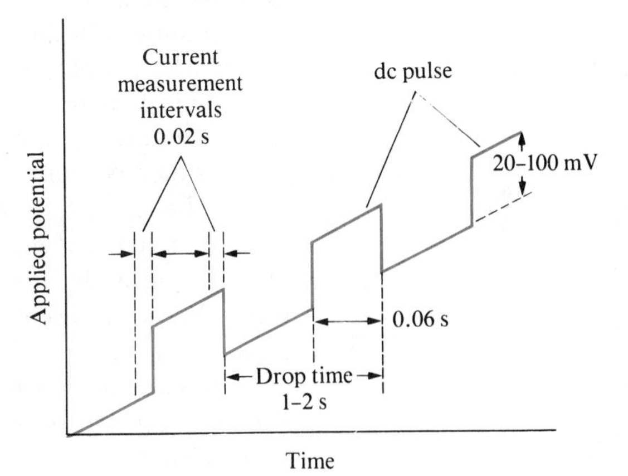

FIGURE 17-15 Voltage program for differential-pulsed polarography.

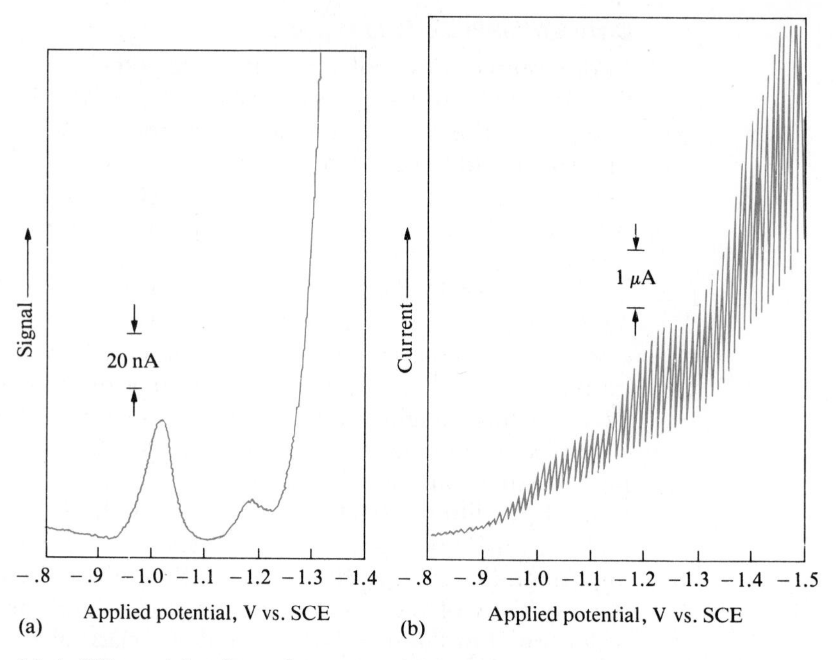

FIGURE 17-16 (a) A differential-pulse polarogram: 0.36 ppm tetracycline·HCl in 0.1 M acetate buffer, pH 4, PAR Model 174 polarographic analyzer, dropping mercury electrode, 50-mV pulse amplitude, 1-s drop. (b) A dc polarogram: 180 ppm tetracycline·HCl in 0.1 M acetate buffer, pH 4, similar conditions. (Reprinted with permission from J. B. Flato, *Anal. Chem.*, **1972,** *44*(11), 75A. Copyright 1972 American Chemical Society.)

diffusion, that is, to the diffusion-controlled current. In classical polarography, the initial surge of current is not observed because the time scale of the measurement is long relative to the lifetime of the momentary current. In contrast, in pulse polarography, the current measurement is made before the surge has completely decayed. Thus, the current measured contains both a diffusion-controlled component and a component that has to do with lowering the surface layer to the concentration to that demanded by the Nernst expression; the total current is typically several times larger than the diffusion current. It should be noted that the solution again becomes homogeneous with respect to the analyte when the drop is detached. Thus, at any given voltage, an identical current surge accompanies each voltage pulse.

When the potential pulse is first applied to the electrode, a surge in the nonfaradaic current also occurs as the charge on the drop increases (page 431). This current, however, decays exponentially with time and approaches zero near the end of the life of a drop when the surface area of the drop is changing only slightly (Figure 17-3). Thus, by measuring currents at this time only, the nonfaradaic residual current is greatly reduced, and the signal-to-noise ratio is larger. Enhanced sensitivity results.

Reliable instruments for pulse polarography are now available at reasonable cost. Differential-pulse polarography in particular has thus become an electroanalytical tool of considerable importance.

17F-3

Stripping Methods

Stripping methods encompass a variety of electrochemical procedures having a common, characteristic initial step.[8] In all these procedures, the analyte is first collected by a constant-electrode-potential deposition at a mercury or a solid microelectrode. After an accurately measured period, the electrolysis is discontinued, and the deposited analyte determined by one of the voltammetric procedures just described. During this second step, the analyte is redissolved, or stripped, from the microelectrode; hence the name attached to these methods. Figure 17-14e illustrates the voltage program followed when linear scanning is used as the second step in an analysis. Often one of the pulse techniques shown in Figure 17-11c and d is used in place of the simple linear scan.

Stripping methods are of prime importance in trace work because the electrolysis step concentrates the analyte and thus permits the determination of minute amounts of analyte with reasonable accuracy. Thus, the analysis of solutions in the 10^{-6} to 10^{-9} M range becomes feasible by methods that are both simple and rapid.

The Electrodeposition Step

Ordinarily, only a fraction of the analyte is deposited during the electrodeposition step; hence, quantitative results depend not only upon control or electrode potential but also upon such factors as electrode size, length of deposition, and stirring rate for both the sample and a standard solution employed for calibration.

Microelectrodes for stripping methods have been formed from a variety of materials, including gold, silver, platinum, mercury, and glassy carbon. The most popular electrode is the *hanging-drop* electrode, which consists of a single drop of mercury in contact with a platinum wire. Figure 17-17 illustrates one method of forming a hanging-drop electrode. An ordinary dropping mercury capillary provides a means of transferring a reproducible quantity of mercury (usually one to three drops) to a Teflon scoop. Note that the dropping mercury capillary does *not* serve as an electrode in this application but merely as a means of dispensing a highly reproducible volume of mercury. The hanging-drop electrode is then formed by rotating the scoop and bringing the mercury into contact with a platinum wire sealed in a glass tube. The drop adheres strongly enough so that the solution can be stirred without displacing the drop; it can be dislodged by tapping the electrode at the completion of the electrolysis, however.

To carry out the determination of a metal ion, a fresh hanging drop is formed, stirring is begun, and a potential a few tenths of a volt more negative than the half-wave potential for the ion of interest is applied. Deposition is allowed to occur for a carefully measured period; 5 min usually suffices for solutions that are 10^{-7} M or greater, 15 min for 10^{-8} M solutions, and 60 min for those that are 10^{-9} M. It should be emphasized that these times seldom result in complete removal of the ion. The electrolysis period is de-

[8]For detailed discussions of stripping methods, see F. Vydra, K. Stulik, and E. Julakova, *Electrochemical Stripping Analysis.* New York: Halsted, 1977; A. M. Bond, *Modern Polarographic Methods in Analytical Chemistry,* Chapter 9. New York: Marcel Dekker, 1980; W. M. Peterson and R. V. Wong, *Amer. Lab.,* **1981,** *13* (11), 116.

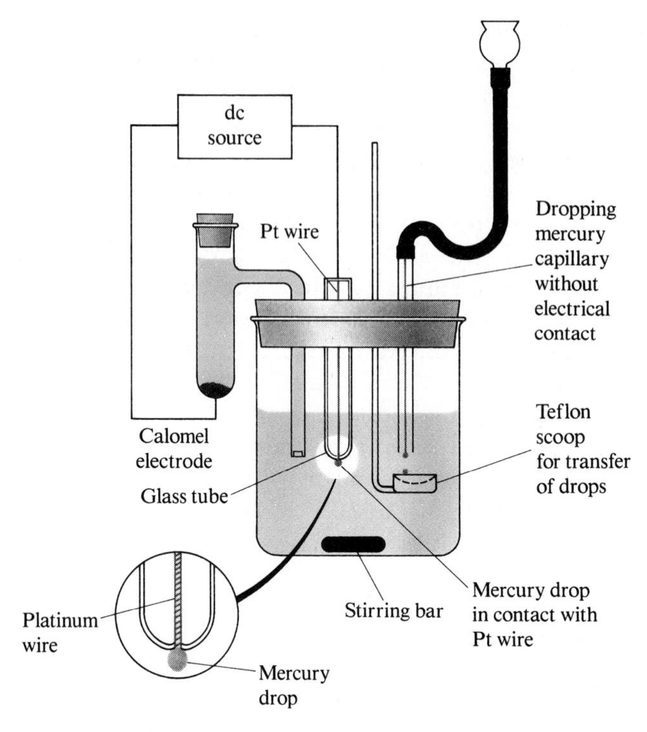

FIGURE 17-17 Apparatus for stripping analysis.

termined by the sensitivity of the method ultimately employed for completion of the analysis.

Voltammetric Completion of the Analysis

The analyte collected in the hanging-drop electrode can be determined by any of several voltammetric procedures. For example, in a linear anodic scan procedure, stirring is discontinued for perhaps 30 s after termination of the deposition. The voltage is then decreased at a linear fixed rate from its original cathodic value, and the resulting anodic current is recorded as a function of the applied voltage. This linear scan produces a curve of the type shown in Figure 17-18. In this experiment, cadmium was first deposited from a 1×10^{-8} M solution by the application of a potential of about -0.9 V (versus

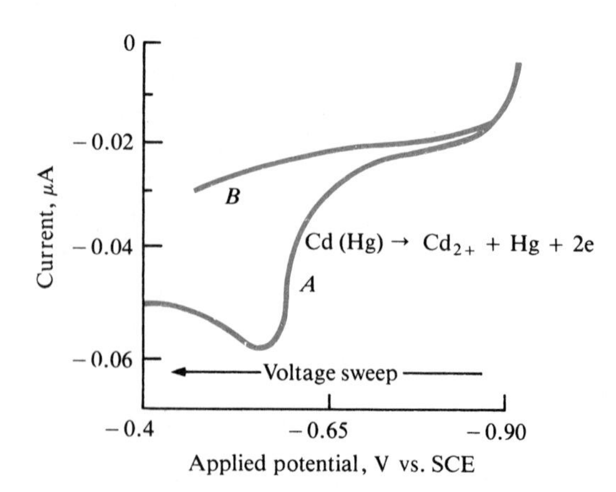

FIGURE 17-18 *A*: Current-voltage curve for the anodic stripping of cadmium. *B*: Residual current curve for blank. (Reprinted with permission and adapted from R. D. DeMars and I. Shain, *Anal. Chem.*, **1957,** *29*, 1826. Copyright 1957 American Chemical Society.)

SCE), which is about 0.3 V more negative than the half-wave potential for this ion. After 15 min of electrolysis, stirring was discontinued; 30 s later, the potential was decreased at a rate of 21 mV/s. A rapid increase in anodic current occurred at about -0.65 V as a result of the reaction

$$Cd(Hg) \longrightarrow Cd^{2+} + Hg + 2e$$

The current decayed after reaching a maximum, owing to depletion of elemental cadmium in the hanging drop. The peak current, after correction for the residual current (curve B in Figure 17-18), was directly proportional to the concentration of cadmium ions over a range of 10^{-6} to 10^{-9} M and inversely proportional to deposition time. This analysis was based on calibration with standard solutions of cadmium ion. With reasonable care, an analytical precision of about 2% relative was obtained.

17F-4 Amperometric Titrations

In an amperometric titration, end points are established by measuring the diffusion or limiting current obtained with a microelectrode as a function of reagent volume (or time for a coulometric titration). In one type of amperometric titration, a microelectrode and counter electrode serve as the indicator system. In a second type, twin microelectrodes are used for end-point determination.

Amperometric titrations are inherently more accurate than conventional voltammetric methods and are less dependent upon such experimental variables as temperature, composition and concentration of supporting electrolyte, and electrode characteristics. Moreover, analytes that are not oxidized or reduced at microelectrodes can be determined, provided the titrant or product of the titration is reactive.

Amperometric Titrations with One Microelectrode

Titration Curves. Typical amperometric titration curves are displayed in Figure 17-19. The curve in Figure 17-19a is characteristic of a titration in which the analyte is reactive at the microelectrode but the titrant is not; the titration of lead(II) with oxalate or sulfate ion is an example. Here, the applied potential is sufficient (say, -1.0 V) to give a diffusion current for lead, which causes a linear decrease in current as lead ions are precipitated from the solution. Beyond equivalence, the current becomes constant at 0 μA. The end point is established by extrapolation of the linear portions of the two curves to their intersection, as shown. Incompleteness of the titration reaction is responsible for the curvature in the equivalence-point region.

The curve in Figure 17-19b is typical of a reaction in which the titrant is reactive but the analyte is not. An example is the titration of magnesium ion with 8-hydroxyquinoline, a reagent that is reduced at -1.6 V (versus SCE). Magnesium ion is unreactive at that potential.

The titration of lead with 8-hydroxyquinoline at -1.6 V gives a curve similar to that shown in Figure 17-19c; both titrant and analyte are reactive at this potential, and the end point corresponds to the minimum in the curve.

In order to obtain plots with linear regions before and following the equivalence point, corrections must be made for changes in volume that occur during the titration. By multiplying the measured diffusion currents by $(V + v)/V$, where V is the original volume of the solution containing the analyte

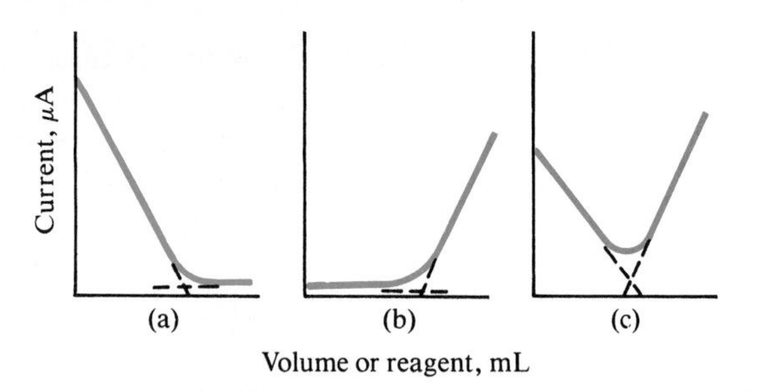

FIGURE 17-19 Typical amperometric titration curves. (a) Analyte is reduced, reagent is not. (b) Reagent is reduced, analyte is not. (c) Both reagent and analyte are reduced.

and v is the volume of titrant, all measured currents are corrected back to the original volume. An alternative is to make v negligibly small by using a titrant solution that is 20 (or more) times as concentrated as the analyte.

Apparatus. A simple manual polarograph is entirely satisfactory for detecting end points in an amperometric titration. The applied potential does not need to be known very accurately (± 0.1 V) since all that is necessary is to select a potential within the diffusion current region of at least one reactant or product in the titration.

The dropping mercury electrode, while useful for many amperometric titrations, cannot be applied to reactions in which one of the reactants is an oxidizing agent because of the ease with which mercury is oxidized. For such analyses, inert microelectrodes manufactured from platinum, graphite, or glassy carbon are used. An example is the *rotating platinum electrode,* which has found widespread use for titrations with such oxidants as bromine, chlorine, and iodine.

Figure 17-20 shows a rotating platinum electrode. It consists of a short length of platinum wire sealed into the side of a soft glass tube. Mercury in the tube provides contact between the wire and the lead to the polarograph. The tube is held in the hollow chuck of a synchronous motor and is rotated in excess of 600 rpm.

FIGURE 17-20 Typical cell arrangement for amperometric titrations with a rotating platinum electrode.

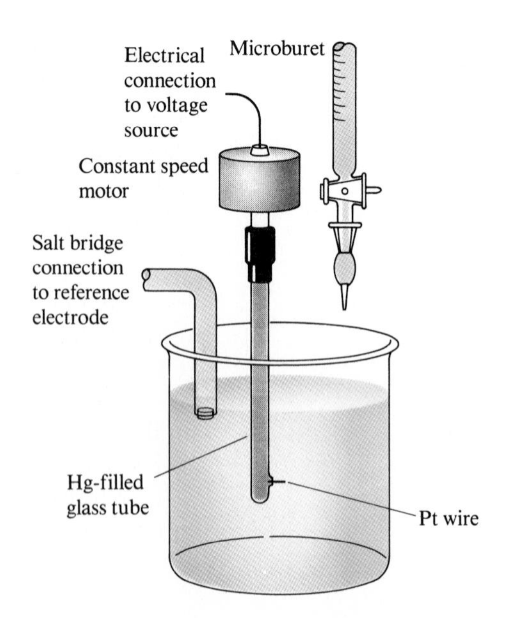

Current-voltage curves with rotating platinum electrodes are similar in appearance to those obtained with the dropping mercury electrode except that current fluctuations due to the growth and detachment of the drop are not observed. Limiting currents tend to be as much as 20 times greater, however, because the reactive species is brought to the surface of the rotating electrode not only by diffusion but by mechanical mixing as well. The rotating electrode also provides steady currents instantaneously, which is in distinct contrast to inert microelectrodes in an unstirred solution, which take several minutes to reach a constant current after each reagent addition.

The low overvoltage of hydrogen severely limits the use of the rotating platinum electrode as a cathode in acidic solutions. Moreover, the high currents produced make the electrode particularly sensitive to traces of dissolved oxygen. Limiting currents obtained with a rotating platinum electrode are influenced in some measure by the previous history of the electrode and are seldom as reproducible as those obtained with the dropping electrode. This lack of reproducibility is not ordinarily a serious problem for most amperometric titrations.

Applications. The amperometric end point has been largely confined to titrations in which the product is a sparingly soluble solid or a stable complex and to certain oxidation/reduction titrations, particularly those in which bromine or iodine is the titrant. Selected applications are shown in Table 17-2.

Amperometric Titrations with Twin Microelectrodes

An amperometric end point can also be obtained with a pair of identical microelectrodes to which a small potential (0.1 to 0.2 V) is applied. The current is then plotted as a function of titrant volume. The end point is marked by a decrease in the current to zero, by a sudden increase in the current from zero, or as a minimum (at zero) in a V-shaped curve.

The use of two polarizable electrodes for end-point detection was first

TABLE 17-2 **Applications of Amperometric Titrations**

Reagent	Reaction Product	Type Electrode*	Substance Determined
K_2CrO_4	Precipitate	DME	Pb^{2+}, Ba^{2+}
$Pb(NO_3)_2$	Precipitate	DME	SO_4^{2-}, MoO_4^{2-}, F^-, Cl^-
8-Hydroxyquinoline	Precipitate	DME	Mg^{2+}, Zn^{2+}, Cu^{2+}, Cd^{2+}, Al^{3+}, Bi^{3+}, Fe^{3+}
Cupferron	Precipitate	DME	Cu^{2+}, Fe^{3+}
Dimethylglyoxime	Precipitate	DME	Ni^{2+}
α-Nitroso-β-naphthol	Precipitate	DME	Co^{2+}, Cu^{2+}, Pd^{2+}
$K_4Fe(CN)_6$	Precipitate	DME	Zn^{2+}
$AgNO_3$	Precipitate	RP	Cl^-, Br^-, I^-, CN^-, RSH
EDTA	Complex	DME	Bi^{3+}, Cd^{2+}, Cu^{2+}, Ca^{2+} and so on
$KBrO_3$, KBr	Substitution, addition, or oxidation	RP	Certain phenols, aromatic amines, olefins; N_2H_4, As(III), Sb(III)

*DME = dropping mercury electrode; RP = rotating platinum electrode.

proposed before 1900. Almost 30 years elapsed, however, before chemists came to appreciate the advantages of the method.[9] The name *dead-stop end point* was used to describe the technique, and the term is still occasionally encountered.

Apparatus. A principal advantage to an amperometric titration with microelectrodes is the simplicity of the equipment. No reference electrode is required, and the only instrumentation needed, beyond the identical microelectrodes, is a simple voltage divider powered by a dry cell and a microammeter for current detection.

Typical Applications. Twin silver microelectrodes can be used to detect the end point for many titrations that involve silver nitrate as titrant (Table 7-1). Consider, for example, what occurs when 0.1 V is applied between two such electrodes immersed in a solution being titrated to determine its bromide concentration. No current is observed before the equivalence point is reached because no easily reduced species is present in the solution; that is, complete cathodic polarization prevents the flow of electricity. Note that the anode is not polarized because the reaction

$$Ag(s) \longrightarrow Ag^{+} + e$$

could occur if a suitable cathodic reactant existed.

Beyond the equivalence point, depolarization of the cathode takes place because silver ions are now in excess and can be reduced at the electrode:

$$Ag^{+} + e \longrightarrow Ag(s)$$

A current therefore develops as a result of these two half-reactions. In common with other amperometric methods, the magnitude of the current is directly proportional to the concentration of excess silver ion, and the titration curve resembles that shown in Figure 17-19b.

A pair of platinum microelectrodes can be used to obtain end points for oxidation/reduction titrations. For example, for the titration of iron(II) with cerium(IV), the reaction is

$$Fe^{2+} + Ce^{4+} \longrightarrow Fe^{3+} + Ce^{3+}$$

At the outset, the cathode is polarized when a small potential is applied to the electrode because the system contains no easily reduced species [the anode is not polarized since oxidation of iron(II) could occur there]. The first addition of cerium(IV) causes the formation of iron(III), which depolarizes the cathode. A current then develops in the cell, the magnitude of which is dependent upon the concentration of iron(III). With further additions of reagent, this current increases and attains a maximum value when the concentrations of iron(II) and iron(III) are equal. The current then decreases as the supply of iron(II) becomes depleted (that is, as the anode starts to become polarized). The cell becomes completely polarized at the equivalence point. Although the

[9] C. W. Foulk and A. T. Bawden, *J. Amer. Chem. Soc.*, **1926**, *48*, 2045. For an excellent analysis of this type of end point, see J. J. Lingane, *Electroanalytical Chemistry*, 2nd ed., pp. 280–294. New York: Interscience, 1958.

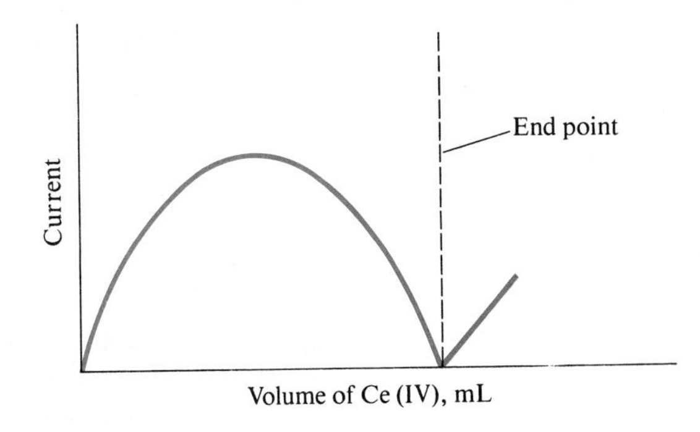

FIGURE 17-21 The amperometric titration of Fe(II) with Ce(IV) with twin microelectrodes.

solution contains a reducible species (Fe^{3+}) and an oxidizable species (Ce^{3+}), a potential of about 0.7 V is required to cause the cell reaction

$$Fe^{3+} + Ce^{3+} \longrightarrow Fe^{2+} + Ce^{4+}$$

to take place. Thus, the cell is completely polarized with an applied potential of only 0.1 V, and the current is zero. With the further addition of cerium(IV), both the cathode and the anode are depolarized as a consequence of the half-reactions

$$Ce^{4+} + e \longrightarrow Ce^{3+} \quad \text{cathode}$$

$$Ce^{3+} \longrightarrow Ce^{4+} + e \quad \text{anode}$$

The current now becomes dependent upon the concentration of excess cerium(IV). Figure 17-21 depicts the curve for this titration.

Questions and Problems

17-1 Make a distinction between
*(a) limiting current, diffusion current, and residual current.
(b) faradaic current and nonfaradaic current.
*(c) voltammetry and polarography.
(d) normal- and differential-pulse polarography.
*(e) classic and current-sampled polarography.
(f) two-electrode and three-electrode polarography.

*17-2 What would you expect to observe if a series of polarograms were obtained for a solution that was 1.0×10^{-3} M in Cd^{2+} and 0.0, 0.001, 0.01, 0.10, and 1.0 M in KNO_3?

17-3 Why is nitrogen bubbled through solutions before polarograms are obtained?

*17-4 Describe the causes of a residual current.

17-5 For voltammetry, what are the advantages of dropping mercury electrodes over other microelectrodes?

*17-6 Why are limiting currents with a rotating platinum electrode larger than with a dropping mercury electrode of a comparable surface area?

17-7 Under what circumstances are stripping methods particularly advantageous?

17-8 The polarogram for 25.0 mL of a solution that is 4.62×10^{-3} M in Pb^{2+} produces a diffusion current for Pb^{2+} of 42.1 μA. Calculate the percent change in Pb^{2+} concentration if the current in the diffusion current region is allowed to continue for
*(a) 5 min (b) 10 min *(c) 20 min (d) 30 min

*17-9 The following data were obtained from polarograms for a series of Pb^{2+} solutions:

Concn Pb^{2+}, mmol/L	Limiting Current, μA	Concn Pb^{2+}, mmol/L	Limiting Current, μA
0.00	1.32	3.06	27.91
0.510	5.65	4.08	36.08
1.02	10.70	5.10	45.82
2.04	19.08		

(a) Plot diffusion current versus concentration.
(b) Use the least-squares method to derive an equation for the best straight line for the plot in part (a).
(c) Calculate the standard deviations for the slope and about regression.
(d) Calculate the concentration of lead and the absolute and relative standard deviations for each of the following analyses:

	i_d	No. of Replicate Measurements
*(1)	2.76	2
(2)	7.75	1
*(3)	7.75	3
(4)	26.32	3
*(5)	40.01	4

*17-10 For the data in Problem 17-9, calculate
(a) a mean value for i_d/C.
(b) the absolute and relative standard deviation for the mean.
(c) a theoretical value for i_d/C assuming the data are for average currents, the diffusion coefficient for $Pb^{2+} = 9.8 \times 10^{-6}$ cm²/s, the reaction product is Pb(Hg), the drop time is 2.88 s, and the mercury flow rate is 2.53 mg/s.
(d) the percent error in the theoretical value for i_d/C.

17-11 The following data for the polarographic reduction of Zn^{2+} to zinc amalgam were obtained with an electrode having a drop time of 3.07 s and a mercury flow rate of 2.13 mg/s:

mmol Zn^{2+}/L	i_d, μA
0.500	4.11
2.00	16.23
3.50	28.16
7.00	57.73

(a) Calculate a mean value for i_d/C for the four data. What is the standard deviation for the set?
(b) What would be the value of i_d/C for similar solutions with an electrode having a drop time of 6.48 s and a flow rate of 0.956 mg/s?
(c) Calculate the zinc concentration of a solution that produced a diffusion current of 31.76 μA with the electrode in part (b).
(d) Calculate the percent error introduced in the zinc determination if a correction for the change in capillaries had not been made.

*17-12 The following data were collected for three dropping electrodes. Complete the data for electrodes A and C:

	A	B	C
Flow rate, mg/s	0.982	3.92	6.96
Drop time, s	6.53	2.36	1.37
i_d/C, μA L mmol^{-1}		4.86	

*17-13 Calculate the diffusion coefficient for the species responsible for the data in part (b) of Problem 17-12 assuming that the reduction involved 2 mol of electrons for each mole of reactant.

*17-14 Electrode C in Problem 17-12 was employed to study the reduction of an organic compound known to have a diffusion coefficient of 6.2×10^{-6} cm²/s. A 4.00×10^{-4} M solution of the compound yielded a diffusion current of 8.41 μA. Calculate n for the reaction.

17-15 In basic solution, a 1.00×10^{-3} M solution of benzaldehyde produced a wave having an average diffusion current of 6.12 μA. The electrode had a drop time of 4.11 s and a mercury flow rate of 1.77 mg/s. In a solution buffered to pH 2.0, the same concentration of benzaldehyde had a diffusion current of 3.05 μA. Suggest what the reduction product was in each case if the diffusion coefficient for benzaldehyde is 7.1×10^{-6} cm²/s.

*17-16 An organic compound underwent a two-electron reduction at dropping electrode A in Problem 17-12. A diffusion current of 7.01 μA was produced by a 9.6×10^{-4} M solution of the compound. Calculate the diffusion coefficient for the compound.

17-17 A standard-addition method for the polarographic determination of Pb^{2+} involved addition of 5.00 mL of 8.00×10^{-3} Pb^{2+} to one of two 25.00-mL aliquots of the sample. Supporting electrolyte and water were then added to each aliquot to give a final volume of 50.00 mL. Calculate the molar Pb^{2+} concentration based upon the following data:

Sample	Diffusion Current, μA	
	Sample Only	Sample Plus Standard Addition
*(a)	78.3	88.7
(b)	13.5	23.9
*(c)	41.0	51.4
(d)	54.6	65.0

*17-18 Shown below is the polarogram for a solution that is 1.0×10^{-4} M in KBr and 0.1 M in KNO_3. Offer an explanation of the wave that occurs at +0.12 V and the rapid change in current that starts at about +0.48 V. Would the wave at +0.12 V have any analytical applications? Explain.

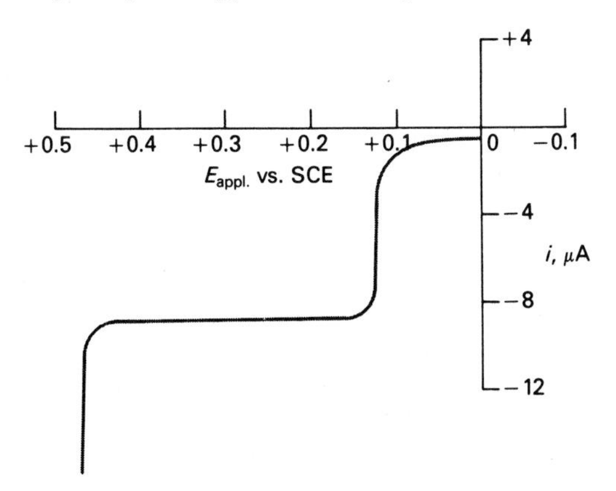

17-19 The reaction

$$Ox + 4H^+ + 4e \rightleftharpoons R$$

is reversible and has a half-wave potential of −0.349 V when carried out at a dropping mercury electrode from a solution buffered to pH 2.5. Predict the half-wave potential at pH (a) 1.0; (b) 3.5; (c) 7.0.

17-20 A standard-addition method for the polarographic determination of zinc calls for the introduction of 2.00 mL of 1.00×10^{-2} M Zn^{2+} to one of two 10.00 mL aliquots of the sample. Both then receive 10.0 mL of a tartrate solution that acts as supporting electrolyte and buffer; finally, both are diluted to 25.00 mL. Calculate the weight in milligrams of zinc in 100 mL of the following solutions:

	Diffusion Current, μA	
Solution	Aliquot Only	Aliquot Plus Standard Addition
*(a)	28.2	37.1
(b)	40.8	50.1
*(c)	79.6	90.7
(d)	60.8	72.2
(e)	19.1	26.0

17-21 The sulfate in a 25.00-mL aliquot was determined amperometrically by titration with 0.220 M $Pb(NO_3)_2$ at −1.20 V (versus SCE). Use the data below to

(a) construct a curve relating the current (corrected for volume change) observed to the volume of reagent.

(b) calculate the molar concentration of sulfate in the solution.

0.220 M Pb^{2+}, mL	i_{obs}, μA
5.00	1.0
10.0	1.4
15.0	2.1
18.0	2.9
21.0	5.0
24.0	20.8
27.0	122
30.0	262
33.0	386

CHAPTER 18

An Introduction to Spectroscopic Methods of Analysis

Historically, the term *spectroscopy* referred to a branch of science in which light (that is, visible radiation) was resolved into its component wavelengths to produce *spectra*. Spectra, of course, played a vital role in the development of modern atomic theory. In addition, spectroscopy has proved to be a powerful tool for qualitative and quantitative analysis.

With the passage of time, the meaning of spectroscopy has become broadened to include studies not only with light but also with other types of electromagnetic radiation, such as X-ray, ultraviolet, infrared, microwave, and radio-frequency radiation. Indeed, current usage extends the meaning of spectroscopic methods still further to include techniques that do not involve electromagnetic radiation. Examples of the last include acoustic, mass, and electron spectroscopy.

The spectroscopic methods described in the next three chapters are largely based upon ultraviolet and visible radiation, although brief mention is given to techniques involving other parts of the electromagnetic spectrum.[1]

18A Properties of Electromagnetic Radiation

Electromagnetic radiation is a type of energy that is transmitted through space at enormous velocities. Many of the properties of electromagnetic radiation are conveniently described by means of a classical wave model that employs such parameters as wavelength, frequency, velocity, and amplitude. In contrast to other wave phenomena, such as sound, electromagnetic radiation requires no supporting medium for its transmission and readily passes through a vacuum.

The wave model fails to account for phenomena associated with the absorption or emission of radiant energy. To treat these properties adequately, electromagnetic radiation must be viewed as a stream of discrete particles of energy called *photons*, with the energy of the photon being proportional to the frequency of the radiation. These dual views of radiation as particles and waves are not mutually exclusive but, rather, complementary. Indeed, the duality is found to apply to the behavior of streams of electrons and other elementary particles as well and is completely rationalized by wave mechanics.

[1]For further study, see E. J. Meehan, in *Treatise on Analytical Chemistry,* 2nd ed., Part I, Vol. 7, Chapters 1–3, P. J. Elving, E. J. Meehan, and I. M. Kolthoff, Eds. New York: Wiley, 1981; J. D. Ingle, Jr., and S. R. Crouch, *Analytical Spectroscopy*. Englewood Cliffs, N.J.: Prentice-Hall, 1988; J. E. Crooks, *The Spectrum in Chemistry*. New York: Academic Press, 1978.

18A-1

Wave Properties

For many purposes, electromagnetic radiation is conveniently pictured as an electric field that undergoes sinusoidal oscillations in space. Figure 18-1 is a two-dimensional representation of a beam of monochromatic (that is, single-wavelength), plane-polarized radiation. The term *plane-polarized* implies that the oscillations of the electric field are in a single plane. The electric field is represented as a vector whose length is proportional to the field strength. The abscissa in this plot is either time as the radiation passes a fixed point in space or distance. Note that the direction in which the field oscillates is perpendicular to the direction in which the radiation is being propagated.

Wave Parameters

In Figure 18-1, the *amplitude* A of the sinusoidal wave is defined as the length of the electrical vector at the maximum in the wave. The time required for the passage of successive maxima (or minima) through a fixed point in space is called the *period* p of the radiation. The *frequency* ν is the number of oscillations of the field per second[2] and is equal to $1/p$.

It is important to realize that *frequency* is determined by the source and remains *invariant* regardless of the medium traversed by the radiation. In contrast, the *velocity* of propagation v_i, the rate at which the wave front moves through a medium, is *dependent* upon both the medium and the frequency; the subscript i is employed to indicate this frequency dependence. Another parameter of interest is the *wavelength* λ_i, which is the linear distance between successive maxima or minima of a wave.[3] Multiplication of the frequency in

FIGURE 18-1

Representation of a beam of monochromatic radiation of wavelength λ and amplitude A. The arrows represent the electrical vector of the radiation.

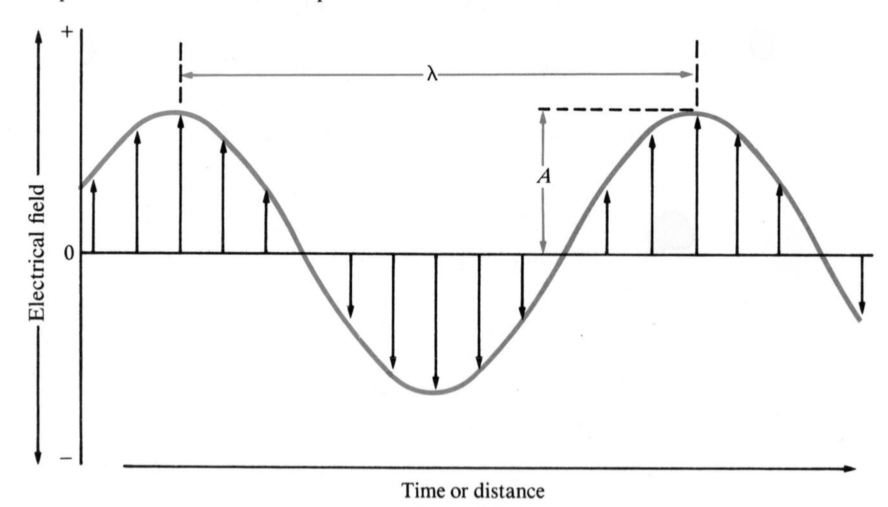

[2]The common unit of frequency is the *hertz* (Hz), which is equal to one cycle, or wave, per second.

[3]The commonly used units for wavelength depend upon the spectral region in question. For example, the angstrom (Å, 10^{-10} m) is convenient for X-ray and short ultraviolet radiation; the nanometer (nm, 10^{-9} m) is employed with visible and ultraviolet radiation; the micrometer (μm, 10^{-6} m) is useful for the infrared region.

waves per second by the wavelength in centimeters gives the velocity of propagation in centimeters per second:

$$v_i = \nu\lambda_i \tag{18-1}$$

In a vacuum, the velocity of radiation propagation becomes independent of wavelength and is at its maximum. This velocity, which is given the symbol c, has been determined to be 2.99792×10^{10} cm/s. The velocity of radiation in air differs only slightly from c (it is about 0.03% less). Thus, to three significant figures, Equation 18-2 is equally applicable in air or vacuum:

$$c = \nu\lambda = 3.00 \times 10^{10} \text{ cm/s} \tag{18-2}$$

The rate of propagation of radiation is less than c in a medium containing matter because the electromagnetic field of the radiation interacts with the electrons in the atoms or molecules of the medium and is slowed as a consequence. Since the radiant frequency is invariant and fixed by the source, the *wavelength of radiation must decrease* as it passes from a vacuum to a medium containing matter (Equation 18-1). This effect is illustrated in Figure 18-2 for a beam of visible radiation. Note that the wavelength shortens nearly 200 nm, or more than 30%, as the radiation passes from air into glass; a reverse change occurs when the radiation again enters air.

The wavenumber σ, defined as the reciprocal of the wavelength in centimeters ($1/\lambda$), is yet another way of describing electromagnetic radiation. Ordinarily, the unit for σ is cm^{-1}, which is the number of waves of a particular frequency in a distance of 1 cm.

Radiant Power and Intensity

The *power* P of radiation is the energy of the beam that reaches a given area per second; the *intensity* I is the power per unit solid angle. These quantities are related to the square of the amplitude A (Figure 18-1). Although it is not strictly correct to do so, power and intensity are often used synonymously.

18A-2 The Particle Properties of Radiation

The Energy of Electromagnetic Radiation

To understand many of the interactions between radiation and matter, it is necessary to postulate that electromagnetic radiation is made up of packets

FIGURE 18-2 The change in wavelength as radiation passes from air into a dense glass and back to air.

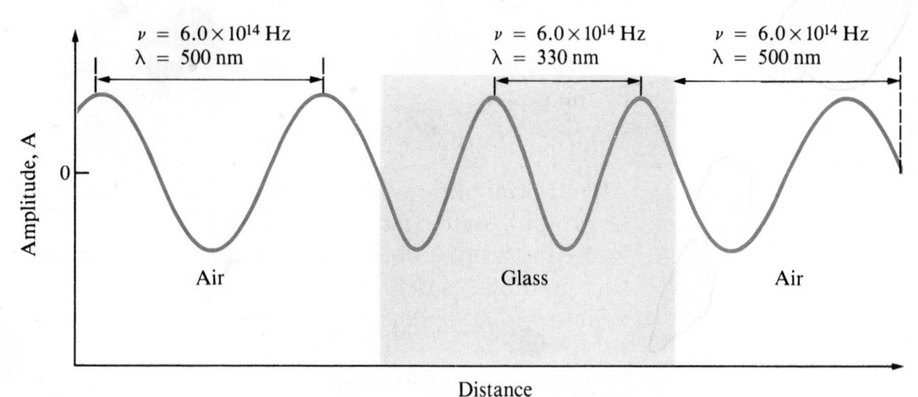

of energy called *photons* (or *quanta*). The energy of a photon depends upon the frequency of the radiation and is given by

$$E = h\nu \tag{18-3}$$

where h is Planck's constant (6.63×10^{-34} J·s). In terms of wavelength and wavenumber,

$$E = \frac{hc}{\lambda} = hc\sigma \tag{18-4}$$

Note that wavenumber, like frequency, is directly proportional to energy.

18B The Electromagnetic Spectrum

The electromagnetic spectrum encompasses an enormous range of wavelengths and energies. For example, an X-ray photon ($\lambda \simeq 10^{-10}$ m) is approximately 10,000 times more energetic than a photon emitted by an incandescent tungsten wire ($\lambda \simeq 10^{-6}$ m) and 10^{11} times more energetic than a photon in the radio-frequency range.

The major divisions of the electromagnetic spectrum are depicted in the color plate located inside the front cover of this book. Note that both the frequency and the wavelength scales are logarithmic; note also that the region to which the human eye is perceptive (the *visible spectrum*) is but a minute part of the whole spectrum. Such diverse radiations as gamma rays or radio waves differ from visible light only in the matter of frequency and, hence, energy.

Figure 18–3 indicates the regions of the spectrum that are useful for analytical spectroscopy and the molecular or atomic transitions responsible for absorption or emission of radiation in each region.

18C Absorption of Radiation

In spectroscopic nomenclature, *absorption* is a process in which a chemical species in a transparent medium selectively *attenuates* (decreases the intensity of) certain frequencies of electromagnetic radiation. According to quantum theory, every elementary particle (atom, ion, or molecule) has a unique set of energy states, the lowest of which is the *ground state;* at room temperature, most elementary particles exist in their ground state. When a photon of radiation passes near an elementary particle, absorption becomes probable if (and only if) the energy of the photon matches *exactly* the energy difference between the ground state and one of the higher energy states of the particle. Under these circumstances, the energy of the photon is transferred to the atom, ion, or molecule, converting it to the higher energy state, which is termed an *excited state.* Excitation of a species M to its excited state M* can be depicted by the equation

$$M + h\nu \longrightarrow M^*$$

After a brief period (10^{-6} to 10^{-9} s), the excited species *relaxes* to its original, or ground, state, transferring its excess energy to other atoms or molecules in the medium. This process, which causes a small rise in temperature of the surroundings, is described by the equation

$$M^* \longrightarrow M + \text{heat}$$

Relaxation may also occur by *photochemical decomposition* of M* to form new

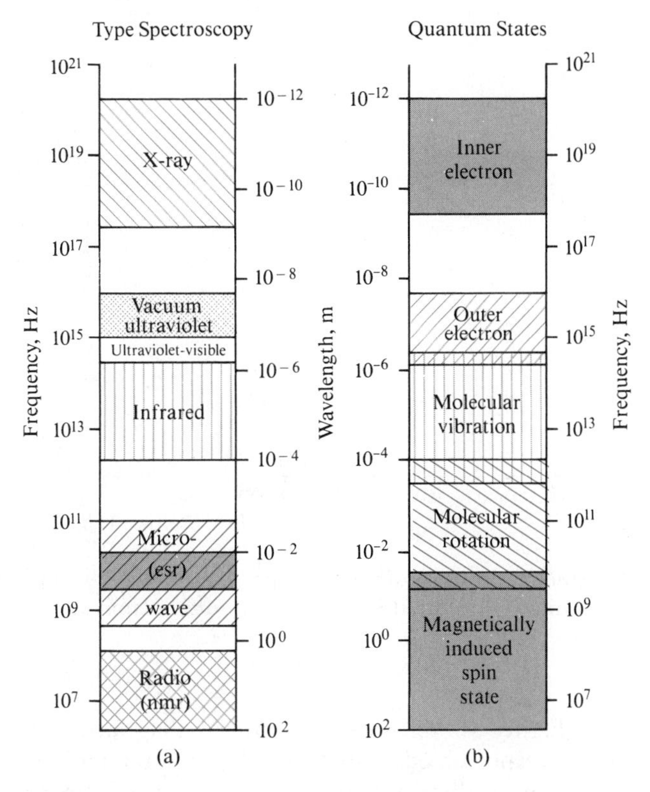

FIGURE 18-3 Parts of the electromagnetic spectrum employed for spectroscopy.

species or by the *fluorescent* or *phosphorescent* reemission of radiation. It is important to note that the lifetime of M* is so very short that its concentration at any instant is ordinarily negligible. Furthermore, the amount of thermal energy released during relaxation is usually so small as to be undetectable. Thus, absorption measurements have the advantage of creating minimal disturbance of the system under study.

The absorbing characteristics of a species are conveniently described by means of an *absorption spectrum,* which is a plot of some function of the attenuation of a beam of radiation versus wavelength, frequency, or wavenumber. Ultraviolet and visible absorption spectra are usually obtained on a gaseous sample of the analyte or on a dilute solution of the analyte in a transparent solvent.

18C-1 Atomic Absorption

When a beam of polychromatic ultraviolet or visible radiation passes through a medium containing gaseous atoms, only a few frequencies are attenuated by absorption, and the spectrum is made up of a number of very narrow (about 0.005 nm) *absorption lines.* Figure 18-4a is an ultraviolet/visible absorption spectrum for gaseous sodium atoms. The ordinate is *absorbance,* which is a measure of the degree of attenuation of the beam (Section 18C-4).[4]

[4]It is noteworthy that at a higher sensitivity, a handful of additional lines would appear. In addition, the three lines shown are in fact split into pairs called *doublets* that differ in wavelength by only a few tenths of a nanometer. The cause of this splitting is beyond the scope of this text.

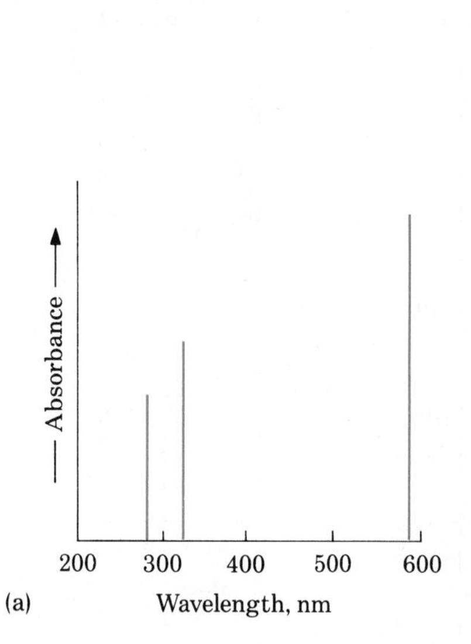

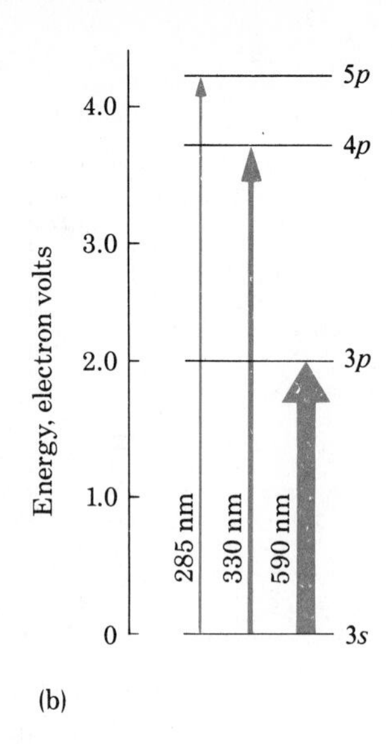

FIGURE 18-4 (a) Absorption spectrum for sodium vapor. (b) Partial energy-level diagram for sodium, showing the transitions resulting from absorption at 590, 330, and 285 nm.

Figure 18-4b is a partial energy diagram for sodium showing the transitions responsible for the three absorption lines in (a). The transitions involve excitation of the single outer electron of sodium from its room temperature or ground state 3*s* orbital to the 3*p*, 4*p*, and 5*p* orbitals. These excitations are brought on by absorption of photons of radiation whose energies *exactly* match the differences in energies between the excited states and the 3*s* ground state.

EXAMPLE 18-1

The energy difference between the 3*p* and the 3*s* orbitals in Figure 18-4b is 2.107 eV. Calculate the wavelength of radiation that would be absorbed in exciting the 3*s* electron to the 3*p* state (1 eV = 1.60×10^{-19} J).

Rearranging Equation 18-4 gives

$$\lambda = hc/E$$

$$= \frac{6.63 \times 10^{-34}\ \text{J s} \times 3.00 \times 10^{10}\ \text{cm s}^{-1} \times 10^{7}\ \text{nm cm}^{-1}}{2.107\ \text{eV} \times 1.60 \times 10^{-19}\ \text{J/eV}} = 590\ \text{nm}$$

Absorption spectra for the alkali metal atoms are a good deal simpler than those of elements with additional outer electrons. The transition metal atomic spectra are particularly rich in lines with some elements exhibiting several thousand lines.

18C-2 Molecular Absorption

Figure 18-5a is a partial energy-level diagram that depicts some of the processes that occur when a polyatomic species absorbs infrared, visible, and ultraviolet radiation. The energies E_1 and E_2, two of the several electronically excited states of a molecule, are shown relative to the energy of its ground state E_0. In addition, the relative energies of a few of the many vibrational states associated with each electronic state are indicated by the lighter horizontal lines.

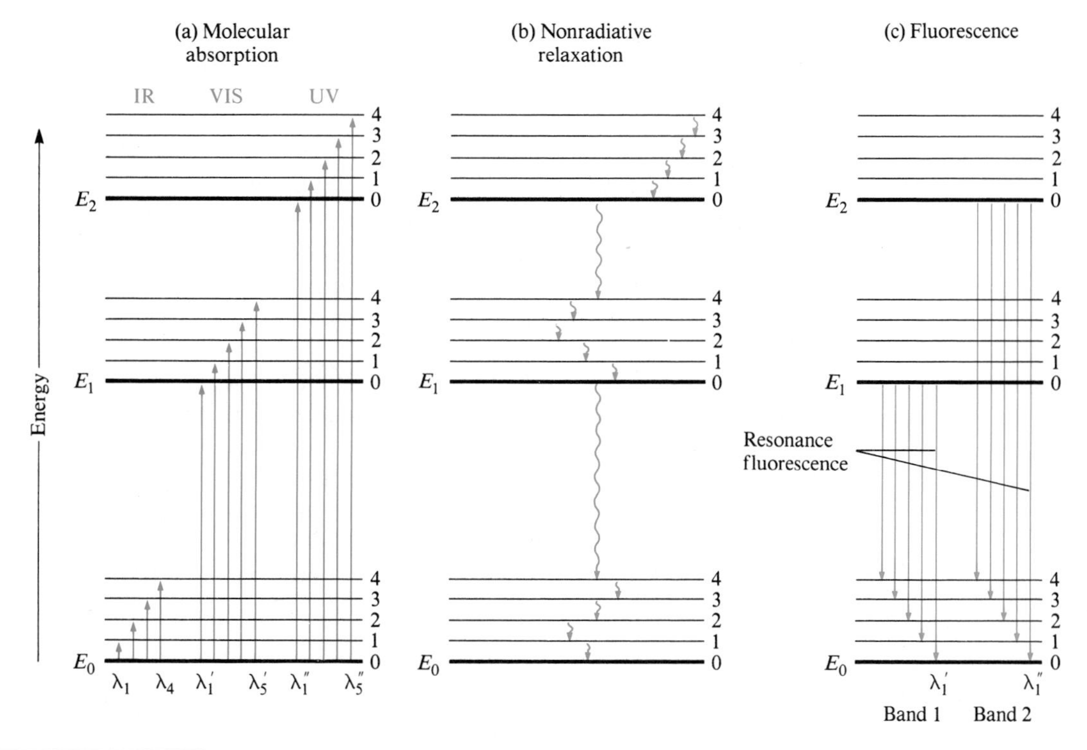

FIGURE 18-5 Energy-level diagram showing some of the energy changes that occur during absorption, nonradiative relaxation, and fluorescence by a molecular species.

An idea as to the nature of vibrational states can be gained by picturing a bond in a molecule as a vibrating spring with atoms attached to both ends. With each vibration, the atoms periodically approach and move away from one another. The potential energy of such a system at any instant depends upon the extent to which the spring is stretched or compressed. For an ordinary spring, the energy of the system varies continuously and reaches a maximum when the spring is fully stretched or fully compressed. In contrast, the energy of a spring system of atomic dimensions can assume only certain discrete energies called vibrational energy levels. Some of the vibrational energy levels associated with each of the electronic states of a molecule are depicted by the lines labeled 1, 2, 3, and 4 in Figure 18-5a (the lowest vibrational levels are labeled 0). Note that the differences in energy among the vibrational states are significantly smaller than among energy levels of the electronic states (typically, an order of magnitude smaller).

Although they are not shown, a set of rotational energy states is superimposed on each of the vibrational states shown in the energy diagram. The energy differences among these states are smaller than those among vibrational states by an order of magnitude.

Infrared Absorption

Infrared radiation generally is not sufficiently energetic to cause electronic transitions but can induce transitions in the vibrational and rotational states associated with *the ground electronic state* of the molecule. Four of these transitions are depicted in the lower left part of Figure 18-5a. For absorption to occur, the source has to emit radiation of frequencies corresponding exactly to the energies indicated by the lengths of the four arrows; infrared radiation of only those four frequencies is then absorbed by a collection of these molecules.

Absorption of Ultraviolet and Visible Radiation

The center arrows in Figure 18-5a suggest that the molecules under consideration absorb visible radiation of five wavelengths, thereby promoting electrons to the five vibrational levels of the excited electronic level E_1. Ultraviolet photons that are more energetic are required to produce the absorption indicated by the five arrows to the right.

As suggested by Figure 18-5a, molecular absorption in the ultraviolet and visible regions consists of absorption *bands* made up of closely spaced lines. (A real molecule has many more energy levels than shown here; thus the typical absorption band consists of a multitude of lines.) In a solution, the absorbing species are surrounded by solvent, and the band nature of molecular absorption often becomes blurred because collisions tend to spread the energies of the quantum states, thus giving smooth and continuous absorption peaks.

Relaxation Processes

As noted earlier, the lifetime of an excited species is brief because several mechanisms exist whereby an excited atom or molecule can give up its excess energy and relax to its ground state. Two of the most important of these mechanisms, nonradiative relaxation and fluorescent relaxation, are illustrated in Figures 18-5b and c.

Two types of nonradiative relaxation are shown in Figure 18-5b. *Vibrational deactivation or relaxation,* depicted by the short wavy arrows between vibrational energy levels, takes place during collisions between excited molecules and molecules of the solvent. During the collisions, the excess vibrational energy is transferred to solvent molecules in a series of steps as indicated in the figure. The gain in vibrational energy of the solvent is reflected in an increase in the temperature of the medium. Vibrational relaxation is such an efficient process that the average lifetime of an excited *vibrational* state is only about 10^{-15} s.

Nonradiative relaxation between the lowest vibrational level of an excited electronic state and the upper vibrational level of another electronic state can also occur. This type of relaxation, depicted by the two longer wavy arrows in Figure 18-5b, is much less efficient than vibrational relaxation so that the average lifetime of an electronic excited state is between 10^{-6} and 10^{-9} s. The mechanisms by which this type of relaxation occurs are not fully understood, but the net effect is again a rise in the temperature of the medium.

Figure 18-5c depicts another relaxation process: fluorescence. Molecular fluorescence is discussed in Sections 18D-2 and 20C.

18C-3 Terms Employed in Absorption Spectroscopy

Table 18-1 lists the common terms and symbols used in absorption spectroscopy. This nomenclature is recommended by the American Society for Testing Materials as well as the American Chemical Society. Column 3 contains alternative symbols encountered in the older literature. Because a standard nomenclature is highly desirable in order to avoid ambiguities, the reader is

TABLE 18-1

Important Terms and Symbols Employed in Absorption Measurement

Term and Symbol*	Definition	Alternative Name and Symbol
Radiant power P, P_0	Energy of radiation (in ergs) impinging on a 1-cm^2 area of a detector per second	Radiation intensity I, I_0
Absorbance A	$\log \frac{P_0}{P}$	Optical density D; extinction E
Transmittance T	$\frac{P}{P_0}$	Transmission T
Path length of radiation† b	—	l, d
Absorptivity† a	$\frac{A}{bc}$	Extinction coefficient k
Molar absorptivity‡ ϵ	$\frac{A}{bc}$	Molar extinction coefficient

*Terminology recommended by the American Chemical Society (*Anal. Chem.*, **1952**, *24*, 1349; **1976**, *48*, 2298).

† c may be expressed in g/L or in other specified concentration units; b may be expressed in cm or in other units of length.

‡ c is expressed in mol/L; b is expressed in cm.

urged to learn and use the recommended terms and symbols and avoid those in column 3.

Transmittance

Figure 18-6 depicts a beam of parallel radiation before and after it has passed through a layer of solution with a thickness b cm and a concentration c of an absorbing species. As a consequence of interactions between the photons and absorbing particles, the power of the beam is attenuated from P_0 to P. The *transmittance* T of the solution is defined as the fraction of incident radiation transmitted by the solution:

$$T = P/P_0 \qquad (18\text{-}5)$$

Transmittance is often expressed as a percentage.

Absorbance

The absorbance of a solution is defined by the equation

$$A = -\log_{10} T = \log \frac{P_0}{P} \qquad (18\text{-}6)$$

FIGURE 18-6

Attenuation of a beam of radiation by an absorbing solution.

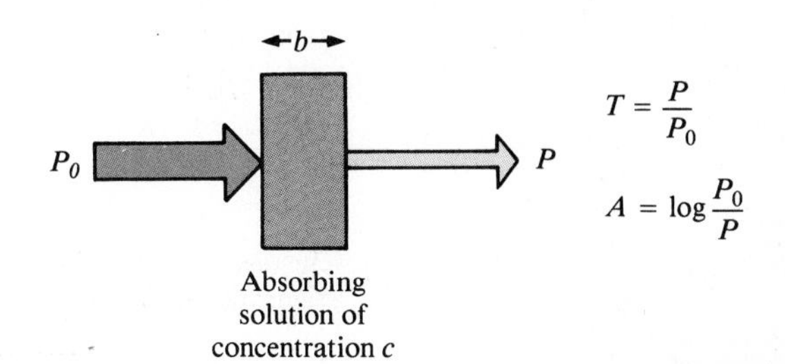

Note that, in contrast to transmittance, the absorbance of a solution increases as the attenuation of the beam becomes greater.

18C-4 The Relationship Between Absorbance and Concentration

The functional relationship between the quantity measured in an absorption method (A) and the quantity sought (the analyte concentration c) is known as *Beer's law* and can be written

$$A = \log (P_0/P) = abc \qquad (18\text{-}7)$$

where a is a proportionality constant called the *absorptivity* and b is the path length of the radiation through the absorbing medium. Since absorbance is a unitless quantity, the absorptivity has units that render the right side of the equation dimensionless.

When concentration in Equation 18-7 is expressed in moles per liter and b is in centimeters, the proportionality constant is called the *molar absorptivity* and is given the special symbol ϵ. Thus,

$$A = \epsilon bc \qquad (18\text{-}8)$$

where ϵ has the units of L cm^{-1} mol^{-1}.

Beer's law can be rationalized by considering the fate of a beam of parallel monochromatic radiation with power P_0 after it enters a layer of absorbing matter (solid, liquid, or gas) at a normal angle, as shown in Figure 18-7.[5] After passing through a length b of the material, which contains n absorbing particles (atoms, ions, or molecules), the power of the radiation is decreased to P as a result of absorption. Consider now a cross section of the layer that has an area S and an infinitesimal thickness dx and that contains dn absorbing particles. Associated with each particle, we can imagine a surface at which photon capture occurs. That is, if a photon reaches one of these areas by chance, absorption follows immediately. The total projected area of these capture surfaces within the section of thickness dx is designated dS. The ratio of the capture area to the total area is then dS/S. On a statistical average, this ratio represents the probability for photon capture within the section.

The power of the beam entering the section P_x is proportional to the number of photons per square centimeter per second, and dP_x represents the quantity removed per second within the section. The fraction absorbed is then $-dP_x/P_x$, and this ratio also equals the average probability for capture. The term is given a minus sign to indicate that P undergoes a decrease. Thus,

$$-\frac{dP_x}{P_x} = \frac{dS}{S} \qquad (18\text{-}9)$$

Recall that dS is the sum of the capture areas for particles within the section, which must be proportional to the number of particles, or

$$dS = adn \qquad (18\text{-}10)$$

where dn is the number of particles and a is a proportionality constant that

[5]The discussion that follows is based on a paper by F. C. Strong (*Anal. Chem.*, **1952**, *24*, 338). For a rigorous derivation of the law, see D. J. Swinehart, *J. Chem. Educ.*, **1972**, *39*, 333.

can be called the *capture cross section.* Combining Equations 18-9 and 18-10 and integrating over the interval between zero and n, we obtain

$$-\int_{P_0}^{P} \frac{dP_x}{P_x} = \int_0^n \frac{a\,dn}{S}$$

When the integrals are evaluated, we find

$$-\ln \frac{P}{P_0} = \frac{an}{S}$$

Upon converting to base-10 logarithms and inverting the fraction to change the sign, we obtain

$$\log \frac{P_0}{P} = \frac{an}{2.303S} \qquad (18\text{-}11)$$

where n is the total number of particles within the block shown in Figure 18-7. The cross-sectional area S can be expressed in terms of the volume of the block V and its length b:

$$S = \frac{V}{b}\ \text{cm}^2$$

Substitution of this quantity into Equation 18-11 yields

$$\log \frac{P_0}{P} = \frac{anb}{2.303V} \qquad (18\text{-}12)$$

Note that n/V has the units of concentration (that is, number of particles per cubic centimeter), which is readily converted to moles per liter:

$$c = \frac{n \text{ particles}}{6.02 \times 10^{23} \text{ particles/mol}} \times \frac{1000 \text{ cm}^3\text{/L}}{V \text{ cm}^3} = \frac{1000n}{6.02 \times 10^{23}\ V} \text{ mol/L}$$

Combining this relationship with Equation 18-12 yields

$$\log \frac{P_0}{P} = \frac{(6.02 \times 10^{23})abc}{2.303 \times 1000}$$

Finally, the constants in this equation can be collected into a single factor ϵ to give

$$\log \frac{P_0}{P} = \epsilon bc = A$$

which is a statement of Beer's law.

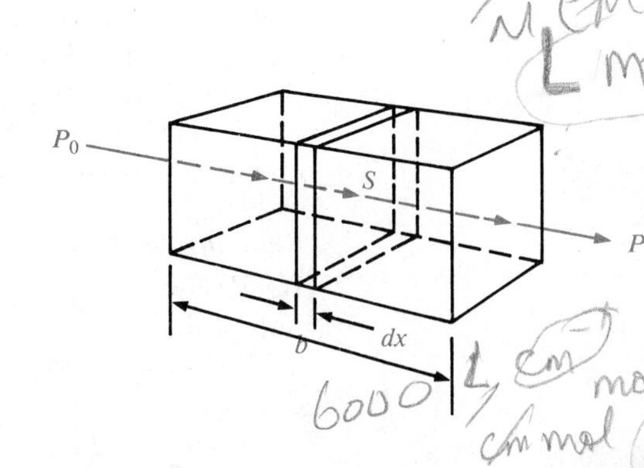

FIGURE 18-7 Attenuation of radiation with initial power P_0 by a solution containing c mol/L of absorbing solute and a path length of b cm ($P < P_0$).

18C-5

The Experimental Measurement of Transmittance and Absorbance

The relationships given by Equations 18-7 and 18-8 are not directly applicable in the laboratory because the power of the attenuated beam P cannot be determined in a straightforward way. The difficulty has its origin in the need to hold the sample whose absorbance is sought in some sort of container with transparent windows. Interaction between the radiation and the windows is inevitable, leading to a loss by reflection at each air-to-window interface as well as each window-to-sample interface. Typical losses can be appreciable. For example, about 4% of a beam is reflected upon vertical passage of visible radiation across each of the four interfaces associated with the two windows of a glass container. Moreover, significant absorption may occur within the container walls. Finally, the beam may suffer a diminution in power during its passage through the solvent as a result of scattering by large molecules or inhomogeneities.

In order to compensate for these effects, the power of the beam transmitted through a cell containing an absorbing solution is generally compared with that of a beam that passes through an identical cell containing only solvent. An experimental absorbance is then defined by the equation

$$A = \log \frac{P_{\text{solvent}}}{P_{\text{solution}}} = \log \frac{P_0}{P} \tag{18-13}$$

Such experimental absorbances obey Beer's law and are presumably good approximations of true absorbances. Henceforth, the term *absorbance* will refer to the ratio defined by Equation 18-13, P_0 being the power of radiation after passage through a cell containing only the solvent, and P being the power after passage through an identical cell containing a solution of the analyte.

18C-6

The Application of Beer's Law to Mixtures

Beer's law also applies to solutions containing more than one kind of absorbing substance. Provided no interaction occurs among the various species, the total absorbance for a multicomponent system is

$$A_{\text{total}} = A_1 + A_2 + \cdots + A_n = \epsilon_1 bc_1 + \epsilon_2 bc_2 + \cdots + \epsilon_n bc_n \tag{18-14}$$

where the subscripts refer to absorbing components 1, 2, . . ., n.

18C-7

Limitations to the Applicability of Beer's Law

The linear relationship between absorbance and path length at a fixed concentration of an absorbing substance is a generalization for which no exceptions are known. Deviations from the direct proportionality between absorbance and concentration at constant b are frequently encountered, however. Some of these deviations are fundamental and represent *real* limitations to the law. Others occur as a consequence of the manner in which the absorbance measurements are made (*instrumental deviations*) or as a result of chemical changes associated with concentration changes (*chemical deviations*).

Real Limitations

Beer's law is successful in describing the absorption behavior of dilute solutions only and in this sense is a *limiting law*. At high concentrations (usually > 0.01 M), the average distances between particles of the absorbing species are diminished to the point where each particle affects the charge distribution of its neighbors. This interaction can alter their ability to absorb a given wavelength of radiation. Because the extent of interaction depends upon concentration, the occurrence of this phenomenon causes deviations from the linear relationship between absorbance and concentration. A similar effect is sometimes encountered in dilute solutions of absorbers that contain high concentrations of other species, particularly electrolytes. The close proximity of ions to the absorber alters the molar absorptivity of the latter by electrostatic interactions, which leads to departures from Beer's law.

While the effect of molecular interactions is ordinarily not significant at concentrations below 0.01 M, some exceptions are encountered among certain large organic ions or molecules. For example, the molar absorptivity at 436 nm for the cation of methylene blue is reported to increase by 88% as the dye concentration is increased from 10^{-5} to 10^{-2} M; even below 10^{-6} M, strict adherence to Beer's law is not observed.

Deviations from Beer's law also arise because molar absorptivity is dependent upon the refractive index of the solution.[6] Thus, if concentration changes cause significant alterations in the refractive index of a solution, departures from Beer's law are observed. In general, this effect is small and rarely significant at concentrations less than 0.01 M.

Chemical Deviations

Apparent deviations from Beer's law are frequently encountered as a consequence of association, dissociation, or reaction of the absorbing species with the solvent. These deviations result from shifts in chemical equilibria and not from changes in molar absorptivities. As shown by the following example, apparent departures from Beer's law are readily predicted from the equilibrium constants for the reactions and the molar absorptivities of the solutes.

EXAMPLE 18-2

A series of solutions containing various concentrations of the acidic indicator HIn ($K_a = 1.42 \times 10^{-5}$) was prepared in 0.1 M HCl and 0.1 M NaOH. In both media, a linear relationship between absorbance and concentration was observed at 430 and 570 nm. From the magnitude of the acid dissociation constant, it is apparent that, for all practical purposes, the indicator is entirely in the undissociated form (HIn) in the HCl solution and completely dissociated as In^- in NaOH. The molar absorptivities at the two wavelengths were found to be

	ϵ_{430}	ϵ_{570}
HIn (HCl solution)	6.30×10^2	7.12×10^3
In^- (NaOH solution)	2.06×10^4	9.60×10^2

Derive absorbance data (1.00-cm cell) at the two wavelengths for unbuffered solutions with indicator concentrations ranging from 2×10^{-5} to 16×10^{-5} M. Plot the data.

Let us calculate the concentration of HIn and In^- in an unbuffered 2.00×10^{-5} M solution of the indicator. From the equation for the dissociation reaction, it is apparent that

$$[H^+] = [In^-]$$

Furthermore,

$$[In^-] + [HIn] = 2.00 \times 10^{-5}$$

[6] G. Kortum and M. Seiler, *Angew. Chem.*, **1939**, *52*, 687.

Substitution of these relationships into the expression for K_a gives

$$\frac{[\mathrm{In}^-]^2}{2.00 \times 10^{-5} - [\mathrm{In}^-]} = 1.42 \times 10^{-5}$$

Rearrangement yields the quadratic expression

$$[\mathrm{In}^-]^2 + (1.42 \times 10^{-5})\,[\mathrm{In}^-] - 2.84 \times 10^{-10} = 0$$

which can be solved to give

$$[\mathrm{In}^-] = 1.12 \times 10^{-5}$$

$$[\mathrm{HIn}] = 2.00 \times 10^{-5} - 1.12 \times 10^{-5} = 0.88 \times 10^{-5}$$

The absorbances at the two wavelengths are found by substituting into Equation 18-14:

$$A_{430} = 6.30 \times 10^2 \times 1.00 \times 0.88 \times 10^{-5} + 2.06 \times 10^4 \times 1.00 \times 1.12 \times 10^{-5} = 0.236$$

$$A_{570} = 7.12 \times 10^3 \times 1.00 \times 0.88 \times 10^{-5} + 9.60 \times 10^2 \times 1.00 \times 1.12 \times 10^{-5} = 0.073$$

The following data were derived in a similar way and are plotted in Figure 18-8:

c_{HIn}	[HIn]	$[\mathrm{In}^-]$	A_{430}	A_{570}
2.00×10^{-5}	0.88×10^{-5}	1.12×10^{-5}	0.236	0.073
4.00×10^{-5}	2.22×10^{-5}	1.78×10^{-5}	0.381	0.175
8.00×10^{-5}	5.27×10^{-5}	2.73×10^{-5}	0.596	0.401
12.00×10^{-5}	8.52×10^{-5}	3.48×10^{-5}	0.771	0.640
16.00×10^{-5}	11.9×10^{-5}	4.11×10^{-5}	0.922	0.887

Figure 18-8 is a plot of the data derived in the foregoing example and illustrates the types of deviations from Beer's law that occur when the absorber is a participant in an association or dissociation equilibrium. Note that the direction of curvature is opposite at the two wavelengths.

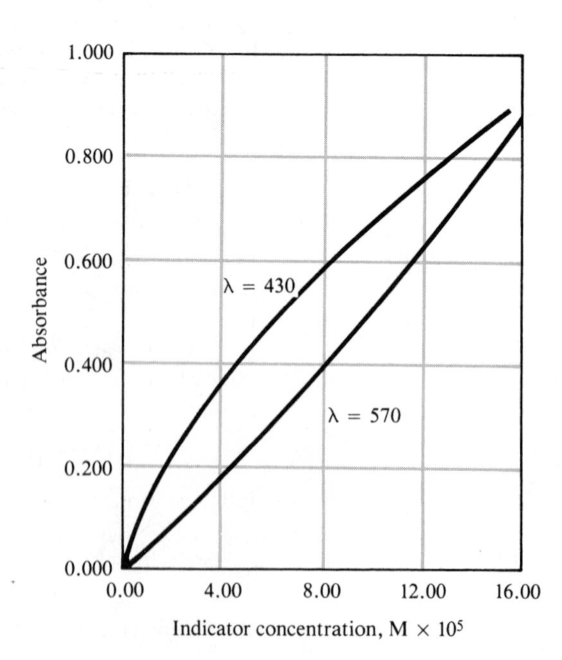

FIGURE 18-8 Chemical deviations from Beer's law for unbuffered solutions of the indicator HIn. For data, see Example 18-2.

Instrumental Deviations with Polychromatic Radiation

Beer's law is also a limiting law in the sense that it applies only when absorbance is measured with monochromatic radiation. Truly monochromatic sources, such as lasers, are not practical for routine analytical instruments, however. Instead, a polychromatic continuous source is employed in conjunction with a grating or a filter that isolates a more or less symmetric band of wavelengths around the desired one (page 486). The following derivation illustrates how such a source may lead to deviations from Beer's law.

Consider a beam made up of just two wavelengths λ' and λ'', and assume that Beer's law applies strictly to each. With this assumption, we can write for radiation λ'

$$A' = \log \frac{P_0'}{P'} = \epsilon' bc$$

$$\frac{P_0'}{P'} = 10^{\epsilon' bc} \quad \text{and} \quad P' = P_0' \, 10^{-\epsilon' bc}$$

Similarly, for λ''

$$\frac{P_0''}{P''} = 10^{\epsilon'' bc} \quad \text{and} \quad P'' = P_0'' \, 10^{-\epsilon'' bc}$$

When an absorbance measurement is made with radiation composed of both wavelengths, the power of the beam emerging from the solution is given by $P' + P''$ and that of the beam emerging from the solvent by $P_0' + P_0''$. Therefore, the measured absorbance is

$$A_m = \log \frac{P_0' + P_0''}{P' + P''}$$

which can be rewritten as

$$A_m = \log \frac{P_0' + P_0''}{P_0' \, 10^{-\epsilon' bc} + P_0'' \, 10^{-\epsilon'' bc}}$$

$$= \log (P_0' + P_0'') - \log (P_0' \, 10^{-\epsilon' bc} + P_0'' \, 10^{-\epsilon'' bc})$$

Now, when $\epsilon' = \epsilon''$, this equation simplifies to

$$A_m = \epsilon' bc$$

and Beer's law is followed. As shown in Figure 18-9, however, the relationship between A_m and concentration is no longer linear when the molar absorptivities differ. Moreover, departures from linearity become greater as the difference between ϵ' and ϵ'' increases. When this treatment is expanded to include additional wavelengths, the effect remains the same.

It is an experimental fact that deviations from Beer's law resulting from the use of a polychromatic beam are not appreciable, provided the radiation used does not encompass a spectral region in which the absorber exhibits large changes in absorbance as a function of wavelength. This observation is illustrated in Figure 18-10.

Instrumental Deviations in the Presence of Stray Radiation

The radiation employed for absorbance measurements is usually contaminated with small amounts of *stray* radiation due to instrumental imperfections. Stray radiation arises from scattering phenomena off the surfaces of prisms, lenses,

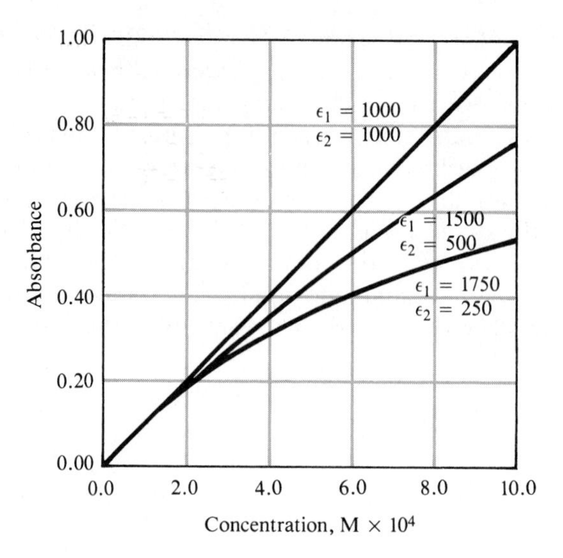

FIGURE 18-9 Deviations from Beer's law with polychromatic light. The absorber has the indicated molar absorptivities at the two wavelengths λ_1 and λ_2.

filters, and windows (page 491). It often differs greatly in wavelength from the principal radiation and, in addition, may not have passed through the sample or solvent.

When measurements are made in the presence of stray radiation, the observed absorbance is given by

$$A' = \log \frac{P_0 + P_s}{P + P_s}$$

where P_s is the power of the stray radiation. Figure 18-11 shows a plot of A' versus concentration for various levels of P_s relative to P_0.

Note that the instrumental deviations illustrated in Figures 18-10 and 18-11 result in absorbances that are smaller than theoretical. It can be shown that instrumental deviations always lead to negative absorbance errors.[7]

FIGURE 18-10 The effect of polychromatic radiation upon Beer's law. Band *A* shows little deviation because ϵ does not change greatly throughout the band. Band *B* shows marked deviation because ϵ undergoes significant changes in this region.

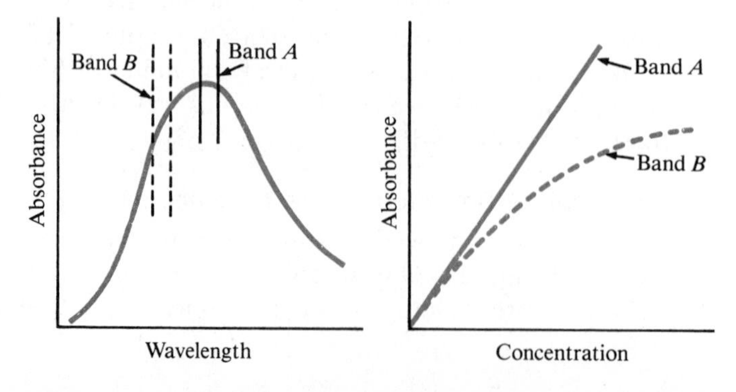

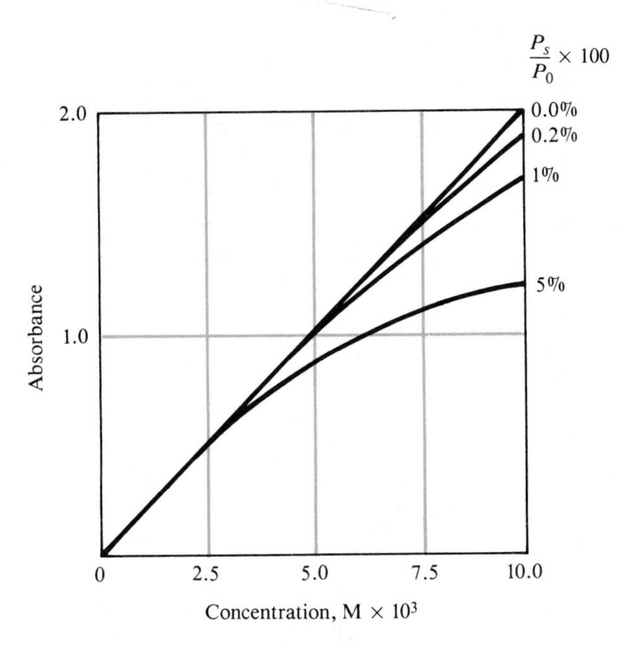

FIGURE 18-11 Apparent deviation from Beer's law caused by various amounts of stray radiation.

18D Emission of Electromagnetic Radiation

Atoms, ions, and molecules can be excited to one or more higher energy levels by any of several processes, including bombardment with electrons or other elementary particles, exposure to a high-voltage ac spark, heat treatment in a flame or arc, or exposure to a source of electromagnetic radiation. The lifetime of an excited species is generally transitory (10^{-6} to 10^{-9} s), and relaxation to a lower energy level or the ground state takes place with a release of the excess energy in the form of electromagnetic radiation, heat, or perhaps both.

18D-1 Emission Spectra

Radiation from a source is conveniently characterized by means of an *emission spectrum,* which usually takes the form of a plot of the relative power of the emitted radiation as a function of wavelength or frequency. Figure 18-12 illustrates a typical emission spectrum, which was obtained by aspirating a brine solution into an oxyhydrogen flame. Three types of spectra are evident in the figure: *line, band,* and *continuous.* The line spectrum is made up of a series of sharp, well-defined peaks arising from the excitation of individual atoms. The band spectrum consists of several groups of lines so closely spaced that they are not completely resolved. The source of the bands are small molecules or radicals. Finally, the continuous spectrum is responsible for the increase in the background that becomes evident above about 350 nm. The line and band spectra are superimposed on this continuum. The source of the continuum is described on page 476.

[7]E. J. Meehan, in *Treatise on Analytical Chemistry,* 2nd ed., P. J. Elving, E. J. Meehan, and I. M. Kolthoff, Eds., Part I, Vol. 7, pp. 71–79. New York: Interscience, 1981.

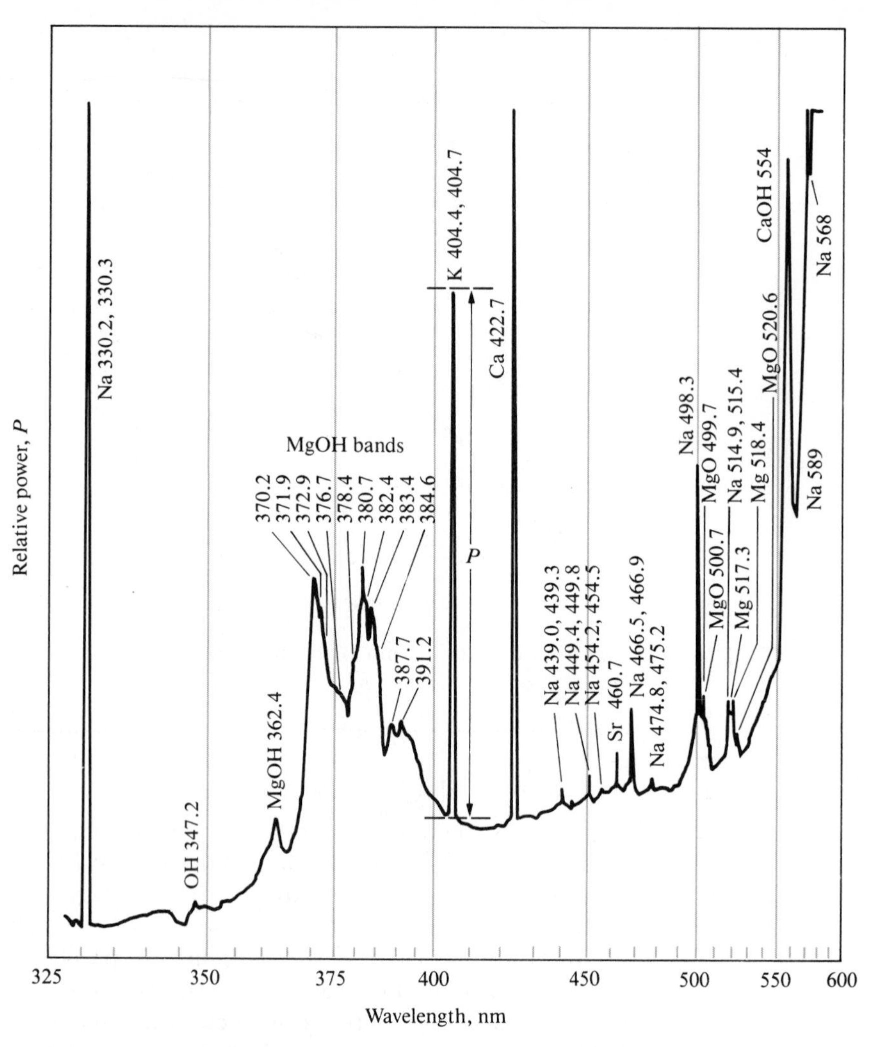

FIGURE 18-12 Emission spectrum of a brine obtained with an oxyhydrogen flame. (R. Hermann and C. T. J. Alkemade, *Chemical Analysis by Flame Photometry*, 2nd ed., p. 484. New York: Interscience, 1963. With permission.)

Line Spectra

Line spectra are encountered when the radiating species are individual atomic particles that are well separated, as in a gas. The individual particles in a gaseous medium behave independently of one another, and the spectrum consists of a series of sharp lines with widths of about 10^{-4} Å. In Figure 18-12, lines for sodium, potassium, strontium, and calcium are identified.

The energy-level diagram in Figure 18-13a shows the source of two of the lines in a typical emission spectrum of an element. The horizontal line labeled E_0 corresponds to the lowest, or ground state, energy of the atom. The horizontal lines labeled E_1 and E_2 are two higher energy electronic levels of the species. For example, the single outer electron in the ground state E_0 for a sodium atom is located in the 3*s* orbital. Energy level E_1 then represents the energy of the atom when this electron has been promoted to the 3*p* state by absorption of thermal, electrical, or radiant energy (see also Figure 18-4). The promotion is depicted by the shorter wavy arrow on the left in Figure

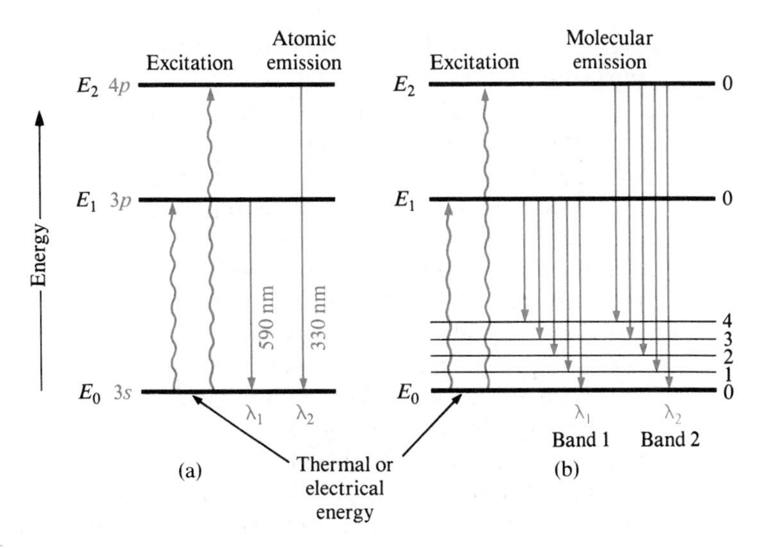

FIGURE 18-13 Energy-level diagrams for a sodium atom and a simple molecule, showing the source of (a) a line spectrum and (b) a band spectrum.

18-13a. After perhaps 10^{-8} s, the atom returns to the ground state, emitting a photon whose frequency and wavelength are given by Equations 18-3 and 18-4:

$$\nu_1 = (E_1 - E_0)/h$$

$$\lambda_1 = hc/(E_1 - E_0)$$

This emission process is illustrated by the shorter straight arrow on the right in Figure 18-13a.

For the sodium atom, E_2 in the figure corresponds to the more energetic $4p$ state; the resulting radiation λ_2 would then appear at a shorter wavelength. The line appearing at about 330 nm in Figure 18-12 results from this transition; the $3p$-to-$3s$ transition provides a line at about 590 nm. It is important to note that the emitted wavelengths *are identical to the wavelengths of the absorption peaks* for sodium (Figure 18-4) because the transitions involved are between the same two states.

Band Spectra

Band spectra are often encountered in spectral sources because of the presence of gaseous radicals or small molecules. For example, in Figure 18-12 bands for OH, MgOH, and MgO are labeled and consist of a series of closely spaced lines not fully resolved by the instrument used to obtain the spectrum. Bands arise from the numerous quantized vibrational levels that are superimposed on the ground-state electronic energy level of a molecule.

Figure 18-13b is a partial energy-level diagram for a molecule showing its ground state E_0 and two of its several excited electronic states, E_1 and E_2. A few of the many vibrational levels associated with the ground state are also shown, but those associated with the two excited states have been omitted because the lifetime of an excited vibrational state is brief compared with that of an electronically excited state (about 10^{-15} s versus 10^{-8} s). A consequence of this tremendous difference in lifetimes is that when an electron is excited to one of the higher vibrational levels of an electronic state, relaxation to the lowest vibrational level of that state occurs before an electronic transition to the ground state can occur. Therefore, the radiation produced by the electrical or thermal excitation of polyatomic species nearly always involves a transition from the *lowest vibrational level of an excited electronic state* to any of the several vibrational levels of the ground state.

The mechanism by which a vibrationally excited species relaxes to the nearest electronic state involves a transfer of its excess energy to other atoms in the system through a series of collisions. As noted, this process takes place at an enormous speed. Relaxation from one electronic state to another can also occur by collisional transfers of energy, but the rate of this process is slow enough that relaxation by photon release is favored.

The energy-level diagram in Figure 18-13b illustrates the mechanism by which two radiation bands consisting of five closely spaced lines are emitted by a molecule excited by thermal or electrical energy. For a real molecule, the number of individual lines is much larger because the ground state contains many more vibrational levels than shown. In addition, a multitude of rotational states would be superimposed on each of the vibrational levels. The differences in energy among the rotational levels is perhaps an order of magnitude smaller than that for vibrational states. Thus, a real molecular band would be made up of many more lines than shown in Figure 18-13b, and these lines would be much more closely spaced.

Continuous Spectra

As shown in Figure 18-14, truly continuous radiation is produced when solids are heated to incandescence. Thermal radiation of this kind, which is called *blackbody radiation,* is more characteristic of the temperature of the emitting surface than of the material of which that surface is composed. Blackbody radiation is produced by the innumerable atomic and molecular oscillations excited in the condensed solid by the thermal energy. Note that the energy peaks in Figure 18-14 shift to shorter wavelengths with increasing temperature. It is clear that very high temperatures are needed to cause a thermally excited source to emit a substantial fraction of its energy as ultraviolet radiation.

As noted earlier, part of the continuous background radiation exhibited in the flame spectrum shown in Figure 18-12 is probably thermal emission

FIGURE 18-14 Blackbody radiation curves.

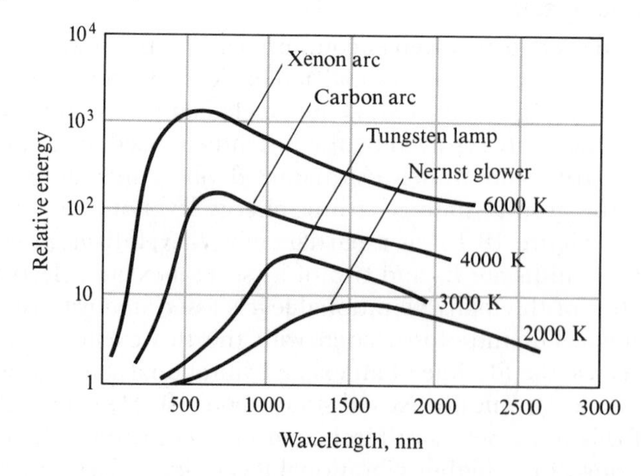

from incandescent particles in the flame. Note that this background decreases rapidly as the ultraviolet region is approached.

Heated solids are important sources of infrared, visible, and longer-wavelength ultraviolet radiation for analytical instruments.

The Effect of Concentration on Line and Band Spectra. The radiant power P of a line or a band depends directly upon the number of excited atoms or molecules, which in turn is proportional to the total concentration c of the species present in the source. Thus, we can write

$$P = kc \tag{18-15}$$

where k is a proportionality constant. This relationship is the basis of quantitative emission spectroscopy.

18D-2 Emission by Fluorescence and Phosphorescence

Fluorescence and phosphorescence are analytically important emission processes in which atoms or molecules are excited by the absorption of a beam of electromagnetic radiation. The excited species then relax to the ground state, giving up their excess energy as photons. Fluorescence takes place much more rapidly than phosphorescence and is generally complete in about 10^{-5} s (or less) from the time of excitation. Phosphorescence emission may extend for minutes or even hours after irradiation has ceased.

Fluorescence is considerably more important than phosphorescence in analytical chemistry. Thus, our discussions focus largely on the former.

Atomic Fluorescence

Gaseous atoms fluoresce when they are exposed to radiation that has a wavelength that exactly matches that of one of the absorption (or emission) lines of the element in question. For example, gaseous sodium atoms are promoted to the excited energy state E_1 shown in Figure 18-13b through absorption of 590 nm radiation. Relaxation may then take place by fluorescent reemission of radiation of the identical wavelength. Fluorescence in which the excitation and emission wavelengths are the same is termed *resonance fluorescence*. Sodium atoms could also exhibit resonance fluorescence when exposed to 330 nm radiation. In addition, however, the element could also produce nonresonance fluorescence by first relaxing to energy level E_1 through a series of nonradiative collisions with other species in the medium. Further relaxation to the ground state can then take place either by the emission of a 590 nm photon or by further collisional deactivation.

Resonance fluorescence is commonly encountered with atoms and to a lesser extent with molecular species.

Molecular Fluorescence

The number of molecules that fluoresce is relatively small because fluorescence requires structural features that slow the rate of the nonradiative relaxation processes illustrated in Figure 18-5b and enhance the rate of fluorescence relaxation shown in Figure 18-5c. Most molecules lack these features and undergo nonradiative relaxation at a rate that is significantly greater than the radiative relaxation rate; thus fluorescence is precluded.

As shown in Figure 18-5c, bands of radiation are produced when molecules fluoresce. Like molecular absorption bands, molecular fluorescence bands are made up of a multitude of closely spaced lines that are often difficult to resolve. Note that the lines that terminate the two fluorescence bands on the short-wavelength, or high-energy, side (λ_1' and λ_1'') are resonance lines. That is, molecular fluorescence bands consist largely of lines that are longer in wavelength than the band of absorbed radiation responsible for their excitation. This shift in wavelength is sometimes called the *Stokes shift.*

To understand the source of Stokes shifts, let us consider what occurs when the molecule under consideration is irradiated by a single wavelength λ_5''. As shown in Figure 18-5a, absorption of this radiation promotes an electron into vibrational level 4 of the second excited electronic state E_2. In 10^{-15} s or less, relaxation to the zero vibrational level occurs (Figure 18-5b). At this point, further relaxation can follow either the nonradiative route depicted in Figure 18-5b or the radiative route shown in Figure 18-5c. If the latter is followed, relaxation to any of the several vibrational levels of the ground state can take place, giving a band of emitted wavelengths as shown. Note that all these lines are lower in energy, or longer in wavelength, than the excitation line λ_5''.

Let us now turn to those molecules in excited state E_2 that undergo nonradiative relaxation to electronic state E_1. As before, further relaxation can take a nonradiative or a radiative route to the ground state. In the latter case, band 1 of fluorescence is produced. Note that in this case the Stokes shift is from ultraviolet radiation to visible. Note also that band 1 can be produced not only by the mechanism just described but also by the absorption of visible radiation of wavelengths λ_1' through λ_5' (Figure 18-5a).

A detailed discussion of the applications of molecular fluorescence is given in Section 20C.

Questions and Problems

18-1 What kind of transitions are responsible for the
- *(a) absorption of ultraviolet radiation?
- (b) absorption of infrared radiation?
- *(c) emission of band spectra?
- (d) fluorescence of atoms?
- *(e) fluorescence of molecules?
- (f) emission of line spectra?

18-2 Define
- *(a) ground state.
- (b) excited electronic state.
- *(c) photon.
- (d) band spectra.
- *(e) continuous spectra.
- (f) line spectra.
- *(g) resonance fluorescence.
- (h) phosphorescence.
- *(i) percent transmittance.
- (j) absorbance.
- *(k) molar absorptivity.
- (l) absorptivity.
- *(m) stray radiation.
- (n) wavenumber.
- *(o) relaxation.
- (p) Stokes shift.

18-3 Calculate the frequency in hertz of
- *(a) an X-ray beam with a wavelength of 2.65 Å.
- (b) an emission line for copper at 211.0 nm.
- *(c) the line at 694.3 nm produced by a ruby laser.
- (d) the output of a CO_2 gas laser at 10.6 μm.
- *(e) an infrared absorption peak at 19.6 μm.
- (f) a microwave beam at 1.86 cm.

18-4 Calculate the wavelength in centimeters of

*(a) an airport tower transmitting at 118.6 MHz.
(b) a VOR (radio navigation aid) transmitting at 114.10 kHz.
*(c) an NMR signal at 105 MHz.
(d) an infrared absorption peak having a wavenumber of 1210.

*18-5 A typical simple infrared spectrophotometer covers a wavelength range from 3 to 15 μm. Express its range (a) in wavenumbers and (b) in hertz.

18-6 A sophisticated ultraviolet/visible/near-IR instrument has a wavelength range of 185 to 3000 nm. What are its wavenumber and frequency ranges?

18-7 Express the following absorbances in terms of percent transmittance:

*(a) 0.064. (d) 0.209.
*(b) 0.765. (e) 0.437.
*(c) 0.318. (f) 0.413.

18-8 Convert the accompanying transmittance data to absorbances:

*(a) 19.4%. (d) 4.51%.
*(b) 0.863. (e) 0.100.
*(c) 27.2%. (f) 79.8%.

18-9 Calculate the percent transmittance of solutions having twice the absorbance of the solutions in Problem 18-7.

18-10 Calculate the absorbances of solutions having half the percent transmittance of those in Problem 18-8.

18-11 Use the data in Table 18-2 to evaluate the missing quantities. Wherever necessary, assume that the molecular weight of the analyte is 250.

TABLE 18-2

	A	$\% T$	ϵ	b, cm	c, M	c, ppm	a, cm^{-1} ppm^{-1}
*(a)	0.416			1.40	1.25×10^{-4}		
(b)		45.5		2.10	8.15×10^{-3}		
*(c)	1.424			0.996			0.137
(d)		19.6	5.42×10^3		2.50×10^{-4}		
*(e)			3.46×10^3	2.50		3.33	
(f)			1.214×10^4	1.25	7.77×10^{-4}		
*(g)		48.3		0.250		6.72	
(h)	0.842		7.73×10^3	2.00			
*(i)		76.3		1.10			0.0631
(j)		6.54	9.82×10^2		8.64×10^{-3}		

*18-12 A solution containing 4.48 ppm $KMnO_4$ has a transmittance of 0.309 in a 1.00-cm cell at 520 nm. Calculate the molar absorptivity of $KMnO_4$.

18-13 A solution containing 3.75 mg/100 mL of A (fw = 220) has a transmittance of 39.6% in a 1.50-cm cell at 480 nm. Calculate the molar absorptivity of A.

*18-14 A solution containing the complex formed between Bi(III) and thiourea has a molar absorptivity of 9.32×10^3 L cm^{-1} mol^{-1} at 470 nm.

(a) What is the absorbance of a 6.24×10^{-5} M solution of the complex at 470 nm in a 1.00-cm cell?
(b) What is the percent transmittance of the solution described in (a)?
(c) What is the molar concentration of the complex in a solution that has the absorbance described in (a) when measured at 470 nm in a 5.00-cm cell?

18-15 At 580 nm, which is the wavelength of its maximum absorption, the complex $FeSCN^{2+}$ has a molar absorptivity of 7.00×10^3 L cm^{-1} mol^{-1}. Calculate

(a) the absorbance of a 2.50×10^{-5} M solution of the complex at 580 nm in a 1.00-cm cell.
(b) the absorbance of a solution in which the concentration of the complex is twice that in (a).
(c) the transmittance of the solutions described in (a) and (b).
(d) the absorbance of a solution that has half the transmittance of that described in (a).

*18-16 A 2.50-mL aliquot of a solution that contains 3.8 ppm iron(III) is treated with an appropriate excess of KSCN and diluted to 50.0 mL. What is the absorbance of the resulting solution at 580 nm in a 2.50-cm cell? See Problem 18-15 for absorptivity data.

18-17 Zinc(II) and the ligand L form a product that absorbs strongly at 600 nm. As long as the molar concentration of L exceeds that of zinc(II) by a factor of 5, the absorbance is dependent only on the cation concentration. Neither zinc(II) nor L absorbs at 600 nm. A solution that is 1.60×10^{-4} M in zinc(II) and 1.00×10^{-3} M in L has an absorbance of 0.464 in a 1.00-cm cell at 600 nm. Calculate

(a) the percent transmittance of this solution.
(b) the percent transmittance of this solution in a 2.50-cm cell.
(c) the light path needed to match the absorbance of solution (a) with a 3.00-cm column that is 4.00×10^{-4} M in the complex.

*18-18 The equilibrium constant for the conjugate acid/base pair

$$HIn + H_2O \rightleftharpoons H_3O^+ + In^-$$

is 8.00×10^{-5}. From the additional information

Species	Absorption Maximum, nm	Molar Absorptivity 430 nm	Molar Absorptivity 600 nm
HIn	430	8.04×10^3	1.23×10^3
In^-	600	0.775×10^3	6.96×10^3

(a) calculate the absorbance at 430 nm and 600 nm for the following indicator concentrations: 3.00×10^{-4} M, 2.00×10^{-4} M, 1.00×10^{-4} M, 0.500×10^{-4} M, and 0.250×10^{-4} M.
(b) plot absorbance as a function of indicator concentration.

18-19 The equilibrium constant for the reaction

$$2CrO_4^{2-} + 2H^+ \rightleftharpoons Cr_2O_7^{2-} + H_2O$$

is 4.2×10^{14}. The molar absorptivities for the two principal species in a solution of $K_2Cr_2O_7$ are

λ	ϵ_1 (CrO_4^{2-})	ϵ_2 ($Cr_2O_7^{2-}$)
345	1.84×10^3	10.7×10^2
370	4.81×10^3	7.28×10^2
400	1.88×10^3	1.89×10^2

Four solutions were prepared by dissolving 4.00×10^{-4}, 3.00×10^{-4}, 2.00×10^{-4}, and 1.00×10^{-4} moles of $K_2Cr_2O_7$ in water and diluting to 1.00 L with a pH 5.60 buffer. Derive theoretical absorbance values (1.00-cm cells) for each solution and plot the data for (a) 345 nm, (b) 370 nm, (c) 400 nm.

CHAPTER 19

Instruments for Optical Spectroscopy

The basic components of analytical instruments for emission, absorption, and fluorescence spectroscopy are remarkably alike in function and general performance requirements regardless of whether the instruments are designed for ultraviolet, visible, or infrared radiation. Because of their similarities, such instruments are frequently referred to as *optical instruments* even when they are applied to spectral regions to which the eye is insensitive. In this chapter, we first examine the characteristics of the components common to all optical instruments and point out those features that are independent of the wavelength region being employed as well as those that are not. We then consider the general design characteristics of typical instruments, particularly those used for absorption spectroscopy.

19A Instrument Components

Most spectroscopic instruments are made up of five components: (1) a stable source of radiant energy, (2) a wavelength selector that permits the isolation of a restricted wavelength region, (3) one or more sample containers, (4) a radiation detector, or transducer, which converts radiant energy to a measurable signal (usually electrical), and (5) a signal processor and readout. Figure 19-1 shows how these components are assembled in instruments for emission, absorption, and fluorescence spectroscopy. Note that the configuration of components 4 and 5 is the same for the three types of instruments.

Emission instruments differ from the other two types in that components 1 and 3 are combined. That is, the sample container is an arc, a spark, a heated surface, or a flame that both holds the sample and causes it to emit characteristic radiation. In contrast, absorption and fluorescence spectroscopy require an external source of radiant energy and a cell to hold the sample. In absorption measurements, the beam from the source passes through the sample after leaving the wavelength selector (in some instruments, the positions of the sample and selector are reversed). In fluorescence, the source induces the sample to emit characteristic radiation, which is usually measured at a 90-deg angle with respect to the beam from the source.

19A-1 The Transmittance of Various Construction Materials

Figure 19-2 shows the spectral-transparency regions for various materials used for constructing the windows, lenses, sample containers, and prisms of spectroscopic instruments. Ordinary silicate glass is widely used for instruments designed for use in the visible region only. Fused silica and quartz extend the range of spectroscopic instruments down to 180 to 200 nm in the ultraviolet. Infrared spectroscopy requires the use of windows and cells fashioned from

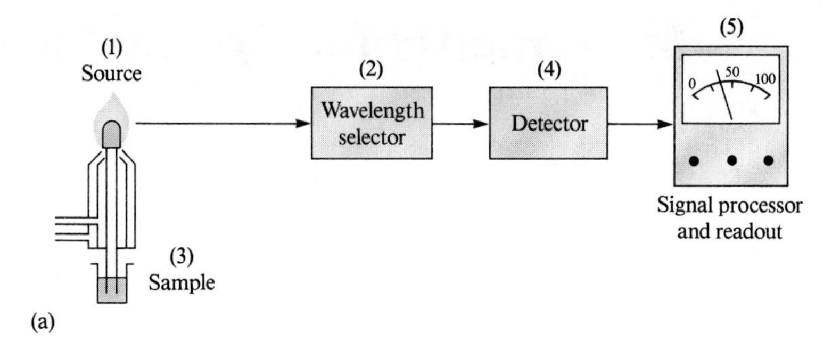

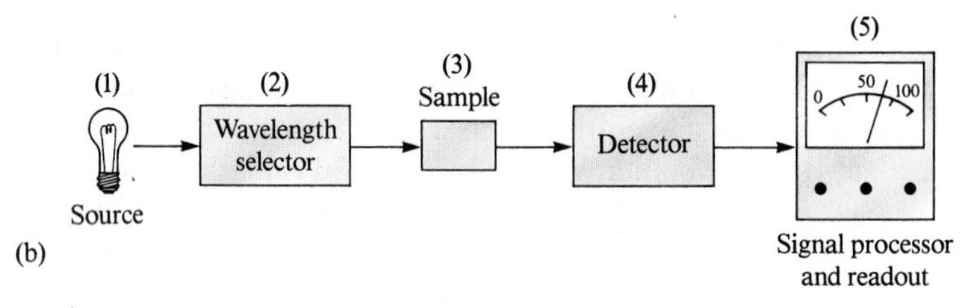

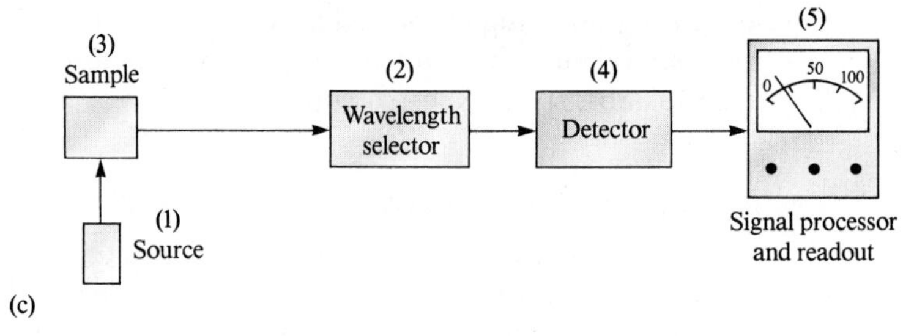

FIGURE 19-1 Components of various types of instruments for optical spectroscopy: (a) emission spectroscopy; (b) absorption spectroscopy; (c) fluorescence and scattering spectroscopy.

FIGURE 19-2 Transmittance range for various construction materials.

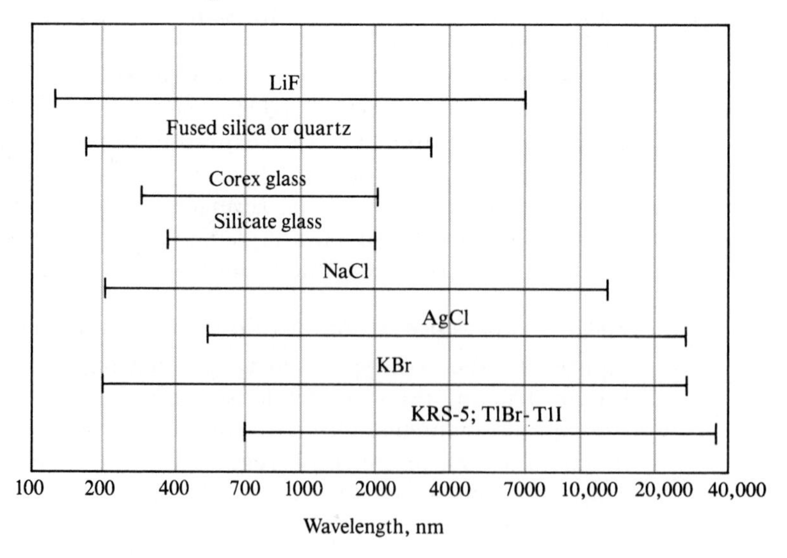

such materials as polished sodium chloride, potassium chloride, or silver chloride. All infrared-transparent materials tend to become fogged as a result of the absorption of moisture and must therefore be polished regularly until clear again.

19A-2

Spectroscopic Sources

Absorption and fluorescence spectroscopy require an external radiation source whose output is constant and intense enough to make detection and measurement easy. Typically, the radiant power of a source varies exponentially with the voltage of its electrical supply. For this reason, voltage regulators are often used to power spectroscopic sources.

The problem of source stability is sometimes circumvented by splitting the output of a source into a reference beam, which passes through the solvent, and a sample beam, which passes through a solution of the analyte. The detector is then illuminated alternately with the two beams, or the beams are monitored by matched detectors. The ratio of the intensities of the two beams then provides an analytical parameter that is largely independent of source fluctuations.

Table 19-1 lists the common sources used in the various types of spectroscopy. Note that both continuous and line sources find application. Details concerning line sources are found in Section 21B-2. In this section, we describe the most common continuous sources of ultraviolet, visible, and infrared radiation.

TABLE 19-1

Sources for Spectroscopy

Source	Wavelength Region, nm	Type of Spectroscopy
Continuous Sources		
Xenon lamp	250–600	Molecular fluorescence; Raman
H_2 and D_2 lamps	160–380	UV molecular absorption
Tungsten/halogen lamp	240–2500	UV/vis/near-IR molecular absorption
Tungsten lamp	350–2200	Vis/near-IR molecular absorption
Nernst glower	400–20,000	IR molecular absorption
Nichrome wire	750–20,000	IR molecular absorption
Globar	1200–40,000	IR molecular absorption
Line Sources		
Hollow cathode lamp	UV/vis	Atomic absorption; atomic fluorescence
Electrodeless discharge lamp	UV/vis	Atomic absorption; atomic fluorescence
Metal vapor lamp	UV/vis	Atomic absorption; molecular fluorescence; Raman
Laser	UV/vis/IR	Raman; molecular absorption; molecular fluorescence

Hydrogen and Deuterium Lamps

A truly continuous spectrum in the ultraviolet region is produced by the electrical excitation of deuterium or hydrogen at low pressure. The mechanism by which a continuum is produced involves the formation of an excited molecule (D_2^* or H_2^*) by absorption of electrical energy. This species then dissociates to give two hydrogen or deuterium atoms plus an ultraviolet photon. The reactions for hydrogen are

$$H_2 + E_e \longrightarrow H_2^* \longrightarrow H' + H'' + h\nu$$

where E_e is the electrical energy absorbed by the molecule. The energy for the overall process is

$$E_e = E_{H_2^*} = E_{H'} + E_{H''} + h\nu$$

where $E_{H_2^*}$ is the *fixed quantized energy* of H_2^* and $E_{H'}$ and $E_{H'}$ are the *kinetic energies* of the two hydrogen atoms. The sum of the latter two energies can vary from zero to $E_{H_2^*}$. Thus, the energy and the frequency of the photon can also vary within this range of energies. That is, when the two kinetic energies are by chance small, $h\nu$ is large, and when the two energies are large, $h\nu$ is small. The consequence is a truly continuous spectrum from about 160 nm to the beginning of the visible region.

Most modern lamps for generating ultraviolet radiation contain deuterium and are of a low voltage type in which an arc is formed between a heated, oxide-coated filament and a metal electrode. The heated filament provides electrons to maintain a direct current at a potential of about 40 V; a regulated power supply is required for constant intensities.

Both deuterium and hydrogen lamps provide a useful continuous spectrum in the region from 160 to 375 nm. The intensity of the deuterium lamp is greater than that of the hydrogen lamp, however, which accounts for the more widespread use of the former. At longer wavelengths (>360 nm), the lamps generate emission lines, which are superimposed on the continuum. For many applications, these lines are a nuisance. They are useful for wavelength calibration of absorption instruments, however.

Tungsten-Filament Lamps

The most common source of visible and near-infrared radiation is the tungsten-filament lamp. The energy distribution of this source approximates that of a blackbody and is thus temperature-dependent. Figure 18-14 illustrates the output of the tungsten filament lamp at 3000 K. In most absorption instruments, the operating filament temperature is about 2900 K; the bulk of the energy is thus emitted in the infrared region. A tungsten-filament lamp is useful for the wavelength region between 320 and 2500 nm. The lower limit is imposed by absorption by the glass envelope that houses the filament.

The energy output of a tungsten lamp in the visible region varies approximately as the fourth power of the operating voltage, thus making close voltage control essential. For this reason, constant-voltage transformers or electronic voltage regulators are employed. As an alternative, the lamp can be operated from a 6-V storage battery, which provides a remarkably stable voltage if it is maintained in good condition.

Tungsten/halogen lamps contain a small quantity of iodine within a quartz envelope that houses the filament. Quartz allows the filament to be operated at a temperature of about 3500 K, which leads to higher intensities and extends the range of the lamp well into the ultraviolet. The lifetime of a tungsten/halogen lamp is more than double that of an ordinary tungsten lamp because the life of the latter is limited by sublimation of the tungsten from the filament. In the presence of iodine, the sublimed tungsten reacts to give gaseous WI_2 molecules, which then diffuse back to the hot filament, where they decompose and redeposit as tungsten atoms. Tungsten/halogen lamps are finding ever-increasing use in modern spectroscopic instruments because of their extended wavelength range, greater intensity, and longer life.

Infrared Sources

Continuous infrared radiation is obtained from hot inert solids. A *Globar* source consists of a 5 by 50 mm silicon carbide rod. Radiation in the region from 1 to 40 μm is emitted when the Globar is heated to about 1500°C by the passage of electricity.

A *Nernst Glower* is a cylinder of zirconium and yttrium oxides having typical dimensions of 2 by 20 mm; it emits infrared radiation when heated to a high temperature by an electric current. Electrically heated spirals of nichrome wire also serve as infrared sources.

19A-3 Wavelength Selectors

Spectroscopic instruments are ordinarily equipped with a device that restricts the radiation being measured to a narrow band that is absorbed or emitted by the analyte. Such devices often greatly enhance both the selectivity and the sensitivity of an instrument. In addition, for absorption measurements, narrow bandwidths increase the likelihood of adherence to Beer's law.

At the outset, it must be understood that no selector is capable of producing radiation of a single wavelength. Instead, the output of such a device is a group of contiguous wavelengths called a band[1]; these wavelengths are distributed more or less symmetrically about a central *nominal wavelength.* As shown in Figure 19-3, the *effective bandwidth,* or *bandwidth,* of a selector is defined as the width of the band in wavelength units at half-peak height. Note that the ordinate in this plot is the percentage of incident radiation of a given wavelength that is transmitted. Bandwidths vary enormously from one wavelength selector to another. For example, a high-quality monochromator for the visible region may have an effective bandwidth of a few tenths of a nanometer or less, whereas an absorption filter in this same region may possess a bandwidth of 200 nm or more.

As shown in Table 19-2, two general types of wavelength selectors, *filters* and *monochromators,* are used to provide narrow bands of radiation. Monochromators have the advantage that the output wavelength can be varied continuously over a considerable spectral range.

Filters

Filters operate by absorbing all but a restricted band of radiation from a continuous source. As shown in Figure 19-3, a filter is generally characterized by its *nominal wavelength,* its maximum percent transmittance, and its effective bandwidth.

[1]Note that the term *band* in this context has a somewhat different meaning from that used in describing types of spectra.

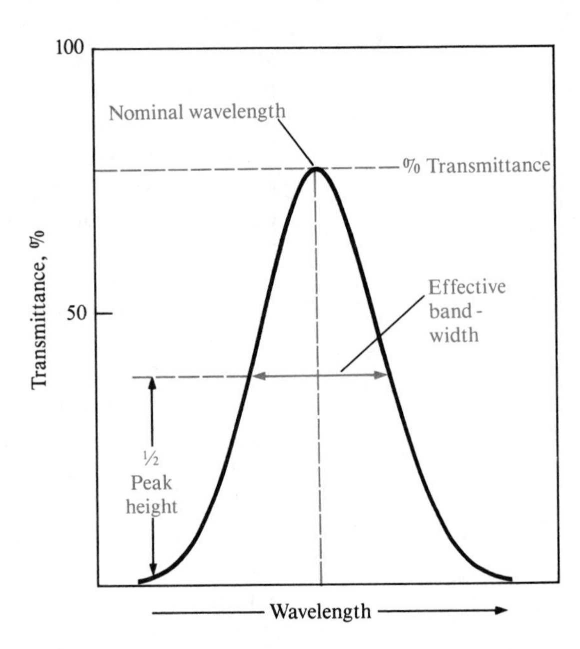

FIGURE 19-3 Output of a typical wavelength selector.

Interference Filters. Interference filters find use with ultraviolet and visible radiation, as well as with wavelengths up to about 14 μm in the infrared region. As the name implies, an interference filter relies on optical interference to provide a relatively narrow band of radiation.

An interference filter consists of a very thin layer of a transparent material (frequently calcium fluoride or magnesium fluoride) coated on both sides with a film of metal that is thin enough so that it transmits approximately half of the radiation striking it and reflects the other half. This array is sandwiched between two glass plates that protect it from the atmosphere. When radiation strikes the central array at a 90-deg angle, approximately half is transmitted by the first metallic layer and the other half reflected. The transmitted radiation undergoes a similar partition when it reaches the second layer of metal. If the reflected portion from the second layer is of the proper wavelength, it is partially reflected from the inner portion of the first layer in phase with the incoming light of the same wavelength. The result is constructive interference of the radiation of this wavelength and destructive removal of most other

TABLE 19-2 **Wavelength Selectors for Spectroscopy**

Type	Wavelength Range, nm	Note
Continuously variable		
Grating	100–40,000	3000 lines/mm for vacuum UV 50 lines/mm for far IR
Prism	120–30,000	See Figure 19-2 for construction materials.
Discontinuous		
Interference filter	200–14,000	
Absorption filter	380–750	

wavelengths. It is readily shown that the nominal wavelength for an interference filter λ_{max} is given by the equation[2]

$$\lambda_{max} = 2tn/\mathbf{n} \qquad (19\text{-}1)$$

where t is the thickness of the central fluoride layer, n is its refractive index, and $\mathbf{n}$ is an integer called the *interference order.* The glass layers of the filter are often selected to absorb all but one of the wavelengths transmitted by the central layer, thus restricting the transmission of the filter to a single order.

Figure 19-4 illustrates the performance characteristics of a typical interference filter. Most filters of this type have bandwidths of better than 1.5% of the nominal wavelength, although this figure is lowered to 0.15% in some narrow-band filters; the latter have a maximum transmittance of about 10%.

Absorption Filters. Absorption filters, which are generally less expensive and more rugged than interference filters, are limited in application to the visible region. This type of filter usually consists of a colored glass plate that removes part of the incident radiation by absorption. Absorption filters have effective bandwidths that range from perhaps 30 to 250 nm. Filters that provide the narrowest bandwidths also absorb a significant fraction of the desired radiation and may have a transmittance of 1% or less at their band peaks. Figure 19-4 contrasts the performance characteristics of a typical absorption filter with its interference counterpart.

Glass filters with transmittance maxima throughout the entire visible region are available from commercial sources. While their performance characteristics are distinctly inferior to those of interference filters, their cost is appreciably less, and they are perfectly adequate for many routine applications.

FIGURE 19-4 Bandwidths for two types of filters.

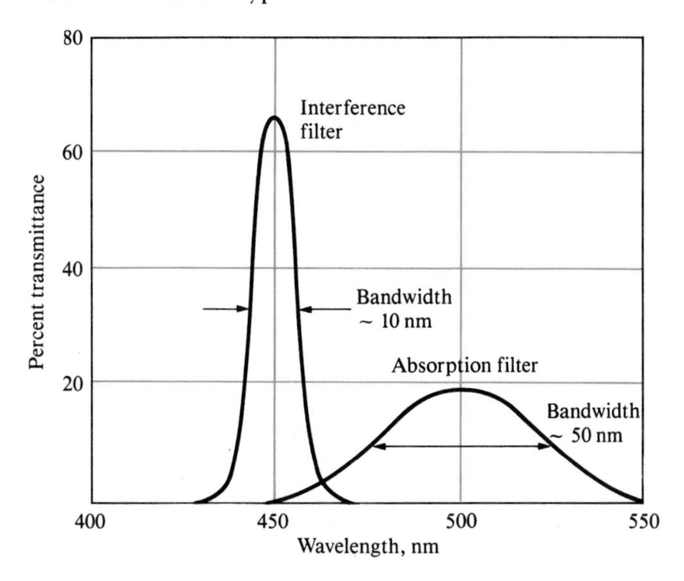

[2]For example, see D. A. Skoog, *Principles of Instrumental Analysis,* 3rd ed., pp. 121–122. Philadelphia: Saunders College Publishing, 1985.

Monochromators

Monochromators for ultraviolet, visible, and infrared radiation are all similar in construction in the sense that they employ slits, lenses, mirrors, windows, and dispersing devices. To be sure, the materials from which these components are fabricated depend upon the wavelength region of intended use (Figure 19-2).

The Components of a Monochromator. Figure 19-5 illustrates the optical elements found in all monochromators: (1) an entrance slit, (2) a collimating lens or mirror to produce a parallel beam, (3) a prism or grating to disperse the radiation into its component wavelengths, and (4) a focusing element that projects a series of rectangular images of the entrance slit upon a planar surface called the *focal plane*. In addition, most monochromators have entrance and exit *windows* to protect the components from dust and corrosive laboratory fumes.

As shown in Figure 19-5, two types of dispersing devices are found in monochromators: reflection gratings and prisms. For purposes of illustration, a beam made up of just two wavelengths, λ_1 and λ_2 ($\lambda_1 > \lambda_2$), is shown. The beam enters the monochromator via a narrow rectangular opening (the *slit*),

FIGURE 19-5 Two types of monochromators: (a) Czerny-Turner grating monochromator; (b) Bunsen prism monochromator. (In both instances, $\lambda_1 > \lambda_2$.)

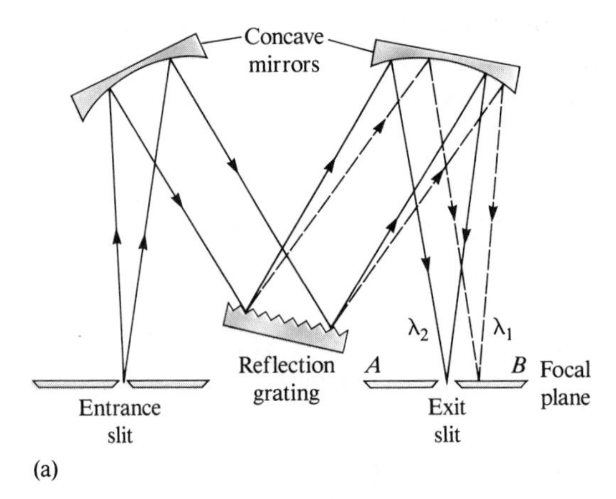

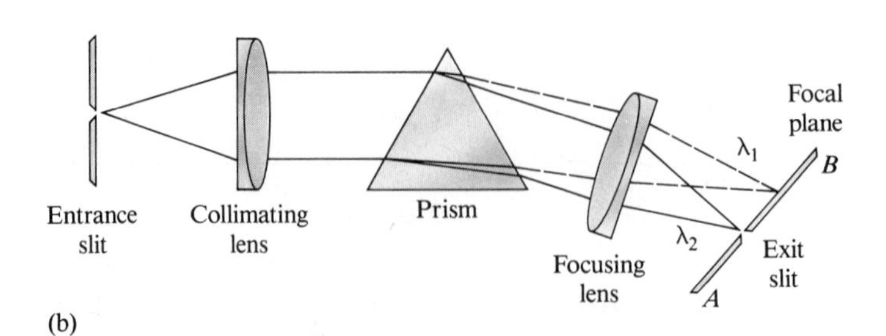

is collimated, and then strikes the surface of the dispersing element at an angle. In the grating monochromator, angular dispersion of the beam into its individual wavelengths results from diffraction at the reflective surface. For the prism instrument, bending, or refraction, of the radiation at the two surfaces leads to dispersion. In either case, the dispersed radiation is focused on the focal plane *AB*, where it appears as two images of the entrance slit (one for each wavelength). These images can be focused on the exit slit by rotating the dispersing element.

Reflection gratings serve as the dispersing element in most modern spectroscopic instruments. Thus, the discussion that follows is restricted to grating monochromators exclusively.

Reflection Gratings. Most reflection gratings are *replica gratings* prepared from a *master grating,* which consists of a large number of parallel and closely spaced grooves (or blazes) ruled on a hard, polished surface with a suitably shaped diamond tool. A magnified cross-sectional view of a few typical grooves is shown in Figure 19-6. A grating for the ultraviolet and visible regions contains from 300 to 2000 grooves/mm, with 1200 to 1400 being most common. For the infrared region, 10 to 200 grooves/mm is common.

Replica gratings are formed by evaporating a film of aluminum onto a master grating after it has been coated with a parting agent that permits ready separation of the aluminum from the master. A glass plate is then cemented to the aluminum, and the grooved film is lifted from the master mold to give a finished grating. In recent times, *holographic gratings* have begun to appear in monochromators. This type of grating is manufactured by sophisticated lithographic techniques based upon the use of a pair of laser beams. The performance of these gratings is far superior to that of replica gratings.

Dispersion by a Reflection Grating. The grating shown in Figure 19-6 is an *echellette* grating, which is grooved, or *blazed,* to have relatively broad faces

FIGURE 19-6 The mechanism of diffraction from an echellette-type grating.

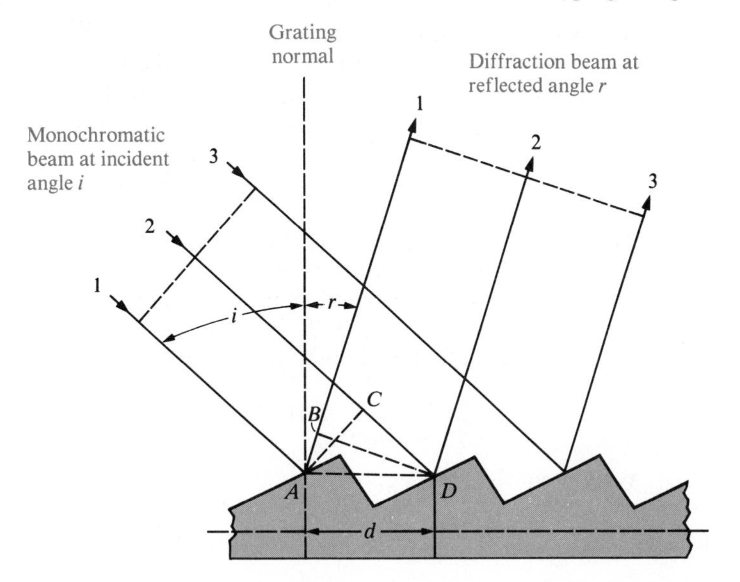

from which reflection occurs and narrow unused faces. This geometry provides highly efficient diffraction. Each broad face can be treated as a point source of radiation, giving reflected beams 1, 2, and 3, which interfere with one another. In order for the interference to be constructive, it is necessary that the path lengths differ by an integral multiple **n** of the wavelength of the incident beam.

In Figure 19-6, parallel beams of monochromatic radiation 1 and 2 strike the grating at an incident angle i to the *grating normal.* Maximum constructive interference is shown as occurring at the reflected angle r. It is evident that beam 2 travels a greater distance than beam 1 and that this difference is equal to $\overline{CD} - \overline{AB}$. For constructive interference to occur, this difference must equal $\mathbf{n}\lambda$:

$$\mathbf{n}\lambda = \overline{CD} - \overline{AB}$$

where **n**, a small whole number, is called the diffraction *order.* Note, however, that angle CAD is equal to angle i and that angle BDA is identical to angle r. Therefore, from trigonometry,

$$\overline{CD} = d \sin i$$

where d is the spacing between the reflecting surfaces. It is also seen that

$$\overline{AB} = -d \sin r$$

The minus sign by convention indicates that the angle of reflection r lies on the opposite side of the grating normal from the incident angle i (as in Figure 19-6); angle r is positive when it is on the same side as i. Substitution of the last two expressions into the first gives the condition for constructive interference:

$$\mathbf{n}\lambda = d(\sin i + \sin r) \quad (19\text{-}2)$$

Equation 19-2 suggests that several values of λ exist for a given diffraction angle r. Thus, if a first-order line ($\mathbf{n} = 1$) of 900 nm is found at r, second-order (450 nm) and third-order (300 nm) lines also appear at this angle. Ordinarily, the first-order line is the most intense; indeed, it is possible to design gratings that concentrate as much as 90% of the incident intensity in this order. The higher-order lines can generally be removed by filters. For example, glass, which absorbs radiation below 350 nm, eliminates the high-order spectra associated with first-order radiation in most of the visible region. The example that follows illustrates these points.

EXAMPLE 19-1

An echellette grating containing 1450 blazes per millimeter was irradiated with a polychromatic beam at an incident angle 48 deg to the grating normal. Calculate the wavelengths of radiation that appear at an angle of reflection +20, +10, and −10 deg.

To obtain d in Equation 19-2, we write

$$d = \frac{1 \text{ mm}}{1450 \text{ blazes}} \times 10^6 \frac{\text{nm}}{\text{mm}} = 689.7 \frac{\text{nm}}{\text{blaze}}$$

When r equals +20 deg,

$$\lambda = \frac{689.7}{\mathbf{n}}(\sin 48 + \sin 20) = \frac{748.4}{\mathbf{n}}$$

and the wavelengths for the first-, second-, and third-order reflections are 748, 374, and 249 nm.

When angle r is -10 deg,

$$\lambda = \frac{689.7}{n}[\sin 48 + \sin(-10)] = 392.8 \text{ nm}$$

Further calculations of a similar kind yield the following data:

	Wavelength, nm		
r, deg	n = 1	n = 2	n = 3
+20	748	374	249
+10	632	316	211
0	513	256	171
−10	393	196	131

In contrast to a prism, a grating disperses radiation linearly along the focal plane of the monochromator, which greatly simplifies monochromator design. A second advantage is that replica gratings are significantly less costly than prisms.

Monochromator Slits. The slits of a monochromator play an important role in determining the quality of the instrument. Slit jaws are formed by carefully machining two pieces of metal to give sharp edges. Care must be taken to ensure that these edges are parallel to one another and in the same plane.

The effective bandwidth of a monochromator depends upon the dispersion of the prism or grating as well as on the width of the entrance and/or exit slit. Most monochromators are equipped with variable slits so that the effective bandwidth can be changed. Narrow slits, and thus narrow bandwidths, lead to higher instrument *resolution,* which is desirable because greater spectral detail is revealed under such conditions (for example, see Figure 20-4). Because the power of the beam exiting from a monochromator falls off rapidly as the slit widths are narrowed, the resolution at which a monochromator can be operated is generally determined by the sensitivity of its radiation detector.

Stray Radiation in Monochromators. The exit beam of a monochromator is usually contaminated with small amounts of radiation having wavelengths far removed from that of the instrument setting. Sources of this unwanted radiation include reflection from various surfaces within the monochromator and scattering by dust particles in the atmosphere or on the surfaces of optical parts. Generally, the effects of spurious radiation are minimized by introducing baffles at appropriate spots in the monochromator and by coating interior surfaces with flat black paint. In addition, the monochromator is sealed with windows over the slits to prevent entrance of dust and fumes. Despite these precautions, however, some spurious radiation still reaches the exit slit of even the best monochromators. The effects of such radiation on absorption measurements are described in Sections 18C-7 and 20A-3.

19A-4 *Radiation Detectors and Transducers*

A *detector* is a device that indicates the existence of some physical phenomenon. Familiar examples of detectors include photographic film for indicating the presence of electromagnetic or radioactive radiation, the pointer of a balance for detecting mass differences, and the mercury level in a thermometer for detecting temperature changes. The human eye is also a detector; it converts visible radiation into an electrical signal that is passed to the brain via the chain of neurons in the optic nerve.

A *transducer* is a special type of detector that converts signals, such as light intensity, pH, mass, and temperature into *electrical* signals that can be subsequently amplified, manipulated, and finally converted into numbers representing the magnitude of the original signal.

Properties of Transducers

The ideal electromagnetic radiation transducer responds rapidly to low levels of radiant energy over a broad wavelength range. In addition, it produces an electrical signal that is easily amplified and has a relatively low noise level.[3] Finally, it is essential that the electrical signal produced by the transducer be directly proportional to the beam power P:

$$G = KP + K' \quad (19\text{-}3)$$

where G is the electrical response of the detector in units of current, resistance, or potential. The proportionality constant K measures the sensitivity of the detector in terms of electrical response per unit of radiant power. Many detectors exhibit a small constant response, known as a *dark current* K', even when no radiation impinges upon their surfaces. Instruments with detectors that have a significant dark-current response are ordinarily equipped with a compensating circuit that permits subtraction of a signal proportional to the dark current to reduce K' to zero. Thus, under ordinary circumstances, we can write

$$G = KP \quad (19\text{-}4)$$

Types of Transducers

As shown in Table 19-3, two general types of transducers are encountered: one responds to photons, the other to heat. All photon detectors are based upon the interaction of radiation with a reactive surface to produce electrons (photoemission) or to promote electrons to energy states in which they can conduct electricity (photoconduction). Only ultraviolet, visible, and near-infrared radiation have sufficient energy to cause these processes to occur; thus photon detectors are limited to wavelengths shorter than about 2 μm.

Generally, infrared radiation is detected by measuring the temperature rise of a blackened material located in the path of the beam. Because the temperature changes resulting from the absorption of the infrared energy are minute, close control of the ambient temperature is required if large errors are to be avoided. It is usually the detector system that limits the sensitivity and precision of an infrared instrument.

[3] Generally, the output from analytical instruments fluctuates in a random way as a consequence of the operation of a large number of uncontrolled variables. These fluctuations, which limit the sensitivity of an instrument, are called *noise*. The terminology is derived from radio engineering, where the presence of unwanted signal fluctuations was recognizable to the ear as static, or noise.

TABLE 19-3 **Detectors for Spectroscopy**

Type	Wavelength Range, nm
Photon Detectors	
Phototubes	150–1000
Photomultiplier tubes	150–1000
Silicon diodes	350–1100
Photoconductors	750–3000
Photovoltaic cells	380–780
Heat Detectors	
Thermocouples	600–20,000
Bolometers	600–20,000
Pneumatic cells	600–40,000
Pyroelectric cells	1000–20,000

Photon Detectors

Four widely used types of photon detectors are described in the paragraphs that follow: (1) phototubes, (2) photomultiplier tubes, (3) silicon photodiodes, and (4) photovoltaic cells.

Phototubes. As shown in Figure 19-7, a phototube consists of a semicylindrical cathode and a wire anode sealed inside an evacuated transparent envelope. The concave surface of the cathode supports a layer of photoemissive material, such as an alkali metal or metal oxide, that tends to emit electrons upon being irradiated. When a potential is applied across the electrodes, the emitted electrons flow to the wire anode, producing a photocurrent that is readily amplified and displayed or recorded.

The number of electrons ejected from a photoemissive surface is directly proportional to the radiant power of the beam striking that surface. With an applied potential of about 90 V, all these electrons reach the anode to give a

FIGURE 19-7 A phototube and accessory circuit. The photocurrent induced by the radiation causes a potential drop in the resistor, which is then amplified to drive a meter or recorder.

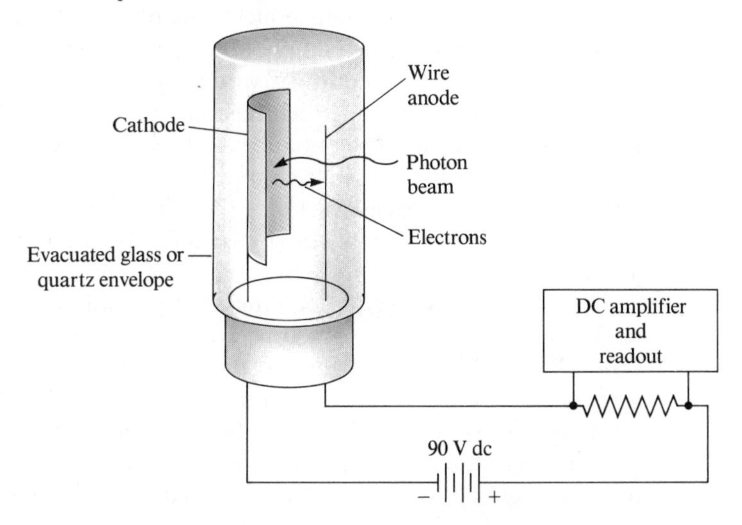

current that is also proportional to radiant power. Phototubes frequently produce a small dark current in the absence of radiation (Equation 19-3) that results from thermally induced electron emission.

Photomultiplier Tubes. The *photomultiplier tube,* shown schematically in Figure 19-8, is similar in construction to the phototube just described but is significantly more sensitive. Its cathode surface is similar in composition to that of a phototube, with electrons being emitted upon exposure to radiation. The emitted electrons are accelerated toward a *dynode* (labeled 1 in the figure) maintained at a potential 90 V more positive than the cathode. Upon striking the dynode surface, each accelerated photoelectron produces several additional electrons, all of which are then accelerated to dynode 2, which is 90 V more positive than dynode 1. Here again, electron amplification occurs. By the time this process has been repeated at each of the remaining dynodes, 10^6 to 10^7 electrons have been produced for each photon; this cascade is finally collected at the anode. The resulting current is then further amplified electronically and measured.

Silicon Photodiodes. Crystalline silicon is a *semiconductor*—that is, a material whose electrical conductivity is less than that of a metal but greater than that of an electrical insulator. Silicon is a Group IV element and thus has four valence electrons. In a silicon crystal, each of these electrons is combined with electrons from four other silicon atoms to form four covalent bonds. At room temperature, sufficient thermal agitation occurs in this structure to liberate an occasional electron from its bonded state leaving it free to move throughout the crystal. Thermal excitation of an electron leaves behind a positively charged region termed a *hole,* which, like the electron, is also mobile. The mechanism of hole movement is stepwise, with a bound electron from a neighboring silicon atom jumping into the electron-deficient region (the hole) and thereby creating another positive hole in its wake. Conduction in a semiconductor involves the movement of electrons and holes in opposite directions.

The conductivity of silicon can be greatly enhanced by *doping,* a process whereby a tiny, controlled amount (approximately 1 ppm) of a Group V or Group III element is distributed homogenously throughout a silicon crystal. For example, when a crystal is doped with a Group V element, such as arsenic, four out of five of the valence electrons of the dopant form covalent bonds with four silicon atoms leaving one electron free to contribute to the conductivity of the crystal. In contrast, when the silicon is doped with a Group III element, such as gallium, which has but three valence electrons, an excess of holes develops, which also enhances conductivity. A semiconductor containing unbonded electrons (*n*egative charges) is termed an *n-type* semiconductor, and one containing an excess of holes (*p*ositive charges) is a *p-type.* In an *n*-type semiconductor, the *majority carrier* is electrons; in a *p*-type, holes are the majority carrier.

Present silicon technology makes it possible to fabricate what is called a *pn junction* or a *pn diode,* which is conductive in one direction and not in the other. Figure 19-9a is a schematic diagram of a silicon diode. The *pn* junction is shown as a dashed line through the middle of the crystal. Electrical wires are attached to both ends of the device. Figure 19-9b shows the junction in its conduction mode, wherein the positive terminal of a dc source is connected to the *p* region and the negative terminal to the *n* region (the diode is said to

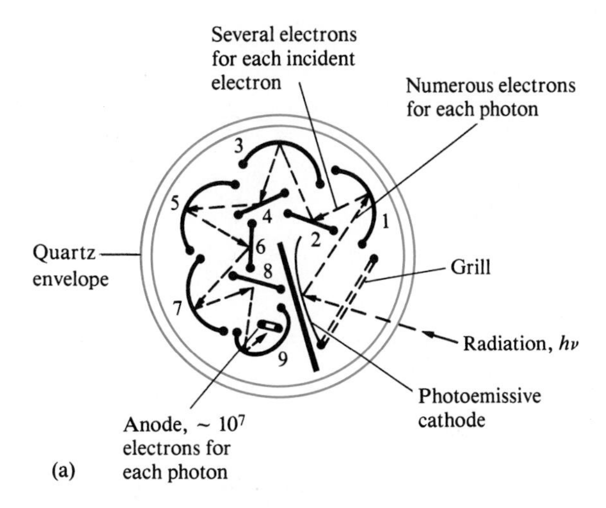

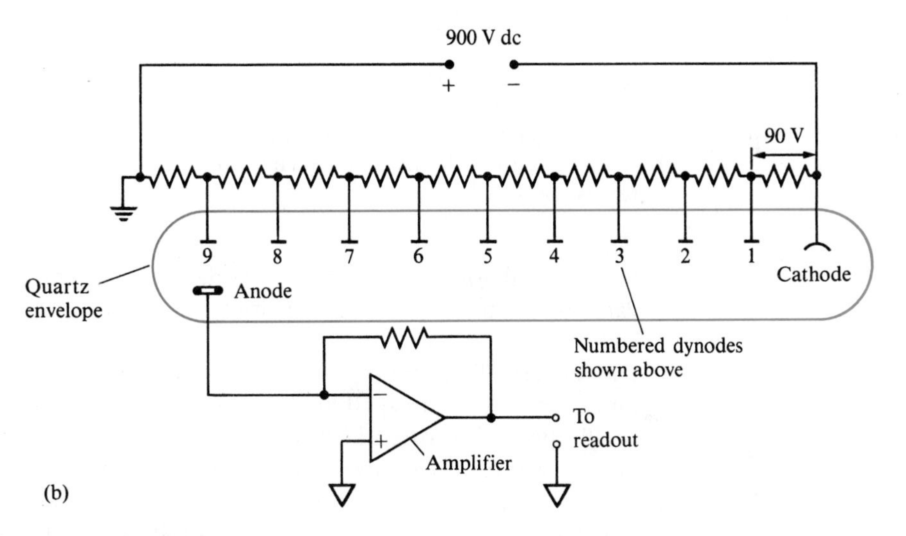

FIGURE 19-8 Photomultiplier tube: (a) cross section; (b) electrical circuit.

be *forward-biased* under these conditions). The excess electrons in the *n* region and the positive holes in the *p* region move toward the junction, where they combine and annihilate each other. The negative terminal of the source injects new electrons into the *n* region, which can continue the conduction process. The positive terminal extracts electrons from the *p* region, thus creating new holes that are free to migrate toward the *pn* junction.

Figure 19-9c illustrates the behavior of a silicon diode under *reverse biasing*. Here, the majority carriers are drawn away from the junction, leaving a nonconductive *depletion layer*. The conductance under reverse bias is only about 10^{-6} to 10^{-8} of that under forward biasing; thus, a silicon diode is a current rectifier.

A reverse-biased silicon diode can serve as a radiation detector because ultraviolet and visible photons are sufficiently energetic to create additional electrons and holes when they strike the depletion layer of a *pn* junction. The resulting increase in conductivity is readily measured and is directly propor-

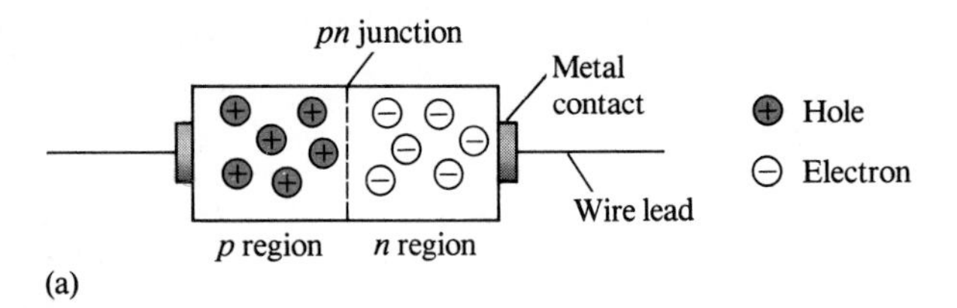

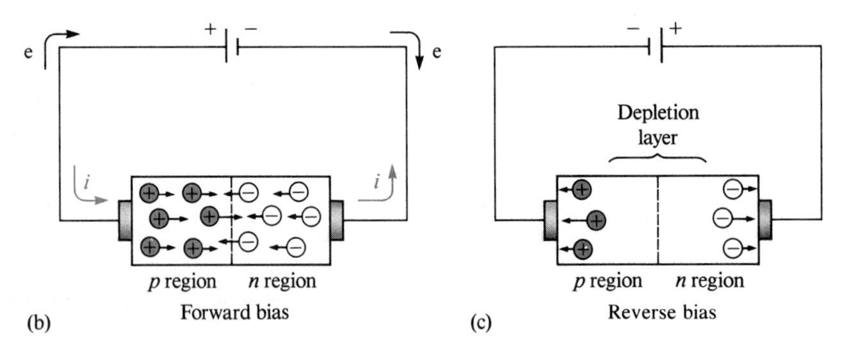

FIGURE 19-9 (a) Schematic of a silicon diode. (b) Flow of electricity under forward bias. (c) Formation of depletion layer that prevents flow of electricity under reverse bias.

tional to radiant power. A silicon-diode detector is more sensitive than a simple vacuum phototube but less sensitive than a photomultiplier tube.

Diode-Array Detectors. Silicon photodiodes have become of notable importance recently because 1000 or more can be fabricated side by side on a single small silicon chip (the width of individual diodes is about 0.02 mm). With one or two of these *diode-array detectors* placed along the length of the focal plane of a monochromator, all wavelengths can be monitored simultaneously, thus making high-speed spectroscopy possible. Multichannel instruments based upon diode arrays are discussed in Section 19B-2.

Photovoltaic Cells. A photovoltaic cell (or photocell), the simplest of all radiation transducers, consists of a flat copper or iron electrode upon which is deposited a layer of a semiconducting material, such as selenium or copper(I) oxide. The outer surface of the semiconductor is coated with a thin, transparent film of gold, silver, or lead, which serves as the second, or collector, electrode. When radiation is absorbed on the surface of the semiconductor, electrons and holes are formed and migrate in opposite directions, thus creating a current. If the two electrodes are connected through a low-resistance external circuit, the current produced is directly proportional to the power of the incident beam. The currents are large enough (10 to 100 μA) to be measured with a simple microammeter without amplification.

The typical photovoltaic cell has maximum sensitivity at about 550 nm, with the response falling off to perhaps 10% of maximum at 350 and 750 nm. The photocell constitutes a rugged, low-cost detector of visible radiation that has the advantage of not requiring an external power source. It is not, however, as sensitive as other detectors and in addition suffers from *fatigue*, which causes its current output to decrease gradually with continued illumination. Despite these disadvantages, photovoltaic cells are quite useful for simple, portable, low-cost filter instruments.

Heat Detectors

The convenient photon detectors discussed in the previous section cannot be used to measure infrared radiation because photons of these frequencies lack the energy to cause photoemission of electrons; as a consequence, thermal detectors must be used. Unfortunately, the performance characteristics of thermal detectors are much inferior to those of phototubes, photomultiplier tubes, silicon diodes, or photovoltaic cells.

A thermal detector consists of a tiny blackened surface that absorbs infrared radiation and increases in temperature as a consequence. The temperature rise is converted to an electrical signal that is amplified and measured. Under the best of circumstances, the temperature changes involved are minuscule, amounting to a few thousandths of a degree Celsius. The difficulty of measurement is compounded by thermal radiation from the surroundings, which is a potential source of uncertainty. To minimize the effects of this background radiation, or noise, thermal detectors are housed in a vacuum and are carefully shielded from their surroundings. To further minimize the effects of this external noise, the beam from the source is chopped by a rotating disk inserted between source and detector. Chopping produces a beam that fluctuates regularly from zero intensity to a maximum. The transducer converts this periodic radiation signal to an alternating electrical current that can be amplified and separated from the dc signal arising from the background radiation. Despite all these measures, infrared measurements are significantly less precise than measurements of ultraviolet and visible radiation.

As shown in Table 19-3, four types of heat detectors are used for infrared spectroscopy. The most widely used is a tiny thermocouple or a group of thermocouples called a *thermopile.* These devices consist of one or more pairs of dissimilar metal junctions that develop a potential difference when their temperatures differ. The magnitude of the potential depends upon the temperature difference.

A bolometer consists of a conducting element whose electrical resistance changes as a function of temperature. Bolometers are fabricated from thin strips of metals, such as nickel or platinum, or from semiconductors consisting of oxides of nickel or cobalt; the latter are called *thermistors.*

A *pneumatic detector* consists of a small cylindrical chamber that is filled with xenon and contains a blackened membrane to absorb infrared radiation and heat the gas. One end of the cylinder is sealed with a window that is transparent to infrared radiation; and the other is sealed with a flexible diaphragm that moves in and out as the gas pressure changes with cooling or heating. The temperature is determined from the position of the diaphragm.

Pyroelectric detectors are manufactured from crystals of a pyroelectric material, such as barium titanate or triglycine sulfate. When a crystal of either of these compounds is sandwiched between a pair of electrodes (one of which is transparent to infrared radiation), a temperature-dependent voltage develops that can be amplified and measured.

Signal Processors and Readouts

A signal processor is ordinarily an electronic device that amplifies the electrical signal from the detector; in addition, it may alter the signal from dc to ac (or the reverse), change the phase of the signal, and filter it to remove unwanted components. The signal processor may also be called upon to perform such mathematical operations on the signal as differentiation, integration, or conversion to a logarithm.

Several types of readout devices are found in modern instruments. Digital meters, scales of potentiometers, recorders, cathode-ray tubes and monitors of microcomputers are some examples.

19A-5 Sample Containers

Sample containers, which are usually called *cells* or *cuvettes,* must have windows fabricated from a material that is transparent in the spectral region of interest. Thus, as shown in Figure 19-2, quartz or fused silica is required for the ultraviolet region (below 350 nm) and may be used in the visible region and to about 3000 nm in the infrared. Because of its lower cost, silicate glass is ordinarily used for the region between 375 and 2000 nm. Plastic containers have also found application in the visible region. The most common window material for infrared studies is crystalline sodium chloride.

The best cells have windows that are normal to the direction of the beam so that reflection losses are minimized. The most common cell length for studies in the ultraviolet and visible regions is 1 cm; matched, calibrated cells of this size are available from several commercial sources. Other path lengths, from shorter than 0.1 cm to 10 cm, can also be purchased. Transparent spacers for shortening the path length of 1-cm cells to 0.1 cm are also available.

For reasons of economy, cylindrical cells are sometimes encountered. Particular care must be taken to duplicate the position of such cells with respect to the beam; otherwise variations in path length and reflection loss at the curved surfaces can cause significant error.

The quality of spectroscopic data is critically dependent upon the way the matched cells are used and maintained. Fingerprints, grease, or other deposits on the walls markedly alter the transmission characteristics of a cell. Thus, thorough cleaning before and after use is imperative, and care must be taken to avoid touching the windows after cleaning is complete. Matched cells should never be dried by heating in an oven or over a flame because this may cause physical damage or a change in path length. Matched cells should be calibrated against each other regularly with an absorbing solution.

19B Spectroscopic Instruments

The components discussed in the previous section have been assembled in various ways to produce dozens of designs for instruments to be used for spectroscopic measurements. These designs run the gamut from remarkably simple to highly sophisticated. Costs also vary widely, from a few hundred dollars to a hundred times that amount or more. No single instrument is best for all purposes, and selection must be determined by the type of work for which the instrument is intended and by the economics of its applications.

19B-1 Types of Spectroscopic Instruments

A *spectroscope* is an instrument for identifying the elements in a sample that have been excited in a flame or other hot medium. It consists of a modified monochromator, such as that shown in Figure 19-5b, in which the focal plane containing the exit slit is replaced by a movable eyepiece that permits visual detection of the emission lines. The wavelength of a line is determined from the angle between the incident beam and the path of the line to the eyepiece.

Strictly speaking, a *colorimeter* is an instrument for absorption measurements in which the human eye serves as the detector. One or more comparison standards are required each time the instrument is used.

A *photometer* is a simple instrument that can be used for absorption, emission, or fluorescence measurements with ultraviolet, visible, or infrared radiation. A photometer is distinguished by its use of absorption or interference filters for wavelength selection and a photoelectric device for measuring radiant power. Instruments used for absorption measurements with visible radiation are sometimes called *photoelectric colorimeters* or even simply *colorimeters*. Use of the latter term can lead to ambiguity. A photometer that is to be employed for fluorescence measurements exclusively is often termed a *fluorometer*.

A *spectrograph* records spectra on a photographic plate or film located along the focal plane of a monochromator. Thus, the monochromators in Figure 19-5 could be converted to spectrographs by replacing the focal plane *AB* by a plate or film holder. The spectra would then appear as a series of black images of the entrance slit. Spectrographs are used primarily for qualitative elemental analysis.

A *spectrometer* is a monochromator equipped with a fixed slit at the focal plane. The two monochromators shown in Figure 19-5 are examples of spectrometers. A spectrometer equipped with a phototransducer is called a *spectrophotometer*. Spectrometers can be used for absorption, emission, and fluorescence measurements. Fluorescence spectrometers are often called *spectrofluorometers*.

19B-2

Instrument Designs

In this section, we consider four general types of spectroscopic instruments: (1) single-beam, (2) double-beam in space, (3) double-beam in time, and (4) multichannel.

Single-Beam Instruments

Figure 19-10a is a schematic diagram of a single-beam instrument for absorption measurements. It consists of one of the radiation sources shown in Table 19-1, a filter or a monochromator for wavelength selection (Table 19-2), matched cells that can be interposed alternately in the radiation beam, one of the detectors listed in Table 19-3, an amplifier, and a readout device.

The measurement of percent transmittance with a manual single-beam instrument involves three steps: (1) the *0% T adjustment,* (2) the *100% T adjustment,* and (3) the *determination of % T for the analyte*. The 0% *T* adjustment is carried out with the shutter imposed between the source and the photodetector. The meter needle is mechanically or electrically adjusted until it points to the 0 graduation. Step 2 is then carried out by placing the solvent in the light path, opening the shutter, and varying the intensity of the radiation until the meter reads 100 (100% *T*). The beam intensity can be varied in several ways, including adjusting the electrical power to the source or the amplification of the electrical signal from the detector. Alternatively, the beam can be attenuated by a diaphragm, an optical wedge, or an optical comb that physically blocks a fraction of the radiation (the amount removed can be varied). In step 3, the solvent cell is replaced by one containing the analyte, and the percent transmittance is read from the scale. Because the transduced signal from a photodetector is linear with respect to the power of the radiation it receives, the scale reading with the sample in the light path is the percent

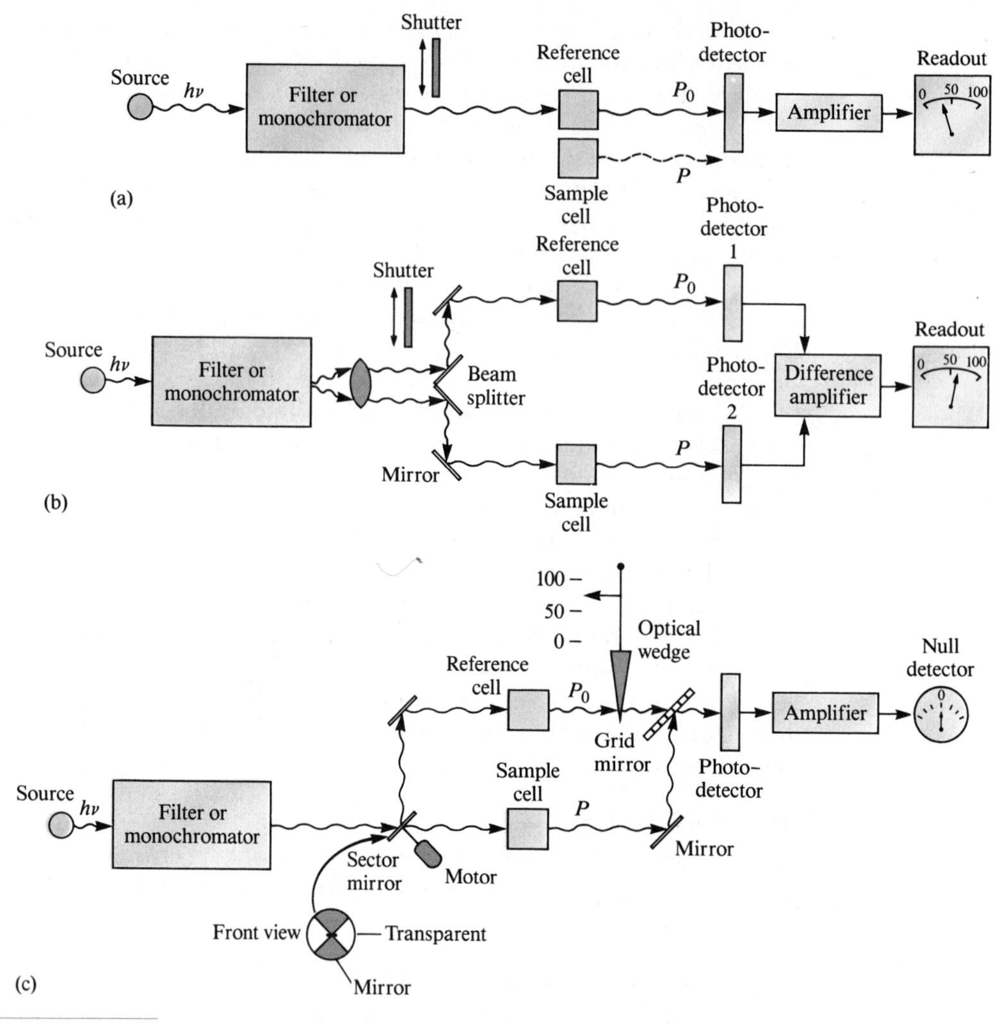

FIGURE 19-10 Instrument designs for photometers and spectrophotometers: (a) single-beam instrument; (b) double-beam instrument with beams separated in space; (c) double-beam instrument with beams separated in time.

transmittance (that is, the percent of full scale). Clearly, a logarithmic scale can be substituted to give the absorbance of the solution directly.

Normally, a single-beam instrument requires a stabilized voltage supply to avoid errors resulting from changes in the beam intensity during the time required to make the 100% T adjustment and determine % T for the analyte.

Single-beam instruments vary widely in their complexity and performance characteristics. The simplest and least expensive (a few hundred dollars or less) consists of a battery-operated tungsten bulb as the source, a set of glass filters for wavelength selection, test tubes for sample holders, a photovoltaic cell as the detector, and a small microammeter as the readout device. At the other extreme are sophisticated, computer-controlled instruments with a range of 200 to 1000 nm or more. These spectrophotometers have interchangeable tungsten/deuterium lamp sources, use rectangular silica cells, and are equipped with a high resolution grating with variable slits. Photomultiplier tubes are used as detectors, and the output is often digitized and stored so that it can be printed out in several forms. An instrument of this type may cost $50,000 or more.

Double-Beam Instruments

Many modern photometers and spectrophotometers are based upon a double-beam design. Figure 19-10b illustrates a double-beam-in-space instrument in which two beams are formed in space by a V-shaped mirror called a beam splitter. One beam passes through the reference solution to a photodetector, and the second simultaneously traverses the sample to a second, matched photodetector. The two outputs are amplified, and their ratio (or the log of their ratio) is determined electronically and displayed by the readout device. With manual instruments, the measurement is a two-step operation involving first the zero adjustment with a shutter in place between selector and beam splitter. In the second step, the shutter is opened and the transmittance or absorbance is read directly from the meter.

The second type of double-beam instrument is illustrated in Figure 19-10c. Here the beams are separated in time by a rotating sector mirror that directs the entire beam from the monochromator first through the reference cell and then through the sample cell. The pulses of radiation are recombined by another sector mirror, which transmits one pulse and reflects the other to the detector. As shown by the insert labeled "front view" in Figure 19-10c, the motor-driven sector mirror is made up of pie-shaped segments, half of which are mirrored and half of which are transparent. The mirrored sections are held in place by blackened metal frames that periodically interrupt the beam and prevent its reaching the detector. The detector circuit is programmed to use these periods to perform the dark-current adjustment.

The instrument shown in Figure 19-10c is a null type, in which the beam passing through the solvent is attenuated until its intensity just matches that of the beam passing through the sample. Attenuation is accomplished in this design with an optical wedge, the transmission of which decreases linearly along its length. Thus, the null point is reached by moving the wedge in the beam until the two electrical pulses are identical as indicated by the null detector. The transmittance (or absorbance) is then read directly from the pointer attached to the wedge.

Double-beam instruments offer the advantage that they compensate for all but the most short-term fluctuations in the radiant output of the source as well as for drift in the detector and amplifier. Furthermore, the double-beam design lends itself well to the continuous recording of transmittance or absorbance spectra. Consequently, most modern ultraviolet and visible recording instruments are double-beam (usually in time). Most infrared spectrophotometers are based on this design.

Multichannel Instruments

During the last decade, a number of multichannel spectrophotometers have become available. Figure 19-11 is a simplified schematic drawing showing the optical design of a type of multichannel spectrometer called a *diode-array spectrometer*. Radiation from a tungsten or deuterium lamp is focused upon the sample or solvent container and then passes into a monochromator with a fixed grating. The dispersed radiation falls on a photodiode-array detector, which, as mentioned earlier, consists of a linear array of several hundred photodiodes that have been formed along the length of a silicon chip. Typi-

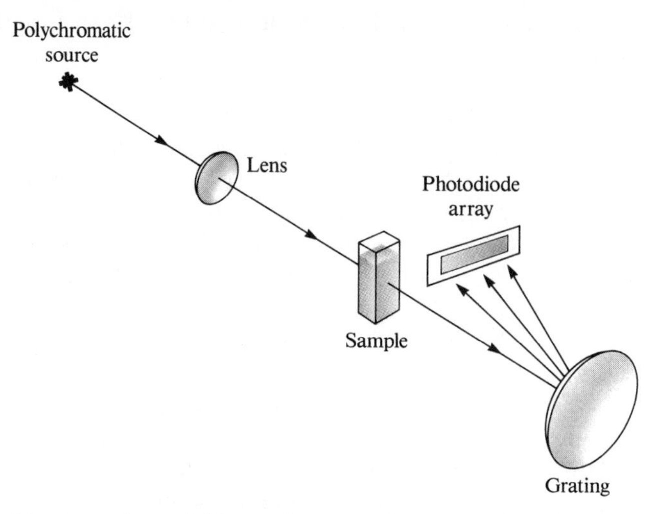

FIGURE 19-11 Diagram of a multichannel spectrophotometer based upon a grating and photodiode detector.

cally, the chips are 1 to 6 cm in length and the widths of the individual diodes are 15 to 50 μm. The chip also contains a capacitor and an electronic switch for each diode. A computer-driven shift register sequentially closes each switch momentarily, which causes each capacitor to be charged to -5 V. Radiation impinging on any diode surface causes partial discharge of its capacitor. This lost charge is replaced during the next switching cycle. The resulting charging currents, which are proportional to the radiant power, are amplified, digitized, and stored in computer memory. The entire cycle is completed in a few milliseconds.

The monochromator slit width of a diode-array instrument is usually made identical to the width of one of the silicon diodes. Thus, the output of each diode corresponds to the radiation of a different wavelength, and a spectrum is obtained by scanning these outputs sequentially. Since the electronic scanning process is remarkably rapid, data for an entire spectrum are accumulated in 1 s or less.

A diode-array instrument is a powerful tool for studying transient intermediates in moderately fast reactions, for kinetic studies, and for the qualitative and quantitative determination of the components exiting from a chromatographic column. The disadvantages of this type of instrument are its somewhat limited resolution (usually 1 to 2 nm) and its moderately high cost.

Questions and Problems

19-1 Define the term *effective bandwidth of a filter*.

*19-2 How many lines per millimeter are required in a grating if the first-order diffraction line at 500 nm is to be observed at a reflection angle of -40 deg when the angle of incidence is 60 deg?

19-3 An infrared grating has 72.0 lines/mm. Calculate the wavelengths of the first- and second-order diffraction spectra at reflection angles of (a) -15 deg, (b) 0 deg, and (c) $+15$ deg. Assume the incident angle is 50 deg.

*19-4 Describe how a spectroscope, a spectrograph, and a spectrophotometer differ from each other.

19-5 Why do quantitative and qualitative analyses often require different monochromator slit widths?

19-6 The Wien displacement law states that the wavelength maximum in micrometers for blackbody radiation is

$$\lambda_{max}T = 2.90 \times 10^3$$

where T is the temperature in kelvins. Calculate the wavelength maximum for a blackbody that has been heated to *(a) 4000 K, (b) 3000 K, *(c) 2000 K, and (d) 1000 K.

19-7 Stefan's law states that the total energy E_t emitted by a blackbody per unit time and per unit area is

$$E_t = \alpha T^4$$

where α is 5.69×10^{-8} W/m^2·K^4. Calculate the total energy output in W/m^2 for the blackbodies described in Problem 19-6.

*19-8 The relationships described in Problems 19-6 and 19-7 may be of help in solving the following.

(a) Calculate the wavelength of maximum emission of a tungsten-filament bulb operated at 2870 K and at 3000 K.
(b) Calculate the total energy output of the bulb in W/cm^2.

19-9 Describe the differences between the following and list any particular advantages possessed by one over the other:

*(a) hydrogen- and deuterium-discharge lamps as sources for ultraviolet radiation.
(b) filters and monochromators as wavelength selectors.
*(c) photovoltaic cells and phototubes as detectors for electromagnetic radiation.
(d) phototubes and photomultiplier tubes.
*(e) photometers and colorimeters.
(f) spectrophotometers and photometers.
*(g) single-beam and double-beam instruments for absorbance measurements.
(h) conventional and diode-array spectrophotometers.

*19-10 A portable photometer with a linear response to radiation registered 73.6 μA with a blank solution in the light path. Replacement of the blank with an absorbing solution yielded a response of 24.9 μA. Calculate

(a) the percent transmittance of the sample solution.
(b) the absorbance of the sample solution.
(c) the transmittance to be expected for a solution in which the concentration of the absorber is one third that of the original sample solution.
(d) the transmittance to be expected for a solution that has twice the concentration of the sample solution.

19-11 A photometer with a linear response to radiation gave a reading of 685 mV with a blank in the light path and 179 mV when the blank was replaced by an absorbing solution. Calculate

(a) the percent transmittance and absorbance of the absorbing solution.
(b) the expected transmittance if the concentration of absorber is one half that of the original solution.
(c) the transmittance to be expected if the light path through the original solution is doubled.

*19-12 Why does a deuterium lamp produce a continuous rather than a line spectrum in the ultraviolet?

19-13 What are the differences between a photon detector and a heat detector?

*19-14 How is the power of infrared radiation measured?

19-15 Why can photomultiplier tubes not be used with infrared radiation?

*19-16 Why is iodine sometimes introduced into a tungsten lamp?

19-17 Describe how an absorption photometer and a fluorescence photometer differ from each other.

*19-18 Describe the basic design difference between a spectrometer for absorption measurements and one for emission studies.

19-19 What data are needed to describe the performance characteristics of an interference filter?

19-20 Define

*(a) dark current.
(b) transducer.
*(c) scattered radiation (in a monochromator).
(d) n-type semiconductor.
*(e) majority carrier.
(f) depletion layer.

*19-21 An interference filter is to be constructed for isolation of the CS_2 absorption band at 4.54 μm.

(a) If the determination is to be based upon first-order interference, how thick should the dielectric layer be (refractive index 1.34)?
(b) What other wavelengths will be transmitted?

CHAPTER 20

Molecular Spectroscopy

Molecular spectroscopy based upon ultraviolet, visible, and infrared radiation is widely used for the identification and determination of myriad inorganic and organic species.[1] Infrared absorption spectroscopy, for example, is one of the most powerful tools available to the chemist for determining the structure of both inorganic and organic compounds. In addition, it is now assuming an important role in quantitative analysis, particularly for determining environmental pollutants.

Molecular ultraviolet/visible absorption spectroscopy is employed primarily for quantitative analysis and is probably more widely used in chemical and clinical laboratories throughout the world than any other single procedure.

Molecular fluorescence methods, while less generally applicable than absorption methods, are of considerable importance because of their high selectivity and extraordinary sensitivity. These methods have proved particularly useful for the quantitative determination of molecules of biological and biochemical interest.

20A Ultraviolet and Visible Absorption Spectroscopy

In this section, we consider the types of molecular species that absorb ultraviolet or visible radiation and can thus be determined by absorption spectroscopy.

20A-1 Absorbing Species

As noted in Section 18C-2, absorption of ultraviolet and visible radiation by molecules generally takes the form of one or more electronic absorption bands, each of which is made up of numerous closely packed but discrete lines. Each line arises from the transition of an electron from the ground state to one of the many vibrational and rotational energy states associated with each excited electronic energy state. Because so many of these vibrational and rotational states exist and because their energies differ only slightly, the number of lines contained in the typical band is large and their displacement from one another minute.

Figure 20-1a, which is part of the visible absorption spectrum for 1,2,4,5-

[1]For further reading, see E. J. Meehan, in *Treatise on Analytical Chemistry,* 2nd ed., P. J. Elving, E. J. Meehan, and I. M. Kolthoff, Eds., Part I, Vol. 7, Chapters 1–3. New York: Wiley, 1981; R. P. Bauman, *Absorption Spectroscopy.* New York: Wiley, 1962; G. F. Lothian, *Absorption Spectrophotometry,* 3rd ed. London: Adam Hilger, 1969.

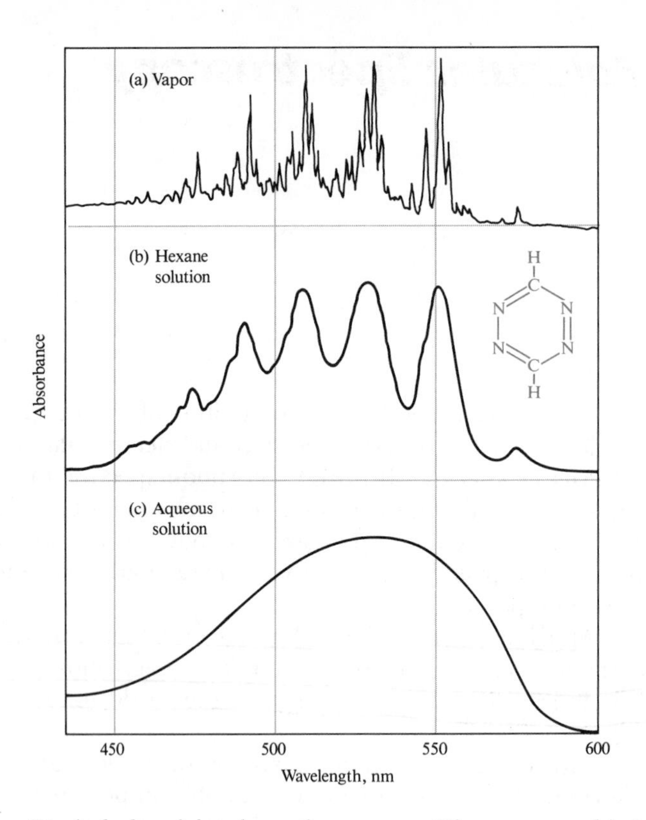

FIGURE 20-1 Typical ultraviolet absorption spectra. The compound is 1, 2, 4, 5 tetrazine. From S. F. Mason, J. Chem. Soc. **1959**, 1265. With permission.

tetrazine vapor, shows the fine structure that arises from the numerous rotational and vibrational levels associated with the excited electronic states of this aromatic molecule. In the gaseous state, the individual tetrazine molecules are sufficiently separated from one another to vibrate and rotate freely, and the many individual absorption lines resulting from the multitude of vibrational rotational energy states is clearly evident. In the condensed state or in solution, however, freedom to rotate is largely lost, and lines due to differences in rotational energy levels are obliterated. Furthermore, in the presence of solvent molecules, energies of the various vibrational levels are modified in an irregular way. Thus, the energy of a given state in an assemblage of molecules takes on a Gaussian distribution; line broadening is the result. This effect is more pronounced in polar solvents, such as water, than in nonpolar hydrocarbon media. This solvent effect is illustrated in Figures 20-1b and c.

Absorption by Organic Compounds

Two types of electrons are responsible for the absorption of ultraviolet and visible radiation by organic molecules: (1) shared electrons that participate directly in bond formation and are thus associated with more than one atom and (2) unshared outer electrons that are largely localized about such atoms as oxygen, the halogens, sulfur, and nitrogen.

The wavelengths at which an organic molecule absorbs depend upon how tightly its various electrons are bound. Thus, the shared electrons in single bonds such as carbon/carbon or carbon/hydrogen, are so firmly held that their excitation requires energies corresponding to wavelengths in the vacuum ultraviolet region (below 180 nm). Single-bond spectra have not been widely exploited for analytical purposes because of the experimental difficulties of working in this region. These difficulties are attributable to the fact that both quartz and atmospheric components absorb radiation below 180 nm, a circumstance that requires the use of evacuated spectrophotometers equipped with lithium fluoride optics.

Organic compounds containing double or triple bonds generally exhibit useful absorption peaks in the readily accessible ultraviolet region because the electrons in unsaturated bonds are relatively loosely held and thus easily excited. Unsaturated organic functional groups that absorb in the ultraviolet and visible regions are termed *chromophores*. Table 20-1 lists common chromophores and the approximate location of their absorption maxima.

Organic compounds containing sulfur, bromine, and iodine also absorb in the ultraviolet region because these elements contain loosely bonded unshared electrons that are more easily excited than the shared electrons of a saturated bond and are thus more readily excited by the absorption of photons.

Absorption by Inorganic Species

The spectra for most absorbing inorganic complex ions and molecules resemble those for organic compounds (Figure 20-1b and c), with broad absorption maxima and little fine structure. The spectra for ions of the lanthanide and actinide series represent an important exception. The electrons responsible for absorption by these elements (4f and 5f, respectively) are shielded from

TABLE 20-1

Absorption Characteristics of Some Common Organic Chromophores

Chromophore	Example	Solvent	λ_{max}, nm	ϵ_{max}
Alkene	$C_6H_{13}CH{=}CH_2$	n-Heptane	177	13,000
Conjugated alkene	$CH_2{=}CHCH{=}CH_2$	n-Heptane	217	21,000
Alkyne	$C_5H_{11}C{\equiv}C{-}CH_3$	n-Heptane	178	10,000
			196	2000
			225	160
Carbonyl	$CH_3\overset{O}{\overset{\|}{C}}CH_3$	n-Hexane	186	1000
			280	16
	$CH_3\overset{O}{\overset{\|}{C}}H$	n-Hexane	180	Large
			293	12
Carboxyl	$CH_3\overset{O}{\overset{\|}{C}}OH$	Ethanol	204	41
Amido	$CH_3\overset{O}{\overset{\|}{C}}NH_2$	Water	214	60
Azo	$CH_3N{=}NCH_3$	Ethanol	339	5
Nitro	CH_3NO_2	Isooctane	280	22
Nitroso	C_4H_9NO	Ethyl ether	300	100
			665	20
Nitrate	$C_2H_5ONO_2$	Dioxane	270	12
Aromatic	Benzene	n-Hexane	204	7900
			256	200

external influences by electrons that occupy orbitals with larger principal quantum numbers. As a consequence, the absorption bands are narrow and relatively unaffected by the nature of the species bonded by the outer electrons.

With few exceptions, the ions and complexes of the 18 elements in the first two transition series are colored in one if not all of their oxidation states. Absorption of visible radiation by these species involves transitions of electrons between filled and unfilled *d* orbitals that differ in energy as a consequence of ligands bonded to the metal ions. The energy differences between *d* orbitals (and thus the position of the corresponding absorption peak) depend upon the oxidation state of the element, its position in the periodic table, and the kind of ligand bonded to its ion.

Charge-Transfer Absorption

For quantitative purposes, *charge-transfer absorption* is particularly important because molar absorptivities are unusually large ($\epsilon_{max} > 10{,}000$), a circumstance that leads to high sensitivity. Many inorganic and organic complexes exhibit this type of absorption and are therefore called *charge-transfer complexes.*

A charge-transfer complex consists of an electron-donor group bonded to an electron acceptor. When this product absorbs radiation, an electron from the donor is transferred to an orbital that is largely associated with the acceptor. The excited state is thus the product of a kind of internal oxidation/reduction process. This behavior differs from that of an organic chromophore in which the excited electron is in a *molecular* orbital that is shared by two or more atoms.

Familiar examples of charge-transfer complexes include the phenolic complex of iron(III), the 1,10-phenanthroline complex of iron(II), the iodide complex of molecular iodine, and the ferro/ferricyanide complex responsible for the color of Prussian blue. The red color of the iron(III)/thiocyanate complex is a further example of charge-transfer absorption. Absorption of a photon results in the transfer of an electron from the thiocyanate ion to an orbital that is largely associated with the iron(III) ion. The product is an excited species involving predominantly iron(II) and the thiocyanate radical SCN. As with other types of electronic excitation, the electron in this complex ordinarily returns to its original state after a brief period. Occasionally, however, an excited complex may dissociate and produce photochemical oxidation/reduction products.

In most charge-transfer complexes involving a metal ion, the metal serves as the electron acceptor. Exceptions are the 1,10-phenanthroline complexes of iron(II) (Section 13B-1) and copper(I), where the ligand is the acceptor and the metal ion the donor. A few other examples of this type of complex are known.

20A-2 Instruments for Ultraviolet and Visible Absorption Spectroscopy

Photometers

Photometers for absorption methods offer the advantages of low cost, simplicity, ruggedness, portability, and ease of maintenance. Moreover, where high spectral purity is not important (and often it is not), the accuracy and precision of measurements made with a photometer can approach those made with a spectrophotometer. The disadvantages of photometers are their lesser versatility, their inability to generate entire spectra, and often their wider effective bandwidths.

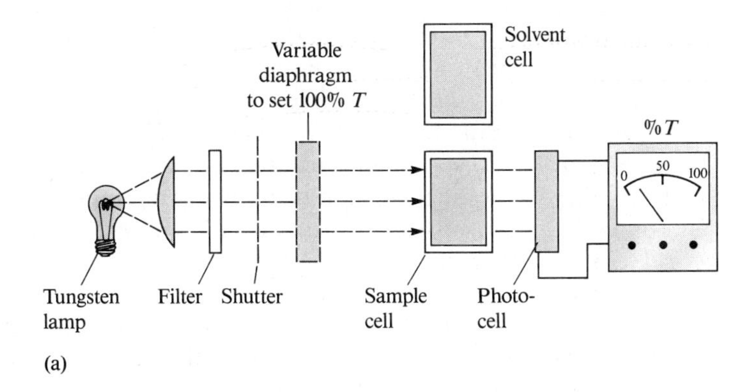

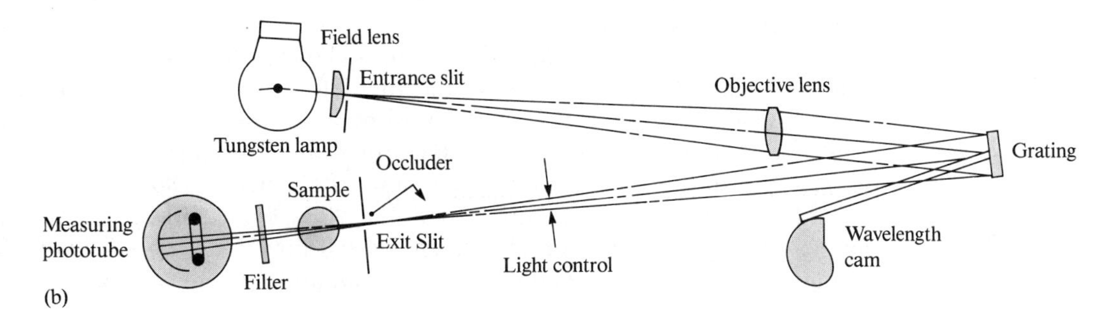

FIGURE 20-2 Single-beam instruments for absorption measurements in the visible region: (a) filter photometer; (b) spectrophotometer. (Courtesy of Milton Roy Company, Analytical Products Division, Rochester, NY.)

Figure 20-2a is a diagram of a simple single-beam photometer used for quantitative measurements in the visible region. Photometers of this kind are generally supplied with several filters, each of which transmits a different portion of the visible spectrum. Generally, a suitable filter is one whose color is the complement of the color of the analyte solution because it is this complementary color that is absorbed by the solution. For example, a liquid appears red because it transmits the red portion of the spectrum while absorbing the green. Because the intensity of the green radiation thus varies with analyte concentration, a green filter should be employed. If several filters possessing the same general hue are available, the one that provides the greatest absorbance (or least transmittance) for a solution of the analyte should be used.

Ultraviolet photometers have become important detectors in high-performance liquid chromatography. In this application, a mercury-vapor lamp serves as the source and the emission line at 254 or 280 nm is isolated by an interference filter. This type of photometer is described briefly in Section 25A-5.

Spectrophotometers

Several dozen models of spectrophotometers for the visible or the ultraviolet/visible region are now marketed by various instrument manufacturers.

Instruments for the Visible Region. Visible-region spectrophotometers are generally single-beam, grating instruments that are relatively inexpensive (less than $1000), rugged, and readily portable. At least one is battery-operated and small enough to be held in the hand. The most common application of these instruments is for fixed-wavelength quantitative analysis, although several produce surprisingly good absorption spectra that are useful for qualitative analysis also.

The instrument shown schematically in Figure 20-2b, the Bausch and Lomb Spectronic 20 spectrophotometer, first appeared on the market in the mid-1950s and a modified version is still being manufactured and widely sold. Undoubtedly, more of these instruments are currently in use throughout the world than any other single spectrophotometer model. As shown in the figure, the Spectronic 20 employs a tungsten-filament light source operated by a stabilized power supply that provides radiation of constant intensity. After diffraction by a simple reflection grating, the radiation passes through the sample or reference cuvettes to a phototube. The amplified electrical signal from the detector then powers a meter with a $5\frac{1}{2}$-in scale calibrated in transmittance and absorbance.

The instrument is equipped with an occluder, which is a vane that automatically falls between the beam and the detector whenever the cuvette is removed from its holder; the 0% T adjustment can then be made. The light-control device shown in Figure 20-2b consists of a V-shaped aperture that is moved in or out of the beam in order to set the meter to 100% T.

The spectral range of the Spectronic 20 is from 340 to 625 nm; an accessory phototube extends this range to 950 nm. Other specifications for the instrument include a bandwidth of 20 nm and a wavelength accuracy of ± 2.5 nm.

Recording Spectrophotometers for the Ultraviolet and Visible Regions. Figure 20-3 shows the optics of a typical double-beam (in time) spectrophotometer designed to operate over a range from about 190 to 750 nm. The instrument is equipped with interchangeable deuterium/tungsten sources, a reflection grating monochromator, and a photomultiplier detector. The beam splitter is a motor-driven circular disk or chopper that is divided into three segments, one of which is transparent, the second reflecting, and the third opaque. With each rotation, the detector receives three signals, the first corresponding to P_0, the second to P, and the third to the dark current. The resulting electrical signals are then processed electronically to give the transmittance or absorbance on a readout device.

An instrument of the kind shown in Figure 20-3 is usually provided with a motor-driven grating that is synchronized with the paper drive of a recorder so that automatic scanning and recording of an entire spectrum becomes possible.

20A-3 *Qualitative Applications of Ultraviolet Spectrophotometry*

Spectrophotometric measurements with ultraviolet radiation are useful for detecting chromophoric groups, such as those shown in Table 20-1.[2] Because large parts of even the most complex organic molecules are transparent to radiation longer than 180 nm, the appearance of one or more peaks in the region from 200 to 400 nm is clear indication of the presence of unsaturated groups or of atoms such as sulfur or halogens. Often, an idea as to the identity

[2]For a detailed discussion of ultraviolet absorption spectroscopy in the identification of organic functional groups, see R. M. Silverstein, G. C. Bassler, and T. C. Morrill, *Spectrometric Identification of Organic Compounds*, 4th ed., Chapter 6. New York: Wiley, 1981.

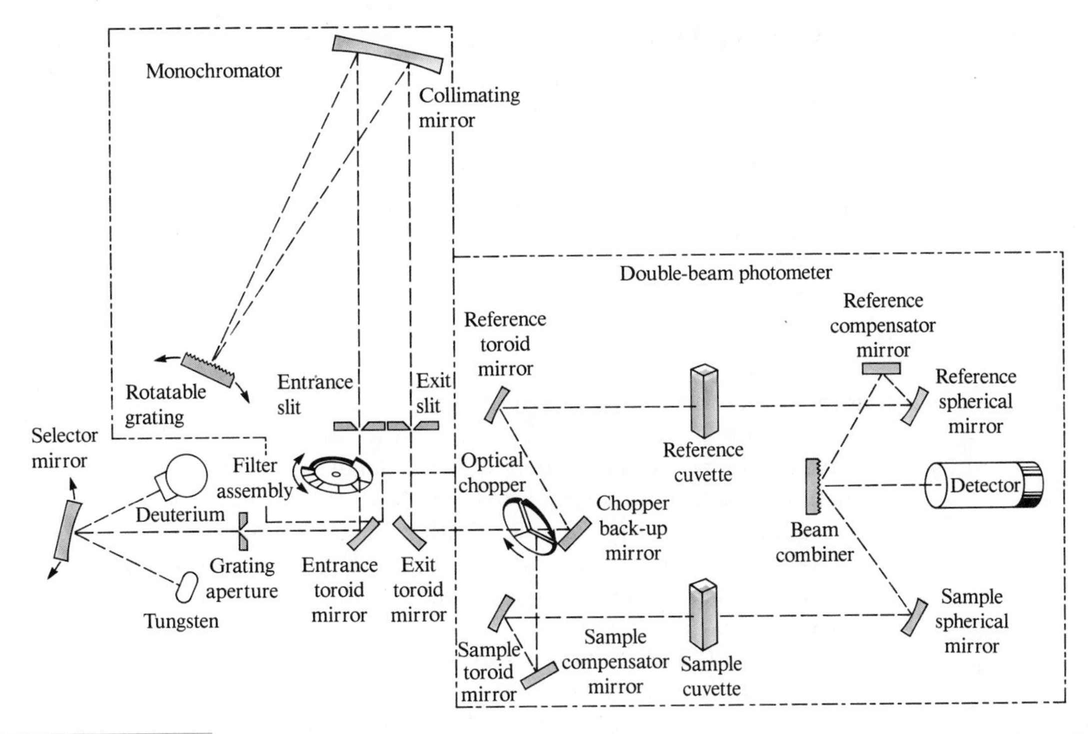

FIGURE 20-3 A double-beam recording spectrophotometer for the ultraviolet and visible regions; the Perkin-Elmer Series. (Courtesy of Coleman Instruments Division, Oak Brook, IL 50421.)

of the absorbing groups can be gained by comparing the spectrum of an analyte with those of simple molecules containing various chromophoric groups.[3] Ordinarily, however, ultraviolet spectra lack sufficient fine structure to allow unambiguous identification of an analyte. Thus, ultraviolet qualitative data must be supplemented with other physical or chemical evidence such as infrared, nuclear magnetic resonance, and mass spectra as well as solubility and melting- and boiling-point information.

Solvents

Ultraviolet spectra for qualitative analysis are most commonly derived for dilute solutions of the analyte. For volatile compounds, however, more useful spectra often result when the sample is examined as a gas (for example, compare Figure 20-1a and 20-1b). Such gas-phase spectra can often be obtained by allowing a drop or two of the pure compound to equilibrate with the atmosphere in a stoppered cuvette.

A solvent for ultraviolet/visible spectroscopy must be transparent throughout this region and should dissolve a sufficient quantity of the sample to give well-defined peaks. Moreover, consideration must be given to possible

[3]Several organizations publish catalogs of spectra, including American Petroleum Institute, *Ultraviolet Spectral Data, A.P.I. Research Project 44.* Pittsburgh: Carnegie Institute of Technology; *Sadtler Ultraviolet Spectra.* Philadelphia: Sadtler Research Laboratories; and American Society for Testing Materials, Committee E-13, Philadelphia.

TABLE 20-2

Solvents for the Ultraviolet and Visible Regions

Solvent	Lower Wavelength Limit, nm	Solvent	Lower Wavelength Limit, nm
Water	180	Carbon tetrachloride	260
Ethanol	220	Diethyl ether	210
Hexane	200	Acetone	330
Cyclohexane	200	Dioxane	320
Benzene	280	Cellosolve	320

interactions with the absorbing species. For example, polar solvents, such as water, alcohols, esters and ketones, tend to obliterate vibrational spectra and should thus be avoided when spectral detail is desired. Nonpolar solvents, such as cyclohexane, often provide spectra that more closely approach that of a gas (compare, for example, the three spectra in Figure 20-1). In addition, the polarity of the solvent often influences the position of absorption maxima. Consequently a common solvent must be employed when comparing spectra for the purpose of identification.

Table 20-2 lists common solvents for studies in the ultraviolet and visible regions and their approximate lower wavelength limits. These limits are strongly dependent upon the purity of the solvent. For example, ethanol and the hydrocarbon solvents are frequently contaminated with benzene, which absorbs below 280 nm.[4]

The Effect of Slit Width

The effect of variation in slit width, and hence effective bandwidth, is illustrated by the spectra in Figure 20-4. Clearly, peak heights and separation are

FIGURE 20-4

Spectra for reduced cytochrome *c* obtained with four spectral bandwidths: (1) 20 nm, (2) 10 nm, (3) 5 nm, and (4) 1 nm. At bandwidths <1 nm, peak heights were the same but instrument noise became pronounced. (Courtesy of Varian Instrument Division, Palo Alto, CA.)

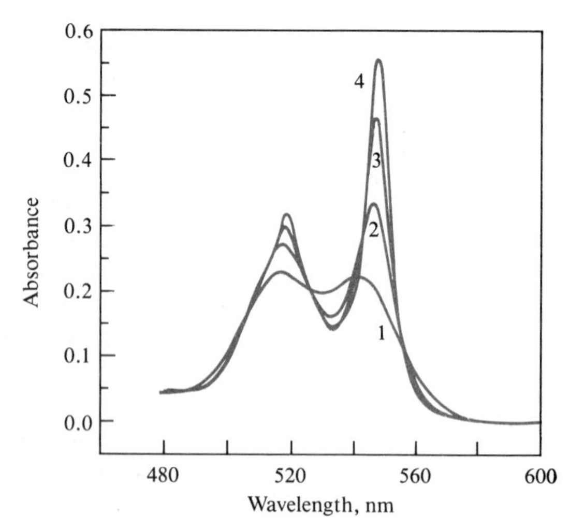

[4]Most major suppliers of reagent chemicals in the United States offer spectrochemical grades of solvents. Spectral-grade solvents have been treated so as to remove absorbing impurities and meet or exceed the requirements set forth in *Reagent Chemicals, American Chemical Society Specifications*, 7th ed. Washington, D.C.: American Chemical Society, 1986.

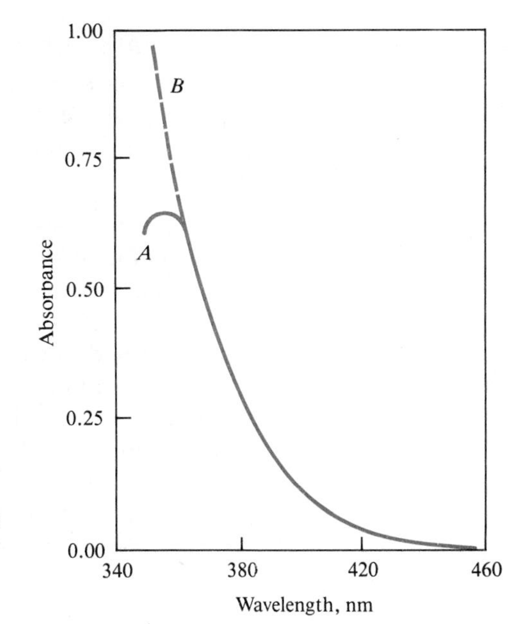

FIGURE 20-5 Spectra of cerium(IV) obtained with a spectrophotometer having glass optics (*A*) and quartz optics (*B*). The false peak in *A* arises from the transmission of stray radiation of longer wavelengths.

distorted at wider bandwidths. For this reason, spectra for qualitative applications are obtained at minimal slit widths.

The Effect of Scattered Radiation at the Wavelength Extremes of a Spectrophotometer

Earlier we demonstrated that scattered radiation may lead to instrumental deviations from Beer's law (page 471). Another undesirable effect of this type of radiation is that it occasionally causes false peaks to appear when a spectrophotometer is being operated at its wavelength extremes. Figure 20-5 shows an example of such behavior. Curve *B* is the true spectrum for a solution of cerium(IV) produced with a research-quality spectrophotometer responsive down to 200 nm or less. Curve *A* was obtained for the same solution with an inexpensive instrument operated with a tungsten source designed for work in the visible region only. The false peak at about 360 nm is directly attributable to scattered radiation, which was not absorbed because it was made up of wavelengths longer than 400 nm. Under most circumstances, such stray radiation has a negligible effect because its power is only a tiny fraction of the total power of the beam exiting from the monochromator. At wavelength settings below 380 nm, however, radiation from the monochromator is greatly attenuated as a result of absorption by the glass optical components and cuvettes. In addition, both the output of the source and the photocell sensitivity fall off dramatically below 380 nm. These factors combine to cause a substantial fraction of the measured absorbance to be due to the scattered radiation of wavelengths to which cerium(IV) is transparent. A false peak results.

This same effect is sometimes observed with ultraviolet/visible instruments when attempts are made to measure absorbances at wavelengths lower than about 190 nm.

20A-4

Quantitative Ultraviolet and Visible Photometry and Spectrophotometry

Absorption spectroscopy based upon ultraviolet and visible radiation is one of the most useful tools available to the chemist for quantitative analysis.[5] The important characteristics of spectrophotometric and photometric methods are

1. *Wide applicability.* Enormous numbers of inorganic, organic, and biochemical species absorb ultraviolet or visible radiation and are thus amenable to direct quantitative determination. Many nonabsorbing species can also be determined after chemical conversion to absorbing derivatives. It has been estimated that over 90% of the analyses performed in clinical laboratories are based upon ultraviolet and visible absorption spectroscopy.
2. *High sensitivity.* Typical detection limits for absorption spectroscopy range from 10^{-4} to 10^{-5} M. This range can often be extended to 10^{-6} or even 10^{-7} M with certain procedural modifications.
3. *Moderate to high selectivity.* Often a wavelength can be found at which the analyte alone absorbs, thus making preliminary separations unnecessary. Furthermore, where overlapping absorption bands do occur, corrections based upon additional measurements at other wavelengths sometimes eliminate the need for a separation step.
4. *Good accuracy.* The relative errors in concentration encountered with a typical spectrophotometric or photometric procedure employing ultraviolet and visible radiation lie in the range from 1 to 5%. Such errors can often be decreased to a few tenths of a percent with special precautions.
5. *Ease and convenience.* Spectrophotometric and photometric measurements are easily and rapidly performed with modern instruments. In addition, the methods readily lend themselves to automation.

Scope

The applications of quantitative absorption methods not only are numerous but also touch upon every field in which quantitative chemical information is required. The reader can gain a notion of the scope of spectrophotometry by consulting a series of review articles published biennially in *Analytical Chemistry*[6] and from monographs on the subject.[7]

Applications to Absorbing Species

The spectrophotometric determination of any organic compound containing one or more of the chromophoric groups listed in Table 20-1 is potentially feasible. In addition, numerous inorganic species absorb ultraviolet or visible radiation and can thus be determined by direct photometric or spectropho-

[5]For a wealth of detailed, practical information on spectrophotometric practices, see *Techniques in Visible and Ultraviolet Spectrometry,* Vol. I, *Standards in Absorption Spectroscopy,* C. Burgess and A. Knowles, Eds. London: Chapman and Hall, 1981; and J. R. Edisbury, *Practical Hints on Absorption Spectrometry.* New York: Plenum Press, 1968.

[6]J. A. Howell and L. G. Hargis, *Anal. Chem.,* 1978, *50,* 243R; 1980, *52,* 306R; 1982, *54,* 171R; 1984, *56,* 225R; 1986, *58,* 108R.

[7]See, for example, E. B. Sandell and H. Onishi, *Colorimetric Determination of Traces of Metals,* 4th ed. New York: Interscience, 1978; *Colorimetric Determination of Nonmetals,* 2nd ed., D. F. Boltz and J. A. Howell, Eds. New York: Wiley, 1978; Z. Marczenko, *Spectrophotometric Determination of Elements.* New York: Halsted Press, 1975; and M. Pisez and J. Bartos, *Colorimetric and Fluorometric Analysis of Organic Compounds and Drugs.* New York: Marcel Dekker, 1974.

tometric procedures. Among these are nitrite and nitrate ions, the oxides of nitrogen, the elemental halogens, ozone, and most of the transition metals in one oxidation state or another.

Applications to Nonabsorbing Species

Many nonabsorbing inorganic and organic analytes can be determined spectrophotometrically by causing them to react with a chromophoric reagent to yield a product that absorbs in the ultraviolet or visible region. The successful application of such reagents usually requires that the reaction between analyte and reagent be forced to near completion.

Typical inorganic color-forming reagents are thiocyanate ion for iron, cobalt, and molybdenum; the anion of hydrogen peroxide for titanium, vanadium, and chromium; and iodide ion for bismuth, palladium, and tellurium. Of even greater importance are organic chelating agents that form stable colored complexes with cations. Examples are diethyldithiocarbamate for the determination of copper, diphenylthiocarbazone for lead, 1,10-phenanthrolene for iron, and dimethylglyoxime for nickel. Figure 20-6 shows the color-forming reactions for the first two of these reagents. The structure of the 1,10-phenanthroline complex of iron(II) is shown on page 326, and the reaction of nickel with dimethylglyoxime to form a red precipitate is described on page 78. In the application of the last reaction to the photometric determination of nickel, an aqueous solution of the cation is extracted with a solution of the chelating agent in an immiscible organic liquid. The absorbance of the resulting bright red organic layer serves as a measure of the concentration of the metal.

Procedural Details

In developing a photometric or spectrophotometric procedure, conditions that yield a reproducible relationship (preferably linear) between concentration and absorbance must be established at the outset.

FIGURE 20-6 Typical chelating reagents for absorption. (a) Diethyldithiocarbamate. (b) Diphenylthiocarbazone.

(a)

$$2\,(C_2H_5)_2N{-}C(=S)SNa + M^{2+} \rightleftharpoons (C_2H_5)_2N{-}C(S_2)M(S_2)C{-}N(C_2H_5)_2 + 2Na^+$$

(b)

$$2\,C_6H_5(H)N{-}N{=}C(SH){-}N{=}N{-}C_6H_5 + Pb^{2+} \rightleftharpoons Pb[S{-}C(=N{-}N(H)C_6H_5){-}N{=}N{-}C_6H_5]_2 + 2H^+$$

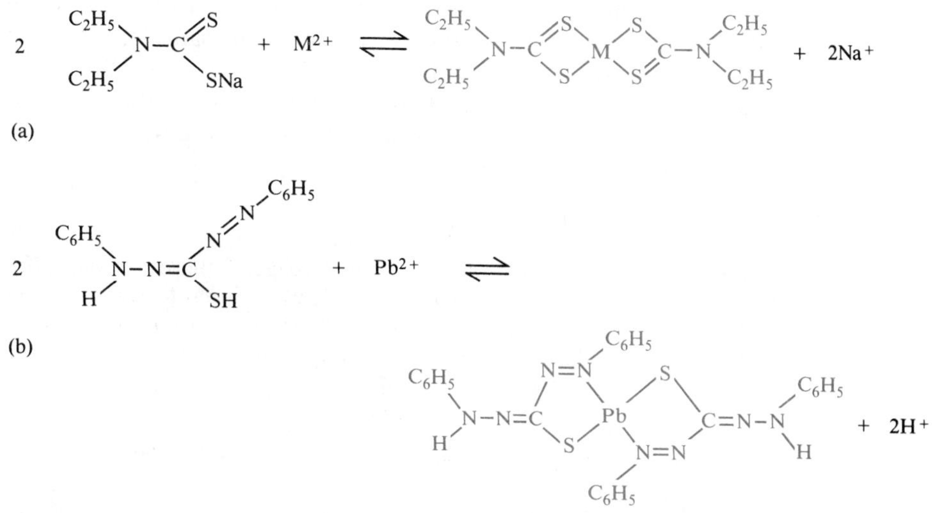

Selection of Wavelength. In order to realize maximum sensitivity, spectrophotometric absorbance measurements are ordinarily made at a wavelength corresponding to an absorption peak because the change in absorbance per unit of concentration is greatest here. In addition, the absorption curve is often relatively flat at its maximum, which leads to good linearity (Figure 18-10, page 472) and less possibility for error from failure to reproduce precisely the wavelength setting of the instrument. For measurements with a photometer, a filter that has a color that is complementary to that of the analyte solution is chosen. Such a choice leads to enhanced sensitivity as well as greater probability of obtaining a linear calibration curve.

Variables That Influence Absorbance. The absorbance of a solution is often influenced by such variables as the nature of the solvent, pH, temperature, electrolyte concentration, reaction time, and presence of interfering substances. The effects of these variables must be studied in order to establish a set of conditions that yield reproducible analytical data.

The Cleaning and Handling of Cells. Accurate spectrophotometric measurements require the use of good-quality matched cells that have been regularly calibrated against one another to detect differences arising from scratches, etching, and wear. Cell cleaning techniques, which are described in Section 31M-1, are equally important.[8]

Determination of the Relationship Between Absorbance and Concentration. The standard solutions for calibration should approximate the overall composition of the samples as closely as possible and should encompass a reasonable range of analyte concentrations. Seldom, if ever, is it safe to assume adherence to Beer's law and use only a single standard to determine the molar absorptivity. It is even less prudent to base an analysis on a literature value for the molar absorptivity because measured molar absorptivities often vary from instrument to instrument even when the same models are being employed.

The Standard-Addition Method. The difficulties that attend the production of a set of standards whose overall composition closely resembles that of the sample can be formidable if not insurmountable. Under these circumstances, a *standard-addition* approach may prove useful. In this procedure, a known quantity of standard is added to an aliquot of the sample, with the absorbance being measured before and after the addition. Provided Beer's law is obeyed (and this must be confirmed experimentally), the analyte concentration can be derived from the two absorbances, the two volumes, and the concentration of the standard.

EXAMPLE 20-1

A 2.00-mL urine specimen was treated with reagents that react with phosphate to produce a color, following which the sample was diluted to 100 mL. Photometric measurement on a 25.0-mL aliquot yielded an absorbance of 0.428. Addition of 1.00 mL of a solution containing 0.0500 mg of phosphate to a second 25.0-mL ali-

[8]See J. O. Erickson and T. Surles, *American Laboratory*, **1976**, *8*(6), 50.

quot resulted in an absorbance of 0.517. Use these data to calculate the number of milligrams of phosphate in each milliliter of the specimen.

The absorbance of the second measurement must be corrected for dilution:

$$\text{Corrected absorbance} = 0.517 \times \frac{26.0}{25.0} = 0.538$$

$$\text{Absorbance caused by 0.0500 mg phosphate} = 0.538 - 0.428 = 0.110$$

$$\text{Weight of phosphate in } \frac{25.0}{100} \text{ of specimen} = \frac{0.428}{0.110} \times 0.0500 = 0.195 \text{ mg}$$

Finally, then,

$$\text{mg phosphate/mL of specimen} = \frac{100}{25.0} \times 0.195 \times \frac{1}{2.00} = 0.390$$

The Analysis of Mixtures. The total absorbance of a solution at any given wavelength is equal to the sum of the absorbances of the individual components in the solution (Equation 18-14). This relationship makes it possible, in principle, to determine the concentrations of the individual components of a mixture even if their spectra totally overlap. For example, Figure 20-7 shows three spectra, one for a solution of species A, the second for a solution of B, and the third for a mixture of the two. It is evident that both components contribute to the absorbance at every wavelength. To analyze this mixture, molar absorptivities for A and B are first determined at wavelengths λ_1 and λ_2 with enough standards to be sure that Beer's law is obeyed over an absorbance range that encompasses the absorbance of the sample. Note that the wavelengths selected are ones at which the two spectra differ significantly. To complete the analysis, the absorbance of the mixture is determined at the same two wavelengths. Example 20-2 demonstrates how the composition of the mixture is derived from data of this kind.

FIGURE 20-7 Absorption spectrum for a two-component mixture.

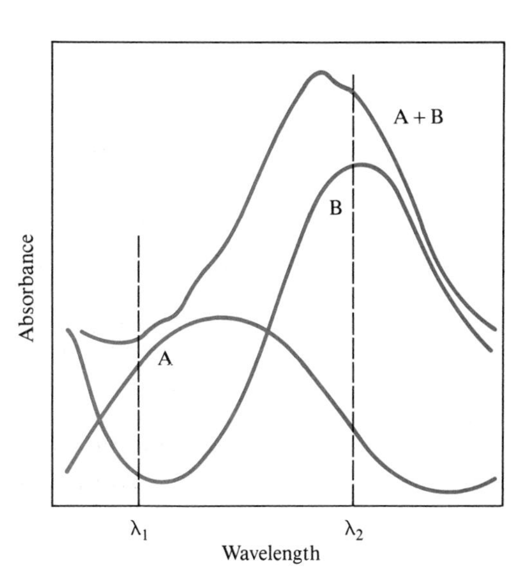

EXAMPLE 20-2

Palladium(II) and gold(III) can be determined simultaneously by complexing the two ions with methiomeprazine ($C_{19}H_{24}N_2S_2$). The absorption maximum for the palladium complex occurs at 480 nm, and that for the gold complex is at 635 nm. Molar absorptivity data at these wavelengths are

	Molar Absorptivity	
	480 nm	635 nm
Pd complex	3.55×10^3	5.64×10^2
Au complex	2.96×10^3	1.45×10^4

A 25.0-mL sample was treated with an excess of methiomeprazine and subsequently diluted to 50.0 mL. Calculate the molar concentrations of Pd(II) and Au(III) if the diluted solution had an absorbance of 0.533 at 480 nm and 0.590 at 635 nm when measured in a 1.00-cm cell.

Letting c_{Pd} and c_{Au} be the molar concentrations of the two ions, we can write for 480 nm (see Equation 18-14)

$$0.533 = 3.55 \times 10^3 \times 1.00 \times c_{\mathrm{Pd}} + 2.96 \times 10^3 \times 1.00 \times c_{\mathrm{Au}}$$

$$c_{\mathrm{Pd}} = \frac{0.533 - 2.96 \times 10^3 \times c_{\mathrm{Au}}}{3.55 \times 10^3}$$

At 635 nm

$$0.590 = 5.64 \times 10^2 \times 1.00 \times c_{\mathrm{Pd}} + 1.45 \times 10^4 \times 1.00 \times c_{\mathrm{Au}}$$

Substitution for c_{Pd} in this expression gives

$$0.590 = \frac{5.64 \times 10^2(0.533 - 2.96 \times 10^3 \times c_{\mathrm{Au}})}{3.55 \times 10^3} + 1.45 \times 10^4 \times c_{\mathrm{Au}}$$

$$= 0.0847 - 4.70 \times 10^2 \times c_{\mathrm{Au}} + 1.45 \times 10^4 \times c_{\mathrm{Au}}$$

$$c_{\mathrm{Au}} = \frac{0.590 - 0.0847}{1.403 \times 10^4} = 3.60 \times 10^{-5}\ \mathrm{M}$$

$$c_{\mathrm{Pd}} = \frac{0.533 - (2.96 \times 10^3)(3.60 \times 10^{-5})}{3.55 \times 10^3} = 1.20 \times 10^{-4}\ \mathrm{M}$$

Since the analysis involved a twofold dilution, the concentrations of Pd(II) and Au(III) in the original sample were 7.20×10^{-5} and 2.40×10^{-4} M, respectively.

Mixtures containing more than two absorbing species can be analyzed, in principle at least, if one additional absorbance measurement is made for each added component. The uncertainties in the resulting data become greater, however, as the number of measurements increases. Some of the newer computerized spectrophotometers are capable of minimizing these uncertainties by overdetermining the system; that is, these instruments use many more data points than unknowns and effectively match the entire spectrum of the unknown as closely as possible by deriving synthetic spectra for various concentrations of the components. The derived spectra are then compared with that of the analyte until a close match is found. Spectra for standard solutions of each component are required, of course.

The Effect of Instrumental Uncertainties[9]

The accuracy and precision of spectrophotometric analyses are often limited by the indeterminate error, or *noise,* associated with the instrument.[10] As pointed out in Section 19B-2, a spectrophotometric absorbance measurement entails three steps: a 0% T adjustment, a 100% T adjustment, and a measurement of % T. The indeterminate errors associated with each of these steps combine to give a net indeterminate error for the final value obtained for T. The relationship between the noise encountered in the measurement of T and the resulting *concentration uncertainty* can be derived by writing Beer's law in the form

$$c = -\frac{1}{\epsilon b}\log T = \frac{-0.434}{\epsilon b}\ln T$$

Taking the partial derivative of this equation while holding ϵb constant leads to the expression

$$\partial c = \frac{-0.434}{\epsilon b T}\,\partial T$$

where ∂c can be interpreted as the uncertainty in c that results from the noise (or uncertainty) ∂T in T. Dividing this equation by the previous one gives

$$\frac{\partial c}{c} = \frac{0.434}{\log T} \times \frac{\partial T}{T} \qquad (20\text{-}1)$$

where $\partial T/T$ is the *relative* indeterminate error in T that arises from the noise in the three measurement steps, and $\partial c/c$ is the resulting relative indeterminate concentration error.

The best and most useful measure of the indeterminate error ∂T is the standard deviation σ_T, which is easily measured for a given instrument by making 20 or more replicate transmittance measurements of an absorbing solution. Substituting σ_T and σ_c for the corresponding differential quantities in Equation 20-1 leads to

$$\frac{\sigma_c}{c} = \frac{0.434}{\log T} \times \frac{\sigma_T}{T} \qquad (20\text{-}2)$$

where σ_c/c and σ_T/T are relative standard deviations.

It is clear from an examination of Equation 20-2 that the uncertainty in a photometric concentration measurement varies in a complex way with the magnitude of the transmittance. The situation is even more complicated than suggested by the equation, however, because the uncertainty σ_T is, under many circumstances, also *dependent upon T.*

In a detailed theoretical and experimental study, Rothman, Crouch, and Ingle[11] described several sources of instrumental indeterminate errors and

[9]For further reading, see J. D. Ingle Jr. and S. R. Crouch, *Analytical Spectroscopy,* Chapter 5. Englewood Cliffs, NJ: Prentice Hall, 1988.

[10]In the context of this discussion, *noise* refers to random variations in the instrument output due not only to electrical fluctuations but also to such other variables as the way the operator reads the meter, the position of the cell in the light beam, the temperature of the solution, and the output of the source.

[11]L. D. Rothman, S. R. Crouch, and J. D. Ingle Jr., *Anal. Chem.,* **1975,** *47,* 1226.

showed the net effect of these errors on the precision of concentration measurements. The errors fall into three categories: those for which the magnitude of σ_T is (1) independent of T, (2) proportional to $\sqrt{T^2 + T}$, and (3) proportional to T. Table 20-3 summarizes information about these sources of uncertainty. When the three relationships for σ_T in the first column are substituted into Equation 20-2, three equations for the relative standard deviation in the concentration are obtained; these are shown in the third column.

Concentration Errors When $\sigma_T = k_1$. For many photometers and spectrophotometers, the standard deviation in the measurement of T is constant and independent of the magnitude of T. This type of indeterminate error is often encountered with direct-reading instruments, and has its origin in the somewhat limited resolution of the meter scale. The size of a typical scale is such that a reading cannot be reproduced to better than a few tenths of a percent of the full-scale reading, and the magnitude of this uncertainty is the same from one end of the scale to the other. For typical inexpensive instruments, standard deviations of about $0.003T$ ($\sigma_T = \pm 0.003T$) are observed.

EXAMPLE 20-3

A spectrophotometric analysis was performed with a manual instrument that exhibited an absolute standard deviation of $\pm 0.003T$ throughout its transmittance scale. Calculate the relative standard deviation in concentration that results from this uncertainty when the analyte solution has an absorbance of (a) 1.000 and (b) 2.000.

(a) To convert absorbance to transmittance, we write

$$\log T = -A = -1.00$$

$$T = \text{antilog}(-1.00) = 0.100$$

For this instrument, $\sigma_T = k_1 = \pm 0.003$ (see first entry in Table 20-3). Substituting this value and $T = 0.100$ into Equation 20-3 yields

$$\frac{\sigma_c}{c} = \frac{0.434}{\log 0.100} \times \frac{\pm 0.003}{0.1000} = \pm 0.013 \text{ or } \pm 1.3\%$$

(b) At $A = 2.000$, $T = \text{antilog}(-2.000) = 0.010$

$$\frac{\sigma_c}{c} = \frac{0.434}{\log 0.010} \times \frac{\pm 0.003}{0.010} = \pm 0.065 \text{ or } \pm 6.5\%$$

The data plotted as curve A in Figure 20-8 were derived from calculations similar to those in Example 20-3. Note that the relative standard deviation in the concentration passes through a minimum at an absorbance of about 0.5 and rises rapidly when the absorbance is less than about 0.1 or greater than approximately 1.5.

Figure 20-9a is a plot of the relative standard deviation for experimentally determined concentrations as a function of absorbance. It was obtained with the inexpensive spectrophotometer shown in Figure 20-2b. The striking similarity between this curve and curve A in Figure 20-8 indicates that the instrument studied is affected by an absolute indeterminate error of about $\pm 0.003T$ and that this error is independent of transmittance. It is probable that the source of this uncertainty lies in the limited resolution of the transmittance scale.

TABLE 20-3

Categories of Instrumental Indeterminate Errors in Transmittance Measurements

Category	Sources	Effect of T on Relative Standard Deviation of Concentration	
$\sigma_T = k_1$	Readout resolution; thermal detector noise; dark current and amplifier noise	$\frac{\sigma_c}{c} = \frac{0.434}{\log T} \times \frac{k_1}{T}$	(20-3)
$\sigma_T = k_2\sqrt{T^2 + T}$	Photon detector shot noise	$\frac{\sigma_c}{c} = \frac{0.434}{\log T} \times k_2\sqrt{1 + \frac{1}{T}}$	(20-4)
$\sigma_T = k_3T$	Cell positioning uncertainty; fluctuation in source intensity	$\frac{\sigma_c}{c} = \frac{0.434}{\log T} \times k_3$	(20-5)

Note: σ_T is the standard deviation of the transmittance measurements, σ_c/c is the relative standard deviation of the concentration measurements, T is transmittance, and k_1, k_2, and k_3 are constants for a given instrument.

Infrared spectrophotometers also exhibit an indeterminate error that is independent of transmittance. With such instruments, the source of this error lies in the thermal detector. Fluctuations in the output of this type of transducer are independent of the output; indeed, fluctuations are observed even in the absence of radiation. An experimental plot of data from an infrared spectrophotometer is similar in appearance to Figure 20-9a. The curve is displaced upward, however, because of the greater standard deviation associated with infrared measurements.

Concentration Errors When $\sigma_T = k_2\sqrt{T^2 + T}$. This type of indeterminate uncertainty is characteristic of the highest-quality spectrophotometers. It has its origin in the so-called *shot noise* that causes the output of photomultipliers and phototubes to fluctuate randomly about a mean value. Equation 20-4 in

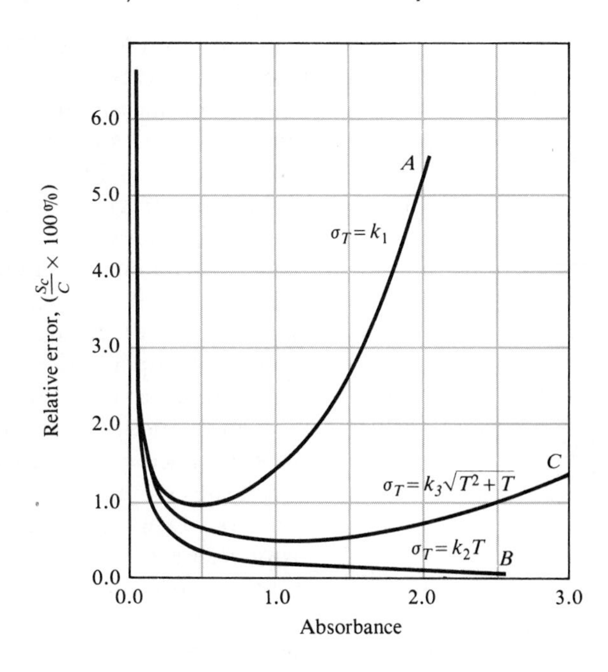

FIGURE 20-8 Error curves for various categories of instrumental uncertainties.

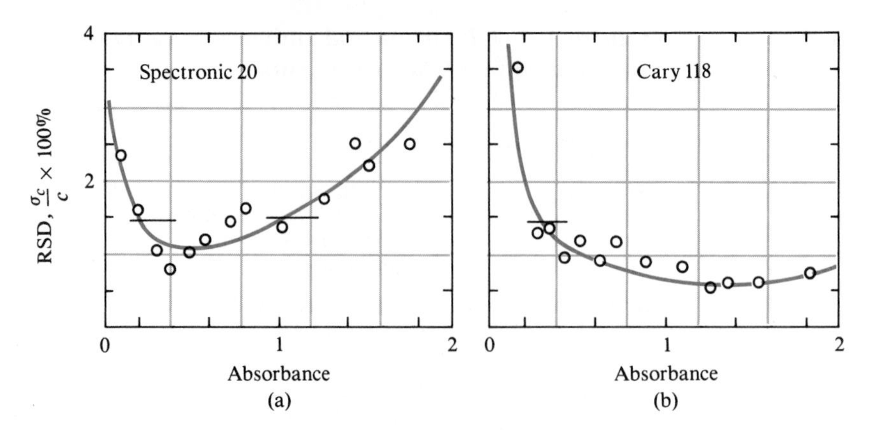

FIGURE 20-9 Experimental curves relating relative concentration uncertainties to absorbance for two spectrophotometers. Data obtained with (a) a Spectronic 20, a low-cost instrument (Figure 20-2a), (b) a Cary 118, a research-quality instrument. (From W. E. Harris and B. Kratochvil, *An Introduction to Chemical Analysis,* p. 384. Philadelphia: Saunders College Publishing, 1981. With permission.)

Table 20-3 describes the effect of shot noise on the relative standard deviation of concentration measurements. A plot of this relationship appears as curve *B* in Figure 20-8. In obtaining these data, k_2 was assumed to be ± 0.003, a typical value for high-quality spectrophotometers.

Figure 20-9b shows an analogous plot of experimental data obtained with an ultraviolet/visible spectrophotometer having performance characteristics similar to those of the instrument shown in Figure 20-3. Note that, in contrast to the less expensive instrument, absorbances of 2.0 or greater can be measured here without serious deterioration in the quality of the data.

Concentration Errors When $\sigma_T = k_3 T$. Substitution of $\sigma_T = k_3 T$ into Equation 20-2 reveals that the relative standard deviation in concentration from this type of uncertainty is inversely proportional to the logarithm of the transmittance (Equation 20-5 in Table 20-3). Curve *C* in Figure 20-8, which is a plot of Equation 20-5, reveals that this type of uncertainty is important at low absorbances (high transmittances) but approaches zero at high absorbances.

At low absorbances, the precision obtained with high-quality double-beam instruments is often described by Equation 20-5. The source of this behavior is failure to position cells reproducibly with respect to the beam during replicate measurements. This position dependence probably arises from small imperfections in the cell windows, which cause reflective losses and transparency to differ from one area of the window to another.

Evaluation of k_3 in Equation 20-5 is possible by comparing the precision of absorbance measurements made in the usual way with measurements in which the cells are left undisturbed at all times with replicate solutions being introduced with a syringe. Experiments of this kind with a high-quality spectrophotometer yielded a value of 0.013 for k_3 (note 11 page 519). Curve *C* in Figure 20-8 was obtained by substituting this numerical value into Equation 20-5. Cell positioning errors affect all types of spectrophotometric measurements in which cells are repositioned between measurements.

Fluctuations in source intensity also yield standard deviations that are described by Equation 20-5. This type of behavior is sometimes encountered in inexpensive single-beam instruments that have unstable power supplies and in infrared instruments.

20A-5 Photometric and Spectrophotometric Titrations

Photometric and spectrophotometric measurements are useful for locating the equivalence points of titrations.[12] This application of absorption measurements obviously requires that one or more of the reactants or products absorb radiation or that an absorbing indicator be present.

Titration Curves

A photometric titration curve is a plot of absorbance (corrected for volume change) as a function of titrant volume. If conditions are chosen properly, the curve consists of two straight-line regions with different slopes, one occurring at the outset of the titration and the other located well beyond the equivalence-point region; the end point is taken as the intersection of extrapolated linear portions of the two lines.

Figure 20-10 shows typical photometric titration curves. Figure 20-10a is the curve for the titration of a nonabsorbing species with an absorbing titrant that is decolorized by the reaction. An example is the titration of thiosulfate ion with triiodide ion. The titration curve for the formation of an absorbing product from colorless reactants is shown in Figure 20-10b; an example is the titration of iodide ion with a standard solution of iodate ion to form triiodide. The remaining figures illustrate the curves obtained with various combinations of absorbing analytes, titrants, and products.

In order to obtain titration curves with linear portions that can be extrapolated, the absorbing system(s) must obey Beer's law. Furthermore, absorbances must be corrected for volume changes by multiplying the observed absorbance by $(V + v)/V$, where V is the original volume of the solution and v is the volume of added titrant.

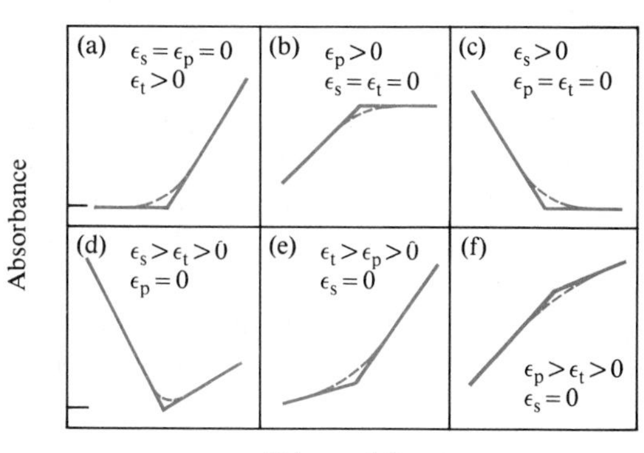

FIGURE 20-10 Typical photometric titration curves. Molar absorptivities of the substance titrated, the product, and the titrant are ϵ_s, ϵ_p, ϵ_t.

[12] For further information, see J. B. Headridge, *Photometric Titrations*. New York: Pergamon Press, 1961.

Instrumentation

Photometric titrations are ordinarily performed with a spectrophotometer or a photometer that has been modified so that the titration vessel is held in the light path.[13] After the instrument is set to a suitable wavelength (or an appropriate filter is inserted), the 0% T adjustment is made in the usual way. With radiation passing through the analyte solution to the detector, the instrument is then adjusted to a convenient absorbance reading by varying the source intensity or the detector sensitivity. Ordinarily, no attempt is made to measure the true absorbance since relative values are perfectly adequate for end-point detection. Titration data are then collected without alteration of the instrument settings. The power of the radiation source and the response of the detector must remain constant during a photometric titration. Cylindrical containers are ordinarily used, and care must be taken to avoid any movement of the vessel that might alter the length of the radiation path.

Both filter photometers and spectrophotometers have been employed for photometric titrations. The latter are preferred, however, because their narrower bandwidths enhance the probability of adherence to Beer's law.

Applications of Photometric Titrations

Photometric titrations often provide more accurate results than a direct photometric determination because the data from several measurements are pooled in determining the end point. Furthermore, the presence of other absorbing species may not interfere since only a change in absorbance is being measured.

One advantage of a photometric end point is that the experimental data are taken well away from the equivalence-point region. Consequently, the equilibrium constants of the reactions need not be as favorable as those required for a titration that depends upon observations near the equivalence point (for example, potentiometric or indicator end points). For the same reason, more dilute solutions may be titrated.

The photometric end point has been applied to all types of reactions.[14] For example, most standard oxidizing agents have characteristic absorption spectra and thus produce photometrically detectable end points. Although standard acids or bases do not absorb, the introduction of acid/base indicators permits photometric neutralization titrations. The photometric end point has also been used to great advantage in titrations with EDTA and other complexing agents. Figure 20-11 illustrates the application of this technique to the successive titration of bismuth(III) and copper(II). At 745 nm, the cations, the reagent, and the bismuth complex formed in the first part of the titration do not absorb but the copper complex does. Thus, the solution exhibits no absorbance until essentially all the bismuth has been titrated. With the first formation of the copper complex, an increase in absorbance occurs. The increase continues until the copper equivalence point is reached. Further reagent additions cause no further absorbance change. Clearly, two well-defined end points result.

The photometric end point has also been adapted to precipitation titra-

[13]Titration flasks and cells for use in the instrument shown in Figure 20-2b are available from the Kontes Manufacturing Corp., Vineland, NJ 08360.

[14]See, for example, the review by A. L. Underwood in *Advances in Analytical Chemistry and Instrumentation,* C. N. Reilley, Ed., Vol. 3, pp. 31–104. New York: Interscience, 1964.

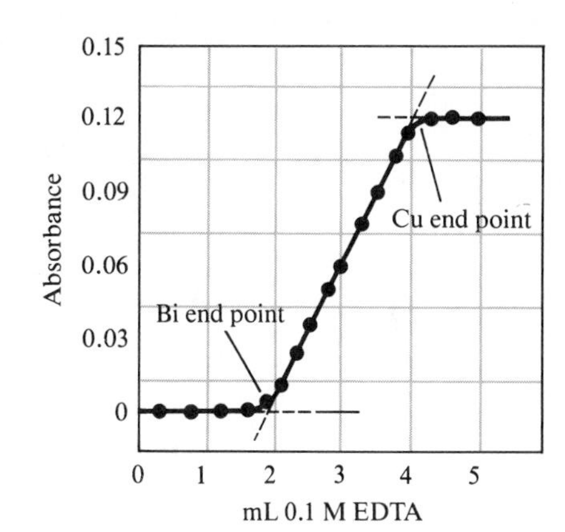

FIGURE 20-11 Photometric titration curve at 745 nm for 100 mL of a solution that was 2.0×10^{-3} M in Bi^{3+} and Cu^{2+}. (A. L. Underwood, *Anal. Chem.*, **1954,** *26,* 1322. With permission of the American Chemical Society.)

tions. The suspended solid product diminishes the radiant power by scattering; titrations are carried to a condition of constant turbidity.

20A-6 Spectrophotometric Studies of Complex Ions

Spectrophotometry is a valuable tool for elucidating the composition of complex ions in solution and for determining their formation constants. The power of the technique lies in the fact that quantitative absorption measurements can be performed without disturbing the equilibria under consideration. Although most spectrophotometric studies of complexes involve systems in which a reactant or a product absorbs, nonabsorbing systems can also be investigated successfully. For example, the composition and formation constant for a complex of iron(II) and a nonabsorbing ligand could probably be determined by measuring the color decreases that occur when solutions of the absorbing iron(II) complex of 1,10-phenanthroline are mixed with various amounts of the ligand. For this approach to be successful, the formation constant and the composition of the 1,10-phenanthroline complex have to be known.

The three most common techniques employed for complex-ion studies are (1) the method of continuous variations, (2) the mole-ratio method, and (3) the slope-ratio method.

The Method of Continuous Variations[15]

In the method of continuous variations, cation and ligand solutions with identical analytical concentrations are mixed in such a way that the total volume and the total moles of reactants in each mixture is constant but the mole ratio of reactants varies systematically (for example, 1:9, 8:2, 7:3, and so forth). The absorbance of each solution is then measured at a suitable wavelength and corrected for any absorbance the mixture might exhibit if no reaction had occurred. The corrected absorbance is plotted against the volume fraction of one reactant, that is, $V_M/(V_M + V_L)$, where V_M is the volume of the cation solution and V_L that of the ligand. A typical plot is shown in Figure 20-12. A maximum (or minimum if the complex absorbs less than the reactants) occurs at a volume ratio V_M/V_L corresponding to the combining ratio of cation and ligand in the complex. In Figure 20-12, $V_M/(V_M + V_L)$ is 0.33 and

[15]See W. C. Vosburgh and G. R. Cooper, *J. Amer. Chem. Soc.*, **1941,** *63,* 437.

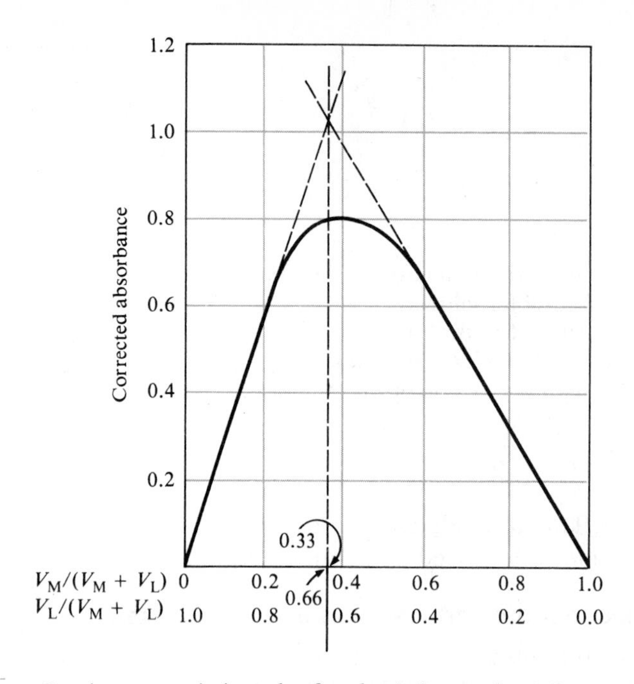

FIGURE 20-12 Continuous-variation plot for the 1:2 complex ML_2.

$V_L/(V_M + V_L)$ is 0.66; thus, V_M/V_L is 0.33/0.66, which suggests that the complex has the formula ML_2.

The curvature of the experimental lines in Figure 20-12 is the result of incompleteness of the complex-formation reaction. A formation constant for the complex can be evaluated from measurements of the deviations from the theoretical straight lines.

The Mole-Ratio Method

In the mole-ratio method, a series of solutions is prepared in which the analytical concentration of one reactant (usually the cation) is held constant while that of the other is varied. A plot of absorbance versus mole ratio of the reactants is then prepared. If the formation constant is reasonably favorable, two straight lines of different slopes that intersect at a mole ratio that corresponds to the combining ratio in the complex are obtained. Typical mole-ratio plots are shown in Figure 20-13. Note that the ligand of the 1:2 complex absorbs at the wavelength selected so that the slope beyond the equivalence point is greater than zero. We deduce that the uncomplexed cation involved in the 1:1 complex absorbs, because the initial point has an absorbance greater than zero.

The formation constant can be evaluated from the data in the curved portion of mole-ratio plots.

EXAMPLE 20-4

Derive sufficient equations to permit calculation of the equilibrium concentrations of all the species involved in the 1:2 complex-formation reaction illustrated in Figure 20-13.

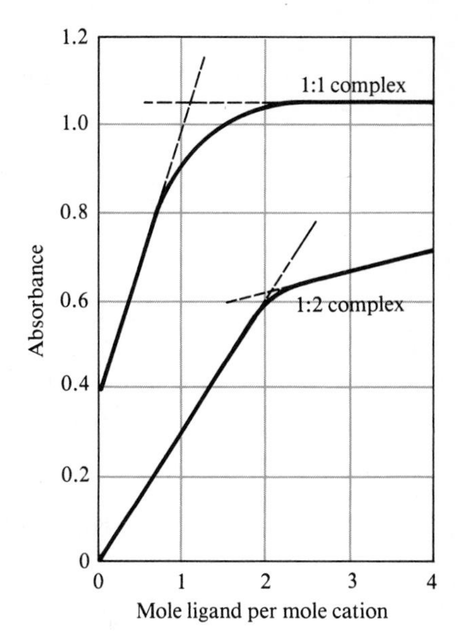

FIGURE 20-13 Mole-ratio plots for a 1:1 and a 1:2 complex. The 1:2 complex is the more stable, as indicated by less curvature near the stoichiometric ratio.

Two mass-balance expressions can be written that are based upon the preparatory data. Thus, for the reaction

$$M + 2L \rightleftharpoons ML_2$$

we can write

$$c_M = [M] + [ML_2]$$

$$c_L = [L] + 2[ML_2]$$

where c_M and c_L are the molar concentrations of M and L before reaction occurs. For 1-cm cells, the absorbance of the solution is

$$A = \epsilon_M[M] + \epsilon_L[L] + \epsilon_{ML_2}[ML_2]$$

From the mole-ratio plot, we see that $\epsilon_M = 0$. Values for ϵ_L and ϵ_{ML_2} can be obtained from the two straight-line portions of the curve. With one or more measurements of A in the curved position of the plot, sufficient data are available to calculate the three equilibrium concentrations and tnus the formation constant.

A mole-ratio plot may reveal the stepwise formation of two or more complexes as successive slope changes, provided the complexes have different molar absorptivities and provided the formation constants are sufficiently different from each other.

The Slope-Ratio Method

This procedure is particularly useful for weak complexes but is applicable only to systems in which a single complex is formed. The method assumes (1) that the complex-formation reaction can be forced to completion by a large excess of either reactant and (2) that Beer's law is followed under these circumstances.

Let us consider the reaction in which the complex M_xL_y is formed by the reaction of x moles of the cation M with y moles of a ligand L:

$$xM + yL \rightleftharpoons M_xL_y$$

Mass-balance expressions for this system are

$$c_M = [M] + x[M_xL_y]$$

$$c_L = [L] + y[M_xL_y]$$

where c_M and c_L are the molar analytical concentrations of the two reactants. We now assume that at very high analytical concentrations of L, the equilibrium is shifted far to the right and $[M] << x[M_xL_y]$. Under this circumstance, the first mass-balance expression simplifies to

$$c_M = x[M_xL_y]$$

If Beer's law obtains,

$$A_1 = \epsilon b[M_xL_y] = \epsilon b c_M/x$$

A plot of absorbance as a function of c_M becomes linear whenever sufficient L is present to satisfy the assumption that $[M] << x[M_xL_y]$. The slope of this plot is $\epsilon b/x$.

When c_M is made very large, we assume that $[L] << y[M_xL_y]$, whereupon the second mass-balance equation reduces to

$$c_L = y[M_xL_y]$$

and

$$A_2 = \epsilon b[M_xL_y] = \epsilon b c_L/y$$

Again, if our assumptions are valid, a linear plot of A_2 versus c_L is observed at high concentrations of M. The slope of this line is $\epsilon b/y$.

The ratio of the slopes of the two straight lines gives the combining ratio between M and L:

$$\frac{\epsilon b/x}{\epsilon b/y} = \frac{y}{x}$$

20B Infrared Absorption Spectroscopy

Infrared spectrophotometry is one of the most powerful tools available to the chemist for identifying pure organic and inorganic compounds because, with the exception of a few homonuclear molecules, such as O_2, N_2, and Cl_2, all molecular species absorb infrared radiation. Furthermore, with the exception of chiral molecules in the crystalline state, each molecular species has a unique infrared absorption spectrum. Thus, an exact match between the spectrum of a compound of known structure and that of an analyte unambiguously identifies the latter.

Infrared spectroscopy is a less satisfactory tool for quantitative analyses than its ultraviolet and visible counterparts because the narrow peaks that characterize infrared absorption usually lead to deviations from Beer's law. Furthermore, infrared absorbance measurements are considerably less precise. Nevertheless, where modest precision suffices, the unique nature of infrared spectra provides a degree of selectivity in a quantitative measurement that may offset these undesirable characteristics.[16]

20B-1 Infrared Absorption Spectra

Vibrational absorption occurs in the infrared region, where the energy of radiation is insufficient to excite electronic transitions. As shown in Figure 20-14, infrared spectra exhibit narrow, closely spaced absorption peaks resulting from transitions among the various vibrational quantum levels. Variations in rotational levels may also give rise to a series of peaks for each vibrational state; with liquid or solid samples, however, rotation is often hindered or prevented, and the effects of these small energy differences are not detected. Thus, a typical infrared spectrum for a liquid, such as that in Figure 20-14, consists of a series of vibrational peaks.

The number of ways a molecule can vibrate is related to the number of atoms, and thus the number of bonds, it contains. For even a simple molecule, the number of possible vibrations is large. For example, *n*-butanal ($CH_3CH_2CH_2CHO$) has 33 vibrational modes, most differing from each other in energy. Not all of these vibrations give rise to infrared peaks; nevertheless, as shown in Figure 20-14, the spectrum for *n*-butanal is relatively complex.

Infrared absorption occurs not only with organic molecules but also with covalently bonded metal complexes, which are generally active in the longer-wavelength infrared region. Infrared spectrophotometric studies have thus provided much useful information about complex metal ions.

20B-2 Instruments for Infrared Spectroscopy

Three types of infrared instruments are found in modern laboratories: dispersive spectrometers (or spectrophotometers), Fourier-transform (FTIR) spectrometers, and filter photometers. The first two are used for obtaining

FIGURE 20-14 Infrared spectrum for *n*-butanal (*n*-butyraldehyde). Note that transmittance rather than absorbance is plotted. [*Catalog of Selected Infared Spectral Data,* Serial No. 225, Thermodynamics Research Center Data Project Thermodynamics Research Center, Texas A&M University, College Station, TX (loose-leaf data sheets extant, 1964).]

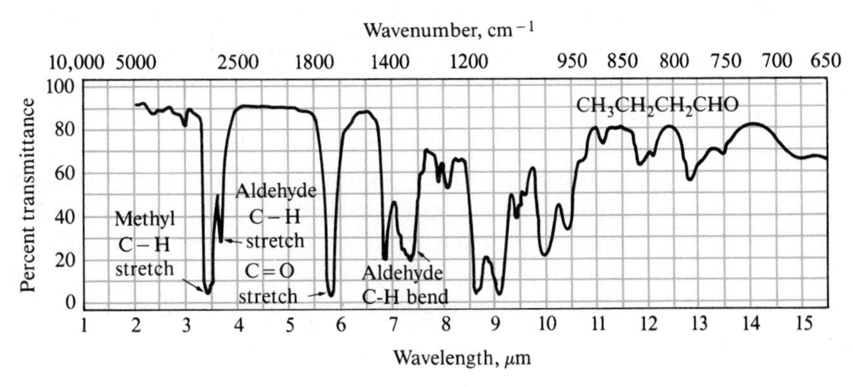

[16]For further reading, see A. L. Smith, in *Treatise on Analytical Chemistry,* 2nd ed., P. J. Elving, E. J. Meehan, and I. M. Kolthoff, Eds., Part I, Vol. 7, Chapter 5. New York: Wiley, 1981.

complete spectra for qualitative identification, whereas filter photometers are designed for quantitative work. Fourier-transform and filter instruments are nondispersive in the sense that neither employs a grating or prism to disperse radiation into its component wavelengths.

Dispersive Instruments

With one difference, dispersive infrared instruments are similar in general design to the double-beam (in time) spectrophotometers shown in Figures 19-10c and 20-3. The difference lies in the location of the cell compartment with respect to the monochromator. In ultraviolet/visible instruments, cells are always located between the monochromator and the detector in order to avoid photochemical decomposition, which may occur if samples are exposed to the full power of an ultraviolet or visible source. Infrared radiation, in contrast, is not sufficiently energetic to bring about photodecomposition; thus the cell compartment can be located between the source and the monochromator. This arrangement is advantageous because any scattered radiation generated in the cell compartment is largely removed by the monochromator.

As shown in Section 19A, the components of infrared instruments differ considerably in detail from those found in ultraviolet and visible instruments. Thus, infrared sources are heated solids rather than deuterium or tungsten lamps, infrared gratings are much coarser than those required for ultraviolet/visible radiation, and infrared detectors respond to heat rather than photons. Furthermore, the optical components of infrared instruments are constructed from polished solids, such as sodium chloride or potassium bromide.

Fourier-Transform Spectrometers

Fourier-transform infrared spectrometers offer the advantages of unusually high sensitivity, resolution, and speed of data acquisition (data for an entire spectrum can be obtained in 1 s or less). Offsetting these advantages are the complexity of the instruments and their high cost, which arise in part because a moderately sophisticated dedicated computer is needed to decode the output data.

Fourier-transform instruments contain no dispersing element, and all wavelengths are detected and measured simultaneously. In order to separate wavelengths, it is necessary to modulate the source signal in such a way that it can subsequently be decoded by a Fourier transformation, a mathematical operation that requires a high-speed computer. The theory of Fourier-transform measurements is beyond the scope of this book.[17]

Filter Photometers

Infrared photometers designed to monitor the concentration of air pollutants, such as carbon monoxide, nitrobenzene, vinyl chloride, hydrogen cyanide, and pyridine, are now being marketed and used to ensure compliance with regulations established by the Occupational Safety and Health Administration (OSHA). Interference filters, each designed for the determination of a specific

[17]For an elementary discussion of the principles of Fourier-transform spectroscopy, see D. A. Skoog, *Principles of Instrumental Analysis*, 3rd ed., pp. 148–157 and 332–336. Philadelphia: Saunders College Publishing, 1985.

pollutant, are available. These transmit narrow bands of radiation in the range of 3 to 14 μm.

20B-3

Qualitative Applications of Infrared Spectrophotometry

An infrared absorption spectrum, even one for a relatively simple compound, often contains a bewildering array of sharp peaks and minima. Peaks useful for the identification of functional groups are located in the shorter-wavelength region of the infrared (from about 2.5 to 8.5 μm), where the positions of the maxima are only slightly affected by the carbon skeleton to which the groups are attached. Investigation of this region of the spectrum thus provides considerable information regarding the overall constitution of the molecule under investigation. Table 20-4 gives the positions of characteristic maxima for some common functional groups.[18]

Identification of the functional groups in a molecule is seldom sufficient to permit positive identification of the compound, and the entire spectrum from 2.5 to 15 μm must be compared with that of known compounds. Collections of spectra are available for this purpose.[19]

20B-4

Quantitative Infrared Photometry and Spectrophotometry

Quantitative infrared absorption methods differ somewhat from their ultraviolet and visible counterparts because of the greater complexity of the spectra, the narrowness of the absorption bands, and the instrumental limitations of infrared photometers and spectrophotometers.[20]

TABLE 20-4

Some Characteristic Infrared Absorption Peaks

Functional Group		Absorption Peaks	
		Wavenumber, cm^{-1}	Wavelength, μm
O—H	Aliphatic and aromatic	3600–3000	2.8–3.3
NH_2	Also secondary and tertiary	3600–3100	2.8–3.2
C—H	Aromatic	3150–3000	3.2–3.3
C—H	Aliphatic	3000–2850	3.3–3.5
C≡N	Nitrile	2400–2200	4.2–4.6
C≡C—	Alkyne	2260–2100	4.4–4.8
COOR	Ester	1750–1700	5.7–5.9
COOH	Carboxylic acid	1740–1670	5.7–6.0
C=O	Aldehydes and ketones	1740–1660	5.7–6.0
$CONH_2$	Amides	1720–1640	5.8–6.1
C=C—	Alkene	1670–1610	6.0–6.2
ϕ—O—R	Aromatic	1300–1180	7.7–8.5
R—O—R	Aliphatic	1160–1060	8.6–9.4

[18]For more detailed information, see N. B. Colthup, *J. Opt. Soc. Amer.*, **1950**, *40*, 397; R. M. Silverstein, G. W. Bassler, and T. C. Morrill, *Spectrometric Identification of Organic Compounds*, 4th ed. New York: Wiley, 1981.

[19]American Petroleum Institute, *Infrared Spectral Data, A.P.I. Research Project 44*. Pittsburgh: Carnegie Institute of Technology; *Sadtler Standard Spectra*. Philadelphia: Sadtler Research Laboratories.

[20]For an extensive discussion of quantitative infrared analysis, see A. L. Smith, in *Treatise on Analytical Chemistry*, 2nd ed., P. J. Elving, E. J. Meehan, and I. M. Kolthoff, Eds., Part I, Vol. 7, pp. 415–456. New York: Wiley, 1981.

Absorbance Measurements

The use of matched cuvettes for solvent and analyte is seldom practical for infrared measurements because of the difficulty in obtaining cells with identical transmission characteristics. Part of this difficulty arises from degradation in the transparency of infrared cell windows (typically polished sodium chloride) with use due to attack by traces of moisture in the atmosphere and in samples. Furthermore, path lengths are hard to reproduce because infrared cells are often less than 1 mm thick. Such narrow cells are required to permit the transmission of measurable intensities of infrared radiation through pure samples or through very concentrated solutions of the analyte. Measurements of dilute analyte solutions, as is done in ultraviolet or visible spectroscopy, are frequently precluded by the lack of good solvents that transmit over appreciable regions of the infrared spectrum.

For the reasons just mentioned, a reference absorber is often dispensed with entirely in qualitative infrared work, and the intensity of the radiation passing through the sample is simply compared with that of the unobstructed beam; alternatively, a salt plate may be placed in the reference beam. Either way, the resulting transmittance is ordinarily less than 100%, even in regions of the spectrum where the sample is totally transparent. This effect is readily seen by examining the spectrum in Figure 20-14.

For quantitative work, two methods are employed to correct for the scattering and absorption by the solvent and the cell. In the *cell in/cell out* procedure, a single cell is used to obtain successive spectra of the solvent and sample with respect to the unobstructed reference beam. The transmittance of each solution versus the reference beam is then determined at an analyte absorption maximum. These transmittances can be written as

$$T_0 = P_0/P_r$$

$$T_s = P/P_r$$

where P_r is the power of the reference beam and T_0 and T_s are the transmittances of the solvent and the sample, respectively, against this reference. If P_r remains constant during the two measurements, the transmittance of the sample with respect to the solvent can be obtained by division of one equation by the other:

$$T = T_s/T_0 = P/P_0$$

An alternative way of obtaining P_0 and T is the *base-line* method, in which the solvent transmittance is assumed to be constant or at least to change linearly between the shoulders of the absorption peak. This technique is demonstrated in Figure 20-15.

Deviations from Beer's Law

Instrumental deviations from Beer's law are common in infrared spectroscopy because the source intensity and detector sensitivity of most infrared instruments are so low that relatively wide slits are required to produce an output signal that can be measured accurately. As a consequence of these limitations, the bandwidths of radiation from a monochromator are frequently of the same order of magnitude as the widths of the typically narrow infrared absorption peak. As noted on page 471, this combination of circumstances usually leads to a nonlinear relationship between absorbance and concentration.

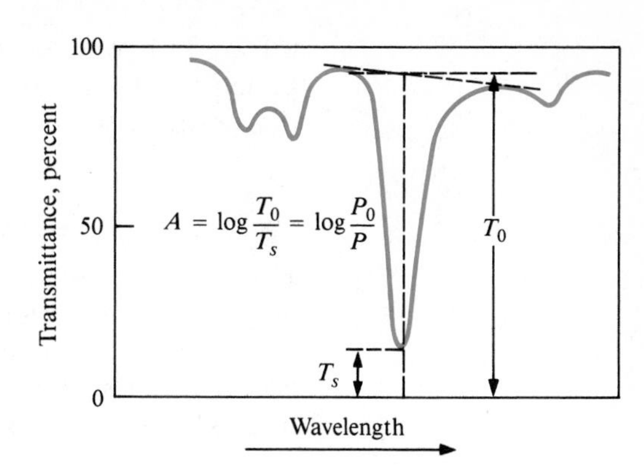

FIGURE 20-15 Base-line method for determination of absorbance.

Calibration curves with significant curvature are therefore common in quantitative infrared spectroscopy.

Applications of Quantitative Infrared Spectroscopy

Infrared spectrophotometry offers the potential for determining an unusually large number of substances because nearly all molecular species absorb in the infrared region. Moreover, the uniqueness of an infrared spectrum provides a degree of specificity that is matched or exceeded by relatively few other analytical methods. This specificity has particular application to the analysis of mixtures of closely related organic compounds.

The recent proliferation of government regulations on atmospheric contaminants has demanded the development of sensitive, rapid, and highly specific methods for a variety of chemical compounds. Infrared absorption procedures appear to meet this need better than any other single analytical tool.

Table 20-5 illustrates the variety of atmospheric pollutants that can be determined with a simple, portable filter photometer equipped with a separate interference filter for each analyte species. Of the more than 400 chemicals for which maximum tolerable limits have been set by OSHA, half or more have absorption characteristics that make them amenable to determination by

TABLE 20-5 **Examples of Infrared Vapor Analysis for OSHA Compliance**

Compound	Allowable Exposure, ppm*	Wavelength, μm	Minimum Detectable Concentration, ppm†
Carbon disulfide	10	4.54	0.5
Chloroprene	10	11.4	4
Diborane	0.1	3.9	0.05
Ethylenediamine	10	13.0	0.4
Hydrogen cyanide	10	3.04	0.4
Methyl mercaptan	0.5	3.38	0.4
Nitrobenzene	1	11.8	0.2
Pyridine	5	14.2	0.2
Sulfur dioxide	2	8.6	0.5
Vinyl chloride	5	10.9	0.3

Courtesy of The Foxboro Company, Foxboro, MA 02035.

*1986 OSHA exposure limits for 8-h weighted average.

†For 20.25-m cell.

infrared photometry or spectrophotometry. Obviously, peak overlaps are to be expected with so many compounds absorbing; nevertheless, the method does provide a moderately high degree of selectivity.

20C Molecular Fluorescence Methods

Fluorometric and spectrofluorometric methods of analysis, which are unusually sensitive and highly selective, have been applied to the determination of many inorganic and organic molecules in trace amounts.[21] Fluorometric procedures are less widely applicable than those based upon absorption, however, because many species do not fluoresce.

20C-1 Fluorescent Species

As pointed out in Section 18D-2, fluorescence is one of several mechanisms by which a molecule returns to the ground state after it has been excited by absorption of radiation. Thus, all absorbing molecules have the potential to fluoresce. Most compounds do not, however, because their structure provides pathways by which relaxation can occur at a greater rate than fluorescent emission.

The *quantum yield,* or quantum efficiency, of molecular fluorescence is simply the ratio of molecules that fluoresce to the total number of excited molecules (or the ratio of photons emitted to photons absorbed). Highly fluorescent molecules, such as fluorescein, have quantum efficiencies that approach unity under some conditions. Nonfluorescent species have efficiencies that are essentially zero.

Fluorescence and Structure

The most intense and most useful molecular fluorescence behavior is found in compounds containing aromatic rings. While certain aliphatic and alicyclic carbonyl compounds as well as highly conjugated double-bonded structures also fluoresce, their numbers are small in comparison with the number of fluorescent compounds containing aromatic systems.

Most unsubstituted aromatic hydrocarbons fluoresce in solution, with the quantum efficiency increasing with the number of rings and their degree of condensation. The simplest heterocyclics, such as pyridine, furan, thiophene, and pyrrole, do not exhibit molecular fluorescence, but fused-ring structures containing these rings often do.

Substitution on an aromatic ring causes shifts in the wavelength of absorption maxima and corresponding changes in the fluorescence peaks. In addition, substitution frequently affects the fluorescence efficiency. For example, the relative fluorescence intensities for aniline, phenol, benzene, and bromobenzene are 20, 18, 10, and 5, respectively.

The Effect of Structural Rigidity

It is found experimentally that fluorescence is particularly favored in rigid molecules. For example, under similar conditions of measurement, the quan-

[21]For detailed discussions of fluorescence methods, see W. R. Seitz, in *Treatise on Analytical Chemistry,* 2nd ed., P. J. Elving, E. J. Meehan, and I. M. Kolthoff, Eds., Part I, Vol. 7, Chapter 4. New York: Wiley, 1981; J. D. Winefordner, S. G. Schulman, and T. C. O'Haver, *Luminescence Spectrometry in Analytical Chemistry.* New York: Wiley-Interscience, 1972; and S. G. Schulman, *Fluorescence and Phosphorescence Spectroscopy.* New York: Pergamon Press, 1977.

tum efficiency of fluorene is nearly 1.0 whereas that of biphenyl is about 0.2. The difference in behavior appears to be

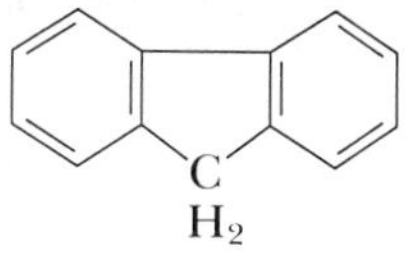

fluorene 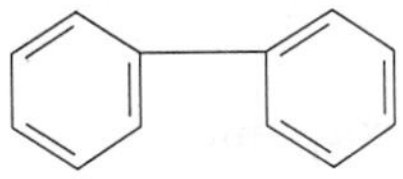 biphenyl

largely a result of the increased rigidity furnished by the bridging methylene group in fluorene. Many similar examples can be cited. In addition, enhanced emission frequently results when fluorescing dyes are adsorbed on a solid surface; here again, the added rigidity provided by the solid may account for the observed effect.

The influence of rigidity has also been invoked to account for the increase in fluorescence of certain organic chelating agents when they are complexed with a metal ion. For example, the fluorescence intensity of 8-hydroxyquinoline is much less than that of the zinc complex:

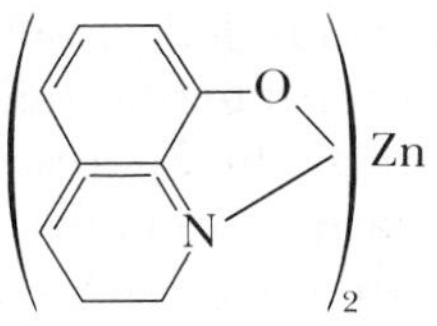

20C-2 Temperature and Solvent Effects

In most molecules, the quantum efficiency of fluorescence decreases with increasing temperature because the increased frequency of collision at elevated temperatures improves the probability of collisional relaxation. A decrease in solvent viscosity leads to the same result.

20C-3 The Effect of Concentration on Fluorescence Intensity

The power of fluorescent radiation F is proportional to the radiant power of the excitation beam absorbed by the system:

$$F = K'(P_0 - P) \qquad (20\text{-}6)$$

where P_0 is the power of the beam incident on the solution and P is its power after it traverses a length b of the medium. The constant K' depends upon the quantum efficiency of the fluorescence. In order to relate F to the concentration c of the fluorescing particle, we write Beer's law in the form

$$\frac{P}{P_0} = 10^{-\epsilon bc} \qquad (20\text{-}7)$$

where ϵ is the molar absorptivity of the fluorescing species and ϵbc is the absorbance A. By substituting Equation 20-7 into Equation 20-6, we obtain

$$F = K'P_0(1 - 10^{-\epsilon bc}) \qquad (20\text{-}8)$$

Expansion of the exponential term in Equation 20-8 leads to

$$F = K'P_0\left[2.3\epsilon bc - \frac{(-2.3\epsilon bc)^2}{2!} - \frac{(-2.3\epsilon bc)^3}{3!} - \cdots\right] \qquad (20\text{-}9)$$

Provided $\epsilon bc = A < 0.05$, all the subsequent terms in the brackets are small with respect to the first and we can write

$$F = 2.3K'\epsilon bcP_0 \tag{20-10}$$

or, at constant P_0,

$$F = Kc \tag{20-11}$$

Thus, a plot of the fluorescence power of a solution versus the concentration of the emitting species should be linear at low concentrations. When c becomes great enough that the absorbance is larger than about 0.05 (or the transmittance is smaller than about 90%), linearity is lost and F lies below an extrapolation of the straight-line plot. Indeed, at very high concentrations, F reaches a maximum and then begins to decrease with increasing concentration.

20C-4 Fluorometers and Spectrofluorometers

Figure 20-16 shows a typical configuration for the components of *fluorometers* and *spectrofluorometers.* These components are identical to the ones described in Section 19A for ultraviolet/visible spectroscopy. A fluorometer, like a photometer, employs filters for wavelength selection. Most spectrofluorometers, in contrast, employ a filter for limiting the excitation radiation and a grating monochromator for dispersing the fluorescence radiation from the sample. A few spectrofluorometers have two monochromator systems, one for the excitation radiation and one for the fluorescence. With such an instrument both *excitation* and fluorescence spectra can be obtained.

As shown in Figure 20-16, fluorescence instruments are usually double-beam in design in order to compensate for fluctuations in the power of the source. The beam to the sample first passes through a primary filter or a primary monochromator, which transmits radiation that excites fluorescence but excludes or limits radiation that corresponds to the fluorescence wavelengths. Fluorescence radiation is propagated from the sample in all directions but is most conveniently observed at right angles to the excitation beam; at other angles, increased scattering from the solution and the cell walls may

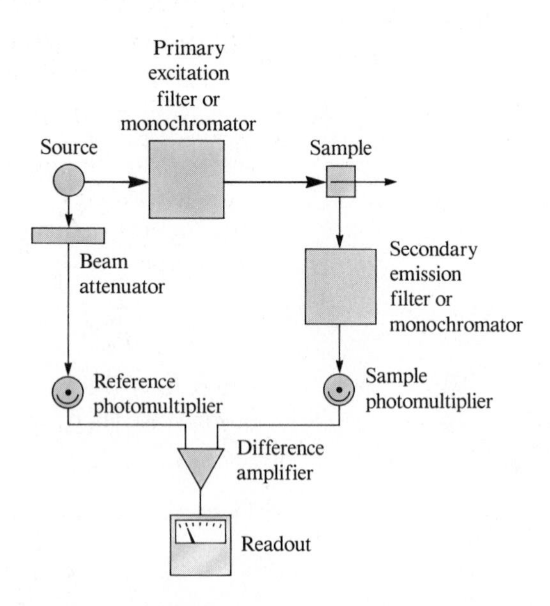

FIGURE 20-16 Components of a fluorometer or a spectrofluorometer.

cause large errors in the intensity measurement. The emitted radiation reaches a photoelectric detector after passing through the secondary filter or monochromator, which isolates a fluorescence peak for measurement.

The reference beam passes through an attenuator to decrease its power to approximately that of the fluorescence radiation (the power reduction is usually by a factor of 100 or more). The signals from the reference and sample phototubes are then processed by a difference amplifier whose output is displayed on a meter or recorder. Many fluorescence instruments are of the null type, this state being achieved by optical or electrical attenuators.

The sophistication, performance characteristics, and cost of fluorometers and spectrofluorometers differ as widely as do the corresponding instruments for absorption measurements. In many regards, filter-type instruments are better suited for quantitative analytical work than are the more elaborate instruments based on monochromators. Generally, fluorometers are more sensitive than spectrofluorometers because filters have a higher radiation throughput than do monochromators. In addition, source and detector can be positioned closer to the sample in the simpler instrument, a factor that enhances sensitivity.

20C-5 Applications of Fluorescence Methods

Fluorescence methods are generally one to three orders of magnitude more sensitive than methods based upon absorption because the sensitivity of the former can be enhanced either by increasing the power of the excitation beam (Equation 20-10) or amplifying the detector signal. Neither of these options improves the sensitivity of methods based upon absorption, however, because the concentration-related parameter in this case is a ratio:

$$c = k \log (P_0/P)$$

where $k = 1/ab$ (Equation 18-7). Increasing the power P_0 increases P proportionately and thus has no effect on sensitivity. Similarly, increasing the amplification of the detector signal affects the two measured quantities in an identical way and leads to no improvement.

Methods for Inorganic Species

Inorganic fluorometric methods are of two types. Direct methods are based upon the reaction of the analyte with a chelating agent to form a complex that fluoresces. In contrast, indirect methods depend upon the diminution, or *quenching,* of fluorescence of a reagent as a result of its reaction with the analyte. Quenching is used primarily for the determination of anions.

The most successful fluorometric reagents for the determination of cations are aromatic compounds having two or more donor functional groups that permit chelate formation with the metal ion. A typical example is 8-hydroxyquinoline, the structure of which is given on page 77. A few other fluorometric reagents and their applications are found in Table 20-6. For a more complete summary, see Meites.[22]

Nonradiation relaxation of transition-metal chelates is so efficient that fluorescence of these species is seldom encountered. It is noteworthy that most transition metals absorb in the ultraviolet or visible region whereas nontran-

[22]L. Meites, *Handbook of Analytical Chemistry,* pp. 6-178 to 6-181. New York: McGraw-Hill, 1963.

TABLE 20-6 **Selected Fluorometric Methods for Inorganic Species**

Ion	Reagent	Wavelength, nm Absorption	Fluorescence	Sensitivity, μg/mL	Interference
Al^{3+}	Alizarin garnet R	470	500	0.007	Be, Co, Cr, Cu, F^-, NO_3^-, Ni, PO_4^{3-}, Th, Zr
F^-	Al complex of Alizarin garnet R (quenching)	470	500	0.001	Be, Co, Cr, Cu, Fe, Ni, PO_4^{3-}, Th, Zr
$B_4O_7^{2-}$	Benzoin	370	450	0.04	Be, Sb
Cd^{2+}	2-(*o*-Hydroxyphenyl)-benzoxazole	365	Blue	2	NH_3
Li^+	8-Quinolinol	370	580	0.2	Mg
Sn^{4+}	Flavanol	400	470	0.1	F^-, PO_4^{3-}, Zr
Zn^{2+}	Benzoin	—	Green	10	B, Be, Sb, colored ions

From A. Weissler and C. E. White, *Handbook of Analytical Chemistry,* L. Meites, Ed., pp. 6-178 to 6-181. New York: McGraw-Hill, 1963. With permission.

sition metal ions do not. For this reason, fluorometry often complements spectrophotometry as a method for the determination of cations.

Methods for Organic Species

The number of applications of fluorometric methods to organic problems is impressive. Weissler and White have summarized the most important of these in several tables.[23] More than 100 entries are found under the heading *Organic and General Biochemical Substances,* including such diverse compounds as adenine, anthranilic acid, aromatic polycyclic hydrocarbons, cysteine, guanidine, indole, naphthols, certain nerve gases, proteins, salicylic acid, skatole, tryptophan, uric acid, and warfarin. Some 50 medicinal agents that can be determined fluorometrically are listed. Included among these are adrenaline, alkylmorphine, chloroquin, digitalis principles, lysergic acid diethylamide (LSD), penicillin, phenobarbital, procaine, and reserpine. Methods for the analysis of ten steroids and an equal number of enzymes and coenzymes are also listed in these tables. Some of the plant products listed are chlorophyll, ergot alkaloids, rauwolfia serpentian alkaloids, flavonoids, and rotenone.

Without question, the most important application of fluorometry is in the analysis of food products, pharmaceuticals, clinical samples, and natural products. The sensitivity and selectivity of the method make it a particularly valuable tool in these fields.

20D Automation of Photometric and Spectrophotometric Methods

The first fully automated instrument for chemical analysis (the Technicon AutoAnalyzer®) appeared on the market in 1957. This instrument was designed to fulfill the needs of clinical laboratories, where blood and urine samples are routinely analyzed for a dozen or more chemical species. The

[23]A. Weissler and C. E. White, in *Handbook of Analytical Chemistry,* L. Meites, Ed., pp. 6-182 to 6-196. New York: McGraw-Hill, 1963.

number of such analyses demanded by modern medicine is enormous[24]; the need to keep their cost at a reasonable level is obvious. These two considerations motivated the development of analytical systems that perform several analyses simultaneously with a minimum input of human labor. The use of automatic instruments has spread from clinical laboratories to laboratories for the control of industrial processes and the routine determination of a wide spectrum of species in air, water, soils, and pharmaceutical and agricultural products.[25] In the majority of these applications, the analyses are completed by a photometric or fluorometric measurement.

20D-1 Types of Automatic Analytical Systems

Automatic analytical instruments are of two types, *discrete* and *continuous,* although combinations of the two are sometimes encountered. In a discrete instrument, individual samples are maintained as separate entities and kept in separate vessels throughout such unit operations as sampling, definition of the sample (measurement of its weight or volume), dilution, reagent addition, mixing, centrifugation, and transportation to the measuring device. Often discrete systems require the use of robots.

In continuous systems, the sample becomes a part of a flowing stream in which the several unit operations just mentioned take place as the sample is carried from the injection point to a flow-through measuring unit and finally to waste.

This section is devoted to continuous flow methods, which most commonly employ photometric and spectrophotometric measurements. Two types of continuous procedures are encountered: *segmented-flow methods,* in which the analytical stream is divided into individual segments by the periodic injection of air bubbles, and *nonsegmented-flow procedures,* in which the analytical stream is unbroken. A nonsegmented-flow analysis is generally termed a *flow-injection analysis.*

20D-2 Segmented-Flow Methods

Figure 20-17 is a diagram of a single-channel AutoAnalyzer® used for the analysis of one of the constituents of blood. Ordinarily, several of these channels are arranged in parallel, each being dedicated to a single type of determination. Multichannel instruments have a sampler module that dilutes the sample and partitions it into aliquots for introduction in each channel.

The Sample and Reagent Transport System

The heart of a continuous-flow instrument, be it segmented or nonsegmented, is the *peristaltic proportionating pump system.* A peristaltic pump is a device in which a fluid (liquid or gas) is squeezed through plastic tubing by metal rollers mounted on a pair of parallel continuous chains or on a spindle. A spring-loaded platen pinches the tubing against one of the rollers at all times, thus forcing a continuous flow of fluid through the tubing. Generally, peristaltic pumps are driven by a constant-speed motor and are capable of delivering

[24]For example, in 1976, about 100 million samples were analyzed in domestic clinical laboratories. See Snyder *et al., Anal. Chem.,* **1976,** *48,* 1942A.

[25]For an extensive treatment of automatic analysis, see J. K. Foreman and P. B. Stockwell, *Automatic Chemical Analysis.* New York: Wiley, 1975.

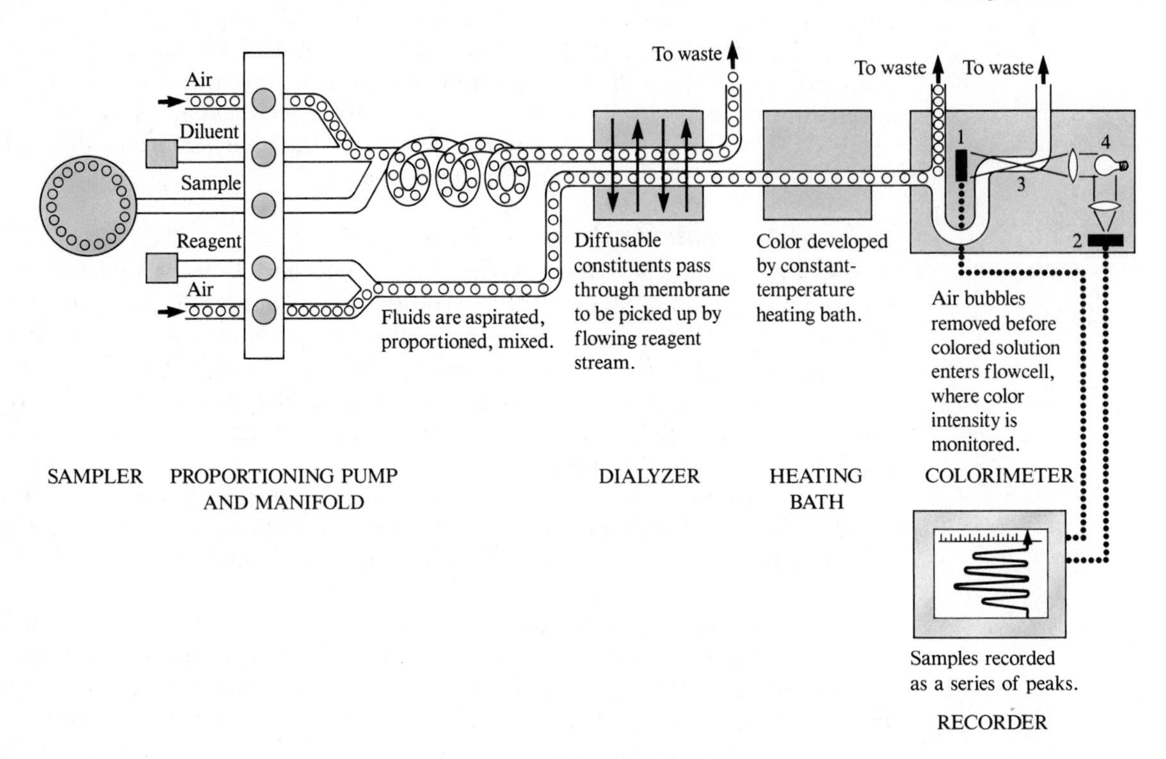
1. Sample photocell
2. Reference photocell
3. Flowcell
4. Light source
To waste
To waste
To waste
Air
Diluent
Sample
Reagent
Air
Fluids are aspirated, proportioned, mixed.
Diffusable constituents pass through membrane to be picked up by flowing reagent stream.
Color developed by constant-temperature heating bath.
Air bubbles removed before colored solution enters flowcell, where color intensity is monitored.
SAMPLER
PROPORTIONING PUMP AND MANIFOLD
DIALYZER
HEATING BATH
COLORIMETER
Samples recorded as a series of peaks.
RECORDER

FIGURE 20-17 A single-channel Technicon AutoAnalyzer® system. (Reproduced with permission of Technicon Instruments Corporation. Technicon, SMA, and AutoAnalyzer are registered trademarks of Technicon Instruments Corporation.)

remarkably reproducible volumes. The volume is controlled by the inside diameter of the tubing and by the pumping time. A wide variety of tube sizes permit flow rates as low as 0.015 mL/min and as high as 3.9 mL/min to be attained.

Ordinarily, the rollers in continuous-flow instruments are long enough to accommodate several tubes (as many as 28) simultaneously.

Segmentation

An important feature of the apparatus shown in Figure 20-17 is that provision is made to pump regularly spaced air bubbles into each stream. The spacing of the bubbles is such that every sample is carried through the system in several successive fluid segments. Before final measurement, the stream passes through a debubbler, which removes the bubbles and recombines the segments. Segmenting tends to maintain a sharp concentration profile at the leading and following edges of each sample. In the absence of bubbles, tailing along the tube walls occurs and enhances the probability of contaminating the following sample.

A second purpose of the bubbles is to promote mixing of sample and reagent. As shown in Figure 20-17, each segment is inverted as it rises and falls through the turns of a mixing coil; maximum mixing efficiency occurs when the length of each segment is less than half the coil diameter.

Separations in Continuous-Flow Analyzers

The instrument shown in Figure 20-17 contains a dialysis module in which the small analyte ions and molecules diffuse through a membrane into a segmented stream of a colorimetric reagent. Larger molecules remain in the original stream and are carried to waste. The membrane is supported between two Lucite or Kel-F plates in which congruent channels have been cut to accommodate the two flows. The transfer of smaller species through this membrane is quite incomplete (less than 50%). Thus, successful quantitative analysis requires close control of temperature and flow rates for both samples and standards.

Another common separation technique used in continuous-flow methods is extraction. Two immiscible liquids are brought together in a mixing coil and then passed into a separator, which is a horizontal glass tube containing an inner tube that partially removes either the lighter or the heavier part of the stream and carries it to the detector module. Separation is again quite incomplete, but the lack of completeness is of no consequence because unknowns and standards are treated in an identical way.

It is important to appreciate that the timing sequences in automatic instruments are sufficiently reproducible so that loss of precision does not accompany incomplete separation or incomplete reactions, as is often the case with manual operations.

Detectors

Various types of detectors have been used with the AutoAnalyzer®, including atomic absorption and emission instruments, fluorometers, electrochemical systems, refractometers, spectrophotometers, and photometers, the last being by far the most common. The detector shown in Figure 20-17 is a twin-beam photometer in which part of the radiation passes through a tubular flow cell to a photoelectric detector; after collimation, the second beam falls on a reference photocell.

Figure 20-18 shows the computer-generated readout from a 12-channel AutoAnalyzer® employed for the routine analysis of blood. The shaded areas show the range of concentrations considered normal for the population. A modern AutoAnalyzer® for routine blood screening is capable of analyzing 200 serum samples[26] per day for 20 constituents. That is, the instrument performs 4000 separate analyses each day. Interspersed with the 200 samples are typically 100 standards for calibration.

Applications of Segmented-Flow Analyzers

As noted earlier, a major area of application of segmented-flow analyzers has been in clinical laboratories, where these instruments are the workhorses for routine clinical tests. In addition, however, there now exists a vast literature having to do with the applications of segmented-flow instruments in environ-

[26]Blood serum is the clear fluid remaining after clotting has occurred.

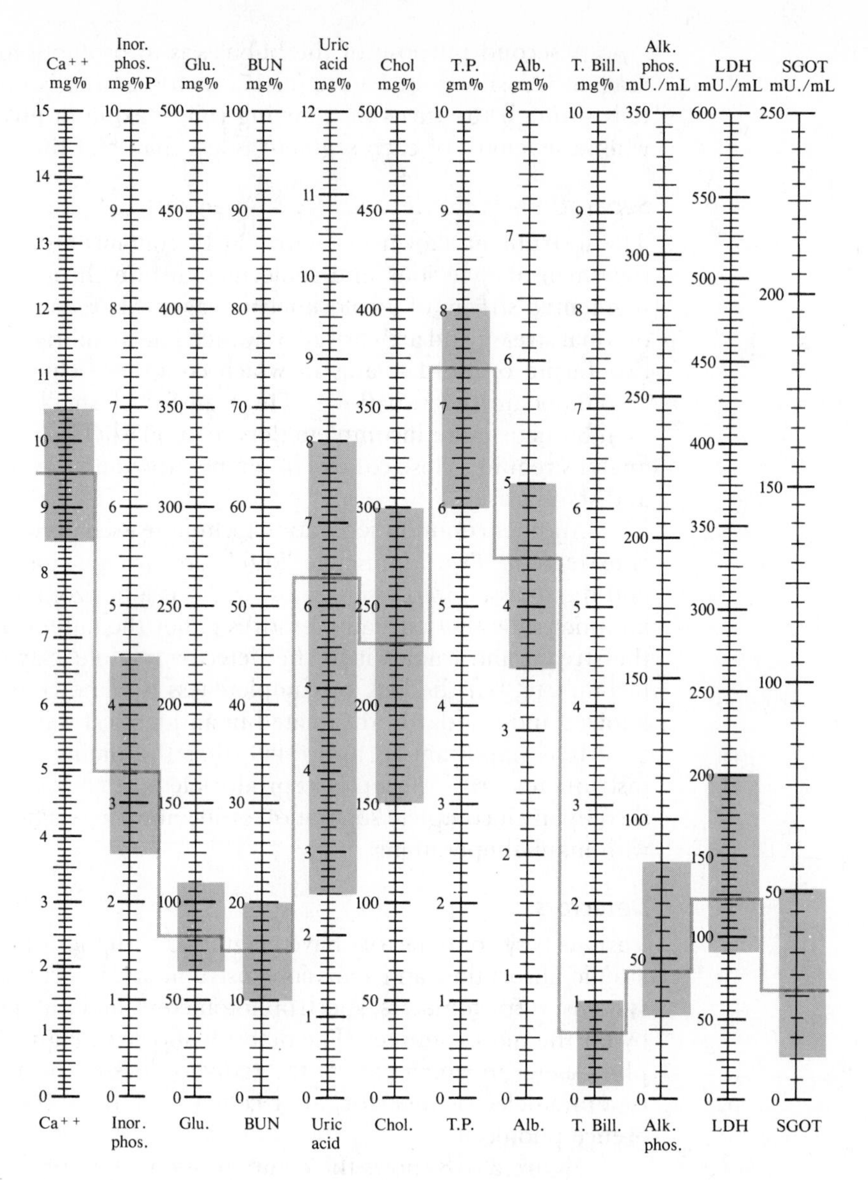

FIGURE 20-18 Readout from a 12-channel Technicon SMA® system. (Reproduced with permission of Technicon Instruments Corporation.)

mental, pharmaceutical, food, agricultural, and metallurgy areas, for both laboratory and process-monitoring uses. Thus, it has been estimated that continuous-flow automated methods are being used in industry for the determination of more than 300 species in more than 1000 types of sample matrices.[27]

20D-3 Flow-Injection Methods

In flow-injection analysis (FIA), the newest continuous-flow method, highly precise sample volumes are introduced into an unsegmented stream. The first

[27] A. Coneta, W. T. Dorsheimer, and M. J. F. DuCros, *Amer. Lab.*, **1981**, *13*, 116.

description of an unsegmented stream as a medium in which to perform analyses appeared more than a decade ago. By now, it has become apparent that flow injection is an important new way of carrying out automatic wet chemical analyses rapidly and efficiently.[28] Equipment for flow-injection analysis is relatively simple and is marketed by several manufacturers.

Figure 20-19a shows the details of a typical flow-injection instrument. As in segmented-flow analyzers, sample and reagents are transported through flexible plastic tubing by the action of a peristaltic pump. The tubing used in flow-injection methods is typically 0.5 mm in diameter, which is narrower than that used in segmented-flow procedures. In some flow-injection instruments, depulsed positive-displacement pumps are used in place of peristaltic models to reduce detector noise.

All of the various flow-type detectors of segmented-flow analyzers are equally applicable to flow-injection measurements. In addition, separations by extraction or dialysis are possible.

FIGURE 20-19 (a) Flow-injection apparatus for determining calcium in water by formation of a colored complex with *o*-cresolphthalein complexone at pH 10. All tubing had an inside diameter of 0.5 mm. *A* and *B* are reaction coils having the indicated lengths. (b) Recorder output. The four sets of curves at left are for duplicate injections of standards containing 5, 10, 15, and 20 ppm calcium. (From E. H. Hansen, J. Ruzicka, and A. K. Ghose, *Anal. Chim. Acta*, 1978, *100*, 151. With permission.)

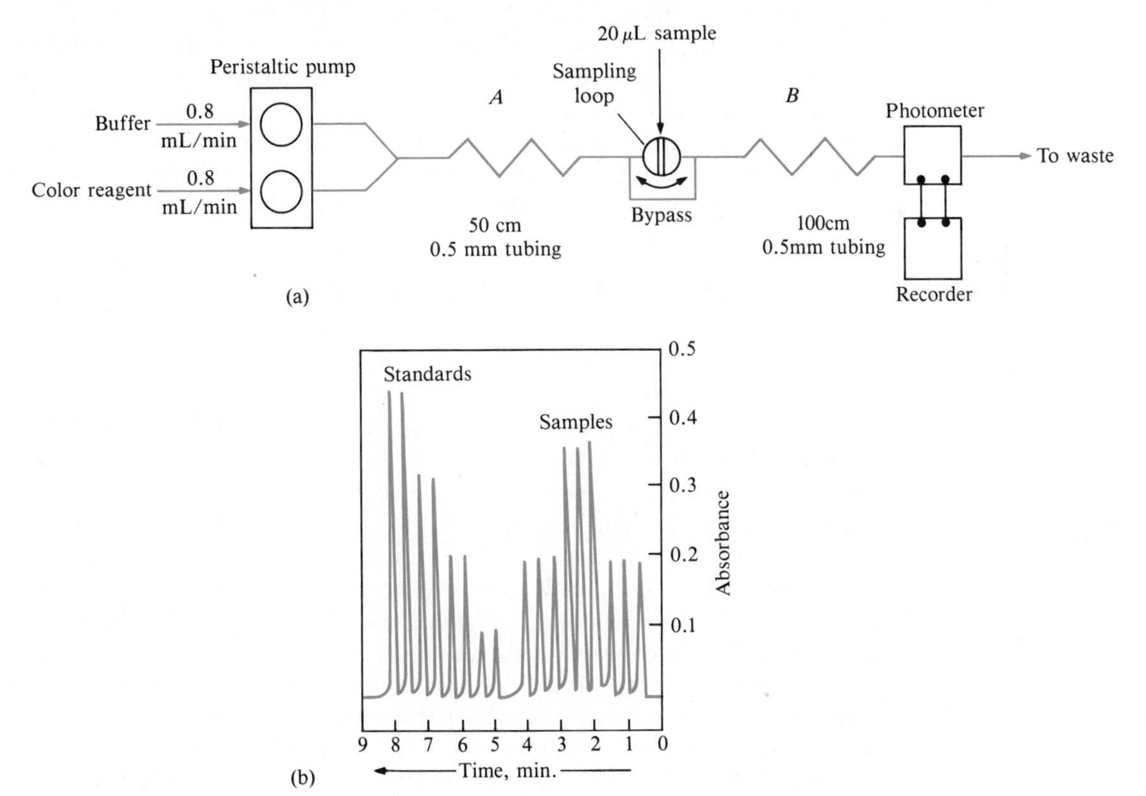

[28]For a monograph dealing with flow-injection analysis, see J. Ruzicka and E. H. Hansen, *Flow Injection Analysis.* New York: Wiley, 1981. For brief reviews of the method, see J. Ruzicka, *Anal. Chem.*, **1983,** *55,* 11041A; K. K. Stewart, *Anal. Chem.*, **1983,** *55,* 931A; C. B. Ranger, *Anal. Chem.*, **1981,** *53,* 20A; and D. Betteridge, *Anal. Chem.*, **1978,** *50,* 832A.

Injectors

Sample sizes for flow-injection procedures range from 5 to 200 μL, with 10 to 30 μL being typical for most applications. For a successful analysis, it is vital that the sample solution be injected rapidly as a pulse, or plug, of liquid; in addition, the injections must not disturb the flow of the carrier stream. The requirements here are significantly more rigorous than is the case with air-segmented flow because the sample is more likely to spread in the absence of bubbles. To date, the most satisfactory injector systems are based upon sampling loops similar to those used in chromatography (see, for example, Figure 25-3). The method of operation of a sampling loop can be seen by reference to Figure 20-19. With the valve of the loop in the position shown, reagents flow through the bypass. When a sample has been injected into the loop and the valve turned 90 deg, the sample enters the flow as a single, well-defined zone. For all practical purposes, flow through the bypass ceases with the valve in this position because the diameter of the sample loop is significantly greater than that of the bypass tubing.

The Principles of Flow-Injection Analysis

Immediately after injection, the sample zone in a flow-injection apparatus has a rectangular concentration profile. As the sample moves through the tubing, band broadening, or *dispersion,* takes place and the zone profile then has the shape shown in Figure 20-20c. The width of the zone depends upon the distance between the injector and the detector as well as on the pumping rate.

Figure 20-20a is a flow diagram of the simplest of all flow-injection systems. A colorimetric reagent for chloride ion that consists of a mixture of mercury(II) thiocyanate and iron(III) ion is pumped directly into the sampling valve and thence through a 50-cm reactor coil, where the reagent diffuses into the sample plug and produces a colored product by the sequence of reactions

$$Hg(SCN)_2(aq) + 2Cl^- \rightleftharpoons HgCl_2(aq) + 2SCN^-$$

$$Fe^{3+} + SCN^- \rightleftharpoons \underset{\text{red}}{Fe(SCN)^{2+}}$$

The recorder output for a series of standards containing from 5 to 75 ppm of chloride are shown in Figure 20-20b. Note that four injections of each standard were made to demonstrate the reproducibility of the system. This process took 23 min, which corresponds to a sampling rate of 130 samples/h.

The two curves in Figure 20-20c are scans of samples containing 30 (R_{30}) and 75 (R_{75}) ppm of chloride. These curves, in contrast to those in Figure 20-20b, were obtained with a high-speed recorder so that their shapes could be studied. The two plots demonstrate that less than 1% of the analyte is present in the flow cell after 28 s, the time of the next injection (S_2). This system has been successfully used for the routine determination of chloride ion in brackish and waste waters as well as in serum samples.

Figure 20-19 illustrates a more complicated system for the colorimetric determination of calcium in serum, milk, and drinking water. A borax buffer and a color-forming reagent are combined in a 50-cm mixing coil (A) prior to sample injection. The recorder output for three samples in triplicate and four standards in duplicate is shown in Figure 20-19b.

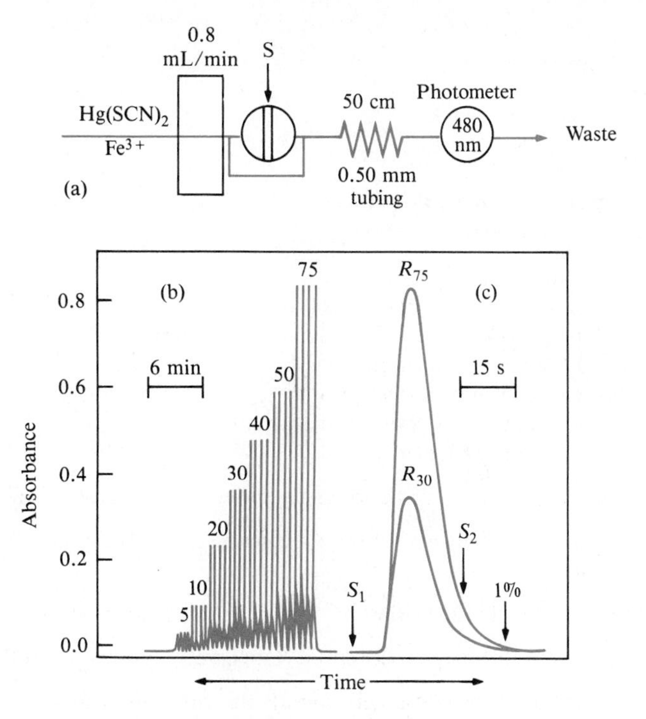

FIGURE 20-20 Flow-injection determination of chloride: (a) flow diagram; (b) recorder readout for quadruplicate runs on standards containing 5 to 75 ppm chloride ion; (c) fast scan of two of the standards to demonstrate the low analyte carryover (less than 1%) from run to run. Note that the point marked 1% corresponds to where the response would just begin for a sample injected at time S_2. (From J. Ruzicka and E. H. Hansen, *Flow Injection Methods,* p. 8. Wiley: New York, 1981. Reprinted by permission.)

Advantages of Flow-Injection Measurements

Flow-injection equipment differs markedly from its segmented-flow counterpart in the respect that no air-bubble segmentation is used. Before the advent of flow-injection methods, it was believed that air bubbles were vital to the success of continuous-flow techniques, serving to prevent excess sample dispersion, to promote turbulent mixing, and to scrub the walls of the conduit and thus prevent cross-contamination between samples. In fact, however, excess dispersion or dilution and cross-contamination are nearly completely avoided in a properly designed system without air bubbles. In addition, the mixing of reagents and sample occurs rapidly, albeit by a different mechanism.

Flow-injection measurements have several important advantages over segmented-flow procedures, including (1) higher analysis rates (typically 100 to 300 samples/h), (2) enhanced response times (often less than 1 min between sample injection and recorder response), (3) lower solvent costs (because of the narrower tubing), (4) much more rapid startup and shutdown times (less than 5 min for each), and (5) except for the injection system, simpler and more flexible equipment. The last two advantages are of particular importance because they make it feasible and economic to apply automated measurements to a relatively few nonroutine samples. In other words, continuous-flow methods are no longer restricted to situations where the number of samples is large and the analytical method highly routine.

Questions and Problems

20-1 Define
- *(a) chromophore.
- (b) standard-addition method.
- *(c) noise in a spectrophotometric measurement.
- (d) photometric titration.
- *(e) charge-transfer absorption.
- (f) the method of continuous variation.
- *(g) quantum efficiency.

*20-2 Explain the difference between a segmented-flow and a flow-injection method.

20-3 Explain the difference between a discrete and a continuous system for automatic analyses.

*20-4 Describe the characteristics of organic compounds that fluoresce.

20-5 What is the difference between a fluorometer and a spectrofluorometer?

*20-6 Why do some absorbing compounds fluoresce and others not?

20-7 Identify the characteristics of infrared absorbance measurements that tend to cause departure from Beer's law.

*20-8 Explain how the composition of an absorbing complex is determined by the mole-ratio method.

20-9 Explain how the composition of an absorbing complex is determined by the method of continuous variation.

*20-10 Sketch a photometric titration curve for Fe(III) with SCN^- ion when a photometer with a green filter is used to collect data. Why is a green filter used?

20-11 Sketch a photometric titration curve for the titration of Sn^{2+} with MnO_4^-. What color filter should be used for this titration? Explain.

*20-12 Why does molecular fluorescence generally occur at wavelengths that are longer than the excitation wavelengths?

20-13 Why cannot the excellent detector systems available for ultraviolet and visible radiation be used with infrared radiation?

*20-14 Why is the sample compartment located between the monochromator and the detector in an ultraviolet spectrophotometer but between the source and the monochromator in an infrared spectrophotometer?

*20-15 A 4.97-g petroleum specimen was decomposed by wet-ashing and subsequently diluted to 500 mL in a volumetric flask. Cobalt was determined by treating 25.00-mL aliquots of this diluted solution as follows:

Reagent Volume			
Co(II), 3.00 ppm	Ligand	H_2O	Absorbance
0.00	20.00	5.00	0.398
5.00	20.00	0.00	0.510

Assume that the Co(II)/ligand chelate obeys Beer's law, and calculate the percentage of cobalt in the original sample.

20-16 A two-tablet sample of vitamin/mineral supplement weighing 6.08 g was wet-ashed to eliminate organic matter and then diluted to 1.00 L. Two 10.00-mL aliquots were then analyzed. Calculate the average weight of iron in each tablet, based upon the following information:

Reagent Volume			
Fe(III), 1.00 ppm	Ligand	H_2O	Absorbance
0.00	25.00	15.00	0.492
15.00	25.00	0.00	0.571

20-17 The standard deviation in transmittance for a particular instrument is 0.006, regardless of the magnitude of the transmittance. Calculate the relative standard deviation in concentration due to this source when

*(a) $T = 0.015$
*(b) $A = 0.334$
*(c) $\% T = 64.8$
(d) $A = 0.921$
(e) $T = 0.804$
(f) $\% T = 50.0$

20-18 The meter of an inexpensive spectrophotometer has a 5-in scale scribed in linear units from 0 to 100% T. The scale, which limits the precision of the instrument, can be read to about $\pm 0.5\%$ T. Calculate the relative precision of concentration determinations for an absorbance of (a) 0.020, (b) 0.050, (c) 0.100, (d) 0.400, (e) 0.800, (f) 1.200, (g) 2.000.

*20-19 Ethylenediaminetetraacetic acid abstracts bismuth(III) from its thiourea complex:

$$Bi(tu)_6^{3+} + H_2Y^{2-} \longrightarrow BiY^- + 6tu + 2H^+$$

Predict the shape of a photometric titration curve based on this process, given that the Bi(III)/thiourea complex is the only species in the system that absorbs at 465 nm, the wavelength selected for the analysis.

20-20 The accompanying data (1.00-cm cells) were obtained for the spectrophotometric titration of 10.00 mL of Pd(II) with 2.44×10^{-4} M Nitroso R (O. W. Rollins and M. M. Oldham, *Anal. Chem.*, **1971**, *43*, 262):

Volume of Nitroso R, mL	A_{500}
0	0
1.00	0.147
2.00	0.271
3.00	0.375
4.00	0.371
5.00	0.347
6.00	0.325
7.00	0.306
8.00	0.289

Calculate the concentration of the Pd(II) solution, given that the ligand-to-cation ratio in the colored product is 2:1.

20-21 Solutions containing species A and B obey Beer's law over an extensive concentration range. Molar absorptivity data for each are as follows:

λ, nm	ϵ_A	ϵ_B
400	893	0.00
420	940	0.00
440	955	0.00
460	936	24.6
480	874	102
500	795	185
520	691	289
540	574	428
560	440	622
580	297	980
600	167	1178
620	39.7	1692
640	3.45	1742
660	0.00	1806
680	0.00	1809
700	0.00	1757

*(a) A solution containing both solutes has an absorbance of 0.360 in a 1.00-cm cell at 540 nm. What is the concentration of B in this solution if [A] = 5.00×10^{-4}M?

*(b) A solution containing both species has an absorbance of 0.510 both at 440 nm and at 600 nm when measured in a 1.00-cm cell. Calculate the concentrations of A and B.

(c) Construct an absorption spectrum for a 5.80×10^{-4} M solution of A in a 1.00-cm cell.

(d) Construct an absorption spectrum for a 3.25×10^{-4} M solution of B in a 1.00-cm cell.

(e) Construct an absorption spectrum for a solution that is 5.80×10^{-4} M in A and 3.25×10^{-4} M in B in a 1.00-cm cell.

20-22 A. J. Mukhedkar and N. V. Deshpande (*Anal. Chem.*, **1963**, *35*, 47) report on a simultaneous determination for cobalt and nickel based upon absorption by their 8-quinolinol complexes. Molar absorptivities are $\epsilon_{Co} = 3529$ and $\epsilon_{Ni} = 3228$ at 365 nm and $\epsilon_{Co} = 428.9$ and $\epsilon_{Ni} = 0$ at 700 nm. Calculate the concentration of nickel and cobalt in each of the following solutions (1.00-cm cells):

Solution	A_{365}	A_{700}
*1	0.724	0.0710
2	0.614	0.0744
3	0.693	0.0460

20-23 Solutions of P and of Q individually obey Beer's law over a large concentration range. Spectral data for these species in 1.00-cm cells are

	Absorbance	
λ, nm	8.55×10^{-5} M P	2.37×10^{-4} M Q
400	0.078	0.550
420	0.087	0.592
440	0.096	0.599
460	0.102	0.590
480	0.106	0.564
500	0.110	0.515

Continued on next page

Continued from previous page

λ, nm	Absorbance 8.55 × 10^{-5} M P	2.37 × 10^{-4} M Q
520	0.113	0.433
540	0.116	0.343
560	0.126	0.255
580	0.170	0.170
600	0.264	0.100
620	0.326	0.055
640	0.359	0.030
660	0.373	0.030
680	0.370	0.035
700	0.346	0.063

(a) Plot an absorption spectrum for a solution that is 8.55 × 10^{-5} M in P and 2.37 × 10^{-4} M in Q.
(b) Calculate the absorbance (1.00-cm cells) at 440 nm of a solution that is 4.00 × 10^{-5} M in P and 3.60 × 10^{-4} M in Q.
(c) Calculate the absorbance (1.00-cm cells) at 620 nm for a solution that is 1.61 × 10^{-4} M in P and 7.35 × 10^{-4} M in Q.

20-24 Use the data in the previous problem to calculate the molar concentration of P and Q in each of the following solutions:

A_{440}	A_{620}	A_{440}	A_{620}
*(a) 0.357	0.803	(d) 0.910	0.338
(b) 0.830	0.448	*(e) 0.480	0.825
*(c) 0.248	0.333	(f) 0.194	0.315

20-25 The indicator HIn has an acid dissociation constant of 4.80 × 10^{-6} at ordinary temperatures. The accompanying absorbance data are for 8.00 × 10^{-5} M solutions of the indicator measured in 1.00-cm cells in strongly acidic and strongly alkaline media.

λ, nm	Absorbance pH 1.00	pH 13.00
420	0.535	0.050
445	0.657	0.068
450	0.658	0.076
455	0.656	0.085
470	0.614	0.116
510	0.353	0.223
550	0.119	0.324
570	0.068	0.352
585	0.044	0.360
595	0.032	0.361
610	0.019	0.355
650	0.014	0.284

Estimate the wavelength at which absorption by the indicator becomes independent of pH (that is, the isosbestic point).

20-26 Calculate the absorbance (1.00-cm cells) at 450 nm of a solution in which the total molar concentration of the indicator described in Problem 20-25 is 8.00 × 10^{-5} and the pH is *(a) 4.92, (b) 5.46, *(c) 5.93, (d) 6.16.

20-27 What is the absorbance at 595 nm (1.00-cm cells) of a solution that is 1.25 × 10^{-4} M in the indicator of Problem 20-25 and has a pH of *(a) 5.30, (b) 5.70, *(c) 6.10?

20-28 Several buffer solutions were made 1.00×10^{-4} M in the indicator of Problem 20-25. Absorbance data (1.00-cm cells) are

Solution	A_{450}	A_{595}
*A	0.344	0.310
B	0.508	0.212
*C	0.653	0.136
D	0.220	0.380

Calculate the pH of each solution.

20-29 Construct an absorption spectrum for an 8.00×10^{-5} M solution of the indicator of Problem 20-25 when measurements are made with 1.00-cm cells and

*(a) $\frac{[HIn]}{[In^-]} = 3$ (b) $\frac{[HIn]}{[In^-]} = 1$ (c) $\frac{[HIn]}{[In^-]} = \frac{1}{3}$

20-30 Molar absorptivity data for the cobalt and nickel complexes with 2,3-quinoxalinedithiol are $\epsilon_{Co} = 36{,}400$ and $\epsilon_{Ni} = 5520$ at 510 nm and $\epsilon_{Co} = 1240$ and $\epsilon_{Ni} = 17{,}500$ at 656 nm.

A 0.425-g sample was dissolved and diluted to 50.0 mL. A 25.0-mL aliquot was treated to eliminate interferences; after addition of 2,3-quinoxalinedithiol, the volume was adjusted to 50.0 mL. This solution had an absorbance of 0.446 at 510 nm and 0.326 at 656 nm in a 1.00-cm cell. Calculate the parts per million of cobalt and nickel in the sample.

*20-31 The chelate ZnQ_2^{2-} exhibits a maximum absorption at 480 nm. When the chelating agent is present in at least a fivefold excess, the absorbance is dependent only upon the molar concentration of Zn(II) and obeys Beer's law over a large range. Neither Zn^{2+} nor Q^{2-} absorbs at 480 nm. A solution that is 2.30×10^{-4} M in Zn^{2+} and 8.60×10^{-3} M in Q has an absorbance of 0.690 in a 1.00-cm cell at 480 nm. Under the same conditions, a solution that is 2.30×10^{-4} M in Zn^{2+} and 5.00×10^{-4} M in Q^{2-} has an absorbance of 0.540. Calculate the numerical value of K_f for the process

$$Zn^{2+} + 2Q^{2-} \rightleftharpoons ZnQ_2^{2-}$$

21-32 The sodium salt of 2-quinizarinsulfonic acid (NaQ) forms a complex with aluminum(III) that absorbs strongly at 560 nm (E. G. Owens and J. H. Yoe, *Anal. Chem.*, **1959**, *31*, 385). Use the accompanying data to establish the combining ratio between cation and ligand. In all solutions, $c_{Al} = 3.7 \times 10^{-5}$ M, and all measurements were made in 1.00-cm cells.

Molar Ligand Concentration	Absorbance
1.00×10^{-5}	0.131
2.00×10^{-5}	0.265
3.00×10^{-5}	0.396
4.00×10^{-5}	0.468
5.00×10^{-5}	0.487
6.00×10^{-5}	0.498
8.00×10^{-5}	0.499
1.00×10^{-4}	0.500

*20-33 The accompanying data (1.00-cm cells) were obtained in a slope-ratio investigation of the product formed between Ni(II) and 1-cyclopentene-1-dithiocarboxylic acid (CDA):

$c_{CDA} = 1.00 \times 10^{-3}$ M		$c_{Ni} = 1.00 \times 10^{-3}$ M	
c_{Ni}, M	A_{530}	c_{CDA}, M	A_{530}
5.00×10^{-6}	0.051	9.00×10^{-6}	0.031
1.20×10^{-5}	0.123	1.50×10^{-5}	0.051
3.50×10^{-5}	0.359	2.70×10^{-5}	0.092
5.00×10^{-5}	0.514	4.00×10^{-5}	0.137
6.00×10^{-5}	0.616	6.00×10^{-5}	0.205
7.00×10^{-5}	0.719	7.00×10^{-5}	0.240

Calculate the ligand-to-cation ratio in this complex.

20-34 Iron(II) forms a chelate with the ligand P. Evaluate the composition of FeP_n from the accompanying data (1.00-cm cells):

Fe(II) = 2.00×10^{-3} M		P = 2.00×10^{-3} M	
Molar Concentration of P	A	Molar Concentration of Fe	A
4.00×10^{-6}	0.025	6.00×10^{-6}	0.113
1.50×10^{-5}	0.094	1.00×10^{-5}	0.189
3.00×10^{-5}	0.189	1.75×10^{-5}	0.330
5.00×10^{-5}	0.314	3.00×10^{-5}	0.566
7.00×10^{-5}	0.440	4.00×10^{-5}	0.754

*20-35 The accompanying absorbance data (1.00-cm cells) were recorded for a continuous-variation study of the colored product formed between cadmium(II) and the complexing reagent R:

	Reactant Volume, mL		
Solution	1.25×10^{-4} M Cd^{2+}	1.25×10^{-4} M R	A_{390}
0	10.00	0.00	0.000
1	9.00	1.00	0.174
2	8.00	2.00	0.353
3	7.00	3.00	0.530
4	6.00	4.00	0.672
5	5.00	5.00	0.723
6	4.00	6.00	0.673
7	3.00	7.00	0.537
8	2.00	8.00	0.358
9	1.00	9.00	0.180
10	0.00	10.00	0.000

(a) Establish the ligand-to-cation ratio in the product.
(b) Determine an average value for the molar absorptivity of the complex; assume that the species in lesser amount is completely incorporated into the complex in the linear portions of the plot.
(c) Evaluate K_f for the complex, using the stoichiometric relationships that exist under conditions of maximum absorbance.

20-36 (a) Use the following absorbance data (1.00-cm cells) to evaluate the ligand-to-cation ratio in the complex formed between copper(II) and the ligand H_2B:

Solution	Reactant Volume, mL		A_{475}
	8.00×10^{-5} M Cu	8.00×10^{-5} M H_2B	
0	10.00	0.00	0.000
1	9.00	1.00	0.104
2	8.00	2.00	0.210
3	7.00	3.00	0.314
4	6.00	4.00	0.419
5	5.00	5.00	0.507
6	4.00	6.00	0.571
7	3.00	7.00	0.574
8	2.00	8.00	0.423
9	1.00	9.00	0.211
10	0.00	10.00	0.000

(b) Evaluate an average molar absorptivity for the complex; assume that the species in lesser amount is completely incorporated into the complex in the linear portions of the plot.
(c) Evaluate K_f for the complex, using the stoichiometric relationships that exist under conditions of maximum absorption.

*20-37 (a) Use the following absorbance data (1.00-cm cells) to evaluate the ligand-to-cation ratio in the complex formed between Co(II) and the bidentate ligand Q:

Solution	Reactant Volume, mL		A_{560}
	9.50×10^{-5} M Co(II)	9.50×10^{-5} M Q	
0	10.00	0.00	0.000
1	9.00	1.00	0.094
2	8.00	2.00	0.193
3	7.00	3.00	0.291
4	6.00	4.00	0.387
5	5.00	5.00	0.484
6	4.00	6.00	0.570
7	3.00	7.00	0.646
8	2.00	8.00	0.585
9	1.00	9.00	0.295
10	0.00	10.00	0.000

(b) Determine an average molar absorptivity for the complex; assume that the species in lesser amount is completely incorporated into the complex in the linear portions of the plot.
(c) Evaluate K_f for the complex, using the stoichiometric relationships that exist under conditions of maximum absorbance.

*20-38 The logarithm of the molar absorptivity for acetone in ethanol is 2.75 at 366 nm. Calculate the range of acetone concentrations that can be used if the percent transmittance is to be greater than 10% and less than 90% with a 1.50-cm cell.

20-39 The logarithm of the molar absorptivity of phenol in aqueous solution is 3.812 at 211 nm. Calculate the range of phenol concentrations that can be used if the absorbance is to be greater than 0.100 and less than 2.000 with a 1.25-cm cell.

*20-40 A standard solution was put through appropriate dilutions to give the concentrations of iron shown below. The iron(II)-1,10-phenanthroline complex was then developed in 25.0-mL aliquots of these solutions, following which each was diluted to 50.0 mL. The following absorbances (1.00-cm cells) were recorded at 510 nm:

Fe(II) Concentration in Original Solutions, ppm	A_{510}
4.00	0.160
10.0	0.390
16.0	0.630
24.0	0.950
32.0	1.260
40.0	1.580

(a) Sketch a calibration curve from these data.
(b) Use the method of least squares to derive an equation relating absorbance and the concentration of iron(II).
(c) Calculate the standard deviation about regression.
(d) Calculate the standard deviation of the slope.

20-41 The method developed in Problem 20-40 was used for the routine determination of iron in 25.0-mL aliquots of ground water. Express the concentration (as ppm Fe) in samples that yielded the accompanying absorbance data (1.00-cm cell). Calculate the relative standard deviation of the result. Repeat the calculation assuming the absorbance data are means of three measurements.

*(a) 0.143 *(c) 0.068 *(e) 1.512
(b) 0.675 (d) 1.009 (f) 0.546

CHAPTER 21

Atomic Spectroscopy Based on Ultraviolet and Visible Radiation

Atomic spectroscopy is used for the qualitative and quantitative determination of perhaps 70 elements. Sensitivities of atomic methods lie typically in the parts-per-million to parts-per-billion range. Additional virtues of these methods are speed, convenience, unusually high selectivity, and moderate instrument costs.

Spectroscopic studies of atoms (or of elementary ions, such as Fe^+, Mg^+, or Al^+) with ultraviolet and visible radiation can be performed only in a gaseous medium in which the individual atoms or ions are well separated from one another. Consequently, the first step in all atomic spectroscopic procedures is *atomization,* a process in which a sample is volatilized and decomposed in such a way as to produce an atomic gas. The efficiency and reproducibility of the atomization step in large measure determine the method's sensitivity, precision, and accuracy; that is, atomization is by far the most critical step in atomic spectroscopy.

As shown in Table 21-1, atomic spectroscopic methods are conveniently categorized on the basis of how the sample is atomized. Note that the temperature of the medium created by atomization varies considerably among the several methods. Note also that atomic methods are based upon absorption, fluorescence, and emission phenomena.[1]

In this chapter we consider spectroscopic methods that are based on the four most common methods of atomization: (1) flame, (2) electrothermal, (3) inductively coupled plasma, and (4) direct-current plasma. Before undertaking this discussion, however, it is worthwhile pointing out the similarities and the differences between atomic spectroscopic methods and the molecular methods described in Chapter 20.

21A A Comparison of Atomic and Molecular Spectroscopic Methods

Atomic spectroscopy is based on absorption, fluorescence, and emission phenomena, whereas only absorption and fluorescence are generally applicable for molecular spectroscopy. The reason that thermal-emission methods are

[1]References that deal with the theory and applications of atomic spectroscopy include R. D. Sacks, A. Syty, and J. W. Robinson, in *Treatise on Analytical Chemistry,* 2nd ed., P. J. Elving, E. J. Meehan, and I. M. Kolthoff, Eds., Part I, Vol. 7, Chapters 6, 7, and 8. New York: Wiley, 1981; C. Th. J. Alkemade *et al., Metal Vapors in Flames.* Elmsford, NY: Pergamon Press, 1982; B. Magyar, *Guidelines to Planning Atomic Spectrometric Analysis.* New York: Elsevier, 1982; J. D. Ingle Jr. and S. R. Crouch, *Spectrochemical Analysis.* Englewood Cliffs, NJ: Prentice-Hall, 1988.

little used for determining molecular species is that most molecules decompose at the temperatures required to excite emission.

There are few conceptual differences between atomic and molecular spectroscopy; the primary difference lies in the types of species studied. Atomic spectroscopy gives information about the identity and concentration of *atoms* in a sample regardless of how these atoms are combined. In contrast, molecular spectroscopy gives qualitative and quantitative information about the *molecules* in a sample regardless of what atoms they contain.

Atomic spectroscopy is limited to ultraviolet, visible, and X-ray frequencies because only such radiation is energetic enough to cause electronic transitions. Molecules, on the other hand, have vibrational and rotational as well as electronic energy states. As a consequence, molecular spectroscopy is based upon ultraviolet, visible, infrared, microwave, and radio-frequency radiation.

Atomic and molecular spectroscopic instruments have a number of common features. For example, both use monochromators or filters for wavelength selection and photon detectors to determine radiation intensity. Instruments for both atomic and molecular *absorption* studies have a reservoir to contain the sample. In molecular instruments, the reservoir is a cell, or cuvette, that contains a liquid or gaseous solution of the sample. The reservoir for atomic spectroscopy is a flame, a plasma, an arc, or a spark containing the gaseous atomic sample. The atom reservoir serves two purposes: it is both the vehicle for atomization and the container of the atomic vapor.

Atomic absorption and fluorescence instruments also have in common a source of radiation. The precise nature of this source is, as we shall see, a major difference between atomic and molecular spectroscopy.

21B Atomic Spectroscopy Based on Flame Atomization

As shown in Table 21-1, three types of atomic methods are based upon flame atomization: (1) atomic absorption spectroscopy (AAS), (2) atomic emission spectroscopy (AES), and (3) atomic fluorescence spectroscopy (AFS).

TABLE 21-1 **Classification of Atomic Spectral Methods**

Atomization Method	Typical Atomization Temperature, °C	Basis for Method	Common Name and Abbreviation of Method
Flame	1700–3150	Absorption	Atomic absorption spectroscopy, AAS
		Emission	Atomic emission spectroscopy, AES
		Fluorescence	Atomic fluorescence spectroscopy, AFS
Electrothermal	1200–3000	Absorption	Electrothermal atomic absorption spectroscopy
		Fluorescence	Electrothermal atomic fluorescence spectroscopy
Inductively coupled argon plasma	6000–8000	Emission	Inductively coupled plasma spectroscopy, ICP
		Fluorescence	Inductively coupled plasma fluorescence spectroscopy
Direct-current argon plasma	6000–10,000	Emission	DC plasma spectroscopy, DCP
Electric arc	4000–5000	Emission	Arc-source emission spectroscopy
Electric spark	40,000(?)	Emission	Spark-source emission spectroscopy

In flame atomization, an aqueous solution of the sample is dispersed (or *nebulized*) as a fine spray and then mixed with gaseous fuel and oxidant that carry it into a burner. The solvent evaporates in the *base region* of the flame, which is located just above the tip of the burner (Figure 21-1). The resulting finely divided solid particles are carried to a region in the center of the flame called the *inner cone.* Here, in this hottest part of the flame, gaseous atoms and elementary ions are formed from the solid particles. Excitation of atomic emission spectra also takes place in this region. Finally, the atoms and ions are carried to the outer edge, or *outer cone,* where oxidation may occur before the atomization products disperse into the atmosphere. Because the velocity of the fuel/oxidant mixture through the flame is high, only a fraction of the sample undergoes all these processes; indeed, a flame is not a very efficient atomizer.

21B-1 Flame Atomizers

Two types of burners are used in flame spectroscopy: *turbulent-flow burners* and *laminar-flow burners.* Figure 21-2a is a diagram of a commercially available turbulent-flow, or *total-consumption,* burner in which the nebulizer and burner are combined in a single unit. The sample is drawn up the capillary and nebulized by the Venturi action caused by the flow of gases around the capillary tip. Typical sample flow rates are 1 to 3 mL/min.

Turbulent-flow burners offer the advantage of introducing a relatively large and representative sample into the flame. Disadvantages include a relatively short path length through the flame and problems with clogging of the tip. In addition, these burners are noisy from both the electronic standpoint and the auditory standpoint. Although sometimes used for emission and fluorescence, turbulent-flow burners find little use in present-day absorption instruments.

Figure 21-2b is a diagram of a typical commercial laminar-flow, or *premix,* burner. The sample is nebulized by the flow of oxidant past a capillary tip. The resulting aerosol is then mixed with fuel and flows past a series of baffles that remove all but the finest droplets. As a result of the baffles, much of the sample collects in the bottom of the mixing chamber, where it drains into a waste container. The aerosol, oxidant, and fuel are then fed into a slotted burner that provides a flame that is usually 5 to 10 cm in length.

Laminar-flow burners provide a relatively quiet flame and a significantly longer path length. These properties tend to enhance sensitivity and reproducibility. Furthermore, clogging is seldom a problem. Disadvantages include

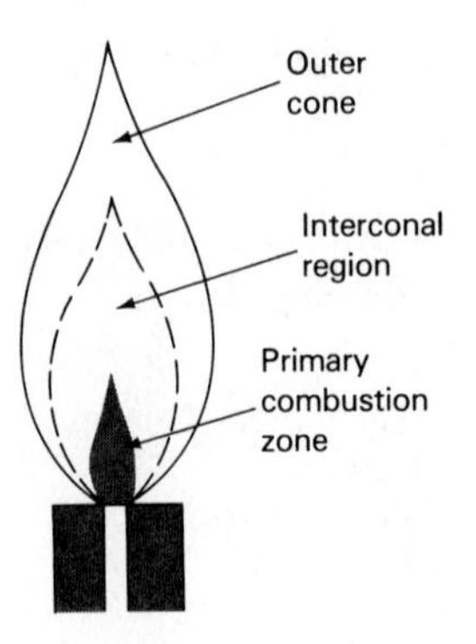

FIGURE 21-1 Regions of an acetylene/air flame.

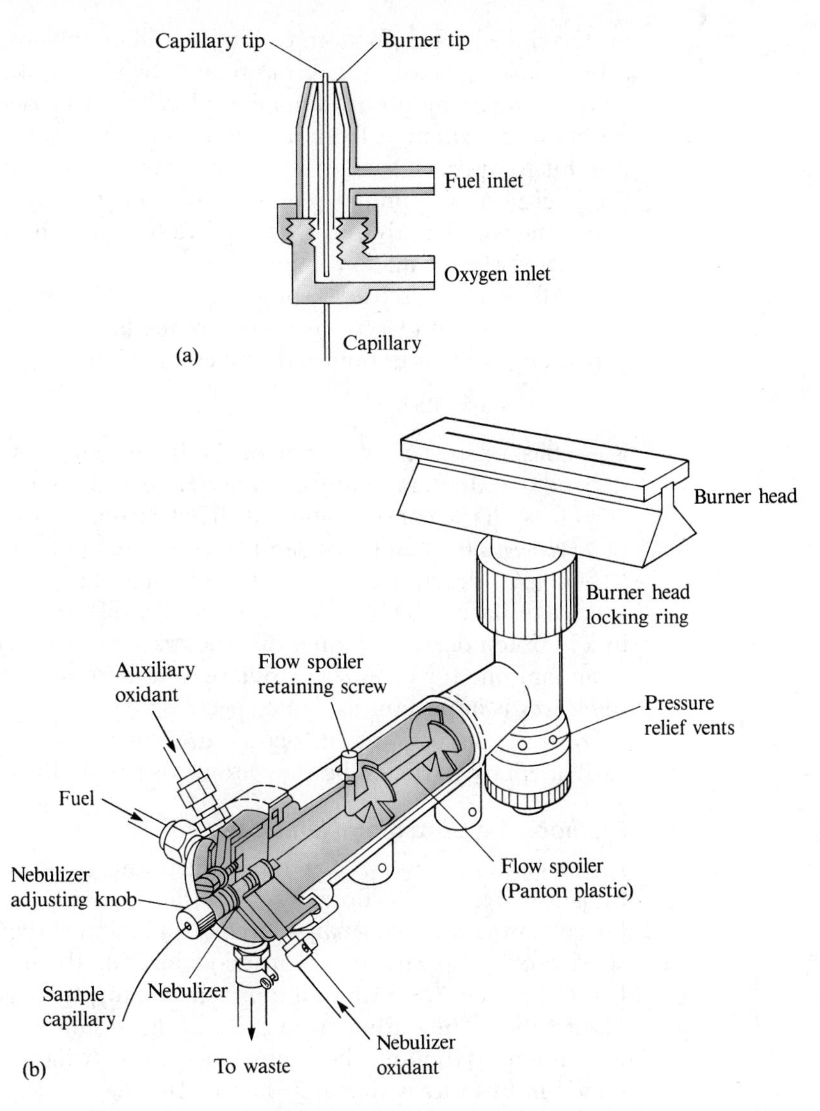

FIGURE 21-2 Diagrams of burners for atomic flame spectroscopy. (a) Turbulent-flow, or total-consumption, burner. (Courtesy of Beckman Instruments, Fullerton, CA.) (b) Laminar (premix) burner. (Courtesy of Perkin-Elmer Corporation, Norwalk, CT.)

a lower rate of sample introduction (which may offset the longer path length advantage) and the possibility of selective evaporation of mixed solvents in the mixing chamber, which can lead to analytical uncertainties. Furthermore, the mixing chamber contains a potentially explosive mixture that can be ignited by a flashback. Note that the burner in Figure 21-2b is equipped with pressure-relief vents for this reason. In addition, the burner head is sometimes held in place by stainless steel cables.

21B-2 *Emission and Absorption Spectra in Flames*

Emission and absorption spectra for both atoms and elementary ions are obtained from flames. As shown in Section 18D-1, atomic emission spectra are produced when an atom or ion excited by the absorption of energy from a hot source relaxes to its ground state by giving off a photon of radiation. In contrast, atomic absorption takes place when a gaseous atom or ion absorbs a photon of radiation from an external source and is thus excited. It is important to appreciate that, when the same electronic transition is involved, the energy of an emitted photon is *identical* to that of an absorbed photon. Thus, the wavelength of the emitted radiation is the same as the wavelength of the absorbed radiation.

All elements ionize to some degree in a flame, which leads to a mixture of atoms, ions, and electrons in the hot medium. For example, when a sample containing barium is atomized, the equilibrium

$$Ba \rightleftharpoons Ba^+ + e$$

is established in the inner cone of the flame. The position of this equilibrium depends on the temperature of the flame and total concentration of barium as well as on the concentration of the electrons produced from the ionization of *all elements* present in the sample. At the temperatures of the hottest flames (>3000 K), nearly half of the barium may be present in ionic form. The spectra of Ba and Ba^+ are, however, totally different from one another. Thus, in a high-temperature flame, two spectra for barium are excited, one for the atom and one for its ion. For this reason (and others), control of flame temperatures is important in flame spectroscopy.

In the discussion that follows, the words *atom* and *atomic* generally refer to elementary particles, be they atoms or simple ions.

21B-3 Methods for Obtaining Flame Spectra

In *flame emission spectroscopy*, the excited analyte ions serve as the source of radiation. That is, in contrast to all other types of spectroscopy we have thus far encountered, no *external* source of radiation is required in flame emission spectroscopy. Spectra are recorded by locating the inner cone of the flame in front of the entrance slit of a monochromator, such as one of those shown in Figure 19-5. The output from the exit slit is then monitored as the spectrum is scanned by rotating the grating or prism. A flame emission spectrum obtained in this way is shown in Figure 18-12.

In *atomic absorption spectroscopy*, the radiation from a special type of external source is passed through the inner cone of the flame, through a monochromator (or sometimes an interference filter), and to the surface of a radiation detector. In contrast to molecular absorption methods, atomic absorption methods *do not employ a continuous source of radiation* but instead use sources that emit *lines* of radiation that have the same wavelength as that of an absorption peak of the analyte. The need for and properties of this type of source are considered in Section 21B-6. A consequence of the use of line sources is that complete spectra seldom appear in the literature dealing with atomic absorption spectroscopy.

For *atomic fluorescence spectroscopy*, fluorescence is excited by a beam of radiation that enters the flame at right angles to the path to the entrance slit of the wavelength selector.

21B-4 Types of Flames

Several combinations of fuel and oxidant are employed in flame spectroscopy. Low-temperature flames (1750 to 1850°C), which are obtained with propane or natural gas as the fuel and air as the oxidant, have sufficient energy to provide satisfactory spectra for the alkali metals but are cool enough to preclude significant ionization and loss of sensitivity from this cause. Furthermore, few other elements are excited in this type of flame, and so spectra are simple and analyte lines readily isolated even with inexpensive glass filters.

Air/acetylene flames, which have temperatures in the 2200 to 2400°C range, are useful for many atomic absorption methods. This mixture is not satisfactory, however, with elements such as aluminum, silicon, the alkaline earths, and vanadium, which form refractory oxides that are incompletely atomized at these temperatures.

To obtain emission spectra for most of the elements, acetylene is employed as the fuel with oxygen or nitrous oxide as oxidant; these mixtures produce flames having temperatures of 2950 to 3050°C.

Air/hydrogen (2100°C) and oxygen/hydrogen (2700°C) flames are useful for observing lines in the shorter-wavelength ultraviolet because they are transparent in this region, whereas acetylene and other hydrocarbon flames are not.

21B-5 The Effects of Flame Temperature

Both emission and absorption spectra are affected in a complex way by variations in flame temperature. One effect common to the two methods is that increases in atomization efficiency and thus in the total atom population of the flame accompany temperature increases; enhanced sensitivity is the consequence. As noted earlier, however, this increase in atom population is more than offset by the number of atoms lost to ionization with certain elements.

Another effect of elevated temperatures is line broadening, which is accompanied by a decrease in peak height.

Flame temperature also determines the relative number of excited and unexcited atoms in a flame. In an air/acetylene flame, for example, the ratio of excited to unexcited magnesium atoms can be computed to be about 10^{-8}, whereas in an oxygen/acetylene flame, which is about 700°C hotter, this ratio is larger by a factor of about 70. Control of temperature is thus of prime importance in flame emission methods. For example, with a 2500°C flame, a temperature increase of 10°C causes the number of sodium atoms in the excited $3p$ state to increase by about 3%. In contrast, the corresponding *decrease* in the much larger number of ground-state atoms is only about 0.002%. Therefore, emission methods, based as they are on the population of *excited atoms,* require much closer control of flame temperature than do absorption procedures, in which the analytical signal depends upon the number of *unexcited atoms.*

The number of unexcited atoms in a typical flame exceeds the number of excited ones by a factor of 10^3 to 10^{10} or more. This fact suggests that absorption methods should be significantly more sensitive than emission methods. In fact, however, several other variables also influence sensitivity, and the two methods tend to complement each other in this regard. Table 21-2 illustrates this point.

21B-6 Atomic Absorption Spectroscopy Based on Flames

Flame atomic absorption spectroscopy is currently the most widely used of all the atomic methods listed in Table 21-1 because of its simplicity, effectiveness, and relatively low cost.

TABLE 21-2

Comparison of Detection Limits for Various Elements by Flame Absorption and Flame Emission Methods

Flame Emission More Sensitive	Sensitivity About the Same	Flame Absorption More Sensitive
Al, Ba, Ca, Eu, Ga, Ho, In, K, La, Li, Lu, Na, Nd, Pr, Rb, Re, Ru, Sm, Sr, Tb, Tl, Tm, W, Yb	Cr, Cu, Dy, Er Gd, Ge, Mn, Mo, Nb, Pd, Rh, Sc, Ta, Ti, V, Y, Zr	Ag, As, Au, B, Be, Bi, Cd, Co, Fe, Hg, Ir, Mg, Ni, Pb, Pt, Sb, Se, Si, Sn, Te, Zn

Adapted with permission from E. E. Pickett and S. R. Koirtyohann, *Anal. Chem.*, **1969,** *41* (14), 42A. Copyright 1969 American Chemical Society.

Atomic Absorption Spectra

Typically, the absorption of radiation from an external radiation source by an atomic species in a flame takes the form of a series of narrow peaks (lines) that are the result of transitions of an electron from the ground state to one of several higher energy levels (Section 18C-1). It is important to emphasize that the wavelength of an absorbed line *is identical to* the wavelength of the line that is emitted when the electron returns from the excited state to the ground state. That is, atomic absorption involves *resonance lines* (page 477).

Atomic Absorption Line Widths

The natural width of an atomic absorption or an atomic emission line is on the order of 10^{-5} nm. Two effects, however, cause observed line widths to be broadened by a factor of 100 (or more). As will become apparent shortly, line broadening is an important consideration in the design of atomic absorption instruments.

Doppler Broadening. Doppler broadening results from the rapid motion of atoms as they emit or absorb radiation. Atoms moving toward the detector emit wavelengths that are slightly shorter than the wavelengths emitted by atoms moving at right angles to the detector. This difference is a manifestation of the well-known Doppler shift; the effect is reversed for atoms moving away from the detector. The net effect is an increase in the width of the emission line. For precisely the same reason, the Doppler effect also causes broadening of absorption lines. This type of broadening becomes more pronounced as the flame temperature increases because of the consequent increased rate of motion of the atoms.

Pressure Broadening. Pressure broadening arises from collisions among atoms that result in slight variations in their ground-state energies and thus slight energy differences between ground and excited states. Like Doppler broadening, pressure broadening becomes greater with increases in temperature. Therefore, broader absorption and emission peaks are always encountered at elevated temperatures.

The Effect of Narrow Line Widths on Absorbance Measurements

Since transition energies for atomic absorption lines are unique for each element, analytical methods based upon atomic absorption have the potential of being highly specific. On the other hand, the narrow lines create a problem in quantitative analysis that is not encountered in molecular absorption. No ordinary monochromator is capable of yielding a band of radiation as narrow as the peak width of an atomic absorption line (0.002 to 0.005 nm). Consequently, the use of radiation that has been isolated from a continuous source by a monochromator inevitably causes instrumental departures from Beer's law (see the discussion of instrument deviations from Beer's law in Section 18C-7). In addition, since the fraction of radiation absorbed from such a beam is small, the detector receives a signal that is only slightly attenuated (that is, $P \rightarrow P_0$) and the sensitivity of the measurement is reduced.

The problem created by narrow absorption peaks has been surmounted by using radiation from a source that emits not only a *line of the same wavelength* as the one selected for absorption measurements but also one that is *narrower.* For example, a mercury vapor lamp is selected as the external radiation source for the determination of mercury. Gaseous mercury atoms electrically excited in such a lamp return to the ground state by *emitting* radiation with wavelengths that are identical to the wavelengths *absorbed* by the analyte mercury atoms in the flame. Since the lamp is operated at a temperature lower than that of the flame, the Doppler and pressure broadening of the mercury emission lines from the lamp is less than the corresponding broadening of the analyte absorption peaks in the hot flame that holds the sample. As a consequence, the effective bandwidths of the lines emitted by the lamp are significantly less than the corresponding bandwidths of the absorption peaks for the analyte in the flame.

Figure 21-3 illustrates the strategy generally used in measuring absorbances in atomic absorption methods. Figure 21-3a depicts four narrow *emission* lines from a typical atomic lamp source. Also shown is how one of these lines is isolated by a filter or monochromator. Figure 21-3b shows the flame *absorption spectrum* for the analyte between the wavelengths λ_1 and λ_2; note that the width of the absorption peak in the flame is significantly greater than the width of the emission line from the lamp. As shown in Figure 21-3c, the intensity of the incident beam P_0 has been decreased to P by passage through the sample. Since the bandwidth of the emission line from the lamp is now significantly less than the bandwidth of the absorption peak in the flame, log P_0/P is likely to be linearly related to concentration.

Line Sources for Atomic Absorption Spectroscopy

Two types of lamps are used in atomic absorption instruments: *hollow-cathode lamps* and *electrodeless discharge lamps.*

Hollow-Cathode Lamps. The most useful radiation source for atomic absorption spectroscopy is the *hollow-cathode lamp,* shown schematically in Figure 21-4. It consists of a tungsten anode and a cylindrical cathode sealed in a glass tube containing an inert gas, such as argon, at a pressure of 1 to 5 torr. The cathode is either fabricated from the analyte metal or else serves as a support for a coating of that metal.

The application of a potential of about 300 V across the electrodes causes ionization of the argon and generation of a current of 5 to 10 mA as the argon cations and electrons migrate to the two electrodes. If the potential is sufficiently large, the argon cations strike the cathode with sufficient energy to dislodge some of the metal atoms and thereby produce an atomic cloud; this

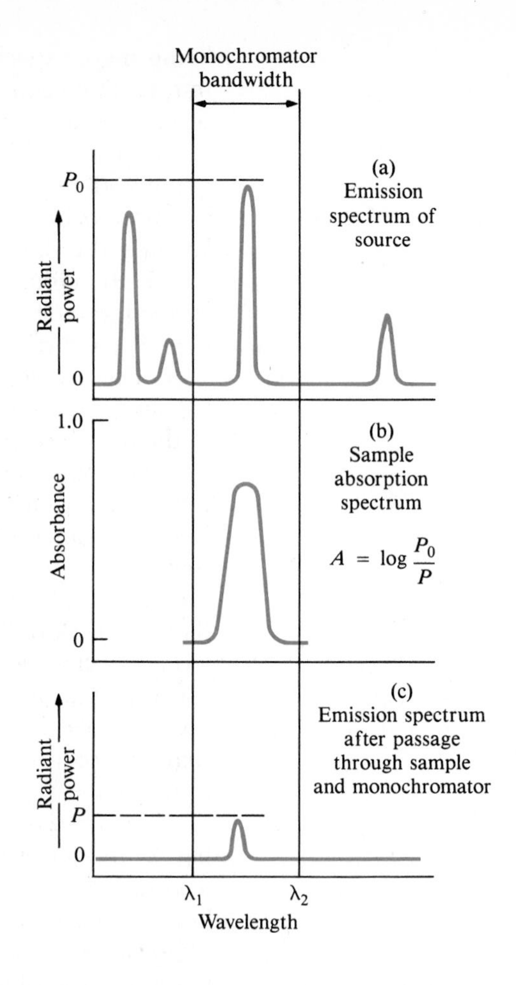

FIGURE 21-3 Atomic absorption of a resonance line.

process is called *sputtering.* A fraction of the sputtered metal atoms is in an excited state and emit their characteristic wavelengths as they return to the ground state. It is important to recall that the atoms producing emission lines in the lamp are at a significantly lower temperature than the analyte atoms in the flame. Thus the emission lines from the lamp are broadened less than the absorption peaks in the flame.

The sputtered metal atoms in a lamp eventually diffuse back to the cathode surface (or to the walls of the lamp) and are deposited.

Hollow-cathode lamps for about 40 elements are available from commercial sources. Some are fitted with a cathode containing more than one element; such lamps provide spectral lines for the determination of several

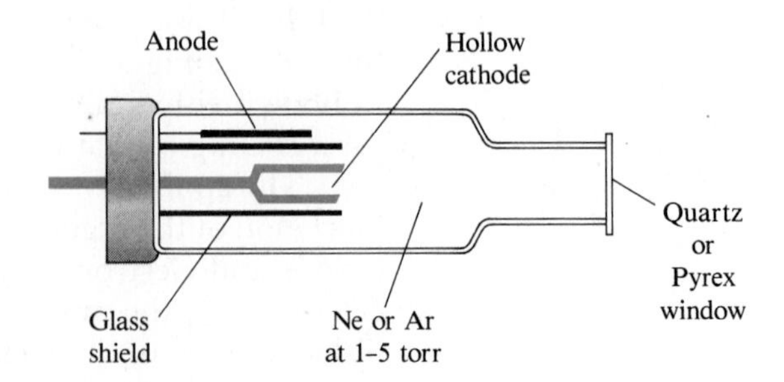

FIGURE 21-4 Diagram of a hollow-cathode lamp.

species. The development of the hollow-cathode lamp is widely regarded as the single most important event in the evolution of *atomic* absorption spectroscopy.

Electrodeless Discharge Lamps. Electrodeless discharge lamps are useful sources of atomic line spectra and provide radiant intensities that are usually one to two orders of magnitude greater than their hollow-cathode counterparts. A typical lamp is constructed from a sealed quartz tube containing an inert gas, such as argon, at a pressure of a few torr and a small quantity of the analyte metal (or its salt). The lamp contains no electrode but instead is energized by an intense field of radio-frequency or microwave radiation. The argon ionizes in this field, and the ions are accelerated by the high-frequency component of the field until they gain sufficient energy to excite (by collision) the atoms of the metal whose spectrum is sought.

Electrodeless discharge lamps are available commercially for several elements. Their performance does not appear to be as reliable as that of the hollow-cathode lamp.

Source Modulation

In an atomic absorption measurement, it is necessary to discriminate between radiation from the hollow-cathode or electrodeless discharge lamp and radiation from the flame. Much of the latter is eliminated by the monochromator, which is always located between the flame and the detector. The thermal excitation of a fraction of the analyte atoms in the flame, however, produces radiation of the wavelength at which the monochromator is set. Since such radiation is not removed, it acts as a potential source of interference.

The effect of analyte emission is overcome by *modulating* the output from the hollow-cathode lamp so that its intensity fluctuates at a constant frequency. The detector thus receives an alternating signal from the hollow-cathode lamp and a continuous signal from the flame and converts these signals into the corresponding types of electric current. A relatively simple electronic system then eliminates the unmodulated dc signal produced by the flame and passes the ac signal from the source to an amplifier and finally to the readout device.

Modulation is most often accomplished by interposing a motor-driven circular chopper *between the source and the flame* (Figure 21-5). Segments of the metal chopper have been removed so that radiation passes through the device half of the time and is reflected the other half. Rotation of the chopper at a constant speed causes the beam reaching the flame to vary periodically from

FIGURE 21-5 Optical paths in a double-beam atomic absorption spectrophotometer.

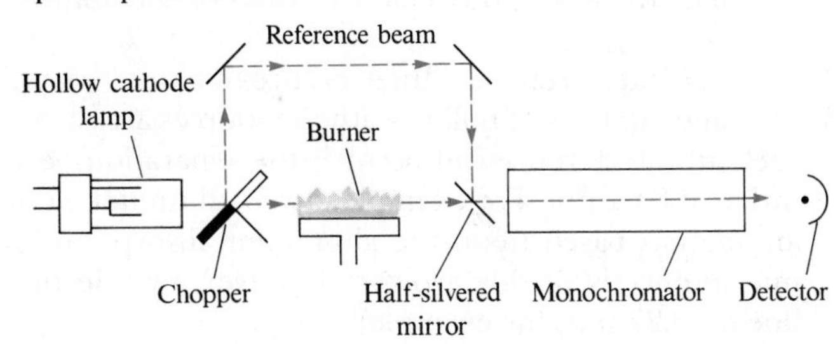

zero intensity to some maximum intensity and then back to zero. As an alternative, the power supply for the source can be designed for intermittent (or ac) operation.

Instruments

An atomic absorption instrument contains the same basic components as an instrument designed for molecular absorption measurements: a source, a sample container (here, a flame reservoir), a wavelength selector, and a detector/readout system. Both single- and double-beam instruments are offered by numerous manufacturers. The range of sophistication and the cost (upward from a few thousand dollars) are both substantial.

Photometers. As a minimum, an instrument for atomic absorption spectroscopy must be capable of providing a sufficiently narrow bandwidth to isolate the line chosen for a measurement from other lines that may interfere with or diminish the sensitivity of the method. A photometer equipped with a hollow-cathode source and glass filters is satisfactory for measuring concentrations of the alkali metals, which have only a few widely spaced resonance lines in the visible region. A more versatile photometer is sold with readily interchangeable interference filters and lamps. A separate filter and lamp are used for each element. Satisfactory results for the analysis of 22 metals are claimed.

Spectrophotometers. Most atomic absorption measurements are made with instruments equipped with an ultraviolet/visible grating monochromator. Figure 21-5 is a schematic of a typical double-beam instrument. Radiation from the hollow-cathode lamp is chopped and mechanically split into two beams, one of which passes through the flame and the other around the flame. A half-silvered mirror returns both beams to a single path, by which they pass alternately through the monochromator and to the detector. The signal processor then separates the ac signal generated by the chopped light source from the dc signal produced by the flame. The logarithm of the ratio of the reference and sample components of the ac signal is then computed and sent to the readout device for display as absorbance.

Interferences

Two types of interference are encountered in atomic absorption methods. *Spectral interferences* occur when particulate matter from the atomization scatters the incident radiation from the source or when the absorption or emission of an interfering species either overlaps or is so close to the analyte wavelength that resolution by the monochromator becomes impossible. *Chemical interferences* result from various chemical processes that occur during atomization and alter the absorption characteristics of the analyte.

Spectral Interferences. Interference due to overlapping lines is rare because the emission lines of hollow-cathode sources are so very narrow. Nevertheless, such an interference can occur if the separation between two lines is on the order of 0.01 nm. For example, a vanadium line at 308.211 nm interferes in an analysis based upon the aluminum absorption line at 308.215 nm. The interference is readily avoided, however, by selecting a different aluminum line (309.27 nm, for example).

Spectral interferences also result from the presence of either *molecular* combustion products that exhibit broad-band absorption or particulate products that scatter radiation. Both diminish the power of the transmitted beam and lead to positive analytical errors. Where the source of these products is the fuel/oxidant mixture alone, corrections are readily obtained from absorbance measurements made with a blank aspirated into the flame.

A much more troublesome problem is encountered when the source of absorption or scattering originates in the sample matrix. In this type of interference, the power of the transmitted beam P is reduced by the matrix components, but the incident beam power P_0 is not; a positive error in absorbance and thus concentration results. A potential matrix interference due to absorption occurs in the determination of barium in alkaline earth mixtures, for example. The wavelength of the barium line used for atomic absorption analysis appears in the center of a broad absorption *band* for molecular CaOH; interference by calcium in a barium analysis results. The net effect is readily eliminated by substituting nitrous oxide for air as the oxidant; the higher temperature decomposes the CaOH and eliminates the absorption band.

Spectral interference due to scattering by products of atomization often occurs when concentrated solutions containing elements such as titanium, zirconium, and tungsten—which form stable oxides—are aspirated into the flame. Metal oxide particles with diameters greater than the wavelength of light appear to be formed and cause scattering of the incident beam.

Fortunately, spectral interferences by matrix products are not widely encountered with flame atomization and usually can be avoided by variations in such analytical parameters as temperature and fuel-to-oxidant ratio. Alternatively, if the source of interference is known, an excess of the interfering substance can be added to both sample and standards; provided the excess is large with respect to the concentration from the sample matrix, the contribution of the latter will become insignificant. The added substance is sometimes called a *radiation buffer.*

The matrix-interference problem is greatly exacerbated with electrothermal atomization and is one of the major causes of the lower accuracy associated with nonflame methods.[2]

Chemical Interferences. Chemical interferences can frequently be minimized by a suitable choice of operating conditions. Perhaps the most common type of chemical interference is by anions that form compounds of low volatility with the analyte and thus decrease the rate at which it is atomized. Low results are the consequence. An example is the decrease in calcium absorbance observed with increasing concentrations of sulfate or phosphate ions, which form nonvolatile compounds with calcium ion.

Interferences due to the formation of species of low volatility can often be eliminated or moderated by use of higher temperatures. Alternatively, *releasing agents,* which are cations that react preferentially with the interference and prevent its interaction with the analyte, can be introduced. For example, the addition of excess strontium or lanthanum ion minimizes interference by

[2]For a discussion of methods for overcoming matrix interference problems, see D. A. Skoog, *Principles of Instrumental Analysis,* 3rd ed., pp. 270–273. Philadelphia: Saunders College Publishing, 1985.

phosphate in the determination of calcium. Here, the strontium or lanthanum replaces the analyte in the nonvolatile compound formed with the interfering species.

Protective agents prevent interference by preferentially forming stable but volatile species with the analyte. Three common reagents for this purpose are EDTA, 8-hydroxyquinoline, and APDC (the ammonium salt of 1-pyrrolidine-carbodithioic acid). For example, the presence of EDTA has been shown to eliminate interference by silicon, phosphate, and sulfate in the determination of calcium.

The ionization of atoms and molecules is usually inconsequential in combustion mixtures that involve air as the oxidant. In high-temperature flames, however, where oxygen or nitrous oxide serves as the oxidant, ionization becomes appreciable, and a significant concentration of free electrons exists as a consequence of the equilibrium

$$M \rightleftharpoons M^+ + e$$

where M represents a neutral atom or molecule and M^+ is its ion. Ordinarily, the spectrum of M^+ is quite different from that of M, so that ionization of the analyte ions leads to low results. It is important to appreciate that treating the ionization process as an equilibrium—with free electrons as one of the products—implies that the degree of ionization of an analyte atom is strongly influenced by the presence of other ionizable metals in the flame. Thus, if the medium contains not only species M but species B as well, and if B ionizes according to the equation

$$B \rightleftharpoons B^+ + e$$

then the degree of ionization of M is decreased by the mass-action effect of the electrons formed from B. The errors caused by analyte ionization can frequently be eliminated by addition of an *ionization suppressor,* which provides a relatively high concentration of electrons to the flame; suppression of analyte ionization results. Potassium salts are frequently used as ionization suppressors because of the low ionization energy of this element.

Applications of Atomic Absorption Spectroscopy

Atomic absorption spectroscopy provides a sensitive means of determining more than 60 elements. The method is well suited for routine measurements by relatively unskilled operators.

Quantitative Techniques. Quantitative atomic absorption methods are usually based on calibration curves, which, in principle, are linear. Departures from linearity occur, however, and analyses should *never* be based on the measurement of a single standard with the assumption that Beer's law is being followed. In addition, the production of an atomic vapor involves sufficient uncontrollable variables to warrant measuring the absorbance of at least one standard solution each time an analysis is performed. Any deviation of the standard from its original calibration value can then be applied as a correction to the analytical results.

The standard-addition method is extensively used in atomic absorption spectroscopy. In this procedure, two or more aliquots of the sample are transferred to volumetric flasks. One is diluted to volume directly, and a known amount of analyte is introduced into the other before dilution to the same volume. The absorbance of each is measured (several different standard additions are recommended if the method is unfamiliar). If a linear relationship exists between absorbance and concentration (and this must be verified experimentally), the following relationships apply:

$$A_x = \frac{kV_xc_x}{V_T}$$

$$A_T = \frac{k(V_xc_x + V_sc_s)}{V_T}$$

where V_x and c_x are the volume and concentration of the analyte solution, V_s and c_s are the same parameters for the standard, and V_T is the total volume; A_x and A_T are the absorbances of the sample alone and the sample plus standard, respectively. These two equations are readily combined to give

$$c_x = \frac{A_x}{A_T - A_x} \times \frac{c_sV_s}{V_x}$$

If several standard additions have been made, A_T can be plotted against c_s. The resulting straight line can then be extrapolated to $A_T = 0$, at which point

$$c_x = \frac{-c_sV_s}{V_x}$$

Use of the standard-addition method tends to compensate for variations caused by physical and chemical interferences in the analyte solution.

Sensitivity and Accuracy. Columns 2 and 3 of Table 21-3 provide information on detection limits for a number of common elements by flame atomic absorption and also by electrothermal methods, which are discussed in Section 21C. For comparison purposes, limits for some of the other atomic procedures are also included. Small differences among the quoted values are not significant. An order of magnitude is probably meaningful, but a factor of 2 or 3 certainly is not.

Under usual conditions, the relative error associated with a flame absorption analysis is of the order of 1 to 2%. With special precautions, this figure can be lowered to a few tenths of one percent. Errors encountered with nonflame atomization usually exceed those for flame atomization by a factor of 5 to 10.

21B-7 *Flame Emission Spectroscopy*

Atomic emission spectroscopy employing a flame (also called flame emission spectroscopy or flame photometry) has found widespread application in elemental analysis. Its most important uses are in the determination of sodium, potassium, lithium, and calcium, particularly in biological fluids and tissues. Because of its convenience, speed, and relative freedom from interferences, flame emission spectroscopy has become the method of choice for these elements, which are otherwise difficult to determine. The method has also been applied, with various degrees of success, to the determination of perhaps half the elements in the periodic table.

In addition to its quantitative applications, flame emission spectroscopy is also useful for qualitative analysis. Complete spectra are readily recorded; identification of the elements present is then based upon the peak wavelengths,

TABLE 21-3

Detection Limits of Atomic Spectroscopy Methods for Selected Elements§

Element	Absorption, Flame*	Absorption, Electrothermal†	Emission, Flame*	Emission, ICP*‡
Al	30	0.005	5	2
As	100	0.02	0.0005	40
Ca	1	0.02	0.1	0.02
Cd	1	0.0001	800	2
Cr	3	0.01	4	0.3
Cu	2	0.002	10	0.1
Fe	5	0.005	30	0.3
Hg	500	0.1	0.0004	1
Mg	0.1	0.00002	5	0.05
Mn	2	0.0002	5	0.06
Mo	30	0.005	100	0.2
Na	2	0.0002	0.1	0.2
Ni	5	0.02	20	0.4
Pb	10	0.002	100	2
Sn	20	0.1	300	30
V	20	0.1	10	0.2
Zn	2	0.00005	0.0005	2

*From V. A. Fassel and R. N. Kniseley, *Anal. Chem.*, **1974**, *46*, 1111A. With permission of the American Chemical Society.

†From C. W. Fuller, *Electrothermal Atomization for Atomic Absorption Spectroscopy*, pp. 65–83. London: The Chemical Society, 1977. With permission of The Royal Society of Chemistry.

‡ICP = inductively coupled plasma.

§All values in nanograms/milliliter = 10^{-3} μg/mL = 10^{-3} ppm.

which are unique for each element. In this respect, flame emission has a clear advantage over flame absorption, which does not provide complete absorption spectra because of the discontinuous nature of the radiation sources that must be used.

Instrumentation

Instruments for flame emission work are similar in design to flame absorption instruments except that in the former the flame acts as the radiation source; a hollow-cathode lamp and chopper are therefore unnecessary. Some instruments can be configured for either emission or absorption measurements. Much of the early work in atomic emission analyses was accomplished with turbulent-flow burners. Laminar-flow burners, however, are becoming more and more widely used.

Photometers and Spectrophotometers. For nonroutine analysis, a recording ultraviolet/visible spectrophotometer with a resolution of perhaps 0.05 nm is desirable. Such an instrument can provide complete emission spectra that are useful for the identification of the elements present in a sample. Figure 18-12 is an example of a typical flame emission spectrum excited by an oxyhydrogen flame. The sample was a brine, and spectral lines and bands for several elements are identified. Figure 21-6 illustrates how background corrections are made with a recorded spectrum.

Simple filter photometers often suffice for routine determinations of the alkali and alkaline earth metals. A low-temperature flame is employed to prevent excitation of most other metals. As a consequence, the spectra are simple, and interference filters can be used to isolate the desired emission line.

Several instrument manufacturers supply flame photometers designed specifically for the analysis of sodium, potassium, and lithium in blood serum and other biological samples. In these instruments, the radiation from the flame is split into three beams of approximately equal power. Each beam then passes into a separate photometric system consisting of an interference filter (which transmits an emission line of one of the elements while absorbing those of the other two), a phototube, and an amplifier. The outputs can be measured separately if desired. Ordinarily, however, lithium serves as an *internal standard* for the analysis. For this purpose, a fixed amount of lithium is introduced into each standard and sample. The output ratio between the sodium transducer and the lithium transducer and the potassium transducer and the lithium transducer then serve as analytical parameters. This system provides improved accuracy because the intensities of the three lines are affected in the same way by most analytical variables, such as flame temperature, fuel flow rate, and background radiation. Clearly, lithium must be absent from the sample.

Automated Flame Photometers

Fully automated photometers based on the segmented-flow system described in Section 20D-2 are now widely used in clinical laboratories for the determination of sodium and potassium. In these instruments, samples are withdrawn sequentially from a sample turntable, dialyzed to remove protein and particulates, diluted with a lithium internal standard, and aspirated into a flame. Calibration is performed automatically after every nine samples.

Interferences

The interferences encountered in flame emission spectroscopy have the same sources as those encountered in atomic absorption methods (Section 21B-6); the severity of any given interference often differs for the two procedures, however.

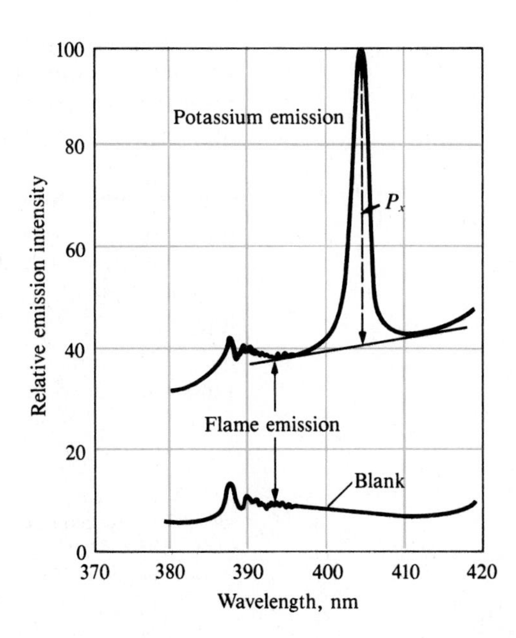

FIGURE 21-6 A portion of the emission spectrum for potassium in a hydrogen/oxygen flame. The potassium emission is superimposed on the background emission from the flame. For clarity, the spectrum for the blank has been displaced downward.

Analytical Techniques

The analytical techniques for flame emission spectroscopy are similar to those described for atomic absorption spectroscopy. Both calibration curves and the standard-addition method are employed. In addition, internal standards can be used to compensate for flame variables.

21C Atomic Absorption Methods with Electrothermal Atomizers

Flame atomization is not very efficient for two reasons. First, a large portion of the sample flows down the drain (laminar burner) or is not completely atomized (turbulent burner). Second, the residence time of individual atoms in the optical path of the flame is brief ($\simeq 10^{-4}$ s). Electrothermal atomizers, which first appeared on the market in the early 1970s, generally provide enhanced sensitivity because the entire sample is atomized in a short period, and the average residence time of the atoms in the optical path is 1 s or more.[3]

Other than the atomizer, the equipment needed for an electrothermal analysis is identical to that used in flame absorption methods. Most modern instruments are designed so that the change from one type of atomizer to the other is a relatively simple matter.

In an electrothermal atomizer, a few microliters of sample are first evaporated at a low temperature and then ashed at a somewhat higher temperature in an electrically heated tube or cup (usually fabricated from carbon). After ashing, the current is rapidly increased to several hundred amperes, which causes the temperature to soar to perhaps 2000 to 3000°C; atomization of the sample occurs in a period of a few milliseconds to seconds. The absorbance of the atomized particles is then determined in the region immediately above the heated conductor.

Figure 21-7a illustrates a commercially available electrothermal atomizer that fits in front of the entrance slit of a monochromator. In this device, the sample is contained in a graphite cup (A) supported between a pair of identical electrodes (B). The cup, which is surrounded by a sheath of inert gas, has holes on both sides to permit radiation from the hollow-cathode lamp to pass over the surface of the sample. The metal mounting (C) is water-cooled.

With the monochromator set at a wavelength at which absorption occurs, the detector output during electrothermal atomization rises to a maximum after a few seconds of ignition and then rapidly decays back to zero as the atomization products escape into the surroundings. The change is rapid enough (often <1 s) to require a high-speed recorder or a computerized data-acquisition system. Quantitative analyses are usually based on peak height, although peak area is also useful.

Figure 21-7b shows the output signals (as a function of time) from an atomic absorption spectrophotometer equipped with a graphite-cup atomizer. The four peaks on the right were recorded at the wavelength of a lead peak as a 2 μL sample of canned orange juice was dried, ashed, and atomized. The peaks produced during drying and ashing were caused by particulate ignition

[3]See R. E. Sturgeon, *Anal. Chem.*, **1977,** *49,* 1255A; S. R. Koirtyohann and M. L. Kaiser, *Anal. Chem.*, **1982,** *54,* 1515A; and C. W. Fuller, *Electrothermal Atomization for Atomic Absorption Spectroscopy.* London: The Chemical Society, 1978.

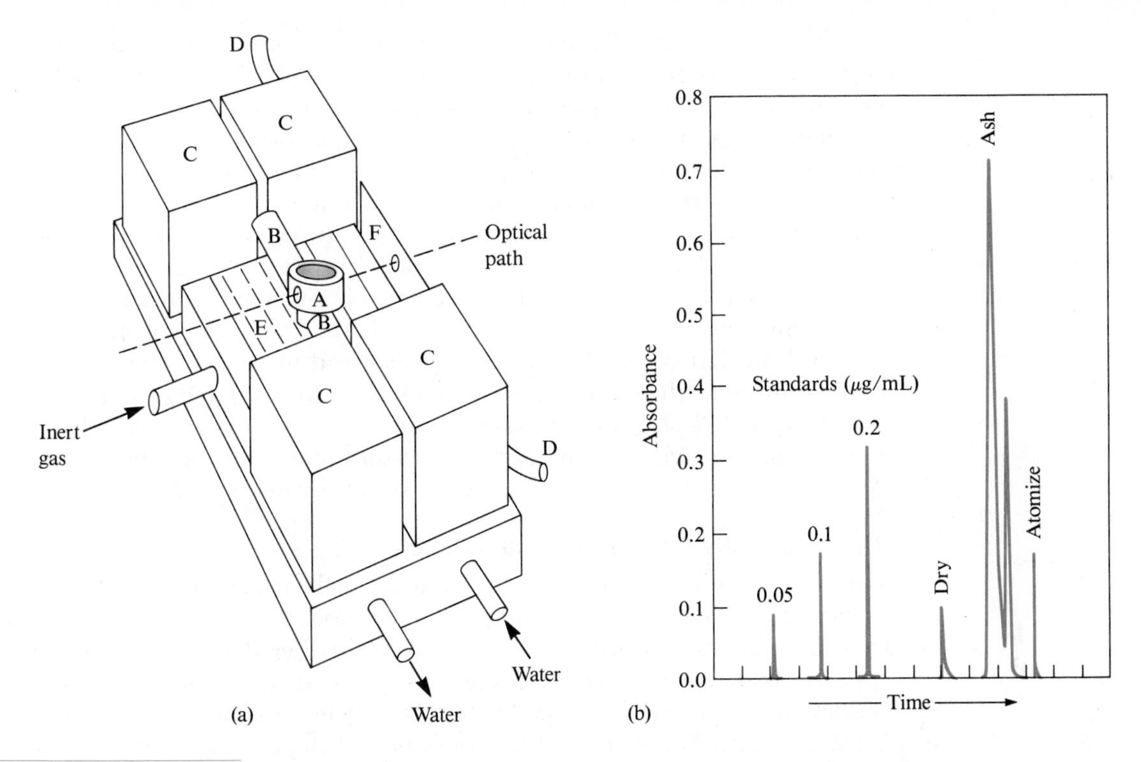

FIGURE 21-7 (a) Diagram of a graphite-cup atomizer. (b) Typical output from a spectrophotometer equipped with an electrothermal atomizer. The time for drying and ashing are 20 and 60 s, respectively. (Courtesy of Varian Instrument Division, Palo Alto, CA.)

products. Comparison of the lead peak on the far right with the peaks recorded for standards indicates a lead concentration of about 0.1 μg/mL of juice.

Electrothermal atomizers offer the advantage of unusually high sensitivity for small volumes of sample. Typically, sample volumes are between 0.5 and 10 μL; under these circumstances, absolute detection limits typically lie in the range from 10^{-10} to 10^{-13} g of analyte. The relative precision of electrothermal methods is generally in the range of 5 to 10%, compared with the 1 to 2% that can be expected for flame atomization. Furthermore, the interference problems discussed in Section 21B-6 tend to be more severe with electrothermal than with flame atomizers.

21D Atomic Emission Methods Based on Atomization in Plasmas

Plasma atomizers, which became available commercially in the mid-1970s, offer several advantages over flame atomizers.[4] Plasma atomization has been used for both thermal emission and fluorescence spectroscopy. It has not been widely used as an atomizer for atomic absorption methods.

[4]For a detailed discussion of the various plasma sources, see P. Tschopel, in *Comprehensive Analytical Chemistry,* G. Svehla, Ed., Vol. IX, Chapter 3. New York: Elsevier, 1979; R. D. Sacks, in *Treatise on Analytical Chemistry,* 2nd ed., P. J. Elving, E. J. Meehan, and I. M. Kolthoff, Eds., Part I, Vol. 7, pp. 516–526. New York: Wiley, 1981; J. D. Ingle Jr., and S. R. Crouch, *Spectrochemical Analysis.* Englewood Cliffs, N.J.: Prentice-Hall, 1988.

By definition, a plasma is a conducting gaseous mixture containing a significant concentration of cations and electrons. In the argon plasma employed for emission analyses, argon ions and electrons are the principal conducting species, although cations from the sample also contribute. Argon ions, once formed in a plasma, are capable of absorbing sufficient power from an external source to maintain the temperature at a level at which further ionization sustains the plasma indefinitely; temperatures as great as 10,000 K are encountered. Three power sources have been employed in argon plasma spectroscopy. One is a dc electrical source capable of maintaining a current of several amperes between electrodes immersed in the argon plasma. The second and third are powerful radio-frequency and microwave-frequency generators through which the argon flows. Of the three, the radio-frequency, or *inductively coupled plasma* (ICP), source appears to offer the greatest advantage in terms of sensitivity and freedom from interference. On the other hand, the *dc plasma source* (DCP) has the virtues of simplicity and lower cost.

21D-1 The Inductively Coupled Plasma Source[5]

Figure 21-8a is a schematic drawing of an inductively coupled plasma source. It consists of three concentric quartz tubes through which streams of argon flow at a total rate of between 11 and 17 L/min. The diameter of the largest tube is about 2.5 cm. Surrounding the top of this tube is a water-cooled induction coil powered by a radio-frequency generator capable of producing 2 kW of energy at about 27 MHz. Ionization of the flowing argon is initiated by a spark from a Tesla coil. The resulting ions, and their associated electrons, then interact with the fluctuating magnetic field (labeled H in Figure 21-8) produced by the induction coil. This interaction causes the ions and electrons within the coil to flow in the closed annular paths depicted in the figure; ohmic heating is the consequence of their resistance to this movement.

The temperature of this plasma is high enough to require that it be thermally isolated from the outer quartz cylinder. Isolation is achieved by flowing argon tangentially around the walls of the tube, as indicated by the arrows in Figure 21-8a. The tangential flow cools the inside walls of the central tube and centers the plasma radially.

Sample Injection

The sample is carried into the hot plasma at the head of the tubes by argon flowing at about 1 L/min through the central quartz tube. The sample can be an aerosol, a thermally generated vapor, or a fine powder.

Plasma Appearance and Spectra

The typical plasma has a very intense, brilliant white, nontransparent core topped by a flamelike tail. The core, which extends a few millimeters above the tube, produces a spectral continuum upon which is superimposed the atomic spectrum for argon. The continuum apparently results when argon and other ions recombine with electrons. In the region 10 to 30 mm above the core, the continuum fades and the plasma is optically transparent. Spectral

[5]For a more complete discussion, see V. A. Fassel, *Science,* **1978,** *202,* 183; V. A. Fassel, *Anal. Chem.,* **1979,** *51,* 1290A; M. Thompson and J. N. Walsh, *Inductively Coupled Plasma Spectrometry.* London: Blackie, 1983; R. M. Barnes, *CRC Crit. Rev. Anal. Chem.,* **1978,** *7,* 203; J. D. Ingle Jr. and S. R. Crouch, *Spectrochemical Analysis,* Chapter 8. Englewood, NJ: Prentice-Hall, 1988.

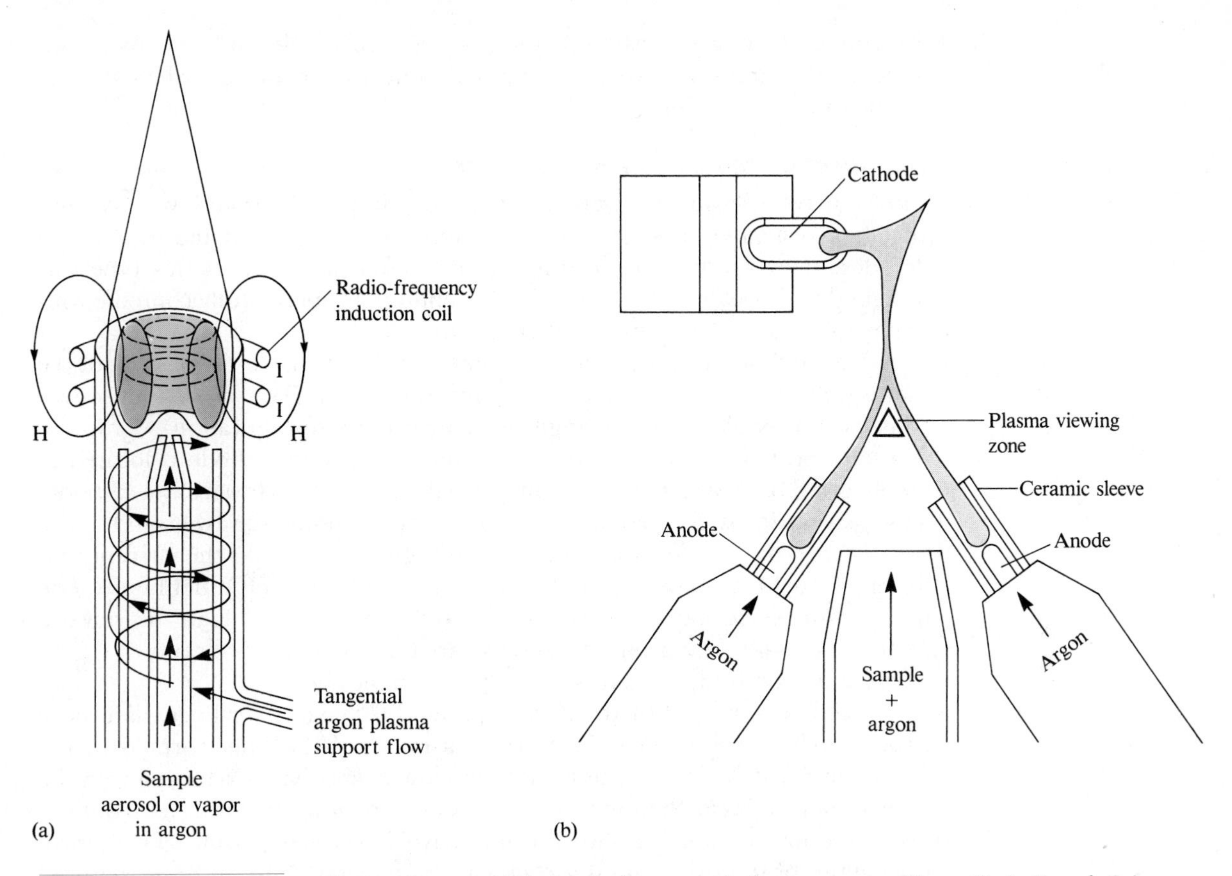

FIGURE 21-8 Plasma sources. (a) Inductively coupled plasma source. (From V. A. Fassel, *Science*, **1978,** *202,* 185. With permission. Copyright 1978 by the American Association for the Advancement of Science.) (b) A three-electrode dc plasma jet. (Courtesy of Applied Research Laboratories, Inc., Sunland, CA.)

observations are generally made 15 to 20 mm above the induction coil. Here, the background radiation is remarkably free of argon lines and is well suited to spectral measurements. Many of the most sensitive analyte lines in this region of the plasma are from ions such as Ca^+, Cd^+, Cr^+, and Mn^+.

Analyte Atomization and Ionization

By the time the sample atoms reach the observation point in the plasma, they have had a residence time of about 2 ms at temperatures ranging from 6000 to 8000 K. These times and temperatures are two to three times as great as those attainable in the hottest combustion flames (acetylene/nitrous oxide). As a consequence, atomization is more nearly complete, and fewer chemical interferences are encountered. Surprisingly, ionization interference effects are small or nonexistent, perhaps because the large concentration of electrons from the ionization of the argon maintains a more or less constant electron concentration in the plasma.

Several other advantages are associated with the plasma source. First, atomization occurs in a chemically inert environment, which should also enhance the lifetime of the analyte. In addition, and in contrast to flame sources, the temperature cross section of the plasma is relatively uniform. As a consequence, calibration curves tend to remain linear over several orders of magnitude of concentration.

21D-2 The Direct-Current Argon Plasma Source

Direct-current plasma jets were first described in the 1920s and were systematically investigated as sources for emission spectroscopy for more than two decades. It was not until recently, however, that a source of this type was designed that provides data reproducible enough to successfully compete with flame and inductively coupled plasma sources.[6]

Figure 21-8b is a diagram of a commercially available dc plasma source that is well suited to the excitation of emission spectra. This plasma-jet source consists of three electrodes arranged in an inverted Y configuration. A graphite anode is located in each arm of the Y, and a tungsten cathode is located at the inverted base. Argon flows from the two anode blocks toward the cathode. The plasma jet is formed when the cathode is momentarily brought into contact with the anodes. Ionization of the argon occurs, and the current that develops ($\simeq$ 14 A) generates additional ions to sustain itself indefinitely. The temperature is perhaps 10,000 K in the arc core and 5000 K in the viewing region. The sample is aspirated into the area between the two arms of the Y, where it is atomized, excited, and its spectrum viewed.

Spectra produced by the plasma jet tend to have fewer lines than those produced by the inductively coupled plasma, and the lines formed in the former are largely from atoms rather than ions. Sensitivities achieved with the dc plasma jet appear to range from an order of magnitude lower to about the same as those obtainable with the inductively coupled plasma. The reproducibilities of the two systems are similar. Significantly less argon is required for the dc plasma, and the auxiliary power supply is simpler and less expensive. On the other hand, the graphite electrodes must be replaced every few hours, whereas the inductively coupled plasma source requires little or no maintenance.

21D-3 Instruments for Plasma Spectroscopy

Several manufacturers offer instruments for plasma emission spectroscopy. In general, these consist of a high-quality grating spectrophotometer for the ultraviolet and visible regions with a photomultiplier detector. Many are automated, so that an entire spectrum can be scanned sequentially. Others have several photomultiplier tubes located in the focal plane, so that the lines for several elements (two dozen or more) can be monitored simultaneously. Such instruments are very expensive.

21D-4 Quantitative Applications of Plasma Sources

Unquestionably, inductively coupled and dc plasma sources yield significantly better quantitative analytical data than other emission sources. The excellence of these results stems from the high stability, low noise, low background, and freedom from interferences of the sources when operated under appropriate experimental conditions. The performance of the inductively coupled plasma

[6]For additional details, see G. W. Johnson, H. E. Taylor, and R. K. Skogerboe, *Anal. Chem.*, **1979**, *51*, 2403; *Spectrochim. Acta, Part B*, **1979**, *34*, 197; J. Reednick, *Amer. Lab.*, **1979**, *11*(3) 53.

source is somewhat better than that of the dc plasma source in terms of detection limits. The latter, however, is less expensive to purchase and operate and is entirely adequate for many applications.

In general, the detection limits with the inductively coupled plasma source appear comparable to or better than those of other atomic spectral procedures. Table 21-3 compares the sensitivity of several of these methods.

21D-5 *Atomic Fluorescence Methods Based on Plasma Atomization*

Atomic fluorescence analyses can be performed with a relatively simple instrument when a hollow-cathode lamp is used to excite analyte atoms. Under this circumstance, no wavelength selector is needed because the only atoms in a sample that are excited are those of the element whose spectrum is emitted by the lamp. For example, the radiation from a cadmium hollow-cathode lamp contains unique lines that can excite only cadmium atoms in a sample. Thus it is possible to determine cadmium selectively with an instrument consisting of just a hollow-cathode lamp, a flame or plasma source, and a photomultiplier tube. The selectivity here is provided not by a monochromator, as is usually the case, but by the single-element nature of the radiation emitted by the hollow-cathode source.

A nondispersive instrument that permits the simultaneous determination of 12 elements was recently described.[7] This instrument is much simpler in design than most multielement instruments. Figure 21-9 is a diagram of one of the 12 source/detector modules that are arranged in a circle around a central inductively coupled plasma source. Each module consists of a hollow-cathode excitation source that emits radiation for one of the elements to be determined, an interference filter (needed to remove background radiation from the plasma), and a photomultiplier detector. Here, in contrast to the usual design, fluorescence is observed at a vertical angle of about 45 deg. The lamp is pulsed

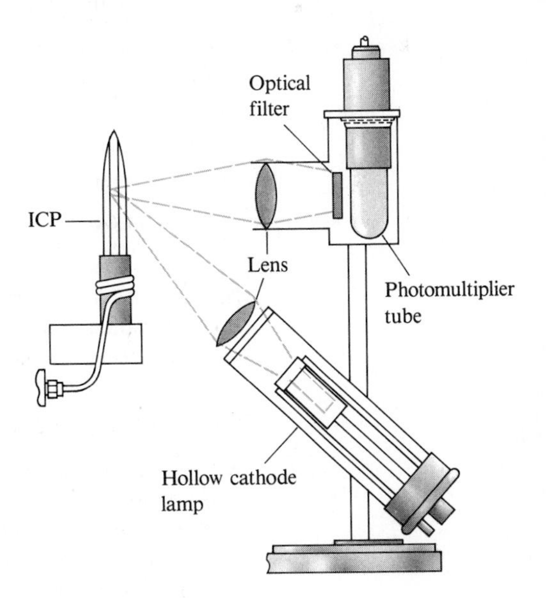

FIGURE 21-9 A source/detector module for a multielement fluorescence instrument based upon inductively coupled plasma atomization. (Courtesy of Baird Corporation, Bedford, MA.)

[7] D. R. Demers and C. D. Allemand, *Anal. Chem.*, 1981, *53*, 1915; D. R. Demers, D. A. Busch, and C. D. Allemand, *Amer. Lab.*, 1982, *14*(3), 167.

at a frequency of about 500 Hz, and the detection system is synchronized to receive these pulses. The timing of the pulses for the 12 lamps is such that at any given instant, only one fluorescence signal is produced and detected.

The instrument is highly versatile because each module can be prealigned to observe fluorescence in a part of the plasma that is optimal for a given element. Rapid sequential measurement of the fluorescence of 12 elements under the best conditions for each is thus possible. Furthermore, modules can be readily interchanged in 1 min or less and are ready to operate after a 10-min warm-up period.

Questions and Problems

*21-1 Describe the basic differences between atomic emission and atomic fluorescence spectroscopy.

21-2 Define *(a) atomization, (b) pressure broadening, *(c) Doppler broadening, (d) turbulent-flow nebulizer, *(e) laminar-flow nebulizer, (f) hollow-cathode lamp, *(g) sputtering, (h) ionization suppressor, *(i) spectral interference, (j) chemical interference, *(k) radiation buffer, (l) releasing agent, *(m) protective agent.

*21-3 Why is atomic emission more sensitive to flame instability than atomic absorption or fluorescence?

21-4 Why is a nonflame atomizer more sensitive than a flame atomizer?

*21-5 Why is source modulation employed in atomic absorption spectroscopy?

21-6 In a hydrogen/oxygen flame, an atomic absorption peak for iron decreased in the presence of large concentrations of sulfate ion.
- (a) Suggest an explanation for this observation.
- (b) Suggest three possible methods for overcoming the potential interference of sulfate in a quantitative determination of iron.

*21-7 Why are the lines from a hollow-cathode lamp generally narrower than the lines emitted by atoms in a flame?

21-8 Assume that the peaks shown in Figure 21-7 were obtained for 2-μL aliquots of standards and sample. Calculate the parts per million of lead in the sample of canned orange juice.

*21-9 Suggest sources of the two peaks in Figure 21-7 that appear during drying and ashing.

21-10 In the concentration range from 500 to 2000 ppm of U, a linear relationship is found between absorbance at 351.5 nm and concentration. At lower concentrations, the relationship becomes nonlinear unless about 2000 ppm of an alkali metal salt is introduced. Explain.

*21-11 What is the purpose of an internal standard in flame emission methods?

*21-12 A 5.00-mL sample of blood was treated with trichloroacetic acid to precipitate proteins. After centrifugation, the resulting solution was brought to pH 3 and extracted with two 5-mL portions of methyl isobutyl ketone containing the organic lead-complexing agent APCD. The extract was aspirated directly into an air/acetylene flame and yielded an absorbance of 0.502 at 283.3 nm. Five-milliliter aliquots of standard solutions containing 0.400 and 0.600 ppm of lead were treated in the same way and yielded absorbances of 0.396 and 0.599. Calculate the parts per million of lead in the sample assuming that Beer's law is followed.

21-13 The sodium in a series of cement samples was determined by flame emission spectroscopy. The flame photometer was calibrated with a series of standards containing 0, 20.0, 40.0, 60.0, and 80.0 μg Na_2O per milliliter. The instrument readings for these solutions were 3.1, 21.5, 40.9, 57.1, and 77.3.

(a) Plot the data.
(b) Derive a least-squares line for the data.
(c) Calculate standard deviations for the slope and about regression for the line in (b).
(d) The following data were obtained for replicate 1.000-g samples of cement dissolved in HCl and diluted to 100.0 mL after neutralization.

	Emission Reading			
	Blank	Sample A	Sample B	Sample C
Replicate 1	5.1	28.6	40.7	73.1
Replicate 2	4.8	28.2	41.2	72.1
Replicate 3	4.9	28.9	40.2	Spilled

Calculate the % Na_2O in each sample. What are the absolute and relative standard deviations for the average of each determination?

*21-14 The chromium in an aqueous sample was determined by pipetting 10.0 mL of the unknown into each of five 50.0-mL volumetric flasks. Various volumes of a standard containing 12.2 ppm Cr were added to the flasks, and the solutions were then diluted to volume.

Unknown, mL	Standard, mL	Absorbance
10.0	0.0	0.201
10.0	10.0	0.292
10.0	20.0	0.378
10.0	30.0	0.467
10.0	40.0	0.554

(a) Plot absorbance as a function of volume of standard V_s.
(b) Derive an expression relating absorbance to the concentrations of standard and unknown (C_s and C_x) and the volumes of the standards and unknown (V_s and V_x) as well as the volume to which the solutions were diluted (V_t).
(c) Derive expressions for the slope and the intercept of the straight line obtained in (a) in terms of the variables listed in (b).
(d) Show that the concentration of the analyte is given by the relationship $C_x = aC_s/bV_x$, where a and b are the slope and the intercept of the straight line in (a).
(e) Determine values for a and b by the method of least squares.
(f) Calculate the standard deviation for the slope and about regression in (e).
(g) Calculate the ppm Cr in the sample using the relationship given in (d).
(h) The relative variance of a result obtained in this way is approximately equal to the sum of the relative variances of the slope and the intercept. Calculate the absolute standard deviation for the result obtained in (g).

CHAPTER 22

Kinetic Methods of Analysis

Kinetic methods of analysis differ in a fundamental way from the equilibrium, or thermodynamic, methods we have dealt with so far. In kinetic methods, measurements are made under *dynamic* conditions in which the concentrations of reactants and products are changing continuously. Here, the rate of appearance of products or disappearance of reactants serves as the analytical parameter. The measurements in thermodynamic methods, in contrast, are performed upon systems that have come to equilibrium so that concentrations are *static*.

The distinction between the two types of methods is illustrated in Figure 22-1, which shows the progress over time of the reaction

$$A + R \rightleftharpoons P \qquad (22\text{-}1)$$

where A represents the analyte, R the reagent, and P the product. Thermodynamic methods operate in the region beyond time t_e where the bulk concentrations of reactants and product have become constant and the chemical system is at equilibrium. In contrast, kinetic methods are carried out during the time interval from 0 to t_e where reactant and product concentrations are varying continuously, and the rate of their disappearance or appearance is conveniently measured.

Selectivity is achieved in kinetic methods by choosing reagents and conditions that amplify differences in the *rates* at which the analyte and potential interferences react. Selectivity in thermodynamic methods is realized by choosing reagents and conditions that amplify differences in *equilibrium constants.*

Kinetic methods greatly extend the number of chemical reactions that can be used for analytical purposes because they permit the use of reactions that are too slow or too incomplete for thermodynamics-based procedures.

Two types of kinetic methods are based upon catalyzed reactions. In the first, an analyte is the catalyst and is determined from its catalytic effect upon an *indicator reaction* involving reactants or products that are easily measured. Such methods are among the most sensitive in the chemist's repertoire. In the second type, a catalyst is introduced to hasten the reaction between analyte and reagent. This approach is often highly selective, or even specific, particularly when an enzyme serves as the catalyst.

Undoubtedly, the most widespread use of kinetic methods is in biochemical and clinical laboratories, where the number of analyses based upon kinetics exceeds those based upon thermodynamics.

22A Rates of Chemical Reactions; Rate Laws

This section provides a brief introduction to the theory of chemical kinetics, which is needed to understand the basis for kinetic methods of analysis.

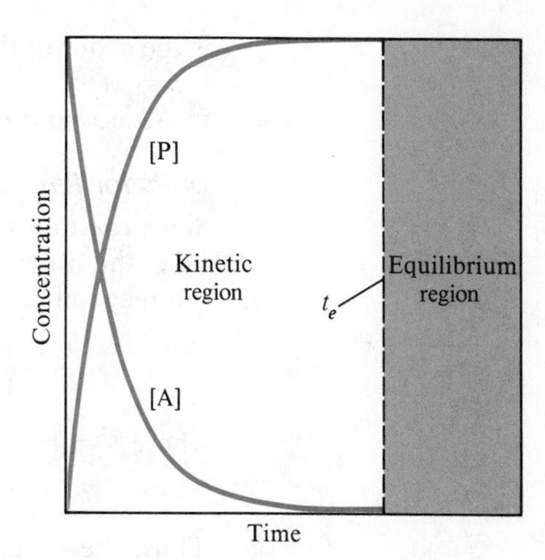

FIGURE 22-1 Change in concentration of analyte [A] and product [P] as a function of time.

22A-1 Some Terms and Symbols Used in Chemical Kinetics

The *mechanism* for a chemical reaction consists of a series of chemical equations describing the individual elementary steps by which products are formed from reactants. Much of what chemists know about mechanisms has been gained from studies in which the rate at which reactants are consumed or products formed is measured as a function of such variables as reactant and product concentration, temperature, pressure, pH, and ionic strength. Such studies lead to an empirical *rate law* that relates the reaction rate to the concentrations of reactants, products, and intermediates at any instant. Mechanisms are derived by postulating a series of elementary steps that are chemically reasonable and consistent with the empirical rate law.

Concentration Terms in Rate Laws

Rate laws are algebraic expressions consisting of concentration terms and constants, which often look somewhat like an equilibrium constant expression (see Equation 22-2). The reader should realize, however, that the square bracketed terms in a rate expression represent molar concentrations *at a particular instant* rather than equilibrium molar concentrations (as in equilibrium constant expressions). This meaning is frequently emphasized by adding a subscript to show the time to which the concentration refers. Thus, $[A]_t$, $[A]_0$, and $[A]_\infty$ indicate the concentration of A at time t, time zero, and infinite time, respectively. Infinite time is regarded as any time greater than that required for the reaction under consideration to come to equilibrium. That is, $t_\infty > t_e$ in Figure 22-1.

Reaction Order

Let us assume that the empirical rate law for the general reaction shown as Equation 22-1 is found by experiment to take the form

$$\text{Rate} = -\frac{d[A]}{dt} = -\frac{d[R]}{dt} = \frac{d[P]}{dt} = k[A]^m[R]^n \tag{22-2}$$

where the rate is the derivative of the concentration of A, R, or P with respect to time. Note that the first two rates carry a negative sign because the concentrations of A and R decrease as the reaction proceeds. In this rate expression, k is the *rate constant,* m is the *order of the reaction* with respect to A, and n

is the order of the reaction with respect to R. The *overall order* of the reaction is $p = m + n$. Thus, if $m = 1$ and $n = 2$, the reaction is said to be first order in A, second order in R, and third order overall.

Units for Rate Constants

Since reaction rates are always expressed in terms of concentration per unit time, the units of the rate constant are determined by the overall order p of the reaction according to the relation

$$\frac{\text{Concentration}}{\text{Time}} = (\text{units of } k)(\text{concentration})^p$$

where $p = m + n$. Rearranging leads to

$$\text{Units of } k = (\text{concentration})^{1-p}\ \text{time}^{-1}$$

Thus, the units for a first-order rate constant are s^{-1}, and the units for a second-order rate constant are $M^{-1}s^{-1}$.

22A-2 The Rate Law for First-Order Reactions

The simplest case in the mathematical analysis of reaction kinetics is that of a spontaneous irreversible decomposition of a species A:

$$A \xrightarrow{k} P \qquad (22\text{-}3)$$

The reaction is first order in A, and the rate is

$$\text{Rate} = \frac{-d[A]}{dt} = k[A] \qquad (22\text{-}4)$$

Pseudo-First-Order Reactions

A first-order decomposition reaction per se is of no use in analytical chemistry because an analysis is ordinarily based upon reactions involving at least two species, an analyte and a reagent. Usually, however, the rate law for a reaction involving two species is sufficiently complex as to be of no use for analytical purposes. In fact, the only kinetic methods that are useful are those that can be performed under conditions that permit the chemist to simplify complex rate laws to a form analogous to Equation 22-4. A higher order reaction that is carried out so that such a simplification is feasible is termed a *pseudo-first-order reaction.* Methods for converting higher-order reactions to pseudo-first-order are dealt with in later sections.

Mathematical Expressions Describing First-Order Behavior

Because essentially all kinetic analyses are performed under pseudo-first-order conditions, it is worthwhile to examine in detail some of the characteristics of reactions having rate laws that approximate Equation 22-4.

By rearranging Equation 22-4, we obtain

$$\frac{d[A]}{[A]} = -kdt \qquad (22\text{-}5)$$

The integral of this equation from time zero, when $[A] = [A]_0$, to time t, when $[A] = [A]_t$, is

$$\int_{[A]_0}^{[A]_t} \frac{d[A]}{[A]} = -k \int_0^t dt$$

Evaluation of the integrals gives

$$\ln \frac{[A]_t}{[A]_0} = -kt \tag{22-6}$$

Finally, by taking the exponential of both sides of Equation 22-6, we obtain

$$\frac{[A]_t}{[A]_0} = e^{-kt} \quad \text{or} \quad [A]_t = [A]_0\, e^{-kt} \tag{22-7}$$

This integrated form of the rate law gives the concentration of A as a function of the initial concentration $[A]_0$, the rate constant k, and the time t. A plot of this relationship is depicted in Figure 22-1.

EXAMPLE 22-1

A reaction is first-order with $k = 0.0370\ s^{-1}$. Calculate the concentration of reactant remaining 18.2 s after initiation of the reaction if its initial concentration is 0.0100 M.

Substituting into Equation 22-7 gives

$$[A]_{18.2} = (0.0100\ M)e^{-(0.0370\ s^{-1})(18.2\ s)} = 0.00510\ M$$

When the rate of a reaction is being followed by the rate of appearance of a product P rather than the rate of disappearance of analyte A, it is useful to modify Equation 22-7 to relate the concentration of P at time t to the initial analyte concentration $[A]_0$. The concentration of A at any time is equal to its original concentration minus the concentration of product (when 1 mol of product forms for 1 mol of analyte). Thus,

$$[A]_t = [A]_0 - [P]_t \tag{22-8}$$

Substitution of this expression for $[A]_t$ into Equation 22-7 gives, after rearrangement,

$$[P]_t = [A]_0(1 - e^{-kt}) \tag{22-9}$$

A plot of this relationship is also shown in Figure 22-1.

The form of Equations 22-7 and 22-9 is that of a pure exponential, which appears widely in science and engineering. A pure exponential in this case has the useful characteristic that equal elapsed times give equal fractional decreases in reactant concentration or increases in product concentration. As an example, consider a time interval $t = \tau = 1/k$, which upon substitution into Equation 22-7 gives

$$[A]_\tau = [A]_0 e^{-k\tau} = [A]_0 e^{-k/k} = (1/e)[A]_0$$

and likewise for a period of $t = 2\tau = 2/k$

$$[A]_{2\tau} = (1/e)^2[A]_0$$

and so on for successive periods, as shown in Figure 22-2.

The period $\tau = 1/k$ is sometimes referred to as the *natural lifetime* of species A. During time τ, the concentration of A decreases to $1/e$ of its original value. A second period, from $t = \tau$ to $t = 2\tau$, produces an equivalent fractional decrease in concentration to $1/e$ of the value at the beginning of the second interval, which is $(1/e)^2$ of $[A]_0$. A more familiar example of this property of exponentials is found in the half-life $t_{1/2}$ of radionuclides. During a period $t_{1/2}$, half of the atoms in a sample of a radioactive element decay to products; a

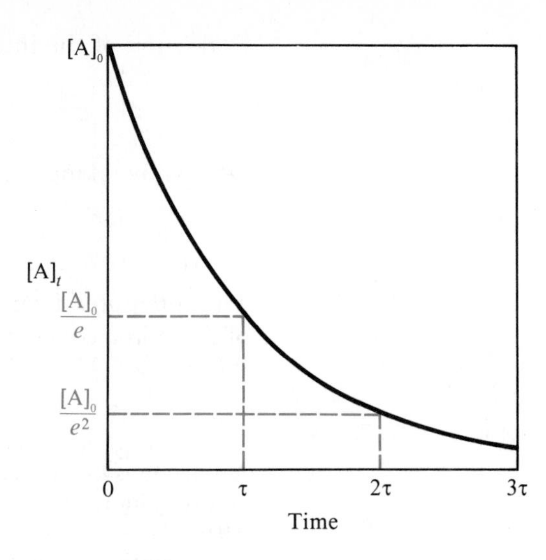

FIGURE 22-2 Rate curve for a first-order reaction showing that equal elapsed times produce equal fractional decreases in analyte concentration.

second period of $t_{1/2}$ reduces the amount of the element to one quarter of its original number, and so on for succeeding periods. Regardless of the time interval chosen, equal elapsed times produce equal fractional decreases in reactant concentration for a first-order process.

EXAMPLE 22-2

Calculate the time required for a first-order reaction with $k = 0.0500\ s^{-1}$ to proceed to 99.0% completion.

For 99.0% completion, $[A]_t/[A]_0 = (100 - 99)/100 = 0.010$; substitution into Equation 22-6 then gives

$$\ln 0.010 = -kt = -(0.0500\ s^{-1})t$$

$$t = -\frac{\ln 0.010}{0.0500\ s^{-1}} = 92\ s$$

22A-3 Rate Laws for Second-Order and Pseudo-First-Order Reactions

Let us consider a typical analytical reaction in which 1 mol of analyte A reacts with 1 mol of reagent B to give a single product P. For the present, we assume the reaction is irreversible and write

$$A + R \xrightarrow{k} P \tag{22-10}$$

If the reaction occurs in a single elementary step, the rate is proportional to the concentration of each of the reactants, and the rate law is

$$-\frac{d[A]}{dt} = k\,[A][R] \tag{22-11}$$

The reaction is first order in each of the reactants and second order overall. If the concentration of R is chosen such that [R] >> [A], the concentration of R changes very little during the course of the reaction and we can write $k\,[R] = \text{constant} = k'$. Equation 22-11 is then rewritten as

$$-\frac{d[A]}{dt} = k'[A] \tag{22-12}$$

which is identical in form to the first-order case of Equation 22-4. Hence the reaction is said to be *pseudo first order* in A.

EXAMPLE 22-3

For a pseudo-first-order reaction in which the reagent is present in 100-fold excess, find the relative error resulting from the assumption that $k\,[R] = k'$ is constant when the reaction is 40% complete.

The initial concentration of the reagent can be expressed as

$$[R]_0 = 100[A]_0$$

At 40% reaction, 60% of A remains. Thus,

$$[A]_{40\%} = 0.60[A]_0$$

$$[R]_{40\%} = [R]_0 - 0.40[A]_0 = 100[A]_0 - 0.40[A]_0 = 99.6[A]_0$$

Assuming pseudo-first-order behavior, the rate at 40% reaction is

$$-\frac{d[A]_{40\%}}{dt} = k\,[R]_0[A]_{40\%} = k\,(100[A]_0)(0.40[A]_0)$$

The true rate at 40% reaction is $k(99.6[A]_0)(0.60[A]_0)$. Thus, the relative error is

$$\frac{k(100[A]_0)(0.60[A]_0) - k(99.6[A]_0)(0.60[A]_0)}{k(99.6[A]_0)(0.60[A]_0)} = 0.004 \text{ (or 0.4\%)}$$

As Example 22-3 clearly illustrates, the error associated with the determination of the rate of a pseudo-first-order reaction with a 100-fold excess of reagent is negligible. A 50-fold reagent excess leads to a 1% error; an error of this magnitude usually deemed acceptable in kinetic methods. Moreover, the error is even less significant at times when the reaction is less than 40% complete.

Reactions are seldom completely irreversible, and a rigorous description of the kinetics of a second-order reaction that occurs in a single step must take into account the reverse reaction. The rate of the reaction is the difference between the forward rate and the reverse rate:

$$-\frac{d[A]}{dt} = k_1[A][R] - k_{-1}[P]$$

where k_1 is the second-order rate constant for the forward reaction and k_{-1} is the first-order rate constant for the reverse reaction. In deriving this equation we have assumed for simplicity that a single product is formed, but more complex cases can be described as well.[1] As long as conditions are maintained such that k_{-1} and/or [P] are relatively small, the rate of the reverse reaction is negligible, and little error is introduced by the assumption of pseudo-first-order behavior.

22A-4 Catalyzed Reactions

Catalyzed reactions, particularly those in which enzymes serve as catalysts, are widely used for the determination of a variety of biological and biochemical species as well as a number of inorganic cations and anions. We shall therefore

[1]See J. H. Espenson, *Chemical Kinetics and Reaction Mechanisms,* pp. 42–48. New York: McGraw-Hill, 1981.

use enzyme-catalyzed reactions to illustrate catalytic rate laws and to show how these rate laws can be reduced to relatively simple algebraic relationships, such as the pseudo-first-order equation shown as Equation 22-12. These simplified relationships can then be used for analytical purposes.

Enzyme-Catalyzed Reactions

Enzymes are high-molecular-weight molecules that catalyze a variety of reactions of biological and biomedical importance. Enzymes are particularly useful as analytical reagents owing to their high degree of selectivity. Consequently, they are widely used in the determination of molecules with which they combine when acting as catalysts. Such molecules are usually designated as *substrates*. In addition to the determination of substrates, enzyme-catalyzed reactions are employed for the determination of activators, inhibitors, and, of course, enzymes.

The behavior of a large number of enzymes is consistent with the general mechanism

$$E + S \underset{k_{-1}}{\overset{k_1}{\rightleftharpoons}} ES \xrightarrow{k_2} P + E \qquad (22\text{-}13)$$

In this equation, the enzyme E is shown as reacting reversibly with the substrate S to form an enzyme-substrate complex ES; this complex then decomposes irreversibly to form the product(s) and the regenerated enzyme. The rate law for this mechanism assumes one of two forms, depending upon the relative rates of the two steps. If the second step is considerably slower than the first (case 1), the reactants and ES essentially achieve equilibrium. In this case, the second step determines the overall rate and is thus termed *the rate-determining step*. When, however, the rates of the two steps are comparable in magnitude (case 2), ES decomposes as rapidly as it is formed, and its concentration can be assumed to be small and relatively constant throughout much of the reaction. In the two sections that follow, we show that in both cases, the reaction conditions can be arranged to yield simple relationships between rate and analyte concentration.

Case 1: Rate Determined by the Rate of Complex Decomposition

A rate law consistent with the mechanism of Equation 22-13 when the second step is rate-determining is derived in the following way. The concentrations of S and E at time t are given by

$$[S]_t = [S]_0 - [ES]_t \quad \text{and} \quad [E]_t = [E]_0 - [ES]_t \qquad (22\text{-}14)$$

Since the equilibrium for the first step is achieved rapidly relative to the rate of the second step, we can equate the forward and reverse rates of the first step and obtain

$$-\frac{d[E]}{dt} = -\frac{d[S]}{dt} \simeq \frac{d[ES]}{dt}$$

At any time t, the rate of the forward reaction is proportional to the product of the concentration of the two reactants ($[E]_t[S]_t$); the rate of the reverse reaction is then proportional to the instantaneous concentration of the complex ($[ES]_t$). Thus,

$$k_1[E]_t[S]_t = k_{-1}[ES]_t$$

Rearranging this equation leads to

$$\frac{k_1}{k_{-1}} = \frac{[ES]_t}{[E]_t[S]_t} = K \tag{22-15}$$

When the rates of the forward and reverse reactions are equal, the system is *at chemical equilibrium.* Thus the concentration terms in Equation 22-15 are *equilibrium concentrations,* and K in this equation is the *equilibrium constant* for the first step in the reaction. Substitution of $[S]_t$ and $[E]_t$ from Equation 22-14 into Equation 22-15 gives after rearranging

$$[ES]_t = K([S]_0 - [ES]_t)([E]_0 - [ES]_t) \tag{22-16}$$

In theory, either the substrate or the enzyme catalyst can be present in considerable excess; however, enzyme reagents are expensive and are thus nearly always maintained at low levels when a substrate molecule is being determined. Furthermore, natural levels of enzyme are very low so that we can generally assume that $[S]_0 >> [ES]_t$. Equation 22-16 then becomes

$$[ES]_t = K[S]_0([E]_0 - [ES]_t) \tag{22-17}$$

When Equation 22-17 is solved for $[ES]_t$, we have

$$[ES]_t = \frac{K[S]_0[E]_0}{1 + K[S]_0} \tag{22-18}$$

and the rate of the reaction is

$$\frac{d[P]}{dt} = k_2[ES]_t = \frac{k_2K[S]_0[E]_0}{1 + K[S]_0} = \frac{k_2[S]_0[E]_0}{(1/K) + [S]_0} \tag{22-19}$$

It is usually possible to adjust conditions such that Equation 22-19 simplifies to a relationship that can be used for the determination of either the catalyst or the substrate. For enzyme determinations, the concentration of the substrate is made large enough so that $[S]_0 >> 1/K$; the denominator of Equation 22-19 is then approximately equal to $[S]_0$, and the rate is

$$\frac{d[P]}{dt} = k_2[E]_0 \tag{22-20}$$

Thus, the rate of the reaction is directly proportional to the initial concentration of the enzyme catalyst and can be used for determining that species.

If, however, the substrate is the species to be determined, its initial concentration is made small enough so that $[S]_0 << 1/K$. Equation 22-19 then reduces to

$$\frac{d[P]}{dt} = k_2K[S]_0[E]_0$$

In order to determine substrate concentrations, measurements are performed at constant enzyme concentration so that

$$\frac{d[P]}{dt} = k'[S]_0 \tag{22-21}$$

where $k' = k_2K[E]_0$. Under these conditions, the rate of the reaction is clearly proportional to substrate concentration. Note that the catalyst is being regenerated continuously (Equation 22-13). Thus $[E]_t \simeq [E]_0$.

Case 2: Rate Determined by the Steady-State Condition

In the event that the reaction rates of the two steps in the mechanism shown by Equation 22-13 are comparable, it is necessary to use what is called the *steady-state approximation* to arrive at a rate law. In this treatment, which is described in many sources[2] and therefore is not reproduced here, the rate of change of the concentration of the ES complex is presumed to be negligible. This assumption leads to the rate law

$$\frac{d[P]}{dt} = \frac{k_2[E]_0[S]_t}{[S]_t + (k_{-1} + k_2)/k_1} = \frac{k_2[E]_0[S]_t}{[S]_t + K_m} \qquad (22\text{-}22)$$

The *Michaelis constant*, $K_m = (k_{-1} + k_2)/k_1$, is essentially an "equilibrium-like" constant analogous to $1/K$ in case 1; it expresses the ratio of the sum of the rates of the two reactions tending to destroy ES to the rate of the reaction favoring ES formation. The constant is usually expressed in units of millimoles/liter (mM) and ranges from 0.01 mM to 100 mM for common enzymes.

As in case 1, the rate equation can be simplified so that the reaction rate is proportional to either enzyme or substrate concentration. For example, if the concentration of substrate is made large enough so that $[S] >> K_m$, Equation 22-22 reduces to

$$\frac{d[P]}{dt} = k_2[E]_0 \qquad (22\text{-}23)$$

Under these conditions, when the rate is independent of substrate concentration, the reaction is said to be *pseudo zero order* in substrate, and the rate is directly proportional to the concentration of enzyme.

When conditions are such that the concentration of S is small or when K_m is relatively large, then $[S]_t << K_m$, and Equation 22-22 simplifies to

$$\frac{d[P]}{dt} = \frac{k_2}{K_m}[E]_0[S]_t = k'[S]_t$$

where $k' = k_2[E]_0/K_m$. Hence, the kinetics are first order in substrate. In order to use this equation for determining analyte concentrations, it is necessary to measure $d\,[P]/dt$ at the beginning of the reaction, where $[S]_t \simeq [S]_0$, so that

$$\frac{d[P]}{dt} \simeq k'[S]_0 \qquad (22\text{-}24)$$

The regions where Equations 22-23 and 22-24 are applicable are illustrated in Figure 22-3, in which the initial rate of an enzyme-catalyzed reaction is plotted as a function of substrate concentration. When substrate concentration is low, Equation 22-24, which is linear in substrate concentration, governs the shape of the curve, and it is this region that is used to determine the amount of substrate present.

If it is desired to determine the amount of enzyme, the region of high substrate concentration is employed, where Equation 22-23 applies, and the rate is independent of substrate concentration. The limiting rate of the reaction at large values of [S] is often referred to as v_{max}, as indicated in the figure. It can be shown that the value of the substrate concentration at exactly $v_{max}/2$ is equal to the Michaelis constant K_m.

[2]For example, see J. H. Espenson, *Chemical Kinetics and Reaction Mechanisms,* pp. 72–80. New York: McGraw-Hill, 1981.

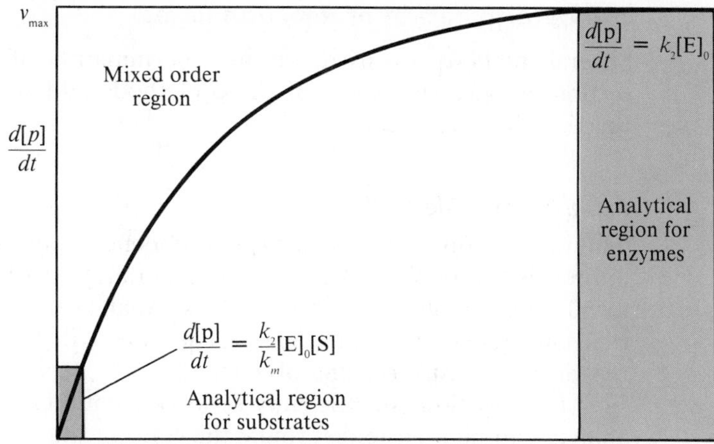

FIGURE 22-3 Change in rate of product formation as a function of substrate concentration, showing the parts of the curve useful for determination of substrate and enzyme.

EXAMPLE 22-4

The enzyme urease, which catalyzes the hydrolysis of urea, is widely used for determining urea in blood. Details of this application are given on page 594. The Michaelis constant for urease at room temperature is 2.0 mM, and $k_2 = 2.5 \times 10^4 \text{ s}^{-1}$ at pH 7.5. (a) Calculate the initial rate of the reaction when the urea concentration is 0.030 mM and the urease concentration is 5.0 μM, and (b) find v_{max}.

(a) From Equation 22-22,

$$\frac{d[P]}{dt} = \frac{k_2[E]_0[S]_t}{[S]_t + K_m}$$

At the beginning of the reaction, $[S]_t = [S]_0$ and

$$\frac{d[P]}{dt} = \frac{(2.5 \times 10^4 \text{ s}^{-1})(5.0 \times 10^{-6} \text{ M})(0.030 \times 10^{-3} \text{ M})}{0.030 \times 10^{-3} \text{ M} + 2.0 \times 10^{-3} \text{ M}}$$

$$= 1.8 \times 10^{-3} \text{ M s}^{-1}$$

(b) Figure 22-3 reveals that $d[P]/dt = v_{max}$ when the concentration of substrate is large and Equation 22-23 applies. Thus,

$$d[P]/dt = v_{max} = k_2[E]_0 = (2.5 \times 10^4 \text{ s}^{-1})(5.0 \times 10^{-6} \text{ M}) = 0.125 \text{ M s}^{-1}$$

Although our discussion thus far has been concerned with enzymatic methods, an analogous treatment for ordinary catalysis gives rate laws that are similar in form to those for enzymes. These expressions often reduce to the first-order case for ease of data treatment, and many examples of kinetic-catalytic methods are found in the literature.[3]

[3]See K. B. Yatsimirskii, *Kinetic Methods of Analysis.* Oxford: Pergamon Press, 1966; H. B. Mark, G. A. Rechnitz, and R. A. Greinke, *Kinetics in Analytical Chemistry.* New York: Wiley-Interscience, 1968.

22B The Determination of Reaction Rates

Several methods are used for the determination of reaction rates. In this section, we describe some of these methods and when in the course of a reaction they are used.

22B-1 Experimental Methods

The way reaction rates are measured depends upon whether the reaction of interest is fast or slow. A reaction is generally regarded as fast if it proceeds to 50% of completion in 10 s or less. Analytical methods based upon fast reactions generally require special equipment that permits rapid mixing of reagents and fast recording of data.

If a reaction is sufficiently slow, conventional equilibrium methods of analysis can be used to determine the concentration of a reactant or product as a function of time. Often, however, the reaction of interest is too rapid for equilibrium measurements—that is, concentrations change appreciably during the measurement process. Under these circumstances, either the reaction must be stopped (*quenched*) while the measurement is made or an instrumental technique that records concentrations continuously as the reaction proceeds must be employed. In the former case, an aliquot is removed from the reaction mixture and rapidly quenched by mixing it with a reagent that combines with one of the reactants to stop the reaction. Alternatively, quenching is accomplished by lowering the temperature rapidly to slow the reaction to an acceptable level for the measurement step. Unfortunately, quenching techniques tend to be laborious and often time-consuming and are thus not widely used for analytical purposes.

The most convenient approach for obtaining kinetic data is to monitor the progress of the reaction continuously by spectrophotometry, conductometry, potentiometry, or some other instrumental technique. With the advent of inexpensive microcomputer technology, instrumental readings proportional to concentrations of reactants and/or products are often recorded directly as a function of time, stored in the computer's memory, and retrieved later for data processing.

In the following sections, we explore some strategies used to determine rates of reactions and thus concentrations from plots of analyte or product concentration as a function of time.

22B-2 Types of Kinetic Methods

Kinetic methods are classified according to the type of relationship that exists between the measured variable and the analyte concentration.

The Differential Method

In the *differential method,* concentrations are computed from reaction rates by means of a differential form of a rate expression. Rates are determined by measuring the slope of a curve relating analyte or product concentration to reaction time. To illustrate, let us substitute $[A]_t$ from Equation 22-7 for [A] in Equation 22-4:

$$\text{Rate} = -\left(\frac{d[A]}{dt}\right)_t = k[A]_t = k[A]_0\, e^{-kt} \tag{22-25}$$

As an alternative, the rate can be expressed in terms of the product concentration. That is,

$$\text{Rate} = \left(\frac{d[\text{P}]}{dt}\right)_t = k[\text{A}]_0 1^{-kt} \qquad (22\text{-}26)$$

Equations 22-25 and 22-26 give the dependence of the rate upon k, t, and, most important, $[\text{A}]_0$, the initial concentration of the analyte. At any fixed time t, the factor ke^{-kt} is a constant, and the rate is directly proportional to the initial analyte concentration.

EXAMPLE 22-5

The rate constant for a pseudo-first-order reaction is $0.156\ \text{s}^{-1}$. Find the initial concentration of the reactant if its rate of disappearance 10.00 s after the initiation of the reaction is $2.79 \times 10^{-4}\ \text{M s}^{-1}$.

The proportionality constant ke^{-kt} is

$$ke^{-kt} = (0.156\ \text{s}^{-1})e^{-(0.156\ \text{s}^{-1})(10.00\ \text{s})} = 3.28 \times 10^{-2}\ \text{s}^{-1}$$

Rearranging Equation 22-25 and substituting numerical values, we have

$$[\text{A}]_0 = \text{Rate}/ke^{-kt}$$

$$= (2.79 \times 10^{-4}\ \text{M s}^{-1})/(3.28 \times 10^{-2}\ \text{s}^{-1})$$

$$= 8.51 \times 10^{-3}\ \text{M}$$

The choice of the time at which a reaction rate is measured is often based upon a variety of factors, including convenience, the existence of interfering side reactions, and the inherent precision of making the measurement at a particular time. It is often advantageous to make the measurement near $t = 0$ because this portion of the exponential curve is nearly linear (see, for example, the initial parts of the curves in Figure 22-1) and the slope is readily estimated from the tangent to the curve. Moreover, if the reaction is pseudo first order, such a small amount of excess reagent is consumed that no error arises from changes in k resulting from changes in reagent concentration. Finally, the relative error in determining the slope is minimal at the beginning of the reaction because the slope is a maximum in this region.

Figure 22-4 illustrates how the differential method is used to determine the concentration of an analyte $[\text{A}]_0$ from experimental rate measurements for the reaction shown as Equation 22-1. The solid curves in Figure 22-4a are plots of the experimentally measured product concentration [P] as a function of reaction time for four standard solutions of A. These curves are then used to prepare the differential calibration plot shown in Figure 22-4b. To obtain the rates, tangents are drawn to each of the curves in 22-4a at a time near zero (dashed lines in part a). The slopes of the tangents are then plotted as a function of [A], giving the straight line shown in 22-4b. Unknowns are treated in the same way, and analyte concentrations are determined from the calibration curve.

Of course, it is not necessary to record the entire rate curve, as has been done in Figure 22-4a, since only a small portion of the plot is utilized for measuring the slope. As long as sufficient data points are collected to determine the initial slope precisely, time is saved and the entire procedure is simplified. More sophisticated data-handling procedures and numerical analysis of the

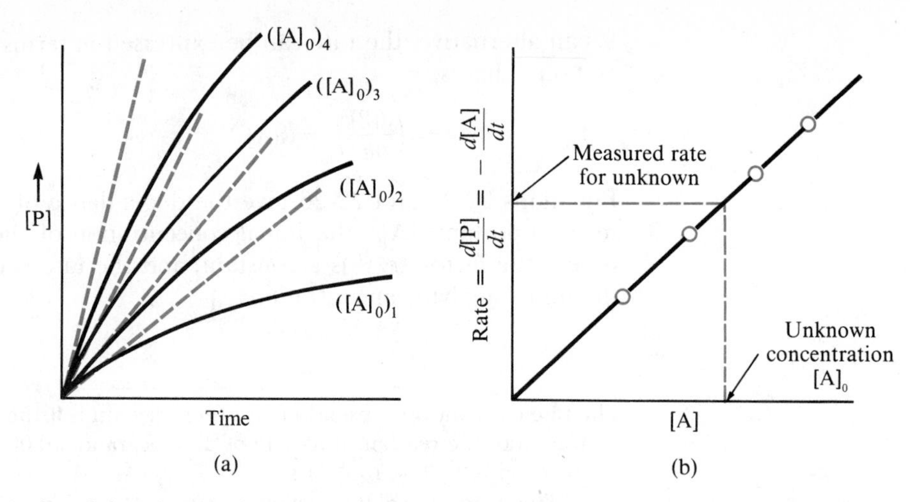

FIGURE 22-4 A plot of data for the determination of A by the differential method. (a) Solid lines are the experimental plots of product concentration as a function of time for four initial concentrations of A. Dashed lines are tangents to the curve at $t \rightarrow 0$. (b) A plot of the slopes obtained from the tangents in (a) as a function of analyte concentration.

data make possible high-precision rate measurements at later times as well; under certain circumstances such measurements are more accurate and precise than those made near $t = 0$.

Integral Methods

In contrast to the differential method, *integral methods* take advantage of integrated forms of rate laws, such as those shown by Equations 22-6, 22-7, and 22-9.

Graphical Methods. Equation 22-6 may be rearranged to give

$$\ln[A]_t = -kt + \ln[A]_0 \tag{22-27}$$

Thus a plot of the natural logarithm of experimentally measured concentrations of A (or P) as a function of time should yield a straight line with a slope of $-k$ and a y intercept of $\ln[A]_0$. Use of this procedure for the determination of nitromethane is illustrated in Example 22-6.

EXAMPLE 22-6

The data in the first two columns of Table 22-1 were recorded for the pseudo-first-order decomposition of nitromethane in the presence of excess base. Find the initial concentration of nitromethane and the pseudo-first-order rate constant for the reaction.

Computed values for the natural logarithms of nitromethane concentrations are shown in the third column of Table 22-1. The data are plotted in Figure 22-5. A least-squares analysis of the data (Section 2E-4) leads to an intercept a of

$$a = \ln[CH_3NO_2]_0 = -5.129$$

which upon exponentiation gives

$$[CH_3NO_2]_0 = 5.92 \times 10^{-3}\ M$$

The least-squares analysis also gives the slope of the line b, which in this case is

$$b = -1.62 = -k$$

and thus

$$k = 1.62 \text{ s}^{-1}$$

Fixed-Time Methods. Fixed-time methods are based upon Equation 22-7 or 22-9. The former can be rearranged to

$$[A]_0 = \frac{[A]_t}{e^{-kt}} \quad (22\text{-}28)$$

The simplest way of employing this relationship is to perform a calibration experiment with a standard solution that has a known concentration $[A]_0$. After a carefully measured reaction time t, $[A]_t$ is determined and used to evaluate e^{-kt} by means of Equation 22-28. Unknowns are then analyzed by measuring $[A]_t$ after exactly the same reaction time and employing the calculated value for e^{-kt} to compute the analyte concentrations.

Equation 22-28 is easily modified for the situation where [P] is measured experimentally rather than $[A]_t$. Equation 22-9 may be rearranged to solve for $[A]_0$. That is,

$$[A]_0 = \frac{[P]_t}{1 - e^{-kt}} \quad (22\text{-}29)$$

A more desirable approach to the uses of Equation 22-28 or 22-29 is to measure [A] or [P] at two times t_1 and t_2. For example, if the product concentration is determined, we can write

$$[P]_{t_1} = [A]_0(1 - e^{-kt_1})$$

$$[P]_{t_2} = [A]_0(1 - e^{-kt_2})$$

Subtracting the first equation from the second yields, upon rearrangement,

$$[A]_0 = \frac{[P]_{t_2} - [P]_{t_1}}{e^{-kt_1} - e^{-kt_2}} = C([P]_{t_2} - [P]_{t_1}) \quad (22\text{-}30)$$

The reciprocal of the denominator is constant for constant t_1 and t_2 and is assigned the symbol C.

The use of Equation 22-30 has the fundamental advantage common to most kinetic methods that the absolute determination of concentration or of a variable proportional to concentration is unnecessary. It is the *difference* between two concentrations that is proportional to the initial concentration of the analyte. This means that factors influencing the long-term stability of the measuring instrument are subtracted out.

TABLE 22-1 **Data for the Decomposition of Nitromethane**

Time, s	$[CH_3NO_2]$, M	$\ln[CH_3NO_2]$
0.25	3.86×10^{-3}	−5.557
0.50	2.59×10^{-3}	−5.956
0.75	1.84×10^{-3}	−6.298
1.00	1.21×10^{-3}	−6.717
1.25	0.742×10^{-3}	−7.206

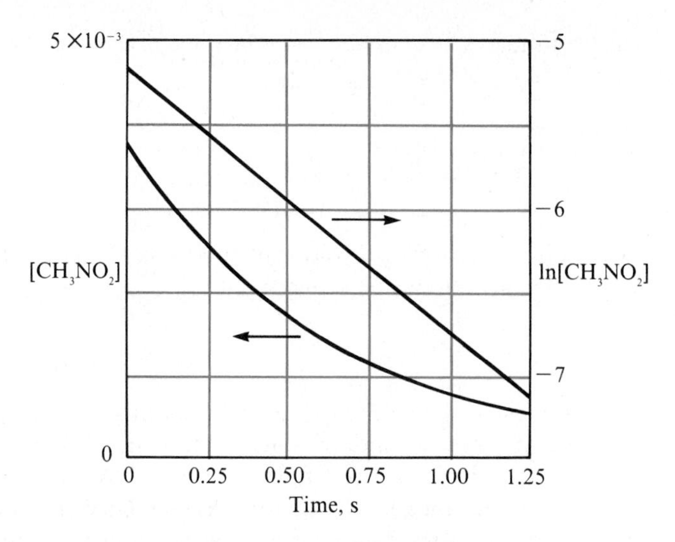

FIGURE 22-5 Plots of nitromethane concentration and the natural logarithm of nitromethane concentration as a function of time. The data are from Example 22-5.

An important example of uncatalyzed methods is the fixed-time method for the determination of thiocyanate ion based upon spectrophotometric measurements of the red iron(III) thiocyanate complex. The reaction in this application is

$$Fe^{3+} + SCN^- \underset{k_{-1}}{\overset{k_1}{\rightleftharpoons}} \underset{\text{red}}{Fe(SCN)^{2+}}$$

Under conditions of excess Fe^{3+}, the reaction is pseudo first order in SCN^- The curves in Figure 22-6a show the increase in absorbance due to the ap-

FIGURE 22-6 (a) Absorbance due to the formation of $FeSCN^{2+}$ as a function of time for five concentrations of SCN^-. (b) A plot of the difference in absorbance ΔA at times t_2 and t_1 as a function of SCN^- concentration.

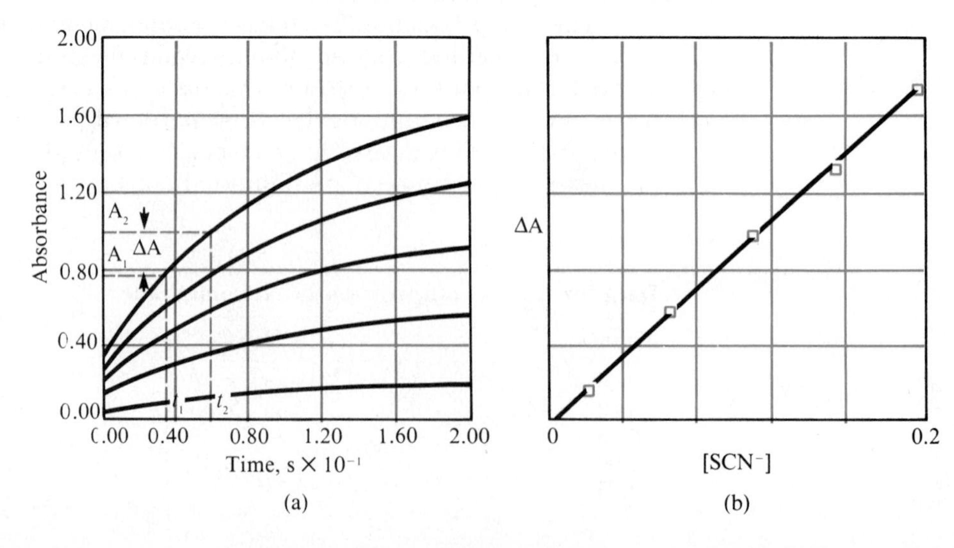

pearance of $Fe(SCN)^{2+}$ versus time following the rapid mixing of 0.100 M Fe^{3+} with various concentrations of SCN^- at pH 2. Since the concentration of $Fe(SCN)^{2+}$ is related to the absorbance by Beer's law, the experimental data can be used directly without conversion to concentration. Thus, the change in absorbance ΔA between times t_1 and t_2 is computed and plotted versus $[SCN^-]_0$, as in Figure 22-6b. Unknown concentrations are then determined by evaluating ΔA under the same experimental conditions and reading the concentration of thiocyanate ion from the linear working curve.

Fixed-time methods are advantageous because the measured quantity is directly proportional to analyte concentration and because measurements can be made *at any time* during the progress of first-order reactions. When instrumental methods are used for monitoring reactions by fixed-time procedures, the precision of the analytical results approaches the precision of the instrument used.

22C Applications of Kinetic Methods

Kinetic methods, which are used for the determination of both organic and inorganic species, fall into two categories, catalyzed and uncatalyzed. As noted earlier, uncatalyzed reactions are not nearly as widely used as catalyzed reactions because the selectivity and sensitivity of the former are often quite inferior to those of the latter. Uncatalyzed reactions are used to advantage when high-speed, automated measurements are required, however, or when the sensitivity of the detection methods is great.[4]

22C-1 Catalytic Methods

The Determination of Inorganic Species

Many inorganic cations and anions catalyze *indicator reactions*—that is, reactions whose rates are readily measured by instrumental methods, such as spectrophotometric, fluorometric, or electroanalytical. Conditions are then employed such that the rate is proportional to the concentration of catalyst, and from the rate data, the concentration of catalyst is determined. An example of this type of method involves the kinetic determination of traces of mercury(II).

The Catalytic Determination of Mercury. The catalytic effect of mercury on the reaction of hexacyanoferrate(II) with nitrosobenzene has been employed to determine trace levels of the metal in the +2 state. The uncatalyzed indicator reaction proceeds as follows:

$$Fe(CN)_6^{4-} + H_2O \underset{}{\overset{\text{slow}}{\rightleftharpoons}} Fe(CN)_5H_2O^{3-} + CN^- \qquad (22\text{-}31)$$

$$Fe(CN)_5H_2O^{3-} + C_6H_5NO \overset{\text{fast}}{\rightleftharpoons} Fe(CN)_5C_6H_5NO^{3-} + H_2O \qquad (22\text{-}32)$$

The presence of Hg^{2+} accelerates the first step, apparently by removing cyanide to form an intermediate complex ion $Hg(CN)^+$, thus forcing the equilibrium to the right. Mercury(II) is then regenerated by reaction of the $Hg(CN)^+$

[4]For reviews of applications of modern kinetic methods, see M. Kopanica and V. Stara, in *Comprehensive Analytical Chemistry,* G. Svehla, Ed., Vol. XVIII, pp. 11–227. New York: Elsevier, 1983; and G. G. Guilbault, in *Treatise on Analytical Chemistry,* 2nd ed., I. M. Kolthoff and P. J. Elving, Eds., Part I, Vol. 1, Chapter 11. New York: Wiley, 1978.

complex with hydrogen ions in the solution:

$$Hg^{2+} + CN^- \longrightarrow Hg(CN)^+$$

$$Hg(CN)^+ + H^+ \longrightarrow HCN + Hg^{2+}$$

Reaction 22-32 is relatively fast and serves as the indicator reaction for monitoring the formation of the product by spectrophotometric measurements at 528 nm. The formation of $Fe(CN)_5C_6H_5NO^{3-}$ is consistent with a general catalyst mechanism similar to that presented in Section 22A-4, so that the rate of the reaction is given by

$$\frac{d[Fe(CN)_5C_6H_5NO^{3-}]}{dt} = k_{cat}[Hg^{2+}]$$

The reaction is slow enough that a fixed-time measurement of absorbance at $t = 1800$ s permits the determination of Hg^{2+} at levels as low as 3×10^{-7} M. The method has been used for the determination of mercury in biological materials, in industrial by-products, and in water.

The Scope of Catalyzed Methods in Inorganic Analysis. Kinetic methods based on catalysis by inorganic analytes are widely applicable. For example, Kopanica and Stara[5] list 40 cations and 15 anions that have been determined by a variety of indicator reactions.

The Determination of Organic Species

Without question, the most important applications of catalyzed kinetic methods to organic analyses involve the use of enzymes as catalysts. These methods have been used for the determination of both enzymes and substrates and serve as the basis for many of the routine and automated screening tests performed by the thousands in clinical laboratories throughout the world. One of these tests is for determining the quantity of urea in blood and is called the blood urea nitrogen (BUN) test (Figure 20-18). A brief description of this test follows.

The Enzymatic Determination of Urea. The determination of urea in blood and urine is frequently carried out by measuring the rate of hydrolysis of urea in the presence of the enzyme urease. The equation for the reaction is

$$CO(NH_2)_2 + 2H_2O + H^+ \xrightarrow{\text{urease}} 2NH_4^+ + HCO_3^-$$

As suggested in Example 22-4, urea can be determined by measuring the initial rate of production of the products of this reaction. The high selectivity of the enzyme permits the use of nonselective detection methods, such as electrical conductance, for initial rate measurements. Commercial instruments, such as the Beckman BUN Analyzer, operate on this principle. The sample is mixed with a small amount of an enzyme-buffer solution in a conductivity cell. The maximum rate of increase in conductance is measured within 10 s of mixing, and the concentration of urea is determined from a calibration curve consisting of a plot of maximum initial rate as a function of urea concentration. The precision of the instrument is on the order of 2 to 5% relative for concentrations in the physiological range of 2 to 10 mM.

[5]M. Kopanica and V. Stara, in *Comprehensive Analytical Chemistry*, G. Svehla, Ed., Vol. XVIII, pp. 192–209. New York: Elsevier, 1983.

Another method for following the rate of urea hydrolysis is based on a specific-ion electrode for ammonium ions (Section 15B-4).

The Scope of Methods Based on Enzyme-Catalyzed Reactions. Enzyme-catalyzed reactions are used for the kinetic determination of a large and diverse group of substrate species. These methods have the twin virtues of remarkable selectivity and good sensitivity. In addition, they are generally readily automated.

A number of inorganic species can be determined by enzyme-catalyzed reactions, including ammonia, hydrogen peroxide, carbon dioxide, hydroxylamine, as well as nitrate, phosphate, and pyrophosphate ions.

Methods have also been developed for the determination of individual components in mixtures of closely related organic compounds. For example, a procedure for the determination of the individual components in a mixture of 21 organic acids without preliminary separation has been described (several enzymes are used). Similarly, methods have been developed for the determination of components in mixtures of alcohols and hydroxy compounds, sugars, amines, and steroid alcohols.

Enzymatic methods have been described for the quantitative determination of several hundred enzymes. In addition, some two dozen inorganic cations and anions are known to decrease the rates of certain enzyme-catalyzed indicator reactions. These *inhibitors* can thus be determined from the decrease in rate brought about by their presence.

Enzyme activators are substances, often inorganic ions, that are required for certain enzymes to become active as catalysts. Activators can thus be determined by their effect on the rates of enzyme-catalyzed reactions. For example, it has been reported that magnesium at concentrations as low as 10 ppb can be determined in blood plasma based on activation by this ion of the enzyme isocitric dehydrogenase.

22C-2 *Noncatalytic Reactions*

As noted earlier, kinetic methods based on uncatalyzed reactions are not nearly as widely used as those in which a catalyst is involved. We have already described two of these methods (page 590 and 592).

Generally, uncatalyzed reactions can be rendered useful when selective reagents are employed in conjunction with sensitive detection methods. For example, the selectivity of complexing agents can be controlled by adjusting the pH of the medium in the determination of metal ions, as discussed in Section 11B-7. Sensitivity can be achieved through the use of spectrophotometric detection to monitor complexing agents that form complexes having large molar absorptivities. The determination of Cu^{2+} presented in Problem 22-13 is an example.

A highly sensitive alternative is to select complexes that fluoresce so that the rate of change of fluorescence is used as a measure of analyte concentration (Problem 22-14).

The precision of both noncatalytic and catalytic kinetic methods depends upon a variety of experimental conditions, such as pH, ionic strength, and temperature. With careful control of these variables, precisions of 1 to 10% relative standard deviation are typical. Automation of kinetic methods and computerized data analysis can often improve the relative precision to 1% or less.

22C-3

The Kinetic Determination of Components in Mixtures

An important application of kinetic methods is in the determination of closely related species in mixtures, such as alkaline earth cations or organic compounds with the same functional groups. For example, suppose two species A and B react with a common excess reagent to form products under pseudo-first-order conditions:

$$A + R \xrightarrow{k_A} P'$$

$$B + R \xrightarrow{k_B} P''$$

Generally, k_A and k_B differ from each other. Thus, if $k_A > k_B$, A is depleted before B. It is possible to show that, if the ratio k_A/k_B is greater than about 500, the consumption of A is approximately 99% complete before 1% of B is used up. Thus, a differential determination of A with no significant interference from B is possible provided the rate is measured shortly after mixing.

When the ratio of the two rate constants is small, determination of both species is still possible by more complex methods of data treatment. These methods are beyond the scope of this text.[6]

Questions and Problems

22-1 Define the following terms as they are used in reaction kinetics:

*(a) order of a reaction
(b) pseudo first order
*(c) enzyme
(d) substrate
*(e) Michaelis constant
(f) differential method
*(g) integral method
(h) indicator reaction

22-2 The analysis of multicomponent mixtures by kinetic methods is sometimes referred to as "kinetic separations." Explain the significance of this term.

*22-3 Explain why pseudo-first-order conditions are utilized in most kinetic methods.

22-4 List three advantages of kinetic methods.

*22-5 Develop an expression for the half-life of the reactant in a first-order process in terms of k.

22-6 Find the natural lifetime in seconds for first-order reactions corresponding to

*(a) $k = 0.497\ s^{-1}$.
(b) $k = 6.62\ h^{-1}$.
*(c) $[A]_0 = 3.16$ M and $[A]_t = 0.496$ M at $t = 3650$ s.
(d) $[P]_\infty = 0.176$ M and $[P]_t = 0.0423$ M at $t = 9.62$ s.
*(e) half-life $t_{1/2} = 32.4$ years.
(f) $t_{1/2} = 0.478$ s.

22-7 Find the first-order rate constant for a reaction that is 66.6% complete in

*(a) 0.0100 s.
(b) 0.100 s.
*(c) 1.00 s.
(d) 5280 s.
*(e) 26.8 μs.
(f) 8.86 ns.

22-8 Calculate the number of lifetimes τ required for a pseudo-first-order reaction to achieve the following levels of completion:

*(a) 10%
(b) 50%
*(c) 90%
(d) 99%
*(e) 99.9%
(f) 99.99%

[6]For examples of the use of kinetic methods for the analysis of multicomponent mixtures, see G. M. Ridder and D. W. Margerum, *Anal. Chem.* **1977**, *49*, 2090.

22-9 Find the number of half-lives $t_{1/2}$ required to reach the levels of completion listed in Problem 22-8.

22-10 Find the relative error associated with the assumption that k' is invariant during the course of a pseudo-first-order reaction under the following conditions.

	Extent of Reaction, %	Excess of Reagent
*(a)	1	5X
(b)	1	10X
*(c)	1	50X
(d)	1	100X
*(e)	5	5X
(f)	5	10X
*(g)	5	100X
(h)	63.2	5X
*(i)	63.2	10X
(j)	63.2	50X
*(k)	63.2	100X

*22-11 Show that for an enzyme reaction obeying Equation 22-22 the substrate concentration for which the rate equals $v_{max}/2$ is equal to K_m.

22-12 Equation 22-22 can be rearranged to produce the equation

$$\frac{1}{d[P]/dt} = \frac{K_m}{v_{max}[S]} + \frac{1}{v_{max}}$$

(a) Suggest a way to employ this equation in the construction of a working curve for the enzymatic determination of substrate.
(b) Describe how the resulting working curve can be used to find K_m and v_{max}.

*22-13 Copper(II) forms a 1:1 complex with the organic complexing agent R in acidic medium. The formation of the complex can be monitored by spectrophotometry at 480 nm. Use the following data collected under pseudo-first-order conditions to construct a calibration curve of rate versus concentration of R. Find the concentration of copper(II) in an unknown whose rate under the same conditions was 6.2×10^{-3} Absorbance·s^{-1}.

$c_{Cu^{2+}}$, ppm	Rate, Absorbance·s^{-1}
3.0	3.6×10^{-3}
5.0	5.4×10^{-3}
7.0	7.9×10^{-3}
9.0	1.03×10^{-2}

22-14 Aluminum forms a 1:1 complex with 2-hydroxy-1-naphthaldehyde *p*-methoxybenzoylhydrazonal that exhibits fluorescence emission at 475 nm. Under pseudo-first-order conditions, a plot of the initial rate of the reaction (emission units/s) versus the concentration of aluminum (in μM) yields a straight line described by the equation

$$\text{Rate} = 1.74c_{Al} - 0.225$$

Find the concentration of aluminum in a solution that exhibits a rate of 0.76 emission units/s under the same experimental conditions.

*22-15 The enzyme amine oxidase catalyzes the oxidation of amines to aldehydes. For tryptamine, K_m for the enzyme is 4.0×10^{-4} M and $v_{max} = k_2[E]_0 = 1.6 \times 10^{-3}$ μM/min at pH 8. Find the concentration of a solution of tryptamine that reacts at a rate of 0.18 μM/min in the presence of amine oxidase under the above conditions. Assume that [tryptamine] $<<$ K_m.

CHAPTER 23

An Introduction to Chromatographic Methods

Chromatography is an analytical method that is widely used for the separation, identification, and determination of the chemical components in complex mixtures, many of which could not otherwise be resolved. Chromatography was invented and named by the Russian botanist Mikhail Tswett shortly after the turn of the century. He employed the technique to separate various plant pigments, such as chlorophylls and xanthophylls, by passing solutions of them through glass columns packed with finely divided calcium carbonate. The separated species appeared as colored bands on the column, which accounts for the name he chose for the method (Greek *chroma* meaning "color" and *graphein* meaning "to write").

The applications of chromatography have grown explosively in the last four decades, owing not only to the development of several new techniques but also to the expanding need of scientists for better methods of separating complex mixtures.[1] The tremendous impact of these methods on science is attested by the 1952 Nobel prize awarded to A.J.P. Martin and R.L.M. Synge for their discoveries in the field. Perhaps more impressive is the list of 12 Nobel prizes awarded between 1937 and 1972 that are based upon work in which chromatography played a vital role.[2]

23A A General Description of Chromatography

The term *chromatography* is difficult to define rigorously, owing to the variety of systems and techniques to which it has been applied. All of these methods, however, have in common the use of a *stationary phase* and a *mobile phase.* Components of a mixture are carried through the stationary phase by the flow of gaseous or liquid mobile phase, separations being based on differences in migration rates among the sample components.

23A-1 Classification of Chromatographic Methods

Chromatographic methods can be categorized in two ways. The first is based upon the physical means by which the stationary and mobile phases are brought into contact with each other. In *column chromatography,* the stationary phase is

[1]General references on chromatography include E. Heftmann, *Chromatography,* 3rd ed. New York: Van Nostrand-Reinhold, 1975; B. L. Karger, L. R. Snyder, and C. Horváth, *An Introduction to Separation Science,* New York: Wiley, 1973; R. Stock and C.B.F. Rice, *Chromatographic Methods,* 3rd ed. London: Chapman & Hall, 1974; and *Chromatographic and Allied Methods,* O. Mikeš, Ed. New York: Wiley, 1979.

[2]See L. S. Ettre, in *High-Performance Liquid Chromatography,* C. Horváth, Ed., Vol. 1, p. 4. New York: Academic Press, 1980.

held in a narrow tube, and the mobile phase is forced through the tube under pressure or by gravity. In *planar chromatography*, the stationary phase is supported on a flat plate or in the pores of a paper. Here the mobile phase moves through the stationary phase by capillary action or under the influence of gravity. Although this discussion focuses largely on column chromatography, it should be noted that the equilibria upon which the two types of chromatography are based are identical and that the theory developed for column chromatography is readily adapted to planar as well.

A more fundamental classification of chromatographic methods is one based upon whether the mobile phase is a liquid (*liquid chromatography*) or a gas (*gas chromatography*). Table 23-1 lists several subdivisions of each method according to the type of equilibrium by which solutes distribute themselves between the mobile and stationary phases. When the stationary phase is a liquid, as it is in three of the methods listed in Table 23-1, a means must be provided to hold it immobilized. Immobilization is often accomplished by adsorbing a thin film of the liquid on the surface of a finely divided inert solid. Alternatively, the liquid may be retained in the pores, or interstices, of solid particles. A liquid may also be immobilized by adsorption or bonding on the inner walls of a capillary tubing. Ideally, the solid plays no direct part in the separation, serving only as a support for the liquid. Often, however, the nature of the solid does have an effect on the separation.

It is also noteworthy that liquid chromatography can be performed in columns and on planar surfaces, but gas chromatography is restricted to column procedures.

23A-2 Elution Chromatography

Figure 23-1 shows schematically how two components A and B are resolved on a column by *elution chromatography*. Elution involves washing a solute through a column by additions of fresh solvent. A single portion of the sample dissolved in the mobile phase is introduced at the head of the column (at time t_0 in Figure 23-1), whereupon components A and B distribute themselves between the two phases. Introduction of additional mobile phase (the *eluent*) forces the dissolved portion of the sample down the column, where further partition between the mobile phase and fresh portions of the stationary phase occurs

TABLE 23-1 **Classification of Column Chromatographic Methods**

General Classification	Specific Method	Stationary Phase	Type of Equilibrium
Liquid chromatography (LC) (mobile phase: liquid)	Liquid-liquid, or partition	Liquid adsorbed on a solid	Partition between immiscible liquids
	Liquid-bonded phase	Organic species bonded to a solid surface	Partition/adsorption
	Liquid-solid, or adsorption	Solid	Adsorption
	Ion-exchange	Ion-exchange resin	Ion-exchange
	Size-exclusion	Liquid in interstices of a polymeric solid	Partition/sieving
Gas chromatography (GC) (mobile phase: gas)	Gas-liquid	Liquid adsorbed on a solid	Partition between gas and liquid
	Gas-bonded phase	Organic species bonded to a solid surface	Partition/adsorption
	Gas-solid	Solid	Adsorption

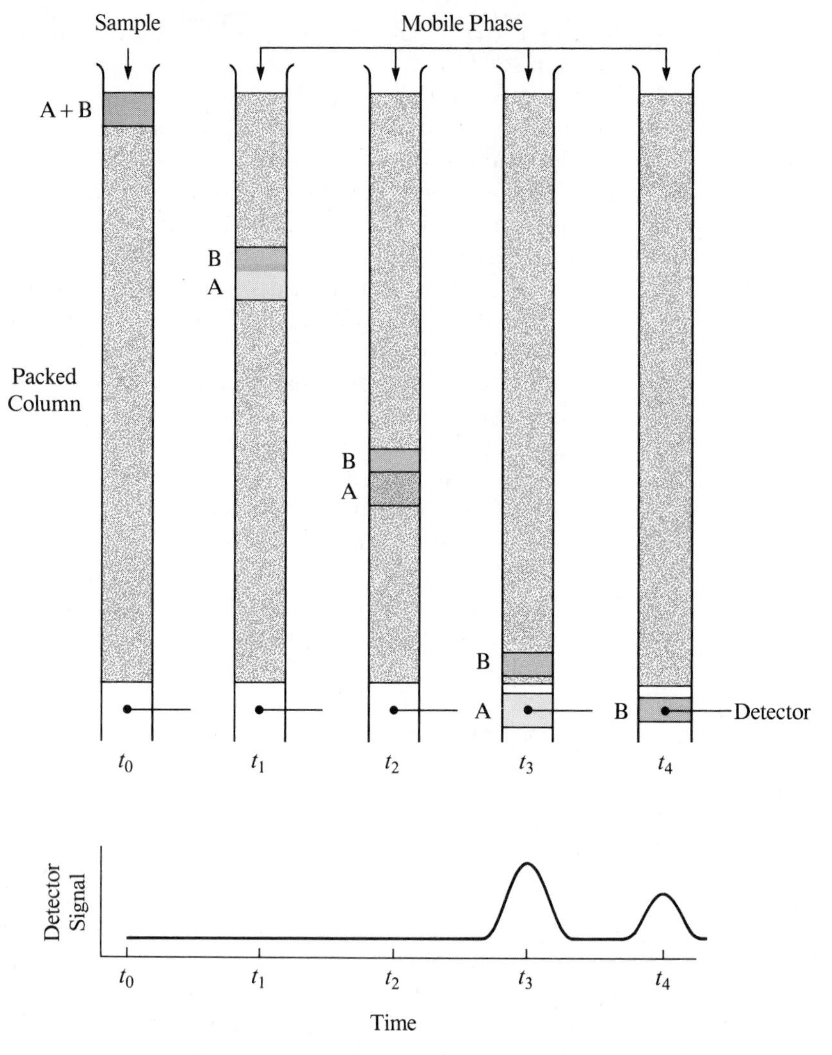

FIGURE 23-1 Diagram showing the separation of a mixture of components A and B by column elution chromatography. The lower figure shows the output of the signal detector at the various stages of elution shown in the upper figure.

(time t_1). Partitioning between the fresh solvent and the stationary phase takes place simultaneously at the site of the original sample.

Further additions of solvent carry solute molecules down the column in a continuous series of transfers between the two phases. Because solute movement can occur only in the mobile phase, the average *rate* at which a solute migrates *depends upon the fraction of time it spends in that phase.* This fraction is small for solutes that are strongly retained by the stationary phase (component B in Figure 23-1, for example) and large where retention in the mobile phase is more likely (component A). Ideally, the resulting differences in rates cause the components in a mixture to separate into *bands,* or *zones,* along the length of the column (see Figure 23-2). Isolation of the separated species is then accomplished by passing a sufficient quantity of mobile phase through the column to cause the individual bands to pass out the end (to be eluted from the column), where they can be collected (times t_3 and t_4 in Figure 23-1).

Chromatograms

If a detector that responds to solute concentration is placed at the end of the column and its signal is plotted as a function of time (or of volume of added mobile phase), a series of symmetric peaks is obtained, as shown in the lower part of Figure 23-1. Such a plot, called a *chromatogram,* is useful for both qualitative and quantitative analysis. The positions of the peaks on the time axis can be used to identify the components of the sample; the areas under the peaks provide a quantitative measure of the amount of each species.

The Effects of Relative Migration Rates and Band Broadening on Resolution

Figure 23-2 shows concentration profiles for the bands containing solutes A and B on the column in Figure 23-1 at time t_1 and at a later time t_2.[3] Because B is more strongly retained by the stationary phase than is A, B lags during the migration. Clearly, the distance between the two increases as they move down the column. At the same time, however, broadening of both bands takes place, which lowers the efficiency of the column as a separating device. While band broadening is inevitable, conditions can often be found where it occurs more slowly than band separation. Thus, as shown in Figure 23-2, a clean resolution of species is possible provided the column is sufficiently long.

Several chemical and physical variables influence the rates of band separation and band broadening. As a consequence, improved separations can often be realized by the control of variables that either increase the rate of band separation or decrease the rate of band spreading. These alternatives are illustrated in Figure 23-3, where the two overlapping bands shown in chromatogram (a) are resolved in chromatogram (b) by altering conditions so that the elution time of the first compound is shortened while that of the second is lengthened. Figure 23-3c shows that resolution can also be achieved by causing the two bands to become narrower.

The variables that influence the relative rates at which solutes migrate

FIGURE 23-2

Concentration profiles for solute bands A and B at two different times in their migration down the column in Figure 23-1. The times t_1 and t_2 are indicated in Figure 23-1.

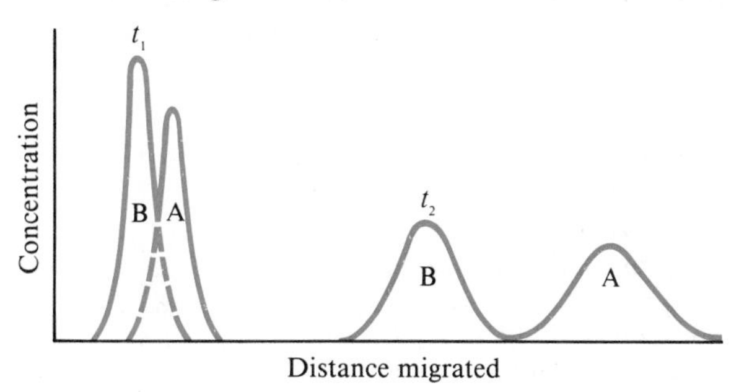

[3]Note that the relative positions of the bands for A and B in the concentration profile in Figure 23-2 appear to be reversed from their positions in the lower part of Figure 23-1. The difference is that the abscissa is distance along the column in Figure 23-2 but time in Figure 23-1. Thus, in Figure 23-1, the *front* of a peak lies to the left and the *tail* to the right; in Figure 23-2, the reverse is true.

through a stationary phase are described in the next section. Following this discussion, we turn to those factors that play a part in zone broadening.

23B Migration Rates of Solutes

The effectiveness of a chromatographic column in separating two solutes depends in part upon the relative rates at which the two species are eluted. These rates are in turn determined by the partition ratios of the solutes between the two phases.

23B-1 Partition Ratios in Chromatography

All chromatographic separations are based upon differences in the extent to which solutes are partitioned between the mobile and the stationary phase. For the solute species A, the equilibrium involved is described by the equation

$$A_{\text{mobile}} \rightleftharpoons A_{\text{stationary}}$$

The equilibrium constant K for this reaction is called a *partition ratio,* or *partition coefficient and is defined as*

$$K = \frac{c_S}{c_M} \tag{23-1}$$

where c_S is the molar analytical concentration of a solute in the stationary phase and c_M is its analytical concentration in the mobile phase. Ideally, the partition ratio is constant over a wide range of solute concentrations; that is, c_S is directly proportional to c_M. At high solute concentrations, however, marked departures from linearity are often encountered. Fortunately, most chromatography is performed under conditions in which Equation 23-1 does apply, which greatly simplifies the derivation of expressions that describe the separation process. Chromatography carried out under conditions in which K is more or less constant is termed *linear chromatography.* The discussions that follow deal exclusively with separations of this type.

23B-2 Retention Times

For the chromatogram shown in Figure 23-4, zero on the time axis corresponds to the instant the sample is injected onto the column and elution is started. The peak at t_M is for a species that is *not* retained by the column; its rate of migration is the same as the average rate of motion of the molecules of the

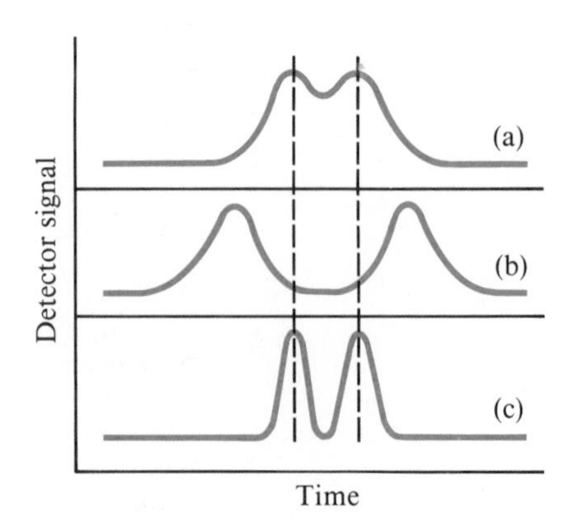

FIGURE 23-3 Two-component chromatograms illustrating two methods of improving separation: (a) original chromatogram with overlapping peaks; improvement brought about by (b) an increase in band separation and (c) a decrease in band width.

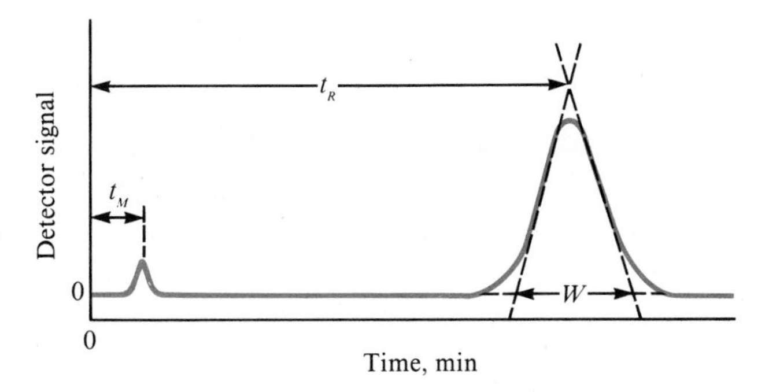

FIGURE 23-4 Determination of the standard deviation τ from a chromatographic peak: $W = 4\tau$, t_R is the retention time for a solute retained by the packing, and t_M is the retention time for one that is not. Thus, t_M is approximately equal to the time required for a molecule of the mobile phase to pass through the column.

mobile phase. The *retention time* t_R for the solute responsible for the second peak is the time for that peak to reach the detector at the end of the column.

The average linear rate of solute migration $\bar{v}$ is

$$\bar{v} = \frac{L}{t_R} \tag{23-2}$$

where L is the length of the column packing. Similarly, the average linear rate of movement u of the molecules of the mobile phase is

$$u = \frac{L}{t_M} \tag{23-3}$$

23B-3 The Relationship Between Retention Time and Partition Ratio

In order to relate the retention time of a solute to its partition ratio, we express its migration rate as a fraction of the velocity of the mobile phase:

$$\bar{v} = u \times \text{fraction of time solute spends in mobile phase}$$

This fraction, however, equals the average number of moles of solute in the mobile phase at any instant divided by the total number of moles of solute in the column:

$$\bar{v} = u \times \frac{\text{moles of solute in mobile phase}}{\text{total moles of solute}}$$

$$\bar{v} = u \times \frac{c_M V_M}{c_M V_M + c_S V_S} = u \times \frac{1}{1 + c_S V_S / c_M V_M}$$

where c_M and c_S are the molar concentrations of the solute in the mobile and stationary phases, respectively, and V_M and V_S are the total volumes of the two phases in the column.

Substitution of Equation 23-1 into this equation gives an expression for the rate of solute migration as a function of its partition ratio as well as a function of the volumes of the stationary and mobile phases:

$$\bar{v} = u \times \frac{1}{1 + K V_S / V_M} \tag{23-4}$$

The two volumes can be estimated from the method by which the column is prepared.

23B-4

The Rate of Migration of Solutes: The Capacity Factor

The *capacity factor* is an important parameter that is widely used to describe the migration rates of solutes on columns. For a solute A, the capacity factor k'_A is defined as

$$k'_A = \frac{K_A V_S}{V_M} \tag{23-5}$$

where K_A is the partition ratio for the species A. Substitution of Equation 23-5 into 23-4 yields

$$\bar{v} = u \times \frac{1}{1 + k'_A} \tag{23-6}$$

In order to show how k'_A can be derived from a chromatogram, we substitute Equations 23-2 and 23-3 into Equation 23-6:

$$\frac{L}{t_R} = \frac{L}{t_M} \times \frac{1}{1 + k'_A} \tag{23-7}$$

This equation rearranges to

$$k'_A = \frac{t_R - t_M}{t_M} \tag{23-8}$$

As shown in Figure 23-4, t_R and t_M are readily obtained from a chromatogram. When the capacity factor for a solute is much less than unity, elution goes on so rapidly that accurate determination of the retention times is difficult. When the capacity factor is larger than perhaps 20 to 30, elution times become inordinately long. Ideally, separations are performed under conditions in which the capacity factors for the solutes in a mixture lie in the range between 1 to 5.

In the chapters that follow, it is shown that capacity factors in gas chromatography can be varied by changing the temperature and the column packing. In liquid chromatography, capacity factors can often be manipulated to give better separations by varying the composition of the mobile phase and the stationary phase.

23B-5

Differential Migration Rates: The Selectivity Factor

The *selectivity factor* α of a column for the two species A and B is defined as

$$\alpha = \frac{K_B}{K_A} \tag{23-9}$$

where K_B is the partition ratio for the more strongly retained species B and K_A is the constant for the less strongly held or more rapidly eluted species A. By this definition, α *is always greater than unity.*

Substitution of Equation 23-5 and the analogous equation for solute B into Equation 23-9 provides after rearrangement a relationship between the selectivity factor for two solutes and their capacity factors:

$$\alpha = \frac{k'_B}{k'_A} \tag{23-10}$$

where k'_B and k'_A are the capacity factors for B and A, respectively. Substitution of Equation 23-8 for the two solutes into Equation 23-10 gives an expression that permits the determination of α from an experimental chromatogram:

$$\alpha = \frac{(t_R)_B - t_M}{(t_R)_A - t_M} \tag{23-11}$$

23C Band Shapes and Band Broadening

A successful theory of column chromatography must account for three phenomena: (1) the differential migration rates of solutes, (2) the Gaussian shapes of chromatographic peaks, and (3) peak broadening. The *kinetic*, or *rate, theory of chromatography* explains all these phenomena in largely quantitative terms.[4]

The discussion in Section 23B addressed differential migration rates. In this section, we consider the shapes and breadths of chromatographic peaks.

23C-1 The Shapes of Chromatographic Peaks

The rate theory of chromatography is capable of describing the shapes of chromatographic peaks in quantitative terms based on a random-walk mechanism for the migration of molecules through a column. For present purposes, however, a qualitative description of the migration process suffices.

Examination of the peaks in typical chromatograms (Figure 23-4) or the concentration profile of bands on a column (Figure 23-2) reveals that their shapes are similar to normal error, or Gaussian, curves (Figure 2-5) obtained when replicate values of a measurement are plotted as a function of their frequency of occurrence. As shown in Section 2D-1, normal error curves are rationalized by assuming that the uncertainty associated with any single measurement is the summation of a much larger number of small, individually undetectable and random uncertainties, each of which has an equal probability of being positive or negative. The most common occurrence is for these uncertainties to cancel one another, thus leading to the mean value. With less likelihood, the summation may cause results to be greater or smaller than the mean. The consequence is a symmetric distribution of data around the mean value shown in Figure 2-5. In a similar way, the typical Gaussian shape of a chromatographic band can be attributed to the additive combination of the random motions of the myriad solute particles as the chromatographic band moves down the column.

It is instructive to consider a single solute molecule as it undergoes many thousands of transfers between the stationary and the mobile phases during its movement through the column. Residence time in either phase is highly irregular. Transfer from one phase to the other requires energy, and the molecule must acquire this energy from its surroundings. Thus, the residence time in a given phase may be transitory in some instances and relatively long in others. Recall that movement can occur *only while the molecule is in the mobile phase*. As a consequence, certain particles travel rapidly by virtue of their accidental inclusion in the mobile phase for a majority of the time whereas others lag because they happen to be incorporated in the stationary phase for

[4]For a detailed presentation of the rate theory, see J. C. Giddings, in *Chromatography*, 3rd ed., E. Heftmann, Ed., Chapter 3. New York: Van Nostrand-Reinhold, 1975. For a shorter presentation, see J. C. Giddings, *J. Chem. Educ.*, **1958**, *35*, 588; **1967**, *44*, 704.

a greater-than-average length of time. The result of these random individual processes is a symmetric spread of velocities around the mean value, which represents the behavior of the average particle.

The breadth of a band increases as it moves down the column because more time is allowed for spreading to occur. Thus, zone breadth is directly related to residence time in the column and inversely related to the flow velocity of the mobile phase.

23C-2 Methods for Describing the Efficiency of Chromatographic Columns

Two related parameters are widely used as measures of the efficiency of chromatographic columns: (1) *number of theoretical plates* N and (2) *plate height* H (or, sometimes, *height equivalent of a theoretical plate HETP*). The two are related by the equation

$$N = L/H \tag{23-12}$$

where L is the length (usually in centimeters) of the column packing.

The efficiency of chromatographic columns increases as the number of plates becomes greater and as the plate height becomes smaller. Enormous differences in efficiencies are encountered in columns, depending upon their type and the kinds of mobile and stationary phases they contain. Efficiencies in terms of plate numbers can vary from a few hundred to several hundred thousand; plate heights ranging from a few tenths to one thousandth of a centimeter or smaller are not uncommon.

Genesis of the Terms Plate and Plate Height

The nomenclature just described for defining column efficiency is unfortunate because it tends to perpetuate the myth that a chromatographic column can be treated as a series of contiguous layers, or plates, within which equilibrium conditions always prevail. This model, which was first developed early in the century to describe the behavior of fractional distillation columns, was modified and applied to column chromatography by Martin and Synge in 1941. Their plate model successfully accounts for the Gaussian shape of chromatographic peaks as well as for factors that influence differences in solute-migration rates. The plate model is totally incapable of accounting for zone broadening, however, because of its basic assumption that equilibrium conditions prevail throughout a column during elution. This assumption can never be valid in the dynamic state that exists in a chromatographic column, where phases are moving past one another at such a pace that sufficient time is not available for equilibration.

Because the plate model is such a poor representation of a chromatographic column, the reader is strongly urged (1) to avoid attaching any real or imaginary significance to the terms *plate* and *plate height* and (2) to view these terms as designators of column efficiency that are retained for historic reasons only and not because they have physical significance. Unfortunately, these terms are so well entrenched in the chromatographic literature that their replacement by more appropriate designations seems unlikely, at least in the near future.

The Relationship Between Plate Height and Band Broadening

As shown in Section 2D-2, the standard deviation σ and the variance σ^2 define the breadth of a Gaussian curve. Because chromatographic bands are also Gaussian and because the efficiency of a column is reflected in the breadth of chromatographic peaks, it is convenient to define column efficiency H in terms of variance per unit length of column:

$$H = \frac{\sigma^2}{L} \tag{23-13}$$

where H, the efficiency parameter, is the plate height. This definition is illustrated in Figure 23-5a, which shows a column having a packing L cm in length. Above this schematic drawing is a plot showing the distribution of molecules along the length of the column at the moment the analyte peak reaches the end of the packing (that is, at the retention time t_R). The curve is Gaussian, and the locations of $L + 1\sigma$ and $L - 1\sigma$ are indicated as broken vertical lines. Note that L carries units of centimeters and σ^2 units of centimeters squared; thus H represents a linear distance in centimeters as well (Equation 23-13). In fact, the plate height can be thought of as the length of column that contains a fraction of the analyte that lies between L and $L - \sigma$. Because the area under a normal error curve bounded by $\pm\sigma$ is about 68% of the total area (page 24), the plate height, as defined, contains 34% of the analyte.

Experimental Evaluation of N and H

Figure 23-4 is a typical chromatogram with time as the abscissa. The *retention time* t_R is defined as the time required after sample injection for the solute peak to appear at the end of the column. The variance of this peak, which can be obtained by a simple graphical procedure, has units of seconds squared, however, and is usually designated as τ^2 to distinguish it from σ^2, which has units of centimeters squared. The two standard deviations τ and σ are related by

$$\tau = \frac{\sigma}{L/t_R} \tag{23-14}$$

where L/t_R is the average linear velocity of the solute in centimeters per second.

Figure 23-4 illustrates a simple means for approximating τ and σ from an experimental chromatogram. Tangents at the inflection points on the two sides of the chromatographic peak are extended to form a triangle with the abscissa. The area of this triangle can be shown to be approximately 96% of the total area under the peak. In Section 2D-2, it was shown that about 96% of the area under a Gaussian peak is included within plus or minus two

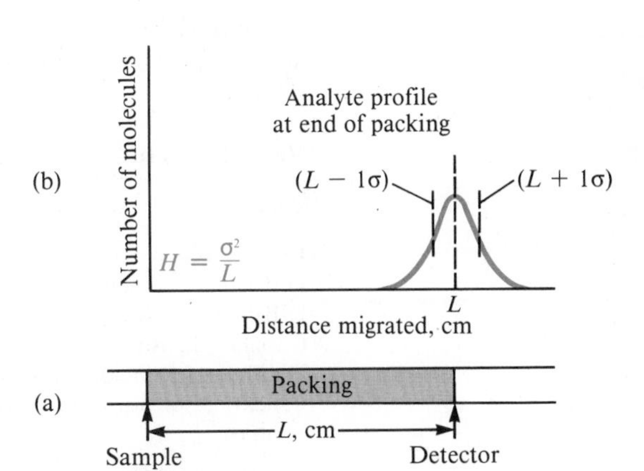

FIGURE 23-5 Definition of plate height $H = \sigma^2/L$.

standard deviations ($\pm 2\sigma$) of its maximum. Thus, the intercepts shown in Figure 23-4 occur at approximately $\pm 2\tau$ from the maximum, and $W = 4\tau$, where W is the magnitude of the base of the triangle. Substituting this relationship into Equation 23-14 and rearranging yield

$$\sigma = \frac{LW}{4t_R}$$

Substitution of this equation for σ into Equation 23-13 gives

$$H = \frac{LW^2}{16t_R^2} \tag{23-15}$$

To obtain N, we substitute into Equation 23-12 and rearrange to get

$$N = 16\left(\frac{t_R}{W}\right)^2 \tag{23-16}$$

Thus, N can be calculated from two time measurements, t_R and W; to obtain H, the length of the column packing L must also be known.

Another method for approximating N, which some workers believe to be more reliable, is to determine $W_{1/2}$, the width of a peak at half its maximum height. The number of theoretical plates is then given by

$$N = 5.54\left(\frac{t_R}{W_{1/2}}\right)^2 \tag{23-17}$$

The two parameters N and H are widely used in the literature and by instrument manufacturers as measures of column performance. For these parameters to be meaningful in comparing two columns, it is essential that they be determined with the *same compound.*

23C-3 Kinetic Variables Affecting Band Broadening

Band broadening is the consequence of the finite rate at which several mass-transfer processes occur during migration of a solute down a column. Some of these rates are controllable by the adjustment of experimental variables, thus permitting improvement in separations. Table 23-2 lists the most important of these variables. Their effects on column efficiency, as measured by plate height H, are described in the paragraphs that follow.

The Effect of Mobile-Phase Flow Rate

The magnitude of kinetic effects on column efficiency clearly depends upon the length of time the mobile phase is in contact with the stationary phase, which in turn depends upon the flow rate of the mobile phase. For this reason,

TABLE 23-2 **Variables That Affect Column Efficiency**

Variable	Symbol	Usual Units
Linear velocity of mobile phase	u	$cm \cdot s^{-1}$
Diffusion coefficient in mobile phase*	D_M	$cm^2 \cdot s^{-1}$
Diffusion coefficient in stationary phase*	D_S	$cm^2 \cdot s^{-1}$
Capacity factor (Equation 23-8)	k'	Unitless
Diameter of packing particle	d_p	cm
Thickness of liquid coating on stationary phase	d_f	cm

*Increases as temperature increases and viscosity decreases.

efficiency studies have generally been carried out by determining H (by means of Equation 23-16 or 23-17 and Equation 23-12) as a function of mobile-phase velocity. The data obtained from such studies are typified by the two plots shown in Figure 23-6, one for liquid chromatography and the other for gas chromatography. While both show a minimum in H (or a maximum in efficiency) at low flow rates, the minimum for liquid chromatography usually occurs at flow rates that are well below those for gas chromatography and often so low that they are not observed under normal operating conditions.

Generally, liquid chromatograms are obtained at lower flow rates than gas chromatograms. Furthermore, as shown in the figure, plate heights for liquid chromatographic columns are an order of magnitude or more smaller than those encountered with gas chromatographic columns. Offsetting this advantage, however, is the fact that it is impractical to employ liquid columns that are longer than about 25 to 50 cm (because of high pressure drops), whereas gas chromatographic columns may be 50 m or more in length. Consequently, the total number of plates, and thus overall column efficiency, are usually superior with gas chromatographic columns.

A Theory of Band Broadening

Over the last 30 years, an enormous amount of theoretical and experimental effort has been devoted to developing quantitative relationships describing the effects of experimental variables on plate heights for various types of columns. Perhaps a dozen or more mathematical expressions for calculating plate height have been put forward and applied with various degrees of success. It is apparent that none of these is entirely adequate to explain the complex physical interactions and effects that lead to zone broadening and thus lower column efficiencies. Some of the equations, though imperfect, have been of considerable use, however, in pointing the way toward improved

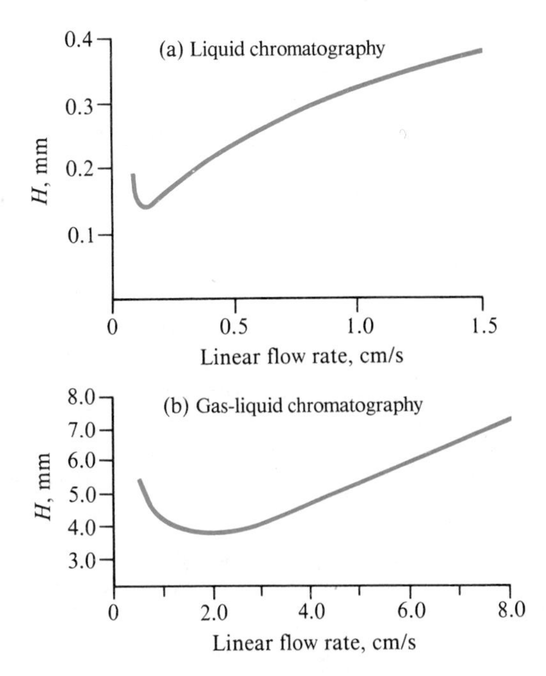

FIGURE 23-6 Effect of mobile-phase flow rates on plate height for (a) liquid chromatography and (b) gas chromatography.

TABLE 23-3 **Kinetic Processes That Contribute to Peak Broadening**

Process	Term in Equation 23-18	Relationship to Column* and Solute Properties	Equation Number
Longitudinal diffusion	B/u	$\frac{B}{u} = \frac{2k_D D_M}{u}$	(23-19)
Mass transfer to and from liquid stationary phase†	$C_S u$	$C_S u = \frac{qk' d_f^2 u}{(1 + k')^2 D_S}$	(23-20)
Mass transfer to and from solid stationary phase†	$C_S u$	$C_S u = \frac{2t_d k' u}{(1 + k')^2}$	(23-21)
Mass transfer in mobile phase	$C_M u$	$C_M u = \frac{f(d_p^2, d_c^2, u)}{D_M} u$	(23-22)

*u, D_M, D_S, d_f, d_p, k' are as defined in Table 23-2.

f: function of.

k_D, q: constants.

t_d: average desorption time of analyte from surface; $t_d = 1/k_d$, where k_d is first-order rate constant for desorption.

d_c: column diameter.

B: coefficient of longitudinal diffusion.

C_S, C_M: coefficients of mass transfer in stationary and mobile phases, respectively.

†Equation 23-20 applies for a liquid stationary phase only. Equation 23-21 applies to a solid stationary phase where adsorption occurs.

column performance.[5] One of these is presented here.

The efficiency of most chromatographic columns can be approximated by the expression

$$H = B/u + C_S u + C_M u \tag{23-18}$$

where H is the plate height in centimeters and u is the linear velocity of the mobile phase in centimeters per second. The quantity B is the *longitudinal diffusion coefficient* while C_S and C_M are *mass transfer coefficients* for the stationary and mobile phases, respectively. The equations given in Table 23-3 reveal the effects of column variables on the three terms in Equation 23-18.

The Longitudinal Diffusion Term B/u. Diffusion is a process in which species migrate from a more concentrated part of a medium to a more dilute. The rate of migration is proportional to the concentration difference between the regions as well as to the *diffusion coefficient* D_M of the species. The latter, which is a measure of the mobility of a substance in a given medium, is a constant equal to the velocity of migration under a unit concentration gradient.

[5]Theoretical studies of zone broadening in the 1950s by Dutch chemical engineers led to the *van Deemter equation*, which can be written in the form

$$H = A + B/u + Cu$$

where the constants A, B, and C are coefficients of eddy diffusion, longitudinal diffusion, and mass transfer, respectively. This equation is of considerable historic interest; its modernized version takes the form of Equation 23-18 (see S. J. Hawkes, *J. Chem. Educ.*, **1983**, *60*, 393).

In chromatography, longitudinal diffusion results in the migration of a solute from the concentrated center of a band to the more dilute regions on either side (that is, toward and opposed to the direction of flow). Longitudinal diffusion is a common source of band broadening in gas chromatography but is of little significance in liquid chromatography because the rate at which molecules diffuse in a gaseous medium is high, whereas the rate in a liquid solvent is relatively low. The magnitude of the B term in Equation 23-18 is largely determined by the diffusion coefficient D_M of the analyte in the mobile phase and is directly proportional to this constant (Equation 23-19).

As shown by Equation 23-18, the contribution of longitudinal diffusion to plate height is inversely proportional to the linear velocity of the eluent. Such a relationship is not surprising inasmuch as the analyte is in the column for a briefer period when the flow rate is high. Thus, diffusion from the center of the band to the two edges has less time to occur.

The initial decreases in H shown in both curves in Figure 23-6 are a direct consequence of the longitudinal diffusion. Note that the effect is much less pronounced in liquid chromatography because of the much lower diffusion rates in a liquid mobile phase. The striking difference in plate heights shown by the two graphs in Figure 23-6 can also be explained by considering the relative rates of longitudinal diffusion in the two mobile phases. That is, diffusion coefficients in gaseous media are orders of magnitude larger than in liquids. Thus band broadening goes on to a much greater extent in gas chromatography than in liquid chromatography.

The Mass-Transfer Coefficients C_S and C_M. The need for the two mass transfer coefficients C_S and C_M in Equation 23-18 arises because the equilibrium between the mobile and the stationary phase is established so slowly that a chromatographic column always operates under nonequilibrium conditions. Consequently, analyte molecules at the front of a band are swept ahead before they have time to equilibrate with the stationary phase and thus be retained. Similarly, equilibrium is not reached at the trailing edge of a band, and molecules are left behind in the stationary phase by the fast moving mobile phase.

Band broadening from mass-transfer effects arises because the many flowing streams of a mobile phase within a column and the layer of immobilized liquid making up the stationary phase both have finite widths. Consequently time is required for solute molecules to diffuse from the interior of these phases to their interface where transfer occurs. This time lag results in the persistence of nonequilibrium conditions along the length of the column. If the rates of mass transfer within the two phases were infinite, broadening of this type would not occur.

Note that the extent of both longitudinal broadening and mass-transfer broadening depend upon the rate of diffusion of analyte molecules but that the direction of diffusion in the two cases is different. Longitudinal broadening arises from the tendency of molecules to move in directions that tend to parallel the flow, whereas mass-transfer broadening occurs from diffusion that tends to be at right angles to the flow. As a consequence, the extent of longitudinal broadening is *inversely* related to flow rate. For mass-transfer broadening, in contrast, the faster the mobile phase moves, the less time there is for equilibrium to be approached. Thus, as shown by the last two terms in Equation 23-18, the mass-transfer effect on plate height is directly proportional to the rate u of movement of the mobile phase.

The Stationary Phase Mass-Transfer Term $C_S u$. When the stationary phase is an immobilized liquid, the mass-transfer coefficient is directly proportional to the square of the thickness of the film on the support particles d_f^2 and inversely proportional to the diffusion coefficient D_S of the solute in the film (Equation 23-20). These effects can be understood by realizing that both reduce the average frequency at which analyte molecules reach the interface where transfer to the mobile phase can occur. That is, with thick films, molecules must on the average travel farther to reach the surface, and with smaller diffusion coefficients, they travel slower. The consequence is a slower rate of mass transfer and an increase in plate height.

When the stationary phase is a solid surface, the mass-transfer coefficient C_S is directly proportional to the time required for a species to be adsorbed or desorbed, which in turn is inversely proportional to the first order rate constant for the processes (Equation 23-21).

The Mobile Phase Mass-Transfer Term $C_M u$. The mass-transfer processes that occur in the mobile phase are sufficiently complex to defy complete and rigorous analysis, at least to date. On the other hand, a good qualitative understanding of the variables affecting zone broadening from this cause exists, and this understanding has led to vast improvements in all types of chromatographic columns.

The mobile-phase mass-transfer coefficient C_M is known to be inversely proportional to the diffusion coefficient of the analyte in the mobile phase D_M and also to be some function of the square of the particle diameter of the packing d_p^2, the square of the column diameter d_c^2, and the flow rate (Equation 23-22).

The contributions of mobile-phase mass-transfer to plate height is the product of the mass transfer coefficient C_M (which is a function of solvent velocity) and the velocity of the solvent. Thus, the net contribution of $C_M u$ to plate height is not linear in u (see the curve labeled $C_M u$ in Figure 23-8) but bears a complex dependency on solvent velocity.

Zone broadening in the mobile phase arises in part from the multitude of pathways by which a molecule (or ion) can find its way through a packed column. As shown in Figure 23-7, the length of these pathways may differ significantly; thus, the residence time in the column for molecules of the same species is also variable. Solute molecules then reach the end of the column over a time interval, which leads to a broadened band. This effect, which is

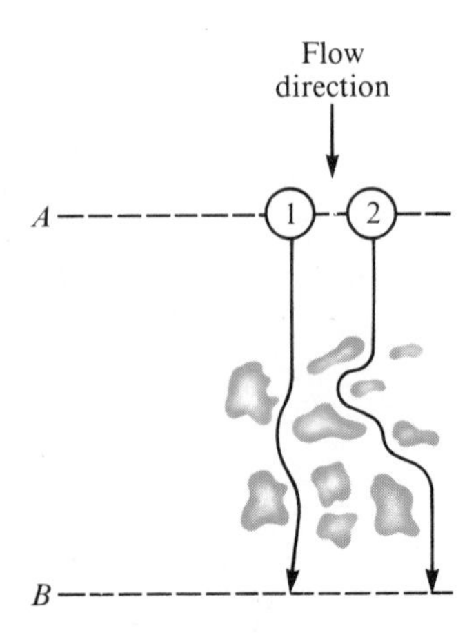

FIGURE 23-7 Typical pathways of two solute molecules during elution. Note that the distance traveled by molecule 2 is greater than that traveled by molecule 1. Thus, molecule 2 arrives at B later than molecule 1.

sometimes called *eddy diffusion,* would be independent of solvent velocity if it were not partially offset by ordinary diffusion, which results in molecules being transferred from a stream following one pathway to a stream following another. If the velocity of flow is very low, a large number of these transfers will occur, and each molecule in its movement down the column samples numerous flow paths, spending a brief time in each. As a consequence, the rate at which each molecule moves down the column tends to approach that of the average. Thus, at low mobile-phase velocities, the molecules are not significantly dispersed by the multiple-path nature of the packing. At moderate or high velocities, however, sufficient time is not available for diffusion averaging to occur, and band broadening due to the different path lengths is observed. At sufficiently high velocities, the effect of eddy diffusion becomes independent of flow rate.

Superimposed upon the eddy diffusion effect is one that arises from stagnant pools of the mobile phase retained in the stationary phase. Thus, when a solid serves as the stationary phase, its pores are filled with *static* volumes of mobile phase. Solute molecules must then diffuse through these stagnant pools before transfer can occur between the *moving* mobile phase and the stationary phase. This situation applies not only to solid stationary phases but also to liquid stationary phases immobilized on porous solids because the immobilized liquid does not usually fully fill the pores.

The presence of stagnant pools of mobile phase slows the exchange process and results in a contribution to the plate height that is directly proportional to the mobile phase velocity and inversely proportional to the diffusion coefficient for the solute in the mobile phase. An increase in particle diameter d_p also has a significant effect because of the increase in internal volume that accompanies increases in particle size.

Effect of Mobile Phase Velocity on Terms in Equation 23-18. Figure 23-8 shows the variation of the three terms in Equation 23-18 as a function of mobile phase velocity. The top curve is the summation of these various effects. Note that an optimum flow rate exists at which the plate height is a minimum and the separation efficiency is a maximum.

Summary of Methods for Reducing Band Broadening. Two important controllable variables that affect column efficiency are the diameter of the particles making up the packing and the diameter of the column. The effect of particle diameter is demonstrated by the data shown in Figures 24-1 and 25-1. To

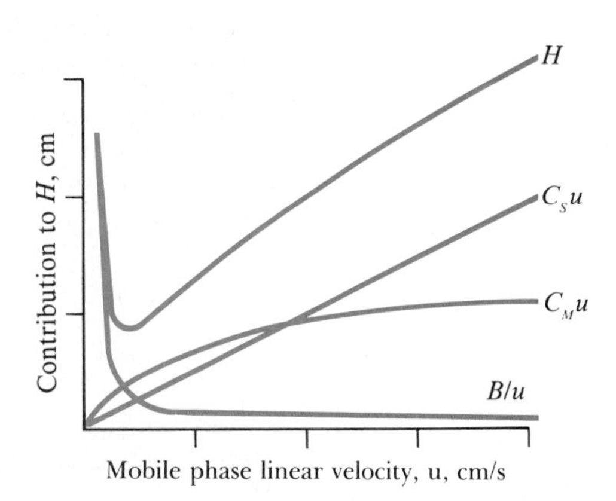

FIGURE 23-8 Contribution of various mass-transfer coefficients to column plate height. $C_S u$ arises from the rate of mass transfer to and from the stationary phase, $C_M u$ comes from a limitation in the rate of mass transfer in the mobile phase, and B/u is associated with longitudinal diffusion.

take advantage of the effect of column diameter, narrower and narrower columns have been used in recent years.

With gaseous mobile phases, the rate of longitudinal diffusion can be reduced appreciably by lowering the temperature and thus the diffusion coefficient D_M. The consequence is significantly smaller plate heights at low temperatures. This effect is usually not noticeable in liquid chromatography because diffusion is slow enough so that the longitudinal diffusion term has little effect on overall plate height.

With liquid stationary phases, the thickness of the layer of adsorbed liquid should be minimized since C_S in Equation 23-18 is proportional to the square of this variable d_f in Equation 23-20.

23D Optimization of Column Performance

A chromatographic separation is optimized by varying experimental conditions until the components of a mixture are separated cleanly with a minimum expenditure of time. Optimization experiments are aimed at either (1) reducing zone broadening or (2) altering relative migration rates of the components. As we have shown in Section 23C, zone broadening is increased by those kinetic variables that increase the plate height of a column. Migration rates, on the other hand, are varied by changing those variables that affect the capacity and selectivity factors of the solutes (Section 23B).

23D-1 Column Resolution

The *resolution* R_s of a column provides a quantitative measure of its ability to separate two analytes. The significance of this term is illustrated in Figure 23-9, which consists of chromatograms for species A and B on three columns with different resolving powers. The resolution of each column is defined as

$$R_s = \frac{2\Delta Z}{W_A + W_B} = \frac{2[(t_R)_B - (t_R)_A]}{W_A + W_B} \tag{23-23}$$

where all of the terms on the right side are as defined in the figure.

It is evident from Figure 23-9 that a resolution of 1.5 gives an essentially complete separation of A and B, whereas a resolution of 0.75 does not. At a resolution of 1.0, zone A contains about 4% B and zone B contains about 4% A. At a resolution of 1.5, the overlap is about 0.3%. The resolution for a given stationary phase can be improved by lengthening the column, thus increasing the number of plates. An adverse consequence of the added plates, however, is an increase in the time required for the resolution.

The Relationship Between Resolution and Properties of the Column and Solute

A useful equation is readily derived that relates the resolution of a column to the number of plates it contains as well as to the capacity and selectivity factors of a pair of solutes on the column. Thus, it is readily shown[6] that for the two solutes A and B in Figure 23-9, the resolution is given by the equation

$$R_s = \frac{\sqrt{N}}{4}\left(\frac{\alpha - 1}{\alpha}\right)\left(\frac{k'_B}{1 + k'_B}\right) \tag{23-24}$$

[6]See D. A. Skoog, *Principles of Instrumental Analysis*, 3rd ed., pp. 741–743. Philadelphia: Saunders College Publishing, 1985.

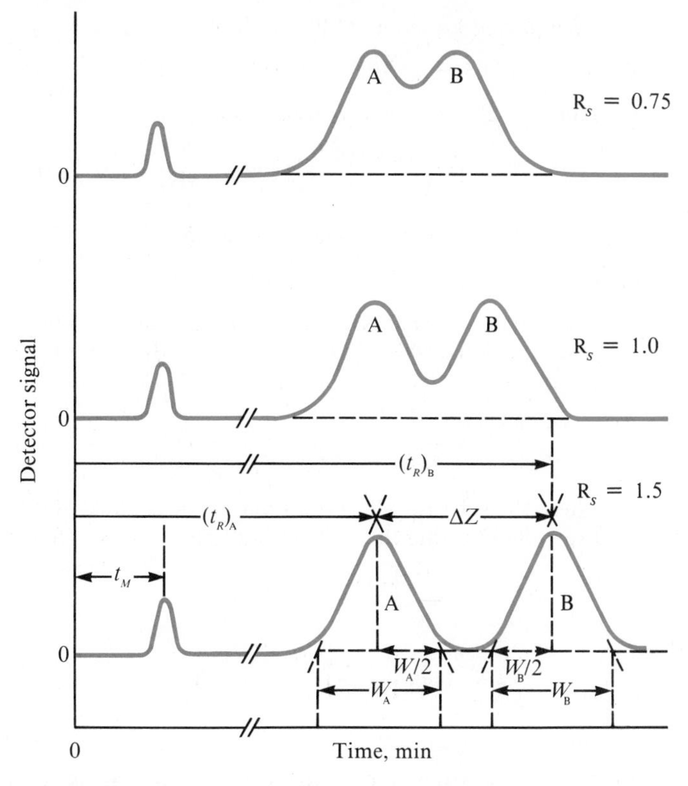

FIGURE 23-9 Separation at three resolutions: $R_s = 2\Delta Z/(W_A + W_B)$.

where k'_B is the capacity factor of the slower-moving species and α is the selectivity factor. This equation can be rearranged to give the number of plates needed to realize a given resolution:

$$N = 16R_s^2 \left(\frac{\alpha}{\alpha - 1}\right)^2 \left(\frac{1 + k'_B}{k'_B}\right)^2 \tag{23-25}$$

The Relationship Between Resolution and Elution Time

As mentioned earlier, the goal in chromatography is the highest possible resolution in the shortest possible elapsed time. Unfortunately, these goals tend to be incompatible, and a compromise between the two is usually necessary. The time $(t_R)_B$ required to elute the two species in Figure 23-9 with a resolution of R_s is

$$(t_R)_B = \frac{16R_s^2 H}{u} \left(\frac{\alpha}{\alpha - 1}\right)^2 \frac{(1 + k'_B)^3}{(k'_B)^2} \tag{23-26}$$

where u is the linear rate of movement of the mobile phase.

EXAMPLE 23-1

Substances A and B have retention times of 16.40 and 17.63 min, respectively, on a 30.0-cm column. An unretained species passes through the column in 1.30 min. The peak widths (at base) for A and B are 1.11 and 1.21 min, respectively. Calculate (a) column resolution, (b) average number of plates in the column, (c) plate height, (d) length of column required to achieve a resolution of 1.5, and (e) time required to elute substance B on the longer column.

Employing Equation 23-23, we find

(a) $R_s = 2(17.63 - 16.40)/(1.11 + 1.21) = 1.06$

(b) Equation 23-16 permits computation of N:

$$N = 16\left(\frac{16.40}{1.11}\right)^2 = 3493 \quad \text{and} \quad N = 16\left(\frac{17.63}{1.21}\right)^2 = 3397$$

$$N_{av} = (3493 + 3397)/2 = 3445 = 3.4 \times 10^3$$

(c) $H = L/N = 30.0/3445 = 8.7 \times 10^{-3}$ cm

(d) k' and α do not change with increasing N and L. Thus, substituting N_1 and N_2 into Equation 23-25 and dividing one of the resulting equations by the other yield

$$\frac{(R_s)_1}{(R_s)_2} = \frac{\sqrt{N_1}}{\sqrt{N_2}}$$

where the subscripts 1 and 2 refer to the original and longer columns, respectively. Substituting the appropriate values for N_1, $(R_s)_1$, and $(R_s)_2$ gives

$$\frac{1.06}{1.5} = \frac{\sqrt{3445}}{\sqrt{N_2}}$$

$$N_2 = 3445\left(\frac{1.5}{1.06}\right)^2 = 6.9 \times 10^3$$

But

$$L = NH = 6.9 \times 10^3 \times 8.7 \times 10^{-3} = 60 \text{ cm}$$

(e) Substituting $(R_s)_1$ and $(R_s)_2$ into Equation 23-26 and dividing yield

$$\frac{(t_R)_1}{(t_R)_2} = \frac{(R_s)_1^2}{(R_s)_2^2} = \frac{17.63}{(t_R)_2} = \frac{(1.06)^2}{(1.5)^2}$$

$$(t_R)_2 = 35 \text{ min}$$

Thus, to obtain the improved resolution, the separation time must be doubled.

23D-2 Optimization Techniques

Equations 23-24 and 23-26 serve as guides in choosing conditions that lead to a desired degree of resolution with a minimum expenditure of time. An examination of these equations reveals that each is made up of three parts. The first describes the efficiency of the column in terms of $\sqrt{N}$ or H. The second, which is the quotient containing α, is a selectivity term that depends on the properties of the two solutes. The third component is the capacity term, which is the quotient containing k'_B; this term depends on the properties of both the solute and the column.

Variation in Plate Height

As shown by Equation 23-24, the resolution of a column improves as the square root of the number of plates it contains increases. Example 23-1e reveals, however, that increasing the number of plates is expensive in terms of time unless the increase is achieved by reducing the plate height and not by increasing column length.

Methods for minimizing plate height are discussed in Section 23C and include reducing the particle size of the packing, the diameter of the column, the column temperature (gas chromatography), and the thickness of the liquid film (liquid chromatography). Optimizing the flow rate of the mobile phase is also helpful.

Variation in the Capacity Factor

Often, a separation can be improved significantly by manipulation of the capacity factor k'_B. Increases in k'_B generally enhance resolution (but at the expense of elution time). To determine the optimum range of values for k'_B, it is convenient to write Equation 23-24 in the form

$$R_s = Q \frac{k'_B}{1 + k'_B}$$

and Equation 23-26 as

$$(t_R)_B = Q' \frac{(1 + k'_B)^3}{(k'_B)^2}$$

where Q and Q' contain the rest of the terms in the two equations. Figure 23-10 is a plot of R_s/Q and $(t_R)_B/Q'$ as a function of k'_B, assuming Q and Q' remain approximately constant. It is clear that values of k'_B greater than about 10 are to be avoided because they provide little increase in resolution but markedly increase the time required for separations. The minimum in the elution-time curve occurs at $k'_B \approx 2$. Often, then, the optimal value of k'_B lies in the range from 1 to 5.

Usually, the easiest way to improve resolution is by optimizing k'. For gaseous mobile phases, k' can often be improved by temperature changes. For liquid mobile phases, changes in the solvent composition often permit manipulation of k' to yield better separations. An example of the dramatic effect that relatively simple solvent changes can bring about is demonstrated in Figure 23-11. Here, modest variations in the methanol/water ratio convert unsatisfactory chromatograms (a and b) to ones with well-separated peaks for each component (c and d). For most purposes, the chromatogram shown in (c) is best since it shows adequate resolution in minimum time.

Variation in the Selectivity Factor

Optimizing k' and increasing N are not sufficient to give a satisfactory separation of two solutes in a reasonable time when α approaches unity. Under this circumstance, a means must be sought to increase α while maintaining k' in the range of 1 to 10. Several options are available; in decreasing order of

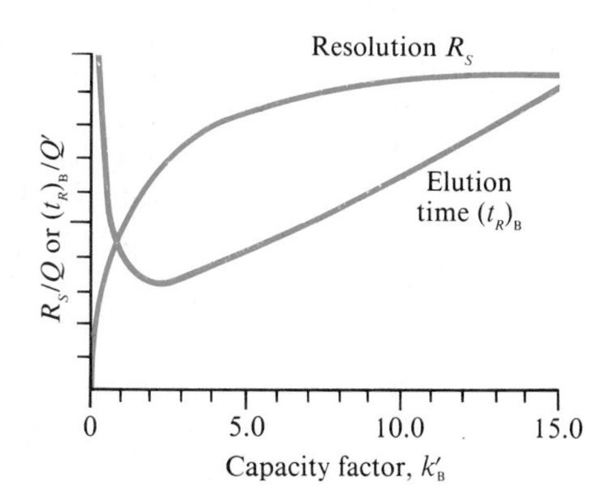

FIGURE 23-10 Effect of capacity factor k'_B on resolution R_s and elution time $(t_R)_B$. It is assumed that Q and Q' remain constant with variations in k'_B.

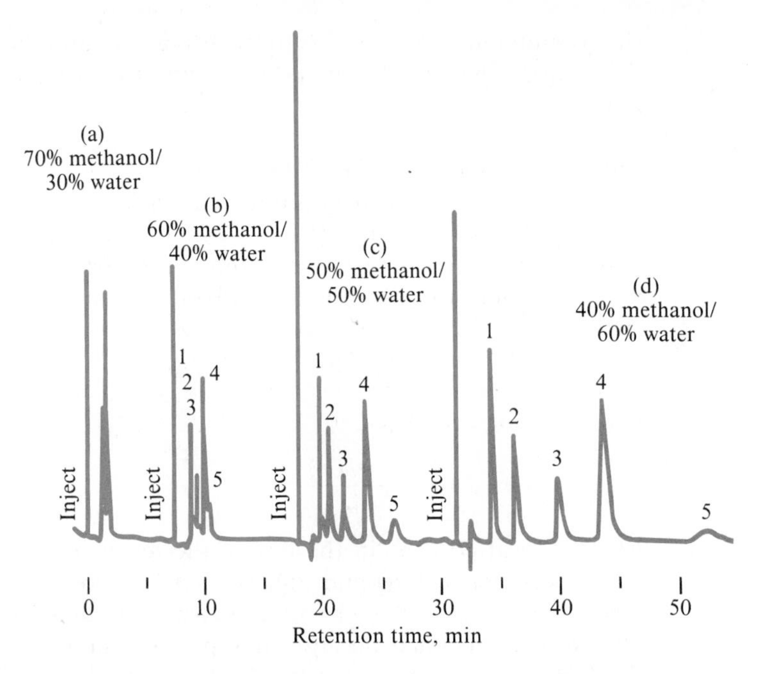

FIGURE 23-11 Effect of solvent variation on chromatograms. Analytes: (1) 9,10-anthraquinone; (2) 2-methyl-9,10-anthraquinone; (3) 2-ethyl-9,10-anthraquinone; (4) 1,4-dimethyl-9,10-anthraquinone; (5) 2-*t*-butyl-9,10-anthraquinone. (Courtesy of DuPont Biotechnology Systems, Wilmington, DE.)

their desirability as determined by promise and convenience, the options are (1) changing the composition of the mobile phase, (2) changing the column temperature, (3) changing the composition of the stationary phase, and (4) using special chemical effects.

An example of the use of option 1 has been reported for the separation of anisole ($C_6H_5OCH_3$) and benzene.[7] With a mobile phase that was a 50% mixture of water and methanol, k' was 4.5 for anisole and 4.7 for benzene, while α was only 1.04. Substitution of an aqueous mobile phase containing 37% tetrahydrofuran gave k' values of 3.9 and 4.7 and an α value of 1.20. Peak overlap was significant with the first solvent system and negligible with the second.

A less convenient but often highly effective method of improving α while maintaining values for k' in their optimal range is to alter the chemical composition of the stationary phase. To take advantage of this option, most laboratories that carry out chromatographic separations frequently maintain several columns that can be interchanged with a minimum of effort.

Increases in temperature usually cause increases in k' but have little effect on α values in liquid-liquid and liquid-solid chromatography. In contrast, with ion-exchange chromatography, temperature effects can be large enough to make exploration of this option worthwhile before resorting to a change in column packing.

[7]L. R. Snyder and J. J. Kirkland, *Introduction to Modern Liquid Chromatography*, 2nd ed., p. 75. New York: Wiley, 1979.

A final method for enhancing resolution is to incorporate into the stationary phase a species that complexes or otherwise interacts with one or more components of the sample. A well-known example of the use of this option arises where an adsorbent impregnated with a silver salt improves the separation of olefins as a consequence of the formation of complexes between the silver ions and unsaturated organic compounds.

23D-3 The General Elution Problem

In Figure 23-12 are hypothetical chromatograms for a six-component mixture made up of three pairs of components with widely different distribution coefficients and thus widely different capacity factors. In chromatogram (a), conditions have been adjusted so that the capacity factors for components 1 and 2 (k'_1 and k'_2) are in the optimal range of 1 to 5. The factors for the other components are far larger than the optimum, however. Thus, the peaks for components 5 and 6 appear only after an inordinate length of time has passed; furthermore, these peaks are so broad that they may be difficult to identify unambiguously.

As shown in chromatogram (b), changing conditions to optimize the separation of components 5 and 6 bunches the peaks for the first four components to the point where their resolution is unsatisfactory. Here, however, the total elution time is ideal.

A third set of conditions, in which k' values for components 3 and 4 are optimal, results in chromatogram (c). Again, separation of the other two pairs is not entirely satisfactory.

The phenomenon illustrated in Figure 23-12 is encountered often enough to be given a name—the *general elution problem.* A common solution to this problem is to change conditions that determine the values of k' as the separation proceeds. These changes can be performed in a stepwise manner or continuously. Thus, for the mixture shown in Figure 23-12, conditions at the

FIGURE 23-12 The general elution problem in chromatography.

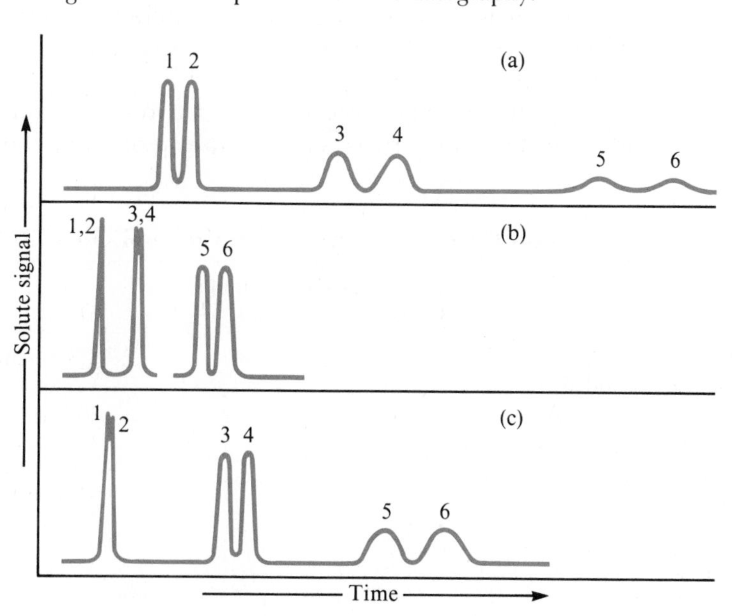

outset could be those producing chromatogram (a). Immediately after the elution of components 1 and 2, conditions could be changed to those that are optimal for separating components 3 and 4 (as in chromatogram c). With the appearance of peaks for these components, the elution could be completed under the conditions used for producing chromatogram (b). Often such a procedure leads to satisfactory separation of all the components of a mixture in minimal time.

For liquid chromatography, variations in k' are brought about by varying the composition of the mobile phase during elution (*gradient elution* or *solvent programming*). For gas chromatography, temperature increases (*temperature programming*) achieve optimal conditions for separations.

23E A Summary of Important Relationships for Chromatography

The number of quantities, terms, and relationships employed in chromatography is large and often confusing. Tables 23-4 and 23-5 summarize the most important definitions and equations used in this text.

23F Applications of Chromatography

Chromatography has grown to be the premiere method for separating closely related chemical species. In addition, it can be employed for the qualitative identification and quantitative determination of separated species.

23F-1 Qualitative Analysis

Chromatography is widely used for recognizing the presence or absence of components in mixtures that contain a limited number of possible species whose identities are known. For example, 30 or more amino acids in a protein hydrolysate can be detected with a reasonable degree of certainty by means of a chromatogram. On the other hand, because a chromatogram provides but a single piece of information about each species in a mixture (the retention time), the application of the technique to the qualitative analysis of complex samples of unknown composition is limited. Nevertheless, chromatography often serves as a first step in a qualitative analysis by various spectroscopic techniques.

It is important to note that while a chromatogram may not lead to positive identification of the species in a sample, it often provides sure evidence of the

TABLE 23-4 **Important Chromatographic Experimental Quantities and Relationships**

Name	Symbol of Experimental Quantity	Determined from
Migration time, nonretained species	t_M	Chromatogram (Figure 23-9)
Retention times, species A and B	$(t_R)_A$, $(t_R)_B$	Chromatogram (Figure 23-9)
Adjusted retention time, species A	$(t'_R)_A$	$(t'_R)_A = (t_R)_A - t_M$
Peak widths, species A and B	W_A, W_B	Chromatogram (Figure 23-9)
Length of column packing	L	Direct measurement
Flow rate	F	Direct measurement
Volume of stationary phase	V_S	Packing preparation data
Concentration of solute in mobile and stationary phases	C_M, C_S	Analysis and preparation data

TABLE 23-5 **Important Derived Quantities and Relationships**

Name	Calculation of Derived Quantities	Relationship to Other Quantities
Linear mobile-phase velocity	$u = L/t_M$	
Volume of mobile phase	$V_M = t_M F$	
Capacity factor	$k' = (t_R - t_M)/t_M$	$k' = \frac{KV_S}{V_M}$
Partition coefficient	$K = \frac{k'V_M}{V_S}$	$K = \frac{C_S}{C_M}$
Selectivity factor	$\alpha = \frac{(t_R)_B - t_M}{(t_R)_A - t_M}$	$\alpha = \frac{k'_B}{k'_A} = \frac{K_B}{K_A}$
Resolution	$R_S = \frac{2[(t_R)_B - (t_R)_A]}{W_A + W_B}$	$R_S = \frac{\sqrt{N}}{4}\left(\frac{\alpha - 1}{\alpha}\right)\left(\frac{k'_B}{1 + k'_B}\right)$
Number of plates	$N = 16\left(\frac{t_R}{W}\right)^2$	$N = 16R_S^2\left(\frac{\alpha}{\alpha - 1}\right)^2\left(\frac{1 + k'_B}{k'_B}\right)^2$
Plate height	$H = L/N$	
Retention time	$(t_R)_B = \frac{16R_S^2H}{u}\left(\frac{\alpha}{\alpha - 1}\right)^2\frac{(1 + k'_B)^3}{(k'_B)^2}$	

absence of species. Thus, failure of a sample to produce a peak at the same retention time as a standard obtained under identical conditions is strong evidence that the compound in question is absent (or present at a concentration below the detection limit of the procedure).

23F-2 Quantitative Analysis

Chromatography owes its enormous growth in part to its speed, simplicity, relatively low cost, and wide applicability as a separating tool. It is doubtful, however, that its use would have become so widespread had it not been for the fact that it can also provide quantitative information about separated species.

Quantitative chromatography is based upon a comparison of either the height or the area of the analyte peak with that of one or more standards. If conditions are properly controlled, both of these parameters vary linearly with concentration.

Analyses Based on Peak Height

The height of a chromatographic peak is obtained by connecting the baselines on the two sides of the peak by a straight line and measuring the perpendicular distance from this line to the peak. This measurement can ordinarily be made with reasonably high precision and yields accurate results, provided variations in column conditions do not alter peak width during the period required to obtain chromatograms for sample and standards. The variables that must be controlled closely are column temperature, eluent flow rate, and rate of sample injection. In addition, care must be taken to avoid overloading the column.

The effect of sample-injection rate is particularly critical for the early peaks of a chromatogram. Relative errors of 5 to 10% due to this cause are not unusual with syringe injection.

Analyses Based on Peak Area

Peak area is independent of broadening effects caused by the variables mentioned in the previous paragraph. From this standpoint, therefore, area is a more satisfactory analytical parameter than peak height. On the other hand, peak heights are more easily measured and, for narrow peaks, more accurately determined.

Many modern chromatographic instruments are equipped with electronic integrators that provide precise measurements of relative peak areas. If such equipment is not available, a manual estimate must be made. A simple method that works well for symmetric peaks of reasonable widths is to multiply peak height by the width at one-half peak height. Alternatively, a planimeter can be used, or the recorded peak can be cut out and weighed, with its weight then being compared with the weight of a known area of recorder paper. McNair and Bonelli measured the precision of these various techniques on chromatograms for ten replicate samples.[8] They reported the following relative standard deviations: electronic integration, 0.44%; mechanical integration, 1.3%; weight of paper, 1.7%; height times width at one-half height, 2.6%; planimeter, 4.1%.

Calibration with Standards

The most straightforward method for quantitative chromatographic analyses involves the preparation of a series of standard solutions that approximate the composition of the unknown. Chromatograms for the standards are then obtained, and peak heights or areas are plotted as a function of concentration. A plot of the data should yield a straight line passing through the origin; analyses are based upon this plot. Frequent restandardization is necessary for highest accuracy.

The most important source of error in analyses by the method based on calibration standards is usually the uncertainty in the volume of sample; occasionally, the rate of injection is also a factor. Samples are ordinarily small ($\simeq 1\ \mu L$), and the uncertainties associated with the injection of a reproducible volume of this size with a microsyringe can amount to several percent relative. The situation is even worse in gas-liquid chromatography, where the sample must be injected into a heated sample port. In this circumstance, evaporation from the needle tip may lead to large variations in the volume injected.

Errors in sample volume can be reduced to perhaps 1 to 2% relative by means of a rotary sample valve such as that shown in Figure 23-13. The sample loop *ACB* in (a) is filled with liquid; rotation of the valve by 45 deg then introduces a reproducible volume of sample (the volume originally contained in *ACB*) into the mobile-phase stream.

[8] H. M. McNair and E. J. Bonelli, *Basic Gas Chromatography*, p. 158. Walnut Creek, CA: Varian Aerograph, 1978.

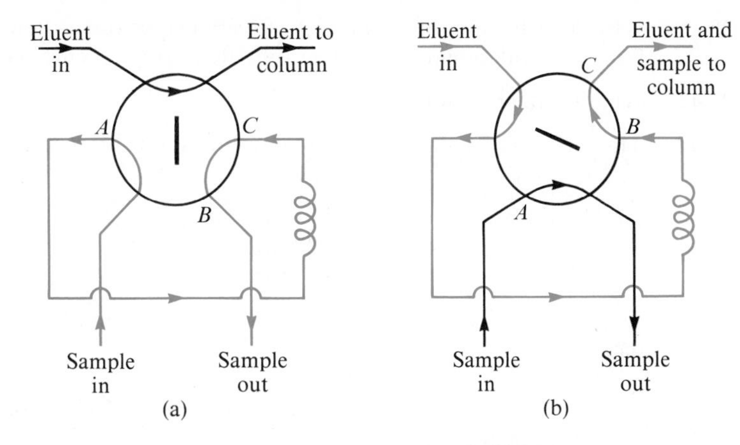

FIGURE 23-13 A rotary sample valve: (a) valve position for filling sample loop *ACB;* (b) valve position for introduction of sample into column.

The Internal-Standard Method

The highest precision for quantitative chromatography is obtained by using internal standards because the uncertainties introduced by sample injection are avoided. In this procedure, a carefully measured quantity of an internal-standard substance is introduced into each standard and sample, and the ratio of analyte peak area (or height) to internal-standard peak area (or height) is the analytical parameter. For this method to be successful, it is necessary that the internal-standard peak be well separated from the peaks of all other components in the sample ($R_s > 1.25$), but it must appear close to the analyte peak, however. With a suitable internal standard, precisions of 0.5 to 1% relative are reported.

Questions and Problems

23-1 Define

*(a) elution	*(g) selectivity factor
(b) mobile phase	(h) plate height
*(c) stationary phase	*(i) longitudinal diffusion
(d) partition ratio	(j) eddy diffusion
*(e) retention time	*(k) column resolution
(f) capacity factor	(l) eluent

*23-2 Describe the general elution problem.

23-3 List the variables that lead to zone broadening.

*23-4 What is the difference between gas-liquid and liquid-liquid chromatography?

23-5 What is the difference between liquid-liquid and liquid-solid chromatography?

*23-6 What variables are likely to affect the α value for a pair of analytes?

23-7 How can the capacity factor for a solute be manipulated?

*23-8 Describe a method for determining the number of plates in a column.

23-9 What are the effects of temperature variation on chromatograms?

*23-10 Why does the minimum in a plot of plate height versus flow rate occur at lower flow rates with liquid chromatography than with gas chromatography?

23-11 What is gradient elution?

*23-12 The following data apply to a column for liquid chromatography:

Length of packing	24.7 cm
Flow rate	0.313 mL/min
V_M	1.37 mL
V_S	0.164 mL

A chromatogram of a mixture of species A, B, C, and D provided the following data:

	Retention Time, min	Width of Peak Base (W), min
Nonretained	3.1	—
A	5.4	0.41
B	13.3	1.07
C	14.1	1.16
D	21.6	1.72

Calculate

(a) the number of plates from each peak.
(b) the mean and the standard deviation for N.
(c) the plate height for the column.

*23-13 From the data in Problem 23-12, calculate for A, B, C, and D

(a) the capacity factor.
(b) the partition coefficient.

*23-14 From the data in Problem 23-12, calculate for species B and C

(a) the resolution.
(b) the selectivity factor.
(c) the length of column necessary to separate the two species with a resolution of 1.5.
(d) the time required to separate the two species with a resolution of 1.5.

*23-15 From the data in Problem 23-12, calculate for species C and D

(a) the resolution.
(b) the length of column required to separate the two species with a resolution of 1.5.

23-16 The following data were obtained by gas-liquid chromatography on a 40-cm packed column:

Compound	t_R, min	$W_{1/2}$, min
Air	1.9	—
Methylcyclohexane	10.0	0.76
Methylcyclohexene	10.9	0.82
Toluene	13.4	1.06

Calculate

(a) an average number of plates from the data.
(b) the standard deviation for the average in (a).
(c) an average plate height for the column.

23-17 Referring to Problem 23-16, calculate the resolution for
(a) methylcyclohexene and methylcyclohexane.
(b) methylcyclohexene and toluene.
(c) methylcyclohexane and toluene.

23-18 If a resolution of 1.5 is desired in separating methylcyclohexane and methylcyclohexene in Problem 23-16,
(a) how many plates are required?
(b) how long must the column be if the same packing is employed?
(c) what is the retention time for methylcyclohexene on the column in Problem 23-16b?

23-19 If V_S and V_M for the column in Problem 23-16 are 19.6 and 62.6 mL, respectively, and a nonretained air peak appears after 1.9 min, calculate the
(a) capacity factor for each compound.
(b) partition coefficient for each compound.
(c) selectivity factor for methylcyclohexane and methylcyclohexene.

*23-20 List the variables that lead to (a) band broadening and (b) band separation.

*23-21 What is the effect on a chromatographic peak of introducing the sample too slowly?

*23-22 From distribution studies, species M and N are known to have water/hexane partition coefficients of 6.01 and 6.20 ($K = [M]_{H_2O}/[M]_{hex}$). The two species are to be separated by elution with hexane in a column packed with silica gel containing adsorbed water. The ratio V_S/V_M for the packing is 0.422.
(a) Calculate the capacity factor for each solute.
(b) Calculate the selectivity factor.
(c) How many plates are needed to provide a resolution of 1.5?
(d) How long a column is needed if the plate height of the packing is 2.2×10^{-3} cm?
(e) If a flow rate of 7.10 cm/min is employed, how long will it take to elute the two species?

23-23 Repeat the calculations in Problem 23-22 assuming $K_M = 5.81$ and $K_N = 6.20$.

CHAPTER 24

Gas-Liquid Chromatography

In *gas chromatography* (GC), an inert *carrier gas* serves as the mobile phase that elutes the components of a mixture from a column containing an immobilized stationary phase. In contrast to liquid chromatography, no interaction occurs between the mobile phase and the analyte in gas chromatography; thus, the rate of movement of the analyte is largely independent of the chemical nature of the mobile phase.

As its name implies, the stationary phase in *gas-solid chromatography* is a solid having a large surface area at which adsorption of the analyte species can take place. Separation occurs because of differences in the positions of adsorption equilibria between the gaseous components of the sample and the solid surface of the stationary phase. Currently, the application of gas-solid chromatography is limited to the separation of low-molecular-weight gaseous species such as carbon monoxide, oxygen, nitrogen, and hydrocarbons. More polar compounds are semipermanently retained on solid surfaces, which precludes the use of solid adsorbents for the gas-chromatographic separation of most species.

Gas-liquid chromatography (GLC), the subject of this chapter, is one of the most important and widely used methods for separating and determining the chemical components of complex mixtures. In this technique, the stationary phase is a liquid that is immobilized on the surface of a solid support by adsorption or by chemical bonding. The rate of movement of an analyte through the column is determined by its distribution ratio between the gaseous and the immobilized liquid phase.

Although the concept of gas-liquid chromatography was first enunciated in 1941 by Martin and Synge, its usefulness as a tool for the separation of closely related species was not demonstrated experimentally for more than a decade. Within three years of this demonstration, however, the first commercial apparatus for gas chromatography appeared on the market. Since that time, the growth in applications of this technique has been phenomenal. For example, by 1965 more than 2000 publications on gas chromatography were appearing annually; by 1985, this number had grown to 5000.[1]

[1]For monographs on GLC, see J. A. Perry, *Introduction to Analytical Gas Chromatography,* New York: Marcel Dekker, 1981; *Modern Practice of Gas Chromatography,* 2nd ed., R. L. Grob, Ed. New York: Wiley, 1985; J. Q. Walker, M. J. Jackson Jr., and J. B. Maynard, *Chromatographic Systems,* 2nd ed., Part II. New York: Academic Press, 1977.

24A

Principles of Gas-Liquid Chromatography

The general principles of chromatography, developed in Chapter 23, and the mathematical relationships summarized in Tables 23-4 and 23-5 are applicable to gas chromatography with only a minor modification to correct for the compressibility of gaseous mobile phases.[2]

Equation 23-18 for plate height is applicable to gas chromatography as well as liquid chromatography. The longitudinal diffusion term (B/u) in this equation is, however, more important where a gas is the mobile phase because diffusion rates in gases are typically 10^4 times greater in this medium than in liquids. As a consequence, the minimum in a plot of plate height versus flow rate is considerably broadened in gas chromatography (Figure 23-6b). This effect is also reflected in the curves shown in Figure 24-1, which illustrate the improvement in plate height that accompanies a reduction in the particle size of the column packing.

24B

Instruments for Gas-Liquid Chromatography

Figure 24-2 shows the basic components of an instrument for performing gas-liquid chromatographic separations.

24B-1

Carrier-Gas Supply

Carrier gases, which must be chemically inert, include helium, argon, nitrogen, and hydrogen, with helium being the most widely used. As is shown later, the choice of gas is often dictated by the type of detector used. Associated with the gas supply are pressure regulators, gauges, and flowmeters. In addition,

FIGURE 24-1 Effect of particle size on plate height. The numbers to the right are particle diameters. (From J. Boheman and J. H. Purnell, in *Gas Chromatography 1958,* D. H. Desty, Ed. New York: Academic Press, 1958. With permission of Butterworths, Stoneham, MA.)

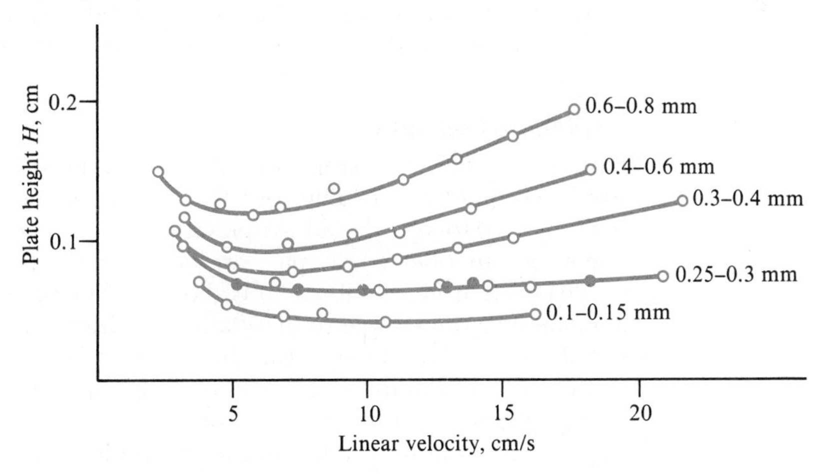

[2]See D. A. Skoog, *Principles of Instrumental Analysis,* 3rd ed., pp. 758–759. Philadelphia: Saunders College Publishing, 1985.

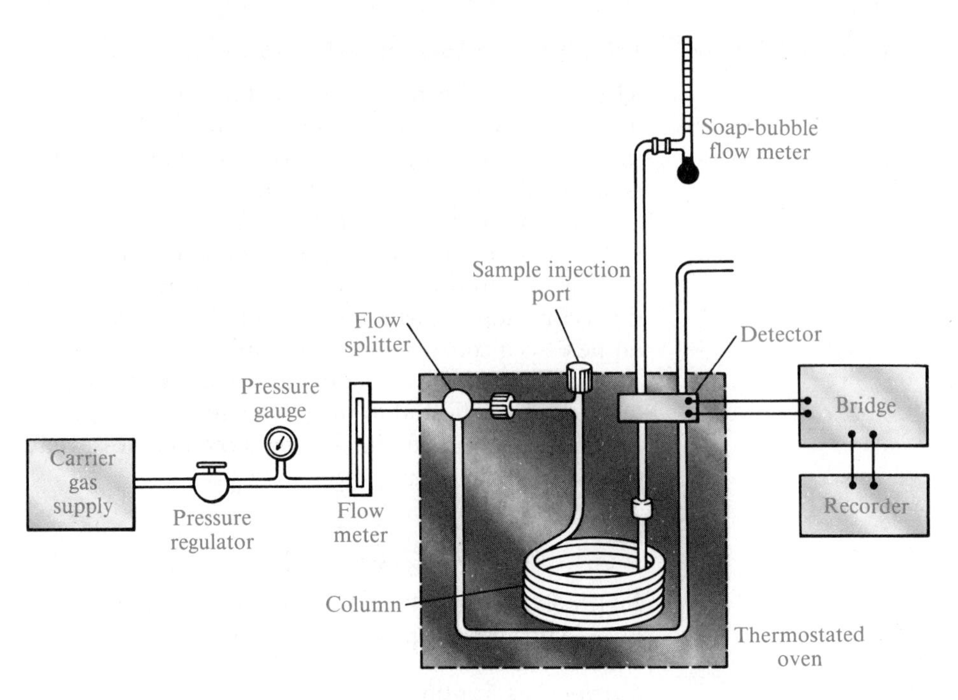

FIGURE 24-2 Diagram of a gas chromatograph.

the carrier-gas system often contains a molecular sieve to remove water and other impurities.

Flow rates are controlled by a pressure regulator. Inlet pressures usually range from 10 to 50 psi (lb/in^2 above room pressure), which leads to flow rates of 25 to 150 mL/min. As shown in Figure 24-2, flow rates are best established by a simple soap-bubble meter located at the end of the column. A soap film is formed in the path of the gas by squeezing a bulb containing an aqueous solution of soap or detergent; the time required for this film to move between two graduations on the buret is measured and converted to flow rate.

24B-2 Sample-Injection System

In order to avoid band spreading, the sample must be minute and must be introduced rapidly as a "plug" of vapor. Most commonly, a calibrated microsyringe is used to inject liquid samples through a rubber or silicone diaphragm or septum into a heated sample port located at the head of the column. The sample port is ordinarily about 50°C above the boiling point of the least volatile component of the sample so that flash vaporization occurs. For ordinary analytical columns, sample sizes vary from a few tenths of a microliter to 20 μL. With capillary columns, which require much smaller samples ($\simeq 10^{-3}$ μL), a sample splitter is employed to deliver only a small fraction of the injected sample to the column head, with the remainder going to waste.

Gas samples are best introduced by means of a sample valve similar to that shown in Figure 23-13. Solid samples are commonly introduced as solutions.

24B-3

Columns

Packed and *tubular* columns are encountered in gas-liquid chromatography. The packed columns can accommodate a much larger sample and are generally more convenient to use. Open tubular columns, however, have the advantage of unparalleled resolution.

Packed Columns

Packed columns are fabricated from glass or metal (stainless steel, copper, or aluminum) tubes that typically have lengths of 2 to 3 m and inside diameters of 2 to 4 mm. The tubes are ordinarily formed as coils having diameters of roughly 15 cm—a configuration that allows them to be fitted into ovens for thermostating.

As shown in Figure 24-1, the efficiency of a gas-liquid column increases rapidly with decreasing particle size of the stationary phase. Particles with diameters much smaller than about 150 μm cannot be accommodated, however, because of the high pressures required to force the carrier gas and sample through such finely divided solids (the required pressure varies inversely as the square of particle diameter). As a result, typical packings are prepared from particles that are 60 to 80 mesh (250 to 170 μm) or 80 to 100 mesh (170 to 149 μm).

Column efficiency increases as the range of particle size decreases. As a consequence, some supports are now being sold with 10-mesh rather than 20-mesh ranges.

Two general types of packing are currently in use in gas chromatography. The first consists of an inorganic support material that has been coated with an organic liquid held in place by adsorption or chemical bonding. The second is a porous organic polymer that requires no additional coating. The former is the more widely used at the present time.

Solid Supports. Solid supports hold the liquid stationary phase in place so that the surface area exposed to the mobile phase is as large as possible. The ideal support consists of small, uniform, spherical particles with good mechanical strength and a specific surface area of at least 1 m^2/g. In addition, the material should be inert at elevated temperatures and uniformly wetted by the liquid phase. No substance that meets all these criteria perfectly is yet available.

The first, and still most widely used, supports for gas chromatography were silicates prepared from naturally occurring diatomaceous earth, which consists of the skeletons of thousands of species of single-celled plants that inhabited ancient lakes and seas. Such plants received their nutrients and disposed of their wastes via molecular diffusion through their pores. As a consequence, their remains make excellent support materials because gas chromatography is based upon the same kind of molecular diffusion.

Two types of supports are derived from diatomaceous earth. The first, which is generally known by the trade name Chromosorb P, is prepared by crushing, blending, and briquetting the diatomaceous earth as it comes from the ground and then heating it to above 900°C. The resulting bricks, which are also used for high-temperature insulation, are then ground and screened according to particle diameter. The second type of support, called Chromosorb W or G, is prepared by mixing diatomaceous earth with a sodium carbonate flux before heating to about 900°C. This product is more rugged than the first and shows less tendency to adsorb solutes. Unfortunately, its specific surface area is only about 1 m^2/g, whereas that of Chromosorb P is 4 m^2/g.

Adsorption on Solid Supports. A problem that has plagued gas chromatography from its inception is the physical adsorption of polar or polarizable analyte species, such as alcohols or aromatic hydrocarbons, on support surfaces. Adsorption results in distorted peaks, which are broadened and often exhibit a tail. It has been established that adsorption is the consequence of silanol groups that form on the surface of diatomaceous silicates by reaction with moisture. Thus, a fully hydrolyzed silicate surface has the structure

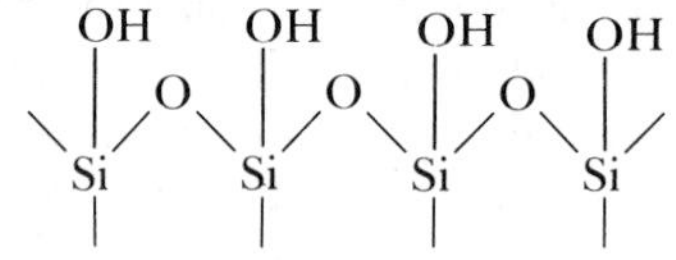

```
 OH     OH     OH     OH
  |  O   |  O   |  O   |
 \| / \  | / \  | / \  |/
  Si     Si     Si     Si
  |      |      |      |
```

The SiOH groups on the support surface have a strong affinity for polar organic molecules and tend to retain them by adsorption.

Support materials can be deactivated by *silanization* with dimethylchlorosilane (DMCS). The reaction is

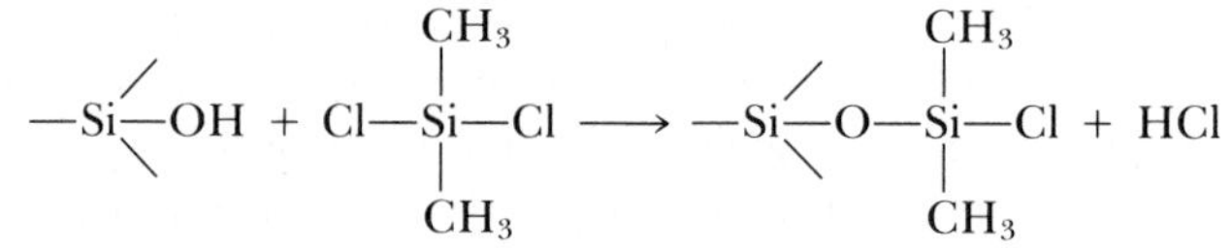

$$\equiv Si{-}OH + Cl{-}\overset{CH_3}{\underset{CH_3}{Si}}{-}Cl \longrightarrow \equiv Si{-}O{-}\overset{CH_3}{\underset{CH_3}{Si}}{-}Cl + HCl$$

Upon washing with methanol, the second chloride is replaced by a methoxy group:

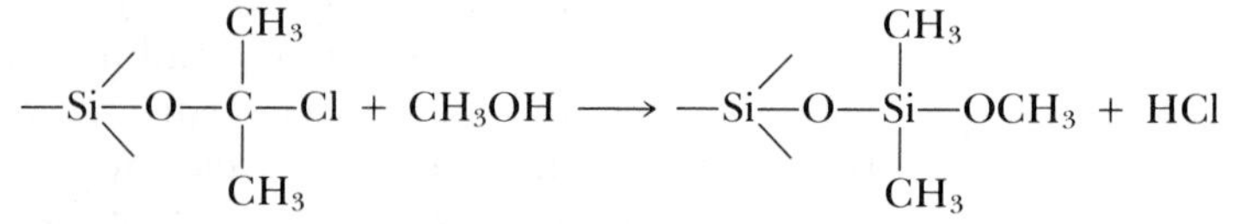

$$\equiv Si{-}O{-}\overset{CH_3}{\underset{CH_3}{C}}{-}Cl + CH_3OH \longrightarrow \equiv Si{-}O{-}\overset{CH_3}{\underset{CH_3}{Si}}{-}OCH_3 + HCl$$

Silanized supports may still show a residual adsorption, which apparently is the result of mineral impurities in the diatomaceous earth. Acid washing prior to silanization removes these impurities. Packings that have been acid-washed and silanized are sold with the designation —AW—DMCS.

Another silanization reagent is hexamethyldisilazane, $(CH_3)_3SiNHSi(CH_3)_3$. Packings treated with this reagent are designated —HMDS.

Liquid Phases. Desirable properties for the immobilized liquid phase in a gas-liquid chromatographic column are (1) *low volatility* (ideally, the boiling point of the liquid should be at least 100° C higher than the maximum operating temperature of the column); (2) *thermal stability;* (3) *chemical inertness;* and (4) *solvent characteristics* such that the capacity and selectivity factors of the solutes to be resolved fall within a suitable range.

Myriad solvents have been proposed as stationary phases in gas-liquid chromatography. By now, only a handful—perhaps a dozen or fewer—suffice for most applications. The proper choice among these solvents is often critical to the success of a separation. Qualitative guidelines exist for making this choice; in the end, however, the best stationary phase must be determined in the laboratory.

The properties of the various common stationary phases are discussed in Section 24C.

Column Preparation. Several techniques are employed for coating support particles with the stationary liquid phase. In one method, a known weight of support is mixed with three to four times its volume of a volatile solvent containing a weight of the stationary phase that is 1 to 10% of the weight of the support particles. The solvent is then evaporated, often in a rotary stripper, to give a dry, free-flowing solid, which is slowly poured into the column with gentle tapping.

A properly prepared column has 750 to 3000 plates per meter and can be used for several hundred analyses. Numerous types of prepacked columns are available from commercial sources.

Chemically Bonded Stationary Phases. Since about 1980, commercial packings that consist of inorganic particles coated with an organic layer held in place by chemical bonding have been available. Chemically bonded packings offer several advantages, one being their uniform and highly reproducible properties. Furthermore, the monomolecular nature of the organic layer leads to rapid equilibration of solutes between the two phases, which, as shown in Section 23C-3, results in improved efficiency (recall that the $C_S u$ term in Equation 23-18 is directly proportional to the square of the thickness of the layer of stationary liquid phase). Another advantage of chemically bonded supports is their thermal stability, which extends to 250°C for some and to 350°C for others. Disadvantages of chemically bonded supports include their limited sample capacity and their sensitivity to degradation by water and oxygen.

Porous-Polymer Packings. Recently, porous organic polymers have found use as packing materials in gas chromatography. These materials, which vary widely in chemical structure, serve as the stationary phase. That is, no additional organic coating is required. Porous polymers have the considerable advantage that they do not "bleed" from the column as do some liquid stationary phases.

Open Tubular Columns

Open tubular, or capillary, columns were first used as early as 1957, when it became apparent from theoretical considerations that such columns should provide separations unprecedented in terms of speed and number of theoretical plates. At that time, it was demonstrated in several laboratories that columns having 300,000 theoretical plates or more were practical. Despite such spectacular performances, capillary columns did not find widespread use until more than two decades after their invention. Several factors contributed to this delay including the small sample capacity and fragile nature of the column, the mechanical problems associated with sample introduction and with connection of the column to the detector, the difficulty in coating the column reproducibly, the short lifetime of poorly prepared columns, the tendency of columns to clog, and patents, which limited commercial development to a single manufacturer (the original patent expired in 1977). By the late 1970s, these problems had become manageable, and several companies were offering capillary columns at a reasonable cost. A major growth in the use of open tubular columns has occurred in the last few years as a consequence.[3]

Types of Columns. Open tubular columns are of three types: *fused-silica* (FSOT), *wall-coated* (WCOT), and *porous-layer* (PLOT), or *support-coated* (SCOT). Table 24-1 compares some important properties of each with those for packed columns.

Wall-coated columns are simply capillary tubes internally coated with a thin layer of the stationary phase. Early columns were constructed of stainless steel, aluminum, copper, or plastic. Subsequently, glass was used. The glass is often etched with gaseous hydrogen chloride, strong aqueous hydrochloric acid, or potassium hydrogen fluoride to give a rough surface that bonds the stationary phase more tightly. Also, the inner surface of glass columns are often treated with silanizing agents (page 630) or other deactivating reagents.

In porous-layer open tubular columns, the inner surface of the capillary is lined with a thin film ($\simeq 30\ \mu m$) of finely divided diatomaceous earth. The presence of this layer permits the column to carry several times as much stationary phase; correspondingly larger samples can thus be accommodated.

The newest capillary columns, which first appeared in 1979, are fashioned from fused silica. Their manufacture is based upon techniques developed for the production of optical fibers. Silica capillaries have much thinner walls than their glass counterparts; outside diameters are about 0.3 to 0.4 mm,

TABLE 24-1 **Properties and Characteristics of Typical Gas Chromatographic Columns**

	Type of Column			
	FSOT*	WCOT†	PLOT‡	Packed
Length, m	10–100	10–100	10–100	1–6
Inside diameter, mm	0.1–0.3	0.25–0.75	0.5	2–4
Efficiency, plates/m	2000–4000	1000–4000	600–1200	500–1000
Sample size, ng	10–75	10–1000	10–1000	10–10^6
Relative pressure	Low	Low	Low	High
Relative speed	Fast	Fast	Fast	Slow
Chemical inertness	Best ⟶			Poorest
Flexible?	Yes	No	No	No

*Fused-silica open tubular column.

†Wall-coated open tubular column.

‡Porous-layer open tubular column (also called support-coated open tubular SCOT).

[3]For a recent monograph on open tubular columns, see M. L. Lee, J. Yang, and K. D. Bartle, *Open Tubular Column Gas Chromatography*. New York: Wiley, 1984.

and inside diameters are about 0.1 to 0.3 mm. The tubes are given added strength by an outside protective polyimide coating that is applied as the capillary tubing is being drawn. The resulting columns are quite flexible and can be bent into coils having diameters of a few inches. Open tubular silica columns are available commercially (as are metal and glass wall-coated columns) and appear to have the important advantages of physical strength, lower reactivity toward solids, and flexibility. They are replacing the older types of open tubular columns for most applications.

Column Coating. The inside surface of an open tubular column is coated with a film of stationary liquid phase that varies in thickness from 0.1 to 1 μm. The liquids used are the same as those employed in packed columns (Section 24C). Several methods have been developed for coating capillary columns. In the dynamic method, a dilute solution of the stationary phase in a volatile solvent is gently driven through the column under a light gas pressure. In the static method, the column is filled with a solution of the stationary phase; with one end of the column capped, the solvent is then slowly removed under vacuum.

Initially, problems were encountered in coating silica columns with the more polar stationary phases. It appears now, however, that these problems have been or will soon be solved, thus providing a full spectrum of wall-coated fused-silica open tubular columns.

24B-4 *Column Thermostating*

Column temperature is an important variable that must be controlled to a few tenths of a degree for precise work. Thus, the column is ordinarily housed in a thermostated oven. The optimum column temperature depends upon the boiling point of the sample and the degree of separation required. Roughly, a temperature equal to or slightly above the average boiling point of a sample results in a reasonable elution time (2 to 30 min). For samples with a broad boiling range, it is often desirable to employ temperature programming so that the column temperature is increased either continuously or in steps as the separation proceeds. Figure 24-3c illustrates the effectiveness of temperature programming in improving resolutions.

In general, optimum resolution is associated with minimum temperature; the cost of lowered temperature, however, is an increase in elution time and therefore an increase in the time required to complete an analysis. Figures 24-3a and b illustrate this principle.

24B-5 *Detectors*

Detection devices for gas-liquid chromatography must respond rapidly to minute concentrations of solutes as they exit the column. The solute concentration in the carrier gas at any instant is no more than a few parts per thousand and is often smaller than this figure by one or two orders of magnitude. Moreover, the time required for a peak to pass the detector is typically 1 s (or less); thus, the device must be capable of exhibiting its full response during this brief period.

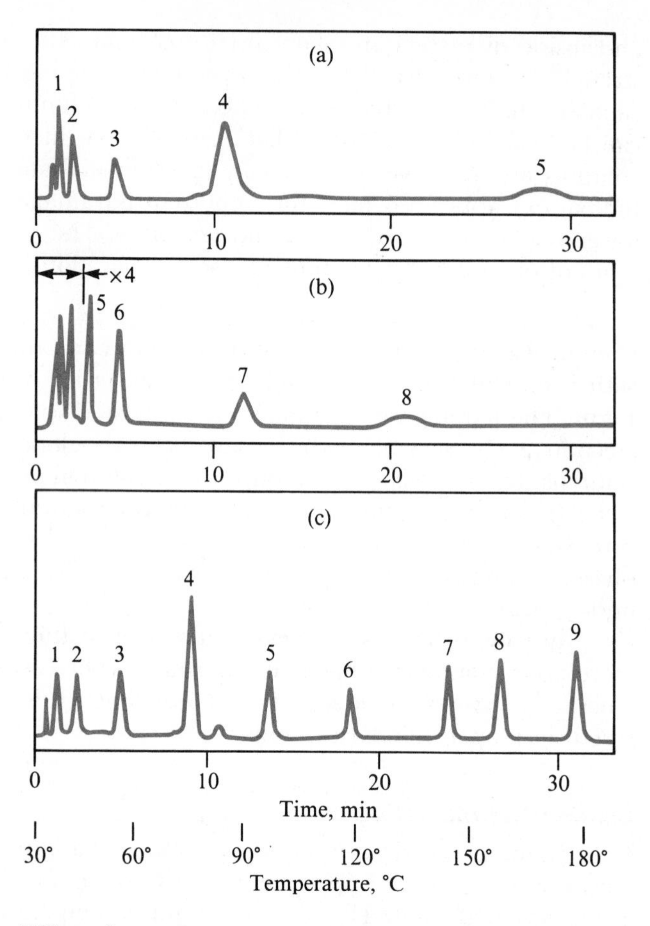

FIGURE 24-3 Effect of temperature on gas chromatograms. (a) Isothermal at 45°C; (b) isothermal at 145°C; (c) programmed at 30 to 180°C. (From W. E. Harris and H. W. Habgood, *Programmed Temperature Gas Chromatography*, p. 10. New York: Wiley, 1966. Reprinted with permission.)

Other desirable properties of a detector are linear response, stability, and uniform response to a wide variety of chemical species or, alternatively, predictable and selective response to one or more classes of solutes. Needless to say, no detector meets all of these desiderata, and it seems unlikely that such a detector will ever be designed. Three of the most widely used detectors for gas chromatography are described briefly in the paragraphs that follow.

Thermal Conductivity Detectors

The thermal conductivity detector, or *katharometer*, was one of the earliest detectors employed in gas chromatographic studies. This device, which still finds widespread use, is based upon changes in the thermal conductivity of the gas stream brought about by the presence of analyte molecules. The sensing device in the thermal conductivity detector consists of an electrically heated element whose temperature at constant electrical power depends upon the thermal conductivity of the surrounding gas. The heated element may be a fine platinum, gold, or tungsten wire or a semiconducting thermistor. The resistance of the wire or thermistor is a measure of its temperature, which depends in part upon the rate at which the surrounding gas molecules conduct thermal energy away from the detector element to the walls of a metal block in which it is housed. Figure 24-4 is a schematic drawing of a typical commercial thermal conductivity detector.

In chromatographic applications, a double detector system is usually employed, with one element being located in the gas stream *ahead* of the sample-injection chamber and the other immediately beyond the column. Alternatively, the gas stream may be split, as shown in Figure 24-2. In either case, the thermal conductivity of the carrier gas is canceled, and the effects of variation in flow rate, pressure, and electrical power are minimized. The resistances of the twin detectors are usually compared by incorporating them into two arms of a simple Wheatstone bridge circuit.

The thermal conductivities of hydrogen and helium are roughly six to ten times greater than those of most organic compounds. Thus, the presence of even small amounts of organic materials causes a relatively large decrease in the thermal conductivity of the column effluent; consequently, the detector undergoes a marked rise in temperature. The conductivities of other carrier gases more closely resemble those of organic constituents; therefore, a thermal conductivity detector dictates the use of hydrogen or helium.

The advantages of a thermal conductivity detector are its simplicity, its large linear dynamic range ($\simeq 10^5$), its general response to both organic and inorganic species, and its nondestructive character, which permits the collection of solutes after detection. A limitation of the thermal conductivity detector is its relatively low sensitivity ($\simeq 10^{-8}$ g solute/mL of carrier gas). Other detectors exceed this sensitivity by factors as large as 10^4 to 10^7. It should be noted that the low sensitivity of thermal detectors precludes their use in conjunction with capillary columns because of the very small samples that can be accommodated by such columns.

Flame Ionization Detectors

Most organic compounds, when pyrolyzed at the temperature of a hydrogen/air flame, produce ionic intermediates and electrons that provide a mechanism by which electricity can be carried through the plasma. With a burner such as that shown in Figure 24-5, the charged species are attracted to and

FIGURE 24-4 A typical thermal conductivity detector. (Courtesy of Varian Instrument Division, Palo Alto, CA.)

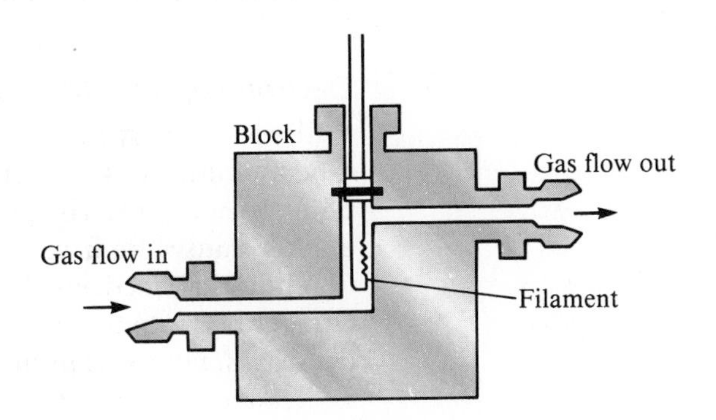

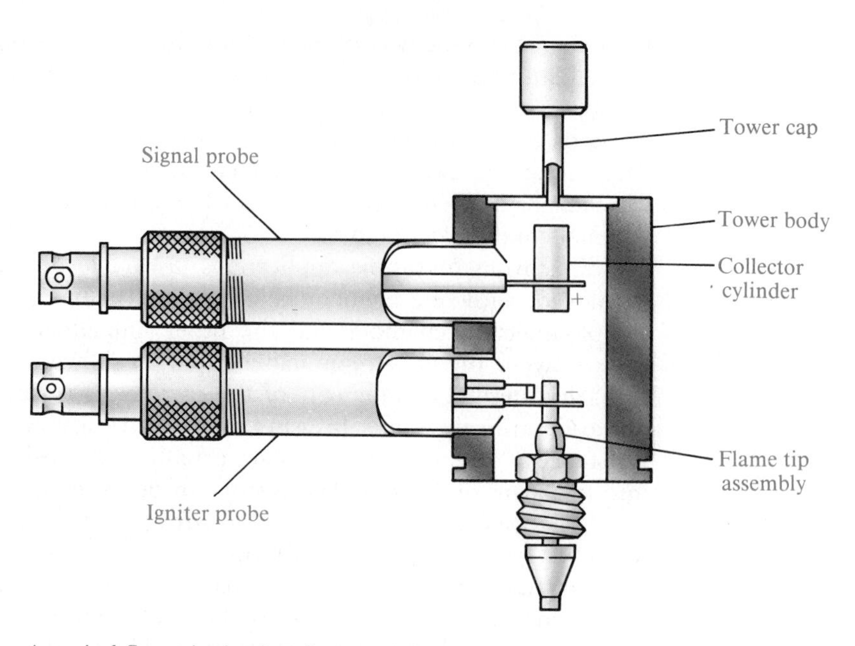

FIGURE 24-5 A typical flame ionization detector. (Courtesy of Varian Instrument Division, Palo Alto, CA.)

captured by a collector; the ion current that results is then amplified and recorded. The electrical resistance of a flame plasma is high (perhaps $10^{12}\ \Omega$), and the resulting currents are therefore minuscule.

The ionization of carbon compounds in a flame is a poorly understood process, although it is observed that the number of ions produced is roughly proportional to the number of *reduced* carbon atoms in the plasma. Functional groups, such as carbonyl, alcohol, halogen, and amine, yield fewer ions or none at all. In addition, the detector is insensitive to noncombustible gases, such as H_2O, CO_2, SO_2, and NO_x. These properties make the flame ionization detector a useful general detector for most organic samples, including those contaminated with water and the oxides of nitrogen and sulfur.

The flame ionization detector is perhaps the most widely used detector because of its high sensitivity ($\simeq 10^{-13}$ g/mL), large linear response range ($\simeq 10^7$), low noise, ruggedness, and convenience. It is normally the detector used with capillary columns. A disadvantage of the flame ionization detector is that it destroys the sample.

Electron-Capture Detectors

With electron-capture detectors, the effluent from the column is passed over a beta emitter (a radioactive compound that emits electrons) such as nickel-63 or tritium (adsorbed on platinum or titanium foil). An electron from the emitter causes ionization of the carrier gas (often nitrogen) and production of a burst of electrons. In the absence of organic species, a constant standing current between a pair of electrodes results from this ionization. The current decreases, however, in the presence of organic molecules that tend to capture electrons. The response is nonlinear unless the potential across the detector is pulsed.

The electron-capture detector is selective in its response and highly sensitive to electronegative functional groups, such as halogens, peroxides, quinones, and nitro groups. It is insensitive to compounds such as amines, alcohols, and hydrocarbons. An important application of the electron-capture detector has been in the detection and determination of chlorinated pesticides.

Electron-capture detectors are highly sensitive and possess the advantage of not consuming the sample to any significant extent (in contrast to flame detectors).

Selective Detectors

Gas chromatography is often coupled with the selective techniques of spectroscopy and electrochemistry. The resulting so-called *hyphenated methods* (for instance, GC-MS and GC-IR) provide the chemist with powerful tools for identifying the components of complex mixtures.[4]

In early hyphenated methods, the eluates from the chromatographic column were collected as separate fractions in a cold trap, a nondestructive, nonselective detector being employed to indicate their appearance. The composition of each fraction was then investigated by nuclear magnetic resonance, infrared, or mass spectroscopy or by electroanalytical measurements. A serious limitation to this approach was the very small (usually micromolar) quantities of solute contained in a fraction; nonetheless, the general procedure proved useful for the qualitative analysis of many multicomponent mixtures.

A second general method, one that now finds widespread use, involves the application of spectroscopic or electroanalytical detectors to monitor the column effluent continuously. Generally, this procedure requires computerized instruments and a large computer memory for storing spectral or electrochemical data for subsequent display as spectra and chromatograms.

24C Liquid Phases for Gas-Liquid Chromatography

Several hundred liquids have been proposed as stationary phases for gas-liquid chromatography because the successful separation of closely related compounds is often critically dependent upon finding a suitable liquid. The retention time for a solute depends upon its partition ratio (Equation 23-1), which in turn is related to the nature of the stationary phase. Clearly, to be useful in gas-liquid chromatography, the immobilized liquid must generate different partition ratios for different solutes. In addition, however, these coefficients must not be extremely large or extremely small because the former leads to prohibitively long retention times and the latter results in such short retention times that separations are incomplete.

To have a reasonable residence time in the column, a species must show at least some degree of compatibility (solubility) with the stationary phase.

[4]For a review on hyphenated methods, see T. Hirschfield, *Anal. Chem.*, **1980,** *52*, 297A; C. L. Wilkens, *Science*, **1983,** *222*, 251.

Thus, the polarities of the two substances should be at least somewhat alike. For example, a stationary phase such as squalene (a high-molecular-weight, nonpolar, saturated hydrocarbon) might be chosen for the separation of members of a nonpolar homologous series such as hydrocarbons or ethers. A more polar stationary phase, such as polyethylene glycol, would be more effective for separating alcohols or amines. For aromatic hydrocarbons, benzyldiphenyl might prove more appropriate.

The elution order for analytes of similar polarity usually correlates with boiling points; where boiling points differ sufficiently, clean separations are feasible. Solutes with nearly identical boiling points but different polarities frequently require a stationary phase that selectively retains one or more of the components by dipole interaction or by adduct formation. Hydrogen bonding is another important interaction that often enhances selectivity. For this type of bonding to be effective, the solute must be a proton donor and the stationary phase must contain a proton acceptor group (such as oxygen, fluorine, or nitrogen), or the converse. Table 24-2 lists a few of the most widely used stationary phases.

24D Applications of Gas-Liquid Chromatography

In evaluating the importance of gas-liquid chromatography, it is necessary to distinguish between the two roles the method plays. The first is as a tool for performing separations; in this capacity, it is unsurpassed when applied to complex organic, metal-organic, and biochemical systems. The second, and distinctly different, function is analysis. Here, retention times are employed for qualitative identification, and peak heights or areas provide quantitative information. Gas-liquid chromatography is much more limited for qualitative purposes than are most spectroscopic methods. As a consequence, an important trend in the field has been in the direction of combining the remarkable

TABLE 24-2 **Some Common Stationary Phases for Gas-Liquid Chromatography**

Trade Name	Chemical Composition	Maximum Temperature, °C	Polarity*	Type of Separation
Squalene	$C_{30}H_{62}$	150	NP	Hydrocarbons
OV-1	Polymethyl siloxane	350	NP	General purpose nonpolar
DC 710	Polymethylphenyl siloxane	300	NP	Aromatics
QF-1	Polytrifluoropropylmethyl siloxane	250	P	Amino acids, steroids, nitrogen compounds
XE-30	Polycyanomethyl siloxane	275	P	Alkaloids, halogenated compounds
Carbowax 20M	Polyethylene glycol	250	P	Alcohols, esters, essential oils
DEG adipate	Diethylene glycol adipate	200	SP	Fatty acids, esters
	Dinonyl phthalate	150	SP	Ketones, ethers, sulfur compounds

*NP = nonpolar; SP = semipolar; P = polar.

fractionation qualities of chromatography with the superior identification properties of such instruments as mass, infrared, and NMR spectrometers.

24D-1

Qualitative Analysis

Gas chromatograms are widely used as criteria of purity for organic compounds. Contaminants are revealed by the appearance of additional peaks. The technique is also useful for evaluating the effectiveness of purification procedures.

In theory, retention times should be useful for the identification of components in mixtures. In fact, however, the applicability of such data is limited by the number of variables that must be controlled in order to obtain reproducible results. Nevertheless, gas chromatography provides an excellent means of confirming the presence or absence of a compound in a mixture, provided an authentic standard sample of the substance is available. The chromatogram of a mixture of the standard and unknown should show no new peaks; enhancement of an existing peak should also be observed. The evidence is particularly convincing if the effect can be duplicated on different columns and at different temperatures.

We have seen (Section 23B-5) that the selectivity factor α for compounds A and B is given by the relationship

$$\alpha = \frac{K_{\mathrm{B}}}{K_{\mathrm{A}}} = \frac{(t_R)_{\mathrm{B}} - t_M}{(t_R)_{\mathrm{A}} - t_M} = \frac{k'_{\mathrm{B}}}{k'_{\mathrm{A}}}$$

If a standard substance is chosen as compound B, then α can provide an index for identification of compound A that is largely independent of column variables other than temperature; that is, numerical tabulations of selectivity factors for pure compounds relative to a common standard can be prepared and then used for the characterization of solutes. The amount of such data available in the literature is presently limited.

24D-2

Quantitative Analysis

The detector signal from a gas-liquid chromatographic column is widely used for quantitative and semiquantitative analyses. An accuracy of 1 to 3% relative is attainable under carefully controlled conditions. As with most analytical tools, reliability is directly related to the control of variables. The nature of the sample also plays a part in determining the potential accuracy.

The general discussion of quantitative chromatographic analysis in Section 23F-2 applies to gas chromatography; therefore, no further consideration of this topic is needed.

24E

Typical Applications of Gas Chromatography

Figure 24-6 illustrates some typical applications of gas chromatography to analytical problems. All are based upon gas-liquid equilibria except Figure 24-6a, where two columns are used in series, the first a gas-liquid column and the second a gas-solid. The gas-liquid column retains only the carbon dioxide and passes the remaining gases without retention. While the carbon dioxide is being eluted from the first column, a switch directs the flow around the second column to avoid permanent adsorption of this gas on the solid packing. After the carbon dioxide signal has returned to zero, the flow is switched back through the second column, thereby permitting separation of the remaining sample components.

Figure 24-6f is particularly noteworthy because it illustrates the power

FIGURE 24-6 Some examples of gas chromatographic separations.

(a) Exhaust mixture: *A*, 35% H_2; *B*, 25% CO_2; *C*, 1% O_2; *D*, 1% N_2; *E*, 1% C_2; *F*, 30% CH_4; *G*, 3% CO; *H*, 1% C_3; *I*, 1% C_4; *J*, 1% *i*-C_5; *K*, 1% *n*-C_5.

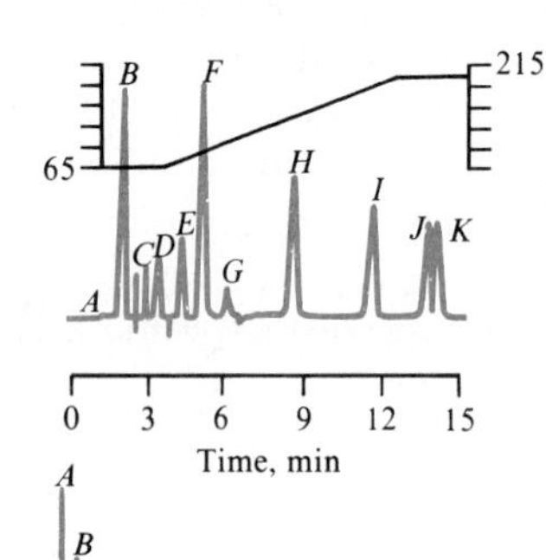

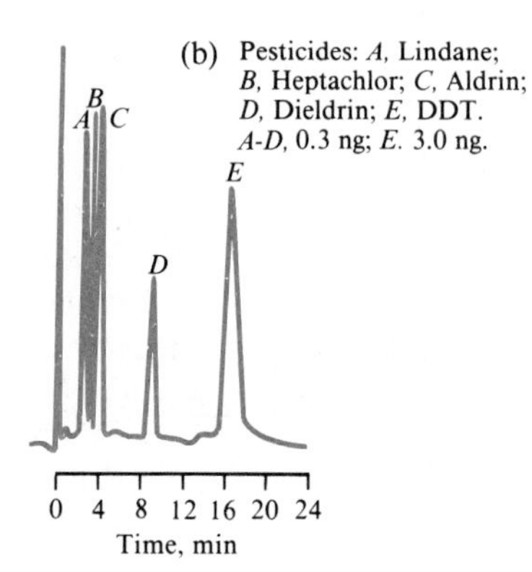

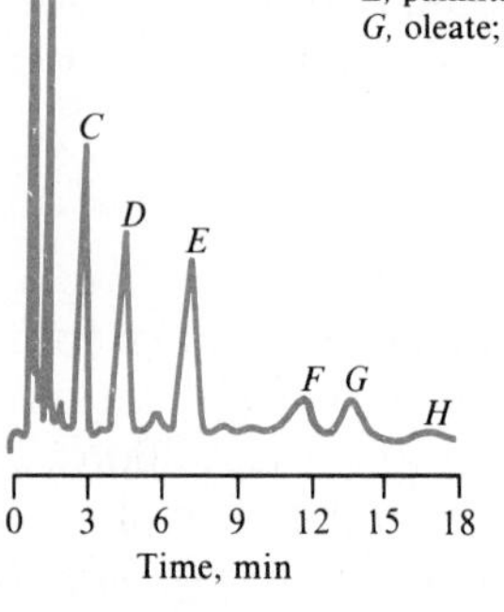

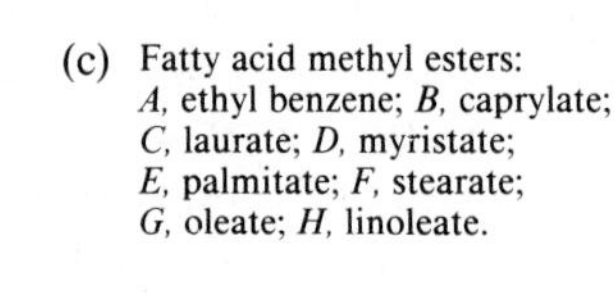

(c) Fatty acid methyl esters: A, ethyl benzene; B, caprylate; C, laurate; D, myristate; E, palmitate; F, stearate; G, oleate; H, linoleate.

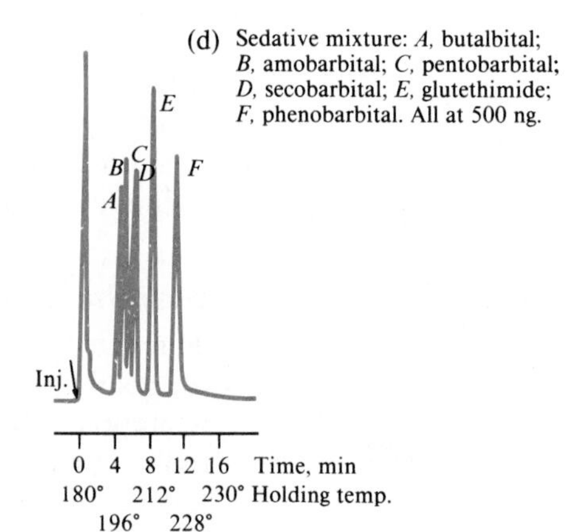

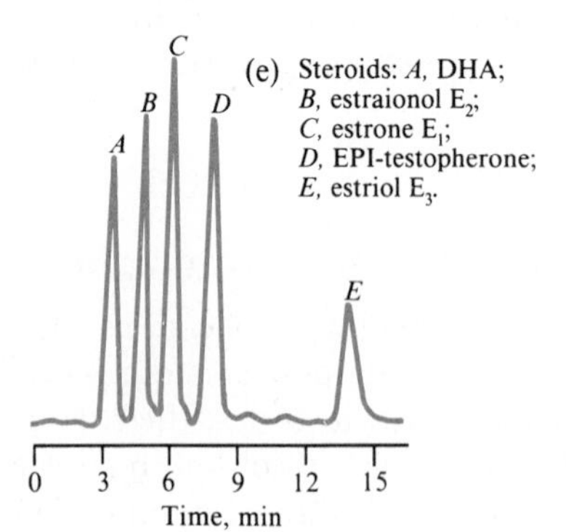

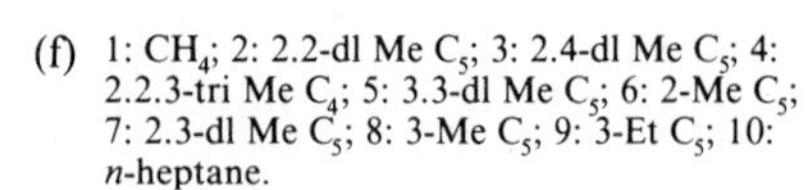

(f) 1: CH4; 2: 2.2-dl Me C5; 3: 2.4-dl Me C5; 4: 2.2.3-tri Me C4; 5: 3.3-dl Me C5; 6: 2-Me C5; 7: 2.3-dl Me C5; 8: 3-Me C5; 9: 3-Et C5; 10: n-heptane.

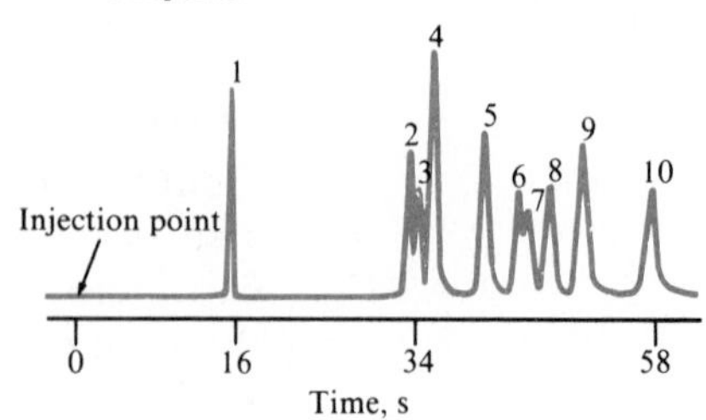

TABLE 24-3

Chromatogram	Column*	Packing	Detector†	Temperature, °C	Carrier	Flow, mL/min
a	10′ × 1/8″ S 5′ × 1/8″ S	Chromosorb 102 Molecular sieve 5A	TCD	65–200	He	30
b	6′ × 1/4″ G	1.5% OV-17 on Chromosorb G	ECD	220		
c	6′ × 1/8″ S	Chromosorb W	TCD	190	He	30
d	6′ × 1/4″ G	1.5% OV-17 on HP Chromosorb G	FID	180–230		
e	6′ × 3.4 mm G	5% OV-210, 2.5% OV-17 on Supelcoport	ECD	260	A	
f	50′ × 0.005″ FSOT	Squalene	FID	20	H_2	

*S = packed stainless steel; G = packed glass; FSOT = fused silica open tubular.

†TCD = thermal conductivity; ECD = electron capture; FID = flame ionization.

and speed of open tubular columns. Here, methane and nine heptane isomers are separated in just under 1 min.

Pertinent data on column conditions for the chromatograms in Figure 24-6 are found in Table 24-3.

Questions and Problems

*24-1 Explain the difference between gas-liquid and gas-solid chromatography.

24-2 Why has gas-solid chromatography been limited in its applications?

*24-3 Describe the differences between an open tubular column and a packed column. What are the advantages and disadvantages of each?

24-4 How do chemically bonded columns differ from conventional gas-liquid columns? What are the advantages and disadvantages of each?

*24-5 What are sample splitters and why must they be used with open tubular columns?

24-6 What are fused-silica columns? What are their advantages and disadvantages?

*24-7 What is temperature programming in gas-liquid chromatography? Why is it used?

24-8 List the desirable properties of a detector for gas-liquid chromatography.

*24-9 Why is helium or hydrogen used as the carrier gas when a thermal conductivity detector is used?

24-10 Why is the flame ionization detector particularly useful for samples containing water, nitrogen, and sulfur?

*24-11 What are hyphenated chromatographic methods?

24-12 A gas-liquid column was labeled —AW—DMCS. What is the meaning of this designation?

*24-13 What is the meaning of silanization when it refers to a column packing?

24-14 Why is it desirable to introduce the sample onto a chromatographic column quickly?

*24-15 The following retention times were found with a 1.10-m gas-liquid chromatographic column: air, 18.0 s; methyl acetate, 1.98 min; methyl propionate, 2.24 min; methyl *n*-butyrate, 7.93 min. The base widths of the last three peaks were 0.19, 0.23, and 0.79 min, respectively. Calculate

(a) k' for each compound.
(b) α for each adjacent pair of compounds.
(c) the average number of theoretical plates and plate height for the column.
(d) the resolution for each adjacent pair of compounds.
(e) the length of column required to achieve a resolution of 1.5 for the first two peaks, assuming the same flow rate.
(f) the time required to elute methyl *n*-butyrate for a column with the dimensions of that in part e.

24-16 One method for the quantitative determination of the constituents in a sample analyzed by gas chromatography is the area-normalization method. In this technique, complete elution of all sample constituents is necessary; the area of each peak is then measured and corrected for differences in detector response to the different eluates. This correction involves multiplication of the area by an empirically determined correction factor. The concentration of the analyte is found from the ratio of its corrected area to the total corrected area of all peaks. For the chromatogram described in Problem 24-15, the areas of the three peaks were 16.4, 45.2, and 30.2 in the order of increasing retention time. Calculate the percentage of each compound if the relative detector responses were 0.60, 0.78, and 0.88, respectively.

*24-17 The stationary liquid in the column described in Problem 24-15 was didecylphthalate, a solvent of intermediate polarity. If a nonpolar solvent such as a silicone oil had been used instead, would the retention times for the three compounds be larger or smaller? Why?

24-18 The following retention times were observed with a 1.25-m chromatographic column: air, 0.395 min; isopropylamine, 4.59 min; and *n*-propylamine, 4.91 min. Peak widths were 0.365 min for isopropylamine and 0.382 min for *n*-propylamine. Calculate

(a) k' for each amine.
(b) an α value.
(c) an average number of theoretical plates in the column.
(d) the plate height for the column.
(e) the resolution for the two amines.
(f) the length of column required to achieve a resolution of 1.5.
(g) the time required to achieve a resolution of 1.5, assuming the original linear flow rate.

24-19 What effect does each of the following have on the plate height of a column? Explain.

*(a) increasing the weight of the stationary phase relative to the packing weight
(b) decreasing the sample-injection rate
*(c) increasing the injection-port temperature
(d) increasing the flow rate
*(e) reducing the particle size of the packing
(f) decreasing the column temperature

*24-20 Peak areas and relative detector responses are to be used to determine the concentration of the five species in a sample. The area-normalization method described in Problem 24-16 is to be used. The relative areas for the five gas chromatographic peaks are given below. Also shown are the relative responses of the detector. Calculate the percentage of each component in the mixture.

Compound	Relative Peak Area	Relative Detector Response
A	32.5	0.70
B	20.7	0.72
C	60.1	0.75
D	30.2	0.73
E	18.3	0.78

24-21 List the variables that lead to (a) band broadening and (b) band separation in gas-liquid chromatography.

CHAPTER 25

High-Performance Liquid Chromatography

Most of this chapter deals with four types of column chromatography in which a liquid serves as the mobile phase: *partition*, or *liquid-liquid, chromatography; adsorption*, or *liquid-solid, chromatography; ion-exchange chromatography;* and *size-exclusion*, or *gel-permeation, chromatography*. Also included at the end of the chapter are short sections on planar liquid chromatography and supercritical-fluid chromatography. In the latter method, a supercritical fluid rather than a liquid serves as the mobile phase.

Early liquid chromatography, including Tswett's original work, was carried out in glass columns with diameters of 1 to 5 cm and lengths of 50 to 500 cm. To ensure reasonable flow rates, the diameters of the particles in the solid stationary phase were usually in the 150 to 200 μm range. Even then, flow rates were only a few tenths of a milliliter per minute. Thus, separation times were long—often several hours. Attempts to speed up the classic procedure by application of vacuum or by pumping were not effective because increases in flow rates increased plate heights beyond the minimum in the typical curve of plate height versus flow rate (Figure 23-8); decreased efficiencies were the result.

Early in the development of liquid chromatography, it was realized that major increases in column efficiency would accompany decreases in the particle size of packings. This effect is demonstrated by the data plotted in Figure 25-1. Note that in none of these plots is the minimum shown in Figure 23-6 reached because this minimum often occurs only at prohibitively low flow rates in liquid chromatography.

It was not until the late 1960s that the technology for producing and using packings with particle diameters as small as 10 μm was developed. This technology required sophisticated instruments that contrasted markedly with the simple glass columns of classic liquid chromatography. The name *high-performance liquid chromatography* (HPLC) is often employed to distinguish these newer procedures from the classic methods, which still find considerable use for preparative purposes. The next six sections of this chapter deal exclusively with HPLC.[1]

[1]Numerous books on high-performance liquid chromatography are available. Among these are L. R. Snyder and J. J. Kirkland, *Introduction to Modern Liquid Chromatography*, 2nd ed. New York: Wiley, 1979; R. J. Hamilton and P. A. Sewell, *Introduction to High Performance Liquid Chromatography*, 2nd ed. New York: Chapman and Hall, 1982; and *High-Performance Liquid Chromatography*, C. Horváth, Ed., Vols. 1 and 2. New York: Academic Press, 1980.

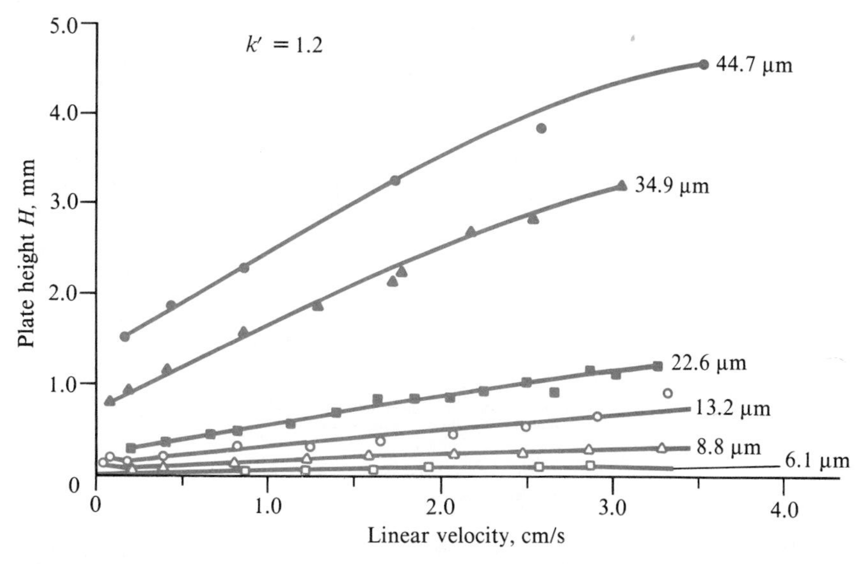

FIGURE 25-1 Effect of packing particle size and flow rate on plate height in liquid chromatography. Column dimensions: 30 cm × 2.4 mm. Solute: N,N-diethyl-*n*-aminoazobenzene. Mobile phase: mixture of hexane, methylene chloride, isopropyl alcohol. (From R. E. Majors, *J. Chromatogr. Sci.*, **1973**, *11*, 92. With permission.)

25A Instruments for High-Performance Liquid Chromatography

Pumping pressures of several hundred atmospheres are required to achieve reasonable flow rates with modern liquid chromatographic packings, which are made up of particles with diameters of 10 μm or less. As a consequence of these high pressures, the equipment for high-performance liquid chromatography tends to be considerably more elaborate and expensive than that encountered in other types of chromatography. Figure 25-2 shows the important components of a typical high-performance chromatograph.

25A-1 Mobile-Phase Reservoirs and Solvent-Treatment Systems

As shown in Figure 25-2, a modern high-performance liquid chromatography apparatus is usually equipped with one or more glass or stainless steel reservoirs, each of which contains 500 mL or more of a solvent. Provisions are often included for removing dissolved gases and particulate matter from the liquids. The gases produce band spreading as a result of bubble formation, while both bubbles and particulates interfere with detector performance. Degassers may consist of a vacuum pumping system, a distillation system, a device for heating and stirring, or, as shown in Figure 25-2, a system for *sparging,* in which the dissolved gases are swept out of solution by fine bubbles of an inert gas that is not soluble in the mobile phase.

The simplest way of performing a liquid chromatographic separation is by *isocratic* elution, in which a single solvent is used to carry the analytes through the column. Often, however, a more satisfactory chromatogram can be obtained by a *gradient elution,* in which two (and sometimes more) solvent systems that differ significantly in polarity are employed. The ratio between volumes of the two solvents is varied in a preprogrammed way, sometimes

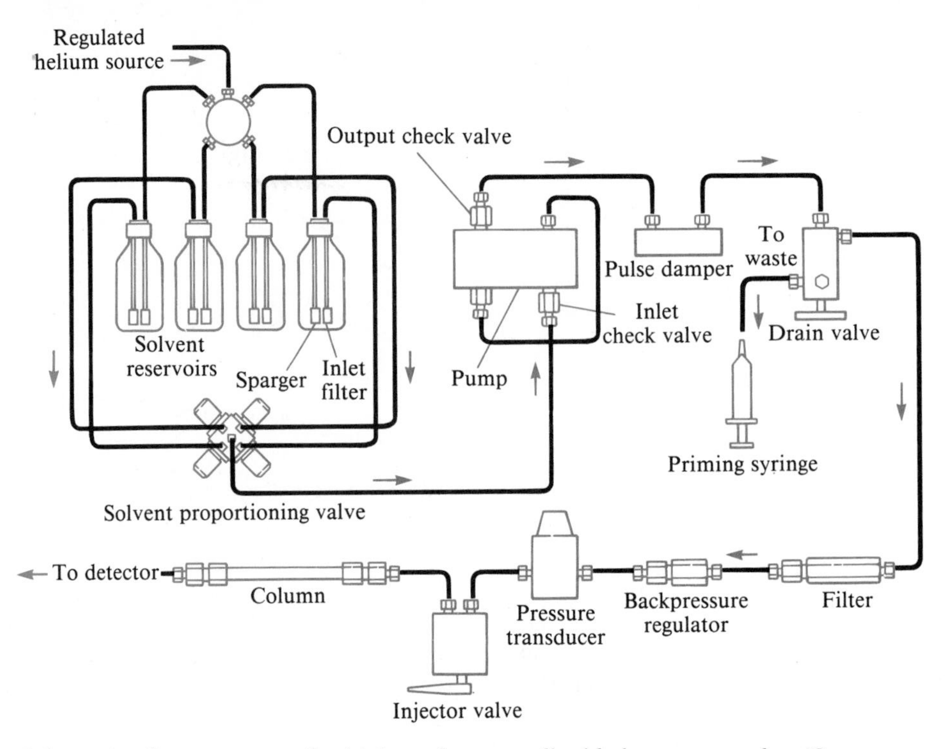

FIGURE 25-2 Schematic of an apparatus for high-performance liquid chromatography. (Courtesy of Perkin-Elmer, Norwalk, CT.)

continuously and sometimes in a series of steps. Gradient elution frequently improves separation efficiency, just as temperature programming helps in gas chromatography. Modern HPLC instruments are often equipped with proportioning valves that introduce liquids from two or more reservoirs at rates that vary continuously (Figure 25-2).

25A-2 Pumping Systems

The requirements for liquid chromatographic pumps are severe: (1) generation of pressures up to 6000 psi (pounds per square inch), (2) pulse-free output, (3) flow rates ranging from 0.1 to 10 mL/min, (4) flow reproducibilities of 0.5% relative or better, and (5) resistance to corrosion by a variety of solvents.

Two types of mechanical pumps are employed: a screw-driven syringe type and a reciprocating pump. The former produces a pulse-free delivery whose flow rate is readily controlled; it suffers, however, from a lack of capacity ($\simeq$ 250 mL) and is inconvenient when solvents must be changed. Reciprocating pumps, which are more widely used, usually consist of a small cylindrical chamber that is filled and then emptied by the back-and-forth motion of a piston. The pumping motion produces a pulsed flow that must be subsequently damped. Advantages of reciprocating pumps include small internal volume, high output pressure (up to 10,000 psi), ready adaptability to gradient elution, and constant flow rates, which are largely independent of column back-pressure and solvent viscosity.

Some instruments use a pneumatic pump, which in its simplest form consists of a collapsible solvent container housed in a vessel that can be pressurized by a compressed gas. Pumps of this type are simple, inexpensive, and pulse-free; they suffer from limited capacity and pressure output, however, and from pumping rates that depend upon solvent viscosity. In addition, they are not adaptable to gradient elution.

It should be noted that the high pressures generated by liquid chromatographic pumps do not constitute an explosion hazard because liquids are not very compressible. Thus, rupture of a component results only in solvent leakage, which may, however, constitute a fire hazard.

25A-3 Sample-Injection Systems

Syringe injection through an elastomeric septum is often used in liquid chromatography owing to its simplicity. This procedure is not very reproducible, however, and is limited to pressures less than about 1500 psi. In *stop-flow* injection, the solvent flow is stopped momentarily, a fitting at the column head is removed, and the sample is injected directly onto the head of the packing by means of a syringe. The most widely used method of sample introduction in liquid chromatography is based upon sampling loops, such as that shown in Figure 25-3. These devices are often an integral part of modern liquid chromatography equipment and have interchangeable loops that provide a choice of sample size ranging from 5 to 500 μL. The precision of injections with a typical sampling loop is a few tenths of a percent relative.

25A-4 Columns for High-Performance Liquid Chromatography

High-performance liquid chromatographic columns are usually constructed from stainless steel tubing, although heavy-walled glass tubing is sometimes employed for lower-pressure applications (<600 psi). Most columns range in length from 10 to 30 cm and have an inside diameter of 4 to 10 mm. Column packings typically have particle sizes of 5 or 10 μm. Columns of this type often contain 40,000 to 60,000 plates/m. Recently, high-performance microcolumns with an inside diameter of 1 to 4.6 mm and a length of 3 to 7.5 cm have become available. These columns, which are packed with 3 or 5 μm particles, contain as many as 100,000 plates/m and have the advantages of speed and minimal solvent consumption.

The most common packing for liquid chromatography is prepared from silica particles, which are synthesized by agglomerating submicrometer silica

FIGURE 25-3 A sampling loop for a liquid chromatograph. (Courtesy of Beckman Instruments, Fullerton, CA.)

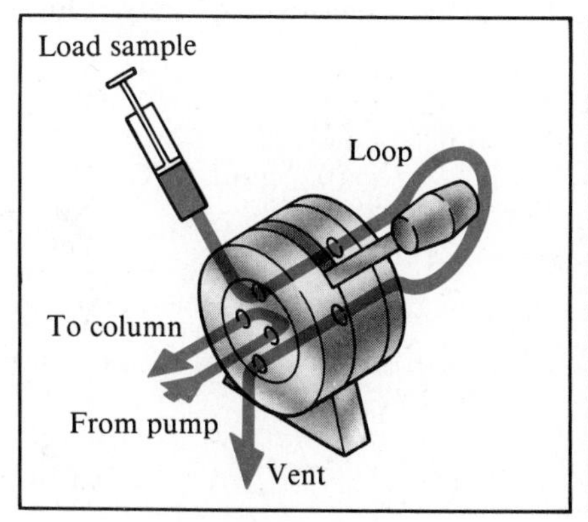

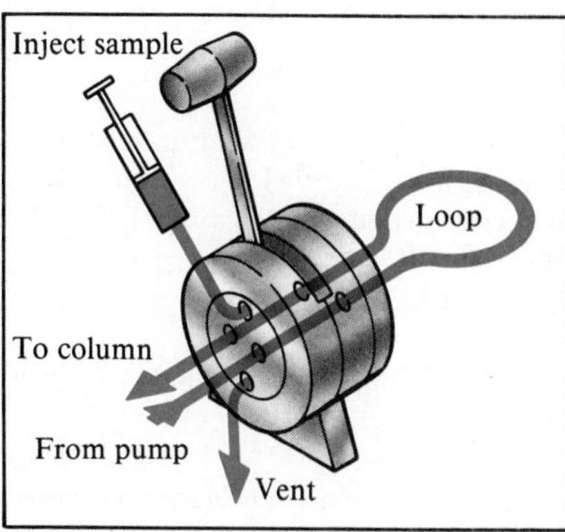

particles under conditions that lead to larger particles with highly uniform diameters. The product is often coated with a thin organic film, which is chemically or physically bonded to the surface. Other packing materials include alumina particles, porous-polymer particles, and ion-exchange resins.

25A-5 Detectors

No highly sensitive universal detector system, such as those used in gas chromatography, is available for high-performance liquid chromatography. Thus, the system used depends upon the nature of the sample. Table 25-1 lists some of the common detectors and their properties.

The most widely used detectors for liquid chromatography are based upon absorption of ultraviolet or visible radiation. Both photometers and spectrophotometers specifically designed for use with chromatographic columns are available from commercial sources. The former often make use of the 254- and 280-nm lines from a mercury source because many organic functional groups absorb in this region (Section 20A-1). Deuterium or tungsten-filament sources with interference filters also provide a simple means of detecting absorbing species. Some modern instruments are equipped with filter wheels that hold several filters that can be rapidly exchanged by rotation of the wheel. Spectrophotometric detectors are considerably more versatile than photometers and are also widely used in high-performance instruments.

Another detector that has found considerable application measures the changes in solvent refractive index caused by the presence of analyte molecules. In contrast to most of the other detectors listed in Table 25-1, the refractive-index detector is general rather than selective and responds to the presence of all solutes. The disadvantage of this type of detector is its somewhat limited sensitivity.

25B High-Performance Partition Chromatography

Partition chromatography, which has become the most widely used of all liquid chromatographic procedures, is conveniently divided into *liquid-liquid* and *bonded-*

TABLE 25-1 **Characteristics of Detectors for High-Performance Liquid Chromatography***

Basis for Detection	Type†	Maximum Sensitivity‡	Flow-Rate Sensitive?	Temperature Sensitivity	Applicable to Gradient Elution?	Available Commercially?
UV absorption	S	2×10^{-10}	No	Low	Yes	Yes
IR absorption	S	10^{-6}	No	Low	Yes	Yes
Fluorometry	S	10^{-11}	No	Low	Yes	Yes
Refractive index	G	1×10^{-7}	No	$\pm 10^{-4}$ g·mL^{-1}/°C	No	Yes
Electrical conductivity	S	10^{-8}	Yes	2%/°C	No	Yes
Mass spectrometry	G	10^{-10}	No	None	Yes	Yes
Electrochemical	S	10^{-12}	Yes	1.5%/°C	No	Yes
Radiochemical	S	—	No	None	Yes	No

*Most data from L. R. Snyder and J. J. Kirkland, *Introduction to Modern Liquid Chromatography*, 2nd ed., p. 162. New York: Wiley-Interscience, 1979. With permission.

†G = general; S = selective.

‡Sensitivity for a favorable sample in grams per milliliter.

phase chromatography. The difference between the two lies in the method by which the stationary phase is held on the packing. In liquid-liquid partition packings, retention is by physical adsorption, while in bonded-phase packings, covalent bonds are involved. Early partition chromatography was exclusively liquid-liquid; now, however, bonded-phase methods predominate, with liquid-liquid separations being relegated to certain special applications.

25B-1

Bonded-Phase Packings

Most bonded-phase packings are prepared by reaction of an organochlorosilane with the —OH groups formed on the surface of silica particles by hydrolysis in hot dilute hydrochloric acid. The product is an organosiloxane. The reaction for one such SiOH site on the surface of a particle can be written as

$$-\overset{|}{\underset{|}{Si}}-OH + Cl-\overset{CH_3}{\underset{CH_3}{Si}}-R \longrightarrow -\overset{|}{\underset{|}{Si}}-O-\overset{CH_3}{\underset{CH_3}{Si}}-R$$

where R is often a straight-chain octyl or octyldecyl group ($C_8H_{17}-$ or $C_{18}H_{37}-$). Other organic functional groups that have been bonded to silica surfaces are aliphatic amines, ethers, nitriles, and aromatic hydrocarbons. Thus, a variety of polarities for the bonded stationary phase are available.

Bonded-phase packings are significantly more stable than packings in which the stationary phase is retained by physical attraction alone. With the latter, periodic recoating of the solid surfaces is required because the stationary phase is gradually dissolved away in the mobile phase. Furthermore, gradient elution is not practical with liquid-liquid packings, again because of losses of the stationary phase by solubility in the mobile phase. The main disadvantage of bonded-phase packings is their somewhat limited sample capacity.

25B-2

Normal- and Reversed-Phase Packings

Two types of partition chromatography are distinguishable based upon the relative polarities of the mobile and stationary phases. Early work in liquid chromatography was based upon highly polar stationary phases, such as triethylene glycol or water; a relatively nonpolar solvent, such as hexane or *i*-propyl ether, served as the mobile phase. For historic reasons, this type of chromatography is now called *normal-phase chromatography*. In *reversed-phase chromatography*, the stationary phase is nonpolar, often a hydrocarbon, and the mobile phase is a relatively polar solvent, such as water, methanol, or acetonitrile.[2] In normal-phase chromatography, the *least* polar component is eluted first; *increasing* the polarity of the mobile phase *decreases* the elution time. In contrast, in the reversed-phase method, the *most* polar component elutes first, and *increasing* the mobile-phase polarity *increases* the elution time.

It has been estimated that more than three quarters of all high-performance liquid chromatography separations are currently performed with reversed-phase, bonded octyl or octyldecyl siloxane packings. With such preparations, the long-chain hydrocarbon groups are aligned parallel to one another

[2]For a detailed discussion of reversed-phase HPLC, see A. M. Krstulovic and P. R. Brown, *Reversed-Phase High-Performance Liquid Chromatography*. New York: Wiley, 1982.

and perpendicular to the surface of the particle, giving a brushlike, nonpolar hydrocarbon surface. The mobile phase used with these packings is often an aqueous solution containing various concentrations of such solvents as methanol, acetonitrile, and tetrahydrofuran.

25B-3 Choice of Mobile and Stationary Phases

Successful partition chromatography requires a proper balance of intermolecular forces among the three participants in the separation process—the solute, the mobile phase, and the stationary phase. These intermolecular forces are described qualitatively in terms of the relative polarities of the reactants. In general, the polarities of common organic functional groups in increasing order are: aliphatic hydrocarbons < olefins < aromatic hydrocarbons < halides < sulfides < ethers < nitro compounds < esters ≃ aldehydes ≃ ketones < alcohols ≃ amine < sulfones < sulfoxides < amides < carboxylic acids < water.

As a rule, most chromatographic separations are achieved by matching the polarity of the analyte to that of the stationary phase; a mobile phase of considerably different polarity is then used. This procedure is generally more successful than one in which the polarities of the analyte and the mobile phase are matched but are different from that of the stationary phase. In the latter situation, the stationary phase often cannot compete successfully for the sample components; retention times then become too short for practical application. At the other extreme is the situation where the polarities of the analyte and stationary phase are too much alike, in which case retention times become inordinately long.

25B-4 Applications

Figure 25-4 shows typical applications of bonded-phase partition chromatography. Table 25-2 further illustrates the variety of samples to which the technique is applicable.

25C High-Performance Adsorption Chromatography

All of the pioneering work in chromatography was based upon liquid-solid adsorption, in which the stationary phase is the surface of a finely divided polar solid. With such a packing, the analyte competes with the mobile phase for sites on the surface of the packing, and retention is the result of adsorption forces.

25C-1 Stationary and Mobile Phases

Finely divided silica and alumina are the only stationary phases that find extensive use in adsorption chromatography. Silica is preferred for most (but not all) applications because of its higher sample capacity and its wider range of useful forms. The adsorption characteristics of the two substances parallel one another. For both, retention times become longer as the polarity of the analyte increases.

In adsorption chromatography, the only variable that affects the partition ratios of analytes is the composition of the mobile phase (in contrast to partition chromatography, where the polarity of the stationary phase is also a factor). Fortunately, enormous variations in capacity factor, selectivity factor, and thus resolution accompany variations in the solvent system; only rarely can a suitable mobile phase not be found.

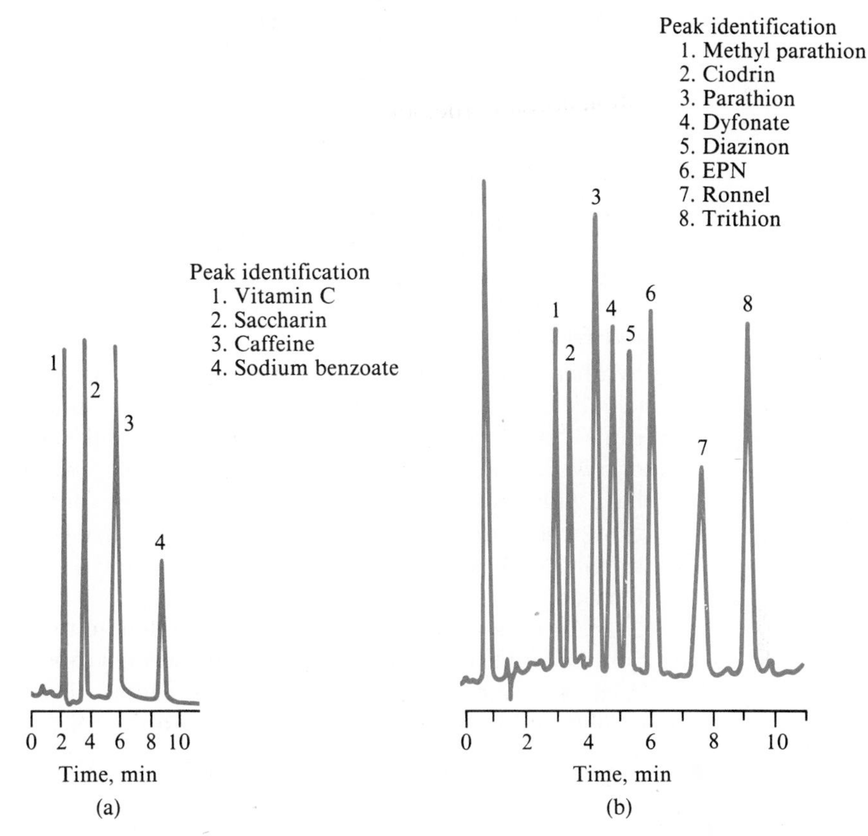

FIGURE 25-4 Typical applications of bonded-phase chromatography. (a) Soft-drink additives. Column: 4.6 mm × 250 mm packed with polar (cyano) bonded-phase packing. Isocratic elution: 6% HOAc/94% H_2O. Flow rate: 1.0 mL/min. (Courtesy of Du Pont Biotechnology Systems, Wilmington, DE.) (b) Organophosphate insecticides. Column: 4.5 mm × 250 mm, packed with 5-μm C_8 bonded-phase particles. Gradient elution: 67% CH_3OH/33% H_2O to 80% CH_3OH/20% H_2O. Flow rate: 2 mL/min. (Courtesy of IBM Instruments, Danbury, CT.)

25C-2 Applications of Adsorption Chromatography

Currently, liquid-solid high-performance liquid chromatography is used extensively for separations of relatively nonpolar, water-insoluble organic compounds with molecular weights below about 5000. One particular strength of adsorption chromatography that is not shared by other methods is its ability

TABLE 25-2 **Typical Applications of High-Performance Partition Chromatography**

Field	Typical Mixtures
Pharmaceuticals	Antibiotics, sedatives, steroids, analgesics
Biochemical	Amino acids, proteins, carbohydrates, lipids
Food products	Artificial sweeteners, antioxidants, aflatoxins, additives
Industrial chemicals	Condensed aromatics, surfactants, propellants, dyes
Pollutants	Pesticides, herbicides, phenols, PCBs
Forensic chemistry	Drugs, poisons, blood alcohol, narcotics
Clinical medicine	Bile acids, drug metabolites, urine extracts, estrogens

to resolve isomeric mixtures, such as meta- and para-substituted benzene derivatives.

25D High-Performance Ion-Exchange Chromatography

Ion-exchange resins are used as the stationary phase in *ion-exchange chromatography,* a separation procedure in which ions of like charge are separated by elution from a column packed with a finely divided resin. This type of chromatography is also called *ion chromatography*.

25D-1 Ion-Exchange Resins

Synthetic ion-exchange resins are high-molecular-weight polymeric materials containing many ionic functional groups per molecule. Cation-exchange resins can be either a strong-acid type containing sulfonic acid groups ($RSO_3^- H^+$) or a weak-acid type containing carboxylic acid groups (RCOOH); the former have wider application. Anion-exchange resins contain basic amine functional groups attached to the polymer molecule. Strong-base exchangers are quaternary amines [$RN(CH_3)_3^+OH^-$]; weak-base types contain secondary or tertiary amines.

An important property of all ion-exchange resins is that they are essentially insoluble in aqueous media. Thus, when a cation exchanger is immersed in an aqueous solution containing the cation M^+, the following exchange equilibrium is quickly established between the solid and solution phases:

$$\underset{\text{solid}}{xRSO_3^- H^+} + \underset{\text{solution}}{M^{x+}} \rightleftharpoons \underset{\text{solid}}{(RSO_3^-)_x M^{x+}} + \underset{\text{solution}}{xH^+}$$

where M^{x+} is a cation and R represents *that part of a resin molecule containing one sulfonic acid group.* The analogous process involving a typical anion-exchange resin can be written as

$$\underset{\text{solid}}{xRN(CH_3)_3^+OH^-} + \underset{\text{solution}}{A^{x-}} \rightleftharpoons \underset{\text{solid}}{[RN(CH_3)_3^+]_x A^{x-}} + \underset{\text{solution}}{xOH^-}$$

where A^{x-} is an anion.

25D-2 Ion-Exchange Equilibria

Ion-exchange equilibria can be treated by the law of mass action. For example, when a dilute solution of calcium ions is brought into contact with a sulfonic acid resin, the following equilibrium develops:

$$Ca^{2+}(aq) + 2H^+(res) \rightleftharpoons Ca^{2+}(res) + 2H^+(aq)$$

Application of the mass law leads to

$$K' = \frac{[Ca^{2+}(res)][H^+(aq)]^2}{[Ca^{2+}(aq)][H^+(res)]^2}$$

where the bracketed terms are molar concentrations (strictly, activities). Note that $[Ca^{2+}(res)]$ and $[H^+(res)]$ are the concentrations of the two ions *in the solid phase.* In contrast to most solids, however, these concentrations can vary from zero to some maximum value at which all the negative sites in the resin are occupied by one species only.

Ion-exchange separations are ordinarily performed under conditions in which one of the ions predominates in *both* phases. For example, in the removal of calcium ions from a dilute and somewhat acidic solution, the calcium ion concentration is much smaller than that of hydrogen ions in both the aqueous and the resin phase. Consequently, the concentration of hydrogen ions does not change significantly as a result of the exchange process. Under these circumstances, $[Ca^{2+}(res)] << [H^+(res)]$ and $[Ca^{2+}(aq)] << [H^+(aq)]$ and the foregoing equilibrium-constant expression can be rearranged to give

$$\frac{[Ca^{2+}(res)]}{[Ca^{2+}(aq)]} = K' \frac{[H^+(res)]^2}{[H^+(aq)]^2} = K \qquad (25\text{-}1)$$

where K is the distribution ratio as defined by Equation 23-1. Note that K in Equation 25-1 represents the affinity of the resin for calcium ion relative to another ion (here, H^+). In general, where K for an ion is large, a strong tendency for the stationary phase to retain that ion exists; where K is small, the opposite is true. Selection of a common reference ion (such as H^+) permits a comparison of distribution ratios for various ions on a given type of resin. Such experiments reveal that polyvalent ions are much more strongly retained than singly charged species. Within a given charge group, differences in K appear to be related to the size of the hydrated ion as well as other properties. Thus, for a typical sulfonated cation-exchange resin, values of K for univalent ions decrease in the order $Ag^+ > Cs^+ > Rb^+ > K^+ > NH_4^+ > Na^+ > H^+ > Li^+$. For divalent cations, the order is $Ba^{2+} > Pb^{2+} > Sr^{2+} > Ca^{2+} > Ni^{2+} > Cd^{2+} > Cu^{2+} > Co^{2+} > Zn^{2+} > Mg^{2+} > UO_2^{2+}$.

25D-3 Applications of Ion-Exchange Resins to Chromatography

In ion chromatography, analyte ions are introduced at the head of a column packed with a suitable ion-exchange resin. Elution is then carried out with a solution that contains an ion that competes with the analyte ions for the charged groups on the resin surface. For example, anions such as chloride, thiocyanate, sulfate, and phosphate can be separated on an anion-exchange resin in its basic form. In this application, the sample is first introduced onto the head of the column, where the anions are retained by the reaction

$$xRN(CH_3)_3^+OH^- + A^{x-} \rightleftharpoons [RN(CH_3)_3^+]_xA^{x-} + xOH^-$$

Elution is then carried out with a dilute solution of a base, which causes the foregoing reaction to be reversed and the anions released. Since the partition ratios for various anions differ from one another, fractionation occurs during the elution.

One of the attractive aspects of ion chromatography is that conductivity measurements provide a convenient general method for detecting and determining the concentration of the eluted species. Ion chromatography with conductometric detection of ions eluted from the column was first described in 1975 and by now has become a powerful and important technique for the quantitative determination of both inorganic and organic charged species.[3]

[3]For a brief review of ion chromatography, see J. S. Fritz, *Anal. Chem.,* **1987,** *59,* 335A. For a detailed description of the method, see F. C. Smith Jr. and R. C. Chang, *The Practice of Ion Chromatography.* New York: Wiley, 1983; D. T. Gjerde and J. S. Fritz, *Ion Chromatography,* 2nd ed. New York: A. Huethig, 1987.

Two types of chromatography based on ion-exchange packings are currently in use: *suppressor-based* and *single-column* ion chromatography. They differ in the method used to prevent the conductivity of the eluting electrolyte from interfering with the measurement of the conductivity of the analytes.

25D-4 Ion Chromatography Based on Suppressors

Conductivity detectors have many of the properties of the ideal detector alluded to in Section 24B-5. They can be highly sensitive, they are universal for charged species, and as a general rule, they respond in a predictable way to concentration changes. Furthermore, such detectors are simple to operate, inexpensive to construct and maintain, easy to miniaturize, and ordinarily give prolonged, trouble-free service. The only limitation to the use of conductivity detectors, which delayed their general application to ion chromatography until the mid-1970s, arises from the high electrolyte concentrations required to elute most analyte ions in a reasonable time. As a consequence, the conductivity from the mobile-phase components tends to swamp that from the analyte ions, thus greatly reducing the detector sensitivity.

In 1975, the problem created by the high conductance of eluents was solved by the introduction of an *eluent suppressor column* immediately following the ion-exchange column.[4] The suppressor column is packed with a second ion-exchange resin that effectively converts the ions of the eluting solvent to a molecular species of limited ionization without affecting the conductivity due to the analyte ions. For example, when cations are being separated and determined, hydrochloric acid is chosen as the eluting reagent, and the suppressor column is an anion-exchange resin in the hydroxide form. The product of the reaction between the eluent and the suppressor is water:

$$H^+(aq) + Cl^-(aq) + resin^+OH^-(s) \longrightarrow resin^+Cl^-(s) + H_2O$$

The analyte cations are of course not retained by this second column.

For anion separations, the suppressor packing is the acid form of a cation-exchange resin and sodium bicarbonate or carbonate is the eluting agent. The reaction in the suppressor is

$$Na^+(aq) + HCO_3^-(aq) + resin^-H^+(s) \longrightarrow resin^-Na^+(s) + H_2CO_3(aq)$$

The largely undissociated carbonic acid does not contribute significantly to the conductivity.

An inconvenience associated with the original suppressor columns was the need to regenerate them periodically (typically, every 8 to 10 h) in order to convert the packing back to the original acid or base form. Recently, however, micromembrane suppressors that operate continuously have become available.[5] For example, where sodium carbonate or bicarbonate is to be removed, the eluent is passed over a series of ultra-thin cation-exchange membranes that separate it from a stream of acidic regenerating solution that flows continuously in the opposite direction. The sodium ions from the eluent exchange with hydrogen ions on the inner surface of the exchanger membrane and then migrate to the other surface for exchange with hydrogen ions from

[4] H. Small, T. S. Stevens, and W. C. Bauman, *Anal. Chem.*, **1975**, *47*, 1801.

[5] For a description of this device, see G. O. Franklin, *Amer. Lab.*, **1985** (3), 71.

the regenerating reagent. Hydrogen ions from the regeneration solution migrate in the reverse direction, thus preserving electrical neutrality.

Figure 25-5 shows two applications of ion chromatography based upon a suppressor column and conductometric detection. In each, the ions were present in the parts-per-million range; the sample size was 50 μL in one case and 20 μL in the other. The method is particularly important for anion analysis because no other rapid and convenient method for handling mixtures of this type now exists.

25D-5 Single-Column Ion Chromatography

Recently, equipment has become available commercially for ion chromatography in which no suppressor column is used. This approach depends upon the small differences in conductivity between the eluted sample ions and the prevailing eluent ions. To amplify these differences, low-capacity exchangers that permit elution with dilute eluent solutions are used. Furthermore, eluents of low equivalent conductance are chosen.[6]

Single-column ion chromatography offers the advantage of not requiring special equipment for suppression. It is, however, a somewhat less sensitive method for determining anions than suppressor column methods.

25E High-Performance Size-Exclusion Chromatography

Size-exclusion chromatography is a recent development in liquid chromatographic procedures. It is a powerful technique that is particularly applicable to high-molecular-weight species.[7]

25E-1 Packings

Packings for size-exclusion chromatography consist of small ($\simeq$ 10 μm) silica or polymer particles containing a network of uniform pores into which solute and solvent molecules can diffuse. While in the pores, molecules are effectively trapped and removed from the flow of the mobile phase. The average residence time of analyte molecules depends upon their effective size. Molecules that are significantly larger than the average pore size of the packing are excluded and thus are not retained; that is, they travel through the column at the rate of the mobile phase. Molecules that are appreciably smaller than the pores can penetrate throughout the pore maze and are thus entrapped for the greatest time; they are last to be eluted. Between these two extremes are intermediate-size molecules whose average penetration into the pores of the packing depends upon their diameter. The fractionation that occurs within this group is directly related to molecular size and, to some extent, molecular shape. Note that size-exclusion separations differ from the other chromatographic procedures in the respect that no chemical or physical interactions occur between analyte and stationary phase. Indeed, every effort is made to avoid such interactions because they impair column efficiency.

[6]See R. M. Becker, *Anal. Chem.*, **1980,** *52,* 1510; J. R. Benson, *Amer. Lab.*, **1985,** (6), 30; and T. Jupille, *Amer. Lab.*, **1986,** (5), 114.

[7]See W. W. Yao, J. J. Kirkland, and D. D. Bly, *Modern Size-Exclusion Liquid Chromatography.* New York: Wiley, 1979.

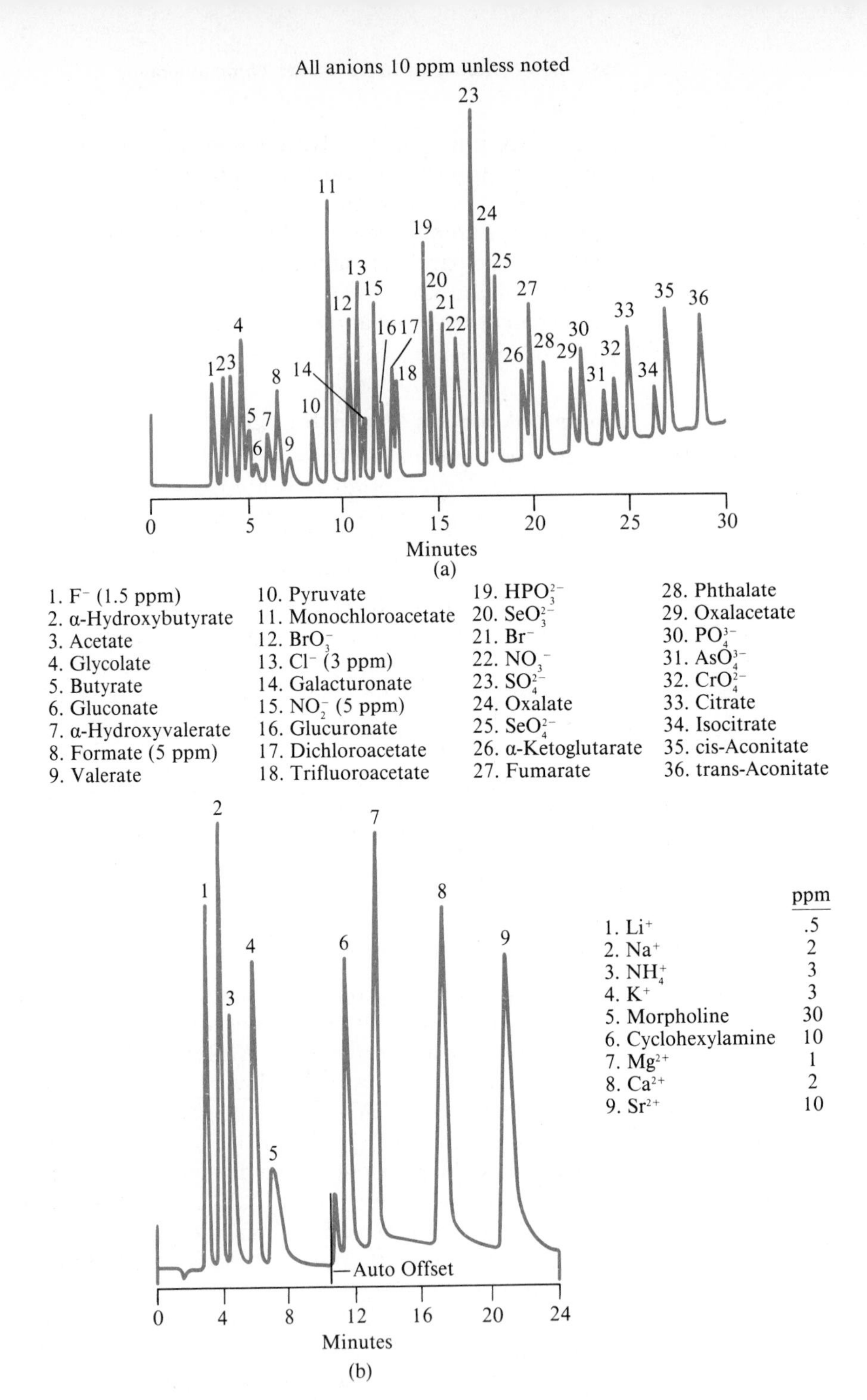

FIGURE 25-5 Applications of ion chromatography. (a) Separation of anions on an anion-exchange column. Gradient eluent: 0.00075 to 0.85 M NaOH. Sample size: 20 μL. (b) Separation of cations on a cation-exchange column. Eluent 1: 0.012 M HCl, 0.00025 M diaminopropionic acid hydrochloride, and 0.00025 M histidine hydrochloride. Eluent 2: 0.048 M HCl, 0.004 M diaminopropionic acid hydrochloride, and 0.004 M histidine hydrochloride. Sample size: 50 μL. Conductivity detector for both samples. (Courtesy of Dionex, Sunnyvale, CA.)

Numerous size-exclusion packings are on the market. Some are hydrophilic for use with aqueous mobile phases; others are hydrophobic and are used with nonpolar organic solvents. Chromatography based on the former is sometimes called *gel filtration,* while techniques based on the latter are termed *gel permeation.* With both types of packing, a wide range of pore diameters is available. The average molecular weight suitable for a given packing may be as small as a few hundred or as large as several million. Ordinarily, a given packing will accommodate a 2- to 2.5-decade molecular weight range.

25E-2 Applications

Figure 25-6a illustrates the application of gel filtration to the determination of three sugars in an aqueous medium. The hydrophilic packing excluded molecular weights greater than 1000. Figure 25-6b is a chromatogram obtained with a hydrophobic packing in which the eluent was tetrahydrofuran. The sample was a commercial epoxy resin in which each monomer unit had a molecular weight of 280.

Another important application of size-exclusion chromatography is in the rapid determination of the molecular weight or the molecular weight distribution of large polymers or natural products. In this method, the elution volumes of the sample are compared with the elution volumes for a series of standard compounds that have the same chemical characteristics.

25F A Comparison of High-Performance Liquid Chromatography and Gas-Liquid Chromatography

A comparison of high-performance liquid chromatography and gas-liquid chromatography is presented in Table 25-3. The former is applicable to nonvolatile substances (including inorganic ions) and thermally unstable materials,

FIGURE 25-6 Applications of size-exclusion chromatography. (a) Gel-filtration determination of glucose (G), fructose (F), and sucrose (S) in canned juices. A 25-cm column was packed with hydrophilic sulfonated polymer particles with an exclusion limit of 1000. (b) Gel-permeation analysis of a commercial epoxy resin (n = number of monomeric units in the polymer). A porous-silica column, 6.2 mm × 250 mm, was used with a mobile phase of tetrahydrofuran. (Courtesy of Du Pont Biotechnology Systems, Wilmington, DE.)

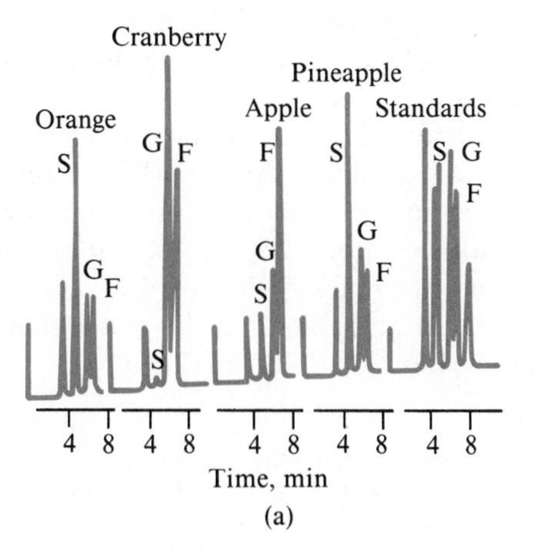

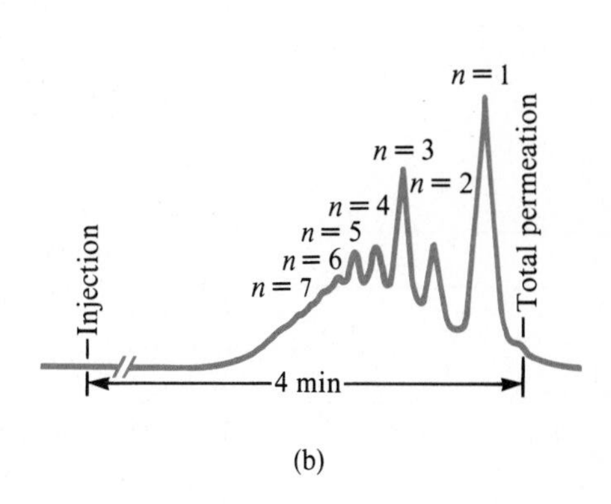

TABLE 25-3 **Comparison of High-Performance Liquid Chromatography and Gas-Liquid Chromatography**

Characteristics of Both Methods
- Efficient, highly selective, and widely applicable
- Only small sample required
- May be nondestructive of sample
- Readily adapted to quantitative analysis

Advantages of HPLC
- Can accommodate nonvolatile and thermally unstable samples
- Generally applicable to inorganic ions

Advantages of GLC
- Simple and inexpensive equipment
- Rapid
- Unparalleled resolution (with capillary columns)
- Easily interfaced with mass spectroscopy

whereas the latter is not. Yet the resolving power of gas-liquid chromatography, particularly of the open-tubular type, is generally superior to that of high-performance liquid chromatography. When both procedures are applicable to a given separation, gas-liquid chromatography offers the advantage of speed and simplicity of equipment. In many situations the methods tend to be complementary.

25G Supercritical-Fluid Chromatography

Supercritical-fluid chromatography (SFC) is a hybrid of gas and liquid chromatography that combines some of the best features of each. While supercritical-fluid chromatography is still in its youth, it appears to be clearly superior to both gas-liquid and high-performance liquid chromatography for certain applications.[8]

25G-1 Important Properties of Supercritical Fluids

A *supercritical fluid* is formed whenever a substance is heated above its *critical temperature*. At the critical temperature, a substance can no longer be condensed into its liquid state through the application of pressure. For example, carbon dioxide becomes a supercritical fluid at temperatures above 31°C. In this state, the molecules of carbon dioxide act independently of one another just as they do in a gas.

As shown by the data in Table 25-4, the physical properties of a substance in the supercritical-fluid state can be remarkably different from the same properties in either the liquid or the gaseous state. For example, the density of a supercritical fluid is typically 200 to 400 times greater than that of the corresponding gas and approaches that of the substance in its liquid state. The properties compared in Table 25-4 are those that are of importance in gas, liquid, and supercritical-fluid chromatography.

[8]For reviews of this technique, see L. G. Randall, *Sep. Sci. Technol.*, **1982**, *17*, 1; D. R. Gere, *Science*, **1983**, *222*, 253; J. C. Fjeldsted and M. L. Lee, *Anal. Chem.*, **1984**, *56*, 619A. For a brief discussion of the supercritical-fluid state, see E. F. Meyer and T. P. Meyer, *J. Chem. Educ.*, **1986**, *63*, 463.

An important property of supercritical fluids and one that is related to their high densities (0.2 to 0.5 g/cm^3), is their ability to dissolve large nonvolatile molecules. For example, supercritical carbon dioxide readily dissolves *n*-alkanes containing from 5 to 22 carbon atoms, di-*n*-alkylphthalates in which the alkyl groups contain 4 to 16 carbon atoms, and various polycyclic aromatic hydrocarbons consisting of several rings.[9]

Critical temperatures for fluids used in chromatography vary widely, from about 30°C to above 200°C. Lower critical temperatures are advantageous from several standpoints. For this reason, much of the work to date has focused on such supercritical fluids as carbon dioxide (31°C), ethane (32°C), and nitrous oxide (37°C). Note that these temperatures, and the pressures at these temperatures, are well within the operating conditions of ordinary high-performance liquid chromatography.

25G-2 Instrumentation and Operating Variables

Instruments for supercritical-fluid chromatography are similar in design to high-performance liquid chromatographs except that provision is made in the former for controlling and measuring the column pressure. Several manufacturers began to offer apparatus for supercritical-fluid chromatography in the mid-1980s.

The Effect of Pressure

The density of a supercritical fluid increases rapidly and nonlinearly with pressure increases. Density increases also alter capacity factors (k') and thus elution times. For example, the elution time for hexadecane is reported to decrease from 25 to 5 min as the pressure of carbon dioxide is raised from 70 to 90 atm. Gradient elution can thus be achieved through linear increases in column pressure or through regulation of pressure to obtain linear density increases. An example of the latter is shown in Figure 25-9a, where the density of the carbon dioxide mobile phase was increased linearly from 0.225 to 0.70 g/mL (by increasing the pressure from 73 to 112 atm).

A clear analogy exists between gradient elution through adjustment of pressure or density of a supercritical fluid and by temperature-gradient elution in gas-liquid chromatography and solvent-gradient elution in liquid chromatography.

TABLE 25-4 **Comparison of Properties of Supercritical Fluids, Liquids, and Gases**

	Gas (STP)	Supercritical Fluid	Liquid
Density, g/cm^3	$(0.6–2) \times 10^{-3}$	0.2–0.5	0.6–1.6
Diffusion coefficient, $cm^2 \cdot s^{-1}$	$(1–4) \times 10^{-1}$	$10^{-3}–10^{-4}$	$(0.2–2) \times 10^{-5}$
Viscosity, $g \cdot cm^{-1} \cdot s^{-1}$	$(1–3) \times 10^{-4}$	$(1–3) \times 10^{-4}$	$(0.2–3) \times 10^{-2}$

All data order of magnitude only.

[9]Certain important industrial processes are based upon the high solubility of organic species in supercritical carbon dioxide. For example, this medium has been employed in extracting caffeine from coffee beans to give decaffeinated coffee and in extracting nicotine from cigarette tobacco.

Columns

Although supercritical-fluid chromatography has been carried out in both packed and open tubular columns, the latter appear to be favored because their length can be greater, a condition that leads to enhanced column efficiency. Typical columns are similar to the fused-silica open tubular (FSOT) columns described in Table 24-1. Because of the low viscosity of supercritical media, columns can be much longer than those used in liquid chromatography, and column lengths of 10 or 20 m and inside diameters of 50 or 100 μm are common. For difficult separations, columns 60 m in length and longer have been used.

Many of the column coatings used in liquid chromatography have been applied to supercritical-fluid chromatography as well. Typically, these are polysiloxanes that are chemically bonded to the inner silica wall of the capillary tubing. Film thicknesses are 0.05 to 0.4 μm.

Mobile Phases

The most widely used mobile phase for supercritical-fluid chromatography is carbon dioxide. It is an excellent solvent for a variety of organic molecules. In addition, it transmits in the ultraviolet and is odorless, nontoxic, readily available, and remarkably inexpensive relative to other chromatographic solvents. Its critical temperature of 31°C and its pressure of 73 atm at the critical temperature permit a wide selection of temperatures and pressures without exceeding the operating limits of modern high-performance liquid chromatography equipment. In some applications, polar organic modifiers, such as methanol, are introduced in small concentrations ($\simeq$ 1%) to modify α values for analytes.

A number of other substances have served as mobile phases in supercritical chromatography, including ethane, pentane, dichlorodifluoromethane, diethyl ether, and tetrahydrofuran.

Detectors

A major advantage of supercritical-fluid chromatography is that the sensitive and universal detectors of gas-liquid chromatography are applicable to this technique as well. For example, the convenient flame ionization detector of gas-liquid chromatography can be applied by simply allowing the supercritical carrier to expand through a restrictor and into a hydrogen flame, where ions formed from the analytes cause variations in the electrical conductivity of the medium.

25G-3 Supercritical-Fluid Chromatography Versus Other Column Methods

The information in Table 25-4, and other data as well, reveal that several physical properties of supercritical fluids are intermediate between the properties of gases and liquids. As a consequence, this new type of chromatography combines some of the characteristics of both gas and liquid chromatography. Thus, like gas chromatography, supercritical-fluid chromatography is inherently faster than liquid chromatography because of the lower viscosity and higher diffusion rates in the mobile phase. High diffusivity, however, leads to longitudinal band spreading (page 610), which is a significant factor with gas but not with liquid chromatography. Thus, the intermediate diffusivities and viscosities of supercritical fluids result in faster separations than are achieved with liquid chromatography accompanied by less zone spreading than is encountered in gas chromatography.

Figure 25-7 compares the performance characteristics of packed columns when elution is performed with supercritical carbon dioxide and with a conventional liquid mobile phase. Note that the supercritical column yields a plate height of about 0.013 mm at a mobile-phase flow rate of 0.6 cm/s, whereas at this same velocity, the plate height of the conventional column is three times as large, or 0.039 mm. Thus, a reduction in peak width by $\sqrt{3}$ should be realized (Equation 23-16). Alternatively, at a plate height corresponding to the minimum in the HPLC curve (0.012 mm) the mobile-phase velocity in the SFC system is four times that of the HPLC system. Thus, a fourfold decrease in analysis time results. These advantages are reflected in the two chromatograms shown in Figure 25-8.

25G-4 *Applications*

Supercritical-fluid chromatography, particularly of the open tubular type, appears to have a potential niche in the spectrum of column-chromatographic methods because it is applicable to a class of compounds that is not readily amenable to either gas-liquid or liquid chromatography. These compounds include species that are nonvolatile or thermally unstable and, in addition, contain no chromophoric groups that can be used for photometric detection. Separation of these compounds is possible with supercritical-fluid chromatography at temperatures below 100°C; furthermore, detection is readily carried out by means of the highly sensitive flame ionization detector. As shown in Figure 25-9, the efficiency of open tubular supercritical columns often approaches that of capillary gas-liquid columns. Note that the supercritical chromatogram was obtained at 40°C, which clearly illustrates the applicability of this technique to thermally labile species.

It is also noteworthy that supercritical columns have the added advantage of being much easier to interface with mass spectrometers than liquid chromatographic columns.

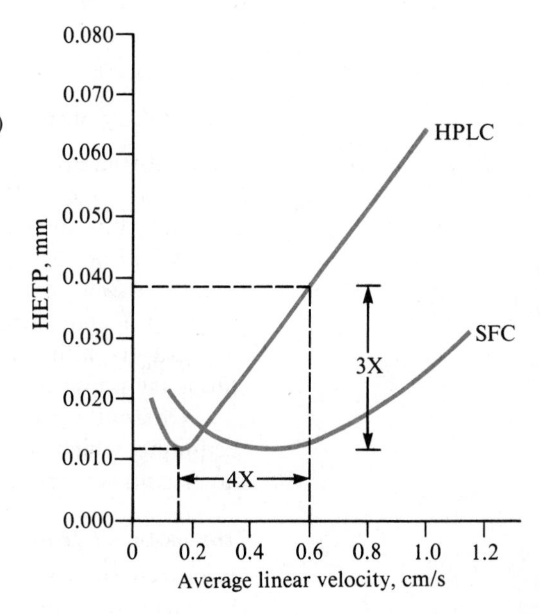

FIGURE 25-7 Performance characteristics of a 5-μm, particle-size packed column when elution is carried out with a conventional mobile phase (HPLC) and with supercritical carbon dioxide (SFC). (From D. R. Gere, *Application Note 800-3,* Hewlett-Packard, 1983. With permission.)

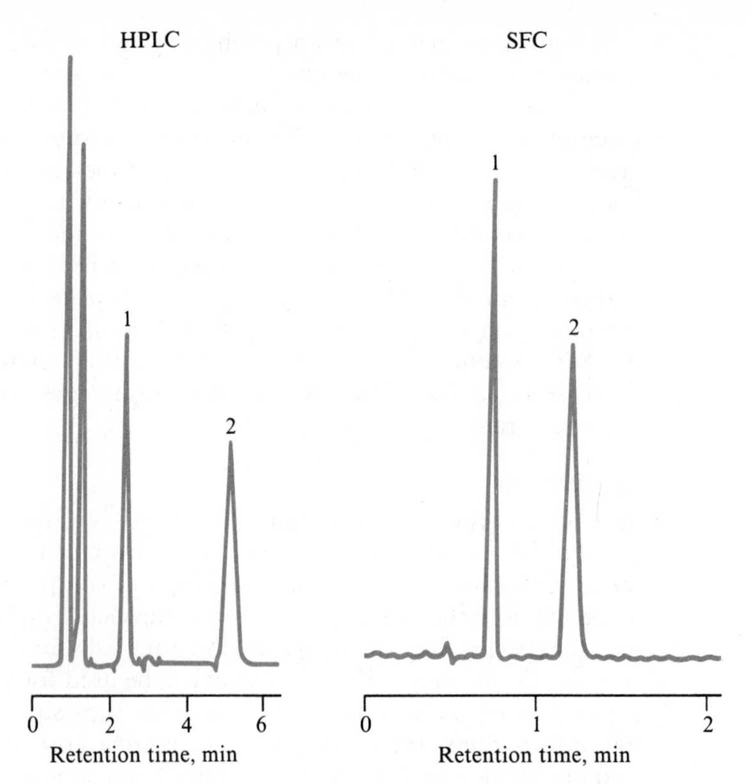

FIGURE 25-8

Comparison of chromatograms obtained by conventional partition chromatography (HPLC) and supercritical-fluid chromatography (SFC). Column: 20 cm × 4.6 mm packed with 10-μm reversed-phase bonded packing. Analytes: 1) biphenyl; 2) terphenyl. For HPLC, mobile phase: 65% CH_3OH/35% H_2O; flow rate: 4 mL/min; linear velocity: 0.55 cm/s; sample size: 10 μL. For SFC, mobile phase: CO_2; flow rate: 5.4 mL/min; linear velocity: 0.76 cm/s; sample size: 3 μL. (From D. R. Gere, T. J. Stark, and T. N. Tweeten, *Application Note 800-4,* Hewlett-Packard, 1983. With permission.)

25H Planar Chromatography

Planar chromatographic methods include *thin-layer chromatography* (TLC), *paper chromatography* (PC), and *electrochromatography*. Each makes use of a flat, relatively thin layer of material that is either self-supporting or is coated on a glass, plastic, or metal surface. The mobile phase moves through the stationary phase by capillary action, sometimes assisted by gravity or an electrical potential. Planar chromatography is sometimes called two-dimensional chromatography, although this description is not strictly correct inasmuch as the stationary phase does have a finite thickness.

Currently, most planar chromatography is based upon the thin-layer technique, which is faster, has better resolution, and is more sensitive than its paper counterpart. This section is devoted to thin-layer methods.

25H-1 The Scope of Thin-Layer Chromatography

In terms of theory, the types of stationary and mobile phases, and applications, thin-layer and liquid chromatography are remarkably similar. In fact, thin-

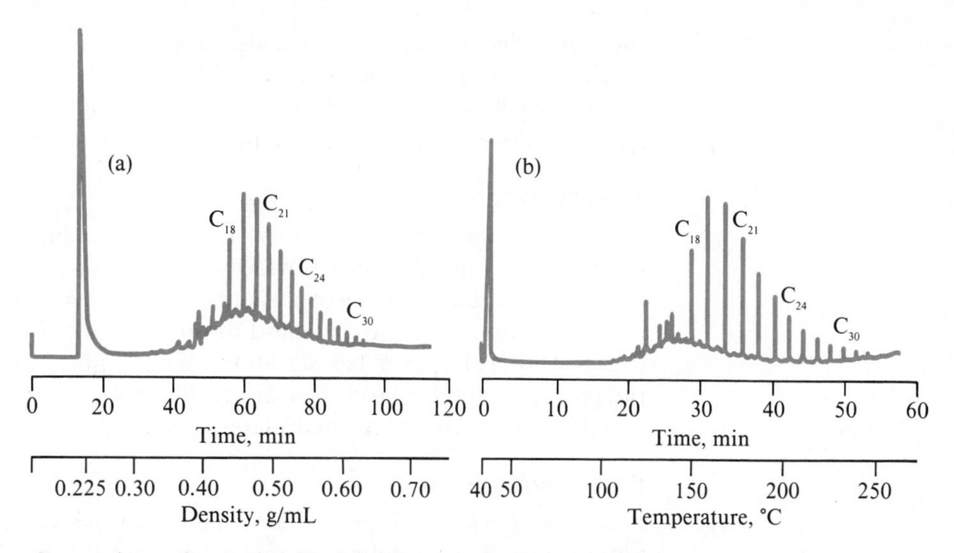

FIGURE 25-9 Comparison of a supercritical-fluid chromatogram and a capillary gas-liquid chromatogram for an aliphatic fraction from a solvent-refined coal fraction. (a) SFC chromatogram; CO_2 mobile phase at 40°C; 34 m × 50 μm inside-diameter fused capillary column coated with a 0.25-μm film of cross-linked SE-54; flame ionization detector; linear density program from 0.225 g/mL (73 atm) to 0.70 g/mL (112 atm). (b) Capillary GLC chromatogram; H_2 carrier gas; 20 m × 300 μm inside-diameter fused-silica column coated with a 0.25-μm film of cross-linked SE-54; flame ionization detector; linear temperature program from 50 to 250°C. (From W. P. Jackson *et al.*, *Ultrahigh Resolution Chromatography,* ACS Symposium Series 250, S. Ahuja, Ed., p. 130. Washington, D.C.: American Chemical Society, 1984. With permission.)

layer plates can be profitably used to develop optimal conditions for separations by column liquid chromatography. The advantages of following this procedure are the speed and low cost of the exploratory thin-layer experiments. Some chromatographers take the position that thin-layer experiments should always precede column experiments.

Thin-layer chromatography has become the workhorse of the drug industry for the all-important determination of product purity. It has also found widespread use in clinical laboratories and is the backbone of many biochemical and biological studies. Finally, it finds widespread use in the industrial laboratories.[10] As a consequence of these many areas of application, it has been estimated that at least as many analyses are performed by thin-layer chromatography as by high-performance liquid chromatography.[11]

25H-2 The Principles of Thin-Layer Chromatography

Typical thin-layer separations are performed on a glass plate that is coated with a thin and adherent layer of finely divided particles; this layer constitutes

[10] Two monographs devoted to the principles and applications of thin-layer chromatography are B. Fried and J. Sherma, *Thin-Layer Chromatography.* New York: Marcel Dekker, 1982; and J. C. Touchstone, *Practice of Thin-Layer Chromatography,* 2nd ed. New York: Wiley, 1983. For briefer reviews, see D. C. Fenimore and C. M. Davis, *Anal. Chem.*, **1981,** *53,* 253A; D. Rogers, *Amer. Lab.*, **1984,** *16* (6), 65.

[11] T. H. Mauch II, *Science,* **1982,** *216,* 161.

the stationary phase. The particles are similar to those described in the discussion of adsorption, normal- and reversed-phase partition, ion-exchange, and size-exclusion column chromatography. Mobile phases are also similar to those employed in high-performance liquid chromatography.

The Preparation of Thin-Layer Plates

A thin-layer plate is prepared by spreading an aqueous slurry of the finely ground solid onto the clean surface of a glass or plastic plate or microscope slide. Often a binder is incorporated into the slurry to enhance adhesion of the solid particles to the glass and to one another. The plate is then allowed to stand until the layer has set and adheres tightly to the surface; for some purposes, it may be heated in an oven for several hours. Several chemical supply houses offer precoated plates of various kinds.

Plate Development

Plate development is the process in which a sample is carried through the stationary phase by a mobile phase; it is analogous to elution in liquid chromatography. The most common way of developing a plate is to place a drop of the sample near one edge of the plate (most plates have dimensions of 5 × 20 or 20 × 20 cm) and mark its position with a pencil. After the sample solvent has evaporated, the plate is placed in a closed container saturated with vapors of the developing solvent. One end of the plate is immersed in the developing solvent, with care being taken to avoid direct contact between the sample and the developer (Figure 25-10). After the developer has traversed one half or two thirds of the length of the plate, the plate is removed from the container and dried. The positions of the components are then determined in any of several ways.

Figure 25-11 illustrates the separation of amino acids in a mixture by development in two directions (*two-dimensional planar chromatography*). The sample was placed in one corner of a square plate, and the plate was developed in the ascending direction with solvent A. This solvent was then removed by evaporation, and the plate was rotated 90 deg, following which ascending development with solvent B was performed. After solvent removal, the positions of the amino acids were determined by spraying with ninhydrin, a

FIGURE 25-10 (a) Ascending-flow developing chamber. (b) Horizontal-flow developing chamber, in which samples are placed on both ends of the plate and developed toward the middle, thus doubling the number of samples that can be accommodated.

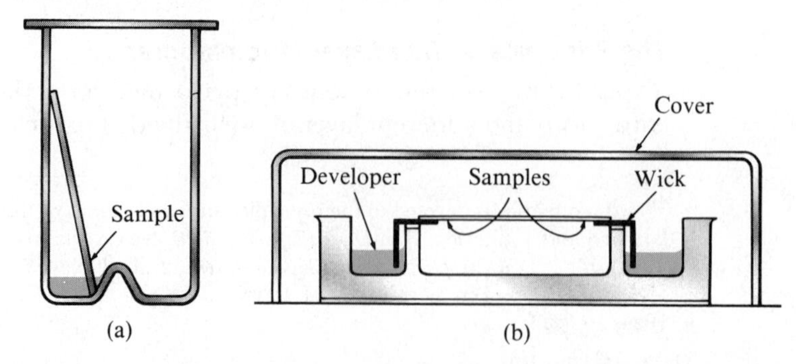

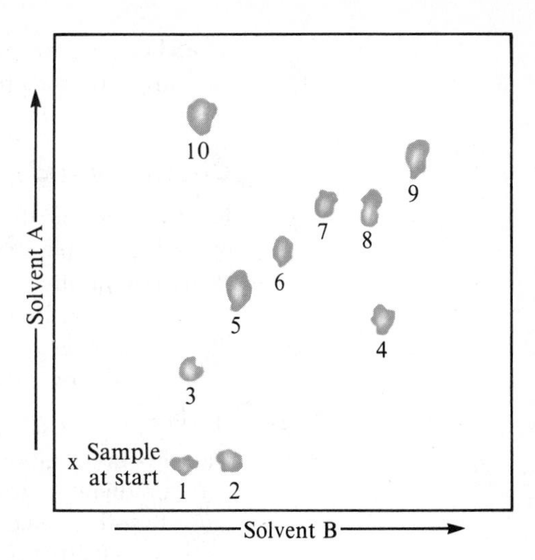

FIGURE 25-11 Two-dimensional thin-layer chromatogram (silica gel) of some amino acids. Solvent A: toluene/2-chloroethanol/pyridine. Solvent B: chloroform/benzyl alcohol/acetic acid. Amino acids: (1) aspartic acid, (2) glutamic acid, (3) serine, (4) β-alanine, (5) glycine, (6) alanine, (7) methionine, (8) valine, (9) isoleucine, (10) cysteine.

reagent that forms a pink to purple product with amino acids. The spots were identified by comparison of their positions with those of standards.

Locating Analytes on the Plate

Several methods are employed to locate sample components after separation. Two common methods, which can be applied to most organic mixtures, involve spraying with a solution of iodine or sulfuric acid, both of which react with organic compounds to yield dark products. Several specific reagents (such as ninhydrin) are also useful for locating separated species.

Another method of detection is based upon incorporating a fluorescent material into the stationary phase. After development, the plate is examined under ultraviolet light. The sample components quench the fluorescence of the material so that all of the plate fluoresces except where the nonfluorescing sample components are located.

Quantitative Analysis

A semiquantitative estimate of the amount of a component present can be obtained by comparing the area of a spot with that of a standard. Better data can be obtained by scraping the spot from the plate, extracting the analyte from the stationary-phase solid, and measuring the analyte by a suitable physical or chemical method. In a third method, a scanning densitometer can be employed to measure the radiation emitted from the spot by fluorescence or reflection.

25H-3 ***Paper Chromatography***

Separations by paper chromatography are performed in the same way as those on thin-layer plates. The papers are manufactured from highly purified cellulose with close control over porosity and thickness. Such papers contain sufficient adsorbed water to make the stationary phase aqueous. Other liquids can be made to displace the water, however, thus providing a different type of stationary phase. For example, paper treated with silicone or paraffin oil permits reversed-phase paper chromatography, in which the mobile phase is a polar solvent. Also available commercially are special papers that contain an adsorbent or an ion-exchange resin, thus permitting adsorption and ion-exchange paper chromatography.

Questions and Problems

25-1 List the types of substances to which each of the following chromatographic methods is most applicable:

*(a) gas-liquid
(b) liquid partition
*(c) ion-exchange
(d) liquid adsorption
*(e) gel permeation
(f) gel filtration
*(g) gas-solid
(h) supercritical fluid

25-2 Define

*(a) isocratic elution.
(b) gradient elution.
*(c) stop-flow injection.
(d) reversed-phase packing.
*(e) normal-phase packing.
(f) ion chromatography.
*(g) eluent-suppressor column.
(h) gel filtration.
*(i) gel permeation.
(j) critical temperature.
*(k) FSOT column.
(l) two-dimensional thin-layer chromatography.
*(m) supercritical fluid.

25-3 List the differences in properties and roles of the mobile phase in gas, liquid, and supercritical-fluid chromatography. How do these differences influence the characteristics of the three methods?

25-4 For a normal-phase separation, predict the order of elution of

*(a) n-hexane, n-hexanol, benzene.
(b) ethyl acetate, diethyl ether, nitrobutane.

25-5 For a reversed-phase separation, predict the order of elution of the solutes in Problem 25-4.

25-6 Describe the physical differences between open tubular and packed columns. What are the advantages and disadvantages of each?

*25-7 Describe the fundamental difference between adsorption and partition chromatography.

25-8 Describe the fundamental difference between ion-exchange and size-exclusion chromatography.

*25-9 What types of species can be separated by high-performance liquid chromatography but not by gas-liquid chromatography?

*25-10 To what types of compounds is supercritical-fluid chromatography particularly applicable?

*25-11 Describe the various kinds of pumps used in high-performance liquid chromatography. What are the advantages and disadvantages of each?

25-12 Describe the difference between single-column and suppressor-column ion chromatography.

*25-13 List the advantages of supercritical fluid chromatography over (a) high-performance liquid chromatography and (b) gas-liquid chromatography.

25-14 Why is there more zone broadening with gas chromatography than with supercritical-fluid chromatography? Why is there less zone broadening with high-performance liquid chromatography than with supercritical-fluid chromatography?

CHAPTER 26

The Analysis of Real Samples

Very early in this text (Section 1C), we pointed out that a quantitative analysis involves a sequence of steps: (1) selecting a method, (2) sampling, (3) preparing a laboratory sample, (4) defining replicate samples (by weight or volume measurements), (5) preparing solutions of the samples, (6) eliminating interferences, (7) completing the analysis by performing measurements that are related in a known way to analyte concentration, and (8) computing the results and estimating their reliability.

Thus far we have focused largely on steps 7 and 8 and to a lesser extent on steps 4 and 6. This emphasis is *not* because the earlier steps are unimportant or easy—quite the opposite. The preliminary steps may be not only more difficult and time-consuming than the two final steps of an analysis but also a greater source of error.

The reasons for postponing a discussion of the preliminary steps to this point are pedagogical. Experience has shown that it is easier to introduce students to analytical techniques by having them first perform measurements on simple materials for which no method selection is required and for which problems with sampling, sample preparation, and sample dissolution are either nonexistent or easily solved. Thus, we have been largely concerned so far with measuring the concentration of analytes in simple aqueous solutions that are uncluttered by interfering species.

The determination of an analyte in a simple solution is usually not particularly difficult once learned because the number of variables that must be controlled is small and the tools available are numerous and easy to use. Furthermore, with simple systems, our theoretical knowledge suffices to allow us to anticipate problems and correct for them. Thus, if chemical analysis involved only determining the concentration of a single species in a simple and readily soluble homogeneous mixture, analytical chemistry could profitably be entrusted to the hands of a skilled technician; certainly, a professional chemist can find more useful and challenging work for mind and hands.

In fact, the materials in which academic and industrial chemists are interested are not as a rule simple. To the contrary, most substances that require analysis are complex, consisting of several species or several tens of species. Such materials are frequently far from ideal in matters of solubility, volatility, stability, and homogeneity, and several steps must precede the final measurement step. Indeed, the final measurement is often anticlimactic in the sense that it is by far easier and less time-consuming than any of the several steps that must precede it.

For example, we showed in earlier chapters that the calcium ion concentration of an aqueous solution is readily determined by titration with a standard EDTA solution or by potential measurements with a specific-ion electrode. Alternatively, the calcium content of a solution can be determined from flame absorption or emission measurements or by the precipitation of calcium oxalate followed by weighing or by titrating with a standard solution of potassium permanganate.

All the aforementioned methods provide a straightforward means of determining the calcium content of a simple salt, such as the carbonate. The chemist is seldom interested in the calcium content of calcium carbonate, however. More likely, what is needed is the percentage of this element in a sample of animal tissue, a silicate rock, or a piece of glass. The analysis thereby acquires new and formidable complexities. For example, none of these materials is soluble in water or dilute aqueous reagents. Before calcium can be determined, therefore, the sample must be decomposed by high-temperature treatment with concentrated reagents. Unless care is taken, this step may cause losses of some of the calcium from the sample or, equally bad, the introduction of that element as a contaminant in the relatively large quantities of reagent usually required to decompose a sample.

Even after the sample has been decomposed to give a solution of calcium ion, the excellent procedures mentioned in the two previous paragraphs cannot ordinarily be applied immediately to complete the analysis, for they are all based upon reactions or properties shared by several elements in addition to calcium. Thus, a sample of animal tissue, silicate rock, or glass almost surely contains one or more components that also react with EDTA, act as a chemical interference in a flame-absorption measurement, or form a precipitate with oxalate ion; furthermore, the high ionic strength resulting from the reagents used for sample decomposition would complicate a direct potentiometric measurement. As a consequence, several additional operations are required before the final measurement to free the solution of interferences.

We have chosen the term *real samples* to describe materials such as those in the preceding illustration. In this context, most of the samples encountered in an elementary quantitative analysis laboratory course definitely are not real, for they are generally homogeneous, usually stable even with rough handling, readily soluble, and, above all, chemically simple. Moreover, well-established and thoroughly tested methods exist for their analysis. There is unquestioned value in introducing analytical techniques with such substances, for they do allow the student to concentrate on the mechanical aspects of an analysis. Once these mechanics have been mastered, however, there is little point in the continued analysis of unreal samples; to do so creates the impression that any chemical analysis involves nothing more than the slavish adherence to a well-defined and narrow path, the end result of which is a number that is accurate to one or two parts per thousand. Unfortunately, all too many chemists retain this view far into their professional lives.

In truth, the pathway leading to knowledge of the composition of real samples is frequently more demanding of intellectual skills and chemical intuition than of mechanical aptitude. Furthermore, a compromise must often be struck between the time that can be afforded and the accuracy that is believed necessary. The chemist is frequently happy to settle for an accuracy of one or two parts per hundred instead of one or two parts per thousand, knowing that the latter may require several hours or even days in additional effort. In fact, with complex materials, even parts-per-hundred accuracy may be unrealistic.

The difficulties encountered in the analysis of real samples stem from their complexity. As a consequence, the literature often contains no well-tested analytical route for the kind of sample under consideration, and so an existing procedure must be modified to take into account compositional differences between the sample in hand and the samples for which the original method was designed. Alternatively, a new analytical method must be developed. In either case, the number of variables that have to be taken into account usually increases exponentially with the number of species contained in the sample.

As an example, let us contrast the problems associated with the flame photometric analysis of calcium carbonate with those for a real calcium-containing sample. In the former, the number of components is small and the variables likely to affect the results are reasonably few. Principal among the variables to be considered are the physical losses of analyte due to the evolution of carbon dioxide when the sample is dissolved in acid, the effect of the anion of the acid and of the fuel-to-air ratio on the intensity of the calcium emission line, the position of the inner cone of the flame with respect to the entrance slit to the photometer, and the quality of the standard calcium solutions used for calibration.

The determination of calcium in a sample, such as a silicate rock, which contains a dozen or more other elements, is far more complex. First of all, the sample can be dissolved only by fusing it at a high temperature with a large excess of a reagent such as sodium carbonate. Physical loss of the analyte is possible during this treatment unless suitable precautions are taken. Furthermore, the introduction of calcium from the excess sodium carbonate or the fusion vessel is of real concern. Following fusion, the sample and reagent are dissolved in acid. With this step, all the variables affecting the calcium carbonate sample are operating, but in addition, a host of new variables are introduced because the dozens of components in the sample matrix. Now, measures must be taken to minimize instrumental and chemical interference brought about by the presence of various anions and cations in the solution being atomized.

The analysis of a real substance is often a challenging problem requiring knowledge, intuition, and experience. The development of a procedure for such materials is not to be taken lightly even by the experienced chemist.

26A Choice of Method for the Analysis of Real Samples

The choice of a method for the analysis of a complex substance requires good judgment and a sound knowledge of the advantages and limitations of the various analytical tools available. In addition, a familiarity with the literature of analytical chemistry is essential. We cannot be too explicit concerning how an analytical method is selected because there is no single best way that applies under all circumstances. We can, however, suggest a systematic approach to the problem and present some generalities that can aid in making intelligent decisions.

26A-1 Definition of the Problem

A first step, which must precede any choice of method, involves a clear definition of the analytical problem. The method of approach selected will be largely governed by the answers to the following questions:

What is the concentration range of the species to be determined?

What degree of accuracy is demanded?
What other components are present in the sample?
What are the physical and chemical properties of the gross sample?
How many samples are to be analyzed?

The concentration range of the analyte may well limit the number of feasible methods. If, for example, the analyst is interested in an element present to the extent of a few parts per million, gravimetric or volumetric methods can generally be eliminated, and spectrometric, potentiometric, and other more sensitive methods become likely candidates. For components in the parts-per-million range, the chemist must guard against even small losses resulting from coprecipitation or volatility, and contamination from reagents and apparatus becomes a major concern. In contrast, if the analyte is a major component of the sample, these considerations become less important, and the classic analytical methods may well be preferable.

The answer to the question of required accuracy is of vital importance in the choice of method and in the way it is performed because the time required to complete an analysis increases, often exponentially, with demands for greater accuracy. Thus, to improve the reliability of analytical results from, say, 2% to 0.2% relative may require a 100-fold or greater increase in operator time. Consequently, the chemist should always spend a few minutes before undertaking an analysis in careful consideration of the degree of accuracy really needed. It is usually foolish to produce physical or chemical data having accuracies that are significantly greater than what is demanded by the use to which the data is to be put.

The demands of accuracy frequently dictate the procedure chosen for an analysis. For example, if the allowable error in the determination of aluminum is only a few parts per thousand, a gravimetric procedure is probably required. On the other hand, if an error of, say, 50 ppt can be tolerated, spectroscopic or electroanalytical procedures are applicable.

The way in which an analysis is carried out is also affected by accuracy requirements. If precipitation with ammonia is chosen for the analysis of a sample containing 20% aluminum, the presence of 0.2% iron is of serious concern where accuracy in the parts-per-thousand range is demanded, and a preliminary separation of the two elements is necessary. If an error of 50 ppt is tolerable, however, the separation of iron is not necessary. Furthermore, this tolerance governs other aspects of the method as well. For example, 1-g samples can be weighed to perhaps 10 mg and certainly no closer than 1 mg. In addition, less care is needed in transferring and washing the precipitate and in other time-consuming operations of the gravimetric method. The intelligent use of shortcuts is not a sign of carelessness, but is instead a recognition of realities in matters of time and effort. The question of accuracy, then, must be settled in clear terms at the outset.

In order to choose a method for the determination of one or more species in a sample, it is necessary to know what other elements and/or compounds are present. Lacking such information, a qualitative analysis must be undertaken in order to identify components that are likely to interfere in the various methods under consideration. As we have noted repeatedly, most analytical methods are based on reactions and physical properties that are shared by several elements or compounds. Thus, measurement of the concentration of a given element by a method that is simple and straightforward in the presence of one group of elements or compounds may require many tedious and time-consuming separations in the presence of others. A solvent suitable for one combination of compounds may be totally unsatisfactory when applied for another. Clearly, a knowledge of the qualitative chemical composition of the sample is a prerequisite for selecting a method for the quantitative determination of one or more of its components.

The chemist must also consider the physical state of the sample in order to determine whether it must be homogenized, whether volatility losses are likely, and whether its composition may change under laboratory conditions due to the absorption of water or to efflorescence. It is also important to determine what treatment is sufficient to decompose or dissolve the sample without loss of analyte. Preliminary tests of one sort or another may be needed to provide this type of information.

Finally, the number of samples to be analyzed is an important criterion in selecting a method. If there are many samples, considerable time can be expended in calibrating instruments, preparing reagents, assembling equipment, and investigating shortcuts, since the cost of these operations can be spread over the large number of samples. If, however, a few samples at most are to be analyzed, a longer and more tedious procedure involving a minimum of these preparatory operations may prove to be the wiser choice from the economic standpoint.

Having answered the preliminary questions, the chemist is now in a position to consider possible approaches to the problem. Sometimes, based upon past experience, the route to be followed is obvious. In other instances, it is not, and the chemist must speculate on problems that are likely to be encountered in the analysis and how they can be solved. By now, some methods probably have been eliminated from consideration and others put on the doubtful list. Ordinarily, however, the chemist turns to the analytical literature in order to profit from the experience of others. This, then, is the next logical step in choosing an analytical method.

26A-2 *Investigation of the Literature*

A list of reference books and journals concerned with various aspects of analytical chemistry appears in Appendix 1. This list is not an exhaustive catalog but rather one that is adequate for most work. It is divided into several categories. In many instances, the division is arbitrary since some works could be logically placed in more than one category.

The chemist usually begins a search of the literature by referring to one or more of the treatises on analytical chemistry or to those devoted to the analysis of specific types of materials. In addition, it is often helpful to consult a general reference work relating to the compound or element of interest. From this survey, a clearer picture of the problem at hand may develop—what steps are likely to be difficult, what separations must be made, what pitfalls must be avoided. Occasionally, all the answers needed or even a set of specific instructions for the analysis may be found. Alternatively, journal references that lead directly to this information may be discovered. On other occasions, the chemist will acquire only a general notion of how to proceed. Several possible methods may appear suitable; others may be eliminated. At this point, it is often helpful to consider reference works concerned with specific substances or specific techniques. Alternatively, the various analytical journals may be consulted. Monographs on methods for completing the analysis are valuable in deciding among several possible techniques.

A major problem in using analytical journals is locating articles pertinent to the problem at hand. The various reference books are useful since most are liberally annotated with references to the original journals. The key to a thorough search of the literature, however, is *Chemical Abstracts* and its *Index Guide*. Such a survey involves the expenditure of a great deal of time, however, and is often made unnecessary by consulting reliable reference works. The advent of computer-aided literature searches promises to minimize the time required for a careful literature survey.

26A-3 Choosing or Deriving a Method

Having defined the problem and investigated the literature for possible approaches, the chemist must next decide upon the route to be followed in the laboratory. If the choice is simple and obvious, analysis can be undertaken directly. Frequently, however, the decision requires the exercise of considerable judgment and ingenuity; experience, an understanding of chemical principles, and perhaps intuition all come into play.

If the substance to be analyzed occurs widely, the literature survey usually yields several alternative methods for the analysis. Economic considerations may dictate which method will yield the desired reliability with the least expenditure of time and effort. As mentioned earlier, the number of samples to be analyzed is often a determining factor in this choice.

Investigation of the literature does not invariably reveal a method designed specifically for the type of sample in question. Ordinarily, however, the chemist will encounter procedures for materials that are at least analogous in composition to the one in question; the decision then has to be made as to whether the variables introduced by differences in composition are likely to have any influence on the results. This judgment is often difficult and fraught with uncertainty; recourse to the laboratory may be the only way of obtaining an unequivocal answer.

If it is decided that existing procedures are not applicable, consideration must be given to modifications that may overcome the problems imposed by the variation in composition. Again, it may be possible to propose only tentative alterations, owing to the complexity of the system; whether these modifications will accomplish their purpose without introducing new difficulties can be determined only in the laboratory.

After giving due consideration to existing methods and their modifications, the chemist may decide that none fits the problem and new procedures must be developed. In doing so, all the facts on the chemical and physical properties of the analyte must be marshalled and given consideration. Several possible ways of performing the desired measurement may become evident from this information. Each possibility must then be examined critically, with consideration given to the influence of the other components in the sample as well as the reagents that must be used for solution or decomposition. At this point, the chemist must try to anticipate sources of error and possible interferences arising from interactions among the components and reagents; methods may have to be derived by which problems of this sort can be circumvented. In the end, it is to be hoped that one or more tentative methods worth testing will have been located. It is probable that the feasibility of some of the steps in the procedure cannot be determined on the basis of theoretical considerations alone; recourse must be made to preliminary laboratory testing of such steps. Certainly, critical evaluation of the entire procedure can come only from careful laboratory work.

26A-4

Testing the Procedure

Once a procedure for an analysis has been selected, a decision must be made as to whether it can be employed directly, without testing, to the problem at hand. The answer to this question is not simple and depends upon a number of considerations. If the method chosen is the subject of a single, or at most a few, literature references, there may be a real point to preliminary laboratory evaluation. With experience, the chemist becomes more and more cautious about accepting claims regarding the accuracy and applicability of a new method. All too often, statements found in the literature are overly optimistic; a few hours spent in testing the procedure in the laboratory may be enlightening.

Whenever a major modification of a standard procedure is undertaken or an attempt is made to apply it to a type of sample different from that for which it was designed, a preliminary laboratory test is advisable. The effects of such alterations simply cannot be predicted with certainty, and the chemist who dispenses with such precautions is sanguine indeed.

Finally, of course, a newly devised procedure must be extensively tested before it is adapted for general use. We must now consider the means by which a new method or a modification of an existing method can be tested for reliability.

The Analysis of Standard Samples

Unquestionably, the best technique for evaluating an analytical method involves the analysis of one or more standard samples whose analyte composition is reliably known. For this technique to be of value, however, it is essential that the standards closely resemble the samples to be analyzed with respect to both the analyte concentration range and the overall composition.

Occasionally, standards suitable for method testing can be synthesized by thoroughly homogenizing weighed quantities of pure compounds. Such a procedure is generally inapplicable, however, when the samples to be analyzed are complex naturally occurring substances, such as rocks, animal tissue, and soils.

As mentioned in Section 2B-4, the National Bureau of Standards has for sale a variety of standard reference materials (SRM) that have been specifically prepared for validation of analytical methods. Most standard reference materials are substances commonly encountered in commerce or in environmental, pollution, clinical, biological, and forensic studies. The concentration of one or more components in these materials is certified by the Bureau based upon measurements using (1) a previously validated reference method, (2) two or more independent, reliable measurements methods, or (3) results from a network of cooperating laboratories, technically competent and thoroughly familiar with the material being tested. More than 800 of these materials are available, including such substances as ferrous and nonferrous metals; ores, ceramics, and cements; environmental gases, liquids, and solids; primary and secondary chemicals; clinical, biological, and botanical samples; fertilizers; and glasses. Several industrial concerns also offer various kinds of standard materials designed for validating analytical procedures.[1]

When standard reference materials are not available (and often they are not), the best the chemist can do is prepare a solution of known concentration whose composition approximates that of the sample after it has been decomposed and dissolved. Obviously, such a standard gives no information at all concerning the fate of the substance being determined during the important decomposition and solution steps.

Analysis by Other Methods

The results of an analytical method can sometimes be evaluated by comparison with data obtained from an entirely different method. Clearly, a second method must exist and in addition should be based on chemical principles that differ considerably from the one under examination. Comparable results from the two serve as presumptive evidence that both are yielding satisfactory results, inasmuch as it is unlikely that the same determinate errors would affect each. Such a conclusion does not apply to those aspects of the two methods that are similar.

Standard Addition to the Sample

When the foregoing approaches are inapplicable, the standard-addition method may prove useful. Here, in addition to being used to analyze the sample, the proposed procedure is tested against portions of the sample to which known amounts of the analyte have been added. The effectiveness of the method can then be established by evaluating the extent of recovery of the added quantity. The standard-addition method may reveal errors arising from the way the sample was treated or from the presence of the other elements and/or compounds in the matrix.

26B The Accuracy Obtainable in the Analysis of Complex Materials

To provide a clear idea of the accuracy that can be expected when a complex material is analyzed with a reasonable amount of effort and care, data on the determination of four elements in a variety of materials are presented in Tables 26-1 to 26-4. These data were taken from a much larger set of results collected by W. F. Hillebrand and G.E.F. Lundell of the National Bureau of Standards and published in the first edition of their excellent book on inorganic analysis.[2]

The materials analyzed were naturally occurring substances and items of commerce; they were especially prepared to give uniform and homogeneous samples and were distributed among chemists who were, for the most part, actively engaged in the analysis of similar materials. The analysts were allowed to use the methods they considered most reliable and best suited for the problem at hand. In most instances, special precautions were taken, and the results are consequently better than can be expected from the average routine analysis.

[1]For literature describing standard reference materials, see the references in footnotes 6 and 7 in Chapter 2 on page 16.

[2]W. F. Hillebrand and G.E.F. Lundell, *Applied Inorganic Analysis*, pp. 874–887. New York: Wiley, 1929.

The numbers in the second column of Tables 26-1 to 26-4 are best values, obtained by the most painstaking analysis for the measured quantity. Each is considered to be the true value for calculation of the absolute and relative errors shown in the fourth and fifth columns. The fourth column was obtained by discarding extremely divergent results, determining the deviation of the remaining individual data from the best value (second column), and averaging these deviations. The fifth column was obtained by dividing the data in the fourth column by the best value (second column) and multiplying by 100%.

The results shown in these tables are typical of the data for 26 elements reported in the original publication. It is to be concluded that analyses reliable to a few tenths of a percent relative are the exception rather than the rule in the analysis of complex mixtures by ordinary methods and that, unless the chemist is willing to invest an inordinate amount of time in the analysis, errors on the order of 1 or 2% must be accepted. If the sample contains less than 1% of the analyte, even larger relative errors are to be expected.

It is clear from these data that the accuracy obtainable in the determination of an element is greatly dependent upon the nature and complexity of the substrate. Thus, the relative error in the determination of phosphorus in two phosphate rocks was 1.1%; in a synthetic mixture, it was only 0.27%. The relative error in an iron determination in a refractory was 7.8%; in a manganese bronze having about the same iron content, it was only 1.8%. Here, the limiting factor in the accuracy is not in the completion step but rather in the dissolution of the samples and the separation of interferences.

The data in these four tables are more than 50 years old, and it is tempting to think that analyses carried out with more modern tools and after additional experience are likely to be significantly better in terms of accuracy and precision. A study by S. Abbey suggests that this assumption is not valid, however.[3] For example, the data in Table 26-5, which was taken from his paper, reveal no significant improvement in silicate analyses of standard reference glass and rock samples in the 43-year period from 1931 to 1974. Indeed, the standard deviation among participating laboratories appears to be larger in later years.

TABLE 26-1

Determination of Iron in Various Materials*

Material	Iron, %	Number of Analysts	Average Absolute Error	Average Relative Error, %
Soda-lime glass	0.064 (Fe_2O_3)	13	0.01	15.6
Cast bronze	0.12	14	0.02	16.7
Chromel	0.45	6	0.03	6.7
Refractory	0.90 (Fe_2O_3)	7	0.07	7.8
Manganese bronze	1.13	12	0.02	1.8
Refractory	2.38 (Fe_2O_3)	7	0.07	2.9
Bauxite	5.66	5	0.06	1.1
Chromel	22.8	5	0.17	0.75
Iron ore	68.57	19	0.05	0.07

*From W. F. Hillebrand and G.E.F. Lundell, *Applied Inorganic Analysis*, p. 878. New York: Wiley, 1929. With permission.

[3]S. Abbey, *Anal. Chem.*, **1981**, *53*, 529A.

From the data in the five tables in this chapter, it is clear that the chemist is well advised to adopt a pessimistic viewpoint regarding the accuracy of an analysis, be it one's own or one performed by someone else.

TABLE 26-2 **Determination of Manganese in Various Materials***

Material	Manganese, %	Number of Analysts	Average Absolute Error	Average Relative Error, %
Ferro-chromium	0.225	4	0.013	5.8
Cast iron	0.478	8	0.006	1.3
	0.897	10	0.005	0.56
Manganese bronze	1.59	12	0.02	1.3
Ferro-vanadium	3.57	12	0.06	1.7
Spiegeleisen	19.93	11	0.06	0.30
Manganese ore	58.35	3	0.06	0.10
Ferro-manganese	80.67	11	0.11	0.14

*From W. F. Hillebrand and G.E.F. Lundell, *Applied Inorganic Analysis*, p. 880. New York: Wiley, 1929. With permission.

TABLE 26-3 **Determination of Phosphorus in Various Materials***

Material	Phosphorus, %	Number of Analysts	Average Absolute Error	Average Relative Error, %
Ferro-tungsten	0.015	9	0.003	20.
Iron ore	0.040	31	0.001	2.5
Refractory	0.069 (P_2O_5)	5	0.011	16.
Ferro-vanadium	0.243	11	0.013	5.4
Refractory	0.45	4	0.10	22.
Cast iron	0.88	7	0.01	1.1
Phosphate rock	43.77 (P_2O_5)	11	0.5	1.1
Synthetic mixtures	52.18 (P_2O_5)	11	0.14	0.27
Phosphate rock	77.56 ($Ca_3(PO_4)_2$)	30	0.85	1.1

*From W. F. Hillebrand and G.E.F. Lundell, *Applied Inorganic Analysis*, p. 882. New York: Wiley, 1929. With permission.

TABLE 26-4 **Determination of Potassium in Various Materials***

Material	Potassium Oxide, %	Number of Analysts	Average Absolute Error	Average Relative Error, %
Soda-lime glass	0.04	8	0.02	50.
Limestone	1.15	15	0.11	9.6
Refractory	1.37	6	0.09	6.6
	2.11	6	0.04	1.9
	2.83	6	0.10	3.5
Lead-barium glass	8.38	6	0.16	1.9

*From W. F. Hillebrand and G.E.F. Lundell, *Applied Inorganic Analysis*, p. 883. New York: Wiley, 1929. With permission.

TABLE 26-5 **Standard Deviation of Silica Results***

Year Reported	Sample Type	Number of Results	Standard Deviation (%, absolute)
1931	Glass	5	0.28†
1951	Granite	34	0.37
1963	Tonalite	14	0.26
1970	Feldspar	9	0.10
1972	Granite	30	0.18
1972	Syenite	36	1.06
1974	Granodiorite	35	0.46

*From S. Abbey, *Anal. Chem.*, **1981**, *53* (4), 529A. With permission.

†0.09 after eliminating one result.

CHAPTER 27

Preparing Samples for Analysis

This chapter deals with two preliminary steps that are important parts of most chemical analyses: (1) sampling and (2) either drying the sample or determining its moisture content. Under some circumstances, neither of these steps is important or necessary, but more commonly, one or both are vital and may limit the accuracy and significance of the analytical data.

27A Sampling

Generally, a chemical analysis is performed on only a small fraction of the material whose composition is of interest. Clearly, the composition of this fraction must reflect as closely as possible the average composition of the bulk of the material if the results are to have value. The process by which a representative fraction is acquired is termed *sampling*. Often, sampling is the most difficult step in the entire analytical process and the step that limits the accuracy of the procedure. This statement is particularly true when the material to be analyzed is a large and inhomogeneous liquid, such as a lake, or an inhomogeneous solid, such as an ore, a soil, or a piece of animal tissue.

The end product of the sampling step is a quantity of homogeneous material weighing a few grams or, at most, a few hundred grams that may constitute as little as one part in 10^7 or 10^8 of the material whose composition is sought. Yet the composition of this minute fraction must, as closely as possible, be identical to the average composition of the total mass. Where, as with the examples just mentioned, the quantity of material to be sampled is large and its composition inherently nonhomogeneous, the task of producing a representative sample is indeed formidable. Clearly, the reliability of the analysis cannot exceed that of the sampling step, and painstaking analysis of a poor sample is wasted effort.

The literature on sampling is extensive;[1] at best we can provide only a brief outline of sampling methods here.

Three steps are involved in sampling bulk materials: (1) identification of the population from which the sample is to be obtained, (2) collection of a *gross sample* that is truly representative of the population being sampled, and (3) reduction of the gross sample to a few hundred grams of a homogeneous *laboratory sample* that is suitable for the analysis to be employed.

[1]See, for example, F. J. Welcher, Ed., *Standard Methods of Chemical Analysis,* 6th ed., Vol. 2, Part A, pp. 21–55. Princeton, NJ: Van Nostrand, 1963. An extensive bibliography of specific sampling information has been compiled by C. A. Bicking, in *Treatise on Analytical Chemistry,* 2nd ed., I. M. Kolthoff and P. J. Elving, Eds., Part I, Vol. 1, p. 299. New York: Interscience, 1978. For a brief review of sampling problems, see B. Kratochvil and J. K. Taylor, *Anal. Chem.,* **1981,** *53,* 924A.

Ordinarily, step 1 is straightforward, with the population being such diverse things as a carload of ore, a carton of bottles containing vitamin pills, a field of wheat, the brain of a rat, or the mud from a stretch of river bottom. Steps 2 and 3 are seldom simple and may require more effort and ingenuity than any other step in the entire analytical scheme.

27A-1 The Effects of Sampling Uncertainties

In Chapter 2, we concluded that both determinate and indeterminate errors in analytical data can be traced to instrument, method, and personal causes. Most determinate errors can be eliminated by exercising care, by calibration, and by the proper use of standards, blanks, and reference materials. Indeterminate errors, which are reflected in the precision of data, can generally be kept at an acceptable level by close control of the variables that influence the measurements. Errors due to invalid sampling are different, however, in the sense that they are not controllable by the use of blanks and standards or by closer control of experimental variables. For this reason, sampling errors are ordinarily treated separately from the other uncertainties associated with an analysis.

For random uncertainties, the overall standard deviation s_o for an analytical measurement is related to the standard deviation of the sampling process s_s and to the standard deviation of the method s_m by the relationship

$$s_o^2 = s_s^2 + s_m^2 \tag{27-1}$$

In many cases, the method variance will be known from replicate measurements on the product of a single laboratory sample. Under this circumstance, s_s can be computed from measurements of s_o for a series of laboratory samples each of which is obtained from several gross samples.

Youden has shown that, once the measurement uncertainty has been reduced to one third or less of the sampling uncertainty (that is, $s_m < s_s/3$), further improvement in the measurement uncertainty is fruitless.[2] As a consequence, if the sampling uncertainty is large and cannot be improved, it is often wise to switch to a less precise but rapid method of analysis so that more samples can be analyzed in a given length of time, thus improving the precision of the average for the lot.

27A-2 The Gross Sample

Ideally, the gross sample is a miniature replica of the entire mass of material to be analyzed. It corresponds to the whole not only in chemical composition but, *equally important,* in particle-size distribution.

The Size of the Gross Sample

From the standpoint of convenience and economy, it is desirable that the gross sample weigh no more than absolutely necessary. Basically, gross sample weight is determined by (1) the uncertainty that can be tolerated between the composition of the gross sample and that of the whole, (2) the degree of heterogeneity of the whole, and (3) the level of particle size at which heterogeneity begins.

This last point warrants amplification. In a well-mixed, homogeneous

[2]W. J. Youden, *J. Assoc. Off. Anal. Chem.*, 1981, *50,* 1007.

solution of a gas or a liquid, heterogeneity exists only on a molecular scale, and the weight of the molecules themselves governs the minimum weight of the gross sample. A particulate solid, such as an ore or a soil, represents the opposite situation. In such materials, the individual pieces of solid can be seen to differ from each other in composition. Here, heterogeneity develops in particles that may have dimensions on the order of a centimeter or more and may weigh several grams. Intermediate between these extremes are colloidal materials and solidified metals. With the former, heterogeneity is first encountered in the particles of the dispersed phase; these typically have diameters in the range of 10^{-5} cm or less. In an alloy, heterogeneity first occurs in the crystal grains.

In order to obtain a truly representative gross sample, a certain number **n** of the particles referred to in (3) must be taken. The magnitude of this number depends upon (1) and (2) and may involve only a relatively few particles, several millions, or even several millions of millions. The need for large numbers of particles is of no great concern for homogeneous gases and liquids since heterogeneity among particles first occurs at the molecular level. Thus, even a very small weight of sample will contain more than the requisite number of particles. However, the individual particles of a particulate solid may weigh a gram or more, which sometimes leads to a gross sample of several tons. Sampling of such material is a costly, time-consuming procedure at best; determination of the smallest weight of material required to provide the needed information minimizes this expense.

The composition of a gross sample removed randomly from a bulk of material is governed by the law of chance; thus, by suitable statistical manipulations, it is possible to predict the probability that a given fraction is similar to the whole. A simple, idealized case serves as an example. A carload of lead ore is made up of just two kinds of particles: galena (lead sulfide) and a gangue containing no lead. All particles have the same size. The car contains, let us say, 100 million particles, and we wish to know how many of these are galena. Since the two components differ in appearance, the composition of the carload could be obtained exactly by counting all the galena particles. This approach would probably involve several lifetimes of work, however. We must therefore settle for the lesser accuracy involved in counting some reasonable fraction of the total number of particles. The number of particles contained in this fraction depends, of course, upon the error we are willing to tolerate in the measurement.

The relationship between the allowable error and the number of particles **n** to be counted can be stated as[3]

$$\mathbf{n} = \frac{1 - p}{p\sigma_r^2} \tag{27-2}$$

where p is the fraction of galena particles, $1 - p$ is the fraction of gangue particles, and σ_r is the allowable relative standard deviation in the count of the galena particles. Thus, for example, if 80% of the particles are galena ($p = 0.8$) and the tolerable standard deviation is 1% ($\sigma_r = 0.01$), a random

[3]For a discussion of the derivation and significance of Equations 27-2 and 27-3, see A. A. Benedetti-Pichler, in *Physical Methods in Chemical Analysis*, W. G. Berl, Ed., Vol. 3, pp. 183–194. New York: Academic Press, 1956; A. A. Benedetti-Pichler, *Essentials of Quantitative Analysis*, Chapter 19. New York: Ronald Press, 1956.

sampling of 2500 particles should be made. A standard deviation of 0.1% requires a sample containing 250,000 particles.

Let us now make the problem more realistic and assume that one of the components in the car contains a higher percentage of lead P_1 and the other component contains a lesser amount P_2. Furthermore, the average density d of the shipment differs from the densities d_1 and d_2 of these components. We are now interested in deciding what number of particles and thus what weight we should take to ensure a sample possessing the overall average percent of lead P with a sampling relative standard deviation of σ_r. Equation 27-2 can be extended to include these stipulations:

$$\mathrm{n} = p(1 - p)\left(\frac{d_1 d_2}{d^2}\right)^2 \left(\frac{P_1 - P_2}{\sigma_r P}\right)^2 \qquad (27\text{-}3)$$

From this equation, we see that the demands of accuracy are costly, in terms of the sample size required, because of the inverse-square relationship between the allowable standard deviation and the number of particles taken. Furthermore, a greater number of particles must be taken as the average percentage P of the element of interest becomes smaller.

The degree of heterogeneity as measured by $P_1 - P_2$ has a profound effect on the number of particles required, with the number increasing as the square of the difference in composition of the two components of the mixture.

The problem of deciding upon the weight of the gross sample for a solid material is ordinarily more difficult than this example because most samples not only contain more than two components but also consist of a range of particle sizes. In most instances, the first of these problems can be met by dividing the sample into an imaginary two-component system. Thus, with an actual lead ore, one component selected might be all the various lead-bearing minerals of the ore and the other all the residual components containing little or no lead. After average densities and percentages of lead are assigned to each part, the system is treated as if it has only two components.

The problem of variable particle size can be handled by calculating the number of particles that would be needed if the sample consisted of particles of a single size. The gross sample weight is then determined by taking into account the particle-size distribution. One approach is to calculate the needed weight by assuming that all particles are the size of the largest. This procedure is not very efficient, however, for it usually calls for removal of a larger weight of material than necessary. Benedetti-Pichler gives alternative methods for computing the weight of gross sample to be chosen.[4]

An interesting conclusion from Equation 27-3 is that the number of particles in the gross sample is *independent* of particle size. The weight of the sample, of course, increases directly as the volume (or as the cube of the particle diameter) so that reduction in the particle size of a given material has a large effect on the weight required in the gross sample.

Clearly, a great deal of information must be known about a substance in order to make use of Equation 27-3. Fortunately, reasonable estimates of the various parameters in the equation can often be made. These estimates

[4]A. A. Benedetti-Pichler, in *Physical Methods in Chemical Analysis,* W. G. Berl, Ed., Vol. 3, p. 192. New York: Academic Press, 1956.

can be based upon a qualitative analysis of the substance, visual inspection, and information from the literature on substances of similar origin. Crude measurements of the density of the various sample components may also be necessary.

EXAMPLE 27-1

Assume that the average particle in the carload of lead ore just considered is judged to be approximately spherical with a radius of about 5 mm. Roughly 4% of the particles appear to be galena ($\simeq$ 70% Pb), which has a density of 7.6 g/cm^3; the remaining particles have a density of about 3.5 g/cm^3 and contain little or no lead. How many pounds of ore should the gross sample contain if the sampling uncertainty is to be kept below 0.5% relative?

We first compute values for the average density and percent lead:

$$d = 0.04 \times 7.6 + 0.96 \times 3.5 = 3.7 \text{ g/cm}^3$$

$$P = \frac{(0.04 \times 7.6 \times 0.70) \text{ g Pb/cm}^3}{3.7 \text{ g sample/cm}^3} \times 100\% = 5.8\% \text{ Pb}$$

Then, substituting into Equation 27-3 gives

$$\text{n} = 0.04\,(1 - 0.04)\left[\frac{7.6 \times 3.5}{(3.7)^2}\right]^2 \left(\frac{70 - 0}{0.005 \times 5.8}\right)^2 = 8.45 \times 10^5 \text{ particles required}$$

$$\text{Wt. of sample} = 8.45 \times 10^5 \text{ particles} \times \frac{4}{3}\pi(0.5)^3 \frac{\text{cm}^3}{\text{particle}} \times \frac{3.7 \text{ g}}{\text{cm}^3} \times \frac{1}{454 \text{ g/lb}}$$

$$= 3.61 \times 10^3 \text{ lb or about 1.8 ton}$$

Sampling Homogeneous Solutions of Liquids and Gases

For solutions of liquids or gases, the gross sample can be relatively small since ordinarily nonhomogeneity first occurs at the molecular level, and even small volumes of sample will contain many more particles than the number computed from Equation 27-3. Whenever possible, the liquid or gas to be analyzed should be stirred well prior to sampling to make sure that the gross sample is homogeneous. With large volumes of solutions, mixing may be impossible; it is then best to sample several portions of the container with a "sample thief," a bottle that can be opened and filled at any desired location in the solution. This type of sampling is important, for example, in determining the constituents of liquids exposed to the atmosphere. Thus, the oxygen content of lake water may vary by a factor of as large as 1000 over a depth difference of a few feet.

Industrial gases or liquids are often sampled continuously as they flow through pipes, with care being taken to ensure that the sample collected represents a constant fraction of the total flow and that all portions of the stream are sampled.

Sampling Particulate Solids

Obtaining a random sample from a bulky particulate material is often difficult. It can best be accomplished while the material is being transferred. For example, randomly chosen shovelfuls or wheelbarrow loads may be consigned to a sample pile, or portions of the material may be intermittently removed from a conveyor belt. Alternatively, the material may be forced through a mechanical device called a riffle or a series of riffles that continuously isolate a fraction of the stream. Mechanical devices of this sort have been developed for handling coals and ores. Details regarding sampling of these materials are beyond the scope of this book.

Sampling Metals and Alloys

Samples of metals and alloys are obtained by sawing, milling, or drilling. In general, it is not safe to assume that chips of the metal removed from the surface are representative of the entire bulk, and solid from the interior must be sampled as well. With billets or ingots of metal, a representative sample can be obtained by sawing across the piece at random intervals and collecting the "sawdust" as the sample. Alternatively, the specimen may be drilled, again at various randomly spaced intervals, and the drillings collected as the sample; the drill should pass entirely through the block or halfway through from opposite sides. The drillings can then be broken up and mixed or melted together in a graphite crucible. A granular sample can often then be produced by pouring the melt into distilled water.

27A-3 Preparation of a Laboratory Sample

For nonhomogeneous solids, the gross sample may weigh several hundred pounds or more, and so reduction of the gross sample to a finely ground and homogeneous laboratory sample, weighing at the most a few pounds, is necessary. This process involves a cycle of operations that includes crushing and grinding, sieving, mixing, and dividing the sample (often into halves) to reduce its weight. During each division, a weight of sample that contains the number of particles computed from Equation 27-3 is retained.

EXAMPLE 27-2

It is desired to reduce the gross sample in Example 27-1 to a laboratory sample that weighs about one pound. How can this be done?

The laboratory sample should contain the same number of particles as the gross sample, or 8.45×10^5. Each particle will weigh on the average

$$\text{Av. wt. of particle} = 1 \text{ lb} \times 454 \frac{\text{g}}{\text{lb}} \times \frac{1}{8.45 \times 10^5 \text{ particles}} = 5.37 \times 10^{-4} \text{ g/particle}$$

The average weight of a particle is related to its radius by the equation

$$\text{Av. wt. of particle} = \frac{4}{3}\pi[r(\text{cm})]^3 \times \frac{3.7 \text{ g}}{\text{cm}^3}$$

Equating these two relationships and solving for r give

$$r = \left(5.37 \times 10^{-4} \text{ g} \times \frac{3}{4\pi} \times \frac{\text{cm}^3}{3.7 \text{ g}}\right)^{1/3} = 3.3 \times 10^{-2} \text{ cm or } 0.3 \text{ mm}$$

Thus the sample should be repeatedly ground, mixed, and divided until the particles are about 0.3 mm in diameter.

Crushing and Grinding of Laboratory Samples

A certain amount of crushing and grinding is ordinarily required to decrease the particle size of solid samples. Because these operations tend to alter the composition of the sample, the particle size should be reduced no more than is required for homogeneity and ready attack by reagents.

Several factors can cause appreciable changes in sample composition as a result of grinding. The heat inevitably generated can cause losses of volatile components. In addition, grinding increases the surface area of the solid and thus increases its susceptibility to reaction with the atmosphere. For example, it has been observed that the iron(II) content of a rock may be decreased by as much as 40% during grinding—apparently a direct result of the iron being oxidized to the +3 state.

Often, the water content of a sample is altered substantially during grinding. Increases are observed as a consequence of the increased specific surface area that accompanies a decrease in particle size (page 69). The increased surface area leads to greater amounts of adsorbed water. For example, the water content of a piece of porcelain changed from 0 to 0.6% when it was ground to a fine powder.

In contrast, decreases in the water content of hydrates often take place during grinding as a result of localized frictional heating. For example, the water content of gypsum ($CaSO_4 \cdot 2H_2O$) decreased from about 21% to 5% when the compound was ground to a fine powder.

Differences in hardness of the component can also introduce errors during crushing and grinding. Softer materials are ground to fine particles more rapidly than are hard ones and may be lost as dust as the grinding proceeds. In addition, flying fragments tend to contain a higher fraction of the harder components.

Intermittent screening often increases the efficiency of grinding. Screening involves shaking the ground sample on a wire or cloth sieve that will pass particles of a desired size. The residual particles are then returned for further grinding; the operation is repeated until the entire sample passes through the screen. The hardest materials, which often differ in composition from the bulk of the sample, are last to be reduced in particle size and are thus last through the screen. Therefore, grinding must be continued until every particle has been passed if the screened sample is to have the same composition as it had before grinding and screening.

A serious error can arise during grinding and crushing as a consequence of sample contamination resulting from the mechanical wear and abrasion of the grinding surfaces. Even though these surfaces are fabricated from hardened steel, agate, or boron carbide, contamination of the sample is nevertheless occasionally encountered. The problem is particularly acute in analyses for minor constituents.

A variety of tools are employed for reducing the particle size of solids, including jaw crushers and disk pulverizers for large samples containing large lumps, ball mills for medium-size samples and particles, and various types of mortars for small amounts of material.

The *ball mill* is a useful device for grinding solids that are not too hard. It consists of a porcelain crock of perhaps 2-L capacity that can be sealed and rotated mechanically. The container is charged with approximately equal volumes of the sample and flint or procelain balls having a diameter of 20 to 50 mm. Grinding and crushing occur as the balls tumble in the rotating container. A finely ground and well-mixed powder can be produced in this way.

The *Plattner diamond mortar* (Figure 27-1) is used for crushing hard, brittle materials. It is constructed of hardened tool steel and consists of a base plate, a removable collar, and a pestle. The sample is placed on the base plate inside the collar. The pestle is then fitted into place and struck several blows with a hammer; this reduces the solid to a fine powder that is collected on glazed paper after the apparatus has been disassembled.

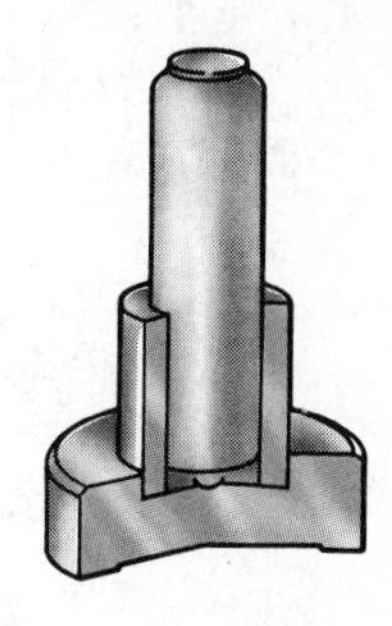

FIGURE 27-1 A Plattner diamond mortar.

Mixing Solid Laboratory Samples

It is essential that solid materials be thoroughly mixed to ensure random distribution of the components in the analytical sample. A common method for mixing powders involves rolling the sample on a sheet of glazed paper. A pile of the substance is placed in the center and mixed by lifting one corner of the paper enough to roll the particles of the sample to the opposite corner. This operation is repeated many times, with the four corners of the sheet being lifted alternately.

Effective mixing of solids is also accomplished by rotating the sample for some time in a ball mill or a twin-shell V-blender. The latter consists of two connected cylinders that form a V-shaped container for the sample. As the blender is rotated, the sample is split and recombined with each rotation, leading to highly efficient mixing.

It is worthwhile noting that, with long standing, finely ground homogeneous materials may segregate on the basis of particle size and density. For example, analyses of layers of a set of student unknowns that had not been used for several years revealed a regular variation in the analyte concentration from top to bottom of the container. Apparently, segregation had occurred as a consequence of vibrations and of density differences in the sample components.

27B Moisture in Samples

Laboratory samples of solids often contain water that is in equilibrium with the atmosphere. As a consequence, unless special precautions are taken, the composition of the sample depends upon the relative humidity and ambient temperature at the time it is analyzed. To cope with this variability in composition, it is common practice to remove moisture from solid samples prior to weighing or, if this is not possible, to bring the water content to some reproducible level that can be duplicated later if necessary. A third alternative involves the determination of the water content at the time the samples are weighed for analysis so that the results can be corrected to a dry basis. In any event, many analyses are preceded by some sort of preliminary treatment designed to take into account the presence of water.

27B-1 Forms of Water in Solids

Essential Water

Essential water forms an integral part of the molecular or crystalline structure of a compound in its solid state. Thus, the *water of crystallization* in a stable solid hydrate (for example, $CaC_2O_4 \cdot 2H_2O$ and $BaCl_2 \cdot 2H_2O$) qualifies as a type of essential water.

Water of constitution is a second type of essential water; it is found in compounds that yield stoichiometric amounts of water when heated or otherwise decomposed. Examples of this type of water are found in potassium hydrogen sulfate and calcium hydroxide, which when heated come to equilibrium with the moisture in the atmosphere as shown by the reactions

$$2KHSO_4(s) \rightleftharpoons K_2S_2O_7(s) + H_2O(g)$$

$$Ca(OH)_2(s) \rightleftharpoons CaO(s) + H_2O(g)$$

Nonessential Water

Nonessential water is retained by the solid as a consequence of physical forces. It is not necessary for characterization of the chemical constitution of the sample and therefore does not occur in any sort of stoichiometric proportion.

Adsorbed water is a type of nonessential water that is retained on the surface of solids by the forces of adsorption. The amount adsorbed is dependent upon humidity, temperature, and the specific surface area of the solid. Adsorption of water occurs to some degree on all solids.

A second type of nonessential water is called *sorbed water* and is encountered with many colloidal substances, such as starch, protein, charcoal, zeolite minerals, and silica gel. In contrast to adsorption, the quantity of sorbed water is often large, amounting to as much as 20% or more of the total weight of the solid. Interestingly enough, solids containing even this amount of water may appear as perfectly dry powders. Sorbed water is held as a condensed phase in the interstices or capillaries of the colloidal solid. The quantity contained in the solid is greatly dependent upon temperature and humidity.

A third type of nonessential moisture is *occluded water,* liquid water entrapped in microscopic pockets spaced irregularly throughout solid crystals. Such cavities often occur in minerals and rocks (and in gravimetric precipitates).

27B-2 The Effect of Temperature and Humidity on the Water Content of Solids

In general, the concentration of water in a solid tends to decrease with increasing temperature and decreasing humidity. The magnitude of these effects and the rate at which they manifest themselves differ considerably according to the manner in which the water is retained.

Compounds Containing Essential Water

The chemical composition of a compound containing essential water is dependent upon temperature and relative humidity.[5] For example, anhydrous barium chloride tends to react with atmospheric moisture to give one of two

[5]Relative humidity is the ratio of the vapor pressure of water in the atmosphere to its vapor pressure in air that is saturated with moisture. At 25°C, the partial pressure of water in saturated air is 23.76 torr. Thus, when air contains water at a partial pressure of 6 torr, the relative humidity is

$$\frac{6.00}{23.76} = 0.253 \text{ or } 25.3\%$$

stable hydrates, depending upon temperature and relative humidity. The equilibria involved are

$$BaCl_2(s) + H_2O(g) \rightleftharpoons BaCl_2{\cdot}H_2O(s)$$

$$BaCl_2{\cdot}H_2O(s) + H_2O(g) \rightleftharpoons BaCl_2{\cdot}2H_2O(s)$$

At room temperature and at a relative humidity between 25 and 90%, $BaCl_2{\cdot}2H_2O$ is the stable species. Since the relative humidity in most laboratories lies well within these limits, the essential-water content of the dihydrate is ordinarily independent of atmospheric conditions. Exposure of either $BaCl_2$ or $BaCl_2{\cdot}H_2O$ to these conditions causes compositional changes that ultimately lead to formation of the dihydrate. On a very dry winter day (relative humidity <25%), however, the situation changes; the dihydrate becomes unstable with respect to the atmosphere, and a molecule of water is lost to form the new stable species $BaCl_2{\cdot}H_2O$. At relative humidities less than about 8%, both hydrates lose water and the anhydrous compound is the stable species. From these remarks, it is apparent that the composition of a sample containing essential water is greatly dependent upon the relative humidity of its environment.

Many hydrated compounds can be converted to the anhydrous condition by oven drying at 100 to 120°C for an hour or two. Such treatment often precedes an analysis of samples containing hydrated compounds.

Compounds Containing Adsorbed Water

Figure 27-2 shows an *adsorption isotherm,* in which the weight of the water adsorbed on a typical solid is plotted against the partial pressure of water in the surrounding atmosphere. It is apparent from the diagram that the extent of adsorption is particularly sensitive to changes in water-vapor pressure at low partial pressures.

The amount of water adsorbed on a solid decreases as the temperature of the solid increases and generally approaches zero when the solid is heated above 100°C. Adsorption or desorption of moisture usually occurs rapidly, with equilibrium often being reached after 5 or 10 min. The speed of the process is often apparent during the weighing of finely divided anhydrous solids, where a continuous increase in weight is observed unless the solid is contained in a tightly stoppered vessel.

FIGURE 27-2 Typical adsorption and sorption isotherms.

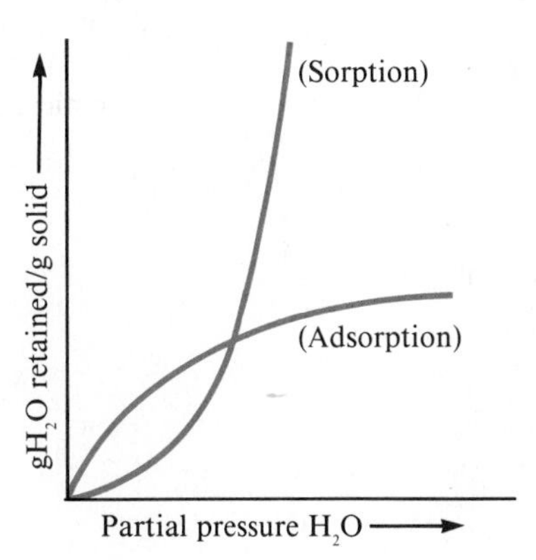

Compounds Containing Sorbed Water

The quantity of moisture sorbed by a colloidal solid varies tremendously with atmospheric conditions, as shown in Figure 27-2. In contrast to the behavior of adsorbed water, however, equilibrium may require days or even weeks for attainment, particularly at room temperature. Moreover, the amounts of water retained by the two processes are often quite different from each other; typically, adsorbed moisture amounts to a few tenths of a percent of the weight of the solid, whereas sorbed water can amount to 10 or 20%.

The amount of water sorbed in a solid also decreases as the solid is heated. Complete removal of this type of moisture at 100°C is by no means a certainty, however, as indicated by the drying curves for an organic compound shown in Figure 27-3. After this material was dried for about 70 min at 105°C, constant weight was apparently reached. It is clear, however, that additional moisture was removed by elevating the temperature. Even at 230°C, dehydration was probably not complete.

Compounds Containing Occluded Water

Occluded water is not in equilibrium with the atmosphere and is therefore insensitive to changes in humidity. Heating a solid containing occluded water may cause a gradual diffusion of the moisture to the surface, where it evaporates. Frequently, heating is accompanied by *decrepitation,* in which the crystals of the solid are suddenly shattered by the steam pressure created from moisture contained in the internal cavities.

27B-3 Drying the Analytical Sample

The methods for dealing with moisture in solid samples depend upon the information desired. When the composition of the material is desired on an as-received basis, the principal concern is that the moisture content not be altered as a consequence of grinding or other preliminary treatment and storage. Where such changes are unavoidable or probable, it is often advantageous to determine the weight loss upon heating at some suitable temperature (say, 105°C) immediately upon receipt of the sample. Then, when the analysis is to be performed, the sample is again dried at this temperature so that the data can be corrected back to the original basis.

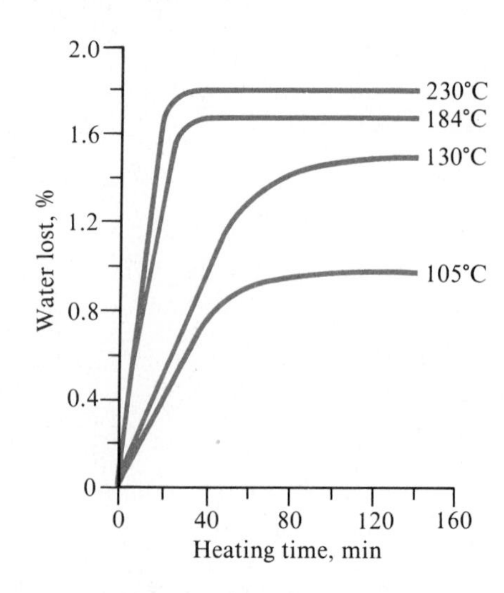

FIGURE 27-3 Removal of sorbed water from an organic compound at various temperatures. (Data from C. O. Willits, *Anal. Chem.*, **1951**, *23*, 1058. With permission of the American Chemical Society.)

Sometimes analytical results are reported on an air-dry basis, where the analysis is performed after the sample has been allowed to equilibrate with the atmosphere. Such a procedure is commonly followed in the analysis of metals and alloys as well as particulate matter that does not adsorb moisture strongly.

We have already noted that the moisture content of some substances is markedly changed by variations in humidity and temperature. Colloidal materials containing large amounts of sorbed moisture are particularly susceptible to the effects of these variables. For example, the moisture content of a potato starch has been found to vary from 10 to 21% as a consequence of an increase in relative humidity from 20 to 70%. With substances of this sort, comparable analytical data from one laboratory to another or even within the same laboratory can be achieved only by carefully specifying a procedure for taking the moisture content into consideration. For example, samples are frequently dried to constant weight at 105°C or at some other specified temperature. Analyses are then performed and results reported on this dry basis. While such a procedure may not render the solid completely free of water, it usually lowers the moisture content to a reproducible level.

27C The Determination of Water in Samples

Often the only satisfactory procedure for obtaining a result on a dry basis requires a separate determination of moisture in a set of samples taken concurrently with the samples that are to be analyzed. Several methods of determining water in solid samples are available. The simplest involves determining the weight loss after the sample has been heated at 100 to 110°C (or some other specified temperature) until the weight of the dried sample becomes constant. Unfortunately, this simple procedure is not at all specific for water, and large positive determinate errors occur in samples that yield volatile decomposition products (other than water) upon being heated. This method can also yield negative errors when applied to samples containing sorbed moisture (for example, see Figure 27-3).

Several highly selective methods have been developed for the determination of water in solid and liquid samples. One of these, the Karl Fischer method, is described in Section 14D-3. Several others are described in a monograph by Mitchell and Smith.[6]

Questions and Problems

*27-1 Describe the steps in a sampling operation.

27-2 What is the object of the sampling step in an analysis?

27-3 Differentiate between
- *(a) sorbed water, adsorbed water, and occluded water.
- (b) water of crystallization and water of constitution.
- *(c) essential water and nonessential water.
- (d) gross sample and laboratory sample.

27-4 What factors determine the weight of a gross sample?

[6]J. J. Mitchell Jr. and D. M. Smith, *Aquametry*, 2nd ed. New York: Wiley, 1977.

*27-5 The following results were obtained for the determination of calcium in a NBS limestone sample: % CaO = 50.38, 50.20, 50.31, 50.22, 50.41. Five gross samples were then obtained for a carload of limestone. The average % CaO values for the gross samples were found to be 49.53, 50.12, 49.60, 49.87, 50.49. Calculate the relative standard deviation associated with the sampling step.

*27-6 The preparation of a heterogeneous catalyst involves coating spherical support particles with a layer of active material. A satisfactory product is entirely coated with a layer of catalyst; the existence of breaks is unacceptable. Complete the accompanying tabulation.

	Number of Particles Counted	Number of Particles Counted That Were		Percent Relative Standard Deviation of the Count	Absolute Standard Deviation of the Count
		Satisfactory	Unsatisfactory		
(a)	325	291			
(b)	325		34		
(c)	800		82		
(d)	675		68		
(e)	1200	1025			

27-7 A coating that weighs at least 3.00 mg is needed to impart adequate shelf life to a pharmaceutical tablet. A random sampling of 250 tablets revealed that 14 failed to meet this requirement.

(a) Use this information to estimate the relative standard deviation for the measurement.
(b) What is the 90% confidence interval for the number of unsatisfactory tablets?
(c) Assuming that the fraction of rejects remains unchanged, how many tablets should be taken to ensure a relative standard deviation of 10% in this measurement?

27-8 Changes in the method used to coat the tablets lowered the percentage of rejects from 5.6% (Problem 27-7) to 2.0%. How many tablets should be taken for inspection if the permissible relative standard deviation in the measurement is to be

*(a) 25%? (b) 10%? *(c) 5%? (d) 1%?

*27-9 The mishandling of a shipping container loaded with 750 cases of wine caused some of the bottles to break. An insurance adjuster proposed to settle the claim at 20.8% of the value of the shipment, based upon a random 250-bottle sample in which 52 were cracked or broken. Calculate

(a) the relative standard deviation of the adjuster's evaluation.
(b) the absolute standard deviation for the 750 cases (12 bottles/case).
(c) the 90% confidence interval for the total number of bottles.
(d) the size of a random sampling needed for a relative standard deviation of 5.0%, assuming a breakage rate of about 21%.

*27-10 Approximately 15% of the particles in a shipment of silver-bearing ore are judged to be argentite, Ag_2S (d = 7.3 $g{\cdot}cm^{-3}$, 87% Ag); the remainder are siliceous (d = 2.6 $g{\cdot}cm^{-3}$) and contain essentially no silver.

(a) Calculate the number of particles that should be taken for the gross sample if the relative standard deviation due to sampling is to be 1% or less.
(b) Estimate the weight of the gross sample, assuming that the particles are spherical and have an average diameter of 4.0 mm.
(c) The sample taken for analysis is to weigh 0.600 g and contain the same number of particles as the gross sample. To what diameter must the particles be ground to satisfy these criteria?

27-11 The average diameter of the particles in a shipment of copper appears to be 5.0 mm. Approximately 5% of the particles are cuprite (80% Cu) with a density of 6 $g{\cdot}cm^{-3}$; the remainder is estimated to have a density of 4 $g{\cdot}cm^{-3}$ and contain 3% Cu.

(a) How many particles of the ore should be sampled if the relative standard deviation due to sampling is to be 4% or less?
(b) What should the weight of the gross sample be?
(c) To what diameter must the particles be ground in order to yield a sample for analysis that weighs 0.500 g and has the same number of particles as the gross sample?

*27-12 The average particle diameter of an ore sample is 2.0 mm. It is estimated that the stibnite content ($d_{Sb_2S_3}$ = 4.5 $g{\cdot}cm^{-3}$, 71.7% Sb) is approximately 2.0%; the remainder has a density of 3.0 $g{\cdot}cm^{-3}$ and contains about 1% Sb.

(a) How many particles of the ore should be taken if the relative standard deviation due to sampling is to be 1% or less?
(b) What should the weight of the gross sample be?
(c) To what diameter must the particles be ground in order to yield a sample for analysis that weighs 0.750 g and has the same number of particles as the gross sample?

27-13 The seller of a mining claim took a random ore sample that weighed approximately 5 lb and had an average particle diameter of 5.0 mm. Inspection revealed that about 1% of the sample was argentite (Problem 27-10), and the remainder had a density of about 2.6 $g{\cdot}cm^{-3}$ and contained no silver. The prospective buyer insisted upon knowing the silver content of the claim with a relative error no greater than 5%. Establish whether the seller provided a sufficiently large sample to permit such an evaluation.

CHAPTER 28

Decomposing and Dissolving the Sample

Most analytical measurements are performed on solutions (usually aqueous) of the analyte. While some samples dissolve readily in water or dilute aqueous solutions of the common acids or bases, others require powerful reagents and rigorous treatment. For example, when sulfur or halogens are to be determined in an organic compound, the sample must be subjected to high temperatures and potent reagents in order to rupture the strong bonds between these elements and carbon. Similarly, drastic conditions are usually required to destroy the silicate structure of a siliceous mineral, thus rendering its cations free for analysis.

The proper choice among the various reagents and techniques for decomposing and dissolving analytical samples can be critical to the success of an analysis, particularly where refractory substances are involved or where the analyte is present in trace amounts. This chapter describes some of the more common methods for obtaining aqueous solutions of samples, particularly those that are difficult to decompose or dissolve.[1]

28A Some General Considerations

Ideally, a reagent should dissolve the sample completely because attempts to leach an analyte quantitatively from an insoluble residue are usually not successful. In choosing a solvent, consideration must also be given to possible interferences introduced during dissolution or decomposition. When trace amounts of an analyte are being determined, the most important considerations are frequently the purity of the reagent used for sample dissolution and the amounts that must be used.

An important concern when dissolving samples is the possibility that some portion of the analyte may volatilize. For example, carbon dioxide, sulfur dioxide, hydrogen sulfide, hydrogen selenide, and hydrogen telluride are generally volatilized when a sample is dissolved in strong acid, whereas ammonia is commonly lost when a basic reagent is employed. Similarly, hydrofluoric acid reacts with silicates and boron-containing compounds to produce volatile fluorides. Strong oxidizing solvents often result in the evolution of chlorine, bromine, or iodine; reducing solvents may lead to the volatilization of such compounds as arsine, phosphine, and stibine.

A number of elements form volatile chlorides that are partially or completely lost from hot hydrochloric acid solutions. Among these are the chlorides

[1]For an extensive discussion of this topic, see R. Bock, *A Handbook of Decomposition Methods in Analytical Chemistry*. New York: Wiley, 1979.

of tin(IV), germanium(IV), antimony(III), arsenic(III), and mercury(II). The oxychlorides of selenium and tellurium also volatilize to some extent from hot hydrochloric acid. The presence of chloride ion in hot concentrated sulfuric or perchloric acid solutions can cause volatilization losses of bismuth, manganese, molybdenum, thallium, vanadium, and chromium.

Boric acid, nitric acid, and the halogen acids are lost from boiling aqueous solutions, and phosphoric acid distills from hot concentrated sulfuric or perchloric acid. Certain volatile oxides can also be lost from hot acidic solutions, including the tetroxides of osmium and ruthenium and the heptoxide of rhenium.

28B Aqueous Reagents for Dissolving or Decomposing Samples

The most common reagents for attacking analytical samples are the mineral acids and, less frequently, aqueous solutions of ammonia and the alkali metal hydroxides.

28B-1 Hydrochloric Acid

Concentrated hydrochloric acid is an excellent solvent for many metal oxides as well as for metals more easily oxidized than hydrogen; often, it is a better solvent for oxides than the oxidizing acids. Concentrated hydrochloric acid is about 12 M, but upon heating, hydrogen chloride is lost until a constant-boiling 6 M solution remains (boiling point about 110°C).

28B-2 Nitric Acid

Hot concentrated nitric acid dissolves all common metals with the exception of aluminum and chromium, which become passive to this reagent as a consequence of surface oxide formation. When alloys containing tin, tungsten, or antimony are treated with the hot reagent, slightly soluble hydrated oxides, such as $SnO_2 \cdot 4H_2O$, form. After coagulation, these colloidal materials can be separated from other metallic species by filtration.

28B-3 Sulfuric Acid

Many materials are decomposed and dissolved by hot concentrated sulfuric acid, which owes part of its effectiveness as a solvent to its high boiling point (about 340°C). Most organic compounds are dehydrated and oxidized at this temperature and are thus eliminated from samples as carbon dioxide and water. Most metals and many alloys are attacked by the hot acid.

28B-4 Perchloric Acid

Hot concentrated perchloric acid, a potent oxidizing agent, attacks a number of iron alloys and stainless steels that are intractable to other mineral acids. Care must be taken in using the reagent, however, because of *its potentially explosive nature.* The cold concentrated acid is not hazardous, nor are heated dilute solutions; *violent explosions occur, however, when hot concentrated perchloric acid comes into contact with organic materials or easily oxidized inorganic substances.* Because of this property, the concentrated reagent should be heated only in special hoods, which are lined with glass or stainless steel, are seamless, and have a fog system for washing down the walls with water. A perchloric-acid hood should always have its own fan system, one that is independent of all other systems. *With proper precautions,*[2] perchloric acid is a safe and useful reagent.

Perchloric acid is marketed as the 60 or 72% acid. A constant-boiling mixture (72.4% $HClO_4$) is obtained at 203°C.

28B-5 Oxidizing Mixtures

More rapid solvent action can sometimes be obtained by the use of mixtures of acids or by the addition of oxidizing agents to a mineral acid. *Aqua regia,* a mixture containing three volumes of concentrated hydrochloric acid and one of nitric acid, is well known. The addition of bromine or hydrogen peroxide to mineral acids often increases their solvent action and hastens the oxidation of organic materials in the sample. Mixtures of nitric and perchloric acid are also useful for this purpose and less dangerous than perchloric acid alone. With this mixture, however, care must be taken to avoid evaporation of all the nitric acid before oxidation of the organic material is complete. *Severe explosions and injuries have resulted from failure to observe this precaution.*

28B-6 Hydrofluoric Acid

The primary use of hydrofluoric acid is for the decomposition of silicate rocks and minerals in the determination of species other than silica. In this treatment, silicon is evolved as the tetrafluoride. After decomposition is complete, the excess hydrofluoric acid is driven off by evaporation with sulfuric acid or perchloric acid. Complete removal is often essential to the success of an analysis because fluoride ion reacts with several cations to form extraordinarily stable complexes that then interfere with the determination of the cations. For example, precipitation of aluminum (as $Al_2O_3 \cdot xH_2O$) with ammonia is quite incomplete if fluoride is present even in small amounts. Frequently, removal of the last traces of fluoride ion from a sample is so difficult and time-consuming as to negate the attractive features of the parent acid as a solvent for silicates.

Hydrofluoric acid finds occasional use in conjunction with other acids in attacking steels that dissolve with difficulty in other solvents.

Because hydrofluoric acid is extremely toxic, dissolution of samples and evaporation to remove excess reagent *should always be carried out in a well-ventilated hood.* Hydrofluoric acid *causes serious damage and painful injury* when brought into contact with the skin. Its effects may not become evident until hours after exposure. If the acid comes into contact with the skin, the affected area should be immediately washed with copious quantities of water. Treatment with a dilute solution of calcium ion, which precipitates fluoride ion, may also be of help.

28C Decomposition of Samples by Fluxes

Many common substances—notably silicates, some mineral oxides, and a few iron alloys—are attacked slowly, if at all, by the aqueous reagents just consid-

[2]See A. A. Schilt, *Perchloric Acid and Perchlorates.* Columbus, Ohio: G. Frederick Smith Chemical Company, 1979.

ered. In such cases, recourse to a fused-salt medium is indicated. In this procedure, the sample is mixed with an alkali metal salt, called the *flux*, and the combination is then fused to form a water-soluble product called the *melt*. Fluxes decompose most substances by virtue of the high temperature required for their use (300 to 1000°C) and the high concentration of reagent brought in contact with the sample.

Where possible, the employment of a flux is avoided, for several dangers and disadvantages attend its use. Among these is the possible contamination of the sample by impurities in the flux. This possibility is exacerbated by the relatively large amount of flux (typically ten times the sample weight) required for a successful fusion. Moreover, the aqueous solution that results when the melt from a fusion is dissolved has a high salt content, which may cause difficulties in the subsequent steps of the analysis. In addition, the high temperatures required for a fusion increase the danger of volatilization losses. Finally, the container in which the fusion is performed is almost inevitably attacked to some extent by the flux; again, contamination of the sample is the result.

For a sample containing only a small fraction of material that dissolves with difficulty, it is common practice to employ a liquid reagent first; the undecomposed residue is then isolated by filtration and fused with a relatively small quantity of a suitable flux. After cooling, the melt is dissolved and combined with the major portion of the sample.

28C-1 Carrying Out a Fusion

The sample in the form of a very fine powder is mixed intimately with perhaps a tenfold excess of the flux. Mixing is usually carried out in the crucible in which the fusion is to be performed. The time required for fusion can range from a few minutes to hours. The production of a clear melt signals completion of the decomposition, although often this condition is not totally obvious.

When the fusion is complete, the mass is allowed to cool slowly; just before solidification, the crucible is rotated to distribute the solid around the walls to produce a thin layer of melt that is easy to dislodge.

28C-2 Types of Fluxes

With few exceptions, the common fluxes used in analysis are compounds of the alkali metals. Alkali metal carbonates, hydroxides, peroxides, and borates are basic fluxes employed to attack acidic materials. The acidic fluxes are pyrosulfates, acid fluorides, and boric oxide. If an oxidizing flux is required, sodium peroxide can be used. As an alternative, small quantities of the alkali nitrates or chlorates can be mixed with sodium carbonate.

The properties of the common fluxes are summarized in Table 28-1.

Sodium Carbonate

Silicates and certain other refractory materials can be decomposed by heating to 1000 to 1200°C with sodium carbonate. This treatment generally converts the cationic constituents of the sample to acid-soluble carbonates or oxides; the nonmetallic constituents are converted to soluble sodium salts.

Carbonate fusions are normally carried out in platinum crucibles.

TABLE 28-1 **Common Fluxes**

Flux	Melting Point, °C	Type of Crucible for Fusion	Type of Substance Decomposed
Na_2CO_3	851	Pt	Silicates and silica-containing samples, alumina-containing samples, sparingly soluble phosphates and sulfates
Na_2CO_3 + an oxidizing agent, such as KNO_3, $KClO_3$, or Na_2O_2	—	Pt (not with Na_2O_2), Ni	Samples requiring an oxidizing environment; that is, samples containing S, As, Sb, Cr, etc.
NaOH or KOH	318 380	Au, Ag, Ni	Powerful basic fluxes for silicates, silicon carbide, and certain minerals (main limitation is purity of reagents)
Na_2O_2	Decomposes	Fe, Ni	Powerful basic oxidizing flux for sulfides; acid-insoluble alloys of Fe, Ni, Cr, Mo, W, and Li; platinum alloys; Cr, Sn, Zr minerals
$K_2S_2O_7$	300	Pt, porcelain	Acidic flux for slightly soluble oxides and oxide-containing samples
B_2O_3	577	Pt	Acidic flux for silicates and oxides where alkali metals are to be determined
$CaCO_3$ + NH_4Cl	—	Ni	Upon heating the flux, a mixture of CaO and $CaCl_2$ is produced; used to decompose silicates for the determination of the alkali metals

Potassium Pyrosulfate

Potassium pyrosulfate is a potent acidic flux that is particularly useful for attacking the more intractable metal oxides. Fusions with this reagent are performed at about 400°C; at this temperature, the slow evolution of the highly acidic sulfur trioxide takes place:

$$K_2S_2O_7 \longrightarrow K_2SO_4 + SO_3(g)$$

Potassium pyrosulfate can be prepared by heating potassium hydrogen sulfate:

$$2KHSO_4 \longrightarrow K_2S_2O_7 + H_2O$$

Other Fluxes

Table 28-1 contains data for several other common fluxes. Noteworthy are boric oxide and the calcium carbonate/ammonium chloride mixture. Both are employed to decompose silicates for the analysis of alkali metals. Boric oxide is removed after solution of the melt by evaporation to dryness with methyl alcohol; methyl borate, $B(OCH_3)_3$, distills.

28D

Decomposition of Organic Compounds for Elemental Analysis[3]

Determination of the elemental composition of an organic sample generally requires drastic treatment to convert the element of interest to a form susceptible to the common analytical techniques. These treatments are usually oxidative and involve conversion of carbon to carbon dioxide and hydrogen to water; occasionally, however, heating the sample with a potent reducing agent is sufficient to rupture the covalent bonds in the compound and free the analyte element from the carbonaceous residue.

Oxidation procedures can be grouped in two categories. *Wet-ashing* (or wet-oxidation) makes use of liquid oxidizing agents, such as sulfuric, nitric, and perchloric acids. *Dry-ashing* usually implies ignition of the organic compound in air or in a stream of oxygen. In addition, oxidations can be carried out in certain fused-salt media; sodium peroxide is the most common flux for this purpose.

In the sections that follow, we consider briefly some of the methods for decomposing organic substances prior to elemental analysis.

28D-1

Wet-Ashing Procedures

Solutions of strong oxidizing agents are often used to decompose organic samples for the determination of metallic constituents. The main problem associated with the use of these reagents is possible losses of the elements of interest by volatilization.

We have already encountered an example of wet-ashing in the Kjeldahl method for the determination of nitrogen in organic compounds (page 238), where concentrated sulfuric acid is the oxidizing agent. This reagent is also frequently employed for the decomposition of organic materials in which metallic constituents are to be determined. Nitric acid can be added periodically to the solution to hasten the oxidation rate.[4] The following elements are volatilized (at least partially) by this procedure, particularly if the sample contains chlorine: arsenic, boron, germanium, mercury, antimony, selenium, tin, the halogens, sulfur, and phosphorus.

A reagent even more effective than a sulfuric/nitric acid mixture is perchloric acid mixed with nitric acid. *Great care must be exercised in the use of this reagent,* however, because of the tendency of hot anhydrous perchloric acid to react explosively with organic material. It is essential to start the oxidation with a mixture in which nitric acid predominates; this reagent attacks the easily oxidized components in the early stages. With continued heating, water and nitric acid are lost by decomposition and evaporation, and the solution becomes a progressively stronger oxidant as the perchloric acid concentration increases. If the solution becomes too concentrated in perchloric acid before most of the oxidation is complete, it will darken or turn black. *Should darkening occur, the mixture should be immediately removed from the heat and diluted with water and nitric acid.* The heating can then be continued. As mentioned on page 693, perchloric acid oxidations should be carried out only in a special hood. If properly performed, oxidations with a mixture of nitric and perchloric acids

[3]For a thorough treatment of this topic, see T. S. Ma and R. C. Rittner, *Modern Organic Elemental Analysis.* New York: Marcel Dekker, 1979.

[4]*Official Methods of Analysis of the AOAC,* 11th ed., p. 400. Washington, D.C.: Association of Official Analytical Chemists, 1970.

are rapid, and losses of metallic ions are negligible.[5] *It cannot be too strongly emphasized that hot concentrated perchloric acid reacts explosively with organic materials and other easily oxidizable substances and that extraordinary care is needed in using this reagent.*

High-Pressure Wet-Ashing

An apparatus has recently been described for automatically wet-ashing various kinds of organic samples preliminary to determination of their metal content by atomic spectroscopy.[6] The oxidation is performed in a closed quartz vessel mounted in an aluminum heating block. The vessel and block are contained in an autoclave that can be pressurized with nitrogen up to 100 atm. Concentrated nitric acid and nitric/hydrochloric acid mixtures are used for the oxidation, and these reagents reach temperatures of 250 to 300°C at the elevated pressures made possible by the autoclave. Organic samples such as foodstuffs, biological materials, petroleum products, and coals are completely oxidized in 15 to 120 min in the apparatus.

Microwave Wet-Ashing

A recent development in wet-ashing at elevated pressures (and thus elevated temperatures) is based upon the microwave decomposition of samples in nitric and other mineral acids contained in sealed Teflon vessels.[7] Heating is performed in a specially designed microwave oven that is now on the market. The rate of wet-ashing and efficiency of decomposition are reported to increase dramatically with this type of equipment. For example, decompositions that take several hours at atmospheric conditions are said to be completed in a few minutes by the microwave technique. A further advantage is that the amount of reagent required for the digestion is much smaller, thus reducing the magnitude of reagent blanks significantly.

28D-2 Dry-Ashing Procedures

The simplest method for decomposing an organic sample prior to determining the cations it contains is to heat the sample over a flame in an open dish or crucible until all carbonaceous material has been oxidized to carbon dioxide. Red heat is often required to complete the oxidation. Analysis of the nonvolatile components follows dissolution of the residual solid. Unfortunately, there is always a great deal of uncertainty about the completeness of recovery of supposedly nonvolatile elements from a dry-ashed sample. Some losses probably result from the entrainment of finely divided particulate matter in the convection currents around the crucible. In addition, volatile metallic compounds may be lost during the ignition. For example, copper, iron, and vanadium are appreciably volatilized when samples containing porphyrin compounds are ashed.

Although dry-ashing is the simplest method for decomposing organic compounds, it is often the least reliable. It should not be employed unless tests have demonstrated its applicability to a given type of sample.

[5] T. T. Gorsuch, *Analyst,* **1959,** *84,* 135; G. F. Smith, *Anal. Chim. Acta,* **1953,** *8,* 397; G. F. Smith, *The Wet Chemical Oxidation of Organic Compositions Employing Perchloric Acid.* Columbus, Ohio: G. F. Smith Chemical Company, 1965.

[6] G. Knapp and A. Grillo, *Amer. Lab.,* **1986,** *18* (3), 76.

[7] *Anal. Chem.,* **1986,** *58,* 1424A; H. M. Kingston and L. B. Jassie, *Anal. Chem.,* **1986,** *58,* 1986.

28D-3

Combustion-Tube Methods

Several common and important elemental components of organic compounds are converted to gaseous products as a sample is pyrolyzed in the presence of oxygen. With suitable apparatus, it is possible to trap these volatile compounds quantitatively, thus making them available for the analysis of the element of interest. The heating is commonly performed in a glass or quartz combustion tube through which a stream of carrier gas is passed. The stream transports the volatile products to parts of the apparatus where they are separated and retained for the measurement; the gas may also serve as the oxidizing agent. Elements susceptible to this type of treatment are carbon, hydrogen, oxygen, nitrogen, the halogens, sulfur, and oxygen. Table 28-2 gives details for some of these methods.

Automated combustion-tube analyzers are now available for the determination of either carbon, hydrogen, and nitrogen or carbon, hydrogen, and oxygen in a single sample. The apparatus requires essentially no attention by the operator, and the analysis is complete in less than 15 min. In one such analyzer, the sample is ignited in a stream of helium and oxygen and passes over an oxidation catalyst consisting of a mixture of silver vanadate and silver tungstate. Halogens and sulfur are removed with a packing of silver salts. A packing of hot copper is located at the end of the combustion train to remove oxygen and convert nitrogen oxides to nitrogen. The exit gas, consisting of a mixture of water, carbon dioxide, nitrogen, and helium, is collected in a glass bulb. The analysis of this mixture is accomplished with three thermal-conductivity measurements (Section 24B-5). The first is made on the intact mix-

TABLE 28-2 **Combustion-Tube Methods for the Elemental Analysis of Organic Substances**

Element	Name of Method	Method of Oxidation	Method of Completion of Analysis
Halogens	Pregl	Sample burned in stream of oxygen over red-hot platinum catalyst; halogens converted primarily to HX and X_2	Gas stream passed through carbonate solution containing SO_3^{2-} (to reduce halogens and oxyhalogens to halides); X^- then determined by usual procedures
	Grote	Sample burned in stream of air over hot silica catalyst; products are HX and X_2	Same as above
Sulfur	Pregl	Similar to halogen determination; combustion products are SO_2 and SO_3	Gas stream passed through aqueous H_2O_2 to convert sulfur oxides to H_2SO_4, which can then be titrated with standard base
	Grote	Similar to halogen determination; products are SO_2 and SO_3	Similar to above
Nitrogen	Dumas	Sample oxidized by hot CuO to CO_2, H_2O, and N_2	Gas stream passed through concentrated KOH solution, leaving only N_2, which is measured volumetrically
Carbon and hydrogen	Pregl	Similar to halogen analysis; products are CO_2 and H_2O	H_2O adsorbed on desiccant and CO_2 on Ascarite; determined gravimetrically
Oxygen	Unterzaucher	Sample pyrolyzed over carbon; oxygen converted to CO; H_2 used as carrier gas	Gas stream passed over I_2O_5 [$5CO + I_2O_5(s) \rightarrow 5CO_2 + I_2(g)$] and liberated I_2 titrated

ture, the second is made on the mixture after water has been removed by passage of the gas through a dehydrating agent, and the third is made on the mixture after carbon dioxide has been removed by an absorbent. The relationship between thermal conductivity and concentration is linear, and the slope of the curve for each constituent is established by calibration with a pure compound such as acetanilide.

28D-4

Combustion with Oxygen in a Sealed Container

A relatively straightforward method for the decomposition of many organic substances involves combustion with oxygen in a sealed container. The reaction products are absorbed in a suitable solvent before the reaction vessel is opened; they are subsequently analyzed by ordinary methods.

A remarkably simple apparatus for performing such oxidations has been suggested by Schöniger (Figure 28-1).[8] It consists of a heavy-walled flask of 300- to 1000-mL capacity fitted with a ground-glass stopper. Attached to the stopper is a platinum gauze basket that holds from 2 to 200 mg of sample. If the substance to be analyzed is a solid, it is wrapped in a piece of low-ash filter paper cut in the shape shown in Figure 28-1. Liquid samples are weighed into gelatin capsules, which are then wrapped in a similar fashion. The paper tail serves as the ignition point.

A small volume of an absorbing solution (often sodium carbonate) is placed in the flask, and the air in the flask is displaced by oxygen. The tail of the paper is ignited, the stopper is quickly fitted into the flask, and the flask is inverted to prevent the escape of the volatile oxidation products. The reaction ordinarily proceeds rapidly, being catalyzed by the platinum gauze surrounding the sample. During the combustion, the flask is shielded to minimize damage in case of explosion.

After cooling, the flask is shaken thoroughly and disassembled, and the inner surfaces are carefully rinsed. The analysis is then performed on the resulting solution. This procedure has been applied to the determination of halogens, sulfur, phosphorus, fluorine, arsenic, boron, carbon, and various metals in organic compounds.

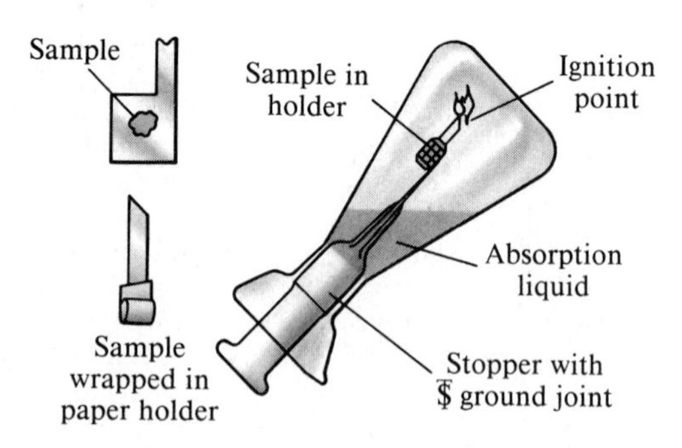

FIGURE 28-1 Schöniger combustion apparatus. (Courtesy Thomas Scientific, Swedesboro, NJ 08085.)

[8] W. Schöniger, *Mikrochim. Acta,* **1955,** 123; **1956,** 869. See also the review articles by A. M. G. MacDonald, in *Advances in Analytical Chemistry and Instrumentation,* C. E. Reilley, Ed., Vol. 4, p. 75. New York: Interscience, 1965; and M. E. McNally and R. L. Grob, *Amer. Lab.,* **1981,** *13* (1), 31.

Questions and Problems

*28-1 Differentiate between wet-ashing and dry-ashing.

28-2 What is a flux? When is its use called for?

*28-3 What fluxes are suitable for the determination of alkali metals in silicates?

28-4 What flux is commonly used for the decomposition of certain refractory oxides?

*28-5 Under what conditions is the use of perchloric acid likely to be dangerous?

28-6 How are organic compounds decomposed for the determination of
*(a) halogens?
(b) sulfur?
*(c) nitrogen?
(d) heavy-metal species?

CHAPTER 29 Eliminating Interferences

Interferences are encountered in chemical analyses whenever the sample matrix contains a species that either produces a signal indistinguishable from that of the analyte or else attenuates the analyte signal. Few analytical measurements are entirely specific and thus free from interference. Consequently, an important step in most analyses is one that eliminates interfering species.

Two general methods are available for coping with interferences. The first makes use of a *masking* agent, which is a substance that reacts with the interference to form a species that no longer contributes to or attenuates the analyte signal. Obviously, the masking agent must not have a significant effect on the behavior of the analyte. An example of a masking agent is fluoride ion in the iodometric determination of copper in ores that contain significant amounts of iron. Fluoride ion does not inhibit the reaction of the analyte with iodide ion to give iodine, which is subsequently titrated. It does, however, prevent iron(III) from undergoing an analogous reaction and thus interfering. The masking effect of the fluoride ion is attributable to the formation of stable iron(III) fluoride complexes that lower the electrode potential of the iron(III) to a point where it is no longer capable of producing iodine from iodide ion.

The second approach to the interference problem is to physically separate the analyte from potential interferences. In Chapters 23 through 25, we considered the chemist's most powerful tool for performing separations: chromatography. In this chapter, we consider other methods, including precipitation, solvent extraction, ion exchange, and distillation.

29A The Nature of the Separation Process

All separation procedures have in common the distribution of the components in a mixture between two phases that subsequently can be separated mechanically. If the ratio of the amount of one particular component in each phase (the *distribution ratio*) differs significantly from that of another, a separation of the two is potentially feasible. To be sure, the complexity of the separation depends upon the magnitude of this difference in distribution ratios. Where the difference is large, a single-stage process suffices. For example, a single precipitation with silver ion is adequate for the isolation of chloride from many other anions because the ratio of the amount of chloride ion in the solid phase to that in the aqueous phase is immense while comparable ratios for, say, nitrate or perchlorate ion approach zero.

A more complex situation prevails when the distribution ratio for one component is essentially zero, as in the foregoing example, but the ratio for the other is not very large. Here, a multistage process is required. For example, uranium(VI) can be extracted into ether from an aqueous nitric acid solution.

Although the distribution ratio approaches unity for a single extraction, uranium(VI) can nevertheless be isolated quantitatively by repeated, or *exhaustive,* extraction of the aqueous solution with fresh portions of ether.

The most complex procedures are required when the distribution ratios of the species to be separated are both greater than zero and approach one another in magnitude; here, multistage *fractionation* techniques, such as chromatography, are necessary. Fractionation, like its simpler counterparts, is based upon differences in the distribution ratios of solutes. Two factors, however, account for the gain in separation efficiency with fractionation. First, the number of times that partitioning occurs between phases is increased enormously; second, distribution occurs between fresh portions of both phases. An exhaustive extraction differs from a fractionation in that fresh portions of only one phase are involved in the former.

29B Separation by Precipitation

Separations by precipitation require large solubility differences between analyte and potential interferences. The theoretical feasibility of this type of separation can be determined by solubility calculations such as those shown in Section 6C. Unfortunately, several other factors may preclude the use of precipitation to achieve a separation. For example, the various coprecipitation phenomena described in Section 3B-4 may cause extensive contamination of a precipitate by an unwanted component, even though the solubility product of the contaminant has not been exceeded. Likewise, the rate of an otherwise feasible precipitation may be too slow to be useful for a separation. Finally, when precipitates form as colloidal suspensions, coagulation may be difficult and slow, particularly when the isolation of a small quantity of a solid phase is attempted.

Many precipitating agents have been employed for quantitative inorganic separations. Some of the most generally useful are described in the sections that follow.

29B-1 Separations Based on Control of Acidity

Enormous differences exist in the solubilities of the hydroxides, hydrous oxides, and acids of various elements. Moreover, the concentration of hydrogen or hydroxide ions in a solution can be varied by a factor of 10^{15} or more and can be readily controlled by the use of buffers. As a consequence, many separation procedures based on pH control are, in theory, available to the chemist. In practice, these separations can be grouped in three categories: (1) those made in relatively concentrated solutions of strong acids, (2) those made in buffered solutions at intermediate pH values, and (3) those made in concentrated solutions of sodium or potassium hydroxide. Table 29-1 lists common separations that can be achieved by control of acidity.

29B-2 Sulfide Separations

With the exception of the alkali and alkaline-earth metals, most cations form sparingly soluble sulfides whose solubilities differ greatly from one another. Because it is relatively easy to control the sulfide ion concentration of an aqueous solution by adjustment of pH (Section 6C-2), separations based on the formation of sulfides have found extensive use. Sulfides can be conveni-

TABLE 29-1

Separations Based upon Control of Acidity

Reagent	Species Forming Precipitates	Species Not Precipitated
Hot concd HNO_3	Oxides of W(VI), Ta(V), Nb(V), Si(IV), Sn(IV), Sb(V)	Most other metal ions
NH_3/NH_4Cl buffer	Fe(III), Cr(III), Al(III)	Alkali and alkaline earths, Mn(II), Cu(II), Zn(II), Ni(II), Co(II)
$HOAc/NH_4OAc$ buffer	Fe(III), Cr(III), Al(III)	Common dipositive ions
$NaOH/Na_2O_2$	Fe(III), most dipositive ions, rare earths	Zn(II), Al(III), Cr(VI), V(V), U(VI)

ently precipitated from homogeneous solution, with the anion being generated by the hydrolysis of thioacetamide (Table 3-1).

A theoretical treatment of the ionic equilibria influencing the solubility of sulfide precipitates was considered in Section 6C-2. Such treatment may fail to provide realistic conclusions regarding the feasibility of separations, however, because of coprecipitation and the slow rate at which some sulfides form. As a consequence, resort must be made to empirical observations.

Table 29-2 shows some common separations that can be accomplished with hydrogen sulfide through control of pH.

29B-3 *Other Inorganic Precipitants*

No other inorganic ion is as generally useful for separations as hydroxide and sulfide ions. Phosphate, carbonate, and oxalate ions are often employed as precipitants for cations, but their behavior is nonselective; therefore separations must ordinarily precede their use.

Chloride and sulfate ions are useful because of their highly selective behavior. The former are used to separate silver from most other metals, and the latter are frequently employed to isolate a group of metals that includes lead, barium, and strontium.

29B-4 *Organic Precipitants*

Selected organic reagents for the isolation of various inorganic ions were discussed in Section 3D-3. Some of these organic precipitants, such as dimethylglyoxime, are useful because of their remarkable selectivity in forming

TABLE 29-2

Precipitation of Sulfides

Elements	Conditions for Precipitation*	Conditions for No Precipitation*
Hg(II), Cu(II), Ag(I)	1, 2, 3, 4	
As(V), As(III), Sb(V), Sb(III)	1, 2, 3	4
Bi(III), Cd(II), Pb(II), Sn(II)	2, 3, 4	1
Sn(IV)	2, 3	1, 4
Zn(II), Co(II), Ni(II)	3, 4	1, 2
Fe(II), Mn(II)	4	1, 2, 3

*1 = 3 M HCl; 2 = 0.3 M HCl; 3 = buffered to pH 6 with acetate; 4 = buffered to pH 9 with $NH_3/(NH_4)_2S$.

precipitates with a few ions. Others, such as 8-hydroxyquinoline, yield slightly soluble compounds with a host of cations. The selectivity of this sort of reagent is owing to the wide range of solubility products among its reaction products and also the fact that the precipitating reagent is ordinarily an anion that is the conjugate base of a weak acid. Thus, separations based on pH control can be realized just as with hydrogen sulfide.

29B-5 The Separation of Constituents Present in Trace Amounts

A problem often encountered in trace analysis is that of isolating the species of interest, which may be present in microgram quantities, from the major components of the sample. Although such a separation is sometimes based on a precipitation, the techniques required differ from those used when the analyte is present in generous amounts.

Several problems attend the quantitative separation of a trace element by precipitation even when solubility losses are not important. Supersaturation often delays formation of the precipitate, and coagulation of small amounts of a colloidally dispersed substance is often difficult. In addition, it is likely that an appreciable fraction of the solid will be lost during transfer and filtration. To minimize these difficulties, a quantity of some other ion that also forms a precipitate with the reagent is often added to the solution. The precipitate from the added ion is called a *collector* and carries the desired minor species out of solution. For example, in isolating manganese as the sparingly soluble manganese dioxide, a small amount of iron(III) is frequently added to the analyte solution before the introduction of ammonia as the precipitating reagent. The basic iron(III) oxide carries down even the smallest traces of the dioxide. Other examples are basic aluminum oxide as a collector of trace amounts of titanium and copper sulfide for collection of traces of zinc and lead. Many other collectors are described by Sandell and Onishi.[1]

Sometimes collectors carry down the trace precipitate by entrainment. Other times, the process must involve coprecipitation in which the minor component is adsorbed on or incorporated into the collector precipitate as the result of mixed-crystal formation. Clearly, the collector must not interfere with the method selected for determining trace component.

29B-6 Separation by Electrolytic Precipitation

Electrolytic precipitation constitutes a highly useful method for accomplishing separations. In this process, the more easily reduced species, be it the wanted or the unwanted component of the mixture, is isolated as a separate phase. The method becomes particularly effective when the potential of the working electrode is controlled at a predetermined level (Section 16C-5).

The mercury cathode (page 404) has found wide application in the removal of many metal ions prior to the analysis of the residual solution. In general, metals more easily reduced than zinc are conveniently deposited in the mercury, leaving such ions as aluminum, beryllium, the alkaline earths, and the alkali metals in solution. The potential required to decrease the concentration of a metal ion to any desired level is readily calculated from polarographic data.

[1]E. B. Sandell and H. Onishi, *Colorimetric Determination of Traces of Metals,* 4th ed., pp. 709–721. New York: Interscience, 1978.

29C Extraction Methods

The extent to which solutes, both inorganic and organic, distribute themselves between two immiscible solvents differs enormously from one species to another, and these differences have been exploited for decades to separate chemical species. This section deals with applications of extractions to analytical separations.

29C-1 Theory

Two terms are employed to describe the distribution of a solute between two immiscible solvents: *distribution coefficient* and *distribution ratio.* It is important to have a clear understanding of the distinction between the two.

The Distribution Coefficient

The distribution coefficient is an equilibrium constant that describes the distribution of a solute between two immiscible solvents. For example, when an aqueous solution of an organic solute A is shaken with an organic solvent, such as hexane, there is quickly established an equilibrium that is described by the equation

$$A(aq) \rightleftharpoons A(org) \tag{29-1}$$

where (aq) and (org) refer to the aqueous and organic phases. Ideally, the ratio of the activities of A in the two phases is constant and independent of the total quantity of A. That is, at any given temperature

$$K_d = \frac{[A(org)]}{[A(aq)]} \tag{29-2}$$

where the equilibrium constant K_d is the distribution coefficient. The terms in brackets are strictly the activities of A in the two solvents, but molar concentrations can frequently be substituted without serious error. Often, K_d is approximately equal to the ratio of the solubility of A in the two solvents.

When the solute exists in different states of aggregation in the two solvents, the equilibrium becomes

$$xA_y(aq) \rightleftharpoons yA_x(org)$$

and the distribution coefficient takes the form

$$K_d = \frac{[A_x(org)]^y}{[A_y(aq)]^x}$$

The Distribution Ratio

The distribution ratio D for an analyte is defined as the ratio of its *analytical* concentration in two immiscible solvents. For a simple system, such as that described by Equation 29-1, the distribution ratio is identical to the distribution coefficient. For more complex systems, however, the two are usually quite different. For example, for the distribution of a fatty acid HA between water and diethyl ether, we can write

$$D = \frac{c_{org}}{c_{aq}} \tag{29-3}$$

where c_{org} and c_{aq} are the molar *analytical* concentrations of HA in the two phases. In the aqueous medium, the analytical concentration of the acid is equal to the sum of the equilibrium concentrations of the weak acid and its conjugate base:

$$c_{aq} = [HA(aq)] + [A^-(aq)]$$

In contrast, no significant dissociation of the acid occurs in the nonpolar organic layer so that the analytical and equilibrium concentrations of HA are identical, and we can write

$$c_{org} = [HA(org)]$$

Substituting the last two relationships into Equation 29-3 gives

$$D = \frac{[HA(org)]}{[HA(aq)] + [A^-(aq)]}$$

In order to relate D to K_d for the *species* HA, we substitute the dissociation-constant expression for HA into this equation:

$$D = \frac{[HA(org)]}{[HA(aq)] + [HA(aq)]\,K_a/[H_3O^+(aq)]}$$

where K_a is the acid dissociation constant for HA. Factoring out [HA(aq)] leads to

$$D = \frac{[HA(org)]}{[HA(aq)]} \times \frac{1}{1 + K_a/[H_3O^+]}$$

Substituting K_d for the ratio of the equilibrium HA concentrations gives, after rearranging,

$$D = \frac{K_d[H_3O^+]}{K_a + [H_3O^+]} = \frac{c_{org}}{c_{aq}} \tag{29-4}$$

This equation can be used to compute the extent of extraction of HA from aqueous solutions having various pH values. (Note the difference between the distribution coefficient K_d and the distribution ratio D.)

EXAMPLE 29-1

The distribution coefficient for a weak acid between water and diethyl ether is found to be 800, and its acid dissociation constant in water is 1.50×10^{-5}. Calculate the analytical concentration of HA remaining in an aqueous solution after 50.0 mL of 0.0500 M HA is extracted with 25.0 mL of ether, assuming the aqueous solution is buffered to a pH of (a) 2.00 and (b) 8.00.

(a) Substituting $[H_3O^+] = 1.00 \times 10^{-2}$ and the two equilibrium constants into Equation 29-4 yields

$$D = \frac{c_{org}}{c_{aq}} = \frac{800 \times 1.00 \times 10^{-2}}{1.50 \times 10^{-5} + 1.00 \times 10^{-2}} = 799 \tag{29-5}$$

The total number of millimoles of the acid contained in the two solvents is equal to the original number of millimoles in the aqueous solution:

$$\text{Total no. mmol HA} = 50.0 \times 0.0500 = 2.50$$

After extraction, the 2.50 mmol of HA is distributed between the two solvents so that

$$50.0c_{aq} + 25.0c_{org} = 2.50$$

where c_{aq} and c_{org} are the analytical concentrations in the two solvents. Rearranging Equation 29-5 shows that

$$c_{org} = 799c_{aq}$$

and so we can write

$$50.0c_{aq} + (25.0)(799c_{aq}) = 2.50$$

$$c_{aq} = 1.25 \times 10^{-4}\ \text{M}$$

(b) When $[H_3O^+] = 1.00 \times 10^{-8}$,

$$D = \frac{800 \times 1.00 \times 10^{-8}}{1.50 \times 10^{-5} + 1.00 \times 10^{-8}} = 0.533$$

Proceeding as in part (a), we have

$$50.0 \times c_{aq} + (25.0)(0.533c_{aq}) = 2.50$$

$$c_{aq} = 3.95 \times 10^{-2}\ \text{M}$$

The Completeness of Multiple Extractions

Distribution coefficients and distribution ratios are useful because they provide guidance as to the most efficient way to perform extractive separations. For example, consider again the extraction of a species HA from an aqueous solution having a pH of 2.00 where Equation 29-5 applies. Suppose that V_{aq} milliliters of water containing a_0 millimoles of HA is extracted with V_{org} milliliters of diethyl ether. At equilibrium, a_1 millimoles of HA remain in the aqueous layer and $a_0 - a_1$ millimoles have been transferred to the organic layer. The analytical concentration of HA in each layer is

$$c_{aq1} = \frac{a_1}{V_{aq}}$$

$$c_{org1} = \frac{a_0 - a_1}{V_{org}}$$

Substitution of these quantities into Equation 29-5 and rearrangement gives

$$a_1 = \left(\frac{V_{aq}}{V_{org}D + V_{aq}}\right) a_0 \qquad (29\text{-}6)$$

The number of millimoles a_2 remaining in the aqueous layer after a second extraction with an identical volume of solvent is, by the same reasoning,

$$a_2 = \left(\frac{V_{aq}}{V_{org}D + V_{aq}}\right) a_1$$

When this expression is multiplied by Equation 29-6, a_1 cancels and we obtain

$$a_2 = \left(\frac{V_{aq}}{V_{org}D + V_{aq}}\right)^2 a_0$$

After n extractions, the number of millimoles of HA remaining in the aqueous layer is

$$a_n = \left(\frac{V_{aq}}{V_{org}D + V_{aq}}\right)^n a_0 \qquad (29\text{-}7)$$

Equation 29-7 can be rewritten in terms of the initial and final analytical concentrations of HA in the water by substituting the relationships

$$a_n = (c_{aq})_n V_{aq} \quad \text{and} \quad a_0 = (c_{aq})_0 V_{aq}$$

where $(c_{aq})_n$ is the analytical concentration of HA in the aqueous phase after n extractions. Substitution of these relationships into Equation 29-7 gives

$$(c_{aq})_n = \left(\frac{V_{aq}}{V_{org}D + V_{aq}}\right)^n (c_{aq})_0 \tag{29-8}$$

As shown in the example that follows, a more efficient extraction is achieved with several small volumes of solvent than a single large one.

EXAMPLE 29-2

The distribution ratio of I_2 between CCl_4 and H_2O is 85. Calculate the concentration of I_2 remaining after 50.0 mL of an aqueous 1.00×10^{-3} M solution of I_2 is extracted with (a) one 50.0-mL portion of CCl_4, (b) two 25.0-mL portions, and (c) five 10.0-mL portions.

(a) Substituting into Equation 29-8 gives

$$(c_{aq})_1 = \left(\frac{50.0}{50.0 \times 85 + 50.0}\right)^1 \times 1.00 \times 10^{-3} = 1.16 \times 10^{-5}\ \text{M}$$

(b) $$(c_{aq})_2 = \left(\frac{50.0}{25.0 \times 85 + 50.0}\right)^2 \times 1.00 \times 10^{-3} = 5.28 \times 10^{-7}\ \text{M}$$

(c) $$(c_{aq})_5 = \left(\frac{50.0}{10.0 \times 85 + 50.0}\right)^5 \times 1.00 \times 10^{-3} = 5.29 \times 10^{-10}\ \text{M}$$

Figure 29-1 demonstrates that the improved efficiency brought about by multiple extractions falls off rapidly as the number of extractions increases. Clearly, little is gained by dividing the extracting solvent into more than five or six portions.

29C-2 Types of Extraction Procedures

Three types of separation procedures are based upon distribution equilibria between immiscible solvents: *simple, exhaustive,* and *countercurrent extractions.*

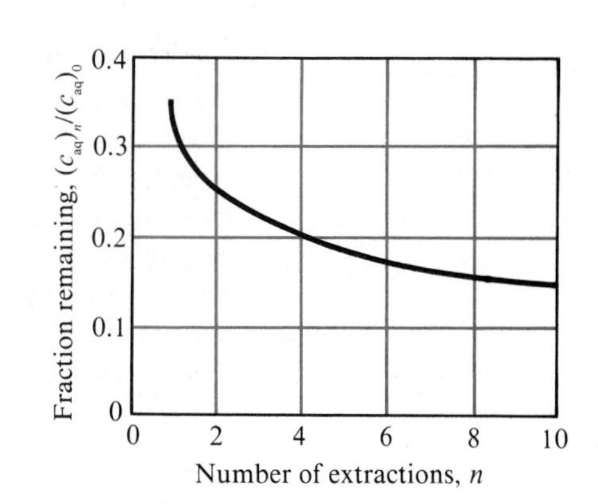

FIGURE 29-1 Plot of Equation 29-8, assuming $K_d = 2$ and $V_{aq} = 100$. The total volume of the organic solvent was also assumed to be 100, so that $V_{org} = 100/n$.

Simple Extractions

When the distribution ratio for one species in a mixture is reasonably favorable (of the order of 5 to 10 or greater) and that for the others is unfavorable (<0.001), an extractive separation can be simple, rapid, and quantitative. The solution containing the analyte is extracted successively with up to six portions of fresh solvent. An ordinary separatory funnel is used with either the original solution or the extract being retained for completing the analysis.

Exhaustive Extractions

Exhaustive extraction permits the separation of components of a mixture that have relatively unfavorable distribution ratios (<1) from those having ratios that approach zero. An apparatus is used in which the organic solvent is automatically distilled, condensed, and caused to pass continuously through the aqueous layer. Thus, the equivalent of several hundred extractions with fresh solvent is accomplished in 1 h or less with equipment that requires no attention.

Countercurrent Fractionation

Automated devices that permit hundreds of automatic successive extractions have been developed. With these instruments, fractionation occurs by a *countercurrent* scheme in which distribution between *fresh* portions of the two phases occurs in a series of discrete steps. An exhaustive extraction differs from the countercurrent technique in that fresh portions of only one phase are introduced in the former.

The countercurrent method permits the separation of components with nearly identical partition ratios. For example, Craig[2] has demonstrated that ten amino acids can be separated by countercurrent extraction even though their partition coefficients differ by less than 0.1.

29D Applications of Extraction Procedures

Extraction is often more attractive than a classic precipitation for separating inorganic species because the equilibration and separation of phases in a separatory funnel are less tedious and time-consuming than precipitation, filtration, and washing. In addition, the problems of coprecipitation and postprecipitation are avoided. Finally, extraction procedures are ideally suited for the isolation of trace quantities of a species.

29D-1 The Extractive Separation of Metal Ions as Chelates

Many organic chelating agents are weak acids that react with metal ions to give uncharged complexes that are highly soluble in organic solvents such as ethers, hydrocarbons, ketones, and chlorinated species (including chloroform and carbon tetrachloride). Distribution ratios for such reagents vary widely among cations and can be controlled by changes in pH and reagent concentration, thus making possible many useful extractive separations.

[2]L. C. Craig, *Anal. Chem.*, **1950**, *22*, 1346.

The Effect of pH and Reagent Concentration on Distribution Ratios

Several chelating agents have proved useful for separations based upon the selective extraction of metal ions from a buffered aqueous solution into a nonaqueous solvent containing these agents. As shown in Figure 29-2, such a process involves several equilibria and several species. Among the latter are the undissociated ligand HL, its conjugate base L^-, the metal-ligand complex ML_n, and metal and hydronium ions. The important equilibria are

$$HL(aq) \rightleftharpoons HL(org) \qquad K_{d1} = \frac{[HL(org)]}{[HL(aq)]} \tag{29-9}$$

$$HL(aq) + H_2O \rightleftharpoons H_3O^+(aq) + L^-(aq) \qquad K_a = \frac{[H_3O^+(aq)][L^-(aq)]}{[HL(aq)]} \tag{29-10}$$

$$M^{n+}(aq) + nL^-(aq) \rightleftharpoons ML_n(aq) \qquad K_f = \frac{[ML_n(aq)]}{[M^{n+}(aq)][L^-(aq)]^n} \tag{29-11}$$

$$ML_n(aq) \rightleftharpoons ML_n(org) \qquad K_{d2} = \frac{[ML_n(org)]}{[ML_n(aq)]} \tag{29-12}$$

Organic chelating agents as well as neutral metal chelates are usually highly soluble in organic liquids so that the distribution coefficients K_{d1} and K_{d2} are generally large numerically. Furthermore, the concentration of M^{n+} in the nonpolar organic layer approaches zero in most cases. The selectivity of the reagent is determined by the relative magnitudes of the formation constants K_f for various cations. As shown by Equation 29-10, the concentration of the active reagent L^- is pH-dependent. Thus, by controlling pH, one can control the concentration of L^- and thus which cations are extracted and which are not.

In order to derive an expression relating the amount of a cation extracted to pH and concentration of chelating agent, we employ the distribution ratio, which for the system shown in Figure 29-2 takes the form

$$D = \frac{c_{org}}{c_{aq}} = \frac{(ML_n(org)]}{[M^{n+}(aq)] + [ML_n(aq)]} \simeq \frac{[ML_n(org)]}{[M^{n+}(aq)]} \tag{29-13}$$

where c_{org} and c_{aq} are the molar analytical concentrations of M^{n+} in the organic and aqueous phases. Ordinarily, the assumption that $[ML_n(aq)] << [M^{n+}(aq)]$ is valid because (1) the chelate is generally not very soluble in water and (2)

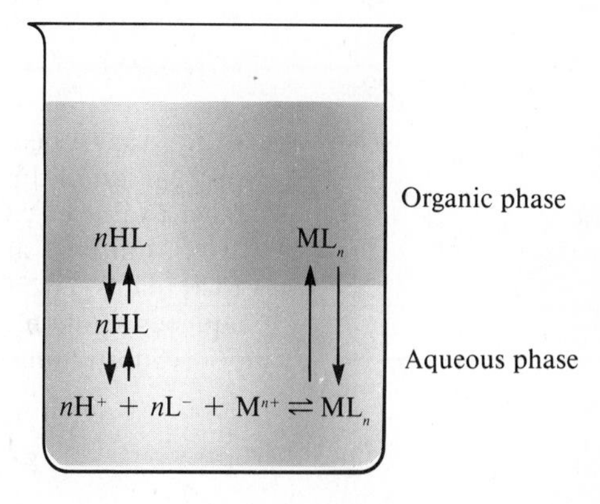

FIGURE 29-2 Equilibria in the extraction of the aqueous cation M^{n+} into an immiscible organic phase containing the organic chelating agent HL.

that which is in solution is largely dissociated. As shown by the following derivation, D is independent of the total amount of metal in the two phases but dependent upon both the concentration of HL in the organic layer and the hydronium ion concentration in the aqueous solution.

If c_L is the original molar concentration of HL in the organic phase, mass balance requires that

$$c_L = [HL(org)] + [HL(aq)] + [L^-(aq)] + n[ML_n(aq)] + n[ML_n(org)]$$

Ordinarily, extractions are carried out with such a large excess of chelating agent that the concentration of the *species* HL in the organic layer far exceeds the concentration of all other species containing L. Thus, the foregoing mass-balance expression simplifies to

$$c_L \simeq [HL(org)] \quad (29\text{-}14)$$

In order to arrive at an expression that relates D for this system to the original concentration of the chelating agent in the organic solution and to the pH of the aqueous solution, let us multiply Equation 29-11 by 29-12 and rearrange, which leads to

$$[ML_n(org)] = K_f K_{d2} [M^{n+}(aq)][L^-(aq)]^n$$

Substituting into Equation 29-13 gives

$$D = \frac{c_{org}}{c_{aq}} = K_f K_{d2} [L^-(aq)]^n \quad (29\text{-}15)$$

Dividing Equation 29-10 by 29-9 allows us to express $[L^-(aq)]$ in terms of $[H_3O^+]$ and $[HL(org)]$:

$$[L^-(aq)] = \frac{K_a}{K_{d1}} \frac{[HL(org)]}{[H_3O^+(aq)]}$$

Substitution of this equation and Equation 29-14 into Equation 29-15 leads to the desired relationship:

$$D = \frac{c_{org}}{c_{aq}} = \frac{K_f K_{d2} K_a^n}{K_{d1}^n} \times \frac{c_L^n}{[H_3O^+(aq)]^n} \quad (29\text{-}16)$$

Combining the four equilibrium constants into a single constant K_{ex} yields

$$D = \frac{c_{org}}{c_{aq}} = \frac{K_{ex} c_L^n}{[H_3O^+(aq)]^n} \quad (29\text{-}17)$$

EXAMPLE 29-3

Lead forms a neutral complex PbL_2 with the ligand L^-. The constant K_{ex} for the distribution of this complex between water and CCl_4 was found by experiment to have a value of 2.0×10^4. A 25.0-mL aliquot of an aqueous solution that is 5.00×10^{-4} M in Pb^{2+} and 0.500 M in $HClO_4$ is extracted with two 10.0-mL portions of CCl_4 that is 0.0250 M in HL. Calculate the percentage of unrecovered Pb^{2+} in the aqueous solution.

Substituting into Equation 29-17 gives

$$D = \frac{(2.0 \times 10^4)(0.0250)^2}{(0.500)^2} = 50.0$$

Substituting into Equation 29-8 for two extractions gives

$$(c_{aq})_2 = \left(\frac{25.0}{10.0 \times 50.0 + 25.0}\right)^2 (5.00 \times 10^{-4}) = 1.13 \times 10^{-6}$$

$$\%\ \text{unextracted Pb}^{2+} = \frac{1.13 \times 10^{-6}}{5.0 \times 10^{-4}} \times 100\% = 0.23\%$$

Some Separations Based on the Extraction of Metal Chelates

Extractions with Diphenylthiocarbazone. Diphenylthiocarbazone, or dithizone, is a useful reagent for separating minute quantities of a dozen or more metal ions. Its reaction with a divalent cation such as Pb^{2+}, can be written as

$$2\ C_6H_5{-}N{=}N{-}C({=}S){-}N(H){-}N(H){-}C_6H_5 + Pb^{2+} \rightleftharpoons \left(C_6H_5{-}N(H){-}N{=}C(S){-}N{=}N{-}C_6H_5\right)_2 Pb + 2H^+$$

green $\qquad\qquad$ red

Both dithizone and its metal chelates are essentially insoluble in water but dissolve readily in solvents such as carbon tetrachloride and chloroform. Solutions of the reagent are deep green, whereas solutions of the metal chelates are intensely red, violet, orange, or yellow and provide a sensitive means for the photometric determination of the separated ions.

The distribution ratio of the various metal dithizonates is described by Equations 29-17 and 29-13. Thus, for lead,

$$D = \frac{c_{org}}{c_{aq}} = \frac{[PbDz_2(org)]}{[Pb^{2+}(aq)]} = \frac{K_{ex}c_{Dz}^2}{[H_3O^+(aq)]^2}$$

Figure 29-3 shows how the percentages of lead and other metal ions extracted into a chloroform solution of dithizone change as the pH of the aqueous medium changes. The data for these curves were derived in a way analogous

FIGURE 29-3 Effect of pH on the extraction of cations with CCl_4 solutions of dithizone.

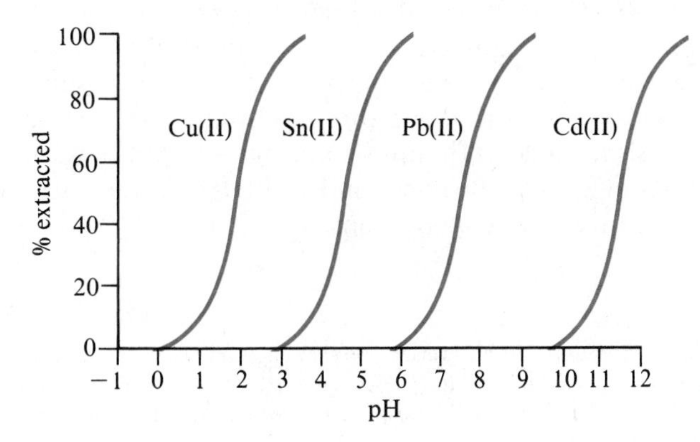

to the calculation shown in Example 29-3. It is apparent from these curves that clean separations of copper, lead, and cadmium can be achieved by pH control. Depending upon its absorption spectrum, tin(II) may interfere slightly in the determination of copper and lead.

Extractions with 8-Hydroxyquinoline. Many of the chelates of 8-hydroxyquinoline (Section 3D-3) are readily extracted into various organic solvents. The extraction process can be formulated as

$$2(HQ)_{org} + (M^{2+})_{aq} \rightleftharpoons (MQ_2)_{org} + 2(H^+)_{aq}$$

where HQ symbolizes the chelating agent. The equilibrium is clearly pH-dependent, which permits the separation of metals through control of the aqueous-phase pH. The method has proved particularly useful for the separation of trace amounts of metals.

Extractions with Other Chelating Agents. Separations that make use of other organic chelating agents are described in several reference works.[3]

29D-2 The Extraction of Metal Chlorides

The data in Table 29-3 indicate that a substantial number of metal chlorides can be extracted into diethyl ether from 6 M hydrochloric acid solution; equally important, a large number of metal ions are either unaffected or extracted only slightly under these conditions. Thus, many useful separations are possible. One of the most important is the separation of iron(III) (99% extracted) from a host of other cations. For example, the greater part of the iron from steel or iron ore samples can be removed by extraction prior to analysis for such trace elements as chromium, aluminum, titanium, and nickel. The species extracted has been shown to be the ion pair $H_3O^+FeCl_4^-$. It has also been shown that the percentage of iron transferred to the organic phase is dependent upon the hydrochloric acid content of the aqueous phase (little is removed from solutions that are below 3 M and above 9 M HCl) and, to some extent, upon the iron content. Unless special precautions are taken, extraction of the last traces of iron is incomplete.

Methyl isobutyl ketone also extracts iron from hydrochloric acid solutions. It is reported to have a somewhat more favorable distribution ratio than diethyl ether and has the added advantage of being less flammable.

29D-3 The Extraction of Nitrates

Certain nitrate salts are selectively extracted by diethyl ether as well as other organic solvents. For example, uranium(VI) is conveniently separated from such elements as lead and thorium by ether extraction of an aqueous solution that is saturated with ammonium nitrate and has a nitric acid concentration of about 1.5 M. Bismuth and iron(III) nitrates are also extracted to some extent under these conditions.

[3]G. H. Morrison and H. Freiser, *Solvent Extraction in Analytical Chemistry*. New York: Wiley, 1957; A. K. De, S. M. Khophar, and R. A. Chalmers, *Solvent Extraction of Metals*. New York: Van Nostrand, 1970; E. B. Sandell and H. Onishi, *Colorimetric Determination of Traces of Metals*, 4th ed. New York: Interscience, 1978.

TABLE 29-3

Diethyl Ether Extraction of Various Chlorides from 6 M Hydrochloric Acid*

Percent Extracted	Elements and Oxidation State
90–100	Fe(III), 99%; Sb(V), 99%**; Ga(III), 97%; Ti(III), 95%**; Au(III), 95%
50–90	Mo(VI), 80–90%; As(III), 80%**†; Ge(IV), 40–60%
1–50	Te(IV), 34%; Sn(II), 15–30%; Sn(IV), 17%; Ir(IV), 5%; Sb(III), 2.5%**
<1 >0	As(V),** Cu(II), In(III), Hg(II), Pt(IV), Se(IV), V(V), V(IV), Zn(II)
0	Al(III), Bi(III), Cd(II), Cr(III), Co(II), Be(II), Fe(II), Pb(II), Mn(II), Ni(II), Os(VIII), Pd(II), Rh(III), Ag(I), Th(IV), Ti(IV), W(VI), Zr(IV)

*From E. H. Swift, *Introductory Quantitative Analysis*, p. 431. Englewood Cliffs, NJ: Prentice-Hall, 1950. With permission.

**Isopropyl ether employed rather than diethyl ether.

†8 M HCl rather than 6 M.

29E Ion-Exchange Separations

Ion-exchange resins have several applications in analytical chemistry. The most important is in high-performance liquid chromatography, described in Section 25D. A brief description of other analytical applications of these useful materials follows.

29E-1 The Separation of Interfering Ions of Opposite Charge

Ion-exchange resins are useful for the removal of interfering ions, particularly where these ions have a charge opposite that of the analyte. For example, iron(III), aluminum(III), and other cations interfere in the gravimetric determination of sulfate by virtue of their tendency to coprecipitate with barium sulfate. Passage of a solution to be analyzed through a column containing a cation-exchange resin results in the retention of all the cations and the liberation of a corresponding number of protons. The sulfate ion passes freely through such a column, however, and the analysis can then be performed on the effluent. In a similar manner, phosphate ion, which interferes in the determination of barium and calcium ions, can be removed by passing the sample through an anion-exchange resin.

29E-2 The Concentration of Traces of an Electrolyte

A useful application of ion exchangers is the concentration of traces of an ion from a very dilute solution. Cation-exchange resins, for example, have been employed to collect traces of metallic elements from large volumes of natural waters. The ions are then liberated by treating the resin with acid; the result is a considerably more concentrated solution for analysis.

29E-3 The Conversion of Salts to Acids or Bases

The total salt content of a sample can be determined by titrating the hydrogen ions released as an aliquot of sample is washed through a cation exchanger

TABLE 29-4

Separation of Some Inorganic Species by Distillation

Analyte	Sample Treatment	Volatile Species	Method of Collection
CO_3^{2-}	Acidification	CO_2	$Ba(OH)_2(aq) + CO_2(g) \rightarrow BaCO_3(s) + H_2O$ or on Ascarite
SO_3^{2-}	Acidification	SO_2	$SO_2(g) + H_2O_2(aq) \rightarrow H_2SO_4(aq)$
S^{2-}	Acidification	H_2S	$Cd^{2+}(aq) + H_2S(g) \rightarrow CdS(s) + 2H^+$
F^-	Addition of SiO_2 and acidification	H_2SiF_6	Basic solution
Si	Addition of HF	SiF_4	Basic solution
H_3BO_3	Addition of H_2SO_4 and methanol	$B(OCH_3)_3$	Basic solution
$Cr_2O_7^{2-}$	Addition of concd HCl	CrO_2Cl_2	Basic solution
NH_4^+	Addition of NaOH	NH_3	Acidic solution
As, Sb	Addition of concd HCl and H_2SO_4	$AsCl_3$ $SbCl_3$	Water
Sn	Addition of HBr	$SnBr_4$	Water

in its acidic form. Similarly, a standard hydrochloric acid solution can be prepared by passing a solution containing a known weight of sodium chloride through a cation-exchange resin in its acidic form. The effluent and washings are collected in a volumetric flask and diluted to volume. In an analogous way, standard sodium hydroxide solutions can be prepared by treatment of an anion-exchange resin with a known quantity of sodium chloride.

29F *The Separation of Inorganic Species by Distillation*

Distillation permits the separation of components whose solution/vapor-phase distribution ratios differ significantly from one another. If one species has a distribution ratio that is large compared with those of the other components of the mixture, the separation is simple. Table 29-4 lists some inorganic species that can be conveniently separated by simple distillation.

Questions and Problems

*29-1 What is a masking agent and how does it function?

29-2 Differentiate between
 (a) an exhaustive extraction and a countercurrent extraction.
 (b) a distribution coefficient and a distribution ratio.

*29-3 What is a collector and when is it used?

29-4 Suggest a precipitation method for separating
 *(a) Fe(III) from Al(III).
 (b) Fe(III) from Cu(II).
 *(c) Fe(III) from Sn(IV).
 (d) As(III) from Cd(II).
 *(e) Hg(II) from Mn(II), Zn(II), and Fe(III).

*29-5 The distribution coefficient for X between chloroform and water is 9.6. Calculate the concentration of X remaining in the aqueous phase after 50.0 mL of 0.150 M X is treated by extraction with the following quantities of chloroform: (a) one 40.0-mL portion, (b) two 20.0-mL portions, (c) four 10.0-mL portions, and (d) eight 5.00-mL portions.

29-6 The distribution coefficient for Z between *n*-hexane and water is 6.25. Calculate the percent of Z remaining in 25.0 mL of water that was originally 0.0600 M in Z after extraction with the following volumes of *n*-hexane: (a) one 25.0-mL portion, (b) two 12.5-mL portions, (c) five 5.00-mL portions, and (d) ten 2.50-mL portions.

*29-7 What volume of $CHCl_3$ is required to decrease the concentration of X in Problem 29-5 to 1.00×10^{-4} M if 25.0 mL of 0.0500 M X is extracted with (a) 25.0-mL portions of $CHCl_3$, (b) 10.0-mL portions of $CHCl_3$, and (c) 2.0-mL portions of $CHCl_3$?

29-8 What volume of *n*-hexane is required to decrease the concentration of Z in Problem 29-6 to 1.00×10^{-5} M if 40.0 mL of 0.0200 M Z is extracted with (a) 50.0-mL portions of *n*-hexane, (b) 25.0-mL portions, (c) 10.0-mL portions?

*29-9 What is the minimum distribution coefficient that permits removal of 99% of a solute from 50.0 mL of water with (a) two 25.0-mL extractions with benzene and (b) five 10.0-mL extractions with benzene?

29-10 If 30.0 mL of water that is 0.0500 M in Q is to be extracted with four 10.0-mL portions of an immiscible organic solvent, what is the minimum distribution coefficient that allows transfer of all but the following percentages of the solute to the organic layer: *(a) 1.00×10^{-4}, (b) 1.00×10^{-2}, (c) 1.00×10^{-3}?

*29-11 A 0.150 M aqueous solution of the weak organic acid HA was prepared from the pure compound, and three 50.0-mL aliquots were transferred to 100-mL volumetric flasks. Solution 1 was diluted to 100 mL with 1.0 M $HClO_4$, solution 2 was diluted to the mark with 1.0 M NaOH, and solution 3 was diluted to the mark with water. A 25.0-mL aliquot of each was extracted with 25.0-mL of *n*-hexane. The extract from solution 2 contained no detectable trace of A-containing species, indicating that A^- is not soluble in the organic solvent. The extract from solution 1 contained no ClO_4^- or $HClO_4$ but was found to be 0.0454 M in HA (by extraction with standard NaOH and back-titration with standard HCl). The extract from solution 3 was found to be 0.0225 M in HA. Assume that HA does not associate or dissociate in the organic solvent, and calculate

(a) the distribution ratio for HA between the two solvents.
(b) the concentration of the *species* HA and A^- in aqueous solution 3 after extraction.
(c) the dissociation constant of HA in water.

29-12 To determine the equilibrium constant for the reaction

$$I_2 + 2SCN^- \rightleftharpoons I(SCN)_2^- + I^-$$

25.0 mL of a 0.0100 M aqueous solution of I_2 was extracted with 10.0 mL of CCl_4. After extraction, spectrophotometric measurements revealed that the I_2 concentration *of the aqueous layer* was 1.12×10^{-4} M. An aqueous solution that was 0.0100 M in I_2 and 0.100 M in KSCN was then prepared. After extraction of 25.0 mL of this solution with 10.0 mL of CCl_4, the concentration of I_2 *in the CCl_4 layer* was found from spectrophotometric measurement to be 1.02×10^{-3} M.

(a) What is the distribution coefficient for I_2 between CCl_4 and H_2O?
(b) What is the formation constant for $I(SCN)_2^-$?

*29-13 The total cation content of natural water is often determined by exchanging the cations for hydrogen ions on a strong-acid ion-exchange resin. A 25.0-mL sample of a natural water was diluted to 100 mL with distilled water, and 2.0 g of a cation-exchange resin was added. After stirring, the mixture was filtered and the solid remaining on the filter paper was washed with three 15.0-mL portions of water. The filtrate and washings required 15.3 mL of 0.0202 M NaOH to give a bromocresol green end point.

(a) Calculate the number of milliequivalents of cation present in exactly 1 L of sample. (Here, the equivalent weight of a cation is its formula weight divided by its charge.)
(b) Report the results in terms of milligrams of $CaCO_3$ per liter.

29-14 An organic acid was isolated and purified by recrystallization of its barium salt. To determine the equivalent weight of the acid, a 0.393-g sample of the salt was dissolved in about 100 mL of water. The solution was passed through a strong-acid ion-exchange resin, and the column was then washed with water; the eluate and washings were titrated with 18.1 mL of 0.1006 M NaOH to a phenolphthalein end point.

(a) Calculate the equivalent weight of the organic acid.
(b) A potentiometric titration curve of the solution resulting when a second sample was treated in the same way revealed two end points, one at pH 5 and the other at pH 9. What is the molecular weight of the acid?

*29-15 Describe the preparation of exactly 2 L of 0.1500 M HCl from primary-standard-grade NaCl using a cation-exchange resin.

29-16 An aqueous solution containing $MgCl_2$ and HCl was analyzed by first titrating a 25.00-mL aliquot to a bromocresol green end point with 18.96 mL of 0.02762 M NaOH. A 10.00-mL aliquot was then diluted to 50.00 mL with distilled water and passed through a strong-acid ion-exchange resin. The eluate and washings required 36.54 mL of the NaOH solution to reach the same end point. Report the molar concentrations of HCl and $MgCl_2$ in the sample.

*29-17 Copper(II) reacts with the chelating agent H_2L to give a complex CuL_2 that is readily soluble in $CHCl_3$. A spectrophotometric study revealed that when a 1.00×10^{-4} aqueous solution of copper(II) was extracted with $CHCl_3$ that was 0.0100 M in H_2L, the analytical concentration of copper in the two phases was identical at pH 5.65.

(a) Write equations describing the equilibria in the system, assuming that dissociation of CuL_2 in the organic phase is negligible.
(b) Calculate K_{ex}.
(c) Calculate the distribution ratio for the system at pH 6.00.
(d) If 50.0 mL of 5.00×10^{-5} M Cu^{2+} in a pH 6.00 buffer were to be extracted with 25.0 mL portions of 0.0100 M H_2L in $CHCl_3$, how many extractions would be required to remove 99% of the copper from the aqueous phase?
(e) Repeat the calculations in part (d) for 99.9% removal.

29-18 Exactly 25.0 mL of a standard solution that was 2.24×10^{-4} M in Ag^+ and buffered to pH 4.30 was extracted with 10.0 mL of a *n*-hexane solution that was 0.025 M in the chelating agent H_2Z (the extracted complex has the formula Ag_2Z). After the phases were separated, the silver concentration of the aqueous phase was found to be 5.41×10^{-6} M.

(a) Calculate the distribution ratio for the system.
(b) Calculate K_{ex} for the system, assuming that the organic phase contained no uncomplexed Ag^+.
(c) How many 10.0-mL extractions would be required to extract 99.5% of the original Ag^+ at pH 4.30?
(d) Repeat the calculations in part (c) for a solution buffered to pH 3.00 and having a silver ion concentration of 5.15×10^{-4} M.

CHAPTER 30

The Chemicals, Apparatus, and Unit Operations of Analytical Chemistry

This chapter is concerned with the practical aspects of the unit operations encountered in an analytical laboratory as well as with the apparatus and chemicals used in these operations.

30A The Selection and Handling of Reagents and Other Chemicals

The purity of reagents has an important bearing upon the accuracy attained in any analysis. It is therefore essential that the quality of a reagent be consistent with the use for which it is intended.

30A-1 The Classification of Commercial Chemicals

Technical Grade

Chemicals labeled *technical grade* are of indeterminate quality and should be used only where purity is not important. A cleaning solution can be prepared from technical-grade potassium dichromate and sulfuric acid. Similarly, technical-grade calcium chloride is adequate for use in a desiccator. These are isolated examples, however, and in general technical-grade chemicals are not appropriate in an analytical laboratory.

USP Grade

Tolerances for USP chemicals are set forth in the *United States Pharmacopoeia.*[1] The specifications are designed to control the amount of contaminants that might be dangerous to health; it is thus conceivable that USP chemicals can contain significant impurities that are not considered to be physiological hazards.

Reagent Grade

Reagent-grade chemicals conform to the minimum standards set forth by the Reagent Chemical Committee of the American Chemical Society[2] and are used wherever possible in analytical work. Some suppliers label their products with the maximum limits of impurity allowed by the ACS specifications, whereas others print actual assays for the various impurities.

[1] U.S. Pharmacopeial Convention, *Pharmacopoeia of the United States of America,* 21st rev. ed. Rockville, MD, 1985.

[2] Committee on Analytical Reagents, *Reagent Chemicals,* 7th ed. Washington, D.C.: American Chemical Society, 1986.

Primary-Standard Grade

The qualities required of a *primary standard*—in addition to extraordinary purity—are set forth in Section 4A-3. Primary-standard reagents, which are available from commercial sources, have been carefully analyzed, and the assay is printed on the container label. The National Bureau of Standards is an excellent source for primary standards. This agency also provides *reference standards,* which are complex substances that have been exhaustively analyzed.[3]

Special-Purpose Reagent Chemicals

Chemicals that have been prepared for a specific application are also available. Included among these are solvents for spectrophotometry and high-performance liquid chromatography and reagents for nonaqueous spectroscopy and electron microscopy. Information pertinent to the intended use is supplied with these reagents. Data provided with a spectrophotometric solvent, for example, might include its absorbance at selected wavelengths and its ultraviolet cutoff wavelength as well as its assay.

30A-2 Rules for Handling Reagents and Solutions

High quality in a chemical analysis requires the availability of reagents and solutions with established purity. A freshly opened bottle of a reagent-grade chemical can ordinarily be used with confidence; whether this same confidence is justified when the bottle is half empty depends entirely on the way it has been handled after being opened. The following rules should be observed to prevent the accidental contamination of reagents and solutions.

1. Select the best grade of chemical available for analytical work. If a choice exists, pick the smallest bottle that will supply the desired quantity.
2. Replace the top of every container *immediately* after the removal of the reagent; do not rely on someone else to do this.
3. Hold the stoppers of reagent bottles between your fingers; never set a stopper on a desk top.
4. *Unless specifically directed to the contrary, never return any excess reagent to a bottle.* The minor financial saving that might accompany the return of an excess is overshadowed by the risk of contaminating the entire bottle.
5. Unless directed otherwise, never insert spatulas, spoons, or knives into a bottle that contains a solid chemical. Instead, shake the capped bottle vigorously or tap it gently against a wooden table to break up any encrustation; then pour out the desired quantity. These measures are occasionally ineffective, and in such cases a clean porcelain spoon should be used.
6. Keep the reagent shelf and the laboratory balance clean and neat. Clean up any spillages immediately, even though someone else is waiting to use the same chemical or reagent.

[3]United States Department of Commerce, *NBS Standard Reference Materials Catalog, 1986–87, NBS Special Publication 260.* Washington, D.C. 20234: Government Printing Office, 1986.

30B

The Cleaning and Marking of Laboratory Ware

Because a chemical analysis is ordinarily performed in duplicate or triplicate, each vessel that holds a sample must be marked so that its contents can always be identified. Flasks, beakers, and some crucibles have small etched areas on which semipermanent markings can be made with a pencil. Special marking inks are available for porcelain surfaces. The marking is baked permanently into the glaze by heating at a high temperature. A saturated solution of iron(III) chloride, while not as satisfactory as the commercial preparation, can also be used for marking.

Every beaker, flask, or crucible that will contain the sample must be thoroughly cleaned before being used. The apparatus should be washed with a hot detergent solution and then rinsed—initially with copious amounts of tap water and finally with several small portions of distilled or deionized water. A properly cleaned object will be coated with a uniform and unbroken film of water. *It is seldom necessary to dry the interior surface of glassware before use;* drying is ordinarily a waste of time at best and a potential source of contamination at worst.

If a grease film persists after thorough cleaning with detergent, a cleaning solution of potassium dichromate in concentrated sulfuric acid may be needed. Extensive rinsing is required after use of this preparation because dichromate ions adhere persistently to glass and porcelain surfaces; a soak in an EDTA solution may be needed to eliminate the last traces. Cleaning solution is most effective when warmed to about 70°C; *at this temperature, it vigorously attacks plant and animal matter and is thus a potentially dangerous preparation.* Any spillages should be promptly diluted with copious volumes of water.

Preparation of Cleaning Solution. Mix 10 to 15 g of $K_2Cr_2O_7$ or Na_2CrO_4 with about 15 mL of H_2O in a 500-mL conical flask. Add concentrated H_2SO_4 *slowly and in small increments;* swirl the flask thoroughly between additions. The contents will become a semisolid red mass; add just enough H_2SO_4 to dissolve this mass. Allow the solution to cool and then transfer it to a glass-stoppered storage bottle. Cleaning solution can be reused until it acquires the green color of Cr(III), at which time it should be discarded; consult with the instructor concerning disposal. *CAUTION: cleaning solution is highly corrosive and must be used with extreme care.*

Chemical suppliers now market alternatives to dichromate cleaning solutions.

30C

The Evaporation of Liquids

It is frequently necessary to decrease the volume of a solution without loss of a nonvolatile solute. Figure 30-1 illustrates how this operation is performed. The ribbed cover glass permits vapors to escape and protects the remaining solution from accidental contamination. Less satisfactory is the use of glass hooks to provide space between the rim of the beaker and a conventional cover glass.

Evaporation is frequently difficult to control, owing to the tendency of

FIGURE 30-1 Arrangement for the evaporation of a liquid.

some solutions to overheat locally. The bumping that results can be sufficiently vigorous to cause partial loss of the solution. The danger of such loss is minimized through careful and gentle heating. Glass beads are also helpful in minimizing bumping.

Some unwanted species can be eliminated during evaporation. For example, chloride and nitrate can be removed from a solution by adding sulfuric acid and evaporating until copious white fumes of sulfur trioxide are observed (this operation must be performed in a hood). Nitrate ion and nitrogen oxides can be eliminated from acidic solutions by adding urea, evaporating to dryness, and gently igniting the residue. The removal of ammonium chloride is best accomplished by adding concentrated nitric acid and evaporating the solution to a small volume. Ammonium ion is rapidly oxidized upon heating; the solution is then evaporated to dryness.

Organic constituents can frequently be eliminated from a solution by adding sulfuric acid and heating to the appearance of sulfur trioxide fumes (hood); this process is known as *wet-ashing*. Nitric acid can be added toward the end of heating to hasten the oxidation of the last traces of organic matter.

30D The Measurement of Mass

Highly reliable weighing data are ordinarily required at one stage or another in the course of a chemical analysis; an *analytical balance* is used to acquire these data. A less precise but more rugged *laboratory balance* suffices for weighings where the demands for reliability are not critical.

30D-1 The Distinction Between Mass and Weight

The reader should appreciate the difference between mass and weight. *Mass*, which is the more fundamental quantity, is an invariant measure of the amount of matter in an object. *Weight* is the force of attraction that exists between an object and its surroundings, principally the earth. Because gravitational attraction varies with geographical location, the weight of an object depends upon where it is weighed. For example, the weight of a crucible is less in Denver than in Atlantic City (both cities are at approximately the same latitude) because the attractive force between crucible and earth is smaller at the higher altitude. Similarly, the crucible weighs more in Seattle than in Panama (both cities are at sea level) because the earth is somewhat flattened at the poles, and the force of attraction increases measurably with latitude. The mass of the crucible remains constant regardless of the location at which it is measured.

Weight and mass are related by the familiar expression

$$W = Mg$$

where W is the weight of an object, M is its mass, and g is the acceleration due to gravity.

A chemical analysis is always based upon mass in order to free the results from a dependence upon locality. A balance is used to compare the weight of an object with the weight of a set of standard masses. Because g affects both unknown and known identically, an equality between their weights indicates an equality in mass.

The distinction between mass and weight tends to be lost in common usage; that is, the operation of comparing masses is ordinarily referred to as *weighing,* and the objects of known mass as well as the results of weighing are called *weights.* It should always be borne in mind, however, that analytical data are based upon mass rather than weight.

30D-2 Types of Analytical Balances

By definition, an *analytical balance* is a weighing instrument that has a maximum capacity that ranges from 1 g to a few kilograms with a precision of at least 1 part in 10^5 at maximum capacity. The precision and accuracy of many modern analytical balances exceed one part in 10^6 at full capacity.

The most commonly encountered analytical balances have a maximum capacity ranging between 160 and 200 g; measurements can be made with a standard deviation of ± 0.1 mg. *Semimicroanalytical balances* have a maximum loading of 10 to 30 g with a precision of ± 0.01 mg. Typical *microanalytical balances* have a capacity of 1 to 3 g and a precision of ± 0.001 mg.

The analytical balance has undergone a dramatic evolution over the past several decades. The traditional analytical balance had two pans attached to either end of a light-weight beam that pivoted about a knife edge located in the center of the beam. The object to be weighed was placed on one pan; sufficient standard weights were then added to the other pan to restore the beam to its original position. Weighing with such an *equal-arm* balance was tedious and time consuming.

The first *single-pan analytical balance* appeared on the market in 1946. The speed and convenience of weighing with this balance were vastly superior to what could be realized with the traditional equal-arm balance. Consequently, the single-pan balance rapidly replaced the latter in most laboratories. The single-pan balance is still extensively used. The design and operation of a single-pan balance are discussed in Section 30D-3.

The *hybrid electronic analytical balance* dates from the early 1970's. This balance retained the beam and knife edge of a single-pan balance but employed a solenoid in place of standard weights to restore the beam to its original position. The current in a solenoid is proportional to the electromagnetic force it creates and is thus proportional to the weight of the object on the opposite end of the beam as well.

The brief era of the hybrid balance ended with the appearance of the true *electronic analytical balance,* which has neither a beam nor a knife edge; this type of balance is discussed briefly in Section 30D-4. It seems likely that both mechanical and hybrid balances will be supplanted by electronic balances in the years to come.

30D-3 The Single-Pan Mechanical Analytical Balance

Components

Although they differ considerably in appearance and performance characteristics, all mechanical balances—equal-arm as well as single-pan—have several common components. Figure 30-2 is a diagram of a typical single-pan balance. Fundamental to this instrument is a light-weight *beam* that is supported on a planar surface by a prism-shaped knife edge labeled A in the figure. Attached to the left end of the beam is a pan for holding the object to be weighed and a full set of weights held in place by hangers. These weights can be lifted from the beam one at a time by a mechanical arrangement that is controlled by a set of knobs on the exterior of the balance case. The right end of the beam holds a counterweight of such a size as to just balance the pan and weights on the left end of the beam.

A second knife edge (B) is located near the left end of the beam and serves to support a second planar surface, which is located in the inner side of a stirrup that couples the pan to the beam. The two knife edges and their planar surfaces are fabricated from extraordinarily hard materials (agate or synthetic sapphire) and form two bearings that permit motion of the beam and pan with a very minimum of friction. The performance of a mechanical balance is critically dependent upon the perfection of these two bearings.

To protect the bearings from damage and wear when an object is placed on the pan or when the balance is not being used, balances are generally equipped with a *beam arrest* and a *pan arrest.* The beam arrest is a mechanical device that raises the beam so that the central knife edge no longer touches its bearing surface and simultaneously frees the stirrup from contact with the outer knife edge. The purpose of the arrest mechanism is to prevent damage

FIGURE 30-2 Modern single-pan analytical balance. (From R. M. Schoonover, *Anal. Chem.*, **1982,** *54,* 973A. Published 1982 American Chemical Society.)

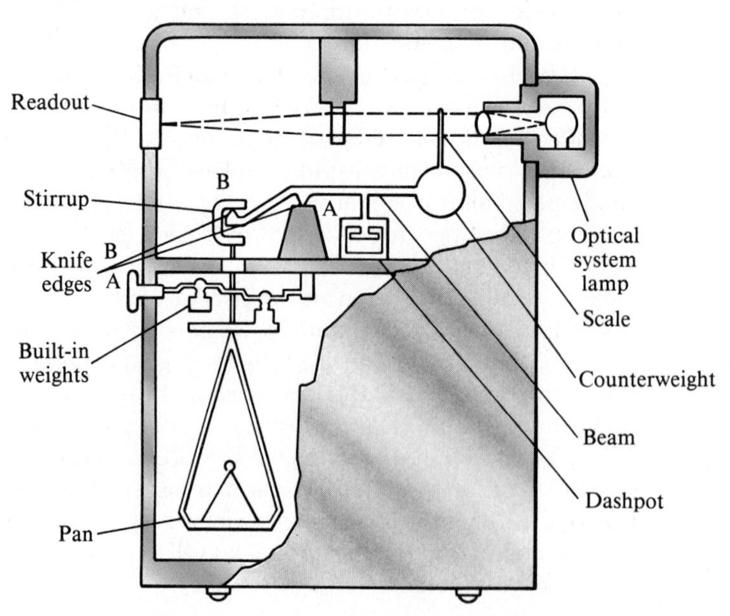

to the bearings of the balance while objects are being placed upon or removed from the pan. When engaged, the pan arrest supports most of the weight of the pan and its contents and thus prevents oscillation. Both arrests are controlled by a lever mounted on the outside of the balance case and should be engaged whenever the balance is not in use.

An *air damper* (sometimes called a *dashpot*) is mounted near the end of the beam opposite the pan. This device consists of a piston that moves within a concentric cylinder attached to the balance case. Air in the cylinder undergoes expansion and contraction as the beam is set in motion; the beam rapidly comes to rest as a result of this opposition to motion.

Protection from air currents is needed to permit discrimination between small differences in weight (<1 mg). An analytical balance is thus always enclosed in a case equipped with doors to permit the introduction or removal of objects.

Weighing with a Single-Pan Balance

With no object on the pan and all of the weights in place, the beam of a properly adjusted balance assumes an essentially horizontal position when the pan and beam arrests are removed so the beam is free to rotate around knife edge *A*. Placement of an object on the pan causes the left end of the beam to be displaced downward. Weights are then removed systematically one by one from the beam until the imbalance is less than 100 mg. The angle of deflection of the beam with respect to its original horizontal position is directly proportional to the milligrams of additional weight that must be removed to restore the beam to its original horizontal position. The optical system shown in the upper part of Figure 30-2 measures this angle of deflection and converts this angle to milligrams. A *reticle,* which is a small transparent screen mounted on the beam, is scribed with a scale that reads 0 to 100 mg. A beam of light passes through the scale to an enlarging lens, which in turn focuses a small part of the enlarged scale onto a frosted glass plate located on the front of the balance. A vernier makes it possible to read this scale to the nearest 0.1 mg.

Stability

An important property of a balance is *stability,* a property that causes the beam of the balance to return to its rest position after being displaced by the momentary application of a slight force to one end. Stability requires the center of gravity of the beam (including the stirrup, pan, and any load) to be below the central knife edge so that the weight of the beam acts as a restoring force to offset the displacement.

Instructions for the Use of a Single-Pan Balance

1. Zero the empty balance. Rotate the arrest control to the full release position. Adjust the vernier so that the scale reading is zero.
2. Arrest the balance again, and place the object to be weighed on the pan. Turn the arrest knob to the partially released position. Rotate the knob controlling the heaviest likely weight for the object until the illuminated scale changes or the instruction "remove weight" appears; then turn the weight knob back one stop. Repeat this procedure with all the other knobs, working systematically through the lighter weights.

3. Turn the arrest control to its fully released position and allow time for the balance to come to equilibrium. The weight of the object is the sum of the weights indicated on the dials or that displayed on the illuminated scale. Use the vernier to establish the weight to the nearest 0.1 mg.

Note. Directions for operation differ somewhat with make and model. Consult the instructor for any modifications to this procedure that may be required.

Summary of Rules for the Use of a Single-Pan Analytical Balance

The single-pan mechanical analytical balance is a delicate instrument. Care must be taken to adhere to the accompanying rules.

To Minimize Wear or to Avoid Damage

1. Be certain that the arresting mechanisms for the beam are engaged whenever the loading is being changed and whenever the balance is not in use.
2. Center the load on the pan insofar as possible.
3. Protect the balance from corrosion. Objects to be placed on the pan should be limited to nonreactive metals, nonreactive plastics, and vitreous materials.
4. Observe special precautions (Section 30E-6) for the weighing of liquids.
5. Consult with the instructor if the balance appears to need adjustment.
6. Keep the balance and its case scrupulously clean. A camel's-hair brush is useful for the removal of spilled material or dust.

To Obtain Reliable Weighing Data

7. Always allow an object that has been heated to return to room temperature before weighing it.
8. Use tongs or finger pads to prevent the uptake of moisture by dried objects.

30D-4 *The Electronic Analytical Balance*[4]

Figure 30-3 is a diagram of an electronic analytical balance. The pan rides above a hollow metal cylinder that is surrounded by a coil and fits over the inner pole of a cylindrical permanent magnet. An electric current in the coil creates a magnetic field that supports or levitates the cylinder, the pan and indicator arm, and whatever load is on the pan. The current is adjusted so that the level of the indicator arm is in the null position when the pan is empty. Placing an object on the pan causes the pan and indicator arm to move downward, which increases the amount of light striking the photocell of the null detector. The increased current from the photocell is amplified and fed into the coil, creating a larger magnetic field, which returns the pan to its original null position. A device such as this, in which a small electric current causes a mechanical system to maintain a null position, is called a *servo system*. The current required to keep the pan and object in the null position is directly

[4]For a more detailed discussion, see R. M. Schoonover, *Anal. Chem.*, **1982,** *54,* 973A; K. M. Lang, *Amer. Lab.*, **1983,** *15* (3), 72.

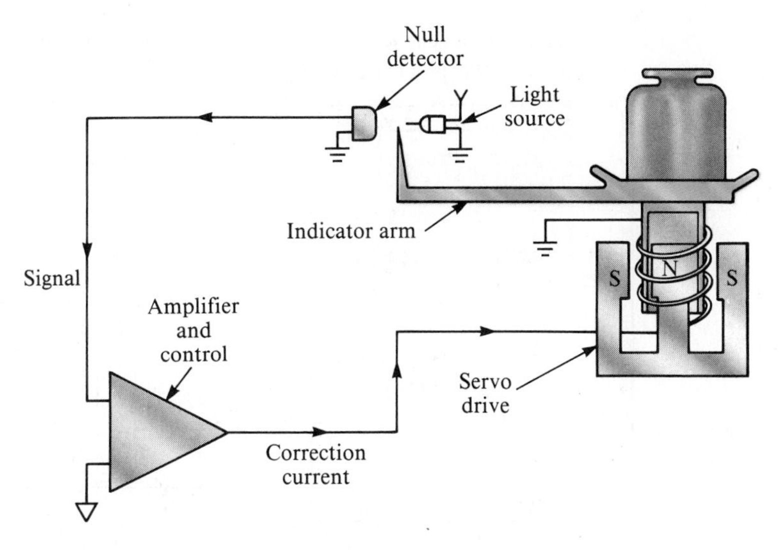

FIGURE 30-3 Electronic analytical balance. (From R. M. Schoonover, *Anal. Chem.*, 1982, *54*, 974A. Published 1982 American Chemical Society.)

proportional to the weight of the object and is readily measured, digitized, and displayed. The calibration of an electronic balance involves the use of a standard mass and adjustment of the current so that the mass of the standard is exhibited on the display.

Figure 30-4 shows the configurations for two electronic analytical balances. In each, the pan is tethered to a system of constraints known collectively as a *cell.* The cell incorporates several *flexures* that permit limited movement of the pan and prevent torsional forces (resulting from off-center loading) from disturbing the alignment of the balance mechanism. At null, the beam is parallel to the gravitational horizon and each flexure pivot is in a relaxed position.

Figure 30-4a shows an electronic balance with the pan located below the cell. Higher precision is achieved with this arrangement than with the top-loading design shown in Figure 30-4b. Even so, top-loading electronic balances have a precision that equals or exceeds that of the best mechanical balances and additionally provide unencumbered access to the pan.

Electronic balances generally feature an automatic *taring control* that causes the display to read zero with a container (such as a boat or weighing bottle) on the pan. Most balances permit taring to 100% of capacity.

Some electronic balances have dual capacities and dual precisions. These features permit the capacity to be decreased from that of a macrobalance to that of a semimicrobalance (30 g) with a concomitant gain in precision to 0.01 mg. Thus, the chemist has effectively two balances in one.

A modern electronic analytical balance provides unprecedented speed and ease of use. For example, one instrument is controlled by touching a single bar at various positions along its length. One position on the bar turns the instrument on or off, another automatically calibrates the balance against a standard mass, and a third zeros the display, either with or without an object on the pan. Reliable weighing data are obtainable with little or no instruction.

A few disadvantages attend the use of an electronic balance. Problems are sometimes encountered in the weighing of ferromagnetic materials. Strong

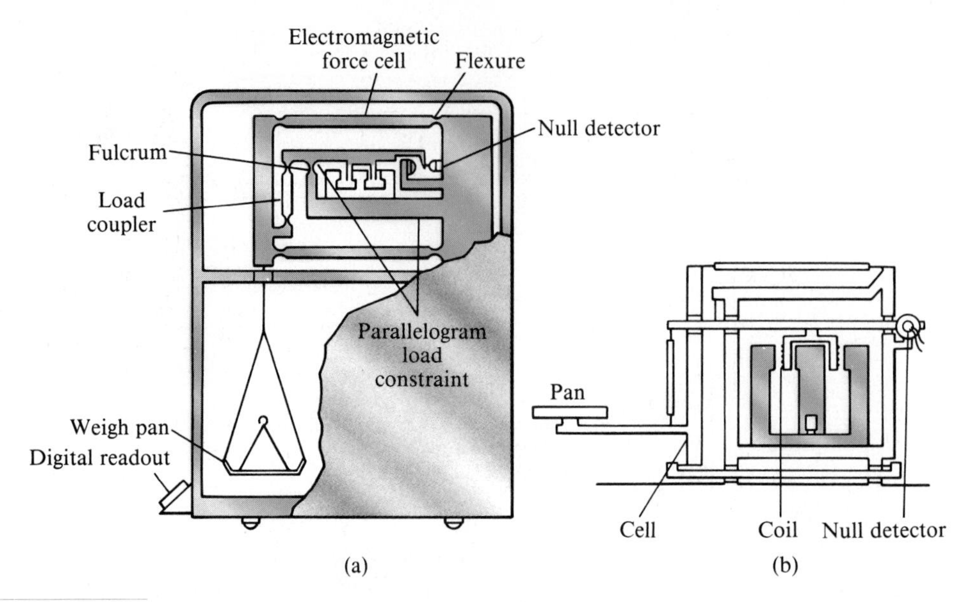

FIGURE 30-4 Electronic analytical balances. (a) Classical configuration with pan beneath the cell. (From R. M. Schoonover, *Anal. Chem.*, **1982,** *54,* 976A. Published 1982 American Chemical Society.) (b) A top-loading design. Note that the mechanism is enclosed in a windowed case. (Reprinted with permission from K. M. Lang, *Amer. Lab.*, **1983,** *15*(3), 72. Copyright 1983 by International Scientific Communications, Inc.)

electromagnetic radiation may also interfere with performance. Finally, the precision and accuracy of these instruments appear to be more susceptible to the effects of dust than do the precision and accuracy of their mechanical counterparts.

30D-5 Sources of Error in Weighing

Correction for Buoyancy[5]

A *buoyancy error* will affect weighing data if the density of the object being weighed differs significantly from that of the standard weights. This error has its origin in the difference in the buoyant force exerted by the medium (air) upon the object and upon the weights. Correction for buoyancy is accomplished with the equation

$$W_1 = W_2 + W_2\left(\frac{d_{air}}{d_{obj}} - \frac{d_{air}}{d_{wts}}\right) \qquad (30\text{-}1)$$

where W_1 is the corrected weight of the object, W_2 is the mass of the standard weights, d_{obj} is the density of the object, d_{wts} is the density of the weights, and d_{air} is the density of the air displaced by them; d_{air} has a value of 0.0012 $g \cdot cm^{-3}$.

The consequences of Equation 30-1 are shown in Figure 30-5, in which the relative error due to buoyancy is plotted against the density of objects weighed in air against stainless steel weights. Note that this error is less than 0.1% for objects that have a density of 2 $g \cdot cm^{-3}$ or greater. It is thus seldom

[5]For further information, see R. Battino and A. G. Williamson, *J. Chem. Educ.*, **1984,** *64,* 51.

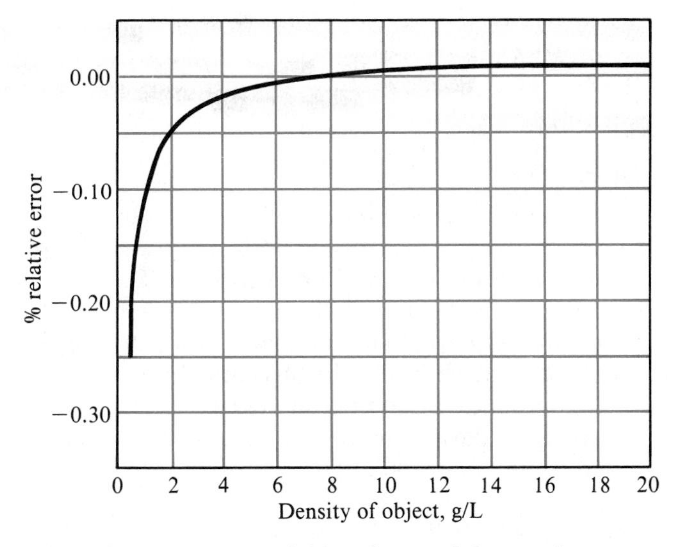

FIGURE 30-5 Effect of buoyancy on weighing data (stainless steel weights). Plot of relative error as a function of the density of the object weighed.

necessary to apply a correction to the weight of most solids. The same cannot be said for low-density solids, liquids, or gases, however; for these, the effects of buoyancy are significant and a correction must be applied.

The density of weights used in single-pan balances ranges from 7.8 to 8.4 $g{\cdot}cm^{-3}$, depending upon the manufacturer. In most cases, use of a density of 8 $g{\cdot}cm^{-3}$ provides a sufficiently accurate correction. If a greater accuracy is required, the specifications for the balance to be used should be consulted for the necessary density data.

EXAMPLE 30-1

A bottle weighed 7.6500 g empty and 9.9700 g after introduction of an organic liquid with a density of 0.92 $g{\cdot}cm^{-3}$. The balance was equipped with stainless steel weights having a density of 8.0 $g{\cdot}cm^{-3}$. Correct the weight of the sample for the effects of buoyancy.

The apparent weight of the liquid is 9.9700 − 7.6500 = 2.3200 g. The same buoyant force acts on the container during both weighings; thus, we need to consider only the force that acts on the 2.3200 g of liquid. Substitution of 0.0012 $g{\cdot}cm^{-3}$ for d_{air}, 0.92 $g{\cdot}cm^{-3}$ for d_{obj}, and 8.0 $g{\cdot}cm^{-3}$ for d_{wts} in Equation 30-1 gives

$$W_1 = 2.3200 + 2.3200\left(\frac{0.0012}{0.92} - \frac{0.0012}{8.0}\right) = 2.3227 \text{ g}$$

Temperature Effects

Attempts to weigh an object whose temperature is different from that of its surroundings will result in a significant error. Failure to allow sufficient time for a heated object to return to room temperature is the commonest source of this problem. Errors due to a difference in temperature have two sources. First, convection currents within the balance case exert a buoyant effect on the pan and object. Second, warm air trapped in a closed container weighs less than the same volume at a lower temperature. Both effects cause the apparent weight of the object to be low. This error can amount to as much as 10 or 15 mg for a typical porcelain filtering crucible or a weighing bottle (Figure 30-6). Heated objects must always be cooled to room temperature before being weighed.

Other Sources of Error

A porcelain or glass object will occasionally acquire a static charge that is sufficient to cause a balance to perform erratically; this problem is particularly serious when the relative humidity is low. Spontaneous discharge frequently occurs after a short period. A low-level source of radioactivity (such as a photographer's brush) in the balance case will provide sufficient ions to relieve the charge. Alternatively, the object can be wiped with a faintly damp chamois.

The optical scale of a single-pan balance should be checked regularly for accuracy, particularly under loading conditions that require the full scale range. A standard 100-mg weight is used for this check.

30D-6 Auxiliary Balances

Balances that are less precise than analytical balances find much use in the analytical laboratory. These offer the advantages of speed, ruggedness, large capacity, and convenience and should be used whenever high sensitivity is not required.

Top-loading auxiliary balances are particularly convenient. A sensitive top-loading balance will accommodate 150 to 200 g with a precision of about 1 mg—an order of magnitude less than a macro analytical balance. Some balances of this type tolerate loads as great as 25,000 g with a precision of ± 0.05 g. Most are equipped with a taring device that brings the balance reading to zero with an empty container on the pan. Some are fully automatic, require no manual dialing or weight handling, and provide a digital readout of the weight. Modern top-loading balances are electronic.

A triple-beam balance with a sensitivity less than that of a typical top-loading auxiliary balance is also useful. This is a single-pan balance with three decades of weights that slide along individual calibrated scales. The precision of a triple-beam balance may be one or two orders of magnitude less than

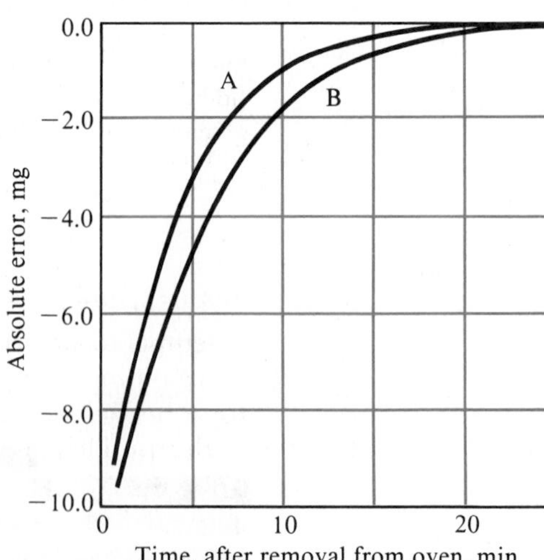

FIGURE 30-6 Effect of temperature on weighing data. Absolute error in weight as a function of time after the object was removed from a 110°C drying oven. A: porcelain filtering crucible. B: weighing bottle containing about 7.5 g of KCl.

that of a top-loading instrument but is adequate for many weighing operations. This type of balance offers the advantages of simplicity, durability, and low cost.

30E The Equipment and Manipulations Associated with Weighing

The weight of many solids changes with humidity, owing to their tendency to absorb weighable amounts of moisture. This effect is especially pronounced when a large surface area is exposed, as with a reagent chemical or a sample that has been ground to a fine powder. The first step in a typical analysis, then, involves drying the sample so that the results will not be affected by the humidity of the surrounding atmosphere.

A sample, a precipitate, or a container is brought to *constant weight* by a cycle that involves heating (ordinarily for 1 h or more) at an appropriate temperature, cooling, and weighing. This cycle is repeated as many times as needed to obtain successive weighings that agree within 0.2 to 0.3 mg of one another. The establishment of constant weight provides some assurance that the chemical or physical processes that occur during the heating (or ignition) are complete.

30E-1 Weighing Bottles

Solids are conveniently dried and stored in *weighing bottles,* two common varieties of which are shown in Figure 30-7. The ground-glass portion of the cap-style bottle shown on the left is on the outside and does not come into contact with the contents; this design eliminates the possibility of some of the sample becoming entrained upon and subsequently lost from the ground-glass surface.

Plastic weighing bottles are available; ruggedness is the principal advantage of these bottles over their glass counterparts.

30E-2 Desiccators and Desiccants

Oven drying is the most common way of removing moisture from solids. To be sure, this approach is not appropriate for substances that decompose or for those from which water is not removed at the ambient temperature of the oven; see Section 27B.

Dried materials are stored in *desiccators* while they cool in order to minimize the uptake of moisture. Figure 30-8 shows the components of a typical desiccator. The base section contains a chemical drying agent, such as anhydrous calcium chloride, calcium sulfate (Drierite[6]), anhydrous magnesium perchlorate (Anhydrone[7] or Dehydrite[8]), or phosphorus pentoxide. The ground-glass surfaces are lightly coated with grease.

The lid of a desiccator is removed or replaced by a sliding motion rather than by a vertical one to minimize the likelihood of disturbing the sample. An airtight seal is achieved by slight rotation and downward pressure upon the positioned lid.

[6]W. A. Hammond Drierite Co.

[7]J. T. Baker Co.

[8]Thomas Scientific Co.

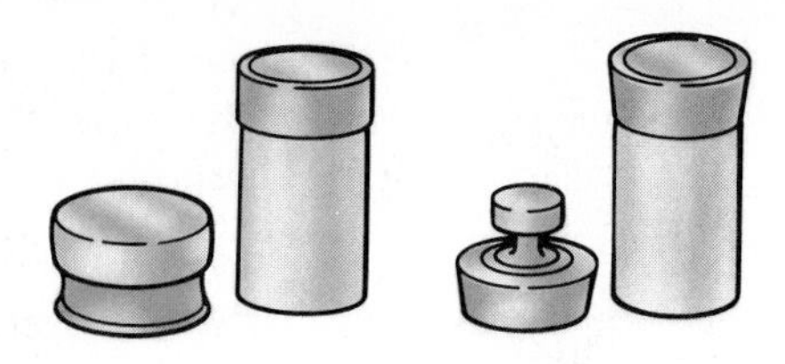

FIGURE 30-7 Typical weighing bottles.

When a heated object is placed in a desiccator, the increase in pressure as the enclosed air is warmed may be sufficient to break the seal between lid and base. Conversely, if the seal is not broken, the cooling of heated objects can cause development of a partial vacuum. Both of these conditions can cause the contents of the desiccator to be physically lost or contaminated. Although it defeats the purpose of the desiccator somewhat, it is advisable to allow some cooling to occur before the lid is seated. It is also helpful to break the seal once or twice during cooling to relieve any excessive vacuum that might otherwise develop. Finally, it is prudent to lock the lid in place with one's thumbs while moving the desiccator from place to place.

Very hygroscopic materials should be stored in containers equipped with snug covers, such as weighing bottles; the covers remain in place while in the desiccator. Most other solids can be safely stored uncovered.

30E-3 The Manipulation of Weighing Bottles

Heating at 105 to 110°C is sufficient to remove the moisture from the surface of most solids. Figure 30-9 depicts the arrangement recommended for drying a sample. The weighing bottle is contained in a labeled beaker covered with a ribbed cover glass. This arrangement protects the sample from accidental contamination and also allows for the free access of air. Crucibles that contain a precipitate that can be freed of moisture by simple drying can be treated similarly. The beaker containing the weighing bottle or crucible to be dried must be carefully marked to permit identification.

FIGURE 30-8 Components of a typical desiccator.

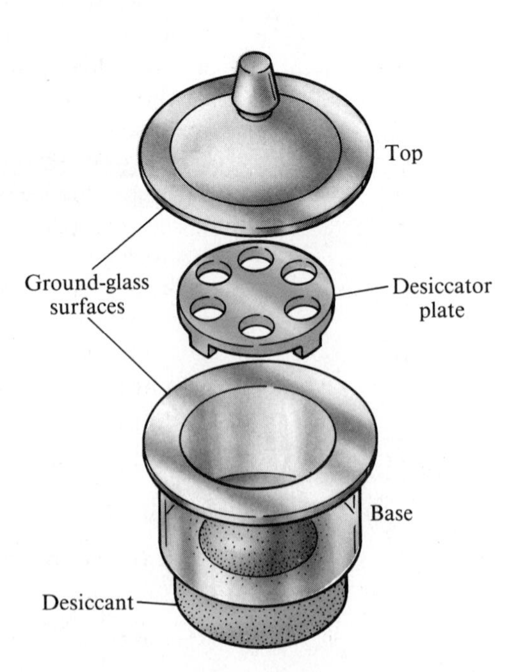

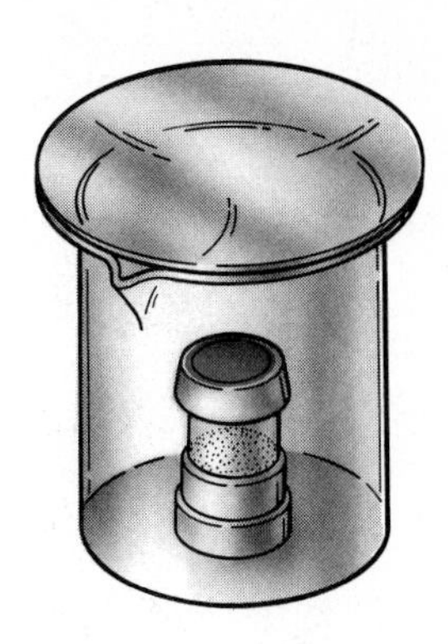

FIGURE 30-9 Arrangement for the drying of samples.

The handling of a dried object with one's fingers can result in the transfer of a weighable amount of water or oil from the skin to the object. Weighing bottles should thus be manipulated with tongs, chamois finger cots, clean cotton gloves, or strips of paper. A weighing bottle being manipulated with strips of paper is shown in Figure 30-10.

30E-4 Weighing by Difference

Weighing by difference is a simple method for determining a series of sample weights. First the bottle and its contents are weighed. One sample is then transferred from the bottle to a container; gentle tapping of the bottle with its top and slight rotation of the bottle provide control over the amount of sample removed. Following transfer, the bottle and its residual contents are weighed. The weight of the sample is the difference between the two weighings. It is essential that all the solid removed from the weighing bottle be transferred to the container without loss.

30E-5 Weighing Hygroscopic Solids

Special precautions are needed for the weighing of hygroscopic solids because of the rapid rate with which they equilibrate with moisture in the atmosphere. The approximate amount of each sample is placed in individual weighing bottles. After the contents have been dried, the bottles are capped, and cooled in a desiccator. Their exact weights are then determined by difference as described above, care being taken to replace the cap of the weighing bottle as quickly as possible after transfer.

FIGURE 30-10 Method for quantitative transfer of a solid sample. Note the use of paper strips to avoid contact between glass and skin.

30E-6 *Weighing Liquids*

The weight of a liquid is always obtained by difference. Liquids that are noncorrosive and relatively nonvolatile can be transferred to previously weighed containers with snugly fitting covers (such as weighing bottles); the weight of the container is subtracted from the total weight.

A volatile or corrosive liquid should be sealed in a weighed glass ampoule. The ampoule is heated, and the neck is then immersed in the sample; as cooling occurs, the liquid is drawn into the bulb. The ampoule is then inverted and the neck sealed off with a small flame. The ampoule and its contents, along with any glass removed during sealing, are cooled to room temperature and weighed. The ampoule is then transferred to an appropriate container and broken. A volume correction for the glass of the ampoule may be needed if the receiving vessel is a volumetric flask.

30F *Weight Titrations*

Weight (or gravimetric) titrimetry is discussed in Section 4E. Directions for the weight titration of chloride ion by the Mohr method are found in Section 31B-3.

A convenient reagent dispenser for a weight titration is a small polyethylene bottle equipped with a fine delivery tip (Figure 30-11). Weighings are normally performed with a top-loading balance having a sensitivity of 1 mg or 0.001 mL. It has been found that the weight of such a bottle does not change significantly when it is handled with bare hands.

30F-1 *Directions for Preparing a Reagent Dispenser*

A dispenser is readily fashioned from a 60-mL polyethylene bottle with a screw cap [such as the Nalge 2-oz ($\simeq$ 60 mL) or 4-oz ($\simeq$ 125 mL) Boston Round Bottle]. With a cork borer, make a hole in the cap that is slightly smaller than the outside diameter of the tip (Note). Carefully force the tip through the hole; apply a bead of epoxy cement to seal the tip to the cap. Apply a pressure-sensitive label, and identify the contents of the bottle with a ballpoint pen.

Note. A tip can be prepared by constricting the opening of an ordinary medicine dropper in a flame. Equally satisfactory are glass tips for conventional Kimax® burets.

30F-2 *Directions for Performing a Weight Titration*

Fill the reagent dispenser with a quantity of the standard titrant, wipe away any liquid spilled on the outside of the container with an absorbing tissue, and tighten the screw cap firmly. Weigh the bottle and its contents to the nearest milligram. Introduce a suitable indicator into the solution of the analyte. Grasp the dispenser in one hand and the titration flask in the other. Tilt the dispenser so that its tip is below the lip of the flask and deliver several increments of the reagent by squeezing the bottle while rotating the flask with the other hand. When it is judged that only a few more drops of reagent are needed, ease the pressure on the bottle so that the flow ceases; then touch the tip to the inside of the flask and further reduce the pressure on the dispenser so that the liquid in the tip is drawn back into the bottle as the tip

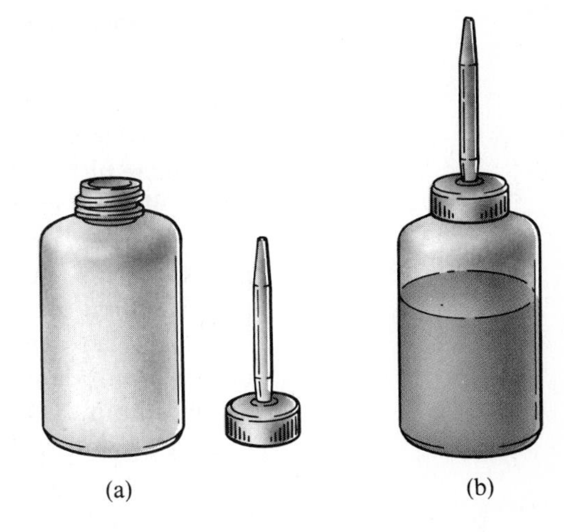

FIGURE 30-11 Reagent dispenser for weight titrations.

is removed from the flask. Set the dispenser on a piece of clean, dry glazed paper and rinse down the inner walls of the flask with a stream of distilled or deionized water (Note 1). Add reagent a drop at a time until the end point is reached (Note 2). Weigh the dispenser and record the data.

Notes

1. Instead of rinsing the walls, the titration flask can be tilted and rotated so that the bulk of the liquid picks up droplets that adhere to the inner surface.
2. Increments smaller than an ordinary drop can be added by forming a partial drop on the tip and then touching the tip to the wall. This partial drop is then combined with the bulk of the solution by rinsing the walls with wash water or tilting the flask until the droplet is incorporated into the bulk.

30G The Equipment and Manipulations for Filtration and Ignition

30G-1 Apparatus

Simple Crucibles

Simple crucibles serve only as containers. The more common types maintain constant weight, within the limits of experimental error, and are used principally to convert precipitates to suitable weighing forms. The solid is first collected on a filter paper. The filter and contents are then transferred to a weighed crucible, and the paper is ignited. Porcelain, aluminum oxide, silica, and platinum are used for the manufacture of these crucibles.

Simple crucibles of nickel, iron, silver, and gold are used as containers for the high-temperature fusion of samples that are not soluble in aqueous reagents. Attack by both the atmosphere and the contents may cause these crucibles to suffer weight changes. Moreover, such attack will contaminate the sample with species derived from the crucible. The chemist selects the crucible whose products will offer the least interference in subsequent steps of the analysis.

Filtering Crucibles

Filtering crucibles serve not only as containers but also as filters. A vacuum is used to hasten the filtration; a tight seal between crucible and filtering flask is accomplished with any of several types of rubber adaptors (Figure 30-12; a complete filtration train is shown in Figure 30-17). Collection of a precipitate with a filtering crucible is frequently less time-consuming than with paper.

Sintered-glass (also called *fritted-glass*) crucibles are manufactured in three porosities: fine, medium, and coarse (marked *f*, *m*, and *c*). The upper temperature limit for a sintered-glass crucible is ordinarily about 200°C. Filtering crucibles made entirely of quartz can tolerate substantially higher temperatures without damage. The same is true for crucibles with unglazed porcelain or aluminum oxide frits. The latter are not as costly as quartz.

A *Gooch crucible* has a perforated bottom that supports a fibrous mat. Asbestos was at one time the filtering medium of choice for a Gooch crucible; current regulations concerning this material have virtually eliminated its use. Small circles of glass matting have now replaced asbestos; they are used in pairs to protect against disintegration during the filtration. Glass mats can tolerate temperatures in excess of 500°C and are substantially less hygroscopic than asbestos.

Filter Paper

Paper is an important filtering medium. Ashless paper is manufactured from cellulose fibers that have been treated with hydrochloric and hydrofluoric acids to remove metallic impurities and silica; ammonia is then used to neutralize the acids. The residual ammonium salts in many filter papers may be sufficient to affect the analysis for nitrogen by the Kjeldahl method (Section 31C-10).

All papers tend to pick up moisture from the atmosphere, and ashless paper is no exception. It is thus necessary to destroy the paper by ignition if the precipitate collected on it is to be weighed. Typically, 9- or 11-cm circles of ashless paper leave a residue that weighs less than 0.1 mg, an amount that

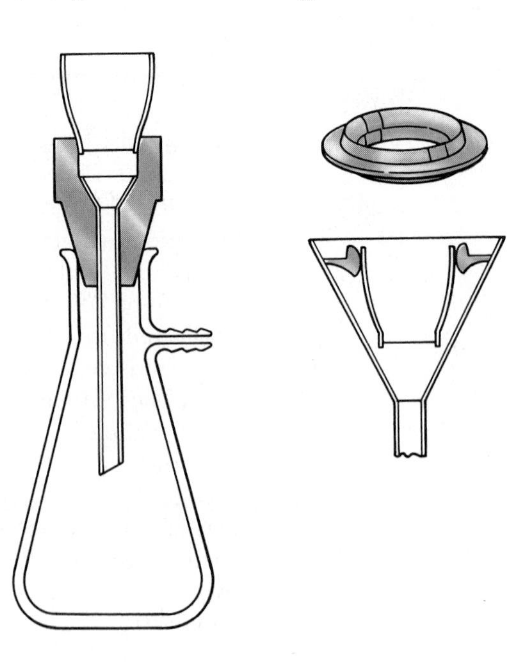

FIGURE 30-12 Adaptors for filtering crucibles.

is ordinarily negligible. Ashless paper can be obtained in several porosities (Appendix 8).

Gelatinous precipitates, such as hydrous iron(III) oxide, clog the pores of any filtering medium. A coarse-porosity ashless paper is most effective for the filtration of such solids, but even here clogging occurs. This problem can be minimized by mixing a dispersion of ashless filter paper with the precipitate prior to filtration. Filter paper pulp is available in tablet form from chemical suppliers; if necessary, the pulp can be prepared by treating a piece of ashless paper with concentrated hydrochloric acid and washing the disintegrated mass free of acid.

Table 30-1 summarizes the characteristics of common filtering media. None satisfies all requirements.

Heating Equipment

Many precipitates can be weighed directly after being brought to constant weight in a low-temperature drying oven. Such an oven is electrically heated and capable of maintaining a constant temperature to within 1°C (or better). The maximum attainable temperature ranges from 140 to 260°C, depending upon make and model; for many precipitates, 110°C is a satisfactory drying temperature. The efficiency of a drying oven is greatly increased by the forced circulation of air. The passage of predried air through an oven designed to operate under a partial vacuum represents an additional improvement.

Microwave laboratory ovens are currently appearing on the market. Where applicable, these greatly shorten drying cycles. For example, slurry samples that require 12 to 16 h for drying in a conventional oven are reported to be dried within 5 to 6 min in a microwave oven.[9]

An ordinary heat lamp can be used to dry a precipitate that has been collected on ashless paper and to char the paper as well. The process is conveniently completed by ignition at an elevated temperature in a muffle furnace.

TABLE 30-1 **Comparison of Filtering Media for Gravimetric Analysis**

		Gooch Crucible				
Characteristic	**Paper**	**Asbestos mat***	**Glass mat**	**Glass Crucible**	**Porcelain Crucible**	**Aluminum Oxide Crucible**
Speed of filtration	Slow	Rapid	Rapid	Rapid	Rapid	Rapid
Convenience and ease of preparation	Troublesome, inconvenient	Troublesome, inconvenient	Convenient	Convenient	Convenient	Convenient
Maximum ignition temperature, °C	None	1200	>500	200–500	1100	1450
Chemical reactivity	Carbon has reducing properties	Inert	Inert	Inert	Inert	Inert
Porosity	Many available	Difficult to control	Several available	Several available	Several available	Several available
Convenience with gelatinous precipitates	Satisfactory	Unsuitable; filter tends to clog	Unsuitable; filter tends to clog	Unsuitable; filter tends to clog	Unsuitable; filter tends to clog	Unsuitable; filter tends to clog
Cost	Low	Low	Low	High	High	High

*The use of asbestos, a confirmed carcinogen, is subject to stringent control in many areas.

[9]D. G. Kuehn, R. L. Brandvig, D. C. Lundean, and R. H. Jefferson, *Amer. Lab.*, **1986**, *18* (7), 31.

Burners are convenient sources of intense heat. The maximum attainable temperature depends upon the design of the burner and the combustion properties of the fuel. Of the three common laboratory burners, the Meker provides the highest temperatures, followed by the Tirrill and Bunsen types.

A heavy-duty electric furnace (*muffle furnace*) is capable of maintaining controlled temperatures of 1100°C or higher. Long-handled tongs and heat-resistant gloves are needed for protection when transferring objects to or from such a furnace.

30G-2 The Manipulations Associated with Filtration and Ignition

Preparation of Crucibles

A crucible used to convert a precipitate to a form suitable for weighing must maintain—within the limits of experimental error—a constant weight throughout the drying or ignition. The crucible is first cleaned thoroughly (filtering crucibles are conveniently cleaned by backwashing on a filtration train) and subjected to the same regimen of heating and cooling as that required for the precipitate. This process is repeated until constant weight (page 731) has been achieved, that is, until consecutive weighings differ by 0.3 mg or less.

Filtration and Washing of Precipitates

Three steps are involved in filtering an analytical precipitate: *decantation, washing,* and *transfer.* In decantation, as much supernatant liquid as possible is passed through the filter while the precipitated solid is kept essentially undisturbed in the beaker where it was formed. This procedure speeds the overall filtration rate by delaying the time at which the pores of the filtering medium become clogged with precipitate. A stirring rod is used to direct the flow of decantate (Figures 30-13). When flow ceases, the drop of liquid at the end of the pouring spout is collected with the stirring rod and returned to the beaker. Wash liquid is next added to the beaker and thoroughly mixed with the precipitate. The solid is allowed to settle, following which this liquid is also decanted through the filter. Several such washings may be required, depending upon the precipitate. Most washing should be carried out *before* the solid is transferred; this results in a more thoroughly washed precipitate and a more rapid filtration.

The transfer process is illustrated in Figures 30-13c and d. The bulk of the precipitate is moved from beaker to filter by suitably directed streams of wash liquid. As in decantation and washing, a stirring rod provides direction for the flow of material to the filtering medium.

The last traces of precipitate that cling to the inside of the beaker are dislodged with a *rubber policeman,* which is a small section of rubber tubing that has been crimped on one end. The open end of the tubing is fitted onto the end of a stirring rod and is wetted with wash liquid before use. Any solid collected with it is combined with the main portion on the filter. Small pieces of ashless paper can be used to wipe the last traces of hydrous oxide precipitates from the wall of the beaker; these papers are ignited along with the paper that holds the bulk of the precipitate.

Many precipitates possess the exasperating property of *creeping,* or spreading over a wetted surface against the force of gravity. Filters are never filled to more than three quarters of capacity, owing to the possibility that some of the precipitate could be lost as the result of creeping. The addition

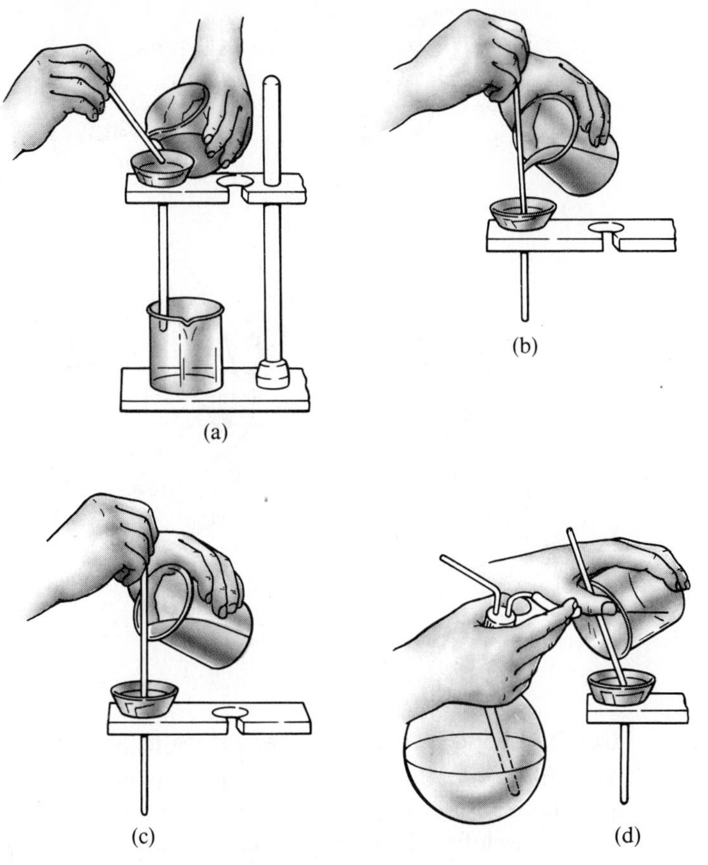

FIGURE 30-13 Steps in filtering. (a and b) Washing by decantation; (c) and (d) Transfer of the precipitate.

of a small amount of nonionic detergent, such as Triton-X-100, to the supernatant liquid or to the wash liquid can be helpful in minimizing creeping.

A gelatinous precipitate must be completely washed before it is allowed to dry because drying causes the solid to shrink and develop cracks. Further additions of wash liquid simply pass through these cracks and accomplish little or no washing.

30G-3 Directions for the Filtration and Ignition of Precipitates

Preparation of a Filter Paper

Figure 30-14 shows the sequence followed in folding a filter paper and seating it in a 58-deg funnel (or 60-deg fluted funnel). The paper is folded exactly in half (a), firmly creased, and folded again (b). A triangular piece from one of the corners is torn off parallel to the second fold (c). The paper is then opened so that the untorn quarter forms a cone (d). The cone is fitted into the funnel, and the second fold is creased (e). Seating is completed by dampening the cone with water from a wash bottle and *gently* patting it with a finger. There is no leakage of air between the funnel and a properly seated cone; in addition, the stem of the funnel will be filled with an unbroken column of liquid.

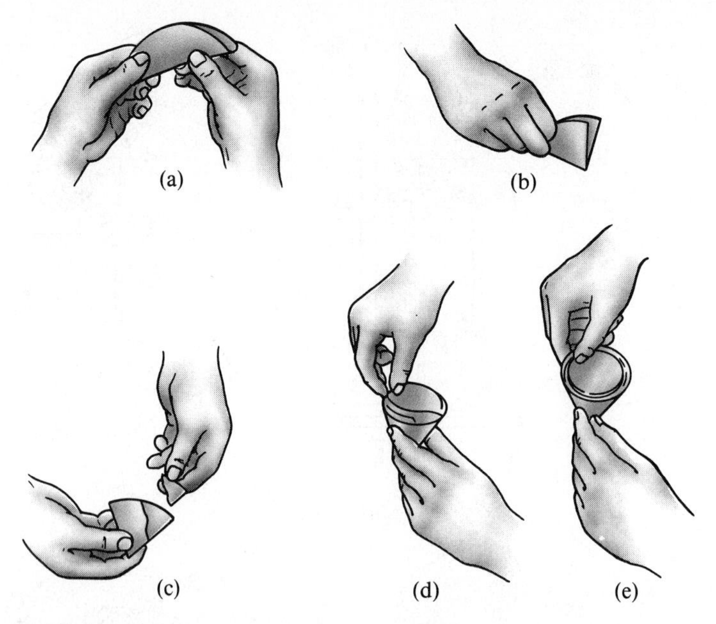

FIGURE 30-14 Folding and seating a filter paper.

The Transfer of Paper and Precipitate to a Crucible

After filtration and washing have been completed, the filter and its contents must be transferred from the funnel to a crucible that has been brought to constant weight. Ashless paper has very low wet strength and must be handled with care during the transfer. The danger of tearing is lessened considerably if the paper is allowed to dry somewhat before it is removed from the funnel.

Figure 30-15 illustrates the transfer process. The triple-thick portion of the filter paper is drawn across the funnel to flatten the cone along its upper edge (a); the corners are next folded inward (b); the top edge is then folded over (c). Finally, the paper and its contents are eased into the crucible (d) so that the bulk of the precipitate is near the bottom.

FIGURE 30-15 Transferring a filter paper and precipitate to a crucible.

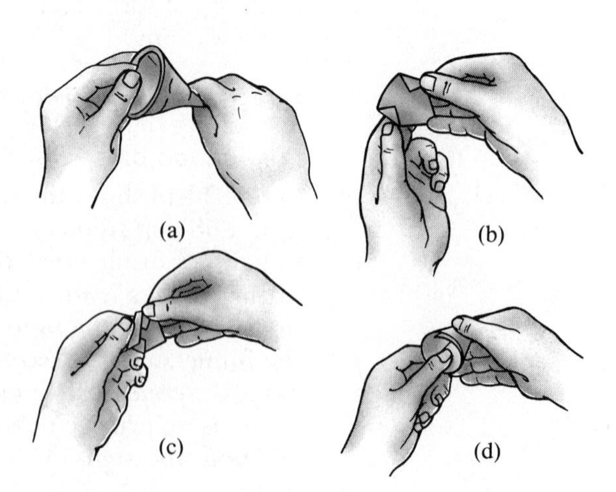

Ashing of a Filter Paper

If a heat lamp is to be used, the crucible is placed on a clean, nonreactive surface, such as a wire screen covered with aluminum foil. The lamp is then positioned about 1 cm above the rim of the crucible and turned on. Charring takes place without further attention. The process is considerably accelerated if the paper is moistened with no more than one drop of concentrated ammonium nitrate solution. Elimination of the residual carbon is accomplished with a burner, as described in the next paragraph.

Considerably more attention must be paid if a burner is used to ash a filter paper. The burner produces much higher temperatures than a heat lamp. The possibility thus exists for the mechanical loss of precipitate if moisture is expelled too rapidly in the initial stages of heating or if the paper bursts into flame. Also, partial reduction of some precipitates can occur through reaction with the hot carbon of the charring paper; such reduction is a serious problem if reoxidation following ashing is inconvenient. These difficulties can be minimized by positioning the crucible as illustrated in Figure 30-16. The tilted position allows for the ready access of air; a clean crucible cover should be available to extinguish any flame that might develop.

Heating is commenced with a small flame. The temperature is gradually increased as moisture is evolved and the paper begins to char. The intensity of heating that can be tolerated can be gauged by the amount of smoke given off. Thin wisps are normal. A significant increase in the amount of smoke indicates that the paper is about to flash and that heating should be temporarily discontinued. Any flame that does appear should be immediately extinguished with a crucible cover. (The cover may become discolored, owing to the condensation of carbonaceous products; these products must ultimately be removed from the cover by ignition to confirm the absence of entrained particles of precipitate.) When no further smoking can be detected, heating is increased to eliminate the residual carbon. Strong heating, as necessary, can then be undertaken.

This sequence ordinarily precedes the final ignition of a precipitate in a muffle furnace, where a reducing atmosphere is equally undesirable.

FIGURE 30-16 Ignition of a precipitate. Proper crucible position for preliminary charring.

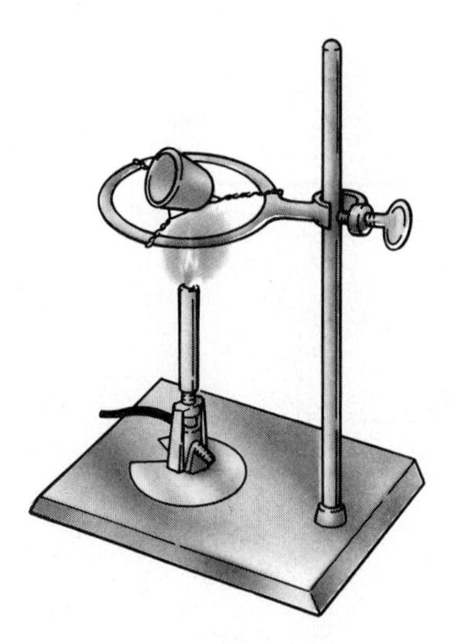

The Use of Filtering Crucibles

A vacuum filtration train (Figure 30-17) is used where a filtering crucible can be used instead of paper. The trap isolates the filter flask from the source of vacuum.

30G-4 Rules for the Manipulation of Heated Objects

Careful adherence to the following rules will minimize the possibility of accidental loss of a precipitate.

1. Practice unfamiliar manipulations before putting them to use.
2. *Never* place a heated object on the benchtop; instead, place it on a wire gauze or a heat-resistant ceramic plate.
3. Allow a crucible that has been subjected to the full flame of a burner or to a muffle furnace to cool momentarily (on a wire gauze or ceramic plate) before transferring it to the desiccator.
4. Keep the tongs and forceps used to handle heated objects scrupulously clean. In particular, do not allow the tips to touch the benchtop.

30H The Measurement of Volume

The precise measurement of volume is as important to many analytical methods as is the precise measurement of mass.

30H-1 Units of Volume

The unit of volume is the *liter* (L), defined as one cubic decimeter. The *milliliter* (mL) is one one-thousandth of a liter and is used where the liter represents an inconveniently large volume unit.

30H-2 The Effect of Temperature on Volume Measurements

The volume occupied by a given mass of liquid varies with temperature; to a lesser extent, so also does the volume of the container that holds the liquid. The accurate measurement of volume may require that both of these effects be taken into account.

Most volumetric measuring devices are made of glass, a material that fortunately has a small coefficient of expansion. The volume of a soft-glass container changes by about 0.003%/°C; the volume of a heat-resistant glass

FIGURE 30-17 Train for vacuum filtration.

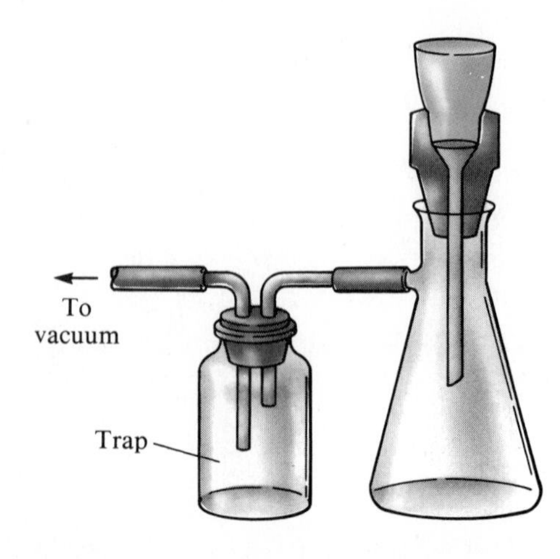

changes by about one third this amount. Clearly, variations in the volume of a glass container with temperature need to be considered only for the most exacting work.

The coefficient of expansion for dilute aqueous solutions (approximately 0.025%/°C) is such that a 5°C change has a measurable effect upon the reliability of ordinary volumetric measurements.

EXAMPLE 30-2

A 40.00-mL sample is taken from an aqueous solution at 5°C; what volume does it occupy at 20°C?

$$V_{20^\circ} = V_{5^\circ} + 0.00025(20 - 5)(40.00) = 40.00 + 0.15 = 40.15 \text{ mL}$$

Volumetric measurements must be referred to some standard temperature; this reference point is ordinarily 20°C. The ambient temperature of most laboratories is sufficiently close to 20°C to eliminate the need for temperature corrections in volume measurements for aqueous solutions. In contrast, the coefficient of expansion for organic liquids may require corrections for temperature differences of 1°C or less.

EXAMPLE 30-3

The coefficient of expansion for an alcoholic KOH solution is 0.11%/°C. Calculate the relative error in the volume of an aliquot delivered at 25.2°C by a 50.00-mL pipet if no correction for temperature is made.

$$V_{20^\circ} = 50.00 - 0.0011(25.2 - 20.0)(50.00) = 49.71 \text{ mL}$$

$$E_{\text{rel}} = \frac{49.71 - 50.00}{50.00} \times 1000 \text{ ppt} = -5.8 \text{ ppt}$$

30H-3 Apparatus for the Precise Measurement of Volume

The reliable measurement of volume is performed with the *pipet*, the *buret*, and the *volumetric flask*.

Volumetric equipment is marked by the manufacturer to indicate not only the manner of calibration (usually TD for "to deliver" or TC for "to contain") but also the temperature at which the calibration strictly applies. Pipets and burets are ordinarily calibrated to deliver specified volumes, whereas volumetric flasks are calibrated on a to-contain basis.

Pipets

Pipets are devices that permit the transfer of accurately known volumes from one container to another. Common types are shown in Figure 30-18; information concerning their use is given in Table 30-2. A *volumetric*, or *transfer*, pipet (Figure 30-18a) delivers a single, fixed volume between 0.5 and 200 mL. Many such pipets are color-coded by volume for convenience in identification and sorting. *Measuring* pipets (Figure 30-18b and c) are calibrated in convenient units to permit delivery of any volume up to a maximum capacity ranging from 0.1 to 25 mL.

Volumetric and measuring pipets are filled to a calibration mark at the outset; the manner in which the transfer is completed depends upon the particular type. Because an attraction exists between most liquids and glass,

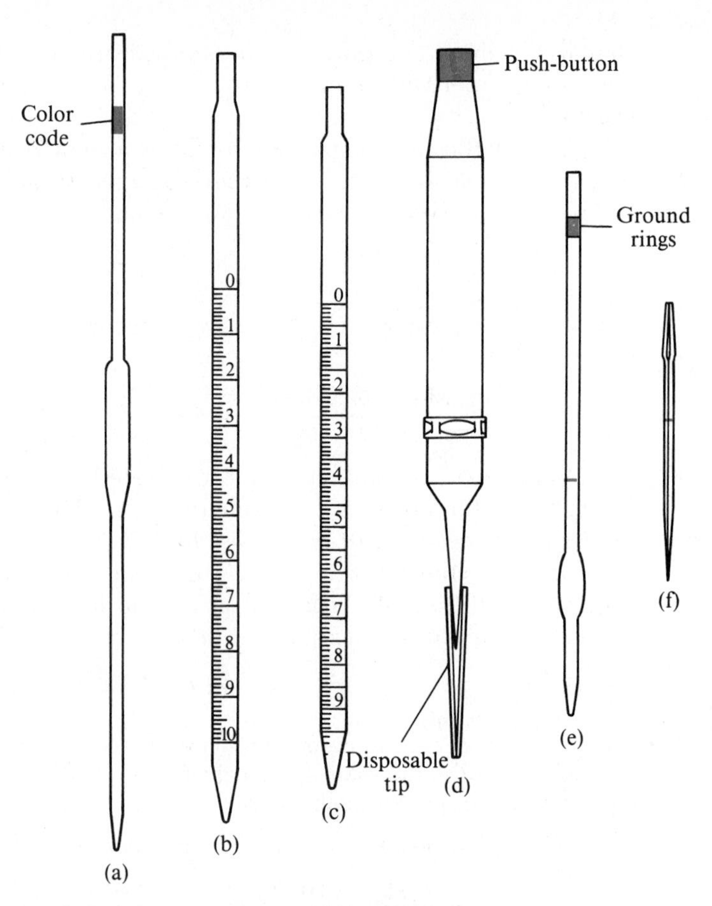

FIGURE 30-18 Typical pipets: (a) volumetric, (b) Mohr, (c) serological, (d) syringe, (e) Ostwald-Folin, (f) lambda.

a small amount of liquid tends to remain in the tip of the pipet after the pipet is emptied. This residual liquid is never blown out of a volumetric pipet or from some measuring pipets; it is blown out of other types of pipets (Table 30-2).

Numerous *automatic* pipets are available for situations that call for the

TABLE 30-2 **Characteristics of Pipets**

Name	Type of Calibration*	Function	Available Capacity, mL	Type of Drainage
Volumetric	TD	Delivery of fixed volume	1–200	Free
Mohr	TD	Delivery of variable volume	1–25	To lower calibration line
Serological	TD	Delivery of variable volume	0.1–10	Blow out last drop†
Serological	TD	Delivery of variable volume	0.1–10	To lower calibration line
Ostwald-Fohn	TD	Delivery of fixed volume	0.5–10	Blow out last drop†
Lambda	TC	Containment of fixed volume	0.001–2	Wash out with suitable solvent
Lambda	TD	Delivery of fixed volume	0.001–2	Blow out last drop†
Syringe	TD	Delivery of variable or fixed volume	0.001–1	Tip emptied by syringe

*TD–to deliver; TC–to contain.

†A frosted ring near the top of recently manufactured pipets indicates that the last drop is to be blown out.

repeated delivery of a particular volume. Hand-held syringe pipets (Figure 30-18d) deliver volumes from 1 to 1000 μL (1 mL). The liquid in this type of pipet is contained in a disposable plastic tip. A volume is drawn into the tip by a spring-operated piston that is activated by a push button at the top of the syringe. The liquid is then delivered by reversing the action of the push button on the piston. Remarkable precision is claimed for these devices (± 0.02 μL for a 1-μL measurement, ± 0.3 μL at 1000 μL).

A motorized, computer-controlled microliter pipet is now available. This device is programmed to function as a pipet, a dispenser of multiple volumes, a buret, and a means for diluting samples. The volume desired is entered on a keyboard and is displayed on a panel. A motor-driven piston dispenses the liquid. Maximum volumes range from 10 to 2500 μL are available.

Burets

Burets, like measuring pipets, enable the analyst to deliver any volume up to their maximum capacity. The precision attainable with a buret is substantially greater than with a pipet.

A buret consists of a calibrated tube to hold titrant plus a valve arrangement by which the flow of titrant is controlled. This valve is the principal source of difference among burets. The simplest valve consists of a close-fitting glass bead inside a short length of rubber tubing that connects the buret and its tip; only when the tubing is deformed does liquid flow past the bead.

A buret equipped with a glass stopcock for a valve relies upon a lubricant between the ground-glass surfaces of stopcock and barrel for a liquid-tight seal. Some solutions, notably bases, cause a glass stopcock to freeze upon long contact; therefore thorough cleaning is needed after each use. Valves made of Teflon are commonly encountered; these are unaffected by most common reagents and require no lubricant.

Volumetric Flasks

Volumetric flasks are manufactured with capacities ranging from 5 mL to 5 L and are usually calibrated to contain a specified volume when filled to a line etched on the neck. They are used for the preparation of standard solutions and for the dilution of samples to a fixed volume prior to taking aliquots with a pipet. Some are also calibrated on a to-deliver basis; these are readily distinguished by two reference lines on the neck. If delivery of the stated volume is desired, the flask is filled to the upper line.

30H-4 General Considerations Concerning the Use of Volumetric Equipment

Volume markings are blazed upon clean volumetric equipment by the manufacturer. An equal degree of cleanliness is needed in the laboratory if these markings are to have their stated meanings. Only clean glass surfaces support a uniform film of liquid. Dirt or oil causes breaks in this film; the existence of breaks is a certain indication of an unclean surface.

Cleaning

A brief soaking in a warm detergent solution is usually sufficient to remove the grease and dirt responsible for water breaks. Prolonged soaking should be avoided because a rough area or ring is likely to develop at a detergent-air interface. This ring cannot be removed and causes a film break that destroys the usefulness of the equipment. If detergent is ineffective, treatment with warm cleaning solution (*Caution!*) usually helps.

After being cleaned, the apparatus must be thoroughly rinsed with tap water and then with three or four portions of distilled or deionized water. It is seldom necessary to dry volumetric ware. As a general rule, calibrated glass equipment should not be heated.

Avoidance of Parallax

The top surface of a liquid confined in a narrow tube (such as a pipet or buret or the neck of a volumetric flask) exhibits a marked curvature, or *meniscus*. It is common practice to use the bottom of the meniscus as the point of reference in calibrating and using volumetric equipment. This minimum can be established more exactly by holding an opaque card or piece of paper behind the graduations (Figure 30-19).

In reading volumes, the eye must be at the level of the liquid surface to avoid an error due to *parallax*, a condition that causes the volume to appear smaller than its actual value if the meniscus is viewed from above and larger if the meniscus is viewed from below (Figure 30-19).

30H-5 Directions for the Use of a Pipet

The following directions pertain specifically to volumetric pipets but can be modified for the use of other types as well.

Liquid is drawn into a pipet through the application of a slight vacuum. *The mouth should never be used for suction because of the possibility of accidentally ingesting the liquid being pipetted.* Instead, a rubber suction bulb or a rubber tube connected to a vacuum source should be used (Figure 30-20a).

Cleaning

Use a rubber bulb to draw detergent (or cleaning) solution to a level 2 to 3 cm above the calibration mark of the pipet. Drain this solution and then rinse the pipet with several portions of tap water. Inspect for film breaks; repeat this portion of the cleaning cycle if necessary. Finally, fill the pipet with distilled water to perhaps one third of its capacity and carefully rotate it so that the entire interior surface is wetted. Repeat this rinsing step at least twice.

Measurement of an Aliquot

Use a rubber bulb to draw a small volume of the liquid to be sampled into the pipet and thoroughly wet all the interior surface. Repeat with *at least* two additional portions. Then carefully fill the pipet to a level somewhat above the graduation mark (Figure 30-20a). Quickly replace the bulb with a *forefinger* to arrest the outflow of liquid (Figure 30-20b). Make certain there are no bubbles in the bulk of the liquid or foam at the surface. Tilt the pipet slightly

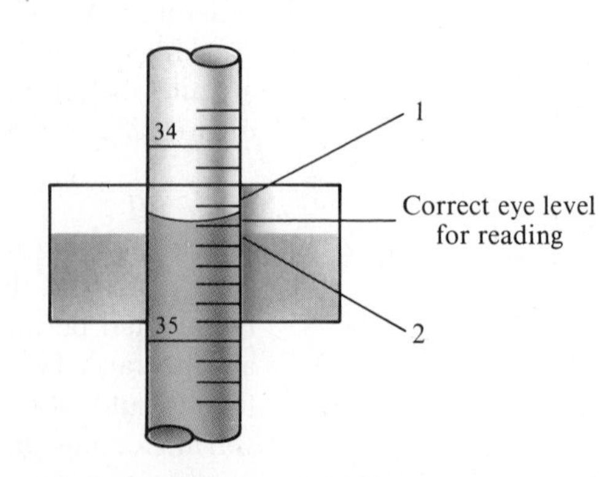

FIGURE 30-19 Method for reading a buret. The eye should be level with the meniscus. The reading shown is 34.39 mL. If viewed from position 1, the reading appears smaller than 34.39 mL; from position 2, it appears larger.

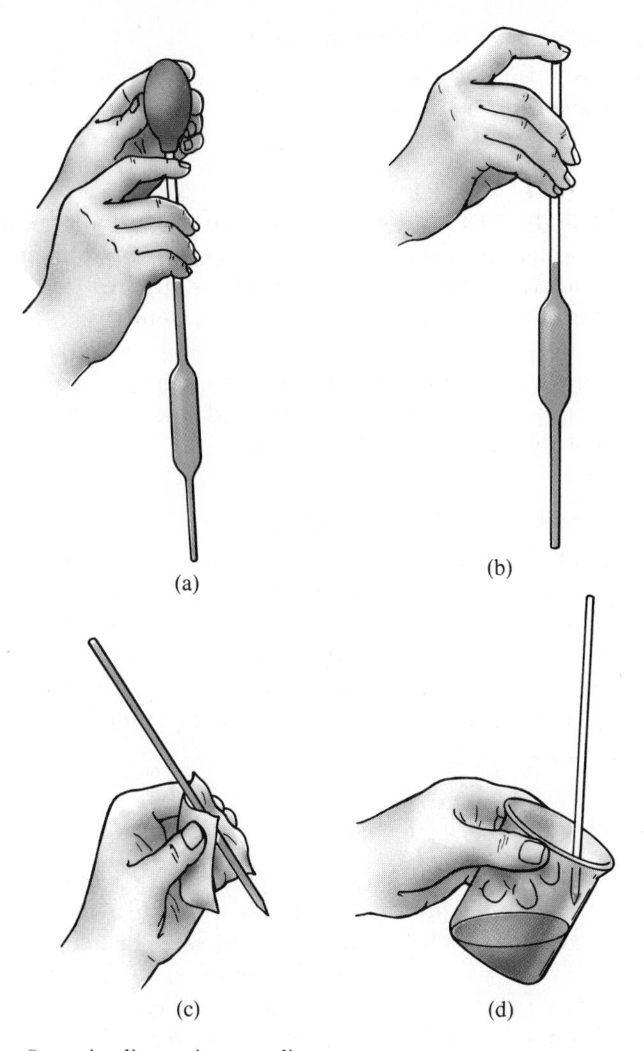

FIGURE 30-20 Steps in dispensing an aliquot.

from the vertical and wipe the exterior free of adhering liquid (Figure 30-20c). Touch the tip of the pipet to the wall of a glass vessel (*not* the container into which the aliquot is to be transferred), and slowly allow the liquid level to drop by partially releasing the forefinger (Note 1). Halt further flow as the bottom of the meniscus coincides exactly with the graduation mark. Then place the pipet tip well within the receiving vessel, and allow the liquid to drain. When free flow ceases, rest the tip against the inner wall of the receiver for a full 10 s (Figure 30-20d). Finally, withdraw the pipet with a rotating motion to remove any liquid adhering to the tip. *The small volume remaining inside the tip of a volumetric pipet should not be blown or rinsed into the receiving vessel* (Note 2).

Notes

1. The liquid can best be held at a constant level if the forefinger is *faintly* moist. Too much moisture makes control impossible.
2. Rinse the pipet thoroughly after use.

30H-6

Directions for the Use of a Buret

Before it is placed in service, a buret must be scrupulously clean; in addition, its valve must be liquid-tight.

Cleaning

Thoroughly clean the tube of the buret with detergent and a long brush. If water breaks persist after rinsing, clamp the buret in an inverted position with the end in a beaker of warm cleaning solution. Use a rubber hose to connect the buret tip with a source of vacuum. Gently pull the cleaning solution into the buret, stopping well short of the stopcock (Note 1). Allow the cleaning solution to stand for about 15 min and then drain. Rinse the buret thoroughly with tap water and then with distilled water. Inspect again for water breaks. Repeat the treatment if necessary.

Lubrication of a Glass Stopcock

Carefully remove all old grease from a glass stopcock and its barrel with a paper towel and dry both parts completely. Lightly grease the stopcock, taking care to avoid the area adjacent to the hole. Insert the stopcock into the barrel and rotate it vigorously with slight inward pressure. A proper amount of lubricant has been used when (1) the area of contact between stopcock and barrel appears nearly transparent, (2) the seal is liquid-tight, and (3) no grease has worked its way into the tip.

Notes

1. Cleaning solution often disperses more stopcock lubricant than it destroys and thus leaves a buret with a heavier grease film than before treatment. This solution should *never* be allowed to come into contact with lubricated stopcock assemblies.
2. Grease films that are unaffected by cleaning solution may yield to such organic solvents as acetone or benzene. Thorough washing with detergent should follow such treatment. The use of silicone lubricants is not recommended; contamination by such preparations is difficult—if not impossible—to remove.
3. So long as the flow of liquid is not impeded, fouling of a buret tip with stopcock grease is not a serious matter. Removal is best accomplished with organic solvents. A stoppage during a titration can be freed by *gentle* warming of the tip with a lighted match.
4. Before a buret is returned to service after reassembly, it is advisable to test for leakage. Simply fill the buret with water and establish that the volume reading does not change with time.

Filling

Make certain the stopcock is closed. Add 5 to 10 mL of the titrant, and carefully rotate the buret to wet the interior completely. Allow the liquid to drain through the tip. *Repeat this procedure at least two more times.* Then fill the buret well above the zero mark. Free the tip of air bubbles by rapidly rotating the stopcock and permitting small quantities of the titrant to pass. Finally, lower the level of the liquid just to or somewhat below the zero mark. Allow for drainage ($\simeq$ 1 min), and then record the initial volume reading, estimating to the nearest 0.01 mL.

Titration

Figure 30-21 illustrates the preferred method for the manipulation of a stopcock; when the hand is held as shown, any tendency for lateral movement by the stopcock will be in the direction of firmer seating. Be sure the tip of the buret is well within the titration vessel (ordinarily a flask). Introduce the titrant in increments of about 1 mL. Swirl (or stir) constantly to ensure thorough mixing. Decrease the size of the increments as the titration progresses; add titrant dropwise in the immediate vicinity of the end point (Note 2). When it is judged that only a few more drops are needed, rinse the walls of the container (Note 3). Allow for drainage (at least 30 s) at the completion of the titration. Then record the final volume, again to the nearest 0.01 mL.

Notes

1. When unfamiliar with a particular titration, many chemists prepare an extra sample. No care is lavished on its titration since its functions are to reveal the nature of the end point and to provide a rough estimate of titrant requirements. This deliberate sacrifice of one sample frequently results in an overall saving of time.
2. Increments smaller than one drop can be taken by allowing a small volume of titrant to form on the tip of the buret and then touching the tip to the wall of the flask. This partial drop is then combined with the bulk of the liquid as in Note 3.
3. Instead of being rinsed toward the end of a titration, the flask can be tilted and rotated so that the bulk of the liquid picks up any drops that adhere to the inner surface.

30H-7 Directions for the Use of a Volumetric Flask

Before being put into use, volumetric flasks should be washed with detergent (and cleaning solution, if necessary) and thoroughly rinsed. Only rarely do they need to be dried. If required, however, drying is best accomplished by clamping the flask in an inverted position. Insertion of a glass tube connected to a vacuum line hastens the process.

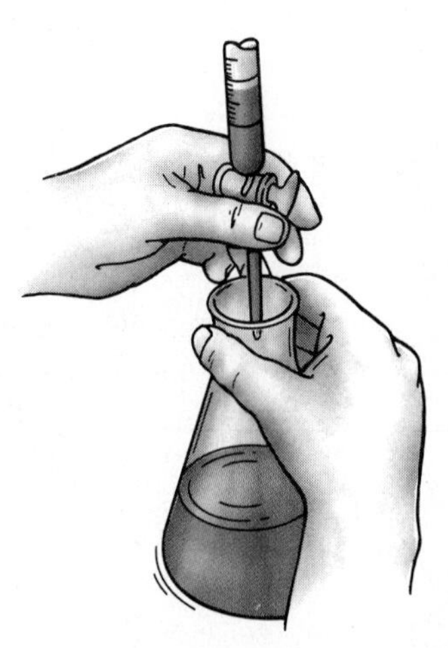

FIGURE 30-21 Recommended method for manipulation of a buret stopcock.

Direct Weighing into a Volumetric Flask

The direct preparation of a standard solution requires the introduction of a known weight of solute to a volumetric flask. Use of a powder funnel minimizes the possibility of loss of solid during the transfer. Rinse the funnel thoroughly; collect the washings in the flask.

The foregoing procedure is inappropriate if heating is needed to dissolve the solute. Instead, weigh the solid into a beaker or flask, add solvent, heat to dissolve the solute, and allow the solution to cool to room temperature. Transfer this solution quantitatively to the volumetric flask, as described in the next section.

The Quantitative Transfer of Liquid to a Volumetric Flask

Insert a funnel into the neck of the volumetric flask; use a stirring rod to direct the flow of liquid from the beaker into the funnel. Tip off the last drop of liquid on the spout of the beaker with the stirring rod. Rinse both the stirring rod and the interior of the beaker with distilled water and transfer the washings to the volumetric flask, as before. Repeat the rinsing process *at least* two more times.

Dilution to the Mark

After the solute has been transferred, fill the flask about half-full and swirl the contents to hasten solution. Add more solvent and again mix well. Bring the liquid level almost to the mark, and allow time for drainage ($\simeq$ 1 min); then use a medicine dropper to make such final additions of solvent as are necessary (Note). Firmly stopper the flask, and invert it repeatedly to ensure thorough mixing. Transfer the contents to a storage bottle that either is dry or has been thoroughly rinsed with several small portions of the solution from the flask.

Note. If, as sometimes happens, the liquid level accidentally exceeds the calibration mark, the solution can be saved by correcting for the excess volume. Use a gummed label to mark the location of the meniscus. After the flask has been emptied, carefully refill to the manufacturer's etched mark with water. With a buret, determine the additional volume needed to fill the flask so that the meniscus is at the gummed-label mark. This volume must be added to the nominal volume of the flask when calculating the concentration of the solution.

30I

The Calibration of Volumetric Ware

The reliability of a volumetric analysis depends upon agreement between the volumes purportedly and actually contained in (or delivered by) the apparatus. Calibration simply verifies this agreement or provides a correction if agreement is lacking. The latter involves the assignment of corrections to the existing volume markings or the striking of new markings that more closely agree with the nominal values.

A calibration consists of determining the mass of a liquid of known density that is contained in (or delivered by) volumetric ware. Although calibration appears to be a straightforward process, a number of important variables must be controlled. Principal among these is temperature, which influences a calibration in two ways. First, and more important, the volume occupied by a given mass of liquid varies with temperature. Second, the volume of the apparatus itself is variable, owing to the tendency of the glass to expand or contract with temperature changes.

We have noted (page 728) that the effect of buoyancy upon weighing data is most pronounced when the density of the object differs significantly from that of the weights. If water is used as the calibration liquid, a correction for buoyancy is generally necessary.

Finally, the liquid selected for calibration requires consideration. Water is the liquid of choice for most work. Mercury is also useful, particularly for small volumes. The volume contained in an apparatus will be identical with that delivered by it because mercury does not wet glass surfaces. A small correction must be applied to account for the convex meniscus of mercury. The magnitude of this correction depends upon the diameter of the apparatus at the graduation mark.

The calculations associated with calibration, while not difficult, are somewhat involved. The raw weighing data are first corrected for buoyancy with Equation 30-1. Next, the volume of the apparatus at the temperature of calibration (T) is obtained by dividing the density of the liquid at that temperature into the corrected weight. Finally, this volume is corrected to the standard temperature of 20°C as in Example 30-2.

Table 30-3 is provided to ease the computational burden of calibration. Corrections for buoyancy with respect to stainless steel or brass weights (the density difference between the two is small enough to be neglected) and for the volume change of water and of glass containers have been incorporated into these data. Multiplication by the appropriate factor from Table 30-3 converts the mass of water at temperature T to (1) the corresponding volume at that temperature and (2) the volume at 20°C.

EXAMPLE 30-4

A 25-mL pipet delivers 24.976 g of water weighed against stainless steel weights at 25°C. Use the data in Table 30-3 to calculate the volume delivered by this pipet at 25 and 20°C.

$$\text{At } 25°\text{C: } V = 24.976 \text{ g} \times 1.0040 \text{ mL/g} = 25.08 \text{ mL}$$

$$\text{At } 20°\text{C: } V = 24.976 \text{ g} \times 1.0037 \text{ mL/g} = 25.07 \text{ mL}$$

30I-1 General Directions for Calibration Work

All volumetric ware should be painstakingly freed of water breaks before being calibrated. Burets and pipets need not be dry; volumetric flasks should be thoroughly drained and dried at room temperature. The water used for calibration should be in thermal equilibrium with its surroundings. This condition is best established by drawing the water well in advance, noting its temperature at frequent intervals, and waiting until no further changes occur.

Although an analytical balance can be used for calibration, weighings to the nearest milligram are perfectly satisfactory except for all but the very smallest volumes. Thus a top-loading balance is more conveniently used. Weighing bottles or small, well-stoppered conical flasks can serve as receivers for the calibration liquid.

Calibration of a Volumetric Pipet

Determine the empty weight of the stoppered receiver to the nearest milligram. Transfer a portion of temperature-equilibrated water to the receiver with the pipet (Section 30H-5), weigh the receiver and its contents (again, to the nearest milligram), and calculate the weight of water delivered from the difference in these weights. Calculate the volume delivered with the aid of Table 30-3. Repeat the calibration several times; calculate the mean volume delivered and its standard deviation.

Calibration of a Buret

Fill the buret with temperature-equilibrated water and make sure that no air bubbles are trapped in the tip. Allow about 1 min for drainage; then lower the liquid level to bring the bottom of the meniscus to the 0.00-mL mark. Touch the tip to the wall of a beaker to remove any adhering drop. Wait 10 min and recheck the volume; if the stopcock is tight, there should be no perceptible change. During this interval, weigh (to the nearest milligram) a 125-mL conical flask fitted with a rubber stopper.

Once tightness of the stopcock has been established, slowly transfer (at about 10 mL/min) approximately 10 mL of water to the flask. Touch the tip to the wall of the flask. Wait 1 min, record the volume that was apparently delivered, and refill the buret. Weigh the flask and its contents to the nearest milligram; the difference between this weight and the initial value gives the

TABLE 30-3 **Volume Occupied by 1.0000 g of Water Weighed in Air Against Stainless Steel Weights**

Temperature, *T* °C	Volume, mL At *T*	Volume, mL Corrected to 20°C
10	1.0013	1.0016
11	1.0014	1.0016
12	1.0015	1.0017
13	1.0016	1.0018
14	1.0018	1.0019
15	1.0019	1.0020
16	1.0021	1.0022
17	1.0022	1.0023
18	1.0024	1.0025
19	1.0026	1.0026
20	1.0028	1.0028
21	1.0030	1.0030
22	1.0033	1.0032
23	1.0035	1.0034
24	1.0037	1.0036
25	1.0040	1.0037
26	1.0043	1.0041
27	1.0045	1.0043
28	1.0048	1.0046
29	1.0051	1.0048
30	1.0054	1.0052

Corrections for buoyancy (stainless steel weights) and change in container volume have been applied.

mass of water delivered. Use Table 30-3 to convert this mass to the true volume. Subtract the apparent volume from the true volume. This difference is the correction that should be applied to the apparent volume to give the true volume. Repeat the calibration until agreement within ± 0.02 mL is achieved.

Starting again from the zero mark, repeat the calibration, this time delivering about 20 mL to the receiver. Test the buret at 10-mL intervals over its entire volume. Prepare a plot of the correction to be applied as a function of volume delivered. The correction associated with any interval can be determined from this plot.

Calibration of a Volumetric Flask with a Single-Pan Balance

Weigh the clean, dry flask to the nearest milligram. Then fill to the mark with equilibrated water and reweigh. Calculate the volume contained with the aid of Table 30-3.

A glass tube that has been drawn out to a tip or a medicine dropper is useful in making final adjustments to the liquid level.

Calibration of a Volumetric Flask Relative to a Pipet

The calibration of a volumetric flask relative to a pipet provides an excellent method for partitioning a sample into aliquots. These directions pertain to a 50-mL pipet and a 500-mL volumetric flask; other combinations are equally convenient.

Carefully transfer ten 50-mL aliquots from the pipet to a dry 500-mL volumetric flask. Mark the location of the meniscus with a gummed label. Cover with a label varnish to ensure permanence. Once the flask is calibrated in this way dilution to the label permits the same pipet to deliver precisely a one-tenth aliquot of the solution in the flask. Note that if the pipet is changed, recalibration with the new pipet is necessary.

30J The Laboratory Notebook

A laboratory notebook is needed to record measurements and observations concerning an analysis. The book should be permanently bound with consecutively numbered pages (if necessary, the pages should be hand-numbered before any entries are made). Most notebooks have more than ample room; there is no need to crowd entries.

The first few pages should be saved for a table of contents that is updated as entries are made.

30J-1 Rules for the Maintenance of a Laboratory Notebook

1. *Record all data and observations directly into the notebook in ink.* Neatness is greatly to be desired. Nevertheless, the pursuit of this quality must not include the transcribing of observations from a sheet of paper to the notebook or from one notebook to another. The risk of misplacing—or incorrectly transcribing—crucial data and thereby ruining an experiment is unacceptable.
2. Supply each entry or series of entries with a heading or label. A series of weighing data for a set of empty crucibles should carry the heading "empty crucible weights" (or something similar), for example, and the weight of each crucible should be identified by the same number or

letter used to label the crucible. The significance of such an entry is obvious when it is recorded but may become unclear with the passage of time.

3. Date each page of the notebook as it is used.
4. *Never* attempt to erase or obliterate an incorrect entry. Instead, cross it out with a single horizontal line and locate the correct entry as nearby as possible. Do not write over incorrect numbers; with time, it may become impossible to distinguish the correct entry from the incorrect one.
5. Never remove a page from the notebook. Draw diagonal lines across any page that is to be disregarded. Provide a brief rationale for disregarding the page.

30J-2

Format

The instructor should be consulted concerning the format to be used in keeping the laboratory notebook.[10] One convention involves using each page consecutively for the recording of data and observations as they occur. The completed analysis is then summarized on the next available page spread (that is, left and right facing pages). As shown in Figure 30-22, the first of these two facing pages should contain the following entries:

1. The title of the experiment ("The Gravimetric Determination of Chloride").
2. A brief statement of the principles upon which the analysis is based.
3. A complete summary of the weighing, volumetric, and/or instrument response data needed to calculate the results.
4. A report of the best value for the set and a statement of its precision.

The second page should contain the following items:

1. Equations for the principal reactions in the analysis.
2. An equation showing how the results were calculated.
3. A summary of observations that appear to bear upon the validity of a particular result or the analysis as a whole. *Any such entry must have been originally recorded in the notebook at the time the observation was made.*

30K

Safety in the Laboratory

Work in a chemical laboratory necessarily involves a degree of risk; accidents can and do happen. Strict adherence to the following rules will go far toward preventing (or minimizing the effect of) accidents.

1. At the outset, learn the location of the nearest eye fountain, fire blanket, shower, and fire extinguisher. Learn the proper use of each, and do not hesitate to use this equipment should the need arise.
2. *WEAR EYE PROTECTION AT ALL TIMES.* The potential for serious and perhaps permanent eye injury makes it mandatory that adequate eye protection be worn at all times by students, instructors, and visitors. Eye protection should be donned before entering the laboratory

[10]See also Howard M. Kanare, *Writing the Laboratory Notebook.* Washington, D.C. 20036: The American Chemical Society, 1985.

Gravimetric Determination of Chloride 08

The Chloride in a soluble sample was precipitated as AgCl and weighed as such

Sample weights	1	2	3
Wt. bottle plus sample, g	27.6115	27.2185	26.8105
– less sample, g	27.2185	26.8105	26.4517
wt. sample, g	0.3930	0.4080	0.3588
Crucible weights, empty	~~20.7925~~	~~22.8311~~	~~21.2488~~
	20.7926	22.8311	~~21.2482~~
			21.2483
Crucible weights, with AgCl, g	~~21.4294~~	~~23.4920~~	~~21.8324~~
	~~21.4297~~	~~23.4914~~	21.8323
	21.4296	23.4915	
Weight of AgCl, g	0.6370	0.6604	0.5840
Percent Cl⁻	40.10	40.04	40.27
Average percent Cl⁻		40.12	
Relative standard deviation	3.0 parts per thousand		

Date Started 1-9-88
Date Completed 1-16-88

FIGURE 30-22 Summary data page of a laboratory notebook.

and should be used continuously until it is time to leave. Serious eye injuries have occurred to people performing such innocuous tasks as computing or writing in a laboratory notebook; such incidents are usually the result of someone else's losing control of an experiment. Regular prescription glasses are not adequate substitutes for eye protection approved by the Office of Safety and Health Administration (OSHA). Contact lenses should never be used in the laboratory because laboratory fumes may react with them and have a harmful effect on the eyes.

3. Most of the chemicals in a laboratory are toxic; some are very toxic, and some—such as concentrated solutions of acids and bases—are highly corrosive. Avoid contact between these liquids and the skin. In the event of such contact, *immediately* flood the affected area with copious quantities of water. If a corrosive solution is spilled on clothing, remove the garmet immediately. Time is of the essence; modesty cannot be a matter of concern.
4. *NEVER* perform an unauthorized experiment. Such activity is grounds for disqualification at many institutions.

5. Never work alone in the laboratory; be certain that someone is always within earshot.
6. Never bring food or beverages into the laboratory. Do not drink from laboratory glassware. Do not smoke in the laboratory.
7. Always use a bulb to draw liquids into a pipet; *NEVER* use the mouth to provide suction.
8. Wear adequate foot covering (no sandals). Confine long hair with a net. A laboratory coat or apron will provide some protection and may be required.
9. Be extremely tentative in touching objects that have been heated; hot glass looks just like cold glass.
10. Always fire-polish the ends of freshly cut glass tubing. *NEVER* attempt to force glass tubing through the hole of a stopper. Instead, make sure that both tubing and hole are wet with soapy water. Protect hands with several layers of towel while inserting glass into a stopper.
11. Use fume hoods whenever toxic or noxious gases are likely to be evolved. Be cautious in testing for odors; use the hand to waft vapors above containers toward the nose.
12. Notify the instructor in the event of an injury.
13. Dispose of solutions and chemicals as instructed. It is illegal to flush solutions containing heavy metal ions or organic liquids down the drain in many localities; alternative arrangements are required for the disposal of such liquids.

CHAPTER 31 Selected Methods of Analysis

This chapter contains detailed directions for performing a variety of chemical analyses. The methods have been chosen with the aim of providing experience with common analytical techniques that are widely used by chemists.

The chances of success in the laboratory greatly improve when time is taken at the outset to read carefully and *understand* each step in an analytical method and to develop a plan for how and when each step is to be performed. For greatest efficiency, such study and planning should take place *before entering the laboratory.*

The recommendations that follow regarding laboratory work will not only help provide an understanding of the various experiments to be undertaken but will also result in more efficient use of the time available for laboratory work.

Background Information

Before an analysis is undertaken, the student should understand the significance of each step in the procedure in order to avoid the pitfalls and potential sources of error present in all analytical methods. Information about these steps can usually be found in the discussion section that precedes most of the procedures, in sections of earlier chapters that are referred to in the discussion section, and in the "Notes" that follow many of the procedures. If these sources fail to explain the need for certain of the recommended steps, the instructor should be consulted before the laboratory work is begun.

The Accuracy of Measurements

In studying a procedure, it is wise to distinguish between those measurements that must be made with maximum precision, and thus with maximum care, and those that can be carried out rapidly with little concern for precision. Generally, measurements that appear in the equation used to compute the results must be performed with maximum precision. The remaining measurements can *and should* be made less carefully to conserve time. Often the words *about* and *approximately* are used to indicate that a measurement does not have to be done carefully. It is a waste of time and effort to measure, let us say, a volume to ± 0.02 mL when an uncertainty of ± 0.5 mL or even ± 5 mL will have no discernible effect on the results.

In some procedures, a statement such as "weigh three 0.5-g samples to the nearest 0.1 mg . . ." is encountered. Here, samples of perhaps 0.4 to 0.6 g are to be taken and weighed to the nearest 0.1 mg. The number of significant figures in the specification of a volume or a weight is also a guide as to the care that should be taken in making a measurement. For example, the statement "add 10.00 mL of a solution to the beaker" indicates that the volume should be measured carefully with a buret or pipet with the aim of limiting the uncertainty to perhaps ±0.02 mL.

Time Utilization

The time requirements of the several unit operations involved in an analysis should be studied before work is started. Such study will reveal operations that require considerable elapsed, or clock, time but little or no operator time—for example, when a sample is dried in an oven, cooled in a desiccator, or evaporated on a hot plate. The experienced chemist plans to use such periods of waiting to perform other operations or perhaps to begin a new analysis. Some workers find it worthwhile to prepare a written time schedule for each laboratory period to avoid periods when no work can be done.

Time planning is also needed to identify places where an analysis can be interrupted for overnight or longer, as well as those operations that must be completed without a break.

Reagents

Directions for the preparation of reagents accompany many of the procedures. Before preparing such reagents, it is wise to check to see if they are already prepared and available for general use.

If a reagent is known to pose a hazard, the student should plan *in advance of the laboratory period* what steps must be taken to minimize injury or damage. Furthermore, rules for the disposal of waste liquids and solids should be known and followed. These rules vary from one part of the country to another and from one laboratory to another.

Water

Some laboratories use deionizers to purify water; others employ stills for this purpose. The terms "distilled water" and "deionized water" are used interchangeably in the directions that follow. Either type is satisfactory for analytical work.

Tap water should be employed only for preliminary cleaning of glassware. The cleaned glassware is then rinsed with at least three small portions of distilled or deionized water.

31A Gravimetric Methods of Analysis

General aspects, calculations, and typical applications of gravimetric analysis are discussed in Chapter 3.

31A-1 The Gravimetric Determination of Chloride in a Soluble Sample

Discussion

The chloride content of a soluble salt can be determined by precipitation of the anion as silver chloride:

$$Ag^+ + Cl^- \longrightarrow AgCl(s)$$

The precipitate is collected in a weighed filtering crucible and washed; its weight is determined after it has been dried to constant weight at 110°C.

The solution containing the sample is kept somewhat acidic during the precipitation to eliminate possible interference from anions of weak acids (such as CO_3^{2-}) that form sparingly soluble silver salts in a neutral environment. A moderate excess of silver ion is needed to diminish the solubility of silver chloride, but an excess is avoided to minimize coprecipitation of silver nitrate.

Silver chloride forms first as a colloid and is subsequently coagulated with heat. Nitric acid and the small excess of silver nitrate promote coagulation by providing a moderately high electrolyte concentration. Nitric acid in the wash solution maintains the electrolyte concentration and eliminates the possibility of peptization during the washing step; the acid subsequently decomposes to give volatile products when the precipitate is dried. See Section 3B-2 for additional information concerning the properties and treatment of colloidal precipitates.

In common with other silver halides, finely divided silver chloride undergoes photodecomposition:

$$2AgCl(s) \longrightarrow 2Ag(s) + Cl_2(g)$$

The elemental silver produced in this reaction is responsible for the violet color that develops in the precipitate. In principle, this reaction leads to low results for chloride ion. In practice, however, its effect is negligible provided direct and prolonged exposure to sunlight is avoided.

If photodecomposition of silver chloride occurs before filtration, the additional reaction

$$3Cl_2(aq) + 3H_2O + 5Ag^+ \longrightarrow 5AgCl(s) + ClO_3^- + 6H^+$$

tends to cause high results.

Some photodecomposition of silver chloride is inevitable as the analysis is ordinarily performed. It is worthwhile to minimize exposure of the solid to intense sources of light as far as possible.

Iodide, bromide, and thiocyanate, if present, precipitate along with silver chloride and cause high results. Additional interference can be expected from tin and antimony, which are likely to precipitate as oxychlorides under the conditions of the analysis.

Because silver nitrate is expensive, any unused reagent should be collected in a storage container; similarly, precipitated silver chloride should be retained after the analysis is complete.[1]

PROCEDURE

Clean three sintered-glass or porcelain filtering crucibles by allowing about 5 mL of concentrated HNO_3 to stand in each for about 5 min. Use a vacuum (Figure 30-17) to draw the acid through the crucible. Rinse each crucible with three portions of tap water, and then discontinue the vacuum. Next, add about 5 mL of 6 M NH_3 and wait for about 5 min before drawing it through the filter. Finally, rinse each crucible with six to eight portions of distilled or deionized water. Provide each crucible with an identifying mark. Bring the crucibles to constant weight by heating at 110°C while the other steps in the analysis are being carried out. The first drying should be for at least 1 h; subsequent heating periods can be somewhat shorter (30 to 40 min).

Transfer the unknown to a weighing bottle and dry it at 110°C (Figure 30-9) for 1 to 2 h; allow the bottle and contents to cool to room temperature in a desiccator. Weigh (to the nearest 0.1 mg) individual samples by difference (page 733) into

[1]Silver can be recovered from silver chloride and from surplus reagent by reduction with ascorbic acid; see J. W. Hill and L. Bellows, *J. Chem. Educ.*, **1986**, *63* (4), 357; see also J. P. Rawat and S. Iqbal M. Kamoonpuri, *ibid.*, **1986**, *63* (6), 537 for recovery (as $AgNO_3$) based on ion exchange.

400-mL beakers (Note 1). Dissolve each sample in about 100 mL of distilled water to which 2 to 3 mL of 6 M HNO_3 has been added.

Slowly, and with good stirring, add 0.2 M $AgNO_3$ to each of the cold sample solutions until AgCl is observed to coagulate (Notes 2, 3); then introduce an additional 3 to 5 mL. Heat almost to boiling, and digest the solids for about 10 min. Add a few drops of $AgNO_3$ to confirm that precipitation is complete. If additional precipitate forms, add about 3 mL of $AgNO_3$, digest, and again test for completeness of precipitation. Pour any unused $AgNO_3$ into a waste container (NOT into the original reagent bottle). Cover each beaker, and store in a dark place for at least 2 h and preferably until the next laboratory period.

Read the instructions for filtration in Section 30G-2. Decant the supernatant liquids through weighed filtering crucibles. Wash the precipitates several times (while they are still in the beaker) with a wash solution consisting of 2 to 5 mL of 6 M HNO_3 per liter of distilled water; decant these washings through the filters. Quantitatively transfer the AgCl from the beakers to the individual crucibles with fine streams of wash solution; use rubber policemen to dislodge any particles that adhere to the walls of the beakers. Continue washing until the filtrates are essentially free of Ag^+ ion (Note 4).

Dry the precipitate at 110°C for at least 1 h. Store the crucibles in a desiccator while they cool. Determine the weight of the crucibles and their contents. Repeat the cycle of heating, cooling, and weighing until consecutive weighings agree to within 0.2 mg. Calculate the percentage of Cl^- in the sample.

Upon completion of the analysis, remove the precipitates by gently tapping the crucibles over a piece of glazed paper. Transfer the collected AgCl to a container for silver wastes. Remove the last traces of AgCl by filling the crucibles with 6 M NH_3 and allowing them to stand.

Notes

1. Consult with the instructor concerning an appropriate sample size.
2. Determine the approximate amount of $AgNO_3$ needed by calculating the volume that would be required if the unknown were pure NaCl.
3. Use a separate stirring rod for each sample and leave it in its beaker throughout the determination.
4. To test the washings for Ag^+, collect a small volume in a test tube and add a few drops of HCl. Washing is judged complete when little or no turbidity develops.

31A-2 The Gravimetric Determination of Tin in Brass

Discussion

Brasses are important alloys. Copper is ordinarily the principal constituent, with lesser amounts of lead, zinc, tin, and possibly other elements being present as well. Treatment of a brass with nitric acid results in the formation of the sparingly soluble "metastannic acid" $H_2SnO_3 \cdot xH_2O$; all other constituents are dissolved. The solid is filtered, washed, and ignited to SnO_2.

The gravimetric determination of tin provides experience in the use of ashless filter paper and is frequently performed in conjunction with a more inclusive analysis of a brass sample.

PROCEDURE

Provide identifying marks on three porcelain crucibles and their covers. During waiting periods in the experiment, bring these to constant weight by ignition at 900°C in a muffle furnace.

Do not dry the unknown. If so instructed, rinse it with acetone to remove any oil or grease. Weigh (to the nearest 0.1 mg) approximately 1 g samples of the unknown into 250-mL beakers. Cover the beakers with watch glasses. Place the beakers in the HOOD, and cautiously introduce a mixture containing about 15 mL of concentrated HNO_3 and 10 mL of H_2O. Digest the samples for at least 30 min; add more HNO_3 if necessary. Rinse the watch glasses; then evaporate the solutions to about 5 mL but not to dryness (Note 1).

Add about 5 mL of 3 M HNO_3, 25 mL of distilled water, and one quarter of a tablet of filter paper pulp to each sample; heat without boiling for about 45 min. Collect the precipitated $H_2SnO_3{\cdot}xH_2O$ on fine-porosity ashless filter papers (Section 30G-3 and Notes 2, 3). Use many small volumes of hot 0.3 M HNO_3 to wash the last traces of copper from the precipitate. Test for completeness of washing with a drop of NH_3 on the top of the precipitate; wash further if the precipitate turns blue.

Remove the filter paper and its contents from the funnels, fold, and place in crucibles that, with their covers, have been brought to constant weight (Figure 30-15). Ash the filter paper at as low a temperature as possible. There must be free access of air throughout the charring (Section 30G-3 and Figure 30-16). Gradually increase the temperature until all the carbon has been removed. Then bring the covered crucibles and their contents to constant weight in a 900°C furnace (Note 4). Calculate the percentage of tin in the unknown.

Notes

1. It is often time-consuming and difficult to redissolve the soluble components of the residue obtained when a sample is evaporated to dryness.
2. The filtration step can be quite time-consuming and once started cannot be interrupted.
3. If the unknown is to be analyzed electrolytically for its lead and copper content (Section 31J-1), collect the filtrates in tall-form beakers. The final volume should be about 125 mL; evaporate to that volume if necessary. If the analysis is for tin only, the volume of washings is not important.
4. Partial reduction of SnO_2 may cause the ignited precipitate to appear gray. In this case, add a drop of nitric acid, cautiously evaporate, and ignite again.

31A-3 The Gravimetric Determination of Nickel in Steel

Discussion

The nickel in a steel sample can be precipitated from a slightly alkaline medium with an alcoholic solution of dimethylglyoxime (page 77). Interference from iron(III) is eliminated by masking with tartaric acid. The product is freed of moisture by drying at 110°C.

The bulky character of nickel dimethylglyoxime limits the weight of nickel that can be accommodated conveniently and thus the sample weight. Care must also be taken to control the excess of alcoholic dimethylglyoxime used. If too much is added, the alcohol concentration becomes great enough to dissolve appreciable amounts of the nickel dimethylglyoxime, which leads to low results. If the alcohol concentration becomes too low, however, some of the reagent may precipitate, giving a positive error.

PREPARATION OF SOLUTIONS

(a) *Dimethylglyoxime, 1% (w/v).* Dissolve 10 g of dimethylglyoxime in 1 L of ethanol. (Sufficient for about 50 precipitations.)

(b) *Tartaric acid, 15% (w/v).* Dissolve 225 g of tartaric acid in sufficient water to give 1500 mL of solution. Filter before use if the solution is not clear. (Sufficient for about 50 precipitations.)

PROCEDURE

Clean and mark three medium-porosity sintered-glass crucibles (Note 1); bring them to constant weight by drying at 110°C for at least 1 h.

Weigh (to the nearest 0.1 mg) samples containing between 30 and 35 mg of nickel into individual 400-mL beakers (Note 2). In the HOOD, dissolve each sample in about 50 mL of 6 M HCl with gentle warming. Carefully add approximately 15 mL of 6 M HNO_3, and boil gently to expel any oxides of nitrogen that may have been produced. Dilute to about 200 mL and heat to boiling. Introduce about 30 mL of 15% tartaric acid and sufficient concentrated NH_3 to produce a faint odor of NH_3 in the vapors over the solutions (Note 3); then add another 1 to 2 mL of NH_3. If the solutions are not clear at this stage, proceed as directed in Note 4. Make the solutions acidic with HCl (no odor of NH_3), heat to 60 to 80°C, and add about 20 mL of the 1% dimethylglyoxime solution. With good stirring, add 6 M NH_3 until a slight excess exists (faint odor of NH_3) plus an additional 1 to 2 mL. Digest the precipitates for 30 to 60 min, cool for at least 1 h, and filter.

Wash the solids with water until the washings are free of Cl^- (Note 5). Bring the crucible and their contents to constant weight at 110°C. Report the percentage of nickel in the sample. The dried precipitate has the composition $Ni(C_4H_7O_2N_2)_2$(fw = 288.93).

Notes

1. Medium-porosity porcelain filtering crucibles or Gooch crucibles with glass pads can be substituted for sintered-glass crucibles in this determination.
2. Use a separate stirring rod for each sample and leave it in the beaker throughout the analysis.
3. The existence or absence of excess NH_3 is readily established by odor; the vapors over the container should be wafted toward the nose with a waving motion of the hand.
4. If $Fe_2O_3 \cdot xH_2O$ forms upon addition of NH_3, acidify the solution with HCl, introduce additional tartaric acid, and neutralize again. Alternatively, remove the solid by filtration. Thorough washing with a hot NH_3/NH_4Cl solution is required; the washings are combined with the solution containing the bulk of the sample.
5. Test the washings for Cl^- by collecting a small portion in a test tube, acidifying with HNO_3, and adding a drop or two of 0.1 M $AgNO_3$. Washing is judged complete when little or no turbidity develops.

31B Precipitation Titrations

As noted in Chapter 7, most precipitation titrations make use of a standard silver nitrate solution as titrant. Directions follow for the volumetric titration of chloride ion using an adsorption indicator and for the weight titration of the same species by the Mohr method, in which chromate ion serves as the indicator.

31B-1 Preparation of a Standard Silver Nitrate Solution

PROCEDURE

Use a top-loading balance to transfer the approximate weight of $AgNO_3$ to a weighing bottle (Note 1). Dry at 110°C for about 1 h but not much longer (Note 2), and then cool to room temperature in a desiccator. Weigh the bottle and contents (to the nearest 0.1 mg). Transfer the bulk of the $AgNO_3$ to a volumetric flask using a powder funnel. Cap the weighing bottle and reweigh it and any solid that remains.

Rinse the powder funnel thoroughly. Dissolve the $AgNO_3$, dilute to the mark with water, and mix well (Note 3). Calculate the molar concentration of this solution.

Notes

1. Consult with the instructor concerning the volume and concentration of $AgNO_3$ to be prepared. The weight of $AgNO_3$ to be taken is as follows:

Silver Ion Concentration, M	Approximate Weight (g) of $AgNO_3$ Needed to Prepare 1000 mL	500 mL	250 mL
0.10	16.9	8.5	4.2
0.05	8.5	4.2	2.1
0.02	3.4	1.8	1.0

2. Prolonged heating causes partial decomposition of $AgNO_3$. Some discoloration may occur, even after only 1 h at 110°C; the effect of this decomposition on the purity of the reagent is ordinarily imperceptible.
3. Silver nitrate solutions should be stored in a dark place when not in use.

31B-2 The Determination of Chloride by Titration with an Adsorption Indicator

Discussion

In this titration, the anionic adsorption indicator dichlorofluorescein is used to locate the end point. With the first excess of titrant, the indicator becomes incorporated in the counter-ion layer surrounding the silver chloride and imparts color to the solid (Section 7A-4). In order to obtain a satisfactory color change, it is desirable to maintain the particles of silver chloride in the colloidal state. Dextrin is added to the solution to stabilize the colloid and prevent its coagulation.

PREPARATION OF SOLUTIONS

Dichlorofluorescein indicator. Dissolve 0.2 g of dichlorofluorescein in a solution prepared by mixing 75 mL of ethanol and 25 mL of water. (Sufficient for several hundred titrations.)

PROCEDURE

Dry the unknown at 110°C for about 1 h; allow it to return to room temperature in a desiccator. Weigh individual samples (to the nearest 0.1 mg) into individual conical flasks, and dissolve them in appropriate volumes of distilled water (Note 1). To each, add about 0.1 g of dextrin and 5 drops of indicator. Titrate (Note 2) with $AgNO_3$ to the first permanent pink color of silver dichlorofluoresceinate. Report the percentage of Cl^- in the unknown.

Notes

1. Use 0.25-g samples for 0.1 M $AgNO_3$ and about half that amount for 0.05 M reagent. Dissolve the former in about 200 mL of distilled water and the latter in about 100 mL. If 0.02 M $AgNO_3$ is to be used, a 0.4-g sample should be weighed into a 500-mL volumetric flask, and 50-mL aliquots should be taken for titration.
2. Colloidal AgCl is sensitive to photodecomposition, particularly in the presence of the indicator; attempts to perform the titration in direct sunlight will fail. If photodecomposition appears to be a problem, establish the approximate end point with a rough preliminary titration, and use this information to estimate the volumes of $AgNO_3$ needed for the other samples. For each subsequent

sample, add the indicator and dextrin only after most of the $AgNO_3$ has been added, and then complete the titration without delay.

31B-3

The Determination of Chloride by a Weight Titration Based on the Mohr Method

Discussion

The Mohr method uses CrO_4^{2-} ion as an indicator in the titration of chloride ion with silver nitrate. The first excess of titrant results in the formation of a red silver chromate precipitate, which signals the end point. See Section 7B-4 for additional details.

Instead of a buret, a balance is employed in this procedure to determine the weight of silver nitrate solution needed to reach the end point. The concentration of the silver nitrate is most conveniently determined by standardization against primary-standard sodium chloride, although direct preparation by weight is also feasible. The reagent concentration is expressed as weight molarity (mmol $AgNO_3$/g of solution).

PREPARATION OF SOLUTIONS

(a) *Silver nitrate, approximately 0.1 mmol/g of solution.* (Sufficient for about ten titrations.) Dissolve about 4.5 g of $AgNO_3$ in about 500 mL of distilled water. Standardize the solution against weighed quantities of reagent-grade NaCl as directed in *Procedure* (Note). Express the concentration as weight molarity (mmol $AgNO_3$/g of solution). When not in use, the solution should be stored in a dark place.

(b) *Potassium chromate, 5%.* (Sufficient for about ten titrations.) Dissolve about 1.0 g of K_2CrO_4 in about 20 mL of distilled water.

Note. Alternatively, standard $AgNO_3$ can be prepared directly by weight. To do so, follow the directions in Section 31B-1 for weighing out a known amount of primary-standard $AgNO_3$. Use a powder funnel to transfer the weighed $AgNO_3$ to a 500-mL polyethylene bottle that has been previously weighed to the nearest 10 mg. Add about 500 mL of water and weigh again. Calculate the weight molarity.

PROCEDURE

Dry the unknown at 110°C for at least 1 h (Note 1). Cool in a desiccator. Consult with the instructor for a suitable sample size. Weigh (to the nearest 0.1 mg) individual samples into 250-mL conical flasks, and dissolve in about 100 mL of distilled water. Add small quantities of $NaHCO_3$ until effervescence ceases. Introduce about 2 mL of K_2CrO_4 solution, and titrate (Note 2) to the first permanent appearance of red Ag_2CrO_4.

Determine an indicator blank by suspending a small amount of chloride-free $CaCO_3$ in 100 mL of distilled water containing 2 mL of K_2CrO_4.

Correct reagent weights for the blank. Report the percentage of Cl^- in the unknown.

Dispose of AgCl and reagents as directed by the instructor.

Notes

1. The $AgNO_3$ is conveniently standardized concurrently with the analysis. Dry reagent-grade NaCl for about 1 h. Cool; then weigh (to the nearest 0.1 mg) 0.25-g portions into conical flasks and titrate as above.
2. Directions for performing a weight titration are given in Section 30F-2.

31C

Neutralization Titrations

Neutralization titrations are performed with standard solutions of strong acids or bases. While a single solution (of either acid or base) is sufficient for the titration of a given type of analyte, it is convenient to have standard solutions of both acid and base available in the event a back-titration is needed to locate end points more exactly. The concentration of one solution is established by titration against a primary standard; the concentration of the other is then determined from the acid/base ratio (that is, the volume of acid needed to neutralize 1.000 mL of the base).

The directions in this section call for volumetric measurements. The methods are readily adapted to the weight titration technique described in Section 30F, however. In the latter case, weight molarity (mmol reagent/g solution) is computed from standardization data and used to calculate analyte concentrations.

31C-1

The Effect of Atmospheric Carbon Dioxide on Neutralization Titrations

Water in equilibrium with the atmosphere is about 1×10^{-5} M in carbonic acid as a consequence of the equilibrium

$$CO_2(g) + H_2O \rightleftharpoons H_2CO_3(aq)$$

At this concentration level, the amount of 0.1 M base consumed by the carbonic acid in a typical titration is negligible. With more dilute reagents (<0.05 M), however, the water used as a solvent for the analyte and in the preparation of reagents must be freed of carbonic acid by boiling for a brief period.

Water that has been purified by distillation rather than by deionization is often supersaturated with carbon dioxide and may thus contain sufficient acid to affect the results of an analysis.[2] The instructions that follow are based upon the assumption that the amount of carbon dioxide in the water supply can be neglected without causing serious error. For further discussion on the effects of carbon dioxide in neutralization titrations, see Section 10A-3.

31C-2

Preparation of Indicator Solutions for Neutralization Titrations

Discussion

The theory of acid/base indicators is discussed in Section 8A-2, and the structures of some common indicators are shown in Section 8G-1. An indicator exists for virtually any pH range between 1 and 13.[3] Directions follow for the preparation of indicator solutions suitable for most neutralization titrations.

PROCEDURE

Stock solutions ordinarily contain between 0.5 and 1.0 g of indicator per liter. (One liter of indicator is sufficient for hundreds of titrations.)

(a) *Methyl orange, methyl red, and sulphonthaleins.* Dissolve the sodium salt directly in distilled or deionized water. (The common sulphonthaleins include bromocresol

[2]Water that is to be used for neutralization titrations can be tested by adding 5 drops of phenolphthalein to a 500-mL portion. Less than 0.2 to 0.3 mL of 0.1 M OH^- should suffice to produce the first faint pink color of the indicator. If a larger volume is needed, the water should be boiled and cooled before it is used to prepare standard solutions or to dissolve samples.

[3]See, for example, J. Beukenkemp and W. Rieman III, in *Treatise on Analytical Chemistry*, I. M. Kolthoff and P. J. Elving, Eds., Part I, Vol. 11, pp. 6987–7001. New York: Wiley, 1974.

green, bromothymol blue, bromophenol blue, thymol blue, cresol red, and phenol red.)

(b) *Phenolphthalein, thymolphthalein.* Dissolve the solid indicator in a solution consisting of 800 mL ethanol and 200 mL of distilled or deionized water.

31C-3

Preparation of Dilute Hydrochloric Acid Solutions

Discussion

The preparation and standardization of acids are considered in Sections 10A-1 and 10A-2.

PROCEDURE

For a 0.1 M solution, add about 8 mL of concentrated HCl to about 1 L of distilled or deionized water (Note). Mix thoroughly, and store in a glass-stoppered bottle.

Note. It is advisable to eliminate CO_2 from the water by a preliminary boiling if very dilute solutions (<0.05 M) are being prepared.

31C-4

Preparation of Carbonate-Free Sodium Hydroxide

Discussion

See Sections 10A-3 and 10A-4 for information concerning the preparation and standardization of bases.

PROCEDURE

If so directed by the instructor, prepare a bottle for protected storage (Figure 10-1; Note 1). Transfer 1 L of distilled or deionized water to the storage bottle (see the Note in Section 31C-3). Decant 4 to 5 mL of 50% NaOH into a small container (Note 2), add it to the water, and *mix thoroughly*. USE EXTREME CARE IN HANDLING 50% NaOH, which is highly corrosive. If the reagent comes into contact with skin, IMMEDIATELY flush the area with *copious* amounts of water.

Protect the solution from unnecessary contact with the atmosphere.

Notes

1. A solution of base that will be used up within two weeks can be stored in a tightly capped polyethylene bottle. After each removal of base, squeeze the bottle while tightening the cap to minimize the air space above the reagent. The bottle will become embrittled after extensive use as a container for bases.
2. Be certain that any solid Na_2CO_3 in the 50% NaOH has settled to the bottom of the container and that the decanted liquid is absolutely clear. If necessary, filter the base through a glass mat in a Gooch crucible; collect the clear filtrate in a test tube inserted in the filter flask.

31C-5

The Determination of the Acid/Base Ratio

Discussion

When both acid and base solutions have been prepared, it is useful to determine their volumetric combining ratio so that the two solutions can be used to establish end points more accurately in any of methods given later in this section. In addition, knowledge of this ratio and the concentration of one solution permits calculation of the molarity of the other.

PROCEDURE

Instructions for placing a buret into service are given in Sections 30H-4 and 30H-6; consult these instructions if necessary. Place a test tube or small beaker over the top of the buret that holds the NaOH solution to minimize contact between the solution and the atmosphere.

Record the initial volumes of acid and base in the burets to the nearest 0.01 mL. Deliver 35 to 40 mL of the acid into a 250-mL conical flask. Touch the tip of the buret to the inside wall of the flask, and rinse down with a little distilled water. Add two drops of phenolphthalein (Note 1) and then sufficient base to render the solution a definite pink. Introduce acid dropwise to discharge the color, and again rinse down the walls of the flask. Carefully add base until the solution acquires a faint pink hue that persists for at least 30 s (Notes 2, 3). Record the final buret volumes (again, to nearest 0.01 mL). Repeat the titration. Calculate the acid/base volume ratio. The ratios for duplicate titrations should agree to within 1 to 2 ppt. Perform additional titrations, if necessary, to achieve this order of precision.

Notes

1. The volume ratio can also be determined with an indicator that has an acidic transition range, such as bromocresol green (page 184). If the NaOH is contaminated with carbonate, the ratio obtained with this indicator will differ significantly from the value obtained with phenolphthalein. In general, the acid/base ratio should be evaluated with the indicator that is to be used in subsequent titrations.
2. Fractional drops can be formed on the buret tip, touched to the wall of the flask, and then rinsed down with a small amount of water.
3. The phenolphthalein end point fades as CO_2 is absorbed from the atmosphere.

31C-6 *Standardization of Hydrochloric Acid Against Sodium Carbonate*

Discussion

See Section 10A-2.

PROCEDURE

Dry a quantity of primary-standard Na_2CO_3 for about 2 h at 110°C (Figure 30-9), and cool in a desiccator. Weigh individual 0.20- to 0.25-g samples (to the nearest 0.1 mg) into 250-mL conical flasks, and dissolve each in about 50 mL of distilled water. Introduce 3 drops of bromocresol green, and titrate with HCl until the solution just begins to change from blue to green. Boil the solution for 2 to 3 min, cool to room temperature (Note 1), and complete the titration (Note 2).

Determine an indicator correction by titrating approximately 100 mL of 0.05 M NaCl and 3 drops of indicator. Boil briefly, cool, and complete the titration. Subtract any volume needed for the blank from the titration volumes. Calculate the concentration of the HCl solution.

Notes

1. The indicator should change from green to blue as CO_2 is removed during heating. If no color change occurs, an excess of acid was added originally. This excess can be back-titrated with base, provided the acid/base combining ratio is known; otherwise, the sample must be discarded.
2. It is permissible to back-titrate with base to establish the end point with greater certainty.

31C-7 Standardization of Sodium Hydroxide Against Potassium Hydrogen Phthalate

Discussion

See Section 10A-4.

PROCEDURE

Dry a quantity of primary-standard potassium hydrogen phthalate (KHP) for about 2 h at 110°C (Figure 30-9), and cool in a desiccator. Weigh individual 0.7- to 0.8-g samples (to the nearest 0.1 mg) into 250-mL conical flasks, and dissolve each in 50 to 75 mL of distilled or deionized water. Add 2 drops of phenolphthalein; titrate with base until the pink color of the indicator persists for 30 s (Note). Calculate the concentration of the NaOH solution.

Note. It is permissible to back-titrate with acid to establish the end point more precisely. The volume of acid must, of course, be recorded so that a correction to the volume of base used in the titration can be computed from the acid/base ratio.

31C-8 The Determination of Potassium Hydrogen Phthalate in an Impure Sample

Discussion

The unknown is a mixture of KHP and a neutral salt. This analysis is conveniently performed concurrently with the standardization of the base.

PROCEDURE

Consult with the instructor concerning an appropriate sample size. Then follow the directions in Section 31C-7.

31C-9 The Determination of the Acid Content of Vinegars and Wines

Discussion

The total acid content of a vinegar or a wine is readily determined by titration with a standard base. It is customary to report the acid content of vinegar in terms of acetic acid, the principal acidic constituent, even though other acids are present. Similarly, the acid content of a wine is expressed in terms of percent tartaric acid, notwithstanding the presence of other acids in the sample. Most vinegars contain about 5% acid (w/v) expressed as acetic acid; wines ordinarily contain somewhat under 1% acid (w/v) expressed as tartaric acid.

PROCEDURE

(a) *If the unknown is a vinegar* (Note 1), pipet 25.00 mL into a 250-mL volumetric flask and dilute to the mark with distilled water. Mix thoroughly, and pipet 50.00-mL aliquots into 250-mL conical flasks. Add about 50 mL of water and 2 drops of phenolphthalein (Note 2) to each, and titrate with standard 0.1 M NaOH to the first permanent (≈30 s) pink color.

Report the acidity of the vinegar as percent (w/v) CH_3COOH (fw = 60.053).

(b) *If the unknown is a wine,* pipet 50.00-mL aliquots into 250-mL conical flasks, add about 50 mL of distilled water and 2 drops of phenolphthalein to each (Note 2), and titrate to the first permanent (≈30 s) pink color.

Express the acidity of the sample as percent (w/v) tartaric acid $C_2H_4O_2(COOH)_2$ (fw = 150.09).

Notes

1. The acidity of bottled vinegar tends to decrease on exposure to air. It is recommended that unknowns be stored in individual vials with snug covers.
2. The amount of indicator used should be increased as necessary to make the color change visible in colored samples.

31C-10

The Determination of Sodium Carbonate in an Impure Sample

Discussion

The titration of sodium carbonate is discussed in Section 10A-2 in connection with its use as a primary standard; the same considerations apply for the determination of carbonate in an unknown that has no interfering contaminants.

PROCEDURE

Dry the unknown at 110°C for 2 h, and then cool in a desiccator. Consult with the instructor on an appropriate sample size. Weigh (to the nearest 0.1 mg) individual samples into 250-mL conical flasks. Dissolve each in 50 to 75 mL of distilled water, add 2 drops of bromocresol green, and titrate with standard acid until the indicator just begins to turn green. Boil the solution for 2 to 3 min, cool to room temperature, and complete the titration. If additional acid is not required after boiling, the solution already contains an excess. Either back-titrate this excess with standard base, or discard the sample if a base solution is not available.

Report the percentage of Na_2CO_3 in the sample.

31C-11

The Determination of Amine Nitrogen by the Kjeldahl Method

Discussion

These directions are suitable for the determination of protein in materials such as blood meal, wheat flour, pasta products, dry cereals, and pet foods. A simple modification permits the analysis of unknowns that contain more highly oxidized forms of nitrogen.[4] The chemistry of the Kjeldahl method is described in Section 10B-1.

PROCEDURE

Preparation of samples

Consult with the instructor on sample size. *If the unknown is powdered* (such as blood meal), weigh samples onto individual 9-cm filter papers (Note 1). Fold the paper around the sample and drop each into a Kjeldahl flask (the paper keeps the samples from clinging to the neck of the flask). *If the unknown is not powdered* (such as breakfast cereals or pasta), the samples can be weighed into the Kjeldahl flasks without the paper.

Add 25 mL of concentrated H_2SO_4, 10 g of powdered K_2SO_4, and the catalyst (Note 2) to each flask.

[4]See *Official Methods of Analysis,* 14th ed., p. 16. Washington, D.C.: Association of Official Analytical Chemists, 1984.

Digestion

Clamp the flasks in a slanted position in a hood or vented digestion rack. Heat carefully to boiling. Discontinue heating briefly if foaming becomes excessive; never allow the foam to reach the neck of the flask. Once foaming ceases and the acid is boiling vigorously, the samples can be left unattended; prepare the distillation apparatus during this time. Continue digestion until the solution becomes colorless or faint yellow; 2 to 3 h may be needed for some materials. If necessary, *cautiously* replace the acid lost by evaporation.

When digestion is complete, discontinue heating, and allow the flasks to cool to room temperature; swirl the flasks if the contents show signs of solidifying. Cautiously add 250 mL of water to each flask and again allow the solution to cool to room temperature. If mercury is used as the catalyst, introduce 25 mL of 4% (w/v) Na_2S solution (Note 3).

Distillation of ammonia

Arrange a distillation apparatus similar to that shown in Figure 10-2b (page 239). Pipet 50.00 mL of standard 0.1 M HCl into the receiver flask (Note 4). Clamp the flask so that the tip of the adapter extends below the surface of the standard acid. Circulate water through the condenser jacket.

Hold the Kjeldahl flask at an angle and gently introduce about 60 mL of 50% (w/v) NaOH solution, taking care to minimize mixing with the solution in the flask. *The concentrated caustic solution is highly corrosive and should be handled with great care* (Note 5). Add several pieces of granulated zinc (Note 6) and a small piece of litmus paper. *Immediately* connect the Kjeldahl flask to the spray trap. Cautiously mix the contents by gentle swirling. The litmus paper should be blue after mixing is complete, indicating that the solution is basic.

Bring the solution to a boil, and distill at a steady rate until one half to one third of the original volume remains. Control the rate of heating to prevent the liquid in the receiver flask from being drawn back into the Kjeldahl flask. After distillation is judged complete, lower the receiver flask to bring the adapter well clear of the liquid. Discontinue heating, disconnect the apparatus, and rinse the inside of the condenser with small portions of distilled water, collecting the washings in the receiver flask. Add 2 drops of bromocresol green to the receiver flask, and titrate the residual HCl with standard 0.1 M NaOH to the color change of the indicator.

Report the percentage of nitrogen and the percentage of protein (Note 7) in the unknown.

Notes

1. If filter paper is used to hold the sample, carry a similar piece through the analysis as a blank. Acid-washed filter paper is frequently contaminated with measurable amounts of ammonium ion and should be avoided if possible.
2. Any of the following catalyze the digestion: a drop of mercury, 0.5 g of HgO, a crystal of $CuSO_4$, 0.1 g of selenium, 0.2 g of $CuSeO_3$. The catalyst can be omitted, if desired.
3. Mercury(II) ions must be precipitated as the sulfide to prevent retention of some of the ammonia as a mercury complex.
4. A modification of this procedure uses about 50 mL of 4% boric acid solution in lieu of the standard HCl in the receiver flask (page 240). After distillation is complete, the ammonium borate produced is titrated with standard 0.1 M HCl, with 2 to 3 drops of bromocresol green as indicator.
5. If any sodium hydroxide solution comes into contact with the skin, the affected area should be washed IMMEDIATELY with copious amounts of water.
6. Granulated zinc (10 to 20 mesh) is added to minimize bumping during the distillation; it reacts slowly with the base to give small bubbles of hydrogen that prevent superheating of the liquid.

7. The percentage of protein in the unknown is calculated by multiplying the % N by an appropriate factor: 5.70 for cereals, 6.25 for meats, and 6.38 for dairy products.

31D

Complex-Formation Titrations with EDTA

See Chapter 11 for a discussion of the analytical uses of EDTA as a chelating reagent. Directions follow for a direct titration of magnesium, a displacement titration of calcium with the magnesium/EDTA complex, and a determination of the hardness of a natural water. Although these procedures are written for volumetric titrimetry, they are readily adapted to weight titrations (Section 30F).

31D-1

Preparation of Solutions

A pH-10 buffer and an indicator solution are needed for these titrations.

(a) *Buffer solution, pH 10.* (Sufficient for 80 to 100 titrations.) Dilute 57 mL of concentrated NH_3 and 7 g of NH_4Cl in sufficient distilled water to give 100 mL of solution.

(b) *Eriochrome Black T indicator.* (Sufficient for about 100 titrations.) Dissolve 100 mg of the solid in a solution containing 15 mL of ethanolamine and 5 mL of absolute ethanol. This solution should be freshly prepared every two weeks; refrigeration slows its deterioration.

(c) *Calmagite indicator.* (Sufficient for about 200 titrations.) Dissolve 0.05 g of the indicator in sufficient distilled water to give 50 mL of solution.

31D-2

Preparation of Standard 0.01 M EDTA Solution

Discussion

See Section 11B-1 for a description of the properties of reagent-grade $Na_2H_2Y \cdot 2H_2O$ and its use in the direct preparation of standard EDTA solutions.

PROCEDURE

Dry about 4 g of the purified dihydrate $Na_2H_2Y \cdot 2H_2O$ (Note 1) at 80°C to remove superficial moisture. Cool to room temperature in a desiccator. Weigh (to the nearest milligram) about 3.8 g into a 1-L volumetric flask (Note 2). Use a powder funnel to ensure quantitative transfer; rinse the funnel well with water before removing it from the flask. Add 600 to 800 mL of water (Note 3) and swirl periodically. Dissolution may take 15 min or longer. When all the solid has dissolved, dilute to the mark with water and mix well (Note 4). In calculating the molarity of the solution, correct the weight of the salt for the 0.3% moisture it ordinarily contains after drying at 80°C.

Notes

1. Directions for the purification of the disodium salt are described by W. J. Blaedel and H. T. Knight, *Anal. Chem.*, **1954**, *26*(4), 741.
2. The solution can be prepared from the anhydrous disodium salt, if desired. The weight taken should be about 3.6 g.
3. Water used in the preparation of standard EDTA solutions must be totally free of polyvalent cations. If any doubt exists concerning its quality, the water should be passed through a cation-exchange resin before use.

4. As an alternative, an EDTA solution that is approximately 0.01 M can be prepared and standardized by direct titration against a Mg^{2+} solution of known concentration (using the directions in Section 31D-3) or against primary-standard $CaCO_3$ by displacement titration (Section 31D-4).

31D-3 The Determination of Magnesium by Direct Titration

Discussion

See Section 11B-6.

PROCEDURE

Transfer the aqueous solution of the unknown to a clean 500-mL volumetric flask, dilute to the mark with water, and mix thoroughly. Transfer 50.00-mL aliquots to 250-mL conical flasks, add 1 to 2 mL of pH-10 buffer and 3 to 4 drops of Erio T or Calmagite indicator to each. Titrate with 0.01 M EDTA until the color changes from red to pure blue (Notes 1, 2).

Express the results as parts per million of Mg^{2+} in the sample.

Notes

1. The color change tends to be slow in the vicinity of the end point. Care must be taken to avoid overtitration.
2. Other alkaline earths, if present, are titrated along with the Mg^{2+}; removal of Ca^{2+} and Ba^{2+} can be accomplished with $(NH_4)_2CO_3$. Most polyvalent cations are also titrated. Precipitation as hydroxides or the use of a masking reagent may be needed to eliminate this source of interference.

31D-4 The Determination of Calcium by Displacement Titration

Discussion

A solution of the magnesium/EDTA complex is useful for the titration of cations, which form stabler complexes than the magnesium complex but for which no indicator is available. Magnesium ions in the complex are displaced by a chemically equivalent quantity of analyte cations. The remaining uncomplexed analyte and the liberated magnesium ions are then titrated; either Eriochrome Black T or Calmagite can serve as indicator. Note that the concentration of the magnesium solution is not important; all that is necessary is that the molar ratio between Mg^{2+} and EDTA be exactly unity in the reagent.

PROCEDURE

Preparation of the magnesium/EDTA complex, 0.1 M

(Sufficient for 90 to 100 titrations.) To 3.72 g of $Na_2H_2Y \cdot 2H_2O$ in 50 mL of distilled water, add an equivalent quantity (2.46 g) of $MgSO_4 \cdot 7H_2O$. Add a few drops of phenolphthalein, followed by sufficient 0.1 M NaOH to turn the solution faintly pink. Dilute to about 100 mL with water. The addition of a few drops of Erio T to a portion of this solution buffered to pH 10 should cause development of a dull violet color. Moreover, a single drop of 0.01 M Na_2H_2Y solution added to the violet solution should cause a color change to blue, and an equal quantity of 0.01 M Mg^{2+} should cause a change to red. The composition of the original solution should be adjusted with additional Mg^{2+} or H_2Y^{2-} until these criteria are met.

Titration

Weigh a sample of the unknown (to the nearest 0.1 mg) into a 500-mL beaker (Note 1). Cover with a watch glass, and carefully add 5 to 10 mL of 6 M HCl. After the sample has dissolved, remove CO_2 by adding about 50 mL of deionized water and boiling gently for a few minutes. Cool, add a drop or two of methyl red, and neutralize with 6 M NaOH until the red color is discharged. Quantitatively transfer the solution to a 500-mL volumetric flask, and dilute to the mark. Take 50.00-mL aliquots of the diluted solution, for titration, treating each as follows: add about 2 mL of pH-10 buffer, 1 mL of Mg/EDTA solution, and 3 to 4 drops of Erio T or Calmagite indicator. Titrate (Note 2) with standard 0.01 M Na_2H_2Y to a color change from red to blue.

Report the number of milligrams of CaO in the sample.

Notes

1. The sample taken should contain 150 to 160 mg of Ca^{2+}.
2. Intereferences with this titration are substantially the same as those encountered in the direct titration of Mg^{2+} and are eliminated in the same way.

31D-5 The Determination of Hardness in Water

Discussion

See Section 11B-8.

PROCEDURE

Acidify 100.0-mL aliquots of the sample with a few drops of HCl, and boil gently for a few minutes to eliminate CO_2. Cool, add 3 to 4 drops of methyl red, and neutralize with 0.1 M NaOH. Introduce 2 mL of pH-10 buffer, 3 to 4 drops of Erio T or Calmagite, and titrate with standard 0.01 M Na_2H_2Y to a color change from red to pure blue (Note).

Report the results in terms of milligrams of $CaCO_3$ per liter of water.

Note. The color change is sluggish if Mg^{2+} is absent. In this event, add 1 to 2 mL of 0.1 M MgY^{2-} (Section 31D-4, Procedure) before starting the titration.

31E Oxidation/Reduction Titrations with Potassium Permanganate

The properties and uses of potassium permanganate are described in Section 14B-1. Directions follow for the determination of iron in an ore and calcium in a limestone. These directions are readily adapted to weight titrimetry as described in Section 30F.

31E-1 Preparation of 0.02 M Potassium Permanganate

Discussion

See page 335 for a discussion of the precautions needed in the preparation and storage of permanganate solutions.

PROCEDURE

Dissolve about 3.2 g of $KMnO_4$ in 1 L of distilled water. Keep the solution at a gentle boil for about 1 h. Cover and let stand overnight. Remove MnO_2 by filtration (Note 1) through a fine-porosity sintered-glass crucible (Note 2) or through a Gooch crucible fitted with glass mats. Transfer the solution to a clean glass-stoppered bottle; store in the dark when not in use.

Notes

1. The heating and filtering can be omitted if the permanganate solution is standardized and used on the same day.

2. Remove the MnO_2 that collects on the fritted plate with 1 M H_2SO_4 containing a few milliliters of 3% H_2O_2, followed by a rinse with copious quantities of water.

31E-2

Standardization of Potassium Permanganate Solutions

Discussion

See page 336 for a discussion of sodium oxalate and other primary standards for permanganate solutions.

PROCEDURE

Dry about 1.5 g of primary-standard-grade $Na_2C_2O_4$ at 110°C for at least 1 h. Cool in a desiccator; weigh (to the nearest 0.1 mg) individual 0.2- to 0.3-g samples into 400-mL beakers. Dissolve each in about 250 mL of 1 M H_2SO_4. Heat each solution to 80 to 90°C, and titrate with $KMnO_4$ while stirring with a thermometer. The pink color imparted by one addition should be permitted to disappear before any further titrant is introduced (Notes 1, 2). Reheat if the temperature drops below 60°C. Take the first persistent (≃30 s) pink color as the end point (Notes 3, 4). Determine a blank by titrating an equal volume of the 1 M H_2SO_4.

Correct the titration data for the blank, and calculate the concentration of the permanganate solution (Note 5).

Notes

1. Promptly wash any $KMnO_4$ that spatters on the walls of the beaker into the bulk of the liquid with a stream of water.
2. Finely divided MnO_2 will form along with Mn^{2+} if the $KMnO_4$ is added too rapidly and will cause the solution to acquire a faint brown discoloration. Precipitate formation is not a serious problem so long as sufficient oxalate remains to reduce the MnO_2 to Mn^{2+}; the titration is simply discontinued until the brown color disappears. The solution must be free of MnO_2 at the end point.
3. The surface of the permanganate solution rather than the bottom of the meniscus can be used to measure titrant volumes. Alternatively, backlighting with a flashlight or a match permits reading of the meniscus in the conventional manner.
4. A permanganate solution should not be allowed to stand in a buret any longer than necessary because partial decomposition to MnO_2 may occur. Freshly formed MnO_2 can be removed from a glass surface with 1 M H_2SO_4 containing a small amount of 3% H_2O_2.
5. As noted on page 337, this procedure yields molarities that are a few tenths of a percent low. For more accurate results, introduce from a buret sufficient permanganate to react with 90 to 95% of the oxalate (about 40 mL of 0.02 M $KMnO_4$ for a 0.3-g sample). Let the solution stand until the permanganate color disappears. Then warm to abut 60°C and complete the titration, taking the first permanent pink (≃30 s) as the end point (Notes 3, 4). Determine a blank by titrating an equal volume of the 1 M H_2SO_4.

31E-3

The Determination of Iron in an Ore

Discussion

The common ores of iron are hematite (Fe_2O_3), magnetite (Fe_3O_4), and limonite ($2Fe_2O_3{\cdot}3H_2O$). Steps in the analysis of these ores are (1) dissolution of the sample, (2) reduction of iron to the divalent state, and (3) titration of iron(II) with a standard oxidant.

The Decomposition of Iron Ores. Iron ores are often decomposed completely in hot concentrated hydrochloric acid. The rate of attack by this reagent is increased by the presence of a small amount of tin(II) chloride, which probably acts by reducing sparingly soluble iron(III) oxides on the surface of the ore to more soluble iron(II) species. The tendency of iron(II) and iron(III) to form chloro complexes accounts for the effectiveness of hydrochloric acid over nitric or sulfuric acid as a solvent for iron ores.

Many iron ores contain silicates that may not be entirely decomposed by treatment with hydrochloric acid. Incomplete decomposition is indicated by a dark residue that remains after prolonged treatment with the acid. A white residue of hydrated silica, which does not interfere in any way, is indicative of complete decomposition.

The Prereduction of Iron. Because part or all of the iron is in the trivalent state after decomposition of the sample, prereduction to iron(II) must precede titration with the oxidant. Any of the methods described in Section 14A-1 can be used. Perhaps the most satisfactory prereductant for iron is tin(II) chloride:

$$2Fe^{3+} + Sn^{2+} \longrightarrow 2Fe^{2+} + Sn^{4+}$$

The only other common species reduced by this reagent are the high-oxidation states of arsenic, copper, mercury, molybdenum, tungsten, and vanadium.

The excess reducing agent is eliminated by the addition of mercury(II) chloride:

$$Sn^{2+} + 2HgCl_2 \longrightarrow Hg_2Cl_2(s) + Sn^{4+} + 2Cl^-$$

The slightly soluble mercury(I) chloride does not reduce permanganate, nor does the excess mercury(II) chloride reoxidize iron(II). Care must be taken, however, to prevent the occurrence of the alternative reaction

$$Sn^{2+} + HgCl_2 \longrightarrow Hg(l) + Sn^{4+} + 2Cl^-$$

Elemental mercury reacts with permanganate and causes the results of the analysis to be high. The formation of mercury, which is favored by an appreciable excess of tin(II), is prevented by careful control of this excess and by the rapid addition of excess mercury(II) chloride. A proper reduction is indicated by the appearance of a small amount of a silky white precipitate after the addition of mercury(II). Formation of a gray precipitate at this juncture indicates the presence of metallic mercury; the total absence of a precipitate indicates that an insufficient amount of tin(II) chloride was used. In either event, the sample must be discarded.

The Titration of Iron(II). The reaction of iron(II) with permanganate is smooth and rapid. The presence of iron(II) in the reaction mixture, however, *induces* oxidation of chloride ion by permanganate, a reaction that does not ordinarily proceed rapidly enough to cause serious error. High results are obtained if this parasitic reaction is not controlled. Its effects can be eliminated through removal of the hydrochloric acid by evaporation with sulfuric acid or by introduction of *Zimmermann-Reinhardt reagent*, which contains manganese(II) in a fairly concentrated mixture of sulfuric and phosphoric acids.

The oxidation of chloride ion during a titration is believed to involve a direct reaction between this species and the manganese(III) ions that form as an intermediate in the reduction of permanganate ion by iron(II). The presence of manganese(II) in the Zimmermann-Reinhardt reagent is believed to inhibit the formation of chlorine by decreasing the potential of the manganese(III)/manganese(II) couple. Phosphate ion is believed to exert a similar effect by forming stable manganese(III) complexes. Moreover, phosphate ions react with iron(III) to form nearly colorless complexes so that the yellow color of the iron(II)/chloro complexes does not interfere with the end point.[5]

PREPARATION OF REAGENTS

The following solutions suffice for about 100 titrations.

(a) *Tin(II) chloride, 0.25 M.* Dissolve 60 g of iron-free $SnCl_2 \cdot 2H_2O$ in 100 mL of concentrated HCl; warm if necessary. After the solid has dissolved, dilute to 1 L with distilled water and store in a well-stoppered bottle. Add a few pieces of mossy tin to help preserve the solution.

(b) *Mercury(II) chloride, 5% (w/v).* Dissolve 50 g of $HgCl_2$ in 1 L of distilled water.

(c) *Zimmermann-Reinhardt reagent.* Dissolve 300 g of $MnSO_4 \cdot 4H_2O$ in 1 L of water. Cautiously add 400 mL of concentrated H_2SO_4, 400 mL of 85% H_3PO_4, and dilute to 3 L.

PROCEDURE

Sample preparation

Dry the ore at 110°C for at least 3 h, and then allow it to cool to room temperature in a desiccator. Consult with the instructor for a sample size that will require from 25 to 40 mL of standard 0.02 M $KMnO_4$. Weigh samples into 500-mL conical flasks. To each, add 10 mL of concentrated HCl and about 3 mL of 0.25 M $SnCl_2$ (Note 1). Cover each flask with a small watch glass or Tuttle flask cover. Heat the flasks in a hood at just below boiling until the samples are decomposed and the undissolved solid—if any—is pure white (Note 2). Use another 1 or 2 mL of $SnCl_2$ to eliminate any yellow color that may develop as the solutions are heated. Heat a blank consisting of 10 mL of HCl and 3 mL of $SnCl_2$ for the same amount of time.

After the ore has been decomposed, remove the excess Sn(II) by the dropwise addition of 0.02 M $KMnO_4$ until the solutions become faintly yellow. Dilute to about 15 mL. Add sufficient $KMnO_4$ solution to impart a faint pink color to the blank; then decolorize with one drop of the $SnCl_2$ solution.

Take samples and blank individually through subsequent steps to minimize air-oxidation of iron(II).

Reduction of iron

Heat the sample solution nearly to boiling, and make dropwise additions of 0.25 M $SnCl_2$ until the yellow color just disappears; then add two more drops (Note 3). Cool to room temperature, and *rapidly* add 10 mL of 5% $HgCl_2$ solution. A small amount of silky white Hg_2Cl_2 should precipitate (Note 4). The blank solution should be treated with the $HgCl_2$ solution.

Titration

Following addition of the $HgCl_2$, wait 2 to 3 min. Then add 25 mL of Zimmermann-Reinhardt reagent and 300 mL of water. Titrate *immediately* with standard 0.02 M $KMnO_4$ to the first faint pink that persists for 15 to 20 s. Do not add the $KMnO_4$ rapidly at any time. Correct the titrant volume for the blank.

Report the percentage of Fe_2O_3 in the sample.

[5]The mechanism by which Zimmermann-Reinhardt reagent acts has been the subject of much study. For a discussion of this work, see H. A. Laitinen, *Chemical Analysis*, pp. 369–372. New York: McGraw-Hill, 1960.

Notes

1. The $SnCl_2$ hastens decomposition of the ore by reducing iron(III) oxides to iron(II). Insufficient $SnCl_2$ is indicated by the appearance of yellow iron(III)/chloride complexes.
2. If dark particles persist after the sample has been heated with acid for several hours, filter the solution through ashless paper, wash the residue with 5 to 10 mL of 6 M HCl, and retain the filtrate and washings. Ignite the paper and its contents in a small platinum crucible. Mix 0.5 to 0.7 g of Na_2CO_3 with the residue and heat until a clear melt is obtained. Cool, add 5 mL of water, and then cautiously add a few milliliters of 6 M HCl. Warm the crucible until the melt has dissolved, and combine the contents with the original filtrate. Evaporate the solution to 15 mL and continue the analysis.
3. The solution may not become entirely colorless but instead may acquire a faint yellow-green hue. Further additions of $SnCl_2$ will not alter this color. If too much $SnCl_2$ is added, it can be removed by adding 0.2 M $KMnO_4$ and repeating the reduction.
4. The absence of precipitate indicates that insufficient $SnCl_2$ was used and that the reduction of iron(III) was incomplete. A gray residue indicates the presence of elemental mercury, which reacts with $KMnO_4$. The sample must be discarded in either event.

31E-4 The Determination of Calcium in a Limestone

Discussion

In common with a number of other cations, calcium is conveniently determined by precipitation with oxalate ion. The solid calcium oxalate is filtered, washed free of excess precipitating reagent, and dissolved in dilute acid. The oxalic acid liberated in this step is then titrated with standard permanganate or some other oxidizing reagent. This method is applicable to samples that contain magnesium and the alkali metals, but most other cations must be absent since they either precipitate or coprecipitate as oxalates and cause positive errors in the analysis.

Factors Affecting the Composition of Calcium Oxalate Precipitates. It is essential that the mole ratio between calcium and oxalate be exactly unity in the precipitate and thus in solution at the time of titration. A number of precautions are needed to ensure this condition. For example, the calcium oxalate formed in a neutral or ammoniacal solution is likely to be contaminated with calcium hydroxide or a basic calcium oxalate, either of which leads to low results. The formation of these compounds is prevented by adding the oxalate to an acidic solution of the sample and slowly forming the precipitate by the dropwise addition of ammonia. The coarsely crystalline calcium oxalate that is produced under these conditions is readily filtered. Losses resulting from the solubility of calcium oxalate are negligible above pH 4, provided washing is limited to freeing the precipitate of excess oxalate.

Coprecipitation of sodium oxalate becomes a source of positive error in the determination of calcium whenever the concentration of sodium in the sample exceeds that of calcium. The error from this source can be eliminated by reprecipitation (Section 3B-4).

Magnesium, if present in high concentration, may precipitate as the oxalate and contaminate the analytical precipitate. An excess of oxalate ion helps prevent this interference through the formation of soluble oxalate complexes of magnesium. Prompt filtration of the calcium oxalate can also help prevent interference because of the pronounced tendency of magnesium oxalate to form supersaturated solutions from which precipitate formation occurs only after an hour or more. For samples containing more magnesium than calcium, these measures do not suffice to give accurate results and reprecipitation of the calcium becomes necessary.

The Composition of Limestones. Limestones are composed principally of calcium carbonate; dolomitic limestones contain large amounts of magnesium carbonate as well. Calcium and magnesium silicates are also present in smaller amounts, along with the carbonates and silicates of iron, aluminum, manganese, titanium, sodium, and other metals.

Hydrochloric acid is an effective solvent for most limestones. Only silica, which does not interfere with the analysis, remains undissolved. Some limestones are more readily decomposed after they have been ignited; a few yield only to a carbonate fusion (Section 28C-2).

The method that follows is remarkably effective for determining calcium in most limestones. Iron and aluminum, in amounts equivalent to that of calcium, do not interfere. Small amounts of manganese and titanium can also be tolerated.

PROCEDURE

Sample preparation

Dry the unknown for 1 to 2 h at 110°C, and cool in a desiccator. If the material is readily decomposed in acid, weigh 0.25- to 0.30-g samples (to the nearest 0.1 mg) into 250-mL beakers. Add 10 mL of water to each sample and cover with a watch glass. Add 10 mL of concentrated HCl dropwise, taking care to avoid losses due to spattering as the acid is introduced. Proceed to the paragraph labeled "Precipitation of calcium oxalate."

If the limestone is not completely decomposed by acid, weigh the sample into a small porcelain crucible and ignite. Raise the temperature slowly to 800 to 900°C and maintain this temperature for about 30 min. After cooling, place the crucible and its contents in a 250-mL beaker, add 5 mL of water, and cover with a watch glass. Introduce 10 mL of concentrated HCl dropwise, and then heat to boiling. Remove the crucible with a stirring rod and rinse it thoroughly with water; combine the washings with the solution containing the sample.

Precipitation of calcium oxalate

Add 5 drops of saturated bromine water to oxidize any iron in the samples and boil gently (HOOD) for 5 min to remove the excess Br_2. Dilute each sample solution to about 50 mL, heat to boiling, and add 100 mL of hot 6% (w/v) $(NH_4)_2C_2O_4$ solution. Add 3 to 4 drops of methyl red, and precipitate CaC_2O_4 by slowly adding 6 M NH_3. When the indicator just begins to change color, add the NH_3 at a rate of one drop every 3 to 4 s. Continue until the solutions turn to the intermediate yellow-orange color of the indicator (pH 4.5 to 5.5). Allow the solutions to stand for no more than 30 min (Note) and filter; medium-porosity filtering crucibles or Gooch crucibles with glass mats are satisfactory. Wash the precipitates with several 10-mL portions of cold water. Rinse the outside of the crucibles to remove residual $(NH_4)_2C_2O_4$, and return them to the beakers in which the CaC_2O_4 was formed.

Titration

Add 100 mL of water and 50 mL of 3 M H_2SO_4 to each of the beakers containing the precipitated calcium oxalate and the crucible. Heat to 80 to 90°C, and titrate with 0.02 M permanganate. The temperature should be above 60°C throughout the titration; reheat if necessary.

Report the percentage of CaO in the unknown.

Note. The period of standing can be longer if the unknown contains no Mg^{2+}.

31F Iodimetric Titrations

The oxidizing properties of iodine, the composition and stability of triiodide solutions, and the applications of this reagent in volumetric analysis are discussed in Section 14B-3. Starch is ordinarily employed as an indicator for iodimetric titrations. The directions that follow are readily adapted to weight titrimetry, which is discussed in Section 30F.

31F-1 Preparation of Reagents

(a) *Iodine approximately 0.05 M.* Weigh about 40 g of KI into a 100-mL beaker. Add 12.7 g of I_2 and 10 mL of water. Stir for several minutes (Note 1). Introduce an additional 20 mL of water, and stir again for several minutes. Carefully decant the bulk of the liquid into a storage bottle containing 1 L of distilled water. It is essential that any undissolved iodine remain in the beaker (Note 2).

Notes

1. Iodine dissolves slowly in the KI solution. Thorough stirring is needed to hasten the process.
2. Any solid I_2 inadvertently transferred to the storage bottle will cause the concentration of the solution to increase gradually. Filtration through a sintered-glass crucible eliminates this potential source of difficulty.

(b) *Starch indicator.* Rub 1 g of soluble starch and 15 mL of water into a paste. Dilute to about 500 mL with boiling water, and heat until the mixture is clear. Cool; store in a tightly stoppered bottle. For most titrations, 3 to 5 mL of the indicator is used. (Sufficient for about 100 titrations.)

The indicator is readily attacked by airborne organisms and should be freshly prepared every few days.

31F-2 Standardization of Iodine Solutions

Discussion

Arsenic(III) oxide, long a favored primary standard for iodine solutions, is now seldom used because of the elaborate federal regulations governing the use of even small amounts of arsenic-containing compounds. Barium thiosulfate monohydrate and anhydrous sodium thiosulfate have been proposed as alternative standards.[6,7] Perhaps the most convenient method for determining the concentration of an iodine solution is the titration of aliquots with

[6]W. M. McNevin and O. H. Kriege, *Anal. Chem.*, **1953,** *25* (5), 767.

[7]A. A. Woolf, *Anal. Chem.*, **1982,** *54* (12), 2134.

a sodium thiosulfate solution that has been standardized against pure potassium iodate. Instructions for this method follow.

PREPARATION OF REAGENTS

(a) *Sodium thiosulfate, 0.1 M.* Follow the directions in Sections 31G-1 and 31G-2 for the preparation and standardization of this solution.

(b) *Starch indicator.* See Section 31F-1.

PROCEDURE

Transfer 25.00-mL aliquots of the iodine solution to 250-mL conical flasks, and dilute to about 50 mL. Introduce approximately 1 mL of 3 M H_2SO_4, and titrate immediately with standard sodium thiosulfate until the solution becomes a faint straw yellow. Add about 5 mL of starch indicator, and complete the titration, taking as the end point the change in color from blue to colorless (Note).

Note. The blue color of the starch/iodine complex may reappear after the titration has been completed, owing to the air-oxidation of iodide ion.

31F-3 The Determination of Antimony in Stibnite

Discussion

The analysis of stibnite, a common antimony ore, is a typical application of iodimetry and is based upon the oxidation of Sb(III) to Sb(V):

$$SbO_3^{3-} + I_2 + H_2O \rightleftharpoons SbO_4^{3-} + 2I^- + 2H^+$$

The position of this equilibrium is strongly dependent upon the hydrogen ion concentration. In order to force the reaction to the right, it is common practice to carry out the titration in the presence of an excess of sodium hydrogen carbonate, which consumes the hydrogen ions as they form.

Stibnite is an antimony sulfide ore containing silica and other contaminants. Provided the material is free of iron and arsenic, the analysis of stibnite for its antimony content is straightforward. Samples are decomposed in hot concentrated hydrochloric acid to eliminate sulfide as gaseous hydrogen sulfide. In order to avoid the loss of antimony chloride by volatilization, excess potassium chloride is added to the solvent to convert the antimony to nonvolatile chloro complexes, such as $SbCl_4^-$ and $SbCl_6^{3-}$.

Sparingly soluble basic antimony salts, such as SbOCl, often form when the excess hydrochloric acid is neutralized; these react incompletely with iodine and cause low results. The difficulty is overcome by adding tartaric acid, which forms a soluble complex ($SbOC_4H_4O_6^-$) from which antimony is rapidly oxidized by the reagent.

PROCEDURE

Dry the unknown at 110°C for 1 h, and allow it to cool in a desiccator. Weigh individual samples (Note 1) into 500-mL conical flasks. Introduce about 0.3 g of KCl and 10 mL of concentrated HCl to each flask. Heat the mixtures (HOOD) just below boiling until only white or slightly gray residues of SiO_2 remain.

Add 3 g of tartaric acid to each sample and heat for an additional 10 to 15 min. Then, with good swirling, add water (Note 2) from a pipet or buret until the volume is about 100 mL. If reddish Sb_2S_3 forms, discontinue dilution and heat further to eliminate H_2S; add more HCl if necessary.

Add 3 drops of phenolphthalein, and neutralize with 6 M NaOH to the first faint pink of the indicator. Discharge the color by the dropwise addition of 6 M HCl, and then add 1 mL in excess. Introduce 4 to 5 g of $NaHCO_3$, taking care to avoid losses of solution by spattering during the addition. Add 5 mL of starch indicator, rinse down the inside of the flask, and titrate with standard 0.05 M I_2 to the first blue color that persists for 30 s.

Report the percentage of Sb_2S_3 in the unknown.

Notes

1. Samples should contain between 1.5 and 2 mmol of antimony; consult with the instructor for an appropriate sample size. Weighings to the nearest milligram are adequate for samples larger than 1 g.
2. The slow addition of water, with efficient stirring, is essential to prevent the formation of SbOCl.

31G *Iodometric Methods of Analysis*

Numerous methods are based upon the reducing properties of iodide ion:

$$2I^- \longrightarrow I_2 + 2e$$

Iodine, the reaction product, is ordinarily titrated with a standard sodium thiosulfate solution, with starch serving as the indicator:

$$I_2 + 2S_2O_3^{2-} \longrightarrow 2I^- + S_4O_6^{2-}$$

A discussion of iodometric methods is found in Section 14C-2. All the procedures described in this section can also be performed conveniently by weight titrimetry (Section 30F).

31G-1 *Preparation of 0.1 M Sodium Thiosulfate*

PROCEDURE

Boil about 1 L of distilled water for 10 to 15 min. Allow the water to cool to room temperature; then add about 25 g of $Na_2S_2O_3 \cdot 5H_2O$ and 0.1 g of Na_2CO_3. Stir until the solid has dissolved. Transfer the solution to a clean glass or plastic bottle, and store in a dark place.

31G-2 *Standardization of Sodium Thiosulfate Against Potassium Iodate*

Discussion

Solutions of sodium thiosulfate are conveniently standardized by titration of the iodine produced when an unmeasured excess of potassium iodide is added to a known volume of an acidified solution of standard potassium iodate solution. The reaction is

$$IO_3^- + 5I^- + 6H^+ \longrightarrow 3I_2 + 3H_2O$$

Note that each formula weight of iodate results in the production of three formula weights of iodine. The procedure that follows is based upon this reaction.

PREPARATION OF SOLUTIONS

(a) *Potassium iodate, 0.0100 M.* Dry about 1.2 g of primary-standard KIO_3 at 110°C for at least 1 h and cool in a desiccator. Weigh (to the nearest 0.1 mg) about 1.1 g into a 500-mL volumetric flask; use a powder funnel to ensure quantitative transfer of the solid. Rinse the funnel well, dissolve the KIO_3 in about 200 mL of distilled water, dilute to the mark, and mix thoroughly.

(b) *Starch indicator.* See Section 31F-1.

PROCEDURE

Pipet 50.00-mL aliquots of standard iodate solution into 250-mL conical flasks. *Treat each sample individually from this point to minimize error resulting from the air-oxidation of iodide ion.* Introduce 2 g of iodate-free KI, and swirl the flask to hasten soution. Add 2 mL of 6 M HCl, and immediately titrate with thiosulfate until the solution becomes pale yellow. Introduce 5 mL of starch indicator, and titrate with constant stirring to the disappearance of the blue color. Calculate the molarity of the iodine solution.

31G-3 Standardization of Sodium Thiosulfate Against Copper

Discussion

Thiosulfate solutions can also be standardized against pure copper wire or foil. This procedure is advantageous when the solution is to be used for the determination of copper because any determinate error in the method tends to be canceled.

Copper(II) is reduced quantitatively to copper(I) by iodide ion:

$$2Cu^{2+} + 4I^- \longrightarrow 2CuI(s) + I_2$$

The importance of CuI formation in forcing this reaction to completion can be seen from the following standard electrode potentials:

$$Cu^{2+} + e \rightleftharpoons Cu^+ \qquad E^0 = 0.15\ V$$

$$I_2 + 2e \rightleftharpoons 2I^- \qquad E^0 = 0.54\ V$$

$$Cu^{2+} + I^- + e \rightleftharpoons CuI(s) \qquad E^0 = 0.86\ V$$

The first two potentials suggest that iodide should have no tendency to reduce copper(II); the formation of CuI, however, favors the reduction. The solution must contain at least 4% excess iodide to force the reaction to completion. Moreover, the pH must be below 4 to prevent the formation of basic copper species that react slowly and incompletely with iodide ion. The acidity of the solution cannot be greater than about 0.3 M, however, because of the tendency of iodide ion to undergo air-oxidation, a process catalyzed by copper salts. Nitrogen oxides also catalyze the air-oxidation of iodide ion. A common source of these oxides is the nitric acid ordinarily used to dissolve metallic copper and other copper-containing solids. Urea is used to scavenge nitrogen oxides from solutions:

$$(NH_2)_2CO + 2HNO_2 \longrightarrow 2N_2(g) + CO_2(g) + 3H_2O$$

The titration of iodine by thiosulfate tends to yield slightly low results when CuI is present, owing to the adsorption of small but measurable quantities of iodine upon the solid. The adsorbed iodine is released only slowly, even when thiosulfate is present; transient and premature end points result.

This difficulty is largely overcome by the addition of thiocyanate ion. The sparingly soluble copper(I) thiocyanate replaces part of the copper iodide at the surface of the solid:

$$CuI(s) + SCN^- \longrightarrow CuSCN(s) + I^-$$

Accompanying this reaction is the release of the adsorbed iodine, which thus becomes available for titration. The addition of thiocyanate must be delayed until most of the iodine has been titrated to prevent interference from a slow reaction between the two species, possibly

$$2SCN^- + I_2 \longrightarrow 2I^- + (SCN)_2$$

PREPARATION OF SOLUTIONS

(a) *Urea, 5% (w/v).* Dissolve about 5 g of urea in sufficient water to give 100 mL of solution. Approximately 10 mL is needed for each titration.

(b) *Starch indicator.* See Section 31F-1.

PROCEDURE

Use scissors to cut copper wire or foil into 0.20- to 0.25-g portions. Wipe the metal free of dust and grease with a filter paper; do not dry it. The pieces of copper should be handled with paper strips, cotton gloves, or tweezers to prevent contamination by contact with the skin.

Use a weighed watch glass or weighing bottle to obtain the weight of individual copper samples by difference (to the nearest 0.1 mg). Transfer each sample to a 250-mL conical flask. Add 5 mL of 6 M HNO_3, cover with a small watch glass, and warm gently (HOOD) until the metal has dissolved. Dilute with about 25 mL of distilled water, add 10 mL of 5% (w/v) urea, and boil briefly to eliminate nitrogen oxides. Rinse the watch glass, collecting the rinsings in the flask. Cool.

Add concentrated NH_3 dropwise and with thorough mixing to produce the intensely blue $Cu(NH_3)_4^{2+}$; the solution should smell faintly of ammonia (Note). Make dropwise additions of 3 M H_2SO_4 until the color of the complex just disappears, and then add 2.0 mL of 85% H_3PO_4. Cool to room temperature.

Treat each sample individually from this point on to minimize the air-oxidation of iodide ion. Add 4.0 g of KI to the sample, and titrate immediately with $Na_2S_2O_3$ until the solution becomes pale yellow. Add 5 mL of starch indicator, and continue the titration until the blue color becomes faint. Add 2 g of KSCN; swirl vigorously for 30 s. Complete the titration, using the disappearance of the blue starch/I_2 color as the end point.

Calculate the molarity of the $Na_2S_2O_3$ solution.

Note. Vapors should not be sniffed directly from the flask but instead should be wafted toward the nose with a waving motion of one's hand.

31G-4 The Determination of Copper in Brass

Discussion

The standardization procedure described in Section 31G-3 is readily adapted to the determination of copper in brass, an alloy that also contains appreciable amounts of tin, lead, and zinc (and perhaps minor amounts of nickel and iron). The method is relatively simple and applicable to brasses with less than 2% iron. A weighed sample is treated with nitric acid, which causes the tin to precipitate as a hydrated oxide of uncertain composition (Section 31A-2).

Evaporation with sulfuric acid to the appearance of sulfur trioxide eliminates the excess nitrate, redissolves the tin compound, and possibly causes the formation of lead sulfate. The pH is adjusted through the addition of ammonia, followed by acidification with a measured amount of phosphoric acid. An excess of potassium iodide is added, and the liberated iodine is titrated with standard thiosulfate. See Section 31G-3 for additional discussion.

PROCEDURE

If so directed, free the metal of oils by treatment with an organic solvent; briefly heat in an oven to drive off the solvent. Weigh (to the nearest 0.1 mg) 0.3-g samples into 250-mL conical flasks, and introduce 5 mL of 6 M HNO_3 into each; warm (HOOD) until solution is complete. Add 10 mL of concentrated H_2SO_4, and evaporate (HOOD) until copious white fumes of SO_3 are given off. Allow the mixture to cool. Cautiously add 30 mL of distilled water, boil for 1 to 2 min, and again cool.

Follow the instructions in the third and fourth paragraphs of the *Procedure* in Section 31G-3.

Report the percentage of Cu in the sample.

31H Titrations with Potassium Bromate

Applications of standard bromate solutions to the determination of organic functional groups are described in Section 14D-1. Directions follow for the determination of phenol in an aqueous solution and for ascorbic acid in vitamin C tablets.

31H-1 Preparation of Solutions

(a) *Potassium bromate, 0.015 M.* Transfer about 1.5 g of reagent-grade potassium bromate to a weighing bottle, and dry at 110°C for at least 1 h. Cool in a desiccator. Weigh approximately 1.3 g (to the nearest 0.1 mg) into a 500-mL volumetric flask; use a powder funnel to ensure quantitative transfer of the solid. Rinse the funnel well, and dissolve the $KBrO_3$ in about 200 mL of distilled water. Dilute to the mark, and mix thoroughly.

Solid potassium bromate can cause a fire if it comes into contact with damp organic material (such as paper toweling in a waste container). Consult with the instructor concerning the disposal of any excess.

(b) *Sodium thiosulfate, 0.05 M.* Follow the directions in Section 31G-1; use about 12.5 g of $Na_2S_2O_3 \cdot 5H_2O$ per liter of solution.

(c) *Starch indicator.* See Section 31F-1.

31H-2 Standardization of Sodium Thiosulfate Against Potassium Bromate

Discussion

Iodine is generated by the reaction between a known volume of standard potassium bromate and an unmeasured excess of potassium iodide:

$$BrO_3^- + 6I^- + 6H^+ \longrightarrow Br^- + 3I_2 + 3H_2O$$

The iodine produced is titrated with the sodium thiosulfate solution.

PROCEDURE

Pipet 25.00-mL aliquots of the $KBrO_3$ solution into 250-mL conical flasks and rinse the interior wall with distilled water. *Treat each sample individually beyond this point.*

Introduce 2 to 3 g of KI and about 5 mL of 3 M H_2SO_4. Immediately titrate with 0.05 M $Na_2S_2O_3$ until the solution is pale yellow. Add 5 mL of starch indicator, and titrate to the disappearance of the blue color.

Calculate the concentration of the thiosulfate solution.

31H-3

The Determination of Phenol by Bromination

Discussion

The phenol content of waste waters from manufacturing processes is conveniently determined by mixing the sample with a measured excess of standard bromate followed by an excess of bromide. The bromine liberated upon acidification reacts with the phenol:

$$BrO_3^- + 5Br^- + 6H^+ \longrightarrow 3Br_2 + 3H_2O$$

$$C_6H_5OH + 3Br_2 \longrightarrow C_6H_2Br_3OH + 3H^+ + 3Br^-$$

After the bromination is complete, the excess bromine is determined employing the procedure used for the standardization.

PROCEDURE

Transfer a sample containing between 1 and 1.5 mmol of phenol to a 250-mL volumetric flask, dilute to the mark with water, and mix well. Pipet 25.00 mL aliquots of the diluted sample into 250-mL conical flasks, and add 25.00 mL aliquots of standard $KBrO_3$ solution. Add about 1 g of KBr and about 5 mL of 3 M H_2SO_4 to each flask. *Stopper each flask immediately after acidification* to prevent the loss of Br_2. Mix and let stand for about 10 min. Introduce 2 to 3 g of KI to each flask, and immediately restopper. Swirl the solutions until the KI has dissolved. Titrate the liberated iodine with standard 0.05 M $Na_2S_2O_3$ until the solution is pale yellow. Add 5 mL of starch indicator, and complete the titration.

Report the number of milligrams of phenol contained in each milliliter of the unknown.

31H-4

The Determination of Ascorbic Acid in Vitamin C Tablets by Titration with Potassium Bromate

Discussion

Ascorbic acid, $C_6H_8O_6$, is cleanly oxidized to dehydroascorbic acid by bromine:

```
       H  H
       |  |
       O  O      H   H                          O  O      H   H
       |  |      |   |                          ‖  ‖      |   |
       C==C      O   O                          C--C      O   O
      /    \     |   |                         /    \     |   |
  O==C      C----C---C--H  +  Br2  ---->  O==C       C----C---C--H  +  2Br-  +  2H+
      \    /|    |   |                         \    /|    |   |
        O   H    H   H                           O   H    H   H
```

An unmeasured excess of potassium bromide is added to an acidified solution of the sample. The solution is titrated with standard potassium bromate to the first permanent appearance of excess bromine; this excess is then determined iodometrically with standard sodium thiosulfate. The entire titration must be performed without delay to prevent air-oxidation of the ascorbic acid.

PROCEDURE

Weigh (to the nearest milligram) 3 to 5 vitamin C tablets (Note 1). Pulverize them thoroughly in a mortar, and transfer the powder to a dry weighing bottle. Weigh individual 0.40- to 0.50-g samples (to the nearest 0.1 mg) into dry 250-mL conical flasks. *Treat each sample individually beyond this point.* Dissolve the sample (Note 2) in 50 mL of 1.5 M H_2SO_4; then add about 5 g of KBr. Titrate immediately with standard $KBrO_3$ to the first faint yellow due to excess Br_2. Record the volume of $KBrO_3$ used. Add 3 g of KI and 5 mL of starch indicator; back-titrate (Note 3) with standard 0.05 M $Na_2S_2O_3$.

Calculate the average weight (in milligrams) of ascorbic acid (fw = 176.13) in each tablet.

Notes

1. This method is not applicable to chewable vitamin C tablets.
2. The binder in many vitamin C tablets remains in suspension throughout the analysis. If the binder is starch, the characteristic color of the complex with iodine appears upon the addition of KI.
3. The volume of thiosulfate needed for the back-titration seldom exceeds a few milliliters.

31I Potentiometric Methods

Potentiometric measurements provide a highly selective method for the quantitative determination of numerous cations and anions. A discussion of the principles and applications of potentiometric measurements is found in Chapter 15. Detailed instructions are given in this section on the use of potentiometric measurements to locate end points in volumetric titrations. In addition, a procedure for the direct potentiometric determination of fluoride ion in drinking water and in toothpaste is described.

31I-1 General Directions for Performing a Potentiometric Titration

The procedure that follows is applicable to the three titrimetric methods described in this section. With the proper choice of indicator electrode, it can also be applied to most of the volumetric and gravimetric titrimetry experiments given in Sections 31B through 31H.

1. Dissolve the sample in 50 to 250 mL of water. Rinse a suitable pair of electrodes with distilled water, and immerse them in the sample solution. Provide magnetic (or mechanical) stirring. Position the buret so that reagent can be delivered without splashing.
2. Connect the electrodes to the meter, commence stirring, and measure and record the initial buret volume and the initial potential (or pH).
3. Measure and record the meter reading and buret volume after each addition of titrant. Introduce fairly large volumes (about 5 mL) at the outset. Withhold a succeeding addition until the meter reading remains constant within 1 to 2 mV (or 0.05 pH unit) for at least 30 s (Note). Judge the volume of reagent to be added by estimating a value for $\Delta E/\Delta V$ after each addition. In the immediate vicinity of the equivalence point, introduce the reagent in 0.1-mL increments. Continue the titration 2 to 3 mL beyond the equivalence point, increasing the volume increments as $\Delta E/\Delta V$ again becomes smaller.

Note: Stirring motors occasionally cause erratic meter readings; it may be advisable to turn off the motor while meter readings are being made.

31I-2 *The Potentiometric Titration of Chloride and Iodide in a Mixture*

Discussion

The potentiometric titration of halide mixtures is discussed in Section 15E-2. The silver indicator electrode can be a commercial billet type or simply a polished wire. A calomel electrode can be used as reference, although diffusion of chloride ion from the salt bridge may cause the results of the titration to be measurably high. This source of error can be eliminated by placing the calomel electrode in a potassium nitrate solution that is in contact with the analyte solution by means of a KNO_3 salt bridge. Alternatively, the analyte solution can be made slightly acidic with several drops of nitric acid; a glass electrode can then serve as the reference electrode because the pH of the solution and thus its potential remain essentially constant throughout the titration.

The titration of I^-/Cl^- mixtures demonstrates how a potentiometric titration can have multiple end points. The potential of the silver electrode is proportional to pAg. Thus, a plot of E_{Ag} against titrant volume yields an experimental curve with the same shape as the theoretical curve shown in Figure 7-4 (the ordinate units will be different, of course).

Experimental curves for the titration of I^-/Cl^- mixtures do not show the sharp discontinuity that occurs at the first equivalence point of the theoretical curve (Figure 7-4). More important, the volume of silver nitrate needed to reach the I^- end point is generally somewhat greater than theoretical; the total volume closely approaches the correct amount, however. This effect is the result of coprecipitation of the more soluble AgCl during formation of the less soluble AgI. An overconsumption of reagent thus occurs in the first part of the titration.

Despite this coprecipitation error, the potentiometric method is useful for the analysis of halide mixtures. With approximately equal quantities of iodide and chloride, relative errors can be kept to within 2%.

PREPARATION OF REAGENTS

(a) *Silver nitrate, 0.05 M.* Follow the instructions in Section 31B-1.

(b) *Potassium nitrate salt bridge.* Bend an 8-mm glass tube into a U-shape with arms that are long enough to extend nearly to the bottom of two 100-mL beakers. Heat 50 mL of water to boiling, and stir in 1.8 g of powdered agar; continue to heat and stir until a uniform suspension is formed. Dissolve 12 g of KNO_3 in the hot suspension. Allow the mixture to cool somewhat. Clamp the U-tube with the openings facing up, and use a medicine dropper to fill it with the warm agar suspension. Cool the tube under a cold-water tap to form the gel. When the bridge is not in use, immerse the ends in 2.5 M KNO_3.

PROCEDURE

Obtain the unknown in a clean 250-mL volumetric flask; dilute to the mark with water, and mix well.

Transfer 50.00 mL of the sample to a clean 100-mL beaker, and add a drop or two of concentrated HNO_3. Place about 25 mL of 2.5 M KNO_3 in a second 100-mL beaker, and make contact between the two solutions with the agar salt bridge. Immerse a silver electrode in the analyte solution and a calomel reference electrode in the second beaker. Titrate with $AgNO_3$ as described in Section 31I-1. Use small increments of titrant in the vicinity of the two end points.

Plot the data, and establish end points for the two analyte ions. Plot a theoretical titration curve, assuming the measured concentrations of the two constituents to be correct.

Report the number of milligrams of I^- and Cl^- in the sample or as otherwise instructed.

31I-3 The Potentiometric Determination of Solute Species in a Phosphate Mixture

Discussion

The use of a glass/calomel electrode system to locate end points in neutralization titrations and to estimate dissociation constants is discussed in Section 15E-4. As a preliminary step to the titrations, the electrode system is standardized against a buffer of known pH.

The unknown is issued as an aqueous solution prepared from one or perhaps two adjacent members of the following series: HCl, H_3PO_4, NaH_2PO_4, Na_2HPO_4, Na_3PO_4, and NaOH. The object is to determine which of these components were used to prepare the unknown as well as the weight percent of each solute.

Most unknowns require a titration with either standard acid or standard base. A few may require separate titrations, one with acid and one with base. The initial pH of the unknown provides guidance concerning the appropriate titrant(s); a study of curve *A* in Figure 9-3 may be helpful in interpreting the data.

PREPARATION OF SOLUTIONS

Standardized 0.1 M HCl and/or 0.1 M NaOH. Follow the directions in Sections 31C-3 through 31C-7.

PROCEDURE

Obtain the unknown in a clean 250-mL volumetric flask. Dilute to the mark and mix well. Transfer a small amount of the diluted unknown to a beaker, and determine its pH. Titrate a 50.00-mL aliquot with standard acid or standard base (or perhaps both). Use the resulting titration curves to select indicator(s) suitable for end-point detection, and perform duplicate titrations with these.

Identify the solute species in the unknown, and report the weight/volume percent of each. Calculate the approximate dissociation constant that can be obtained for any phosphate-containing species from the titration data.

31I-4 The Potentiometric Titration of Copper with EDTA

Discussion

Mercury serves as an electrode of the second kind (Section 15B-1) for the titration of many cations with EDTA. A small volume of the mercury(II)/EDTA complex is added to a solution of the sample. The formation constant of

HgY^{2-} is so favorable that $[HgY^{2-}]$ remains essentially constant throughout the titration; the electrode is thus responsive to $[Y^{4-}]$ and indirectly to the concentration of the cation being titrated (in this experiment, Cu^{2+}). See Section 15E-3 for additional information.

PREPARATION OF SOLUTIONS

(a) *Copper(II), 0.02 M.* Weigh (to the nearest 0.1 mg) about 0.32 g of copper into a 150-mL beaker. Cover with a watch glass, and dissolve in a minimum volume of dilute HNO_3 (HOOD); warm gently, if necessary, to hasten solution. Rinse the underside of the watch glass and the inner wall of the beaker. Transfer the solution quantitatively to a 250-mL volumetric flask, dilute to the mark, and mix well.

(b) *EDTA, about 0.02 M.* Weigh about 1.9 g (to the nearest 0.1 mg) of purified $Na_2H_2Y \cdot 2H_2O$ into a 250-mL volumetric flask and dissolve in water, following the instructions given in Section 31D-2.

(c) *Acetate buffer.* (Sufficient for 40 to 50 titrations.) Dilute about 7 mL of glacial acetic acid to about 100 mL. With a pH meter, adjust the pH to 4.6 through the addition of 6 M NaOH. Dilute to about 250 mL with water.

(d) *Mercury(II), 0.02 M.* (Sufficient for several hundred titrations.) Dissolve 0.40 g of Hg in a minimum volume of HNO_3 (HOOD). Transfer quantitatively to a 100-mL volumetric flask, dilute to the mark, and mix well.

PROCEDURE

Adjust the meter to read in millivolts. Connect the mercury electrode and a calomel electrode to the meter. Transfer a 25.00-mL aliquot of the copper solution to a 250-mL beaker. Add about 5 mL of acetate buffer and one drop of a solution prepared by mixing equal volumes of the EDTA and Hg(II) solutions. Add water, if needed, to cover the ends of the electrodes (Note 1). Titrate (Note 2) with EDTA according to Section 31I-1.

Compare the milligrams of copper found with the amount taken. Calculate the relative error in the titration.

Notes

1. Sufficient chloride ion may leak from the calomel electrode to interfere with the functioning of the mercury indicator electrode. If necessary, place the reference electrode in a beaker that is connected by a KNO_3 bridge to the analyte container; see Section 31I-2.
2. The titration must be performed slowly in the vicinity of the end point.

31I-5 The Direct Potentiometric Determination of Fluoride Ion

Discussion

The solid-state fluoride electrode (Section 15B-6) has found extensive use in the determination of fluoride in a variety of materials. Directions follow for the determination of this ion in drinking water and in toothpaste. A total ionic strength adjustment buffer (TISAB) is used to adjust all unknowns and standards to essentially the same ionic strength; when this is done, the concentration of fluoride, rather than its activity, is measured. The pH of the buffer is about 5, a level at which F^- is the predominant fluorine-containing species. The buffer also contains cyclohexylaminedinitrilotetraacetic acid, which forms stable chelates with iron(III) and aluminum(III), thus freeing fluoride ion from its complexes with these cations.

Before undertaking these experiments, a review of Sections 15B and 15D is suggested.

PREPARATION OF SOLUTIONS

(a) *Total ionic strength adjustment buffer (TISAB).* This solution is marketed commercially under the trade name TISAB.[8] Sufficient buffer for 15 to 20 determinations can be prepared by mixing (with stirring) 57 mL of glacial acetic acid, 58 g of NaCl, 4 g of cyclohexylaminedinitrilotetraacetic acid, and 500 mL of distilled water in a 1-L beaker. Cool the contents in a water or ice bath, and carefully add 6 M NaOH until a pH of 5.0 to 5.5 is reached. Dilute to 1 L with water, and store in a plastic bottle.

(b) *Standard fluoride solution, 100 ppm.* Dry a quantity of NaF at 110°C for 2 h. Cool in a desiccator; then weigh (to the nearest milligram) 0.22 g into a 1-L volumetric flask. (CAUTION! NaF IS HIGHLY TOXIC. *Immediately* wash any skin touched by this compound with copious quantities of water.) Dissolve in water, dilute to the mark, mix well, and store in a plastic bottle. Calculate the exact concentration of fluoride in parts per million.

A standard F^- solution can be purchased from commercial sources.

PROCEDURE

The apparatus for this experiment consists of a solid-state fluoride electrode, a saturated calomel electrode, and a pH meter. A sleeve-type calomel electrode is needed for the toothpaste determination because the measurement is made on a suspension that tends to clog the liquid junction. The sleeve must be loosened momentarily to renew the interface after each series of measurements.

Determination of fluoride in drinking water

Transfer 50.00-mL portions of the water to 100-mL volumetric flasks, and dilute to the mark with TISAB solution.

Prepare a 5-ppm F^- solution by diluting 25.0 mL of the 100-ppm standard to 500 mL in a volumetric flask. Transfer 5.00-, 10.0-, 25.0-, and 50.0-mL aliquots of the 5-ppm solution to 100-mL volumetric flasks, add 50 mL of TISAB solution, and dilute to the mark. (These solutions correspond to 0.5, 1.0, 2.5, and 5.00 ppm F^- in the sample.)

After thorough rinsing and drying with paper tissue, immerse the electrodes in the 0.5-ppm standard. Stir mechanically for 3 min; then measure and record the potential. Repeat with the remaining standards and samples.

Plot the measured potential against the log of the concentration of the standards. Use this plot to determine the concentration in parts per million of fluoride in the unknown.

Determination of fluoride in toothpaste[9]

Weigh (to the nearest milligram) 0.2 g of toothpaste into a 250-mL beaker. Add 50 mL of TISAB solution, and boil for 2 min with good mixing. Cool and then transfer the suspension quantitatively to a 100-mL volumetric flask, dilute to the mark with distilled water, and mix well. Follow the directions for the analysis of drinking water, beginning with the second paragraph.

Report the parts per million of F^- in the sample.

[8]Orion Research, Boston, MA.

[9]From T.S. Light and C. C. Cappuccino, *J. Chem. Educ.*, 1975, *52*, 247.

31J Electrogravimetric Methods

A convenient example of an electrogravimetric method of analysis is the simultaneous determination of copper and lead in a sample of brass. Additional information concerning electrogravimetric methods is found in Section 16C.

31J-1 The Electrogravimetric Determination of Copper and Lead in Brass

Discussion

This procedure is based upon the deposition of metallic copper on a cathode and of lead as PbO_2 on an anode. As a first step, the hydrous oxide of tin ($SnO_2{\cdot}xH_2O$) that forms when the sample is treated with nitric acid must be removed by filtration. Lead dioxide is deposited quantitatively at the anode from a solution with a high nitrate ion concentration; copper is only partially deposited on the cathode under these conditions. It is therefore necessary to eliminate the excess nitrate after deposition of the PbO_2 is complete. Removal is accomplished through the addition of urea:

$$6NO_3^- + 6H^+ + 5(NH_2)_2CO \longrightarrow 8N_2(g) + 5CO_2(g) + 13H_2O$$

Copper then deposits quantitatively from the solution after the nitrate ion concentration has been decreased.

PROCEDURE

Preparation of electrodes

Immerse the platinum electrodes in hot 6 M HNO_3 for about 5 min (Note 1). Wash them thoroughly with distilled or deionized water, rinse with several small portions of acetone or ethanol, and dry in an oven at 110°C for 2 to 3 min. Cool and weigh both anodes and cathodes to the nearest 0.1 mg.

Preparation of samples

It is not necessary to dry the unknown. A rinse with acetone is recommended if there is evidence of oil on the surface of the metal. Weigh (to the nearest 0.1 mg) 1-g samples into 250-mL beakers. Cover the beakers with watch glasses. Cautiously add about 35 mL of 6 M HNO_3 (HOOD). Digest for at least 30 min; add more acid if necessary. Evaporate to about 5 mL but never to dryness (Note 2).

To each sample, add 5 mL of 3 M HNO_3, 25 mL of water, and one quarter of a tablet of filter paper pulp; digest without boiling for about 45 min. Filter off the $SnO_2{\cdot}xH_2O$, using a fine-porosity filter paper (Note 3); collect the filtrates in tall-form electrolysis beakers. Use many small washes with hot 0.3 M HNO_3 to remove the last traces of copper; test for completeness with a few drops of NH_3. The final volume of filtrate and washings should be between 100 and 125 mL; either add water or evaporate to attain this volume.

Electrolysis

With the current switch off, attach the cathode to the negative terminal and the anode to the positive terminal of the electrolysis apparatus. Briefly turn on the stirring motor to be sure the electrodes do not touch. Cover the beakers with split watch glasses and commence the electrolysis. Maintain a current of 1.3 A for 35 min.

Rinse the cover glasses and add 10 mL of 3 M H_2SO_4 followed by 5 g of urea to each beaker. Maintain a current of 2 A until the solutions are colorless. To test for completeness of the electrolysis, remove one drop of the solution with an eye dropper, and mix it with a few drops of NH_3 in a small test tube. If the mixture turns blue, rinse the contents of the tube back into the electrolysis vessel, and continue the electrolysis for an additional 10 min. Repeat the test until no blue $Cu(NH_3)_4^{2+}$ is produced.

When electrolysis is complete, discontinue stirring but leave the current on. Rinse the electrodes thoroughly with water as they are removed from the solution. After rinsing is complete, turn off the electrolysis apparatus (Note 4), disconnect the electrodes, and dip them in acetone. Dry the cathodes for about 3 min and the anodes for about 15 min at 110°C. Allow the electrodes to cool in air, and then weigh them.

Report the percentages of lead (Note 5) and copper in the brass.

Notes

1. Alternatively, grease and organic materials can be removed by heating platinum electrodes to redness in a flame. Electrode surfaces should not be touched with the fingers after cleaning because grease and oil cause nonadherent deposits that can flake off during washing and weighing.
2. Chloride ion must be totally excluded from this determination because it attacks the platinum anode during electrolysis. This reaction is not only destructive but also causes positive errors in the analysis by codepositing platinum with copper on the cathode.
3. If desired, the tin content can be determined by ignition of the $SnO_2 \cdot xH_2O$. See Section 31A-2.
4. It is important to maintain a potential between the electrodes until they have been removed from the solution and washed. Some copper may redissolve if this precaution is not observed.
5. Experience has shown that a small amount of moisture is retained by the PbO_2 and that better results are obtained if 0.8643 is used instead of 0.8662, the stoichiometric factor.

31K Coulometric Titrations

In a coulometric titration, the "reagent" is a constant direct current of exactly known magnitude. The time required for this current to quantitatively oxidize or reduce the analyte (directly or indirectly) is measured. See Section 16D-5 for a discussion of this electroanalytical method.

31K-1 The Coulometric Titration of Cyclohexene

Discussion[10]

Many olefins react sufficiently rapidly with bromine to permit their direct titration. The reaction is carried out in a largely nonaqueous environment with mercury(II) as a catalyst. A convenient way of performing this titration is to add excess bromide ion to a solution of the sample and generate the bromine at an anode that is connected to a constant-current source. The electrode processes are

$$2Br^- \longrightarrow 2Br_2 + 2e \qquad \text{anode}$$

$$2H^+ + 2e \longrightarrow H_2(g) \qquad \text{cathode}$$

The hydrogen produced does not react with bromine rapidly enough to interfere. The bromine reacts with an olefin, such as cyclohexene, to give the addition product:

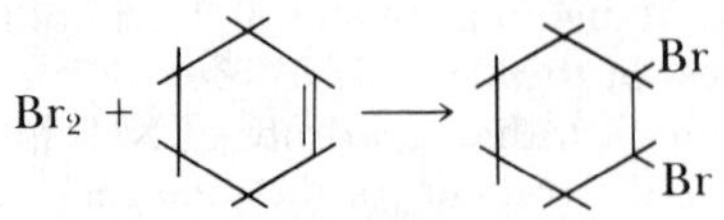

Br_2 + [cyclohexene] ⟶ [1,2-dibromocyclohexane: Br, Br]

[10]This procedure was described by D. H. Evans in *J. Chem. Educ.*, **1968,** *45* (1), 88.

The amperometric method with twin-polarized electrodes (Section 17F-4) provides a convenient way to detect the end point in this titration. A potential difference of 0.2 to 0.3 V is maintained between two small electrodes. This potential is not sufficient to cause the generation of hydrogen at the cathode. Thus, short of the end point, the cathode is polarized and no current is observed. At the end point, the first excess of bromine depolarizes the cathode and produces a current. The electrode reactions at the twin indicator electrodes are

$$2Br^- \longrightarrow Br_2 + 2e \quad \text{anode}$$

$$Br_2 + 2e \longrightarrow 2Br^- \quad \text{cathode}$$

The current is proportional to the bromine concentration and is readily measured with a microammeter.

A convenient way to perform several analyses is to initially generate sufficient bromine in the solvent to give a readily measured current, say, 20 μA. An aliquot of the sample is then introduced, whereupon the current immediately decreases and approaches zero. Generation of bromine is again commenced, and the time needed to regain a current of 20 μA is measured. A second aliquot of the sample is added to the same solution, and the process is repeated. Several samples can thus be analyzed without changing the solvent.

The procedure that follows is for the determination of cyclohexene in a methanol solution. Other olefins can be determined as well.

PREPARATION OF SOLVENT

Dissolve about 9 g of KBr and 0.5 g of mercury(II) acetate (Note 1) in a mixture consisting of 300 mL of glacial acetic acid, 130 mL of methanol, and 65 mL of water. (Sufficient for about 35 mmol of Br_2.) (CAUTION! Mercury compounds are highly toxic and the solvent is a skin irritant. If inadvertent contact occurs, flood the affected area with copious quantities of water.)

PROCEDURE

Obtain the unknown in a 100-mL volumetric flask; dilute to the mark with methanol, and mix well. The temperature of the methanol should be between 18 and 20°C (Note 2).

Add sufficient acetic acid/methanol solvent to cover the indicator and generator electrodes in the electrolysis vessel. Apply about 0.2 V to the indicator electrodes. Activate the generator electrode system, and generate bromine until a current of about 20 μA is indicated on the microammeter. Stop the generation of bromine, record the indicator current to the nearest 0.1 μA, and set the timer to zero. Transfer 10.00 mL of the unknown to the solvent; the indicator current should decrease to almost zero. Resume bromine generation. Produce bromine in smaller and smaller increments by activating the generator for shorter and shorter periods as the indicator current rises and approaches the previously recorded value. Read and record the time needed to reach the original indicator current. Reset the timer to zero, introduce a second aliquot of sample (make the volume larger if the time needed for the first titration was too short, and conversely), and repeat the process. Titrate several aliquots.

Report the number of milligrams of cyclohexene in the unknown.

Notes

1. Mercury(II) ions catalyze the addition of bromine to olefinic double bonds.
2. The coefficient of expansion for methanol is 0.11%/°C; thus, significant volumetric errors result if the temperature is not controlled.

31L Voltammetry

Various aspects of polarographic and amperometric methods are considered in Chapter 17. Two examples that illustrate these methods are described in this section. Enormous diversity exists in the instrumentation available for these determinations. It will thus be necessary for the reader to consult the manufacturer's operating instructions concerning the details of operation for the particular instrument used.

31L-1 The Polarographic Determination of Copper and Zinc in Brass

Discussion

The percentage of copper and zinc in a sample of brass can be determined from polarographic measurements. The method is particularly useful for rapid, routine analyses; in return for speed, however, the accuracy is considerably lower than that obtained with volumetric or gravimetric methods.

The sample is dissolved in a minimum amount of nitric acid. It is not necessary to remove the $SnO_2{\cdot}xH_2O$ produced. Addition of an ammonia/ammonium chloride buffer causes the precipitation of lead as a basic oxide. A polarogram of the supernatant liquid has two copper waves. The one at about -0.2 V (versus SCE) corresponds to the reduction of copper(II) to copper(I), and the one at about -0.5 V represents further reduction to the metal. The analysis is based upon the total diffusion current of the two waves. The zinc concentration is determined from its wave at -1.3 V. For instruments that permit current offset, the copper waves are measured at the highest feasible sensitivity. These waves are then suppressed by the offset control of the instrument, and the zinc wave is obtained, again at the highest possible sensitivity setting.

PREPARATION OF SOLUTIONS

(a) *Copper(II) solution, 2.5×10^{-2} M.* Weigh (to the nearest milligram) 0.4 g of copper wire. Dissolve in 5 mL of concentrated HNO_3 (HOOD). Boil briefly to remove oxides of nitrogen; then cool, dilute with water, transfer quantitatively to a 250-mL volumetric flask, dilute to the mark with water, and mix thoroughly.

(b) *Zinc(II) solution, 2.5×10^{-2} M.* Dry reagent-grade ZnO for 1 h at 110°C, cool in a desiccator, and weigh 0.5 g (to the nearest milligram) into a small beaker. Dissolve in a mixture of 25 mL of water and 5 mL of concentrated HNO_3. Transfer to a 250-mL volumetric flask, and dilute to the mark with water.

(c) *Gelatin, 0.1%.* Add about 0.1 g of gelatin to 100 mL of boiling water.

(d) *Ammonia/ammonium chloride buffer.* (Sufficient for about 15 polarograms.) Mix 27 g of NH_4Cl and 35 mL of concentrated ammonia in sufficient distilled water to give about 500 mL. This solution is about 1 M in NH_3 and 1 M in NH_4^+.

PROCEDURE

Preparation of calibration standards

Use a buret to transfer 0-, 1-, 8-, and 15-mL portions of standard Cu(II) solution to 50-mL volumetric flasks. Add 5 mL of gelatin solution and 30 mL of buffer to each. Dilute to the mark, and mix well. Prepare an identical series of Zn(II) solutions.

Rinse the polarographic cell three times with small portions of a copper(II) solution; then fill the cell. Bubble nitrogen through the solution for 10 to 15 min to remove oxygen. Apply a potential of about −1.6 V, and adjust the sensitivity to cause the detector to give a response that is essentially full scale. Obtain a polarogram, scanning from 0 to −1.5 V (versus SCE). Measure the limiting current at a potential just beyond the second wave. Obtain a diffusion current by subtracting the current for the blank (Note) at this same potential. Calculate i_d/C.

Repeat the foregoing with the other two copper solutions and with the zinc solutions.

Sample preparation

Weigh 0.10- to 0.15-g (to the nearest 0.5 mg) samples of brass into 50-mL beakers, and dissolve in 2 mL of concentrated HNO_3 (HOOD). Boil briefly to eliminate oxides of nitrogen. Cool, add 10 mL of distilled water, and transfer each quantitatively to a 50-mL volumetric flask; dilute to the mark with water, and mix well.

Transfer 10.00 mL of the diluted sample to another 50.0-mL volumetric flask, add 5 mL of gelatin and 30 mL of buffer, dilute to the mark with water, and mix well.

Analysis

Follow the directions in the second paragraph of *Preparation of calibration standards*. Evaluate the diffusion currents for copper and zinc.

Calculate the percentage of Cu and of Zn in the brass sample.

Note. The polarogram for the blank (that is, 0 mL of standard) should be obtained at the sensitivity setting used for the standard with the lowest metal-ion concentration.

31L-2 The Amperometric Titration of Lead

Discussion

Amperometric titrations are discussed in Section 17F-4. In the procedure that follows, the lead concentration of an aqueous solution is determined by titration with a standard potassium dichromate solution. The reaction is

$$Cr_2O_7^{2-} + 2Pb^{2+} + H_2O \longrightarrow 2PbCrO_4(s) + 2H^+$$

The titration can be performed with a dropping mercury electrode maintained at either 0 or −1.0 V (versus SCE). At 0 V, the current remains near zero short of the end point but rises rapidly immediately thereafter, owing to the reduction of dichromate, which is now in excess. At −1.0 V, both lead ion and dichromate ion are reduced. The current thus decreases in the region short of the end point (reflecting the decreases in the lead ion concentration as titrant is added), passes through a minimum at the end point, and rises as dichromate becomes available. The end point at −1.0 V should be the easier of the two to locate exactly. Removal of oxygen is unnecessary, however, for the titration at 0 V.

PREPARATION OF SOLUTIONS

(a) *Supporting electrolyte.* Dissolve 10 g of KNO_3 and 8.2 g of sodium acetate in about 500 mL of distilled water. Add glacial acetic acid to bring the pH to 4.2 (pH meter); about 20 mL of the acid will be required. (Sufficient for about 20 titrations.)

(b) *Gelatin, 0.1%.* Add 0.1 g of gelatin to 100 mL of boiling water.

(c) *Potassium dichromate, 0.01 M.* Use a powder funnel to weigh about 0.75 g (to the nearest 0.5 mg) of primary-standard $K_2Cr_2O_7$ into a 250-mL volumetric flask. Dilute to the mark with distilled water, and mix well.

(d) *Potassium nitrate salt bridge.* See Section 31I-2, *Preparation of Reagents.*

PROCEDURE

Obtain the unknown (Note) in a clean 100-mL volumetric flask; dilute to the mark with distilled water, and mix well. Perform the titration in a 100-mL beaker. Locate a saturated calomel electrode in a second 100-mL beaker, and provide contact between the solutions in the two containers with the KNO_3 salt bridge.

Transfer a 10.00-mL aliquot of the unknown to the titration beaker. Add 25 mL of the supporting electrolyte and 5 mL of gelatin. Insert a dropping mercury electrode in the sample solution. Connect both electrodes to the polarograph. Measure the current at an applied potential of 0 V. Add 0.01 M $K_2Cr_2O_7$ in 1-mL increments; record the current and volume after each addition. Continue the titration to about 5 mL beyond the end point. Correct the currents for volume change (page 449), and plot the data. Evaluate the end-point volume.

Repeat the titrations at -1.0 V. Here, it is necessary to bubble nitrogen through the solution for 10 to 15 min before the titration and after each addition of reagent. The flow of nitrogen must, of course, be interrupted during the current measurements. Again, correct the currents for volume change, plot the data, and determine the end-point volume.

Report the number of milligrams of Pb in the unknown.

Note. A stock 0.0400 M solution can be prepared by dissolving 13.5 g of reagent-grade $Pb(NO_3)_2$ in 10 mL of 6 M HNO_3 and diluting to 1 L with water. Unknowns should contain 15 to 25 mL of this solution.

31M Methods Based on the Absorption of Radiation

Molecular absorption methods are discussed in Chapter 20. Directions follow for (1) the use of a calibration curve for the determination of iron in water, (2) the use of a standard-addition procedure for the determination of manganese in steel, and (3) a spectrophotometric determination of the pH of a buffer solution.

31M-1 The Cleaning and Handling of Cells

The accuracy of spectrophotometric measurements is critically dependent upon the availability of good-quality matched cells. These should be calibrated against one another at regular intervals to detect differences resulting from scratches, etching, and wear. Equally important is the proper cleaning of the exterior sides (the *windows*) just before the cells are inserted into a photometer or spectrophotometer. The preferred method is to wipe the windows with a lens paper soaked in methanol; the methanol is then allowed to evaporate, leaving the windows free of contaminants. It has been shown that this method is far superior to the usual procedure of wiping the windows with a dry lens paper, which tends to leave a residue of lint and a film on the window.[11]

[11] For further information, see J. O. Erickson and T. Surles, *Amer. Lab.*, **1976,** *8* (6), 50.

31M-2

The Determination of Iron in a Natural Water

Discussion

The red-orange complex that forms between iron(II) and 1,10-phenanthroline (orthophenanthroline) is useful for determining iron in water supplies. The reagent is a weak base that reacts to form phenanthrolinium ions, $PhenH^+$, in acidic media. Complex formation with iron is thus best described by the equation

$$Fe^{2+} + 3PhenH^+ \rightleftharpoons Fe(Phen)_3^{2+} + 3H^+$$

The formation constant for this equilibrium is 2.5×10^6 at 25°C. Iron(II) is quantitatively complexed in the pH range between 3 and 9. A pH of about 3.5 is ordinarily recommended to prevent precipitation of iron salts, such as phosphates.

An excess of a reducing reagent, such as hydroxylamine or hydroquinone, is needed to maintain iron in the +2 state. The complex, once formed, is very stable.

This determination can be performed with a spectrophotometer set at 508 nm or with a photometer equipped with a green filter.

PREPARATION OF SOLUTIONS

(a) *Standard iron solution, 0.01 mg/mL.* Weigh (to the nearest 0.2 mg) 0.0702 g of reagent-grade $Fe(NH_4)_2(SO_4)_2 \cdot 6H_2O$ into a 1-L volumetric flask. Dissolve in 50 mL of water that contains 1 to 2 mL of concentrated sulfuric acid; dilute to the mark, and mix well.

(b) *Hydroxylamine hydrochloride.* Dissolve 10 g of $H_2NOH \cdot HCl$ in about 100 mL of distilled water. (Sufficient for 80 to 90 measurements.)

(c) *Orthophenanthroline solution.* (Sufficient for 80 to 90 measurements.) Dissolve 1.0 g of orthophenanthroline monohydrate in about 1 L of water. Warm slightly if necessary. Each milliliter is sufficient for no more than about 0.09 mg of Fe. Prepare no more reagent than needed; it darkens on standing and must then be discarded.

(d) *Sodium acetate, 1.2 M.* (Sufficient for 80 to 90 measurements.) Dissolve 166 g of $NaOAc \cdot 3H_2O$ in 1 L of distilled water.

PROCEDURE

Preparation of a calibration curve

Transfer 25.00 mL of the standard iron solution to a 100-mL volumetric flask and 25 mL of distilled water to a second 100-mL volumetric flask. Add 1 mL of hydroxylamine, 10 mL of sodium acetate, and 10 mL of orthophenanthroline to each flask. Allow the mixtures to stand for 5 min; dilute to the mark and mix well.

Clean a pair of matched cells for the instrument. Rinse each cell with at least three portions of the solution it is to contain. Determine the absorbance of the standard with respect to the blank.

Repeat the above procedure with at least three other volumes of the standard iron solution; attempt to encompass an absorbance range between 0.1 and 1.0. Plot a calibration curve.

Determination of iron

Transfer 10.00 mL of the unknown to a 100-mL volumetric flask; treat in exactly the same way as the standards, measuring the absorbance with respect to the blank. Alter the volume of unknown taken to obtain absorbance measurements for replicate samples that are within the range of the calibration curve.

Report the parts per million of iron in the unknown.

31M-3 The Determination of Manganese in Steel

Discussion

Small quantities of manganese are readily determined photometrically by the oxidation of Mn(II) to the intensely colored permanganate ion. Potassium periodate is an effective oxidizing reagent for this purpose. The reaction is

$$5IO_4^- + 2Mn^{2+} + 3H_2O \longrightarrow 5IO_3^- + 2MnO_4^- + 6H^+$$

Permanganate solutions that contain an excess of periodate are quite stable.

Interferences to the method are few. The presence of most colored ions can be compensated for with a blank. Cerium(III) and chromium(III) are exceptions; these yield oxidation products with periodate that absorb to some extent at the wavelength used for the measurement of permanganate.

The method given here is applicable to steels that do not contain large amounts of chromium. The sample is dissolved in nitric acid. Any carbon in the steel is oxidized with peroxodisulfate. Iron(III) is eliminated as a source of interference by complexation with phosphoric acid. The standard-addition method (page 516) is used to establish the relationship between absorbance and amount of manganese in the sample.

A spectrophotometer set at 525 nm or a photometer with a green filter can be used for the absorbance measurements.

PREPARATION OF SOLUTIONS

(a) *Standard manganese(II) solution.* (Sufficient for several hundred analyses.) Weigh 0.1 g (to the nearest 0.1 mg) of manganese into a 50-mL beaker, and dissolve in about 10 mL of 6 M HNO_3 (HOOD). Boil gently to eliminate oxides of nitrogen. Cool; then transfer the solution quantitatively to a 1-L volumetric flask. Dilute to the mark with water, and mix thoroughly. The manganese in 1 mL of the standard solution, after being converted to permanganate, causes a volume of 50 mL to increase in absorbance by about 0.09.

PROCEDURE

The unknown does not require drying. If there is evidence of oil, rinse with acetone and dry briefly. Weigh (to the nearest 0.1 mg) duplicate samples (Note 1) into 150-mL beakers. Add about 50 mL of 6 M HNO_3, and boil gently (HOOD); heating for 5 to 10 min should suffice. Cautiously add about 1 g of ammonium peroxodisulfate, and boil gently for an additional 10 to 15 min. If the solution is pink or has a deposit of MnO_2, add 1 mL of NH_4HSO_3 (or 0.1 g of $NaHSO_3$) and heat for 5 min. Cool; transfer quantitatively (Note 2) to 250.0-mL volumetric flasks. Dilute to the mark with water, and mix well. Use a 20.00-mL pipet to transfer three aliquots of each sample to individual beakers. Treat as follows:

Aliquot	Volume of 85% H_3PO_4, mL	Volume of standard Mn, mL	Weight of KIO_4, g
1	5	0.00	0.4
2	5	5.00 (Note 3)	0.4
3	5	0.00	0.0

Boil each solution gently for 5 min, cool, and transfer quantitatively to a 50-mL volumetric flask. Mix well. Measure the absorbance of aliquots 1 and 2 using aliquot 3 as the blank (Note 4).

Report the percentage of manganese in the unknown.

Notes

1. The sample size depends upon the manganese content of the unknown; consult with the instructor.
2. If there is evidence of turbidity, filter the solutions as they are transferred to the volumetric flasks.
3. The volume of the standard addition may be dictated by the absorbance of the sample. It is useful to obtain a rough estimate by generating permanganate in about 20 mL of sample, diluting to about 50 mL, and measuring the absorbance.
4. A single blank can be used for all measurements, provided the samples weigh within 50 mg of one another.

31M-4 The Spectrophotometric Determination of pH

Discussion

The pH of an unknown buffer is determined by addition of an acid/base indicator and spectrophotometric measurement of the absorbance of the resulting solution. Because overlap exists between the spectra for the acid and base forms of the indicator, it is necessary to evaluate individual molar absorptivities for each form at two wavelengths. See page 517 for further discussion.

The relationship between the two forms of bromocresol green in an aqueous solution is described by the equilibrium

$$HIn + H_2O \rightleftharpoons H_3O^+ + In^-$$

for which

$$K_a = \frac{[H_3O^+][In^-]}{[HIn]} = 1.6 \times 10^{-5}$$

The spectrophotometric evaluation of $[In^-]$ and $[HIn]$ permits the calculation of $[H_3O^+]$.

PREPARATION OF SOLUTIONS

(a) *Bromocresol green, 1.0×10^{-4} M.* (Sufficient for about five determinations.) Dissolve 40.0 mg (to the nearest 0.1 mg) of the sodium salt of bromocresol green (fw = 720) in water, and dilute to 500 mL in a volumetric flask.

(b) *HCl, 0.5 M.* Dilute 4 mL of concentrated HCl to approximately 100 mL with water.

(c) *NaOH, 0.4 M.* Dilute 7 mL of 6 M NaOH to about 100 mL with water.

PROCEDURE

Determination of individual absorption spectra

Transfer 25.00-mL aliquots of the bromocresol green indicator solution to two 100-mL volumetric flasks. To one add 25 mL of 0.5 M HCl; to the other add 25 mL of 0.4 M NaOH. Dilute to the mark and mix well.

Obtain the absorption spectra for the acid and conjugate-base forms of the indicator between 400 and 600 nm, using water as a blank. Record absorbance values at 10-nm intervals routinely and at closer intervals as needed to define maxima and minima. Evaluate the molar absorptivity for HIn and In^- at wavelengths corresponding to their absorption maxima.

Determination of the pH of an unknown buffer

Transfer 25.00 mL of the stock bromocresol green indicator to a 100-mL volumetric flask. Add 50.0 mL of the unknown buffer, dilute to the mark, and mix well. Measure the absorbance of the diluted solutions at the wavelengths for which absorptivity data were calculated.

Report the pH of the buffer.

31N Molecular Fluorescence

The phenomenon of fluorescence and its analytical applications are discussed in Section 20C. Directions follow for the determination of quinine in beverages, in which the quinine concentration is typically between 25 and 60 ppm.

31N-1 The Determination of Quinine in Beverages[12]

Discussion

Solutions of quinine fluoresce strongly when excited by radiation at 350 nm. The relative intensity of the fluorescent peak at 450 nm provides a sensitive method for the determination of quinine in beverages. Preliminary measurements are needed to define a concentration region in which fluorescent intensity is either linear or nearly so. The unknown is then diluted as necessary to produce readings within this range.

PREPARATION OF REAGENTS

(a) *Sulfuric acid, 0.05 M.* Add about 17 mL of 6 M H_2SO_4 to 2 L of distilled water.

(b) *Quinine sulfate standard, 1 ppm.* Weigh (to the nearest 0.5 mg) 0.100 g of quinine sulfate into a 1-L volumetric flask, and dilute to the mark with 0.05 M H_2SO_4. (Sufficient for 60 to 70 analyses.) Transfer 10.00 mL of this solution to another 1-L volumetric flask, and again dilute to the mark with 0.05 M H_2SO_4. This latter solution contains 1 ppm of quinine; it should be prepared daily and stored in the dark when not in use.

[12]These directions were adapted from J. E. O'Reilly, *J. Chem. Educ.*, **1975,** *52* (9), 610.

PROCEDURE

Determination of a suitable concentration range
To find a suitable working range, measure the relative fluorescent intensity of the 1 ppm standard at 450 nm (or with a suitable filter that transmits in this range). Use a graduated cylinder to dilute 10 mL of the 1 ppm solution with 10 mL of 0.05 M H_2SO_4; again measure the relative fluorescence. Repeat this dilution and measurement process until the relative intensity approaches that of a blank consisting of 0.05 M H_2SO_4. Make a plot of the data, and select a suitable range for the analysis (that is, a region within which the plot is linear).

Preparation of a calibration curve
Use volumetric glassware to prepare three or four standards that span the linear region; measure the fluorescence intensity for each. Plot the data.

Analysis
Obtain an unknown. Make suitable dilutions with 0.05 M H_2SO_4 to bring its fluorescence intensity within the calibration range.

Calculate the parts per million of quinine in the unknown.

310 Atomic Spectroscopy

Several methods of analysis based upon atomic spectroscopy are discussed in Chapter 21. One such application is atomic absorption, which is demonstrated in the experiment that follows.

310-1 The Determination of Lead in Brass by Atomic Absorption Spectroscopy

Discussion

Brasses and other copper-based alloys contain from 0 to about 10% lead as well as tin and zinc. Atomic absorption spectroscopy permits the quantitative estimation of these elements. The accuracy of this procedure is not as great as that obtainable with gravimetric or volumetric measurements, but the time needed to acquire the analytical information is considerably less.

A weighed sample is dissolved in a mixture of nitric and hydrochloric acids, the latter being needed to prevent the precipitation of tin as metastannic acid, $SnO_2 \cdot xH_2O$. After suitable dilution, the sample is aspirated into a flame, and the absorption of radiation from a hollow cathode lamp is measured.

PREPARATION OF SOLUTIONS

(a) *Standard lead solution, 100 mg/L.* Dry a quantity of reagent-grade $Pb(NO_3)_2$ for about 1 h at 110°C. Cool; weigh (to the nearest 0.1 mg) 0.17 g into a 1-L volumetric flask. Dissolve in a solution of 5 mL water and 1 to 3 mL of concentrated HNO_3. Dilute to the mark with distilled water, and mix well.

PROCEDURE

Weigh duplicate samples of the unknown (Note 1) into 150-mL beakers. Cover with watch glasses, and then dissolve (HOOD) in a mixture consisting of about 4 mL of concentrated HNO_3 and 4 mL of concentrated HCl (Note 2). Boil gently to remove oxides of nitrogen. Cool; transfer the solutions quantitatively to a 250-mL volumetric flasks, dilute to the mark with water, and mix well.

Use a buret to deliver 0-, 5-, 10-, 15-, and 20-mL portions of the standard lead solution to individual 50-mL volumetric flasks. Add 4 mL of concentrated HNO_3 and 4 mL of concentrated HCl to each, and dilute to the mark with water.

Transfer 10.00-mL aliquots of each sample to 50-mL volumetric flasks, add 4 mL of concentrated HCl and 4 mL of concentrated HNO_3, and dilute to the mark with water.

Set the monochromator at 283.3 nm, and measure the absorbance for each standard and the sample at that wavelength. Take at least three—and preferably more—readings for each measurement.

Plot the calibration data. Report the percentage of lead in the brass.

Notes

1. The weight of sample depends on the lead content of the brass and on the sensitivity of the instrument used for the absorption measurements. A sample containing 6 to 10 mg of lead is reasonable. Consult with the instructor.
2. Brasses that contain a large percentage of tin require additional HCl to prevent the formation of metastannic acid. The diluted samples may develop some turbidity on prolonged standing; a slight turbidity has no effect on the determination of lead.

31O-2 The Determination of Sodium, Potassium, and Calcium in Mineral Waters by Atomic Emission Spectroscopy

Discussion

A convenient method for the determination of alkali and alkaline earth metals in water and in blood serum is based upon the characteristic spectra these elements emit upon being aspirated into a natural gas/air flame. The accompanying directions are suitable for the analysis of the three elements in water samples. Radiation buffers (Section 21B-6) are used to minimize the effect of each element upon the emission intensity of the others.

PREPARATION OF SOLUTIONS

(a) *Standard calcium solution, 500 ppm.* Dry a quantity of $CaCO_3$ for about 1 h at 110°C. Cool in a desiccator; weigh (to the nearest milligram) 1.25 g into a 600-mL beaker. Add about 200 mL of distilled water and about 10 mL of concentrated HCl; cover the beaker with a watch glass during the addition of the acid to avoid loss due to spattering. After reaction is complete, transfer the solution quantitatively to a 1-L volumetric flask, dilute to the mark and mix well.

(b) *Standard potassium solution, approximately 500 ppm.* Dry a quantity of KCl for about 1 h at 110°C. Cool; weigh (to the nearest milligram) about 0.95 g into a 1-L volumetric flask. Dissolve in distilled water, and dilute to the mark.

(c) *Standard sodium solution, approximately 500 ppm.* Proceed as in (b), using 1.25 g (to the nearest milligram) of dried NaCl.

(d) *Radiation buffer for the determination of calcium.* Prepare about 100 mL of a solution that has been saturated with NaCl, KCl, and $MgCl_2$, in that order.

(e) *Radiation buffer for the determination of potassium.* Prepare about 100 mL of a solution that has been saturated with NaCl, $CaCl_2$, and $MgCl_2$, in that order.

(f) *Radiation buffer for the determination of sodium.* Prepare about 100 mL of a solution that has been saturated with $CaCl_2$, KCl, and $MgCl_2$, in that order.

PROCEDURE

Preparation of working curves

Add 5.00 mL of the appropriate radiation buffer to each of a series of 100-mL volumetric flasks. Add volumes of standard that will produce solutions that range from 0 to 10 ppm in the cation to be determined. Dilute to the mark with water, and mix well.

Measure the emission intensity for each solution, taking at least three readings for each. Aspirate distilled water between each set of measurements. Correct the average values for background luminosity, and prepare a working curve from the data.

Repeat for the other two cations.

Analysis of a water sample

Prepare duplicate aliquots of the unknown as directed for preparation of working curves. If necessary, use a standard to calibrate the response of the instrument to the working curve; then measure the emission intensity of the unknown. Correct the data for background. Determine the cation concentration in the unknown by comparison with the working curve.

31P Separation of Cations by Ion Exchange

The application of ion-exchange resins to the separation of ionic species of opposite charge is discussed in Section 29E-1. Directions follow for the ion-exchange separation of nickel(II) from zinc(II) based upon converting zinc ions to the negatively charged chloride complexes. After separation, each of the cations is determined by EDTA titration.

31P-1 The Separation of Nickel(II) and Zinc(II) by Ion Exchange

Discussion

The separation of the two cations is based upon differences in their tendency to form anionic complexes. Stable chlorozincate(II) complexes (such as $ZnCl_3^-$ and $ZnCl_4^{2-}$) are formed in 2 M hydrochloric acid and retained on an anion-exchange resin. In contrast, nickel(II) is not complexed appreciably in this medium and passes rapidly through such a column. After separation is complete, elution with water effectively decomposes the chloro complexes and permits removal of the zinc.

PREPARATION OF ION-EXCHANGE COLUMNS

A typical ion-exchange column is a cylinder 25 to 40 cm in length and 1 to 1.5 cm in diameter. A stopcock at the lower end permits adjustment of the flow of liquid through the column. A buret makes a convenient column. It is recommended that two columns be prepared to permit the simultaneous treatment of duplicate samples.

Insert a plug of glass wool to retain the resin particles. Then introduce sufficient strong-base anion-exchange resin (Note) to give a 10- to 15-cm column. Wash the column with about 50 mL of 6 M NH_3, followed by 100 mL of water and 100 mL of 2 M HCl. At the end of this cycle, the flow should be stopped so that the liquid level remains about 1 cm above the resin column. *At no time should the liquid level be allowed to drop below the top of the resin.*

PROCEDURE

Obtain the unknown, which should contain between 2 and 4 mmol of Ni^{2+} and Zn^{2+}, in a clean 100-mL volumetric flask. Add 16 mL of 12 M HCl, dilute to the mark with distilled water, and mix well. The resulting solution is approximately 2 M in acid. Transfer 10.00 mL of the diluted unknown onto the column. Place a 250-mL conical flask beneath the column, and slowly drain until the liquid level is barely above the resin. Rinse the interior of the column with several 2- to 3-mL portions of the 2 M HCl; lower the liquid level to just above the resin surface after each washing. Elute the nickel with about 50 mL of 2 M HCl at a flow rate of 2 to 3 mL/min. After elution is complete, evaporate the collected liquid just to dryness on a hot plate or steam bath.

Elute the Zn(II) by passing about 100 mL of water through the column, using the same flow rate; collect the liquid in a 500-mL conical flask.

Note. Amberlite® CG 400 or its equivalent can be used.

31P-2 The Titration of Nickel and Zinc with EDTA

Discussion

Both nickel and zinc are determined by titration with standard EDTA at pH 10. A weight titration is recommended. Eriochrome Black T is the indicator for the zinc titration; bromopyrogallol or murexide is used for the nickel titration.

PREPARATION OF REAGENTS

(a) *Standard EDTA.* See Section 31D-2. Express the concentration in terms of mmol EDTA/g of solution.

(b) *pH-10 buffer.* See Section 31D-1.

(c) *Eriochrome Black T indicator.* See Section 31D-1.

(d) *Bromopyrogallol indicator.* Dissolve 0.5 g of the solid indicator in 100 mL of 50% (v/v) ethanol. (Sufficient for 100 titrations.)

(e) *Murexide indicator.* The solid is approximately 0.2% indicator by weight in NaCl. Approximately 0.2 g is needed for each titration. The solid preparation is used because solutions of the indicator are quite unstable.

PROCEDURE

(a) *Titration of nickel.* Evaporate the solution containing the nickel to dryness to eliminate excess HCl. Avoid overheating; the residual $NiCl_2$ must not be permitted to decompose to NiO. Dissolve the residue in 100 mL of distilled water, and add 10 to 20 mL of pH-10 buffer. Add 15 drops of bromopyrogallol indicator or 0.2 g of murexide. Titrate to the color change (blue to purple for bromopyrogallol, yellow to purple for murexide).

Calculate the weight in milligrams of nickel in the unknown.

(b) *Titration of zinc.* Add 10 to 20 mL of pH-10 buffer and 1 to 2 drops of Eriochrome Black T to the eluate. Titrate with standard EDTA solution to a color change from red to blue.

Calculate the weight in milligrams of zinc in the unknown.

31Q

Gas-Liquid Chromatography

As noted in Chapter 24, gas-liquid chromatography enables the analyst to resolve the components of complex mixtures. The accompanying directions are for the determination of ethanol in beverages.

31Q-1

The Gas Chromatographic Determination of Ethanol in Beverages[13]

Discussion

Ethanol is conveniently determined in aqueous solutions by means of gas chromatography. The method is readily extended to measurement of the *proof* of alcoholic beverages. By definition, the proof of a beverage is two times its volume percent of ethanol at 60°F.

The operating instructions pertain to a 1/4-in. (o.d.) × 0.5-m Poropack column containing 80- to 100-mesh packing. A thermal conductivity detector is needed. (Flame ionization is not satisfactory because of its insensitivity to water.)

The analysis is based upon a calibration curve in which the ratio of the area under the ethanol peak to the area under the ethanol-plus-water peak is plotted as a function of the volume percent of ethanol:

$$\text{vol \% EtOH} = \frac{\text{vol EtOH}}{\text{vol soln}}$$

This relationship is not strictly linear. At least two reasons can be cited to account for the curvature. First, the thermal conductivity detector responds linearly to mass ratios rather than volume ratios. Second, at the high concentrations involved, the volumes of ethanol and water are not strictly additive, as would be required for linearity. That is,

$$\text{vol EtOH} + \text{vol } H_2O \neq \text{vol soln}$$

PREPARATION OF STANDARDS

Use a buret to measure 10.00, 20.00, 30.00, and 40.00 mL of absolute ethanol into separate 50-mL volumetric flasks (Note). Dilute to volume with distilled water, and mix well.

Note. The coefficient of thermal expansion for ethanol is approximately five times that for water. It is thus necessary to keep the temperature of the solutions used in this experiment to ±1°C during volume measurements.

PROCEDURE

The following operating conditions have yielded satisfactory chromatograms for this analysis:

Column temperature	100°C
Detector temperature	130°C
Injection-port temperature	120°C
Bridge current	100 mA
Flow rate	60 mL/min

[13]Adapted from J. J. Leary, *J. Chem. Educ.*, 1983, *60*, 675.

Inject a 1-μL sample of the 20% (v/v) standard, and record the chromatogram. Obtain additional chromatograms, adjusting the recorder speed until the water peak has a width of about 2 mm at half-height. Then vary the volume of sample injected and the attenuation until peaks with a height of at least 40 mm are produced. Obtain chromatograms for the remainder of the standards (including pure water and pure ethanol) in the same way. Measure the area under each peak, and plot $\text{area}_{\text{EtOH}}/(\text{area}_{\text{EtOH}} + \text{area}_{\text{H}_2\text{O}})$ as a function of the volume percentage of ethanol.

Obtain chromatograms for the unknown. Report the volume percentage of ethanol.

APPENDIX 1 Selected References to the Literature of Analytical Chemistry

Treatises

As used here, the term *treatise* means a comprehensive presentation of one or more broad areas of analytical chemistry.

N. H. Furman and F. J. Welcher, Eds., *Standard Methods of Chemical Analysis,* 6th ed. New York: Van Nostrand, 1962–66. In five parts; largely devoted to specific applications.

I. M. Kolthoff and P. J. Elving, Eds., *Treatise on Analytical Chemistry.* New York: Wiley, 1959–86. Part I (12 volumes) is devoted to theory; Part II (17 volumes) deals with analytical methods for the elements; Part III (4 volumes) treats industrial analytical chemistry. Early volumes of the second edition of Part I of this monumental work began to appear in 1978.

G. Svehla, C. L. Wilson, and D. W. Wilson, Eds., *Comprehensive Analytical Chemistry.* New York: Elsevier, 1959– . To 1986, 20 volumes of this work have appeared.

A. Weissberger, Ed., *Techniques of Chemistry,* Volume I, *Physical Methods of Chemistry,* 4th ed. New York: Interscience, 1971– . This work consists of a large number of individually bound books dealing with various instruments employed for chemical measurements.

Official Methods of Analysis

These publications are often single volumes that provide a useful source of analytical methods for the determination of specific substances in articles of commerce. The methods have been developed by various scientific societies and serve as standards in arbitration as well as in the courts.

Standard Methods for the Examination of Water and Wastewater, 14th ed. New York: American Public Health Association, 1976.

ASTM Book of Standards. Philadelphia: American Society for Testing Materials. This 48-volume work is revised annually and contains methods for both physical testing and chemical analysis. Volume 12, *Chemical Analysis of Metals and Metal-Bearing Ores,* is one of the useful volumes for the chemist.

N. H. Hanson, *Official, Standardized and Recommended Methods of Analysis,* 2nd ed. London: Society for Analytical Chemistry, 1973.

Official Methods of Analysis, 14th ed. Washington, D.C.: Association of Official Analytical Chemists, 1984. This is a very useful source of methods for the analysis of such materials as drugs, food, pesticides, agricultural materials, cosmetics, vitamins, and nutrients. It is revised every five years.

Review Serials

Analytical Chemistry, Fundamental Reviews. These reviews appear biennially as a supplement to the April issue of *Analytical Chemistry* in even-numbered years. Most of the significant developments occurring in the past two years in 30 or more areas of analytical chemistry are covered.

Analytical Chemistry, Application Reviews. These reviews appear as part of *Analytical Chemistry* biennially in odd-numbered years. The articles are devoted to recent analytical work in 18 specific areas, such as water analysis, clinical chemistry, petroleum products, and air pollution.

Critical Reviews in Analytical Chemistry. This publication appears quarterly and provides in-depth reviews of various aspects of analytical chemistry.

D. Glick, Ed., *Methods of Biochemical Analysis.* This annual publication consists of a series of review articles covering the newest developments in the analysis of biochemical substances.

Tabular Compilations

E. Hoegfeldt and D. D. Perrin, *Stability Constants of Metal-Ion Complexes.* London: The Chemical Society, 1979 and 1981. Two volumes.

L. Meites, Ed., *Handbook of Analytical Chemistry.* New York: McGraw-Hill, 1963.

G. Milazzo, S. Caroli, and V. K. Sharma, *Tables of Standard Electrode Potentials.* New York: Wiley, 1978.

Advanced Analytical and Instrumental Textbooks

J. N. Butler, *Ionic Equilibrium: A Mathematical Approach.* Reading, MA: Addison-Wesley, 1964.

G. D. Christian and J. E. O'Reilly, *Instrumental Analysis,* 2nd ed. Boston: Allyn and Bacon, 1986.

H. A. Laitinen and W. E. Harris, *Chemical Analysis,* 2nd ed. New York: McGraw-Hill, 1975.

E. D. Olsen, *Modern Optical Methods of Analysis.* New York: McGraw-Hill, 1975.

D. A. Skoog, *Principles of Instrumental Analysis,* 3rd ed. Philadelphia: Saunders College Publishing, 1985.

H. Strobel, *Chemical Instrumentation,* 2nd ed. Boston: Addison-Wesley, 1973.

H. H. Willard, L. L. Merritt Jr., J. A. Dean, and F. A. Settle, *Instrumental Methods of Analysis,* 6th ed. New York: Van Nostrand, 1981.

Monographs

Hundreds of monographs devoted to limited areas of analytical chemistry are available. In general, these are authored by experts and are excellent sources of information. Representative monographs in various areas are listed below.

Gravimetric and Titrimetric Methods

M.R.F. Ashworth, *Titrimetric Organic Analysis.* New York: Interscience, 1965. Two volumes.

L. Erdey, *Gravimetric Analysis.* Oxford: Pergamon, 1965.

J. S. Fritz, *Acid-Base Titrations in Nonaqueous Solvents.* Boston: Allyn and Bacon, 1973.

W. F. Hillebrand, G.E.F. Lundell, H. A. Bright, and J. I. Hoffman, *Applied Inorganic Analysis,* 2nd ed. New York: Wiley, 1953.

I. M. Kolthoff, V. A. Stenger, and R. Belcher, *Volumetric Analysis.* New York: Interscience, 1942–57. Three volumes.

T. S. Ma and R. C. Rittner, *Modern Organic Elemental Analysis.* New York: Marcel Dekker, 1979.

W. Wagner and C. J. Hull, *Inorganic Titrimetric Analysis.* New York: Marcel Dekker, 1971.

Organic Analysis

S. Siggia and J. G. Hanna, *Quantitative Organic Analysis via Functional Groups,* 4th ed. New York: Wiley, 1979.

F. T. Weiss, *Determination of Organic Compounds: Methods and Procedures.* New York: Wiley-Interscience, 1970.

Spectrometric Methods

D. F. Boltz and J. A. Howell, *Colorimetric Determination of Nonmetals,* 2nd ed. New York: Wiley-Interscience, 1978.

J. A. Dean and T. C. Rains, Eds., *Flame Emission and Atomic Absorption Spectroscopy.* New York: Marcel Dekker, 1974. Three volumes.

J. D. Ingle and S. R. Crouch, *Analytical Spectroscopy.* Englewood Cliffs, NJ: Prentice-Hall, 1988.

W. J. Price, *Analytical Absorption Spectrometry.* London: Heyden, 1972.

E. B. Sandell and H. Onishi, *Colorimetric Determination of Traces of Metals,* 4th ed. New York: Interscience, 1978.

F. D. Snell, *Photometric and Fluorometric Methods of Analysis.* New York: Wiley, 1978. Two volumes.

Electroanalytical Methods

A. J. Bard and L. R. Faulkner, *Electrochemical Methods.* New York: Wiley, 1980.

P. T. Kissinger and W. R. Heineman, Eds., *Laboratory Techniques in Electroanalytical Chemistry.* New York: Marcel Dekker, 1984.

J. J. Lingane, *Electroanalytical Chemistry,* 2nd ed. New York: Interscience, 1954.

D. T. Sawyer and J. L. Roberts Jr., *Experimental Electrochemistry for Chemists.* New York: Wiley, 1974.

Analytical Separations

E. Heftmann, *Chromatography,* 3rd ed. New York: Van Nostrand-Reinhold, 1975.

B. L. Karger, L. R. Snyder, and C. Horváth, *An Introduction to Separation Science.* New York: Wiley, 1973.

O. Mikeš, Ed., *Chromatographic and Allied Methods.* New York: Wiley, 1979.

J. M. Miller, *Separation Methods in Chemical Analysis.* New York: Wiley, 1975.

Miscellaneous

R. G. Bates, *Determination of pH: Theory and Practice,* 2nd ed. New York: Wiley, 1973.

R. Bock, *Decomposition Methods in Analytical Chemistry.* New York: Wiley, 1979.

G. H. Morrison, Ed., *Trace Analysis.* New York: Interscience, 1965.

D. D. Perrin, *Masking and Demasking Chemical Reactions.* New York: Wiley, 1970.

M. Pinta, *Modern Methods for Trace Element Analysis.* Ann Arbor, MI: Ann Arbor Science, 1978.

W. Rieman and H. F. Walton, *Ion Exchange in Analytical Chemistry.* Oxford: Pergamon, 1970.

W. J. Williams, *Handbook of Anion Determination.* London: Butterworths, 1979.

Periodicals

Numerous journals are devoted to analytical chemistry; these are primary sources of information in the field. Some of the best-known titles are listed below. (The boldface portion of the title is the *Chemical Abstracts* abbreviation for the journal.)

Amer*ican* **Lab***oratory*
Analyst, *The*
Anal*ytical* **Biochem***istry*
Anal*ytical* **Chem***istry*
Anal*ytica* **Chim***ica* **Acta**
Anal*ytical* **Instrum***entation*
Anal*ytical* **Lett***ers*
Appl*ied* **Spectrosc***opy*
Clin*ical* **Chem***istry*
J*ournal of the* **Assoc***iation of* **Off***icial* **Anal***ytical* **Chem***ists*
J*ournal of* **Chromatogr***aphic* **Sci***ence*
J*ournal of* **Chromatogr***aphy*
J*ournal of* **Electroanal***ytical* **Chem***istry and* **Interfac***ial* **Electrochem***istry*
Microchem*ical* **J***ournal*
Separ*ation* **Sci***ence*
Spectrochim*ica* **Acta**
Talanta
Z*eitschrift für* **Anal***ytische* **Chem***ie*

APPENDIX 2 Some Standard and Formal Electrode Potentials

Half-Reaction	E^0, V*	Formal Potential, V[†]
Aluminum		
$Al^{3+} + 3e \rightleftharpoons Al(s)$	−1.662	
Antimony		
$Sb_2O_5(s) + 6H^+ + 4e \rightleftharpoons 2SbO^+ + 3H_2O$	+0.581	
Arsenic		
$H_3AsO_4 + 2H^+ + 2e \rightleftharpoons H_3AsO_3 + H_2O$	+0.559	0.577, 1 M HCl, $HClO_4$
Barium		
$Ba^{2+} + 2e \rightleftharpoons Ba(s)$	−2.906	
Bismuth		
$BiO^+ + 2H^+ + 3e \rightleftharpoons Bi(s) + H_2O$	+0.320	
$BiCl_4^- + 3e \rightleftharpoons Bi(s) + 4Cl^-$	+0.16	
Bromine		
$Br_2(l) + 2e \rightleftharpoons 2Br^-$	+1.065	1.05, 4 M HCl
$Br_2(aq) + 2e \rightleftharpoons 2Br^-$	+1.087[‡]	
$BrO_3^- + 6H^+ + 5e \rightleftharpoons \frac{1}{2}Br_2(l) + 3H_2O$	+1.52	
$BrO_3^- + 6H^+ + 6e \rightleftharpoons Br^- + 3H_2O$	+1.44	
Cadmium		
$Cd^{2+} + 2e \rightleftharpoons Cd(s)$	−0.403	
Calcium		
$Ca^{2+} + 2e \rightleftharpoons Ca(s)$	−2.866	
Carbon		
$C_6H_4O_2$ (quinone) $+ 2H^+ + 2e \rightleftharpoons C_6H_4(OH)_2$	+0.699	0.696, 1 M HCl, $HClO_4$, H_2SO_4
$2CO_2(g) + 2H^+ + 2e \rightleftharpoons H_2C_2O_4$	−0.49	
Cerium		
$Ce^{4+} + e \rightleftharpoons Ce^{3+}$		+1.70, 1 M $HClO_4$; +1.61, 1 M HNO_3; +1.44, 1 M H_2SO_4
Chlorine		
$Cl_2(g) + 2e \rightleftharpoons 2Cl^-$	+1.359	
$HClO + H^+ + e \rightleftharpoons \frac{1}{2}Cl_2(g) + H_2O$	+1.63	
$ClO_3^- + 6H^+ + 5e \rightleftharpoons \frac{1}{2}Cl_2(g) + 3H_2O$	+1.47	
Chromium		
$Cr^{3+} + e \rightleftharpoons Cr^{2+}$	−0.408	
$Cr^{3+} + 3e \rightleftharpoons Cr(s)$	−0.744	
$Cr_2O_7^{2-} + 14H^+ + 6e \rightleftharpoons 2Cr^{3+} + 7H_2O$	+1.33	

Continued on next page

Continued from previous page

Half-Reaction	E^0, V*	Formal Potential, V[†]
Cobalt		
$Co^{2+} + 2e \rightleftharpoons Co(s)$	−0.277	
$Co^{3+} + e \rightleftharpoons Co^{2+}$	+1.808	
Copper		
$Cu^{2+} + 2e \rightleftharpoons Cu(s)$	+0.337	
$Cu^{2+} + e \rightleftharpoons Cu^{+}$	+0.153	
$Cu^{+} + e \rightleftharpoons Cu(s)$	+0.521	
$Cu^{2+} + I^{-} + e \rightleftharpoons CuI(s)$	+0.86	
$CuI(s) + e \rightleftharpoons Cu(s) + I^{-}$	−0.185	
Fluorine		
$F_2(g) + 2H^{+} + 2e \rightleftharpoons 2HF(aq)$	+3.06	
Hydrogen		
$2H^{+} + 2e \rightleftharpoons H_2(g)$	0.000	−0.005, 1 M HCl, $HClO_4$
Iodine		
$I_2(s) + 2e \rightleftharpoons 2I^{-}$	+0.5355	
$I_2(aq) + 2e \rightleftharpoons 2I^{-}$	+0.615[‡]	
$I_3^{-} + 2e \rightleftharpoons 3I^{-}$	+0.536	
$ICl_2^{-} + e \rightleftharpoons \frac{1}{2}I_2(s) + 2Cl^{-}$	+1.056	
$IO_3^{-} + 6H^{+} + 5e \rightleftharpoons \frac{1}{2}I_2(s) + 3H_2O$	+1.196	
$IO_3^{-} + 6H^{+} + 5e \rightleftharpoons \frac{1}{2}I_2(aq) + 3H_2O$	+1.178[‡]	
$IO_3^{-} + 2Cl^{-} + 6H^{+} + 4e \rightleftharpoons ICl_2^{-} + 3H_2O$	+1.24	
$H_5IO_6 + H^{+} + 2e \rightleftharpoons IO_3^{-} + 3H_2O$	+1.601	
Iron		
$Fe^{2+} + 2e \rightleftharpoons Fe(s)$	−0.440	
$Fe^{3+} + e \rightleftharpoons Fe^{2+}$	+0.771	0.700, 1 M HCl; 0.732, 1 M $HClO_4$; 0.68, 1 M H_2SO_4
$Fe(CN)_6^{3-} + e \rightleftharpoons Fe(CN)_6^{4-}$	+0.36	0.71, 1 M HCl; 0.72, 1 M $HClO_4$, H_2SO_4
Lead		
$Pb^{2+} + 2e \rightleftharpoons Pb(s)$	−0.126	−0.14, 1 M $HClO_4$; −0.29, 1 M H_2SO_4
$PbO_2(s) + 4H^{+} + 2e \rightleftharpoons Pb^{2+} + 2H_2O$	+1.455	
$PbSO_4(s) + 2e \rightleftharpoons Pb(s) + SO_4^{2-}$	−0.350	
Lithium		
$Li^{+} + e \rightleftharpoons Li(s)$	−3.045	
Magnesium		
$Mg^{2+} + 2e \rightleftharpoons Mg(s)$	−2.363	
Manganese		
$Mn^{2+} + 2e \rightleftharpoons Mn(s)$	−1.180	
$Mn^{3+} + e \rightleftharpoons Mn^{2+}$		1.51, 7.5 M H_2SO_4
$MnO_2(s) + 4H^{+} + 2e \rightleftharpoons Mn^{2+} + 2H_2O$	+1.23	
$MnO_4^{-} + 8H^{+} + 5e \rightleftharpoons Mn^{2+} + 4H_2O$	+1.51	
$MnO_4^{-} + 4H^{+} + 3e \rightleftharpoons MnO_2(s) + 2H_2O$	+1.695	
$MnO_4^{-} + e \rightleftharpoons MnO_4^{2-}$	+0.564	
Mercury		
$Hg_2^{2+} + 2e \rightleftharpoons Hg(l)$	+0.788	0.274, 1 M HCl; 0.776, 1 M $HClO_4$; 0.674, 1 M H_2SO_4

Continued on next page

Continued from previous page

Half-Reaction	E^0, V*	Formal Potential, V[†]
$2Hg^{2+} + 2e \rightleftharpoons Hg_2^{2+}$	+0.920	0.907, 1 M $HClO_4$
$Hg^{2+} + 2e \rightleftharpoons 2Hg(l)$	+0.854	
$Hg_2Cl_2(s) + 2e \rightleftharpoons 2Hg(l) + 2Cl^-$	+0.268	0.244, sat'd KCl; 0.282 1 M KCl; 0.334, 0.1 M KCl
$Hg_2SO_4(s) + 2e \rightleftharpoons 2Hg(l) + SO_4^{2-}$	+0.615	
Nickel		
$Ni^{2+} + 2e \rightleftharpoons Ni(s)$	−0.250	
Nitrogen		
$N_2(g) + 5H^+ + 4e \rightleftharpoons N_2H_5^+$	−0.23	
$HNO_2 + H^+ + e \rightleftharpoons NO(g) + H_2O$	+1.00	
$NO_3^- + 3H^+ + 2e \rightleftharpoons HNO_2 + H_2O$	+0.94	0.92, 1 M HNO_3
Oxygen		
$H_2O_2 + 2H^+ + 2e \rightleftharpoons 2H_2O$	+1.776	
$HO_2^- + H_2O + 2e \rightleftharpoons 3OH^-$	+0.88	
$O_2(g) + 4H^+ + 4e \rightleftharpoons 2H_2O$	+1.229	
$O_2(g) + 2H^+ + 2e \rightleftharpoons H_2O_2$	+0.682	
$O_3(g) + 2H^+ + 2e \rightleftharpoons O_2(g) + H_2O$	+2.07	
Palladium		
$Pd^{2+} + 2e \rightleftharpoons Pd(s)$	+0.987	
Platinum		
$PtCl_4^{2-} + 2e \rightleftharpoons Pt(s) + 4Cl^-$	+0.73	
$PtCl_6^{2-} + 2e \rightleftharpoons PtCl_4^{2-} + 2Cl^-$	+0.68	
Potassium		
$K^+ + e \rightleftharpoons K(s)$	−2.925	
Selenium		
$H_2SeO_3 + 4H^+ + 2e \rightleftharpoons Se(s) + 3H_2O$	+0.740	
$SeO_4^{2-} + 4H^+ + 2e \rightleftharpoons H_2SeO_3 + H_2O$	+1.15	
Silver		
$Ag^+ + e \rightleftharpoons Ag(s)$	+0.799	0.228, 1 M HCl; 0.792, 1 M $HClO_4$; 0.77, 1 M H_2SO_4
$AgBr(s) + e \rightleftharpoons Ag(s) + Br^-$	+0.073	
$AgCl(s) + e \rightleftharpoons Ag(s) + Cl^-$	+0.222	0.228, 1 M KCl
$Ag(CN)_2^- + e \rightleftharpoons Ag(s) + 2CN^-$	−0.31	
$Ag_2CrO_4(s) + 2e \rightleftharpoons 2Ag(s) + CrO_4^{2-}$	+0.446	
$AgI(s) + e \rightleftharpoons Ag(s) + I^-$	−0.151	
$Ag(S_2O_3)_2^{3-} + e \rightleftharpoons Ag(s) + 2S_2O_3^{2-}$	+0.017	
Sodium		
$Na^+ + e \rightleftharpoons Na(s)$	−2.714	
Sulfur		
$S(s) + 2H^+ + 2e \rightleftharpoons H_2S(g)$	+0.141	
$H_2SO_3 + 4H^+ + 4e \rightleftharpoons S(s) + 3H_2O$	+0.450	
$SO_4^{2-} + 4H^+ + 2e \rightleftharpoons H_2SO_3 + H_2O$	+0.172	
$S_4O_6^{2-} + 2e \rightleftharpoons 2S_2O_3^{2-}$	+0.08	
$S_2O_8^{2-} + 2e \rightleftharpoons 2SO_4^{2-}$	+2.01	

Continued on next page

Continued from previous page

Half-Reaction	E^0, V*	Formal Potential, V†
Thallium		
$Tl^+ + e \rightleftharpoons Tl(s)$	−0.336	−0.551, 1 M HCl; −0.33, 1 M $HClO_4$, H_2SO_4
$Tl^{3+} + 2e \rightleftharpoons Tl^+$	+1.25	0.77, 1 M HCl
Tin		
$Sn^{2+} + 2e \rightleftharpoons Sn(s)$	−0.136	−0.16, 1 M $HClO_4$
$Sn^{4+} + 2e \rightleftharpoons Sn^{2+}$	+0.154	0.14, 1 M HCl
Titanium		
$Ti^{3+} + e \rightleftharpoons Ti^{2+}$	−0.369	
$TiO^{2+} + 2H^+ + e \rightleftharpoons Ti^{3+} + H_2O$	+0.099	0.04, 1 M H_2SO_4
Uranium		
$UO_2^{2+} + 4H^+ + 2e \rightleftharpoons U^{4+} + 2H_2O$	+0.334	
Vanadium		
$V^{3+} + e \rightleftharpoons V^{2+}$	−0.256	−0.21, 1 M $HClO_4$
$VO^{2+} + 2H^+ + e \rightleftharpoons V^{3+} + H_2O$	+0.359	
$V(OH)_4^+ + 2H^+ + e \rightleftharpoons VO^{2+} + 3H_2O$	+1.00	1.02, 1 M HCl, $HClO_4$
Zinc		
$Zn^{2+} + 2e \rightleftharpoons Zn(s)$	−0.763	

*G. Milazzo, S. Caroli, and V. K. Sharma, *Tables of Standard Electrode Potentials*. London: Wiley, 1978. Reprinted by permission of John Wiley & Sons, Ltd.

†E. H. Swift and E. A. Butler, *Quantitative Measurements and Chemical Equilibria*. New York: W. H. Freeman and Company, 1972.

‡These potentials are hypothetical because they correspond to solutions that are 1.00 M in Br_2 or I_2. The solubilities of these two compounds at 25°C are 0.18 M and 0.0020 M, respectively. In saturated solutions containing an excess of $Br_2(\ell)$ or $I_2(s)$, the standard potentials for the half-reaction $Br_2(l) + 2e \rightleftharpoons 2Br^-$ or $I_2(s) + 2e \rightleftharpoons 2I^-$ should be used. In contrast, at Br_2 and I_2 concentrations less than saturation, these hypothetical electrode potentials should be employed.

APPENDIX 3 Solubility-Product Constants

Substance	Formula	K_{sp}
Aluminum hydroxide	$Al(OH)_3$	2×10^{-32}
Barium carbonate	$BaCO_3$	5.1×10^{-9}
Barium chromate	$BaCrO_4$	1.2×10^{-10}
Barium iodate	$Ba(IO_3)_2$	1.57×10^{-9}
Barium manganate	$BaMnO_4$	2.5×10^{-10}
Barium oxalate	BaC_2O_4	2.3×10^{-8}
Barium sulfate	$BaSO_4$	1.3×10^{-10}
Bismuth oxide chloride	$BiOCl$	7×10^{-9}
Bismuth oxide hydroxide	$BiOOH$	4×10^{-10}
Cadmium carbonate	$CdCO_3$	2.5×10^{-14}
Cadmium hydroxide	$Cd(OH)_2$	5.9×10^{-15}
Cadmium oxalate	CdC_2O_4	9×10^{-8}
Cadmium sulfide	CdS	2×10^{-28}
Calcium carbonate	$CaCO_3$	4.8×10^{-9}
Calcium fluoride	CaF_2	4.9×10^{-11}
Calcium oxalate	CaC_2O_4	2.3×10^{-9}
Calcium sulfate	$CaSO_4$	2.6×10^{-5}
Copper(I) bromide	$CuBr$	5.2×10^{-9}
Copper(I) chloride	$CuCl$	1.2×10^{-6}
Copper(I) iodide	CuI	1.1×10^{-12}
Copper(I) thiocyanate	$CuSCN$	4.8×10^{-15}
Copper(II) hydroxide	$Cu(OH)_2$	1.6×10^{-19}
Copper(II) sulfide	CuS	6×10^{-36}
Iron(II) hydroxide	$Fe(OH)_2$	8×10^{-16}
Iron(II) sulfide	FeS	6×10^{-18}
Iron(III) hydroxide	$Fe(OH)_3$	4×10^{-38}
Lanthanum iodate	$La(IO_3)_3$	6.2×10^{-12}
Lead carbonate	$PbCO_3$	3.3×10^{-14}
Lead chloride	$PbCl_2$	1.6×10^{-5}
Lead chromate	$PbCrO_4$	1.8×10^{-14}
Lead hydroxide	$Pb(OH)_2$	2.5×10^{-16}
Lead iodide	PbI_2	7.1×10^{-9}
Lead oxalate	PbC_2O_4	4.8×10^{-10}
Lead sulfate	$PbSO_4$	1.6×10^{-8}
Lead sulfide	PbS	7×10^{-28}
Magnesium ammonium phosphate	$MgNH_4PO_4$	3×10^{-13}
Magnesium carbonate	$MgCO_3$	1×10^{-5}
Magnesium hydroxide	$Mg(OH)_2$	1.8×10^{-11}
Magnesium oxalate	MgC_2O_4	8.6×10^{-5}
Manganese(II) hydroxide	$Mn(OH)_2$	1.9×10^{-13}
Manganese(II) sulfide	MnS	3×10^{-13}
Mercury(I) bromide	Hg_2Br_2	5.8×10^{-23}
Mercury(I) chloride	Hg_2Cl_2	1.3×10^{-18}
Mercury(I) iodide	Hg_2I_2	4.5×10^{-29}
Silver arsenate	Ag_3AsO_4	1×10^{-22}

Continued on next page

Continued from previous page

Substance	Formula	K_{sp}
Silver bromide	$AgBr$	5.2×10^{-13}
Silver carbonate	Ag_2CO_3	8.1×10^{-12}
Silver chloride	$AgCl$	1.82×10^{-10}
Silver chromate	Ag_2CrO_4	1.1×10^{-12}
Silver cyanide	$AgCN$	7.2×10^{-11}
Silver iodate	$AgIO_3$	3.0×10^{-8}
Silver iodide	AgI	8.3×10^{-17}
Silver oxalate	$Ag_2C_2O_4$	3.5×10^{-11}
Silver sulfide	Ag_2S	6×10^{-50}
Silver thiocyanate	$AgSCN$	1.1×10^{-12}
Strontium oxalate	SrC_2O_4	5.6×10^{-8}
Strontium sulfate	$SrSO_4$	3.2×10^{-7}
Thallium(I) chloride	$TlCl$	1.7×10^{-4}
Thallium(I) sulfide	Tl_2S	1×10^{-22}
Zinc hydroxide	$Zn(OH)_2$	1.2×10^{-17}
Zinc oxalate	ZnC_2O_4	7.5×10^{-9}
Zinc sulfide	ZnS	4.5×10^{-24}

From R. P. Frankenthal, in *Handbook of Analytical Chemistry*, L. Meites, Ed., pp. 1-13–1-19. New York: McGraw-Hill, 1963. With permission.

APPENDIX 4 Dissociation Constants for Acids

Acid	Formula	Dissociation Constant at 25°C		
		K_1	K_2	K_3
Acetic	CH_3COOH	1.75×10^{-5}		
Arsenic	H_3AsO_4	6.0×10^{-3}	1.05×10^{-7}	3.0×10^{-12}
Arsenous	H_3AsO_3	6.0×10^{-10}	3.0×10^{-14}	
Benzoic	C_6H_5COOH	6.14×10^{-5}		
Boric	H_3BO_3	5.83×10^{-10}		
1-Butanoic	$CH_3CH_2CH_2COOH$	1.51×10^{-5}		
Carbonic	H_2CO_3	4.45×10^{-7}	4.7×10^{-11}	
Chloroacetic	$ClCH_2COOH$	1.36×10^{-3}		
Citric	$HOOC(OH)C(CH_2COOH)_2$	7.45×10^{-4}	1.73×10^{-5}	4.02×10^{-7}
Ethylenediamine-tetraacetic	H_4Y	1.0×10^{-2}	2.1×10^{-3}	6.9×10^{-7}
				$K_4 = 5.5 \times 10^{-11}$
Formic	$HCOOH$	1.77×10^{-4}		
Fumaric	*trans*-$HOOCCH{:}CHCOOH$	9.6×10^{-4}	4.1×10^{-5}	
Glycolic	$HOCH_2COOH$	1.48×10^{-4}		
Hydrazoic	HN_3	1.9×10^{-5}		
Hydrogen cyanide	HCN	2.1×10^{-9}		
Hydrogen fluoride	H_2F_2	7.2×10^{-4}		
Hydrogen peroxide	H_2O_2	2.7×10^{-12}		
Hydrogen sulfide	H_2S	5.7×10^{-8}	1.2×10^{-15}	
Hypochlorous	$HOCl$	3.0×10^{-8}		
Iodic	HIO_3	1.7×10^{-1}		
Lactic	$CH_3CHOHCOOH$	1.37×10^{-4}		
Maleic	*cis*-$HOOCCH{:}CHCOOH$	1.20×10^{-2}	5.96×10^{-7}	
Malic	$HOOCCHOHCH_2COOH$	4.0×10^{-4}	8.9×10^{-6}	
Malonic	$HOOCCH_2COOH$	1.40×10^{-3}	2.01×10^{-6}	
Mandelic	$C_6H_5CHOHCOOH$	3.88×10^{-4}		
Nitrous	HNO_2	5.1×10^{-4}		
Oxalic	$HOOCCOOH$	5.36×10^{-2}	5.42×10^{-5}	
Periodic	H_5IO_6	2.4×10^{-2}	5.0×10^{-9}	
Phenol	C_6H_5OH	1.00×10^{-10}		
Phosphoric	H_3PO_4	7.11×10^{-3}	6.34×10^{-8}	4.2×10^{-13}
Phosphorous	H_3PO_3	1.00×10^{-2}	2.6×10^{-7}	
o-Phthalic	$C_6H_4(COOH)_2$	1.12×10^{-3}	3.91×10^{-6}	
Picric	$(NO_2)_3C_6H_2OH$	5.1×10^{-1}		
Propanoic	CH_3CH_2COOH	1.34×10^{-5}		
Pyruvic	$CH_3COCOOH$	3.24×10^{-3}		
Salicylic	$C_6H_4(OH)COOH$	1.05×10^{-3}		
Sulfamic	H_2NSO_3H	1.03×10^{-1}		
Sulfuric	H_2SO_4	Strong	1.20×10^{-2}	
Sulfurous	H_2SO_3	1.72×10^{-2}	6.43×10^{-8}	
Succinic	$HOOCCH_2CH_2COOH$	6.21×10^{-5}	2.32×10^{-6}	
Tartaric	$HOOC(CHOH)_2COOH$	9.20×10^{-4}	4.31×10^{-5}	
Trichloroacetic	Cl_3CCOOH	1.29×10^{-1}		

From L. Meites, *Handbook of Analytical Chemistry*, p. 1-21. New York: McGraw-Hill, 1963. With permission.

APPENDIX 5 Dissociation Constants for Bases

Base	Formula	Dissociation Constant at 25°C
Ammonia	NH_3	1.76×10^{-5}
Aniline	$C_6H_5NH_2$	3.94×10^{-10}
1-Butylamine	$CH_3(CH_2)_2CH_2NH_2$	4.0×10^{-4}
Dimethylamine	$(CH_3)_2NH$	5.9×10^{-4}
Ethanolamine	$HOC_2H_4NH_2$	3.18×10^{-5}
Ethylamine	$CH_3CH_2NH_2$	4.28×10^{-4}
Ethylenediamine	$NH_2C_2H_4NH_2$	$K_1 = 8.5 \times 10^{-5}$
		$K_2 = 7.1 \times 10^{-8}$
Hydrazine	H_2NNH_2	1.3×10^{-6}
Hydroxylamine	$HONH_2$	1.07×10^{-8}
Methylamine	CH_3NH_2	4.8×10^{-4}
Piperidine	$C_5H_{11}N$	1.3×10^{-3}
Pyridine	C_5H_5N	1.7×10^{-9}
Trimethylamine	$(CH_3)_3N$	6.25×10^{-5}

From L. Meites, *Handbook of Analytical Chemistry*, p. 1-21. New York: McGraw-Hill, 1963. With permission.

APPENDIX 6 Stepwise Formation Constants

Ligand	Cation	log K_1	log K_2	log K_3	log K_4	log K_5	log K_6
CH_3COO^-	Ag^+	0.4	−0.2				
	Cd^{2+}	1.3	1.0	0.1	−0.4		
	Cu^{2+}	2.2	1.1				
	Hg^{2+}	$\log K_1K_2 = 8.4$					
	Pb^{2+}	2.7	1.5				
NH_3	Ag^+	3.3	3.8				
	Cd^{2+}	2.6	2.1	1.4	0.9	−0.3	−1.7
	Co^{2+}	2.1	1.6	1.0	0.8	0.2	−0.6
	Cu^{2+}	4.3	3.7	3.0	2.3	−0.5	
	Ni^{2+}	2.8	2.2	1.7	1.2	0.8	0.0
	Zn^{2+}	2.4	2.4	2.5	2.1		
Br^-	Ag^+	$AgBr(s) + Br^- \rightleftharpoons AgBr_2^-$			$\log K_{s2} = -4.7$		
		$AgBr_2^- + Br^- \rightleftharpoons AgBr_3^{2-}$			$\log K_3 = 0.7$		
	Hg^{2+}	9.0	8.3	1.4	1.3		
	Pb^{2+}	1.2					
Cl^-	Ag^+	$AgCl(s) + Cl^- \rightleftharpoons AgCl_2^-$			$\log K_{s2} = -4.7$		
		$AgCl_2^- + Cl^- \rightleftharpoons AgCl_3^{2-}$			$\log K_3 = 0.0$		
	Bi^{3+}	2.4	2.0	1.4	0.4	0.5	
	Cd^{2+}	1.5	0.4	0.4			
	Cu^+	$Cu^+ + 2Cl^- \rightleftharpoons CuCl_2^-$			$\log K_1K_2 = 4.9$		
	Fe^{2+}	0.4	0.0				
	Fe^{3+}	1.5	0.6	−1.0			
	Hg^{2+}	6.7	6.5	0.9	1.0		
	Pb^{2+}	1.6	$Pb^{2+} + 3Cl^- \rightleftharpoons PbCl_3^-$		$\log K_1K_2K_3 = 1.7$		
	Sn^{2+}	1.1	0.6	0.0			
CN^-	Ag^+	$Ag^+ + 2CN^- \rightleftharpoons Ag(CN)_2^-$			$\log K_1K_2 = 21.1$		
	Cd^{2+}	5.5	5.1	4.6	3.6		
	Hg^{2+}	18.0	16.7	3.8	3.0		
	Ni^{2+}	$Ni^{2+} + 4CN^- \rightleftharpoons Ni(CN)_4^-$			$\log K_1K_2K_3K_4 = 22$		
EDTA	See Table 11-1 (page 261)						
F^-	Al^{3+}	6.1	5.0	3.8	2.7	1.6	0.5
	Fe^{3+}	5.3	4.0	2.8			
OH^-	Al^{3+}	8.9	$Al(OH)_3(s) + OH^- \rightleftharpoons Al(OH)_4^-$			$\log K_{s4} = 1.0$	
	Cd^{2+}	2.3					
	Cu^{2+}	6.5					
	Fe^{2+}	3.9					
	Fe^{3+}	11.1	10.7				
	Hg^{2+}	10.3					
	Ni^{2+}	4.6					
	Pb^{2+}	6.2	$Pb(OH)_2(s) + OH^- \rightleftharpoons Pb(OH)_3^-$			$\log K_{s3} = -1.3$	
	Zn^{2+}	4.4	$Zn(OH)_2(s) + 2OH^- \rightleftharpoons Zn(OH)_4^{2-}$			$\log K_{s4} = -0.9$	
I^-	Cd^{2+}	2.4	1.6	1.0	1.1		
	Cu^+		$CuI(s) + I^- \rightleftharpoons CuI_2^-$		$\log K_{s2} = -3.1$		

Continued on next page

Continued from previous page

Ligand	Cation	log K_1	log K_2	log K_3	log K_4	log K_5	log K_6
	Hg^{2+}	12.9	11.0	3.8	2.3		
	Pb^{2+}	1.3	$PbI_2(s) + I^- \rightleftharpoons PbI_3^-$		log K_{s3} = −4.7		
			$PbI_3^- + I^- \rightleftharpoons PbI_4^{2-}$		log K_4 = −3.8		
$C_2O_4^{2-}$	Al^{3+}	log K_1K_2 = 13		3.8			
	Fe^{3+}	9.4	6.8	4.0			
	Mg^{2+}	3.4	1.0				
	Mn^{2+}	3.9	1.9				
	Pb^{2+}	$Pb^{2+} + 2C_2O_4^{2-} \rightleftharpoons Pb(C_2O_4)_2^{2-}$			log K_1K_2 = 6.5		
SO_4^{2-}	Al^{3+}	3.2	1.9				
	Cd^{2+}	2.3					
	Cu^{2+}	2.4					
	Fe^{3+}	3.0	1.0				
SCN^-	Ag^+		$AgSCN(s) + SCN^- \rightleftharpoons Ag(SCN)_2^-$			log K_{s2} = −7.2	
	Cd^{2+}	1.0	0.7	0.6	1.0		
	Co^{2+}	2.3	0.7	−0.7	0.0		
	Cu^{2+}		$CuSCN(s) + SCN^- \rightleftharpoons Cu(SCN)_2^-$			log K_{s2} = −3.4	
	Fe^{3+}	2.1	1.3				
	Hg^{2+}	log K_1K_2 = 17.3		2.7	1.8		
	Ni^{2+}	1.2	0.5	0.2			

From L. Meites, *Handbook of Analytical Chemistry*, p. 1-39, New York: McGraw-Hill, 1963. With permission.

APPENDIX 7 Designations and Porosities for Filtering Crucibles

Crucible Type	Designation				
Glass, Pyrex†		C (60)*	M (15)	F (5.5)	
Glass, Kimax‡	EC (170–220)	C (40–60)	M (10–15)	F (4–4.5)	VF (2–2.5)
Porcelain, Coors U.S.A.§	Medium (40)	Fine (5)	VF (1.2)		
Porcelain, Selas‖	XF (100)	XFF (40)	#10 (8.8)	#01 (6)	
Aluminum oxide, ALUNDUM¶	Extra Coarse (30)	Coarse (20)	Medium (5)	Fine (0.1)	

*The numbers in parentheses are nominal maximum pore diameters in micrometers.

†Corning Glass Works, Corning, NY

‡Owens-Illinois, Toledo, OH.

§Coors Porcelain, Golden, CO.

‖Selas Corporation of America, Dresher, PA.

¶Norton, Worcester, MA.

APPENDIX 8 Designations Carried by Ashless Filter Papers

Manufacturer	Fine Crystals	Moderately Fine Crystals	Coarse Crystals		Gelatinous Precipitates	
Schleicher and Schuell*	507, 590 589 blue ribbon	589 white ribbon	589 green ribbon	589 black ribbon	589 black ribbon	589-1H
Munktell†	OOH	OK OO	OOR		OOR	
Whatman‡	42	44, 40	41		41	41H
Eaton-Dikeman§	90	80	60		50	

Manufacturers' literature should be consulted for more complete specifications. Tabulated are manufacturer designations of papers suitable for filtration of the indicated type of precipitate.

*Schleicher and Schuell, Inc. Keene, NH

†E. H. Sargent and Company, Chicago, IL, agents.

‡H. Reeve Angel and Company, Inc., Clifton, NJ, agents.

§Eaton-Dikeman Company, Mount Holly Springs, PA.

APPENDIX 9 Volumetric Calculations Using Normality and Equivalent Weight

The *normality* of a solution expresses the number of equivalents of a solute contained in 1 L or the number of milliequivalents in 1 mL. The equivalent and milliequivalent, like the mole and millimole, are units for describing the amount of a chemical species. The former are defined, however, in such a way that it is possible to state that, at the equivalence point in *any* titration,

$$\text{no. meq of analyte present} = \text{no. meq standard reagent added} \qquad (A9\text{-}1)$$

or

$$\text{no. eq analyte present} = \text{no. eq standard reagent added} \qquad (A9\text{-}2)$$

As a consequence, stoichiometric factors such as those described in Section 4C-4 need not be derived every time a volumetric calculation is performed. Instead, the stoichiometry is taken into account by the way equivalent or milliequivalent weight is defined.

A9-1 The Definition of Equivalent and Milliequivalent

In contrast to the mole, the amount of a substance contained in one equivalent can vary from reaction to reaction. Consequently, the weight of one equivalent of a compound can never be computed *without reference to a chemical reaction* in which that compound is, directly or indirectly, a participant. Similarly, the normality of a solution can never be specified *without knowledge about how the solution will be used.*

Equivalents in Neutralization Reactions

One equivalent weight of a substance participating in a neutralization reaction is that amount of a substance (molecule, ion, or paired ion such as NaOH) that either contributes or consumes 1 mol of hydrogen ions *in that reaction.*[1] A milliequivalent is simply 1/1000 of an equivalent.

The relationship between equivalent weight (eqw) and formula weight is straightforward for strong acids or bases and for other acids or bases that contain a single reactive hydrogen or hydroxide ion. For example, the equivalent weights of potassium hydroxide, hydrochloric acid, and acetic acid are

[1] An alternative definition, proposed by the International Union of Pure and Applied Chemistry, is as follows: an equivalent is "that amount of a substance, which, in a specified reaction, releases or replaces that amount of hydrogen that is combined with 3 g of carbon-12 in methane $^{12}CH_4$" (see *Information Bulletin* No. 36, International Union of Pure and Applied Chemistry, August 1974). This definition applies to acids. For other types of reactions and reagents, the amount of hydrogen referred to may be replaced by the equivalent amount of hydroxide ions, electrons, or cations. The reaction to which the definition is applied must be specified.

equal to their formula weights because each has but a single reactive hydrogen ion or hydroxide ion. Barium hydroxide $Ba(OH)_2$, which contains two identical hydroxide ions, reacts with two hydrogen ions in any acid/base reaction, and its equivalent weight is one half its formula weight:

$$\text{eqw } Ba(OH)_2 = \text{fw } Ba(OH)_2/2$$

The situation becomes more complex for acids or bases that contain two or more reactive hydrogen or hydroxide ions with different tendencies to dissociate. With certain indicators, for example, only the first of the three protons in phosphoric acid is titrated:

$$H_3PO_4 + OH^- \longrightarrow H_2PO_4^- + H_2O$$

With certain other indicators, a color change occurs only after two hydrogen ions have reacted:

$$H_3PO_4 + 2OH^- \longrightarrow HPO_4^{2-} + 2H_2O$$

For a titration involving the first reaction, the equivalent weight of phosphoric acid is equal to its formula weight; for the second, the equivalent weight is one half its formula weight. (Because it is not practical to titrate the third proton, an equivalent weight that is one third the formula weight is not generally encountered for H_3PO_4.) If it is not known which of these reactions is involved, an unambiguous definition of the equivalent weight for phosphoric acid *cannot be made.*

Equivalent Weight in Oxidation/Reduction Reactions

The equivalent weight of a participant in an oxidation/reduction reaction is that weight which directly or indirectly produces or consumes 1 mol of electrons. The numerical value for the equivalent weight is conveniently established by dividing the formula weight of the substance of interest by the change in oxidation number associated with its reaction. As an example, consider the oxidation of oxalate ion by permanganate ion:

$$5C_2O_4^{2-} + 2MnO_4^- + 16H^+ \longrightarrow 10CO_2 + 2Mn^{2+} + 8H_2O \qquad (A9\text{-}3)$$

The change in oxidation number for manganese in this reaction is 5 because the element passes from the +7 to the +2 state; the equivalent weights for MnO_4^- and Mn^{2+} are therefore one fifth their formula weights. Each carbon atom in the oxalate ion is oxidized from the +3 to the +4 state, leading to the production of two electrons by that species. Therefore, the equivalent weight of sodium oxalate is one half its formula weight. It is also possible to assign an equivalent weight to the carbon dioxide produced by the reaction. Since this molecule contains but a single carbon atom and since that carbon undergoes a change in oxidation number of 1, the formula weight and equivalent weight of the two are identical.

It is important to note that in evaluating the equivalent weight of a substance, *only* its change in oxidation number during the titration is considered. For example, suppose the manganese content of a sample containing Mn_2O_3 is to be determined by a titration based upon the reaction given in Equation A9-3. The fact that each manganese in the Mn_2O_3 has an oxidation number of +3 plays no part in determining its equivalent weight. That is, we must assume that by suitable treatment, all the manganese is oxidized to the +7 state before the titration is begun. Each manganese from the Mn_2O_3 is then reduced from the +7 to the +2 state in the titration step. The equivalent weight is thus the formula weight of Mn_2O_3 divided by $2 \times 5 = 10$.

As in neutralization reactions, the equivalent weight for a given oxidizing or reducing agent is not invariant. Potassium permanganate, for example, reacts under some conditions to give MnO_2:

$$MnO_4^- + 3e + 2H_2O \longrightarrow MnO_2(s) + 4OH^-$$

The change in the oxidation state of manganese in this reaction is from +7 to +4, and the equivalent weight of potassium permanganate is now equal to its formula weight divided by 3 (instead of 5 as in the earlier example).

Equivalent Weights in Precipitation and Complex-Formation Reactions

The equivalent weight of a participant in a precipitation or a complex-formation reaction is that weight which reacts with or provides one mole of the *reacting* cation if it is univalent, one-half a mole if it is divalent, one-third of a mole if it is trivalent, and so on. It is important to note that the cation referred to in this definition is always *the cation directly involved in the analytical reaction* and not necessarily the cation contained in the compound whose equivalent weight is being defined.

EXAMPLE A9-1

Define equivalent weights for $AlCl_3$ and $BiOCl$ if the two compounds are determined by a precipitation titration with $AgNO_3$:

$$Ag^+ + Cl^- \longrightarrow AgCl(s)$$

In this instance, the equivalent weight is based upon the number of moles of *silver ions* consumed in the titration of each compound. Since 1 mol of Ag^+ reacts with 1 mol of Cl^- that is provided by one-third mole of $AlCl_3$, we can write

$$\text{eqw } AlCl_3 = \text{fw } AlCl_3/3$$

Because each mole of BiOCl reacts with only 1 Ag^+ ion,

$$\text{eqw BiOCl} = \text{fw BiOCl}/1$$

Note that the fact that Bi^{3+} (or Al^{3+}) is trivalent has no bearing because the definition is based *upon the cation involved in the titration:* Ag^+.

A9-2 The Definition of Normality

The normality c_N of a solution expresses the number of milliequivalents of solute contained in 1 mL of solution or the number of equivalents contained in 1 L. Thus, a 0.20 N hydrochloric acid solution contains 0.20 meq of HCl in each milliliter of solution or 0.20 eq in each liter.

The normal concentration of a solution is defined by a pair of equations analogous to Equation 4-1a and 4-1b. Thus, for a solution of the species A, the normality $c_{N(A)}$ is given by the equations

$$c_{N(A)} = \frac{\text{no. meq A}}{\text{vol soln, mL}} \qquad (A9\text{-}4a)$$

$$c_{N(A)} = \frac{\text{no. eq A}}{\text{vol soln, L}} \qquad (A9\text{-}4b)$$

A9-3

Some Useful Algebraic Relationships

Two pairs of algebraic equations, analogous to Equation 4-2a and 4-2b as well as 4-3a and 4-3b, apply when normal concentrations are being used:

$$\text{amount of A} = \text{no. meq A} = \frac{\text{wt A (g)}}{\text{meqw A (g/meq)}} \tag{A9-5a}$$

$$\text{amount of A} = \text{no. eq A} = \frac{\text{wt A (g)}}{\text{eqw A (g/eq)}} \tag{A9-5b}$$

$$\text{amount of A} = \text{no. meq A} = \text{vol (mL)} \times c_{N(A)} \text{ (meq/mL)} \tag{A9-6a}$$

$$\text{amount of A} = \text{no. eq A} = \text{vol (L)} \times c_{N(A)} \text{ (eq/L)} \tag{A9-6b}$$

A9-4

Calculation of the Normality of Standard Solutions

Example A9-2 shows how the normality of a standard solution is computed from preparatory data. Note the similarity between this example and Example 4-3 (page 89).

EXAMPLE A9-2

Describe the preparation of 5.000 L of 0.1000 N Na_2CO_3 from the primary-standard solid, assuming the solution is to be used for titrations in which the chemical reaction is

$$CO_3^{2-} + 2H^+ \longrightarrow H_2O + CO_2$$

Applying Equation A9-6b gives

$$\text{amount of } Na_2CO_3 = \text{vol soln (L)} \times c_{N(Na_2CO_3)} \text{ (eq/L)}$$

$$= 5.000 \text{ L} \times 0.1000 \text{ eq/L} = 0.5000 \text{ eq } Na_2CO_3$$

Rearranging Equation A9-5b gives

$$\text{wt } Na_2CO_3 = \text{no. eq } Na_2CO_3 \times \text{eqw } Na_2CO_3$$

But 2 eq of Na_2CO_3 are contained in each mole of the compound; therefore,

$$\text{wt } Na_2CO_3 = 0.5000 \text{ eq } Na_2CO_3 \times \frac{105.99 \text{ g } Na_2CO_3}{2 \text{ eq } Na_2CO_3} = 26.50 \text{ g}$$

Therefore, dissolve 26.50 g in water and dilute to 5.000 L.

It is worth noting that, when the carbonate ion reacts with two protons, the weight of sodium carbonate required to prepare a 0.10 N solution is just one half that required to prepare a 0.1 M solution.

A9-4

The Treatment of Titration Data When Normality Is Used

In Section 4C-4 we introduced a systematic way of solving problems involving titration data based on molarities. This same technique can be applied equally well to problems in which concentrations are given in units of normality, provided it is understood that the stoichiometric ratio is always equal to unity when equivalents or milliequivalents are substituted for moles or millimoles. That is, the stoichiometric ratio of the number of *equivalents* of the standard or analyte to the number of equivalents of the titrant must always be unity because of the way the equivalent is defined.

Calculation of Normalities from Titration Data

Examples A9-3 and A9-4 illustrate how normality is computed from standardization data. Note that these examples are similar to Examples 4-7 and 4-8.

EXAMPLE A9-3

Exactly 50.00 mL of an HCl solution required 29.71 mL of 0.03926 N $Ba(OH)_2$ to give an end point with bromocresol green indicator. Calculate the normality of the HCl.

Note that the molarity of $Ba(OH)_2$ is one half its normality. That is,

$$c_{Ba(OH)_2} = 0.03926 \frac{meq}{mL} \times \frac{1 \text{ mmol}}{2 \text{ meq}} = 0.01963 \text{ M}$$

We now proceed by the systematic method outlined in Section 4C-2.

Step 1. Because milliequivalents are the units employed, we write

$$\text{stoichiometric ratio} = \frac{1 \text{ meq HCl}}{1 \text{ meq Ba(OH)}_2}$$

Step 2. The number of milliequivalents of standard is obtained by substituting into Equation A9-6a:

$$\text{no. meq Ba(OH)}_2 = 29.71 \text{ mL Ba(OH)}_2 \times 0.03926 \frac{\text{meq Ba(OH)}_2}{\text{mL Ba(OH)}_2}$$

Step 3. To obtain the number of milliequivalents of HCl, we multiply the result from step 2 by the ratio derived in step 1:

$$\text{no. meq HCl} = (29.71 \times 0.03926) \text{ meq Ba(OH)}_2 \times \frac{1 \text{ meq HCl}}{1 \text{ meq Ba(OH)}_2}$$

Step 4. Equating the result from step 3 to Equation A9-6a yields

$$\text{no. meq HCl} = 50.00 \text{ mL HCl} \times c_{N(HCl)} = (29.71 \times 0.03926 \times 1) \text{ meq HCl}$$

$$c_{N(HCl)} = \frac{(29.71 \times 0.03926 \times 1) \text{ meq HCl}}{50.00 \text{ mL HCl}} = 0.02333 \text{ N}$$

EXAMPLE A9-4

A 0.2121-g sample of pure $Na_2C_2O_4$ (fw = 134.00) was titrated with 43.31 mL of $KMnO_4$. What was the normality of the $KMnO_4$ solution? The chemical reaction is

$$2MnO_4^- + 5C_2O_4^{2-} + 16H^+ \longrightarrow 2Mn^{2+} + 10CO_2 + 8H_2O$$

As in the previous example, at the equivalence point in the titration,

$$\text{no. meq Na}_2\text{C}_2\text{O}_4 = \text{no. meq KMnO}_4$$

Substituting Equations A9-6a and A9-5a into this relationship gives

$$\text{vol KMnO}_4 \times c_{N(KMnO_4)} = \frac{\text{wt Na}_2\text{C}_2\text{O}_4 \text{ (g)}}{\text{meqw Na}_2\text{C}_2\text{O}_4 \text{ (g/meq)}}$$

$$43.31 \text{ mL KMnO}_4 \times c_{N(KMnO_4)} = \frac{0.2121 \text{ g Na}_2\text{C}_2\text{O}_4}{0.13400 \text{ g Na}_2\text{C}_2\text{O}_4/2 \text{ meq}}$$

$$c_{N(KMnO_4)} = \frac{0.2121 \text{ g Na}_2\text{C}_2\text{O}_4}{43.31 \text{ mL KMnO}_4 \times 0.1340 \text{ g Na}_2\text{C}_2\text{O}_4/2 \text{ meq}}$$

$$= 0.073093 \text{ meq/mL KMnO}_4 = 0.07309 \text{ N}$$

Note that the normality found here is five times the molarity computed in Example 4-8.

Calculation of the Quantity of Analyte from Titration Data

The examples that follow illustrate how analyte concentrations are computed when normalities are involved.

EXAMPLE A9-5

A 0.8040-g sample of an iron ore was dissolved in acid. The iron was then reduced to Fe^{2+} and titrated with 47.22 mL of 0.1121 N (0.02242 M) $KMnO_4$ solution. Calculate the results of this analysis in terms of (a) % Fe (fw = 55.847) and (b) % Fe_3O_4 (fw = 231.54). The reaction of the analyte with the reagent is described by the equation

$$MnO_4^- + 5Fe^{2+} + 8H^+ \longrightarrow Mn^{2+} + 5Fe^{3+} + 4H_2O$$

(a) At the equivalence point, we know that (Equation A9-1)

$$\text{no. meq } KMnO_4 = \text{no. meq } Fe^{2+} = \text{no. meq } Fe_3O_4$$

Substituting Equations A9-6a and A9-5a leads to

$$\text{vol } KMnO_4 \text{ (mL)} \times c_{N(KMnO_4)} \text{ (meq/mL)} = \frac{\text{wt } Fe^{2+} \text{ (g)}}{\text{meqw } Fe^{2+} \text{ (g/meq)}}$$

Substituting numerical data into this equation gives, after rearranging,

$$\text{wt } Fe^{2+} = 47.22 \text{ mL } KMnO_4 \times 0.1121 \frac{\text{meq}}{\text{mL } KMnO_4} \times \frac{0.055847 \text{ g}}{1 \text{ meq}}$$

Note that the milliequivalent weight of the Fe^{2+} is equal to its milliformula weight. The percentage of iron is

$$\% \; Fe^{2+} = \frac{(47.22 \times 0.1121 \times 0.055847) \text{ g } Fe^{2+}}{0.8040 \text{ g sample}} \times 100\% = 36.77\%$$

(b) Here,

$$\text{no. meq } KMnO_4 = \text{no. meq } Fe_3O_4$$

and

$$\text{vol } KMnO_4 \text{ (mL)} \times c_{N(KMnO_4)} = \frac{\text{wt } Fe_3O_4 \text{ (g)}}{\text{meqw } Fe_3O_4 \text{ (g/meq)}}$$

Substituting numerical data and rearranging give

$$\text{wt } Fe_3O_4 = 47.22 \text{ mL} \times 0.1121 \frac{\text{meq}}{\text{mL}} \times 0.23154 \frac{\text{g } Fe_3O_4}{3 \text{ meq}}$$

Note that the milliequivalent weight of Fe_3O_4 is one third its formula weight because each Fe^{2+} undergoes a one-electron change and the compound is converted to 3 Fe^{2+} before titration. The percentage of Fe_3O_4 is then

$$\% \; Fe_3O_4 = \frac{(47.22 \times 0.1121 \times 0.23154/3) \text{ g } Fe_3O_4}{0.8040 \text{ g sample}} \times 100\% = 50.81\%$$

Note that the answers to this example are identical to those in Example 4-9.

EXAMPLE A9-6

A 0.4755-g sample containing $(NH_4)_2C_2O_4$ and inert compounds was dissolved in water and made alkaline with KOH. The liberated NH_3 was distilled into 50.00 mL of 0.1007 N (0.05035 M) H_2SO_4. The excess H_2SO_4 was back-titrated with 11.13 mL of 0.1214 N NaOH. Calculate the percentage of N (fw = 14.007) and of $(NH_4)_2C_2O_4$ (fw = 124.10) in the sample.

At the equivalence point, the number of milliequivalents of acid and base are equal. In this titration, however, two bases are involved: NaOH and NH_3. Thus,

$$\text{no. meq } H_2SO_4 = \text{no. meq } NH_3 + \text{no. meq NaOH}$$

After rearranging,

$$\text{no. meq } NH_3 = \text{no. meq N} = \text{no. meq } H_2SO_4 - \text{no. meq NaOH}$$

Substituting Equations A9-5a and A9-6a for the number of milliequivalent of N and H_2SO_4, respectively, yields

$$\frac{\text{wt N (g)}}{\text{meqw N (g/meq)}} = 50.00 \text{ mL } H_2SO_4 \times 0.1007 \frac{\text{meq}}{\text{mL } H_2SO_4} - 11.13 \text{ mL NaOH} \times 0.1214 \frac{\text{meq}}{\text{mL NaOH}}$$

$$\text{wt N} = (50.00 \times 0.1007 - 11.13 \times 0.1214) \text{ meq} \times 0.014007 \text{ g N/meq}$$

$$\% \text{ N} = \frac{(50.00 \times 0.1007 - 11.13 \times 0.1214) \times 0.014007 \text{ g N}}{0.4755 \text{ g sample}} \times 100\% = 10.85\%$$

The number of milliequivalents of $(NH_4)_2C_2O_4$ is equal to the number of milliequivalents of NH_3 and N, but the milliequivalent weight of the $(NH_4)_2C_2O_4$ is equal to one half its formula weight. Thus,

$$\text{wt } (NH_4)_2C_2O_4 = (50.00 \times 0.1007 - 11.13 \times 0.1214) \text{ meq} \times 0.12410 \text{ g/(2 meq)}$$

$$\% \text{ } (NH_4)_2C_2O_4 = \frac{(50.00 \times 0.1107 - 11.13 \times 0.1214) \times 0.06205 \text{ g } (NH_4)_2C_2O_4}{0.4755 \text{ g sample}} \times 100\% = 48.07\%$$

Note that the results obtained here are identical to those obtained in Example 4-11.

EXAMPLE A9-7

The CO concentration in a gas sample was obtained by passing 20.3 L of the gas over iodine pentoxide heated to 150°C. The reaction is

$$I_2O_5(s) + 5CO(g) \longrightarrow 5CO_2(g) + I_2(g)$$

The iodine distilled at this temperature and was collected in an absorber containing 8.25 mL of 0.01101 N (0.01101 M) $Na_2S_2O_3$:

$$I_2(aq) + 2S_2O_3^{2-}(aq) \longrightarrow 2I^-(aq) + S_4O_6^{2-}(aq)$$

The excess $Na_2S_2O_3$ was back-titrated with 2.16 mL of 0.01894 N (0.00947 M) I_2. Calculate the number of milligrams of CO (fw = 28.01) per liter of sample.

Note that the $Na_2S_2O_3$ reacted with the I_2 produced by the reaction of CO as well as with that contained in the standard solution. Therefore, we write

$$\text{no. meq } Na_2S_2O_3 = \text{no. meq CO} + \text{no. meq } I_2$$

where the number of milliequivalents of I_2 is the amount of I_2 from the standard I_2 solution. The milliequivalent weight of CO is its milliformula weight divided by its change in oxidation number during the volumetric reaction. Since 5 mol of CO produces 1 mol of I_2, which undergoes a two-electron change, each CO is responsible for a change in oxidation number of 2/5 and

$$\text{eqw CO} = \frac{\text{fw CO}}{2/5} = \frac{5 \text{ fw CO}}{2}$$

Rearranging the first equation and substituting numerical data give

$$\text{no. meq CO} = 8.25\text{ mL Na}_2\text{S}_2\text{O}_3 \times 0.01101\,\frac{\text{meq}}{\text{mL}} - 2.16\text{ mL I}_2 \times 0.01894\,\frac{\text{meq}}{\text{mL I}_2}$$

$$\text{wt CO} = (8.25 \times 0.01101 - 2.16 \times 0.01894)\text{ meq CO} \times \frac{28.01\text{ mg CO}}{5/2\text{ meq CO}}$$

$$\frac{\text{wt CO}}{\text{vol sample}} = \frac{(0.04992 \times 28.01 \times 5/2)\text{ mg CO}}{20.36\text{ L}} = 0.172\,\frac{\text{mg CO}}{\text{L}}$$

Note that this result is identical to that found in Example 4-12.

APPENDIX 10

The Potentiometer

The potentiometer is a null-type instrument that permits the accurate measurement of potentials while drawing a minimum of electricity from the source under study.

Figure A10-1 shows a circuit diagram for the simplest form of a potentiometer. It consists of a *linear voltage divider AB,* a dc source V_B, a standard cell having a precisely known potential V_S, and a current detector A. The voltage divider consists of a uniform resistance wire mounted on a meter stick. A sliding contact, or wiper, permits variation of the output voltage V_{AC} between points A and C. In more sophisticated instruments, the divider is a precision wire-wound resistor formed into a helical coil. A continuously variable wiper that can be moved from one end of the helix to the other provides the variable voltage. Ordinarily, a divider is powered by a line-operated dc power supply or by mercury batteries, which provide a potential V_B that is somewhat larger than that which is to be measured.

A10-1 Voltage Measurements with a Potentiometer

To determine an unknown potential V_X with the instrument shown in Figure A10-1, two null-point measurements are required. Switch P is first closed to provide a current in the slide wire and a potential drop across AC. The standard cell is next placed in the circuit by moving switch SW to position 1. When the tapping key K is closed momentarily, a current is indicated by the current detector A unless the output from the standard cell is identical with the potential drop V_{AC} between A and C. The position of contact C is then varied until a null condition is achieved, as indicated by the detector when K is closed.

This process is repeated with SW in position 2 so that the unknown cell is in the circuit. For both adjustments, the magnitude of the potential V_{AC} between points A and C is given by

$$V_{AC} = V_B \frac{R_{AC}}{R_{AB}} = \frac{\not{k}AC}{\not{k}AB} = \frac{AC}{AB}$$

where AC and AB are the linear distances from A to C and B, respectively, and k is the proportionality constant relating length of the conductor to its resistance. When SW is at position 1, we can write

$$V_S = V_B \frac{AC_S}{AB}$$

where AC_S represents the linear distance between A and C as shown on the scale. At position 2,

$$V_X = V_B \frac{AC_X}{AB}$$

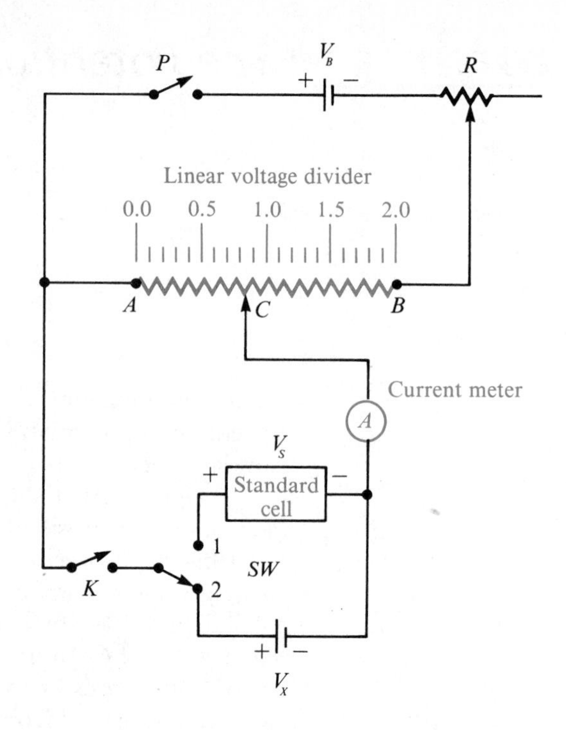

FIGURE A10-1 Circuit diagram for a laboratory potentiometer for measuring V_X.

where AC_X is the new distance between A and C. Dividing these equations and rearranging give

$$V_X = V_S \frac{AC_X}{AC_S} \tag{A10-1}$$

Thus, V_X is obtained from the potential of the standard cell and the two slide-wire settings.

The slide wire shown in Figure A10-1 is readily calibrated to read directly in volts, thus avoiding the necessity of calculating V_X each time. Calibration is accomplished by initially setting the slide-wire reading to the voltage of the standard cell. Then, with that cell in the circuit, the potential across AB is adjusted by means of the variable resistor R until a null reading is achieved. The slide-wire setting then gives V_X directly when C is adjusted to the null position with the unknown voltage.

A10-2 Current Measurements with a Potentiometer

The null method is readily applied to the determination of current. A small precision resistor is placed in series in the circuit, and the potential drop across the resistor is measured with a potentiometer. The current can then be calculated by means of Ohm's law.

APPENDIX 11

The Method of Successive Approximations

The solution of quadratic, cubic, or higher-order equations is often necessary in the study of chemical equilibria. As an example, consider the calculation of the hydrogen ion concentration in a solution of a weak acid. The expression that must be solved is

$$K_a = [H^+]^2/(c_{HA} - [H^+]) = x^2/(c_{HA} - x)$$

where $x = [H^+]$. When the simplifying assumption that $c_{HA} \gg x$ is invalid or when an answer accurate to several significant figures is desired, the quadratic equation is employed to solve for x. Another approach that is often simpler and less prone to error, particularly when a calculator or computer is used, is *the method of successive approximations,* which is summarized as follows:

1. Rearrange the equation of interest so that $x_1 = f(x)$.
2. Estimate a value for x_1, and substitute x_1 into $f(x)$ so that $x_2 = f(x_1)$.
3. Compare x_2 with x_1. If the difference between x_2 and x_1 is sufficiently small, then x_2 is the solution. If x_1 and x_2 are not close enough, repeat steps 2 and 3 by calculating $x_3 = f(x_2)$, $x_4 = f(x_3)$, . . . , and so on, until successive values of x are equal within the desired level of accuracy.

The procedure is best illustrated by example. Suppose that, for a weak acid, $K_a = 2.0000 \times 10^{-3}$ and $c_{HA} = 0.10000$, and we wish to determine $[H^+] = x$ to five significant figures.

Step 1: We arrange the equation to obtain

$$x_1 = \sqrt{K_a(c_{HA} - x)} = \sqrt{(2.0000 \times 10^{-3})(0.10000 - x)}$$

Step 2: As the first approximation, we guess that $x_1 = 0$, and so

$$x_2 = \sqrt{(2.0000 \times 10^{-3})(0.10000 - x_1)}$$
$$= \sqrt{(2.0000 \times 10^{-3})(0.10000)} = 1.4142 \times 10^{-2}$$

Step 3: Since x is considerably greater than zero, we repeat step 2:

$$x_3 = \sqrt{(2.0000 \times 10^{-3})(0.10000 - x_2)}$$
$$= \sqrt{(2.0000 \times 10^{-3})(0.10000 - 1.4142 \times 10^{-2})} = 1.3014 \times 10^{-2}$$

and so on for successive values of x_n:

$$x_4 = 1.3183 \times 10^{-2}$$

$$x_5 = 1.3177 \times 10^{-2}$$

$$x_6 = 1.3177 \times 10^{-2}$$

Note that, in a single cycle of the process, the answer x_2 obtained is accurate to two significant figures. Furthermore, each successive cycle adds another significant figure to the answer. When a calculator with three memories to hold K_a, c_{HA}, and x is used, each cycle requires only a few keystrokes. This relatively simple example illustrates the process, but it is in the solution of very complicated equations that the method of successive approximations is indispensable. Let us consider a final example:

$$x^3 + 2x^2 + 8.43x - 18.5 = 0$$

Step 1: $x_1 = f(x) = 18.5/(x^2 + 2x + 8.43)$

Step 2: Let us make an initial guess of $x_1 = 2$, so that we have

$$x_2 = 18.5/(2^2 + 2 \times 2 + 8.43) = 1.126$$

$$x_3 = 18.5/(1.126^2 + 2 \times 1.126 + 8.43) = 1.548$$

$$x_4 = 1.329$$

$$x_5 = 1.439$$

$$x_6 = 1.383$$

$$x_7 = 1.411$$

$$x_8 = 1.397$$

$$x_9 = 1.404$$

$$x_{10} = 1.400$$

Note that in this example, which would be virtually impossible to solve algebraically, an answer good to three significant figures is obtained in four cycles. If the process does not appear to converge to a final answer in a few cycles, it may be necessary to rearrange the equation to a different form.

A very large body of information has been amassed on the iterative solution of various types of equations by successive approximation and other artful techniques. Textbooks on *numerical analysis* serve as useful guides to the solution of equations that cannot be solved analytically.[1]

[1]For example, see A. C. Norris, *Computational Chemistry*, p. 62. New York: Wiley, 1981.

APPENDIX 12

Compounds Recommended for the Preparation of Standard Solutions of Some Common Elements*

Element	Compound	FW	Solvent**	Notes
Aluminum	Al metal	26.98	Hot dil HCl	a
Antimony	$KSbOC_4H_4O_6 \cdot \frac{1}{2}H_2O$	333.93	H_2O	c
Arsenic	As_2O_3	197.84	dil HCl	i,b,d
Barium	$BaCO_3$	197.35	dil HCl	
Bismuth	Bi_2O_3	465.96	HNO_3	
Boron	H_3BO_3	61.83	H_2O	d,e
Bromine	KBr	119.01	H_2O	a
Cadmium	CdO	128.40	HNO_3	
Calcium	$CaCO_3$	100.09	dil HCl	i
Cerium	$(NH_4)_2Ce(NO_3)_6$	548.23	H_2SO_4	
Chromium	$K_2Cr_2O_7$	294.19	H_2O	i,d
Cobalt	Co metal	58.93	HNO_3	a
Copper	Cu metal	63.55	dil HNO_3	a
Fluorine	NaF	41.99	H_2O	b
Iodine	KIO_3	214.00	H_2O	i
Iron	Fe metal	55.85	HCl, hot	a
Lanthanum	La_2O_3	325.82	HCl, hot	f
Lead	$Pb(NO_3)_2$	331.20	H_2O	a
Lithium	Li_2CO_3	73.89	HCl	a
Magnesium	MgO	40.31	HCl	
Manganese	$MnSO_4 \cdot H_2O$	169.01	H_2O	g
Mercury	$HgCl_2$	271.50	H_2O	b
Molybdenum	MoO_3	143.94	1 M NaOH	
Nickel	Ni metal	58.70	HNO_3, hot	a
Phosphorus	KH_2PO_4	136.09	H_2O	
Potassium	KCl	74.56	H_2O	a
	$KHC_8H_4O_4$	204.23	H_2O	i,d
	$K_2Cr_2O_7$	294.19	H_2O	i,d
Silicon	Si metal	28.09	NaOH, concd	
	SiO_2	60.08	HF	
Silver	$AgNO_3$	169.87	H_2O	a
Sodium	NaCl	58.44	H_2O	i
	$Na_2C_2O_4$	134.00	H_2O	i,d
Strontium	$SrCO_3$	147.63	HCl	a
Sulfur	K_2SO_4	174.27	H_2O	
Tin	Sn metal	118.69	HCl	
Titanium	Ti metal	47.90	H_2SO_4, 1:1	a
Tungsten	$Na_2WO_4 \cdot 2H_2O$	329.86	H_2O	h
Uranium	U_3O_8	842.09	HNO_3	d
Vanadium	V_2O_5	181.88	HCl, hot	
Zinc	ZnO	81.37	HCl	a

*The data in this table were taken from a more complete list assembled by B. W. Smith and M. L. Parsons, *J. Chem. Educ.*, **1973,** *50,* 679. Unless otherwise specified, compounds should be dried to constant weight at 110°C.

**Unless otherwise specified, acids are concentrated analytical grade.

Continued on next page

Continued from previous page

[a]Conforms well to the criteria listed in Section 4B-1 and approaches primary standard quality.

[b]Highly toxic.

[c]Loses $\frac{1}{2}$ H_2O at 110°C. After drying, fw = 324.92. The dried compound should be weighed quickly after removal from the desiccator.

[d]Available as a primary standard from the National Bureau of Standards.

[e]H_3BO_3 should be weighed directly from the bottle. It loses 1 H_2O at 100°C and is difficult to dry to constant weight.

[f]Absorbs CO_2 and H_2O. Should be ignited just before use.

[g]May be dried at 110°C without loss of water.

[h]Loses both waters at 110°C. fw = 293.82. Keep in desiccator after drying.

[i]Primary standard.

Answers to Questions and Problems

Chapter 2

2-1. (a) Accuracy is the closeness of an experimental result to the accepted value or to what is believed to be the true value for the result. Precision is the closeness of a result to the mean of two or more results measured in exactly the same way.

(c) The confidence limit is the interval around an experimental mean $\bar{x}$ within which the true mean μ can be expected to lie with a certain probability. The confidence level is a probability usually expressed as a percentage.

(e) The absolute error E is

$E = x_i - x_t$

where x_i is an experimental result and x_t is the accepted value for the result.

The relative error E_r is

$E_r = (x_i - x_t)/x_t$

Often the numerator of this equation is multiplied by 100 or 1000.

(g) The sample mean $\bar{x}$ is

$$\bar{x} = \frac{\sum_{i=0}^{N} x_i}{N}$$

where x_i are the individual values of x making up a finite set of N values. The population mean μ is given by the same equation when N approaches infinity.

2-2. (a) The range of a set of data is the numerical difference between the largest and the smallest value in the set.

(c) A histogram is a bar graph that is obtained by dividing a set of data into equally spaced cells and plotting the percentage of measurements falling into each cell as a function of the measured quantity.

(e) The standard error of a mean σ_m is

$$\sigma_m = \frac{\sigma}{\sqrt{N}}$$

where σ is the standard deviation of a sample of data made up of N measurements.

(f) The Q test is a statistically based test designed to determine whether an outlying result should be rejected or retained; Q_{exp} is obtained from the equation

$Q_{exp} = |x_q - x_n|/w$

where x_q is the questionable result, x_n is its nearest neighbor, and w is the spread of the set. If Q_{exp} is greater than the statistically derived Q_{crit}, the questionable result can be rejected with a certain level of confidence.

(i) The significant figures in a number are all the certain digits plus one uncertain digit.

2-3. The three types of determinate error are instrumental error, method error, and personal error.

2-5. Constant determinate errors can be detected by varying sample size because the effect of the error becomes smaller as the sample size increases.

2-7. The quantity s is the standard deviation for a sample of data, whereas σ is the standard deviation for a population of data. As the number of data N making up the sample increases, s approaches σ more and more closely. Ordinarily, s can be assumed to be equal to σ when N is about 20 to 30.

2-9.

	A	C	E
(a) Mean	2.08	0.09180	69.53
(b) Median	2.1	0.0885	69.635
(c) Spread	0.9	0.0116	0.44
(d) Standard deviation	0.35	0.0055	0.22
(e) Coefficient of variation	17%	6.0%	0.31%

2-10. For A, $E = 0.08$; $E_r = 40$ ppt
For C, $E = -0.0012$; $E_r = -13$ ppt
For E, $E = 0.48$; $E_r = 7.0$ ppt

2-11. For A, 95% CI = 2.08 ± 0.44 or 2.1 ± 0.4
For C, 95% CI = 0.0918 ± 0.0087 or 0.092 ± 0.009
For E, 95% CI = 69.53 ± 0.35 or 69.5 ± 0.4
In each case, the 95% confidence interval is the interval around the mean $\bar{x}$ wherein the population mean μ can be expected to lie 95 times out of 100.

2-12. For A, 95% CI = 2.08 ± 0.18 or 2.1 ± 0.2
For C, 95% CI = 0.0918 ± 0.0069 or 0.092 ± 0.007
For E, 95% CI = 69.53 ± 0.15 or 69.5 ± 0.2

2-13. For A, (a) $Q_{exp} = 0.67$, $Q_{crit} = 0.73$; ∴ retain
(b) T_n(expt) = 1.66, T_n(crit) = 1.67; ∴ retain
For C, (a) $Q_{exp} = 0.84$, $Q_{crit} = 0.85$; ∴ retain
(b) T_n(expt) = 1.49, T_n(crit) = 1.46; ∴ reject
For E, (a) $Q_{exp} = 0.95$, $Q_{crit} = 0.85$; ∴ reject
(b) T_n(expt) = 1.45, T_n(crit) = 1.46; ∴ retain

2-14. $s_{pooled} = 0.173$; $(\bar{x}_B - \bar{x}_E)_{expt} = 0.354$
(a) $\bar{x}_B - \bar{x}_E$ from Equation 2-19 = 0.267
Since the experimental difference is greater than the computed value, a difference is suggested at the 90% confidence level.
(b) At the 95% confidence level, computed $\bar{x}_B - \bar{x}_E = 0.53$. Thus, a difference has not been demonstrated at this confidence level.

2-15. $F = \dfrac{(0.215)^2}{(0.0757)^2} = 8.07$; $F_{crit} = 19.16$
Since the experimental value of F is well below its critical value, no precision difference has been demonstrated.

2-17. $F = \dfrac{(0.349)^2}{(0.277)^2} = 1.59$; $F_{crit} = 6.39$
No difference in precision has been demonstrated.

2-18. (a) −0.04%, $c = 0.3\%$

2-19. (a) 13 g, $c = 3$ g

2-20.

Sample	s, %K	Sample	s, %K
1	0.095	4	0.10
2	0.12	5	0.10
3	0.11		

$s_{pooled} = 0.11\%$ K^+

2-22. $s_{pooled} = 0.29\%$ heroin

2-24. (a) 80% CI = 18.5 ± 1.29 × $2.4/\sqrt{1}$ = 18.5 ± 3.1 μg/mL
95% CI = 18.5 ± 1.96 × $2.4/\sqrt{1}$ = 18.5 ± 4.7 μg/mL

(b) 80% CI = 18.5 ± 1.29 × $2.4/\sqrt{2}$ = 18.5 ± 2.2 μg/mL
95% CI = 18.5 ± 1.96 × $2.4/\sqrt{2}$ = 18.5 ± 3.3 μg/mL

(c) 80% CI = 18.5 ± 1.29 × $2.4/\sqrt{4}$ = 18.5 ± 1.5 μg/mL
95% CI = 18.5 ± 1.96 × $2.4/\sqrt{4}$ = 18.5 ± 2.4 μg/mL

2-26. 10 and 17 measurements

2-28. $\bar{x} = 3.22$, $s = 0.061$

(a) 95% CI = 3.22 ± 4.30 × $0.061/\sqrt{3}$ = 3.22 ± 0.15

(b) 95% CI = 3.22 ± 1.96 × $0.05/\sqrt{3}$ = 3.22 ± 0.06

2-30. 12 measurements

2-31. (a) $(\bar{x} - \mu)_{actual} = -0.006$
$\bar{x} - \mu = 2.36 \times 0.0087/\sqrt{8} = 0.0073$
No determinate error demonstrated.

(b) $(\bar{x} - \mu)_{actual} = -0.03$
$\bar{x} - \mu = 2.78 \times 0.0198/\sqrt{5} = 0.025$
Determinate error suggested.

(c) $(\bar{x} - \mu)_{actual} = +0.03$
$\bar{x} - \mu = 2.45 \times 0.042/\sqrt{7} = 0.039$
No determinate error demonstrated.

2-33. (a) $\bar{x} - \mu = 1.96 \times 0.094/\sqrt{4} = 0.092$; $(\bar{x} - \mu)_{actual} = 0.11$
Determinate error suggested.

(b) $\bar{x} - \mu = 3.18 \times 0.085/\sqrt{4} = 0.14$
No error demonstrated.

2-35.

Element	$(\bar{x}_1 - \bar{x}_2)_{actual}$	$\pm z\sigma\sqrt{\frac{N_1 + N_2}{N_1 N_2}}$	Difference Suggested?
As	+10	±20	No
Co	−0.07	±0.053	Yes
La	+0.40	±0.42	No
Sb	+0.04	±0.53	No
Th	−0.12	±0.090	Yes

The Co and Th results suggest that a difference does exist, and the defendant's claim appears valid.

2-37. $s_b = 0.19$
(a) 0.58 mg (b) 0.41 mg (c) 0.35 mg

2-38. (a) $F_{expt} = 1.44$; $F_{crit} = 4.00$
No difference in precision demonstrated.

(c) $F_{expt} = 1.96$; $F_{crit} = 2.60$
No difference in precision demonstrated.

2-39.

	s	CV, %	Rounded Result
(a)	0.030	5.2	0.57(±0.03)
(b)	0.089	0.42	21.3(±0.1)
(c)	0.14×10^{-16}	2.0	$6.9(\pm 0.1) \times 10^{-16}$
(d)	1.4×10^{3}	0.77	$1.84(\pm 0.01) \times 10^{5}$
(e)	0.51×10^{-2}	8.5	$6.0(\pm 0.5) \times 10^{-2}$ or $7(\pm 1) \times 10^{-2}$
(f)	0.11×10^{-3}	1.3	$8.1(\pm 0.1) \times 10^{-3}$
(g)	55	6.2	$8.8(\pm 0.6) \times 10^{2}$ or $9(\pm 1) \times 10^{2}$
(h)	0.061×10^{-6}	1.3	$4.69(\pm 0.06) \times 10^{-6}$ or $4.7(\pm 0.1) \times 10^{-6}$

2-41.

	s	CV, %	Rounded Result
(a)	0.0075	3.2	0.238(±0.008) or 0.24(±0.01)
(c)	0.0014	0.0061	23.780(±0.001)
(e)	5.8×10^{5}	1.2×10^{2}	$5(\pm 6) \times 10^{5}$

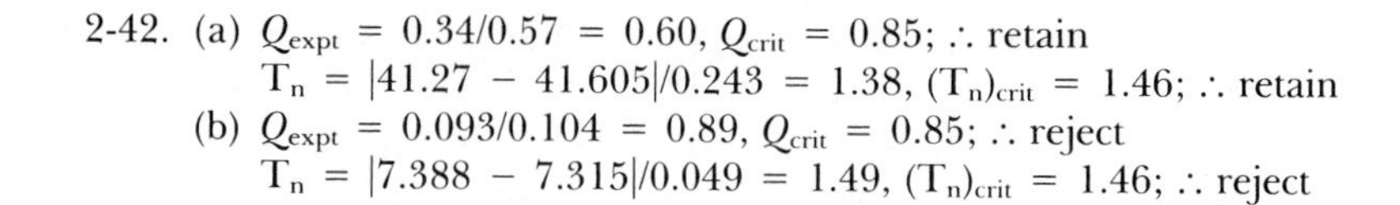

2-42. (a) $Q_{expt} = 0.34/0.57 = 0.60$, $Q_{crit} = 0.85$; ∴ retain
$T_n = |41.27 - 41.605|/0.243 = 1.38$, $(T_n)_{crit} = 1.46$; ∴ retain
(b) $Q_{expt} = 0.093/0.104 = 0.89$, $Q_{crit} = 0.85$; ∴ reject
$T_n = |7.388 - 7.315|/0.049 = 1.49$, $(T_n)_{crit} = 1.46$; ∴ reject

Chapter 3

3-1. (a) The individual particles of a colloid are smaller than about 10^{-5} mm in diameter, while those of a crystalline precipitate are larger. As a consequence, crystalline precipitates settle out of solution relatively rapidly, whereas colloidal particles do not unless they can be caused to agglomerate.
(c) Precipitation is the process by which a solid phase forms and is carried out of solution when the solubility product of a species is exceeded. Coprecipitation is the process in which a *normally soluble* species is carried out of solution during the formation of a precipitate.
(e) Occlusion is a type of coprecipitation in which an impurity is entrapped in a pocket formed by a rapidly growing crystal. Mixed-crystal formation is a type of coprecipitation in which a foreign ion is incorporated into a growing crystal in a lattice position that is ordinarily occupied by one of the ions of the precipitate. The contaminant has the same charge and is about the same size as the ion it replaces.

3-2. (a) Digestion is a process for improving the purity and filterability of a precipitate by heating the solid in contact with the solution from which it is formed (the mother liquor).
(c) In reprecipitation, a precipitate is filtered, washed, redissolved, and then reformed from the new solution. Because the concentration of contaminant is lower in this new solution than in the original, the second precipitate contains less coprecipitated impurity.
(e) A gravimetric factor is a stoichiometric ratio used to convert the weight of a compound to the weight of a second compound that is chemically equivalent to the first:

$$\text{Gravimetric factor} = \frac{\text{fw of substance sought}}{\text{fw of substance weighed}} \times \frac{a}{b}$$

where a and b are small whole numbers that have values such that the number of formula weights in the numerator and denominator are chemically equivalent.
(g) The counter-ion layer is a layer of solution surrounding a colloidal particle that contains a sufficient excess of ions of opposite charge to just equal the number of primarily adsorbed ions on the particle.

3-3. A chelating agent is an organic compound that contains two electron-donor groups located in such a configuration that a five- or six-membered ring is formed when both groups complex a cation.

3-5. (a) plus (b) adsorbed Ag^+ (c) NO_3^-

3-7. Peptization is the process in which a coagulated colloid returns to its original dispersed state as a consequence of a reduction in the electrolyte concentration of the solution in contact with the precipitate. Peptization during the washing of a coagulated colloid can be avoided by washing with an electrolyte solution rather than with pure water.

3-9. (a) fw SO_3/fw $BaSO_4$
(c) 2 × fw In/fw In_2O_3
(e) 2 × fw CuO/fw $Cu_2(SCN)_2$
(g) fw Pb_3O_4/3 × fw PbO_2
(i) fw $Na_2B_4O_7 \cdot 10H_2O$/2 × fw B_2O_3

3-10. 95.36% Cl

3-12. 0.6622 g $Cu(IO_3)_2$

3-14. 0.124 g AgI

3-16. 14.03% Mg

3-18. 18.99% C
3-20. 38.82% Hg_2Cl_2
3-22. 46.40% NH_3
3-24. Minimum weight = 0.617 g
Weight of precipitate = 0.825 g
3-26. (a) 0.239 g sample (b) 0.494 g AgCl (c) 0.407 g sample
3-28. % I^- = 27.05 % Cl^- = 4.72
3-30. 0.395 g CO_2
3-32. (a) 0.369 g $Ba(IO_3)_2$ (b) 0.0148 g $BaCl_2 \cdot 2H_2O$
3-34. (a) 0.488 g Ag_2CrO_4 (b) 0.0142 g K_2CrO_4

Chapter 4

4-1. (a) The millimole is a unit of amount of an elementary species, such as an atom, an ion, a molecule, or an electron. One millimole contains 6.02×10^{20} particles of the species.
(b) The milliformula weight of a species is the weight in grams of one millimole.
4-2. The equivalence point in a titration is that point at which sufficient titrant has been added so that stoichiometrically equivalent amounts of analyte and titrant are present. The end point in a titration is the point at which an observable physical change that signals the equivalence point occurs.
4-4. A primary standard is a highly purified substance that serves as the basis for a titrimetric method. It is used either (1) to prepare a standard solution directly by weight or (2) to standardize a solution to be used in a titration.
4-6. (a) 0.119 mol or 119 mmol Mn_3O_4
(c) 0.277 mol or 277 mmol $Na_2B_4O_7$
(e) 2.23×10^{-5} mol or 2.23×10^{-2} mmol SO_2
4-7. (a) 1.05×10^{-4} mol or 0.105 mmol $Mg_2P_2O_7$
(c) 7.77 mol or 7.77×10^3 mmol NaCl
(e) 1.50×10^{-5} mol or 0.0150 mmol ascorbic acid
4-8. (a) 88.0 g CO_2
(c) 1.60×10^3 g NaOH
(e) 1.48×10^3 g HCl
4-9. (a) Dilute 60.0 mL of 4.00 M H_2SO_4 to 3.00 L.
(b) Dilute 181 g of 13.0% solution to 3.00 L.
(c) Dilute 13.5 mL of the concentrated reagent to 3.00 L.
4-11. 0.09151 M
4-13. 0.0461 M
4-15. 0.2970 M $HClO_4$; 0.3258 M NaOH
4-17. 5.471% As_2O_3
4-19. 7.317% $(NH_2)_2CS$
4-20. (a) pNa = pBr = 2.000; pH = pOH = 7.00
(c) pBa = 2.46; pOH = 2.15, pH = 11.85
(e) pCa = 2.28; pBa = 2.44; pCl = 1.75; pH = pOH = 7.00
4-21. (a) 2.1×10^{-9} M (e) 2.09 M
(c) 9.2×10^{-1} M (g) 9.9×10^{-1} M
4-22. (a) 9.36×10^{-3} M (b) 1.9×10^{-5} M
(c) Rel error = −3.0 ppt; abs error = -2.8×10^{-5} M
4-24. 3.03 M H_2SO_4
4-26. (a) Dissolve 20.0 g glucose in water and dilute to 200 mL.
(b) Dissolve 20.0 g glucose in 80.0 g water.
(c) Dilute 20.0 mL of CH_3OH to 200 mL.
4-28. 346 ppm S
4-30. (a) 0.02053 mmol $AgNO_3$/g soln = 0.02053 M_w
(b) 0.01938 M_w KSCN
(c) 4.572% $BaCl_2 \cdot H_2O$

Chapter 5

5-1. The activity of a species is a concentration-dependent parameter that provides a measure of the effect of the species on chemical equilibria that is independent of the ionic strength of the media. The activity coefficient of a species is an ionic-strength-dependent parameter that converts the molar concentration of the species to its activity.

5-2. (a) Equilibrium is shifted to the left.
(c) Equilibrium is shifted to the left.

5-3. (a) A salt is the product (other than water) of an acid/base reaction.
(c) The hydronium ion is the product of the reaction of a proton with one molecule of water. Its formula is H_3O^+.
(e) A base is a proton acceptor.
(g) The Le Châtelier principle states that the position of a chemical equilibrium is always shifted in such a direction as to relieve the effect of an applied stress.
(h) An amphiprotic solvent is one that undergoes self-ionization to form a pair of ionic species. For example,
$2H_2O \rightleftharpoons H_3O^+ + OH^-$
$2C_2H_5OH \rightleftharpoons C_2H_5OH_2^+ + C_2H_5O^-$

5-4. Water is a leveling solvent toward mineral acids because all of these acids are completely dissociated in that medium and are thus equal in strength. In ethanol, the degree of dissociation of the various mineral acids differs significantly. Thus, ethanol is termed a differentiating solvent because it differentiates among the strengths of the acids.

5-6. This conclusion is invalid because the units of K_{sp} for AgCl are mol²/L² whereas the units for K_{sp} for Ag_2CO_3 are mol³/L³. Thus,
$s_{AgCl} = (1.8 \times 10^{-10})^{1/2} = 1.3 \times 10^{-5}$
$s_{Ag_2CO_3} = (8.1 \times 10^{-12}/4)^{1/3} = 1.3 \times 10^{-4}$
and Ag_2CO_3 is the more soluble.

5-7. (a) $\beta_2 = 1 \times 10^5$ $\beta_4 = 8 \times 10^7$ $\beta_6 = 5 \times 10^8$
$\beta_3 = 5 \times 10^6$ $\beta_5 = 5 \times 10^8$

5-9. Greatest discrepancies occur at high ionic strengths with reactants and/or products that have multiple charges.

5-10. (a) $K_b = \dfrac{[C_2H_5NH_3^+][OH^-]}{[C_2H_5NH_2]} = 4.28 \times 10^{-4}$

(c) $K_a = \dfrac{[C_5H_5N][H_3O^+]}{[C_5H_5NH^+]} = \dfrac{K_w}{K_b} = \dfrac{1.00 \times 10^{-14}}{1.7 \times 10^{-9}} = 5.9 \times 10^{-6}$

(e) $K_f = [AgCl_2^-]/[Cl^-] = \text{antilog } -4.7 = 2 \times 10^{-5}$

(g) $\dfrac{[Cd(NH_3)_2^{2+}]}{[Cd^{2+}][NH_3]^2} = K_f = K_1K_2 = 3.98 \times 10^2 \times 1.26 \times 10^2 = 5 \times 10^4$

(i) $\dfrac{[H_3O^+]^3[AsO_4^{3-}]}{[H_3AsO_4]} = K_1K_2K_3 = 6.0 \times 10^{-3} \times 1.05 \times 10^{-7} \times 3.0 \times 10^{-12}$
$= 1.9 \times 10^{-21}$

5-11. K_{sp} = (a) 4.0×10^{-16} (c) 2.0×10^{-2} (e) 3.5×10^{-10}

5-12. (a) 0.056 g/100 mL (c) 0.014 g/100 mL
(b) 8.2×10^{-4} g/100 mL (d) 0.093 g/100 mL

5-14. (a) 2.18×10^{-7} M (b) 0.98 M

5-16. (a) 0.025 M (c) 1.9×10^{-3} M
(b) 8.9×10^{-4} M (d) 7.6×10^{-7} M

5-18.

	Quadratic Method		Approximate Method	
	$[H_3O^+]$	$[OH^-]$	$[H_3O^+]$	$[OH^-]$
(a)	2.4×10^{-5}	4.1×10^{-10}	2.4×10^{-5}	4.1×10^{-10}
(c)	1.05×10^{-12}	9.6×10^{-3}	1.02×10^{-12}	9.8×10^{-3}
(e)	5.0×10^{-11}	2.0×10^{-4}	5.0×10^{-11}	2.0×10^{-4}
(g)	3.05×10^{-4}	3.28×10^{-11}	3.06×10^{-4}	3.27×10^{-11}

5-19. $[H_3O^+] = [OH^-] = 3.38 \times 10^{-8}$

5-20. (a) 1.10×10^{-2} (quadratic); 1.17×10^{-2} (approximate)
(b) 1.17×10^{-8} (quadratic); 1.17×10^{-8} (approximate)
(e) 1.47×10^{-4} (quadratic); 1.59×10^{-4} (approximate)

5-21. (a) $[Hg^{2+}] = 0.0050$; $[SCN^-] = 2.2 \times 10^{-9}$
(b) $[Hg^{2+}] = 2.3 \times 10^{-7}$; $[SCN^-] = 4.6 \times 10^{-7}$
(c) $[Hg^{2+}] = 5.0 \times 10^{-16}$; $[SCN^-] = 0.010$

5-22. (a) 8×10^{-7} M (b) 8×10^{-6} M (c) 8×10^{-5} M

5-24. (a) $PbI_2(1.2 \times 10^{-3}) > TlI(2.6 \times 10^{-4}) > BI_3(1.3 \times 10^{-5}) > AgI$ (9.1×10^{-9})
(b) $PbI_2(7.1 \times 10^{-7}) > TlI(6.5 \times 10^{-7}) > AgI(8.3 \times 10^{-16}) > BiI_3$ (8.1×10^{-16})
(c) $PbI_2(1.3 \times 10^{-4}) > BiI_3(6.7 \times 10^{-7}) > TlI(6.5 \times 10^{-7}) > AgI$ (8.3×10^{-16})

5-26. (a) (1) 1.4×10^{-6} M (2) 1.1×10^{-6} M
(b) (1) 2.0×10^{-3} M (2) 1.2×10^{-3} M
(c) (1) 3.1×10^{-5} M (2) 1.1×10^{-5} M
(d) (1) 1.4×10^{-5} M (2) 2.0×10^{-6} M

5-28. (a) (1) 3.0×10^{-3} M (2) 2.9×10^{-3} M
(b) (1) 3.6×10^{-3} M (2) 2.9×10^{-3} M

5-29. (a) (1) 5.4×10^{-6} M (2) 5.4×10^{-6} M
(c) (1) 8.5×10^{-13} M (2) 2.5×10^{-13} M
(e) (1) 1.4×10^{-11} M (2) 3.1×10^{-12} M

Chapter 6

6-1. (a) $0.10 = [H_3PO_4] + [H_2PO_4^-] + [HPO_4^{2-}] + [PO_4^{3-}]$
(c) $0.100 + 0.0500 = [HNO_2] + [NO_2^-]$
$[Na^+] = c_{NaNO_2} = 0.0500$
(e) $0.10 = [Na^+] = [OH^-] - 2[Zn(OH)_4^{2-}]$
(g) $[Ca^{2+}] = \frac{1}{2}([F^-] + [HF])$
$[HF] = [OH^-]$
(i) $0.0100 = [NH_3] + [NH_4^+] + [Cd(NH_3)^{2+}] + 2[Cd(NH_3)_2^{2+}] +$
$3[Cd(NH_3)_3^{2+}] + 4[Cd(NH_3)_4^{2+}] + 5[Cd(NH_3)_5^{2+}] +$
$6[Cd(NH_3)_6^{2+}]$
$[OH^-] = 2[Cd^{2+}] + [NH_4^+]$

6-2. (a) $[H_3O^+] = [OH^-] + [H_2PO_4^-] + 2[HPO_4^{2-}] + 3[PO_4^{3-}]$
(c) $[H_3O^+] + [Na^+] = [OH^-] + [NO_2^-]$
(e) $[Na^+] + [H_3O^+] + 2[Zn^{2+}] = [OH^-] + 2[Zn(OH)_4^{2-}]$
(g) $2[Ca^{2+}] + [H_3O^+] = [F^-] + [OH^-]$
(i) $[H_3O^+] + 2([Cd^{2+}] + [Cd(NH_3)^{2+}] \cdots [Cd(NH_3)_6^{2+}] = [OH^-]$

6-3. (a) $K_{sp} = [Ag^+][SCN^-]$ (e) $K_{sp} = [Bi^{3+}][I^-]^3$
(c) $K_{sp} = [Pb^{2+}][CrO_4^{2-}]$ (g) $K_{sp} = [Pb^{2+}][Cl^-][F^-]$

6-4. (a) $K_{sp} = S^2$ (e) $K_{sp} = 27S^4$
(c) $K_{sp} = S^2$ (g) $K_{sp} = S^3$

6-5. (a) 5.81×10^{-19} (c) 1.30×10^{-20} (e) 1.30×10^{-22}

6-6. (a) 3.40×10^{-6} (c) 8.10×10^{-19} (e) 3.00×10^{-13}

6-7. 8.01×10^{-16}

6-9. 6.2×10^{-12}

6-11. (a) 9.1×10^{-9} M (b) 2.4×10^{-3} M (c) 3.9×10^{-5} M

6-13. (a) 0.63 g (b) 0.080 g (c) 0.60 g

6-15. 8.4×10^{-2} M

6-17. (a) 3.2×10^{-7} M (b) 1.6×10^{-2} M

6-19. (a) 2.6×10^{-3} M (c) 7.4×10^{-6} M
(b) 0.100 M (d) 3.3×10^{-3} M

6-21. (a) $[Mg^{2+}] = 0.0102$ M; $[Cl^-] = 0.0204$ M; $[OH^-] = 4.2 \times 10^{-5}$ M; $[H_3O^+] = 2.4 \times 10^{-10}$ M
(b) $[Mg^{2+}] = 0.0686$ M; $[Cl^-] = 0.204$ M; $[H_3O^+] = 6.68 \times 10^{-2}$ M; $[OH^-] = 1.50 \times 10^{-13}$ M

6-23. (a) $Be(OH)_2$ (b) $Hf(OH)_4$

6-25. (a) 8.3×10^{-11} M (b) 1.6×10^{-11} (c) 1.3×10^4 (d) 1.3×10^4

6-27. (a) Not feasible (b) Feasible (c) Feasible (d) Not feasible

6-29. (a) 5.2×10^{-3} M (c) 3.6×10^{-4} M

6-30. (a) 1.5×10^{-4} M (c) 7.4×10^{-5} M

6-31. (a) 1.7×10^{-12} (c) 4.7×10^{-11}

6-32. (b) 5.3×10^{-5} M

6-33. (a) 1.01×10^{-2} M; % $CaSO_4(aq) = 50$
(b) 7.6×10^{-3} M; % $CaSO_4(aq) = 66$

6-34. (a) 1.8×10^{-6} M (b) 2.5×10^{-13} M (c) 2.3×10^{-6} M

6-36. (a) 3×10^{-8} M (b) 3×10^{-11} M

6-38. (a) 0.7 M (b) 2×10^{-3} M

6-40. (a) 8.7×10^{-2} M (b) 8.7×10^{-3} M (c) 1.0×10^{-3} M (d) 9.0×10^{-4} M (e) 1.2×10^{-3} M

6-42. 8.3×10^{-2} M

Chapter 7

7-1. In the Volhard method for Cl^-, it is necessary to filter or otherwise remove the AgCl from the solution before back-titration with KSCN because AgCl is more soluble than AgSCN. As a consequence, SCN^- ion present during the back-titration reacts with the precipitated AgCl and gives a fading end point:
$AgCl(s) + SCN^- \rightleftharpoons AgSCN(s) + Cl^-$
The filtration step is unnecessary when Br^- ion is being determined because AgBr is less soluble than AgSCN and thus does not react with SCN^- ion.

7-2. (a) Before the equivalence point in a Br^- titration, the colloidal particles of AgBr are negatively charged because of the adsorption of excess Br^- ions. This charge attracts InH^+ ions to the surface of the precipitate, thus imparting a color to it. Beyond the equivalence point, the solid particles develop a positive charge due to the adsorption of Ag^+. Now the indicator ions are repelled into the solution and give a color to the liquid phase.
(b) Here, the reverse situation obtains. Before the equivalence point, the InH^+ ions are repelled from the surface of the solid by adsorbed Ag^+ ions and impart a color to the solution. Beyond the equivalence point, Cl^- ions are adsorbed, which attract the indicator ions into the counter-ion layer, thus coloring the solid.

7-3. Cationic adsorption indicators are used advantageously with precipitation titrations that must be carried out in acidic solutions. In such solutions, the indicator is in its cationic form and is thus attracted to or repelled from the charged surface of the precipitate. In contrast, an anionic adsorption indicator exists largely as the uncharged HIn form in acidic solution and is not adsorbed or repelled by the solid. Therefore HIn has no indicator function.

7-4. $2Ag^+ \text{ (excess)} + C_2O_4^{2-} \xrightarrow{\text{neutral soln}} Ag_2C_2O_4(s)$ (solid removed by filtration)
$Ag^+ + SCN^- \xrightarrow{\text{acidic soln}} AgSCN(s)$ (titration of excess $AgNO_3$ with standard KSCN)
$Fe^{3+} + SCN^- \rightleftharpoons FeSCN^{2+}$ (indicator reaction)
Since the titration of excess Ag^+ must be carried out in the presence of acid

to prevent precipitation of Fe^{3+} ions, the $Ag_2C_2O_4$ must be filtered to prevent its redissolving in the acidic medium.

7-5. In the presence of adsorbed fluoresceinate ions, the rate of photodecomposition of AgCl is enhanced to the point where errors are incurred. Thus, the titration must be carried out in subdued light.

7-6. In the Mohr method, it is desirable to standardize the $AgNO_3$ against NaCl in order to correct for the $AgNO_3$ reagent required to produce an observable amount of red Ag_2CrO_4.

7-7. Potassium is determined by precipitation with an excess of a standard solution of sodium tetraphenyl boron. An excess of standard $AgNO_3$ is then added to precipitate the excess tetraphenyl boron ion. The excess $AgNO_3$ is then titrated with a standard solution of SCN^-. The reactions are

$K^+ + B(C_6H_5)_4^- \rightleftharpoons KB(C_6H_5)_4(s)$ (measured excess $B(C_6H_5)_4^-$)

$B(C_6H_5)_4^- + Ag^+ \rightleftharpoons AgB(C_6H_5)_4(s)$ (measured excess $AgNO_3$)

The excess $AgNO_3$ is then determined by a Volhard titration with KSCN.

7-8. $Pb^{2+} + Cl^- + F^- \rightleftharpoons PbClF(s)$ (neutral solution)

$PbClF(s) + H^+ \rightleftharpoons Pb^{2+} + Cl^- + HF$ (acidic solution)

$Ag^+ + Cl^- \rightleftharpoons AgCl(s)$ (excess standard $AgNO_3$)

$Ag^+ + SCN^- \rightleftharpoons AgSCN(s)$ (standard KSCN)

$Fe^{3+} + SCN^- \rightleftharpoons FeSCN^+$ (indicator reaction)

7-9. (a) 0.08368 M (c) 0.1558 M (e) 0.3130 M

7-11. 28.30%

7-13. 32.79%

7-15. 116.7 mg

7-16. 44.41%

7-18. 15.60 mg/tablet

7-20. One of the seven chlorine atoms is titrated.

7-23. 18.9 ppm

7-25. 0.4348%

7-26. 10.60% Cl^-, 55.65% ClO_4^-

7-28. 73.22% KCl, 26.78% K_2SO_4

7-29. (a)

Vol NH_4SCN, mL	$[Ag^+]$	$[SCN^-]$	pAg
30.00	9.1×10^{-3}	1.2×10^{-10}	2.04
40.00	3.8×10^{-3}	2.9×10^{-10}	2.42
49.00	3.4×10^{-4}	3.2×10^{-9}	3.47
50.00	1.05×10^{-6}	1.05×10^{-10}	5.98
51.00	3.3×10^{-9}	3.3×10^{-4}	8.48
60.00	3.7×10^{-10}	2.9×10^{-3}	9.43
70.00	2.1×10^{-10}	5.3×10^{-3}	9.68

(c)

Vol NaCl, mL	$[Ag^+]$	$[Cl^-]$	pAg
10.00	3.75×10^{-2}	4.85×10^{-9}	1.43
20.00	1.50×10^{-2}	1.21×10^{-8}	1.82
29.00	1.27×10^{-3}	1.43×10^{-7}	2.90
30.00	1.35×10^{-5}	1.35×10^{-5}	4.87
31.00	1.48×10^{-7}	1.23×10^{-3}	6.83
40.00	1.70×10^{-8}	1.07×10^{-2}	7.77
50.00	9.71×10^{-9}	1.88×10^{-2}	8.01

(e)

Vol Na_2SO_4, mL	$[Ba^{2+}]$	$[SO_4^{2-}]$	pBa
0.00	2.50×10^{-2}	0.0	1.60
10.00	1.00×10^{-2}	1.3×10^{-8}	2.00
19.00	8.48×10^{-4}	1.5×10^{-7}	3.07
20.00	1.1×10^{-5}	1.1×10^{-5}	4.94
21.00	1.6×10^{-7}	8.20×10^{-4}	6.80
30.00	1.8×10^{-8}	7.14×10^{-3}	7.74
40.00	1.0×10^{-8}	1.25×10^{-2}	7.98

7-30.

Vol $AgNO_3$, mL	$[Ag^+]$	pAg
5.00	1.63×10^{-11}	10.79
40.00	7.21×10^{-7}	6.14
45.00	2.63×10^{-3}	2.58

7-31.

Vol NaCl, mL	$[Hg_2^{2+}]$	pHg_2
10.00	3.33×10^{-2}	1.48
30.00	9.09×10^{-3}	2.04
40.00	6.9×10^{-7}	6.16
50.00	5.5×10^{-15}	14.26

Chapter 8

8-1. The initial pH of the NH_3 solution is less than that of the solution containing NaOH. With the first addition of titrant, the pH of the NH_3 solution decreases rapidly and then levels off and becomes nearly constant throughout the middle part of the titration. In contrast, additions of standard acid to the NaOH solution cause the pH of the NaOH solution to decrease gradually and nearly linearly until the equivalence point is approached. The equivalence-point pH for the NH_3 solution is well below 7, whereas that for the NaOH solution is exactly 7. Beyond the equivalence point, the two curves are identical.

8-3. Physiological fluids, such as blood, must remain more or less constant in pH for an animal to survive. Acidic and basic compounds are constantly being introduced into these fluids as a consequence of the intake of food and air. The buffers in the fluids maintain a constant pH despite these additions.

8-5. Mixture (a) has the greater buffer capacity because the concentration of buffering reagents is greater.

8-7. The equivalence point is the theoretical point in a titration at which the amount of reagent added is chemically equivalent to the amount of analyte. The end point is that point where an observable physical change related to the condition of chemical equivalence occurs.

8-9. The standard reagents in neutralization titrations are always strong acids or strong bases because the reactions with these types of reagents are more complete than are the reactions with their weaker counterparts. Sharper end points are the consequence of this difference.

8-11. 3.25

8-12. (a) NaOCl (c) hydrazine

8-13. (a) Formic acid/sodium formate (c) NH_3/NH_4Cl

8-14. 7.47 at 0°C; 6.63 at 50°C; 6.16 at 100°C

8-16. −1.12

8-18. (a) 1.05 (b) 1.05 (c) 1.81 (d) 1.81 (e) 12.60

8-20. 6.62

8-22. (a) $[H_3O^+] = 0.0500$; pH = 1.30 (b) $a_{H_3O^+} = 0.430$; pH = 1.37

8-24. (a) 1.94 (b) 2.45 (c) 3.52

8-26. (a) 0.273 (b) 1.03 (c) 2.96

8-27. (a) 11.12 (b) 10.61 (c) 9.53

8-28. (a) 5.12 (b) 5.62 (c) 6.62

8-30. (a) 10.26 (b) 9.76 (c) 8.75

8-32. (a) 2.41 (b) 8.35 (c) 12.35 (d) 3.85

8-34. (a) 3.85 (b) 4.06 (c) 2.64 (d) 2.08

8-36. (a) 0.00 (b) 1.00 (c) −1.00 (d) −0.509 (e) 0.500 (f) −0.003 (g) 0.000

8-37. (a) −5.00 (b) −0.079 (c) −0.097 (d) −1.124 (e) −3.273 (f) −0.176 (g) −0.017

8-38. (a) 5.00 (b) 0.097 (c) 0.079 (d) 1.032 (e) 3.37 (f) 0.176 (g) 0.017

8-40. 15.2 g

8-42. 196 mL

8-44. Equiv-point pH = 8.14; cresol purple

8-46.

Vol, mL	pH	Vol, mL	pH
0.00	13.00	49.00	11.00
10.00	12.82	50.00	7.00
25.00	12.52	51.00	3.00
40.00	12.05	55.00	2.32
45.00	11.72	60.00	2.04

8-47.

	(a)	(c)
Vol, mL	pH	pH
0.00	2.16	3.12
5.00	2.49	4.28
15.00	2.95	4.86
25.00	3.31	5.23
40.00	3.90	5.83
45.00	4.25	6.18
49.00	4.99	6.92
50.00	8.00	8.96
51.00	11.00	11.00
55.00	11.68	11.68
60.00	11.96	11.96

8-48.

Vol, mL	pH	Vol, mL	pH
0.00	11.12	49.00	7.56
5.00	10.19	50.00	5.27
15.00	9.61	51.00	3.00
25.00	9.25	55.00	2.32
40.00	8.64	60.00	2.04
45.00	8.29		

8-49.

	(a)	(c)
Vol, mL	pH	pH
0.00	2.80	4.26
5.00	3.65	6.57
15.00	4.23	7.15
25.00	4.60	7.52
40.00	5.20	8.12
49.00	6.29	9.21
50.00	8.65	10.11
51.00	11.00	11.00
55.00	11.68	11.68
60.00	11.96	11.96

Chapter 9

9-1. An amphiprotic substance is one that is capable of behaving as both an acid and a base.

9-3. (a) Very slightly acidic (e) Basic
(c) Basic (g) Acidic

9-5. Equiv-point pH $\simeq$ 9.2; $\therefore$ phenolphthalein

9-7. $\alpha_0 = [H_3O^+]^3/D$ $\quad \alpha_2 = K_1K_2[H_3O^+]/D$
$\alpha_1 = K_1[H_3O^+]^2/D$ $\quad \alpha_3 = K_1K_2K_3/D$
where $D = [H_3O^+]^3 + K_1[H_3O^+]^2 + K_1K_2[H_3O^+] + K_1K_2K_3$

9-8. (a) $\dfrac{[H_3AsO_4][HAsO_4^{2-}]}{[H_2AsO_4^-]^2} = \dfrac{K_2}{K_1} = \dfrac{1.05 \times 10^{-7}}{6.0 \times 10^{-3}} = 1.8 \times 10^{-5}$

9-9. $K = \dfrac{[NH_3][HOAc]}{[NH_4^+][OAc^-]} = \dfrac{K_w}{K_aK_b} = 3.25 \times 10^{-5}$

9-10. (a) Equiv-point pH ≃ 8.3; cresol purple
(c) Equiv-point pH ≃ 8.4; cresol purple
(e) Equiv-point pH ≃ 5.8; methyl red or bromocresol purple
(g) Equiv-point pH ≃ 9.8; phenolphthalein or thymolphthalein

9-11. (a) 4.23 (b) 1.16 (c) 2.07 (d) 12.78 (e) 11.35 (f) 8.52

9-13. (a) 9.92 (b) 2.95 (c) 4.56 (d) 8.39

9-15. (a) 1.54 (b) 1.99 (c) 12.07 (d) 12.01

9-17. (a) $[HSO_3^-]/[SO_3^{2-}] = 15.6$
(b) $[HCit^{2-}]/[Cit^{3-}] = 2.5$
(c) $[HM^-]/[M^{2-}] = 0.498$
(d) $[HT^-]/[T^{2-}] = 0.0232$

9-19. (a) 2.06 (b) 7.20 (c) 10.62 (d) 2.09 (e) 2.01

9-21. (a) 2.11 (b) 7.37

9-23. Mix 442 mL of 0.300 M Na_2CO_3 with 558 mL of 0.2000 M HCl.

9-26. 50.3 g

9-28.

mL Reagent	(a) pH	(c) pH
0.00	11.65	0.96
12.50	10.32	1.26
24.00	8.95	1.61
25.00	8.35	1.64
26.00	7.73	1.67
37.50	6.35	2.14
45.00	5.75	2.65
49.00	4.97	3.40
50.00	3.83	7.31
51.00	2.70	11.30
60.00	1.74	12.26

9-29.

Vol $HClO_4$, mL	pH	Vol $HClO_4$, mL	pH
0.00	13.00	35.00	8.11
10.00	12.70	44.00	6.84
20.00	12.16	45.00	4.74
24.00	11.43	46.00	2.68
25.00	10.42	50.00	2.00
26.00	9.39		

9-31.

	pH	α_0	α_1	α_2	α_3
(a)	2.00	0.899	0.101	3.94×10^{-3}	
	6.00	1.82×10^{-4}	0.204	0.796	
	10.00	2.28×10^{-12}	2.56×10^{-5}	1.000	
(c)	2.00	0.931	6.93×10^{-2}	1.20×10^{-4}	4.82×10^{-9}
	6.00	5.32×10^{-5}	3.96×10^{-2}	0.685	0.275
	10.00	1.93×10^{-16}	1.44×10^{-9}	2.49×10^{-4}	1.000
(e)	2.00	0.500	0.500	1.30×10^{-5}	
	6.00	7.94×10^{-5}	0.794	0.206	
	10.00	3.84×10^{-12}	3.84×10^{-4}	1.000	

Chapter 10

10-1. When carbon dioxide is dissolved in water, it is largely present as CO_2 and H_2CO_3, which upon heating decomposes to give CO_2 (and H_2O). The CO_2 is not strongly bonded by water molecules, however, and thus is readily volatilized from aqueous media. Gaseous HCl molecules, on the other hand, are fully dissociated into H_3O^+ and Cl^- when dissolved in water; neither of these species is volatile.

10-3. Primary-standard Na_2CO_3 can be obtained by heating primary-standard-grade $NaHCO_3$ for about 1 h at 270 to 300°C. The reaction is
$NaHCO_3(s) \longrightarrow Na_2CO_3(s) + H_2O(g) + CO_2(g)$

10-5. The carbonate error arises in acid/base titrations whenever the pH range of the indicator used is different from that of the indicator used for reagent standardization. For example, if an indicator with an acidic range is used in a standardization of carbonate containing base, each carbonate ion present reacts with two hydronium ions. If this solution is then used with an indicator with a basic range of color change, the carbonate present consumes but a single proton. Thus the molarity of the base has been effectively lowered by the change in indicator. This type of carbonate error is avoided by using carbonate-free solutions of base or by using the same indicator for standardization and analyses. When very dilute reagent solutions are being employed, the equilibrium concentration of carbon dioxide in water may provide sufficient titratable protons to affect the results of acid/base titrations. This type of carbonate error is avoided by boiling and cooling the water to be used for preparing reagents and dissolving samples.

10-7. Treat an acidic solution of the sample with a good reducing agent, which converts the HNO_3 to NH_4^+. Then carefully make the solution basic as in the Kjeldahl method, and distill the NH_3 into a standard solution of acid. Back-titrate the excess acid with standard base.

10-9. (1) The autoprotolysis constant; the smaller, the more complete the reaction. (2) The acidic or basic properties; when a weak acid is being titrated, a basic solvent leads to more complete reaction. When a weak base is being titrated, an acidic solvent is better. (3) The dielectric solvent; the higher, the more complete the reaction.

10-11. For the strong acid, pH = 2.00; for the strong base, pH = 13.30.

10-13. (a) Autoprotolysis is the process in which an amphiprotic species undergoes self-ionization or self-dissociation. As an example,
$NH_3 + NH_3 \rightleftharpoons NH_4^+ + NH_2^-$
(c) An acidic amphiprotic solvent is one that can behave as both an acid and a base. Its acidic properties are considerably stronger than its basic properties, however. Examples are anhydrous acetic acid, formic acid, and sulfuric acid.
(e) The dielectric constant of a liquid provides a measure of its ability to separate ions of opposite charge. The dielectric constant of water is very large (78.5); thus, many salts are completely dissociated in this medium. In contrast, the dielectric constant of acetic acid is small (6.2), and few if any species are completely dissociated in this solvent.

10-14. (a) $2H_2O \rightleftharpoons H_3O^+ + OH^-$ $\quad K_S = K_w = [H_3O^+][OH^-]$
(c) $2H_2SO_4 \rightleftharpoons H_3SO_4^+ + HSO_4^-$ $\quad K_S = [H_3SO_4^+][HSO_4^-]$
(e) $2HCOOH \rightleftharpoons HCOOH_2^+ + HCOO^-$ $\quad K_S = [HCOOH_2^+][HCOO^-]$
(g) $2HCN \rightleftharpoons H_2CN^+ + CN^-$ $\quad K_S = [H_2CN^+][CN^-]$

10-15. (a) 0.1388 M (b) 0.1500 M

10-17. (a) Dissolve 17 g KOH in water and dilute to about 2.0 L.
(b) Dissolve 9.5 g $Ba(OH)_2 \cdot 8H_2O$ in water and dilute to about 2.0 L.
(c) Dilute 120 mL of the reagent to about 2.0 L.
(d) Dilute 53.99 g of the HCl to about 2.00 L.

10-19. (a) 0.09961 (b) 0.00049

10-21. (a) 0.28 to 0.36 g (c) 0.85 to 1.1 g (e) 0.17 to 0.22 g

10-22. 0.05950 M

10-24. 0.1217 g H_2T/100 mL

10-26. 76.97 g/equiv

10-28. 7.079%
10-30. 19.0%
10-32. 3.35×10^3 ppm
10-34. 1.37×10^{-2} M
10-36. 13.86%
10-38. 0.7065% acetaldehyde, 0.6453% ethyl acetate
10-40. 6.333%
10-41. 69.84% KOH, 21.04% K_2CO_3, 9.12% H_2O
10-43. (a) 18.15 mL (b) 36.30 mL (c) 38.28 mL (d) 12.27 mL
10-45. (a) 4.314 mg NaOH/mL
(b) 7.985 mg Na_2CO_3/mL, 4.358 mg $NaHCO_3$/mL
(c) 4.396 mg NaOH/mL, 3.455 mg Na_2CO_3/mL
(d) 8.215 mg Na_2CO_3/mL
(e) 13.46 mg $NaHCO_3$/mL
10-47. (a) 2.422 mg HCl/mL, 2.296 mg H_3PO_4/mL
(b) 1.390 mg HCl/mL
(c) 9.658 mg NaH_2PO_4/mL
(d) 6.270 mg H_3PO_4/mL
10-49. (a)

Vol, mL	pH	Vol, mL	pH
0.00	1.30	25.0	9.55
12.5	1.70	25.1	15.22
24.0	2.87	26.0	16.22
24.9	3.87	30.0	16.89

(b) In ethanolic solution, $\Delta pH = 11.35$; in water, $\Delta pH = 6.25$.
10-51. (a) pH (water) = 8.30; pH (ethanol) = 11.40
(b) pH (water) = 4.90; pH (ethanol) = 6.00
(c) pH (water) = 3.31; pH (ethanol) = 3.85
10-53.

mL base	pH		mL base	pH	
	C_2H_5OH	H_2O		C_2H_5OH	H_2O
0.00	3.35	2.80	49.90	8.40	7.29
10.00	5.10	4.00	50.00	11.75	8.65
25.00	5.70	4.60	50.10	15.10	10.00
40.00	6.30	5.20	51.00	16.09	11.00
49.00	7.40	6.29	60.00	17.06	11.96

Chapter 11

11-1. (a) A chelate is a cyclic complex consisting of a metal ion and a reagent that contains two or more electron donor groups located in such a position that they can bond with the metal ion to form a heterocyclic ring.
(c) A ligand is a species that contains one or more electron-pair donor groups that tend to form bonds with metal ions.
(e) A conditional formation constant is an equilibrium constant for the reaction between a metal ion and a complexing agent that applies only when the pH and/or the concentration of other complexing ions is carefully specified.
(g) The hardness of water is the concentration of calcium carbonate that is equivalent to the total concentration of all multivalent metal carbonates in the water.

11-2. Three general methods for performing EDTA titrations are (1) direct titration, (2) back-titration, and (3) displacement titration. Method 1 is simple and rapid and requires but one standard reagent. Method 2 is advantageous for those metals that react so slowly with EDTA as to make direct titration inconvenient. In addition, this procedure is useful for cations for which a satisfactory indicator is not available. Finally, it is useful for analyzing samples that contain anions that form sparingly soluble precipitates with the analyte under the analytical conditions. Method 3 is particularly useful in situations where no satisfactory indicators are available for direct titration.

11-4. Multidentate ligands offer the advantage that they usually form more stable complexes than do unidentate ligands. Furthermore, they often form but a single complex with the cation, which simplifies their titration curves and makes end-point detection easier.

11-5. (a) $Ag^+ + S_2O_3^{2-} \rightleftharpoons AgS_2O_3^-$ $\qquad \beta_1 = \dfrac{[AgS_2O_3^-]}{[Ag^+][S_2O_3^{2-}]}$

$AgS_2O_3^- + S_2O_3^{2-} \rightleftharpoons Ag(S_2O_3)_2^{3-}$ $\qquad \beta_2 = \dfrac{[Ag(S_2O_3)_2^{3-}]}{[AgS_2O_3^-][S_2O_3^{2-}]}$

(c) $Cd^{2+} + NH_3 \rightleftharpoons CdNH_3^{2+}$ $\qquad \beta_1 = \dfrac{[CdNH_3^{2+}]}{[Cd^{2+}][NH_3]}$

$CdNH_3^+ + NH_3 \rightleftharpoons Cd(NH_3)_2^{2+}$ $\qquad \beta_2 = \dfrac{[Cd(NH_3)_2^{2+}]}{[CdNH_3^{2+}][NH_3]}$

$Cd(NH_3)_2^{2+} + NH_3 \rightleftharpoons Cd(NH_3)_3^{2+}$ $\qquad \beta_3 = \dfrac{[Cd(NH_3)_3^{2+}]}{[Cd(NH_3)_2^{2+}][NH_3]}$

$Cd(NH_3)_3^{2+} + NH_3 \rightleftharpoons Cd(NH_3)_4^{2+}$ $\qquad \beta_4 = \dfrac{[Cd(NH_3)_4^{2+}]}{[Cd(NH_3)_3^{2+}][NH_3]}$

11-6. The overall formation constant is equal to the product of the individual stepwise constants. Thus the overall constant for the formation of $Cd(NH_3)_3^{2+}$ in Problem 11-5c is

$$K_3 = \beta_1\beta_2\beta_3 = \frac{[Cd(NH_3)_3^{2+}]}{[Cd^{2+}][NH_3]^3}$$

which is the equilibrium constant for the reaction

$Cd^{2+} + 3NH_3 \rightleftharpoons Cd(NH_3)_3^{2+}$

11-7. The MgY^{2-} is added to ensure a sufficient analytical concentration of Mg^{2+} to provide a sharp end point with Eriochrome Black T indicator.

11-8. (a) 0.0452 M (b) 2.44 mg Ag^+ mL (c) 1.62 mg Cu_2O/mL

11-10. (a) 0.0193 M (b) 0.776 mg MgO/mL (c) 1.54 mg Fe_2O_3/mL

11-12. 0.01032 M

11-14. 0.01005 M

11-16. 323.9 mg Ca, 256.4 mg Mg^{2+}

11-18. 94.40%

11-20. 31.48% NaBr, 48.57% $NaBrO_3$

11-22. 0.08432 M

11-24. 7.515% Pb; 9.456% Zn; 4.304% Mg

11-26. 8.517% Pb; 24.85% Zn; 64.07% Cu; 2.56% Sn

11-27. (a) 1.4×10^9 (b) 3.3×10^{11} (c) 2.2×10^{13}

11-29. (a) 8.15×10^{12} (c) 4.93×10^9
(b) 1.33×10^{14} (d) 8.06×10^{10}

11-31. $K'_{Sr} = 3.66 \times 10^8$

Vol, mL	pSr	Vol, mL	pSr
0.00	2.00	25.00	5.37
10.00	2.30	25.10	6.16
24.00	3.57	26.00	7.16
24.90	4.57	30.00	7.86

11-33. $K''_{Co} = 5.74 \times 10^{13}$

Vol, mL	pCo	Vol, mL	pCo
0.00	2.66	20.00	8.96
5.00	2.86	21.00	13.72
10.00	3.10	30.00	14.72
19.00	4.20		

Chapter 12

12-1. (a) Oxidation is the process in which electrons are abstracted from a species, thus converting it to a higher oxidation state.

(c) A galvanic cell is an electrochemical cell that is operated in such a way as to produce an electrical current. It consists of two conductors immersed in an electrolyte medium.

(e) An anode is one of the two electrodes in an electrochemical cell. It is the electrode at which chemical oxidation takes place, thus introducing electrons into the exterior conductor that connects the anode to the cathode.

(g) A liquid junction in an electrochemical cell is the interface between two dissimilar electrolyte solutions. A potential develops at this interface.

(i) The standard hydrogen electrode consists of a piece of platinized platinum immersed in a solution that is kept saturated with hydrogen gas (the gas is bubbled continuously through the solution); the hydrogen ion activity of this solution is exactly unity, and the gas is at a pressure of 1.00 atm. The potential of the standard hydrogen electrode is assigned a value of unity at all temperatures.

(k) The standard electrode potential for a half-reaction is the potential of a *cell* consisting of a cathode at which that half-reaction is occurring and a standard hydrogen electrode behaving as the anode. The activities of all participants in the half-reaction are specified as having a value of unity. The additional specification that the standard hydrogen electrode is the anode implies that the standard potential for a half-reaction is always a *reduction potential.*

(m) The formal potential of a half-reaction is the potential of the system (measured against the SHE) when the concentration of each solute participating in the half-reaction is exactly 1 M and the concentrations of all other constituents of the solution are carefully specified.

12-3. The standard hydrogen electrode is never used in the laboratory because it is impossible to prepare solutions that have a hydrogen activity of exactly unity. At ionic strengths that approach unity, the activity of the hydronium ion is not known and cannot be computed from theory. Thus, practical hydrogen electrodes employ considerably lower hydronium ion concentrations, where the activity coefficient is known. Potential data with such electrodes can then be extrapolated to give potentials against the *hypothetical* standard hydrogen electrode.

12-5. The potential in the presence of base is more negative because the nickel concentration in this solution is far less than 1 M. Consequently, the driving force for the reduction of Ni(II) to the metallic state is also far less, and the electrode potential is significantly more negative. (In fact, the electrode potential for the reaction $Ni(OH)_2(s) + 2e \rightleftharpoons Ni(s) + 2OH^-$ is -0.72 V.)

12-7. The first standard potential is for a solution that is saturated with I_2, which has an $I_2(aq)$ activity significantly less than unity. The second potential is for a *hypothetical* half-cell in which the $I_2(aq)$ activity is unity. Such a half-cell, if it existed, would have a greater potential since the driving force for the reduction would be greater at the higher I_2 concentration. The second half-cell potential, although hypothetical, is nevertheless useful for calculating electrode potentials for solutions that are not saturated in I_2.

12-9. $SCE||CuSO_4(1.00 \times 10^{-4}\ M),NH_3(0.100\ M)|Cu$

$$\log K_f = 2\left(E^0_{Cu^{2+}} - E_{cell} - E_{SCE} + \frac{0.0592}{2}\log\frac{c_{CuSO_4}}{c^2_{NH_3}}\right)/0.0592$$

12-10. (a) $Tl^{3+} + 2Ag(s) + 3Br^- \rightleftharpoons TlBr(s) + 2AgBr(s)$

(b) $2Fe^{2+} + UO_2^{2+} + 4H^+ \rightleftharpoons 2Fe^{3+} + U^{4+} + 2H_2O$

(c) $N_2(g) + 2H_2(g) + H^+ \rightleftharpoons N_2H_5^+$
(d) $Cr_2O_7^{2-} + 9I^- + 14H^+ \rightleftharpoons 2Cr^{3+} + 3I_3^- + 7H_2O$
(e) $H_2O_2 + 2Ce^{4+} \rightleftharpoons 2Ce^{3+} + O_2(g) + 2H^+$
(f) $IO_3^- + 5I^- + 6H^+ \rightleftharpoons 3I_2(s) + 3H_2O$

12-12. (a) Oxidizing agent Tl^{3+}; $Tl^{3+} + Br^- + 2e \rightleftharpoons TlBr(s)$
Reducing agent Ag; $Ag + Br^- \rightleftharpoons AgBr(s) + e$
(b) Oxidizing agent UO_2^{2+}; $UO_2^{2+} + 4H^+ + 2e \rightleftharpoons U^{4+} + 2H_2O$
Reducing agent Fe^{2+}; $Fe^{2+} \rightleftharpoons Fe^{3+} + e$
(c) Oxidizing agent N_2; $N_2 + 5H^+ + 4e \rightleftharpoons N_2H_5^+$
Reducing agent H_2; $H_2 \rightleftharpoons 2H^+ + 2e$
(d) Oxidizing agent $Cr_2O_7^{2-}$; $Cr_2O_7^{2-} + 14H^+ + 6e \rightleftharpoons 2Cr^{3+} + 7H_2O$
Reducing agent I^-; $3I^- \rightleftharpoons I_3^- + 2e$
(e) Oxidizing agent Ce^{4+}; $Ce^{4+} + e \rightleftharpoons Ce^{3+}$
Reducing agent H_2O_2; $H_2O_2 \longrightarrow O_2 + 2H^+ + 2e$
(f) Oxidizing agent IO_3^-; $IO_3^- + 6H^+ + 5e \rightleftharpoons \frac{1}{2}I_2(s) + 3H_2O$
Reducing agent I^-; $I^- \longrightarrow \frac{1}{2}I_2(s) + e$

12-14. (a) $1.5(\pm 1) \times 10^{46}$
(b) $1.7(\pm 0.1) \times 10^{-15}$
(c) $3(\pm 4) \times 10^{-16}$
(d) $3(\pm 7) \times 10^{80}$
(e) $4(\pm 3) \times 10^{25}$
(f) $6.1(\pm 1) \times 10^{55}$ (In 1M H_2SO_4)

12-16. (a) 0.813 V (b) 0.747 V (c) 0.351 V (d) 0.664 V

12-18. (a) 0.795 V (b) −0.693 V (c) 0.600 V (d) 0.39 V (e) 0.437 V (f) 0.171 V

12-20. (a) Anode; 0.235 V (b) Cathode; 0.163 V (c) Anode; 0.015 V (d) Anode; 0.012 V (e) Cathode; 0.310 V (f) Cathode; 0.359 V

12-22. (a) To the right (b) To the left (c) To the left (d) To the right (e) To the right (f) To the right

12-24. (a) 0.490 V; galvanic (b) −0.117 V; electrolytic (c) −0.077 V; electrolytic (d) 0.806 V; galvanic (e) −0.591 V; electrolytic (f) 1.23 V; galvanic

12-26. 0.390 V

12-28. $1.3(\pm 0.1) \times 10^{-6}$

12-30. −1.25 V

12-32. $1.62(\pm 0.06) \times 10^{-14}$

12-34. $4.5(\pm 0.4) \times 10^{16}$

12-36. $4.2(\pm 0.3) \times 10^{14}$

12-38. 1.9×10^{-8}

Chapter 13

13-1. The half-reactions for most indicators involve hydrogen ions. Therefore, their electrode potentials are pH-dependent. Thus, the range of potentials over which a color change occurs is also pH-dependent.

13-3. Symmetric titration curves are observed when the stoichiometric ratio of oxidant to reductant is unity and the hydrogen ion concentration is held constant.

13-5. The titration curve would exhibit three equally spaced end points. The potentials before the first end point would be somewhat negative. Between the first and second end points, the potentials would lie in the range from +0.3

to +0.4 V. Between the second and third end points, they would be in a range around +1.0 V.

13-6. (a) 0.452 to 0.511 V (c) 0.215 to 0.274 V

13-7. (a) Diphenylamine or *p*-ethoxychrysoidine
(c) Methylene blue

13-8. (a) −0.020 V (b) 1.25 V (c) 1.25 V

13-10. (a) $4.7(\pm 0.4) \times 10^{17}$
(b) $6(\pm 5) \times 10^{9}$
(c) $2(\pm 7) \times 10^{96}$

13-12.

	E, V		
Vol, mL	(a)	(c)	(e)
10.00	−0.292	0.32	0.316
25.00	−0.256	0.36	0.334
49.00	−0.156	0.46	0.384
49.90	−0.097	0.52	0.414
50.00	+0.017	0.95	1.17
50.10	+0.074	1.17	1.48
51.00	+0.104	1.20	1.49
60.00	+0.133	1.23	1.50

Chapter 14

14-1. Only in the presence of Cl^- ion is Ag a good enough reducing agent to be very useful for prereductions. In the presence of Cl^- ion, the half-reaction in the Walden reductor is
$Ag(s) + Cl^- \rightleftharpoons AgCl(s) + e$
The excess HCl increases the tendency of this reaction to occur by the common-ion effect.

14-3. $UO_2^{2+} + 2Ag(s) + 4H^+ + 2Cl^- \rightleftharpoons U^{4+} + 2AgCl(s) + 2H_2O$

14-5. Standard $KMnO_4$ solutions are seldom used to titrate solutions containing HCl because of the tendency of MnO_4^- to oxidize Cl^-, thus causing an overconsumption of MnO_4^-.

14-7. $2MnO_4^- + 3Mn^{2+} + 2H_2O \longrightarrow 5MnO_2(s) + 4H^+$

14-9. Solutions of $K_2Cr_2O_7$ are used extensively for the back-titration of solutions of Fe^{2+} when the latter is being used as a standard reductant for the determination of oxidizing agents.

14-11. The solution becomes stronger because of air-oxidation of the excess I^- present, which increases the I_3^- concentration:
$6I^- + O_2(g) + 4H^+ \longrightarrow 2I_3^- + 2H_2O$

14-13. Standard solutions of reductants find somewhat limited use because of their susceptibility to air-oxidation.

14-15. $I_2 + H_3AsO_3 + H_2O \rightleftharpoons 2I^- + H_3AsO_4 + 2H^+$
In order for this reaction to go to completion, it is necessary to remove H^+ ions as they are formed by buffering the solution to a basic pH.

14-17. Starch is decomposed in the presence of high concentrations of iodine to give products that do not behave as satisfactory indicators. This reaction is prevented by not adding the starch until the iodine concentration is very small.

14-19. $2MnO_4^- + 3Mn^{2+} + 2H_2O \longrightarrow \underset{\text{brown}}{5MnO_2(s)} + 4H^+$

14-21. $BrO_3^- + 6I^- + 6H^+ \longrightarrow Br^- + 3I_2 + 3H_2O$
$I_2 + 2S_2O_3^{2-} \longrightarrow 2I^- + S_4O_6^{2-}$

14-23. $H_5IO_6 + 2I^- + H^+ \longrightarrow I_2 + IO_3^- + 3H_2O$ (in slightly basic solution)
$H_5IO_6 + 7I^- + 7H^+ \longrightarrow 4I_2 + 6H_2O$ (in acidic solution)
$I_2 + 2S_2O_3^{2-} \longrightarrow 2I^- + S_4O_6^{2-}$

14-25. The sample is refluxed with anhydrous methanol until the water has been removed. The resulting solution is then titrated with the reagent.

14-27. (a) $2Mn^{2+} + 5S_2O_6^{2-} + 8H_2O \longrightarrow 10SO_4^{2-} + 2MnO_4^- + 16H^+$
(b) $NaBiO_3(s) + 2Ce^{3+} + 4H^+ \longrightarrow BiO^+ + 2Ce^{4+} + 2H_2O + Na^+$
(c) $H_2O_2 + U^{4+} \longrightarrow UO_2^{2+} + 2H^+$
(d) $V(OH)_4^+ + Ag(s) + Cl^- + 2H^+ \longrightarrow VO^{2+} + AgCl(s) + 2H_2O$
(e) $2MnO_4^- + 5H_2O_2 + 6H^+ \longrightarrow 5O_2 + 2Mn^{2+} + 8H_2O$
(f) $ClO_3^- + 6I^- + 6H^+ \longrightarrow 3I_2 + Cl^- + 3H_2O$

14-29. (a) 4 mol $H\overset{\overset{\displaystyle O}{\|}}{C}OH$ + 2 mol $H\overset{\overset{\displaystyle O}{\|}}{C}H$
(b) 2 mol $H\overset{\overset{\displaystyle O}{\|}}{C}OH$
(c) 1 mol $H\overset{\overset{\displaystyle O}{\|}}{C}OH$ + 1 mol $CH_3\overset{\overset{\displaystyle O}{\|}}{C}H$

14-31. Dissolve 25.74 g $K_2Cr_2O_7$ in water and dilute to 2.500 L.
14-33. Dissolve 3.131 g $KBrO_3$ in water and dilute to 750.0 mL.
14-35. 16.77 mg Fe_2O_3/mL
14-37. 0.06711 M
14-39. 36.48%
14-41. (a) 14.72% (b) 20.54%
14-43. 0.7013%
14-45. 0.5554 g
14-47. (a) 15.88% (b) 7.377%
14-49. (a) $MnO_4^- + 4Mn^{2+} + 20F^- + 8H^+ \longrightarrow 5MnF_4^- + 4H_2O$
(b) 35.00%
14-51. % Fe = 69.07; % Cr = 21.07
14-53. 52.81%
14-55. 12.42%
14-57. 0.371 mg H_2S/L; 0.317 mg SO_2/L
14-59. 2.635%
14-61. 19.9 mg/L
14-63. 8.260%
14-65. 3.644%

Chapter 15

15-1. An electrode of the first kind is a metal electrode used to determine the concentration of the ion derived from that metal. For example, a copper electrode is an electrode of the first kind for determining the concentration of Cu^{2+} ions in a solution. An electrode of the second kind is a metal electrode that responds to the concentration of an anion that forms a precipitate of limited solubility or a stable complex with the cation derived from that metal. For example, a silver electrode is an electrode of the second kind for Br^- ions. In this application, the analyte solution is saturated with AgBr before measurements are made.

15-2. (b) The boundary potential of a membrane electrode is the potential that develops when the membrane separates two solutions that have different concentrations of a cation or anion that the membrane binds selectively. For example, if a component of the membrane AM equilibrates with A^+ analyte ions in an aqueous solution, the following equilibria develop when the membrane is positioned between two solutions for A^+:

$$\underset{\text{membrane}_1}{A^+M^-} \rightleftharpoons \underset{\text{soln}_1}{A^+} + \underset{\text{membrane}_1}{M^-}$$

$$\underset{\text{membrane}_2}{A^+M^-} \rightleftharpoons \underset{\text{soln}_2}{A^+} + \underset{\text{membrane}_2}{M^-}$$

where the subscripts refer to the two sides of the membrane. A potential develops across this membrane if one of these equilibria proceeds farther to the right than the other, and this potential is the boundary potential. For example, if the concentration of A^+ is greater in solution 1 than in solution 2, the negative charge on side 1 of the membrane is less than that on side 2 because the equilibrium on side 1 lies farther to the left. Thus, a greater fraction of the negative charge on side 1 is neutralized by A^+.

(d) The membrane in a solid-state electrode for F^- is crystalline LaF_3, which when immersed in aqueous solution dissociates according to the equation

$LaF_3 \rightleftharpoons La^{3+} + 3F^-$

Thus, a boundary potential develops across this membrane when it separates two solutions of different F^- ion concentration. The source of this potential is described in part (b) of this answer.

15-4. The alkaline error in a glass electrode develops in solutions having a low concentration of H_3O^+ and a high concentration of an alkali metal ion. Under these circumstances, the electrode begins to respond to the alkali ion concentration as well as to the concentration of H_3O^+. A negative pH error results.

15-6. The equilibrium involved is

$$C_2H_5NH_2 + H_2O \rightleftharpoons C_2H_5NH_3^+ + OH^- \qquad K_b = \frac{[OH^-][C_2H_5NH_3^+]}{[C_2H_5NH_2]}$$

At half-neutralization,

$[C_2H_5NH_3^+] = [C_2H_5NH_2]$

and the base dissociation constant expression simplifies to

$K_b = [OH^-]$

Taking the negative logarithm of this equation yields

$-\log K_b = -\log [OH^-] = pOH = 14.00 - pH$

$K_b = \text{antilog}\,(pH - 14.00)$

15-7. (a) -0.327 V

(b) SCE||SCN^-(xM),CuSCN(sat'd)|Cu

(c) $pSCN = (E_{cell} + 0.571)/0.0592$

(d) 8.36

15-9. (a) SCE||Hg_2Cl_2(sat'd),Cl^-(xM)|Hg

$pCl = (E_{cell} - 0.024)/0.0592$

(b) SCE||Ag_2CO_3(sat'd),CO_3^{2-}(xM)|Ag

$pCO_3 = 2(E_{cell} - 0.227)/0.0592$

(c) SCE||Sn^{2+}(1.00×10^{-4} M),Sn^{4+}(xM)|Pt

$pSn(IV) = 2(0.028 - E_{cell})/0.0592$

15-11. $pCrO_4 = 6.76$

15-13. (a) 0.157 V (b) -0.026 V (c) 0.054 V

15-15. (a) 12.629 (b) 5.579

(c) 12.596 to 12.663; 5.545 to 5.612

15-17. $2.0(\pm 0.2) \times 10^{-6}$

15-19.

Vol, mL	E(vs. SCE), V	Vol, mL	E(vs. SCE), V
5.00	0.58	49.00	0.66
10.00	0.59	50.00	0.80
15.00	0.60	51.00	1.10
25.00	0.61	55.00	1.14
40.00	0.63	60.00	1.15

Chapter 16

16-1. (a) A coulometric titration is an electroanalytical method in which a constant current of known magnitude is used to generate a reagent that reacts with the analyte (in some instances, the analyte itself may react partially at the working electrode). The time required to complete the reaction between the electrogenerated reagent and the analyte is measured.

(c) Concentration polarization is encountered when the current in an electrochemical cell is limited by the rate at which reactants are brought to or removed from the surface of one or both electrodes. Under this circumstance, the current in the cell is no longer directly proportional to the cell potential.

(e) The faraday is the quantity of charge (electricity) that produces one equivalent of chemical change at an electrode. It consists of 1 mol, or 6.02×10^{23}, electrons.

(g) A controlled-cathode-potential electrolysis is one in which the cathode potential is monitored continuously against a reference electrode and the cell potential is adjusted to maintain the cathode at a fixed, predetermined level.

(i) A cathode depolarizer is a chemical species introduced into an electrochemical cell in excess to prevent undesired concentration polarization. For example, nitrate ion is often used as a cathode depolarizer to keep the cathode potential low enough to prevent hydrogen evolution.

(k) A working electrode is the electrode in an electrochemical cell at which the analyte reacts directly or indirectly.

16-2. In amperostatic coulometry, the current in the electroanalytical cell is held constant. In potentiostatic coulometry, the potential of the working electrode is maintained constant.

16-4. (1) Migration, which results from electrostatic attraction between the ions and the electrode, (2) diffusion, which results from concentration differences between the film of solution at an electrode surface and the bulk of the solution, and (3) convection, which brings ions to the electrode surface mechanically.

16-6. Kinetic polarization is often encountered when the reaction product is a gas, particularly when the electrode is a softer metal, such as mercury, zinc, or copper. It is likely to occur at low temperatures and high current densities.

16-8. In a coulometric titration, the reagent is an electric current of known strength; in a volumetric titration, the reagent is a standard solution. In the former, time is measured; in the latter, volume is determined. Both require an end point. In a coulometric method, increments of reagent are introduced by means of a switch that connects the current source to an electrochemical cell. In a volumetric titration, the flow of reagent is controlled by the stopcock of a buret.

16-10. A constant-cathode-potential electrolysis is performed by employing a reference electrode to monitor the potential of a working cathode. The potential applied across the working cathode and an anodic counter electrode is then varied in such a way that the potential of the cathode remains constant at a desired level throughout the electrolysis.

16-11. (a) -1.381 V (b) -3.46 V (c) -5.70 V (d) -5.79 V

16-13. (a) $E_1^0 - E_2^0 = 0.077$ V (c) $E_1^0 - E_2^0 = 0.223$ V

16-14. (a) 4.25×10^{-6} M (b) -0.425 V

16-16. (a) Ag deposits first.

(b) -2.316 V

(c) $E_{cathode} = 0.200$ to 0.054 V

16-19. (a) not feasible (b) feasible (c) +0.099 to −0.014 V
16-19. (a) 0.433 V (b) 0.277 V (c) −0.127 V
16-22. (a) 28.4 min (b) 27.80 min
16-24. 13.3% Cd; 5.43% Zn
16-26. 79.5 ppm
16-28. 4.062%
16-30. 2.47% CCl_4; 1.85% $CHCl_3$
16-32. 4.68%
16-34. 0.445
16-36. 2.73×10^{-4} g

Chapter 17

17-1. (a) A limiting current is a voltammetric current that is constant and independent of applied potential. Its magnitude is limited by the rate at which reactant is brought to or removed from the surface of one or both electrodes as a consequence of diffusion, migration, and convection. A diffusion current is a limiting current obtained under conditions in which migration and convection have been eliminated as mechanisms of reactant transport. The current is then limited solely by the rate of diffusion. The residual current in polarography is the small current observed in the absence of analyte. It arises from the presence of impurities and from the flow of electrons needed to charge each mercury drop.

(c) Voltammetry is an analytical method based upon current/voltage curves obtained under conditions that lead to concentration polarization. Generally, the working electrode in voltammetry is a microelectrode having a surface area of a few square millimeters. Polarography is a type of voltammetry in which the microelectrode is a dropping mercury electrode.

(e) In classical polarography, the current is monitored continuously with a device that damps the wide current fluctuations resulting from the formation and detachment of mercury drops. In current-sampled polarography, the current is measured only during the last few milliseconds of the lifetime of the drops. In order to make such measurements, it is necessary to have a drop-knocker that causes detachment of the drop at fixed time intervals. The currents obtained with each drop are then essentially constant.

17-2. The limiting current for the reduction of Cd^{2+} ion decrease as the electrolyte concentration increases; ultimately, it becomes constant in the 0.1 and 1.0 M solutions of KNO_3. The current for the 0.0 M KNO_3 solution is made up of a migration component and a diffusion component. Increasing the electrolyte concentration reduces the attractive force between the cathode and Cd^{2+}. Thus, the migration current decreases and approaches zero at the higher KNO_3 concentrations.

17-4. A residual current develops in part because of the presence of small concentrations of reducible contaminants. In addition, a small current arises as a result of the flow of electrons required to charge each drop as it forms and detaches.

17-6. Limiting currents with a rotating platinum electrode are greater than those with a dropping mercury electrode because in the former case reactants are brought to the electrode surface by convection as well as by diffusion.

17-8. (a) −0.057% (c) −0.23%

17-9. (a)

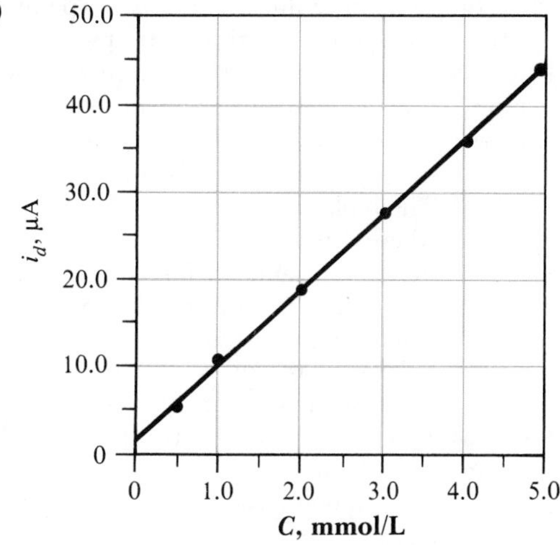

(b) $i_d = 1.437 + 8.634C$

(c) $s_r = 0.381$, $s_b = 8.2 \times 10^{-2}$

(d)

	C, mmol/L	s_c, mmol/L	$(s_c)_r$, %
(1)	0.15	0.041	27
(3)	0.73	0.034	4.6
(5)	4.47	0.035	0.8

17-10. (a) 8.72 μA mmol^{-1} L
(b) $s = 0.25$; $s_r = 2.9\%$
(c) 8.42 μA mmol^{-1} L
(d) -3.5%

17-12. For electrode A, $i_d/C = 2.29$; for electrode C, $i_d/C = 6.51$

17-14. $n = 3.6 = 4$

17-16. 2.0×10^{-5} cm^2/s

17-17. (a) 1.20×10^{-2} M (c) 6.31×10^{-3} M

17-18. The wave at +0.12 V is due to the anodic reaction
$2Hg(l) + 2Br^- \rightleftharpoons Hg_2Br_2(s) + 2e$
The wave that starts at 0.48 V is due to the anodic reaction
$Hg(l) \rightleftharpoons Hg^{2+} + 2e$
The first wave is useful for the determination of Br^-.

17-20. (a) 41.4 mg Zn/100 mL (c) 93.8 mg Zn/100 mL

Chapter 18

18-1. (a) Transitions among various electronic energy levels.
(c) Transitions from the lowest vibrational levels of the various excited electronic states to the various vibrational states of the ground state.
(e) As in part (c).

18-2. (a) The ground state is the lowest energy state of an atom, ion, or molecule and the one occupied by most of the atoms, ions, or molecules of a species at room temperature.
(c) A photon is a particle of radiant energy whose magnitude is given by $h\nu$, where h is Planck's constant and ν is the frequency of a radiation.
(e) A continuous spectrum is one produced by a heated body. Even at highest resolution, a continuous spectrum cannot be dispersed into lines.

(g) Resonance fluorescence is a type of fluorescence in which the emitted radiation is identical in frequency to the excitation frequency.

(i) $\% T = (P/P_0) \times 100$, where T is the transmittance, P_0 is the power of a beam of radiation incident upon an absorbing medium, and P is the power of the beam after it has passed through a layer of the medium having a thickness of b cm.

(j) The molar absorptivity ϵ of a species is defined by the equation $\epsilon = A/bc$, where A is the absorbance of a solution that is c molar in the absorbing species, and b is the path length in centimeters of the radiation used to measure A. The absorbance is given by $A = \log (P_0/P) = \log (1/T)$, where the terms used in defining A are found in the answer to part (i).

(k) Relaxation is a process whereby an excited species loses energy and returns to a lower energy state.

18-3. (a) 1.13×10^{18} Hz (c) 4.32×10^{14} Hz (e) 1.53×10^{13} Hz

18-4. (a) 252.75 cm (c) 286 cm

18-5. (a) 3333 to 666.7 cm^{-1} (b) 9.99×10^{13} to 1.99×10^{13}

18-7. (a) 86.3% (b) 17.2% (c) 48.1%

18-8. (a) 0.712 (b) 0.064 (c) 0.565

18-9. (a) 74.5% (b) 3.0% (c) 23.1%

18-10. (a) 1.013 (b) 0.365 (c) 0.866

18-11. (a) $\% T = 38.4$; $\epsilon = 2.38 \times 10^3$; $c = 31.2$ ppm; $a = 0.0133$
(c) $\% T = 3.77$; $\epsilon = 3.42 \times 10^4$; $c = 4.18 \times 10^{-5}$ M; $c = 10.44$ ppm
(e) $A = 0.115$; $\% T = 76.7$; $c = 1.33 \times 10^{-5}$ M; $a = 0.0138$
(g) $A = 0.316$; $\epsilon = 4.70 \times 10^4$; $c = 2.69 \times 10^{-5}$ M; $a = 0.188$
(i) $A = 0.1175$; $\epsilon = 1.58 \times 10^4$; $c = 6.76 \times 10^{-6}$ M; $c = 1.69$ ppm

18-12. 1.80×10^4

18-14. (a) 0.582 (b) 26.2% (c) 1.25×10^{-5} M

18-16. 0.0595

18-18. (a)

c_{Ind}, M	A_{430}	A_{600}
3.00×10^{-4}	1.54	1.06
2.00×10^{-4}	0.935	0.777
1.00×10^{-4}	0.383	0.455
0.500×10^{-4}	0.149	0.261
0.250×10^{-4}	0.0557	0.145

(b)

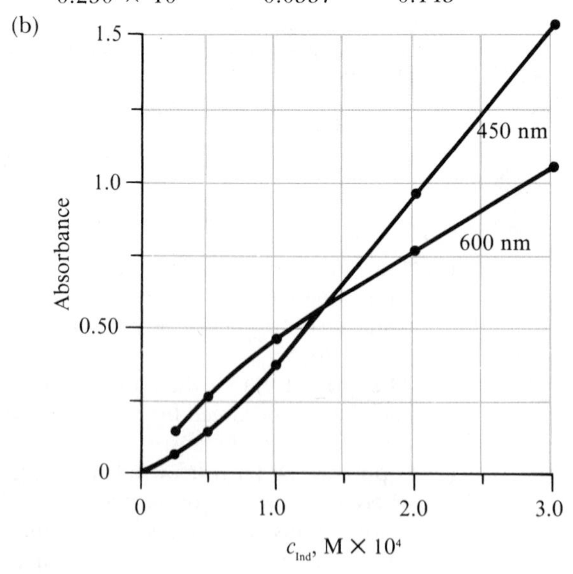

Chapter 19

19-2. (a) 446 lines/mm

19-4. A spectroscope consists of a monochromator that has been modified so that the focal plane contains a movable eyepiece that permits visual detection of the various emission lines of the elements. A spectrograph is a monochromator equipped with a photographic film or plate holder located along its focal plane; spectra are recorded photographically. A spectrophotometer is a monochromator equipped with a photoelectric detector that is located behind the exit slit.

19-6. (a) 725 nm (c) 1.45 μm

19-7. (a) 1.46×10^7 W/m^2 (c) 9.10×10^5 W/m^2

19-8. (a) 1.01 μm, 0.967 μm (b) 3.86×10^6 W/m^2, 4.61×10^6 W/m^2

19-9. (a) Hydrogen and deuterium lamps differ only in the gases they contain. The latter produces radiation of somewhat higher intensity.

(c) A phototube is a vacuum tube equipped with a photoemissive cathode. It has a high electrical resistance and requires a potential of 90 V or more to produce a photocurrent. The currents are generally small enough to require considerable amplification before they can be measured. A photovoltaic cell consists of a photosensitive semiconductor sandwiched between two electrodes. A current is generated between the electrodes when radiation is absorbed by the semiconducting layer. The current is generally large enough to be measured directly with a microammeter. The advantages of a phototube are greater sensitivity and wavelength range as well as better reproducibility. The advantages of the photocell are its simplicity, low cost, and general ruggedness. In addition, it does not require an external power supply or elaborate electronic circuitry. Its use is limited to visible radiation.

(e) A photometer is an instrument for absorption measurements that consists of a source, a filter, and a photoelectric detector. A colorimeter differs from a photometer in the respect that the human eye serves as the detector in the former. The photometer offers the advantage of greater precision and the ability to discriminate between colors provided they are not too much alike. The main advantages of a colorimeter are simplicity, low cost, and the fact that no power supply is needed.

(g) A single-beam spectrophotometer employs a fixed beam of radiation that irradiates first the solvent and then the analyte solution. In a double-beam instrument, the solvent and solution are irradiated simultaneously or nearly so. The advantages of the double-beam instrument are freedom from problems arising from fluctuations in the source intensity and from drift in electronic circuits; in addition, it is more easily adapted to automatic spectral recording. The single-beam instrument offers the advantages of simplicity and lower cost.

19-10. (a) 33.8% (b) 0.471 (c) 0.697

19-12. In a deuterium lamp, the input energy from the power source produces an excited deuterium *molecule* that dissociates into two atoms in the ground state and a photon of radiation. As the excited deuterium molecule relaxes, its quantized energy is distributed between the energy of the photon and the kinetic energies of the two deuterium atoms. The latter can vary from nearly zero to the original energy of the excited molecule. Therefore, the energy of the radiation, which is the difference between the quantized energy of the excited molecule and the kinetic energies of the atoms, can also vary continuously over the same range. Consequently, the emission spectrum is continuous.

19-14. The power of an infrared beam is measured with a heat detector, which consists of a tiny blackened surface that is warmed as a consequence of radiation absorption. This surface is attached to a transducer, which converts the heat signal to an electrical one. Heat transducers are of three types: (1) a thermopile, which consists of several dissimilar metal junctions that develop a potential that depends upon the difference in temperature of the junctions; (2) a bolometer, which is fashioned from a conductor whose electrical resistance depends upon its temperature; and (3) a pneumatic detector whose internal pressure is temperature-dependent.

19-16. Tungsten/halogen lamps contain a small amount of iodine in the evacuated quartz envelope that holds the tungsten filament. The iodine prolongs the life of the lamp and permits it to operate at a higher temperature. The iodine combines with gaseous tungsten that sublimes from the filament and causes the metal to be redeposited, thus adding to the life of the lamp.

19-18. The basic difference between an absorption spectrometer and an emission spectrometer is that the former requires a separate radiation source and a sample compartment that holds containers for the sample and its solvent. With an emission spectrometer, the sample container is a hot flame, a heated surface, or an electric arc or spark that also serves as the radiation source.

19-20. (a) The dark current is the small current that develops in a radiation transducer in the absence of radiation.

(c) Scattered radiation in a monochromator is unwanted radiation that reaches the exit slit as a result of reflection and scattering. Its wavelength is usually different from that of the radiation reaching the slit directly from the dispersing element.

(e) A majority carrier in a semiconductor can be electrons or holes, depending which is present in the greater amount. The majority carrier, as the name implies, conducts a greater fraction of current than does the minority carrier.

19-21. (a) 1.694 μm (b) (4.54/2) μm, 4.54/3 μm, and so forth

Chapter 20

20-1. (a) A chromophore is an unsaturated functional group in an organic molecule that absorbs characteristic wavelengths of ultraviolet or visible radiation.

(c) Noise in a spectrophotometric measurement refers to the random variations in the output of a spectrophotometer due to uncontrollable fluctuations in electronic circuits, power supplies, sources, and readout devices. Spectrophotometric noise also arises from cell-positioning uncertainties and from uncontrolled variables that affect the chemical behavior of the system under study.

(e) Charge-transfer absorption arises from the radiation-induced transfer of an electron from a donor group in a complex to an acceptor group. It is characterized by a very large molar absorptivity.

(g) Quantum efficiency in fluorescence refers to the ratio of the number of quanta absorbed by a sample to the number of photons emitted as fluorescence.

20-2. In segmented-flow methods, solutions are segmented by the introduction of regularly spaced air bubbles into the flowing streams of sample and reagent. In flow-injection analysis, analyte and reagents are introduced into continuously flowing streams of sample and reagent that contain no air bubbles.

20-4. Fluorescing compounds contain aromatic, carbonyl, or highly conjugated double-bonded functional groups. Molecules with rigid structures are particularly likely to fluoresce.

20-6. Compounds that fluoresce generally have structural features that lower the rate at which an electronically excited species relaxes to the ground state by the radiationless transfer of excitation energy to the surroundings. As a consequence, relaxation by fluorescence has time to occur.

20-8. In the mole-ratio method for studying complexes, solutions are prepared in which the concentration of one of the reactants is held constant while the concentration of the other is varied over a considerable range. A plot of absorbance as a function of the mole ratio of the reactants exhibits a change in slope at the ratio that corresponds to the combining ratio of the reactants in the complex.

20-10.

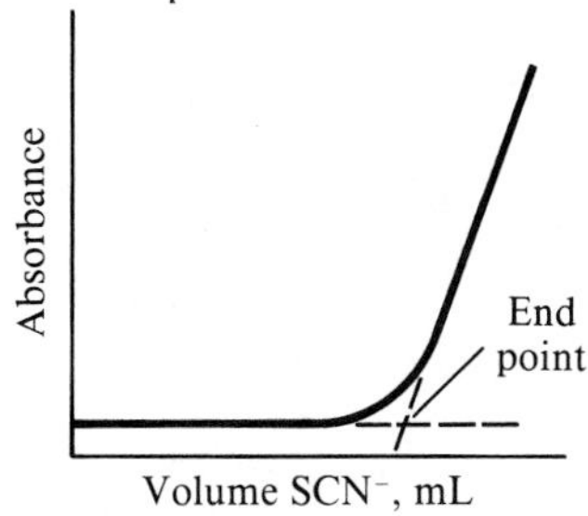

20-12. The rate at which a molecule relaxes from an excited vibrational state to the lowest vibrational state of an electronic state is generally orders of magnitude greater than the rate at which a molecule relaxes from one electronic state to another. Thus the vibrational energy of an excited molecule is lost before fluorescence has time to occur, and the wavelength of fluorescence is consequently less than that of the radiation that caused the excitation.

20-14. The sample compartment in an ultraviolet/visible instrument is located between the monochromator and the detector in order to reduce the possibility of photochemical decomposition of samples that might occur if samples were exposed to the full power of the source. Infrared radiation, in contrast, is not energetic enough to cause photochemical problems but suffers instead from a problem caused by the relatively high percentage of radiation scattered by the sample and its container. With the monochromator located between the sample and the detector, much of this scattered radiation is removed and does not reach the detector.

20-15. 0.0215%

20-17. (a) ±9.5% (b) ±1.7% (c) ±2.1%

20-19.

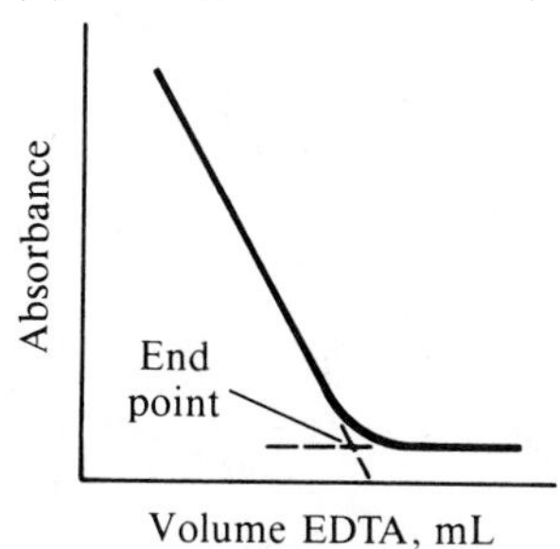

20-21. (a) 1.71×10^{-4} (b) $c_A = 5.34 \times 10^{-4}$ M, $c_B = 3.57 \times 10^{-4}$ M

20-22. (a) $c_{Co^{2+}} = 1.65 \times 10^{-4}$ M, $c_{Ni^{2+}} = 4.33 \times 10^{-5}$ M

		c_P, M	c_Q, M
20-24.	(a)	2.08×10^{-4}	4.90×10^{-5}
	(c)	8.36×10^{-5}	6.10×10^{-5}
	(e)	2.11×10^{-4}	9.64×10^{-5}

20-26. (a) 0.492 (c) 0.190

20-27. (a) 0.302 (c) 0.491

20-28. A, 5.60; C, 4.80

20-29.

Absorbance

0.0 0.1 0.2 0.3 0.4 0.5

400 450 500 550 600 650

Wavelength, nm

20-31. 1.8×10^8

20-33. Ligand-to-cation ratio = 3

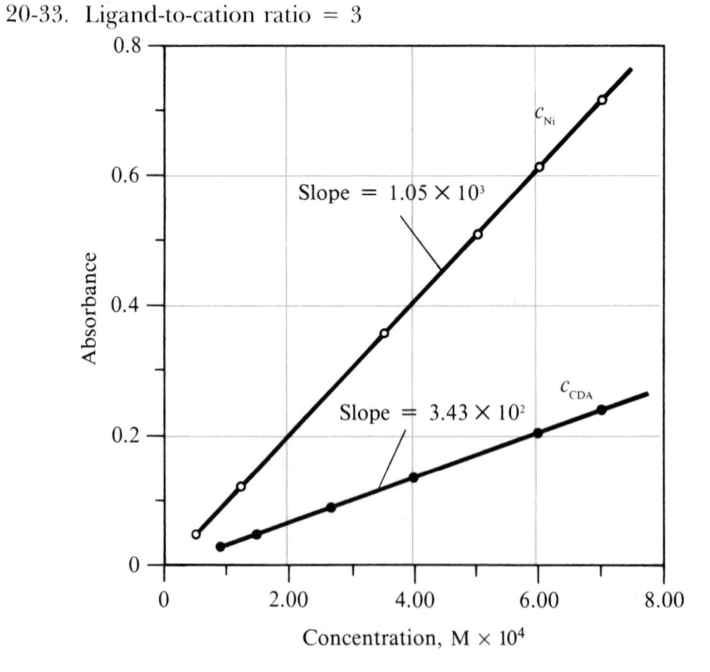

20-35. (a) CdR^{2+}
(b) $1.42(\pm 0.02) \times 10^4$
(c) 9.3×10^3

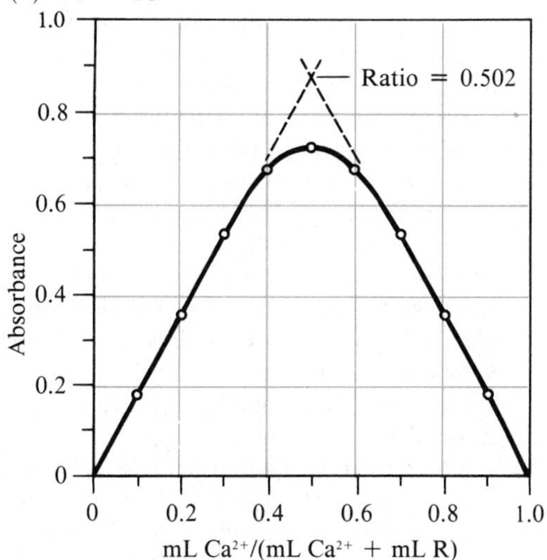

20-38. 5.4×10^{-3} to 1.2×10^{-3} M

20-40. (a)

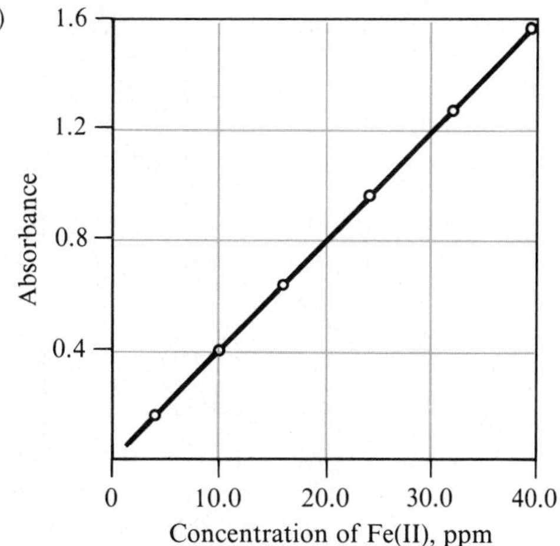

(b) $A = 3.95 \times 10^{-2}C_{Fe} + 1.01 \times 10^{-3}$
(c) $s_r = 3.3 \times 10^{-3}$
(d) $s_b = 1.1 \times 10^{-4}$

20-41. (a) $C_{Fe} = 3.65$ ppm; $(s_1)_r = \pm 2.8\%$; $(s_3)_r = 2.1\%$
(c) $C_{Fe} = 1.75$ ppm; $(s_1)_r = \pm 6.1\%$; $(s_3)_r = 4.6\%$
(e) $C_{Fe} = 3.83$ ppm; $(s_1)_r = \pm 0.3\%$; $(s_3)_r = 0.2\%$

Chapter 21

21-1. In atomic emission, the analyte emits radiation as a consequence of excitation in a flame, arc, spark, or plasma. In atomic fluorescence, the analyte emits radiation as a consequence of being excited by a beam of electromagnetic radiation of the wavelength it absorbs. Fluorescence methods require an external source, which is usually located at right angles to the radiation path to the detector. Atomic emission requires no external source of radiation.

21-3. In atomic emission spectroscopy, the analytical signal is produced by *excited* atoms or ions, whereas in atomic absorption, the signal results from the absorption by *unexcited* species. Typically, the number of unexcited species exceeds the number of excited species by several orders of magnitude. The ratio of unexcited to excited atoms in a hot medium varies exponentially with temperature. Thus a small change in temperature brings about a large relative change in the number of excited atoms. The relative number of unexcited atoms changes very little because they are present in an enormous excess. Therefore, emission spectroscopy is more sensitive to temperature changes than is absorption spectroscopy.

21-5. In atomic absorption spectroscopy, the source radiation must be modulated to an ac signal to prevent radiation emitted by the analyte from interfering with the absorption signal. The detector is then made to reject the dc signal from the flame and measure the modulated signal from the source.

21-7. The temperature in a hollow cathode tube is significantly lower than that in a flame. As a consequence, Doppler broadening is less pronounced in the former and narrower lines result.

21-9. These peaks are caused by light scattering by particulate matter generated during sample ashing.

21-11. When an internal standard is used in emission methods, the ratio of the intensity of the analyte line to the intensity of an internal standard line serves as the analytical parameter. This procedure tends to compensate for indeterminate errors arising from fluctuations in flame temperature.

21-12. 0.504 ppm

21-14. (a)

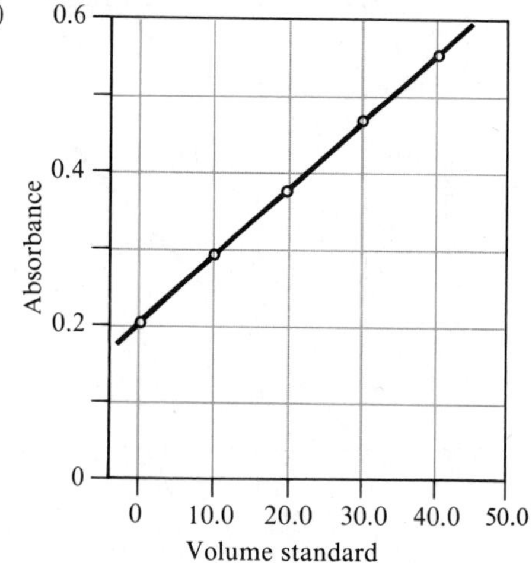

(b) If Beer's law is followed, the absorbance is

$$A = \frac{\epsilon b C_x V_x}{V_t} + \frac{\epsilon b C_s V_s}{V_t}$$

(c) The slope of a plot of A versus V_s is

$$b = \frac{\epsilon b C_s}{V_t}$$

and the intercept is

$$a = \frac{\epsilon b C_x V_x}{V_t}$$

(d) Dividing the second equation in (c) by the first leads to

$$\frac{a}{b} = \frac{\cancel{\epsilon b} C_x V_x / \cancel{V_t}}{\epsilon b C_s / \cancel{V_t}}$$

$$C_x = \frac{a C_s}{b V_x}$$

(e) $b = 8.81 \times 10^{-3}$

$a = 0.202$

(f) $s_b = 4.1 \times 10^{-5}$
$s_r = 1.3 \times 10^{-3}$
(g) 28.0 ppm Cr
(h) 0.22 ppm Cr

Chapter 22

22-1. (a) The order of a reaction is the numerical sum of the exponents of the concentration terms in the rate law for the reaction.
(c) Enzymes are high-molecular-weight organic molecules that catalyze reactions of biochemical importance.
(e) The Michaelis constant is an equilibrium-like constant defined by the equation $K_m = (k_{-1} + k_2)/k_1$, where k_{-1} and k_1 are the rate constants for the forward and reverse reactions in the formation of an enzyme/substrate complex, which is the intermediate in an enzyme-catalyzed reaction. The term k_2 is the rate constant for the decomposition of the complex to give products.
(g) Integral methods use integrated forms of rate equations to compute concentrations from kinetic data.

22-3. Pseudo-first-order conditions are used in kinetic methods because under these conditions the reaction rate is directly proportional to the concentration of the analyte.

22-5. $t_{1/2} = 0.693/k$

22-6.	(a) 2.01 s	(c) 410 s	(e) 1.47×10^9 s
22-7.	(a) 40.5 s^{-1}	(c) 0.405 s^{-1}	(e) 1.51×10^4 s^{-1}
22-8.	(a) 0.105	(c) 2.3	(e) 6.9
22-9.	(a) 0.15	(c) 3.3	(e) 10.0
22-10.	(a) 0.2%	(c) 0.02%	(e) 1.0%
	(g) 0.05%	(i) 6.8%	(k) 0.64%

22-11. $\frac{dP}{dt} = \frac{k_2[E]_0[S]}{[S] + K_m}$ (Equation 22-22)

At v_{max}, $\frac{dP}{dt} = k_2[E]_0$. Thus at $v_{max}/2$, we can write

$$\frac{dP}{dt} = \frac{\cancel{k_2[E]_0}}{2} = \frac{\cancel{k_2[E]_0}[S]}{[S] + K_m}$$

$$[S] + K_m = 2[S]$$

$$K_m = [S]$$

22-13.

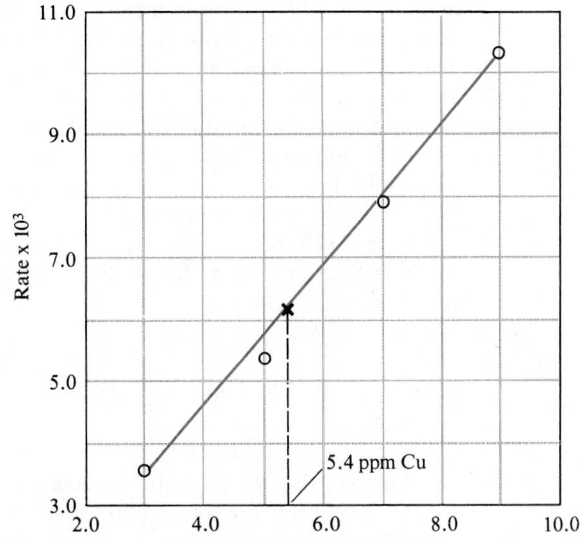

22-15. 4.5×10^{-2} μmol/L

Chapter 23

23-1. (a) Elution is a process in which species are washed through a chromatographic column by additions of fresh solvent.

(c) The stationary phase in a chromatographic column is a solid or liquid that is fixed in place. A mobile phase passes over or through the stationary phase.

(e) The retention time for an analyte is the time interval between its injection onto a column and the appearance of its peak at the other end of the column.

(g) The selectivity factor α of a column toward two species is given by the equation $\alpha = K_B/K_A$, where K_B is the partition ratio of the more strongly held species B and K_A is the corresponding ratio for the less strongly held solute A.

(i) Longitudinal diffusion is a source of band broadening in a column that occurs when a solute diffuses from the concentrated center of a band to the more dilute regions on either side. This movement is thus toward and opposed to the direction of flow of the mobile phase.

(k) The resolution R_s of a column toward two species A and B is given by the equation $R_s = 2\Delta Z/(W_A + W_B)$, where ΔZ is the distance (in units of time) between the peaks for the two species and W_A and W_B are the widths (also in units of time) of the peaks at their bases.

23-2. The general elution problem arises whenever chromatograms are obtained on samples that contain species with widely different partition ratios. When conditions are such that good separation of the more strongly held species is realized, lack of resolution among the weakly retained species is observed. Conversely, when conditions are chosen to give satisfactory separation of the weakly retained compounds, severe band broadening and long retention times are encountered for the strongly bound species. The general elution problem is often solved in liquid chromatography by gradient elution; temperature programming serves the same purpose in gas chromatography.

23-4. In gas-liquid chromatography, the mobile phase is a gas, whereas in liquid-liquid chromatography it is a liquid.

23-6. Variables that affect α values include composition of the mobile phase, column temperature, composition of the stationary phase, and chemical interaction between the stationary phase and one of the solutes being separated.

23-8. The number of plates in a column can be determined by measuring the retention time t_r and width of a peak at its base W. The number of plates N is then given by the equation $N = 16(t_r/W)^2$.

23-10. The minima observed in plots of plate height versus flow rate are caused by longitudinal diffusion, which, in contrast to other broadening sources, goes on to a greater extent at low flow rates than at high. The rate of longitudinal diffusion is orders of magnitude larger in a gaseous mobile phase than in a liquid, however. Thus, the phenomenon becomes noticeable at higher flow rates in gases than in liquids.

23-12. (a) $N_A = 2775$; $N_B = 2472$; $N_C = 2364$; $N_D = 2523$

(b) $\bar{N} = 2534(\pm 174) = 2.5(\pm 0.2) \times 10^3$ plates

(c) $H = 0.0097$ cm

23-13.

	A	B	C	D
(a) k'	0.74	3.3	3.5	6.0
(b) K	6.2	28	29	50

23-14. (a) 0.72 (b) 1.1 (c) 105 cm (d) 60 min

23-15. (a) 5.2 (b) 2.1 cm

23-20. (a) Variables that lead to band broadening: (1) very low or very high flow rates; (2) high viscosity; (3) low temperature; (4) large particle size for packing; (5) for liquid stationary phases, thick films; (6) long columns; (7) slow introduction of sample; (8) large samples.

(b) Variables that lead to band separation: (1) packings that produce partition coefficients that differ significantly; (2) increased packing length; (3) variation in solvent composition; (4) optimum temperature; (5) change in mobile-phase pH; (6) incorporation of species in the stationary phase that selectively complex certain analytes.

23-21. Slow sample introduction leads to band broadening.

23-22. (a) $k'_M = 2.54$; $k'_N = 2.62$
(b) $\alpha_{N,M} = 6.20/6.01 = 1.03$
(c) 8.1×10^4 plates
(d) 1.8×10^2 cm
(e) 91 min

Chapter 24

24-1. In gas-liquid chromatography, the stationary phase is a liquid immobilized on a solid. Retention of sample constituents involves equilibria between a gaseous and a liquid phase. In gas-solid chromatography, the stationary phase is a solid surface that retains analytes by physical adsorption. Here, separations involve adsorption equilibria.

24-3. In open tubular columns, the stationary phase is retained by adsorption or chemical bonding on the inner surface of a capillary tubing. In packed columns, the stationary phase is a liquid immobilized on uniform particles of an inert solid contained in a narrow glass or metal tube. The main advantage of open tubular columns is their remarkable resolution. Packed columns are easier and more convenient to use, however, and in addition accommodate much larger amounts of sample.

24-5. Sample splitters are mechanical devices that deliver a fixed small fraction of a sample to the head of a column; the remainder of the sample goes to waste. Capillary columns require such small samples that reproducible volumes cannot be introduced with a syringe. This problem is overcome by delivering easily measured volumes to the splitter, which then reduces the volume delivered to the column in a precise way.

24-7. Temperature programming involves increasing the temperature of a gas-chromatographic column in a programmed way as a separation is being carried out. This technique is particularly useful for samples that contain constituents that boil over a large temperature range. The lower boiling constituents are separated initially at temperatures that provide good resolution. As the separation proceeds, the column temperature is increased so that the higher boiling constituents come off the column with good resolution and at reasonable lengths of time.

24-9. Hydrogen or helium is used as the carrier gas with thermal-conductivity detectors because the conductivities of these gases are several times greater than those of most organic compounds. Thus, the change in conductivity brought about by the presence of organic species in the effluent from a column is larger than it would be with gases having lower thermal conductivities.

24-11. In hyphenated methods, chromatographic columns are coupled with highly selective detectors such as ultraviolet, infrared, and mass spectrometers and various kinds of electroanalytical instruments. These hyphenated methods provide a powerful tool for identifying the components of complex mixtures.

24-13. The label -AW-DMCS stands for acid-washed dimethylchlorosilane column packing.

24-15. (a) $k'_1 = 5.60$; $k'_2 = 6.46$; $k'_3 = 25.4$
(b) $\alpha_{2,1} = 1.16$; $\alpha_{3,2} = 3.93$
(c) 1.6×10^3 plates; 0.068 cm/plate

(d) $(R_s)_{2,1} = 1.2$; $(R_s)_{3,2} = 11$
(e) 2.5×10^3 plates, or 1.7-m column
(f) 12 min

24-16. 13.7% methyl acetate; 49.2% methyl propionate; 37.1% methyl *n*-butyrate

24-17. Retention times would be less because the nonpolar stationary phase would have less tendency to retain the polar analyte molecules.

24-19. (a) This would lead to a thicker layer of liquid stationary phase, which increases band broadening and thus increases plate heights.
(c) Increasing the port temperature leads to greater band broadening owing to a greater longitudinal diffusion; larger plate heights are the result.
(e) Reducing the particle size of the packing improves efficiency and thus reduces plate height.

Chapter 25

25-1. (a) Substances that are somewhat volatile and are thermally stable.
(c) Substances that are ionic.
(e) High-molecular-weight compounds that are soluble in nonpolar solvents.
(g) Relatively nonpolar, water-insoluble compounds with molecular weights lower than 5000. Good for separating isomers.

25-2. (a) In an isocratic elution, the solvent composition is held constant throughout the elution.
(c) In a stop-flow injection, the flow of solvent is stopped, a fitting at the head of the column is removed, and the sample is injected directly onto the head of the column. The fitting is then replaced and pumping is resumed.
(e) In a normal-phase packing, the stationary phase is quite polar and the mobile phase is relatively nonpolar.
(g) An eluent-suppressor column is located after the ion-exchange column in ion chromatography. It converts the ionized species used to elute analyte ions to largely undissociated molecules that do not interfere with conductometric detection.
(i) Gel-permeation chromatography is a type of size-exclusion chromatography in which the packings are hydrophobic and the eluents are nonaqueous. It is used for separating high-molecular-weight nonpolar species.
(k) FSOT columns are fused silica open tubular columns used in gas chromatography.
(m) A supercritical fluid is a substance that is heated above its critical temperature so that it cannot be condensed into a liquid no matter how great the pressure.

25-4. (a) *n*-hexane, benzene, *n*-hexanol

25-5. (a) *n*-hexanol, benzene, *n*-hexane

25-7. In adsorption chromatography, the sample components are selectively retained on the surface of a solid stationary phase by adsorption. In partition chromatography, selective retention occurs in a liquid or liquid-like stationary phase.

25-9. Nonvolatile and thermally unstable compounds.

25-11. The simplest type of pump for liquid chromatography is a pneumatic pump, which consists of a collapsible solvent container housed in a vessel that can be pressurized by a compressed gas. This type of pump is simple, inexpensive, and pulse-free. It has limited capacity and pressure output, it is not adaptable to gradient elution, and its pumping rate depends upon the viscosity of the solvent. A screw-driven syringe pump consists of a large syringe in which the piston is moved in or out by means of a motor-driven screw. It also is pulse-free, and the rate of delivery is easily varied. It suffers from lack of capacity and is inconvenient to use when solvents must be changed. The most versatile and widely used pump is the reciprocating pump, which usually consists of a small cylindrical chamber that is filled and then emptied by the back-and-forth motion of a piston. Advantages of the reciprocating pump include small internal volume, high output pressures, adaptability to gradient elution, and flow rates that are constant and independent of viscosity and back pressure. The main disadvantage is the pulsed output that must subsequently be damped.

25-13. (a) The advantages over HPLC are (1) that it can be used with general detectors, such as the flame ionization detector, and (2) that it is inherently faster or gives better resolution for the same speed.
(b) The advantage over GLC is that it can be used for nonvolatile and thermally unstable samples.

Chapter 27

27-1. The steps in sampling are (1) identification of the population from which the sample is to be drawn, (2) collection of a gross sample, and (3) reduction of the gross sample to a small quantity of homogeneous material for analysis.

27-3. (a) Sorbed water is that held as a condensed liquid phase in the capillaries of a colloid. Adsorbed water is that retained by adsorption on the surface of a finely ground solid. Occluded water is that held in cavities distributed irregularly throughout a crystalline solid.
(c) Essential water is chemically bound water that occurs as an integral part of the molecular or crystalline structure of a compound in its solid state. Nonessential water is that retained by a solid as a consequence of physical forces.

27-5. For NBS sample, mean = 50.30% CaO; s = 0.09% CaO; s_r = 1.8 ppt
For gross samples, mean = 49.92% CaO; s = 0.39% CaO; s_r = 7.8 ppt
Relative variance of sampling = $(7.8)^2 - (1.8)^2 = 5.8$
Relative standard deviation of sampling = 7.6 ppt

27-6.

	σ_{rel}, %	σ_{abs}, particles
(a)	1.9	6
(b)	16	6
(c)	10	9
(d)	12	8
(e)	1.2	13

27-8. (a) 784 tablets (c) 2.0×10^4 tablets

27-9. (a) $\sigma_{rel} = 12\%$
(b) $\sigma_{abs} = 1.1 \times 10^3$ bottles
(c) 9000 ± 1800 bottles
(d) 1.5×10^3 bottles

27-10. (a) 3.5×10^4 particles (b) 3.9×10^3 g (8.5 lb) (c) 0.22 mm

27-12. (a) 2.2×10^5 particles (b) 2.8×10^3 g (6.1 lb) (c) 0.13 mm

Chapter 28

28-1. Dry-ashing is carried out by igniting the sample in air or sometimes oxygen. Wet-ashing is carried out by heating the sample in an aqueous medium containing such oxidizing agents as H_2SO_4, $HClO_4$, HNO_3, H_2O_2, or some combination of these.

28-3. B_2O_3 or $CaCO_3/NH_4Cl$

28-5. When hot concentrated $HClO_4$ comes in contact with organic materials or other oxidizable species, explosions are highly probable.

28-7. (a) Samples for halogen determination can be decomposed in a Schöniger combustion flask, combusted in a tube furnace in a stream of oxygen, or fused in a peroxide bomb.
(c) Samples for nitrogen determination are decomposed in hot concentrated H_2SO_4 in a Kjeldahl flask or oxidized by CuO in a tube furnace in the Dumas method.

Chapter 29

29-1. A masking agent is a complexing reagent that reacts selectively with one or more components of a solution to prevent them from interfering in an analysis.

29-3. A collector is a species that is added to a sample when traces of an analyte are to be precipitated. The collector forms a precipitate with the reagent used to precipitate the analyte and prevents the small amount of analyte precipitate from becoming physically lost.

29-4. (a) NaOH. The basic oxide of iron precipitates, whereas the aluminum remains in solution as the aluminate ion.
(c) Hot concentrated HNO_3 precipitates Sn(IV) as $SnO_2 \cdot xH_2O$.
(e) Precipitate Hg^{2+} as the sulfide from 3 M HCl.

29-5. (a) 1.73×10^{-2} M (b) 6.40×10^{-3} M
(c) 2.06×10^{-3} M (d) 6.89×10^{-4} M

29-7. (a) 75 mL (b) 40 mL (c) 22 mL

29-9. (a) 18.0 (b) 7.56

29-11. (a) $K_D = 1.53$ (b) $[HA] = 0.0147$; $[A^-] = 0.0378$ (c) $K_a = 9.7 \times 10^{-2}$

29-12. (a) 12.4 meq Ca^{2+}/L (b) 1.25×10^3 mg $CaCO_3$/L

29-15. Dissolve 17.53 g of NaCl in about 100 mL of water and pass the solution through a column packed with a cation exchange resin in its acid form. Wash the column with several hundred milliliters of water, collecting the liquid from the original solution and the washings in a 2-L volumetric flask. Dilute to the mark and mix well.

29-17. (a) $H_2L(aq) \leftrightharpoons H_2L(org)$
$H_2L(aq) + H_2O \leftrightharpoons H_3O^+ + HL^-(aq)$
$HL^-(aq) + H_2O \leftrightharpoons H_3O^+ + L^{2-}(aq)$
$Cu^{2+}(aq) + L^{2-}(aq) \leftrightharpoons CuL(aq)$
$CuL(aq) \leftrightharpoons CuL(org)$
(b) $K_{ex} = 5.01 \times 10^{-8}$
(c) At pH 6.00, $K = 5.01$
(d) Four extractions
(e) Six extractions

Index

A **boldface entry** refers to specific laboratory directions; *t* refers to a table.

International Atomic Weights

Element	Symbol	Atomic number	Atomic weight
Actinium	Ac	89	(227)
Aluminum	Al	13	26.9815
Americium	Am	95	(243)
Antimony	Sb	51	121.75
Argon	Ar	18	39.948
Arsenic	As	33	74.9216
Astatine	At	85	(210)
Barium	Ba	56	137.34
Berkelium	Bk	97	(247)
Beryllium	Be	4	9.0122
Bismuth	Bi	83	208.980
Boron	B	5	10.811
Bromine	Br	35	79.909
Cadmium	Cd	48	112.40
Calcium	Ca	20	40.08
Californium	Cf	98	(249)
Carbon	C	6	12.01115
Cerium	Ce	58	140.12
Cesium	Cs	55	132.905
Chlorine	Cl	17	35.453
Chromium	Cr	24	51.996
Cobalt	Co	27	58.9332
Copper	Cu	29	63.54
Curium	Cm	96	(245)
Dysprosium	Dy	66	162.50
Einsteinium	Es	99	(254)
Erbium	Er	68	167.26
Europium	Eu	63	151.96
Fermium	Fm	100	(252)
Fluorine	F	9	18.9984
Francium	Fr	87	(223)
Gadolinium	Gd	64	157.25
Gallium	Ga	31	69.72
Germanium	Ge	32	72.59
Gold	Au	79	196.967
Hafnium	Hf	72	178.49
Helium	He	2	4.0026
Holmium	Ho	67	164.930
Hydrogen	H	1	1.00797
Indium	In	49	114.82
Iodine	I	53	126.9044
Iridium	Ir	77	192.2
Iron	Fe	26	55.847
Krypton	Kr	36	83.80
Lanthanum	La	57	138.91
Lawrencium	Lw	103	(257)
Lead	Pb	82	207.19
Lithium	Li	3	6.942
Lutetium	Lu	71	174.97
Magnesium	Mg	12	24.312
Manganese	Mn	25	54.9380
Mendelevium	Mv	101	(256)
Mercury	Hg	80	200.59
Molybdenum	Mo	42	95.94
Neodymium	Nd	60	144.24
Neon	Ne	10	20.183
Neptunium	Np	93	(237)
Nickel	Ni	28	58.70
Niobium	Nb	41	92.906
Nitrogen	N	7	14.0067
Nobelium	No	102	(254)
Osmium	Os	76	190.2
Oxygen	O	8	15.9994
Palladium	Pd	46	106.4
Phosphorus	P	15	30.9738
Platinum	Pt	78	195.09
Plutonium	Pu	94	(244)
Polonium	Po	84	(210)
Potassium	K	19	39.102
Praseodymium	Pr	59	140.907
Promethium	Pm	61	(145)
Protactinium	Pa	91	(231)
Radium	Ra	88	(226)
Radon	Rn	86	(222)
Rhenium	Re	75	186.2
Rhodium	Rh	45	102.905
Rubidium	Rb	37	85.47
Ruthenium	Ru	44	101.07
Samarium	Sm	62	150.35
Scandium	Sc	21	44.956
Selenium	Se	34	78.96
Silicon	Si	14	28.086
Silver	Ag	47	107.870
Sodium	Na	11	22.9898
Strontium	Sr	38	87.62
Sulfur	S	16	32.064
Tantalum	Ta	73	180.948
Technetium	Tc	43	(99)
Tellurium	Te	52	127.60
Terbium	Tb	65	158.924
Thallium	Tl	81	204.37
Thorium	Th	90	232.038
Thulium	Tm	69	168.934
Tin	Sn	50	118.69
Titanium	Ti	22	47.90
Tungsten	W	74	183.85
Uranium	U	92	238.03
Vanadium	V	23	50.942
Xenon	Xe	54	131.30
Ytterbium	Yb	70	173.04
Yttrium	Y	39	88.905
Zinc	Zn	30	65.37
Zirconium	Zr	40	91.22

Numbers in parentheses indicate mass number of most stable known isotope.